327 (MCR

Michael Klagsbrun
Surgical Research G-1008

10643684

Proteases
and Biological
Control

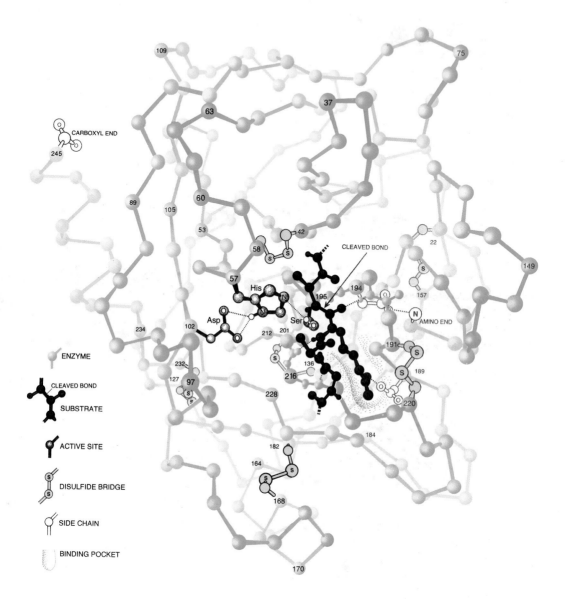

CLEAVED BOND

CLEAVED BOND

CARBOXYL END

AMINO END

His

Asp

Ser

ENZYME

CLEAVED BOND

SUBSTRATE

ACTIVE SITE

DISULFIDE BRIDGE

SIDE CHAIN

BINDING POCKET

Proteases
and Biological
Control

edited by

E. Reich
The Rockefeller University

D. B. Rifkin
The Rockefeller University

E. Shaw
Brookhaven National Laboratory

**COLD SPRING HARBOR CONFERENCES ON CELL PROLIFERATION
VOLUME 2**

Cold Spring Harbor Laboratory
1975

Proteases and Biological Control

© 1975 by Cold Spring Harbor Laboratory
All rights reserved

International Standard Book Number 0-87969-114-X
Library of Congress Catalog Card Number 75-18635

Printed in the United States of America

Cover and book design by Emily Harste

Frontispiece

This illustration represents the structure of the trypsin molecule determined by R. Stroud, L. Kay and R. Dickerson. The blue lines follow the path in which the polypeptide chain is folded. Spheres along this chain represent the α-carbon positions of amino acid residues in the sequence numbered by comparison with the homologous sequence of chymotrypsinogen A. The atoms in red indicate probable positions for atoms in a polypeptide substrate oriented close to the active center of the enzyme. This model was derived by combining the models of benzamidine trypsin and of the pancreatic trypsin inhibitor determined by R. Huber and his collaborators. The normal side chain of lysine 15 on the inhibitor was replaced by an arginine side chain, and the arginine side chain was overlaid onto the position found for benzamidine in benzamidine trypsin. This model therefore embodies an enzyme conformation like that induced by side-chain binding and a substrate conformation which evolved to bind tightly to the enzyme. The structure of the true complex formed by pancreatic trypsin inhibitor and trypsin has been solved by Huber and coworkers and shown to exist as a tetrahedral intermediate. The active center contains the three side chains of aspartic acid 102, histidine 57 and serine 195, whose association was first identified by D. Blow and his collaborators in studies of chymotrypsin. Several hydrogen bonds are formed between the substrate and enzyme which orient this susceptible peptide bond close to the active center. The side-chain binding pocket in enzymes with trypsinlike specificity contains an aspartic acid residue 189, which forms a salt bridge to the substrate side chain. (Drawing by Irving Geis, from *Scientific American*, July 1974. © 1974 by I. Geis, R. M. Stroud and R. E. Dickerson.)

Contents

III. COMPLEMENT AND KININS

IV. FIBRINOLYSIS

VII. PROTEINASES AND REPRODUCTION

VIII. CELL GROWTH AND PROTEASES

Preface

The study of proteases has yielded some of the most enduring triumphs of biochemistry. Proteases were among the first enzymes to be highly purified and crystallized; modern concepts of enzyme specificity were established when Bergmann and Fruton developed the use of synthetic protease substrates, and the resulting insight provided the sine qua non for determining the amino acid sequence of proteins. Most of our understanding of enzyme catalysis and the role of specific amino acid residues in enzyme mechanisms emerged from the study of proteases, especially the serine proteases. Crystallographic analysis of several proteases has ordered this knowledge in three dimensions and deepened our feeling for the structure and functions of proteins and enzymes in general. All of these achievements were milestones in the quest for a chemical understanding of biological phenomena; the experimentally useful proteases during this period were all simple digestive enzymes with little or no specificity for protein substrates.

A second era in protease research was initiated shortly after the first, when Kunitz reported that the trypsin catalyzed activation of chymotrypsinogen occurred without detectable proteolytic degradation. The significance of this finding was greatly extended when Lorand subsequently discovered that the catalytic effect of thrombin in blood clotting was due to limited proteolysis of its substrate, fibrinogen. It has since become clear that specific, limited proteolysis is a general mechanism in many important physiological functions, such as those mediated by the enzyme cascades of blood coagulation, complement activation, fibrinolysis and kinin generation. One remarkable property of these dynamic systems is their stability under normal conditions; this is accomplished by exquisite regulation that permits rapid activation and yet confines enzyme function in time and space. Exposing the nature of this regulation is a tantalizing challenge for the future. Although less clearly defined than in the case of circulating enzyme cascades, limited proteolysis is certainly involved in intracellular zymogen activation and in hormone biosynthesis and activation. Still less defined, but far more interesting and significant, are the tentative suggestions of protease action as essential elements in more complex biological phenomena. These include gamete formation, follicle rupture and fertilization; cell migration and tumor invasiveness; intra- and extracellular protein turnover; morphogenesis and metamorphosis, and many more.

This conference was initiated at the suggestion of J. D. Watson, who surmised that the full significance of protease function in biological processes was insufficiently appreciated. His view was endorsed by most of the participants, whose enthusiasm and discussions made the conference a success. Their contributions, as recorded in this volume, will hopefully attract other workers to the field.

A debt of gratitude is also owed to the National Institutes of Health for their financial support which allowed us the flexibility in arranging the meeting we otherwise might not have enjoyed.

E. Reich
D. B. Rifkin
E. Shaw

Unifying Concepts among Proteases

Kenneth A. Walsh

Department of Biochemistry, University of Washington
Seattle, Washington 98195

Proteolytic enzymes were among the first recognized enzymes. Clues to their involvement in biological control mechanisms can be found in the literature around the turn of the century. It was known then that acidic chyme promoted the conversion of prosecretin to secretin, that secretin and pancreozymin stimulated the production of pancreatic juice, and that trypsinogen in this juice was converted to active trypsin by intestinal enterokinase (Bayliss and Starling 1902). The gastrointestinal proteases in higher animals have counterparts not only in simpler organisms (even bacteria) but also in blood plasma, sperm and a diverse assortment of tissues where proteolytic action appears to serve a control function rather than a nutritional purpose. For years these various proteases were isolated, named according to their origin or physiological role, and classified on the basis of their specificity. In 1960 Hartley proposed that a simpler classification could be based on the similarities in their enzymatic mechanisms as revealed by susceptibility to inhibitors. Thus in Table 1, a minimum of five sets of proteases can be defined, where enzymes are grouped into mechanistic sets on the basis of evidence of detailed analogy in their mechanisms of action. A single set of enzymes (e.g., the serine proteases) may serve diverse biological functions.

Homologous Families of Proteases

More recently, amino acid sequence analysis of members of these mechanistic sets has revealed that extensive structural similarities are often found among several members of a single set (Neurath, Walsh and Winter 1967; Stroud, Kay and Dickerson 1971; Titani et al. 1972b). The patterns of similarity or *homology* are simply explained if the various members of the set have arisen by a process of divergent evolution from a single ancestral prototype. Although observation of homology is suggestive of detailed analogy in enzymatic mechanism, as will be discussed later, lack of homology does not necessarily rule out placement of two enzymes in a single mechanistic set.

1

Table 1
Examples of Proteases, Subdivided in Mechanistic Sets

Set	Identification		Examples	Function
	feature	inhibitor		
Serine protease	"active" serine	fluorophosphates	trypsin	digestion
			thrombin	blood coagulation
			plasmin	lysis of blood clots
			coccoonase	mechanical
			subtilisin	digestion
			acrosin	sperm penetration
Metalloexopeptidase	Zn^{++}	o-phenanthroline	carboxypeptidase	digestion
Sulfhydryl protease	CySH	iodoacetate	papain	digestion
			strept. proteinase	digestion
			cathepsin B	intracellular digestion
Acid protease	acidic pH optimum	diazoketones	pepsin	denaturing attack
			chymosin	milk coagulation
Metalloendopeptidase	Zn^{++}, Ca^{++}	o-phenanthroline	thermolysin	digestion

As suggested by Hartley (1960).

2

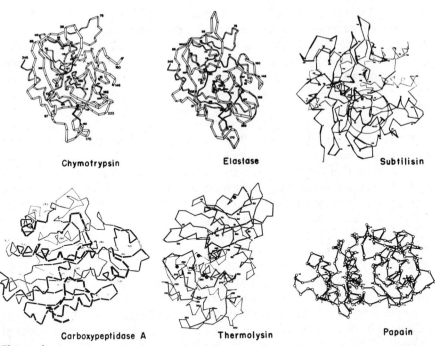

Chymotrypsin   Elastase   Subtilisin

Carboxypeptidase A   Thermolysin   Papain

Figure 1
The conformation of the peptide chains of six proteases from four mechanistic sets. The upper three proteins are all serine proteases, but they belong to two different subsets. In one subset, bovine chymotrypsin and porcine elastase have homologous sequences and similar chain folding. (Reproduced, with permission, from Hartley and Shotton 1971.) In the other subset, subtilisin is not homologous, is folded quite differently, but has an analogous mechanism of action. (Reproduced, with permission, from Wright, Alden and Kraut 1969.) The folding of each of the other three proteases, carboxypeptidase A, thermolysin and papain, is quite different from that of the serine proteases and from each other. (Reproduced, with permission, from Lipscomb et al. 1969, Matthews et al. 1972, and Drenth et al. 1971, respectively.) Their amino acid sequences are not homologous with the serine proteases or with each other.

The striking similarity in the three-dimensional structure of bovine chymotrypsin and porcine elastase (Fig. 1) demonstrated that these two homologous members of a mechanistic set are similarly folded and have analogous assemblies of components of the active site (Shotton and Watson 1970). Stroud, Kay and Dickerson (1971) have since shown that bovine trypsin has an almost identical conformation. Thus it is surprising to find that alignment of any two of their amino acid sequences reveals only 40–50% identity. Evidently the crucial sequences that control their conformations are preserved during evolution. The three enzymes differ in their specificity, and a rational explanation of these differences is provided by changes among only three of the residues which shape their binding sites (Hartley 1970). Two important conclusions can be drawn from these similarities in structure: (1) Homologous sequences are likely to have similar three-dimensional structures and analogous mechanisms of action. (2) Exploring molecular detail of one member of a set may be informative about all members of that set. That is, the detailed structure and mechanism of one member of a set provides a model for understanding the structure, mechanism and specificity of other members of the same set. The clues to these simplifying concepts had actually been first observed by Perutz and Kendrew in comparing the three-dimensional structures of myoglobin and hemoglobin, where homologous sequences were seen to be associated with similar subunit conformation (Perutz 1960).

Hypothetical family trees can be drawn for the evolution of proteolytic enzymes (Neurath, Walsh and Winter 1967). Sequence data indicate that trypsins from cow, pig, sheep, dogfish, lungfish and streptomyces are homologous; biological, paleontological and structural data are consistent with a pattern of mutation of a single gene and selection of the various contemporary trypsins during organismic evolution. But how can the homologous set comprising chymotrypsin, trypsin and thrombin all coexist in one animal unless there are three different genes? Apparently an ancestral gene must have duplicated; one copy evolved as a trypsin and other copies evolved in specificity as chymotrypsin, thrombin or elastase. Thus the potential functions of the organism were expanded without loss of the original function. The complexity and diversity of higher organisms may well be an expression of such expansion of productive loci in the primordial genome. Changes in specificity may be explained by alterations as simple as the single significant difference between trypsin and chymotrypsin (Asp-189–Ser-189), where a negative charge and an hydroxyl group are interchanged at the bottom of the binding pockets (Hartley 1970). In elastase, the specificity pocket is virtually closed by Val-216 and Thr-226, whereas access to the pocket is provided by glycine residues in trypsin and chymotrypsin (Shotton and Watson 1970).

Molecular evolution of enzymes may include changes in the binding (specificity) site, trivial "neutral" changes in the phenotype and changes in the catalytic apparatus (Fig. 2). Examples of the latter case have not been documented, but it seems logical that either the binding site or the catalytic apparatus might be preserved while the other mutates to produce new function.

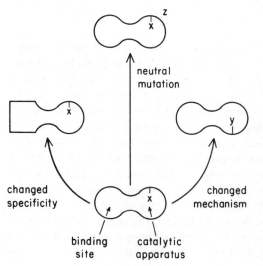

Figure 2
Diagrammatic illustration of three types of change in molecular evolution. A "neutral" mutation (Z) may cause no measurable change in function. But a change in either binding site (*left*) or catalytic apparatus (-X to -Y) would alter the specificity or catalytic function, respectively.

Perhaps in this way a hydrolytic enzyme could be the starting point for evolution of a group transfer enzyme or a synthetase.

The pattern of molecular evolution of sets of proteins has shown surprisingly good correlation between the time of presumed divergence of organisms and the percent change in sequence of their constituent molecules (Dickerson 1971). One can think of reasons why this would not be expected (generation time, lethal mutations, etc.), but the facts speak for themselves. Different rates of divergence are explained by Dickerson (1971) in terms of the proportion of each molecule that is intimately involved in interactions with its molecular environment.

Mechanistic Sets

Returning to the proteases, we should review the five well-characterized mechanistic sets. The active sites of the serine proteases (chymotrypsin, elastase and trypsin) lie in similar crevices in each molecule (Matthews et al. 1967; Shotton and Watson 1970; Stroud, Kay and Dickerson 1971). In each case, the stereochemistry is consistent with a concerted attack on a peptide bond by His-57 and Ser-195, assisted by Asp-102 (chymotrypsinogen numbering). A likely mechanism has been proposed in which the serine, rendered nucleophilic by interaction with the Asp-His pair, attacks the carbonyl carbon to yield a tetrahedral intermediate, stabilized in an "oxyanion hole" (Blow, Birktoft and Hartley 1969; Robertus et al. 1972; Hunkapillar et al. 1973). Transfer of a proton from His-57 releases the amine to yield an acyl enzyme. A water molecule replaces the amine, and the process is reversed to deacylate the enzyme. Entirely analogous mechanisms are thought to operate in the homologous enzymes, thrombin, coccoonase, etc., where the binding pockets would specify different substrates.

The metalloexopeptidase prototype is illustrated by the three-dimensional structure of carboxypeptidase A (Fig. 1), where a very different sequence (Bradshaw et al. 1969) and conformation (Lipscomb et al. 1969) provides a dead-end pocket rather than an open ravine. In proposed mechanisms, a hydrophobic pocket and Arg-145 bind the C-terminal amino acid. A chelated zinc atom is thought to act as an electron sink to polarize the susceptible peptide bond. The peptide bond is attacked in a concerted manner by a general acid (Tyr-248) and a general base (Glu-270). K. Titani in our laboratory has recently shown that carboxypeptidases A and B are homologous, and Schmid, Lattman and Herriott (1974) have shown that the conformation of carboxypeptidase B resembles that of carboxypeptidase A as much as trypsin resembles chymotrypsin. In fact, the specificity difference is accounted for by the interchange of an isoleucine and aspartic acid at position 255, as predicted earlier by Reeck et al. (1971).

The metalloendopeptidases (e.g., thermolysin) have a crevice with a hydrophobic pocket as in chymotrypsin, but a zinc atom is chelated as in carboxypeptidase. The sequence (Titani et al. 1972a) and conformation (Colman, Jansonius and Matthews 1972) are quite different from those of the two previous sets (Fig. 1). In the active site, nine residues are implicated by proximity arguments. Chemical evidence supports the involvement of a histidyl residue, and His-231 is a likely candidate (Burstein, Walsh and Neurath 1974). The

Table 2
Proposed Components of Catalytic Apparati

Set	Example	Nucleophile	Gen. base	Gen. acid	Inhibitor
Serine protease	chymotrypsin	Ser-195	His-57	His-57	DFP
Metalloexopeptidase	carboxypeptidase A		Glu-270	Tyr-248	o-phenanthroline
Metalloendopeptidase	thermolysin		Glu-143?	His-231?	o-phenanthroline
SH-protease	papain	Cys-25	His-159	—	iodoacetate
Acid protease	pepsin	Asp-32?		Asp-215?	diazoketones

Specific references can be found in Boyer (1971) and Colman et al. (1972).

binding site for competitive inhibitors is identified, but the mechanism of catalysis is not clear. This particular enzyme is a thermostable protein from a thermophilic organism. A neutral protease from the mesophile *B. subtilis* appears to be homologous, but it lacks the thermal stability of thermolysin (Pangburn 1974). We are studying neutral protease to identify the structural elements responsible for these differences in thermal stability.

Papain, the prototype of the sulfhydryl proteases, has the conformation illustrated in Figure 1. The nucleophilic Cys-25 is in a crevice near His-159, and these two residues are thought to act in concert to attack susceptible bonds (Drenth et al. 1971). The conformation is quite unlike that of any of the previous three sets.

The fifth set, the acid proteases, is characterized by its dependence on a strongly acidic environment. The sequence of pepsin has recently been described by Tang et al. (1973), and mechanistic studies implicate aspartyl residues 32 and 215. But a detailed mechanism awaits resolution of the conformation. Rennin (chymosin) and certain mold proteases appear to be homologous with pepsin and therefore are likely to operate by analogous mechanisms (Foltmann 1970; Harris et al. 1972).

Mechanistic features of the five sets of proteases are summarized in Table 2. The conservatism of biology is evident. There appears to be a limited number of ways in which residues of one protein attack another. At this point, one must avoid overconfidence. While we do recognize five fundamental sets of proteases and have inklings about their mechanisms, the list of sets is surely not complete and exceptions to apparent sets are already known. For example, chymotrypsin, subtilisin and a carboxypeptidase from yeast (Kuhn 1973; Hayashi, Moore and Stein 1973) are each inhibited by diisopropyl-fluorophosphate (DFP) and therefore might be considered to be serine proteases. Yet the sequence around the active serine is quite different in the three cases. The structure of subtilisin has been solved by Markland and Smith (1967) and by Alden, Wright and Kraut (1970), and it is clear that chymotrypsin and subtilisin differ both in sequence and in conformation (Fig. 1). However, the antiparallel binding of the substrate to the active site of chymotrypsin has its precise counterpart in the active site of subtilisin (Segal et al. 1971). Furthermore, subtilisin and chymotrypsin possess identical constellations of chemical groups for catalysis, supported on quite different architectural matrices. Thus while the location of critical residues in the linear sequence is dissimilar in the two enzymes (Table 3), the substrate is apparently exposed to similar attack. Mechanistically these enzymes are both serine proteases and belong in the same set, but structurally they represent two different families or "subsets." Yeast carboxypeptidase may well represent a third subset. Thus the primary criterion for choosing the mechanistic set for a new protease is the detail of its catalytic mechanism; subsets subdivide enzymes with common mechanisms into homologous families.

There are surely other sets or subsets comprising other proteases which are as yet insufficiently described. Where do we place the aminopeptidases, collagenase, cathepsins, kininases, etc.? Will detailed knowledge of one of these enzymes provide a model for understanding another? For example, is cathepsin B (Mycek 1970) in the same set as papain? Both require thiols, but are the two mechanisms analogous, or is cathepsin B the prototype of a new set?

Table 3

Location of Residues in the Active Sites of
Chymotrypsin and Subtilisin

	Chymotrypsin	*Subtilisin*
"Charge relay system"	Asp-102	Asp-32
	His-57	His-64
	Ser-195	Ser-221
Peptide binding	P_1 Ser-214	Ser-125
	P_3 Gly-216	Gly-127
"Oxyanion hole"	Gly-193	Asn-155
	Ser-195	Ser-221

Data from Robertus et al. (1972).

Physiological Roles of Proteases

Proteases serve well-known degradative functions in nutrition. They also serve important control functions by limited proteolysis. Since there is no known biological repair mechanism for a broken peptide bond, limited proteolysis is an operationally irreversible process. Neurath (this volume) discusses the implications of this statement more fully and draws attention to the potential of proteolysis in unidirectional control processes. We should also not ignore the problem of controlling proteases after their synthesis. How can a protease be synthesized without destruction of the cytological microenvironment, including the very enzymes which are involved in its synthesis?

The tissue locale of protease action is summarized in Figure 3. *Extracellular* digestion is the most familiar process and has been studied since Spallanzani's observations of gastric juice in the eighteenth century. *Intracellular* degradation has been demonstrated by lysosomes. Limited proteolysis in the cytoplasm generates hormones, enzymes and self-assembling proteins as scheduled events in the cell cycle and as scheduled responses to physiological stimuli.

One theme of this volume is *intercellular* communication involving limited proteolysis. Some of these controlling functions are indicated in Figure 3. Conversion of procollagen to collagen in the extracellular matrix avoids intracellular fibrosis. Trypsinogen remains inert in the pancreas and is activated to trypsin in the protected environment of the duodenum. Acrosin from the sperm attacks the *zona pellucida* of eggs; coccoons are produced by one set of cells in moths to be later digested by "coccoonase" from other cells. Factors from wounded tissues cause zymogens from the liver to yield enzymes in the plasma which induce fibrin clots. Neural and chemical signals may be transduced to enzymatic and hormonal effectors by limited proteolysis.

These diverse purposes—nutrition and control—have one feature in common, the cleavage of peptide bonds. The various sets of specific proteases are tailored to these physiological roles by evolutionary modulations of primordial success stories. We are probably only aware of the tip of the iceberg. Due to their self-digesting nature, proteases are difficult to isolate. Furthermore, when

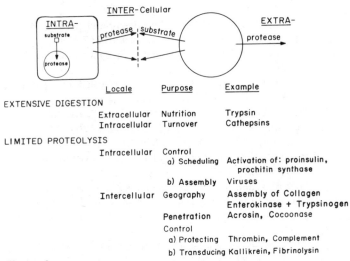

	Locale	Purpose	Example
EXTENSIVE DIGESTION			
	Extracellular	Nutrition	Trypsin
	Intracellular	Turnover	Cathepsins
LIMITED PROTEOLYSIS			
	Intracellular	Control	
		a) Scheduling	Activation of: proinsulin, prochitin synthase
		b) Assembly	Viruses
	Intercellular	Geography	Assembly of Collagen Enterokinase + Trypsinogen
		Penetration	Acrosin, Cocoonase
		Control	
		a) Protecting	Thrombin, Complement
		b) Transducing	Kallikrein, Fibrinolysin

Figure 3

The locale of proteolytic action is dictated by the physiological role of each protease. Thus *extracellular* digestion is largely nutritional; *intracellular* digestion may be extensive (e.g., in lysosomes) or limited (as in control of scheduling and assembly). *Intercellular* digestion serves a variety of purposes which facilitate localization of molecular assembly and zymogen activation, mechanical penetration and transduction of a physiological signal to a chemical response. References to specific examples can be found in Neurath (this volume).

biochemists disrupt the native organization of cells to seek the normal substrate in control processes, great care must be taken to prevent premature proteolysis. The titles of subsequent chapters indicate that these problems are not insuperable.

In recent years it has been thought that proteases lack sophisticated controls, such as allostery, but now it appears that proteases may themselves act as control agents which transform the function of other proteins. What signals release these control agents, how are these signals translated to effect proteolysis, and how does proteolysis control metabolism? In the blood coagulation scheme, a pattern of stepwise control is evident, but what activates the triggering Hageman factor? What causes the release of pancreatic zymogens and the production of enterokinase that triggers the activation of these zymogens? What causes the release or activation of acrosin or coccoonase on schedule? How many zymogen activation events remain undiscovered? There are more questions than answers as we try to fit fragmentary chemical control data into a physiological context.

Acknowledgments

This work was supported by research grants from the National Institutes of Health (GM 15731) and the American Cancer Society (BC91P).

REFERENCES

Alden, R. A., C. S. Wright and J. Kraut. 1970. A hydrogen-bond network at the active site of subtilisin BPN'. *Phil. Trans. Roy. Soc. London* B **257**:119.

Bayliss, W. M. and E. H. Starling. 1902. The mechanism of pancreatic secretion. *J. Physiol.* **28**:325.

Blow, D. M., J. J. Birktoft and B. S. Hartley. 1969. Role of a buried acid group in the mechanism of action of chymotrypsin. *Nature* **221**:337.

Boyer, P. D., ed. 1971. *The Enzymes,* 3rd ed., vol. 3. Academic Press, New York.

Bradshaw, R. A., L. H. Ericsson, K. A. Walsh and H. Neurath. 1969. The amino acid sequence of bovine carboxypeptidase A. *Proc. Nat. Acad. Sci.* **63**:1389.

Burstein, Y., K. A. Walsh and H. Neurath. 1974. Evidence of an essential histidine residue in thermolysin. *Biochemistry* **13**:205.

Colman, P. M., J. N. Jansonius and B. W. Matthews. 1972. The structure of thermolysin: An electron density map at 2.3 Å resolution. *J. Mol. Biol.* **70**:701.

Dickerson, R. E. 1971. The structure of cytochrome *c* and the rates of molecular evolution. *J. Mol. Evol.* **1**:26.

Drenth, J., J. N. Jansonius, R. Koekoek and B. G. Wolthers. 1971. Papain, X-ray structure. In *The Enzymes* (ed. P. Boyer), 3rd ed., vol. 3, p. 484. Academic Press, New York.

Foltmann, B. 1970. The N-terminal and C-terminal amino acid sequence of calf rennin. *Phil. Trans. Roy. Soc. London* B **257**:147.

Harris, C. I., A. Kurosky, L. Rao and T. Hofmann. 1972. Amino acid sequences in penicillo-pepsin: Evidence for homology with porcine pepsin and chymosin. *Biochem. J.* **127**:34P.

Hartley, B. S. 1960. Proteolytic enzymes. *Annu. Rev. Biochem.* **29**:45.

———. 1970. Homologies in serine proteinases. *Phil. Trans. Roy. Soc. London* B **257**:77.

Hartley, B. S. and D. M. Shotton. 1971. Pancreatic elastase. In *The Enzymes* (ed. P. Boyer), 3rd ed., vol. 3, p. 323. Academic Press, New York.

Hayashi, R., S. Moore and W. H. Stein. 1973. Carboxypeptidase from yeast. Large scale preparation and the application to COOH-terminal analysis of peptides and proteins. *J. Biol. Chem.* **248**:2296.

Hunkapillar, M. W., S. H. Smallcombe, D. R. Whitaker and J. H. Richards. 1973. Carbon nuclear magnetic resonance studies of the histidine residue in α-lytic protease. Implications for the catalytic mechanism of serine proteases. *Biochemistry* **11**:4293.

Kuhn, R. W. 1973. Isolation and characterization of an acid carboxypeptidase from yeast. Ph.D. thesis, University of Washington, Seattle.

Lipscomb, W. N., J. A. Hartsuck, F. A. Quiocho and G. N. Reeke, Jr. 1969. The structure of carboxypeptidase A. IX. The X-ray diffraction results in the light of the chemical sequence. *Proc. Nat. Acad. Sci.* **64**:28.

Markland, F. S. and E. L. Smith. 1967. Subtilisin BPN'. VII. Isolation of cyanogen bromide peptides and the complete amino acid sequence. *J. Biol. Chem.* **242**:5198.

Matthews, B. W., P. B. Sigler, R. Henderson and D. M. Blow. 1967. Three-dimensional structure of tosyl-α-chymotrypsin. *Nature* **214**:652.

Matthews, B. W., J. N. Jansonius, P. M. Colman, B. P. Schoenborn and D. Dupourque. 1972. Three-dimensional structure of thermolysin. *Nature New Biol.* **238**:37.

Mycek, M. J. 1970. Cathepsins. In *Methods in Enzymology* (ed. G. E. Perlmann and L. Lorand), vol. 19, p. 285. Academic Press, New York.

Neurath, H., K. A. Walsh and W. P. Winter. 1967. Evolution of structure and function of proteases. *Science* **158**:1638.

Pangburn, M. K. 1974. Structure and function of neutral metalloendopeptidases. Ph.D. thesis, University of Washington, Seattle.

Perutz, M. 1960. Structure of hemoglobin. *Brookhaven Symp. Biol.* **13**:165.

Reeck, G. R., K. A. Walsh, M. A. Hermodson and H. Neurath. 1971. New forms of carboxypeptidase B and their homologous relationships to carboxypeptidase A. *Proc. Nat. Acad. Sci.* **68**:1226.

Robertus, J. D., J. Kraut, R. A. Alden and J. J. Birktoft. 1972. Subtilisin: A stereochemical mechanism involving transition state stabilization. *Biochemistry* **11**:4293.

Schmid, M., E. E. Lattman and J. R. Herriott. 1974. The structure of bovine carboxypeptidase B: Results at 5.5 Å resolution. *J. Mol. Biol.* **84**:97.

Segal, D. M., J. C. Powers, G. H. Cohen, D. R. Davies and P. E. Wilcox. 1971. Substrate binding site in bovine chymotrypsin A_γ. A crystallographic study using peptide chloromethylketones as site-specific inhibitors. *Biochemistry* **10**:3728.

Shotton, D. M. and H. C. Watson. 1970. Three-dimensional structure of tosyl-elastase. *Nature* **225**:811.

Stroud, R. M., L. M. Kay and R. E. Dickerson. 1971. The crystal and molecular structure of DIP-inhibited bovine trypsin at 2.7 Å resolution. *Cold Spring Harbor Symp. Quant. Biol.* **36**:125.

Tang, J., P. Sepulveda, J. Marciniszyn, Jr., K. C. S. Chen, W-Y. Huang, N. Tao, D. Liu and J. P. Lanier. 1973. Amino acid sequence of porcine pepsin. *Proc. Nat. Acad. Sci.* **70**:3437.

Titani, K., M. A. Hermodson, L. H. Ericsson, K. A. Walsh and H. Neurath. 1972a. Amino acid sequence of thermolysin. *Nature New Biol.* **238**:35.

Titani, K., M. A. Hermodson, K. Fujikawa, L. H. Ericsson, K. A. Walsh, H. Neurath and E. W. Davie. 1972b. Bovine factor X_{1a} (activated Stuart factor). Evidence of homology with mammalian serine proteases. *Biochemistry* **11**:4899.

Wright, C. S., R. A. Alden and J. Kraut. 1969. Structure of subtilisin BPN′ at 2.5 Å resolution. *Nature* **221**:235.

Structure-Function Relationships in the Serine Proteases

Robert M. Stroud, Monty Krieger, Roger E. Koeppe II, Anthony A. Kossiakoff and John L. Chambers

Norman W. Church Laboratory of Chemical Biology
California Institute of Technology, Pasadena, California 91109

Of the many ways available to control the biological activity of proteins, e.g., induction or repression of their synthesis at the translational (Jacob and Monod 1961) or transcriptional levels (Tomkins et al. 1969), specific modification or destruction are the most direct. Many biological systems are controlled by methods such as these, and the serine protease family of enzymes plays a major role in many of these systems (Stroud 1974). The pancreatic serine proteases are digestive enzymes which show optimal activity around the neutral pH region. Their function in hydrolyzing peptide bonds and the systems of physiological control over their activity have close homology in many other biological processes, e.g., blood clotting (Owren and Stormorken 1973; Magnusson 1971), bacterial sporulation (Leighton et al. 1973), fertilization (Stambaugh, Brackett and Mastroianni 1969), etc. Many of the enzymes of biological control have been recognized as serine proteases, which in nearly all documented cases have amino acid sequence homology to the pancreatic serine proteases. It is therefore to be expected that these enzymes will have closely homologous tertiary structures and will share the same catalytic mechanism of action. The mechanisms by which such enzymes are activated or inhibited will also share many common features with the digestive serine proteases. In many cases the degrees to which these principles can be extended may be predicted by recognition of the chemical and structural features of the pancreatic serine proteases which appear to define their properties. In this article we will discuss recent advances in the understanding of aspects of the structures and functions of the mammalian serine proteases.

THE PANCREATIC DIGESTIVE ENZYMES: TRYPSIN, CHYMOTRYPSIN, ELASTASE

Intrinsic to the process of digestion in mammals is the breakdown of dietary protein by the pancreatic serine proteases. These pancreatic digestive enzymes are among the most thoroughly studied of all enzymes, principally because they are extracellular enzymes that are easily separated and purified in large quantities (Kunitz and Northrop 1935). They originate in the pancreas as inactive precursors, or proenzymes, which are secreted into the duodenum. There they are activated (Kunitz and Northrop 1936; Northrop, Kunitz and Herriot 1948; Maroux, Baratti and Desnuelle 1971) by the cleavage of one critical peptide bond near the amino-terminal end of the polypeptide chain (Davie and Neurath 1955). This cleavage in turn permits a conformational change (Neurath and Dixon 1957; Sigler et al. 1968) which converts the pro-enzymes to active enzymes. Once activated, these enzymes catalyze the break-down of proteins, first into fragments and ultimately into individual amino acids.

Kinetic studies on a variety of amide and ester substrates have shown that the mechanism of serine protease catalysis (Eq. 1) involves a number of intermediates (Zerner, Bond and Bender 1964; Oppenheimer, Labouesse and Hess 1966; Caplow 1969; Hess et al. 1970; Fersht and Requena 1971a; Fastrez and Fersht 1973a,b; Fersht and Renard 1974):

$$E + S \rightleftharpoons ES \rightleftharpoons ES^T \rightleftharpoons EP_2^a \rightleftharpoons EP_2^T \rightleftharpoons EP_2 \rightleftharpoons E + P_2 . \tag{1}$$
$$+$$
$$P_1$$

Here, E represents free enzyme; S, the substrate; ES and EP_2, enzyme sub-strate and product complexes; ES^T and EP_2^T, tetrahedral intermediates; EP_2^a, an acyl enzyme; P_1, the amino or alcohol portion of the product; and P_2, the carboxylic acid portion of the product. For amides, the rate-determining step is generally acylation, $E + S \rightarrow EP_2^a$, whereas deacylation, $EP_2^a \rightarrow E + P_2$, is usually rate-determining for esters (Zerner and Bender 1964). The character-istic differences between each of the digestive serine proteases—trypsin, chymotrypsin and elastase—lie in their specificity for hydrolyzing the peptide bonds between different amino acids in the protein substrate. Trypsin, the most sharply specific of the digestive enzymes, hydrolyzes those peptide bonds that immediately follow either of the two basic amino acids, lysine or arginine. Chymotrypsin hydrolyzes peptide bonds that follow several of the amino acids with larger hydrophobic side chains, and elastase binds the small side chains of glycine, alanine or serine at the equivalent binding site (Naughton and Sanger 1961; Brown, Kauffman and Hartley 1967; Sampath Narayanan and Anwar 1969). The complete amino acid sequences and three-dimen-sional molecular structures have now been worked out for chymotrypsin (Sigler et al. 1968) and its proenzyme (Freer et al. 1970), elastase (Shotton and Watson 1970), and DIP (diisopropylphosphoryl)-trypsin (Stroud, Kay and Dickerson 1971, 1974) and the proenzyme (Kossiakoff, Kay and Stroud, unpubl.). These structures, along with that of the bacterial serine protease subtilisin (Wright, Alden and Kraut 1969; Alden, Wright and Kraut 1970), have been uniquely valuable in developing an understanding of how these enzymes bind a substrate and how they catalyze the subsequent chemical reaction.

Activating the Proenzyme

The first key to activation of the pancreatic proenzymes is enterokinase, an enzyme secreted in small amounts by the mucous membrane of the stomach. Its prime function is to convert some trypsinogen to trypsin, which then activates all of the proenzymes (including more trypsinogen) (Kunitz and Northrop 1936; Northrop, Kunitz and Herriot 1948; Maroux, Baratti and Desnuelle 1971). In each case, activation involves the cleavage of a few amino acid residues from the amino-terminal end of the proenzyme (Davie and Neurath 1955; Neurath and Dixon 1957).

With the formation of the new amino terminus at Ile-16[1] (Oppenheimer, Labouesse and Hess 1966), the protein undergoes conformational changes (Neurath, Rupley and Dreyer 1956) leading to a catalytically active configuration. A comparison of the high-resolution structure of chymotrypsinogen with that of chymotrypsin (Freer et al. 1970; Wright 1973) and of the high-resolution structure of DIP-trypsin with the recently determined high-resolution structure of trypsinogen (Kossiakoff, Kay and Stroud, unpubl.) helps us to understand the exact nature of these conformational changes. (A detailed description of the trypsinogin structure will be published later.) In both cases,

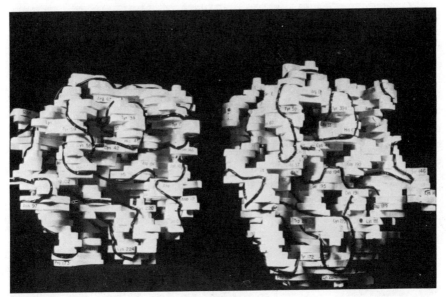

Figure 1
A comparison of the 5-Å models of trypsinogen (*left*) and DIP-trypsin (*right*) shows close structural homology in most areas of the molecule. Striking structural differences are observed in only two areas of the molecule. The first is in the binding pocket region, which is formed by residues 214–220 and 189–192, and the second is along a loop of chain containing residues 140–151 located on the right-hand side of the molecule. The small difference in size of the models is due to a difference in scale and orientation.

[1] The numbering system referred to is that of chymotrypsin, which will be adopted here as a standard for comparison of sequences.

clipping the proenzyme tail permits the new, positively charged α-amino terminus at residue 16 to fold into the interior of the globular structure and form an ion pair with the negative carboxyl group of Asp-194 (Matthews et al. 1967; Sigler et al. 1968). While this change is accompanied by movements in the region of the specificity binding pocket, there appears to be little change in the interaction between Asp-102, His-57 and Ser-195 at the catalytic site (Freer et al. 1970; see also Fig. 1). Thus the arrangement of these catalytic residues is preformed in the proenzyme. One major factor which contributes to the relative inactivity of the zymogen is that the binding of the normal substrates is impaired (Kassell and Kay 1973; Gertler, Walsh and Neurath 1974).

Enzyme Specificity and Substrate Binding

The serine proteases differ in their specificities because of differences in their substrate binding sites. Trypsin, chymotrypsin and elastase all have specific side-chain binding pockets on the surface of the protein close to the catalytic site (see Fig. 2). This pocket is lined by residues 214–220 and 189–192 and

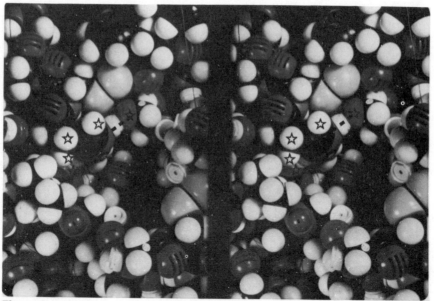

Figure 2
Stereoscopic photograph of a space-filling model showing the active site and specificity pocket in trypsin (see Stroud, Kay and Dickerson 1971, 1974). The imidazole side chain of His-57 is visible; however, the carboxylic acid side chain of Asp-102 is hidden from view by several amino acid residues. The viewing direction is approximately the same as that of Figures 4 and 7. The side-chain binding pocket is located beneath and to the right of the catalytic site.

Stars are placed near the active site as markers. They are located (reading from left to right) at the following positions: the Asp-102/His-57 hydrogen bond, the β-carbon protons of His-57, one of the ring protons on His-57, and the Ser-195 γ-hydroxyl.

defines the primary specificity toward substrate side chains immediately prior to the peptide bond which is to be cleaved. Cysteine residues 220 and 191 are linked by a disulfide bond. In trypsin, residue 189 is an aspartic acid, and its negatively charged carboxyl group ($pK_a = 4.6$) (East and Trowbridge 1968) lies at the bottom of the pocket (Stroud, Kay and Dickerson 1974). Trypsin has primary specificity for basic amino acids because their positively charged side chains bind tightly in this pocket (Mares-Guia and Shaw 1965; Ruhlmann et al. 1973; Blow, Janin and Sweet 1974; Sweet et al. 1974; Krieger, Kay and Stroud 1974). In an attempt to determine the manner in which amino acid side chains bind, we determined the structure of benzamidine trypsin. Benzamidine is a competitive, specific and reversible inhibitor of trypsin. Figure 3 shows how benzamidine, an amino acid side chain analog, binds in the specificity binding pocket (Krieger, Kay and Stroud 1974). In the case of trypsin, there is evidence that when side chains are bound in this pocket, they induce small conformational changes in the enzyme-substrate complex which help to accelerate catalysis (Inagami and Murachi 1964; Inagami and York 1968). In chymotrypsin, residue 189 is a serine (Hartley 1964). The pocket is now relatively hydrophobic and uncharged at neutral pH's, thus explaining chymotrypsin's specificity. In both trypsin and chymotrypsin, residue 216, a glycine, lies at the entrance to the binding pocket. In elastase, valine replaces glycine at position 216 (Shotton and Hartley 1970). The larger hydrophobic side chain blocks the entrance to the pocket and only allows the binding of amino acids with small side chains at the primary binding site (Shotton and Watson 1970).

Other parts of the enzyme are involved in binding other parts of the substrate molecule as well as the side chain, so that the susceptible substrate bond is aligned appropriately on the surface. Secondary specificity toward other

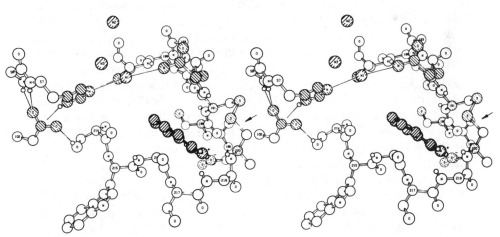

Figure 3
The figure shows the structure of the trypsin binding pocket in benzamidine-trypsin. (The phenyl amidinium is depicted by heavy shading.) This view is approximately the same as that of Figure 2. (Reprinted, with permission, from Krieger, Kay and Stroud 1974.)

side chains in the physiological substrate molecules can also be correlated with enzyme structure (Fersht, Blow and Fastrez 1973), although the role of secondary specificity in the digestive enzymes is clearly much less significant than it is in highly specific enzymes of biological control.

Ideally, one would like to study the three-dimensional structure of an enzyme-substrate complex by X-ray crystallography and in so doing, gain new insights into the mechanism of serine protease catalysis. Unfortunately, this has not yet been possible because the catalyzed reaction takes place almost immediately after the substrate is bound and the system becomes an enzyme-plus-product complex. Data for a three-dimensional structure analysis cannot generally be collected in so short a time. Fersht and Renard (1974) have pointed out, however, that it may be possible to use equilibrium methods to trap intermediates in the reaction pathway and study their structures. Until the structures of such intermediates have been determined, crystallographers will be limited to studying the binding of substrate analogs and inhibitors, which in some limited respects resemble true substrates. Nevertheless, from such studies inferences can be drawn about the structural transformations which occur during the catalyzed reaction.

Among the best analogs to true trypsin substrates are the naturally occurring trypsin inhibitors. They have evolved in parallel with the enzymes so that they bind extremely tightly to the active site. Such protein inhibitors are crucial to physiological control of the serine proteases (Tschesche 1974). For example, if pancreatic secretory trypsin inhibitors were not synthesized along with the pancreatic serine proenzymes, one prematurely activated molecule of trypsin could start an autocatalytic chain reaction which would activate the other serine proenzymes and destroy any nearby proteins. Inhibitors are present to prevent such catastrophes and to control physiological processes mediated by proteolytic enzymes. The structure of an intracellular 6500 molecular weight trypsin inhibitor (PTI) isolated from bovine pancreas and other organs was determined by R. Huber et al. (1970, 1971). Chemical modifications had already shown that Lys-15 of this inhibitor was involved in the trypsin-PTI association (Chauvet and Acher 1967; Kress and Laskowski 1967; Fritz et al. 1969). By combining models of PTI with the known structures of trypsin and chymotrypsin, substrate binding models (Fig. 4) were developed by us (Stroud, Kay and Dickerson 1971; Krieger, Kay and Stroud 1974) and independently by Huber et al. (1971) and Blow et al. (1972). High-resolution structures of the PTI-trypsin complex (Ruhlmann et al. 1973) and of a soybean trypsin inhibitor-trypsin complex (Blow, Janin and Sweet 1974; Sweet et al. 1974) have since been determined. Several hydrogen bonds and stereochemical complementarity between enzyme and inhibitor orient the susceptible bond at the active site. There is one very important difference between the predicted substrate binding models and the structures of the trypsin-inhibitor complexes; the inhibitor and trypsin are covalently bound together via an oxygen-carbon bond between the hydroxyl of Ser-195 and the carbonyl group of Lys-15 in the trypsin-PTI complex and between the serine hydroxyl and the carbonyl carbon of Arg-63 in the trypsin-STI (soybean trypsin inhibitor) complex. These complexes have been shown to exist as tetrahedral adducts which probably resemble normal intermediates (ES^T) in serine protease catalysis.

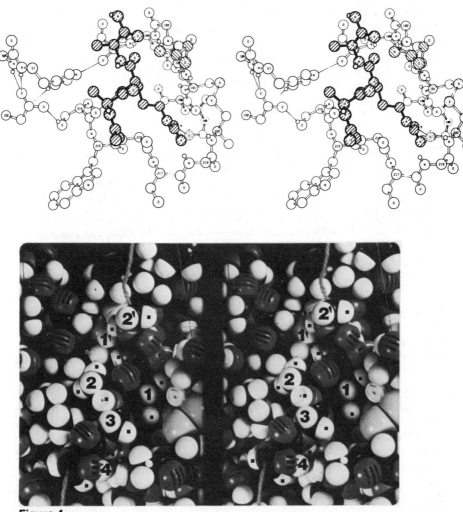

Figure 4

(*Top*) A model for the binding of a portion of the bovine pancreatic trypsin inhibitor (heavy shading) to trypsin in which the side chain of the inhibitor's Lys-15 has been replaced by an arginine side chain. The model was constructed so that the "Arg-15" side chain fitted the electron density for benazmidine in benzamidine-trypsin. This model for enzyme-substrate interaction embodies a substrate conformation that evolved to bind tightly to the enzyme and an enzyme conformation which is presumably like that induced by the binding of specific substrate side chains. (Reprinted, with permission, from Krieger, Kay and Stroud 1974.)

(*Bottom*) Stereoscopic view of a space-filling model of the trypsin-substrate complex described in the text. This view (approximately as above) should be compared with the same region of the enzyme alone shown in Figure 2. This figure beautifully demonstrates the intimate stereochemical compatibility between enzyme and substrate. The woolen threads indicate the end points of the portion of the oligopeptide substrate chain which is included in the model. Protons in the substrate are labeled with square dots, and atoms attached to the α-carbon atoms of the substrate fragment are labeled 4-3-2-1-1'-2' (along the sequence in the substrate). The peptide bond between 1 and 1' is the one to be hydrolyzed.

These and other crystallographic and chemical studies have produced detailed models for the association of enzyme and substrate prior to catalysis. Although based on inferences drawn from substrate analogs or inhibitor binding studies, such models do suggest ways that chemical groups on the enzyme may participate in catalysis.

The Active Center: pH-Activity Correlations

The X-ray crystal structures of the serine proteases have shown that their active sites are almost identical. The catalytic site of all serine proteases is characterized by a serine hydroxyl group (residue 195). Diisopropylfluorophosphate (DFP) (Jansen, Nutting and Balls 1949; Hartley 1960) and phenyl methane sulfonyl fluoride (PMSF) (Fahrney and Gold 1963; Kallos and Rizok 1964) react with this hydroxyl and irreversibly inhibit serine proteases, regardless of their substrate specificity. In the free enzyme, this hydroxyl is hydrogen bonded to the N-ϵ_2 of His-57. N-δ_1 of His-57 is hydrogen bonded to the carboxyl group of Asp-102, which in turn is not directly accessible to solvent (Blow, Birktoft and Hartley 1969; Wright, Alden and Kraut 1969; Alden, Wright and Kraut 1970; Birktoft and Blow 1972; Stroud, Kay and Dickerson 1974). The direct participation of these three groups in catalysis has been established, and chemical modification of any of them can greatly diminish or abolish catalysis (Jansen, Nutting and Balls 1949; Hartley 1960; Fahrney and Gold 1963; Kallos and Rizok 1964; Ong, Shaw and Schoellmann 1964; Shaw, Mares-Guia and Cohen 1965; Henderson 1971; Martinek, Savin and Berezin 1971; Chambers et al. 1974).

The pH-activity profiles for hydrolysis of peptides, amides or esters by trypsin or chymotrypsin are bell shaped and reflect maximal enzymatic activity at about pH 8. The high pH limb of the curve depends on an apparent pK_a of 8.8 for α-chymotrypsin (Fersht and Requena 1971b) or 10.1 for trypsin (Spomer and Wootton 1971). Fersht and Requena have demonstrated that this ionization, which controls enzyme conformation and substrate binding (K_m), is directly associated with titration of the α-amino terminus of Ile-16. The internal salt bridge formed between this amino group and the side chain of Asp-194 in the active enzyme is broken at high pH, where deprotonation of the amino group was shown to favor an alternate conformation for the enzyme.

The low pH limb of the profile depends on a single group of pK_a about 6.7 in both enzymes; protonation of this group adversely affects both acylation and deacylation (Bender and Kezdy 1964). It has often been assumed that this group corresponds to His-57. Jencks (1969), however, has pointed out that this group need not be His-57, but might be some other group on the enzyme either controlling conformation or effecting a change in rate-determining step near this pH. The aspartic-histidine-serine system *as a whole* has been shown to take up a single proton as the pH is lowered below 7.0. Fersht and Renard (1974) have also demonstrated that for the hydrolysis of acetyl phenylalanine-p-nitrophenyl ester by δ-chymotrypsin, k_{cat} or k_{cat}/K_m depends on a single ionization between pH 9.0 and pH 2.0. Thus it would seem that only one group at the active center has a pK_a in the range pH 2–9 which can be detected kinetically.

Richards and his colleagues (Hunkapiller et al. 1973) have shown by nuclear magnetic resonance studies that the imidazole of His-57 in α-lytic protease (a bacterial homolog of the pancreatic serine proteases) does not ionize until the pH is lowered below 4.0. Their research led them to propose that the group of pK_a 7.0 was Asp-102.

In order to study the ionization of Asp-102 in trypsin (Koeppe and Stroud, unpubl.), we monitored infrared absorbance arising from the carboxyl C=O stretch at 1570 cm^{-1} and 1710 cm^{-1} (Timasheff and Rupley 1972) as a function of pH. To diminish the number of titratable carboxyls, accessible carboxyl groups in trypsin were modified with semicarbazide (Fersht and Sperling 1973). The modified enzyme was found to contain 2.5 molar equivalents of free carboxyl groups. These were identified as Asp-102, Asp-194 (1.0 molar equivalent each) and the α-carboxyl terminus of Asn-245 (0.5 equivalent).

The spectrum shown in Figure 5 indicates differential absorbance at 1600 cm^{-1} and at 1680 cm^{-1} between semicarbazide-trypsin solutions of pD 6.50 and 7.13. Each peak is shifted toward the other by about 30 cm^{-1} from the value found for other carboxyls—a result which is to be expected for hydrogen-bonded carboxyls (Susi 1972) such as Asp-102. The titration curve of semicarbazide-trypsin compiled from a series of infrared difference spectra is shown in Figure 6. Based on the number of free carboxyl groups, we assume that the low pH titration of average pK_a 2.9 corresponds to titration of 1.5 carboxyl groups, while the titration of average pK_a 6.8 corresponds to one carboxyl. The gradient of the low pH titration is approximately 1.5 times that of the upper one, which is consistent with the assumption. Both titrations, however, appear sharper than expected for single or noninteracting groups.

Binding of Cu^{++} ions displaces the upper limb of the titration downward from pH 6.8. Martinek et al. (1969, 1971) have shown that Ag$^+$ ions are powerful competitive inhibitors of trypsin, and that Cu^{++} and Ag$^+$ compete with each other in inhibiting chymotrypsin. Either species competes with pro-

Figure 5

Infrared difference spectrum for semicarbazide-trypsin. The path length was 0.150 mm. The sample cell was at pD 7.13, and the reference cell at pD 6.50. pD values in all cases correspond to uncorrected pH meter readings. Concentrations were 1.5 mM enzyme, 6 mM NaNO$_3$, and 12 mM benzamidine. The peak positions at 1680 cm^{-1} for C═O in COOD and at 1600 cm^{-1} for C—O in COO$^-$ are closer together than for other trypsin carboxyls due to the effect of hydrogen bonding. The detected difference in species concentration judged from peak heights corresponds to about 0.4 carboxyl equivalent.

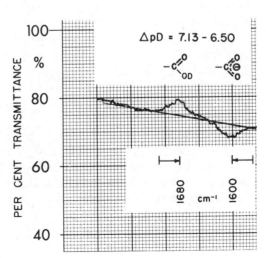

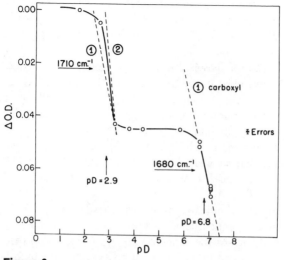

Figure 6

Titration curve for free carboxyls in semicarbazide-trypsin obtained by plotting differential absorbance at 1710 cm^{-1} and 1680 cm^{-1} from a series of infrared difference spectra. Reference solutions were at pD 3.2, 3.8, 4.4 and 6.6. The gradient of the titration at low pD was 1.5–2.0 times greater than the gradient for the group of pKa 6.8 and corresponds to titration of about 1.5 carboxyls.

tons for a site on the enzyme with $pK_{app} \simeq 7.0$. Since we have shown that Ag$^+$ binds specifically between Asp-102 and His-57 in the orthorhombic crystal form of DIP-trypsin (Chambers et al. 1974 and Fig. 7), in trypsinogen and in trigonal DIP-trypsin (Kossiakoff, Kay and Stroud, unpubl.), the pK_a of 6.8 has tentatively been assigned to the carboxyl of Asp-102. (A detailed description of the experimental procedures and results will be published elsewhere.) Copper ion binding shows that Asp-194 and Asn-245 cannot be responsible for this pK_a. These data suggest that the average apparent pK_a of Asp-194 and Asn-245 is 2.9.

Control experiments eliminate the possibility that imidazole-stretching frequencies could account for the infrared bands at 1680 cm^{-1} and 1600 cm^{-1} in Figure 5. However, imidazole titration may perturb a neighboring carbonyl and thereby conceivably be responsible for the infrared difference peaks observed around pD 6.8. This possibility is the subject of continuing investigations in our laboratory.

There are two arguments against this possibility and in support of the assignment of a pK_a of about 7.0 to Asp-102. First, using an average extinction coefficient for the other seven carboxyl groups in β-trypsin, derived from infrared difference spectra of the unmodified enzyme, the low pH limb of the titration shown in Figure 6 corresponds to 1.5 carboxyl equivalents. This implies that the remaining carboxyl group must titrate outside of the range pH 2–pH 5. Second, in the presence of Cu^{++} ions, there is no differential

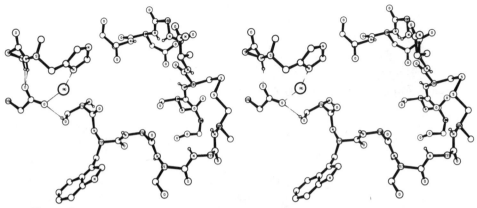

Figure 7
A stereo ORTEP drawing of the catalytic site of silver DIP-trypsin, shown in the same orientation as Figure 1. The DIP group has been omitted for clarity. The γ-oxygen of Ser-195 is in the position found for DIP-trypsin, close to that found in tetrahedral intermediates or in acyl enzymes (see text). (Reprinted, with permission, from Chambers et al. 1974.)

absorbance in the region of pH 7.0, which suggests that either of the two imidazoles at His-40 or His-91 which may titrate around pH 7.0 do not induce changes in their neighboring carbonyl groups which would be detected by the technique.

The Mechanism of Hydrolysis by Serine Proteases

Following their determination of the active-site structure of chymotrypsin, Blow, Birktoft and Hartley (1969) first proposed that proton transfers between His-57 and Asp-102 were important in catalysis. The studies of the microscopic pK_a's of His-57 and Asp-102 referred to in the previous section are consistent with this proposal. From their studies of the histidine ionization in α-lytic protease, Hunkapiller et al. (1973) explained the sequence of proton transfers between Asp-102 and His-57, discussed here and included in Figure 9 (below), in terms of pK_a's.

In this discussion we assume (see previous section) that the pK_a of Asp-102 is 6.8, and that the imidazole of His-57 is essentially neutral above pH 4.0. This leads to the ionization of the active center around pH 7.0 (Fig. 8). The mechanistic importance of these assignments is that the aspartate ion of residue 102 can act as a chemical base which can readily accept a proton from the histidine side chain during catalysis (Hunkapiller et al. 1973). Together, Asp-102 and His-57 shuttle protons back and forth from enzyme to substrate, and so the mechanism can best be described as nucleophilic attack with general base catalysis by His-57 (Bender and Kezdy 1964; Inward and Jencks 1965) and Asp-102. The important differences between this reaction and a nonenzymatic hydrolysis are the binding to the enzyme and the efficient proton shuttle.

Interfering with this shuttle inhibits catalysis. For example, by methylating the N-ϵ_2 of His-57 in chymotrypsin, the shuttle can no longer operate normally

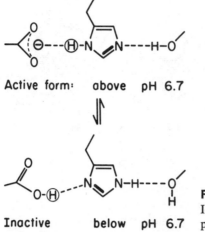

Active form: above   pH 6.7

Inactive   below   pH 6.7

Figure 8
Ionization of the active center in the range pH 4.0–8.0 as discussed in the text.

and the rate of catalysis drops by a factor of 5000 to 200,000 for specific substrates (Henderson 1971). Silver ions bind specifically to trypsin between Asp-102 and His-57 (Chambers et al. 1974). By adding silver, the shuttle is blocked and catalysis is inhibited (Martinek, Savin and Berezin 1971).

The mechanistic scheme shown in Figure 9 is consistent with most experimental data relating to hydrolysis of peptides, esters or amides by trypsin or chymotrypsin. In the first step (I), substrate and enzyme form a Michaelis complex. Nucleophilic attack by the hydroxyl group of Ser-195 follows. As the reaction proceeds, the hydroxyl twists around the C_α—C_β bond and forms a covalent chemical bond to the substrate carbon at step I–II (Steitz, Henderson and Blow 1969). Concerted with this, a proton is transferred from the serine hydroxyl group to the N-ϵ_2 of His-57. From there, it is eventually delivered to the nitrogen of the peptide bond in the substrate. As a result of this proton transfer, the proton previously bound to the N-δ_1 of His-57 is transferred to the carboxyl group of Asp-102, which acts as a base in this reaction (Hunkapiller et al. 1973). Although these proton transfers are rapid, deuterium isotope effects show that proton transfer is involved in the rate-determining step of the catalysis (Bender et al. 1964; Pollock, Hogg and Schowen 1973).

Whether the Asp-His-Ser proton shuttle is concerted or stepwise remains in question. If the mechanism is concerted, the negative charge of Asp-102 would be neutralized while negative charge develops on the carbonyl oxygen of the substrate. The imidazole would remain neutral throughout the reaction; thus unstable intermediates due to charge separation would be avoided (Jencks 1971; Hunkapiller et al. 1973). In fact, charge development in the transition state in chymotrypsin catalysis does appear to be small (Jencks 1971). The shuttle may be stepwise if the energy requirements of charge separation (negative charges on the substrate and Asp-102 and a positive charge on His-57) are offset by a more favorable entropy of activation in a two-step process (Jencks 1972).

One might favor the concerted mechanism because it might be expected that the precise alignment of the shuttle, which has been observed in all serine protease structures, evolved so that the entropic advantage of the

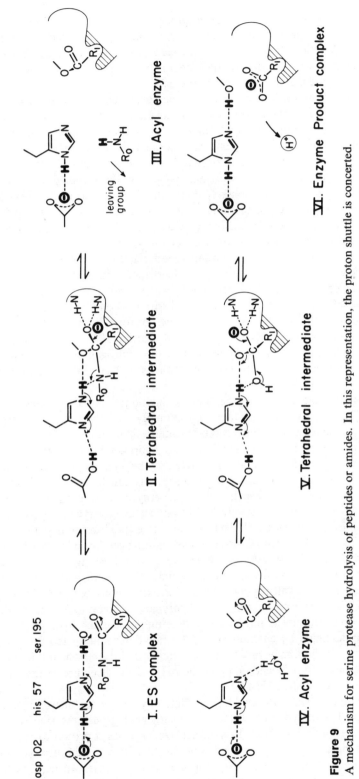

Figure 9

A mechanism for serine protease hydrolysis of peptides or amides. In this representation, the proton shuttle is concerted.

25

two-step process over the concerted process was minimized. Thus the enzyme could exploit for increased reaction rate the energy saved in eliminating charge separation. If this were not the case, it would seem unnecessary to use both an Asp and an His for the general base catalysis. The Asp could be eliminated and the His could act as the base.

After the attack by the serine on the substrate, a short-lived tetrahedral intermediate is formed (II). This intermediate is stabilized by the covalent bond to the enzyme and by a number of hydrogen bonds. The following structural features of the tetrahedral intermediate are primarily based on the crystallographic determination of many different protease-inhibitor structures.

The negatively charged substrate oxygen in the tetrahedral intermediate is stabilized by hydrogen bonds from the amide nitrogens of residues 195 and 193. The importance of the amides of Ser-195 and Gly-193 was first noted by Steitz, Henderson and Blow (1969), and their role in transition-state stabilization was postulated by Henderson (1970) and by Robertus et al. (1972). Another hydrogen bond forms between the carbonyl group of Ser-214 and the α-N of the substrate (Steitz, Henderson and Blow 1969; Segal et al. 1971). Comparison of the kinetics of hydrolysis of specific trypsin and chymo-trypsin substrates with and without the hydrogen bonding capacity of the α-N suggests that the Ser-214—α-N bond may not form in the Michaelis complex (Ingles and Knowles 1968; Caplow and Harper 1972; Kobayashi and Ishi 1974). These results show, however, that this hydrogen bond does play a role in the transition states between intermediates and possibly in the tetrahedral and acyl enzyme intermediates.

One explanation for the exceptional catalytic powers of enzymes is that enzymes have evolved so that they can optimally bind the transition-state structures in the reactions they catalyze rather than the substrates themselves (Pauling 1946; Wolfenden 1972). The hydrogen-bonded structure in the serine protease-substrate transition state may be an example of transition-state stabilization, for the oriented hydrogen bonds can help to speed up the reaction by smoothing down the highest energy barriers between intermediate states. The stability of the tetrahedral adduct in the trypsin-trypsin inhibitor complexes is consistent with the transition-state stabilization hypothesis.

At step II–III, the now unstable carbon-nitrogen bond is broken, and the first product of hydrolysis, an amine, is free to diffuse away, taking with it a proton from the enzyme. At the same time, the bound part of the substrate rearranges to a covalently modified acyl enzyme intermediate (III). At pH 8, N^{14}/N^{15} kinetic isotope effects (O'Leary and Kluetz 1972) show that the C—N bond rupture is partially rate-determining for the hydrolysis of acetyl tryptophanamide by chymotrypsin. The rate-determining step for amide hydrolysis, however, may vary from the formation of the tetrahedral intermediate to its breakdown, depending upon the pH and the structure of the substrates (Fastrez and Fersht 1973a).

The breakdown of the acyl intermediate (IV–VI) is the microscopic reverse of steps I–III; this time water is the attacking group. At step V–VI, the second product is formed. It is an acid which loses a proton to the solution and becomes negatively charged. For the first time (if the proton shuttle is concerted), there are two charges in the system. These two negative charges repel each other and so help to dissociate the second product from the enzyme (Johnson and Knowles 1966), regenerating free enzyme.

The presence of a carboxyl group of high pK_a and a neutral side chain of His-57 with a low pK_a would suggest two compelling evolutionary reasons why the Asp-His-Ser arrangement should be universal to serine proteases (Hunkapiller et al. 1973). First, by neutralizing a negative charge on Asp-102, rather than generating a positive charge on His-57, during formation of the tetrahedral intermediate, there would be no unfavorable charge separation. This would contribute to reducing transition-state internal energies, and thus to rate enhancement (Jencks 1971). Second, if the charged Asp-102 is to be a proton acceptor at physiological pH values, its pK_a must be raised and it must have access to the proton donor. The imidazole of His-57 is ideally suited both to insulate Asp-102 from solvent (so raising the pK_a of the buried carboxyl group) and to serve as a proton conductor, transferring charge from the carboxyl group to the substrate. It is also important to note that both the reverse separation of the pK_a values of Asp-102 and His-57 and the struc-ture of trypsin at pH 7 and pH 8 (Stroud, Kay and Dickerson 1974; Krieger, Kay and Stroud 1974; Huber et al. 1974; Sweet et al. 1974), which shows a symmetric interaction between the charge on Asp-102 and His-57 (see Fig. 3), are unlike the situation expected in aqueous solution and reflect a unique microenvironment for these groups.

As far as we know, all the serine proteases use the same three chemical groups to hydrolyze peptide bonds. They, like trypsin, are active catalysts only when the aspartic acid is negatively charged. Against the active site, the reaction goes on in an unique way as the enzyme smoothes the transition from one intermediate state to another. This emphasizes the importance of exact stereochemical fit and correct orientation (Koshland 1958) of the substrate as the reaction takes place, rather than simply the generation of an especially reactive site. After all, a very reactive site could react in many less specific ways. It is better to have a moderately efficient catalytic site coupled with a very selective binding requirement (Fersht and Sperling 1973). With this in mind, the subtle differences between serine proteases involved in biological control can be understood in terms of differences in their specific substrate binding properties.

Acknowledgments

This is contribution number 4990 from the Norman W. Church Laboratory of Chemical Biology. We would like to acknowledge our indebtedness to the National Institutes of Health for their support by means of grants GM-19984 and GM-70469. R.M.S. is the recipient of an NIH Career Development Award, M.K. is the recipient of a Danforth Foundation Fellowship, R.K. and J.C. are recipients of NIH Predoctoral Traineeships, and A.K. is the recipient of an NIH Postdoctoral Fellowship.

REFERENCES

Alden, R. A., C. S. Wright and J. Kraut. 1970. A hydrogen-bond network at the active site of subtilisin BPN'. *Phil. Trans. Roy. Soc. London* B **257**:119.

Bender, M. L. and F. J. Kezdy. 1964. The current state of the α-chymotrypsin mechanism. *J. Amer. Chem. Soc.* **86**:3704.

Bender, M. L., G. E. Clement, F. J. Kezdy and H. D'A. Heck. 1964. The correlation of the pH (pD) dependence of the stepwise mechanism of α-chymotrypsin-catalyzed reactions. *J. Amer. Chem. Soc.* **86:**3680.

Birktoft, J. J. and D. M. Blow. 1972. Structure of crystalline α-chymotrypsin. V. The atomic structure of tosyl-α-chymotrypsin at 2 Å resolution. *J. Mol. Biol.* **68:**187.

Blow, D. M., J. J. Birktoft and B. S. Hartley. 1969. Role of a buried acid group in the mechanism of action of chymotrypsin. *Nature* **221:**337.

Blow, D. M., J. Janin and R. M. Sweet. 1974. Mode of action of soybean trypsin inhibitor (Kunitz) as a model for specific protein-protein interactions. *Nature* **249:**54.

Blow, D. M., C. S. Wright, D. Kukla, A. Ruhlmann, W. Steigemann and R. Huber. 1972. A model for the association of bovine pancreatic trypsin inhibitor with chymotrypsin and trypsin. *J. Mol. Biol.* **69:**137.

Brown, J. R., D. L. Kauffman and B. S. Hartley. 1967. The primary structure of porcine pancreatic elastase: The N-terminus and disulphide bridges. *Biochem. J.* **103:**497.

Caplow, M. 1969. Chymotrypsin catalysis. Evidence for a new intermediate. *J. Amer. Chem. Soc.* **91:**3639.

Caplow, M. and C. Harper. 1972. Discrete effects of the acylamino proton in a chymotrypsin substrate on different processes in catalysis. *J. Amer. Chem. Soc.* **94:**6508.

Chambers, J. L., G. G. Christoph, M. Krieger, L. Kay and R. M. Stroud. 1974. Silver ion inhibition of serine proteases: Crystallographic study of silver trypsin. *Biochem. Biophys. Res. Comm.* **59:**70.

Chauvet, J. and R. Acher. 1967. The reactive site of the basic trypsin inhibitor of pancreas. *J. Biol. Chem.* **242:**4274.

Davie, E. W. and H. Neurath. 1955. Identification of a peptide released during autocatalytic activation of trypsinogen. *J. Biol. Chem.* **212:**515.

East, E. J. and C. G. Trowbridge. 1968. Binding of benzamidine and protons to trypsin as measured by difference spectra. *Arch. Biochem. Biophys.* **125:**334.

Fahrney, D. E. and A. M. Gold. 1963. Sulfonyl fluorides as inhibitors of esterases. *J. Amer. Chem. Soc.* **85:**997.

Fastrez, J. and A. R. Fersht. 1973a. Mechanism of chymotrypsin. Structure, reactivity, and nonproductive binding relationships. *Biochemistry* **12:**1067.

———. 1973b. Demonstration of the acyl-enzyme mechanism for the hydrolysis of peptides and anilides by chymotrypsin. *Biochemistry* **12:**2025.

Fersht, A. R. and M. Renard. 1974. pH dependence of chymotrypsin catalysis. *Biochemistry* **13:**1416.

Fersht, A. R. and Y. Requena. 1971a. Mechanism of the α-chymotrypsin hydrolysis of amides. pH dependence of k_c and K_m. Kinetic detection of an intermediate. *J. Amer. Chem. Soc.* **93:**7079.

———. 1971b. Equilibrium and rate constants for the interconversion of two conformations of α-chymotrypsin. *J. Mol. Biol.* **60:**279.

Fersht, A. R. and J. Sperling. 1973. The charge relay system in chymotrypsin and chymotrypsinogen. *J. Mol. Biol.* **74:**137.

Fersht, A. R., D. M. Blow and J. Fastrez. 1973. Leaving group specificity in the chymotrypsin catalyzed hydrolysis of peptides. A stereochemical interpretation. *Biochemistry* **12:**2035.

Freer, S. T., J. Kraut, J. D. Robertus, H. T. Wright and Ng.- H. Xuong. 1970. Chymotrypsinogen: 2.5 Å crystal structure, comparison with α-chymotrypsin, and implications for zymogen activation. *Biochemistry* **9:**1997.

Fritz, H., H. Schult, R. Meister and E. Werlo. 1969. Herstellung und eigenschaften von derivaten des Trypsin-Kallikrein-Inhibitors aus Rinderorganen. *Z. Physiol. Chem.* **350**:1531.

Gertler, A., K. A. Walsh and H. Neurath. 1974. Catalysis by chymotrypsinogen. Demonstration of an acyl-zymogen intermediate. *Biochemistry* **13**:1302.

Hartley, B. S. 1960. Proteolytic enzymes. *Annu. Rev. Biochem.* **29**:45.

———. 1964. Amino acid sequence of bovine chymotrypsinogen A. *Nature* **201**:1284.

Henderson, R. 1970. Structure of crystalline α-chymotrypsin. IV. Structure of indoleacryloyl-α-chymotrypsin and its relevance to the hydrolytic mechanism of the enzyme. *J. Mol. Biol.* **54**:341.

———. 1971. Catalytic activity of α-chymotrypsin in which histidine-57 has been methylated. *Biochem. J.* **124**:13.

Hess, G. P., J. McConn, E. Ku and G. McConkey. 1970. Studies of the activity of chymotrypsin. *Phil. Trans. Roy. Soc. London* B **257**:89.

Huber, R., D. Kukla, A. Ruhlmann and W. Steigemann. 1971. Pancreatic trypsin inhibitor: Structure and function. *Cold Spring Harbor Symp. Quant. Biol.* **36**:141.

Huber, R., D. Kukla, A. Ruhlmann, O. Epp and H. Formanek. 1970. The basic trypsin inhibitor of bovine pancreas. I. Structure analysis and conformation of the polypeptide chain. *Naturwissenschaften* **57**:389.

Huber, R., D. Kukla, W. Steigemann, J. Deisenhofer and A. Jones. 1974. Structure of the complex formed by bovine trypsin and bovine pancreatic trypsin inhibitor. In *Bayer Symposium V: Proteinase Inhibitors* (ed. H. Fritz et al.), p. 497. Springer Verlag, Heidelberg.

Hunkapiller, M. W., S. H. Smallcombe, D. R. Whitaker and J. H. Richards. 1973. Carbon nuclear magnetic resonance studies of the histidine residue in α-lytic protease. *Biochemistry* **12**:4732.

Inagami, T. and T. Murachi. 1964. The mechanism of the specificity of trypsin catalysis. *J. Biol. Chem.* **239**:1395.

Inagami, T. and S. S. York. 1968. The effect of alkylguanidines and alkylamines on trypsin catalysis. *Biochemistry* **7**:4045.

Ingles, D. W. and J. R. Knowles. 1968. The stereospecificity of α-chymotrypsin. *Biochem. J.* **108**:561.

Inward, P. W. and W. P. Jencks. 1965. The reactivity of nucleophilic reagents with furoyl-chymotrypsin. *J. Biol. Chem.* **240**:1986.

Jacob, F. and J. Monod. 1961. Genetic regulatory mechanisms in the synthesis of proteins. *J. Mol. Biol.* **3**:318.

Jansen, E. F., M. D. F. Nutting and A. K. Balls. 1949. Mode of inhibition of chymotrypsin by diisopropylfluorophosphate. I. Introduction of phosphorous. *J. Biol. Chem.* **179**:201.

Jencks, W. P. 1969. *Catalysis in Chemistry and Enzymology*, p. 218. McGraw-Hill, New York.

———. 1971. Reactivity correlations and general acid-base catalysis in enzymic transacylation reactions. *Cold Spring Harbor Symp. Quant. Biol.* **36**:1.

———. 1972. General acid-base catalysis of complex reactions in water. *Chem. Rev.* **72**:705.

Johnson, C. H. and J. R. Knowles. 1966. The binding of inhibitors to α-chymotrypsin. *Biochem. J.* **101**:56.

Kallos, J. and D. Rizok. 1964. Heavy atom labeling of the serine of the active center of chymotrypsin: Pipsyl-chymotrypsin. *J. Mol. Biol.* **9**:255.

Kassell, B. and J. Kay. 1973. Zymogens of proteolytic enzymes. *Science* **180**:1022.

Kobayashi, R. and S. Ishi. 1974. The trypsin-catalyzed hydrolysis of some L-α-amino-lacking substrates. *J. Biochem.* **75**:825.

Koshland, D. E., Jr. 1958. Application of a theory of enzyme specificity to protein synthesis. *Proc. Nat. Acad. Sci.* **44**:98.

Kress, L. F. and M. Laskowski. 1967. The basic trypsin inhibitor of bovine pancreas. *J. Biol. Chem.* **242**:4925.

Krieger, M., L. M. Kay and R. M. Stroud. 1974. Structure and specific binding of trypsin: Comparison of inhibited derivatives and a model for substrate binding. *J. Mol. Biol.* **83**:209.

Kunitz, M. and J. H. Northrop. 1935. Crystalline chymo-trypsin and chymo-trypsinogen. I. Isolation, crystallization, and general properties of a new proteolytic enzyme and its precursor. *J. Gen. Physiol.* **18**:433.

————. 1936. Isolation.from beef pancreas of cystalline trypsinogen, trypsin, a trypsin inhibitor, and an inhibitor-trypsin compound. *J. Gen. Physiol.* **19**:991.

Leighton, T. J., R. H. Doi, R. A. J. Warren and R. A. Kelln. 1973. The relationship of serine protease activity to RNA polymerase modifications and sporulation in *Bacillus subtilis*. *J. Mol. Biol.* **76**:103.

Magnusson, S. 1971. Thrombin and prothrombin. In *The Enzymes* (ed. P. D. Boyer), vol. 3, p. 277. Academic Press, New York.

Mares-Guia, M. and E. Shaw. 1965. Studies on the active center of trypsin: The binding of amidines and guanidines as models of the substrate chain. *J. Biol. Chem.* **240**:1579.

Maroux, S., J. Baratti and P. Desnuelle. 1971. Purification and specificity of porcine enterokinase. *J. Biol. Chem.* **246**:5031.

Martinek, K., Y. V. Savin and I. V. Berezin. 1971. Kinetic manifestations of trypsin's active center during inhibition of its enzymatic activity by Ag^+ ions. *Biokhimiya* **36**:806.

Martinek, K., Kh. Vill', Z. A. Strel'tsova and I. V. Berezin. 1969. Kinetic manifestations of the active-center structure of α-chymotrypsin during inhibition of enzyme activity by Ag^+ ions. *Mol. Biol.* (U.S.S.R.) **3**:554.

Matthews, B. W., P. B. Sigler, R. Henderson and D. M. Blow. 1967. Three-dimensional structure of tosyl-α-chymotrypsin. *Nature* **214**:652.

Naughton, M. A. and F. Sanger. 1961. Purification and specificity of pancreatic elastase. *Biochem. J.* **78**:156.

Neurath, H. and G. H. Dixon. 1957. Structure and activation of trypsinogen and chymotrypsinogen. *Fed. Proc.* **16**:791.

Neurath, H., J. A. Rupley and W. J. Dreyer. 1956. Structural changes in the activation of chymotrypsinogen and trypsinogen. Effects of urea on chymotrypsinogen and delta-chymotrypsinogen. *Arch. Biochem. Biophys.* **65**:243.

Northrop, J. H., M. Kunitz and R. M. Herriot. 1948. *Crystalline Enzymes*, p. 96. Columbia University Press, New York.

O'Leary, M. H. and M. D. Kluetz. 1972. Nitrogen isotope effects on the chymotrypsin catalyzed hydrolysis of *N*-acetyl-L-tryptophanamide. *J. Amer. Chem. Soc.* **94**:3585.

Ong, E. B., E. Shaw and G. Schoellmann. 1964. An active center histidine peptide of α-chymotrypsin. *J. Amer. Chem. Soc.* **86**:1271.

Oppenheimer, H. L., B. Labouesse and G. P. Hess. 1966. Implication of an ionizing group in the control of conformation and activity of chymotrypsin. *J. Biol. Chem.* **241**:2720.

Owren, P. A. and H. Stormorken. 1973. The mechanism of blood coagulation. *Ergeb. Physiol. Biol. Chem. Exp. Pharmacol.* **68**:1.

Pauling, L. 1946. Molecular architecture and biological reactions. *Chem. Eng. News* **24**:1375.

Pollack, E., J. L. Hogg and R. L. Schowen. 1973. One-proton catalysis in the deacetylation of acetyl-α-chymotrypsin. *J. Amer. Chem. Soc.* **95**:968.

Robertus, J. D., J. Kraut, R. A. Alden and J. J. Birktoft. 1972. Subtilisin; a stereochemical mechanism involving transition-state stabilization. *Biochemistry* **11:**4293.

Ruhlmann, A., D. Kukla, P. Schwager, K. Bartels and R. Huber. 1973. Structure of the complex formed by bovine trypsin and bovine pancreatic trypsin inhibitor. *J. Mol. Biol.* **77:**417.

Sampath Narayanan, A. and R. A. Anwar. 1969. The specificity of purified porcine pancreatic elastase. *Biochem. J.* **114:**11.

Segal, D. M., J. C. Powers, G. H. Cohen, D. R. Davies and P. E. Wilcox, 1971. Substrate binding site in bovine chymotrypsin Aγ. A crystallographic study using peptide chloromethyl ketones as site-specific inhibitors. *Biochemistry* **10:**3728.

Shaw, E., M. Mares-Guia and W. Cohen. 1965. Evidence for an active-center histidine in trypsin through use of a specific reagent, 1-chloro-3-tosylamido-7-amino-2-heptanone, the chloromethyl ketone derived from N_a-tosyl-L-lysine. *Biochemistry* **4:**2219.

Shotton, D. M. and B. S. Hartley. 1970. Amino-acid sequence of porcine pancreatic elastase and its homologies with other serine proteases. *Nature* **225:**802.

Shotton, D. M. and H. C. Watson. 1970. The three-dimensional structure of crystalline porcine pancreatic elastase. *Phil. Trans. Roy. Soc. London* B **257:**111.

Sigler, P. W., D. M. Blow, B. W. Matthews and R. Henderson. 1968. Structure of crystalline α-chymotrypsin. II. A preliminary report including a hypothesis for the activation mechanism. *J. Mol. Biol.* **35:**143.

Spomer, W. E. and J. F. Wootton. 1971. The hydrolysis of a α-N-benzoyl-L-argininamide catalyzed by trypsin and acetyl trypsin. Dependence on pH. *Biochim. Biophys. Acta* **235:**164.

Stambaugh, R., B. Brackett and L. Mastroianni. 1969. Inhibition of *in vitro* fertilization of rabbit ova by trypsin inhibitors. *Biol. Reprod.* **1:**223.

Steitz, T. A., R. Henderson and D. M. Blow. 1969. Structure of crystalline α-chymotrypsin. III. Crystallographic studies of substrates and inhibitors. *J. Mol. Biol.* **46:**337.

Stroud, R. M. 1974. A family of protein-cutting proteins. *Sci. Amer.* **231:**74.

Stroud, R. M., L. M. Kay and R. E. Dickerson. 1971. The crystal and molecular structure of DIP-inhibited bovine trypsin at 2.7 Å resolution. *Cold Spring Harbor Symp. Quant. Biol.* **36:**125.

————. 1974. The structure of bovine trypsin: Electron density maps of the inhibited enzyme at 5 Å and 2.7 Å resolution. *J. Mol. Biol.* **83:**185.

Susi, H. 1972. The strength of hydrogen bonding: Infrared spectroscopy. In *Methods in Enzymology* (ed. S. P. Colowick and N. O. Kaplan), vol. 26, p. 381. Academic Press, New York.

Sweet, R. M., H. T. Wright, J. Janin, C. H. Chothia and D. M. Blow. 1974. Crystal structure of the complex of porcine trypsin with soybean trypsin inhibitor (Kunitz) at 2.6 Å resolution. *Biochemistry* **13:**4212.

Timasheff, S. N. and J. A. Rupley. 1972. Infrared titration of lysozyme carboxyls. *Arch. Biochem. Biophys.* **150:**318.

Tomkins, G. M., T. D. Gelehrter, D. Granner, D. Martin, Jr., H. H. Samuels and E. B. Thompson. 1969. Control of specific gene expression in higher organisms. *Science* **166:**1474.

Tschesche, H. 1974. Biochemistry of natural proteinase inhibitors. *Angew. Chem. Int. Ed. Eng.* **13:**10.

Wolfenden, R. 1972. Analog approaches to the structure of the transition state in enzyme reactions. *Acc't. Chem. Res.* **5:**10.

Wright, C. S., R. A. Alden and J. Kraut. 1969. Structure of subtilisin BPN′ at 2.5 Å resolution. *Nature* **221:**235.

Wright, H. T. 1973. Comparison of the crystal structures of chymotrypsinogen-A and α-chymotrypsin. *J. Mol. Biol.* **79:**1.

Zerner, B. and M. L. Bender. 1964. The kinetic consequences of the acyl-enzyme mechanism for the reactions of specific substrates with chymotrypsin. *J. Amer. Chem. Soc.* **86:**3669.

Zerner, B., R. P. M. Bond and M. L. Bender. 1964. Kinetic evidence for the formation of acyl-enzyme intermediates in the α-chymotrypsin-catalyzed hydrolysis of specific substrates. *J. Amer. Chem. Soc.* **86:**3674.

The Specificity of Proteinases toward Protein Substrates

Joseph S. Fruton

Kline Biology Tower, Yale University
New Haven, Connecticut 06520

Until recently, the specificity of proteinases was largely described in terms of the particular amino acid residues whose participation in a peptide bond of a substrate rendered that bond most sensitive to the enzyme under study. This view came from the discovery during the 1930's of the first synthetic substrates for well-defined proteinases such as pepsin, trypsin, chymotrypsin and papain (Bergmann aand Fruton 1941) and the much more extensive work done after World War II (Neurath and Schwert 1950; Fruton 1970). The question then arose whether the conclusions drawn from such specificity studies with simple synthetic compounds could be applied to the action of proteinases on more complex substrates such as the B chain of insulin or various proteins. Even if allowance was made for the doubtful purity of enzyme preparations or for the use of large enzyme concentrations and long incubation periods, apparent discrepancies were observed between the results of experiments with the two types of substrates (Hill 1965; Neil, Niemann and Hein 1966). Although the action of pancreatic trypsin on protein substrates accorded well with the conclusion that its action was limited to arginyl or lysyl bonds, the situation for other enzymes was more ambiguous, and proteinases like pepsin were judged to be relatively nonspecific.

Secondary Enzyme-Substrate Interactions

Part of the explanation of these apparent discrepancies is that proteinases appear to have extended active sites, whose interaction with oligopeptide substrates is a multipoint cooperative process. This had been appreciated qualitatively for many years, but a more precise description became possible only recently through X-ray crystallographic studies of enzymes and of complexes of enzymes with substrate analogs. The decisive change came, of course, with the elucidation of the three-dimensional structure of lysozyme (Blake et al. 1967) and the recognition that its extended active-site cleft could accommodate at least six monosaccharide units. In the case of the proteinases for which

33

models based on X-ray diffraction are available, such as chymotrypsin (Birktoft et al. 1970), elastase (Shotton and Watson 1970) and papain (Drenth
et al. 1971), analogous extended active sites have been inferred, in part from
studies on the effect of structural changes in oligopeptide substrates on the
kinetics of their cleavage by the dissolved enzymes. It has been suggested
that the active sites of papain (Schechter and Berger 1967) and elastase
(Atlas and Berger 1973) might correspond in length to about seven amino
acid residues of the polypeptide chain in their substrates, and a similar estimate has been made for pepsin (Sampath-Kumar and Fruton 1974), for
which no three-dimensional model is available.

It is now customary, therefore, to distinguish between so-called "primary"
and "secondary" interactions in describing the specificity of proteinases in
their action on oligopeptide substrates. We define the primary enzyme-
substrate interactions as those involving the enzymatic groups directly concerned with the bond-breaking or bond-making step (such as Ser-195 and
His-57 of chymotrypsin), as well as the immediately adjacent structural features of the enzyme (such as the so-called tosyl pocket of chymotrypsin)
that can bind a preferred amino acid residue in a way that promotes catalysis
at a sensitive bond in which that residue participates. Table 1 indicates features, to the extent that they are known, of the primary specificity of several
of the better known proteinases. In addition to the preferred residues at the
sensitive peptide bond, alternative residues (usually less favorable to catalysis)
are also noted in parentheses. There are, of course, the numerous serine proteinases related to pancreatic trypsin, with their preference for an arginyl or
lysyl residue at the site of cleavage. Then follow serine proteinases such as
chymotrypsin and related enzymes (subtilisin, elastase) with a primary specificity for peptide bonds involving a hydrophobic amino acid residue. It should
be noted that in its action on acetyl amino acid amides (Ac-X-NH$_2$), chymotrypsin exhibits a striking preference for substrates in which X = Trp, Tyr or

Table 1

Apparent Primary Specificity of Some Proteinases
toward Small Oligopeptide Substrates

Enzyme	Preferred cleavage site	
Trypsin	-Arg(Lys)-↓Y-	(Y ≠ Pro)
Clostripain	-Arg-↓Y-	(Y may be Pro)
Chymotrypsin	-Trp(Phe, Tyr, Leu)-↓Y-	
Subtilisin BPN′	-Tyr(Leu,Ala)-↓Y-	
Elastase	-Ala(Ser)-↓Y-	
Papain, streptococcal proteinase	-Phe(Val, Ile, Leu)-↓X-Trp(Phe, Leu)-	
B. subtilis neutral protease	-↓X-Leu	
Pepsin	-Phe-↓Trp(Phe, Tyr)-	

Phe; the compounds with aliphatic amino acid residues (Leu, Met, etc.) are hydrolyzed much more slowly or not at all under the usual experimental conditions. In the case of elastase, the preference for a smaller side chain has been correlated with a difference in the size of the so-called tosyl hole of chymotrypsin (Shotton and Watson 1970). For the cysteine proteinases papain and streptococcal proteinase, there is a preference for hydrophobic amino acid residues on either side of the residue donating the carbonyl group to the sensitive bond. The so-called neutral protease of *B. subtilis* appears to prefer a hydrophobic residue on the imino side of the cleavage site, while pepsin (and related acid proteinases) acts best on peptide bonds linking two hydrophobic (preferably aromatic) amino acid residues.

The secondary enzyme-substrate interactions become evident when we replace a simple substrate containing a single preferred amino acid residue with a series of oligopeptides in which the same kind of peptide bond (e.g., Tyr-Gly) is broken, but vary the chain length of the substrate and the nature of the amino acids flanking the sensitive dipeptidyl unit. We will consider shortly some items of experimental evidence relating to secondary enzyme-substrate interactions at the extended active sites of several proteinases.

The problem of the importance of secondary interactions comes to the fore in relation to the growing number of serine proteinases which resemble pancreatic trypsin in their primary specificity, at least as judged by the cleavage of benzoyl-L-arginine ethyl ester or of tosyl-L-arginine methyl ester, but which exhibit specificity with respect to protein substrates. Among these substrates must be included the various trypsin inhibitors isolated from plant seeds and animal tissues or fluids, since it is now known that the inhibition is in fact an enzyme-substrate interaction in which the products do not leave the active site or do so very slowly (Laskowski and Sealock 1971; Tschesche 1974). In Table 2 are indicated some of the striking differences between

Table 2
Apparent Specificity of Trypsinlike Serine Proteinases toward Natural Substrates

Enzyme	Preferred natural substrate (inhibitor)	Site of cleavage
Bovine trypsin	many proteins, peptides	-X-Arg(Lys)-↓Y
Human thrombin	fibrinogen	-Phe-$(X)_7$-Arg-↓Gly-Pro-Arg-
Enterokinase	trypsinogen	Val-$(Asp)_4$-Lys-↓Ile
Kininogen (kallikrein)	plasma α-globulin	-Arg-Pro-Pro-Gly-Pro-Ser-Pro-Phe-Arg-↓Y-
	pancreatic inhibitor (Kunitz)	-Pro-Pro-Tyr-Thr-Gly-Pro-Cys-Lys-↓Ala-

pancreatic trypsin, which appears to cleave arginyl and lysyl peptide bonds in a wide variety of proteins and to interact with the Arg-X or Lys-X unit of many natural inhibitors, and such relatively specific serine proteinases as thrombin or enterokinase. In the case of thrombin, there appears to be a preference for a Phe residue at a distance from the site of catalytic action (Magnusson 1971), and with enterokinase, the aspartyl residues at the amino terminus of trypsinogen appear to play a special role (Maroux, Baratti and Desnuelle 1971). Moreover, with trypsinlike enzymes, such as the kallikreins (Schachter 1969) or some of the plasminogen activators, there appears to be an especially pronounced species specificity comparable to that observed in immunochemical reactions. This list of serine proteinases that resemble trypsin in their primary specificity could, of course, be extended to include plasmin (fibrinolysin), urokinase, cocoonase, acrosin, streptokinase, etc. There are, in addition, various less well characterized serine proteinases that hydrolyze tosyl-L-arginine methyl ester. Among them is the enzyme involved in the initial stages of the sporulation of *B. subtilis;* one of the possible roles of this enzyme may be the proteolytic modification of the B subunit of RNA polymerase (Doi 1973).

Accessibility of Sensitive Peptide Bonds

One factor that can play a role in such specificity toward proteins is the relative accessibility of potentially sensitive dipeptidyl units in the substrate to the active site of a proteinase whose primary specificity permits it to cleave such units. It has long been known that the random coils of denatured globular proteins are more susceptible to proteolytic attack than the native proteins. As Linderstrøm-Lang pointed out in 1939, for those proteins that can undergo reversible denaturation under a given set of conditions, the more rapid cleavage of the denatured form will pull the equilibrium toward denaturation. A case that has been studied extensively from this point of view is the cleavage of bovine serum albumin by pepsin at pH values below 4, where bovine serum albumin undergoes a well-known expansion of the molecule, the so-called N-F transition. The available data suggest that pepsin preferentially attacks the expanded protein at flexible segments that link smaller globular portions of the substrate (Weber and Young 1964; Wilson and Foster 1971). Of course in addition to the contribution of the unfolding of a globular protein substrate to the accessibility of sensitive dipeptidyl units, there is also the matter of the location of such units in the native protein. For an enzyme like trypsin, it may be expected that most of the Arg-Y or Lys-Y units will be on the periphery of the globular protein, but for an enzyme like chymotrypsin, which prefers units involving hydrophobic residues, a significant proportion of potentially sensitive units may be in the interior of the native protein. Obviously, a proteinase with a relatively broad specificity (such as chymotrypsin or papain) will attack exposed peptide bonds on the protein surface or in flexible loops more readily than interior bonds, even though the latter may be more sensitive to attack in a small oligopeptide substrate or fully unfolded protein. Also, when oligomeric proteins are substrates, the potentially sensitive dipeptidyl units which donate amino acid residues that serve as contact points between subunits will also be relatively less accessible to proteolytic attack.

In addition to the factor of accessibility, there is also an important thermo-dynamic difference in the interaction of a polypeptide segment in a folded globular protein with the active site of a proteinase and that of the same poly-peptide in a random coil. If the polypeptide segment of the protein fits into a complementary active-site region, the binding may not involve a significant change in internal degrees of freedom. On the other hand, the randomly coiled polypeptide may be expected to lose internal degrees of freedom in binding to the active site, with an unfavorable entropic contribution to the free energy of binding. Consequently, where better binding is accompanied by better ca-talysis, the same peptide bond may be cleaved much more rapidly when it is located in an accessible part of a globular protein than when it is in a small polypeptide with the same amino acid sequence as in the protein. Conversely, even if a potentially sensitive peptide bond is on the surface of a native pro-tein, but is located in a relatively rigid peptide segment that does not fit into the extended active site, the hydrolysis of that bond may not occur until the protein has been denatured.

Kinetics of Proteinase Action

Having stated these general propositions, the fact remains that it is difficult to study quantitatively the specificity of proteinases on protein substrates, and valuable knowledge can still be gained from the examination of the kinetic parameters in the hydrolysis of relatively simple oligopeptides. In considering such kinetic data as they bear on our problem, several points should be borne in mind.

The first point relates to the use of the relative magnitudes of K_m (substrate concentration at half-maximal initial velocity) as a measure of binding speci-ficity and of k_{cat} (maximal initial velocity per unit enzyme concentration) as a measure of kinetic specificity. These parameters are usually determined under steady-state conditions, where the substrate concentration is much greater than the enzyme concentration and Michaelis-Menten kinetics are ob-served. The minimum reaction sequence in the action of a proteinase E on an oligopeptide substrate A-B, in which the bond between A and B is hydro-lyzed and the products derived from A and B are released sequentially, may be written as follows:

$$\mathrm{E + A\text{-}B} \underset{k_{-1}}{\overset{k_1}{\rightleftharpoons}} \mathrm{E(AB)} \overset{k_2}{\to} \underset{+\mathrm{B}}{\mathrm{E(A)}} \overset{k_3}{\to} \mathrm{E + A.} \tag{1}$$

The first step is the productive association of the substrate with the active site to form an enzyme-substrate complex. The relative magnitude of the dissociation constant K_s (k_{-1}/k_1) for a series of substrates may then be taken as a measure of binding specificity. Where data are available for the interaction of relatively small oligopeptide substrates with proteinases, the values of K_s range between 10^{-3} and 10^{-6} M, although in some instances tighter binding has been found. The relation between K_m and K_s is given by the expression (Gutfreund and Sturtevant 1956):

$$K_m = \frac{k_{-1} + k_2}{k_1} \cdot \frac{k_3}{k_2 + k_3}. \tag{2}$$

When $k_{-1} >> k_2$, $K_m = k_3 K_s / (k_2 + k_3)$, and K_m approximates K_s when k_2 is rate limiting ($k_3 >> k_2$). It should be noted that for a value of about 10^{-4} M for K_s, if the magnitude of k_1 is about 10^7 M^{-1} sec^{-1} (a usual value for the interaction of small substrates with proteinases), k_{-1} is a relatively large number (about 1000 sec^{-1}). However, if in the interaction of a proteinase with a large substrate there are conformational changes in the reactants that make the value of k_1 (for example) 10^4 M^{-1} sec^{-1}, the magnitude of k_{-1} may be in the range of the values for k_2, and the assumption that $k_{-1} >> k_2$ becomes invalid.

The reaction scheme written above is a minimum sequence of steps in the overall catalytic process. There is increasing evidence that the scheme should be expanded to include a kinetically significant tetrahedral intermediate in the conversion of the initial enzyme-substrate complex to the species that loses the first product (Hirohata, Bender and Stark 1974 and references cited therein). Under these circumstances, k_2 (as measured by the rate of release of B) will only reflect the rate of conversion of the initial enzyme-substrate complex when the rate of the formation of the tetrahedral intermediate is much slower than the succeeding step.

Regarding the use of k_{cat} as a measure of kinetic specificity, it should be noted that for reaction scheme (1), $k_{cat} = k_2 k_3 / (k_2 + k_3)$, and k_{cat} approximates k_2 when the release of product B is rate-limiting or k_3 when the release of product A is rate-limiting. These are the extremes; cases are known in which k_2 and k_3 are of similar magnitude (Mole and Horton 1973). Obviously, if either k_2 or k_3 represents a step that includes a rate-limiting conformational change, the values found for a series of substrates may reflect the influence of alterations in the structure of the substrate on the conformational change and not on the release of the product associated with that step. In any case, when values of k_{cat} for a series of substrates are used to define kinetic specificity, one usually makes the assumption that we are comparing processes in which the ratio of k_2 to k_3 is relatively invariant.

These well-known limitations on the use of K_m and k_{cat} for the definition of the specificity of proteinases require emphasis in the present discussion because one may not be justified in many instances in transferring assumptions that are valid for the cleavage of small synthetic substrates to the action of proteinases on oligopeptides and proteins, where secondary enzyme-substrate interactions may play a significant role. Possibly the situation will be clearer when values are available for the individual rate constants k_1, k_{-1}, k_2 and k_3 in the cleavage of a series of oligopeptide substrates by a well-defined proteinase.

A second point that needs consideration relates to the ordered release of the products in the cleavage of an oligopeptide by a proteinase. In reaction sequence (1), the product that leaves first is denoted B, but there is no statement that it is necessarily derived from the amine portion of the substrate. Among the questions at issue is whether in the intermediate E-A there is an obligatory covalent link between E and A. For the action of chymotrypsin on simple synthetic substrates such as esters of acetyl-L-tryptophan, especially where B is a good leaving group (e.g., the *p*-nitrophenoxy group), the experimental evidence is overwhelming for the intermediate formation of the acyl enzyme Ac-Trp-chymotrypsin (Bender and Kezdy 1964). In this case, k_{cat}

equals k_3, since the acylation reaction is much faster than the hydrolysis of the acyl enzyme. With small peptide substrates of chymotrypsin, the evidence is more circumstantial, but strongly persuasive, that the products are released in the same order as with ester substrates, but now k_2 is rate-limiting and is approximated by the experimentally determined value of k_{cat} (Fastrez and Fersht 1973). It may well be that this conclusion applies to the action of chymotrypsin and other proteinases on oligopeptides and protein substrates, but consideration should also be given to the possibility that the relatively large A and B fragments formed upon hydrolysis may have sufficient affinity for the extended active site so that the release of one of them becomes rate-limiting. To my knowledge, there are at present no data either to support or to rule out this suggestion.

A third point to be considered in relation to the action of proteinases on a series of oligopeptide substrates is the possibility of alternative modes of enzyme-substrate interaction, so as to position a different peptide bond at the catalytic site. In these circumstances, comparisons of k_{cat} and K_s values for different substrates may give misleading information about kinetic specificity and productive binding specificity. A limiting case is the one in which an oligopeptide has been bound at the active site in a manner that positions at the catalytic groups a peptide bond that is resistant to enzymatic attack. This kind of nonproductive binding becomes important in specificity studies when the value of K_s for the resistant enzyme-substrate complex is much lower than that of the productive mode. Here the nonproductive interaction may be considered to be equivalent to substrate inhibition of the catalysis (Hein and Niemann 1962). If the dissociation constant of the nonproductive complex is denoted K_s^i, then both K_{cat} and K_m are changed by the same factor $[1 + (K_s/K_s^i)]$. Thus the ratio k_{cat}/K_m is independent of nonproductive binding and is a valuable parameter in the characterization of the specificity of proteinases (Bender and Kezdy 1965). Moreover, $k_{cat}/K_m = k_2/K_s$, so that if either k_2 or K_s can be determined by independent methods, additional information can be obtained about the kinetic importance of factors that determine the specificity of catalysis.

In view of the limitations and uncertainties inherent in the k_{cat} and K_m values for the action of proteinases on oligopeptide substrates, it could rightly be argued that few conclusions can be drawn from such data about the factors that influence specificity. Since each of the steps in reaction sequence (1) may have its own specificity, one needs to know the influence of changes in substrate structure on the individual rate constants. Unfortunately, such data are not yet available, although not unattainable. The best one can do at present is to use k_{cat} and K_m values and to see what can be discerned about the effects of secondary enzyme-substrate interactions on the cleavage of oligopeptides by proteinases.

Enzyme Kinetics and Secondary Interactions

Selected data for the action of five proteinases on pairs of substrates that differ only by an Ala-Ala unit in the acyl component are presented in Table 3. Because of the paucity of strictly comparable data in the literature, Table 3 includes values for the hydrolysis of both methyl esters and of peptide bonds.

Table 3

Secondary Interactions of Proteinases

Enzyme	Substrates	k_{cat} (sec^{-1})	K_m (mM)	Relative k_{cat}/K_m
Trypsin[a]	↓ Z-Lys-OMe	101	0.23	1
(pH 7.5, 30°C)	Z-(Ala)$_2$-Lys-OMe	106	0.08	3
α-Chymotrypsin[b]	↓ Ac-Tyr-Gly-NH$_2$	0.6	23	1
(pH 7.9, 25°C)	Ac-(Ala)$_2$-Tyr-Gly-NH$_2$	10	2	190
Subtilisin BPN'[c]	↓ Ac-Ala-OMe	9	36	1
(pH 7.5, 30°C)	Ac-(Ala)$_2$-Ala-OMe	256	0.45	2300
Elastase[d]	↓ Ac-Ala-OMe	6.7	153	1
(pH 8.0, 25°C)	Ac-(Ala)$_2$-Ala-OMe	73	0.43	3900
Pepsin[e]	↓ Z-Phe-Phe-OP4P	0.7	0.2	1
(pH 3, 37°C)	Z-(Ala)$_2$-Phe-Phe-OP4P	282	0.04	1900

[a] Morihara and Oka (1973a). [b] Baumann et al. (1973). [c] Morihara and Oka (1973b). [d] Gertler and Hofmann (1970). [e] Sachdev and Fruton (1970).

Clearly in the case of the esters, k_2 is not rate-limiting, whereas for the amides it may or may not be rate-limiting. However, if one makes the not unreasonable assumption (Segal 1972) that for a given pair of comparable substrates, the k_2/k_3 ratio is invariant, then the relative values of k_{cat} may be taken as a measure of relative catalytic efficiency, and the relative values of K_m may be taken as a measure of relative binding affinity. In some cases, the validity of this assumption has been demonstrated experimentally, as in the cleavage of acetyl-L-tyrosine ethyl ester and of furoylacryloyl-L-tyrosine ethyl ester by α-chymotrypsin (Himoe et al. 1969).

In view of the uncertainties in comparisons such as those made in Table 3, one should look only at gross differences. With this in mind, and at the risk of oversimplification, I think one can discern three types of effect on the kinetic parameters of the introduction of an Ala-Ala unit. The first type, exemplified by trypsin, is characterized by relative invariance of both k_{cat} and K_m (Morihara and Oka 1973a). In other words, the primary specificity of the enzyme for Arg or Lys bonds appears to be so stringent that, with small oligopeptide substrates, secondary interactions have relatively little effect on either binding affinity or catalytic efficiency. Clearly this conclusion may not apply to other proteinases (such as thrombin) that resemble trypsin in its primary specificity. The second type, exemplified by chymotrypsin (Segal 1972; Baumann, Bizzozero and Dutler 1973), subtilisin (Morihara and Oka 1973b) and elastase (Gertler and Hofmann 1970; Thompson and Blout 1973; Atlas and Berger 1973), is characterized by striking increases in both binding affinity and catalytic efficiency, and it would seem that in these cases there is at least qualitative evidence for the conclusion that better binding means better catalysis. This is an extension of the view expressed by Knowles (1965) for the primary specificity of chymotrypsin. The third type is that exemplified by pepsin, where a very large increase in catalytic efficiency is associated with a rather small change in binding affinity (Fruton 1970). This type of specificity relationship is also evident with other acid proteinases related to pepsin (Voynick and Fruton 1971; Raymond et al. 1972; Ferguson et al. 1973). It is not limited, however, to this class of enzymes. The data of Morihara and Tsuzuki (1970) on thermolysin fit this pattern, and this may also be the situation for enterokinase, whose high catalytic efficiency toward trypsinogen as a protein substrate is reflected in its action on synthetic substrates related to Val-Asp-Asp-Lys-Ile-Val-Gly (Maroux, Baratti and Desnuelle 1971).

What, if anything, do data of this sort indicate about the specificity of individual proteinases in their action on particular protein substrates? Perhaps the case of rennin (chymosin) may suggest the kind of tentative inferences that may be drawn. At pH 7 this enzyme cleaves its natural substrate, κ-casein, at a single peptide linkage, the Phe-Met bond in the sequence -Pro-His-Leu-Ser-Phe-Met-Ala-Ile-. Now in its action on oligopeptide substrates, including the B chain of insulin, rennin has a primary specificity similar to that of pepsin, but there is an important difference: In all cases studied thus far, the K_m for the action of rennin on a given pepsin substrate is an order of magnitude greater than for pepsin and k_{cat} is much lower. A few representative data are given in Table 4. In the synthetic substrates tested by Raymond et al. (1972, 1973), the Phe-Met unit of casein was replaced by a Phe(NO$_2$)-Nle unit. It is evident that a modification of the B portion by the introduction of a Leu residue enormously increases k_{cat}, without affecting K_m, and a modification of the A portion by the addition of the Pro-His unit greatly increases binding without markedly affecting the catalytic efficiency. If one assumes that structural changes in the A and B groups of the casein segment A-Phe-Met-B have the same effect as on the hydrolysis of oligopeptides such as A-Phe(NO$_2$)-Nle-B, it would seem that a particular polypeptide segment of casein is accessible to the active site of the enzyme, and that the ready cleavage of the Phe-Met bond is primarily a consequence of the contributions of the secondary interactions of the A and B portions of the segment with the extended active site.

Table 4

Secondary Interactions of Acid Proteinases

Enzyme	Substrate	k_{cat} (sec^{-1})	K_m (mM)
Pepsin (pH 4, 37°C)	Phe-Gly-His-Phe(NO$_2$)-Phe-OMe[a]	0.1	0.4
	Phe-Gly-His-Phe(NO$_2$)-Phe-Ala-Ala-OMe[a]	28	0.2
Rennin (pH 4.7, 30°C)	Phe-Gly-His-Phe(NO$_2$)-Phe-Ala-Ala-OMe[b]	0.07	3.6
	Leu-Ser-Phe(NO$_2$)-Nle-Ala-OMe[c]	0.11	1.0
	Leu-Ser-Phe(NO$_2$)-Nle-Ala-Leu-OMe[c]	12	1.1
	Pro-His-Leu-Ser-Phe(NO$_2$)-Nle-Ala-Leu-OMe[d]	18	0.09

[a] Medzihradszky et al. (1970); Phe(NO$_2$) = p-nitro-L-phenylalanyl. [b] Voynick and Fruton (1971); pH 4, 37°C. [c] Raymond et al. (1972); Nle = L-norleucyl. [d] Raymond et al. (1973).

Aside from the question of the role of secondary enzyme-substrate inter-actions in the cleavage of accessible, but relatively resistant, peptide bonds, there is also the matter of the possible conformational flexibility of the active site in response to such secondary interactions. Does the binding of a poly-peptide segment of a protein substrate cause conformational changes at the active site so as to alter the catalytic efficiency of the enzyme? Clearly the most direct approach to this question is through X-ray crystallography. The availability of three-dimensional models based on high-resolution (2–2.5 Å) X-ray diffraction data has invited attempts to fit models of polypeptide seg-ments into the active site region and to discern individual hydrogen bonds and hydrophobic interactions that might be involved in the productive binding of a substrate. In the case of the proteinases, this has perhaps been carried farthest with the structurally related enzymes chymotrypsin, trypsin and elas-tase, as in the studies of Segal et al. (1971) on the interaction of chymotrypsin with a series of peptidyl-L-phenylalanyl chloromethanes or in the work of Blow, Janin and Sweet (1974) on the interaction of trypsin with the soy-bean trypsin inhibitor (Kunitz). The apparent concordance of crystallographic data with the results of kinetic studies suggests that the active sites of these three serine proteinases are relatively rigid structures. Thus tighter productive binding of the substrate leads to better catalysis, and the principal conforma-tional change appears to be at the sensitive CO-NH group of the substrate, especially in the formation of a tetrahedral intermediate that approaches a stabilized transition state essential for catalysis (Wolfenden 1972; Leinhard 1973). For chymotrypsin, trypsin and elastase, therefore, it would seem that there is no need to invoke large-scale changes in the conformation of their active sites. The question is still open whether this will prove to be the case with those serine proteinases for which the contribution of secondary interac-tions is to enhance catalytic efficiency without a comparable increase in bind-ing affinity.

It was noted above that the kinetic data on the cleavage of oligopeptide substrates by the so-called acid proteinases related to pepsin suggest that secondary enzyme-substrate interactions may greatly enhance catalytic effi-ciency without markedly affecting the apparent dissociation constant of the rate-limiting enzyme-substrate complex. This raises the possibility that the con-tribution of the secondary interactions to binding energy is used to distort or strain the sensitive peptide bond so that the transition state is stabilized more effectively. Without wishing to deny this possibility, I think that some recent data on pepsin also suggest that in this enzyme the active site itself may have conformational flexibility. If the active site is not a rigid structure but in-stead can undergo conformational change in response to its interaction with the substrate, a portion of the binding energy could be used to achieve a transition state in which the active site is strained or distorted, and catalysis would be favored by the tendency of the enzyme to return to its normal state.

Action of Pepsin on Oligopeptide Substrates

Table 5 presents a set of data for pepsin substrates of the type A-Phe-Phe-B, in which the Phe-Phe bond is the only one cleaved, the B group is held con-stant, and the A group is varied. Without going into the details of these data,

Table 5

Cleavage of Pepsin Substrates

Substrate	k_{cat} (sec^{-1})	K_m (mM)	k_{cat}/K_m $(sec^{-1}\ mM^{-1})$
Z-Phe-Phe-OP4P[a]	0.7	0.2	3.7
Z-Gly-Phe-Phe-OP4P	3.1	0.4	7.8
Z-(Gly)$_2$-Phe-Phe-OP4P	71.8	0.4	180
Z-(Gly)$_3$-Phe-Phe-OP4P	4.5	0.4	10.1
Z-(Gly)$_4$-Phe-Phe-OP4P	2.1	0.7	3.0
Z-Gly-Ala-Phe-Phe-OP4P	409	0.11	3720
Z-Gly-Sar-Phe-Phe-OP4P[b]	1.0	0.4	2.5
Z-Gly-Pro-Phe-Phe-OP4P	0.06	0.14	0.4

pH 3.5, 37°C (Sachdev and Fruton 1970; Sampath-Kumar and Fruton 1974).
[a] Z = benzyloxycarbonyl; OP4P = 3-(4-pyridyl)propyl-1-oxy.
[b] Sar = sarcosyl.

it would seem that the variation of k_{cat} is not paralleled by comparable increases in binding affinity, as the changes in K_m are relatively small. Direct measurement, by gel filtration, of the affinity of pepsin for the Phe-Phe unit in a series of compounds gave values for K_s near 0.2 mM at pH 4 and 37°C (Raju, Humphreys and Fruton 1972), so that it may be assumed that $K_m \cong K_s$.

These data indicate that for the series of substrates Z-(Gly)$_n$-Phe-Phe-OP4P, the placement of the hydrophobic benzyloxycarbonyl group in relation to the sensitive Phe-Phe unit and the number of CO-NH groups in the peptide has a significant effect on catalysis, with an apparent maximum at Z-Gly-Gly-Phe-Phe-OP4P. However, if one considers only peptides of this length, there are very large differences among the substrates in which A = Z-Gly-Gly, Z-Gly-Ala and Z-Gly-Pro, for which the k_{cat} values range over four orders of magnitude. This kind of result is consistent with the view that the secondary interaction of the A unit of A-Phe-Phe-B with the extended active site may enhance catalysis by the utilization of the potential binding energy in the enzyme-substrate interaction to lower the free energy of activation in the catalytic process.

These data also suggest that the Z-dipeptidyl group interacts with pepsin as a unit and that the hydrophobic benzyloxycarbonyl group participates in the productive interaction of a substrate such as Z-Gly-Gly-Phe-Phe-OP4P with pepsin, although it does not appear to contribute significantly to an increase in total binding energy in the formation of the enzyme-substrate complex. It seemed possible that the amino-terminal hydrophobic group had been drawn into the extended active site of pepsin by the strong affinity of the sensitive Phe-Phe unit for the catalytic site. To test this possibility directly, the amino-terminal benzyloxycarbonyl group was replaced by either a dansyl or mansyl group (Sachdev et al. 1972, 1973, 1975). Dansyl or mansyl compounds are essentially nonfluorescent in aqueous solution, but fluoresce strongly when transferred to a solvent of lower polarity or when bound to protein. This substitution in the pepsin substrates (listed in Table 5) did not alter significantly the relative susceptibility of the Phe-Phe bond.

When a dansyl or mansyl group is located at the amino terminus of a pepsin substrate of this type, the fluorescence of the probe is markedly enhanced when mixed with pepsin in 1:1 molar proportion (Table 6). The data were obtained with compounds that are cleaved slowly by pepsin. By measurement of the change in fluorescence intensity of substrates such as Mns-Gly-Phe-Phe-OP4P as a function of protein concentration, the dissociation constant (K_D) of the pepsin-ligand complex was estimated to be 0.03 mM; the same value was found for K_m in the enzymatic hydrolysis of this substrate. This provides additional evidence for the view that the kinetically determined value of K_m approximates K_s and that the rate-limiting step in the catalysis is the conversion of the initial enzyme-substrate complex. It should be noted in Table 6 that the binding of the compounds with a Gly-Phe or Phe-Gly unit is stronger than that with Mns-Gly-Gly-OP4P but weaker than that with Mns-Phe-Phe-OP4P.

To distinguish between the intrinsic binding affinity of pepsin for the mansyl group and the effect of the Phe-Phe unit in drawing the mansyl group attached to it into the active site, a comparison was made with the binding of Mns-Gly-Gly-OP4P, which is not cleaved by pepsin. Also, use was made of the powerful pepsin inhibitor pepstatin (Umezawa 1972), which forms a tight 1:1 noncovalent complex with pepsin (K_i ca. 10^{-10} M). It is evident (Table 6) that the fluorescence of Mns-Gly-Gly-OP4P is slightly enhanced by equimolar pepsin, but this enhancement is not altered significantly by the addition of equimolar pepstatin. A similar result was obtained with mansylamide. On the other hand, the addition of pepstatin to the mixture of pepsin and a substrate such as Mns-Phe-Phe-OP4P caused a reduction in fluorescence to the level observed with Mns-Gly-Gly-OP4P or mansylamide. The value of K_D for the

Table 6

Interaction of Mansyl Compounds with Pepsin

Mansyl compound	Fluorescence[a]		K_D (mM)
	− pepstatin	+ pepstatin[b]	
Mns-Gly-Gly-OP4P[c]	0.22 (450)	0.22 (450)	0.24
Mns-Gly-Phe-OP4P	0.47 (445)	0.18 (445)	0.15
Mns-Phe-Gly-OP4P	0.40 (450)	0.20 (450)	0.17
Mns-Phe-Phe-OP4P	1.6 (435)	0.20 (460)	0.07
Mns-Gly-Phe-Phe-OP4P	2.9 (445)	0.32 (465)	0.03
Gly-Gly-Phe-Phe-NHNH-Mns	1.4 (435)	0.16 (440)	0.07

Concentrations of mansyl compound and of pepsin, 10 µM; pH 2.35, 25°C (Sachdev, Brownstein and Fruton 1975).

[a] Expressed in arbitrary fluorescence units (FU) at maximum (given in nanometers in parentheses) of uncorrected emission spectrum (excitation at 330 nm), normalized with 1.9 µM quinine sulfate in 0.1 N H_2SO_4 as the standard (1.0 FU at 448 nm). At pH 2.35, the fluorescence of the mansyl compounds in the absence of added pepsin is 0.03–0.05 FU, with a broad emission band centered in the range 445–480 nm.

[b] Concentration of pepstatin, 10 µM.

[c] Mns = 6-(N-methylanilino)-2-naphthalenesulfonyl; OP4P, 3-(4-pyridyl)propyl-1-oxy.

interaction of Mns-Gly-Gly-OP4P or mansylamide with pepsin is about an order of magnitude greater than for Mns-Gly-Phe-Phe-OP4P, and consequently, the intrinsic affinity of the enzyme for the mansyl group is less than for the same group attached to the Phe-Phe unit. Taken together with the effect of pepstatin on the fluorescence of the pepsin-ligand complexes, these results suggest that the active site has a relatively low intrinsic affinity for the mansyl group, but that the group is drawn into the active site by the interaction of the Phe-Phe unit with the enzyme; this interaction is prevented by pepstatin. In addition, there appears to be a separate binding site for the mansyl group (as in Mns-Gly-Gly-OP4P) distinct from the active site; the mansyl group of Mns-Phe-Phe-OP4P can interact with this distinct mansyl binding site when the substrate is excluded by pepstatin from its preferred interaction with the active site.

Further evidence in favor of this interpretation of our fluorescence data was provided by studies on the change in the change in the fluorescence of mansylamide or Mns-Phe-Phe-OP4P in the presence of pepsinogen that is undergoing activation. It had been known from the work of Wang and Edelman (1971) that the fluorescence of 6-p-toluidino-2-napthalenesulfonate (closely related structurally to mansic acid) is *decreased* about 50% upon autocatalytic conversion of pepsinogen to pepsin; they concluded that the ligand was bound at a locus distinct from the active site of the enzyme. A similar result was obtained with mansylamide, but with Mns-Phe-Phe-OP4P (and other substrates of this type), the fluorescence of the probe group was markedly *increased* (at least 3-fold) during the activation process (Sachdev, Brownstein and Fruton 1973). It would appear, therefore, that during the autocatalytic conversion of pepsinogen to pepsin, the conformational change in the binding site for mansylamide (or Mns-Gly-Gly-OP4P) is distinct from the change in the locus for interaction with a mansyl group that is drawn into the active site by the Phe-Phe unit of a peptide substrate.

These results suggest considerable flexibility at the active site of pepsin and raise the possibility that a substrate binding cleft is formed through conformational changes arising from the interaction of a suitable peptide substrate or inhibitor with the extended active site of the enzyme. Such a cleft may be present in pepsinogen, but may be occupied by the amino-terminal portion of the zymogen that is removed by proteolysis during the activation process. Upon conversion to pepsin, the cleft may be narrowed or closed after the release of the activation peptides and thus made inaccessible to the mansyl group of mansylamide. The specific interaction of a suitable peptide substrate with the active site of the enzyme, in widening the cleft, thus may allow an attached mansyl group to enter an enzymatic region of lower polarity. From this point of view, the intrinsic proteinase activity of pepsinogen (Bustin and Conway-Jacobs 1971; McPhie 1974) may therefore be a consequence of the competition between a protein substrate and the amino-terminal portion of pepsinogen for the extended active site of pepsin.

The fluorescence studies discussed thus far involved the interaction of pepsin with substrates labeled at the amino terminus with a fluorescent probe group. It was noted in Table 4 that the introduction of hydrophobic amino acids on the carboxyl side of the sensitive peptide bond can have a large effect on the catalytic efficiency of pepsin, and several compounds have been prepared in which a fluorescent probe group is located at the carboxyl terminus of suitable peptide substrates, among which are peptide hydrazides such as Gly-Gly-Phe-Phe-NHNH-Mns which is cleaved by pepsin at the Phe-Phe bond. The fluorescence of the mansyl group is greatly enhanced upon the addition of equimolar pepsin and is reduced to the low level observed with mansylhydrazide by the further addition of equimolar pepstatin (Sachdev, Brownstein and Fruton 1975). Substrates of this kind, as well as those of the type Mns-$(Gly)_n$-Phe-Phe-OP4P, are useful for stopped-flow kinetic studies of pepsin and related acid proteinases. Upon mixing the substrate with the enzyme, there is a rapid fluorescence increase (k_1 = ca. 10^7 M^{-1}sec^{-1}), followed by a slower, first-order decrease associated with the hydrolytic process (G. P. Sachdev and J. S. Fruton, unpubl.).

The data obtained thus far in studies of the interaction of pepsin with peptide substrates bearing a fluorescent group give no evidence for the existence of a defined "subsite" that binds the fluorescent probe. Since the three-dimensional structure of pepsin is still unknown, more detailed speculation on this question is premature. However, some general comment seems necessary regarding recent efforts to "map" the extended active sites of proteinases by denoting the enzyme area that binds a single amino acid residue a "subsite." In such mapping it is implicitly assumed that the active site of an enzyme is a relatively rigid structure, and when three-dimensional models based on X-ray diffraction studies are available, attempts have been made to identify the subsites by fitting substrate models to the rigid enzyme model. Perhaps this is justified to some extent in the case of chymotrypsin, but where the conformational flexibility of the active site is greater, as appears to be the case with pepsin, such mapping of the complete binding area in terms of subsites is of doubtful validity.

Action of Papain on Oligopeptide Substrates

Since the mapping of the extended active sites of proteinases may be said to have been begun with the work of Schechter and Berger (1967, 1968) on papain, it is perhaps appropriate that we consider this enzyme next. On the basis of kinetic data obtained with Ala-Ala-Ala-Ala, Ala-Ala-Ala-Ala-Ala and Ala-Ala-Ala-Ala-Ala-Ala, in which one or two L-alanyl residues were replaced by the D-form, they formulated a graphic scheme in which they divided the active site into subsites S_1, S_2, S_3, etc., on the acyl side of the sensitive bond and S_1', S_2', S_3', etc., on the amino side of the site of cleavage. These were identified on the basis of the effect of introducing a D-alanyl residue on the apparent first-order rate constant per unit enzyme concentration. To determine these values, they examined by electrophoresis the products formed at a single enzyme concentration and a single substrate concentration. Serious doubt may be raised about the reliability of this procedure for the quantitative estimation of rate constants, especially since most of the alanine peptides were cleaved at more than one peptide bond. There are other problems as well, such as the effect of the location of the terminal α-amino group of the substrate in relation to the site of cleavage.

We have recently tried to approach the problem of the active site of papain in a manner analogous to that just described for pepsin. To meet the specificity requirements of papain, a series of substrates was designed in which the Glu-

Leu bond was the only one cleaved, and this dipeptidyl unit was flanked on the amino side by a valyl residue (to provide a hydrophobic amino acid and to take advantage of the fact that the Val-Y bond is resistant to papain action) and by a glycyl residue on the carboxyl side (to remove the C-terminal carboxyl group from adjacence to the sensitive bond). The resulting tetrapeptide derivative, Z-Val-Glu-Leu-Gly, was then modified by the addition of one or two glycyl residues between the benzyloxycarbonyl and valyl groups, and the Z group was also replaced by the mansyl group for fluorescence studies (Lowbridge and Fruton 1974).

Although the kinetic data for the cleavage of the Glu-Leu bond in the series of substrates with the amino-terminal benzyloxycarbonyl group were consistent with the view that papain possesses an extended active site, no inferences could be drawn about the effect of the hydrophobic Z group on enzyme catalysis. In the face of the current uncertainties about the mechanism of papain action on peptide substrates (Mole and Horton 1973), this is perhaps not surprising.

With the mansyl compounds, fluorescence measurements showed that neither active papain nor papain inactivated by blockage of the active-site sulfhydryl group has appreciable affinity for the mansyl group of mansic acid, Mns-Gly-Gly or Mns-Gly-Val-Glu. However, with substrates of the type A-Val-Glu-Leu-Gly, where A = Mns, Mns-Gly or Mns-Gly-Gly, the mansyl group is drawn, by peptide-protein interaction, into an environment of lower polarity, and the apparent binding affinity increases with increasing chain length of the mansyl peptide. Although much more work is needed along this line, the available data suggest considerable conformational flexibility at the active site of papain. We have recently initiated stopped-flow fluorescence experiments on the interaction of papain with Mns-Gly-Gly-Val-Glu-Leu-Gly. Our initial data under conditions where $E>>S$ give a surprisingly low second-order rate constant for the rate of formation of the ES complex, namely, 10^5 $M^{-1}sec^{-1}$ (pH 6.5, 25°C), suggesting that conformational changes in the enzyme may be involved in the enzyme-substrate interaction. The initial rise in fluorescence, due to the formation of the enzyme-ligand complex, is then followed by a single first-order decrease in fluorescence, which we attribute to the cleavage of the Glu-Leu bond and the release of Mns-Gly-Gly-Val-Glu, for which the enzyme has little affinity.

Although the studies on the interaction of pepsin and papain with substrates bearing a fluorescent probe group suggest that there may be kinetically significant conformational changes at the active sites of these two proteinases, much additional work will be needed before we can affirm that such changes do in fact occur. Moreover, it should be added that the few experiments of this kind that have been performed on the interaction of chymotrypsin with peptides of the type DNS-(Gly)$_n$-Phe-OEt (M. A. Johnston and J. S. Fruton, unpubl.) do not give evidence of a drawing of the fluorescent probe group into an hydrophobic region of this enzyme. This is, of course, consistent with the apparent absence of a well-defined cleft in the three-dimensional model based on the X-ray crystallographic study of chymotrypsin. It would seem, therefore, that among the proteinases there may be a considerable variety of extended active sites, whose topography and conformational flexibility depend on the structure of the enzyme protein as a whole.

This attempt to discuss the factors that may determine the specificity of proteinase action on protein substrates has offered more questions than answers. At the present stage of development of the problem, that is perhaps only to be expected.

Acknowledgments

The recent research of my laboratory described here was aided by grants from the National Institutes of Health (GM-18172 and AM-15682) and from the National Science Foundation (GB 37871X).

REFERENCES

Atlas, D. and A. Berger. 1973. Size and stereospecificity of the active site of porcine elastase. *Biochemistry* **12:**2573.

Baumann, W. K., S. A. Bizzozero and H. Dutler. 1973. Kinetic investigation of the α-chymotrypsin-catalyzed hydrolysis of peptide substrates. *Eur. J. Biochem.* **39:**381.

Bender, M. L. and F. J. Kezdy. 1964. The current status of the α-chymotrypsin mechanism. *J. Amer. Chem. Soc.* **86:**3704.

————. 1965. Mechanism of action of proteolytic enzymes. *Annu. Rev. Biochem.* **34:**49.

Bergmann, M. and J. S. Fruton. 1941. The specificity of proteinases. *Adv. Enzymol.* **1:**63.

Birktoft, J. J., D. M. Blow, R. Henderson and T. A. Steitz. 1970. The structure of α-chymotrypsin. *Phil. Trans. Roy. Soc. London* B **257:**67.

Blake, C. C. F., G. A. Mair, A. C. T. North, D. C. Phillips and V. R. Sarma. 1967. On the conformation of the hen egg-white lysozyme molecule. *Proc. Roy. Soc.* (London) B **167:**365.

Blow, D. M., J. Janin and R. M. Sweet. 1974. Mode of action of soybean trypsin inhibitor (Kunitz) as a model for specific protein-protein interaction. *Nature* **249:**54.

Bustin, M. and A. Conway-Jacobs. 1971. Intramolecular activation of porcine pepsinogen. *J. Biol. Chem.* **246:**615.

Doi, R. H. 1973. Role of proteases in sporulation. *Current Topics in Cellular Regulation* (ed. B. L. Horecker and E. R. Stadtman), vol. 6, pp. 1–20. Academic Press, New York.

Drenth, J., J. N. Jansonius, R. Koekoek and B. G. Wolters. 1971. The structure of papain. *Adv. Protein Chem.* **25:**79.

Fastrez, J. and A. R. Fersht. 1973. Demonstration of the acyl-enzyme mechanism for the hydrolysis of peptides and anilides by chymotrypsin. *Biochemistry* **12:**2025.

Ferguson, J. B., J. R. Andrews, I. M. Voynick and J. S. Fruton. 1973. The specificity of cathepsin D. *J. Biol. Chem.* **248:**6701.

Fruton, J. S. 1970. The specificity and mechanism of pepsin action. *Adv. Enzymol.* **33:**401.

Gertler, A. and T. Hofmann. 1970. Acetyl-L-alanyl-L-alanyl-alanine methyl ester: A new highly specific elastase substrate. *Can. J. Biochem.* **48:**384.

Gutfreund, H. and J. M. Sturtevant. 1956. The mechanism of the reaction of chymotrypsin with *p*-nitrophenyl acetate. *Biochem. J.* **63:**656.

Hein, G. E. and C. Niemann. 1962. Steric course and specificity of α-chymotrypsin-catalyzed reactions. II. *J. Amer. Chem. Soc.* **84:**4495.

Hill, R. L. 1965. Hydrolysis of proteins. *Adv. Protein Chem.* **20**:37.

Himoe, A., K. G. Brandt, R. J. deSa and G. P. Hess. 1969. Investigations of the chymotrypsin-catalyzed hydrolysis of specific substrates. IV. Pre-steady state kinetic approaches to the investigation of the catalytic hydrolysis of esters. *J. Biol. Chem.* **244**:3483.

Hirohata, H., M. L. Bender and R. S. Stark. 1974. Acylation of α-chymotrypsin by oxygen and sulfur esters of specific substrates: Kinetic evidence for a tetrahedral intermediate. *Proc. Nat. Acad. Sci.* **71**:1643.

Knowles, J. R. 1965. Enzyme specificity: α-Chymotrypsin. *J. Theoret. Biol.* **9**:213.

Laskowski, M., Jr. and R. W. Sealock. 1971. Protein proteinase inhibitors—Molecular aspects. In *The Enzymes* (ed. P. D. Boyer), 3rd ed., vol. 3, pp. 375–473. Academic Press, New York.

Lienhard, G. E. 1973. Enzymatic catalysis and transition-state theory. *Science* **180**:149.

Linderstrøm-Lang, K. 1939. Globular proteins and proteolytic enzymes. *Proc. Roy. Soc.* (London) **B 127**:17.

Lowbridge, J. and J. S. Fruton. 1974. Studies on the extended active site of papain. *J. Biol. Chem.* **249**:6754.

Magnusson, S. 1971. Thrombin and prothrombin. In *The Enzymes* (ed. P. D. Boyer), 3rd ed., vol. 3, pp. 277–321. Academic Press, New York.

Maroux, S., J. Baratti and P. Desnuelle. 1971. Purification and specificity of porcine enterokinase. *J. Biol. Chem.* **246**:5031.

McPhie, P. 1974. Pepsinogen: Activation by a unimolecular mechanism. *Biochem. Biophys. Res. Comm.* **56**:789.

Medzihradszky, K., I. M. Voynick, H. Medzihradszky-Schweiger and J. S. Fruton. 1970. Effect of secondary enzyme-substrate interactions on the cleavage of synthetic peptides by pepsin. *Biochemistry* **9**:1154.

Mole, J. E. and H. R. Horton. 1973. Kinetics of papain-catalyzed hydrolysis of α-N-benzoyl-L-arginine-p-nitroanilide. *Biochemistry* **12**:816.

Morihara, K. and T. Oka. 1973a. Effect of secondary interaction on the enzymatic activity of trypsin-like enzymes from *Streptomyces*. *Arch. Biochem. Biophys.* **156**:764.

———. 1973b. Effect of secondary interaction on the enzymatic activity of subtilisin BPN': Comparison with α-chymotrypsin, trypsin, and elastase. *FEBS Letters* **33**:54.

Morihara, K. and H. Tsuzuki. 1970. Thermolysin: Kinetic study with oligopeptides. *Eur. J. Biochem.* **15**:374.

Neil, G. L., C. Niemann and G. E. Hein. 1966. Structural specificity of α-chymotrypsin: Polypeptide substrates. *Nature* **210**:903.

Neurath, H. and G. W. Schwert. 1950. The mode of action of the crystalline pancreatic proteolytic enzymes. *Chem. Rev.* **46**:69.

Raju, E. V., R. E. Humphreys and J. S. Fruton. 1972. Gel filtration studies on the binding of peptides to pepsin. *Biochemistry* **11**:3533.

Raymond, M. N., J. C. Mercier and E. Bricas. 1973. New oligopeptide substrates of chymosin (rennin). In *Peptides 1972* (ed. H. Hanson and H. D. Jakubke), pp. 380–382. North-Holland, Amsterdam.

Raymond, M. N., J. Garnier, E. Bricas, S. Cilianu, M. Blasnic, A. Chaix and P. Lefrancier. 1972. Studies on the specificity of chymosin (rennin). I. Kinetic parameters of the hydrolysis of synthetic oligopeptide substrates. *Biochimie* **54**:145.

Sachdev, G. P. and J. S. Fruton. 1970. Secondary enzyme-substrate interactions and the specificity of pepsin. *Biochemistry* **9**:4465.

Sachdev, G. P., A. D. Brownstein and J. S. Fruton. 1973. *N*-Methyl-2-anilino-naphthalene-6-sulfonyl peptides as fluorescent probes for pepsin-substrate interaction. *J. Biol. Chem.* **248**:6292.

———. 1975. Fluorescence studies on the active sites of porcine pepsin and *Rhizopus*-pepsin. *J. Biol. Chem.* **250**:501.

Sachdev, G. P., M. A. Johnston and J. S. Fruton. 1972. Fluorescence studies on the interaction of pepsin with its substrates. *Biochemistry* **11**:1080.

Sampath-Kumar, P. S. and J. S. Fruton. 1974. Studies on the extended active sites of acid proteinases. *Proc. Nat. Acad. Sci.* **71**:1070.

Schachter, M. 1969. Kallikreins and kinins. *Physiol. Rev.* **49**:509.

Schechter, I. and A. Berger. 1967. On the size of the active site in proteases. I. Papain. *Biochem. Biophys. Res. Comm.* **27**:157.

———. 1968. On the active site of proteases. III. Mapping the active site of papain: Specific peptide inhibitors of papain. *Biochem. Biophys. Res. Comm.* **32**:898.

Segal, D. M. 1972. A kinetic investigation of the crystallographically deduced binding subsites of bovine chymotrypsin A$_\gamma$. *Biochemistry* **11**:349.

Segal, D. M., J. C. Powers, G. H. Cohen, D. R. Davies and P. E. Wilcox. 1971. Substrate binding site in chymotrypsin A$_\gamma$. A crystallographic study using peptide chloromethyl ketones as site-specific inhibitors. *Biochemistry* **10**:3728.

Shotton, D. M. and H. C. Watson. 1970. The three-dimensional structure of crystalline porcine pancreatic elastase. *Phil. Trans. Roy. Soc. London* B **257**:111.

Thompson, R. C. and E. R. Blout. 1973. Dependence of the kinetic parameters for elastase-catalyzed amide hydrolysis on the length of peptide substrates. *Biochemistry* **12**:57.

Tschesche, H. 1974. Biochemistry of natural proteinase inhibitors. *Angew. Chem.* **13**:10.

Umezawa, H. 1972. *Enzyme Inhibitors of Microbial Origin,* pp. 34–50. University of Tokyo Press, Tokyo.

Voynick, I. M. and J. S. Fruton. 1971. The comparative specificity of acid proteinases. *Proc. Nat. Acad. Sci.* **68**:257.

Wang, J. L. and G. M. Edelman. 1971. Fluorescent probes for conformational states of proteins. IV. The pepsinogen-pepsin conversion. *J. Biol. Chem.* **246**:1185.

Weber, G. and L. B. Young. 1964. Fragmentation of bovine serum albumin by pepsin. *J. Biol. Chem.* **239**:1415.

Wilson, W. D. and J. F. Foster. 1971. Conformation-dependent limited proteolysis of bovine plasma albumin by an enzyme present in commercial preparations. *Biochemistry* **10**:1772.

Wolfenden, R. 1972. Analog approaches to the structure of the transition state in enzyme reactions. *Acct. Chem. Res.* **5**:10.

Limited Proteolysis
and Zymogen Activation

Hans Neurath

Department of Biochemistry, University of Washington
Seattle, Washington 98195

A number of physiological events are initiated by the selective cleavage of peptide bonds, a process also referred to as "limited proteolysis." This term was first introduced by Linderstrøm-Lang and Ottesen (1949) to describe the restricted peptide bond cleavage which results from the conversion of ovalbumin to plakalbumin due to a protease of *B. subtilis*. Limited proteolysis is a "one-by-one" process in which peptide bonds are cleaved sequentially. In contrast, the kinetic behavior of protein degradation is that of an "all-or-none" process, which proceeds as if all susceptible bonds were cleaved simultaneously (Linderstrøm-Lang 1952). Numerous examples of limited proteolysis have been described in the literature, but the nature of the phenomenon can only be understood by considering the specificity of the protease and the structure of the protein substrate.

Kinetic studies with model substrates have led to the delineation of the specificity of proteases in terms of the amino acid side chains which surround the susceptible peptide bond (see Walsh; Fruton; both this volume). In the case of large protein substrates, an additional requirement must be met, namely, that the susceptible peptide bond be accessible to the attacking protease and that the topography of its environment fit the active site of the proteolytic enzyme. It is to be expected, therefore, that, in general, limited proteolysis is directed toward surface loops of the protein substrate, rather than toward regions of the peptide chain that lie within the internal domain of the protein molecule (Naslin, Spyridakis and Labeyrie 1973). This expectation is fully borne out by experimental observations. Typical examples are the limited hydrolysis of ribonuclease by subtilisin (Kartha, Bello and Harker 1967; Wyckoff et al. 1967; Carlysle et al. 1974), the conversion of chymotrypsinogen to chymotrypsin (Sigler et al. 1968), and the limited cleavage of peptide bonds of bovine carboxypeptidase B by enzymes of autolyzed pancreatic juice (Reeck et al. 1971; Herriott and Schmid 1975). In each case, the susceptible peptide bonds are in flexible regions of polypeptide chains on the surface of the protein molecules. This "topographical specificity" is responsible for the fact that, in general, compact native protein molecules are more resistant to proteolytic degradation than are denatured ones.

It is almost self-evident that limited proteolysis will convert a single-chain protein into a two- or multichain structure. For instance, whereas chymotrypsinogen is a single-chain zymogen, π-chymotrypsin, the first product of activation, is composed of two chains held together by disulfide bonds, and α-chymotrypsin, the final product, contains three chains (Neurath 1957). Similarly, the two chains of insulin are derived from a single-chain precursor by removal of an internal chain segment, the so-called C peptide (Steiner et al. 1969). The precursors of the blood-clotting proteases, factors IX and prothrombin, are composed of single chains, whereas the active forms, factor IX_a and thrombin, consist of two chains held together by disulfide bonds (Davie et al. 1974; Enfield et al. 1974). It stands to reason that peptide bond cleavage serves to bring distant functional amino acid residues into a three-dimensional relationship that cannot be exactly achieved in the precursor.

Regulation by Limited Proteolysis

The very factors which limit enzymatic proteolysis render this process eminently suitable for controlling physiological processes. These are (1) the specificity of enzymatic peptide bond cleavage and (2) its irreversibility in aqueous solutions. These factors singly and together introduce a permanent change in the molecular environment which, in turn, can initiate biological functions and regulate physiological reactions.

The initiation of functions as well as their regulation can be attributed to a process in which a protein precursor is converted to an active protein. These reactions are essentially irreversible because the free energy of peptide bond cleavage in aqueous solution is a large negative number. The reaction is specific because, as previously discussed, the topography of the susceptible peptide bond has to fit the active site of the attacking protease. This control is different in kind from the reversible control mechanisms that involve allosteric transitions and phosphorylation or adenylylation. Figure 1 intends to

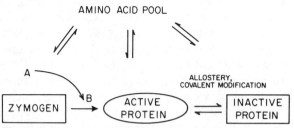

Figure 1
Control mechanisms which mediate the formation of an active protein from inactive precursors. Irreversible, limited proteolysis of a zymogen (*left*), triggered by a specific protease, is contrasted with reversible transitions (*right*). The reversible arrows from the amino acid pool to active protein, zymogen and inactive protein are shorthand abbreviations for control processes on the transcriptional and translational levels.

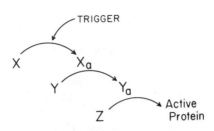

Figure 2

Consecutive enzyme-catalyzed reactions (cascade) involving zymogen to enzyme conversions. For details, see the text.

place these two types of conversions of an inactive protein to the active form within the larger context of various control mechanisms involved in protein metabolism. The control by limited proteolysis can be more finely regulated by the interposition of additional zymogen-enzyme conversions.

The term "cascade" or waterfall reaction was introduced some ten years ago by Davie and Ratnoff (1964) to describe such a sequence of enzymatic inductions in which the product of one reaction is a catalyst for the next. Cascade processes have several features which render them eminently suited for control. These are summarized in Figure 2, which illustrates three consecutive enzyme-catalyzed reactions. The system can be said to be *poised* and only needs to be *triggered* to set the consecutive reactions in motion. There exists a reciprocal relationship between enzyme and zymogen, such that each zymogen contains a site that is recognized by the preceding enzyme and a site which after activation recognizes the succeeding zymogen.

The classical example of a two-stage cascade process is the activation of pancreatic zymogens (Fig. 3). Here the trigger is intestinal enterokinase, whose secretion is under hormonal control. The first stage in the cascade is the conversion of pancreatic trypsinogen to trypsin. The transfer of the zymogen from the zymogen granules in the pancreas to the lumen of the small intestine regulates communication between zymogen and trigger. The product of the first stage, active trypsin, in turn catalyzes the conversion of a number of pancreatic zymogens to their respective enzymes (Neurath 1964) (Fig. 3, right). The dotted arrow indicates that trypsin itself can activate trypsinogen autocatalytically. However, whereas trypsin will potentially cleave any lysyl or arginyl bond in the protein substrate, enterokinase seems to be completely specific for that lysyl bond which is preceded by a series of aspartyl residues

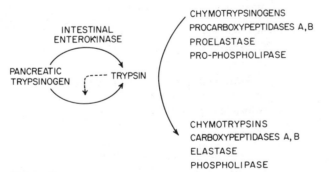

Figure 3

A two-stage cascade involved in the activation of pancreatic zymogens, triggered by enterokinase.

(Maroux, Baratti and Desnuelle 1971). This restricted specificity renders enterokinase resistant to self-digestion during storage in the intestinal cell, in contrast to trypsin which also for its own protection needs to be stored as a zymogen.

Another feature of consecutive zymogen-enzyme conversion reactions (cascades) is that of *enzymatic amplification*. Thus one molecule of trigger might yield 10^3 molecules of X_a, 10^6 molecules of Y_a and 10^9 molecules of active protein. The sequence of events is determined by the specificity of each active enzyme. Finally, an element of control can be introduced if the trigger and the precursor originate in different cells or tissues. In such instances, the production of active protein is regulated also by the *timing of transport* of the components from two or more different cell systems.

Physiological Control Systems

I should like to illustrate briefly the importance of the factors just enumerated for some of the physiological processes which involve limited proteolysis in a single reaction or in cascades. A representative, but by no means exhaustive, list of such systems is given in Table 1.

The most prominent defense reactions are those of the blood-clotting and the complement systems. Both of these reaction systems are poised and depend upon enzyme amplification. The series of reactions which precede the conversion of fibrinogen to fibrin is discussed in greater detail by Davie et al. (this volume).

The complement system that participates in the immune reaction comprises 11 proteins, some of which are members of the group of serine proteases

Table 1

Physiological Systems Controlled by Limited Proteolysis

Physiological system	*Example*
Defense reactions	blood coagulation fibrinolysis complement reaction
Hormone production	proinsulin $\longrightarrow$ insulin angiotensinogen $\longrightarrow$ angiotensin
Assembly	bacteriophage procollagen $\longrightarrow$ collagen fibrinogen $\longrightarrow$ fibrin clot
Development	prochitin synthetase $\longrightarrow$ chitin synthetase prococoonase $\longrightarrow$ cocoonase fertilization (pro-acrosin $\longrightarrow$ acrosin)
Digestion	zymogen $\longrightarrow$ enzyme
Tissue injury	impairment of cell contact inhibition prekallikrein $\longrightarrow$ kallikrein kininogen $\longrightarrow$ kinin

(Müller-Eberhard 1975). The final event is the rupture of the cell membrane followed by cell lysis. The fibrinolysis system is yet another series of defense reactions that are controlled by proteolytic enzymes (Guest 1966).

The initiation of physiological function is best exemplified by the production of protein hormones on one hand and the activation of pancreatic zymogens on the other. With regard to the protein hormones, it is worthy of note that the existence of proinsulin had been overlooked for many years because the zymogen-hormone conversion occurs so rapidly following disruption of the cell structure (Steiner et al. 1969). Both the zymogen and the activating enzymes are products of the same type of cells, but control mechanisms exist both to trigger the reaction and to prevent continuous hormone production. Since these hormones are specific fragments of larger protein precursors, one may wonder whether other fragments of the precursor perform additional physiological functions (Tager and Steiner 1974). Stated in more general terms, one wonders whether the activation peptides are merely throw-away fragments, or whether they serve physiological functions of their own. Two instances can be cited for a physiological action of activation peptides: (1) the inhibitory effect of a fragment (propiece II) derived from bovine prothrombin on the activation of this zymogen (Jesty and Esnouf 1973), and (2) the hormonelike suppression of gastric secretion by the activation peptide of pancreatic trypsinogen (Abita et al. 1973). The question of the physiological role of activation peptides is important and deserving of greater attention.

Limited proteolysis is also involved in reactions which yield self-assembling proteins. Thus components of bacteriophage can be accumulated as precursors until their interlocked mates are prepared. The whole assembly line is then turned on, à la Henry Ford, by limited proteolysis of one or more components (Wood et al. 1968). Similarly, collagen and fibrinogen are secreted by cells in a soluble precursor form which then can be programmed to form irreversible collagen fibrils (Bornstein 1974) or fibrin clots (Davie et al., this volume).

Reference has already been made by Walsh (this volume) to the role of proteolytic enzymes in fertilization and development. The irreversibility of proteolysis and the poised nature of the zymogen activation make these enzymes eminently suited for those purposes. In budding yeast, chitin synthetase is activated from its zymogen when needed for septum formation (Cabib and Farkas 1971), and cocoonase, a trypsinlike protease used by certain silk moths for their escape from the cocoon, is secreted as a precursor and activated in response to developmental programming (Kramer, Felsted and Law 1973). Similarly, acrosin, the protease in the head of mammalian sperm, is secreted as a precursor, but when activated, becomes a loaded gun for penetration of the ovum (Polakoski 1974).

The role of proteolytic enzymes in tissue injury is exemplified here by the effect of protease inhibitors on cell contact inhibition of malignant cells, as discussed by Reich (this volume). Another example is the kallikrein-kinin system (Back 1966), which is activated in certain pathological states associated with increased fibrinolysis (Back 1966).

The release of the hypotensive kinin peptides from plasma kininogens is catalyzed by kallikrein, a trypsinlike enzyme, which in turn is the product of activation of its precursor prokallikrein (Bagdasarian, Lahiri and Colman

1973). This system, too, may be considered to be a cascade akin to the blood coagulation and the fibrinolysis systems. Both steps of the cascade are also controlled by a variety of extrinsic tissue inhibitors (Back 1966).

Zymogen Activation

In all known zymogen-enzyme conversions, the specific peptide bond cleaved is in a region of a polypeptide chain which is amino terminal relative to the active center. At first this may seem to be a curious phenomenon, but could it be any other way if the product of the activation reaction serves a physiological role? To put it simply, if the peptide removed during activation were at the carboxyl terminus, rather than at the amino terminus, trypsin would be synthesized before trypsinogen, fibrin before fibrinogen, collagen before procollagen. By synthesizing an inactivation prefix before synthesizing the active portion of the protein molecule, premature physiological function is avoided. An alternative mechanism on the posttranslational level for protecting an enzyme from expressing its function would be the formation of an enzyme-inhibitor complex. These two mechanisms are fundamentally different, as is illustrated in Figure 4. The enzyme inhibitor is complementary to the structure of the active site and, as shown by Laskowski and coworkers (Laskowski and Sealock 1971), has the characteristics of a pseudosubstrate. In fact, only a few residues of inhibitor interact with the active site of the enzyme, whereas the bulk of the protein inhibitor acts mainly as a supporting structure. Zymogens, on the other hand, are inactive due to their having a slightly different conformation than the active enzyme, and the structure of the activation peptide is unrelated to the structure of the active site. There is some evidence that enzyme-inhibitor complexes preceded the evolution of zymogen activation since, by and large, zymogens are rare among lower forms of life, whereas enzyme inhibitors are not. There is also presumptive evidence that the enzyme evolved prior to the zymogen, and that the additional peptide segment has been added later as a new message between initiation and translation of the enzyme.

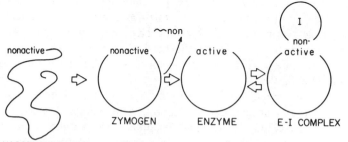

Figure 4
Schematic illustration of the control of enzyme activity by zymogen activation as contrasted with the dissociation of an enzyme-inhibitor complex.

Evolution

It is well recognized that the mammalian serine proteases are a set of homologous proteins, and that during evolution, those amino acid residues have been preserved which are essential for function or for conformation. These include, in particular, the peptide sequences around the amino-terminal residue and around the serine and histidine of the active site (Walsh, this volume). An abbreviated presentation of these homologies in pancreatic trypsin, chymotrypsin and elastase as well as in plasma proteases involved in blood clotting, such as thrombin and factor X_a, are shown in Table 2. These homologies provide evidence that the pancreatic and hepatic serine proteases have evolved from an ancestral gene by gene duplication and divergent evolution. At first sight, the same does not appear to apply to the prefixes or activation peptides of the corresponding zymogens. It is true that the activation peptides of all trypsinlike enzymes are homologous, and so are the activation peptides of all chymotrypsinlike enzymes (Table 3). However, there is no proof that the sequences characteristic of the trypsinogen activation peptides are homologous to those characteristic of the activation peptides of chymotrypsinogens. As a third class of homologous activation peptides, we may consider those of the blood-clotting zymogens: factor IX, factor X and prothrombin. The amino-terminal sequences of these zymogens are homologous (Enfield et al. 1974), but we do not know as yet the sequences of the overlap regions connecting the activation peptides to the zymogens. If, in fact, the activation peptides of these three groups of serine proteases are different, one must conclude that the enzymes arose first and that the prefixes were added later as several independent events following gene duplication and divergence of function. Alternatively, one might argue that the activation peptides of the serine proteases represent distant homologous sequences that have undergone rapid mutations as, for instance, in the case of fibrinopeptides (Doolittle et al. 1971). Indeed, the activation peptides of the serine proteases differ appreciably in length (Fig. 5) as well as in composition and structure. However, as large as these differences appear to be when, for instance, trypsinogen and prothrombin are compared to each other, the possibility cannot be excluded a priori that they are products of a rapidly mutating ancestral gene. It is of

Table 2
Homologous Peptide Sequences of Functional Amino Acid Residues of Serine Proteases

Enzyme	N-terminus	His-57	Ser-195
Trypsin	Ile-Val-Gly-	Ala-Ala-His-Cys-Tyr-	Gly-Asp-Ser-Gly-Gly-
Cocoonase	Ile-Val-Gly		Gly-Asp-Ser-Gly-Gly-
Chymotrypsin A	Ile-Val-Asn	Ala-Ala-His-Cys-Gly-	Gly-Asp-Ser-Gly-Gly-
Elastase	Val-Val-Gly-	Ala-Ala-His-Cys-Val-	Gly-Asp-Ser-Gly-Gly-
Thrombin	Ile-Val-Glu-	Ala-Ala-His-Cys-Leu-	Gly-Asp-Ser-Gly-Gly-
Plasmin	Val-Val-Gly-	Ala-Ala-His-Cys-Leu-	Gly-Asp-Ser-Gly-Gly-
Factor X_a	Ile-Val-Gly-	Ala-Ala-His-Cys-Leu-	Gly-Asp-Ser-Gly-Gly-

For more extensive sequence comparison, see Dayhoff (1973).

Table 3

Zymogen Activation Peptides of Homologous Serine Proteases

I.	Trypsinogen group	
	bovine trypsinogen	Val-Asp-Asp-Asp-Asp-Lys
	porcine trypsinogen	Phe-Pro-Thr-Asp-Asp-Asp-Asp-Lys
	lungfish trypsinogen	Phe-Pro-Ile-Glu-Glu-Asp-Lys
	pro-cocoonase	Lys-Lys-Thr-Pro-Gln-Arg-Thr-Gln-Asp-Asp-Gly-Gly-Lys
II.	Chymotrypsinogen group	
	bovine chtg A	Cys-Gly-Val-Pro-Ala-Ile-Gln-Pro-Val-Leu-Ser-Gly-Leu-Ser-Arg
	bovine chtg B	Cys-Gly-Val-Pro-Ala-Ile-Gln-Pro-Val-Leu-Ser-Gly-Leu-Ala-Arg
	bovine subunit II	Cys-Gly-Ala-Pro-Ile-Phe-Gln-Pro-Asn-Ser- — - — - — -Ala-Arg
	lungfish proelastase	Cys-Gly-Val-Pro-Ser-Tyr-Pro-Pro-His- — - — - — - — -Ala-Arg
III.	Prothrombin group	
	prothrombin	Ala-Asn-Lys-Gly-Phe-Leu-Glx-Glx- — -Val-Arg-Lys-Gly-Asn-Leu
	factor IX	Tyr-Asn-Ser-Gly-Lys-Leu-Glu-Glu-Phe-Val-Arg- — -Gly-Asn-Leu
	factor X	Ala-Asn-Ser- — -Phe-Leu-Glu-Glu- — -Val-Lys-Gln-Gly-Asn-Leu

interest in this connection that computer analysis by de Haen and Teller in our laboratory (de Haen, Neurath and Teller 1975) has led to the proposal that the ancestral serine protease probably had a structure similar to that of thrombin, and that all other serine proteases arose later by divergent evolution. If this is the case, one would expect that the cascade process leading to the formation of the fibrin clot has evolved in the direction from thrombin to factor IX_a, i.e., that gene duplication and mutation have redirected the specificity of the ancestral protease so as to introduce an additional amplifying mechanism into the cascade.

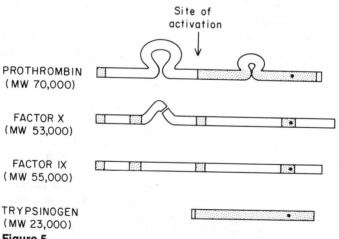

Figure 5

Comparison of the lengths of polypeptide chains of the zymogen forms of homologous serine proteases. The stipled segments are homologous. The dot in the enzyme segment (*right*) indicates the reactive serine.

Mechanism

As a final discussion point, I should like to raise the question of the nature of the structural difference between the enzyme and its zymogen and inquire in what manner the cleavage of a specific peptide bond in the zymogen could generate enzymatic function.

Does limited proteolysis merely remove an obstruction from the active site or does it induce a conformational change in the zymogen? Is such a change of major proportion or is it subtle and perhaps beyond detection by current techniques? Does zymogen activation involve the de novo assembly of the catalytic apparatus or does it merely improve the preexisting arrangement? In the best-studied case, namely, the X-ray comparison of chymotrypsinogen and chymotrypsin, the differences were found to be so subtle that it was not clear which structural changes were crucial to zymogen activation and which were only incidental (Freer et al. 1970). A new insight into the mechanism of zymogen activation has been recently gained by the observation that several well-known zymogens, including procarboxypeptidases A and B, trypsinogen, chymotrypsinogen and pepsinogen, reveal a weak intrinsic activity prior to activation, generally several orders of magnitude smaller than that of the enzyme. If we assume that the activities of the zymogen and of the enzyme are mediated by the same functional groups, the relative ineffectiveness of the zymogen as a catalyst could be ascribed to a less effective binding of the substrate, to an imperfectly developed catalytic apparatus, or to a combination of both.

Diisopropylfluorophosphate, commonly referred to as DFP, is an active-site titrant that reacts specifically with the serine residue of the active site of various serine proteases. The product is inactive DIP enzyme. DFP also reacts with corresponding zymogens, but as shown in Table 4, the second-order rate constant of the reaction with the zymogen is several orders of magnitude lower than that with the corresponding enzyme (Morgan et al. 1972). In each case, the reaction involves the active serine, as evidenced by the fact that one gram atom of radiophosphorus is incorporated per mole of protein. Treatment of DIP trypsinogen, as of trypsinogen, with trypsin cleaves the activation peptide, but the product is inactive DIP-trypsin instead of trypsin. The zymogen

Table 4
Second-order Rate Constants of
Reactions with DFP

Protein	$k_2(l \cdot mole^{-1}min^{-1})$
Chymotrypsinogen	0.15
Trypsinogen	0.04
Kallikrein	8.0
Trypsin	300.0
Chymotrypsin	2700.0
Acetylcholine-esterase	13000.0

pH 7.0 for chymotrypsinogen and trypsinogen and pH 7.2, ionic strength 0.1, 25°C for all others.

chymotrypsinogen, like the enzyme chymotrypsin, can also catalyze the hydrolysis of an acyl ester, such as p-nitrophenyl-p'-guanidinobenzoate, forming a transient acyl zymogen, analogous to the acyl enzyme. The intermediate, guanidinobenzoyl-chymotrypsinogen, is stable at pH 5 and therefore can be isolated, but at pH 7 it undergoes spontaneous hydrolysis to regenerate the zymogen and guanidinobenzoate (Gertler, Walsh and Neurath 1974). A combination of kinetic data derived from the reaction of chymotrypsinogen with this ester and from inhibition studies with the homologous zymogen trypsinogen indicates that the low catalytic activity of these zymogens is primarily due to an imperfectly developed binding site and only secondarily to an impaired catalytic apparatus (Table 5). The overall reaction rates of these zymogens are about 10^5–10^6 times lower than those of the corresponding enzymes. This factor of approximately one million can be thought of as being the product of two factors, one relating to binding and the other relating to bond breaking. The binding ratios of enzyme to zymogen are approximately 10^4, but the catalytic rates of decomposition of the acyl enzyme intermediates are of the order of 10^2 (Table 5). Thus in the case of the serine proteases at least, the zymogen is a poor catalyst because it has an imperfectly developed binding site, whereas the bond-breaking mechanism by and large is preexistent in the zymogen. These conclusions, based primarily on kinetic evidence, agree with crystallographic data which show that in chymotrypsinogen, the substrate binding pocket is imperfectly developed but becomes fully established by rotation of a segment of the polypeptide chain, which moves a methionine residue from the interior of the molecule toward the exterior (Freer et al. 1970). However, the catalytic apparatus (usually referred to as the charge-relay system) appears to be present in the zymogen as well as in the enzyme, a

Table 5

Reactions of Enzymes and Zymogens

Reaction[a]	Relative rates (enzyme/zymogen) (approx.)
Inactivation	
$E/Z + iPr_2P\text{-}F \rightarrow iPrP\text{-}Z/E$[b]	7×10^4
$E/Z + CH_3SO_2F \rightarrow CH_3SO_2\text{-}Z/E$[c]	5×10^1
Acylation	
$E/Z + NBGB \rightarrow GB\text{-}Z/E$[d]	10^6
Deacylation	
$GB\text{-}E/Z \xrightarrow{H_2O} GB + Z/E$[d]	8×10^1
Competitive inhibition	
$E/Z + PABA \rightleftharpoons (Z/E\text{-}PABA)$[d]	$6 \times 10^3 \ (1/K_i)$

[a] The following symbols are used: E = enzyme (chymotrypsin or trypsin); Z = zymogen (chymotrypsinogen or trypsinogen); E/Z means reaction with E or with Z. $iPr_2P\text{-}F$ = diisopropylphosphorofluoridate; CH_3SO_2F = methane sulfonyl fluoride; NPGB = p-nitrophenyl-p'-guanidinobenzoate; GB = guanidinobenzoic acid; PABA = p-amino-benzamidine.

[b] Data of Robinson, Neurath and Walsh (1973) and Morgan et al. (1972).

[c] Data of Morgan, Walsh and Neurath (1974).

[d] Data of Gertler, Walsh and Neurath (1974).

conclusion which has been recently corroborated by Robillard and Shulman (1974) using high-resolution proton nuclear magnetic resonance to assess the presence of the hydrogen bond between histidine 57 and aspartic acid 102 in both chymotrypsinogen and chymotrypsin.[1] We cannot tell whether the intrinsic, though weak, catalytic function of zymogens is of physiological significance or whether it is simply a useful experimental tool. In either event, it provides a probe for studying the refolding of a polypeptide chain during zymogen-enzyme conversion.

CONCLUSIONS

While the conversion of a chymotrypsinogen to chymotrypsin is by far the best understood example of the induction of biological activity by limited proteolysis, it is important to emphasize that it is only one of a great variety of physiological functions that can be induced by limited peptide bond cleavage of precursor proteins. The product of activation may be an enzyme, a hormone, a pharmacologically active peptide or a structural component of tissues, etc., but the trigger is a specific protease in every instance. Proteases can generate functions, but they can also destroy them, and in this sense it may be useful to think of limited proteolysis as a control element that can turn reactions on and off. The on and off reactions are controlled by different switches, so to speak, because proteolysis is an irreversible process, and as pointed out by Walsh (this volume), proteases are not endowed with repair functions.

It seems probable that limited proteolysis precedes protein degradation and that an element of specificity is thereby introduced to regulate protein breakdown. Evidence in support of this idea has recently been advanced by Katunuma and coworkers (Katunuma 1973), who have demonstrated the presence in a variety of mammalian tissues of "group-specific proteases" which recognize the coenzyme binding site of pyridoxal enzymes (such as ornithine transaminase, phosphorylase or homoserine deaminase), attack the apoenzyme more rapidly than the holoenzyme, and condition it for subsequent degradation by lysosomal proteases. The proposed mechanism is illustrated in Figure 6.

The conversion of fibrinogen to fibrin is turned on by thrombin and is regulated by the blood-clotting cascade. The off reaction involves a separate control system which regulates plasmin or fibinolysin, the enzymes involved in the dissolution of the fibrin clot. The growing realization that proteolytic enzymes participate in a great variety of physiological reactions undoubtedly prompted the organizers to convene the conference which resulted in this book. An intrinsic difficulty in the study of these activation reactions is the elusive

---

[1] According to the X-ray models of chymotrypsin and trypsin, the isopropyl group of DFP fits in the substrate binding pocket of the enzyme, which does not exist as such in the zymogen. If this structural element of the active site is, in fact, involved in binding, one would expect that a compound lacking this element would react more nearly equally with enzyme and zymogen. Methane sulfonyl fluoride (MSF) is such a compound. It reacts with the active serine, but the small methyl group does not provide a handle for binding. In fact, as shown in Table 5, the ratio of the second-order rate constants of the enzyme to zymogen in this case is not of the order of 10^6, but only 10^2 (Morgan, Walsh and Neurath 1974).

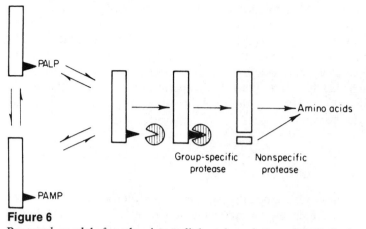

Figure 6
Proposed model for the intracellular degradation, by limited proteolysis, of pyridoxal-P-enzymes by group-specific proteases. (Reprinted, with permission, from Katunuma 1973.)

nature of the zymogens, since they tend to undergo spontaneous activation as soon as the cell structure is destroyed and the proteolytic enzyme which acts as trigger becomes exposed to the zymogen substrate. However, the recognition of this problem contains the elements of its solution since the activation reaction can be inhibited by suitable procedures once the molecular and enzymatic properties of the activating protease are known.

To summarize, limited proteolysis is a unique mechanism for controlling physiological functions. It can amplify signals and respond to them rapidly and irreversibly. Detailed investigation of the specific systems that generate, control and destroy physiologically active proteins by limited proteolysis should be a fruitful endeavor for future research.

Acknowledgments

Investigations described here, which recently originated in our laboratory, were supported by the National Institutes of Health (GM 15731) and by the American Cancer Society (BC91P).

REFERENCES

Abita, J. P., A. Moulin, M. Lazdunski, G. Hage, G. Palasciano, A. Brasca and O. Tiscornia. 1973. A physiological inhibitor of gastric secretion. The activation peptide of trypsinogen. *FEBS Letters* **34**:251.

Back, N. 1966. Fibrinolysin system and kinins. *Fed. Proc.* **25**:77.

Bagdasarian, A., B. Lahiri and R. W. Colman. 1973. Origin of the high molecular weight activator of prekallikrein. *J. Biol. Chem.* **248**:7742.

Bornstein, P. 1974. The biosynthesis of collagen. *Annu. Rev. Biochem.* **43**:567.

Cabib, E. and V. Farkas. 1971. The control of morphogenesis: An enzymatic mechanism for the initiation of septum formation in yeast. *Proc. Nat. Acad. Sci.* **68**:2052.

Carlysle, C. H., R. A. Palmer, S. K. Mazumdar, B. A. Gorinsky and D. G. R. Yeates. 1974. The structure of ribonuclease at 2.5 Å resolution. *J. Mol. Biol.* **85**:1.

Davie, E. W. and O. D. Ratnoff. 1964. Waterfall sequence for intrinsic blood clotting. *Science* **145**:1310.

Dayhoff, M. O., ed. 1973. *Atlas of Protein Sequence and Structure,* vol. 5. National Biomedical Research Foundation, Washington, D.C.

de Haen, C., H. Neurath and D. C. Teller. 1975. The phylogeny of trypsin-related serine proteases and their zymogens. New methods for the investigation of distant evolutionary relationships. *J. Mol. Biol.* **95**:225.

Doolittle, R. F., G. L. Wooding, Y. Lin and M. Riley. 1971. Hominoid evolution as judged by fibrinopeptide structures. *J. Mol. Evol.* **1**:74.

Enfield, D. L., L. H. Ericsson, K. Fujikawa, K. A. Walsh and H. Neurath. 1974. Bovine factor IX (Christmas factor). Further evidence of homology with factor X (Stuart factor) and prothrombin. *FEBS Letters* **47**:1.

Freer, S. T., J. Kraut, J. D. Robertus, H. T. Wright and Ng. H. Xuong. 1970. Chymotrypsinogen: 2.5 Å crystal structure, comparison with α-chymotrypsin, and implications for zymogen activation. *Biochemistry* **9**:1997.

Gertler, A., K. A. Walsh and H. Neurath. 1974. Catalysis by chymotrypsinogen. Demonstration of an acyl zymogen intermediate. *Biochemistry* **13**:1302.

Guest, M. M. 1966. Functional significance of the fibrinolytic enzyme system. *Fed. Proc.* **25**:73.

Herriott, J. R. and M. F. Schmid. 1975. Bovine carboxypeptidase B at 2.8 Angstrom resolution. *Proc. Nat. Acad. Sci.* (in press).

Jesty, J. and M. P. Esnouf. 1973. The preparation of activated factor X and its action on prothrombin. *Biochem. J.* **131**:791.

Kartha, G., J. Bello and D. Harker. 1967. Tertiary structure of ribonuclease. *Nature* **213**:862.

Katunuma, N. 1973. Enzyme degradation and its regulation by group-specific proteases in various organs of rats. In *Current Topics in Cellular Regulation* (ed. B. L. Horecker and E. R. Stadtman), vol. 7, p. 175. Academic Press, New York.

Kramer, K. J., R. L. Felsted and J. H. Law. 1973. Cocoonase. V. Structural studies on an insect serine protease. *J. Biol. Chem.* **248**:3021.

Laskowski, M., Jr. and R. W. Sealock. 1971. Protein proteinase inhibitors. Molecular aspects. In *The Enzymes* (ed. P. D. Boyer), vol. 3, p. 375. Academic Press, New York.

Linderstrøm-Lang, K. U. 1952. Proteins and enzymes. In *Lane Medical Lectures,* vol. 6, p. 1. Stanford University Publications, Stanford, California.

Linderstøm-Lang, K. U. and M. Ottesen. 1949. Formation of plakalbumin from ovalbumin. *Comp. Rend. Trav. Lab. Carlsberg* **26**:403.

Maroux, S., J. Baratti and P. Desnuelle. 1971. Purification and specificity of porcine enterokinase. *J. Biol. Chem.* **246**:5031.

Morgan, P. H., K. A. Walsh and H. Neurath. 1974. Inactivation of trypsinogen by methane sulfonyl fluoride. *FEBS Letters* **41**:108.

Morgan, P. H., N. C. Robinson, K. A. Walsh and H. Neurath. 1972. Inactivation of bovine trypsinogen and chymotrypsinogen by diisopropylphosphorofluoridate. *Proc. Nat. Acad. Sci.* **69**:3312.

Müller-Eberhard, H. J. 1975. Patterns of complement activation. *Prog. Immunol.* (in press).

Naslin, L., A. Spyridakis and F. Labeyrie. 1973. A study of several bonds hypersensitive to proteases in a complex flavohemoenzyme, yeast cytochrome b_2. *Eur. J. Biochem.* **34:**268.

Neurath, H. 1957. The activation of zymogens. *Adv. Protein Chem.* **12:**320.

————. 1964. Mechanism of zymogen activation. *Fed. Proc.* **23:**1.

Polakoski, K. L. 1974. Partial purification and characterization of proacrosin from boar sperm. *Fed. Proc.* **33:**1308.

Reeck, G. R., K. A. Walsh, M. A. Hermodson and H. Neurath. 1971. New forms of carboxypeptidase B and their homologous relationship to carboxypeptidase A. *Proc. Nat. Acad. Sci.* **68:**1226.

Robillard, G. and R. G. Shulman. 1974. High resolution nuclear magnetic resonance studies of the active site of chymotrypsin. II. Polarization of histidine 57 by substrate analogues and competitive inhibitors. *J. Mol. Biol.* **86:**541.

Robinson, N. C., H. Neurath and K. A. Walsh. 1973. The relation of the α-amino group of trypsin to enzyme function and zymogen activation. *Biochemistry* **12:**420.

Sigler, P. B., D. M. Blow, B. W. Matthews and R. Henderson. 1968. Structure of crystalline α-chymotrypsin. II. A preliminary report including a hypothesis for the activation mechanism. *J. Mol. Biol.* **35:**143.

Steiner, D. F., J. L. Clark, C. Nolan, A. H. Rubenstein, E. Margoliash, B. Aten and P. E. Oyer. 1969. Proinsulin and the biosynthesis of insulin. *Recent Prog. Hormone Res.* **25:**207.

Tager, H. S. and D. F. Steiner. 1974. Peptide hormones. *Annu. Rev. Biochem.* **43:**509.

Wood, W. B., R. S. Edgar, J. King, I. Lielausis and M. Henninger. 1968. Bacteriophage assembly. *Fed. Proc.* **27:**1160.

Wyckoff, H. W., K. D. Hardman, N. M. Allewell, T. Inagami, L. N. Johnson and F. M. Richards. 1967. The structure of ribonuclease-S at 3.5 Å resolution. *J. Biol. Chem.* **242:**3984.

Role of Proteases in Blood Coagulation

Earl W. Davie, Kazuo Fujikawa, Mark E. Legaz and Hisao Kato

Department of Biochemistry, University of Washington
Seattle, Washington 98195

The cascade mechanism for blood coagulation involves the interaction of nearly a dozen glycoproteins, which eventually gives rise to fibrin formation. Most of these proteins occur in plasma in a precursor or inactive form. When the coagulation process is initiated, many of these proteins are converted to enzymes in a series of stepwise reactions (Davie and Ratnoff 1964; Macfarlane 1964). Nearly half of these enzymes have been identified as serine endopeptidases, which exert their effect on the coagulation process by highly selective proteolysis of a specific protein substrate. In this paper, we will review the reactions which lead to fibrin formation in the intrinsic coagulation pathway, with particular emphasis on the mechanisms by which the serine proteases are formed from their plasma precursor.

In the initial step of the intrinsic coagulation scheme, factor XII is converted to factor XII$_a$ by its interaction with collagen or vascular basement membranes (Niewiarowski, Bankowski and Rogowicka 1965; Cochrane et al. 1972) or a surface such as glass or kaolin (Ratnoff and Rosenblum 1958; Biggs et al. 1958; Nossel 1964) (Fig. 1). Factor XII$_a$ then activates factor XI, converting it to a serine endopeptidase (Kingdon, Davie and Ratnoff 1964). These initial reactions are greatly accelerated by the participation of another plasma protein called prekallikrein or Fletcher factor (Cochrane et al. 1972; Wuepper 1972; Weiss, Gallin and Kaplan 1974). Fletcher factor deficiency was first observed by Hathaway and coworkers, who reported a prolonged kaolin-activated partial thromboplastin time in several patients (Hathaway, Belhasen and Hathaway 1965; Hathaway and Alsever 1970). These patients lacked prekallikrein (Wuepper 1972), a protein that functions in kinin formation. In the presence of factor XII$_a$, prekallikrein is converted to kallikrein (Kaplan and Austen 1971; Cochrane and Wuepper 1971; Komiya, Nagasawa and Suzuki 1972). Furthermore, kallikrein is capable of activating factor XII (Wuepper 1972). Thus the initial activation of traces of factor XII by collagen can trigger an amplification reaction in the early or contact phase of intrinsic blood coagulation.

66   E. W. Davie et al.

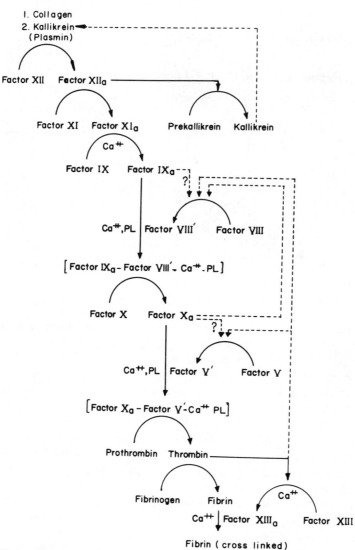

Figure 1

Tentative mechanism for the initiation of blood clotting in mammalian plasma in the intrinsic system. PL refers to phospholipid. (Reprinted with permission, from Davie and Fujikawa, 1975.)

The molecular mechanism for the surface activation of factor XII is not known. The activation of factor XII by kallikrein and plasmin, however, presumably involves proteolysis, since these enzymes are well-known endopeptidases. Cochrane et al. (1972), however, have not observed any difference in rabbit factor XII and factor XII$_a$ by SDS gel electrophoresis in the presence and absence of 2-mercaptoethanol, suggesting that any proteolysis that may occur during the activation reaction is minor. With human factor XII, the molecule is cleaved into fragments during the activation by plasmin and kallikrein (Cochrane, Revak and Wuepper 1973).

In the next reaction, factor XII$_a$ converts factor XI to an enzyme (Soulier, Prou-Wartelle and Menache 1958; Hardisty and Margolis 1959; Ratnoff, Davie and Mallett 1961) (Fig. 1). Factor XI has been highly purified from human and bovine plasma in several different laboratories (Ratnoff and Davie 1962; Kingdon, Davie and Ratnoff 1964; Wuepper 1972; Kato, Fujikawa and Legaz 1974; Schiffman and Lee 1974). It is a glycoprotein with a molecular weight of 160,000 and is composed of two similar or identical polypeptide chains held together by disulfide bonds (Wuepper 1972; H. Kato, M. E. Legaz and E. W. Davie, unpubl.). The polypeptide chains have a molecular weight of 80,000. The activation of factor XI by factor XII$_a$ or trypsin does not appear to change the molecular weight of the protein. Cleavage of internal peptide bonds in the two chains of factor XI can be demonstrated, however, by SDS gel electrophoresis following reduction of factor XI$_a$. In this case, two polypeptide chains can be identified with molecular weights of about 50,000 and 30,000 (Wuepper 1972; H. Kato, M. E. Legaz and E. W. Davie, unpubl.). The catalytic site in factor XI$_a$ is located in the 30,000 subunit, as indicated by [^{32}P]diisopropylfluorophosphate (DFP) labeling experiments (H. Kato, M. E. Legaz and E. W. Davie, unpubl.). The nature of the peptide bonds that are split during this reaction has not been established.

In the next reaction, factor XI$_a$ converts factor IX to factor IX$_a$ (Waaler 1959; Yin and Duckert 1961; Ratnoff and Davie 1962; Schiffman, Rapaport and Patch 1963; Cattan and Denson 1964; Fujikawa et al. 1974b) (Fig. 1). This reaction, which requires a divalent cation (Ratnoff and Davie 1962; Kingdon and Davie 1965), occurs in two steps (Fujikawa et al. 1974b). Factor IX isolated from bovine plasma is a single-chain glycoprotein with an amino-terminal tyrosine and a molecular weight of 54,500 (Fujikawa et al. 1973). Its partial structure is shown in Figure 2 (Fujikawa et al. 1974c; Enfield et al. 1974). In the first step of the activation reaction, factor IX is cleaved by hydrolysis of an internal peptide bond (arrow 1, Fig. 2). This gives rise to a two-chain structure held together by a disulfide bond(s). The

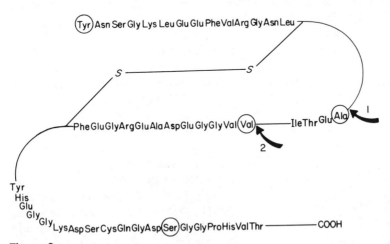

Figure 2

Partial structure for bovine factor IX. (Data from Fujikawa et al. 1974b and Enfield et al. 1974.)

light chain of this intermediate has a molecular weight of about 16,000, and the heavy chain has a molecular weight of 38,000. The amino-terminal group of the light chain is tyrosine and the amino-terminal group of the heavy chain is alanine. This intermediate does not have enzymatic activity. In the second step of the activation reaction, factor XI_a cleaves a glycopeptide from the amino-terminal end of the heavy chain (arrow 2, Fig. 2). The activation peptide that is released has a molecular weight of about 9000. This step leads to the formation of factor IX_a, a protein with a molecular weight of 46,500. The new amino-terminal sequence on the heavy chain of factor IX_a is Val-Val-Gly-Gly- and its molecular weight is 27,300. This chain is homologous with the heavy chains of thrombin and factor X_a (Magnusson 1971; Titani et al. 1972; Fujikawa et al. 1974c; Enfield et al. 1974). Furthermore, the heavy chains of all three proteins are homologous with trypsin. Thus it appears likely that the critical event in the activation of factor IX, as well as prothrombin and factor X, is the formation of the new amino-terminal valine (or isoleucine) which probably forms an ion pair with aspartic acid. This would be analogous to the requirement for a free amino group for ion pair formation with other proteases such as elastase and trypsin (Scrimger and Hofmann 1967; Gertler and Hofmann 1967; Magnusson and Hofmann 1970; Rao and Hofmann 1970; Shotton and Watson 1970; Robinson, Neurath and Walsh 1973; Karibian et al. 1974).

In the next series of reactions, factor IX_a interacts with factor VIII in the presence of calcium and phospholipid to form a complex, and it is this complex which converts factor X to factor X_a (Hougie, Denson and Biggs 1967; Hemker and Kahn 1967; Østerud and Rapaport 1970; Chuang, Sargeant and Hougie 1972) (Fig. 1). Under physiological conditions, the platelets provide the phospholipid for this reaction. In these reactions, factor VIII appears to be a regulatory protein and factor IX_a is the catalyst. Indeed, factor IX_a alone will convert factor X to factor X_a (M. E. Legaz, K. Fujikawa and E. W. Davie, unpubl.). The activation of factor X by factor IX_a is accelerated several thousandfold in the presence of factor VIII. Factor VIII, however, may not participate directly in this reaction and may require activation by a protease such as thrombin. Indeed, it was shown a number of years ago by Rapaport and coworkers (Rapaport et al. 1963; Rapaport, Hjort and Patch 1965) and others (Macfarlane et al. 1964; Ozge-Anwar, Connell and Mustard 1965) that thrombin would dramatically increase the activity of factor VIII (Fig. 1). This increase in factor VIII activity has also been demonstrated with highly purified preparations (Legaz et al. 1973, 1975; Shapiro et al. 1973). The nature of this thrombin modification reaction is not known. No changes have been noted in the molecule by SDS gel electrophoresis before and after activation. Other proteases, such as factor X_a, will also activate factor VIII (circles, Fig. 3). The activated factor VIII formed in the presence of factor X_a is very stable, in contrast to that formed in the presence of thrombin (triangles, Fig. 3). Factor VIII activated by thrombin is partially stabilized, however, by calcium ions and phospholipid (squares, Fig. 3), an observation previously made by Thompson (1971). Whether factor IX_a itself can modify factor VIII prior to the formation of the complex has not been established.

Factor VIII has been extensively purified in a number of different laboratories from bovine and human sources (Schmer et al. 1972; Marchesi, Shul-

Figure 3

Activation of bovine factor VIII by thrombin and factor X_a. The incubation mixture contained 0.1 ml factor VIII (40 μg), 0.03 ml phospholipid (0.4% Centrolex-P), 0.02 ml 0.03 M $CaCl_2$ and 0.05 ml thrombin (10 ng) (squares) or 0.05 ml factor X_a (35 ng) (circles). The triangles represent an experiment with thrombin where the phospholipid and $CaCl_2$ were replaced by 0.05 M Tris-HCl pH 8.0 in the incubation mixture. The reaction temperature was 37°C. At various intervals, 0.01-ml aliquots were diluted 10^2–10^4-fold in 0.05 M Tris-HCl pH 8.0 and assayed for factor VIII activity in factor VIII-deficient plasma.

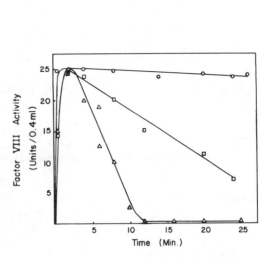

man and Gralnick 1972; Legaz et al. 1973, 1975; Shapiro et al. 1973). It is a glycoprotein with a molecular weight of 1.1 million as determined by sedimentation equilibrium (Schmer et al. 1972; Legaz et al. 1973). In the presence of reducing agents, similar or identical subunits are formed with a molecular weight of about 200,000 (Schmer et al. 1972; Legaz et al. 1973; Shapiro et al. 1973). These preparations have both coagulant and platelet-aggregating activity (Forbes and Prentice 1973; Legaz et al. 1975). A number of investigators have reported that factor VIII coagulant activity can be dissociated from the high molecular weight protein in the presence of 0.25 M $CaCl_2$ or 1.0 M NaCl (Weiss and Kochwa 1970; Weiss, Phillips and Rosner 1972; Owen and Wagner 1972; Cooper, Griggs and Wagner 1973; Griggs et al. 1973; Donati, de Gaetano and Vermylen 1973; Rick and Hoyer 1973). Furthermore, factor VIII coagulant activity appears to be of low molecular weight of around 25,000 (Owen and Wagner 1972) or perhaps 100,000 (Cooper, Griggs and Wagner 1973). This low molecular weight factor VIII is not activated by thrombin (Owen and Wagner 1972). A simple interpretation of the data dealing with high molecular weight versus low molecular weight factor VIII is not evident at the present time.

The molecular basis for the activation of factor X by the factor IX_a-factor VIII complex has recently been clarified (Fujikawa et al. 1974a). This reaction is analogous to the activation of factor X by a protease from Russell's viper venom (Fujikawa, Legaz and Davie 1972). Factor X is a glycoprotein with a molecular weight of about 55,000 (Esnouf and Williams 1962; Jackson, Johnson and Hanahan 1968; Jackson and Hanahan 1968; Aronson, Mustafa and Mushinski 1969; Fujikawa, Legaz and Davie 1972; Jackson 1972; Esnouf, Lloyd and Jesty 1973; Bajaj and Mann 1973) (Fig. 4). It is composed of a light chain with an amino-terminal alanine and a heavy chain

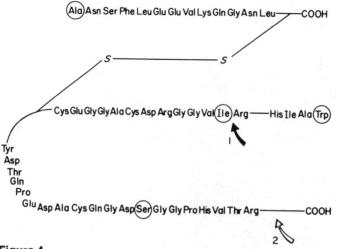

Figure 4

Partial structure for bovine factor X. (Data from Titani et al. 1972 and Fujikawa et al. 1974a,c.)

with an amino-terminal tryptophan (Fujikawa, Legaz and Davie 1972; Jackson 1972). These two chains are held together by a disulfide bond(s). During the activation of factor X by factor IX_a and factor VIII, a glycopeptide with a molecular weight of 11,000 is split from the amino-terminal end of the heavy chain (arrow 1, Fig. 4). This reduces the molecular weight of factor X from 55,000 to 44,000 and gives rise to factor $X_{a\alpha}$, an enzyme with endopeptidase activity. This protein has a new amino-terminal sequence of Ile-Val-Gly-Gly- on the heavy chain. Factor $X_{a\alpha}$ can be converted to factor $X_{a\beta}$ by limited proteolysis in which a glycopeptide(s) with a molecular weight of about 4000 is cleaved from the carboxyl end of the heavy chain (arrow 2, Fig. 4). This leads to the formation of factor $X_{a\beta}$, which has coagulant activity equivalent to factor $X_{a\alpha}$.

In the next reaction of the coagulation process, factor $X_{a\alpha}$ or factor $X_{a\beta}$ interacts with factor V to form a complex that acts as a catalyst in the activation of prothrombin (Papahadjopoulos and Hanahan 1964; Milstone 1964; Barton, Jackson and Hanahan 1967; Jobin and Esnouf 1967; Hemker et al. 1967; Jesty and Esnouf 1973) (Fig. 1). This reaction also requires calcium ions and phospholipid, the latter being supplied by the platelets. Factor X_a is the enzyme in this complex and factor V is a regulatory protein. The activity of factor V is greatly increased by its interaction with thrombin (Papahadjopoulos, Hougie and Hanahan 1964; Newcomb and Hoshida 1965; Barton and Hanahan 1967). Thus the role of factor V in blood coagulation appears to be very similar to that of factor VIII.

The complex of factor X_a and factor V initially converts prothrombin to thrombin in a two-step reaction (Stenn and Blout 1972; Esmon, Owen and Jackson 1974). Prothrombin is a glycoprotein with a molecular weight of about 68,000–72,000 and with an amino-terminal alanine (Magnusson 1971, 1972; Stenflo 1973; Heldebrant et al. 1973a; Fujikawa et al. 1974c) (Fig. 5).

The first step in the activation of prothrombin is the cleavage of a large glycopeptide (molecular weight about 33,000) from the amino-terminal end of the molecule by factor X_a (arrow 1, Fig. 5). This gives rise to a single-chain intermediate with an amino-terminal threonine and a molecular weight of about 38,000. In the second step of the activation reaction, the intermediate is cleaved by factor X_a to form thrombin (arrow 2, Fig. 5). Thrombin is composed of a heavy and a light chain held together by a disulfide bond. The origin of the light chain differs from that of factor IX_a and factor X_a. In these enzymes, the amino-terminal region of the precursor molecule remains attached to the enzyme (Figs. 2 and 4). In thrombin, the light chain comes from the internal portion of the precursor.

The activation of prothrombin proceeds in a three-step reaction after the first traces of thrombin are formed. In the first step, thrombin reacts with prothrombin to form a small fragment (molecular weight about 23,000) and another intermediate (molecular weight about 50,000) with an amino-terminal serine (arrow 3, Fig. 5). This intermediate is then cleaved by factor X_a in a second and third step to form thrombin (Aronson and Menache 1966; Lanchantin, Friedmann and Hart 1968; Stenn and Blout 1972; Heldebrant et al. 1973b; Heldebrant and Mann 1973; Fass and Mann 1973; Kisiel and Hanahan 1973; Benson, Kisiel and Hanahan 1973; Engel and Alexander 1973; Jesty and Esnouf 1973; Owen, Esmon and Jackson 1974; Esmon, Owen and Jackson 1974) (arrows 1 and 2, Fig. 5).

Once thrombin is formed, it plays a multiple role in coagulation in that it (1) converts fibrinogen to fibrin, (2) converts factor XIII to factor XIII_a, and (3) activates factor V and factor VIII. Thrombin also participates in the platelet release reaction in which substances such as ADP, ATP and 5-hydroxytryptamine are secreted to the extracellular phase (Mustard and Packham 1970). This reaction is an essential step in the formation of the dense aggregates of platelets which play an important role in thrombosis.

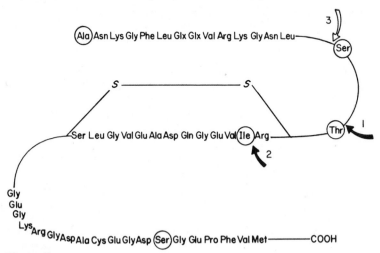

Figure 5
Partial structure for bovine prothrombin. (Data from Magnusson 1971 and Fujikawa et al. 1974c.)

In summary, blood coagulation can be viewed as a series of cascade reactions, many of which involve the formation of highly specific proteases. These proteases participate in coagulation by limited proteolysis of a specific protein substrate(s). In vivo, these enzymes are short-lived since the activated clotting factors are diluted in flowing blood, inactivated by proteolysis and protease inhibitors (Lanchantin et al. 1966; Rimon, Shamash and Shapiro 1966; Abildgaard 1967; Yin, Wessler and Stoll 1971; Heimburger, Haupt and Schwick 1971; Rosenberg and Damus 1973), and rapidly cleared by the liver (Deykin et al. 1968). Thus the coagulation proteins are poised to participate in the arrest of bleeding from a ruptured blood vessel but remain in an inactive form elsewhere in the vascular system where the blood must remain fluid to carry out its physiological functions.

Acknowledgment

This work was supported in part by research grants GM 10793 and HL 11857 from the National Institutes of Health.

REFERENCES

Abildgaard, U. 1967. Purification of two progressive antithrombins of human plasma. *Scand. J. Clin. Lab. Invest.* **19**:190.

Aronson, D. L. and D. Menache. 1966. Chromatographic analysis of the activation of human prothrombin with human thrombokinase. *Biochemistry* **5**:2635.

Aronson, D. L., A. J. Mustafa and J. F. Mushinski. 1969. Purification of human factor X and comparison of peptide maps of human factor X and prothrombin. *Biochim. Biophys. Acta* **188**:25.

Bajaj, S. P. and K. G. Mann. 1973. Simultaneous purification of bovine prothrombin and factor X. *J. Biol. Chem.* **248**:7729.

Barton, P. G. and D. J. Hanahan. 1967. The preparation and properties of a stable factor V from bovine plasma. *Biochim. Biophys. Acta* **133**:506.

Barton, P. G., C. M. Jackson and D. J. Hanahan. 1967. Relationship between factor V and activated factor X in the generation of prothrombinase. *Nature* **214**:923.

Benson, B. J., W. Kisiel and D. J. Hanahan. 1973. Calcium binding and other characteristics of bovine factor II and its activation intermediates. *Biochim. Biophys. Acta* **329**:81.

Biggs, R., A. A. Sharp, J. Margolis, R. M. Hardisty, J. Stewart and W. M. Davidson. 1958. Defects in the early stages of blood coagulation: A report of four cases. *Brit. J. Haematol.* **4**:177.

Cattan, A. D. and K. W. E. Denson. 1964. The interaction of contact product and factor IX. *Thromb. Diath. Haemorrh.* **11**:155.

Chuang, T. F., R. B. Sargeant and C. Hougie. 1972. The intrinsic activation of factor X in blood coagulation. *Biochim. Biophys. Acta* **273**:287.

Cochrane, C. G. and K. D. Wuepper. 1971. The first component of the kinin-forming system in human and rabbit plasma: Its relationship to clotting factor XII (Hageman factor). *J. Exp. Med.* **134**:986.

Cochrane, C. G., S. D. Revak and K. D. Wuepper. 1973. Activation of Hageman factor in solid and fluid phases. *J. Exp. Med.* **138**:1564.

Cochrane, C. G., S. D. Revak, B. S. Aikin and K. D. Wuepper. 1972. The structural characteristics and activation of Hageman factor. In *Inflammation: Mechanisms and Control* (ed. I. H. Lepow and P. A. Ward), p. 119. Academic Press, New York.

Cooper, H. A., T. R. Griggs and R. H. Wagner. 1973. Factor VIII recombination after dissociation by CaCl$_2$. *Proc. Nat. Acad Sci.* **70:**2326.

Davie, E. W. and K. Fujikawa. 1975. Basic mechanisms in blood coagulation. *Annu. Rev. Biochem.* (in press).

Davie, E. W. and O. D. Ratnoff. 1964. Waterfall sequence for intrinsic blood clotting. *Science* **145:**1310.

Deykin, D., F. Cochios, G. DeCamp and A. Lopez. 1968. Hepatic removal of activated factor X by the perfused rabbit liver. *Amer. J. Physiol.* **214:**414.

Donati, M. B., G. de Gaetano and J. Vermylen. 1973. Evidence that bovine factor VIII, not bovine fibrinogen, aggregates human platelets. *Thromb. Res.* **2:**97.

Enfield, D. L., L. H. Ericsson, K. Fujikawa, K. Titani, K. A. Walsh and H. Neurath. 1974. Bovine factor IX (Christmas factor). Further evidence of homology with factor X (Stuart factor) and prothrombin. *FEBS Letters* **47:**1.

Engel, A. M. and B. Alexander. 1973. Molecular changes during prothrombin activation. *Biochim. Biophys. Acta* **320:**687.

Esmon. C. T., W. G. Owen and C. M. Jackson. 1974. The conversion of prothrombin to thrombin. II. Differentiation between thrombin- and factor X$_a$-catalyzed proteolyses. *J. Biol. Chem.* **249:**606.

Esnouf, M. P. and W. J. Williams. 1962. The isolation and purification of a bovine-plasma protein which is a substrate for the coagulation fraction of Russell's viper venom. *Biochem. J.* **84:**62.

Esnouf, M. P., P. H. Lloyd and J. Jesty. 1973. A method for the simultaneous isolation of factor X and prothrombin from bovine plasma. *Biochem. J.* **131:**781.

Fass, D. N. and K. G. Mann. 1973. Activation of fluorescein-labeled prothrombin. *J. Biol. Chem.* **248:**3280.

Forbes, C. D. and C. R. M. Prentice. 1973. Platelet aggregation—bovine and porcine antihaemophilia factor aggregates human platelets. *Nature New Biol.* **241:**149.

Fujikawa, K., M. M. E. Legaz and E. W. Davie. 1972. Bovine factors X$_1$ and X$_2$ (Stuart factor). Isolation and characterization. *Biochemistry* **11:**4882.

Fujikawa, K., M. H. Coan, M. E. Legaz and E. W. Davie. 1974a. The mechanism of activation of bovine factor IX (Christmas factor) by bovine factor XI$_a$ pathways. *Biochemistry* **13:**5290.

Fujikawa, K., M. E. Legaz, H. Kato and E. W. Davie. 1974b. The mechanism of activation of bovine factor IX (Christmas factor) by bovine factor XI$_a$ (activated plasma thromboplastin antecedent). *Biochemistry* **13:**4508.

Fujikawa, K., A. R. Thompson, M. E. Legaz, R. G. Meyer and E. W. Davie. 1973. Isolation and characterization of bovine factor IX (Christmas factor). *Biochemistry* **12:**4938.

Fujikawa, K., M. H. Coan, D. L. Enfield, K. Titani, L. H. Ericsson and E. W. Davie. 1974c. A comparison of bovine prothrombin, factor IX (Christmas factor), and factor X (Stuart factor). *Proc. Nat. Acad. Sci.* **71:**427.

Gertler, A. and T. Hofmann. 1967. The involvement of the amino-terminal amino acid in the activity of pancreatic proteases. I. The effects of nitrous acid on elastase. *J. Biol. Chem.* **242:**2522.

Griggs, T. R., H. A. Cooper, W. P. Webster, R. H. Wagner and K. M. Brinkhous. 1973. Plasma aggregating factor (bovine) for human platelets: A marker for study of antihemophilic and von Willebrand factors. *Proc. Nat. Acad. Sci.* **70:**2814.

Hardisty, R. M. and J. Margolis. 1959. The role of Hageman factor in the initiation of blood coagulation. *Brit. J. Haematol.* **5**:203.

Hathaway, W. E. and J. Alsever. 1970. The relation of Fletcher factor to factors XI and XII. *Brit. J. Haematol.* **18**:161.

Hathaway, W. E., L. P. Belhasen and H. S. Hathaway. 1965. Evidence for a new plasma thromboplastin factor. I. Case report, coagulation studies and physico-chemical properties. *Blood* **26**:521.

Heimburger, N., H. Haupt and H. G. Schwick. 1971. Proteinase inhibitors of human plasma. In *Proceedings of the International Research Conference on Protein Inhibitors* (ed. H. Fritz and H. Tschesche), p. 1. Walter de Gruyter, Berlin.

Heldebrant, C. M. and K. G. Mann. 1973. The activation of prothrombin. I. Isolation and preliminary characterization of intermediates. *J. Biol. Chem.* **248**:3642.

Heldebrant, C. M., R. J. Butkowski, S. P. Bajaj and K. G. Mann. 1973b. The activation of prothrombin. II. Partial reactions, physical and chemical charac-terization of the intermediates of activation. *J. Biol. Chem.* **248**:7149.

Heldebrant, C. M., C. Noyes, H. S. Kingdon and K. G. Mann. 1973a. The acti-vation of prothrombin. III. The partial amino acid sequences at the amino terminal of prothrombin and the intermediates of activation. *Biochem. Biophys. Res. Comm.* **54**:155.

Hemker, H. C. and M. J. P. Kahn. 1967. Reaction sequence of blood coagulation. *Nature* **215**:1201.

Hemker, H. C., M. P. Esnouf, P. W. Hemker, A. C. W. Swart and R. G. Macfar-lane. 1967. Formation of prothrombin converting activity. *Nature* **215**:248.

Hougie, C., K. W. E. Denson and R. Biggs. 1967. A study of the reaction product of factor VIII and factor IX by gel filtration. *Thromb. Diath. Haemorrh.* **18**:211.

Jackson, C. M. 1972. Characterization of two glycoprotein variants of bovine factor X and demonstration that the factor X zymogen contains two polypep-tide chains. *Biochemistry* **11**:4873.

Jackson, C. M. and D. J. Hanahan. 1968. Studies on bovine factor X. II. Charac-terization of purified factor X. Observations on some alterations in zone electro-phoresis and chromatographic behavior occurring during purification. *Biochem-istry* **7**:4506.

Jackson, C. M., T. F. Johnson and D. J. Hanahan. 1968. Studies on bovine factor X. I. Large-scale purification of the bovine plasma protein possessing factor X activity. *Biochemistry* **7**:4492.

Jesty, J. and M. P. Esnouf. 1973. The preparation of activated factor X and its action on prothrombin. *Biochem. J.* **131**:791.

Jobin, F. and M. P. Esnouf. 1967. Studies on the formation of the prothrombin-converting complex. *Biochem. J.* **102**:666.

Kaplan, A. P. and K. F. Austen. 1971. A prealbumin activator of prekallikrein. II. Derivation of activators of prekallikrein from active Hageman factor by digestion with plasmin. *J. Exp. Med.* **133**:696.

Karibian, D., C. Jones, A. Gertler, K. J. Dorrington and T. Hofmann. 1974. On the reaction of acetic and maleic anhydrides with elastase. Evidence for a role of the NH_2-terminal valine. *Biochemistry* **13**:2891.

Kato, H., K. Fujikawa and M. E. Legaz. 1974. Isolation and activation of bovine factor XI (plasma thromboplastin antecedent) and its interaction with factor IX (Christmas factor). *Fed. Proc.* **33**:1505.

Kingdon, H. S. and E. W. Davie. 1965. Further studies on the activation of factor IX by activated factor XI. *Thromb. Diath. Haemorrh.* (Suppl.) **17**:15.

Kingdon, H. S., E. W. Davie and O. D. Ratnoff. 1964. The reaction between activated plasma thromboplastin antecedent and diisopropylphosphofluoridate. *Biochemistry* **3**:166.

Kisiel, W. and D. J. Hanahan. 1973. The action of factor Xa, thrombin and trypsin on human factor II. *Biochim. Biophys. Acta* **329**:221.

Komiya, M., S. Nagasawa and T. Suzuki. 1972. Bovine prekallikrein activator with functional activity as Hageman factor. *J. Biochem.* **72**:1205.

Lanchantin, G. F., J. A. Friedmann and D. W. Hart. 1968. On the occurrence of polymorphic human prothrombin. *J. Biol. Chem.* **243**:476.

Lanchantin, G. F., M. L. Plesset, J. A. Friedmann and D. W. Hart. 1966. Dissociation of esterolytic and clotting activities of thrombin by trypsin-binding macroglobulin. *Proc. Soc. Exp. Biol. Med.* **121**:444.

Legaz, M. E., G. Schmer, R. B. Counts and E. W. Davie. 1973. Isolation and characterization of human factor VIII (antihemophilic factor). *J. Biol. Chem.* **248**:3946.

Legaz, M. E., M. J. Weinstein, C. M. Heldebrant and E. W. Davie. 1975. Isolation, subunit structure, and proteolytic modification of bovine factor VIII. *Ann. N. Y. Acad. Sci.* (in press).

Macfarlane, R. G. 1964. An enzyme cascade in the blood clotting mechanism and its function as a biochemical amplifier. *Nature* **202**:498.

Macfarlane, R. G., R. Biggs, B. J. Ash and K. W. E. Denson. 1964. The interaction of factors VIII and IX. *Brit. J. Haematol.* **10**:530.

Magnusson, S. 1971. Thrombin and prothrombin. In *The Enzymes* (ed. P. D. Boyer), vol. 3, p. 277. Academic Press, New York.

————. 1972. On the primary structure of bovine thrombin. *Folia Haematol.* (Leipzig) **98**:4, S. 385.

Magnusson, S. and T. Hofmann. 1970. Inactivation of bovine thrombin by nitrous acid. *Can. J. Biochem.* **48**:432.

Marchesi, S. L., N. R. Shulman and H. R. Gralnick. 1972. Studies on the purification and characterization of human factor VIII. *J. Clin. Invest.* **51**:2151.

Milstone, J. H. 1964. Thrombokinase as prime activator of prothrombin: Historical perspectives and present status. *Fed. Proc.* **23**:742.

Mustard, J. F. and M. A. Packham. 1970. Factors influencing platelet function: Adhesion, release, and aggregation. *Pharmacol. Rev.* **22**:97.

Newcomb, T. F. and M. Hoshida. 1965. Factor V and thrombin. *Scand. J. Clin. Lab. Invest.* (Suppl. 84) **17**:61.

Niewiarowski, B., E. Bankowski and I. Rogowicka. 1965. Studies on the adsorption and activation of the Hageman factor (factor XII) by collagen and elastin. *Thromb. Diath. Haemorrh.* **14**:387.

Nossel, H. L., ed. 1964. *The Contact Phase of Blood Coagulation*. F. A. Davis, Philadelphia.

Østerud, B. and S. I. Rapaport. 1970. Synthesis of intrinsic factor X activator, inhibition of the function of formed activator by antibodies to factor VIII and to factor IX. *Biochemistry* **9**:1854.

Owen, W. G. and R. H. Wagner. 1972. Antihemophilic factor: Separation of an active fragment following dissociation by salts or detergents. *Thromb. Diath. Haemorrh.* **27**:502.

Owen, W. G., C. T. Emson and C. M. Jackson. 1974. The conversion of prothrombin to thrombin. I. Characterization of the reaction products formed during the activation of bovine prothrombin. *J. Biol. Chem.* **249**:594.

Ozge-Anwar, A. H., G. E. Connell and J. F. Mustard. 1965. The activation of factor VIII by thrombin. *Blood* **26**:500.

Papahadjopoulos, D. and D. J. Hanahan. 1964. Observations on the interaction of phospholipids and certain clotting factors in prothrombin activator formation. *Biochim. Biophys. Acta* **90**:436.

Papahadjopoulos, D., C. Hougie and D. J. Hanahan. 1964. Purification and properties of bovine factor V: A change of molecular size during blood coagulation. *Biochemistry* **3**:264.

Rao, L. and T. Hofmann. 1970. The reaction of 2,4,6-trinitrobenzene sulfonic acid with pancreatic elastase. *Can. J. Biochem.* **48**:1249.

Rapaport, S. I., P. F. Hjort and M. J. Patch. 1965. Further evidence that thrombin-activation of factor VIII is an essential step in intrinsic clotting. *Scand. J. Clin. Lab. Invest.* (Suppl. 84) **17**:88.

Rapaport, S. I., S. Schiffman, M. J. Patch and S. B. Ames. 1963. The importance of activation of antihemophilic globulin and proaccelerin by traces of thrombin in the generation of intrinsic prothrombinase activity. *Blood* **21**:221.

Ratnoff, O. D. and E. W. Davie. 1962. The activation of Christmas factor (factor IX) by activated plasma thromboplastin antecedent (activated factor XI). *Biochemistry* **1**:677.

Ratnoff, O. D. and J. M. Rosenblum. 1958. Role of Hageman factor in the initiation of clotting by glass: Evidence that glass frees Hageman factor from inhibition. *Amer. J. Med.* **25**:160.

Ratnoff, O. D., E. W. Davie and D. L. Mallett. 1961. Studies on the action of Hageman factor: Evidence that activated Hageman factor in turn activates plasma thromboplastin antecedent. *J. Clin. Invest.* **40**:803.

Rick, M. E. and L. W. Hoyer. 1973. Immunologic studies of antihemophilic factor (AHF, factor VIII). V. Immunologic properties of AHF subunits produced by salt dissociation. *Blood* **42**:737.

Rimon, A., Y. Shamash and B. Shapiro. 1966. The plasmin inhibitor of human plasma. IV. Its action on plasmin, trypsin, chymotrypsin, and thrombin. *J. Biol. Chem.* **241**:5102.

Robinson, N. C., H. Neurath and K. A. Walsh. 1973. The relation of the α-amino group of trypsin to enzyme function and zymogen activation. *Biochemistry* **12**:420.

Rosenberg, R. D. and P. S. Damus. 1973. The purification and mechanism of action of human antithrombin-heparin cofactor. *J. Biol. Chem.* **248**:6490.

Schiffman, S. and P. Lee. 1974. Preparation, characterization, and activation of a highly purified factor XI: Evidence that a hitherto unrecognized plasma activity participates in the interaction of factor XI and XII. *Brit. J. Haematol.* **27**:101.

Schiffman, S., S. I. Rapaport and M. J. Patch. 1963. The identification and synthesis of activated plasma thromboplastin component (PTC). *Blood* **22**:733.

Schmer, G., E. P. Kirby, D. C. Teller and E. W. Davie. 1972. The isolation and characterization of bovine factor VIII (antihemophilic factor). *J. Biol. Chem.* **247**:2512.

Scrimger, S. T. and T. Hofmann. 1967. The involvement of the amino-terminal amino acid in the activity of pancreatic proteases. II. The effects of nitrous acid on trypsin. *J. Biol. Chem.* **242**:2528.

Shapiro, G. A., J. C. Anderson, S. V. Pizzo and P. A. McKee. 1973. The subunit structure of normal and hemophilic factor VIII. *J. Clin. Invest.* **52**:2198.

Shotton, D. M. and H. C. Watson. 1970. Three-dimensional structure of tosyl-elastase. *Nature* **225**:811.

Soulier, J.-P., O. Prou-Wartelle and D. Menache. 1958. Caractères différentiels des facteurs Hageman et P.T.A. Rôle du contact dans la phase initiale de la coagulation. *Rev. Fr. Etud. Clin. Biol.* **3**:263.

Stenflo, J. 1973. Normal and dicoumarol-induced prothrombin: Structure and properties. M.D. thesis, University of Malmo, Malmo, Sweden.

Stenn, K. S. and E. R. Blout. 1972. Mechanism of bovine prothrombin activation by an insoluble preparation of bovine factor X_a (thrombokinase). *Biochemistry* **11**:4502.

Thompson, A. R. 1971. Intrinsic factor X activation. Ph.D. thesis, University of Washington, Seattle, Washington.

Titani, K., M. A. Hermodson, K. Fujikawa, L. H. Ericsson, K. A. Walsh, H. Neurath and E. W. Davie. 1972. Bovine factor X_{1a} (activated Stuart factor). Evidence of homology with mammalian serine proteases. *Biochemistry* **11**:4899.

Waaler, B. A. 1959. Contact activation in the intrinsic blood clotting system. *Scand. J. Clin. Lab. Invest.* (Suppl. 37) **11**:1.

Weiss, A. S., J. I. Gallin and A. P. Kaplan. 1974. Fletcher factor deficiency. A diminished rate of Hageman factor activation caused by absence of prekallikrein with abnormalities of coagulation, fibrinolysis, chemotactic activity, and kinin generation. *J. Clin. Invest.* **53**:622.

Weiss, H. J. and S. Kochwa. 1970. Molecular forms of antihaemophilic globulin in plasma, cryoprecipitate and after thrombin activation. *Brit. J. Haematol.* **18**:89.

Weiss, H. J., L. L. Phillips and W. Rosner. 1972. Separation of subunits of antihemophilic factor (AHF) by agarose gel chromatography. *Thromb. Diath. Haemorrh.* **27**:212.

Wuepper, K. D. 1972. Biochemistry and biology of components of the plasma kinin-forming system. In *Inflammation: Mechanisms and Control* (ed. I. H. Lepow and P. A. Ward), p. 93. Academic Press, New York.

Yin, E. T. and F. Duckert. 1961. The formation of intermediate product I in a purified system. The role of factor IX or of its precursor and of a Hageman factor-PTA fraction. *Thromb. Diath. Haemorrh.* **6**:224.

Yin, E. T., S. Wessler and P. J. Stoll. 1971. Rabbit plasma inhibitor of the activated species of blood coagulation factor X. *J. Biol. Chem.* **246**:3694.

Controls in the Clotting of Fibrinogen

Laszlo Lorand

Department of Biochemistry and Molecular Biology, Northwestern University
Evanston, Illinois 60201

The finding that thrombin acted as a proteolytic enzyme by attacking specific peptide bonds near the N-terminal regions of two of the three chains in fibrinogen may have actually been the first demonstration that hydrolysis of proteins could serve a physiological function other than the mere digestion of foodstuff (Bailey et al. 1951; Lorand 1951, 1952; Lorand and Middlebrook 1952; Bettelheim and Bailey 1952; Blomback and Yamashina 1958). Inasmuch as the thrombin-catalyzed removal of fibrinopeptides greatly alters the solubility of the parent protein and induces it to form ordered aggregates, this may also have been the first example for the enzyme-controlled assembly of proteins occurring at the posttranslational level. Later it became evident that through the functioning of the fibrin-stabilizing system, yet another enzymatic step occurs which brings about the final strengthening of the clot by the covalent ligation of fibrin units with γ-glutamyl-ϵ-lysine bridges (see Lorand 1972; Roberts, Lorand and Mockros 1973; Mockros, Roberts and Lorand 1974). Linderstrøm-Lang in 1952 anticipated the existence of such processes of protein "remodeling" when he suggested that certain enzymes "may contribute by embroidering upon the canvas woven by hard-working metabolic systems of the cell."

If the reactions of the fibrinogen molecule leading to a cross-linked fibrin network are viewed in terms of a biosynthetic sequence of further tailoring a protein substrate, thrombin is seen to exercise a most interesting and dual regulatory function. It controls production of the clotting intermediate, fibrin, as well as the rate of generation of the transamidating enzyme, which, in turn, must act on this fibrin. In Figure 1, these events are placed into the framework of the situation found in plasma.

In normal blood, the explosive cascade of sequential activation of clotting factors is accomplished within a few minutes and requires the participation of tissue, platelet and plasma components. Excess thrombin is neutralized to a large extent by forming an inactive stoichiometric complex with an inhibitory protein, antithrombin. Lysis of fibrin by plasmin constitutes another con-

79

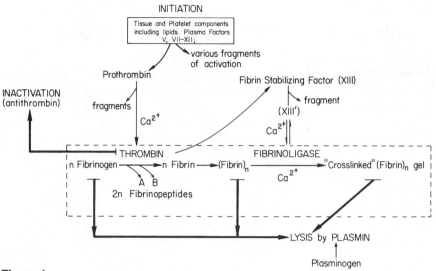

Figure 1
Outline of the clotting reaction in the plasma of vertebrates. Following the initial
interaction of tissue, platelet and plasma factors (initiation), prothrombin is activated to thrombin, a hydrolytic enzyme of great specificity. It brings about the
conversion of fibrinogen to fibrin by the removal of fibrinopeptide fragments and
it also regulates the rate of formation of fibrinoligase, a transamidating enzyme
which cross-links fibrin (broken rectangle). Excess thrombin is eliminated by an inactivating mechanism. Fibrinogen-fibrin deposits which may arise in the circulation
are removed by plasmin, a trypsinlike enzyme generated from plasminogen by a
variety of activators. (Heavy arrows outline the "shut-off" mechanisms.) The
enzymatically cross-linked fibrin gel is a relatively poor substrate for plasmin.

trolling factor, and if one further extends the metabolic analogy and thinks
of clot formation as the anabolic phase, lysis, with its different pathways,
would represent the catabolism of fibrin products. As is discussed elsewhere
in this volume, a number of the factors that promote clotting and lysis in
plasma have already been isolated to a high degree of purity, and a great deal
of biochemical knowledge is available regarding the order and the nature of
activation (mostly proteolytic) of several zymogens. Many of the activated
factors themselves (VII_a, IX_a, X_a, thrombin and plasmin) belong to the
serine-OH type of proteases of the trypsin class.

Fibrinoligase is a notable exception, being the only enzyme in blood
clotting with a cysteine-SH active center functional group (Curtis et al. 1973,
1974a). Conversion of this enzyme from its zymogen (factor XIII) also
presents some rather unique aspects of control. The zymogen has an (*ab*)
protomeric structure (Bohn 1970; Schwartz et al. 1971), and removal of an
N-terminal peptide fragment by thrombin from the *a* subunit constitutes the
first step in activation (Mikuni, Iwanaga and Konishi 1973; Takagi and Doolittle 1974). However, this modified zymogen (*a′b*) is still devoid of catalytic
power, and formation of the active center in the enzyme requires a consecutive step in which the protein reacts with calcium ions. These ions, augmented
by general ionic strength, specifically disrupt the heterologous association of

subunits and, in a coupled event, also cause the unmasking of the catalytically important cysteine side chain buried within a' (Curtis et al. 1973, 1974a; Holbrook, Cooke and Kingston 1973; Cooke and Holbrook 1974; Chung, Lewis and Folk 1974):

$$\begin{array}{ccccc} & \text{thrombin} & & \text{Ca}^{2+} & \\ \text{zymogen} & \longrightarrow & \text{zymogen}' & \rightleftarrows & \text{enzyme} + \text{noncatalytic subunit.} \\ (ab) & \Big\downarrow & (a'b) & \text{EDTA} \quad (a^*) & (b) \\ & \text{fragment} & & & \end{array}$$

Though there is some indication that γ-glutamine residues in the noncatalytic b subunit are altered as enzyme activity is generated in a^* (Gray and Lorand 1974), nevertheless, removal of calcium leads to a reassociation of the two subunit types, reburying the active center cysteine and neutralizing catalytic potency. It may be of particular physiological significance that at the calcium concentration and general ionic strength existing in plasma, only a slow and limited formation of a^* can take place (Curtis et al. 1974a). Thus the chance of generating dangerous transamidase activity in the circulation, which, for example, could lead to the antigenic modification of proteins by the incorporation of drugs such as isoniazid or hydralazine (Lorand, Campbell and Robertson 1972), would be reduced even if activation of the zymogen by thrombin or by another trypsinlike enzyme did accidentally occur—as if some hesitancy was built into the plasma system for generating the last enzyme in the blood coagulation cascade, and as if an additional control was provided so that only the disruption of platelets or other cells could create the necessary condition for the sudden formation of fibrinoligase by causing a local increase in calcium ion concentration. As soon as the enzyme drifted from this ion-enriched environment back into the general plasma milieu, its active-center cysteine would become buried again, and by reassociating with the noncatalytic b carrier protein, which is in excess in blood, it would be rendered inactive.

There is ample kinetic evidence with synthetic substrates that both the thrombin- (Kezdy, Lorand and Miller 1965; Chase and Shaw 1969) and the fibrinoligase-catalyzed reactions (Lorand et al. 1968; Chung and Folk 1972; Curtis et al. 1974b) proceed through intermediary acylation-deacylation steps. Hence the representation given in Figure 1 is only an oversimplified outline of the pathway pertaining to the clotting of fibrinogen, and the existence of specific acyl enzyme intermediates must be postulated. For example, interaction of the serine hydroxyl group of thrombin (HO-E) with the fibrinopeptide A (A-COOH) moiety of the α chain in fibrinogen (α-F) would comprise the following intermediates:

$$\begin{array}{c} \text{A-CONH-}\alpha\text{-F} \ + \ \text{HO-E} \ \rightleftarrows \ (\text{A-CONH-}\alpha\text{-F; HO-E}) \\ \Big\downarrow \\ \text{A-COO-E} \ + \ \text{H}_2\text{N-}\alpha\text{-F} \\ \Big\downarrow \ \text{H}_2\text{O} \\ \text{A-COOH} \ + \ \text{HO-E.} \end{array}$$

The reaction of thrombin with the β chain of fibrinogen and with the a subunit of the factor XIII zymogen would proceed through similar steps, all involving arginyl-O-ester acyl enzyme intermediates. By contrast, the reaction between the active-center cysteine of fibrinoligase (HS-E′) and fibrin (H_2N-F-$CONH_2$; showing only an ϵ-amino and a γ-glutamine functionality) would generate a γ-glutamyl-S-ester intermediate which could undergo aminolysis by reaction either with a lysine residue of a neighboring fibrin molecule (F′) to yield cross-linked protein (P_1), or with a synthetic amine (H_2NR) to produce a uniquely modified fibrin species (P_2):

$$H_2N\text{-}F\text{-}CONH_2 \;+\; HS\text{-}E'$$

$$\downarrow \uparrow$$

$$(H_2N\text{-}F\text{-}CONH_2;\; HS\text{-}E)$$

$$\downarrow$$

$$H_2N\text{-}F\text{-}COS\text{-}E' \;+\; NH_3$$

$$H_2N\text{-}F'\text{-}CONH_2 \qquad\qquad H_2NR$$

$$H_2N\text{-}F\text{-}CONH\text{-}F\text{-}CONH_2 + HS\text{-}E' \qquad\qquad H_2N\text{-}F\text{-}CONHR + HS\text{-}E'.$$
$$(P_1) \qquad\qquad\qquad\qquad\qquad\qquad (P_2)$$

Thus the presence of the synthetic compound specifically inhibits the enzyme-catalyzed formation of covalent bridges between fibrin molecules by modifying the reactive γ-glutamyl functions on the protein. Clotting as such (n fibrin $\rightarrow$ [fibrin]$_n$) is not interfered with even when amines with rather bulky substituents (e.g., dansylcadaverine; Lorand et al. 1968) are used. This indicates that a large degree of flexibility exists in the contact domains between individual fibrin particles in the clot. The viscoelastic properties of the modified gel structures, represented by units of P_2, approximate those of simple fibrin, and both of these are much less elastic than crosslinked fibrin (Roberts, Lorand and Mockros 1973; Mockros, Roberts and Lorand 1974). It is important to note that the covalently ligated (P_1) structures are far more resistant to proteolytic digestion than the inhibitor-modified (P_2) ones even in plasma or in whole blood thrombi (Lorand, Bruner-Lorand and Pilkington 1966; Lorand and Nilsson 1972), a circumstance which offers possibilities for aiding therapeutic thrombolysis.

One of the remarkable aspects of amine incorporation with fibrinoligase (in contrast, for example, to the related transglutaminase from liver) is the very high selectivity of this enzyme for the fibrin substrate. Reactivity with fibrinogen is about tenfold slower, and it has been established that the removal of fibrinopeptide A opens up the protein for the enzyme (Lorand, Chenoweth and Gray 1972). The thrombin-controlled unmasking of the cross-linking sites on fibrinogen is of obvious physiological regulatory significance, since it seems to serve the purpose of coordinating events of clotting with the generation of fibrinoligase activity (Fig. 1).

This thrombin-dependent control of opening up sites for future transamidation by a process of limited proteolysis on the parent protein appears to be a characteristic feature in vertebrate blood only. We found that the clotting of lobster fibrinogen is brought about by transamidation alone, and that synthetic amines completely prevent clot formation in Homarus (Lorand et al.

1963). From the evolutionary point of view, one may surmise that transamidation might have been the fundamental enzymatic reaction in clotting, and that the thrombin-catalyzed steps were added on as extra controls. By contrast to the vertebrate enzyme, the Homarus transamidase does not exist in the form of a zymogen; the active enzyme is sequestered away in the amoebocytes and can come into contact with fibrinogen only after release from these cells upon injury. But perhaps the formation of the copulation plug in rodents is the most interesting example of clotting by transamidation (Williams-Ashman et al. 1972). Here a small basic protein from the secretion of the seminal vesicle forms a clot which is cross-linked on the average by at least one γ-glutamyl-ϵ-lysine bridge for each 23-amino acid residue distance, representing the highest frequency of this peptide cross-link in any system. The enzyme is secreted by a separate anatomical entity (coagulation gland) and is brought into contact with its protein substrate during ejaculation.

Acknowledgments

Research pertaining to this work was supported by a USPHS Research Career Award and by grants from the National Heart and Lung Institute, National Institutes of Health (HL-02212 and 16346).

REFERENCES

Bailey, K., F. R. Bettelheim, L. Lorand and W. R. Middlebrook. 1951. Action of thrombin in the clotting of fibrinogen. *Nature* **167**:233.

Bettelheim, F. R. and K. Bailey. 1952. The products of the action of thrombin on fibrinogen. *Biochim. Biophys. Acta* **9**:578.

Blomback, B. and I. Yamashina. 1958. On the N-terminal amino acids in fibrinogen and fibrin. *Arkiv. Kemi* **12**:299.

Bohn, H. 1970. Isolierung und Charakerisierung des Fibrin stabilisierenden Factors aus menschlichen Thrombozyten. *Thromb. Diath. Haemorrh.* **23**:455.

Chase, T., Jr. and E. Shaw. 1969. Comparison of the esterase activities of trypsin, plasmin, and thrombin on guanidinobenzoate esters. Titration of the enzymes. *Biochemistry* **8**:2212.

Chung, S. I. and J. E. Folk. 1972. Kinetic studies with transglutaminases. The human blood enzymes (activated coagulation factor XIII) and the guinea pig hair follicle enzyme. *J. Biol. Chem.* **247**:2798.

Chung, S. I., M. S. Lewis and J. E. Folk. 1974. Relationships of the catalytic properties of human plasma and platelet transglutaminases (activated blood coagulation factor XIII) to their subunit structures. *J. Biol. Chem.* **249**:940.

Cooke, R. D. and J. J. Holbrook. 1974. The calcium-induced dissociation of human plasma clotting factor XIII. *Biochem. J.* **141**:79.

Curtis, C. G., P. Stenberg, C. H. J. Chou, A. Gray, K. L. Brown and L. Lorand. 1973. Titration and subunit localization of active center cysteine in fibrinoligase (thrombin-activated fibrin stabilizing factor). *Biochem. Biophys. Res. Comm.* **52**:51.

Curtis, C. G., K. L. Brown, R. B. Credo, R. A. Domanik, A. Gray, P. Stenberg and L. Lorand. 1974a. Calcium dependent unmasking of active center cysteine during activation of fibrin stabilizing factor. *Biochemistry* **13**:3774.

Curtis, C. G., P. Stenberg, K. L. Brown, A. Baron, K. Chen, A. Gray, I. Simpson and L. Lorand. 1974b. Kinetics of transamidating enzymes. Production of thiol in the reactions of thiol esters with fibrinoligase. *Biochemistry* **13**:3257.

Gray, A. and L. Lorand. 1974. Selective modification of the *b* subunit of fibrin stabilizing factor during activation of the zymogen. *Fed. Proc.* **33**:1474.

Holbrook, J. J., R. D. Cooke and I. B. Kingston. 1973. The amino acid sequence around the reactive cysteine residue in human plasma factor XIII. *Biochem. J.* **135**:901.

Kezdy, F. J., L. Lorand and K. D. Miller. 1965. Titration of active centers in thrombin solutions. Standardization of the enzyme. *Biochemistry* **4**:2302.

Linderstrøm-Lang, K. U. 1952. Proteins and enzymes. In *Lane Medical Lectures, Medical Sciences,* vol. 6, p. 114. Stanford University Press, Stanford, California.

Lorand, L. 1951. "Fibrino-peptide." New aspects of the fibrinogen-fibrin transformation. *Nature* **167**:992.

———. 1952. Fibrino-peptide. *Biochem. J.* **52**:200.

———. 1972. Fibrinoligase. The fibrin stabilizing factor system of blood plasma. *Ann. N.Y. Acad. Sci.* **202**:6.

Lorand, L. and W. R. Middlebrook. 1952. The action of thrombin on fibrinogen. *Biochem. J.* **52**:196.

Lorand, L. and J. L. G. Nilsson. 1972. Molecular approach for designing inhibitors to enzymes involved in blood clotting. In *Drug Design,* (ed. E. J. Ariens), vol. 3, p. 415. Academic Press, New York.

Lorand, L., J. Bruner-Lorand and T. R. E. Pilkington. 1966. Inhibitors of fibrin crosslinking. Relevance for thrombolysis. *Nature* **210**:1273.

Lorand, L., L. K. Campbell and B. Robertson. 1972. Enzymatic coupling of isoniazid to proteins. *Biochemistry* **11**:434.

Lorand, L., D. Chenoweth and A. Gray. 1972. Titration of the acceptor sites in fibrin. *Ann. N.Y. Acad. Sci.* **202**:15.

Lorand, L., R. F. Doolittle, K. Konishi and S. K. Riggs. 1963. A new class of blood coagulation inhibitors. *Arch. Biochem. Biophys.* **102**:171.

Lorand, L., N. G. Rule, H. H. Ong, R. Furlanetto, A. Jacobsen, J. Downey, N. Oner and J. Bruner-Lorand. 1968. Amine specificity in transpeptidation. Inhibition of fibrin crosslinking. *Biochemistry* **7**:1214.

Mikuni, Y., S. Iwanaga and K. Konishi. 1973. A peptide released from plasma fibrin stabilizing factor in the conversion to the active enzyme by thrombin. *Biochem. Biophys. Res. Comm.* **54**:1393.

Mockros, L. F., W. W. Roberts and L. Lorand. 1974. Viscoelastic properties of ligation-inhibited fibrin clots. *Biophys. Chem.* **2**:164.

Roberts, W. W., L. Lorand and L. F. Mockros. 1973. Viscoelastic properties of fibrin clots. *Biorheology* **10**:29.

Schwartz, M. L., S. V. Pizzo, R. L. Hill and P. A. McKee. 1971. The subunit structures of human plasma and platelet factor XIII (fibrin stabilizing factor). *J. Biol. Chem.* **246**:5851.

Takagi, T. and R. F. Doolittle. 1974. Amino acid sequence studies on factor XIII and the peptide released during its activation by thrombin. *Biochemistry* **13**: 750.

Williams-Ashman, H. G., A. C. Notides, S. S. Pabalan and L. Lorand. 1972. Transamidase reactions involved in the enzymatic coagulation of semen: Isolation of γ-glutamyl-ε-lysine dipeptide from clotted guinea pig seminal vesicle secretion protein. *Proc. Nat. Acad. Sci.* **69**:2322.

The Structural and Enzymatic Properties of the Components of the Hageman Factor-activated Pathways

Richard J. Ulevitch, Charles G. Cochrane, Susan D. Revak, David C. Morrison and Alan R. Johnston

Department of Immunopathology, Scripps Clinic and Research Foundation
La Jolla, California 92037

A series of enzymes of the Hageman factor-activated pathways are responsible for the production of biologically active kinins and for blood coagulation and fibrinolysis. Central to the initiation of these changes is the activation of Hageman factor (HF). The relationship of the initial components of this series is shown in Figure 1. Studies in this laboratory (Cochrane, Revak and Wuepper 1973; Revak et al. 1974) and in others (Margolis 1958; Kaplan, Schreiber and Austen 1971) indicate that activation may proceed by at least two distinct mechanisms: (1) *nonenzymatic activation* by substances bearing negative charges, such as lipopolysaccharide (LPS), kaolin and glomerular basement membrane fragments, where activation is achieved without any change in the molecular weight of the HF molecule, and (2) *enzymatic activation* by proteolytic enzymes, where activation is accompanied by changes in the molecular weight of the precursor HF. Both mechanisms produce an active enzyme capable of initiating all of the HF-activated pathways. Activation of each of the HF pathways proceeds by the HF-catalyzed conversion of a proenzyme to an enzymatically active molecule. In addition, it has been demonstrated that both kallikrein and active factor XI are capable of activating precursor Hageman factor by a reciprocal activation mechanism.

This paper will summarize recent findings in this laboratory which describe the enzymatic and physical properties of the precursor and activated molecules of HF, prekallikrein and factor XI (plasma thromboplastin antecedent—PTA) isolated from human plasma.

Physical Properties of Precursor and Activated Hageman Factor

Hageman factor has been purified in precursor form from normal human plasma by a series of standard chromatographic techniques yielding a homogeneous protein (Revak et al. 1974). A radial immunodiffusion assay has been utilized to quantitate the amount of HF in normal human plasma; concentration varies from between 29–43 μg/ml in the individuals tested. Iodination of the purified protein with ^{125}I permitted analysis of the structural features

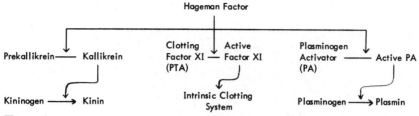

Figure 1

The HF-activated pathways of kinin formation, coagulation and fibrinolysis.

of precursor and activated HF using sodium dodecyl sulfate-polyacrylamide gel electrophoresis (SDS-PAGE).

SDS-PAGE of the purified protein indicated that the native molecule has a molecular weight of approximately 80,000. After reduction and alkylation in urea, no changes in the molecular weight were observed, suggesting that precursor HF exists as a single polypeptide chain. Amino acid analysis provided evidence that all of the available cystines were completely reduced and alkylated (Revak et al. 1974).

Since the native molecule is a single polypeptide chain, it was of interest to determine the effect of various activators on the structure of HF. Two types of activators were chosen for study: (1) soluble polymeric molecules or insoluble particles which contain regions of negative charges, such as bacterial lipopolysaccharide (LPS) or kaolin, and (2) proteolytic enzymes, initially concentrating on trypsin and then extending these studies to biologically important proteolytic enzymes such as kallikrein, plasmin and clotting factor XI.

Activation of HF by Contact with Negatively Charged Substances

The presence of densely positioned negative charges on an activating agent, such as LPS, kaolin or glass, appears to be important in the activation of HF. Kaolin, pretreated with hexadimethrene bromide to neutralize negative charges, binds but does not activate HF, and when LPS fractions containing different amounts of phosphate are compared for their ability to activate HF, fractions with the highest phosphate are found to be the best activators (Morrison and Cochrane 1974). When LPS or kaolin was utilized to activate HF, no changes in molecular weight were observed after electrophoresis of the active molecule in SDS (Fig. 2, top panel). Thus nonenzymatic activation is characterized by interaction of HF with the activator, followed by a presumed conformational change in the protein to yield an active enzyme.

Enzymatic Activation of HF

By comparison with the above activators, trypsin catalyzes the cleavage of HF into three separate fragments of molecular weight 52,000, 40,000 and 28,000. The 28,000 MW fragment possesses enzymatic activity (Fig. 2, bottom panel). Activation of HF with both plasmin and kallikrein produces fragments of HF similar to those produced by trypsin. With both negatively charged activators and enzymes, reduction and alkylation of the activated HF had no additional effect on the electrophoretic patterns of the ^{125}I-labeled proteins.

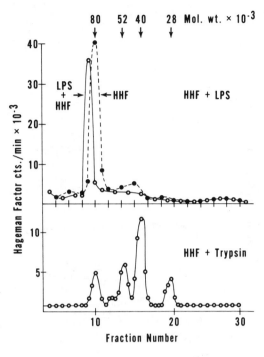

Figure 2
SDS-acrylamide gel electrophoresis of activated HF. Samples of ^{131}I-labeled human HF were activated by either LPS or trypsin. After electrophoresis, gels were sectioned into 1-mm slices and the radioactivity in each section determined.

The question was raised as to which portion of the HF molecule is responsible for binding the molecule to negatively charged activators. To test this, a preparation of trypsin-treated, ^{125}I-labeled Hageman factor was incubated with kaolin for 20 minutes, the suspension centrifuged, and the kaolin precipitate and supernatant analyzed for HF activity and for fragments. The kaolin pellets were placed on the top of an SDS gel and electrophoresis was initiated. All of the bound material is eluted from the kaolin by this procedure. The 52,000 and 40,000 MW fragments were bound to the kaolin, whereas almost all of the 28,000 MW material remained in the supernatant (Fig. 3). The enzymatic activity, as determined by the ability to activate prekallikrein, was associated entirely with the supernatant fraction. Thus it appears that the HF molecule contains at least two distinct regions: (1) the active-site region, which, after enzymatic cleavage, forms a fragment of molecular weight 28,000, and (2) regions involved in the binding of the activated molecule to negatively charged polymeric structures such as kaolin, contained in the 52,000 and 40,000 MW fragments.

Enzymatic Properties of Activated Hageman Factor

Hageman factor activated by both the enzymatic and nonenzymatic mechanisms is characterized by the conversion of a proenzyme to an active enzyme with proteolytic and esterolytic activities. Two methods have been employed for the measurement of the enzymatic activity of activated HF: (1) the activation of a measured amount of prekallikrein followed by quantitation of the kallikrein produced by measurement of the hydrolysis of benzoyl arginine ethyl ester (BAEE) at 253 nm (Wuepper and Cochrane 1972), and (2) the

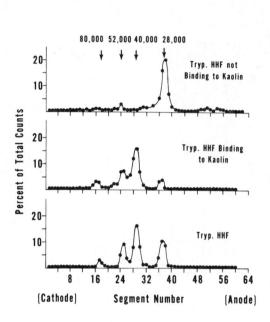

Figure 3

SDS-acrylamide gel electrophoresis of trypsin-activated HF and the interaction of the fragments with kaolin. ^{125}I-labeled HF was added to 4 μg of unlabeled HF, incubated with 0.2 μg trypsin for 15 minutes, 37°C. After the addition of 50 μg ovomucoid trypsin inhibitor, an aliquot was removed for SDS gel electrophoresis and the remainder incubated with 1 μg of kaolin for 20 minutes, 22°C. The supernatant fraction was removed, and both the pellet and supernatant fraction were analyzed for enzymatic activity and for radioactivity on SDS gels. (*Top*) Supernatant fraction; (*middle*) counts bound to kaolin after 20 minutes; (*bottom*) trypsinized HF.

hydrolysis of N-α-acetyl glycine lysine ester (AGLME), with subsequent quantitation of the methanol produced (Ulevitch, Letchford and Cochrane 1974). Evidence supporting the hypothesis that the activation of HF is characterized by the conversion of a proenzyme to an active enzyme derives from a number of experiments, as will be discussed below.

A preparation of precursor HF which possessed no spontaneous prekallikrein-activating activity and which was unable to hydrolyze AGLME was activated with kaolin. When the kaolin-activated HF was incubated with AGLME solutions, simple saturation kinetics for methanol production were observed at substrate concentrations less than 0.04 M AGLME with a TN = 2.5 min^{-1} and K_m = 0.009 M. AGLME concentrations greater than 0.04 M were inhibitory, probably due to substrate inhibition, although this effect has not been investigated in detail. AGLME also acts as a competitive inhibitor of prekallikrein activation, providing additional evidence that the esterolytic site and the proteolytic site of HF are the same. Furthermore, after activation of HF with kaolin, both prekallikrein activation and AGLME hydrolysis were inhibited with diisopropylfluorophosphate (DFP) in a similar manner; K_i = 5.6 × 10^{-4} M (Fig. 4). HF activated with trypsin also acquires esterolytic activity (Ulevitch, Letchford and Cochrane 1974).

Additional data supporting the concept that HF activation represents the conversion of a proenzyme to an active enzyme is found in the following experiment. Lima bean trypsin inhibitor (LBTI), a known inhibitor of active HF (Wuepper and Cochrane 1971), was insolubilized on Sepharose (Seph-4B) with cyanogen bromide (CNBr). A solution containing ^{125}I-labeled precursor HF and ^{131}I-labeled bovine serum albumin (BSA) was added to a 1-ml column of LBTI-Seph 4B and the effluent was assayed for active or

Figure 4

DFP inhibition of kaolin-activated HF. Pellets (500 μg) containing approximately 4.5 μg HF were incubated with either AGLME ($\blacksquare$) or prekallikrein ($\bigcirc$) in the presence or absence of varying concentrations of DFP. The curve is a calculated curve for $K_i = 5.6 \times 10^{-4}$ M from the following equation: percent initial activity $= 1/(1 + I/K_i) \times 100\%$.

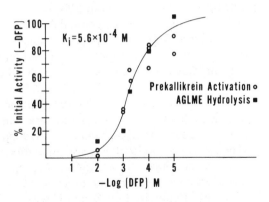

activatable HF and radioactivity. Greater than 90% of the added ^{131}I and ^{125}I were recovered, as well as trypsin-activatable HF. When an identical sample was first activated with trypsin and then placed on the LBTI-Seph 4B columns, 97% of the added BSA, but only 75% of the ^{125}I counts (HF), was recovered in the column eluate. The counts remaining on the LBTI-Seph 4B (approximately 25% of the added ^{125}I-labeled HF) correspond to the percentage of total counts added which would be expected to be associated with the 28,000 MW fragment.

Active HF can also activate factor XI and the plasminogen activator (Kaplan and Austen 1972) by proteolytic cleavage of the precursor molecules, as will be considered in the following sections.

In summary, HF activated with both enzymatic and nonenzymatic activators has been shown to be an active enzyme by its capacity to activate the initial components of the HF-activated pathways, by the hydrolysis of AGLME, and by inhibition with various protease inhibitors, among which are DFP and lima bean trypsin inhibitor.

Initiation of Hageman Factor Pathways
Kinin-forming, Intrinsic-clotting and Fibrinolytic Systems

The activation of both prekallikrein and clotting factor XI is characterized by the conversion of a proenzyme to an active enzyme (Wuepper and Cochrane 1972; Wuepper 1972a). The mechanism of these activations involves proteolytic cleavage, which may be accomplished by activated HF as well as by trypsin. Although the mechanism of activation of the plasminogen proactivator (Kaplan and Austen 1972) has not been well characterized, preliminary evidence suggests that this also represents the conversion of a proenzyme to an active enzyme. We should like to present a summary of the present knowledge of the structure, activation and enzymatic properties of the initial enzymes of the HF-activated pathways of coagulation and kinin formation.

Prekallikrein

Human prekallikrein has been isolated from normal human plasma by standard preparative techniques. Its approximate molecular weight is 107,000, and the proenzyme exists as a single polypeptide chain. After activation of ^{125}I-labeled prekallikrein with either HF or trypsin, kinin-forming and esterolytic activity are observed (Wuepper and Cochrane 1971). Studies of the activated molecule by SDS gel electrophoresis indicate no fragments of the unreduced parent molecule are produced by activation. However after reduction and alkylation, fragments are observed in SDS gels, indicating proteolytic cleavage (Wuepper 1972b).

Kallikrein also utilizes substituted arginine and lysine esters as substrates and is inhibited by a variety of protease inhibitors, among which are DFP, Trasylol and soybean trypsin inhibitor (Wuepper and Cochrane 1972).

The use of insolubilized protease inhibitors has provided additional evidence for the concept that the activation of prekallikrein and factor XI involves the conversion of a proenzyme to an active enzyme. Soybean trypsin inhibitor (STI) was insolubilized on Seph-4B with CNBr. When solutions containing only prekallikrein were added to STI-Seph 4B, all of the activatable prekallikrein was recovered in the effluent, whereas, by comparison, fully active kallikrein was adsorbed on the insoluble STI-Seph 4B (Table 1).

Recripocal Activation of HF

Purified kallikrein has been shown to enhance the clot-promoting and pre-kallikrein-activating activities of precursor HF, suggesting a reciprocal activation mechanism for the HF pathways. The importance of this reaction has been underlined by the observation of individuals whose plasma is deficient in the ability to form plasmin, kinin and active factor XI after exposure to glass (Fletcher trait). These individuals have normal concentrations of HF (Hathaway, Belhasen and Hathaway 1965; Hattersby and Hayse 1970). Data from our laboratory (Wuepper 1973), which have been confirmed and extended by Weiss, Gallin and Kaplan (1972), indicate that these individuals are deficient in prekallikrein, and that the addition of prekallikrein to Fletcher trait plasma restores the deficiencies. Thus besides the kinin-forming properties, the proteolytic activity of kallikrein appears to be central to the amplification of the activation of the HF pathways.

Plasma Thromboplastin Antecedent (Factor XI)

Factor XI has been purified from normal human plasma by column chromatography and Pevikon block electrophoresis (Wuepper 1972a). Electrophoresis of ^{125}I-labeled factor XI in SDS indicated that the precursor molecule has a molecular weight of approximately 160,000. After reduction and alkylation, two subunits of molecular weight 80,000 were observed. When either trypsin or active HF was utilized for activation of factor XI, the active molecule appeared indistinguishable from precursor XI prior to reduction and alkylation (Wuepper 1972a,b). However after reduction, fragments of 47,000 and 27,000 were observed, thus indicating that XI consists of two subunits, each of which is cleaved upon activation.

Active XI possesses clot-promoting activity. This activity is inhibited by DFP, Trasylol and soybean trypsin inhibitor, but not by lima bean or ovomucoid trypsin inhibitors. No detailed studies have been performed on the esterolytic capacity of factor XI, although active XI was found to slowly hydrolyze BAEE and AGLME.

Insolubilized soybean trypsin inhibitor is also useful in separating active and

Table 1

Effect of STI-Seph 4B on Prekallikrein and Kallikrein

Sample	STI-Seph 4B	Δ 253 nm/min/mg protein	Bradykinin generation
1. Prekallikrein, active HF	—	3.1	+
2. Prekallikrein, active HF	+	0.02	—
3. Prekallikrein	—	0.05	—
4. Prekallikrein	+	0.03	—
5. Sample 4, active HF *after* STI-Seph 4B	—	2.5	+

Solutions of prekallikrein were added to columns of STI-Seph 4B prior to and after activation with HF. Effluents were tested for active or activatable enzyme with benzoyl arginine ethyl ester (Wuepper and Cochrane 1972).

Table 2

Effect of STI-Seph 4B on Precursor and Active PTA

Sample	Clotting time (min)
1. Recalcified plasma	14.0
2. PTA, *prior* to STI-Seph 4B	6.1
3. PTA(#2), trypsinized	1.7
4. PTA(#2), *after* STI-Seph 4B	12.0
5. PTA(#4), trypsinized	2.1
6. PTA(#5), *after* STI-Seph 4B	11.0

Clotting tests were performed by adding samples of precursor or active factor XI to 100 microliters of fresh rabbit plasma containing 50 μl of B and A cephalin, followed by 50 mM $CaCl_2$. The solution was gently mixed and the time required at room temperature to form a firm clot recorded.

precursor factor XI, thus providing additional evidence that the activation of XI is characterized by the conversion of a proenzyme to an active enzyme. The data are summarized in Table 2.

Plasminogen Proactivator

The plasminogen proactivator has been isolated from human plasma (Kaplan and Austen 1972). This molecule has an approximate molecular weight of 100,000. The only reported activator for this molecule is active HF. The plasminogen activator isolated by Kaplan and Austen (1972) is most probably identical to a molecule described by Ogston et al. (1969), termed the Hageman factor cofactor. The activated molecule is inhibited by DFP. Preliminary studies in this laboratory also support the concept that the activated molecule is an enzyme, since the activated molecule is adsorbed on STI-Seph 4B, whereas the precursor molecule is not.

At the present time, the mechanism of activation of the plasminogen proactivator by Hageman factor is unknown, but like the other substrates of the Hageman factor pathways, activation most likely proceeds by limited proteolytic cleavage.

Acknowledgments

This work was supported by United States Public Health Service Grant AI-07007, the American Heart Association, and the Council for Tobacco Research.

REFERENCES

Cochrane, C. G., S. D. Revak and K. D. Wuepper. 1973. The activation of Hageman factor in solid and fluid phases: A critical role of kallikrein. *J. Exp. Med.* **138**:1564.

Hathaway, W. E., L. P. Belhasen and H. S. Hathaway. 1965. Evidence for a new thromboplastin factor. I. Case report, coagulation studies and physiocochemical properties. *Blood* **26:**521.

Hattersby, P. G. and D. Hayse. 1970. Fletcher deficiency: A report of three unrelated cases. *Brit. J. Haematol.* **18:**411.

Kaplan, A. P. and K. F. Austen. 1972. The fibrinolytic pathway of human plasma. Isolation and characterization of the plasminogen proactivator. *J. Exp. Med.* **136:**1378.

Kaplan, A. P., A. D. Schreiber and K. F. Austen. 1971. A prealbumin activator of prekallikrein. II. Derivation of activators of prekallikrein from active Hageman factor by digestion with plasmin. *J. Exp. Med.* **133:**696.

Margolis, J. 1958. Activation of plasma by contact with glass: Evidence of a common reaction which releases plasma kinin and initiates coagulation. *J. Physiol.* **144:**1.

Morrison, D. C. and C. G. Cochrane. 1974. A direct evidence for Hageman factor (factor XII) activation by bacterial lipopolysaccharide. *J. Exp. Med.* **140:**797.

Ogston, D., C. M. Ogston, O. D. Ratnoff and C. D. Forbes. 1969. Studies in a complex mechanism for the activation of plasminogen by kaolin and by chloroform. The participation of Hageman factor and additional cofactors. *J. Clin. Invest.* **48:**1786.

Revak, S. D., C. G. Cochrane, A. Johnston and T. Hugli. 1974. Structural changes accompanying activation of human Hageman factor. *J. Clin. Invest.* **54:**619.

Ulevitch, R. J., D. Letchford and C. G. Cochrane. 1974. A direct enzymatic assay for the esterolytic activity of activated Hageman factor. *Thromb. Diath. Haemorrh.* **31:**30.

Weiss, A. S., J. I. Gallin and A. P. Kaplan. 1972. Fletcher factor deficiency: Abnormalities of coagulation, fibrinolysis, chemotactic activity, and kinin generation attributable to absence of prekallikrein. *Fed. Proc.* **32:**353a.

Wuepper, K. D. 1972a. Precursor plasma thromboplastin antecedent (PTA, clotting factor XI). *Fed. Proc.* **31:**624.

———. 1972b. Biochemistry and biology of components of the plasmin kinin-forming system. In *Inflammation: Mechanisms and Control* (ed. I. H. Lepow and P. A. Ward), p. 93. Academic Press, New York.

———. 1973. Prekallikrein deficiency in man. *J. Exp. Med.* **138:**1345.

Wuepper, K. D. and C. G. Cochrane. 1971. Plasma prokinogenase (prekallikrein): Isolation and activation. *J. Clin. Invest.* **50:**100a.

———. 1972. Plasma prekallikrein: Isolation, characterization, and mechanism of activation. *J. Exp. Med.* **135:**1.

The Activation of Bovine Prothrombin

Craig M. Jackson, Charles T. Esmon* and Whyte G. Owen†

Department of Biological Chemistry, Division of Biology and Biomedical Sciences
Washington University, St. Louis, Missouri 63110

Prothrombin activation is catalyzed by an enzyme complex (Straub and Duckert 1961; Milstone 1964; Seegers 1964; Papahadjopoulos and Hanahan 1964; Barton, Jackson and Hanahan 1967; Seegers et al. 1972; Hemker et al. 1967; Deggler and Vreeken 1969) which consists of four components. These are factor X_a, a serine protease (Jackson and Hanahan 1968; Leveson and Esnouf 1969; Radcliffe and Barton 1972; Fujikawa, Legaz and Davie 1972) with trypsinlike bond specificity (Esnouf and Williams 1962; Milstone et al. 1963; Aronson and Menache 1968; Smith 1973); factor V_a, a protein which binds both factor X_a (Papahadjopoulos and Hanahan 1964) and prothrombin (Esmon et al. 1973); phospholipid, which as bilayer aggregates provides a surface onto which all the protein components adsorb (Papahadjopoulos and Hanahan 1964; Cole, Koppel and Olwin 1964, 1965; Jobin and Esnouf 1967; Deggler and Vreeken 1969); and Ca^{++} ions. The Ca^{++} ions are required for binding factor X_a (Papahadjopoulos and Hanahan 1964; Cole, Koppel and Olwin 1964; Jobin and Esnouf 1967) and prothrombin (Barton and Hanahan 1969; Jobin and Esnouf 1967; Bull, Jevons and Barton 1972; Gitel et al. 1973) to the phospholipid surface and for maintaining the stability of factor V_a (Blömback and Blömback 1963; Day and Barton 1972; Dombrose et al. 1972; Dombrose and Seegers 1974). (Factor V_a is a thrombin-modified form of factor V which, as a consequence of the modification, is capable of binding prothrombin and intermediate 1 [Esmon et al. 1973].)

The investigation of prothrombin activation in our laboratory has been separated into three distinct stages. The first stage has consisted of complete identification of the products of prothrombin activation (Owen, Esmon and Jackson 1974; Esmon, Owen and Jackson 1974a). This was necessary since it had been known for a long time that thrombin accounts for only one-half of

Present addresses: *Department of Biochemistry, College of Agricultural and Life Sciences, University of Wisconsin, Madison, Wisconsin 53706; †Department of Pathology, College of Medicine, University of Iowa, Iowa City, Iowa 52242.

the mass of the prothrombin molecule (Magnusson 1971). The second stage has been concerned with determining the function of the nonthrombin-forming half of prothrombin (Gitel et al. 1973; Esmon and Jackson 1974a), and the third stage has been concerned with the elucidation of the activation pathway itself (Esmon and Jackson 1974a, b; Esmon, Owen and Jackson 1974b).

METHODS

All experimental procedures and methods for the isolation of the proteins and phospholipids are given in Jackson, Johnson and Hanahan (1968) and Jackson and Hanahan (1968) for factor X_a, Esmon (1973) for factor V_a, Esmon, Owen and Jackson (1974a) and Owen, Esmon and Jackson (1974) for prothrombin and the partial activation products, and Gitel et al. (1973) for the phospholipids.

RESULTS AND DISCUSSION

Identification and Characterization of Prothrombin Activation Products

After complete activation of prothrombin with $[X_a, V_a,$ Phospholipid, $Ca^{++}]$,[1] thrombin and two nonthrombin-forming fragments are found in the reaction mixture (Fig. 1A) (Owen, Esmon and Jackson 1974). These three activation products account for the total mass of the prothrombin molecule. This conclusion is based on the recovery of protein from QAE-Sephadex ion exchange columns, molecular weight analyses by SDS gel electrophoresis and equilibrium ultracentrifugation, and on amino acid and carbohydrate composition analyses. Designation of the first nonthrombin activation product eluted from the columns as fragment 1 and the second as fragment 2 was based on both historical precedent (Lanchantin, Friedmann and Hart 1965; Aronson and Menache 1966) and the simplification which this designation creates when all of the structural information regarding the prothrombin molecule is available (Morita et al. 1973; Seegers et al. 1974). As prothrombin activation by factor X_a results in the formation of another protease, thrombin, it was important to determine which of the activation products arise as a consequence of factor X_a-catalyzed proteolysis and which as a consequence of thrombin-catalyzed proteolysis. Because thrombin is readily inhibited by diisopropyl-fluorophosphate (iPr$_2$P-F) (Gladner and Laki 1956), in contrast to factor X_a which is only inactivated by very high concentrations of iPr$_2$P-F (Jackson and Hanahan 1968; Leveson and Esnouf 1969; Fujikawa, Legaz and Davie 1972), differentiation between factor X_a-catalyzed and thrombin-catalyzed proteolysis was readily achieved (Stenn and Blout 1972; Esmon, Owen and Jackson 1974a). When prothrombin activation was carried out by the same activator as in the experiment of Figure 1A, but in the presence of iPr$_2$P-F, a previously

---

[1] Brackets are used to indicate that the enclosed materials are prothrombin activator components.

Figure 1

Isolation of prothrombin activation products by QAE-Sephadex chromatography. A 3-ml sample was applied to a column (0.9 × 24 cm) of QAE-Sephadex Q-50 on top of which was a 2-cm bed of soybean trypsin inhibitor-Sepharose. The column was equilibrated in 0.02 M Tris-HCl pH 7.5, 0.1 M NaCl and was eluted with a linear gradient, 45 ml/gradient chamber, 0.1–0.6 M NaCl in the Tris-HCl buffer. Elution was performed at room temperature at 6 ml/hr and 1.0-ml fractions were collected.

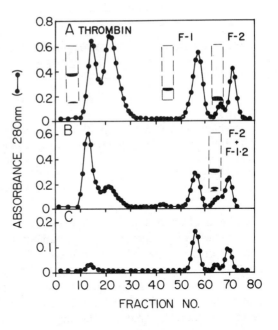

(*A*) Prothrombin, 9 mg in 3 ml of 0.02 M Tris-HCl pH 7.5, 0.1 M NaCl, was activated with X_a, 20 μg; V_a, 28 μg; phospholipid, 200 μg; and $CaCl_2$, 10 mM final concentration for 12 minutes at 23°C. Absorbance, 280 nm (●——●). Active thrombin is found in fractions 10–30. SDS gels are from left to right: thrombin, fragment 1 and fragment 2. In this and all subsequent experiments, thrombin activity is measured from its ability to clot fibrinogen (Owen, Esmon and Jackson 1974).

(*B*) Prothrombin, 9 mg in 3 ml of 0.03 M Tris-HCl pH 7.8, 0.1 M NaCl, 10 mM iPr$_2$P-F, was activated with X_a, 20 μg; V_a, 28 μg; phospholipid, 200 μg; and $CaCl_2$, 10 mM final concentration for 12 minutes at 23°C. The SDS gel is from fraction 71. Column elution conditions are as in *A*.

(*C*) Chromatography of the products from thrombin-catalyzed cleavage of fragment 1·2 on QAE-Sephadex. Fragment 1·2, 1 mg in 1 ml was incubated with thrombin (30 μg) in 0.02 M Tris-HCl pH 7.5, 0.1 M NaCl for 1.35 hours at 23°C. The reaction mixture was chromatographed on a 0.9 × 24-cm column of QAE-Sephadex Q-50. Column equilibration and elution conditions are as given for *A*.

undetected product which cochromatographs with fragment 2 was isolated (Fig. 1B). SDS gel electrophoresis showed that the new product had a molecular weight equal to the sum of the molecular weights of fragments 1 and 2. After isolation, this new product could be cleaved by thrombin to form fragment 1 and fragment 2 (Fig. 1C). On this basis, the new product was designated fragment 1·2. Amino acid analyses confirmed the identification of the new product. Interestingly, neither factor X_a alone nor the complete activator [X_a,V_a,Phospholipid,Ca^{++}] cleaved fragment 1·2, demonstrating that fragments 1 and 2 arise exclusively as a result of a thrombin-catalyzed reaction (Esmon, Owen and Jackson 1974a; Kisiel and Hanahan 1974). On the basis of (1) earlier work indicating that thrombin was derived from the car-

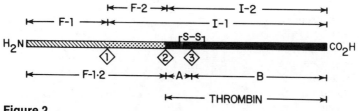

Figure 2

A schematic model for the prothrombin molecule. The diamond
symbol represents the peptide bonds in prothrombin which are
cleaved during activation: factor X_a cleaves bonds 2 and 3, thrombin
cleaves bond 1.

boxyl end of prothrombin (Magnusson 1971), (2) amino-terminal amino
acid analysis of prothrombin and fragment 1·2 (Esmon, Owen and Jackson
1974a) and (3) partial sequence data (Heldebrant et al. 1973; Reuterby
et al. 1974), the unambiguous location of the nonthrombin half of prothrombin was established. This structural information is summarized by the schematic diagram shown in Figure 2. The peptide bonds that are cleaved in
prothrombin are designated in the diagram by 1, 2 and 3 within diamonds.

In addition to cleaving bond 1 in fragment 1·2, thrombin can also catalyze
proteolysis of this bond in prothrombin to form fragment 1 and intermediate 1. Intermediate 1 is a single polypeptide chain precursor of thrombin
which, as seen on the diagram, includes the fragment 2 region. Cleavage of
bond 2 in prothrombin by factor X_a gives rise to intermediate 2 and fragment
1·2. Intermediate 2 has been isolated from incomplete activation mixtures and
consists of a single polypeptide chain with the same amino acid and carbohydrate composition as thrombin. Cleavage of bond 3 in intermediate 2 by
factor X_a yields thrombin which consists of two disulfide-linked polypeptide
chains. From this structural data, it is clear that the activation of prothrombin
by factor X_a must involve cleavage of two peptide bonds. The possible significance of the thrombin-catalyzed cleavage of bond 1 is discussed below.

Function of Fragment 1·2 Region (Propiece)
of Prothrombin

Preliminary experiments in our laboratory and an examination of the literature on prothrombin activation (Baker and Seegers 1967; Seegers et al. 1972)
suggested that the catalytic function of phospholipid and factor V_a might be
mediated by the fragment 1 and fragment 2 regions, respectively. As the two
precursors of thrombin, intermediate 1 and intermediate 2 can be isolated
from incomplete activation mixtures; the general approach to testing this structure-function hypothesis has been to employ these polypeptides as alternative
substrates with different combinations of activator components. Specifically,
the function of the fragment 1 region of prothrombin was investigated by comparing the activation of prothrombin and intermediate 1, and the function of
the fragment 2 region, by comparing the activation of intermediate 1 and
intermediate 2.

Function of Fragment 1 Region: Phospholipid and Calcium Binding

The function of the fragment 1 region in binding prothrombin to phospholipid was first suggested when it was observed that prothrombin activation is markedly accelerated in the presence of phospholipid and calcium (top curve Fig. 3A), whereas intermediate 1 activation is not accelerated by phospholipid under exactly the same experimental conditions (bottom dashed curve) (Gitel et al. 1973). From the curve of Figure 3B, it can also be seen that prothrombin and intermediate 1 activate at identical rates in the absence of phospholipid, indicating that it is the contribution of phospholipid to the process which is being observed. (It should be noted that because of a higher factor X_a concentration, the thrombin formation rate is somewhat greater in the experiment of Fig. 3B than in the experiment shown by the bottom curve of Fig. 3A.) Since it has been shown by other investigators (Papahadjopoulos and Hanahan 1964; Cole, Koppel and Olwin 1964, 1965; Jobin and Esnouf

Figure 3

Phospholipid enhancement of the rate of prothrombin activation.

(*A*) Prothrombin (0.10 mg/ml) or intermediate 1 (0.067 mg/ml) in 0.02 M Tris-HCl, 0.1 M NaCl, pH 7.5, was incubated with [X_a, 0.45 μg/ml; CaCl$_2$, 10 mM] at 37°C in the absence and presence of phospholipid (25 μg/ml). Phospholipid is bovine brain cephalin (Gitel et al. 1973). In the experiment in which the reaction mixture also contains fragment 1, the molar ratio of fragment 1 to prothrombin is 3.2. (■——■) Prothrombin plus phospholipid; (□——□) prothrombin plus phospholipid plus fragment 1; and (– – – –) intermediate 1 plus phospholipid.

(*B*) Prothrombin (0.28 mg/ml) or intermediate 1 (0.19 mg/ml) in the same buffer as in *A* was incubated with [X_a, 5 μg/ml; CaCl$_2$, 10 mM] in the presence and absence of phospholipid (25 μg/ml) at room temperature. Phospholipid was an equimolar mixture of dioleyl phosphatidylcholine and dioleyl phosphatidylglycerol (Gitel et al. 1973. (■——■) Prothrombin in the absence of phospholipid; intermediate 1 in the absence (●——●) and presence (○——○) of phospholipid.

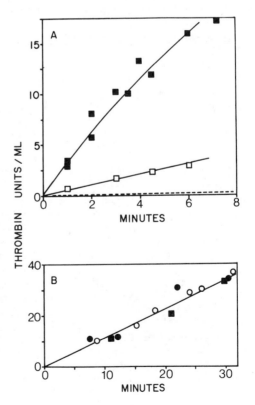

1967; Barton and Hanahan 1969; Bull, Jevons and Barton 1972) that prothrombin and factor X_a bind to phospholipid vesicles, the simplest explanation for the ability of lipid to accelerate activation of prothrombin but not of intermediate 1 is that the fragment 1 region is responsible for binding prothrombin to the lipid vesicle surface. Direct, qualitative evidence for Ca^{++}-mediated binding of fragment 1 to phospholipid vesicles has been published (Gitel et al. 1973). Recent results from both kinetic and quantitative binding experiments (Jackson et al. 1974a,b) also indicate that prothrombin and fragment 1 bind to the lipid surface with the same stoichiometry and affinity.

Since fragment 1 binds to the phospholipid surface, under appropriate conditions it should inhibit prothrombin activation by competing with prothrombin for the lipid surface. The middle curve of Figure 3A shows this inhibition and confirms the prediction. Experiments designed to determine if phospholipid enhancement of intermediate 1 activation could be obtained by adding fragment 1 to the activation mixture showed no increase whatsoever in the activation rate. Work from this and other laboratories (Benson, Kisiel and Hanahan 1973; Stenflo 1973; Stenflo and Ganrot 1973) indicates that virtually all Ca^{++} binding to prothrombin occurs via the fragment 1 region, consistent with the Ca^{++} requirement for phospholipid binding.

Prothrombin activation by factor X_a can also be accelerated by factor V_a in the absence of phospholipid. Since $[X_a, Ca^{++}]$ activates prothrombin and intermediate 1 at the same rate in the absence of phospholipid (Fig. 3B), it was similarly possible to determine whether factor V_a action might be mediated via the fragment 1 region. In Figure 4, the activation of prothrombin and intermediate 1 is investigated as a function of factor V_a concentration. In contrast to the case with phospholipid, with all factor V_a concentrations these two substrates activate at the same rate, indicating clearly that the fragment 1

Figure 4

Comparison of the activation of prothrombin and intermediate 1 with $[X_a, V, Ca^{++}]$. The rates of activation of prothrombin (0.125 mg/ml) and intermediate 1 (0.094 mg/ml) by $[X_a, 4.7$ μg/ml; V_a, variable; Ca^{++}, 0.01 M] are shown as a function of factor V_a concentration (top curve, 16 units V_a/ml; middle curve, 3.2 units of V_a/ml; bottom curve, 0.3 units of V_a/ml). All solution conditions are the same as in Figure 3. Prothrombin was also activated under the same conditions except in the presence of fragment 1 (0.31 mg/ml), which corresponds to a molar ratio of 7.4. (■) Prothrombin; (▲) prothrombin plus fragment 1; (●) intermediate 1. The unit of factor V_a activity is defined in Esmon et al. (1973).

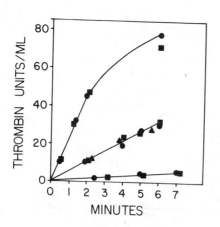

region is not essential for factor V_a action. It should be noted that the activation rate in the presence of factor V_a is 300-fold greater than in its absence. From the middle curve of Figure 4 it can also be seen that fragment 1 does not inhibit prothrombin activation by $[X_a, V_a, Ca^{++}]$. Taken altogether, the results of these experiments imply that during the activation process, the fragment 1 region functions primarily, if not exclusively, via interaction with Ca^{++} and phospholipid.

Function of Fragment 2 Region: Activation in the Presence of Factor V_a

The participation of the fragment 2 region during activation has been investigated by comparing the rate of activation of intermediate 1 with that of intermediate 2 (Fig. 5). In Figure 5A it can be seen that when $[X_a, Ca^{++}]$ alone

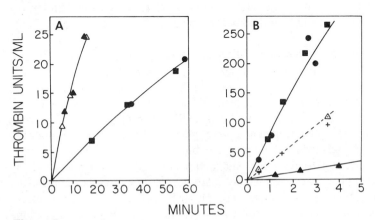

Figure 5
Factor V_a reverses the relative rates of activation of intermediate 1 and intermediate 2.

(*A*) A comparison of the activation of prothrombin, intermediate 1 and intermediate 2 by $[X_a, Ca^{++}]$. All solution conditions are given in the legend to Figure 3. Equimolar concentrations (5.4 μM) of prothrombin (0.39 mg/ml), intermediate 1 (0.27 mg/ml) or intermediate 2 (0.21 mg/ml) were incubated with $[X_a, 3.3\ \mu g/ml; Ca^{++}, 10\ mM$ final conc.]. Phospholipid is *absent* from all reaction mixtures. In one experiment, fragment 2 (0.02 mg/ml) was added to intermediate 2 in order to assess the effect of fragment 2 on the X_a-catalyzed activation of intermediate 2.

(*B*) The effect of factor V_a: A comparison of the activation of prothrombin, intermediate 1 and intermediate 2 by $[X_a, V_a, Ca^{++}]$. All reactant concentrations are the same as in *A*, except that V_a is present at a final concentration of 12.5 U/ml. In one experiment, fragment 2 (0.02 mg/ml) was added to intermediate 2, and in a second experiment, fragment 1 (0.03 mg/ml) was added in addition to fragment 2. (Note that the scales on both abscissa and ordinate are different in *A* and *B*.) (■——■) Prothrombin; (●——●) intermediate 1; (▲——▲) intermediate 2; (△——△) intermediate 2 plus fragment 2; (+——+) intermediate 2 plus fragment 1 plus fragment 2.

is the activator, the rate of intermediate 2 activation is four- to sixfold greater than that of intermediate 1. Upon addition of factor V_a (Fig. 5B), the rate of activation of intermediate 1 is increased 300-fold, whereas that of intermediate 2 is increased only fourfold. As a result, a complete reversal of the status of these two substrates occurs in the presence of factor V_a, and intermediate 1 is now the preferred or more rapidly activated substrate. Clearly, the fragment 2 region is essential for optimal factor V_a action. Moreover from the data of Figure 5, it can also be seen that in these experiments, which do not include phospholipid as part of the activator, intermediate 1 and prothrombin are equivalent substrates. This latter result again implies, as did the data of Figure 4, that the fragment 1 region is not related to factor V_a function.

The middle curve of Figure 5B shows that when fragment 2 is added to intermediate 2 in the presence of factor V_a, a marked stimulation of intermediate 2 activation occurs. In contrast to this latter situation, the addition of fragment 2 to intermediate 2 in the absence of factor V_a (Fig. 5A) is without effect on the activation rate.

The mechanism by which stimulation of intermediate 2 activation occurs has been investigated, and the results of two types of experiments are shown in Figure 6. First, the increase in the rate of thrombin formation from intermediate 2 as a function of both fragment 2 and fragment 1·2 concentrations has been studied. In the experiment shown in Figure 6A, the rate of activation of intermediate 2 plus fragment 2 or fragment 1·2 is equal in the plateau region to that of intermediate 1 or prothrombin at the same molar concentrations. It is interesting to note that maximum stimulation is reached when either of the fragments and intermediate 2 are present in equimolar concentrations, suggesting that a stoichiometric complex with properties of either intermediate 1 or prothrombin may be formed as a result of noncovalent association between the fragments and intermediate 2. Gel electrophoresis at pH 7.5 demonstrates that relatively tight complexes are formed. Panel B shows that intermediate 2 and fragment 2 associate to form a product with electrophoretic characteristics indistinguishable from intermediate 1. Panel C shows that intermediate 2 and fragment 1·2 likewise associate to form a product with electrophoretic characteristics comparable to those of prothrombin. A slight excess of fragment 1·2 was included in the experiment of Figure 6C and can be seen along with traces of fragment 1 and "reconstituted intermediate 1." The presence of fragment 1 is a result of fragment 1·2 cleavage by thrombin which contaminates the intermediate 2.

On the basis of all the data, it can be concluded that the propiece of prothrombin contains two regions with specialized functions. The fragment 1 region functions via interaction with Ca^{++} and phospholipid, whereas the fragment 2 region functions via interaction with intermediate 2 and factor V_a.

Pathway of Prothrombin Activation

Because factor X_a cleaves bonds 2 and 3 in forming thrombin (see Fig. 2), only two initial reaction pathways need to be considered for prothrombin activation. One of these involves cleavage of bond 3 prior to bond 2 to form an intermediate in which both the propiece and the thrombin piece of prothrombin remain covalently linked. As attempts to detect this product in

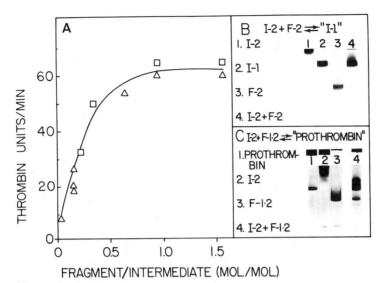

Figure 6

Intermediate 2 activation by $[X_a, V_a, Ca^{++}]$: a comparison of the effects of fragment 2 and fragment 1·2.

(*A*) Intermediate 2 (70 μg/ml, 2 μM) was incubated with $[X_a,$ 14 μg/ml; V, 32 U/ml; CA^{++}, 10 mM final conc.] and either fragment 2 or fragment 1·2 at the molar ratios to intermediate shown on the abcissa of the figure. The initial rate of thrombin formation was determined from plots of the concentration of thrombin versus time. All solution conditions are as given in Figure 3. (△) Intermediate 2 plus fragment 2; (□) intermediate 2 plus fragment 1·2.

(*B*) Noncovalent interaction between intermediate 2 and fragment 2. Acrylamide gel electrophoresis at pH 7.5 was performed as described by Williams and Reisfeld (1964), except that stacking gels were not used. Samples (50 μl each in 0.1 M NaCl, 0.02 M Tris-HCl pH 7.5, and containing 5% glycerol) were layered under the upper buffer, directly over the gel. Electrophoresis was carried out at 3 mA per column and at room temperature, with the gel completely immersed in the lower buffer. Samples are, from left to right: intermediate 2, 13 μg; intermediate 1, 18 μg; fragment 2, 4 μg; intermediate 2, 13 μg plus fragment 2, 4 μg,

(*C*) Noncovalent interaction between intermediate 2 and fragment 1·2. Acrylamide gel electrophoresis at pH 7.5 was performed as in *B*. The samples are, from left to right: prothrombin, 20 μg; intermediate 2, 15 μg; fragment 1·2, 15 μg; intermediate 2, 15 μg plus fragment 1·2, 15 μg. The two components seen in the gel on the far right which are not identified by samples in the other gels are "intermediate 1" (nearest the top), i.e., intermediate 2 plus fragment 2, and a trace of fragment 1 (nearest the bottom).

prothrombin activation mixtures have been unsuccessful, this possibility will not be considered further. The second alternative (Fig. 7), which begins with cleavage of bond 2 to form intermediate 2 and fragment 1·2, accounts for all

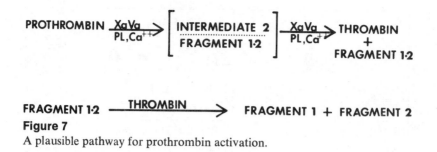

Figure 7
A plausible pathway for prothrombin activation.

the experimental data available and thus will be discussed in detail. The task
of demonstrating that the reaction sequence in Figure 7 can be the pathway of
prothrombin activation involves the following: First, intermediate 2 must be
formed directly from prothrombin; second, intermediate 2 formation must
be accelerated by phospholipid and factor V_a; and third, intermediate 2 must
be converted to thrombin at least as rapidly as prothrombin. Evidence demonstrating that the first two requirements are met was obtained by using radiolabeled prothrombin and determining product distributions in SDS electrophoresis gel as a function of time. This evidence can be found in the primary
reports of this work (Owen, Esmon and Jackson 1974; Esmon, Owen and
Jackson 1974b; Esmon and Jackson 1974a,b). Comparison of activation rates
with different combinations of activator components and different substrates
most clearly illustrates fulfillment of the third requirement (Table 1). From
entries 1 and 2 and Figure 5A it can be seen that intermediate 2 alone is
converted to thrombin more rapidly than to prothrombin by $[X_a, Ca^{++}]$, and
thus for reaction in free solution, the fragment 1·2 region is not required.
Upon addition of phospholipid to $[X_a, Ca^{++}]$ (entries 3, 4 and 5), prothrombin activation is accelerated approximately 50-fold. However with this activator, intermediate 2 is converted to thrombin much too slowly to qualify as
the kinetic intermediate in the activation process. Inclusion of fragment 1·2
with intermediate 2, however, results in intermediate 2 being converted to
thrombin about twice as fast as to prothrombin, and thus the necessary criterion
is now met. In entries 6, 7 and 8 it can be seen that upon addition of factor
V_a to $[X_a, Ca^{++}]$, the prothrombin activation rate is increased approximately
350-fold. Again, as previously shown (Fig. 5), intermediate 2 alone is not a
satisfactory substrate. In this case, addition of fragment 1·2 yields an activation
rate which is nearly equal to that of prothrombin, but clearly not unambiguously sufficient. Most important, however, is the situation with the complete
prothrombin activator $[X_a, V_a, Phospholipid, Ca^{++}]$ (entries 9, 10 and 11).
Here (entry 9) prothrombin activation is nearly 20,000 times faster than it
was with $[X_a, Ca^{++}]$ alone. Once again, however, intermediate 2 alone is not
a satisfactory substrate, but intermediate 2 plus fragment 1·2 is activated more
rapidly than prothrombin, and the criterion for designating intermediate 2 as
an activation intermediate is satisfied. Entry 12 demonstrates that factor X_a
is necessary for generation of thrombin from intermediate 2 and fragment 1·2.

Even though it is clear that intermediate 2 and fragment 1·2 are the products which are formed initially by $[X_a, V_a, Phospholipid, Ca^{++}]$ in this pathway

Table 1

Relative Rates of Thrombin Formation from Prothrombin and
Intermediate 2 with Different Activator Mixtures

Activator[a]	Substrate mixture composition	Activation rate[b]	Relative rate[b]
1. $[X_a, Ca^{++}]$	prothrombin (0.4 mg/ml)	0.042	1
2. $[X_a, Ca^{++}]$	intermediate 2 (0.2 mg/ml)	0.25	6
3. $[X_a, PL, Ca^{++}]$	prothrombin	—	50[c]
4. $[X_a, PL, Ca^{++}]$	intermediate 2	—	5–10[c, d]
5. $[X_a, PL, Ca^{++}]$	intermediate 2 + fragment 1·2	—	95[c]
6. $[X_a, V_a, Ca^{++}]$	prothrombin (0.4 mg/ml)	15.0	350
7. $[X_a, V_a, Ca^{++}]$	intermediate 2 (0.2 mg/ml)	1.5	36
8. $[X_a, V_a, Ca^{++}]$	intermediate 2 (0.2 mg/ml) + fragment 1·2 (0.2 mg/ml)	1.3	310
9. $[X_a, V_a, PL, Ca^{++}]$	prothrombin (0.4 mg/ml)	80	19,000[e]
10. $[X_a, V_a, PL, Ca^{++}]$	intermediate 2 (0.2 mg/ml)	32	7600[e]
11. $[X_a, V_a, PL, Ca^{++}]$	intermediate 2 (0.2 mg/ml) + fragment 1·2 (0.2 mg/ml)	600	140,000[e]
12. $[\_\_, V_a, PL, Ca^{++}]$	intermediate 2 (0.2 mg/ml) + fragment 1·2 (0.2 mg/ml)	0[f]	0

Prothrombin or intermediate 2 plus or minus fragment 1·2 was incubated at room
temperature in 0.02 M Tris-HCl, 0.10 M NaCl, 0.010 M CaCl$_2$, pH 7.5, with different
combinations of activator components. Activator component concentrations are the final
concentrations in the reaction mixture. The molar concentration of the substrates,
prothrombin, intermediate 2 or intermediate 2 plus fragment 1·2 is the same in all
experiments. Thrombin was determined by its ability to clot fibrinogen.

[a] Activator component concentrations are: Ca^{++}, CaCl$_2$, 10 mM in all reactions;
X_a, 3 μg/ml in the reactions of entries 1–8 inclusive and 0.3 μg/ml in the reactions of
entries 9–11; phospholipid, 15 μg/ml in the reactions of entries 3–12; and V_a, 25 μg/ml
(40 U/ml) in the reactions of entries 6–12.

[b] Activation rate is NIH units of thrombin/ml/min. Relative rate is the quotient of
the particular reaction rate divided by the rate of entry 1. (See footnote e for entries
9–11.)

[c] Relative rates are determined by interpolation from data obtained in another set of
experiments at a different X_a concentration.

[d] A slight acceleration of intermediate 2 activation by $[X_a, Phospholipid, Ca^{++}]$ is
observed along with floculation of the phospholipid.

[e] Relative rate is multiplied by 10 since the proteolytic enzyme concentration was decreased tenfold in order to obtain a sufficient number of time points to estimate the
initial rate.

[f] No thrombin (<0.001 NIH U/ml) was detectable after 30 minutes of incubation.

(Fig. 7) and that fragment 1·2 is required for fulfilling the necessary kinetic
requirements, it is also necessary to determine if fragment 1 plus fragment 2
can replace fragment 1·2. Figure 8 (top curve) shows thrombin formation
from intermediate 2 plus fragment 1·2 at a concentration of fragment 1·2
which gives one-half the maximal rate of activation under these conditions.
With an equimolar mixture of fragment 1 and fragment 2 at the same concentration as fragment 1·2 (middle curve), very little acceleration is obtained,
and clearly fragment 1·2 and only fragment 1·2 is capable of functioning
satisfactorily.

All available evidence indicates that the pathway of Figure 7 can be a suffi-

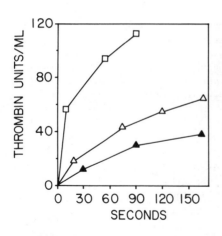

Figure 8

A comparison of the effectiveness of fragments 1 and 2 with fragment 1·2 in intermediate 2 activation. Intermediate 2 (0.075 mg/ml) was activated with [X_a, 0.83 μg/ml; V_a, 12.5 units (8 μg/ml); Phospholipid, 96 μg/ml; Ca^{++}, 10 mM final conc.]. The reaction was performed (▲——▲) without added fragments; (□——□) in the presence of fragment 1·2 (1.8 μg/ml); and (△——△) in the presence of both fragment 2 (0.6 μg/ml) and fragment 1 (1.2 μg/ml). All other conditions are as in the experiments of Table 1.

cient pathway for prothrombin activation. In order to account for the kinetic data, fragment 1·2 must be an integral part of the activation intermediate. Furthermore, because only intact fragment 1·2 possesses the properties which enable intermediate 2 to be activated rapidly enough to meet the kinetic requirements for this activation process, it appears likely that the thrombin-catalyzed cleavage of prothrombin which destroys this capacity functions in shutting down prothrombin activation.

Acknowledgments

This work was supported by grants from the National Institutes of Health (HL 12820, HL 14147 and GM 01311) and the Missouri and American Heart Associations. C. M. J. is the recipient of an Established Investigatorship of the American Heart Association.

REFERENCES

Aronson, D. and D. Menache. 1966. Chromatographic analysis of the activation of human prothrombin with human thrombokinase. *Biochemistry* **5**:2635.

————. 1968. Action of human thrombokinase on human prothrombin and *p*-tosyl-L-arginine methyl ester. *Biochim. Biophys. Acta* **167**:378.

Baker, W. J. and W. H. Seegers. 1967. The conversion of prothrombin to thrombin. *Thromb. Diath. Haemorrh.* **17**:205.

Barton, P. G. and D. J. Hanahan. 1969. Some lipid-protein interactions involved in prothrombin activation. *Biochim. Biophys. Acta* **187**:319.

Barton, P. G., C. M. Jackson and D. J. Hanahan. 1967. Relationship between factor X in the generation of prothrombinase. *Nature* **214**:923.

Benson, B. J., W. Kisiel and D. J. Hanahan. 1973. Calcium binding and other characteristics of bovine factor II and its activation intermediates. *Biochim. Biophys. Acta* **329**:81.

Blömback, B. and M. Blömback. 1963. Purification and stabilization of factor V. *Nature* **198**:886.

Bull, R. K., S. Jevons and P. G. Barton. 1972. Complexes of prothrombin with calcium ions and phospholipids. *J. Biol. Chem.* **247:**2747.

Cole, E. R., J. L. Koppel and J. H. Olwin. 1964. Interaction of bovine auto-prothrombin C with phospholipids and cations. *Can. J. Biochem.* **42:**1595.

————. 1965. Phospholipid-protein interaction in the formation of prothrombin activator. *Thromb. Diath. Haemorrh.* **14:**431.

Day, W. C. and P. G. Barton. 1972. Studies on the stability of bovine plasma factor V. *Biochim. Biophys. Acta* **261:**457.

Deggler, K. K. and J. Vreeken. 1969. The human prothrombin-activating enzyme. *Thromb. Diath. Haemorrh.* **22:**45.

Dombrose, F. A. and W. H. Seegers. 1974. Purification of bovine Ac-globulin (factor V) in physiology and biochemistry of prothrombin conversion. *Thromb. Diath. Haemorrh.* Suppl. 57, p. 241.

Dombrose, F. A., T. Yasui, Z. Roubal, A. Roubal and W. H. Seegers. 1972. Ac-globulin (factor V): Preparation of a practical product. *Prep. Biochem.* **2:**381.

Esmon, C. T. 1973. The function of factor V in prothrombin activation. Ph.D. thesis, Washington University, St. Louis.

Esmon, C. T. and C. M. Jackson. 1974a. The conversion of prothrombin to thrombin. III. The factor Xa catalyzed activation of prothrombin. *J. Biol. Chem.* **249:**7782.

————. 1974b. The conversion of prothrombin to thrombin. IV. The function of the fragment 2 region during activation in the presence of factor V. *J. Biol. Chem.* **249:**7791.

Esmon, C. T., W. G. Owen and C. M. Jackson. 1974a. The conversion of prothrombin to thrombin. II. Differentiation between thrombin- and factor Xa-catalyzed proteolysis. *J. Biol. Chem.* **249:**606.

————. 1974b. The conversion of prothrombin to thrombin. V. The activation of prothrombin by factor Xa in the presence of phospholipid. *J. Biol. Chem.* **249:** 7798.

Esmon, C. T., W. G. Owen, D. L. Duiguid and C. M. Jackson. 1973. The action of thrombin on blood clotting factor V: Conversion of factor V to a prothrombin-binding protein. *Biochim. Biophys. Acta* **310:**289.

Esnouf, M. P. and W. J. Williams. 1962. The isolation of a bovine plasma protein which is a substrate for the coagulant fraction of Russel's viper venom. *Biochem. J.* **84:**62.

Fujikawa, K., M. E. Legaz and E. W. Davie. 1972. Bovine factor X_1 (Stuart factor). Mechanism of activation by a protein from Russell's viper venom. *Biochemistry* **11:**4892.

Gitel, S. N., W. G. Owen, C. T. Esmon and C. M. Jackson. 1973. A polypeptide region of bovine prothrombin specific for binding to phospholipids. *Proc. Nat. Acad. Sci.* **70:**1344.

Gladner, J. and K. Laki. 1956. The inhibition of thrombin by diisopropyl phosphorofluoridate. *Arch. Biochem. Biophys.* **62:**501.

Heldebrant, C. M., C. Noyes, H. S. Kingdon and K. G. Mann. 1973. The activation of prothrombin. III. The partial amino acid sequence at the amino terminal of prothrombin and the intermediates of activation. *Biochem. Biophys. Res. Comm.* **54:**155.

Hemker, H. C., M. P. Esnouf, P. W. Hemker, A. C. W. Swart and R. G. Macfarlane. 1967. Formation of prothrombin converting activity. *Nature* **215:**248.

Jackson, C. M. and D. J. Hanahan. 1968. Studies on bovine factor X. II. Characterization of purified factor X and some observations on alterations in zone electrophoretic and chromatographic behavior occurring during purification. *Biochemistry* **8:**4506.

Jackson, C. M., T. F. Johnson and D. J. Hanahan. 1968. Studies on bovine factor X. I. Large scale purification of the bovine plasma protein possessing factor X activity. *Biochemistry* **7:**4492.

Jackson, C. M., W. G. Owen, S. N. Gitel and C. T. Esmon. 1974a. Chemical role of lipids in prothrombin conversion. *Thromb. Diath. Haemorrh.* Suppl. 57, p. 273.

Jackson, C. M., C. T. Esmon, S. N. Gitel, W. G. Owen and R. A. Henrikson. 1974b. The conversion of prothrombin to thrombin: The function of the pro-piece of prothrombin. In *Prothrombin and Related Coagulation Factors* (ed. H. C. Hemker and J. J. Veltkamp). Leiden University Press, Leiden, The Netherlands.

Jobin, F. and M. P. Esnouf. 1967. Formation of the prothrombin-converting complex. *Biochem. J.* **102:**666.

Kisiel, W. and D. J. Hanahan. 1973. The action of factor Xa, thrombin and trypsin on human factor II. *Biochim. Biophys. Acta* **329:**221.

———. 1974. Proteolysis of human factor II by factor Xa in the presence of hirudin. *Biochem. Biophys. Res. Comm.* **59:**570.

Lanchantin, G. F., J. Friedmann and D. W. Hart. 1965. The conversion of human prothrombin to thrombin by sodium citrate. Analysis of the activation mixture. *J. Biol. Chem.* **240:**3276.

Leveson, J. E. and M. P. Esnouf. 1969. The inhibition of factor X with diiso-propylfluorophosphate. *Brit. J. Haematol.* **17:**173.

Magnusson, S. 1971. Thrombin and prothrombin. In *The Enzymes* (ed. P. Boyer), 3rd ed., vol. 3, p. 278. Academic Press, New York.

Milstone, J. H. 1964. Thrombokinase as prime activator of prothrombin: Historical perspectives and present status. *Fed. Proc.* **23:**742.

Milstone, J. H., N. Oulianoff and V. K. Milstone. 1963. Outstanding characteristics of thrombokinase isolated from bovine plasma. *J. Gen. Physiol.* **47:**315.

Morita, T., S. Iwanaga, T. Suzuki and K. Fujikawa. 1973. Characterization of amino-terminal fragment liberated from bovine prothrombin by activated factor X. *FEBS Letters* **36:**313.

Owen, W. G., C. T. Esmon and C. M. Jackson. 1974. The conversion of prothrombin to thrombin. I. Characterization of the reaction products formed during the activation of bovine prothrombin. *J. Biol. Chem.* **249:**594.

Papahadjopoulos, D. and D. J. Hanahan. 1964. Observation on the interaction of phospholipids and certain clotting factors in prothrombin activator formation. *Biochim. Biophys. Acta* **90:**436.

Radcliffe, R. D. and P. G. Barton. 1972. The purification and properties of activated factor X: Bovine factor X activated with Russell's viper venom. *J. Biol. Chem.* **247:**7735.

Reuterby, J., D. A. Walz, L. E. McCoy and W. H. Seegers. 1974. Amino acid sequence of O fragment of bovine prothrombin. *Thromb. Res.* **4:**885.

Seegers, W. H. 1964. Enzyme theory of blood clotting. *Fed. Proc.* **23:**749.

Seegers, W. H., D. A. Walz, J. Reuterby and L. E. McCoy. 1974. Isolation and some properties of thrombin-E and other prothrombin derivatives. *Thromb. Res.* **4:**829.

Seegers, W. H., N. Sakuragawa, L. E. McCoy, I. A. Sedensky and F. A. Dombrose. 1972. Prothrombin activation: Ac-globulin, lipid, platelet membrane and autoprothrombin C (factor Xa) requirements. *Thromb. Res.* **1:**293.

Smith, R. L. 1973. Titration of activated bovine factor X. *J. Biol. Chem.* **248:**2418.

Stenflo, J. 1973. Vitamin K and the biosynthesis of prothrombin. III. Structural comparison of an NH$_2$-terminal fragment from normal and from dicoumarol-induced bovine prothrombin. *J. Biol. Chem.* **248:**6325.

Stenflo, J. and P. O. Ganrot. 1973. Binding of Ca^{2+} to normal and dicoumarol-induced prothrombin. *Biochem. Biophys. Res. Comm.* **50:**98.

Stenn, K. S. and E. R. Blout. 1972. Mechanism of bovine prothrombin activation by an insoluble preparation of bovine factor Xa (thrombokinase). *Biochemistry* **11:**4502.

Straub, W. and F. Duckert. 1961. The formation of the extrinsic prothrombin activator. *Thromb. Diath. Haemorrh.* **5:**402.

Williams, D. E. and R. A. Reisfeld. 1964. Disc electrophoresis in polyacrylamide gels: Extension to new conditions of pH and buffer. *Ann. N.Y. Acad. Sci.* **121:**373.

Structure of a Vitamin K-dependent Portion of Prothrombin

Johan Stenflo and Per Fernlund

Department of Clinical Chemistry, University of Lund
Malmö General Hospital, Malmö, Sweden

Peter Roepstorff

The Danish Institute of Protein Chemistry
Horsholm, Denmark

The only known function of vitamin K in man and higher animals is to maintain an adequate concentration of biologically active prothrombin and three other coagulation factors in the blood, namely, factors VII, IX and X. Vitamin K deficiency, as well as administration of the vitamin K antagonist dicoumarol, suppresses the concentrations of these coagulation factors, as judged from determinations of their biological activity. Experiments in rats with the protein synthesis inhibitor cycloheximide have corroborated the opinion that the vitamin does not regulate de novo synthesis of these coagulation factors, but exerts its effect at a posttranslational site, presumably by converting precursors of the coagulation factors to their biologically active forms (Hill et al. 1968; Bell and Matschiner 1969; Suttie 1970; Shah and Suttie 1971).

The vitamin K-dependent coagulation factors, which are glycoproteins, can be adsorbed to barium citrate or barium sulfate. Such adsorption is utilized in most purification procedures. Until recently, prothrombin has received more attention than the other vitamin K-dependent coagulation factors because it is more readily available in large amounts.

Immunochemical measurements of the prothrombin concentration in blood plasma during treatment of patients with dicoumarol have shown that the decreased prothrombin activity is not accompanied by a decrease in the immunochemically measurable concentration of the protein (Niléhn and Ganrot 1968; Josso et al. 1968). Ganrot and Niléhn have also used an immunochemical method, crossed immunoelectrophoresis (Laurell 1965; Ganrot 1972), and demonstrated the biosynthesis of biologically inactive prothrombin during dicoumarol treatment. The abnormal prothrombin occurs in the blood plasma in man (Ganrot and Niléhn 1968) as well as in the bovine species (Stenflo 1970). Unlike normal prothrombin, the abnormal, dicoumarol-induced prothrombin does not bind Ca^{++}, is not adsorbed to barium citrate, and has no prothrombin activity. However, it has the same main antigenic determinants as normal prothrombin (Fig. 1). In the rat, a protein apparently correspond-

112   J. Stenflo, P. Fernlund and P. Roepstorff

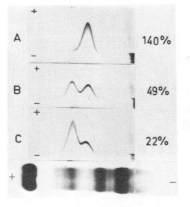

Figure 1
Crossed immunoelectrophoresis pattern of bovine plasma using rabbit antibovine prothrombin. Agarose gel electrophoresis in 0.075 M barbital buffer, 2 mM in calcium lactate pH 8.6. (*A*) Precipitate obtained before administration of dicoumarol; (*B, C*) precipitates given by plasma samples obtained 4 and 7 days after the beginning of dicoumarol administration. The one-stage prothrombin activity is given in percent of an arbitrary standard. The electrophoretic pattern of a bovine plasma is included as a reference.

ing to the dicoumarol-induced prothrombin found in human and bovine plasma is retained within the liver cells (Shah, Suttie and Grant 1973; Morrissey, Jones and Olson 1973). It has recently been demonstrated in cows that administration of dicoumarol leads not only to the biosynthesis of abnormal prothrombin, but also to abnormal factor IX and X, which, like dicoumarol-induced prothrombin, do not bind Ca^{++} (Reekers et al. 1973).

Activation of prothrombin, which results in the formation of thrombin, requires active factor X, factor V, phospholipid and Ca^{++}. Prothrombin and active factor X seem to be bound to the phospholipids via Ca^{++} ions (Bull, Jevons and Barton 1972; Gitel et al. 1973). This affinity of the reactants for the phospholipid results in high local concentrations of them. The dicoumarol-induced prothrombin is presumably not activated because of its deficient Ca^{++} binding.

Knowledge of the structural difference between normal and dicoumarol-induced prothrombin was thought to be essential for a better understanding of the mode of action of vitamin K and for an identification of the Ca^{++} binding structures in prothrombin. This paper summarizes studies designed to locate the differences between normal and dicoumarol-induced prothrombin and to characterize chemically the vitamin K-dependent Ca^{++} binding structures in normal prothrombin. Three aspects of the work will be discussed: (1) the results of a comparison of the composition and properties of purified normal and dicoumarol-induced prothrombin; (2) the degradation of the two molecules and the identification of the vitamin K-dependent structures in normal prothrombin; and (3) the results of structural studies of the vitamin K-dependent Ca^{++} binding portions of normal prothrombin.

Comparison of Composition and Properties of Normal and Dicoumarol-induced Prothrombin

To secure an adequate supply of the purified normal and dicoumarol-induced prothrombins, bovine plasma was used as starting material for the purification. Two similar purification procedures based on conventional chromatographic methods have been devised (Stenflo 1970; Stenflo and Ganrot 1972; Nelsestuen and Suttie 1972a). The purified proteins retained their characteristic electrophoretic and immunochemical properties earlier observed in unfractionated plasma. Thus the anodal electrophoretic mobility of normal prothrombin at pH 8.6 was higher in EDTA-containing buffer than in Ca^{++}-containing buffer, indicating Ca^{++} binding, whereas the electrophoretic mobility of the abnormal prothrombin was independent of the Ca^{++} concentration.

The purified dicoumarol-induced prothrombin had no biological activity, as judged by one-stage prothrombin bioassay, nor did it interfere with the activation of purified normal prothrombin.

The amino acid and carbohydrate compositions of the purified normal and abnormal prothrombin were identical, within the range of experimental error. Likewise, determination of the amino- and carboxyl-terminal amino acids gave identical results. Peptide maps of tryptic digests of the completely reduced and aminoethylated proteins appeared identical. Though these results were obtained with crude methods, they did suggest that there were only minor structural differences between the two prothrombins (Stenflo 1972). The conclusions arrived at by Nelsestuen and Suttie (1972a), who had independently performed similar experiments, were identical. However, these authors also reported that when activated by nonphysiological activators, such as trypsin or the venom from *Echis carinatus,* normal and abnormal prothrombin gave rise to identical amounts of thrombin. This was in agreement with earlier results obtained by Josso et al. (1968) and meant that the carboxyl-terminal thrombin parts of the two prothrombins were identical.

The Ca^{++} binding properties of normal and dicoumarol-induced prothrombin were studied with the rate dialysis method of Colowick and Womack (1969). From a Scatchard plot (Scatchard 1949) of the binding data, the number of Ca^{++} binding sites was found to be 10 to 12 in normal prothrombin, whereas the dicoumarol-induced prothrombin bound at most one Ca^{++} with comparable binding strength (Stenflo and Ganrot 1973) (Fig. 2). The binding curve for normal prothrombin was convex upwards and therefore indicated that some of the high-affinity binding sites, probably three, exhibited positive cooperativity (Koshland 1970). The results obtained by Nelsestuen and Suttie (1972b) in a similar investigation suggested the existence of four high-affinity binding sites in normal prothrombin and less than one in dicoumarol-induced prothrombin. Recent data support the view that the total num-

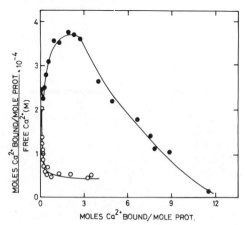

Figure 2
Scatchard plot of Ca^{++} binding of normal and dicoumarol-induced prothrombin. (•) Normal prothrombin; (○) dicoumarol-induced prothrombin.

ber of Ca^{++} binding sites in normal prothrombin is 10 to 12 (Benson, Kisiel and Hanahan 1973; Bajaj, Butkowski and Mann 1974).

The cooperative Ca^{++} binding of normal prothrombin implied a ligand-induced conformation change. This was verified in quantitative immunoprecipitation experiments using antibovine prothrombin. Thus normal prothrombin had Ca^{++}-dependent antigenic determinants, which led to the precipitation of approximately twice as much material in the equivalence zone when the incubation mixture contained Ca^{++} than when it contained EDTA. On the other hand, Ca^{++} did not significantly influence the shape of the precipitation curves obtained with dicoumarol-induced prothrombin and the same antiserum (Stenflo 1972). With the use of circular dichroism as a conformational probe, it was confirmed that Ca^{++} induced a conformational transition. It was expressed only in the aromatic region of the spectrum and is therefore presumably only a local change in the neighborhood of some aromatic amino acid (Björk and Stenflo 1973). Observations pertinent to these results are that fragment 1^1, an amino-terminal fragment (MW approx. 25,000) from prothrombin, harbors all of the Ca^{++} binding sites present in intact normal prothrombin, and that Ca^{++} induces a conformational change in fragment 1 (Benson, Kisiel and Hanahan 1973; Stenflo 1973).

Isolation of Vitamin K-dependent Structures in Normal Prothrombin

Prothrombin consists of a single polypeptide chain. Thrombin is derived from the carboxyl-terminal part of the chain (Magnusson 1971). The activation experiments on normal and abnormal prothrombin with the venom from *Echis carinatus* showed that the defect in dicoumarol-induced prothrombin was in the nonthrombin part of the molecule (Nelsestuen and Suttie 1972a). Both in normal and dicoumarol-induced prothrombin, thrombin cleaves only one peptide bond, which results in two fragments, called fragment 1^1 (MW approx. 25,000) and intermediate 1^1 (MW approx. 47,000). The two fragments are readily separated by chromatography on DEAE-cellulose or QAE-Sephadex or gel filtration on Sephadex G-100 (Heldebrant et al. 1973a; Stenflo 1973; Owen, Esmon and Jackson 1974). Amino acid sequence analysis of fragment 1 verified that it was derived from the amino-terminal part of the prothrombin molecule (Fig. 3). Intermediate 1 from the two prothrombins had identical

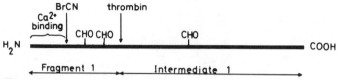

Figure 3

Structure of prothrombin. The points of cleavage by thrombin and cyanogen bromide are indicated by arrows; CHO stands for carbohydrate.

---

[1] The nomenclature of Gitel et al. (1973) is used. In the nomenclature of Heldebrant et al. (1973a, b), fragment 1 is called intermediate 3.

electrophoretic and immunochemical properties and did not bind Ca^{++}. Fragment 1 from normal prothrombin bound Ca^{++}, but that from dicoumarol-induced prothrombin did not. Although the fragments from the two prothrombins had identical amino acid compositions after acid hydrolysis, differences were found between peptide maps prepared from thermolysin digests of them. Fragment 1 contains one methionine residue and was therefore cleaved by cyanogen bromide into two fragments, which were separated by chromatography on Sephadex G-100. The smaller cyanogen bromide fragment (approx. 75 amino acid residues) from normal prothrombin bound Ca^{++}, whereas the corresponding fragment from dicoumarol-induced prothrombin did not. The amino acid compositions in acid hydrolysates of each of the two cyanogen bromide fragments from normal prothrombin were identical with those of the corresponding fragments from dicoumarol-induced prothrombin. Peptide maps of tryptic digests of the completely reduced and S-carboxymethylated large cyanogen bromide fragments were identical, whereas there were obvious differences between the small cyanogen bromide fragments (Fig. 4; Stenflo 1974).

Most of the tryptic peptides from the small Ca^{++}-binding cyanogen bromide fragment from normal prothrombin and the corresponding peptides from the dicoumarol-induced prothrombin were isolated on a preparative scale by high-voltage electrophoresis and gel chromatography on Sephadex G-25 superfine. The amino acid sequences of the peptides were determined to the extent that it was possible to ascertain their positions in the prothrombin sequence (Stenflo 1974). Peptides from normal prothrombin are designated N. . . , and those from abnormal prothrombin, A. . . (Fig. 5). Both peptide N2 and A2 stained faintly with ninhydrin. Both contained residues 4–10 in normal and abnormal prothrombin, respectively. Assuming glutamic acid in positions 7 and 8, peptide A2 had the expected anodal electrophoretic mobility, as judged from Offord's diagram (Offord 1966). However, peptide N2 had an anodal electrophoretic mobility that was too high to be explained by

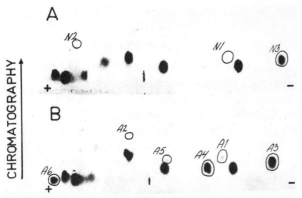

Figure 4
Peptide maps of amino-terminal cyanogen bromide fragments of (*A*) normal and (*B*) dicoumarol-induced prothrombin. Electrophoresis was run at pH 6.5. Peptides from the part of the prothrombin sequence shown in Figure 5 are encircled.

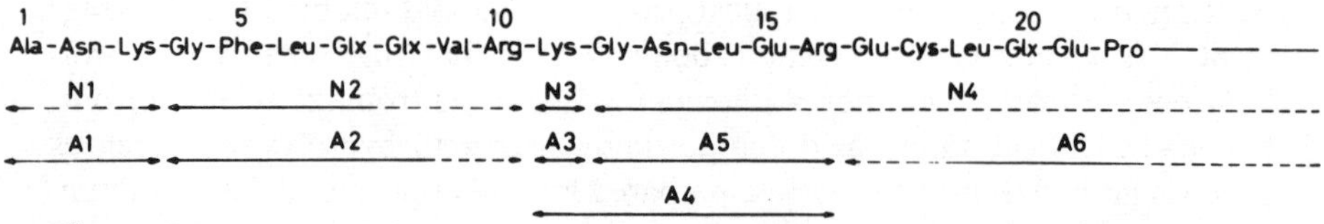

```
1             5              10              15              20
Ala-Asn-Lys-Gly-Phe-Leu-Glx -Glx -Val-Arg-Lys-Gly-Asn-Leu-Glu-Arg-Glu-Cys-Leu-Glx-Glu-Pro —— —— ——

     N1                  N2            N3                           N4
  ◄————————————·  ·————————————————·  ·————————·  ·————————————————————————————————————————
     A1                  A2        A3      A5                        A6
  ◄————————————·  ·————————————·  ·——·  ·————————————·  ·————————————————————————————————
                                         A4
                                   ·————————————————·
```

Figure 5

Amino-terminal sequence of prothrombin. N, peptides from normal prothrombin; A, peptides from dicoumarol-induced prothrombin. (——) Residues positioned by sequence studies on the peptides; (– – – –) residues by composition only. The positions of the peptides were determined by comparison with the sequence given by Heldebrant et al. (1973b), except for residues 10, 16 and 20, which were determined in this investigation, and residue 18 (S. Magnusson, pers. comm.).

its amino acid composition alone (Magnusson 1973; Stenflo 1974). At pH 6.5, the anodal electrophoretic mobility of peptide N2 was lower in Ca^{++}-containing buffer than in buffer without Ca^{++}.

Peptide N4 was isolated from a thermolysin digest of a tryptic peptide. It contained two arginine residues inaccessible to trypsin. In the abnormal prothrombin, trypsin cleaved those peptide bonds that were inaccessible to it in peptide N4. This gave rise to peptides A4 to A6. The anodal electrophoretic mobility at pH 6.5 of peptide N4 was greatly reduced when Ca^{++} was included in the buffer. The Ca^{++} binding of the peptide and its amino acid composition suggest that it is derived from the same part of the prothrombin sequence as the tryptic peptide isolated by Nelsestuen and Suttie (1973) by adsorption to barium citrate.

The results of the comparison of the peptides from normal and dicoumarol-induced prothrombin indicated that peptides N2 and N4 contained the vitamin K-dependent structures. The Ca^{++} binding properties of peptide N4 and its resistance to digestion with proteolytic enzymes suggested that it contained more than one modified residue. Peptide N2 was considered to be suitable for further structural studies.

Structure of a Vitamin K-dependent Portion in Prothrombin

Peptide N2 was thoroughly digested with aminopeptidase M and carboxypeptidase B, whereby glycine, phenylalanine and arginine were quantitatively removed. The resulting tetrapeptide Leu-Glx-Glx-Val (residues 6 to 9) was isolated by gel filtration on Sephadex G-25. It was compared with synthetic Leu-Glu-Glu-Val. Although the native peptide had a far higher anodal electrophoretic mobility at pH 6.5 than the synthetic one, amino acid analysis of acid hydrolysates gave the composition Glu-2, Leu-1, Val-1 for both peptides.

The proton NMR (nuclear magnetic resonance) spectra of the synthetic and native peptides dissolved in H_2O were recorded. The NMR spectrum of the native peptide revealed no resonances that were not present in the spectrum of the synthetic peptide. The spectra showed that the native peptide had no protons on the γ-carbons of its two glutamic acid residues, whereas the synthetic peptide had four. Therefore, both hydrogens on each of the γ-carbons of the two glutamic acid residues must be replaced by substituents, or by

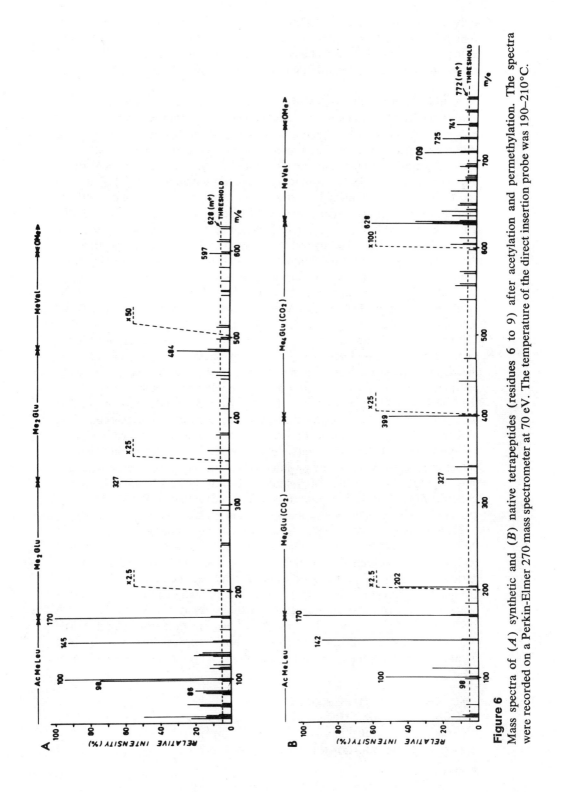

Figure 6

Mass spectra of (*A*) synthetic and (*B*) native tetrapeptides (residues 6 to 9) after acetylation and permethylation. The spectra were recorded on a Perkin-Elmer 270 mass spectrometer at 70 eV. The temperature of the direct insertion probe was 190–210°C.

only one provided that the substituent is electron withdrawing enough to render the remaining proton exchangeable (Stenflo et al. 1974). Any substituent must contain no inexchangeable hydrogen atoms and be labile enough to be lost under the standard conditions of acid hydrolysis (110°C in vacuo for 24 hr).

A mass spectrum (Fig. 6A) of the synthetic tetrapeptide after acetylation and permethylation (Vilkas and Lederer 1968) contained sequence peaks at m/e 170, 327, 484 and 597 and a molecular peak at m/e 628, corresponding to the expected sequence. The presence of leucine and valine was also confirmed by peaks at m/e 100 and 86, respectively, and glutamic acid by an intense peak at m/e 98 (the amine fragment of pyroglutamic acid). The mass spectrum (Fig. 6B) of the native tetrapeptide after acetylation and permethylation shows N-terminal leucine, indicated by peaks at m/e 100, 142 and 170. The sequence peak corresponding to Leu-Glu at m/e 327, as well as the peak at m/e 98, is of rather low intensity, compared with the synthetic peptide, and the sequence peak corresponding to Leu-Glu-Glu is missing entirely. However, new intense peaks are present at m/e 399 and 628, the latter followed by peaks at m/e 741 and 772, corresponding to the chain termination by valine methyl ester. The mass difference between the peaks at m/e 170 and 399 and the difference between the peaks at m/e 399 and 628 are identical (229 amu), which means that both glutamic acid residues are modified in the same way. The mass difference between a permethylated glutamic acid and the permethylated modified glutamic acid residue is 72. This corresponds to one methyl-esterified carboxyl group and one additional methyl group. From the NMR spectra, it was evident that the carboxyl group must be on the γ-carbon atom, which renders the remaining proton on this carbon atom exchangeable, i.e., permethylation leads to C-methylation of this carbon atom. The presence of the modified glutamic acid residues is also confirmed by their amine fragments, as shown by an intense peak at m/e 202. The tetrapeptide thus contains two residues of γ-carboxyglutamic acid (3-amino-1,1,3-propanetricarboxylic acid), a hitherto unidentified amino acid (Stenflo et al. 1974). The structure of the peptide is shown in Figure 7.

High-voltage electrophoresis of the native tetrapeptide at pH 6.5 showed that it was contaminated by a small amount of material with an electrophoretic mobility between that of the main fraction and that of the synthetic tetrapeptide. This can probably be ascribed to loss of one of the extra carboxyl groups during preparation of the peptide. The partially decarboxylated tetrapeptide is revealed in the mass spectrum by the peaks at m/e 98 and 327.

The γ-carboxyglutamic acid presumably has the properties characteristic of malonic acid, i.e., in the acid form it should be decarboxylated by heat treatment but resistant to alkaline hydrolysis. This explains the finding of

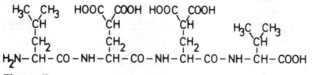

Figure 7
Structure of the tetrapeptide isolated from normal prothrombin.

Figure 8

Pattern obtained after high-voltage electrophoresis at pH 6.5 of acid (A, B) and alkaline (C, D) hydrolysates of synthetic (A, C) and native (B, D) tetrapeptides (residues 6 to 9). The paper was stained with ninhydrin and the anode was to the left. The reference compounds were from right to left on the paper: glycine, glutamic acid, aspartic acid and γ-carboxyglutamic acid. The synthetic procedure for γ-carboxyglutamic acid will be described elsewhere (Fernlund, Roepstorff and Stenflo, in prep.).

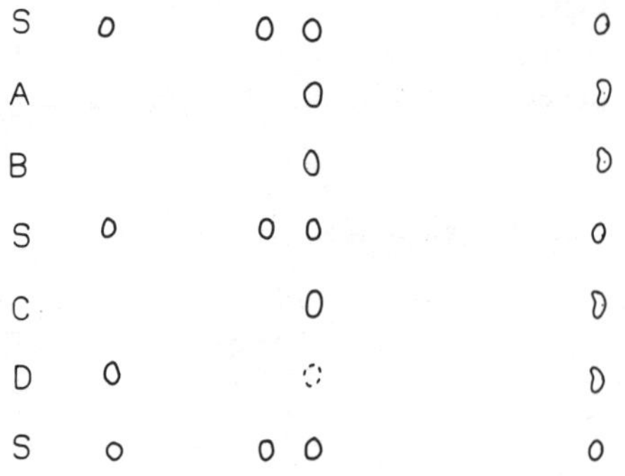

glutamic acid after acid hydrolysis. Furthermore, heating of the peptide to 150°C resulted in decarboxylation and therefore an electrophoretic mobility indistinguishable from that of the synthetic tetrapeptide (Stenflo et al. 1974). In alkaline hydrolysates, γ-carboxyglutamic acid was identified (Fig. 8).

DISCUSSION

The peptide containing residues 4 to 10 in normal prothrombin had an electrophoretic mobility at pH 6.5 that was too high to be explained by its amino acid composition alone, whereas the peptide from abnormal prothrombin had an electrophoretic mobility compatible with its amino acid composition. The finding of two residues of the hitherto unidentified amino acid γ-carboxyglutamic acid fully explains this difference in electrophoretic mobility. Peptide N4 must contain several γ-carboxyglutamic acid residues since it has two arginyl bonds inaccessible to trypsin, it binds Ca^{++}, and is susceptible to heat decarboxylation, as judged from its reduced anodal electrophoretic mobility at pH 6.5 after heat treatment. This is corroborated by Howard and Nelsestuen (1974) and Girardot, Delaney and Johnson (1974). Recently Magnusson et al. (1974) proposed that in addition to residues 7 and 8, residues 15, 17, 20, 21, 26, 27, 30 and 33 are γ-carboxy-glutamic acid residues.

The γ-carboxyglutamic acid residues bind Ca^{++} far better than the glutamic acid residues in the corresponding positions in dicoumarol-induced prothrombin. A requirement for rapid activation of prothrombin in vivo is that it be bound to phospholipids via Ca^{++} ions (Bull, Jevons and Barton 1972; Gitel et al. 1973). The γ-carboxyglutamic acid residues are apparently necessary for such activation, whereas the dicoumarol-induced prothrombin, which has glutamic acid instead of γ-carboxyglutamic acid, does not bind Ca^{++} and is not activated. The function of vitamin K is apparently to mediate the conversion of a number of glutamic acid residues in a prothrombin precursor to γ-carboxyglutamic acid residues by introducing additional carboxyl groups.

The amino acid sequence of the amino-terminal part of factor IX and of

the light chain of factor X was recently determined and found to be homologous with the amino-terminal part of prothrombin (Fujikawa et al. 1974). Administration of dicoumarol results in the synthesis of abnormal forms of these coagulation factors with deficient Ca^{++} binding (Reekers et al. 1973). It therefore seems likely that also the normal forms of these vitamin K-dependent coagulation factors contain γ-carboxyglutamic acid.

Acknowledgment

These investigations were supported by a grant from the Swedish Medical Research Council (Project No. B74-13X-581-10C).

REFERENCES

Bajaj, S. P., R. J. Butkowski and K. G. Mann. 1974. Activation of bovine prothrombin: Calcium binding and activation kinetics. *Fed. Proc.* **33**:1415.

Bell, R. G. and J. T. Matschiner. 1969. Synthesis and destruction of prothrombin in the rat. *Arch. Biochem. Biophys.* **135**:152.

Benson, B. J., W. Kisiel, and D. J. Hanahan. 1973. Calcium binding and other characteristics of bovine factor II and its activation intermediates. *Biochim. Biophys. Acta* **329**:81.

Björk, I. and J. Stenflo. 1973. A conformational study of normal and dicoumarol-induced prothrombin. *FEBS Letters* **32**:343.

Bull, R. K., S. Jevons and P. G. Barton. 1972. Complexes of prothrombin with calcium ions and phospholipids. *J. Biol. Chem.* **247**:2747.

Colowick, S. P. and F. C. Womack. 1969. Binding of diffusible molecules by macromolecules: Rapid measurement by rate of dialysis. *J. Biol. Chem.* **244**:774.

Fujikawa, K., M. H. Coan, D. L. Enfield, K. Titani, L. H. Ericsson and E. W. Davie. 1974. A comparison of bovine prothrombin, factor IX (Christmas factor), and factor X (Stuart factor). *Proc. Nat. Acad. Sci.* **71**:427.

Ganrot, P. O. 1972. Crossed immunoelectrophoresis. *Scand. J. Clin. Lab. Invest.* **29**:39.

Ganrot, P. O. and J.-E. Niléhn. 1968. Plasma prothrombin during treatment with dicoumarol. II. Demonstration of an abnormal prothrombin fraction. *Scand. J. Clin. Lab. Invest.* **22**:23.

Girardot, J. M., R. Delaney and B. C. Johnson. 1974. Carboxylation, the completion step in prothrombin biosynthesis. *Biochem. Biophys. Res. Comm.* **59**:1197.

Gitel, S. N., W. G. Owen, C. T. Esmon and C. M. Jackson. 1973. A polypeptide region of bovine prothrombin specific for binding to phospholipids. *Proc. Nat. Acad. Sci.* **70**:1344.

Heldebrant, C. M., R. J. Butkowski, S. P. Bajaj and K. G. Mann. 1973a. The activation of prothrombin. II. Partial reactions, physical and chemical characterization of the intermediates of activation. *J. Biol. Chem.* **248**:7149.

Heldebrant, C. M., C. Noyes, H. S. Kingdon and K. G. Mann. 1973b. The activation of prothrombin. III. The partial amino acid sequences at the amino terminal of prothrombin and the intermediates of activation. *Biochem. Biophys. Res. Comm.* **54**:155.

Hill, R. B., S. Gaetani, A. M. Paolucci, P. B. Ramarao, R. Alden and G. S.

Ranhotra. 1968. Vitamin K and biosynthesis of protein and prothrombin. *J. Biol. Chem.* **243**:3930.

Howard, J. B. and G. L. Nelsestuen. 1974. Properties of a Ca^{2+} binding peptide from prothrombin. *Biochem. Biophys. Res. Comm.* **59**:757.

Josso, F., J. M. Lavergne, M. Gouault, O. Prou-Wartelle and J. P. Soulier. 1968. Différents états moléculaires du facteur II (prothrombine). Leur étude à l'aide de la staphylocoagulase et d'anticorps antifacteur II. I. Le facteur II chez les sujets traités par les antagonistes de la vitamine K. *Thromb. Diath. Haemorrh.* **20**:88.

Koshland, D. E. 1970. The molecular basis for enzyme regulation. In *The Enzymes: Structure and Control* (ed. P. D. Boyer), vol. 1, p. 341. Academic Press, New York.

Laurell, C.-B. 1965. Antigen-antibody crossed electrophoresis. *Anal. Biochem.* **10**:358.

Magnusson, S. 1971. Thrombin and prothrombin. In *The Enzymes* (ed. P. D. Boyer), vol. 3, p. 277. Academic Press, New York.

————. 1973. Primary structure studies on thrombin and prothrombin. *Thromb. Diath. Haemorrh.* (Suppl.) **54**:31.

Magnusson, S., L. Sottrup-Jensen, T. E. Petersen, H. Morris and A. Dell. 1974. Primary structure of the vitamin K-dependent part of prothrombin. *FEBS Letters* **44**:189.

Morrissey, J. J., J. P. Jones and R. E. Olson. 1973. Isolation and characterization of isoprothrombin in the rat. *Biochem. Biophys. Res. Comm.* **54**:1075.

Nelsestuen, G. L. and J. W. Suttie. 1972a. The purification and properties of an abnormal prothrombin protein produced by dicoumarol-treated cows. *J. Biol. Chem.* **247**:8176.

————. 1972b. Mode of action of vitamin K. Calcium binding properties of bovine prothombin. *Biochemistry* **11**:4961.

————. 1973. The mode of action of vitamin K. Isolation of a peptide containing the vitamin K-dependent portion of prothrombin. *Proc Nat. Acad. Sci.* **70**:3366.

Niléhn, J.-E. and P. O. Ganrot. 1968. Plasma prothrombin during treatment with dicoumarol. I. Immunochemical determination of its concentration in plasma. *Scand. J. Clin. Lab. Invest.* **22**:17.

Offord, R. E. 1966. Electrophoretic mobilities of peptides on paper and their use in the determination of amide groups. *Nature* **6**:591.

Owen, W. G., C. T. Esmon and C. M. Jackson. 1974. The conversion of prothrombin to thrombin. I. Characterization of the reaction products formed during the activation of bovine prothrombin. *J. Biol. Chem.* **249**:594.

Reekers, P. P. M., M. J. Lindhout, B. H. M. Kop-Klaassen and H. C. Hemker. 1973. Demonstration of three anomalous plasma proteins induced by a vitamin K antagonist. *Biochim. Biophys. Acta* **317**:559.

Scatchard, G. 1949. The attractions of proteins for small molecules and ions. *Ann. N.Y. Acad. Sci.* **51**:660.

Shah, D. V. and J. W. Suttie. 1971. Mechanism of action of vitamin K: Evidence for the conversion of a precursor protein to prothrombin in the rat. *Proc. Nat. Acad. Sci.* **68**:1653.

Shah, D. V., J. W. Suttie and G. A. Grant. 1973. A rat liver protein with potential thrombin activity: Properties and partial purification. *Arch. Biochem. Biophys.* **159**:483.

Stenflo, J. 1970. Dicoumarol-induced prothrombin in bovine plasma. *Acta Chem. Scand.* **24**:3762.

————. 1972. Vitamin K and the biosynthesis of prothrombin. II. Structural comparison of normal and dicoumarol-induced bovine prothrombin. *J. Biol. Chem.* **247:**8167.

————. 1973. Vitamin K and the biosynthesis of prothrombin. III. Structural comparison of an NH_2-terminal fragment from normal and from dicoumarol-induced bovine prothrombin. *J. Biol. Chem.* **248:**6325.

————. 1974. Vitamin K and the biosynthesis of prothrombin. IV. Isolation of peptides containing prosthetic groups from normal prothrombin and the corresponding peptides from dicoumarol-induced prothrombin. *J. Biol. Chem.* **249:**5527.

Stenflo, J. and P.-O. Ganrot. 1972. Vitamin K and the biosynthesis of prothrombin. I. Identification and purification of a dicoumarol-induced abnormal prothrombin from bovine plasma. *J. Biol. Chem.* **247:**8160.

————. 1973. Binding of Ca^{2+} to normal and dicoumarol-induced prothrombin. *Biochem. Biophys. Res. Comm.* **50:**98.

Stenflo, J., P. Fernlund, W. Egan and P. Roepstorff. 1974. Vitamin K-dependent modifications of glutamic acid residues in prothrombin. *Proc. Nat. Acad. Sci.* **71:**2730.

Suttie, J. 1970. The effect of cycloheximide administration on vitamin K-stimulated prothrombin formation. *Arch. Biochem. Biophys.* **141:**571.

Vilkas, E. and E. Lederer. 1968. N-methylation de peptides par la methode de Hakomori. Structure du mycoside C_{b1}. *Tetrah. Letters,* 3089.

Complete Primary Structure of Prothrombin: Isolation, Structure and Reactivity of Ten Carboxylated Glutamic Acid Residues and Regulation of Prothrombin Activation by Thrombin

Staffan Magnusson, Torben E. Petersen,
Lars Sottrup-Jensen and Hendrik Claeys*

Department of Molecular Biology, University of Aarhus
Aarhus, Denmark

The B chain of thrombin shows a high degree of amino acid sequence homology with trypsin, elastase and the chymotrypsins A and B from the pancreas, which are the major serine proteases in the digestive tract. Since the primary and tertiary structures of these pancreatic enzymes have been solved and their catalytic sites, substrate binding sites and mechanism of activation have been studied in great detail, in the light of this structural knowledge, one might assume that nothing much would be learned from studying the structures of thrombin and prothrombin. When we decided in 1967 to determine the primary structure of prothrombin, the main reason was that despite what was known about the structures and function of the digestive serine proteases, we could not answer any of the following questions in meaningful terms.

1. *Why is the substrate specificity of thrombin towards protein substrates so very restricted?* Although thrombin has essentially the same specificity as trypsin towards small synthetic substrates and catalyzes the splitting of arginine and lysine esters with a blocked α-amino group, its specificity towards proteins is limited to very few substrates. The best studied of these is fibrinogen, which is the "classical" substrate. Thrombin splits only two of its approximately 150 arginyl and lysyl bonds. Two of the other known thrombin substrates are also coagulation factors, namely, prothrombin and factor XIII; in each, thrombin splits only one single peptide bond concerned with regulation and activation, respectively.

2. *What is the function of the "pro" part of the prothrombin structure?* In the case of trypsinogen, the activation of the zymogen involves the cleavage of one peptide bond, with consequent loss of a hexapeptide, and in chymotrypsinogen, the cleavage of one to four peptide bonds with a loss of at most two dipeptides. In contrast, the activation of prothrombin involves

---

* Present address: Department of Internal Medicine, Academic Hospital, University of Leuven, Leuven, Belgium.

the loss of about half of the molecule (274 amino acid residues with ¾ of the carbohydrate).

3. *What is the function of vitamin K in the biosynthesis of prothrombin and what is the structural modification that it causes?*

4. *What is the structural background for the rapid activation of prothrombin?* It is remarkable that although both prothrombin and the coagulation factors associated with its activation are present in very low concentrations in blood plasma, the generation of thrombin as a result of tissue or cell damage literally takes place in only seconds. As a first step toward solving the structure of prothrombin, we have now completed the determination of its primary structure.

MATERIALS AND METHODS

Bovine Prothrombin and Thrombin

Bovine prothrombin was purified according to Magnusson (1965b, 1970a). The specific activity was 1400–1450 NIH U/mg in a two-stage system where highly purified fibrinogen fraction I-4 (Blombäck and Blombäck 1956) was used as substrate and NIH standard thrombin was used for calibrating the standard curve. This specific activity corresponds to 2350–2450 U/mg in the two-stage method of Ware and Seegers (1949), which is based on the 15-second unit standard and uses acacia gum to stabilize the fibrinogen solution. Bovine thrombin preparations (Magnusson 1965a, 1970a) had a specific activity of 2100 NIH U/mg dry weight.

Activation of prothrombin: Prothrombin dissolved at high concentration (70–200 mg/ml) at pH 7.5 in a Tris buffer with 20–50 mM EDTA was found to be completely converted in 20–30 minutes to a mixture of activation products (R. H. Saundry, S. Magnusson and B. S. Hartley, unpubl.) which have been separated on DEAE-Sephadex A-50 (Magnusson et al. 1975). The nomenclature used for the different activation products is shown in Figure 1.

Methods for Amino Acid Sequence Determination

Isolated peptides were screened for purity and semiquantitative amino acid composition by subjecting 6-M HCl hydrolysates to high-voltage electrophoresis at pH 2.1 and detecting the amino acids with collidine-ninhydrin. Peptides judged to be pure by this technique were then analyzed quantitatively by the standard Moore and Stein method using Locarte amino acid analyzers. Amino acid sequences were determined by the DNS (1-dimethylamino naphthalene-5-sulfonyl [dansyl])-Edman method, sequential degradation with phenyl*iso*thiocyanate, the N-terminal amino acid at each step being identified by two-dimensional chromatography of its DNS derivative on polyamide layers in four different solvent systems. Glycopeptides were detected because of the brown or black color arising as a result of "humin" formation during acid hydrolysis and also because they contained hexosamine. The only hexosamine that we have found on the amino acid analyzer during this work is glucosamine. This agrees with earlier results of carbohydrate analysis of prothrombin (Magnusson 1965b).

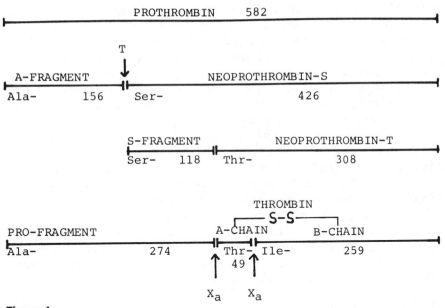

Figure 1

Prothrombin is a single polypeptide chain of 582 amino acid residues with N-terminal Ala. Thrombin (T) cleaves at Arg-156 giving rise to A fragment (residues 1–156) and neoprothrombin-S (residues 157–582). Factor X_a (X_a) cleaves at Arg-274 and Arg-323 to produce Pro fragment (residues 1–274) and thrombin, which is a two-chain protein with the A chain (residues 275–323) being connected to the B chain (residues 324–582) by the disulfide bridge 296–442 (Fig. 2). S fragment (residues 157–274) and neoprothrombin-T (residues 275–582) are both single-chain proteins.

The mass spectrometric analyses referred to have been performed in collaboration with Howard Morris and Anne Dell at the University Chemical Laboratory, Cambridge, England.

Overlap Peptides Connecting A Fragment to S Fragment
and S Fragment to Neoprothrombin-T

Prothrombin (980 mg) was dissolved in 100 ml 1 mM HCl; 0.1 M HCl was used to adjust pH from 4.1 to 3.0. This solution was then adjusted to pH 4.65 with 0.1 M NaOH. The resulting isoelectric precipitate was collected by centrifugation and dissolved with 15 ml 0.5 M Tris-HCl buffer pH 8.65 which was 8 M in freshly deionized urea and 3.65 mM in Na_2EDTA. Dithiothreitol (DTT, 1.68 mmoles) was added and reduction allowed to continue for 2 hours. $ICH_2 \cdot COONa$ (10.08 mmoles) was added, and alkylation was allowed to proceed for 1 hour, after which the mixture was desalted on a 2 x 120-cm column of Sephadex G-50, equilibrated, and eluted with .25 M NH_4HCO_3/NH_3 pH 9.0. The void volume peak was freeze-dried and yielded 850 mg material, which was dissolved in 300 ml 0.15 M NH_4HCO_3 containing 10 mg soybean trypsin inhibitor (STI) (Worthington) then digested with 21 mg α-chymotrypsin (Worthington) pretreated with 1 mg STI in 5 ml. After digesting for 4¾ hours (35–40°C), 1 mg diphenylcarbamyl chloride

(DPCC) in MeOH was added. The digest was freeze-dried and then dissolved at 0°C in 40 ml 50 mM sodium citrate. Ten ml 1.0 M BaCl$_2$ was added, and the calcium-binding fragment (residues 1–35) was adsorbed to the resulting barium citrate precipitate. The supernatant containing the remaining fragments was separated into seven fractions on a 5 x 90-cm column of Sephadex G-50 in 0.25 M NH$_4$HCO$_3$ pH 8.3. The separation was analyzed by paper electrophoresis at pH 6.5, using ninhydrin, Ehrlich, Pauly and phenanthraquinone stains to detect the peptides. The two overlapping peptides connecting the A fragment to the S fragment (residues 129–167) and the S fragment to the A chain of thrombin (residues 234–283), respectively, were isolated from pool II by gradient elution (20 mM–1.0 M NH$_4$HCO$_3$ pH 8.3) on a 1.5 x 60-cm column of DEAE-cellulose followed by high-voltage paper electrophoresis at pH 6.5. The A- to S-fragment overlap (170 nmoles) was subdigested with 0.5 nmole thermolysin (Sigma) for 4 hours, 45°C, pH 6.45, and the peptide (153–167) was isolated by high-voltage electrophoresis at pH 6.5. The peptide (269–277) was isolated from a partial acid hydrolysate (30 mM HCl, 110°C, 8 hr) of 400 nmoles of peptide 234–283 by high-voltage paper electrophoresis at pH 6.5 and pH 2.1.

Isolation of Gla

A small fragment corresponding to residues 1–35 was isolated from a chymotryptic digest of native prothrombin by adsorption to barium citrate, as described above, followed by elution with 1 M Na$_2$SO$_4$, and desalted on a column of Sephadex G-25. Gla (γ-carboxyglutamic acid) was obtained in about 10% yield by partial acid hydrolysis of the peptide 1–35 in 6 M HCl at 45°C for 24 hours. After drying in vacuo, the material was dissolved in 2 M NH$_3$ and subjected to high-voltage electrophoresis at pH 6.5 and pH 2.1, affording separation of Gla from Gla-Gla.

RESULTS AND DISCUSSION

During the course of this work, we have published brief accounts of the details of the sequence work on the A chain of thrombin (Magnusson 1968), most of the B chain of thrombin (Magnusson 1970b, 1971), the completed sequence of the B chain of thrombin and of the A and S fragments and also evidence for the A-chain to B-chain overlap from neoprothrombin-T (Magnusson et al. 1975). Therefore in the present paper, we have chosen to limit the description of sequence evidence to those chymotryptic peptides from reduced, carboxymethylated prothrombin which have provided the evidence that the C terminal of the A fragment overlaps directly to the N terminal of the S fragment and the C terminal of the S fragment overlaps to the N terminal of the A chain of thrombin.

Overlap from A Fragment to S Fragment

One of the thermolytic peptides obtained from the chymotryptic peptide (129–167) that overlapped the A and S fragments was found to have an amino acid composition corresponding to the last four residues in the A fragment plus the first eleven in the S fragment. The sequence Val-Ile-Pro-Arg-

Ser-Gly-Gly- (Ser$_3$, Thr$_2$, Gln, Pro, Leu) of residues 153–167 was obtained. This means that in prothrombin the potential A and S fragments are directly connected without any intervening peptide.

Overlap from S Fragment to A Chain of Thrombin

One of the peptides obtained from a partial acid hydrolysate of the chymotryptic peptide (234–283) overlapping the S fragment to the A chain of thrombin had a composition corresponding to the last six residues of the S fragment and the first three of the A chain of thrombin. The sequence Ala-Ala-Ile-Glu-Gly-Arg-Thr-Ser-Glu was determined, proving that there is a direct overlap in this case as well.

Amino Acid Sequence of Prothrombin

The identification of the last two overlaps completes the amino acid sequence determination of prothrombin. Figure 1 shows the origin of the different activation products in the native prothrombin structure. The term neoprothrombin has been adopted from the chymotrypsin field and indicates a proteolytically modified prothrombin without thrombin activity (Magnusson and Murano 1974). The letters S, T and A refer to the N-terminal amino acid residues by the one-letter code (S = serine, T = threonine, A = alanine). Figure 2 shows the amino acid sequence of prothrombin (containing the sequences of the potential activation products). The amino acid compositions calculated from the sequences of prothrombin and of the different activation products are given in Table 1 and are in good agreement with the found compositions.

The B Chain of Thrombin

The B chain of thrombin (residues 324–582) is extensively homologous with chymotrypsin and the other serine proteases from the pancreas (Magnusson et al. 1975). We have constructed a tentative model of its tertiary structure using the X-ray coordinates for α-chymotrypsin (Birktoft and Blow 1972) and our sequence data for the thrombin B chain. In our model, the catalytic site contains residues homologous with Ser-195, His-57 and Asp-102 in chymotrypsin; there is a salt bridge like the Ile-16 to Asp-194 bridge in chymotrypsin. The thrombin model has a pocket like the side-chain binding pocket of chymotrypsin. At the bottom of this pocket is Asp-522, corresponding to Asp-189 and Ser-189 in trypsin and chymotrypsin, respectively. This is in keeping with the fact that thrombin is a trypsinlike enzyme in the sense that it cleaves at Arg and Lys bonds. There is an entrance to the pocket because thrombin has Gly residues in positions 551 and 561, corresponding to Gly-216 and Gly-226 in chymotrypsin. Thrombin also has the sequence Ser-549-Trp-550-Gly-551 corresponding to Ser-214-Trp-215-Gly-216 in chymotrypsin. On the basis of X-ray studies of the binding of various acetyl(aminoacyl$_{1-3}$-phenylalanine chloromethyl ketones to chymotrypsin Aγ, Segal et al. (1971) showed that alanyl residues in positions P$_{2-4}$ bind to the sequence -Ser-Trp-Gly-214-216, forming an extended, antiparallel, β-type configuration. Therefore, it seems fair to expect that thrombin substrates (e.g., fibrinogen) will bind to thrombin with residues P$_{2-4}$ in a similar configuration.

Relative to the sequence of chymotrypsin, the sequence of the B chain of

Ala-Asn-Lys-Gly-Phe-Leu-GLA-GLA-Val-Arg-Lys-Gly-Asn-Leu-GLA-Arg-GLA-Cys-Leu-GLA-GLA-

-Pro-Cys-Ser-Arg-GLA-GLA-Ala-Phe-GLA-Ala-Leu-GLA-Ser-Leu-Ser-Ala-Thr-Asp-Glu-Ala-Phe-Trp-

-Ala-Lys-Tyr-Thr-Ala-Cys-Glu-Ser-Ala-Arg-Asn-Pro-Arg-Glu-Lys-Leu-Asn-Glu-Cys-Leu-Glu-

-Gly-Asn-Cys-Ala-Glu-Gly-Val-Gly-Met-Asn-Tyr-Arg-Gly-Asn-Val-Ser-Val-Thr-Arg-Ser-Gly-

-Ile-Glu-Cys-Gln-Leu-Trp-Arg-Ser-Arg-Tyr-Pro-His-Lys-Pro-Glu-Ile-Asn-Ser-Thr-Thr-His- (CHO over Asn)

-Pro-Gly-Ala-Asp-Leu-Arg-Glu-Asn-Phe-Cys-Arg-Asn-Pro-Asp-Gly-Ser-Ile-Thr-Gly-Pro-Trp-

-Cys-Tyr-Thr-Thr-Ser-Pro-Thr-Leu-Arg-Arg-Glu-Glu-Cys-Ser-Val-Pro-Val-Cys-Gly-Gln-Asp-

-Arg-Val-Thr-Val-Glu-Val-Ile-Pro-Arg-Ser-Gly-Gly-Ser-Thr-Thr-Ser-Gln-Ser-Pro-Leu-Leu-

-Glu-Thr-Cys-Val-Pro-Asp-Arg-Gly-Arg-Glu-Tyr-Arg-Gly-Arg-Leu-Ala-Val-Thr-Thr-Ser-Gly-

-Ser-Arg-Cys-Leu-Ala-Trp-Ser-Ser-Glu-Gln-Ala-Lys-Ala-Leu-Ser-Lys-Asp-Gln-Asp-Phe-Asn-

-Pro-Ala-Val-Pro-Leu-Ala-Asn-Phe-Cys-Arg-Asn-Pro-Asp-Gly-Asn-Pro-Asp-Gly-Gly-Ala-Trp-

-Cys-Tyr-Val-Ala-Asp-Gln-Pro-Gly-Asp-Phe-Glu-Tyr-Cys-Asn-Leu-Asn-Tyr-Cys-Glu-Glu-Pro-

-Val-Asp-Gly-Asp-Leu-Gly-Asp-Arg-Leu-Gly-Glu-Asp-Pro-Asp-Pro-Asp-Ala-Ala-Ile-Glu-Gly-

-Arg-Thr-Ser-Glu-Asp-His-Phe-Gln-Pro-Phe-Phe-Asn-Glu-Lys-Thr-Phe-Gly-Ala-Gly-Glu-Ala-

-Asp-Cys-Gly-Leu-Arg-Pro-Leu-Phe-Glu-Lys-Lys-Ser-Leu-Glu-Asp-Lys-Thr-Glu-Arg-Glu-Leu-Leu-

-Phe-Glu-Ser-Tyr-Ile-Glu-Gly-Arg-Ile-Val-Glu-Gly-Gln-Asp-Ala-Glu-Val-Gly-Leu-Ser-Pro-

-Trp-Gln-Val-Met-Leu-Phe-Arg-Lys-Ser-Pro-Gln-Glu-Leu-Leu-Cys-Gly-Ala-Ser-Leu-Ile-Ser-

                                                                                                        CHO
                                                                                                         |
-Asp-Arg-Trp-Val-Leu-Thr-Ala-Ala-His-Cys-Leu-Leu-Tyr-Pro-Pro-Trp-Asx-Lys-Asn-Phe-Thr-

-Val-Asp-Asp-Leu-Leu-Val-Arg-Ile-Gly-Lys-His-Ser-Arg-Thr-Arg-Tyr-Glu-Arg-Lys-Val-Glu-

-Lys-Ile-Ser-Met-Leu-Asp-Lys-Ile-Tyr-Ile-His-Pro-Arg-Tyr-Asn-Trp-Lys-Glu-Asn-Leu-Asp-

-Arg-Asp-Ile-Ala-Leu-Leu-Lys-Leu-Lys-Arg-Pro-Ile-Glu-Leu-Ser-Asp-Tyr-Ile-His-Pro-Val-

-Cys-Leu-Pro-Asp-Lys-Gln-Thr-Ala-Ala-Lys-Leu-Leu-His-Ala-Gly-Phe-Lys-Gly-Arg-Val-Thr-

-Gly-Trp-Gly-Asn-Arg-Arg-Glu-Thr-Trp-Thr-Thr-Ser-Val-Ala-Glu-Val-Gln-Pro-Ser-Val-Leu-

-Gln-Val-Val-Asn-Leu-Pro-Leu-Val-Glu-Arg-Pro-Val-Cys-Lys-Ala-Ser-Thr-Arg-Ile-Arg-Ile-

-Thr-Asn-Asp-Met-Phe-Cys-Ala-Gly-Tyr-Lys-Pro-Gly-Glu-Gly-Lys-Arg-Gly-Asp-Ala-Cys-Glu-

-Gly-Asp-Ser-Gly-Gly-Pro-Phe-Val-Met-Lys-Ser-Pro-Tyr-Asn-Asn-Arg-Trp-Tyr-Gln-Met-Gly-

-Ile-Val-Ser-Trp-Gly-Glu-Gly-Cys-Asp-Arg-Asn-Gly-Lys-Tyr-Gly-Phe-Tyr-Thr-His-Val-Phe-

-Arg-Leu-Lys-Lys-Trp-Ile-Gln-Lys-Val-Ile-Asp-Arg-Leu-Gly-Ser

Figure 2

Amino acid sequence of bovine prothrombin (21 residues per line). CHO = carbohydrate, attached to Asn-77, Asn-101 and Asn-376. GLA = γ-carboxyglutamic acid residue (positions 7, 8, 15, 17, 20, 21, 26, 27, 30 and 33). Underlined sequences: -Glu-Asn-Phe-Cys-Arg-Asn-Pro-Asp-Gly- (112–120 and 217–225), identical nonapeptide sequences in A and S fragments; -Val-Ile-Pro-Arg (153–156), sequence at the thrombin-specific cleavage site at C termini of A fragment; -Ile-Glu-Gly-Arg (271–274 and 320–323), sequences at factor X_a-specific cleavage sites at C termini of S fragment and thrombin A chain; -Ile-Val-Glu-Gly- (324–327), sequence at N terminal of serine protease chain (B chain) in thrombin. -Thr-Ala-Ala-His-Cys- (363–367), -Asp- (422) and -Gly-Asp-Ser-Gly-Gly-Pro- (526–531) contain residues (His, Asp and Ser) in sequences that are homologous with His-57, Asp-102 and Ser-195 in active site of chymotrypsin. Asp-522 corresponds to Ser-189 in chymotrypsin and explains why thrombin is a trypsinlike enzyme, specific for Arg and Lys bonds.

Table 1
Amino Acid Compositions Calculated from Sequences of Prothrombin and of Different Activation Products

	Prothrombin	A fragment	S fragment	Thrombin A chain	Thrombin B chain	Thrombin and neopro-thrombin-T	Neopro-thrombin-S	Pro fragment
Asp	34	4	13	3	14	17	30	17
Asn	25	10	5	1	9	10	15	15
Asx	1				1	1	1	
Glu	43	11	11	8	13	21	32	22
Gln	18	2	4	4	8	12	16	6
Gla	10	10	—	—	—	—	—	10
Gly	48	11	12	4	21	25	37	23
Ala	34	10	10	2	12	14	24	20
Val	35	9	5	1	20	21	26	14
Met	6	1	—	—	5	5	5	1
Ile	20	4	1	1	14	15	16	5
Leu	46	10	9	3	24	27	36	19

	Prothrombin (1–582)	A fragment (1–156)	S fragment (157–274)	A chain of thrombin (275–323)	B chain of thrombin (324–582)	Thrombin / Neoprothrombin-T (275–582)	Neoprothrombin-S (157–582)	Pro fragment (1–274)
Pro	35	10	9	2	14	16	25	19
½ Cys	24	10	6	1	7	8	14	16
Trp	14	3	2	—	9	9	11	5
Lys	31	5	2	4	20	24	26	7
His	9	2	—	1	6	7	7	2
Arg	45	15	8	2	20	22	30	23
Ser	36	11	9	2	14	16	25	20
Thr	29	10	5	3	11	14	19	15
Phe	20	4	3	6	7	13	16	7
Tyr	19	4	4	1	10	11	15	8
Total	582	156	118	49	259	308	426	274
MW	66098 +3 CHO	17973 +2 CHO	12775	5721	29683 +CHO	35386 +CHO	48143 +CHO	30730 +2 CHO

Amino acid compositions calculated from the amino acid sequences for prothrombin (residues 1–582), A fragment (1–156), S fragment (157–274), the A chain of thrombin (275–323), the B chain of thrombin (324–582), thrombin and neoprothrombin-T (275–582), neoprothrombin-S (157–582) and Pro fragment (1–274). Molecular weights were obtained by adding integral residue weights for constituent amino acid residues and do not include carbohydrate. The molecular weight of thrombin is 35,404 daltons (excluding the carbohydrate), that of neoprothrombin-T is given in the table.

thrombin has a single deletion, corresponding to residue 218 in chymotrypsin, and single insertions in positions 36, 84, 99, 184 and 221. Longer insertions occur at positions 65 (9 residues with the carbohydrate attached), 128 (3 residues), 147 (5 residues), 188 (4 residues), 203 (2 residues) and 245 (2 residues).

In the model, the two insertions at residues 65 (-Lys-Asn^{CHO}-Phe-Thr-Val-Asp-Asp-Leu-Leu) and at 147 (-Trp-Thr-Thr-Ser-Val-) are not very far from the active site and could very well influence the binding of thrombin substrates at secondary binding sites. The other insertions are situated much further from the active site and therefore seem to be less likely candidates. In addition to the insertions, there is also a large number of substitutions in the B-chain sequence, and the structural background for the restricted specificity of thrombin is not immediately apparent.

The A Chain of Thrombin

As is the case with the A chain in chymotrypsin, the A chain of thrombin is attached to the B chain at Cys-122 (442 in prothrombin). This means that relative to the size of the chymotrypsin A chain, the thrombin A chain has 13 additional residues between its half-cystine and the C-terminal Arg, which in the zymogen is bound to Ile-16. Therefore, it seems likely that there is an extra "loop" in this part of the structure, compared to chymotrypsinogen. The N-terminal end of the A chain, residues 1–21 in the A chain (275–295 in prothrombin), has no counterpart in chymotrypsinogen. The model of the B chain of thrombin has a shallow groove or furrow, lined mostly by residues with hydrophobic side chains, which runs from Cys-122 (Cys-442 in prothrombin) almost straight up along the back of the model, over the top, and comes down slightly to the left of the catalytic site area. It is likely that the N-terminal end of the A chain is situated in this furrow. This means that the N-terminal end of the A chain is a possible candidate for restricting access to the active site in thrombin. The C-terminal end of the A chain is also sufficiently big that it could be wrapped around the right front "corner" of the model and reach the active site region after the activation cleavage of Arg-323-Ile-324 in prothrombin has taken place. A simpler hypothesis might be to assume that the presence of the "extra loop" (the 13-residue insertion in the A chain) in thrombin relative to chymotrypsin causes a difference in the ratio of in-form to out-form (with regard to the N-terminal Ile) from that of chymotrypsin (Fersht 1971) such that thrombin is not very active towards substrates generally, but is "activated" by increasing the ratio of in-form to out-form as a result of interaction with specific structures in thrombin substrates. This hypothesis is now being investigated.

Structure of the Vitamin K-dependent Part of Prothrombin

Dam (1935) discovered vitamin K and found that it was required for maintaining normal prothrombin activity in the blood plasma. Josso et al (1968) and Ganrot and Niléhn (1968) demonstrated that persons treated with dicoumarol, a vitamin K antagonist, have in their plasma a protein which precipitates antibodies to normal prothrombin. This observation indicated that the function of vitamin K in prothrombin biosynthesis was not in the actual synthesis of the glycoprotein itself, but was more likely to be in some kind of postsynthetic modification. The first clue to the chemical identity of this vitamin K-dependent modification came from the discovery (Magnusson 1972, 1973) that a tryptic digest of the A fragment contained two peptides (residues 4–10 and 4–11), Gly-Phe-Leu-Glx-Glx-Val-Arg and Gly-Phe-Leu-Glx-Glx-Val-Arg-Lys, both of which were free of carbohydrate but had electrophoretic mobilities at pH 6.5 which corresponded to a net charge of −3 and −2, respectively. This could only be explained by an unknown constituent having at least two negative charges. Since it was found that these extra negative charges were essentially neutralized at pH 2.1 and no phosphorus had been found in prothrombin (Magnusson 1965b), and since 6-M HCl hydrolysis gave rise to normal glutamic acid residues (on the amino acid analyzer and on amino acid analysis at pH 2.1 in paper electrophoresis), we concluded (Magnusson et al. 1974a) that the extra negative charges were probably due to either or both of the Glx residues having a substituent giving them one or more extra carboxyl groups depending on the size of the substituent. The adsorption of a different tryptic peptide to barium citrate (Nelsestuen and Suttie 1973) indicated that this peptide had also been modified by vitamin K. Stenflo (1974) recently confirmed our results regarding the extra charges on Glx-7 and -8 in normal prothrombin and found normal Glu residues in these positions in dicoumarol prothrombin. We have performed mass spectrometric analyses of peptides 4–10, 5–11, 14–31 (split at 25/26 during the derivatization procedure), 14–21 and 31–35 using an acetylation step followed by a permethylation step for derivatization and proved conclusively that all ten Glx residues (7, 8, 15, 17, 20, 21, 26, 27, 30 and 33) were in fact γ-carboxyglutamic acid residues (Fig. 3) (Magnusson et al. 1974b). The derivatization procedure caused the formation of the fully tetramethylated derivative of the new amino acid with 229 mass units, a decarboxylated product (171 mass units), a cyclized derivative (corresponding to pyroglutamic acid) with 198 mass units, and a cyclized, decarboxylated derivative (140 mass units) that easily loses CO to form a derivative with 112 mass units. High resolution mass measurements accurate to within 1–5 ppm have confirmed the composition of the derivatives.

Reactivity of γ-Carboxyglutamic Acid

We have investigated the extent of methylation of Gla residues in the two derivatization steps by using deuterated methanol (CD_3OD) in the "acetylation" step and in other experiments by using deuterated methyl iodide (CD_3I) in the permethylation step. The combined evidence obtained in these experiments proved that the "acetylation" step causes esterification of the two γ-carboxyl groups in γ-carboxyglutamic acid. The permethylation introduces one methyl group on the α-amino nitrogen and one on the γ-carbon. The esterification of the two γ-carboxyl groups under "acetylation" conditions in acetic anhydride/methanol is unusual for carboxyl groups and shows that the γ-carboxyglutamic acid residues are unusually reactive. Whether this reactivity is utilized physiologically for purposes other than the binding of calcium ions (e.g., in transcarboxylation) is not yet known. The methylation of the α-amino nitrogen occurs normally in the permethylation step. The fact that the γ-carbon is also methylated is not surprising in view of the malonyl character conferred on this carbon by the substitution of the second carboxyl group. The claim by Stenflo et al. (1974) of having obtained dimethyl derivatives of γ-carboxyglutamic acid from the peptide Phe-Leu-Gla-Gla (residues 5–8)

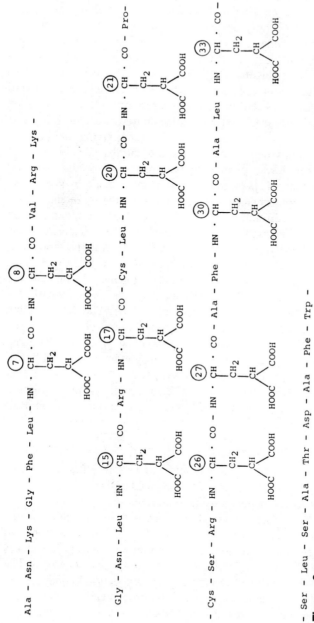

Figure 3

Amino acid sequence of the calcium-binding, vitamin K-dependent part of prothrombin (residues 1–42). Normal amino acid residues in the usual three-letter code. Gla residues (γ-carboxyglutamic acid) in positions 7, 8, 15, 17, 20, 21, 26, 27, 30 and 33. (Reprinted, with permission, from Magnusson et al. 1974b.)

134

using the same derivatization procedure is not compatible with the structure and is difficult to understand unless for some reason the permethylation step did not work in their experiments. Their NMR (nuclear magnetic resonance) evidence is compatible with the γ-carboxyglutamic acid structure, but it is also compatible with other structures and therefore not in itself sufficient to prove the structure.

Detection of Gla-containing Peptides and Isolation of Free Gla

One or both of the extra negative charges on peptides 4–10 and 4–11 were found to be lost if the electrophoresis papers were heated during drying (Magnusson et al. 1975) due to decarboxylation. This property is utilized in a "thermal diagonal" electrophoresis method for finding peptides containing γ-carboxyglutamic acid (Magnusson et al. 1974b) (Fig. 4) in protein digests. The free γ-carboxyglutamic acid has been isolated from both bovine

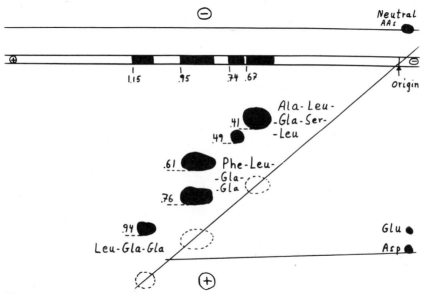

Figure 4
"Thermal-diagonal" electrophoresis of some Gla-containing peptides: Leu-Gla-Gla (residues 19–21), Phe-Leu-Gla-Gla (5–8) and Ala-Leu-Gla-Ser-Leu (31–35). In the first dimension, the peptide mixture was applied at the "origin" on the narrow horizontal paper strip and subjected to 3 kV, 40 minutes at pH 6.5. The black areas marked 1.15, .95, .74 and .67 indicate the positions and mobilities relative to aspartic acid of the peptides on a stained guide strip. After drying and heating for 30 minutes at 150°C, the main strip was stitched to a fresh sheet of paper and subjected to the same electrophoretic conditions, but turned 90° as indicated by the + and − signs. Black ovals indicate actual positions (with new mobilities as a result of decarboxylation) of the peptides after electrophoresis in the second dimension. Dashed ovals on the diagonal indicate the positions the peptides would have migrated to had they not lost one or two negative charges by heat decarboxylation. "Neutral" amino acids "Glu" and "Asp" indicate positions of the respective free amino acids run in the second dimension only.

and human prothrombin by partial acid hydrolysis. From 20 mg of the purified chymotryptic peptide (1–35) from the A fragment of prothrombin, a yield of 1300 nmoles Gla and 300 nmoles Gla-Gla was obtained. The isolated γ-carboxyglutamic acid has an electrophoretic mobility of 1.35 relative to Asp at pH 6.5 and 0.41 relative to Ser at pH 2.1. It shows the same reactivity on "acetylation" and gives the same derivatives on mass spectrometry as the peptide-bound Gla residues (Fig. 5). On field desorption analysis, a single peak of mass number 192 was obtained corresponding to a proton plus the mass number of 191 for the new amino acid (H. R. Morris, A. Dell, T. E. Petersen, L. Sottrup-Jensen and S. Magnusson, in prep.), confirming its identity. Acid hydrolysis of peptides containing Gla in 6 м HCl at 110°C leads in most cases to nearly complete decarboxylation of Gla to glutamic acid. The half-time for this decarboxylation was found to be about 20–30 minutes. In some instances (Magnusson et al. 1974a), a considerable amount of Gla survives the 20-hour hydrolysis. The reason for this is not known. On the amino acid analyzer, Gla is eluted in the same position as Cm-Cys. In hydrolysates where the conversion to Glu had not been completed, there was

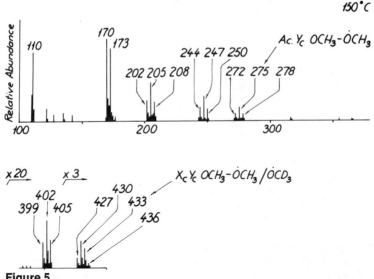

Figure 5

Mass spectrum of mixture of 90% γ-carboxyglutamic acid (Gla) and 10% Gla-Gla isolated from residues 1–35 in bovine prothrombin. "Acetylation" performed in 1:1 mixture of $CH_3OH:CD_3OD$; permethylation with CH_3I. The peaks indicated in the top half of the figure are derived from Gla, the ones in the lower half probably from Gla-Gla. The single peak (mass number 110) has no carboxyl group. The double peak (170, 173) has one. The triplet peaks, with a 1:2:1 abundance distribution (202, 205, 208; 244, 247, 250; 272, 275, 278), have two carboxyl groups each which react in the acetylation step as shown by the distribution of deuterium label. Ac = acetyl, Y_c = core of the N- and γ-methylated γ-carboxyglutamic acid ($C_8H_9NO_3$ of mass number 167), X_c = cyclized Y_c.

also a peak that has not been identified emerging very close to the position of Asp.

On alkaline hydrolysis (4 M NaOH, 20 hr, 110°C in Beckman polyallomer tubes), Gla is recovered in yields of about 50–70%, assuming a normal ninhydrin color value. No significant conversion to Glu was observed on alkaline hydrolysis.

Carbohydrate Attachment Sites

There are three carbohydrate substituents in prothrombin. They are all attached to asparagine residues, namely Asn-77 and Asn-101 in the A fragment (Fig. 6) and Asn-376 in the B chain of thrombin. All three amino acid sequences at the carbohydrate attachment sites fit the general pattern -Asn-X-Ser- and -Asn-X-Thr- for glycoproteins containing glucosamine as the only hexosamine (Marshall 1972). The only hexosamine found in significant amounts in prothrombin was glucosamine (Magnusson 1965b). About 75% of the total carbohydrate in prothrombin (12–13% of the weight) is attached to the A fragment, the remaining 25% being attached to the B chain of thrombin. During the course of the present structure work, glycopeptides from all three sites have been obtained in numerous variants. We have no reason to believe that the three carbohydrate substituents have identical structures. When the amino acid compositions of the "glycopeptides" isolated from bovine prothrombin determined by Nelsestuen and Suttie (1972) are checked against our sequence of prothrombin, it becomes apparent that each of their five fractions contain a mixture of peptide fragments from all three carbohydrate sites. Consequently, no significant conclusion can be drawn from their work regarding the relative composition of the different substituents.

Disulfide Bridges in Prothrombin

There are twelve disulfide bridges in prothrombin. The A fragment has five bridges connecting half-cystines 18–23, 48–61, 66–144, 87–127 and 115–139. The S fragment has three (171–249, 192–232 and 220–244). One bridge connects the A and B chains of thrombin (296–442) and three are internal in the B chain of thrombin (351–367, 496–510 and 524–554).

Internal Homology

The nonapeptide sequence -Glu-Asn-Phe-Cys-Arg-Asn-Pro-Asp-Gly- occurs twice in prothrombin, i.e., in the A fragment (residues 112–120) and in the S fragment (residues 217–225). When the sequences of the A and S fragments were aligned (Fig. 6) on the basis of this identity, it was found that in two regions of 83 residues each (residues 62–144 in the A fragment and residues 167–249 in the S fragment), 31 positions have identical amino acid residues, meaning that a partial gene duplication of this area has occurred during the evolution of prothrombin. The identities include all three disulfide bridges in the S fragment and three of the five bridges in the A fragment (Figs. 7 and 8). The bridges in both homologous areas connect half-cystines 1 to 6, 2 to 4 and 3 to 5, forming "kringlelike"[1] structures. The N-terminal region of the A fragment which contains all ten Gla residues is not duplicated in the S frag-

[1] Kringle: Referring to a classical shape of Scandinavian cake.

A: Ala-Asn-Lys-Gly-Phe-Leu-Gla-Gla-Val-Arg-Lys-Gly-Asn-Leu-Gla-Arg-Gla-Cys-Leu-Gla-

S:

A: -Gla-Pro-Cys-Ser-Arg-Gla-Gla-Ala-Phe-Gla-Ala-Leu-Gla-Ser-Leu-Ser-Ala-Thr-Asp-Ala-

S:

A: -Phe-Trp-Ala-Lys-Tyr-Thr-Ala-Cys-Glu-Ser-Ala-Arg-Asn-Pro-Arg-Glu-Lys-Leu-Asn-Glu-

S: Ser-Gly-Gly-Ser-Thr-Thr-Ser-Gln-Ser-
                                        CHO

A: -Cys-Leu-Glu-Gly-Asn-Cys-Ala-Glu-Gly-Val-Gly-Met-Asn-Tyr-Arg-Gly-ASN-Val-Ser-Val-
62

S: -Pro-Leu-Glu-Glu-Thr-Cys-Val-Pro-Asp-Arg-Gly-Arg-Glu-Tyr-Arg-Gly-Arg-Leu-Ala-Val-
11

A: -Thr-Arg-Ser-Gly-Ile-Glu-Cys-Gln-Leu-Trp-Arg-Ser-Arg-Tyr-Pro-His-Lys-Pro-Glu-Ile-

S: -Thr-Thr-Ser-Gly-Ser-Arg-Cys-Leu-Ala-Trp-Ser-Ser-Glu-Gln-Ala-Lys-Ala-Leu-Ser-Lys-
CHO

A: -ASN-Ser-Thr-His-Pro-Gly-Ala-Asp-Leu-Glu-Asn-Phe-Cys-Arg-Asn-Pro-Asp-Gly-

S: -Asp-GlN-Asp-Phe-Asn-Pro-Ala-Val-Pro-Leu-Ala-Glu-Asn-Phe-Cys-Arg-Asn-Pro-Asp-Gly-

```
A:     -Ser-Ile-Thr-Gly-Pro-Trp-Cys-Tyr-Thr-Thr-Ser-Pro-Thr-Leu-Arg-Arg-Glu-Glu-Cys-Ser-

S:     -Asp-Glu-Glu-Ala-Trp-Cys-Tyr-Val-Ala-Asp-Gln-Pro-Gly-Asp-Phe-Glu-Tyr-Cys-Asn-

PSTI:      -Pro-Val-Cys-Gly-Thr-Asp-Gly-Val-Thr-

A:     -Val-Pro-Val-Cys-Gly-Gln-Asp-Arg-Val-Thr-Val-Glu-Val-Ile-Pro-Arg

S:     -Leu-Asn-Tyr-Cys-Glu-Pro-Val-Asp-Gly-Asp-Arg-Leu-Gly-Glu-Asp-Pro-

A:

S:     -Asp-Pro-Asp-Ala-Ala-Ile-Glu-Gly-Arg
```

Figure 6

Alignment of A fragment (A) and S fragment (S) from prothrombin. Residues 22–30 of the bovine
pancreatic secretory trypsin inhibitor (PSTI) aligned with residues 142–150 in the A fragment is the
only homology with "nonclotting" proteins so far observed. CHO above the line indicates carbohydrate attached to Asn. The 83-residue homologous regions start at Leu-62 in the A fragment and
Leu-11 in the S fragment. Identical residues are underlined. In addition, the two homologous
sequences -Pro-Val-Cys-Gly-Thr-Asp-Gly-Val-Thr- in PSTI and -Pro-Val-Cys-Gly-Gln-Asp-Arg-Val-
Thr- in the A fragment have been underlined.

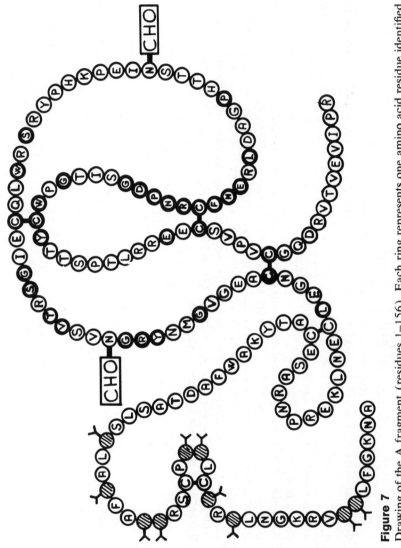

Figure 7

Drawing of the A fragment (residues 1–156). Each ring represents one amino acid residue identified by the one-letter code. A hatched ring with fork represents a Gla residue. A fat line connecting two Cys residues indicates a disulfide bridge. CHO stands for carbohydrate.

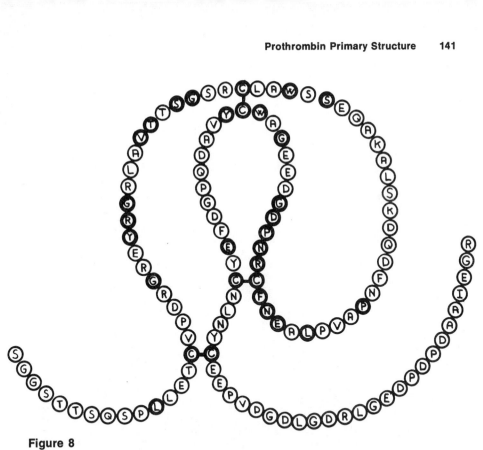

Figure 8

Drawing of the S fragment (residues 157–274). See legend to Figure 7.

ment. The S fragment has an extra C-terminal piece of 13 residues relative to the A fragment. The two carbohydrate attachment sites in the A fragment, at positions 77 and 101, occur exactly halfway between half-cystines 66 and 87 and 87 and 115, respectively. The fact that the corresponding positions in the S fragment have sequences -Arg-Leu-Ala- and -Asp-Gln-Asp- (instead of -Asn-Val-Ser- and -Asn-Ser-Thr- which are not known to code for the attachment of carbohydrate) indicates that whereas the carbohydrate attachment sites in the A fragment are almost certainly to be found on the surface in the tertiary structure of prothrombin, the corresponding parts of the S fragment may be hidden.

Activation of Prothrombin by Factor X_a: Specificity Requirements of Factor X_a

When prothrombin is activated to thrombin by factor X_a under conditions where the resulting thrombin is inhibited by the presence of hirudin (Kisiel and Hanahan 1974) or by a concentration of diisopropyl-fluorophosphate (DFP) sufficient to inhibit thrombin but not factor X_a (Owen, Esmon and Jackson 1974), only two products were obtained (Figs. 1 and 9); namely, Pro fragment (residues 1–274) and active thrombin (residues 275–323 disulfide bridged to 324–582). This means that the two peptide bonds 274–275 and 323–324 (see Figs. 1 and 2) are cleaved selectively by factor X_a. These two cleavage sites have the amino acid sequences -Ile-Glu-

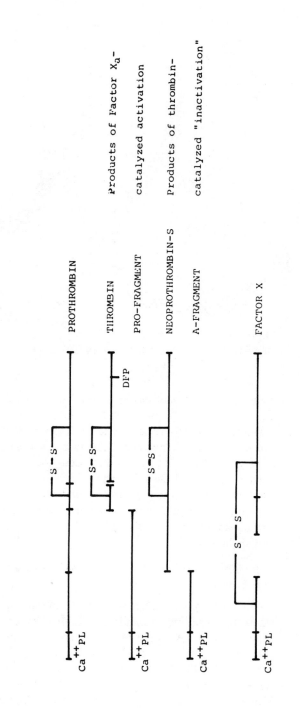

142

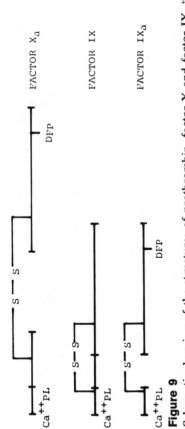

Figure 9

Schematic drawing of the structures of prothrombin, factor X and factor IX. "DFP" indicates approximate positions of the active center -Gly-Asp-Ser-Gly-Gly-Pro- sequences in the three active serine proteases—thrombin, factor X_a and factor IX_a. Only the (potential) interchain disulfide bridges are indicated. "Ca++PL" indicates the calcium-phospholipid binding domain comprising the N-terminal 33 residues in prothrombin and presumably a domain of comparable size in factor X and factor IX. The number of amino acid residues in the chains corresponds approximately to the length of the line in the drawing.

Gly-Arg-Thr-Ser-Glu- and -Ile-Glu-Gly-Arg-Ile-Val-Glu- which are identical in positions P_{1-4} and P'_3 relative to the bond cleaved. A derivative of this sequence has been synthesized, and we are investigating its possible use as a specific substrate/inhibitor for factor X_a.

The synthetic substrate Tos-L-Ile-L-Glu-Gly-L-Arg-pNA (p-nitro-anilide) (Novo, Copenhagen, Denmark) is split by highly purified bovine factor X_a (gift from Dr. P. Esnouf, Oxford, England). This substrate is cleaved rapidly by trypsin. The rate of splitting by thrombin is only about 5% of that by factor X_a under the conditions used (T. E. Petersen, L. Sottrup-Jensen and S. Magnusson, in prep.). Plasmin, urokinase, and streptokinase-plasminogen complex did not split the substrate.

A Thrombin-dependent Control Mechanism for the Regulation of Prothrombin Activation

Neoprothrombin-S, previously called "TEAE peak I material," was first isolated and characterized in 1961 with respect to amino acid and carbohydrate composition, gel electrophoretic mobility and activation characteristics (Magnusson 1962, 1965c). Neoprothrombin-S was essentially inactive in the two-stage test system for prothrombin activity, but it could be activated very slowly if the test system was supplied with an excess of serum. Similar observations have been made by other groups (Asada et al. 1961; Shulman and Hearon 1963; Papahadjopoulos, Hougie and Hanahan 1964; Lechner and Deutsch 1965; Aronson and Ménaché 1966). Seegers et al. (1967) showed that a product with similar activation characteristics could be obtained by thrombin-catalyzed cleavage of prothrombin. This has been confirmed by Esmon, Owen and Jackson (1974). In view of the highly restricted substrate specificity of thrombin, it seems unlikely that a thrombin-specific cleavage site would have been conserved during the evolution of prothrombin unless it had physiological advantage over other structures. We therefore recently proposed (Magnusson et al. 1975) that the "inactivation" of prothrombin to neoprothrombin-S catalyzed by thrombin is an important regulatory mechanism for limiting the extent of prothrombin activation to thrombin by factor X_a. This mechanism is consistent with the following facts:

1. Factor X_a cleaves only at positions 274 and 323, *not* at 156.
2. Thrombin cleaves *only* at 156, not at 274 and 323.
3. Neoprothrombin-S has lost its vitamin K-dependent calcium-phospholipid binding domain and therefore cannot be activated in the fast (phospholipid-dependent) system, despite the fact that it still contains the two factor X_a-sensitive bonds 274 and 323. Therefore, it is not functionally a zymogen for thrombin, although it contains all the structural elements of the thrombin molecule.
4. It has been known for a long time that when a blood or plasma sample is clotted via the extrinsic or intrinsic pathways, the prothrombin activity is quickly depleted, but only a few percent of it can be recovered as thrombin. It has not been clear what happens to the rest of the prothrombin. Since neoprothrombin-S would be detected neither as prothrombin nor as thrombin, it seems likely that it is the major net product of the system.

A consequence of such a regulation mechanism is that inhibitors specific against thrombin would inhibit not only the clotting of fibrinogen as expected, but could also cause the formation of more thrombin by delaying the onset of the "inactivation" of prothrombin to neoprothrombin-S, thus allowing the factor X_a-catalyzed activation to proceed without competition for a longer time.

Comparison between Prothrombin and the Other Vitamin K-dependent Coagulation Factors

Very little is known about the structure of factor VII. The other two vitamin K-dependent coagulation proteins, factor IX and factor X, are known to be zymogens of serine proteases like prothrombin. Their primary structures are being elucidated by the Seattle group (Fujikawa et al. 1974; Fujikawa, Legaz and Davie 1972; Davie et al., this volume), and we know from their work that factor X (the zymogen) consists of a light chain and a heavy chain (Fig. 9). Activation to factor X_a, catalyzed by factor IX_a, removes an activation peptide from the N-terminal end of the heavy chain, which acquires serine protease activity in the process (Davie et al., this volume).

Factor IX (the zymogen) is a single-chain protein. Its activation to factor IX_a (the serine protease) is catalyzed by factor XI_a and involves cleavage of at least two peptide bonds. The resulting serine protease (factor IX_a) consists of two chains. It is not surprising that the heavy chains of both factors X_a and IX_a carrying the DFP-binding active sites are homologous with the pancreatic serine proteases and with the B chain of thrombin. The N-terminal 10–12 residues of the light chains of factors X_a and IX_a have been found by the Seattle group (Fujikawa et al. 1974) to be very homologous with the N-terminal part of prothrombin. We have recently investigated the sequence of the barium citrate adsorbable chymotryptic peptide from human prothrombin by mass spectrometry. Human prothrombin contained peptides with Glx residues corresponding to the same ten positions as in bovine prothrombin (7, 8, 15, 17, 20, 21, 26, 27, 30 and 33). All except number 15 have been unequivocally identified as γ-carboxyglutamic acid residues. The human prothrombin preparation used (Magnusson 1965d) turned out to be contaminated with factor X, and small amounts of the peptic peptide Leu-Gla-Gla-Thr-Lys-Gln-Gly-Asn-Leu were identified (T. E. Petersen, A. Dell, L. Sottrup-Jensen, S. Magnusson and H. R. Morris, in prep.). The strong homology of this peptide with the sequence Leu-Glu-Glu-Val-Lys-Gln-Gly-Asn-Leu obtained from positions 5–13 in bovine factor X (Fujikawa et al. 1974) indicates very strongly that not only prothrombin but also factor X contains γ-carboxyglutamic acid residues corresponding to at least the first two of the ten Gla positions in prothrombin.

Function of Vitamin K in Blood Coagulation

Ganrot and Niléhn (1968) found that normal prothrombin changes its electrophoretic mobility in the presence of calcium ions, indicating the formation of a calcium-prothrombin complex. Dicoumarol prothrombin does not bind calcium strongly enough to be activated in the calcium-phospholipid-dependent test systems. Gitel et al. (1973) demonstrated that the A fragment (residues 1–156) is responsible for the phospholipid-binding capacity of prothrombin, but only in the presence of calcium ions. If these findings and our present knowledge of the structure are rationalized, we can see that the function of vitamin K in coagulation is to cause a postribosomal modification of ten glutamic acid residues in prothrombin, and presumably a similar number in the other vitamin K-dependent factors, such that these generally hydrophilic proteins become endowed with a special affinity for calcium ions and consequently for phospholipids. It is something of a paradox that lipid binding is achieved in this way not by adding a hydrophobic structure, as one might have expected, but by making it more hydrophilic by adding ten extra negative charges.

There are two calcium-phospholipid-dependent reactions involved in coagulation, namely (1) the activation of factor X to factor X_a catalyzed by factor IX_a in the presence of factor VIII and (2) the activation of prothrombin to thrombin catalyzed by factor X_a in the presence of factor V. The net effect of vitamin K in the clotting system is apparently to enable *both* the *substrate and* the *enzyme* in these two activation reactions to be concentrated on the surface of phospholipid micelles. The high local concentrations thus obtained help explain why the generation of thrombin in the presence of phospholipids is extremely fast despite the very low plasma concentrations of prothrombin and factors IX and X. Whereas all three zymogens (factor IX, factor X and prothrombin) contain the vitamin K-dependent binding site for calcium and phospholipids, only two of the active serine proteases (factors IX_a and X_a) have a disulfide bridge connecting their heavy chains to the calcium-phospholipid-binding light chains (Fig. 9). Thrombin has no such disulfide bridge and is therefore a "plasma" enzyme (the thrombin-fibrinogen reaction does not require phospholipids), whereas factors IX_a and X_a stay bound to the phospholipid surface even after they have been activated. Another indirect consequence of the role of vitamin K is that it becomes easier to understand the importance of the thrombin-catalyzed "inactivation" of prothrombin to neoprothrombin-S when one realizes that neoprothrombin-S has lost the calcium-phospholipid binding domain with the A fragment and therefore will be physically separated from the phospholipid-bound factor X_a, thereby explaining why it is not further activated to thrombin despite the fact that it contains the two cleavage sites for factor X_a. A further consequence is that both the A fragment (1–156) and the Pro fragment (1–274) contain the calcium-phospholipid binding domain and can therefore be expected to inhibit prothrombin activation, at least when the phospholipid concentration is limiting.

Function of "Kringle" Structures

The simplest explanation for the existence of the two homologous "kringle" regions in the A and S fragments is to assume that some or all of their identical structure is used together with the two -Ile-Glu-Gly-Arg- sequences to form two independent and quite large binding sites for the simultaneous binding of *two factor X_a molecules* to *one prothrombin molecule*. Such a mechanism is supported by the fact that in activation mixtures with phospholipids, mainly a mixture of thrombin, neoprothrombin-S, A fragment and S fragment is obtained, whereas very little neoprothrombin-T (formed as a result of a single factor X_a split at residue 274) and no "intact thrombin" (1–323 disulfide-bound to 324–582, resulting from a single factor X_a split at 323) has been observed. A situation where the two factor X_a-catalyzed activation splits occur simultaneously would also have obvious advantages for our postulated thrombin-dependent regulation mechanism. If we consider the alternative situation, i.e., allowing the bond 274–275 to be split first, this would result in neo-prothrombin-T being released from the phospholipid and thus not further being activated to thrombin. If the bond 323–324 were to be split first, we would end up having an active thrombin molecule that is phospholipid-bound and which therefore might cause the split at 156–157 (which "inactivates" the remaining prothrombin molecules) to occur before sufficient "plasma" thrombin had been formed. Thus it seems logical to conclude that the activation of prothrombin is caused by two molecules of factor X_a splitting the bonds at Arg-274 and Arg-323 simultaneously. This mechanism confers some restrictions on the tertiary structure of prothrombin. It is interesting to note in this context that the two insertions at the C-terminal end of the S fragment (relative to the A fragment) and at the C-terminal end of the thrombin A chain (relative to the chymotrypsin A chain) are exactly the same size (13 amino acid residues). The prothrombin-thrombin system is likely to be only the first example of systems involving "sophisticated" serine proteases used in extracellular regulation mechanisms.

Acknowledgments

We wish to thank Laila Brøns, Lene Kristensen, Lene Christensen and Margit Skriver for technical assistance. Financial support has been given by the Danish Science Research Council and the U.S. National Heart and Lung Institute (NIH, grant number 1 RO 1 HL 16238-01 HEM). Prothrombin and thrombin used in this investigation were prepared by S. M. at the Chemistry Department II, Karolinska Institutet, Stockholm, Sweden (1956–1965), at the New York State Department of Health, Division of Laboratories and Research, Albany, N.Y. (1965–1966), and at the M.R.C. Laboratory of Molecular Biology, Cambridge, England (1966–1970) with support from the Swedish and British Medical Research Councils and the U.S. National Heart and Lung Institute. L. S. and T. P. were recipients of fellowships from the Danish Science Research Council.

REFERENCES

Aronson, D. L. and D. Ménaché. 1966. Chromatographic analysis of the activation of human prothrombin with human thrombokinase. *Biochemistry* **5**:2635.

Asada, T., Y. Masaki, K. Kitahara, R. Nagayama, T. Hatashita and I. Yanagisawa. 1961. Conversion of prothrombin into thrombin. I. DEAE-cellulose chromatography of prothrombin and isolation of a prothrombin derivative. *J. Biochem.* **49**:721.

Birktoft, J. J. and D. M. Blow. 1972. Structure of crystalline α-chymotrypsin. V. The atomic structure of tosyl-α-chymotrypsin at 2 Å resolution. *J. Mol. Biol.* **68**:187.

Blombäck, B. and M. Blombäck. 1956. Purification of human and bovine fibrinogen. *Arkiv Kemi* **10**:415.

Dam, H. 1935. The antihaemorrhagic vitamin of the chick. Occurrence and chemical nature. *Nature* **135**:652.

Esmon, C. T., W. G. Owen and C. M. Jackson. 1974. The conversion of pro-thrombin to thrombin. II. Differentiation between thrombin and factor X_a-cata-lyzed proteolyses. *J. Biol. Chem.* **249**:606.

Fersht, A. R. 1971. Conformational equilibria and the salt bridge in chymotryp-sin. *Cold Spring Harbor Symp. Quant. Biol.* **36**:71.

Fujikawa, K., M. E. Legaz and E. W. Davie. 1972. Bovine factor X_1 (Stuart factor). Mechanism of activation by a protein from Russell's viper venom. *Biochemistry* **11**:4892.

Fujikawa, K., M. H. Coan, D. L. Enfield, K. Titani, L. H. Ericsson and E. W. Davie. 1974. A comparison of bovine prothrombin, factor IX (Christmas factor), and factor X (Stuart factor). *Proc. Nat. Acad. Sci.* **71**:427.

Ganrot, P. O. and J. E. Niléhn. 1968. Plasma prothrombin during treatment with dicoumarol. II. Demonstration of an abnormal prothrombin fraction. *Scand. J. Clin. Lab. Invest.* **22**:23.

Gitel, S. N., W. G. Owen, C. T. Esmon and C. M. Jackson. 1973. A polypeptide region of bovine prothrombin specific for binding to phospholipids. *Proc. Nat. Acad. Sci.* **70**:1344.

Josso, F., J. M. Lavergne, M. Gouault, O. Prou-Wartelle and J.-P. Soulier. 1968. Differents états moléculaires du facteur II (prothrombine). Leur étude à l'aide de la staphylocoagulase et d'anticorps anti-facteur II. *Thromb. Diath. Hae-morrh.* **20**:88.

Kisiel, W. and D. J. Hanahan. 1974. Proteolysis of human factor II by factor X_a in the presence of hirudin. *Biochem. Biophys. Res. Comm.* **59**:570.

Lechner, K. and E. Deutsch. 1965. Activation of factor X. *Thromb. Diath. Haemorrh.* **13**:314.

Magnusson, S. 1962. Fractionation of prothrombin preparations. *Thromb. Diath. Haemorrh.* (Suppl.) **7**:229.

————. 1965a. Preparation of highly purified bovine thrombin (E. C. 3.4.4. 13) and determination of its N-terminal amino acid residues. *Arkiv Kemi* **24**:349.

————. 1965b. Preparation and carbohydrate analysis of bovine prothrombin. *Arkiv Kemi* **23**:285.

————. 1965c. Fractionation of bovine prothrombin preparations by gradient chromatography on TEAE-cellulose columns. *Arkiv Kemi* **24**:217.

————. 1965d. Purification of prothrombin from human citrated plasma frac-tion II + III (Cohn's method 6). *Arkiv Kemi* **24**:367.

————. 1968. Homologies between thrombin and other serine proteinases. *Bio-chem. J.* **110**:25P.

————. 1970a. Bovine prothrombin and thrombin. In *Methods in Enzymology* (ed. S. P. Colowick and N. O. Kaplan), vol. 19, p. 157. Academic Press, New York.

————. 1970b. Structural aspects of thrombin and prothrombin. In *Structure-Function Relationships of Proteolytic Enzymes* (ed. P. Desnuelle, H. Neurath and M. Ottesen), p. 138. Munksgaard, Copenhagen.

————. 1971. Thrombin and prothrombin. In *The Enzymes* (ed. P. D. Boyer), 3rd ed., vol. 3, p. 277. Academic Press, New York.

————. 1972. On the primary structure of bovine thrombin. *Folia Haematol.* **98**:385.

————. 1973. Primary structure studies on thrombin and prothrombin. *Thromb. Diath. Haemorrh.* (Suppl.) **54**:31.

Magnusson, S. and G. Murano. 1974. Report of the task force on nomencla-ture of thrombin and thrombin-like enzymes, their peptide chains and zymogens thereof. In *Thrombosis: Pathogenesis and Clinical Trials* (ed. E. Deutsch et al.), p. 279. Schattauer, Stuttgart.

Magnusson, S., L. Sottrup-Jensen, T. E. Petersen and H. Claeys. 1975. The primary structure of prothrombin, the role of vitamin K in blood coagulation and a thrombin-catalyzed "negative feed-back" control mechanism for limiting the activation of prothrombin. In *Prothrombin and Related Coagulation Factors* (ed. H. C. Hemker and J. Veltkamp), p. 25. Leiden Universitaire Pers, Leiden, The Netherlands.

Magnusson, S., L. Sottrup-Jensen, T. E. Petersen, P. Klemmensen and E. Kouba. 1974a. Studies on the primary structure of prothrombin. *Thromb. Diath. Haemorrh.* (Suppl.) **57:**153.

Magnusson, S., L. Sottrup-Jensen, T. E. Petersen, H. R. Morris and A. Dell. 1974b. Primary structure of the vitamin K-dependent part of prothrombin *FEBS Letters* **44:**189.

Marshall, R. D. 1972. Glycoproteins. *Annu. Rev. Biochem.* **41:**673.

Nelsestuen, G. L. and J. W. Suttie. 1972. The carbohydrate of bovine prothrombin. *J. Biol. Chem.* **247:**6096.

―――. 1973. The mode of action of vitamin K. Isolation of a peptide containing the vitamin K-dependent portion of prothrombin. *Proc. Nat. Acad. Sci.* **70:** 3366.

Owen, W. G., C. T. Esmon and C. M. Jackson. 1974. The conversion of prothrombin to thrombin. I. Characterization of the reaction products formed during the activation of bovine prothrombin. *J. Biol. Chem.* **249:**594.

Papahadjopoulos, D., C. Hougie and D. J. Hanahan. 1964. Purification and properties of bovine factor V: A change of molecular size during blood coagulation. *Biochemistry* **3:**264.

Seegers, W. H., E. Marciniak, R. K. Kipfer and K. Yasunaga. 1967. Isolation and some properties of prethrombin and autoprothrombin III. *Arch. Biochem. Biophys.* **121:**372.

Segal, D. M., G. H. Cohen, D. R. Davies, J. C. Powers and P. E. Wilcox. 1971. The stereochemistry of substrate binding to chymotrypsin Aγ. *Cold Spring Harbor Symp. Quant. Biol.* **36:**85.

Shulman, N. R. and J. Z. Hearon. 1963. Kinetics of conversion of prothrombin to thrombin by biological activators. *J. Biol. Chem.* **238:**155.

Stenflo, J. 1974. Vitamin K and the biosynthesis of prothrombin. IV. *J. Biol. Chem.* **249:**5527.

Stenflo, J., P. Fernlund, W. Egan and P. Roepstorff. 1974. Vitamin K-dependent modifications of glutamic acid residues in prothrombin. *Proc. Nat. Acad. Sci.* **71:**2730.

Ware, A. G. and W. H. Seegers. 1949. Two-stage procedure for the quantitative determination of prothrombin concentration. *Amer. J. Clin. Pathol.* **19:**471.

The Vitamin K–dependent Incorporation of $H^{14}CO_3^-$ into Prothrombin

John W. Suttie, Charles T. Esmon and James A. Sadowski

Department of Biochemistry, University of Wisconsin
Madison, Wisconsin 53706

Vitamin K is required for the synthesis of four blood-clotting zymogens: prothrombin, factor X, factor IX and factor VII. The vitamin appears to function posttranslationally (Suttie 1973) by modifying a precursor protein which has recently been isolated from rat liver (Esmon, Grant and Suttie 1975). The precursor is inactive in prothrombin bioassay system, but is activated to thrombin by several snake venoms, suggesting that the vitamin K-dependent modification is involved in the physiological activation to thrombin rather than in the activity of the thrombin generated. It has been shown (Stenflo 1970) that a biologically inactive form of prothrombin appears in the plasma of bovine administered the vitamin K antagonist dicoumarol. Unlike prothrombin, this protein does not bind Ca^{++} ions (Nelsestuen and Suttie 1972; Stenflo and Ganrot 1973), and this defect is presumably responsible for its failure to activate in the bioassay. Stenflo et al. (1974) have postulated that the amino-terminal region of prothrombin contains several modified glutamic acid residues (γ-carboxyglutamic acid or 3-amino-1,1,3-propanetricarboxylic acid) which are not found in the abnormal prothrombin. These observations suggest that vitamin K is involved in the γ-carboxylation of some of the glutamic acid residues of the liver prothrombin precursor.

We have recently described an in vitro system (Shah and Suttie 1974) which converts the microsomal precursor protein to biologically active prothrombin in response to the addition of vitamin K. This system should serve to test the hypothesis that the vitamin K-dependent, posttranslational modification of the precursor involves the carboxylation of glutamic acid residues.

If this hypothesis is correct, at least four different criteria should be met when using this in vitro system:

1. There should be a rapid, vitamin K-dependent incorporation of $H^{14}CO_3^-$ into protein, even when de novo protein synthesis is blocked.
2. A significant amount of this radioactivity should be associated with the vitamin K-dependent clotting factors, primarily prothrombin.

151

3. The radioactivity incorporated into prothrombin should be located exclusively in the amino-terminal activation fragment (fragment 1) of prothrombin.
4. Following acid hydrolysis, the remaining radioactivity should be associated with glutamic acid residues.

METHODS AND RESULTS

To investigate the vitamin K-dependent, posttranslational incorporation of $H^{14}CO_3^-$ into protein, postmitochondrial supernatants prepared from the livers of severely vitamin K-deficient rats were incubated in the presence or absence of vitamin K_1 and in the presence of both vitamin K and an antagonist of the vitamin, chloro-K (Table 1). All incubations were carried out in the presence of cycloheximide. The data indicate that addition of vitamin K significantly increased the incorporation of radioactivity into a Triton X-100 extract of the microsomal fraction. Chloro-K inhibited both prothrombin synthesis and bicarbonate incorporation.

One unusual property of the four vitamin K-dependent blood-clotting pro-

Table 1

Incorporation of $H^{14}CO_3^-$ into Protein

	Triton extract			$BaSO_4$ eluate		
Treatment	Pro (U/ml)[a]	cpm/ml	cpm/mg[b]	Pro (U/ml)[a]	cpm/ml	cpm/mg[b]
No addition	0.8	68	4.5	0.4	<5	—[c]
Vitamin K	4.1	800	53.3	11.6	910	36,400
Vitamin K and chloro-K	1.5	297	19.8	1.4	191	—[c]

Livers from vitamin K-deficient rats were homogenized in 0.25 M sucrose. 0.025 M imidazole-HCl pH 7.2 (2 ml/g liver) and the homogenate centrifuged at 12,800g for 10 minutes to obtain a postmitochondrial supernatant which was then incubated under the conditions previously described (Shah and Suttie 1974) for the in vitro synthesis of prothrombin. Cycloheximide (100 µg/ml) and $H^{14}CO_3^-$ (5 µCi/ml) were included in the incubation medium, and prothrombin synthesis was initiated by the addition of vitamin K_1 (20 µg/ml). After incubation for 15 minutes at 37°C, the microsomes were removed by centrifugation at 105,000g for 60 minutes. This pellet was extracted with calcium-free Krebs-Ringer bicarbonate buffer containing 0.015 M potassium oxalate and 0.25% Triton X-100, and remaining debris was removed by centrifugation as described above. The Triton extract was adsorbed with $BaSO_4$ (25 mg/ml). The $BaSO_4$ was removed by centrifugation, and this pellet was washed and eluted as described earlier (Shah and Suttie 1972). Radioactivity incorporated into protein was based on the ^{14}C remaining following two repetitive precipitations with TCA.

[a] Prothrombin determined by two-stage assay.

[b] Protein was estimated assuming an $E_{1\,cm}^{1\%} = 10.0$.

[c] Insufficient protein in the eluate to determine a specific activity.

teins is their ability to bind to barium salts. Preliminary evidence that the radioactivity was incorporated into these proteins was obtained by $BaSO_4$ adsorption of the Triton extract. In the experiment shown in Table 1, 58% of the radioactivity was adsorbed with the $BaSO_4$, whereas less than 5% of the protein was adsorbed. In this experiment, no detectable radioactivity was associated with the $BaSO_4$ eluate when vitamin K was omitted from the reaction. In other experiments, radioactivity was associated with this fraction, but the ratio of ^{14}C in the +K eluate to ^{14}C in the −K eluate was always in excess of 25:1.

The high percentage of radioactivity associated with the barium eluate supports the conclusion that much of the $[^{14}H]CO_3^-$ was incorporated into the vitamin K-dependent clotting proteins. Additional evidence was obtained by ion exchange chromatography of the proteins in the barium eluate on QAE-Sephadex Q-50. The column was developed with a linear gradient of NH_4Cl, and 95% of the radioactivity was found to cochromatograph with the prothrombin activity.

Still further evidence of specificity was obtained by SDS-acrylamide gel electrophoresis of the radioactive barium eluate. Rat prothrombin has a molecular weight of 85,000 by SDS gel electrophoresis (Li and Olson 1967), and most of the radioactivity (Fig. 1) was associated with a product of molecular weight 75–85,000. More important, complete activation of the prothrombin with $[X_a,V,Phospholipid,Ca^{++}]$, Taipan snake venom or thrombin (data not shown) resulted in most of the radioactivity being transferred to a protein with a molecular weight of 20–30,000 daltons. This corresponds to the molecular weight of the amino-terminal activation peptide (fragment 1) of prothrombin, which has been shown (Nelsestuen and Suttie 1973; Stenflo 1974) to contain the vitamin-dependent modification and which is the only prothrombin activation product migrating in this region of the SDS gel. The ability of these specific proteases to generate a peptide of molecular weight

Figure 1

The proteins in the $BaSO_4$ eluate (see Table 1) were dialyzed against 0.05 M NaCl, 0.05 M imidazole-HCl pH 7.8. One sample was activated with $[X_a$, 0.2 μg/ml; V, 2.5 μg/ml; Phospholipid, 20 μg/ml; Ca^{++}, 10 mM]. A separate sample was activated with Taipan snake venom, 25 μg/ml in 10 mM Ca^{++}. The reactions were stopped immediately after complete prothrombin activation by making the mixture 1% in SDS. Electrophoresis was performed by the method of Laemmli (1970). The distribution of radioactivity was determined following combustion of the dried gel slices. The mobility of fragment 1 (F-1) and prothrombin (Pro) was determined with purified proteins run on separate gels.

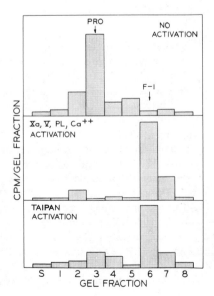

similar, if not identical, to fragment 1 strengthens the conclusion that the bicarbonate is incorporated into prothrombin, and specifically into the fragment-1 region.

The γ-carboxyglutamic acid residues are decarboxylated by strong acid, and the two γ-carboxyl groups should be equally labile to acid hydrolysis. Therefore, 50% of the radioactivity in the in vitro-synthesized prothrombin should be recovered in glutamic acid following acid hydrolysis, and the remainder should be released as $^{14}CO_2$. Acid hydrolysis of radioactive prothrombin resulted in the loss of approximately 50% of the radioactivity. Further, when this hydrolysate was applied to a standard amino acid analyzer column, all of the recovered radioactivity ($>99\%$) was in the glutamic acid peak.

CONCLUSIONS

The data presented here are consistent with the localization of the radioactivity in carboxyl groups of γ-carboxylglutamate residues in vitamin K-dependent coagulation factors. The in vitro system described should prove useful in elucidating the function of vitamin K in this carboxylation, as well as the mechanism of action of various vitamin K antagonists.

Acknowledgments

Previously unpublished research reported here was supported by the College of Agricultural and Life Sciences, University of Wisconsin, Madison, in part by a grant (AM-14881) from the National Institutes of Health, and in part by a National Institutes of Health training grant No. GM-00236 BCH. C. T. E. is a National Institutes of Health Postdoctoral Fellow (1-F22-HL00856).

REFERENCES

Esmon, C. T., G. A. Grant and J. W. Suttie. 1975. Purification of an apparent rat liver prothrombin precursor: Characterization and comparison to normal rat prothrombin. *Biochemistry* (in press).

Laemmli, U. K. 1970. Cleavage of structural proteins during the assembly of the head of bacteriophage T4. *Nature* 227:680.

Li, L. F. and R. E. Olson. 1967. Purification and properties of rat prothrombin. *J. Biol. Chem.* 242:5611.

Nelsestuen, G. L. and J. W. Suttie. 1972. Mode of action of vitamin K. Calcium binding properties of bovine prothrombin. *Biochemistry* 11:4961.

———. 1973. The mode of action of vitamin K. Isolation of a peptide containing the vitamin K-dependent portion of prothrombin. *Proc. Nat. Acad. Sci.* 70:3366.

Shah, D. V. and J. W. Suttie. 1972. The effect of vitamin K and warfarin on rat liver prothrombin concentrations. *Arch. Biochem. Biophys.* 150:91.

———. 1974. The vitamin K dependent, *in vitro* production of prothrombin. *Biochem. Biophys. Res. Comm.* 60:1397.

Stenflo, J. 1970. Dicoumarol-induced prothrombin in bovine plasma. *Acta Chem. Scand.* 24:3762.

———. 1974. Vitamin K and the biosynthesis of prothrombin. IV. Isolation of peptides containing prosthetic groups from normal prothrombin and the corresponding peptides from dicoumarol-induced prothrombin. *J. Biol. Chem.* **249:**5527.

Stenflo, J. and P. Ganrot. 1973. Binding of Ca^{2+} to normal and dicoumarol-induced prothrombin. *Biochem. Biophys. Res. Comm.* **50:**98.

Stenflo, J., P. Fernlund, W. Egan and P. Roepstorff. 1974. Vitamin K dependent modifications of glutamic acid residues in prothrombin. *Proc. Nat. Acad. Sci.* **71:**2730.

Suttie, J. W. 1973. Vitamin K and prothrombin synthesis. *Nutr. Rev.* **31:**105.

Blood Coagulation Factor XIII: Relationship of Some Biological Properties to Subunit Structure

J. E. Folk and Soo Il Chung

Laboratory of Biochemistry, National Institute of Dental Research
National Institutes of Health, Bethesda, Maryland 20014

Intermolecular polymerization of fibrin through formation of $\epsilon(\gamma\text{-glutamyl})$-lysine cross-links is considered to be the last of the many steps that constitute the "cascade" of blood coagulation in higher mammals. To date, it is the only reaction in hemostasis and thrombosis that is known to involve the formation of a covalent bond.

The subject has received much attention in the last few years as reflected in a number of review articles (Doolittle 1973a,b; Dukert and Beck 1968; Finlayson 1974; Folk and Chung 1973; Loewy 1970; Mandel 1971) covering areas on the nature of the cross-link, properties of the enzymes involved, characteristics of enzyme-catalyzed cross-linking, and the biological significance of the reaction.

Establishment of the enzymatic nature of the factor in plasma responsible for fibrin stabilization (Loewy et al. 1961b), together with the elucidation of the specificity (for reviews, see Finlayson 1974; Folk and Chung 1973) and catalytic mechanism (Chung and Folk 1972) of the factors from both human blood plasma and platelets, served to define them as members of a group of enzymes termed transglutaminases. Protransglutaminases, the inactive zymogen forms of these enzymes, exist in placenta (Bohn and Schwick 1971; Chung 1972), uterus (Chung 1972) and prostate (Chung, unpubl.), as well as in blood plasma and platelets. The transglutaminases catalyze a Ca^{++}-dependent acyl transfer reaction in which the γ-carboxamide groups of peptide-bound glutamine residues are acyl donors and the primary amino groups in a variety of compounds (e.g., the ϵ-amino group of peptide-bound lysine) may act as acyl acceptors. Ammonium ion is liberated during enzyme acylation. Deacylation to amine results in formation of monosubstituted γ-amides of peptide-bound glutamic acid (a γ-glutamyl-lysine cross-link in the case where the ϵ-amino group of lysine is the acyl acceptor). Water can act as an acyl acceptor at less than saturating levels of amine or in the absence of amine to give peptide-bound glutamic acid. These reactions may be denoted respectively as:

157

$$\underset{\text{-Glu}}{\overset{\displaystyle \ulcorner \text{NH}_2}{}} + \text{R-NH}_2 \rightleftharpoons \underset{\text{-Glu-}}{\overset{\displaystyle \ulcorner \overset{\displaystyle \text{H}}{\underset{\displaystyle |}{\text{N-R}}}}{}} + \text{NH}_3 \qquad (1)$$

$$\underset{\text{-Glu}}{\overset{\displaystyle \ulcorner \text{NH}_2}{}} + \text{HOH} \rightarrow \text{-Glu-} + \text{NH}_3. \qquad (2)$$

During the past several years, work in this laboratory has been directed toward defining the molecular and catalytic features of transglutaminases. The intent of this presentation is to review and extend present knowledge of coagulation factor XIII from human blood plasma and platelets and of the transglutaminases formed from these proenzymes. Special emphasis will be placed on the relationships of subunit structures to catalytic properties.

ZYMOGEN SUBUNIT STRUCTURES AND ACTIVATION

The protransglutaminase of plasma has a molecular weight of approximately 300,000 (Loewy et al. 1961b). This zymogen is a tetramer composed of two apparently identical catalytic subunits, termed *a* chains, of molecular weight about 75,000 and two identical noncatalytic subunits, *b* chains, of approximately 80,000 molecular weight (Chung 1972; Schwartz et al. 1972). The protransglutaminase of platelets is composed of two *a* chains only and has a molecular weight of 150,000 (Bohn and Schwick 1971; Schwartz et al. 1972). The *a* subunits of platelet zymogen are indistinguishable by numerous criteria from those of the plasma zymogen (Chung and Folk 1972; Chung, Lewis and Folk 1974; Schwartz et al. 1972). Among these criteria are immunochemical reactivity, amino acid composition, carbohydrate content, electrophoretic mobility, molecular weight and change in molecular weight upon treatment with thrombin.

During the course of thrombin-catalyzed conversion of each of these proenzymes to the active enzymes, a single peptide is released from the amino terminus of each of the *a* chains to form *a'* chains. Cleavage occurs at an arginylglycine bond (Tagaki and Doolittle 1974). The molecular weight of the *a'* chain is approximately 4000 less than that of the *a* chains, as estimated by polyacrylamide gel electrophoresis in sodium dodecyl sulfate. This is accounted for by the size of the activation peptide of 36 amino acid residues, the sequence of which has been determined (Takagi and Doolittle 1974). The proteolytic activation of the isolated zymogens does not require calcium ion (Tyler 1970). Ca^{++} is, however, essential for enzymatic activities of the transglutaminases (for review, see Folk and Chung 1973).

Effects of Ca^{++}

Enzyme Subunit Structures

Figure 1 shows the gel filtration patterns obtained on 6% agarose with the plasma zymogen and with the transglutaminase formed upon its activation by thrombin. When gel filtration was carried out in the absence of Ca^{++}, the protransglutaminase (Fig. 1A) and the enzyme (Fig. 1B) emerged from the column in almost identical positions. This is understandable since the loss of a 4000 molecular weight peptide from each *a* chain represents a relatively small

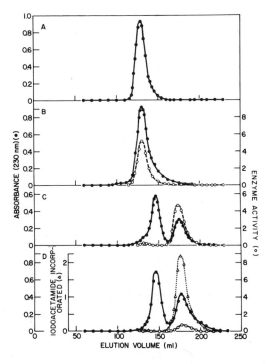

Figure 1

Gel filtration of the protransglutaminase (*A*) and transglutaminase (*B*, *C*, *D*) of human plasma on 6% agarose. In *A* and *B*, gel filtration was carried out without Ca^{++}; in *C* and *D*, $CaCl_2$ was present at 10 mM throughout; and in *D*, the enzyme was partially inactivated with [^{14}C]iodoacetamide.

change in the size of the tetramer. In the presence of 10 mM Ca^{++}, the position of the zymogen was unchanged from that in the absence of the metal ion (Fig. 1A). The enzyme, gel filtered with 10 mM Ca^{++}, however, showed two peaks (Fig. 1C), both of which emerged later than the zymogen or the enzyme without Ca^{++}. Only the last peak showed enzymatic activity. Reversibility of this Ca^{++} effect was evidenced by the finding that the enzyme to which $CaCl_2$ was added at levels of 10–100 mM gave a single peak similar to that in Figure 1B when gel-filtered on a column without Ca^{++}. Plasma transglutaminase was partially inactivated with [^{14}C]iodoacetamide in the presence of Ca^{++}. Figure 1D shows the pattern obtained upon gel filtration of this labeled material in the presence of the metal ion. Radioactivity appeared only in the region corresponding to the catalytically active peak. Analysis of each of the peaks from gel filtration of the enzyme with Ca^{++} (Fig. 1C) by the use of SDS-polyacrylamide gel electrophoresis showed that the first peak to emerge contained only *b* chains, whereas the second peak contained only *a'* chains.

Platelet protransglutaminase and the enzyme formed upon activation of this zymogen with thrombin were gel-filtered in the presence and absence of $CaCl_2$. Figure 2 A and B show a single peak obtained for the proenzyme without Ca^{++} and for the enzyme with Ca^{++}, respectively. The zymogen with Ca^{++} and the enzyme without Ca^{++} occupied this same position on gel filtration. The enzyme was labeled by inactivation with [^{14}C]iodoacetamide. Labeled material also appeared in the same area upon gel filtration (Fig. 2C). Comparison with the gel filtration pattern obtained for plasma transglutaminase in the presence of Ca^{++} (Fig. 2D) showed that the platelet zymogen and enzyme, with and without Ca^{++}, appeared in a position identical to that of the catalytically active *a'* chain component of plasma enzyme.

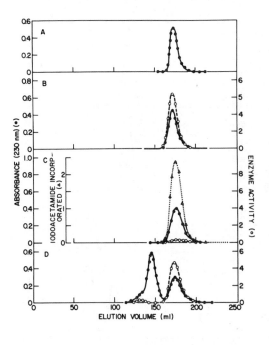

Figure 2
Gel filtration of the protransglutaminase (*A*) and transglutaminase (*B, C*) of human platelets on 6% agarose. In *A*, gel filtration was carried out without Ca++; in *B* and *C*, CaCl₂ was present at 10 mM throughout; and in *C*, the enzyme was inactivated with [¹⁴C]-iodoacetamide. *D* is a reproduction of Figure 1C presented for comparative purposes.

As pointed out above, platelet protransglutaminase (molecular weight 150,000) is composed of two 75,000 molecular weight *a* subunits. Furthermore, the *a* chains of plasma and platelet proenzymes appear to be identical. On the basis of this information and the gel filtration results, we are led to the following tentative conclusions: (1) platelet transglutaminase is a dimer,

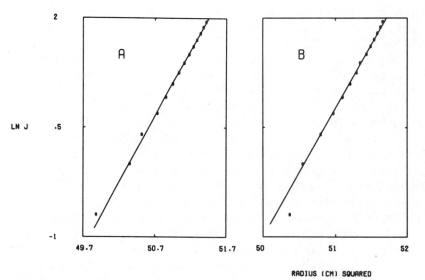

Figure 3
Sedimentation equilibrium of the protransglutaminase (*A*) and transglutaminase (*B, C*) of plasma. In *A* and *B*, centrifugation was carried out without CaCl₂.

a'_2; (2) Ca^{++} does not dissociate this dimer; (3) plasma transglutaminase is a tetramer, a'_2b_2, in the absence of Ca^{++}; and (4) in the presence of Ca^{++}, the a' and b subunits dissociate from one another to give an a' dimer and a b chain component.

Evidence that this b chain component is also a dimer, b_2, and that, indeed, the above conclusions are correct is the following. Ultracentrifugal analyses were carried out on the plasma and platelet zymogens and enzymes. Figure 3 shows some of these results. For plasma proenzyme (Fig. 3A), identical results and a molecular weight of 300,000 were obtained with and without Ca^{++}. Figure 3B is the pattern for the thrombin-activated enzyme in the absence of Ca^{++}. A small difference in molecular weight derived from these data, approximately 10,000 less than with the zymogen, is accounted for by the release of activation peptide from a chains. The curvature of the plot for the enzyme in 10 mM $CaCl_2$ (Fig. 3C) shows marked heterogeneity with respect to molecular weight. The molecular weight estimated by the value of the limiting slope at the lowest concentration was 170,000. This value is only a reasonable approximation for the major components. However, there was no indication of any material of a molecular weight less than 150,000. This is strong evidence that both a' and b components of plasma transglutaminase exist as dimers in the presence of Ca^{++}. Ultracentrifugal analyses of the platelet zymogen and enzyme, either in the presence or absence of Ca^{++}, revealed molecular weights in the 150,000–170,000 range, evidence for dimeric structure in each case.

The results of chemical cross-linking with dimethyl suberimidate (Fig. 4) are consistent with these conclusions. The SDS-polyacrylamide gel electrophoresis was carried out under conditions that do not separate the a or a'

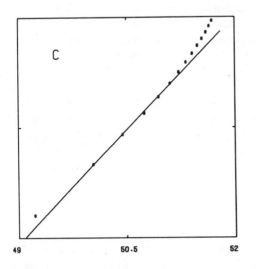

Figure 3 (*continued*)
In *C*, $CaCl_2$ was present at 10 mM. The figure is a reproduction of a computer program graphic display.

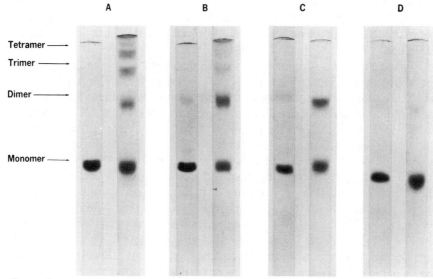

Figure 4
SDS gel electrophoretic patterns of dimethyl suberimidate cross-linked: (*A*) plasma protransglutaminase, (*B*) plasma transglutaminase, (*C*) *b*-chain component of plasma enzyme isolated by gel filtration and (*D*) guinea pig liver transglutaminase. Cross-linking in *A* and *D* was carried out in the absence of $CaCl_2$ and in *B* and *C,* in the presence of 10 mM $CaCl_2$. In each case, the gel on the left is a control incubated without dimethyl suberimidate.

chains from *b* chains. Thus the pattern for plasma proenzyme (Fig. 4A) is typical of a protein molecule composed of four identical subunits. The thrombin-activated plasma enzyme cross-linked in the absence of Ca^{++} showed a similar pattern, again indicative of a four-subunit structure. When the active enzyme was treated with suberimidate in the presence of 10 mM $CaCl_2$, however, the resulting pattern (Fig. 4B) was very different. In this case, the two major bands, indicative of dimer structure, probably arise from the two dimers, a'_2 and b_2, of similar molecular weight. Figure 4C shows a two-band pattern for the *b*-chain material isolated from plasma enzyme by gel filtration in $CaCl_2$. This is in agreement with a dimer structure for this component. Guinea pig liver transglutaminase exists as a monomer (Connellan et al. 1971). The single band obtained with this monomer as a control is as expected (Fig. 4D).

Conformational Changes

The dissociation of the tetrameric plasma enzyme into a'-chain dimers and *b*-chain dimers is not the only effect of Ca^{++}. This metal ion also induces a conformational alteration in the a' chains. Whether this change causes dissociation of a' dimers from *b* dimers is not known at present. Evidence for this conformational change is summarized in Table 1. There are no -SH groups available to titration with 5,5'-dithiobis (2-nitrobenzoic acid) (DTNB) in the plasma or platelet proenzymes, either in the presence or absence of Ca^{++}.

Table 1

Reactivity toward DTNB and Total -SH and -S-S- Contents of Zymogens

Time (min)	-SH reactive with DTNB (groups/molecule)			
	plasma		platelet	
	zymogen ± Ca++ or enzyme − Ca++	enzyme + Ca++	zymogen ± Ca++ or enzyme − Ca++	enzyme + Ca++
2	0	1.6	0	1.5
10	0	2.8	0	2.3
30	0	3.9	0	4.0
90	0	5.8	0	6.0

Zymogen	Total -SH (groups/molecule)	Total -S-S- (groups/molecule)
Plasma		
in a chains	12	32–34
in b chains	12	0
	0	32–34
Platelet		
in a chains	12	0

Further, neither thrombin-activated enzyme is reactive toward DTNB in the absence of Ca++. However, in the presence of 10 mM CaCl$_2$, the active enzymes of both plasma and platelets react with this -SH reagent, and the rates at which they react are identical. It is sulfhydryl groups of the a′ chains of the plasma enzyme that react with DTNB. This follows from the finding that the b subunits of this zymogen contain no -SH groups (Table 1). It is apparent that Ca++ induces a change in the a′ chains of platelet enzymes that results in exposure of -SH groups to reaction with DTNB. Since the a and a′ chains of plasma and platelet zymogens and enzymes, respectively, are indistinguishable by numerous physical, chemical and enzymatic criteria, it seems unlikely that the -SH groups of plasma enzyme a′ subunits are simply masked by b subunits in the tetramer form of the enzyme, i.e., in the absence of Ca++. Rather, one is led to conclude that these a′ chains per se undergo a Ca++-induced alteration that, as in the case of the platelet enzyme, exposes -SH groups.

Findings with the alkylating agents iodoacetamide (Fig. 5) and α-bromo-4-hydroxy-3-nitroacetophenone (Fig. 6) support these conclusions. These agents, which cause enzyme inactivation by reaction with an essential -SH group in an a′ chain of each enzyme, were found to react only with the active enzymes and only in the presence of Ca++. The results summarized in Figure 5 show that total inactivation of both the plasma and platelet enzymes results from incorporation of 1 mole of [14C]iodoacetamide per molecule of enzyme, i.e., 1 mole of agent per 2 a′ chains. Similarly, reaction of 1 mole of α-bromo-4-hydroxy-4-nitroacetophenone per mole of plasma enzyme, or 1 mole per 2 a′ chains, results in total inactivation (Fig. 6). A number of enzymes have been shown to exhibit an "half-of-the-sites reactivity" (for re-

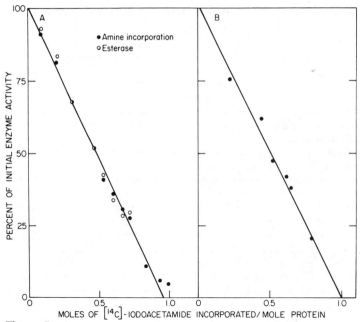

Figure 5

Inactivation of plasma transglutaminase (*A*) and platelet transglutaminase (*B*) by [^{14}C]iodoacetamide. Reactions with iodoacetamide were carried out in the presence of $CaCl_2$.

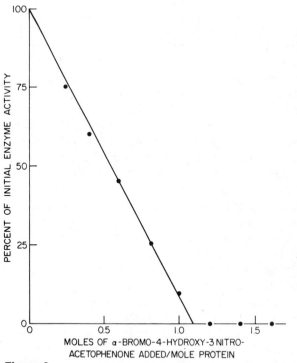

Figure 6

Inactivation of plasma transglutaminase by α-bromo-4-hydroxy-3-nitroacetophenone. Reaction with reagent was carried out in the presence of $CaCl_2$.

164

view, see Levitzki, Stallcup and Koshland 1971) in which only half of the apparently identical subunits in a polymeric protein are expressed in a reaction with a given substrate or inhibitor. The total inactivation of each of these transglutaminases by reaction of alkylating reagent with an -SH group in only one of their two apparently identical catalytic subunits tentatively defines these enzymes as ones of this type (Chung and Folk 1972; Chung, Lewis and Folk 1974).

The sequence of amino acids surrounding the essential cysteine, with which iodoacetamide reacts in the plasma enzyme, has recently been found to be Gly-Gln-Cys(SH)-Trp (Holbrook, Cooke and Kingston 1973). This sequence is identical to a portion of that surrounding the active site cysteine in guinea pig liver transglutaminase (Folk and Cole 1966).

Role of *b* Subunits of Plasma Zymogen

There has been some speculation regarding the role of *b* subunits in plasma protransglutaminase. To date, there is no experimental evidence to support the suggestions that *b* chains may aid in secretion of zymogen from its site of synthesis (Schwartz et al. 1971) or that *b* chains may function to stabilize *a* chains in vivo (Bohn and Schwick 1971; Schwartz et al. 1971). It is true, however, that plasma zymogen $(a_2 b_2)$ is significantly more stable to in vitro manipulations than is the platelet proenzyme (a_2) (Chung, Lewis and Folk 1974; Schwartz et al. 1972).

Kinetic and specificity comparisons of the plasma and platelet transglutaminases have shown that the catalytic properties of these enzymes are identical (Chung and Folk 1972). Further, addition of *b*-chain dimer (isolated from plasma enzyme by gel filtration with Ca^{++}) to platelet enzyme does not affect its enzymatic properties (Chung, Lewis and Folk 1974). The findings presented in Figure 7 show that *b* subunits do have a distinct influence on the rate of zymogen activation by thrombin and suggest a physiological control function for these subunits. The upper curve is that for thrombin activation of platelet protransglutaminase. The open circles on the lower curve are for activation of plasma zymogen under the same experimental conditions. Clearly, activation of the *b*-chain-containing tetrameric plasma proenzyme is slower. When isolated *b*-chain dimer is added to platelet proenzyme (a_2), a tetrameric zymogen $(a_2 b_2)$ indistinguishable from that of plasma is obtained (Chung, Lewis and Folk 1974; Schwartz et al. 1971). Activation of this zymogen by thrombin proceeds at a rate identical to that for the plasma zymogen (open triangles, Fig. 7). The same results were obtained both in the presence and in the absence of Ca^{++}. Thus, noncovalent association of *b* subunits reduces the rate of thrombin-catalyzed conversion of *a* chains to *a′* chains. When the zymogen prepared from isolated *b*-chain dimer and platelet zymogen was activated with thrombin and gel filtered without Ca^{++}, it emerged in a position identical to that of the plasma enzyme without Ca^{++}, evidence for a tetrameric enzyme structure. This, together with the finding that addition of isolated *b*-chain dimer to platelet enzyme in the absence of Ca^{++} yielded the same tetrameric structure, is further strong support for identity of the *a* subunits of plasma and platelet zymogen (Chung, Lewis and Folk 1974).

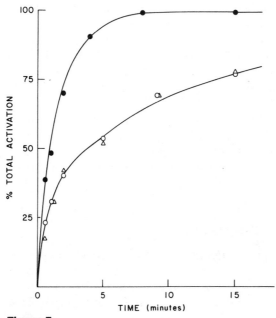

Figure 7
Thrombin activation of protransglutaminases. The
closed circles designate platelet zymogen; the open
circles, plasma zymogen; the open triangles, a zymogen prepared by combining isolated *b*-chain dimer
with platelet zymogen followed by gel filtration. The
initial concentrations of the zymogens were adjusted
so that each activation solution was the same concentration in *a* chains. The thrombin concentration
was the same with each zymogen.

Binding of *a′* Chains of Plasma Transglutaminase to Fibrin

The present in vitro observations on enzyme subunit dissociation and enzyme
and zymogen subunit recombinations serve to focus our attention on the possible states of these molecules in plasma and serum. Difficulties encountered
in separation of plasma protransglutaminase from fibrinogen during early purifications of zymogen led to an assumption that these two proteins were bound
in a complex (Loewy et al. 1961a). Recent careful studies with proenzyme-
rich fibrinogen preparations and with purified zymogen and purified fibrinogen
show that this is not the case (Bohn 1971; Chung 1972; Chung, Finlayson
and Folk 1971). Indeed, there is no evidence of binding.

During the course of clot formation in whole blood or in plasma, transglutaminase activity largely disappears (Dvilansky, Britten and Loewy 1970).
It has been suggested that the enzyme is removed from serum through association with the fibrin clot (Israels, Paraskevase and Israels 1973; Lorand
1950). However, noncatalytic *b* chains have been identified in serum by the
use of an *antiserum* to this material (Bohn 1971).

We have employed two procedures to study the binding of plasma transglutaminase to fibrin. In one of these (Fig. 8), a column was prepared with

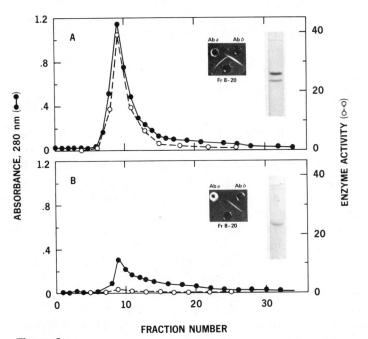

Figure 8

Elution profiles of the protransglutaminase (*A*) and transglutaminase (*B*) of plasma from a fibrin column in the presence of Ca^{++}. The inserts in *A* and *B* represent immunodiffusion and SDS-polyacrylamide gel electrophoretic patterns, respectively, for combined fractions 8 through 20.

fibrin formed from 95% clottable human fibrinogen. After washing well in order to remove thrombin and unclotted material, the column was equilibrated with a Ca^{++}-containing buffer and a sample of plasma protransglutaminase or plasma enzyme was applied. The zymogen passed directly through the column (Fig. 8A). Enzymatic activity was measured after thrombin activation and corresponded exactly with the protein peak. Immunochemical and SDS-polyacrylamide gel electrophoretic analyses (inserts in Fig. 8A) showed evidence of equal amounts of *a* and *b* chains in the pooled peak. Under the experimental conditions employed, the plasma proenzyme appeared to be largely unbound to fibrin and remained in an undissociated form. When thrombin-activated plasma enzyme (Fig. 8B) was applied to the fibrin column, a trailing peak of protein, identified as only *b* chain (inserts in Fig. 8B), was eluted. Negligible enzymatic activity was found in the eluate. In a second procedure, plasma zymogen was added to a solution of fibrinogen in the presence of Ca^{++} and a column of fibrin clot was formed by addition of thrombin. The *a'* chains of the enzyme, formed through thrombin action, remained bound to the fibrin even after extensive washing, whereas a large percentage of the *b* chain was eluted. These findings support the suggestion that the disappearance of transglutaminase activity during clot formation is a result of tight association of catalytic *a'* chains with fibrin clot. Whether this is a result of a covalent or a noncovalent reaction is not known at present.

It seems reasonable to assume that this reaction is the mechanism by which enzyme activity is localized at coagulation sites and by which general intravascular covalent cross-linking is averted.

SUMMARY

Figure 9 summarizes present knowledge of the subunit compositions and interactions of the proenzymes and enzymes from human blood plasma and platelets. Plasma protransglutaminase, a_2b_2 (upper left), upon activation by thrombin in the absence of Ca^{++}, forms a tetrameric enzyme, a'_2b_2, through release of an activation peptide from each of its a subunits. Ca^{++} induces both a conformational change in the a' dimer (as denoted by the double border) and dissociation of tetrameric enzyme into a' dimer and b dimer. The a' chains bind to fibrin; b chains do not. Catalytically active a'_2 may be inactivated by alkylating agents through reaction of an -SH group in one of its subunits, thus in an half-of-the-sites fashion. Platelet protransglutaminase, a_2 (lower left), is activated by thrombin, also through release of a peptide from each of its a subunits. This occurs at a significantly faster rate than with plasma zymogen or with tetrameric zymogen formed by association of platelet zymogen with isolated b-chain dimer. Platelet enzyme, a'_2, will combine with isolated b_2 in the absence of Ca^{++} to form a tetrameric enzyme, a'_2b_2, that is indistinguishable from the plasma enzyme. Platelet enzyme undergoes a Ca^{++}-induced conformational change to give a species with enzymatic properties identical to those of the plasma enzyme.

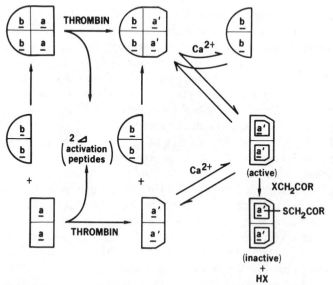

Figure 9

Schematic representation of plasma and platelet protransglutaminase and transglutaminase subunit compositions and interactions.

REFERENCES

Bohn, H. 1971. Immunochemical studies on the fibrin stabilizing factors from human plasma and platelets. *Blut* **22**:237.

Bohn, H. and H. G. Schwick. 1971. Isolierung und Charakterisierung ein fibrin-stabilisierenden Factors aus menschlichen Plazenten. *Arzneimittel-Forschung* **21**:1432.

Bohn, H., H. Haupt and T. Kranz. 1972. Die molekulare Struktur der fibrin-stabilisierenden Factoren des Menschen. *Blut* **25**:235.

Chung, S. I. 1972. Comparative studies on tissue transglutaminase and factor XIII. *Ann. N.Y. Acad. Sci.* **202**:240.

Chung, S. I. and J. E. Folk. 1972. Kinetic studies with transglutaminases. The human blood enzymes (activated coagulation factor XIII) and the guinea pig hair follicle enzyme. *J. Biol. Chem.* **247**:2798.

Chung, S. I., J. S. Finlayson and J. E. Folk. 1971. Tissue transglutaminase and factor XIII. *Fed. Proc.* **30**:1075.

Chung, S. I., M. S. Lewis and J. E. Folk. 1974. Relationship of the catalytic properties of human plasma and platelet transglutaminases (activated blood coagulation factor XIII) to their subunit structures. *J. Biol. Chem.* **249**:940.

Connellan, J. M., S. I. Chung, N. K. Whetzel, L. M. Bradley and J. E. Folk. 1971. Structural properties of guinea pig liver transglutaminase. *J. Biol. Chem.* **246**:1093.

Doolittle, R. F. 1973a. Structural aspects of the fibrinogen to fibrin conversion. *Adv. Prob. Chem.* **27**:1.

———. 1973b. Structural details of fibrin stabilization: Implication for fibrinogen structure and initial fibrin formation. *Thromb. Diath. Haemorrh.* (Suppl.) **54**:83.

Duckert, F. and E. A. Beck. 1968. Clinical disorders due to the deficiency of factor XIII (fibrin-stabilizing factor, fibrinase). *Semin. Hematol.* **5**:83.

Dvilansky, A., A. F. H. Britten and A. G. Loewy. 1970. Factor XIII assay by an isotope method. I. Factor XIII (transamidase) in plasma, serum, leucocytes, erythrocytes and platelets and evaluation of screening tests of clot solubility. *Brit. J. Haematol.* **18**:399.

Folk, J. E. and S. I. Chung. 1973. Molecular and catalytic properties of transglutaminases. *Adv. Enzymol.* **38**:109.

Folk, J. E. and P. W. Cole. 1966. Identification of a functional cysteine essential for the activity of guinea pig liver transglutaminase. *J. Biol. Chem.* **241**:3238.

Finlayson, J. S. 1974. Crosslinking of fibrin. *Semin. Thromb. Hemostas.* **1**:33.

Holbrook, J. J., R. D. Cooke and I. B. Kingston. 1973. The amino acid sequence around the reactive cysteine residue in human plasma factor XIII. *Biochem. J.* **135**:901.

Israels, E. D., F. Paraskevase and L. G. Israels. 1973. Immunochemical studies of coagulation factor XIII. *J. Clin. Invest.* **52**:2398.

Levitzki, A., W. B. Stallcup and D. E. Koshland, Jr. 1971. Half-of-the-sites reactivity and the conformational states of cytidine triphosphate synthetase. *Biochemistry* **10**:3371.

Loewy, A. G. 1970. Mechanism of fibrin crosslinkage. Some historical remarks of uncertain objectivity. *Thromb. Diath. Haemorrh.* (Suppl.) **39**:103.

Loewy, A. G., A. Dahlberg, K. Dunathan, R. Kriel and H. L. Wolfinger, Jr. 1961a. Fibrinase II: Some physical properties. *J. Biol. Chem.* **236**:2634.

Loewy, A. G., K. Dunathan, R. Kriel and H. L. Wolfinger, Jr. 1961b. Fibrinase I: Purification of substrate and enzyme. *J. Biol. Chem.* **236**:2634.

Lorand, L. 1950. Fibrin clots. *Nature* **166**:694.

Mandel, E. E. 1971. The fibrin-stabilizing factor. A decade of progress. *Amer. Clin. Lab. Sci.* **1**:92.

Schwartz, M. L., S. V. Pizzo, R. L. Hill and P. A. McKee. 1971. The subunit structures of human plasma and platelet factor XIII (fibrin stabilizing factor). *J. Biol. Chem.* **246**:5857.

———. 1972. Human factor XIII from plasma and platelets. Molecular weights, subunit structures, proteolytic activation and crosslinking of fibrinogen and fibrin. *J. Biol. Chem.* **248**:1395.

Takagi, T. and R. F. Doolittle. 1974. Amino acid sequence studies on factor XIII and the peptide released during its activation by thrombin. *Biochemistry* **13**:750.

Tyler, H. M. 1970. Studies on the activation of purified human factor XIII. *Biochim. Biophys. Acta* **222**:396.

Initiation and Control of the Extrinsic Pathway of Blood Coagulation

**Jolyon Jesty, James R. Maynard, Robert D. Radcliffe,
Sidonie A. Silverberg, Frances A. Pitlick and Yale Nemerson**

Departments of Internal Medicine and Molecular Biophysics and Biochemistry
Yale University School of Medicine, New Haven, Connecticut 06510

In healthy animals, blood does not clot except in response to damage to a blood vessel. However when blood is collected into glass tubes, it clots in 5 to 10 minutes. This activation of the clotting mechanism by glass, or physiological substances such as collagen and elastin, occurs via the intrinsic pathway of coagulation. This pathway involves the activation of factor XII (Hageman factor), factor XI and factor IX. Activated factor IX, in the presence of factor VIII (antihemophilic factor), then activates factor X to factor X_a. In contrast, on addition of tissue extracts, blood clots in about 12 seconds, owing to the activation of the extrinsic, or tissue factor, pathway of coagulation. This involves the activation of factor X by factor VII, in the form of a complex with tissue factor, a lipoprotein present in many tissues (Nemerson and Esnouf 1973). Thus factor X can be activated by either pathway. Factor X_a then activates prothrombin to thrombin in the presence of factor V, a plasma cofactor. The activations of both factor X and prothrombin are proteolytic conversions of a zymogen to an active proteolytic enzyme (Fig. 1).

All the components of the extrinsic pathway are present in plasma except the initiator, i.e., tissue factor; it is tissue factor that is responsible for the shortening of the clotting time from "infinity" to 12 seconds. The appearance of tissue factor may be regarded as the initial point of control in that its appearance must be a function of the extent of injury. However if the sequence of reactions that comprise the extrinsic pathway were to proceed unchecked, a maximum amount of thrombin would be produced regardless of the size of the initial stimulus. In the past it has been assumed that the major means of control of the pathway is through protease inhibitors present in the plasma, a major one being antithrombin III (heparin cofactor) (Macfarlane 1972). This inhibitor irreversibly inactivates factor X_a and thrombin, but if heparin is absent, inactivation is measured in minutes rather than seconds (Yin, Wessler and Stoll 1971; Rosenberg and Damus 1973). As there are no data on plasma heparin levels, it is not clear how important this control would be in regulating the pathway in vivo.

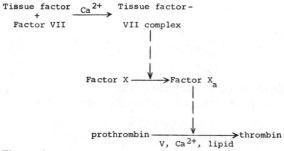

Figure 1
The extrinsic pathway of coagulation.

Mertz, Seegers and Smith (1939a) showed that in the absence of plasma inhibitors, the activation of prothrombin via the extrinsic pathway is subject to an inherent control. Their important results (Fig. 2) show that in activations by tissue extracts of prothrombin preparations containing factors VII and X, the yield of thrombin is related to the amount of initiator used. Using highly purified systems, we have studied the reactions of the pathway separately and have confirmed that in the activations of factor X and prothrombin by the physiological activators, the reactions do not proceed to completion, the yield of the products being a function of the activator concentrations.

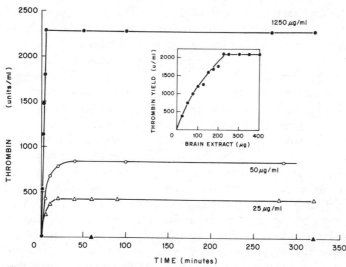

Figure 2
The yield of thrombin by the extrinsic pathway. Partially purified prothrombin (2300 U/ml) was activated with 0, 25, 50, and 1250 μg bovine lung thromboplastin in the presence of calcium ions, and thrombin was measured as a function of time. The inset shows the relationship between the yield of thrombin and the amount of thromboplastin added. (After the results of Mertz, Seegers and Smith 1939a.)

Furthermore, we have gone some way towards elucidating the mechanisms of these controls and have shown that they are intrinsic to the proteolytic reactions involved in each activation.

On the addition of tissue factor to factor VII, a complex is formed in the presence of calcium ions, the activity of the complex depending on the concentrations of both tissue factor and factor VII. The enzymatic activity of the complex, which can be inhibited by diisopropylfluorophosphate (DFP) (Nemerson and Esnouf 1973), is in factor VII; tissue factor appears to function as an essential cofactor. In plasma, the concentration of factor VII is constant (at about 1 μg/ml), so that the activity of the complex will be determined by the concentration of tissue factor. In vivo, tissue factor could be locally in excess of factor VII, but we imagine that if tissue factor is localized on the walls of damaged vessels, it will tend to recruit factor VII from the flowing plasma (Nemerson and Pitlick 1972). Therefore, we suggest that whatever the extent of appearance of tissue factor, the activity of the complex will be a function of this and not of the factor VII concentration.

METHODS

Clotting factors and tissue factor were purified and assayed as previously described (Esnouf, Lloyd and Jesty 1973; Radcliffe and Nemerson 1974), except factor V, which was partially purified by a modification of the method of Dombrose et al. (1972).

WISH amnion cells (American Type Culture Collection) were grown in Eagle's minimum essential medium (supplemented with 10% fetal calf serum, glutamine, nonessential amino acids, streptomycin and penicillin) in flat bottles or roller bottles or on plastic dishes (Falcon). Twenty hours before an experiment, monolayers were treated with trypsin and transferred to the same medium with serum that had been adsorbed with calcium phosphate to remove the vitamin K-dependent clotting factors. Cells were also cultured in suspension in minimum essential medium, supplemented with spinner salts, and dextran sulfate to a concentration of 10 μg/ml (Bremerskov 1973).

Cells grown in monolayers were routinely lifted with trypsin (2.5 mg/ml) for 1 minute; the treatment was stopped by the addition of equimolar amounts of soybean trypsin inhibitor (STI). The cells were then centrifuged (600g, 5 min), washed with saline A (20 mM HEPES buffer pH 7.4), and finally resuspended in this buffered saline at approximately 10^6 cells/ml. Cell viability was measured by trypan-blue exclusion.

RESULTS

Appearance of Tissue Factor Activity in Cultured Cells

Several lines of evidence suggest that tissue factor is a membrane-bound protein. First, detergent is required to extract the protein from tissues, and once extracted and purified, the protein must be recombined with certain types of phospholipid for full activity to be restored (Nemerson 1968). Second, antibodies have been raised against the apoprotein and coupled to horseradish peroxidase to permit histochemical localization. The antibody neutralizes the coagulant activity of tissue factor and precipitates the immunizing protein. When these antibodies are incubated with sections from many tissues, the plasma membranes of endothelial cells appear to be the most heavily stained (Zeldis et al. 1972; Stemerman and Pitlick 1974).

This observation raises the question of how tissue factor is protected from the circulation if it is present at the plasma membrane of endothelial cells. Ashford and Freiman (1968) reported experiments in which the femoral vein of a live rat was gently separated from the surrounding tissue. They found that this manipulation resulted in no change in the morphology of the endothelial cells, but it did provoke the appearance of fibrin strands near the cells. Thus coagulation can be triggered by the most minimal cell damage. Other workers have shown that cultured mammalian cells express tissue factor activity on complete disruption (Green et al. 1971; Zacharski, Hoyer and McIntyre 1973), but there have been no studies on the minimal trauma required to elicit activity, nor on correlating activity with the extent of damage.

It has been our aim to study these two areas. We decided to use cultured cells as a model system and to first determine whether in the undisturbed state they express tissue factor activity. We have shown in two ways that the initial activity is indeed very low. First, the other components of the extrinsic pathway were added to a monolayer of cells and the release of thrombin into the medium was measured. When undisturbed and disrupted cells were compared, it was found that disruption causes at least a tenfold increase in the rate of thrombin formation in the system, showing that undisturbed cells have low tissue factor activity. Second, this result was confirmed by growing (amnion) cells in suspension. This enabled us to measure by direct assay their activity in the undisturbed state. Compared to disrupted cells, untreated cells show less than 5% of the tissue factor activity. Thus it was demonstrated that tissue factor exists normally in a protected state in cultured cells.

Next, we used trypsin treatment as a means of controlled trauma to study the localization of tissue factor. Table 1 gives the results of an experiment where already suspended cells were treated further with more dilute trypsin (25 μg/ml) and the appearance of tissue factor activity and the release of cellular enzymes into the medium were measured. During the course of the

Table 1

The Appearance of Tissue Factor and Cellular Enzymes on Trypsin Treatment

Min	Tissue factor	5'-Nucle-otidase	Alkaline phosphatase	G-6-P dehydrogenase
0	15	1.2	0.5	2.5
5	36	17	6.0	3.0
10	40	23	7.5	3.0
15	43	21	10.5	5.2

Fibroblasts already suspended were treated with trypsin (25 μg/ml), and the activities of tissue factor and cellular enzymes were measured in the medium after centrifugation of the cells. All results are expressed as a percentage of the respective levels obtained with cells disrupted by freezing and thawing.

trypsin treatment, the viability remained at more than 95%; this is confirmed by the fact that the levels of intracellular enzymes, such as glucose-6-phosphate dehydrogenase, do not rise. Concomitant with the appearance of tissue factor activity, however, is a less marked rise in the levels of two membrane-bound enzymes, 5'-nucleotidase and alkaline phosphatase. All the activities are expressed as a percentage of the maximum obtained by cellular disruption. These results clearly show that tissue factor can be expressed without any significant release of intracellular material, but they also suggest that trypsin treatment affects the cell membrane.

The tissue factor activity that appears on trypsin treatment is released into the medium and can thus be separated from the cells by low-speed centrifugation. It can also be centrifuged down at high speed (10^5g for 1 hr) and then examined by electron microscopy (Maynard et al. 1975). Using this technique, it has been shown that on trypsin treatment, a large amount of material that stains positively with ruthenium red is released from the cell surface. The fact that trypsin releases material without affecting cell viability is in accord with the results of Glick and Buck (1973). While the pellet obtained at high speed, which contains tissue factor activity, consists largely of fine fibrils that stain intensely with ruthenium red, it also contains vesicular structures. The presence of the latter is in agreement with the release of membrane-bound enzymes from the cells, and remembering that tissue factor is a lipoprotein, it is likely that tissue factor is also associated with these structures.

Activation and Inactivation of Factor VII

Tissue factor forms a complex with factor VII in the presence of calcium ions, and as mentioned earlier, the activity of this complex may be largely dependent on the concentration of tissue factor.

Factor VII occurs in bovine plasma as a single-chain protein and can be isolated in this form by including 25 mM benzamidine hydrochloride—a potent inhibitor of trypsinlike enzymes—in all the buffers used during the purification (Radcliffe and Nemerson 1975). Since the single-chain material has a lower specific activity than the double-chain material originally described (Jesty and Nemerson 1974), we studied the effect of some proteolytic enzymes on the activity and structure of factor VII. Trypsin, thrombin and factor X_a all increased the activity of the single-chain protein, and at about the same rate. However when lipid and Ca^{++} were added to factor X_a, the rate of activation was increased about 400-fold. This activation results in an increase in activity of up to 80-fold and is accompanied by the formation of the double-chain form originally described (Jesty and Nemerson 1974).

It might be argued that the single-chain form is a zymogen and that the activity of the single-chain preparation can be attributed to contamination by the highly active double-chain form. However, the single-chain form incorporates DFP at the same rate, and to the same extent, as the double-chain form. Furthermore, inhibition occurs in both cases, indicating that DFP is incorporated into the active site. Other protease zymogens have been shown to incorporate this inhibitor into the active site, but in the case of trypsinogen and chymotrypsinogen, the rates of incorporation are about 10^4-fold slower than with the respective enzymes (Morgan et al. 1972). Factor VII is therefore very unusual, if we regard it as a true zymogen, in that there is no change in reactivity towards DFP on activation. The N-terminal sequence of the single-chain form, Ala-Asx-Gly-Phe-, shows close homology with those of pro-thrombin, factor IX and the light chain of factor X (Fujikawa et al. 1974). On activation, this N-terminal sequence is retained in the light chain of double-chain factor VII. The N-terminal sequence of the heavy chain, Ile-Val-Gly-Gly-, shows homology with the comparable regions of trypsin, thrombin and factor X_a (Titani et al. 1972).

Determination of the activity of the single-chain form is at present pre-cluded for two reasons: First, both forms behave identically in purification procedures, so we cannot purify one free of the other. Second, even if the single-chain form could be isolated free of the double-chain one, the results of an assay would be somewhat ambiguous since the assay necessarily produces factor X_a, which is known to cause activation.

The activated form we describe does not correspond with that described by Østerud et al. (1972), in that our double-chain form still requires tissue factor for proteolytic activity on factor X.

At concentrations of the same order as exist in plasma (factor VII, 180 U/ml; factor X_a, 5 U/ml, corresponding to 5% conversion of plasma factor X), the activation rate is initially about 20-fold per minute. Under the same conditions, the double-chain form is inactivated, but at a rate some 20 times slower (Fig. 3). When studied in much more concentrated mixtures by SDS gel electrophoresis, this inactivation is accompanied by the cleavage of the heavy chain into two pieces, the smaller of which still incorporates DFP and therefore contains the active site (Fig. 4). By comparison with a nonreduced

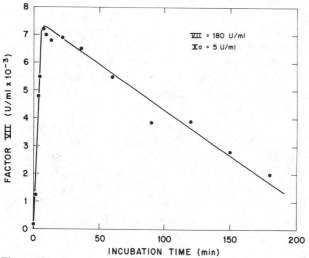

Figure 3

The activation and inactivation of factor VII. Single-chain factor VII (180 U/ml) was incubated with factor X_a (5 U/ml = 5% conversion plasma factor X) in the presence of lipid (0.1 mg/ml mixed brain lipids) and 5 mM $CaCl_2$. Samples were diluted into ovalbumin solutions before assay of factor VII activity.

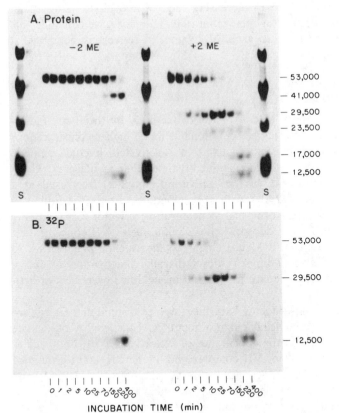

Figure 4
Electrophoresis of the activation and inactivation of factor
VII. Factor VII (0.2 mg/ml) was incubated with factor
X_a (0.07 μg/ml) in the presence of phospholipid (0.1 mg/
ml) and 5 mM $CaCl_2$ at 37°C. Aliquots were removed and
added to 1/20 vol 40 mM [^{32}P]DFP. After 3 hours at 0°C,
the reaction was stopped by the addition of an equal volume
of SDS-urea before gel electrophoresis (Jesty and Nemerson
1974). Samples on the right were reduced with 5% 2-
mercaptoethanol. The gel was stained for protein (A),
dried and autoradiographed to locate the DFP. The molecular weight standards were bovine serum albumin, ovalbumin, α-chymotrypsinogen and cytochrome c. (Reprinted,
with permission, from Radcliffe and Nemerson 1975).

gel (Fig. 4, left), we see that the larger piece is still bound to the light chain.

When double-chain factor VII (at approximately plasma concentration) is
incubated with factor X_a at a concentration that corresponds to complete
conversion of plasma factor X, the rate of inactivation is initially about 30%
per minute, or 2000 U/ml/min of factor VII. We also note in these
experiments that even though the molar concentration of factor X_a is probably tenfold greater than that of factor VII, the rate of inactivation is dependent on the concentration of factor X_a. Thus we predict that in the
activation of factor X, the continuing generation of factor X_a will result first

in activation of factor VII at an increasing rate. Moreover, the subsequent
inactivation may also occur at a substantial rate. Therefore, we conclude that
the process will tend to produce a pulse of factor X-converting activity that
can quickly disappear.

Activation of Factor X

The study of the mechanism of conversion of factor X by the tissue factor-
factor VII complex has necessarily been done at much higher concentrations
of factor X than exist in plasma: factor X was activated, by the complex,
at a concentration of 0.36 mg/ml (about 40 times that in plasma). However, double-chain factor VII (as the activating complex) was used at a
maximum concentration of 240 U/ml, corresponding to only 2.4-fold activation of plasma factor VII. At this high concentration of factor X, we observe
that the yield of factor X_a is reduced when the initial factor VII concentration
is less than about 150 U/ml. Measurement of the rate of inactivation of
factor VII (as the active complex) shows that the inactivation of factor
VII by factor X_a can fully account for the fall in yield at lower rates of activation (Fig. 5).

In the case of activation of factor X in the presence of lipid and Ca^{++}, we
observe two cleavages of the heavy chain of factor X. The cleavage of a single
N-terminal peptide results in the appearance of factor X_a activity and is catalyzed by the tissue factor-factor VII complex and by the coagulant protein of
Russell's viper venom (RVV). Second, factor X_a, in the presence of lipid and
Ca^{++}, cleaves a C-terminal peptide from the heavy chain, giving a new
C-terminal Arg residue. We observe that in the activation of factor X in the

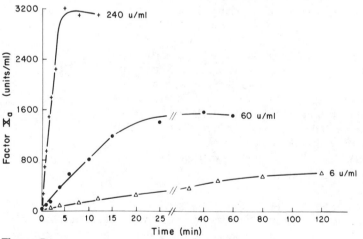

Figure 5

Activation of factor X by the tissue factor-factor VII complex. Factor
X (0.36 mg/ml) was activated with factor VII (as the active complex) at concentrations of 240, 60 and 6 U/ml. Samples were removed into citrate-ovalbumin solution and assayed as described (Jesty
and Nemerson 1974).

presence of lipid there are two alternative pathways to the final product, β-X$_a$; this is because the two cleavages of the heavy chain can occur in either order (Fig. 6).

Because one pathway (reactions 3 + 4, Fig. 6) is initiated by the product, factor X$_a$, it is favored at lower rates of activation. Thus we find that as the rate of activation falls, the pathway initiated by factor X$_a$ becomes more important, as can be seen in Figure 7B. In fast activations, the first product is α-X$_a$, which is formed by the removal of the N-terminal activation peptide; this reaction is catalyzed by the tissue factor-factor VII complex or RVV. The α-X$_a$ is then converted autocatalytically to β-X$_a$ by the removal of the C-terminal peptide. We have shown that this reaction is not catalyzed by the tissue factor-factor VII complex by using DFP-inhibited α-X$_a$ as a substrate; this is cleaved very rapidly by noninhibited α-X$_a$ or β-X$_a$, but not by the activating complex (Jesty, Spencer and Nemerson 1974).

In slower activations, we observe to a greater extent the pathway initiated by the action of factor X$_a$ on factor X to cleave the identical C-terminal peptide (reaction 3, Fig. 6; Fig. 7). This reaction is also lipid-dependent and therefore not seen to a significant extent in activations by RVV in the absence of lipid. We cannot show that it is not catalyzed by the tissue factor-factor VII complex, as these conditions result in activation; but we imagine that as this cleavage is solely a result of factor X$_a$ action in the conversion of α-X$_a$ to β-X$_a$, this is also true of the formation of the intermediate (I) from factor X.

We have isolated this intermediate and shown that it is activated by the tissue factor-factor VII complex or RVV directly to β-X$_a$, and at the same rates as factor X. This result is confirmed by clotting assay, by which we can show that their activities are identical. Similarly, we have isolated α-X$_a$ and β-X$_a$ and shown by assay that their clotting activities are identical. So it appears that the loss of the C-terminal peptide has no effect on the clotting activity of either the zymogen or the enzyme. Thus although the factor X$_a$-initiated pathway predominates in slower activations, this has no effect per se on the final yield of factor X$_a$ activity. However on account of the inactivation of factor VII by factor X$_a$, the yield of factor X$_a$ is reduced in slower activations and therefore correlates with a buildup of the intermediate (I), which

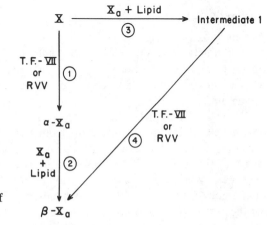

Figure 6
The pathways of activation of factor X.

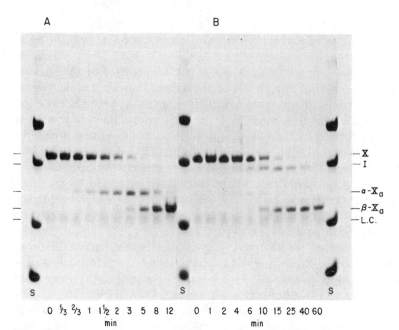

A   B

—X
—I

—α-X$_a$

—β-X$_a$
—L.C.

S   S   S

0 ⅓ ⅔ 1 1½ 2 3 5 8 12   0 1 2 4 6 10 15 25 40 60
        min   min

Figure 7

Electrophoresis of factor X activation by the tissue factor-factor VII
complex. Factor X (0.36 mg/ml) was incubated with factor VII (as
the active complex) at concentrations of (A) 200 and (B) 20 U/ml
in the presence of 5 mM CaCl$_2$ at 37°C. Samples were removed at the
times shown into SDS-urea to stop the reaction and then electrophoresed as described. The mixture of standards (S) contained bovine
serum albumin, ovalbumin, α-chymotrypsinogen and myoglobin. (Reprinted, with permission, from Jesty, Spencer and Nemerson 1974.)

can still be formed by the action of factor X$_a$ on any remaining factor X. The
apparent lack of any control mechanism associated with the alternative pathways of factor X activation is in marked contrast to what we find in the activation of prothrombin.

Activation of Prothrombin

We have found that only at very high levels of activator can prothrombin be
quantitatively activated to thrombin and have studied this apparent control
mechanism in the context of recent evidence showing that prothrombin can be
cleaved in vitro by two pathways, one initiated by factor X$_a$ and the other by
thrombin.

We mentioned earlier that Mertz, Seegers and Smith (1939a) showed that
the extent of thrombin formation from partially purified prothrombin is a
function of the rate of activation (Fig. 2). These authors also studied the
effect of thrombin itself on prothrombin and found that the zymogen rapidly
lost activity, as measured by activation with the same system (Mertz, Seegers
and Smith 1939b). Seegers and McClaughry (1949) showed that inactivation
of the prothrombin was accompanied by a change in its electrophoretic mobility; the inactive material was subsequently isolated by Marciniak and Seegers

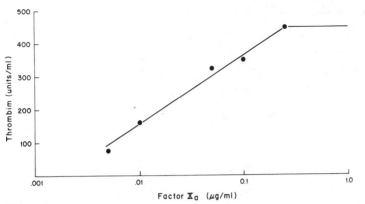

Figure 9

The yield of thrombin as a function of factor X_a concentration. Prothrombin (0.5 mg/ml) was activated in the presence of factor V at 14 U/ml (normal plasma contains 100 U/ml), phospholipid (0.1 mg/ml) and 10 mM $CaCl_2$. The final yield of thrombin (cf. Fig. 2) in each activation, determined by assay, is plotted against log[factor X_a].

thrombin production decreased, but the final yield of thrombin is also progressively less (Fig. 9). SDS gel electrophoresis shows that this is accompanied by the accumulation of P_2 until, at the lowest levels of factor X_a used, P_2 becomes the predominant product (Fig. 8b).

The fact that P_2 is comparatively inert has been ascribed to the loss of the Ca^{++} and lipid binding sites from the prothrombin molecule (Stenflo 1973; Gitel et al. 1973). Whether or not P_2 can be a source of thrombin is not really at issue, since we know that P_2 appears transiently in mixtures which go on to yield thrombin quantitatively (Fig. 8a). However, it is necessary to explain why P_2 accumulates under some conditions and not others. One explanation, not as yet studied in detail, is that the transient species seen during rapid activations is not the same as P_2 that accumulates under slower conditions. Our present experiments have been directed toward determining whether the "enzyme" (one or more components of the activating complex) loses activity once activation has begun, for example, by product inhibition. In this regard, it has already been shown that F_1 inhibits activation of prothrombin by the converting complex (Fig. 10) (Jesty and Esnouf 1973; Heldebrandt et al. 1973; Benson, Kisiel and Hanahan 1973). This inhibition is not large in vitro, but it is worth noting that the ratio of F_1 to prothrombin can quickly become large as thrombin acts on prothrombin. In addition to this inhibition of the activating complex, the complex could be inactivated proteolytically. It has been known for some time that factor V prepared under some conditions gains, and then loses, activity when exposed to thrombin (Rapaport et al. 1963; Colman 1969). The preparation of factor V used in these experiments was made from bovine plasma by a modification of the method of Dombrose et al. (1972) and can be "activated" tenfold. We have studied the behavior of this factor V in two-step activations of prothrombin. Activation is first initiated with the whole complex; further enzyme (factor X_a), cofactor (factor V) or substrate

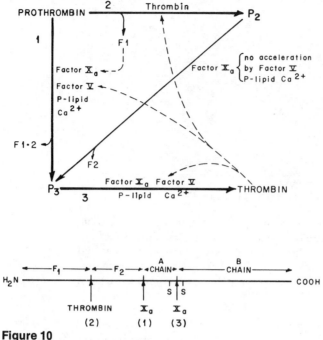

Figure 10

The pathways of prothrombin activation and the sites of feedback control. The numbers beside the reactions correspond with the sites of cleavage of the polypeptide chain, shown at the bottom. The dotted lines represent the apparent sites of control by thrombin.

(prothrombin), or the permutations of these, are then added in an attempt to deduce the state of the various components of the mixture.

We found that after a slow activation (initiated with low factor X_a but high factor V levels), which has produced largely P_2 and F_1, the addition of more factor X_a alone, even in large amounts, has no effect. The factor X_a must be accompanied by the addition of factor V for any increase in thrombin activity to occur. We conclude that factor V in large amounts is essential for the production of thrombin at reasonable rates from P_2, and further, that the factor V initially present in the slow activation is no longer active.

We also found that the amount of factor X_a needed to give a second burst of thrombin is manyfold greater than is required to produce an initial burst of similar size. This demonstrates that a mixture of P_2 and F_1 formed in slow activations is refractory, but does not exclude the further possibility that if components of the activating complex are added later, they come under an additional constraint. Evidence for such a constraint was obtained by adding further prothrombin to a slow activation, as well as more factor X_a and factor V. When, in the second addition, the concentrations of all the components are double (equivalent to initiating a second slow activation in the presence of all the products of the first), there is a negligible increase in the level of thrombin. At present we do not know if further thrombin formation is prevented simply because the second aliquot of prothrombin is cleaved almost at once to P_2, or if inhibitors are present in the mixture as well.

These results lead to the proposal that the conversion of prothrombin to P_2 by thrombin does not merely provide an alternative route for thrombin production. We suggest that under most conditions, the formation of P_2 is nonproductive, because additional feedback reactions operate to inactivate the factor X_a-factor V complex. If the initial concentration of the activating complex is sufficiently high, however, any P_2 formed can be cleaved to P_3, and thence thrombin, before the complex is inactivated.

Thus we consider that the damping of the reaction we observe is a result of two feedback actions of thrombin: first, the inactivation of the cofactor, factor V, and second, the partial "inactivation" of the substrate, prothrombin, in the formation of P_2. When presented with a small stimulus, such a system tends to damp it by the formation of intermediates with no proteolytic activity. On the other hand, a stimulus above a certain size will evoke a response which overrides the usual damping mechanisms.

DISCUSSION

The availability of tissue factor can be regarded as the initial point of control of the extrinsic system of blood coagulation in that factor VII has an absolute requirement for tissue factor in order to activate factor X. We have demonstrated this on a cellular level where, when the other components of the pathway are present, undisturbed cells will not significantly initiate the sequence, whereas cells subjected to mechanical or enzymatic injury cause the rapid generation of thrombin.

The mechanism by which tissue factor controls the activity of factor VII is unknown, but it is almost certain that it is not enzymatic, the active site responsible for the proteolytic activation of factor X being in factor VII (Nemerson and Esnouf 1973). In this regard, the interaction between tissue factor and factor VII may be compared with that between collagen and factor XII or between streptokinase and plasminogen. In all three cases, the "zymogen" becomes enzymatically active upon the formation of a complex with the appropriate effector. Although blood coagulation is the result of a proteolytic cascade, it is noteworthy that the initiating step in each pathway is the nonproteolytic "activation" of an "enzyme"-factor XII or factor VII complex. There are clear advantages to an organism of this type of "activation." If the initiating reaction involved the proteolysis of a plasma zymogen by a tissue protease, as in the cases of the pancreatic zymogens and the other clotting zymogens, an active protease would have to be widely distributed so as to be available at all potential sites of injury. This would require either an enzyme of absolute specificity or an elaborate control mechanism that would protect all other potential substrates. Tissue factor and collagen, however, appear to be inert towards other clotting proteins and have no known proteolytic activity. Thus they can be widely distributed without risk of nonspecific proteolysis and must only be protected from accidental contact with factors VII and XII.

We have investigated the cellular mechanism by which tissue factor might be prevented from interaction with factor VII. As noted, undisturbed cells grown in monolayers or in suspension express very little, or no, tissue factor activity. But after subjection to some forms of trauma, they become able to initiate the extrinsic pathway.

Once activated, the extrinsic pathway is still under control. As early as 1939 it was known that in a system containing all the components of the pathway, except perhaps factor V, the yield of thrombin is a function of the concentration of tissue factor, and we emphasize that this behavior would not be predicted unless the system were constrained by some sort of feedback control. Examining each step of the pathway separately, we have found that various types of control exist by which the initial control, the extent of appearance of tissue factor, may be retained through the activations of factor X and thrombin, and thus be reflected in the amount of thrombin formed.

At the step of activation of factor X, we find that the enzyme, factor VII, is first rapidly activated, and subsequently inactivated, by the product of activation, factor X_a. In the activation of prothrombin by the converting complex of factor X_a, factor V, lipid and Ca^{++}, we find three types of feedback control by the product, thrombin: first, inhibition of activation by F_1, the peptide produced by the action of thrombin on prothrombin; second, the partial "inactivation" of the substrate, prothrombin, by the action of thrombin, producing P_2; and third, the inactivation by thrombin of the cofactor, factor V.

It has been emphasized that these activations have been studied in the absence of plasma inhibitors, notably antithrombin III (heparin cofactor), that inactivate factor X_a and thrombin (Yin, Wessler and Stoll 1971; Rosenberg and Damus 1973). The control mechanisms that we have observed function somewhat differently in that they directly limit the yield of factor X_a or thrombin produced in response to a stimulus, rather than remove the enzymes once they are formed. In vivo, these self-damping controls and the irreversible inhibitors may function in concert to limit coagulation, but the relative importance of each will depend to a large extent on the availability of heparin at the site of the stimulus. In the absence of heparin, the rates of inactivation of factor X_a and thrombin by antithrombin III are measured in minutes and are slow compared to the rates of factor X and prothrombin activation at which we observe control operating. In the presence of heparin, however, the rates of inactivation are very high, and this means of control would become important; for instance, in the case of a constant rate of production of enzyme, its concentration will depend on the rate of inactivation. But at present, very little is known about the availability of heparin at sites of vessel damage, although we note that some cell surfaces contain sulfated mucopolysaccharides of heparan type (Kraemer 1971) that could be involved in the control of coagulation.

Regarding the complete system of coagulation, we note that the activation of prothrombin is common to both the extrinsic and intrinsic pathways. Furthermore, the intrinsic system is probably under kinetic control, as factor VIII, the cofactor of factor IX_a, is activated and inactivated by thrombinlike factor V. This could result in control of a similar type whereby the yield of factor X_a in this activation would be a function of the rate.

Lastly, considering the activation and inactivation of factor VII by factor X_a, we note that factor X_a can be produced by either pathway, and the product is apparently the same in both activations (Radcliffe and Barton 1973). Thus activation of factor X by the intrinsic pathway could lead to activation, and even the subsequent inactivation, of factor VII. Thus initiation of the intrinsic pathway could in effect prime the extrinsic pathway, resulting in a higher initial rate of factor X activation upon the appearance of tissue factor. With these considerations in mind, it is clear that the self-damping mechanisms we observe are not, in fact, restricted to the extrinsic pathway of coagulation and are probably of general importance in ensuring that the yield of thrombin reflects the size of the initial stimulus.

Acknowledgments

We wish to thank Dr. Carol Heckman for allowing us to include her electron microscopy results and Dr. Konigsberg for his welcome criticism. This work was supported by grants from the John A. Hartford Foundation and the National Institutes of Health (HL 16126). S. A. S. holds a Postdoctoral Fellowship from the Connecticut Heart Association.

REFERENCES

Ashford, R. P. and D. G. Freiman. 1968. Platelet aggregation at sites of minimal endothelial injury. *Amer. J. Pathol.* **53**:599.

Benson, B. J., W. Kisiel and D. J. Hanahan. 1973. Calcium binding and other characteristics of bovine factor II and its intermediates. *Biochim. Biophys. Acta* **329**:81.

Bremerskov, V. 1973. Dextran sulphate inhibits cell adhesion in tissue culture. *Nature New Biol.* **246**:174.

Colman, R. W. 1969. The effects of proteolytic enzymes on bovine factor V. *Biochemistry* **8**:1438.

Dombrose, F. A., T. Yasui, Z. Roubal, A. Roubal and W. H. Seegers. 1972. Ac-globulin (factor V): Preparation of a practical product. *Prep. Biochem.* **2**:381.

Esmon, C. T., W. G. Owen and C. M. Jackson. 1974. The conversion of prothrombin to thrombin. *J. Biol. Chem.* **249**:606.

Esnouf, M. P., P. H. Lloyd and J. Jesty. 1973. A method for the simultaneous preparation of bovine factor X and prothrombin. *Biochem. J.* **131**:781.

Fujikawa, K., M. H. Coan, D. L. Enfield, K. Titani, L. H. Ericsson and E. W. Davie. 1974. A comparison of bovine prothrombin, factor IX (Christmas factor), and factor X (Stuart factor). *Proc. Nat. Acad. Sci.* **71**:427.

Gitel, S. N., W. G. Owen, C. T. Esmon and C. M. Jackson. 1973. A polypeptide region of bovine prothrombin specific for binding to phospholipids. *Proc. Nat. Acad. Sci.* **70**:1344.

Glick, M. C. and C. A. Buck. 1973. Glycoproteins from the surface of metaphase cells. *Biochemistry* **12**:85.

Green, D., C. Ryan, N. Malandruccolo and H. L. Nadler. 1971. Characterization of the coagulant activity of cultured human fibroblasts. *Blood* **37**:47.

Heldebrandt, C. M., R. J. Butkowski, S. P. Bajaj and K. G. Mann. 1973. The activation of prothrombin. *J. Biol. Chem.* **248**:7149.

Jesty, J. and M. P. Esnouf. 1973. The preparation of activated factor X and its action on prothrombin. *Biochem. J.* **131**:791.

Jesty, J. and Y. Nemerson. 1974. Purification of factor VII from bovine plasma. *J. Biol. Chem.* **249**:509.

Jesty, J., A. K. Spencer and Y. Nemerson. 1974. The mechanism of activation of factor X. *J. Biol. Chem.* **249**:5614.

Kisiel, W. and D. J. Hanahan. 1974. Proteolysis of human factor II by factor X_a in the presence of hirudin. *Biochem. Biophys. Res. Comm.* **59:**570.

Kraemer, P. M. 1971. Heparan sulfates of cultured cells. *Biochemistry* **10:**1437.

Macfarlane, R. G. 1972. The theory of blood coagulation. In *Human Blood Coagulation, Haemostasis and Thrombosis* (ed. R. Biggs), 4th ed., p. 23. Blackwell, Oxford.

Marciniak, E. 1970. Functional and steric characteristics of modified thrombin zymogen. *Thromb. Diath. Haemorrh.* **24:**361.

Marciniak, E. and W. H. Seegers. 1966. Prethrombin as a new sub-unit of prothrombin. *Nature* **209:**621.

Maynard, J. R., C. A. Heckman, F. A. Pitlick and Y. Nemerson. 1975. Association of tissue factor activity with the surface of cultured cells. *J. Clin. Invest.* (in press).

Mertz, E. T., W. H. Seegers and H. P. Smith. 1939a. Prothrombin, thromboplastin, and thrombin: Quantitative interrelationships. *Proc. Soc. Exp. Biol. Med.* **42:**604.

————. 1939b. Inactivation of prothrombin by purified thrombin solutions. *Proc. Soc. Exp. Biol. Med.* **41:**657.

Morgan, P. H., N. C. Robinson, K. A. Walsh and H. Neurath. 1972. Inactivation of bovine trypsinogen and chymotrypsinogen by di-isopropylphosphorofluoridate. *Proc. Nat. Acad. Sci.* **69:**3312.

Nemerson, Y. 1968. The phospholipid requirement of tissue factor in blood coagulation. *J. Clin. Invest.* **47:**72.

Nemerson, Y. and M. P. Esnouf. 1973. Activation of a proteolytic system by a membrane lipoprotein: The mechanism of action of tissue factor. *Proc. Nat. Acad. Sci.* **70:**310.

Nemerson, Y. and F. A. Pitlick. 1972. The tissue factor pathway of blood coagulation. In *Progress in Hemostasis and Thrombosis* (ed. T. H. Spaet), p. 1. Grune and Stratton, New York.

Østerud, B., A. Berre, A. Otnaess, E. Bjørklid and H. Prydz. 1972. Activation of the coagulation factor VII by tissue thromboplastin and calcium. *Biochemistry* **11:**2853.

Owen, W. G., C. T. Esmon and C. M. Jackson. 1974. The conversion of prothrombin to thrombin. *J. Biol. Chem.* **249:**594.

Radcliffe, R. D. and P. G. Barton. 1973. Comparisons of the molecular forms of activated bovine factor X. *J. Biol. Chem.* **248:**6788.

Radcliffe, R. D. and Y. Nemerson. 1975. Activation and control of factor VII by activated factor X and thrombin. *J. Biol. Chem.* **250:**388.

Rapaport, S. I., S. Schiffman, M. J. Patch and S. B. Ames. 1963. The importance of activation of antihemophilic globulin and proaccelerin in the generation of intrinsic prothrombinase activity. *Blood* **21:**221.

Rosenberg, R. D. and P. S. Damus. 1973. The purification and mechanism of action of human antithrombin-heparin cofactor. *J. Biol. Chem.* **248:**6490.

Seegers, W. H. and R. I. McClaughry. 1949. Production of an inactive derivative of purified prothrombin by means of purified thrombin. *Proc. Soc. Exp. Biol. Med.* **72:**247.

Stemerman, M. B. and F. A. Pitlick. 1974. Tissue factor antigen and blood vessels. *Thromb. Diath. Haemorrh.* **60:**71.

Stenflo, J. 1973. Vitamin K and the biosynthesis of prothrombin. *J. Biol. Chem.* **248:**6325.

Stenn, K. S. and E. R. Blout. 1972. Mechanism of bovine prothrombin activation by an insoluble preparation of bovine factor X_a (thrombokinase). *Biochemistry* **11:**4502.

Titani, K., M. A. Hermodson, K. Fujikawa, L. H. Ericsson, K. A. Walsh, H. Neurath and E. W. Davie. 1972. Bovine factor X_{1a} (activated Stuart factor). Evidence of homology with mammalian serine proteases. *Biochemistry* **11**:4899.

Yin, E. T., S. Wessler and P. J. Stoll. 1971. Biological properties of the naturally occurring plasma inhibitor of activated factor X. *J. Biol. Chem.* **246**:3703.

Zacharski, L. R., L. W. Hoyer and O. R. McIntyre. 1973. Immunologic identification of tissue factor (thromboplastin) synthesized by cultured fibroblasts. *Blood* **41**:671.

Zeldis, S. M., Y. Nemerson, F. A. Pitlick and T. L. Lentz. 1972. Tissue factor (thromboplastin): Localization to plasma membranes by peroxidase-conjugated antibodies. *Science* **175**:766.

Disseminated Intravascular Coagulation: Proteases and Pathology

Scott H. Goodnight, Jr.

Department of Medicine, University of Oregon Medical School
Portland, Oregon 97201

Diffuse intravascular coagulation is the result of pathologic activation of blood coagulation and fibrinolysis. When fully expressed, it is a catastrophic clinical syndrome manifested by bleeding due to depletion of platelets and coagulation factors, organ ischemia consequent to deposition of fibrin in the microcirculation, and hemolytic anemia caused by fibrin-induced traumatic red cell destruction.

The defects in hemostasis stem primarily from the action of thrombin. Thrombocytopenia may reflect thrombin-induced platelet aggregation, entrapment of platelets in a fibrin meshwork, or traumatic disruption of the platelets flowing across fibrin strands. Fibrinogen and factors VIII and V may virtually disappear from the circulating blood. Lysis of the intravascular fibrin yields circulating fibrin degradation products which may inhibit platelet aggregation, thrombin action and fibrin polymerization; the hemostatic failure is thereby enhanced.

Intravascular deposition of fibrin may lead to severe organ dysfunction such as the renal cortical necrosis seen in the generalized Shwartzman reaction in the rabbit (Hjort and Rapaport 1965). In this classic experiment, two spaced doses of intravenous bacterial endotoxin induce intravascular clotting and the deposition of fibrin in the glomerular vasculature. Similar fibrin deposition may occur in human kidneys and adrenal glands during meningococcemia (McGehee, Rapaport and Hjort 1967) or other generalized infections.

Fibrin also appears to be the cause of the hemolytic component of intravascular coagulation which is characterized by fragmented red cells or schistocytes in the peripheral blood. One explanation for these striking morphologic abnormalities has been put forth by Bull et al. (1968). The erythrocytes are thought to "hang up" upon a strand of fibrin, seal off, and then tear away, thereby producing fragmentation, spherocytes and hemolysis.

Triggers of Intravascular Clotting

Some of the mechanisms that "trigger" intravascular clotting are: (1) tissue thromboplastin: brain tissue, granulocytes, amniotic fluid (intrauterine death), erythrocytes; (2) Stuart factor (X): mucus, amniotic fluid (normal); (3) Hageman factor; (4) fibrinogen; and (5) platelets. Various tissue thromboplastins, such as components of brain tissue, granulocytes, amniotic fluid after intrauterine death and red cell membranes, may react with factor VII and lead to diffuse intravascular clotting. Alternatively, factor X, factor XII, fibrinogen or platelets may be directly involved.

Investigators have produced experimental intravascular coagulation by infusing diluted brain tissue thromboplastin intravenously into animals (e.g., Astrup and Albrechtsen 1969). Such infusions lead to venous and pulmonary thrombosis, but arterial lesions and renal glomerular fibrin deposits do not occur. As expected, circulating levels of fibrinogen, factor V, factor VIII and platelets fall. Interestingly, Rapaport et al. (1966) and Evensen and Jeremic (1970) have shown that factor IX (an intrinsic system clotting factor) as well as factor VII of the extrinsic system and factor X are diminished following a tissue thromboplastin infusion. Platelets are not essential for this reaction since rabbits made thrombocytopenic by prior busulfan administration have similar changes in coagulation factors (Evensen and Jeremic 1970). There may well be a human counterpart to this animal experiment.

Recently, Goodnight et al. (1974) studied a group of patients who had received gunshot wounds or other blunt trauma to their heads, leading to massive brain tissue destruction. Many of these patients had intravascular clotting and serious bleeding. Initial levels of fibrinogen, factor V and factor VIII were reduced. Fibrin degradation products were elevated, and the protamine test for fibrin monomer and/or early fibrin degradation products was invariably positive. As in the tissue thromboplastin infusion experiments, the vitamin K-dependent factors (II, VII, IX and X) were reduced. Moreover, factors XII and XI, which are contact activated, were also depressed. These patients did not develop renal insufficiency, and only a few had an occasional fragmented red cell in their peripheral blood. Most likely, the sudden disruption of the normal blood-brain barrier allows mixing of plasma coagulation factors with the macerated brain tissue. Components of the mixture may then gain entrance into the systemic venous circulation, thus producing intravascular clotting. The reason for reduction in the activities of coagulation factors XII, XI and IX of the intrinsic system is not clear.

Under appropriate circumstances, granulocytes may be thromboplastic and cause intravascular clotting. Lerner, Goldstein and Cummings (1971) have shown that granulocytes may develop thromboplastic activity when incubated in plasma or, more importantly, with endotoxin. Factor VII is necessary for this reaction. Niemetz and Fani (1971) have induced intravascular clotting and renal cortical necrosis in rabbits injected with endotoxin-treated granulocytes. Moreover, granulocytes are a necessary reactant in the generalized Shwartzman reaction (Lerner, Rapaport and Spitzer 1968).

In man, acute progranulocytic leukemia is very frequently associated with diffuse intravascular clotting (Rosenthal 1963; Goodnight 1974). One prominent feature of the progranulocytes in these cases is intense cytoplasmic granulation. Quigley (1967) and Gralnick and Abrell (1973) have shown that the lysosomal granules of such cells have potent clot-promoting activity and that this activity requires the presence of factor VII.

Following death of the fetus in utero, amniotic fluid is thromboplastic in vitro and may be a cause of the defibrination seen in the "dead baby" syndrome (Courtney and Allington 1972). Normal amniotic fluid also has a weak clot-promoting effect, but it has been shown to bypass factor VII and act directly on factor X (Courtney and Allington 1972; Phillips and Davidson 1972).

Lastly, infusions of red blood cell stroma into monkeys result in intravascular coagulation (Birndorf and Lopas 1970). Similarly, severe intravascular hemolysis from incompatible blood transfusions may be associated with disseminated clotting in both animals and man (Lopas and Birndorf 1971).

Other clot-promoting substances may activate factor X directly. Patients with metastatic mucus-producing adenocarcinomas may have chronic intravascular coagulation, recurrent venous thromboses, microangiopathic hemolytic anemia and marantic endocarditis with embolization of fibrin into the arterial circulation. Pineo et al. (1973) have purified mucus and shown that it acts on factor X. Furthermore, a purified fraction of mucus injected into rabbits has caused intravascular clotting with accelerated fibrinogen catabolism (G. Pineo, pers. comm.).

Hageman factor has long been considered a likely site for the initiation of disseminated intravascular coagulation. Botti and Ratnoff (1964) have shown that infusions of ellagic acid (an activator of factor XII) into animals leads to shortening of the whole blood clotting time. However, to induce glomerular fibrin deposition, concurrent administration of the fibrinolytic inhibitor ϵ-aminocaproic acid and an α-adrenergic agent were necessary (McKay, Müller-Berghaus and Cruse 1969).

Skjørten and Evensen (1973) injected both endotoxin and liquoid (an acid polymer causing severe endothelial damage) into factor XII-deficient fowl and were unable to detect pulmonary fibrin deposits, whereas tissue thromboplastin infusions cause fibrin deposition with regularity. In contrast, Shen, Rapaport and Feinstein (1973) were not able to block endotoxin-induced intravascular clotting nor the generalized Shwartzman reaction in rabbits depleted of factor VIII by infusion of a factor VIII antibody.

A more dramatic example of the induction of disseminated intravascular coagulation is that of a young female drug addict who tried to extract morphine from an antidiarrheal compound that contained kaolin. She injected an impure extract intravenously and developed overwhelming intravascular clotting, presumably due to factor XII activation by the kaolin (Beresford 1971).

Intravascular clotting can be induced with regularity by snake venoms such as Ankrod or Reptilase, which act by directly splitting fibrinopeptide A from fibrinogen (Ewart et al. 1970; Blömback 1958). Perhaps of more interest, however, is the role of the platelet in intravascular clotting, and particularly the role of complement. It his been shown that platelet-rich plasma from rabbits congenitally deficient in the C6 component of complement clots normally upon the addition of kaolin, but fails to do so upon the addition of inulin or endotoxin (Zimmerman, Arroyave and Müller-Eberhard 1971; Zimmerman and Müller-Eberhard 1971). Inulin and endotoxin activate the alternate complement pathway. Furthermore, Siraganian (1972) has shown that these substances fail to promote platelet-release reactions in C6-deficient rabbit plasma. Therefore, at least in rabbits, endotoxin-induced augmentation of clotting may proceed in part by a complement and platelet pathway. In support of this concept, preliminary work by Stafford, Shen and Rapaport (1973) has shown that endotoxin-induced clotting may be blunted in intact C6-deficient rabbits. However, a note of caution should be sounded. In a recently studied human family with inherited C6 deficiency, absence of C6 had little or no effect on hemostasis (Heusinkveld et al. 1974). It may be that species differences govern the susceptibility of platelets to complement activation.

Thrombin

Regardless of the trigger mechanism, the final common path in most instances involves thrombin. As a result of the action of thrombin, fibrin monomer is formed consequent to removal from fibrinogen of fibrinopeptides A and B. In patients with intravascular clotting, the presence of soluble fibrin monomer complexed with fibrinogen or fibrin degradation products (Sasaki, Page and Shainoff 1966; Lipinski et al. 1967) serves as the basis of a useful clinical test. The addition of protamine sulfate to a plasma containing fibrin monomer results in the formation of an insoluble web or gel. Positive protamine or other "paracoagulation" tests have been found in a large proportion of patients with other clinical and laboratory evidence of intravascular coagulation (Kidder et al. 1972).

Fibrin monomer does not affect platelet function (Müller-Berghaus and Heinrich 1972). However in rabbits, fibrin monomer induced by an Ankrod infusion can be precipitated in the glomerulus by a concomitant endotoxin injection (Müller-Berghaus and Hocke 1972).

Recently, Nossel et al. (1971) have developed an immunoassay for fibrinopeptide A and have noted increased levels of this peptide in a variety of clinical conditions associated with intravascular coagulation (Nossel, Canfield and Butler 1973). The turnover rate of fibrinopeptide A is remarkably short (several minutes), and this peptide is not found in the urine of dogs with thromboplastin-induced systemic clotting (Teger-Nilsson and Gröndahl 1974).

Markedly increased levels of thrombin will destroy factors VIII and V, whereas smaller amounts may activate these clotting factors (Rapaport et al. 1963). Niemetz and Nossel (1969) have observed evidence of factor VIII activation in a variety of patients with chronic intravascular coagulation. Factor·VIII activity measured by a one-stage assay was much higher than levels obtained with a two-stage technique involving aluminum hydroxide adsorption. Similar activation of factor V has been described by Fedder, Prakke and Vreeken (1972) in rabbits infused with low doses of thrombin.

Other activated clotting factors, such as factor X_a or thrombin, are difficult to identify in vivo during intravascular clotting, most likely because of rapid reticuloendothelial clearance and neutralization by plasma inhibitors.

Plasmin

As a result of the deposition of fibrin in the microvasculature, local endothelial plasminogen activator diffuses into the interstices of the fibrin deposits.

Resultant plasmin action leads to circulating soluble fibrin degradation products. As with thrombin, excessive systemic lysis is apparently prevented by specific inhibitors and by clearance of activated enzymes by the liver. Circulating fibrin fragments may inhibit platelet aggregation, polymerization of fibrin monomer and thrombin action, and thereby intensify the hemorrhagic diathesis (for review, see Sherry 1972).

Conclusions

Disseminated intravascular coagulation in animals and man provides an important in vivo model in which to examine some pathologic effects of protease action. Hopefully such studies will lead to a better understanding of this clinically important process and ultimately to improved patient care.

REFERENCES

Astrup, T. and O. K. Albrechtsen. 1969. Serum effects following tissue thromboplastin infusion. *Thromb. Diath. Haemorrh.* **21:**117.

Beresford, H. R. 1971. Coagulation defects after I. V. kaolin. *New Eng. J. Med.* **285:**522.

Birndorf, N. and H. Lopas. 1970. Intravascular coagulation in cynomologus monkeys produced by red cell stroma. *Clin. Res.* **28:**398.

Blömback, B. 1958. Studies on the action of thrombic enzymes on bovine fibrinogen as measured by N-terminal analysis. *Arkiv Kemi* **12:**321.

Botti, R. E. and O. D. Ratnoff. 1964. Studies on the pathogenesis of thrombosis: An experimental "hypercoagulable" state induced by the intravenous injection of ellagic acid. *J. Lab. Clin. Med.* **64:**358.

Bull, B. S., M. L. Rubenberg, J. V. Dacie and M. C. Brain. 1968. Microangiopathic haemolytic anaemia: Mechanisms of red cell fragmentation: *In vitro* studies. *Brit. J. Haematol.* **14:**643.

Courtney, L. D. and M. Allington. 1972. Effect of amniotic fluid on blood coagulation. *Brit. J. Haematol.* **22:**353.

Evensen, S. A. and M. Jeremic. 1970. Do platelets enhance thromboplastin-induced intravascular clotting? *Scand. J. Haematol.* **7:**413.

Ewart, M. R., M. W. C. Hatton, J. M. Basford and K. S. Dodgson. 1970. The proteolytic action of Arvin on human fibrinogen. *Biochem. J.* **118:**603.

Fedder, G., E. M. Prakke and J. Vreeken. 1972. On the early detection of intravascular coagulation. An experimental study. *Thromb. Diath. Haemorrh.* **27:**365.

Goodnight, S. H. 1974. Bleeding and intravascular clotting in malignancy: A review. *Ann. N.Y. Acad. Sci.* **230:**271.

Goodnight, S. H., G. Kenoyer, S. I. Rapaport, M. J. Patch, J. A. Lee and T. Kurze. 1974. Defibrination after brain-tissue destruction. A serious complication of head injury. *New Eng. J. Med.* **290:**1043.

Gralnick, H. R. and E. Abrell. 1973. Studies of the procoagulant and fibrinolytic activity of promyelocytes in acute promyelocytic leukemia. *Brit. J. Haematol.* **24:**89.

Heusinkveld, R. S., J. P. Leddy, M. R. Klemperer and R. T. Breckenridge. 1974. Hereditary deficiency of the sixth component of complement in man. *J. Clin. Invest.* **53:**554.

Hjort, P. F. and S. I. Rapaport. 1965. The Shwartzman reaction: Pathogenetic mechanisms and clinical manifestations. *Annu. Rev. Med.* **16:**135.

Kidder, W. R., L. J. Logan, S. I. Rapaport and M. J. Patch. 1972. The plasma protamine paracoagulation test: Clinical and laboratory evaluation. *Amer. J. Clin. Pathol.* **58**:675.

Lerner, R. G., R. Goldstein and G. Cummings. 1971. Stimulation of human leukocyte thromboplastic activity by endotoxin. *Proc. Soc. Exp. Biol. Med.* **138**:145.

Lerner, R. G., S. I. Rapaport and J. M. Spitzer. 1968. Endotoxin-induced intravascular clotting: The need for granulocytes. *Thromb. Diath. Haemorrh.* **20**:430.

Lipinski, B., Z. Wegrzynowicz, A. Z. Budzynski, M. Kopec, Z. S. Latallo and E. Kowalski. 1967. Soluble unclottable complexes formed in the presence of fibrinogen degradation products (FDP) during the fibrinogen-fibrin conversion and their potential significance in pathology. *Thromb. Diath. Haemorrh.* **17**:65.

Lopas, H. and N. I. Birndorf. 1971. Haemolysis and intravascular coagulation due to incompatible red cell transfusion in isoimmunized monkeys. *Brit. J. Haematol.* **21**:399.

McGehee, W. G., S. I. Rapaport and P. F. Hjort. 1967. Intravascular coagulation in fulminant meningococcemia. *Annu. Intern. Med.* **67**:250.

McKay, D. G., G. Müller-Berghaus and V. Cruse. 1969. Activation of Hageman factor by ellagic acid and the generalized Swartzman reaction. *Amer. J. Pathol.* **54**:393.

Müller-Berghaus, G. and D. Heinrich. 1972. Fibrin monomer and platelet aggregation *in vitro* and *in vivo*. *Brit. J. Haematol.* **23**:177.

Müller-Berghaus, G. and M. Hocke. 1972. Effect of endotoxin on the formation of microthrombi from circulating fibrin monomer complexes in the absence of thrombin generation. *Thromb. Res.* **1**:541.

Niemetz, J. and K. Fani. 1971. Role of leukocytes in blood coagulation and the generalized Shwartzman reaction. *Nature New Biol.* **232**:247.

Niemetz, J. and H. L. Nossel. 1969. Activated coagulation factors: *In vivo* and *in vitro* studies. *Brit. J. Haematol.* **16**:337.

Nossel, H. L., R. E. Canfield and V. P. Butler, Jr. 1973. Plasma fibrinopeptide A concentration as an index of intravascular coagulation. *4th International Congress on Thrombosis and Haemostasis* (Vienna), p. 202. G. Gistel & Cie, Vienna.

Nossel, H. L., L. R. Younger, G. D. Wilner, T. Procupez, R. E. Canfield and V. P. Butler, Jr. 1971. Radioimmunoassay of human fibrinopeptide A. *Proc. Nat. Acad. Sci.* **68**:2350.

Phillips, L. L. and E. C. Davidson. 1972. Procoagulant properties of amniotic fluid. *Amer. J. Obstet. Gynecol.* **113**:911.

Pineo, G. F., E. Regoeczi, M. W. C. Hatton and M. C. Brain. 1973. The activation of coagulation by extracts of mucus: A possible pathway of intravascular coagulation accompanying adenocarcinomas. *J. Lab. Clin. Med.* **82**:255.

Quigley, H. J. 1967. Peripheral leukocyte thromboplastin in promyelocytic leukemia. *Fed. Proc.* **26**:648.

Rapaport, S. I., P. F. Hjort, M. J. Patch and M. Jeremic. 1966. Consumption of serum factors and prothrombin during intravascular clotting in rabbits. *Scand. J. Haematol.* **3**:59.

Rapaport, S. I., S. Schiffman, M. J. Patch and S. B. Ames. 1963. The importance of activation of antihemophilic globulin and proaccelerin by traces of thrombin in the generation of intrinsic prothrombinase activity. *Blood* **21**:221.

Rosenthal, R. L. 1963. Acute promyelocytic leukemia associated with hypofibrinogenemia. *Blood* **21**:495.

Sasaki, T., I. H. Page and J. R. Shainoff. 1966. Stable complex of fibrinogen and fibrin. *Science* **152**:1069.

Shen, S. M-C., S. I. Rapaport and D. I. Feinstein. 1973. Intravascular clotting after endotoxin in rabbits with impaired intrinsic clotting produced by a factor VIII antibody. *Blood* **42:**523.

Sherry, S. 1972. Mechanisms of fibrinolysis. In *Hematology* (ed. W. J. Williams et al.), pp. 1105–1114. McGraw-Hill, New York.

Siraganian, R. P. 1972. Platelet requirement in the interaction of the complement and clotting systems. *Nature New Biol.* **239:**208.

Skjørten, F. and S. A. Evensen. 1973. Induction of disseminated intravascular coagulation in the factor XII-deficient fowl. *Thromb. Diath. Haemorrh.* **30:**25.

Stafford, B. T., S. M-C. Shen and S. I. Rapaport. 1973. Endotoxin-induced clotting and the generalized Shwartzman reaction in C6 deficient rabbits. *Blood* **42:**1020.

Teger-Nilsson, A-C. and N. J. Gröndahl. 1974. A search for fibrinopeptides in urine during experimental intravascular coagulation in dogs. *Thromb. Res.* **4:**131.

Zimmerman, T. S. and H. J. Müller-Eberhard. 1971. Blood coagulation initiation by a complement-mediated pathway. *J. Exp. Med.* **134:**1601.

Zimmerman, T. S., C. M. Arroyave and H. J. Müller-Eberhard. 1971. A blood coagulation abnormality in rabbits deficient in the sixth component of complement (C6) and its correction by purified C6. *J. Exp. Med.* **134:**1591.

Chemistry and Biology
of the Kallikrein-Kinin System

John J. Pisano

Section on Physiological Chemistry, Hypertension-Endocrine Branch
National Heart and Lung Institute, NIH, Bethesda, Maryland 20014

It is the purpose of this short review to introduce the kallikrein-kinin system and present new developments. Several reviews have been published which deal with different aspects of the system (Erdös 1970; Rocha e Silva 1970, 1974; Eisen 1970; Wilhelm 1971; Colman et al. 1971; Sander and Huggins 1972; Maxwell and Acheson 1973). In order to limit the size of the bibliography, only the most recent article on a topic is cited in many instances.

The known components of the kallikrein-kinin system include the kallikreins, prekallikrein activator, kallikrein inhibitors, kininogens, kinins, kinin-converting aminopeptidase and kininases.

Kallikreins

Kallikrein (EC 3.4.21.8) denotes those serine proteinases which liberate kinins from kininogen by limited proteolysis and which have little or no proteolytic activity on other proteins. Another system in which an enzyme acts on a precursor protein to produce an extremely potent, biologically active peptide is the renin-angiotensin system (Fig. 1).

All mammals so far studied have two kallikrein types: plasma kallikrein and glandular kallikrein, which is found in exocrine glands and their secretions. Plasma and glandular kallikreins differ in (1) physicochemical properties, (2) rates of reaction with kininogens and synthetic substrates, (3) kinins produced, and (4) response to a variety of synthetic and natural inhibitors.

Glandular Kallikrein

Glandular kallikreins have been purified by several investigators (Table 1). They are acidic glycoproteins having reported pI's near 4.0 and molecular weights from 24,000 to 43,600. Multiple forms of kallikrein from several sources have been found by chromatography (ion exchange and hydroxylapatite), polyacrylamide gel electrophoresis and isoelectric focusing. Heterogeneity is due at least in part to varying carbohydrate and amino acid amide content.

199

200   J. J. Pisano

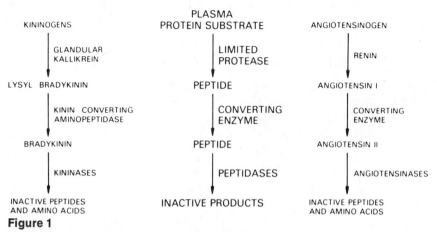

Figure 1
Similarity of the kallikrein-kinin and renin-angiotensin systems.

The pancreas (Fiedler, Hirschauer and Werle 1970) and intestine (Seki, Nakajima and Erdös 1972; Zeitlin 1972) contain prekallikrein, but only the active enzyme has been detected in salivary glands and kidney. It is premature to conclude that the preenzyme is absent in kidney, salivary and other glands, since most of the kallikrein in glands is found in sedimentable granules and activation could occur when they are disrupted by extended homogenization, freezing and thawing, or treatment with deoxycholate or hypotonic medium.

Several investigators have looked for glandular kallikrein in plasma because it has been detected in perfusates of cat salivary glands (Hilton and Lewis 1955) and rat kidney (Roblero et al. 1973). In addition, pancreatic kallikrein entering the intestine may be absorbed (Moriwaki et al. 1973). However, glandular kallikrein has not been found in normal plasma. Perhaps the rate of absorption or entry from the lymph (if it occurs) is too slow to be detected in the presence of inhibitors. On the other hand, glandular kallikrein has been observed in plasma in certain pathological states, such as pancreatitis and postgastrectomy dumping syndrome, and during carcinoid flushes.

Plasma Kallikrein

Only prekallikrein is normally found in plasma because, once activated, kallikrein is rapidly inhibited by several plasma proteinase inhibitors which may also inhibit glandular kallikreins. Plasma prekallikrein has been purified from ox, rabbit and human plasma (Table 1). It is a basic glycoprotein, MW ~ 100,000. Recently, the activated enzyme was isolated (Sampaio, Wong and Shaw 1974) from a human plasma protein fraction using affinity absorbents consisting of immobilized soybean trypsin inhibitor and aminobenzamidine (Fritz and Förg-Brey 1972).

The only known plasma activator is Hageman factor or factor XII (Nagasawa et al. 1968) and its fragments (Kaplan and Austen 1970, 1971; Kaplan, Kay and Austen 1972; Movat, Poon and Takeuchi 1971; Wuepper and Cochrane 1972; Treloar et al. 1972). Prekallikrein is activated by a process of limited proteolysis (Wuepper, Tucker and Cochrane 1970; Wuepper and Cochrane 1972; Takahashi, Nagasawa and Suzuki 1972b) in which

Table 1
Recent Purifications of Kallikreins

Kallikrein	Reference
Glandular	
Porcine pancreas	Fiedler, Hirschauer and Werle 1970; Kutzbach and Schmidt-Kastner 1972; Fritz and Förg-Brey 1972; Zuber and Sache 1974
Guinea pig coagulating gland	Moriwaki et al. 1974
Human saliva	Fujimoto, Moriya and Moriwaki 1973; Fujimoto, Moriwaki and Moriya 1973
Human urine	Pierce 1970; Pierce and Nustad 1972; Moriya et al. 1973; Porcelli et al. 1974; Hial, Diniz and Mares-Guia 1974
Horse urine	Prado, Brandi and Katchburian 1963
Rat urine	Porcelli and Croxatto 1971; Nustad and Pierce 1974; Silva, Diniz and Mares-Guia 1974
Rat intestine	Zeitlin 1972
Human, monkey, pig, dog and rat intestine	Seki, Nakajima and Erdös 1972
Plasma	
Human	Pierce 1970; Kaplan, Kay and Austen 1972; Fritz, Wunderer and Dittmann 1972; Sampaio, Wong and Shaw 1974
Bovine	Takahashi, Nagasawa and Suzuki 1972a
Rabbit	Wuepper and Cochrane 1972

an activation peptide (MW $\sim$ 10,000) is released (Wuepper and Cochrane 1972). Trypsin and certain other proteases also activate plasma and glandular prekallikreins.

Properties of Kallikreins

Human plasma contains about 15 μg prekallikrein per ml, pig pancreas, 75 μg/g fresh tissue, and a 24-hour human urine sample, 400 μg of the activated enzyme. The kallikreins are limited proteases; they hydrolyze two bonds in kininogen to release kinins located inside the polypeptide chain. Plasma kallikrein also hydrolyzes bonds in activated Hageman factor (Ulevitch et al., this volume) and in α_2-macroglobulin (Harpel, Mosesson and Cooper, this volume). Glandular kallikreins have not been tested with Hageman factor or α_2-macroglobulin. The kallikreins have virtually no proteolytic activity on other proteins (Moriwaki et al. 1974; Zuber and Sache 1974) except that high levels of horse urinary kallikrein (Prado, Brandi and Katchburian 1963), porcine pancreatic kallikrein (Zuber and Sache 1974) and perhaps other glandular kallikreins hydrolyze some arginyl bonds in salmine and polyarginine; they do not hydrolyze polylysine.

All plasma kallikreins tested release bradykinin (Rocha e Silva, Beraldo and Rosenfeld 1949) from kininogens (Fig. 2); glandular kallikreins release lysyl-bradykinin (Pierce and Webster 1961; Werle, Trautschold and Leysath

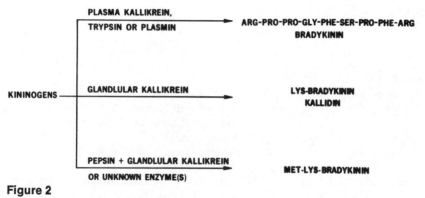

Figure 2

Formation of mammalian kinins.

1961). Trypsin and plasmin also produce bradykinin (Suzuki et al. 1972). A third mammalian kinin, Met-Lys-bradykinin, discovered in incubates of bovine plasma (Elliott and Lewis 1965) and also found free in human urine (Miwa, Erdös and Seki 1969) along with bradykinin and Lys-bradykinin, is generated by unknown enzymes. Attention is currently focused on the enzyme(s) in lysosomes of human polymorphonuclear leukocytes (Movat et al. 1973). Porcine pancreatic kallikrein can form Met-Lys-bradykinin along with Lys-bradykinin when it acts on kininogen fragments instead of native kininogen (Hochstrasser and Werle 1967). Horse urinary kallikrein also generates Met-Lys-bradykinin when acting on several synthetic substrates (E. S. Prado, pers. comm.).

Like trypsin, the kallikreins hydrolyze N-α-substituted esters of arginine and lysine and are inhibited by diisopropylfluorophosphate (DFP) and many of the natural and synthetic trypsin inhibitors. However, there are interesting differences between trypsin and glandular and plasma kallikrein. For example, one trypsin active-site inhibitor, N-α-carbobenzoxy-L-lysine chloromethyl ketone (ZLCK), inhibits human plasma kallikrein (Sampaio, Wong and Shaw 1974), but another, tosyl-L-lysine chloromethyl ketone (TLCK) inhibits none of the kallikreins (Mares-Guia and Diniz 1967; Colman, Mattler and Sherry 1969; Zuber and Sache 1974; Silva, Diniz and Mares-Guia 1974; Hial, Diniz and Mares-Guia 1974; Sampaio, Wong and Shaw 1974). A stable p-guanidinobenzoyl enzyme is formed by reacting p-nitro-p'-guanidinobenzoate (NPGB) with porcine pancreatic (Fiedler, Müller and Werle 1972), rat urinary (Silva, Diniz and Mares-Guia 1974) and human plasma kallikreins (Sampaio, Wong and Shaw 1974). Human urinary kallikrein is unique since it is only slowly inhibited by DFP, is not titratable by NPGB, and its kinin-liberating activity is not inhibited by benzamidine (Hial, Diniz and Mares-Guia 1974).

The kallikreins have a much stronger preference for esters of arginine than for those of lysine (Fiedler, Leysath and Werle 1973), and with the exception of rat urinary kallikrein (Silva, Diniz and Mares-Guia 1974), they barely hydrolyze N-α-substituted amides of these amino acids.

Soybean, but not lima bean, trypsin inhibitor inhibits plasma kallikrein; neither inhibits glandular kallikrein. Kunitz bovine trypsin inhibitor acts on human, porcine and bovine plasma and glandular kallikreins, but not on dog, cat, mouse or guinea pig glandular kallikreins (Vogel, Trautschold and Werle 1968; Vogel and Werle 1970). Porcine urinary (kidney) kallikrein has an unexpectedly weaker affinity for this inhibitor than porcine pancreatic or salivary kallikreins (Werle, Fiedler and Fritz 1973). Pancreatic secretory trypsin inhibitor (Kazal type) does not inhibit any of the kallikreins so far tested (Fritz et al. 1974).

Russell's viper venom contains two powerful broad spectrum inhibitors which inactivate all kallikreins and are structurally and functionally similar to the Kunitz pancreatic trypsin inhibitor (Takahashi, Iwanaga and Suzuki 1974). This type of inhibitor also has been found in sea anemones, snails (*Helix pomatia*) and bovine colostrum (Fritz et al. 1974).

Kinetic and X-ray crystallographic studies show that the affinity of a polypeptide inhibitor for these proteinases is determined by the tertiary structure of the reactive sites of both reactants (Fritz et al. 1974).

Immunochemical identity of human pancreatic, salivary, sweat and urinary kallikreins was found by Ouchterlony double diffusion against monospecific antibody to human urinary kallikrein (T. Kaizu, J. Pierce and J. Pisano, unpubl.). Plasma and glandular kallikreins do not cross-react. Also, no species cross-reactivity between the kallikreins has been noted (Vogel and Werle 1970). However, human and monkey kininogens and plasma and urinary kallikreins cross-react. Nevertheless, the kallikreins from different glands appear to be distinguishable, e.g., human urinary and pancreatic kallikreins (J. Pierce, pers. comm.).

The interspecies enzymatic activity of the kallikreins with kininogens is variable (Webster 1970).

Other Kininogenases

Numerous proteolytic enzymes such as Nargase and ficin can form kininlike peptides when incubated with plasma or purified kininogens (Prado 1970). The activity of trypsin (about half as active as the kallikreins on native kininogen) and the enzymes in snake venom led to the discovery of bradykinin (Rocha e Silva, Beraldo and Rosenfeld 1949). Of the several kininogenases, plasmin (Beraldo 1950) is of greatest interest since it occurs in plasma with kininogen.

Inhibitors of Kallikrein in Plasma

Plasma kallikrein is inhibited by at least four different plasma proteins. Two of these inhibitors, α_2-macroglobulin (Harpel 1970, 1971; McConnell 1972) and C$\overline{1}$ inactivator (Ratnoff et al. 1969; Gigli et al. 1970), react immediately with human and swine plasma kallikreins (Fritz et al. 1972). The other inhibitors, α_1-antitrypsin (Fritz et al. 1972) and antithrombin III (Lahiri et al. 1974), react slowly and progressively. The interaction of plasma kallikrein and other plasma proteases with α_2-macroglobulin and C$\overline{1}$ inhibitor is authoritatively presented by Harpel, Mosesson and Cooper (this volume).

Kininogens

Kininogens are acidic glycoproteins which have been purified from human, bovine and pig plasma (Habermann 1970; Pierce 1970). The occurrence of at least two functionally different kininogens in plasma was suggested by experiments showing that treatment of plasma with glandular kallikrein releases more kinin than can be obtained by glass-activation of the kallikrein in the plasma itself (Vogt 1966). The chromatographic separation from several mammalian plasmas of two kininogens differing both in molecular weight (197,000 and 57,000) and in susceptibility to kallikreins was achieved at the same time in independent studies (Jacobsen 1966a,b; Jacobsen and Kritz 1967). Substrate 1, the high molecular weight (HMW) kininogen form, is a good substrate for both plasma and glandular kallikreins, whereas substrate 2, the low molecular weight (LMW) form, is a good substrate only for glandular kallikreins. These findings were confirmed in other mammals (Yano et al. 1967a,b; Henriques et al. 1967). With low levels of plasma kallikrein, it is possible to exhaust HMW kininogens in human plasma, leaving most if not all, of the LMW kininogen intact (J. Guimarães, pers. comm.). Two human LMW kininogens, types I and II (MW 50,000), have been isolated (Pierce and Webster 1966; Pierce 1968). The kinin moiety appears to be at the C terminus of type I, but inside the polypeptide chain of type II. These LMW kininogens are immunochemically identical. The further purification and characterization of human HMW and LMW kininogens has been achieved. In aqueous media, the estimated molecular weights of purified HMW and LMW kininogens are 200,000 and 50,000, respectively (Habal, Movat and Burrowes 1974). In 6 M guanidine, the estimated molecular weights are 108,000 and 52,000, respectively (H. Z. Movat, pers. comm.). Guimarães et al. (1974) found four distinct, immunochemically identical human kininogens, one of which appears to be the natural substrate for human plasma kallikrein and therefore to correspond to HMW kininogen. On the other hand, one laboratory has found only one human kininogen, MW 70,000 (Spragg, Haber and Austen 1970; Spragg and Austen 1971), and another laboratory found only one rabbit kininogen, MW 79,000 (Wuepper and Cochrane 1971). Bovine kininogens have been extensively studied (Komiya, Kato and Suzuki 1974a,b,c). HMW kininogen, MW 76,000, contains 12.6% carbohydrate; LMW kininogen, MW 48,000, 19.8% carbohydrate. Tryptic digestion of HMW and LMW kininogens gave 43 and 30 peptides, respectively, 28 of which were identical. The HMW and LMW kininogens appear to be immunochemically identical.

All kininogen species so far examined release bradykinin, Lys-bradykinin or Met-Lys-bradykinin when treated with the appropriate enzymes. The amount of kinin released by kininogen is almost universally determined by the method of Diniz and Carvalho (1963) which employs trypsin to release bradykinin from kininogen in plasma acid-treated and heated to destroy kininases. Liberated bradykinin is bioassayed using the isolated rat uterus or guinea pig ileum (Trautschold 1970). About 10 μg bradykinin is reported from kininogens in 1 ml of plasma (Habermann 1970). However, this value is undoubtedly high because trypsin also generates peptides in the sample which potentiate kinin activity. In this author's laboratory, J. Guimarães (unpubl.) has found about 3 μg bradykinin/ml plasma when human or rat urinary kallikrein was used to liberate kinin from kininogen. That this value reflected all the kininogen was evident from the fact that addition of trypsin did not release more kinin.

Kinin-converting Aminopeptidase

An aminopeptidase in rabbit brain (Comargo, Ramalho-Pinto and Greene 1972), human plasma (Guimarães et al. 1973) and human liver (Borges, Prado and Guimarães 1974) rapidly converts Lys-bradykinin and Met-Lys-bradykinin to the limit peptide, bradykinin. Studies involving lung-bypass perfusion of exsanguinated rats with oxygenated Tyrodes-dextran containing Lys-bradykinin or Met-Lys-bradykinin revealed a rapid conversion of these peptides to bradykinin, which was partially recovered in the perfusion medium. Thus kinin conversion can occur at a faster rate than kinin destruction (J. L. Prado, pers. comm.).

Naturally Occurring Kinins

The pharmacological actions of the three mammalian kinins are qualitatively similar, but in quantitative tests with isolated guinea pig ileum, bradykinin was three and ten times more potent than Lys-bradykinin and Met-Lys-bradykinin, respectively. The order was reversed in vascular permeability tests (Reis, Okino and Rocha e Silva 1971). Normal plasma contains less than 3 ng kinin per ml, a value which is of questionable significance. Some authorities believe that kinin destruction is so efficient in plasma and lung that plasma assays are of little value as a measure of kinin production. In this author's laboratory, V. Hial (unpubl.) has found that in a 24-hour sample normal human urine contains about 3–10 μg of each of the three kinins.

The kallikrein-kinin system is widespread in nature, but there are phylogenetic differences in the number, properties and interaction of the components (Seki et al. 1973). For example, plasmas of snakes, frogs and fish contain only kininases; avian plasma contains kallikrein, kininogen and kininase, but lacks an activator like Hageman factor. Turtle and alligator plasmas have all of the components and appear similar to mammalian plasma.

Hymenoptera (wasp, yellowjacket and hornet) venoms (Pisano 1968) and especially amphibian skin, as shown by the outstanding pioneering work from Erspamer's laboratory (Erspamer and Melchiorri 1973), are rich sources of kinins and related peptides (Table 2). Venom from one insect contains a few micrograms of kinin, but the skin of certain amphibians, including our common *Rana pipiens,* contains about 250 μg of bradykinin/g fresh skin. The kinin isolated from yellowjacket venom (Yoshida and Pisano 1974) may be the first vasoactive glycopeptide. The carbohydrate moiety consists entirely of galactose and N-acetylgalactosamine.

Kininases

Plasma and tissue contain a variety of kininases (Erdös and Yang 1970) which can account for the low kinin level of < 3 ng/ml normal plasma (Talamo, Haber and Austen 1969). Hydrolysis of any one of the peptide bonds inactivates bradykinin. The lung normally plays a major role in kinin destruction, since about 80% of administered kinin is inactivated in one pass through this organ (Ferreira and Vane 1967). Ryan, Roblero and Stewart (1968) have shown that no less than five peptide bonds are broken (Arg[1]-Pro[2], Pro[3]-Gly[4], Gly[4]-Phe[5], Ser[6]-Pro[7], Pro[7]-Phe[8]) in one circulation of brady-

Table 2
Naturally Occurring Kinins

Common name	Sequence	Reference
Bradykinin	Arg-Pro-Pro-Gly-Phe-Ser-Pro-Phe-Arg	Rocha e Silva, Beraldo and Rosenfeld 1949; Elliott, Lewis and Horton 1960; Anastasi, Erspamer and Bertaccini 1965
Kallidin	Lys-bradykinin	Pierce and Webster 1961; Werle, Trautschold and Leysath 1961
Methionyllysylbradykinin	Met-Lys-bradykinin	Elliott and Lewis 1965; Dunn and Perks 1970
Thr^6-bradykinin	Arg-Pro-Pro-Gly-Phe-Thr-Pro-Phe-Arg	Watanabe, Yasuhara and Nakajima 1974; Ishikawa et al. 1974
Val^1,Thr^6-bradykinin	Val-Pro-Pro-Gly-Phe-Thr-Pro-Phe-Arg	Nakajima 1968a
Polisteskinin	Pyroglu-Thr-Asn-Lys-Lys-Leu-Arg-Gly-bradykinin	Pisano 1968
	carbohydrate carbohydrate	
Vespulakinin	Thr-Ala-Thr-Thr-Arg-Arg-Arg-Gly-bradykinin	Yoshida and Pisano 1974
Polisteskinin-R	Ala-Arg-Thr^6-bradykinin	Watanabe, Yasuhara and Nakajima 1974
Phyllokinin	Bradykinyl-Ile-Tyr(SO$_3$H)	Anastasi, Bertaccini and Erspamer 1966
Ranakinin N	Bradykinyl-Val-Ala-Pro-Ala-Ser	Nakajima 1968b
Ranakinin O	Bradykinyl-Gly-Lys-Phe-His	Yasuhara et al. 1973
Ranakinin R	Thr^6-bradykinyl-Ile-Ala-Pro-Gly-Ile-Val	Ishikawa et al. 1974

kinin through bloodfree rat lung. Bradykinin delivered into the pulmonary artery of patients undergoing cardiac catheterization is also > 80% destroyed in one pass through the lung. A major catabolite was the C-terminal dipeptide Phe-Arg. This indicates that kininase II (angiotensin-converting enzyme, peptidyldipeptide hydrolase [EC 3.4.15.1]; Erdös 1970, 1975; Nakajima et al. 1973; Oshima, Gecse and Erdös 1974) can account for most of the kinin inactivation in human lung (unpublished observations from this author's laboratory).

The half-life of kinins in cat blood is about 17 seconds (Ferreira and Vane 1967), which is somewhat longer than the circulation time of blood through lung (2–10 sec). Nonetheless, it is obvious that the plasma kininase activity is not insignificant. Virtually all the activity in plasma can be attributed to two enzymes, kininase II and kininase I (carboxypeptidase N, anaphylatoxin inactivator, arginine carboxypeptidase [EC 3.4.12.7]; Oshima, Kato and Erdös 1974). Another kininase has been described in lysed red cells, kidney cortex (Erdös 1970) and lung (Ryan, Roblero and Stewart 1968) which splits the Arg^1-Pro^2 bond. This enzyme may be identical to prolidase (Erdös 1970). Brain contains a kininase which splits the Phe^5-Ser^6 bond of bradykinin (Comargo, Shapanka and Greene 1973). Kininases are undoubtedly present in all organs and probably inactivate locally generated kinins. Because the lung receives the entire output of the right ventricle, it is the main site of inactivation of kinins in blood. It should be noted that kininases are not specific for kinins, but hydrolyze a variety of polypeptides containing the appropriate peptide bonds.

Kinin Inhibitors and Potentiators

While specific drugs are available for blocking histamine, serotonin, catecholamines and acetylcholine, none has been found which specifically blocks the action of kinins. Little progress has been made in the development of potentially useful specific inhibitors of kinin action despite the preparation of many structural analogs (Stewart and Woolley 1966; Stewart 1968; Schröder 1970). Analgesic and anti-inflammatory agents, such as aspirin and indomethacin, can effectively reduce the pain and inflammatory action of kinins, but they probably act to block synthesis of prostaglandins which are the mediators of many kinin actions. Other compounds which block histamine and/or serotonin in isolated organ assays also inhibit bradykinin. They are noncompetitive inhibitors and probably do not act on bradykinin receptors. This work has been reviewed by Rocha e Silva (1970).

Numerous kinin potentiators have been described, including chelating agents, particularly thiol compounds, naturally occurring peptides and peptide derivatives containing alkylating agents.

Chelating agents such as EDTA, 1,10-phenanthroline, cysteine, 2,3-dimercaptopropanol and 2-mercaptoethanol probably act by inhibiting kininases which are metalloenzymes (Erdös and Yang 1970). However, a direct action of these substances on tissue receptors has not been ruled out.

Bovine fibrinopeptide B and human fibrinopeptide A (called human β peptide in Gladner 1966) were the first reported potentiators of kinins (Gladner 1966; Faber 1969). These peptides potentiate the contractile activity of bradykinin on the isolated smooth muscles, but they have not been tested in vivo where their action could be of pathophysiologic significance. Several investigators have since reported on the formation of potentiating peptides when plasma is incubated with trypsin (Aarsen 1968; Hamberg, Elg and Stellwagen 1968; Weyers et al. 1972). These trypsin-generated peptides probably inhibit the kininases (Hamberg, Elg and Stellwagen 1968).

The most interesting peptide potentiators of kinin action in vivo and in vitro have been isolated from the venoms of the crotalid snakes *Bothrops jararaca* (Ferreira, Bartelt and Greene 1970; Ondetti et al. 1971) and *Agkistrodon halys blomofii* (Kato et al. 1973). Of the 14 described to date, the largest is a tridecapeptide, the smallest, a pentapeptide, and the most potent, the nonapeptide Pyr-Glu-Trp-Pro-Arg-Pro-Gln-Ile-Pro-Pro (BPP$_{9a}$, SQ20, 881) found in *Bothrops jararaca* venom. This peptide, at 3×10^{-8} M, doubled the response of the isolated guinea pig ileum to a standard dose of 5×10^{-9} M bradykinin. Thus as few as six molecules of potentiator peptide produce an effect equivalent to that of one molecule of bradykinin. However, the potentiating peptides have no intrinsic agonist activity. More potent potentiators may be present in *Crotalus viridi viridi* (Ferreira 1966) and *C. horridus horridus* venoms (Sander, West and Huggins 1972). These venom peptides potentiate the action of bradykinin by inhibition of kininases and by a direct action on tissues. An important cause of potentiation in vivo appears to be the inhibition of kininases of lung (Freer and Stewart 1973). The inhibition of kininase II, the angiotensin-converting enzyme (EC 3.4.15.1) by BPP$_{9a}$ has been used to detect renovascular hypertension (Krieger et al. 1971), a condition in which there is an overproduction of renin and angiotensin I. The venom peptide blocks the conversion of angiotensin I to angiotensin II which is hypertensive. This block results in a lowering of blood pressure in these hypertensive subjects. The venom peptides also may be useful in the treatment of shock (Errington and Rocha e Silva, Jr. 1973; Erdös et al. 1974). It is of interest also to mention another naturally occurring and more potent peptide inhibitor, the fatty acid-containing pentapeptide, pepstatin (isovaleric acid-valine-valine-4-amino-3-hydroxy-6-methylheptanoic acid-alanine-4-amino-3-hydroxy-6-methylheptanoic acid) found in culture filtrates of actinomycetes. This peptide strongly inhibits pepsin, several other proteases (Umezawa 1972; Aoyagi and Umezawa, this volume) and renin. The K_i for renin is so low, 1.3×10^{-10} M, that it is possible to determine the molarity of renin in only partially purified preparations (McKown, Workman and Gregerman 1974). Mammalian peptide inhibitors of renin also have been described, but they are much less potent than pepstatin (Workman, McKown and Gregerman 1974). Pepstatin inhibition of the kallikreins has not been reported.

Potentiators of bradykinin have been prepared by the attachment of the nitrogen mustard alkylating agent chlorambucil to certain fragments of bradykinin. These derivatives also appear to act by kininase inhibition (Freer and Stewart 1973; Stewart, Freer and Ferreira 1973).

Leukokinin System

Another active peptide(s) is generated from a precursor protein(s) through the action of specific enzyme(s). The leukokinins are thought to play a role in inflammation and in fluid accumulation in malignancy. The leukokinins

(20–25 amino acid residues) are produced by cathepsin D- and E-like enzymes in leukocytes which act on leukokininogen in plasma. Conversion of proleukokininogen to leukokininogen by an unknown mechanism is the rate-limiting step (Greenbaum, this volume).

PATHOPHYSIOLOGIC SIGNIFICANCE OF THE KALLIKREIN-KININ SYSTEM

Kinins are the most potent mammalian vasodilator (hypotensive) peptides known. They also increase vascular permeability and lymph flow, are bron-choconstrictors, contract most isolated smooth muscles, and produce peripheral and visceral pain (Armstrong 1970). Kallikreins rapidly generate kinins which are rapidly destroyed by plasma kininases or by the very efficient kininases in lung. For this reason, kinins are often considered local hormones since they must be utilized where they are formed.

From this list of impressive biologic actions, it would seem that the kallikrein-kinin system is surely involved in certain physiologic and pathologic processes. Such an involvement has been difficult to demonstrate because of the inability to specifically block kallikrein or kinins. Nonetheless, alterations in the levels of components of the system (i.e., kinins, kininogen, prekallikrein and its inhibitors) have been closely associated with several physiologic and pathologic processes, including endotoxin (Hirsch et al. 1974) and anaphylactic shock, edema, including hereditary angioedema (Vogel and Zickgraf-Rüdel 1970), inflammation (Lewis 1970), pancreatitis (Ofstad 1970), acute arthritides (Webster et al. 1972), carcinoid flush (Colman, Mason and Sherry 1969), postgastrectomy dumping syndromes (Zeitlin 1970; Wong et al. 1974) and angina associated with coronary ischemia (Pitt et al. 1969). The cardiovascular effects of kallikrein (Brecher and Brobmann 1970) and kinins (Haddy et al. 1970) are numerous. Kinins also have been implicated in the pathogenesis of migraine headache and some clinical phenomena associated with intracranial vascular accidents (Sicuteri 1970). It is the vasodilating, permeability-increasing and pain-producing actions of kinins and the leukotactic action of plasma kallikrein which seems to be manifest in these pathological conditions. (See some of the reviews cited at the beginning of this chapter for comprehensive discussions of the pathophysiological role of the kallikrein-kinin system.)

Physiological Actions

Glandular Secretion

One of the earliest proposals for a physiological role for the kallikrein-kinin system stemmed from the facts that kallikrein is abundant in exocrine glands and that kinins are potent vasodilators. Based upon the release of kallikrein and kinin and the consumption of kininogen upon appropriate stimulation, it has been proposed that kinins are functional vasodilators in salivary glands, the pancreas, sweat glands and glands in the tongue (Hilton 1970). This view has been challenged because effective vasodilation and salivary secretion can occur without any involvement of the kallikrein-kinin system (Schachter 1970; Schachter, Barton and Karpinski 1973). In recent impressive experiments, it has been shown that in the cat, at least two different mechanisms regulate functional vasodilation, one involving vasodilator nerves and the other the kallikrein-kinin system (Gautvik et al. 1974).

Kinins, Kidney Function and Hypertension

Urinary kallikrein is synthesized in the kidney (Nustad, Pierce and Vaaje 1975). The urinary excretion of kallikrein is analogous to the secretion of kallikrein by the pancreas, salivary and other glands, and, as mentioned above for the other glands, a functional vasodilator role for the kallikrein-kinin system in the kidney has been suggested by many investigators. Infusion of kinins into the renal artery causes a marked vasodilation, increased renal blood flow and an accompanying natriuresis (Jacobson 1970; Wills et al. 1969). Kinins also affect tubular sodium reabsorption (Stein et al. 1972; Alzamora and Capelo 1973). The positive correlation between urinary kallikrein and sodium reported by some investigators (Marin-Grez, Cottone and Carretero 1972; Adetuyibi and Mills 1972) has not been observed by others (Croxatto and San Martín 1970; Geller et al. 1972; Margolius et al. 1974a,b).

Altered urinary kallikrein excretion has been noted in hypertensive patients (Margolius et al. 1974a,b) and animals (Croxatto and San Martín 1970; Geller et al. 1972). Those with essential hypertension excrete about half as much kallikrein as normotensive controls, whereas subjects with primary aldosteronism excrete about twice as much (Margolius et al. 1974a). There is a positive correlation between the levels of circulating sodium-retaining steroid and kallikrein excretion in subjects with primary aldosteronism and in subjects in whom the endogenous steroid level is increased by a low sodium diet (Margolius et al. 1974a,b). The urine of normotensive children with above average blood pressure is less concentrated in kallikrein than age-, sex- and race-matched children with below average blood pressures (H. Margolius, pers. comm.).

An important advance in our understanding of the biological role of the kallikrein-kinin system occurred when Wuepper (1973) reported that highly purified prekallikrein corrected a previously recognized coagulation defect in plasma from the Fletcher family (Hathaway, Belhasen and Hathaway 1965). Shortly thereafter, it was shown that Fletcher plasma was also deficient in fibrinolytic (Weiss, Gallin and Kaplan 1974; Saito, Ratnoff and Donaldson 1974) and chemotactic (Weiss, Gallin and Kaplan 1974) activities which were also restored by addition of purified kallikrein. The manner in which kallikrein promotes coagulation and fibrinolysis is shown in Figure 3. The three systems are linked by coagulation factor XII (Hageman factor). Kallikrein accelerates the activation of factor XII bound to a negatively charged surface into XII_A (Ulevitch et al., this volume). The fragment(s) (XIIf) are about as active as XII_A on prekallikrein and plasminogen proactivator, but clearly less potent in activation of factor XII. Plasma kallikrein functions by a positive feedback mechanism to optimally promote coagulation, fibrinolysis, chemotaxis and kinin generation. Plasma kallikrein and plasminogen activator also possess chemotactic activity for human neutrophils (Kaplan, Kay and Austen 1972; Kaplan, Goetzl and Austen 1974).

It was of great interest to learn that subjects with Fletcher trait are generally healthy, as are subjects with Hageman trait. It has been argued that a

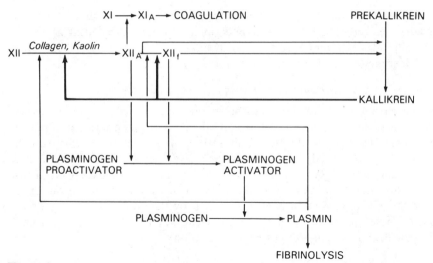

Figure 3
Interrelationship of coagulation, fibrinolysis and the kallikrein-kinin system.

deficiency of Hageman factor is not life threatening because there are alternate enzymes which can perform the important functions of Hageman factor. The same may be true for plasma kallikrein.

The only other kinin-generating enzyme known in plasma is plasmin. Future studies, especially with subjects having Fletcher trait, will determine the physiologic significance of plasma kallikrein and other plasma kininogenases such as plasmin.

Kallikrein-Kinin System and the Gastrointestinal Tract

Bradykinin is a powerful dilator of vessels in both the stomach and gut (Jacobson 1970). A prekallikrein, claimed to be primarily of the plasma type, has been isolated from human and animal colonic mucosa and submucosa (Seki, Nakajima and Erdös 1972). However, the prekallikrein from perfused (and presumably blood-free) rat intestinal wall was only of the glandular type (Zeitlin 1972). Pancreatic kallikrein can be absorbed by isolated rat intestine (Moriwaki et al. 1973) and must be accounted for in studies of intestinal kallikrein. Several laboratories are investigating the role of kallikrein in intestinal absorption (Meng and Haberland 1973; Moriwaki et al. 1973; Fasth and Hultěn 1973a,b).

Cellular Proliferation

The injection of hog pancreatic kallikrein into rats stimulates DNA synthesis and mitosis in the thymus and bone marrow through a calcium-dependent, cyclic AMP-mediated process (Rixon and Whitfield 1973). In vitro studies with rat thymocytes or bone marrow cells show that the effect is due to kinin. Hence, a direct action of kallikrein on regulatory nuclear proteins is not indicated in this study. However, in other studies, long-term treatment with porcine pancreatic kallikrein caused a higher [3H]thymidine labeling of bone marrow and epithelial basal cells of the tongue (Löbbecke 1973), stimulated mitotic activity in the gastrointestinal tract of rats (Rohen and Peterhoff

1973), and promoted regeneration of bone marrow, survival after whole body irradiation, and recovery from radiodermatitis produced by local irradiation (Mandel, Rodesch and Mantz 1973). Kallikreins and kinins are short acting, and their mode of action over a long period is not clear.

Conversion from Intrauterine to Extrauterine Circulation

Kinins have an uncharacteristic vasoconstrictor action on the umbilical artery and vein of human and lamb fetuses and the fetal ductus arteriosus of the calf and lamb (Melmon et al. 1968; Campbell et al. 1968). Pulmonary vascular resistance is decreased and blood flow increased. These effects are enhanced by the increasing oxygen tension of blood at birth. Increased oxygen also promotes kinin formation by granulocytes, which are more numerous in the newborn. Kinin formation is also promoted by the cooling of cord blood at birth. These observations make it highly plausible that kinins have a role in the conversion from pre- to postnatal circulation.

Prostaglandins, Modulators and Mediators of the Kinin Response

Some, but not all, of the responses to bradykinin are partially or completely antagonized by aspirinlike drugs which are inhibitors of prostaglandin synthesis. Bradykinin releases prostaglandins from isolated guinea pig lung (Palmer, Piper and Vane 1973), dog kidney (McGiff et al. 1972; McGiff, Itskovitz and Terragno 1975) and dog spleen (Ferreira, Moncada and Vane 1973). Prostaglandins were not released by other vasodilator substances (eledoisin tested in the kidney and isoprenaline in the spleen), indicating that bradykinin had a specific action. In a most interesting report (Terragno et al. 1975), it was shown that bradykinin added to bovine mesenteric vessel slices stimulated prostaglandin synthesis, predominantly an E prostaglandin by arteries and a F prostaglandin by veins.

Kinins stimulate the synthesis of cyclic GMP and cyclic AMP in guinea pig lung slices (Stoner, Manganiello and Vaughan 1973). The rise in cyclic AMP appears to be prostaglandin mediated, as it is blocked by aspirin and indomethacin. Since these drugs did not lower the accumulation of cyclic GMP, it is suggested that cyclic GMP is required for the synthesis of prostaglandins, which in turn increase cyclic AMP synthesis.

Aspirin antagonizes bradykinin-induced bronchoconstriction in guinea pigs, hypotension in guinea pigs and rabbits, and vasodilation in dogs (Collier 1969, 1970). The increased vascular permeability caused by bradykinin is potentiated by low concentrations of prostaglandins (Moncada, Ferreira and Vane 1973; Thomas and West 1974; Williams and Morley 1973).

Prostaglandins have been shown to sensitize nerve endings to the pain-producing ability of bradykinin in the spleen (Ferreira, Moncada and Vane 1973), in the knee joints of dogs (Moncada, Ferreira and Vane 1973) and in the skin of man (Ferreira and Vane 1974).

CONCLUDING REMARKS

Even though recent results were stressed in this review, two pioneers in the kallikrein-kinin field should be recognized: Professor Eugen Werle, Institute of Clinical Chemistry, University of Munich, Germany; and Professor M. Rocha e Silva, Department of Pharmacology, Faculty of Medicine, Ribeirao Preto, Sao Paulo, Brazil. They discovered most of the components of the system, coined terms, and continue to make important contributions. Many of the leaders in the field today studied in their laboratories or came under their influence at some time in their careers.

Authoritative reviews on all aspects of the chemistry and biology of the kallikrein-kinin system may be found in Pisano and Austen (1975).

REFERENCES

Aarsen, P. N. 1968. Sensitization of guinea pig ileum to the action of bradykinin by trypsin hydrolysate of ox and rabbit plasma. *Brit. J. Pharmacol.* **32**:453.

Adetuyibi, A. and I. H. Mills. 1972. Relation between urinary kallikrein and renal function, hypotension, and excretion of sodium and water in man. *Lancet* **2**:203.

Alzamora, F. and L. R. Capelo. 1973. Increase of kinin in urine after partial occlusion of the renal vein and the effect of bradykinin on renal sodium excretion. *Agents and Actions* **3**:366.

Anastasi, A., G. Bertaccini and V. Erspamer. 1966. Pharmacological data on phyllokinin (bradykinyl-isoleucyl-tyrosine O-sulfate and bradykinyl-isoleucyl-tyrosine). *Brit. J. Pharmacol.* **27**:479.

Anastasi, A., V. Erspamer and G. Bertaccini. 1965. Occurrence of bradykinin in the skin of *Rana temporaria*. *Comp. Biochem. Physiol.* **14**:43.

Armstrong, D. 1970. Pain. In *Handbook of Experimental Pharmacology. Bradykinin, Kallidin and Kallikrein* (ed. E. Erdös), vol. 25, p. 434. Springer, New York.

Beraldo, W. T. 1950. Bradykinin formation by the action of peptone and fibrinolysin on plasma globulius in vitro. *Ciencia Cult.* (*S. Paulo*) **2**:300.

Borges, D. R., J. L. Prado and J. A. Guimarães. 1974. Characterization of a kinin-converting arylaminopeptidase from human liver. *Naunyn-Schmiedeberg's Arch. Pharmacol.* **281**:403.

Brecher, G. A. and G. F. Brobmann. 1970. Effect of kallikrein on the cardiovascular system. In *Handbook of Experimental Pharmacology. Bradykinin, Kallidin, and Kallikrein* (ed. E. Erdös), vol. 25, p. 351. Springer, New York.

Campbell, A. G. M., G. S. Dawes, A. P. Fishman, A. I. Hyman and A. M. Perks. 1968. The release of a bradykinin-like pulmonary vasodilator substance in foetal and newborn lambs. *J. Physiol.* **195**:83.

Collier, H. O. J. 1969. A pharmacological analysis of aspirin. *Adv. Pharmacol. Chemother.* **7**:333.

———. 1970. Kinins and ventilation of the lungs. In *Handbook of Experimental Pharmacology. Bradykinin, Kallidin and Kallikrein* (ed. E. Erdös), vol. 25, p. 409. Springer, New York.

Colman, R. W., J. W. Mason and S. Sherry. 1969. The kallikreinogen-kallikrein enzyme system of human plasma: Assay of components and observations in disease states. *Ann. Intern. Med.* **71**:763.

Colman, R. W., L. Mattler and S. Sherry. 1969. Studies on the prekallikrein (kallikreinogen)-kallikrein enzyme system of human plasma. II. Evidence relating to the kaolin-activated arginine esterase to plasma kallikrein. *J. Clin. Invest.* **48**:23.

Colman, R. W., G. J. D. Girey, R. Zacest and R. C. Talamo. 1971. The human plasma kallikrein-kinin system. In *Progress in Hematology* (ed. E. B. Brown and C. V. Moore), vol. 7, p. 255. Grune and Stratton, New York.

Comargo, A. C. M., F. J. Ramalho-Pinto and L. J. Greene. 1972. Brain peptidases: Conversion and inactivation of kinin hormones. *J. Neurochem.* **19**:37.

Comargo, A. C. M., R. Shapanka and L. J. Greene. 1973. Preparation, assay and partial characterization of a neutral endopeptidase from rabbit brain. *Biochemistry* **12**:1838.

Croxatto, H. R. and M. San Martín. 1970. Kallikrein-like activity in the urine of renal hypertensive rats. *Experientia* **26**:1216.

Diniz, C. R. and I. V. Carvalho. 1963. A micromethod for the determination of bradykininogen under several conditions. *Ann. N.Y. Acad. Sci.* **104**:77.

Dunn, R. S. and A. M. Perks. 1970. A new plasma kinin in the turtle, *Psedemys scripta elegans*. *Experientia* **26**:1220.

Eisen, V. 1970. Formation and function of kinin. In *The Immunochemistry and Biochemistry of Connective Tissue and Its Disease States: Rheumatology*, vol. 3, p. 103. S. Karger, New York.

Elliott, D. F. and G. P. Lewis. 1965. Methionyl-lysyl-bradykinin, a new kinin from ox blood. *Biochem. J.* **95**:437.

Elliott, D. F., G. P. Lewis and E. W. Horton. 1960. The structure of bradykinin— A plasma kinin from ox blood. *Biochem. Biophys. Res. Comm.* **3**:87.

Erdös, E., ed. 1970. *Handbook of Experimental Pharmacology. Bradykinin, Kallidin and Kallikrein,* vol. 25. Springer, New York.

———. 1975. Angiotensin I converting enzyme. *Circ. Res.* **36**:247.

Erdös, E. and H. Y. T. Yang. 1970. Kininases. In *Handbook of Experimental Pharmacology. Bradykinin, Kallidin and Kallikrein* (ed. E. Erdös), vol. 25, p. 289. Springer, New York.

Erdös, E. G., W. H. Massion, D. R. Downs and A. Gecse. 1974. Effect of inhibitors of angiotensin I converting enzyme in endotoxin and hemorrhagic shock. *Proc. Soc. Exp. Biol. Med.* **145**:948.

Errington, M. L. and M. Rocha e Silva, Jr. 1973. The effect of a bradykinin potentiating nonapeptide (BPP$_{9a}$) on prolonged hemorrhagic shock in dogs. *J. Physiol.* **233**:46p.

Erspamer, V. and P. Melchiorri. 1973. Active polypeptides of the amphibian skin and their synthetic analogues. *Pure Appl. Chem.* **35**:463.

Faber, D. B. 1969. Pharmacology of peptides which enhance the action of bradykinin. *Acta Physiol. Pharmacol. Neerl.* **15**:101.

Fasth, S. and L. Hultén. 1973a. The effect of bradykinin on intestinal motility and blood flow. *Acta Chir. Scand.* **139**:699.

———. 1973b. The effect of bradykinin on the consecutive vascular sections of the small and large intestine. *Acta. Chir. Scand.* **139**:707.

Ferreira, S. H. 1966. Bradykinin-potentiating factor. In *Hypotensive Peptides* (ed. E. G. Erdös, N. Back and F. Sicuteri), p. 356. Springer, New York.

Ferreira, S. H. and J. R. Vane. 1967. The disappearance of bradykinin and eledoisin in the circulation and vascular beds of the cat. *Brit. J. Pharmacol.* **30**:417.

———. 1974. New aspects of the mode of action of nonsteroid anti-inflammatory drugs. *Annu. Rev. Pharmacol.* **14**:57.

Ferreira, S. H., D. C. Bartelt and L. J. Greene. 1970. Isolation of bradykinin-potentiating peptides from *Bothrops jararaca* venom. *Biochemistry* **9**:2583.

Ferreira, S. H., S. Moncada and J. R. Vane. 1973. Prostaglandins and the mechanism of analgesia produced by aspirin-like drugs. *Brit. J. Pharmacol.* **49**:86.

Fiedler, F., B. Hirschauer and E. Werle. 1970. Anreicherung von präkallikrein B aus schweinepankreas und eigenschaften verschiedener formen des pankreaskallikreins. *Hoppe-Seyler's Z. Physiol. Chem.* **351**:225.

Fiedler, F., G. Leysath and E. Werle. 1973. Hydrolysis of amino-acid esters by pig-pancreatic kallikreins. *Eur. J. Biochem.* **36**:152.

Fiedler, F., B. Müller and E. Werle. 1972. Active site titration of pig pancreatic kallikrein with *p*-nitrophenyl-*p'*-guanidinobenzoate. *FEBS Letters* **24**:41.

Freer, R. J. and J. M. Stewart. 1972. Alkylating analogs of peptide hormones. Synthesis and properties of *p*-[N,N-BIS(2-chloroethyl)amino]phenylbutyryl derivatives of bradykinin and bradykinin-potentiating factor. *J. Med. Chem.* **15**:1.

————. 1973. Inhibitors of the pulmonary destruction of bradykinin in the rat. In *Protides of the Biological Fluids* (ed. H. Peeters), p. 331. Pergamon, Oxford.

Fritz, H. and B. Förg-Brey. 1972. Zur isolierung von organ und harnkallikrein durch affinitätschromatographie. *Hoppe-Seyler's Z. Physiol. Chem.* **353**:901.

Fritz, H., G. Wunderer and B. Dittmann. 1972. Zur isolierung von schweine und human serum kallikrein durch affinitätschromatographie. Spezifische bindung an wasserunlösliche Kunitz-sojabahnen-inhibitor-cellulosen und dissoziation mit kompetitiven hemmstoffen (benzamidin). *Hoppe-Seyler's Z. Physiol. Chem.* **353**:893.

Fritz, H., H. Tschesche, L. J. Greene and E. Trautscheit, eds. 1974. *Proteinase Inhibitors—Second International Research Conference. Bayer Symposium* V. Springer, New York.

Fritz, H., G. Wunderer, K. Kummer, N. Heimburger and E. Werle. 1972. α_1-Antitrypsin und $\overline{CI}$ inaktivator: Progressiv-inhibitoren für serumkallikreine von mensch und schwein. *Hoppe-Seyler's Z. Physiol. Chem.* **353**:906.

Fujimoto, Y., C. Moriwaki and H. Moriya. 1973. Studies on human salivary kallikrein. II. Properties of purified salivary kallikrein. *J. Biochem.* **74**:247.

Fujimoto, Y., H. Moriya and C. Moriwaki. 1973. Studies on human salivary kallikrein. I. Isolation of human salivary kallikrein. *J. Biochem.* **74**:239.

Gautvik, K. M., M. Kriz, K. Lund-Larsen and K. Nustad. 1974. Control of kallikrein secretion and salivary glands. In *Secretory Mechanisms of Exocrine Glands. Alfred Benzon Symposium VIII* (ed. N. A. Thorn and O. H. Pettersen). Munksgaard, Copenhagen.

Geller, R. G., H. S. Margolius, J. J. Pisano and H. R. Keiser. 1972. Effects of mineralocorticoids, altered sodium intake and adrenalectomy on urinary kallikrein in rats. *Circ. Res.* **31**:857.

Gigli, I., J. W. Mason, R. W. Colman and K. F. Austen. 1970. Interaction of plasma kallikrein with the $\overline{CI}$ inhibitor. *J. Immunol.* **104**:574.

Gladner, J. A. 1966. Potentiation of the effect of bradykinin. In *Hypotensive Peptides* (ed. E. G. Erdös, N. Back and F. Sicuteri), pp. 344–348. Springer, New York.

Guimarães, J. A., D. R. Borges, E. S. Prado and J. L. Prado. 1973. Kinin-converting aminopeptidase from human serum. *Biochem. Pharmacol.* **22**:3157.

Guimarães, J. A., R. Chen Lu, M. E. Webster and J. V. Pierce. 1974. Multiple forms of human plasma kininogen. *Fed. Proc.* **33**:641.

Habal, F. M., H. Z. Movat and C. E. Burrowes. 1974. Isolation of two functionally different kininogens from human plasma: Separation from proteinase inhibitors and interaction with plasma kallikrein. *Biochem. Pharmacol.* **23**:2291.

Habermann, E. 1970. Kininogens. In *Handbook of Experimental Pharmacology. Bradykinin, Kallidin and Kallikrein* (ed. E. Erdös), vol. 25, p. 250. Springer, New York.

Haddy, F. J., T. E. Emerson, Jr., J. B. Scott and R. M. Daughterty, Jr. 1970. The effect of the kinins on the cardiovascular system. In *Handbook of Experimental*

Pharmacology. Bradykinin, Kallidin and Kallikrein (ed. E. Erdös), vol. 25, p. 362. Springer, New York.

Hamberg, U., P. Elg and P. Stellwagen. 1968. On the mechanism of bradykinin potentiation. In *Pharmacology of Hormonal Polypeptides and Proteins* (ed. N. Back, L. Martini and R. Paoletti), p. 626. Plenum, New York.

Harpel, P. C. 1970. Human plasma alpha-2-macroglobulin. An inhibitor of plasma kallikrein. *J. Exp. Med.* **132**:329.

————. 1971. Separation of plasma thromboplastin antecedent from kallikrein by plasma α_2-macroglobulin, kallikrein inhibitor. *J. Clin. Invest.* **50**:2084.

Hathaway, W. E., L. P. Belhasen and H. S. Hathaway. 1965. Evidence for a new plasma thromboplastin factor. I. Case report coagulation studies and physico-chemical properties. *Blood* **26**:521.

Henriques, O. B., N. Kauritcheva, N. Kuznetsova and M. Astrakan. 1967. Substrates of kinin-releasing enzymes isolated from horse plasma. *Nature* **215**:1200.

Hial, V., C. R. Diniz and M. Mares-Guia. 1974. Purification and properties of human urinary kallikrein (kininogenase). *Biochemistry* **13**:4311.

Hilton, S. M. 1970. The physiological role of glandular kallikreins. In *Handbook of Experimental Pharmacology. Bradykinin, Kallidin and Kallikrein* (ed. E. Erdös), vol. 25, p. 389. Springer, New York.

Hilton, S. M. and G. P. Lewis. 1955. The mechanism of the functional hyperaemia in the submandibular salivary gland. *J. Physiol.* **129**:253.

Hirsch, E. F., T. Nakajima, G. Oshima, E. Erdös and C. M. Herman. 1974. Kinin system responses in sepsis after trauma in man. *J. Surg. Res.* **17**:147.

Hochstrasser, K. V. and E. Werle. 1967. Über kininliefernde peptide aus pepsinver-dauten rinderplasmaproteinen. *Hoppe-Seyler's Z. Physiol. Chem.* **348**:177.

Ishikawa, O., T. Yasuhara, T. Nakajima and S. Tachibana. 1974. On the biological active peptide in the skin of *Rana rugosa. The 12th Symposium on Peptide Chemistry,* Kyoto.

Jacobsen, S. 1966a. Substrates for plasma kinin-forming enzymes in human, dog and rabbit plasmas. *Brit. J. Pharmacol.* **26**:403.

————. 1966b. Substrates for plasma kinin-forming enzymes in rat and guinea pig plasma. *Brit. J. Pharmacol.* **28**:64.

Jacobsen, S. and M. Kritz. 1967. Some data on two purified kininogens from human plasma. *Brit. J. Pharmacol.* **29**:25.

Jacobson, E. D. 1970. Effects of bradykinin on the kidney and gastrointestinal organs. In *Handbook of Experimental Pharmacology. Bradykinin, Kallidin and Kallikrein* (ed. E. Erdös), vol. 25, p. 385. Springer, New York.

Kaplan, A. P. and K. F. Austen. 1970. A prealbumin activator of prekallikrein. *J. Immunol.* **105**:802.

————. 1971. A prealbumin activator of prekallikrein. II. Derivation of activators of prekallikrein from active Hageman factor by digestion with plasmin. *J. Exp. Med.* **133**:696.

Kaplan, A. P., E. J. Goetzl and K. F. Austen. 1974. The fibrinolytic pathway of human plasma. II. The generation of *chemotactic* activity by activation of *plasminogen proactivator. J. Clin. Invest.* **52**:2591.

Kaplan, A. P., A. B. Kay and K. F. Austen. 1972. A prealbumin activator of prekallikrein. III. Appearance of chemotactic activity for human neutrophils by the conversion of human prekallikrein to kallikrein. *J. Exp. Med.* **135**:81.

Kato, H., T. Suzuki, K. Okada, T. Kimura and S. Sakakibara. 1973. Structure of potentiator A, one of the five bradykinin potentiating peptides from the venom of *Agkistrodon halys blomhoffii. Experientia* **29**:574.

Komiya, M., H. Kato and T. Suzuki. 1974a. Bovine plasma kininogens. I. Further purification of high molecular weight kininogen and its physicochemical properties. *J. Biochem.* **76:**811.

———. 1974b. Bovine plasma kininogens. II. Microheterogeneities of high molecular weight kininogens and their structural relationships. *J. Biochem.* **76:**823.

———. 1974c. Bovine plasma kininogens. III. Structural comparison of high molecular weight and low molecular weight kininogens. *J. Biochem.* **76:**833.

Krieger, E. M., H. C. Salgado, C. J. Assan, L. J. Greene and S. H. Ferreira. 1971. Potential screening test for detection of overactivity of renin-angiotensin system. *Lancet* **1:**269.

Kutzbach, C. and G. Schmidt-Kastner. 1972. Kallikrein from pig pancreas. *Hoppe-Seyler's Z. Physiol. Chem.* **353:**1099.

Lahiri, B., R. Rosenberg, R. C. Talamo, B. Mitchell, A. Bagdasarian and R. W. Colman. 1974. Antithrombin III, an inhibitor of human plasma kallikrein. *Fed. Proc.* **33:**642.

Lewis, G. P. 1970. Kinins in inflammation and tissue injury. In *Handbook of Experimental Pharmacology. Bradykinin, Kallidin and Kallikrein* (ed. E. Erdös), vol. 25, p. 516. Springer, New York.

Löbbecke, E. A. 1973. Effect of kallikrein on the proliferation of various cell systems. In *Kininogenases. Kallikreins* (ed. G. L. Haberland and J. W. Rohen), p. 161. K. F. Schattauer, New York.

Mandel, P., J. Rodesch and J. M. Mantz. 1973. The treatment of experimental radiation lesions by kallikrein. In *Kininogenases. Kallikrein* (ed. G. L. Haberland and J. W. Rohen), p. 171. K. F. Schattauer, New York.

Mares-Guia, M. and C. R. Diniz. 1967. Studies on the mechanism of rat urinary kallikrein catalysis and its relation to catalysis by trypsin. *Arch. Biochem. Biophys.* **121:**750.

Margolius, H. S., D. Horwitz, J. J. Pisano and H. R. Keiser. 1974a. Urinary kallikrein in hypertension: Relationships to sodium intake and sodium-retaining steriods. *Circ. Res.* **35:**820.

Margolius, H. S., D. Horwitz, R. G. Geller, R. W. Alexander, J. R. Gill, Jr., J. J. Pisano and H. R. Keiser. 1974b. Urinary kallikrein in normal subjects: Relationships to sodium intake and sodium-retaining steroids. *Circ. Res.* **35:**812.

Marin-Grez, M., P. Cottone and O. A. Carretero. 1972. Evidence for an involvement of kinins in regulation of sodium excretion. *Amer. J. Physiol.* **223:**794.

Maxwell, R. A. and G. H. Acheson, eds. 1973. *Proceedings of the 5th International Congress on Pharmacology: Cellular mechanisms,* vol. 5, p. 250. S. Karger, New York.

McConnell, D. J. 1972. Inhibitors of kallikrein in human plasma. *J. Clin. Invest.* **51:**1611.

McGiff, J. C., H. D. Itskovitz and N. A. Terragno. 1975. The action of bradykinin in the canine isolated kidney; relationships to prostaglandins. *Clin. Sci.* (in press).

McGiff. J. C., N. A. Terragno, K. U. Malik and A. J. Lonigro. 1972. Release of a prostaglandin E-like substance from canine kidney by bradykinin. *Circ. Res.* **31:**36.

McKown, M. M., R. J. Workman and R. I. Gregerman. 1974. Pepstatin inhibition of human renin. Kinetic studies and estimation of enzyme purity. *J. Biol. Chem.* **249:**7770.

Melmon K. L., M. J. Cline, T. Hughes and A. S. Nies. 1968. Kinins: Possible mediators of neonatal circulatory changes in man. *J. Clin. Invest.* **47:**1295.

Meng, K. and G. L. Haberland. 1973. Influence of kallikrein on glucose transport in the isolated rat intestine. In *Kininogenases. Kallikrein* (ed. G. L. Haberland and J. W. Rohen), p. 75. F. K. Schattauer, New York.

Miwa, I., E. G. Erdös and T. Seki. 1969. Separation of peptide components of urinary kinin (substance Z). *Proc. Soc. Exp. Biol. Med.* **131:**768.

Moncada, S., S. H. Ferreira and J. R. Vane. 1973. Prostaglandins, aspirin-like drugs and the oedema of inflammation. *Nature* **246:**217.

————. 1974. Sensitization of pain receptors of dog knee joint by prostaglandins. In *Prostaglandin Synthetase Inhibitors* (ed. H. Robinson and J. R. Vane), p. 189. Raven Press, New York.

Moriwaki, C., N. Watnuki, Y. Fujimoto and H. Moriya. 1974. Further purification and properties of kininogenase from the guinea pig's coagulating blood. *Chem. Pharm. Bull.* **22:**628.

Moriwaki, C., H. Moriya, Y. Yamaguchi, K. Kizuki and H. Fujimori. 1973. Intestinal absorption of pancreatic kallikrein and some aspects of its physiological role. In *Kininogenases. Kallikrein* (ed. G. L. Haberland and J. W. Rohen), p. 57. F. K. Schattauer, New York.

Moriya, H., Y. Matsuda, Y. Fujimoto, Y. Hojima and C. Moriwaki. 1973. Some aspects of multiple components on the kallikrein: Human urinary kallikrein (HUK). In *Kininogenases. Kallikrein* (ed. G. L. Haberland and J. W. Rohen), p. 37. K. F. Schattauer, New York.

Movat, H. Z., H. C. Poon and Y. Takeuchi. 1971. The kinin-system of human plasma. I. Isolation of a low molecular weight activator of prekallikrein. *Int. Arch. Allergy* **40:**89.

Movat, H. Z., S. G. Steinberg, F. M. Habal and N. Ranadive. 1973. Demonstration of a kinin-generating enzyme in the lysosomes of human polymorphonuclear leucocytes. *Lab. Invest.* **29:**669.

Nagasawa, S., H. Takahashi, M. Koida, T. Suzuki and J. G. G. Schoenmakers. 1968. Partial purification of bovine plasma kallikreinogen: Its activation by Hageman factor. *Biochem. Biophys. Res. Comm.* **32:**644.

Nakajima, T. 1968a. Occurrence of a new active peptide on smooth muscle and bradykinin in the skin of *Rana nigromaculata* Hallowell. *Chem. Pharm. Bull.* **16:**769.

————. 1968b. On the third active peptide on smooth muscle in the skin of *Rana nigromaculata* Hallowell. *Chem. Pharm. Bull.* **16:**2088.

Nakajima, T., G. Oshima, H. S. J. Yeh, R. P. Igic and E. G. Erdös. 1973. Purification of the angiotensin I-converting enzyme of the lung. *Biochim. Biophys. Acta* **315:**430.

Nustad, K. and J. V. Pierce. 1974. Purification of rat urinary kallikreins and their specific antibody. *Biochemistry* **13:**2312.

Nustad, K., J. V. Pierce and K. Vaaje. 1975. Synthesis of kallikreins by rat kidney slices. *Brit. J. Pharmacol.* (in press).

Ofstad, E. 1970. Formation and destruction of plasma kinins during experimental acute hemorrhagic pancreatitis in dogs. *J. Gastroent.* (Suppl.) **5:**1.

Ondetti, M. A., N. J. Williams, E. F. Sabo, J. Pluscec, E. R. Weaver and O. Kocy. 1971. Angiotensin-converting enzyme inhibitors from the venom of *Bothrops jararaca. Biochemistry* **10:**4033.

Oshima, G., A. Gecse and E. G. Erdös. 1974. Angiotensin I converting enzyme of the kidney cortex. *Biochem. Biophys. Acta* **350:**26.

Oshima, G., J. Kato and E. G. Erdös. 1974. Subunits of human plasma carboxypeptidase N (kininase I; anaphylatoxin inactivator). *Biochim. Biophys. Acta* **365:**344.

Palmer, M. A., P. J. Piper and J. R. Vane. 1973. Release of rabbit aorta contracting substance (RCS) and prostaglandins induced by chemical or mechanical stimulation of guinea pig lungs. *Brit. J. Pharmacol.* **49:**226.

Pierce, J. V. 1968. Structural features of plasma kinins and kininogens. *Fed. Proc.* **27**:52.

————. 1970. Purification of mammalian kallikreins, kininogens and kinins. In *Handbook of Experimental Pharmacology. Bradykinin, Kallidin and Kallikrein* (ed. E. Erdös), vol. 25, p. 21. Springer, New York.

Pierce, J. V. and K. Nustad. 1972. Purification of human and rat urinary kallikreins. *Fed. Proc.* **31**:623.

Pierce, J. V. and M. E. Webster. 1961. Human plasma kallidins: Isolation and chemical studies. *Biochem. Biophys. Res. Comm.* **5**:353.

————. 1966. The purification and some properties of two different kallidinogens from human plasma. In *Hypotensive Peptides* (ed. E. G. Erdös, N. Back and F. Sicuteri), p. 130. Springer, New York.

Pisano, J. J. 1968. Vasoactive peptides in venoms. *Fed. Proc.* **27**:58.

Pisano, J. J. and K. F. Austen, eds. 1975. *Chemistry and Biology of the Kallikrein-Kinin System in Health and Disease*. Fogarty International Center Proceedings No. 27. U.S. Government Printing Office, Washington, D.C. (in press).

Pitt, B., J. Mason, C. R. Conti and R. W. Colman. 1969. Activation of the plasma kallikrein system during myocardial ischemia. *Trans. Assn. Amer. Physicians* **82**:98.

Porcelli, G. and H. R. Croxatto. 1971. Purification on kininogenase from rat urine. *Ital. J. Biochem.* **20**:66.

Porcelli, G., G. B. Marini-Bertòlo, H. R. Croxatto and M. diIorio. 1974. Purification and chemical studies on human urinary kallikrein. *Ital. J. Biochem.* **23**:44.

Prado, E. S., C. M. W. Brandi and A. V. Katchburian. 1963. Some properties of highly purified horse urinary kallikrein. *Ann. N.Y. Acad. Sci.* **104**:186.

Prado, J. L. 1970. Proteolytic enzymes as kininogenases. In *Handbook of Experimental Pharmacology. Bradykinin, Kallidin and Kallikrein* (ed. E. Erdös), vol. 25, p. 156. Springer, New York.

Ratnoff, O. D., J. Pensky, D. Ogston and G. B. Naff. 1969. The inhibition of plasmin, plasma kallikrein, plasma permeability factor, and the C1r subcomponent of the first component of complement by serum C'1 esterase inhibitor. *J. Exp. Med.* **129**:315.

Reis, M. L., L. Okino and M. Rocha e Silva. 1971. Comparative pharmacological actions of bradykinin and related kinins of larger molecular weights. *Biochem. Pharmacol.* **20**:2935.

Rixon, R. H. and J. F. Whitfield. 1973. Kallikrein, kinin and cell proliferation. In *Kininogenases. Kallikrein* (ed. G. L. Haberland and J. W. Rohen), p. 131. K. F. Schattauer, New York.

Roblero, J. S., H. R. Croxatto, J. H. Corthorn, R. L. Garcia and E. deVito. 1973. Kininogenase activity in urine and perfusion fluid of isolated rat kidney. *Acta Physiol. Lat. Amer.* **23**:566.

Rocha e Silva, M. 1970. *Kinin Hormones*. Charles C. Thomas, Springfield, Illinois.

————. 1974. Present trends in kinin research. *Life Sci.* **15**:7.

Rocha e Silva, M., W. T. Beraldo and G. Rosenfeld. 1949. Bradykinin, a hypotensive and smooth muscle stimulating factor released from plasma by snake venoms and by trypsin. *Amer. J. Physiol.* **156**:261.

Rohen, J. W. and I. Peterhoff. 1973. Stimulation of mitotic activity by kallikrein in the gastrointestinal tract of rats. In *Kininogenases. Kallikreins* (ed. G. L. Haberland and J. W. Rohen), p. 147. K. F. Schattauer, New York.

Ryan, J. W., J. Roblero and J. M. Stewart. 1968. Inactivation of bradykinin in the pulmonary circulation. *Biochem. J.* **110**:795.

Saito, H., O. D. Ratnoff and V. H. Donaldson. 1974. Defective activation of clotting, fibrinolytic and permeability-enhancing systems in human Fletcher trait plasma. *Circ. Res.* **34**:641.

Sampaio, C., S. C. Wong and E. Shaw. 1974. Human plasma kallikrein. *Arch. Biochem. Biophys.* **165**:133.

Sander, G. E. and C. G. Huggins. 1972. Vasoactive peptides. *Annu. Rev. Pharmacol.* **12**:227.

Sander, G. E., D. W. West and C. G. Huggins. 1972. Inhibitors of the pulmonary angiotensin I-converting enzyme. *Biochim. Biophys. Acta* **289**:392.

Schachter, M. 1970. Vasodilatation in the submaxillary gland of the cat, rabbit and sheep. In *Handbook of Experimental Pharmacology. Bradykinin, Kallidin and Kallikrein* (ed. E. Erdös), vol. 25, p. 400. Springer, New York.

Schachter, M., S. Barton and E. Karpinski. 1973. Sympathetic vasodilatation in the submaxillary gland and its enhancement after chronic parasympathetic denervation. *Experientia* **29**:1498.

Schröder, E. 1970. Structure activity relationships of kinins. In *Handbook of Experimental Pharmacology. Bradykinin, Kallidin and Kallikrein* (ed. E. Erdös), vol. 25, p. 324. Springer, New York.

Seki, T., T. Nakajima and E. G. Erdös. 1972. Colon kallikrein, its relation to the plasma enzyme. *Biochem. Pharmacol.* **21**:1227.

Seki, T., T. Miwa, T. Nakajima and E. G. Erdös. 1973. Plasma kallikrein-kinin system in nonmammalian blood: Evolutionary aspects. *Amer. J. Physiol.* **224**:1425.

Sicuteri, F. 1970. Bradykinin and intracranial circulation in man. In *Handbook of Experimental Pharmacology. Bradykinin, Kallidin and Kallikrein* (ed. E. Erdös), vol. 25, p. 482. Springer, New York.

Silva, E., C. R. Diniz and M. Mares-Guia. 1974. Rat urinary kallikrein: Purification and properties. *Biochemistry* **13**:4304.

Spragg, J. and K. F. Austen. 1971. The preparation of human kininogen. II. Further characterization of purified human kininogen. *J. Immunol.* **107**:1512.

Spragg, J., E. Haber and K. F. Austen. 1970. The preparation of human kininogen and elicitation of antibody for use in a radial immunodiffusion assay. *J. Immunol.* **104**:1348.

Stein, J. H., R. C. Congbalay, D. L. Karsh, R. W. Osgood and T. F. Ferris. 1972. The effect of bradykinin or proximal tubular sodium reabsorption in the dog: Evidence for functional nephron heterogeneity. *J. Clin. Invest.* **51**:1709.

Stewart, J. M. 1968. Structure-activity relationships in bradykinin analogues. *Fed. Proc.* **27**:63.

Stewart, J. M. and D. W. Woolley. 1966. The search for peptides with specific antibradykinin activity. In *Hypotensive Peptides* (ed. E. G. Erdös, N. Back and F. Sicuteri), p. 23. Springer, New York.

Stewart, J. M., R. J. Freer and S. H. Ferreira. 1973. Two mechanisms for potentiation of the response of smooth muscles to bradykinin. *Fed. Proc.* **32**:846.

Stoner, J., V. C. Manganiello and M. Vaughan. 1973. Effects of bradykinin and indomethacin on cyclic GMP and cyclic AMP in lung slices. *Proc. Nat. Acad. Sci.* **70**:3830.

Suzuki, T., H. Takahashi, M. Komiya, K. Horiuchi and S. Nagasawa. 1972. Protein components which relate to the kinin-releasing system in bovine plasma. In *Advances in Experimental Medicine and Biology* (ed. N. Back and F. Sicuteri), vol. 21, p. 77. Plenum Press, New York.

Takahashi, H., S. Iwanaga and T. Suzuki. 1974. Snake venom proteinase inhibitors. I. Isolation and properties of two inhibitors of kallikrein, trypsin and α-chymotrypsin from the venom of Russell's viper (*Vipera russelli*). *J. Biochem.* **76**:709.

Takahashi, H., S. Nagasawa and T. Suzuki. 1972a. Studies on prekallikrein of bovine plasma. I. Purification and properties. *J. Biochem.* **71**:471.

————. 1972b. Conversion of bovine prekallikrein to kallikrein. Evidence of limited proteolysis of prekallikrein by bovine Hageman factor (factor XII). *FEBS Letters* **24**:98.

Talamo, R. C., E. Haber and K. F. Austen. 1969. A radioimmunoassay for bradykinin in plasma and synovial fluid. *J. Lab. Clin. Med.* **74**:816.

Terragno, D. A., K. Crowshaw, N. A. Terragno and J. C. McGiff. 1975. Prostaglandin synthesis by bovine mesenteric arteries and veins. *Circ. Res.* (in press).

Thomas, G. and G. B. West. 1974. Prostaglandins, kinin and inflammation in the rat. *Brit. J. Pharmacol.* **50**:231.

Trautschold, I. 1970. Assay methods in the kallikrein system. In *Handbook of Experimental Pharmacology. Bradykinin, Kallidin and Kallikrein* (ed. E. Erdös), vol. 25, p. 52. Springer, New York.

Treloar, M. P., H. A. Pyle, P. J. Fuller and H. Z. Movat. 1972. Guinea pig prekallikrein activator. In *Vasopeptides: Chemistry, Pharmacology and Pathophysiology* (ed. N. Back and F. Sicuteri), p. 61. Plenum Press, New York.

Umezawa, H. 1972. *Enzyme Inhibitors of Microbial Origin.* University of Tokyo Press, Japan.

Vogel, R. and E. Werle. 1970. Kallikrein inhibitors. In *Handbook of Experimental Pharmacology. Bradykinin, Kallidin and Kallikrein* (ed. E. Erdös), vol. 25, p. 213. Springer, New York.

Vogel, R. and G. Zickgraf-Rüdel. 1970. Evaluation of the role of kinins in experimental, pathological and clinical conditions: The therapeutic use of kallikrein inhibitors. In *Handbook of Experimental Pharmacology. Bradykinin, Kallidin and Kallikrein* (ed. E. Erdös), vol. 25, p. 550. Springer, New York.

Vogel, R., I. Trautschold and E. Werle. 1968. *Natural Proteinase Inhibitors.* Academic Press, New York.

Vogt, W. 1966. Demonstration of the presences of two separate kinin-forming systems in human and other plasma. In *Hypotensive Peptides* (ed. E. Erdös, N. Back and F. Sicuteri), p. 185. Springer, New York.

Watanabe, M., T. Yasuhara and T. Nakajima. 1974. New bradykinin analogues in wasp (*Polistes rothneyi iwatai*) venom. *1974 International Symposium on Animal, Plant and Microbial Toxins,* Tokyo.

Webster, M. 1970. Kallikreins in glandular tissues. In *Handbook of Experimental Pharmacology. Bradykinin, Kallidin and Kallikrein* (ed. E. Erdös), vol. 25, p. 131. Springer, New York.

Webster, M. E., H. M. Maling, M. H. Zweig, M. A. Williams and W. Anderson, Jr. 1972. Urate crystal induced inflammation in the rat: Evidence for the combined actions of kinins, histamine and components of complement. *Immunol. Comm.* **1**:185.

Weiss, A. S., J. I. Gallin and A. P. Kaplan. 1974. Fletcher factor deficiency. A diminished rate of Hageman factor activation caused by absence of prekallikrein with abnormalities of coagulation, fibrinolysis, chemotactic activity and kinin generation. *J. Clin. Invest.* **53**:622.

Werle, E., F. Fiedler and H. Fritz. 1973. Recent studies on kallikrein and kallikrein inhibitors. In *Proceedings of the 5th International Congress on Pharmacology. Cellular Mechanisms* (ed. R. A. Maxwell and G. H. Acheson), p. 284. S. Karger, New York.

Werle, E., I. Trautschold and G. Leysath. 1961. Isolierung und structure des kallidins. *Hoppe-Seyler's Z. Physiol. Chem.* **326**:174.

Weyers, R., D. Hagel, B. C. Das and C. van der Meer. 1972. Tryptic peptides from rabbit albumin enhancing the effect of bradykinin. *Biochim. Biophys. Acta* **279**:331.

Wilhelm, D. L. 1971. Kinins in human disease. *Annu. Rev. Med.* **22**:63.

Williams, T. J. and J. Morley. 1973. Prostaglandin as potentiators of increased vascular permeability in inflammation. *Nature* **246**:215.

Wills, C. R., J. H. Ludens, J. B. Hook and H. E. Williamson. 1969. Mechanism of natriuretic action of bradykinin. *Amer. J. Physiol.* **217**:1.

Wong, P. Y., R. C. Talamo, B. M. Babior, G. G. Raymond and R. W. Colman. 1974. Kallikrein-kinin system in postgastrectomy dumping syndrome. *Ann. Intern. Med.* **80**:577.

Workman, R. J., M. M. McKown and R. I. Gregerman. 1974. Renin inhibition by proteins and peptides. *Biochemistry* **13**:3029.

Wuepper, K. D. 1973. Prekallikrein deficiency in man. *J. Exp. Med.* **138**:1345.

Wuepper, K. D. and C. G. Cochrane. 1971. Isolation and mechanism of activation of components of the plasma kinin-forming system. In *Biochemistry of the Acute Allergic Reaction* (ed. K. F. Austen and E. L. Becker), p. 299. Blackwell Scientific, Oxford.

————. 1972. Plasma prekallikrein: Isolation, characterization and mechanism of action. *J. Exp. Med.* **135**:1.

Wuepper, K. D., E. S. Tucker, III and C. G. Cochrane. 1970. Plasma kinin system. Proenzyme components. *J. Immunol.* **105**:1307.

Yano, M., H. Kato, S. Nagasawa and S. Suzuki. 1967a. An improved method for the purification of kininogen II from bovine plasma. *J. Biochem.* **62**:386.

————. 1967b. Separation of a new substrate, kininogen I for plasma kallikrein in bovine plasma. *J. Biochem.* **62**:504.

Yasuhara, T., M. Hira, T. Nakajima, N. Yanihara, C. Yanihara, T. Hashimoto, N. Sakura, S. Tachibana, K. Araki, M. Bessho and T. Yamanaka. 1973. Active peptides on smooth muscle in the skin of *Bombina orientalis* Boulenger and characterization of a new bradykinin analogue. *Chem. Pharm. Bull.* **21**:1388.

Yoshida, H. and J. J. Pisano. 1974. Vespula kinin: A new carbohydrate-containing bradykinin analogue. *The 4th International Symposium on Animal, Plant and Microbial Toxins,* Tokyo.

Zeitlin, I. J. 1970. Kinin release associated with the gastrointestinal tract. In *Bradykinin and Related Kinins: Cardiovascular Biochemical and Neural Actions* (ed. F. Sicuteri, M. Rocha e Silva and N. Back), p. 329. Plenum Press, New York.

————. 1972. Rat intestinal kallikrein. In *Vasopeptides* (ed. N. Back and F. Sicuteri), p. 289. Springer, New York.

Zuber, M. and E. Sache. 1974. Isolation and characterization of porcine pancreatic kallikrein. *Biochemistry* **13**:3098.

Cathepsin D-generated, Pharmacologically Active Peptides (Leukokinins) and Their Role in Ascites Fluid Accumulation

Lowell M. Greenbaum

Department of Pharmacology, College of Physicians & Surgeons
Columbia University, New York, New York 10032

Drs. Walsh, Fruton and Neurath (this volume) have mentioned the importance of products of limited proteolysis in biochemical systems. An important example of the results of limited proteolysis by plasma and cellular enzymes is clearly demonstrated by the pathological responses caused by *kinins*. Kinins are peptides which are exquisitely active in causing increased vascular permeability, changes in smooth muscle tone, hypotension and even pain. There are two systems liberating kinins in the body (Fig. 1). The bradykinin or kallikrein-kinin system is a plasma system which rapidly generates bradykinin following an injury; all components are present in the plasma. In this system, kallikrein is liberated from its zymogen, prekallikrein, following activation by blood-clotting enzymes; kallikrein cleaves the plasma α_2-globulin known as bradykininogen to yield the nonapeptide bradykinin (Arg-Pro-Pro-Gly-Phe-Ser-Pro-Phe-Arg) (see Pisano, this volume).

A second system—which one can consider a chronic or a more slowly developing system—is the leukokinin-generating system. Leukokinins are high molecular weight peptides (21–25 amino acids) which are liberated by an acid protease(s) from cells or tissues acting on a protein, leukokininogen (Greenbaum 1972). Unlike bradykininogen, leukokininogen is not found in normal plasma, but it is found in certain pathological fluids, such as ascites fluid resulting from neoplastic disease (Johnston and Greenbaum 1973). Leukokininogen has not been well purified, but our results indicate that it probably results from a plasma protease of unknown specificity cleaving a proleukokininogen protein to the substrate leukokininogen at an appropriate time.

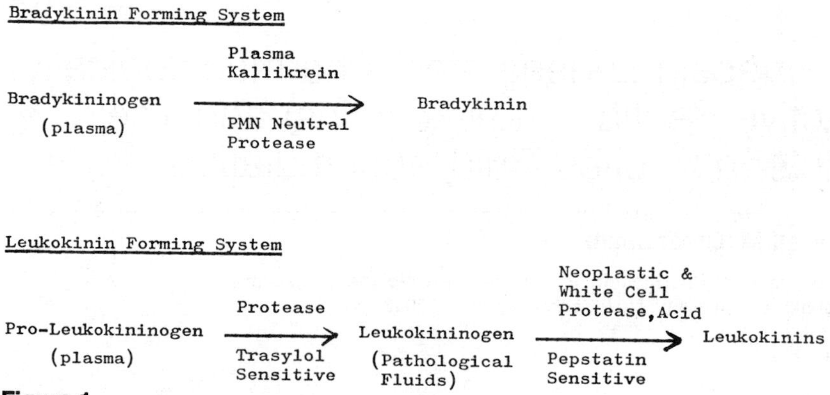

Bradykinin Forming System

```
                        Plasma
                        Kallikrein
Bradykininogen     ─────────────────>     Bradykinin
   (plasma)             PMN Neutral
                        Protease
```

Leukokinin Forming System

```
                                              Neoplastic &
                                              White Cell
                        Protease              Protease,Acid
Pro-Leukokininogen  ──────────>  Leukokininogen  ──────────>  Leukokinins
   (plasma)           Trasylol     (Pathological   Pepstatin
                      Sensitive        Fluids)     Sensitive
```

Figure 1
The bradykinin- and leukokinin-generating systems.

Leukokinin-forming Enzymes

Some of the sources that we have found for leukokinin-forming enzymes are:

1. white cells
     polymorphonuclear leukocytes (rabbit and human)
     macrophages (rabbit)
        alveolar ± BCG
        peritoneal
     lymphocytes ± PHA (human and rat)
2. cancer cells
     leukemic; L-1210 (mouse)
     lymphosarcoma; Meth-A (mouse)
     mastocytoma; P-815Y (mouse)
3. neoplastic ascites fluids
4. tissues
     spleen (bovine)
     liver (dog)

It may be noted that white cells of all types, tissues and neoplastic cells, contain these enzymes. It may also be noted that ascites fluid of neoplastic origin contain these enzymes, no doubt liberated from neoplastic cells present in the fluid. As will be discussed later, these enzymes solubilized in the fluid may play a significant role in the formation of this fluid itself.

For several years we have been investigating the properties of these enzymes. We have particularly directed our attention toward their catalysis of pharmacologically active peptides. We do this by bioassay of the products on isolated rat uterus or guinea pig ileum following incubation with leukokininogen (Freer, Chang and Greenbaum 1972). Since we are dealing in nanogram quantities of materials, bioassay provides a consistent and reasonable way of quantitating leukokinin formation.

The cellular location of these enzymes (determined by breaking up the cell by usual combinations of freezing and thawing and changes in medium tonicity and differential centrifugation) indicated that the most stable activity is found

Table 1
Pepstatin Inhibition of Kinin-forming Enzymes

Enzyme source	Substrate	Pepstatin (moles)	Inhibition %
Macrophage	leukokininogen (plasma)	2×10^{-9}	25
		2×10^{-8}	50
		2×10^{-7}	100
L-1210 cells	leukokininogen (plasma)	2×10^{-9}	100
Mastocytoma	leukokininogen (plasma)	2×10^{-9}	100
Ascites fluid			
murine	leukokininogen (ascites)	2×10^{-9}	80
human	leukokininogen (ascites)	2×10^{-10}	90
Kallikrein			
urinary	bradykininogen	1×10^{-4}	0
plasma	bradykininogen	1×10^{-4}	0

in the "low-speed" (250g) fraction. While the highest specific activity is found in the lysosomal fraction, this activity has proved unstable for purification and storage. The pH optimum for both leukokinin formation and hemoglobin splitting is about 4.0. Recent findings that pepstatin is a potent inhibitor of this enzyme (Table 1) would indicate that we are dealing with cathepsin D or cathepsin D isozymes.

Properties of Leukokinins

The pharmacologically active peptides (leukokinins) that are formed by the action of these enzymes have been under intense investigation by our laboratory. We have isolated at least three different leukokinins and have carried out amino acid analysis and comparisons with the well-known nonapeptide bradykinin (Greenbaum 1972). The leukokinins are much larger than bradykinin, having 21–25 amino acids as compared to the nonapeptide. Moreover, the unexpected finding that the bradykinin sequence is not part of the molecule (only one phenylalanine is present in the leukokinins rather than the two present in bradykinin) is further evidence that the protein substrate from which the leukokinins are cleaved differs from that from which bradykinin is cleaved.

The pharmacological activity of these peptides has demonstrated that on test systems they are potent agents in causing increased vascular permeability. The significance of this is that fluid accumulation, a direct result of increased vascular permeability, occurs in pathological conditions where intracellular proteinases may be released from white cells, tissues, etc. One pathological condition in which we have obtained strong evidence that leukokinin formation may play an important role in fluid accumulation is in neoplastic disease where peritoneal ascites fluid accumulation occurs. Our laboratory has found (Johnston and Greenbaum 1973) that when acidified to pH 3.5–4.5, cell-free

murine ascites fluid (formed in mice inoculated intraperitoneally with mastocytoma cells) produced huge quantities of leukokinins (Fig. 2). This means
that both substrate and enzyme were present in the fluid. The enzymes were
no doubt liberated from neoplastic cells bathed in the fluid. Whether this
liberation is by secretion mechanisms (discussed by Barrett, this volume) or
simply due to destruction of cells, we do not know. We believe that the substrate, leukokininogen, is formed in the fluid by a complex mechanism from
a protein in plasma, although we are not sure of the triggering mechanism.

As can be seen, the amounts of leukokinin formed were in *microgram* quantities, whereas kinins are pharmacologically active in *nanogram* concentrations. Because of the permeability activity of the leukokinins and the presence
of the leukokinin-forming system in the fluid, we began to suspect that this
system might play a major role in ascites fluid accumulation in these mice.

To test this hypothesis, we used the potent inhibitor of this system, pepstatin, in vivo. AKD_2F_1 female mice were injected intraperitoneally with
$6\text{--}8 \times 10^6$ mastocytoma P-815Y cells. Within seven days these animals
accumulate several milliliters of peritoneal fluid and increase the cell count to
10^9 in the peritoneal cavity. When pepstatin was administered to a series of
these mice, either subcutaneously or intraperitoneally, four days after tumor
inoculation, the cell count still rose to 10^9, but the ascites fluid volume was
dramatically and significantly reduced (Greenbaum 1973). We have verified
this finding in pepstatin-treated DBF_2 male mice carrying L-1210 cells. This
means that pepstatin is not blocking ascites fluid accumulation by preventing
cell growth, but is probably involved in blocking leukokinin formation and
consequently an increase in vascular permeability.

Ascites fluid accumulation in humans occurs in neoplastic diseases such as
ovarian carcinoma. In vitro analyses of such fluids have typically demonstrated
that upon acidification, leukokinins are formed. This formation is abolished
when pepstatin is added to the incubating fluid in minute amounts. Thus it

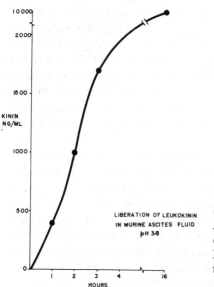

LIBERATION OF LEUKOKININ
IN MURINE ASCITES FLUID
pH 3-8

Figure 2
Ascites fluid was centrifuged to remove
cells and debris and adjusted to pH 4.0.
Leukokinin formation was carried out by
the methods of Johnston and Greenbaum
(1973).

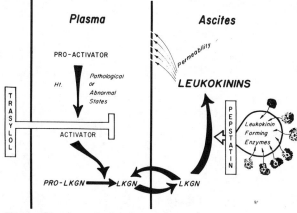

Figure 3
Leukokinin formation and its relationship to ascites
fluid accumulation and protease inhibitors. Invading
neoplastic cells release cathepsin D-like enzymes into
the peritoneum. The enzymes catalyze the formation of
leukokinins, pharmacologically active peptides, which
cause an increase in the permeability of the rich network of blood vessels in the peritoneal membranes resulting in ascites fluid. Pepstatin blocks the enzyme
action on the leukokininogen substrate. The substrate
itself is believed formed by a neutral plasma protease
which is Trasylol (pancreatic trypsin inhibitor) sensitive.

would appear that *ascites fluids from both humans and mice have similar
leukokinin-generating systems.*

Figure 3 is a graphic representation of our ideas and the facts as we know
them concerning leukokinin formation and its relationship to ascites fluid
accumulation and pepstatin inhibition of this accumulation. Invading or inoculated neoplastic cells release cathepsin D-like enzymes into the peritoneum.
The enzymes come into contact with the leukokininogen substrate, whose
formation is triggered by a Trasylol (pancreatic trypsin inhibitor)-sensitive
protease. Once the pathological substrate leukokininogen presents itself in the
peritoneal fluid, the enzymatic formation of quantities of leukokinins ensues,
increasing the permeability of the vasculature perfusing the peritoneal membranes. This results in increasing fluid accumulation in the peritoneal cavity
and, in our hypothesis, is a major factor in ascites formation. Since pepstatin
blocks leukokinin formation, it dramatically retards fluid accumulation.

While the acid pH optimum of the cathepsin D-like enzymes raises some
questions, the fact that pepstatin inhibits leukokinin formation in vitro and
in vivo would lend strong credence to the argument for the leukokinin system
proposed.

Finally, the role of cathepsin D in this system, i.e., forming pharmacologically active peptides by limited proteolysis, broadens the role indicated by
Barrett of a degradative protease in cartilage disease (Barrett and Dingle
1972), as shown in Figure 4.

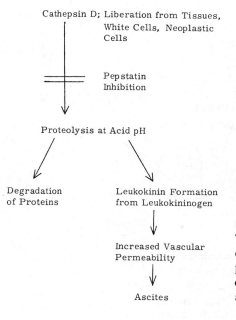

Figure 4
The dual actions of cathepsin D in degrading proteins and in limited proteolysis resulting in the formation of pharmacologically active peptides and pathological fluids.

Acknowledgments

These investigations were supported by grants from USPHS (CA-13061) and the American Cancer Society (DT-6). L. M. G. is a career scientist of the Health Research Council of New York City.

REFERENCES

Barrett, A. J. and J. T. Dingle. 1972. The inhibition of tissue acid proteinases by pepstatin. *Biochem. J.* **127:**439.

Freer, R., J. Chang and L. M. Greenbaum. 1972. Studies on leukokinins. III. Pharmacological activities. *Biochem. Pharmacol.* **21:**3107.

Greenbaum, L. M. 1972. Leukocyte kininogenases and leukokinins from normal and malignant cells. *Amer. J. Pathol.* **68:**613.

Greenbaum, L. M. 1973. The leukokinin system: Its role in fluid accumulation in malignancy and inflammation. *Agents and Actions* **3:**332.

Johnston, M. and L. M. Greenbaum. 1973. Leukokinin-forming system in the ascitic fluid of a murine mastocytoma. *Biochem. Pharmacol.* **22:**1386.

Initiation of Membrane Attack by Complement: Assembly and Control of C3 and C5 Convertase

Hans J. Müller-Eberhard

Department of Molecular Immunology, Scripps Clinic and Research Foundation
La Jolla, California 92037

Activation of complement results in either membrane damage and cell death or in activation of specialized cell functions. I will confine my comments primarily to a discussion of the molecular events that precede membrane damage.[1]

One of the unusual characteristics of complement proteins is their inherent ability to undergo transition from soluble molecules to peripheral or possibly integral membrane constituents. This ability depends on the generation of binding regions by enzymatic removal of activation peptides. The responsible enzymes have a complex quaternary structure, are indigenous to the system, and represent examples of highly specialized proteases. Their action results in the assembly of distinct multimolecular complexes from components which in their native state exhibit stereochemical affinity for each other. Complement-dependent cell membrane damage is accomplished by a multisubunit complex which impairs membrane function probably by physicochemical attack.

The Proteins

The eleven complement proteins differ widely in their physical parameters and in their concentrations in serum (Table 1). The description of their properties will be limited to a few points of general interest. As far as is known, all are glycoproteins and none is a lipoprotein. C1r, C1s and C2 are zymogens, C2, C3, C4 and C5 are particularly susceptible to tryptic attack, and C1q, C4 and C8 are composed of three distinct polypeptide chains, a composition that is rare among proteins. Reactive SH groups are present in C2, C3 and C5 and are essential for the function of these components (Polley and Müller-Eberhard, unpubl.).

C1q, the recognition factor of complement, is unusual in that it is collagen-like (Calcott and Müller-Eberhard 1972). Per 100 amino acid residues it contains five residues of hydroxyproline, two residues of hydroxylysine and

---

[1] For detailed bibliography, see Müller-Eberhard (1975).

Table 1
Proteins of the Classical Human Complement System

Protein	Serum concentration ($\mu g/ml$)	Sedimentation coefficient (S)	MW	Relative electrophoretic mobility	No. of chains
C1q	180	11.1	400,000	γ_2	6 x 3
C1r	—	7.5	180,000	β	2
C1s	110	4.5	86,000	α	1
C2	25	4.5	117,000	β_1	—
C3	1600	9.5	180,000	β_2	2
C4	640	10.0	206,000	β_1	3
C5	80	8.7	180,000	β_1	2
C6	75	5.5	110,000	β_2	1
C7	55	5.5	95,000	β_2	1
C8	80	8.0	163,000	γ_1	3
C9	230	4.5	79,000	α	—

18 residues of glycine. Glucose and galactose are linked as disaccharide units to the hydroxyl group of hydroxylysine. This carbohydrate moiety as well as hydroxylated lysine and proline have not been reported to occur in any other plasma protein. C1q recognizes immune complexes and thereby initiates the complement reaction. One C1q molecule is endowed with six binding sites for IgG. By treatment with sodium dodecyl sulfate (SDS), the molecule may be dissociated into six subunits, each of which is believed to carry one immunoglobulin binding site. On the basis of their work, Reid and Porter (1975) propose that (1) each subunit contains three similar, but not identical, chains joined by disulfide bonds; (2) each chain contains an intermediate region with a collagenlike sequence; and (3) the collagenlike regions of three chains form a triple-helical strand, and the C-terminal, noncollagenous regions, a random-coil globular arrangement. The six subunits associate noncovalently near the end of their helical portions and thus give rise to a bouquetlike structure as visualized by electron microscopy (Shelton, Yonemasu and Stroud 1972).

C8 is noteworthy because it has an unusual chain structure and because the accumulated evidence strongly suggests that complement-dependent membrane impairment is caused by a portion of the C8 molecule. SDS dissociates nonreduced C8 into two subunits of 93,000 and 70,000 daltons. Reduction in the presence of SDS gives rise to three subunits: C8α ($\sim$83,000 daltons), C8β (70,000 daltons), and C8γ ($\sim$10,000 daltons). Accordingly, the small γ chain is linked by disulfide bonds to the α chain, and the β chain is noncovalently bound to the α-γ subunit (Kolb and Müller-Eberhard, in prep.).

Transfer of Soluble Proteins to Three Topologically Distinct Target Membrane Sites

When complement is activated by antibody-coated cells, it undergoes a self-assembling process, in the course of which the entire set of molecules transfers from solution to the solid phase of the target cell surface. All eleven proteins are soluble in aqueous solutions. None has an apparent affinity for membranes in its unaltered, native form. Yet the potential of entering into direct and firm contact with membrane constituents is resident in C3, C4 and C5,6,7. There is strong reason to believe that C8 also possesses this potential (Kolb and Müller-Eberhard, in prep.). In contrast, C1, C2 and C9 bind to sites which are not originally part of the target membrane. Three different tactics are operative in the transfer: (1) reversible ionic interaction, which applies to C1; (2) enzymatic activation of binding sites, which is characteristic for C2, C3, C4 and C5; and (3) adsorption, which applies to C6, C7, C8 and C9.

The most elaborate of these tactics is the enzymatic activation of binding sites (Fig. 1). The respective activating enzyme cleaves a peptide bond of a native C2, C3, C4 or C5 molecule, and this event is followed by dissociation or dislocation of an activation peptide. The molecule is thereby converted to an activated state in which a previously concealed binding region is now exposed. Upon collision with the target membrane surface, it has the potential of entering into strong, probably hydrophobic, interaction with acceptors of unknown chemical nature. The membrane-bound molecule is enabled to fulfill its characteristic function in the cytolytic reaction. Failing collision with a membrane acceptor within fractions of a second after enzymatic activation, the binding site decays, and the molecule remains in cytolytically inactive form in the fluid phase. We assume that the activated state corresponds to a thermodynamically unfavorable conformation, and that decay is the result of transition of this conformation to a stable form.

The polypeptide chain composition is known for C3, C4 and C5 (Fig. 2). Since in each protein the largest chain (α) is attacked by the activating enzyme (Nilsson and Mapes 1973; Schreiber and Müller-Eberhard 1974; Bokisch, Dierich and Müller-Eberhard 1975), the labile binding site appears to be a function of the α chains. This has been shown to be the case for bound C3 (Bokisch, Dierich and Müller-Eberhard 1975) and bound C4 (Cooper 1975a).

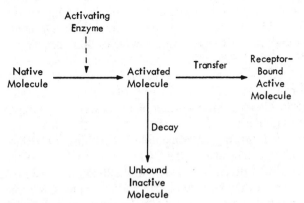

Figure 1
Mechanism of binding of C2, C3, C4 and C5 to the target cell membrane: enzymatic activation of labile binding sites.

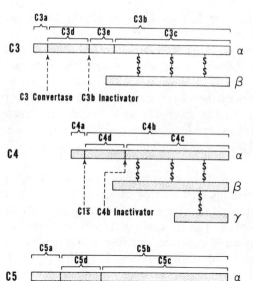

Figure 2

Schematic representation of proposed polypeptide chain structure of C3, C4 and C5. The topological relationships between chains, fragments, functional sites and enzymatic regions are indicated as presently envisaged (Müller-Eberhard 1975).

Figure 3 depicts a three-site model of complement transfer from solution to the solid phase of the target cell surface. It proposes that at completion of the process, the functional units have assumed stable supramolecular organizations at three topologically distinct sites, herein designated SI, SII and SIII. Through ionic interactions, C1 binds via its Clq subunit to target-cell-bound antibody molecules (SI), where the C1 complex undergoes internal activation. This reaction might be triggered by a conformational change of C1q, imposed by the interaction with antibody molecules. The result is the conversion of the zymogen C1r to the active form $C\overline{1r}$ and of C1s to $C\overline{1s}$ (bar denotes enzyme activity). The zymogen C1s is cleaved by $C\overline{1r}$ into a light and a heavy chain, which remain associated through disulfide bonding (Valet and Cooper 1974a). The active site of $C\overline{1s}$ resides in the light chain (Sakai and Stroud 1974). The enzyme is a serine esterase with a primary amino acid sequence around the active-site serine similar to that occurring in trypsin, plasmin and thrombin (Fothergill, unpubl.). From SI, active $C\overline{1}$ catalyzes the assembly of the $C\overline{4,2}$ complex (C3 convertase), which binds to SII and converts itself to $C\overline{4,2,3}$ (C5 convertase). From SII, this enzyme initiates the assembly of the C5b-9 complex which binds to SIII and proceeds to convert this site to a membrane lesion.

The model is based on the following experimental observations: (1) Binding of C1 to a cell is antibody-dependent. (2) C4 can bind to a cell in the absence of antibody and C1 provided $C\overline{1s}$ is present in the fluid phase. (3) C4 can then proceed to bind C2 in the presence of $C\overline{1s}$ and thereafter C3 in the absence of $C\overline{1s}$. (4) Binding of C5-9 to cell *A* can be catalyzed by $C\overline{4,2,3}$ located at the surface of cell *B*. Cells carrying C5-9 undergo lysis irrespective of whether earlier acting components are on their surface (Götze and Müller-Eberhard 1970; Müller-Eberhard 1975).

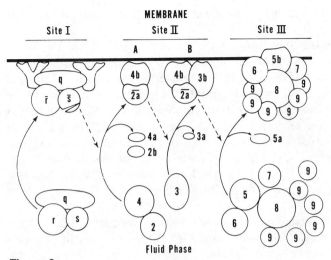

Figure 3
Pictorial representation of the three-site model of complement
transfer from solution to the solid phase of the target cell surface. The C1 complex is reversibly bound to antibody molecules (SI) through its C1q subunit. An internal reaction leads
to activation of C1s. C1s̄ activates labile binding sites in C2
and C4 by enzymatic removal of activation peptides. The C4,2
enzyme (C3 convertase) is thereby enabled to assemble at a
membrane site (SII) which is topologically distinct from SI.
C4̄,2 converts itself to C4̄,2,3 (C5 convertase) by cleavage of
C3 and assimilation of C3b. In cleaving C5, C4̄,2,3 initiates
the self-assembly of the C5b-9 complex which attacks the
membrane at a third, topologically distinct site (SIII) and
kills the cell (Müller-Eberhard 1975).

Thus four proteases are operative in the classical complement reaction,
three of which are instrumental in the activation of labile binding sites: C1s̄
acting on C2 and C4, and C3 and C5 convertase on C3 and C5, respectively.

The molecular economy of the assembly process is determined by the mode
of binding that is operative at a given stage. Within the C1 complex, apparently one molecule of C1r̄ is limited to activating one or perhaps two molecules of C1s (Valet and Cooper 1974b). However, one C1̄ complex can
generate multiple C4̄,2 complexes, and one C4̄,2,3 complex can assemble
multiple C5b-9 complexes. Further, one C4̄,2 gives rise to only one C4̄,2,3.
Yet one C4̄,2 may turn over many molecules of C3, thereby causing individual C3b molecules to attach to membrane acceptors in their immediate
environment. Such clusters of C3b are active in producing immune adherence
(Müller-Eberhard, Dalmasso and Calcott 1966).

C3 Convertase

C3 and C5 convertase depend, with respect to their functionality in the complement system, on a defined, complex subunit structure. By comparison, the
C1 complex does not constitute a complex enzyme because C1r̄ and C1s̄ ful-

fill their characteristic enzymatic functions even after dissociation from the complex.

The precursors of C3 convertase are the serum proteins C2 and C4. These two proteins exhibit stereochemical affinity for each other as evidenced by the formation of a reversible association product in free solution. Action of $\overline{\text{C1}}$ on C4 cleaves an 8000-dalton fragment (C4a) from the α chain of the molecule, thereby producing a site through which C4b (198,000 daltons) can bind to membrane receptors (Fig. 2). Activation of C2 by $\overline{\text{C1}}$ results in cleavage of the molecule and dissociation of the 37,000-dalton fragment C2b. The other portion, C2a (80,000 daltons), has the transient ability to bind firmly to C4b, thus forming the bimolecular complex $\overline{\text{C4b,2a}}$, which is endowed with C3 convertase activity (Müller-Eberhard, Polley and Calcott 1967) (Table 2).

The nature of the enzyme was elucidated when it could be formed in cell-free solution. A mixture of isolated C2 and C4 treated with isolated $\overline{\text{C1s}}$ gave rise to C3 cleaving activity. We concluded that C3 convertase represents a complex of C2 and C4 in which a stable product of C4 serves as acceptor and modulator of activated C2. This conclusion was based on the finding that C3 cleaving activity (1) did not appear on treatment of either C2 or C4 alone with $\overline{\text{C1s}}$, (2) did not appear on addition of C4 to a previously incubated mixture of C2 and $\overline{\text{C1s}}$, but (3) was generated on addition of C2 to a pre-

Table 2

The Cytolytic Complement Reaction: Functional Units and Membrane Sites

---

First site: Activation of recognition unit

1.   SI A + C1q $\xrightarrow[\text{s}]{\text{r}}$ SI A-C1q $\begin{array}{c}\overline{\text{r}}\\\downarrow\\\overline{\text{s}}\end{array}$

Second site: Assembly of activation unit

2. $\text{C4} \xrightarrow{\text{SI A } \overline{\text{C1}}} \text{C4a} + \text{C4b*}$

3. $\text{C2} \xrightarrow{\text{SI A } \overline{\text{C1}}} \text{C2a*} + \text{C2b}$

4. $\text{SII} + \text{C4b*} + \text{C2a} \longrightarrow \text{SII } \overline{\text{C4b,2a}}$

5. $\text{C3} \xrightarrow{\text{SII } \overline{\text{C4b,2a}}} \text{C3a} + \text{C3b*}$

6. $\text{SII } \overline{\text{C4b,2a}} + \text{C3b*} \longrightarrow \text{SII } \overline{\text{C4b,2a,3b}}$

Third site: Assembly of membrane attack mechanism

7. $\text{C5} \xrightarrow{\text{SII } \overline{\text{C4b,2a,3b}}} \text{C5a} + \text{C5b*}$

8. $\text{SIII} + \text{C5b*} + \text{C6} + \text{C7} + \text{C8} + \text{C9} \longrightarrow \text{SIII C5b,6,7,8,9}$

Data from Müller-Eberhard (1975).
SI, SII, SIII: Topographically distinct sites on target cell surface.
A: Antibody to cell surface constituent. Bar denotes active enzyme;
(*) denotes enzymatically activated, labile binding site.

viously incubated mixture of C4 and C1s. Initial attempts to demonstrate the postulated enzymatically active complex after zone ultracentrifugation met with failure. Such demonstration was accomplished, however, using oxidized C2 (see below). C3 convertase was shown to have a sedimentation velocity of 11S and a molecular weight of approximately 300,000 daltons. This molecular size is indicative of a bimolecular complex of C2a and C4b. Assuming that C2a constitutes a single chain, the quaternary structure of the enzyme comprises at least four polypeptide chains. The three chains of C4b (α', 85,000; β, 78,000; and γ, 33,000 daltons) are linked by S–S bonds (Schreiber and Müller-Eberhard 1974) and C2a is linked to C4b by unknown forces.

Assembly and function of C3 convertase are controlled in the cytolytic reaction by (1) rapid loss of binding sites of activated C2 and C4 in the fluid phase, (2) abolition of the acceptor function of C4b by the serum enzyme C4b inactivator (Fig. 2) (Cooper 1975a), and (3) decay of the enzyme with a half-life of 10 minutes at 37°C by dissociation into inactive subunits (Mayer 1970).

The active site of C3 convertase could be topologically assigned to its C2a subunit. In addition to acting on its natural substrate, the enzyme hydrolyzes acetyl-glycyl-lysine methyl ester (AGLMe). Whereas C3 cleaving activity is lost upon decay of the enzyme, the ability to hydrolyze AGLMe is retained. This hydrolytic activity was shown to reside in the C2a subunit ($K_m = 1.8 \times 10^{-2}$ M), and cleavage of C3 by intact C3 convertase ($K_m = 1.8 \times 10^{-6}$ M) was found to be competitively inhibited by AGLMe (Cooper 1975b).

C2 contains two SH groups which are positioned in close proximity to each other and which upon oxidation with iodine form an intramolecular disulfide bond. Through this chemical modification, C2 acquires a 20-fold enhancement of its hemolytic activity (Polley and Müller-Eberhard 1967 and unpubl.). Treatment of the protein with p-chloromercuribenzoate (p-CMB) results in loss of its hemolytic activity. Since 2 moles of radioactive p-CMB were bound per mole of C2, it was concluded that the molecule contains two free, reactive SH groups. In exploring the functional relevance of the SH groups and the nature of the chemical modification induced by iodine treatment, the following observations were made:

1. Iodine treatment led to iodine uptake, but the latter did not correlate with enhancement of activity.
2. Iodine-treated C2 could not be inactivated by p-CMB and did not bind it.
3. [^{14}C]p-CMB-treated C2, upon exposure to iodine, not only was reactivated, but underwent a 10- to 20-fold enhancement of hemolytic activity and released all of the previously bound radioactivity.
4. Upon mild reduction of iodine-treated C2, the 20-fold-enhanced hemolytic activity was lowered to the original level and both SH groups again became available for binding of p-CMB.
5. Iodine treatment did not influence the molecular weight of C2.

Oxidation of C2 with iodine augments three different C2 functions: the binding of activated C2 to C4b, the catalytic function of $\overline{C4,2}$, and the stability of the enzyme (Cooper, Polley and Müller-Eberhard 1970). The half-life at 37°C is 10 minutes for $\overline{C4,2}$ and 200 minutes for $\overline{C4,2}$ containing oxidized C2 (Polley and Müller-Eberhard 1967). The fact that three functional parameters of the $\overline{C4,2}$ complex are affected by the presence of the iodine-induced intramolecular S–S bond in C2 strongly suggests that the location of this bond is extraneous to the active site. Instead, it appears to be located in a region of the molecule where it can influence the function of both the enzymatic as well as the C4b-contact region.

C3 convertase represents a highly specialized protease, C3 being its only known protein substrate. Present evidence indicates that the enzyme attacks a single peptide bond which involves the carboxyl group of an arginyl residue within the amino-terminal portion of the α chain (Budzko, Bokisch and Müller-Eberhard 1971; Nilsson and Mapes 1973; Bokisch, Dierich and Müller-Eberhard 1975). Hydrolysis of this bond leads to dissociation of C3a, which has a molecular weight of 8900 daltons and is composed of 77 amino acid residues (Hugli, this volume).

C5 Convertase

C5 convertase is a derivative of C3 convertase. Addition of C3 to cell-bound $\overline{C4,2}$ not only results in turnover of C3 and binding of C3b to the cell surface (Müller-Eberhard, Dalmasso and Calcott 1966), but also results in the generation of a cell-bound enzyme which turns over C5 (Cooper and Müller-Eberhard 1970). We postulated, therefore, that a product of C3 modulates C3 convertase such that the substrate binding region of the enzyme becomes adapted to C5. The modulation is envisaged as resulting from a physical association of C3b with the $\overline{C4,2}$ complex (Fig. 3, Table 2). Preliminary evidence suggests that C5 convertase has little or no substrate affinity for C3.

Further exploration of the enzyme necessitated its assembly in cell-free solution. Consumption of C5 hemolytic activity was indeed observed upon treatment of C5 with a mixture of $\overline{C4,^{oxy}2}$ (containing C2 in oxidized form) and C3 (Fig. 4). C5 consumption was dependent upon the presence of both $\overline{C4,^{oxy}2}$ and C3 (Fig. 5) and increased with increasing amounts of $\overline{C4,2,3}$. It was also dependent on time, temperature and substrate concentration. Upon zone ultracentrifugation of a preincubated mixture of $\overline{C4,^{oxy}2}$ and C3, C5 consuming activity could be detected in fractions corresponding to 14–16S material (Fig. 6). These data suggest that a relatively firm complex is formed between $\overline{C4,^{oxy}2}$ and nascent C3b, which is endowed with C5 convertase activity. Although it is presently difficult to estimate, it appears that the turnover number of the enzyme is relatively low.

C5 convertase consists of a minimum of three subunits or six polypeptide chains. The catalytic site is probably the same as that of C3 convertase and is located in the C2 subunit of the enzyme.

Assembly and function of C5 convertase are controlled by three mechanisms: (1) rapid decay of the binding sites of the activated subunits during the assembly process, (2) time- and temperature-dependent dissociation of the C2a subunit from the complex, and (3) attack of the C3b subunit by the serum enzyme C3b inactivator (Gigli 1974).

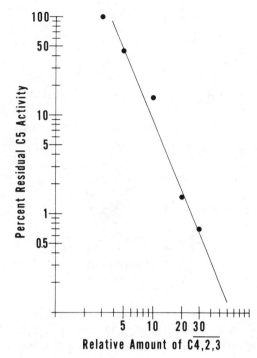

Figure 4
Consumption of C5 activity by $\overline{C4,^{oxy}2,3}$ in cell-free solution. $\overline{C4,^{oxy}2}$ was incubated with isolated C3 for 20 minutes at 37°C; incubation was continued with different volumes of the enzyme solution and constant amounts of C5 for 45 minutes at 37°C. Subsequently, residual C5 activity was determined by standard procedures.

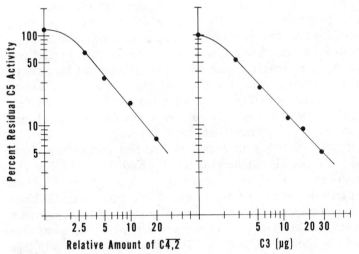

Figure 5
Consumption of C5 activity in cell-free solution: requirement of $\overline{C4,2}$ and C3. C5 was treated with (*left*) a constant amount of C3 and increasing amounts of $\overline{C4,^{oxy}2}$ and (*right*) a constant amount of $\overline{C4,^{oxy}2}$ and increasing amounts of C3.

238   H. J. Müller-Eberhard

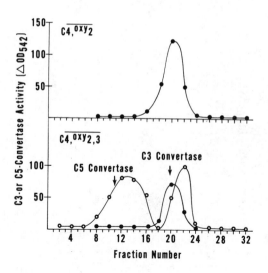

Figure 6
Sucrose density gradient ultracentrifugation of the $\overline{C4,^{oxy}2}$
enzyme (*top*) and of the preincubated mixture of $\overline{C4,^{oxy}2}$ and
C3 (*bottom*). Direction of sedimentation is to the left. Conditions: 5–20% sucrose gradient
in phosphate buffer pH 6, ionic
strength 0.05; 50,000 rpm for
6 hours at 2°C. The velocity of
the heavy C5 consuming component is 14–16S, compared to
11S for C3 convertase. The
slowly sedimenting C5 consuming activity may be generated by
an overlap of free $\overline{C4,^{oxy}2}$ and
C3b (9S).

C5 convertase cleaves an arginyl bond in C5 and produces the fragments
C5a (17,000 daltons) and C5b (163,000 daltons). In its nascent state, C5b
constitutes the biochemical signal that initiates self-assembly of C5b-9.

Assembly of the Membrane Attack System by C5 Convertase

Attack of C5 by the $\overline{C4,2,3}$ enzyme appears to be the final enzymatic event
in the complement reaction (Kolb and Müller-Eberhard 1975). With
cleavage of the low molecular weight activation peptide C5a (Cochrane and
Müller-Eberhard 1968) from the α chain of C5 (Nilsson and Mapes 1973)
(Fig. 2), the resulting C5b fragment acquires the transient ability to bind C6
and C7 and to attach as the trimolecular C5b,6,7 complex to the target cell
membrane. This complex constitutes the molecular arrangement for adsorp-
tive binding of one C8 molecule. The C8 molecule of the tetramolecular
C5b-8 complex functions as binding region for maximally six C9 molecules.
The decamolecular assembly has a cumulative molecular weight of one
million. A clay model of the complex is depicted in Figure 7.

The complex was deduced, with respect to size and composition, from
experiments with cell-bound, radiolabeled C5b-9 (Kolb et al. 1972). The
following were the early experimental results: (1) C5, C6, C7 and C8 are
bound to target cells in equimolar amounts and at saturation of all C9 binding sites, six C9 molecules are bound per one C8 molecule. (2) A minimum
of three C9 molecules are required for the production of one membrane
lesion. (3) Antibodies to either C5, C6 or C7 inhibited binding of C8 to cells
bearing C5b,6,7 sites, and antibody to C8 strongly inhibited binding of C9 to
cells bearing C5b,6,7,8 sites.

The general validity of the proposed model could be examined when it was
found that a soluble, stable C5-9 complex accumulates as a by-product in
the fluid phase during the cytolytic complement reaction (Kolb and Müller-
Eberhard 1973). The isolated complex has a molecular weight of 1,040,000,

Figure 7

Photograph of a model of the C5b-9 membrane attack complex of complement showing three stages of its assembly. The spheres were made of modeling clay, the relative weights being proportional to the molecular weights of the proteins. The numerals refer to the corresponding complement components. (*a*) Model of the membrane-bound C5b,6,7 trimolecular complex displaying triangular geometry and constituting the proposed binding site for C8. (*b*) Model of the tetramolecular complex C5b,6,7,8 having the geometry of a tetrahedron. (*c*) Model of the fully assembled decamolecular C5b-9 complex exhibiting two C9 trimers bound in triangular arrangement to the C8 portion of the tetrahedron (Kolb et al. 1972).

the electrophoretic mobility of an α-globulin, and is cytolytically inactive. It can be dissociated into its subunits by SDS. Subjected to SDS gel electrophoresis, the complex gives rise to seven protein bands. The corresponding proteins could be identified, in order of molecular size, as C5b, C6, C7, the 93,000-dalton subunit of C8, an unknown protein, C9 and the 70,000-dalton subunit of C8. Estimates of the relative concentrations of the components in the complex revealed that C5b, C6, C7 and C8 are represented in equimolar amounts, whereas C9 and the unknown protein are present in multiples thereof (Kolb and Müller-Eberhard 1975).

The accumulated evidence strongly suggests that C8 represents the molecule that executes the cytolytic function of the C5b-9 complex. The assembly of the complex at the surface of a target cell may be experimentally dissected into several reaction steps. Using isolated complement proteins, an intermediate complex may be prepared in which the cell bears C5b,6,7 sites. Whereas the membrane of this cell is completely intact, it becomes leaky upon attachment of C8. In the absence of C9, highly purified C8 causes the cell to undergo protracted low-grade lysis (Stolfi 1968). Lipid bilayers of cholesterol and sphingomyelin also can be impaired by activated complement; and as in the case of cells, impairment commences with C8 action (Haxby et al. 1969). The function of C9 may be regarded as an enhancement mechanism for the expression of C8 activity. We are presently investigating the manner in which the remarkable polypeptide chain composition of C8 might relate to its membrane-impairing function.

CONCLUSION

The complement system constitutes a telling example for the generation of biological activity by protein-protein interaction. In the course of the cytolytic complement reaction, such interactions eventuate in the assembly of two related, multisubunit proteases from apparently inert precursors. Fusion of different protein molecules, initiated by the enzymatic removal of activation peptides, gives rise to the enzymes C3 convertase and C5 convertase. Both constitute highly specialized proteases with respect to substrate and bond specificity. The conversion of C3 convertase to C5 convertase by a product of the substrate of the former enzyme is noteworthy. Whereas these enzymes can attach themselves firmly to the target cell surface, they do not compromise membrane structure or function. However, C5 convertase, in acting on C5, provides the biochemical signal for the nonenzymatic assembly of the multi-molecular membrane attack complex of complement, C5b-9. This complex, with respect to formation, structure and function, is without analogy in other biological systems.

Acknowledgments

This work was supported by United States Public Health Service Grant AI-07007. H. M–E. is a Cecil H. and Ida M. Green Investigator in Medical Research, Scripps Clinic and Research Foundation. This is publication No. 957.

REFERENCES

Bokisch, V. A., M. P. Dierich and H. J. Müller-Eberhard. 1975. Third component of complement (C3): Structural properties in relation to functions. *Proc. Nat. Acad. Sci.* **72**:6.

Budzko, D. B., V. A. Bokisch and H. J. Müller-Eberhard. 1971. A fragment of the third component of human complement with anaphylatoxin activity. *Biochemistry* **10**:1166.

Calcott, M. A. and H. J. Müller-Eberhard. 1972. C1q protein of human complement. *Biochemistry* **11**:3443.

Cochrane, C. G. and H. J. Müller-Eberhard. 1968. The derivation of two distinct anaphylatoxin activities from the third and fifth components of human complement. *J. Exp. Med.* **127**:371.

Cooper, N. R. 1975a. Isolation and analysis of the mechanism of action of an inactivator of C4b in normal human serum. *J. Exp. Med.* **141**:890.

———. 1975b. Enzymatic activity of the second component of complement. *Biochemistry* (in press).

Cooper, N. R. and H. J. Müller-Eberhard. 1970. The reaction mechanism of human C5 in immune hemolysis. *J. Exp. Med.* **132**:775.

Cooper, N. R., M. J. Polley and H. J. Müller-Eberhard. 1970. The second component of human complement (C2): Quantitative molecular analysis of its reactions in immune hemolysis. *Immunochemistry* **7**:341.

Gigli, I. 1974. Control mechanisms of the classical and alternate complement sequence. *Transpl. Proc.* **6**:9.

Götze, O. and H. J. Müller-Eberhard. 1970. Lysis of erythrocytes by complement in the absence of antibody. *J. Exp. Med.* **132**:898.

Haxby, J. A., O. Götze, H. J. Müller-Eberhard and S. C. Kinsky. 1969. Release of trapped marker from liposomes by the action of purified complement components. *Proc. Nat. Acad. Sci.* **64**:290.

Kolb, W. P. and H. J. Müller-Eberhard. 1973. The membrane attack mechanism of complement. Verification of a stable C5-9 complex in free solution. *J. Exp. Med.* **138**:438.

―――. 1975. The membrane attack mechanism of complement: Isolation and subunit composition of the C5b-9 complex. *J. Exp. Med.* **141**:724.

Kolb, W. P., J. A. Haxby, C. M. Arroyave and H. J. Müller-Eberhard. 1972. Molecular analysis of the membrane attack mechanism of complement. *J. Exp. Med.* **135**:549.

Mayer, M. M. 1970. Highlights of complement research during the past 25 years. *Immunochemistry* **7**:485.

Müller-Eberhard, H. J. 1975. Complement. In *Annual Review of Biochemistry* (ed. E. E. Snell), vol. 44, p. 697. Annual Reviews, Palo Alto, California.

Müller-Eberhard, H. J., A. P. Dalmasso and M. A. Calcott. 1966. The reaction mechanism of β_{1C}-globulin (C′3) in immune hemolysis. *J. Exp. Med.* **123**:33.

Müller-Eberhard, H. J., M. J. Polley and M. A. Calcott. 1967. Formation and functional significance of a molecular complex derived from the second and the fourth component of human complement. *J. Exp. Med.* **125**:359.

Nilsson, U. and J. Mapes. 1973. Polyacrylamide gel electrophoresis (PAGE) of reduced and dissociated C3 and C5: Studies of the polypeptide chain (PPC) subunits and their modifications by trypsin (Try) and C̄4,2-C̄4,2,3. *J. Immunol.* **111**:293.

Polley, M. J. and H. J. Müller-Eberhard. 1967. Enhancement of the hemolytic activity of the second component of human complement by oxidation. *J. Exp. Med.* **126**:1013.

Reid, K. B. M. and R. R. Porter. 1975. The structure and mechanism of activation of the first component of complement. *Biochem. J.* (in press).

Sakai, K. and R. M. Stroud. 1974. The activation of C1s with purified C1r. *Immunochemistry* **11**:191.

Schreiber, R. D. and H. J. Müller-Eberhard. 1974. Fourth component of human complement: Description of a three polypeptide chain structure. *J. Exp. Med.* **140**:1324.

Shelton, E., K. Yonemasu and R. M. Stroud. 1972. Ultrastructure of the human complement component C1q. *Proc. Nat. Acad. Sci.* **69**:65.

Stolfi, R. L. 1968. Immune lytic transformation: A state of irreversible damage generated as a result of the reaction of the eighth component in the guinea pig complement system. *J. Immunol.* **100**:46.

Valet, G. and N. R. Cooper. 1974a. Isolation and characterization of the proenzyme form of the C1s subunit of the first complement component. *J. Immunol.* **112**:339.

―――. 1974b. Isolation and characterization of the proenzyme form of the C1r subunit of the first complement component. *J. Immunol.* **112**:1667.

A Model for the Lytic Action
of Complement

Stephen C. Kinsky and Howard R. Six

Departments of Pharmacology and Biological Chemistry
Washington University School of Medicine, St. Louis, Missouri 63110

Classically, complement is defined as a multicomponent system of nine functional serum proteins that react in a specific sequence (C1,C4,C2,C3,C5,C6, C7,C8,C9) to produce lysis of target cells—particularly those recognized as "foreign" due to the presence of certain antigen-antibody complexes on the membrane surface. It is beyond the scope of this article to discuss the intricate details of this sequence (for review, see Müller-Eberhard 1972), and we need mention only briefly that activation of the early components (C1 through C5) is now known to occur by generation of proteolytic enzyme activities. Thus upon interaction of C1 with the membrane-localized immune complex, the first component is converted into a protease capable of cleaving C4 and C2; fragments from each of these components then combine to form an enzyme utilizing C3 as substrate. The next step in the complement cascade is initiated by another enzyme (consisting of the appropriate fragments derived from C4, C2 and C3) which cleaves C5; as a consequence, one of the C5 products becomes attached to the cell membrane along with C6 and C7. This sets the stage for the participation of the terminal components, C8 and C9, resulting in cell lysis by a mechanism that still is poorly understood. However, by analogy with the earlier steps in the complement sequence, it has long been presumed that activation of the terminal components involves the appearance of an enzyme that may directly attack some membrane constituent; alternatively, it has been suggested that activated C8 and/or C9 may stimulate an endogenous autolytic enzyme present in the cell membrane. The purpose of this paper is to review some of the evidence, which we have accumulated in the past six years, indicating that neither of these hypotheses may be correct.

EXPERIMENTAL PROCEDURE

Essentially, this evidence is based on the observation that a variety of liposomal model membranes release trapped markers when incubated with an appropriate antibody and complement source (for review, see Kinsky 1972a, b). To prepare such immunologically sensitive liposomes, the requisite lipid mixture is deposited as a film, coating the walls of a small conical flask. The dried lipids are then dispersed in an isotonic solution of the desired marker, and any marker which is not contained within the aqueous liposomal compartments is subsequently removed by dialysis against isotonic saline or Sepharose chromatography. In most cases, we have employed glucose as the trapped marker because alterations in the permeability of multicompartment liposomes can be followed by a convenient spectrophotometric assay. This assay is based on the increased absorbancy at 340 nm occurring when any released glucose is rapidly oxidized in cuvettes containing hexokinase, ATP, Mg^{++}, glucose-6-phosphate dehydrogenase and TPN. However, it should be noted at the outset that the effect of antibody-complement is not limited to the release of a compound as small as glucose (MW 180), but has been observed with some liposomes using enzymes as large as β-galactosidase (MW 518,000) as trapped marker. Specific experimental details may be found in each of the references cited.

RESULTS AND DISCUSSION

In our initial experiments (Haxby, Kinsky and Kinsky 1968), we employed liposomes that were made with the total lipid extract of sheep erythrocyte membranes containing all of the phospholipid and cholesterol, as well as a major portion of the Forssman antigen, present in the cells. In this regard, it should be emphasized that Forssman antigen is an amphipathic lipid leading to the anticipation that the nonpolar (ceramide) region would anchor the compound in the liposomal bilayers in such a way that the polar oligosaccharide region (which bears the antigenic determinants) remains accessible to antibody; immune complexes formed on the liposomal surface would therefore trigger the complement sequence. This prediction was validated by the fact that these sheep lipid liposomes released approximately 65% of their trapped glucose when incubated with rabbit antisheep erythrocyte antiserum (as a source of anti-Forssman antibodies) and guinea pig serum (as a source of complement).

Subsequent experiments revealed that the liposomes did indeed bind anti-Forssman antibodies and, in their presence, fixed (i.e., consumed) hemolytic complement activity (Alving et al. 1969; Alving and Kinsky 1971). Most importantly, in collaboration with Drs. O. Götze and H. J. Müller-Eberhard, we were able to implicate the same sequence of reactions that are involved in the immune lysis of cells. Thus using purified human complement components and specific anticomponent antibodies as inhibitors, glucose release from the liposomes was shown to be absolutely dependent on the presence of C2 and C8 and markedly stimulated by C9 (Haxby et al. 1969). These results justified the continued exploitation of liposomes to examine the molecular basis of complement-dependent membrane damage.

The chloroform extract employed in the generation of these liposomes did not contain any detectable protein (Haxby, Kinsky and Kinsky 1968), and this prompted the preparation of immunologically sensitive liposomes from defined lipid mixtures. Attainment of this objective was realized in several studies showing release of glucose, in the presence of an appropriate antibody and guinea pig serum, from liposomes made with four constituents (Kinsky et al. 1969; Inoue, Kataoka and Kinsky 1971; Kataoka et al. 1971a, b). These were a phospholipid (either lecithin or sphingomyelin), a sterol (such as cholesterol), a charged amphiphile (dicetyl phosphate for negative liposomes, stearylamine for positive liposomes), and a suitable antigen whose amphipathic properties would favor proper insertion into the lipid bilayers. Two classes of naturally occurring lipid antigens met this requirement: mammalian ceramides (such as Forssman, globoside I and galactocerebroside) or bacterial lipopolysaccharides and lipid A (isolated from S and R forms of Salmonella).

On the basis of these investigations, it was concluded that the direct lytic action of the complement system does not involve the action of a proteolytic enzyme because lipids in bilayer configuration can alone serve as "substrate" (Fig. 1). Accordingly, we tested the possibility that activation of the terminal components resulted in the formation of an enzyme that could degrade phospholipids. The other liposomal constituents were excluded as putative substrates for the following reasons: (1) Sterols are not present in all natural membranes that are susceptible to the action of complement; specifically, they are absent in gram-negative bacteria. (2) Neither of the charged amphiphiles (dicetyl phosphate or stearylamine) are found in any natural membrane. Furthermore, it seemed extremely unlikely that any single enzyme would possess such broad specificity permitting attack on compounds whose structures were so different. (3) This same argument applies to the antigens because obviously numerous substances of diverse structure can trigger the immune lysis of cells and, as just mentioned, immune damage of liposomes.

Thus we were left with phospholipid as the only plausible candidate and therefore prepared liposomes, sensitized with either Forssman or globoside I, using ^{32}P-labeled lecithin or sphingomyelin isolated from rat liver (Inoue

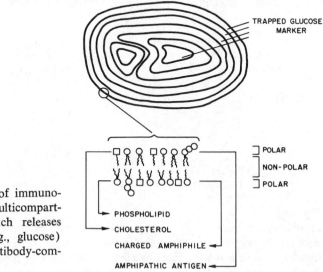

Figure 1
Schematic diagram of immunologically sensitive multicompartment liposome which releases trapped marker (e.g., glucose) in the presence of antibody-complement.

and Kinsky 1970). To our surprise, thin-layer chromatographic analysis revealed no new radioactive compound in reaction mixtures containing liposomes that had released up to 70% of their trapped glucose after incubation with antibody-complement. In fact, at least 98% of the recovered radioactivity was present in the form in which it had been originally incorporated into liposomes, i.e., as either lecithin or sphingomyelin. Although based on negative findings, these results led to the tentative conclusion that the lytic action of complement did not involve enzymatic degradation of membrane phospholipids. Positive supportive evidence was obtained with the demonstration that immunologically sensitive liposomes could be prepared with certain unique phosphonyl and phosphinyl analogs of lecithin (Kinsky et al. 1971; Kinsky 1972b); we should now like to present some recent experiments that illustrate this point.

For this purpose, reference must be made to the structures of two of the compounds which, together with cholesterol and dicetyl phosphate, were used to generate the liposomes. As the phospholipid, we employed the phosphinyl analog (Fig. 2, right), which was kindly donated by Dr. A. F. Rosenthal of the Long Island Jewish Medical Center. This compound differs from natural lecithin (Fig. 2, left) in the following important points: (1) The aliphatic side chains are saturated and hence not liable to peroxidation. (2) The aliphatic side chains are attached to the glycerol backbone via ether, instead of ester, linkages and therefore not susceptible to hydrolysis by either phospholipase A_1, A_2 or B. (3) Finally, the analog lacks the C-O-P and the P-O-C bonds that are cleaved, respectively, by phospholipases C and D.

To render the liposomes immunologically sensitive, we no longer have recourse to naturally occurring lipid antigens, but incorporate the synthetic amphipathic haptens illustrated in Figure 3. These are the dinitrophenylated derivative of phosphatidylethanolamine (shown at the top and abbreviated Dnp-PE) and the dinitrophenylated derivative of aminocaproylphosphatidyl-

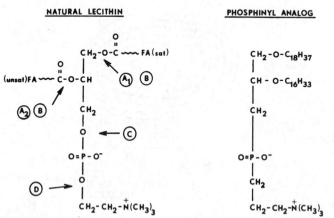

Figure 2
Structural comparison of natural lecithin and synthetic phosphinyl analog; FA designates either saturated or unsaturated long-chain fatty acids attached to the glycerol portion of phosphatidylcholine; arrows denote the bonds hydrolyzed by phospholipases A, B, C and D.

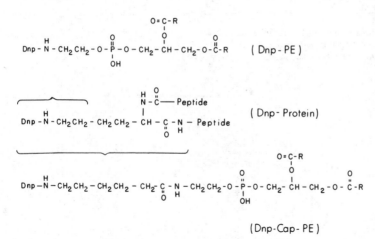

Figure 3

Structural comparison of dinitrophenylated protein (*middle*) and dinitrophenylated phospholipids (*top, bottom*). R designates long-chain fatty acids attached to the glycerol portion of phosphatidyl-ethanolamine.

ethanolamine (shown at the bottom and abbreviated Dnp-Cap-PE). Such compounds possess several advantages. (1) They can be prepared in large quantities. (2) They react with highly purified and well-characterized anti-Dnp antibodies raised in rabbits immunized with dinitrophenylated proteins (structure depicted schematically in Fig. 3, middle). (3) Homologous compounds can be synthesized in which either one or both of the fatty acids is missing.

With these derivatives, it has been possible to investigate such subtle phenomena as the effect of structure on the active vs. passive sensitization of liposomes to antibody-complement (Uemura and Kinsky 1972) and the effect of immunoglobulin class and affinity on immune damage of liposomes (Six, Uemura and Kinsky 1973). Unfortunately, lack of space does not permit discussion of these results, but mention should be made of one unexpected property of Dnp-Cap-PE. We have recently reported that liposomes containing this low molecular weight hapten-phospholipid conjugate are immunogenic in guinea pigs, giving rise to anti-Dnp antibodies. Indeed, the available evidence suggests that liposomes, actively sensitized with N-substituted phosphatidylethanolamine derivatives, may constitute a useful alternative to more conventional methods of antibody production which depend on prior covalent linkage of a hapten to a high molecular weight water-soluble carrier such as protein (Uemura et al. 1974).

To return to the main theme, Figure 4 illustrates the extremely rapid kinetics of glucose release occurring upon the addition of excess complement to phosphinyl liposomes that have been preincubated with excess anti-Dnp antibodies. This experiment also demonstrates that liposomal immune damage requires incorporation of the amphipathic antigen because marker loss was not observed from unsensitized liposomes prepared in the absence of Dnp-Cap-PE.

The dependence of this phenomenon on antibody concentration is depicted in Figure 5. Although not indicated, it should be emphasized that little glucose

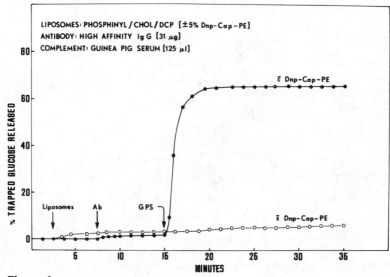

Figure 4

Kinetics of glucose release from phosphinyl liposomes generated in the
presence and absence of Dnp-Cap-PE. In this experiment (as well as those
described in Figs. 5 and 6), actively sensitized multicompartment liposomes were prepared and glucose release assayed by the procedures
described by Uemura and Kinsky (1972).

release occurred (less than 5%) when the anti-Dnp antibodies were replaced
with either an immunoglobulin fraction from normal serum or with antibodies
that do not interact with the dinitrophenyl group.

The effect of complement concentration on the extent of glucose release is
depicted in Figure 6. In this regard, it should be noted that guinea pig serum
(whose hemolytic complement activity had been destroyed by heating at 56°C
for 30 min) was unable to produce marker loss even in the presence of excess
anti-Dnp antibodies.

Such experiments with phosphinyl liposomes strongly suggest that com-
plement-induced lysis does not involve any known phospholipid degradative
enzyme. Coupled with previous arguments against the direct role of any
proteolytic enzyme, we therefore proposed the following alternative (Fig. 7):
Activation of the terminal components (C8 and/or C9) results in transient
exposure of a hydrophobic region in these proteins (or release of a hydrophobic fragment); insertion of the latter into lipid bilayers would presumably
disrupt the noncovalent bonds responsible for maintaining lipids in stable
bilayer configuration.

This concept of complement "detergency" is currently accepted by most
investigators in the field, and indeed, Müller-Eberhard's laboratory has gone
one step further. Their experiments (Kolb et al. 1972, 1973; Kolb and Müller-
Eberhard 1973) suggest that erythrocyte lysis is produced not by a single
small molecule, but rather by a firm, 1 million molecular weight complex consisting of one molecule each of C5, C6, C7 and C8 plus six molecules of C9.

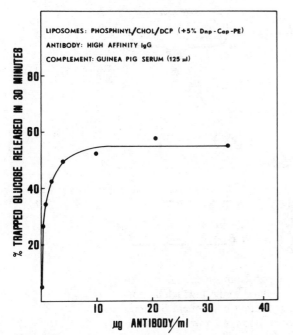

Figure 5

Effect of anti-Dnp antibody (high-affinity rabbit IgG) concentration on glucose release from multicompartment phosphinyl liposomes actively sensitized with Dnp-Cap-PE.

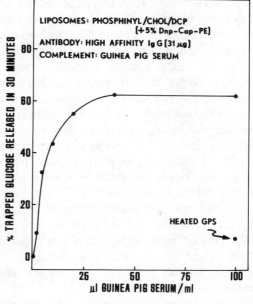

Figure 6

Effect of guinea pig serum (complement) concentration on glucose release from multicompartment phosphinyl liposomes actively sensitized with Dnp-Cap-PE.

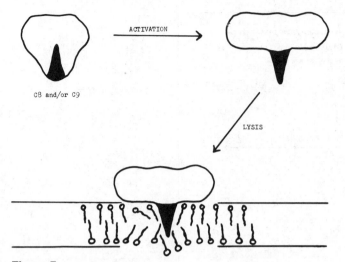

Figure 7

Hypothetical "detergent" action of complement. A hydrophobic region (represented by a dark wedge) which is normally buried within the terminal protein component becomes transiently exposed (or released) upon activation; insertion into the lipid bilayer of cell membranes or liposomes produces "disorganization" leading to lysis.

However at the present time, this hypothesis is difficult to reconcile with certain aspects of liposomal response that must be discussed briefly.

Up to now, we have only referred to data obtained with multicompartment liposomes which, in the presence of excess antibody-complement, release up to 70% of their trapped glucose. This implies that an appreciable portion probably originates from the deeper aqueous compartments, because it can be calculated (Kinsky 1972a) that, for an "average" liposome of 30 concentric bilayers, approximately 7 of the latter must be affected to achieve even 50% release. For a number of years, we have assumed that this could proceed in successive stages: namely, immune damage to the external bilayer permits the entrance of antibody molecules and complement components; the antibodies then combine with antigens in the next bilayer; the resulting immune complexes initiate the complement sequence that destroys the second bilayer, resulting in the exposure of the third bilayer, etc. However, this scheme is no longer tenable in view of the fact that passively sensitized liposomes, which contain antigen only in the outermost bilayer, release as much marker as actively sensitized liposomes in which the antigen is distributed in all of the bilayers (Uemura and Kinsky 1972; Six, Uemura and Kinsky 1973). Furthermore, we have shown that sphingomyelin liposomes, in contrast to lecithin liposomes, release 50% of their trapped glucose under conditions in which little (if any) trapped β-galactosidase appears in the medium (Katoaka, Williamson and Kinsky 1973). Accordingly, it seems likely that if such a big protein molecule cannot get out, equally large (or larger) proteins or complexes may not get in.

In an attempt to resolve this dilemma, we have recently prepared single-

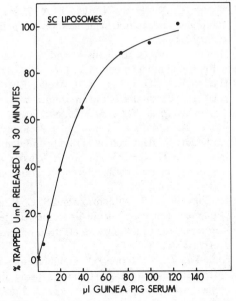

Figure 8
Effect of guinea pig serum (complement) concentration on umbelliferone phosphate (UmP) release from single-compartment liposomes actively sensitized with Dnp-Cap-PE. See Six et al. (1974) for experimental details.

compartment liposomes that respond to anti-Dnp antibodies and guinea pig serum (Fig. 8). These efforts were contingent on the development of a sensitive fluorometric assay for following changes in liposomal permeability that relies on the use of a fluorogenic substrate, umbelliferone phosphate (abbreviated UmP), as trapped marker (Six et al. 1974). It should be noted that a preparation of single-compartment liposomes releases all of its trapped marker when incubated with sufficient complement. This is consistent with our earlier contention that failure to obtain 100% glucose release from multi-compartment liposomes is due to an inability of cytolytically active complement components to reach the innermost lipid bilayers and the fact that, by definition, hydrophobic regions have a limited survival time in water. These single-compartment liposomes should reveal on their surface the postulated large complement component complex upon examination in the electron microscope. Moreover, the single-compartment liposomes are ideal objects for investigation by physical techniques such as differential scanning calorimetry and nuclear magnetic and electron spin resonance spectroscopy. The latter are obviously relevant in future studies concerned with the proposed model for the lytic action of complement.

Acknowledgment

Support for these studies was provided by U.S. Public Health Service Research Grant AI-09319 and Training Grant NS-05221.

REFERENCES

Alving, C. R. and S. C. Kinsky. 1971. Preparation and properties of liposomes in the LA and LAC states. *Immunochemistry* **8:**325.

Alving, C. R., S. C. Kinsky, J. A. Haxby and C. B. Kinsky. 1969. Antibody binding and complement fixation by a liposomal model membrane. *Biochemistry* **8**:1582.

Haxby, J. A., C. B. Kinsky and S. C. Kinsky. 1968. Immune response of a liposomal model membrane. *Proc. Nat. Acad. Sci.* **61**:300.

Haxby, J. A., O. Götze, H. J. Müller-Eberhard and S. C. Kinsky. 1969. Release of trapped marker from liposomes by the action of purified complement components. *Proc. Nat. Acad. Sci.* **64**:290.

Inoue, K. and S. C. Kinsky. 1970. Fate of phospholipids in liposomal model membranes damaged by antibody and complement. *Biochemistry* **9**:4767.

Inoue, K., T. Kataoka and S. C. Kinsky. 1971. Comparative responses of liposomes prepared with different ceramide antigens to antibody and complement. *Biochemistry* **10**:2574.

Kataoka, T., J. R. Williamson and S. C. Kinsky. 1973. Release of macromolecular markers (enzymes) from liposomes treated with antibody and complement. An attempt at correlation with electron microscopic observations. *Biochim. Biophys. Acta* **298**:158.

Kataoka, T., K. Inoue, C. Galanos and S. C. Kinsky. 1971a. Detection and specificity of lipid A antibodies using liposomes sensitized with lipid A and bacterial lipopolysaccharides. *Eur. J. Biochem.* **24**:123.

Kataoka, T., K. Inoue, O. Lüderitz and S. C. Kinsky. 1971b. Antibody and complement dependent damage to liposomes sensitized with bacterial lipopolysaccharides. *Eur. J. Biochem.* **21**:80.

Kinsky, S. C. 1972a. Antibody-complement interaction with lipid model membranes. *Biochim. Biophys. Acta* **265**:1.

———. 1972b. Immune damage to a lipid model membrane. *Ann. N.Y. Acad. Sci.* **195**:429.

Kinsky, S. C., P. P. M. Bonsen, C. B. Kinsky, L. L. M. van Deenen and A. F. Rosenthal. 1971. Preparation of immunologically responsive liposomes with synthetic phosphonyl and phosphinyl analogs of lecithin. *Biochim. Biophys. Acta* **233**:815.

Kinsky, S. C., J. A. Haxby, D. A. Zopf, C. R. Alving and C. B. Kinsky. 1969. Complement-dependent damage to liposomes prepared from pure lipids and Forssman hapten. *Biochemistry* **8**:4149.

Kolb, W. P. and H. J. Müller-Eberhard. 1973. The membrane attack mechanism of complement. Verification of a stable C5-9 complex in free solution. *J. Exp. Med.* **138**:438.

Kolb, W. P., J. A. Haxby, C. M. Arroyave and H. J. Müller-Eberhard. 1972. Molecular analysis of the membrane attack mechanism of complement. *J. Exp. Med.* **135**:549.

———. 1973. The membrane attack mechanism of complement. Reversible interactions among the five native components in free solution. *J. Exp. Med.* **138**:428.

Müller-Eberhard, H. J. 1972. The molecular basis of the biological activities of complement. In *The Harvey Lectures,* Series 66, pp. 75–104. Academic Press, New York.

Six, H. R., K. Uemura and S. C. Kinsky. 1973. Effect of immunoglobulin class and affinity on the initiation of complement dependent damage to liposomal model membranes sensitized with dinitrophenylated phospholipids. *Biochemistry* **12**:4003.

Six, H. R., W. W. Young, Jr., K. Uemura and S. C. Kinsky. 1974. Effect of antibody-complement on multiple vs. single compartment liposomes. Application of a fluorometric assay for following changes in liposomal permeability. *Biochemistry* **13**:4050.

Uemura, K. and S. C. Kinsky. 1972. Active vs. passive sensitization of liposomes toward antibody and complement by dinitrophenylated derivatives of phosphatidylethanolamine. *Biochemistry* **11**:4085.

Uemura, K., R. A. Nicolotti, H. R. Six and S. C. Kinsky. 1974. Antibody formation in response to liposomal model membranes sensitized with N-substituted phosphatidylethanolamine derivatives. *Biochemistry* **13**:1572.

Proteases of the Properdin System

Otto Götze

Department of Molecular Immunology, Scripps Clinic and Research Foundation
La Jolla, California 92037

Work performed in a number of laboratories during the last few years has led to the recognition of an alternative pathway of complement (C) activation (C3 activator system, complement bypass, C3 shunt) operative in the absence of immune antibodies and independent of the early acting complement components C1, C2 and C4[1] (Gewurz, Shin and Mergenhagen 1968; Sandberg et al. 1970; Frank et al. 1971; Götze and Müller-Eberhard 1971). These observations are now firmly linked to the properdin system and have confirmed and extended the work of Pillemer and coworkers (Pillemer et al. 1954; Blum, Pillemer and Lepow 1959; Pensky et al. 1959; Lepow 1961). Although our understanding of the biology of the properdin system and the molecular interactions of its proteins is still incomplete, considerable progress has recently been made in defining factors of this serum system and in delineating their functions. The properdin system can be defined as that group of proteins which, in the absence of the classical C components C1, C2 and C4, will mediate the proteolytic activation of C3 and C5 when serum is incubated with various polysaccharides and lipopolysaccharides. These activating substances include gram-negative endotoxins, yeast cell walls (zymosan), inulin particles, agar and dextrans. In addition, aggregated IgA myeloma proteins and guinea pig γ_1-antibodies (Sandberg et al. 1971), as well as certain polyanionic substances (Hadding et al. 1973; König et al. 1974), can mediate the activation of C3 through this system. The activation of C3 and C5 results in the generation of biological activities associated with the small fragments of C3 and C5 (the anaphylatoxins C3a and C5a), the chemotactic C5b67 complex and the potentially cytolytic C5b6789 complex, as discussed in this volume by Drs. Müller-Eberhard and Hugli. So far, no differences in the biological activities generated or in the chemistry of fragments and complexes have been discerned when properdin system-mediated processes and products

---

[1] The terminology for the classical complement components conforms to the recommendations of the World Health Organization Committee on Complement Nomenclature (Bulletin World Health Organization, vol. 39, p. 939, 1968).

Table 1

Synonyms of Components of the Human Properdin System

Precursor	*Activated form*
Initiating factor (IF)	$\overline{\text{IF}}$ (C3NeF ?)
Properdin (P)	$\overline{\text{P}}$
C3	C3b
Hydrazine sensitive factor (HSF)	HSFa
Factor A	
C3 proactivator (C3PA)	C3A
Glycine-rich β-glycoprotein (GBG)	GGG β_2-glycoprotein II
Factor B	$\overline{\text{B}}$
C3PA convertase (C3PAse)	$\overline{\text{C3PAse}}$
GBGase	
Factor D	$\overline{\text{D}}$
C3b inactivator (C3bINA)	
Conglutinogen-activating factor (KAF)	
Factor C	

were compared with those resulting from activation of the classical, $\overline{\text{C42}}$-dependent pathway of complement activation. Because of the way in which our knowledge about the properdin system developed and for reasons of differences in experimental approach, a number of synonyms, listed in Table 1, have been used for components of this enzyme system.

CONSTITUENTS OF THE PROPERDIN SYSTEM

At present, five serum proteins have clearly been recognized to be factors of the alternative pathway of C activation. These are a factor related to the C3 nephritic factor (C3NeF), properdin (P), the classical third component of complement (C3), C3 proactivator (C3PA) and C3 proactivator convertase (C3PAse). Some of the properties of these components are listed in Table 2. In addition to these factors, a regulatory enzyme, C3b inactivator (C3bINA), has been described.

C3 nephritic factor was detected in the sera of some patients with a form of chronic hypocomplementemic nephritis. Serum of such patients was found to have C3-activating (cleaving) activity when combined with normal serum proteins (Spitzer et al. 1969; Vallota et al. 1970), one of which was later identified as C3PA (Ruley et al. 1973). C3NeF, a non-Ig 7S γ-globulin, has been obtained in highly purified form and has been shown to depend on C3PA and C3PAse for its capacity to initiate C3 activation in vitro. Furthermore, an immune absorbant made from an antiserum to purified C3NeF could be used to remove a component from normal serum which is essential for the activation of C3 by inulin, suggesting that C3NeF is an activated form of a normal constituent of the properdin system (Vallota et al. 1974). The

Table 2
Properties of Human Properdin System Proteins

Property	C3NeF (IF?)	P̄	C3	C3b	C3PA	C3A	C3PAse
MW	150,000	184,000	185,000	176,000	93,000	63,000	24,000
s-Rate	7S	5.4S	9.5S	9S	5–6S	3.5–4.5S	2.5–3S
Electrophoretic mobility	γ	γ2	β	α2	β	γ	α
Approx. serum concentration (μg/ml)	ND	25	1200	—	200	—	ND
Thermolability (52°C, 30 min)	—	—	+	—	+	ND	—
Hydrazine sensitivity	—	—	+	—	—	ND	—
Complex formation with CVF	ND	ND	ND	ND	(+)	+	?

ND, not determined.

dependence of C3NeF on the presence of C3PA and C3PAse points to a role for C3NeF (or its physiological analog) during the early steps of the reaction. It is therefore conceivable that C3NeF is the active form of that protein which, after activation by contact with polysaccharides, initiates the reaction sequence of the properdin system. This factor has been termed the initiating factor (IF). It is not yet known if C3NeF is an enzyme, nor is the mechanism of its action understood, although the experimental data are compatible with the conclusion that C3NeF, alone or in combination with an additional factor, participates in the activation of properdin.

Properdin was first obtained in highly purified form in 1968 (Pensky et al. 1968). The method of preparation was based on the fact that P becomes bound to zymosan when this substance is used to activate the properdin system in normal serum. Properdin can then easily be dissociated from the zymosan particles and can be further purified. Another purification method using conventional column chromatography of human serum in EDTA-containing buffers was recently described (Götze and Müller-Eberhard 1974). Both methods yield properdin which appears to be in an activated state ($\overline{P}$) as evidenced by its capacity to initiate activation of C3 and C3PA in genetically C2-deficient and in normal human serum in the absence of any of the known activating substances. Activation of C3 by $\overline{P}$ was shown to require the presence of C3PA and C3PAse, and activation of C3PA required the presence of C3 and of C3PAse (Götze and Müller-Eberhard 1974). These results firmly linked P to the inulin-mediated activation of C3PA and C3 and confirmed the identity of the C3 activator system or alternative pathway of complement activation (Götze and Müller-Eberhard 1971) and the properdin system.

Activated properdin is one of the most basic proteins of human serum (Fig. 1). It is a glycoprotein containing 9.8% carbohydrate and has a molecular weight of 184,000 daltons. Analysis of its subunit structure revealed that it is composed of four noncovalently linked subunits, each having an approximate molecular weight of 46,000 daltons (Minta and Lepow 1974). Properdin prepared recently in this laboratory from fresh plasma without the use of zymosan and under conditions that prevent the activation of the coagulation and the kinin-forming systems was found to have a subunit molecular weight of 53,000 daltons by sodium dodecyl sulfate-polyacrylamide gel electrophoresis (SDS-PAGE) (Fig. 2). A side-by-side comparison of zymosan-purified properdin and properdin purified from fresh plasma is now in progress in our laboratory. It aims at the identification of differences in the properties and biological activities of the two differently prepared proteins.

The mechanism of action of $\overline{P}$ is still largely unknown. It was reported that P is able to activate precursor C3PAse to the active enzyme, and that this ability is not abolished by treatment with 5×10^{-3} M diisopropylfluorophosphate (DFP) or 10^{-2} M tosyl-L-lysine-chloromethyl ketone (TLCK) (Fearon, Austen and Ruddy 1974a). Since $\overline{P}$ is required for activation of isolated C3PA and C3 in the presence of active C3PAse (Götze and Müller-Eberhard 1974), the results of Fearon and coworkers imply that $\overline{P}$ may have several functions. A factor controlling the activity of zymosan-purified properdin was detected in the euglobulins of human serum by Minta (1974). This protein, termed factor F, has the capacity to form a complex with isolated $\overline{P}$

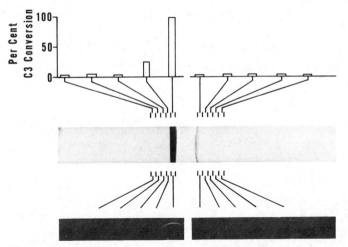

Figure 1

Analysis of isolated $\overline{P}$ by polyacrylamide gel electrophoresis. Two gels were run in identical fashion. Eighty μg $\overline{P}$ in 10% sucrose was layered on top of each 6% running gel. Another 6% running gel was then polymerized on top of each liquid sample. Electrophoresis was performed in 2-amino-2-methyl 1,3-propanediol at pH 9.5 as described by Tamura and Ki (1972). After the run (anode at the right), the two parts of one gel were fixed and stained with amido black. The two parts of the other gel were sliced as indicated into 2-mm sections which were eluted overnight at 4°C with 0.1 ml phosphate-buffered 0.3 M NaCl containing 0.1% gelatin. The eluates were then assayed (1) for C3-converting activity by incubation with fresh serum followed by immunoelectrophoretic analysis of the degree of C3 conversion and (2) for the presence of properdin by double diffusion in agarose using a monospecific antiserum. (Reprinted, with permission, from Götze and Müller-Eberhard 1974.)

in which properdin activity can no longer be detected. Concomitantly, a shift of the properdin line from a γ_2 to a β_2 position can be demonstrated by immunoelectrophoresis. Whether properdin also exists complexed to factor F in fresh plasma or serum and whether dissociation of such a complex precedes activation of properdin has not yet been resolved.

C3 proactivator was the first component of the properdin system to be recognized as an enzyme or, more specifically, the precursor of a protease. C3PA is identical to the glycine-rich β-globulin (GBG) (Boenisch and Alper 1970a) and to factor B (Goodkofsky and Lepow 1971). A protein closely resembling the active fragment of C3PA, i.e., C3 activator (C3A), was isolated in 1965 and termed β_2-glycoprotein II (Haupt and Heide 1965). It appears to be identical to the glycine-rich γ-glycoprotein of Boenisch and Alper (1970b). C3PA was first described as a biologically active molecule by Müller-Eberhard, who found that it is required for the activation of C3 which results from the addition of a cobra venom glycoprotein to normal human serum (Müller-Eberhard 1967). It was subsequently demonstrated that C3PA

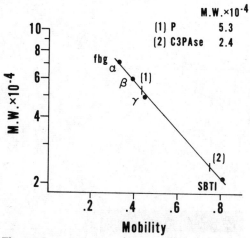

Figure 2

Determination of the molecular weights of the subunits of properdin and of C3PA convertase. SDS gels (7%) were run with 5 μg of each isolated protein and approximately 10 μg of the marker proteins as described by Weber and Osborn (1969). Before layering in 25% glycerol on top of the gels, all samples were incubated at 37°C for 60 minutes with an equal volume of 10 M urea containing 1.4×10^{-2} M DTT and 2% SDS. The proteins used as molecular weight markers were human fibrinogen (fbg) provided by Dr. P. C. Harpel and soybean trypsin inhibitor (STI), obtained from Worthington. Their molecular weights in daltons are respectively: fbg A α-chain, 70,900; fbg B β-chain, 60,400 and γ-chain, 49,400 (Mosesson et al. 1972); and STI, 21,000 (Wu and Scherega 1962). Virtually identical results were obtained with P and C3PAse which had not been reduced in urea-DTT prior to electrophoresis.

is able to enter into a complex with cobra venom factor (CVF) to form an enzyme capable of cleaving C3 in typical fashion (Müller-Eberhard and Fjellström 1971). C3PA is a 5–6S β_1-globulin having a molecular weight of 93,000 daltons. It is composed of a single polypeptide chain, as evidenced by SDS-PAGE of the reduced or performic acid oxidized protein (Fig. 3). Upon activation, C3PA is cleaved into two fragments, a basic larger one and a more acidic smaller fragment, as can be demonstrated by experiments using gel permeation chromatography, sucrose density gradient ultracentrifugation or immunoelectrophoretic analysis (Fig. 4). The molecular weights of these fragments as established by SDS-PAGE are 63,000 daltons and 30,000 daltons, respectively (Fig. 5). After purification by DEAE-cellulose and sub-

Figure 3
Analysis of the structure of isolated C3PA by SDS-polyacrylamide gel electrophoresis. Three 7% SDS gels were
run as described by Weber and Osborn (1969). Purified
C3PA (2–10 μg) was treated prior to electrophoresis with
an equal volume of 2% SDS (*left*) or 10 M urea containing 1.4×10^{-2} M DTT and 2% SDS (*middle*), or was first
oxidized with performic acid (Hirs 1967) and then treated
with urea-SDS (*right*).

C3PA C3PA C3PA
— DTT Perf.
Acid

sequent gel permeation chromatography on Sephadex G-100, the large basic
fragment still retained some of the C3-cleaving activity of the molecule. It
was therefore termed C3 activator (C3A) (Götze and Müller-Eberhard
1971). Furthermore, Cooper (1971; Cooper and Götze, unpubl.) was
able to show that both purified C3PA and C3A cleave N-α-acetyl-glycyl-L-
lysine-methyl ester (AGLMe), which is also a substrate for C2. These results
are supported by the finding that the enzymatically highly active CVF complex is composed of CVF and C3A (Cooper 1973; Vogt et al. 1974). The
biological activity of the preformed CVF-C3A complex is not abolished by

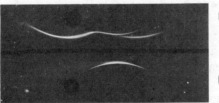

C3PA After
Activation

C3PA Control

Figure 4
Demonstration of the cleavage of C3PA by immunoelectrophoresis. Isolated C3PA (6 μg) was incubated
for 30 minutes at 37°C with $\overline{P}$ (0.4 μg), native C3 (13
μg) and C3PAse (.08 μg) in the presence of 10^{-3} M
Mg^{++}. The control was set up in identical fashion except that $\overline{P}$ was omitted from the reaction mixture.
Immunoelectrophoresis was performed in 1% agarose
and the pattern developed with an anti-C3PA serum.
The anode was at the right. (Adapted from Götze and
Müller-Eberhard 1974.)

262   O. Götze

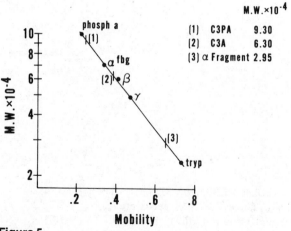

Figure 5
Determination of the molecular weights of C3PA and
C3A. The 7% SDS gel electrophoresis was performed
as described in the legend to Figure 2. The additional
marker proteins were phosphorylase a (phosph a),
100,000 daltons (Cohen, Duewer and Fischer 1972),
and trypsin (tryp), 23,300 daltons (Weber and Osborn
1969). Prior to electrophoresis, C3PA was treated for
5 minutes at 37°C with 5% (w/w) trypsin. The reaction
was stopped by the addition of a threefold excess of STI,
and the samples were boiled for 5 minutes in urea-SDS.
Analysis of the trypsin-treated C3PA by immunoelectrophoresis (not shown) revealed complete cleavage of
C3PA into C3A and the fragment with α-globulin mobility (α-fragment).

treatment with DFP up to a concentration of 10^{-2} M, nor is p-nitrophenyl
p'-guanidinobenzoate (Chase and Shaw 1967) a titrant or substrate for the
active complex (Götze, unpubl.). Using a different, C3b-dependent assay
for C3A, Fearon, Austen and Ruddy (1974b) were also unable to inhibit
the enzyme with 10^{-3} M DFP. These results may indicate that C3A is not a
protease of the serine esterase type, but it is more likely that they are an expression of the high specificity of this enzyme which activates C3 in a manner
seemingly identical to that of trypsin.

C3PA convertase is the second factor of the properdin system which was
recognized as a protease. Its existence was first postulated after the proteolytic
activation mechanism of C3PA had been established. Expression of C3PAse
activity requires the presence of C3b, the major fragment of C3 (Müller-Eberhard and Götze 1972), or of CVF (Cooper 1973; Vogt et al. 1974; Hunsicker, Ruddy and Austen 1973). C3PAse appears to exist both in a precursor
and an active form in fresh serum. The active enzyme (C3PAse) can be
derived from its precursor by treatment with trypsin or with $\overline{P}$ (Fearon, Austen and Ruddy 1974a,b). C3PAse is a 2.5–3.0S α-globulin (Müller-Eberhard and Götze 1972). It is one of the trace proteins of human serum, since
only 20–50 μg of the purified enzyme can be obtained from 800–1000 ml of

fresh starting material. After treatment with dithiothreitol (DTT) in urea, isolated C3PAse exhibits in SDS-PAGE a single, stainable band with a molecular weight of 24,000 daltons (Fig. 2). The molecular weight of precursor C3PAse which is activatable by trypsin was determined to be 25,000 daltons using analytical Sephadex G-75 gel permeation chromatography (Fearon, Austen and Ruddy 1974b). According to the same authors, treatment with 5×10^{-3} M DFP completely inactivates C3PAse, whereas 10^{-2} M TLCK has no effect on the enzyme. In our own experience, DFP at a concentration of 10^{-2} M inactivated 25–55% of the biological activity of a highly purified preparation of C3PAse. In these experiments, the capacity of C3PAse to mediate formation of the CVF-C3A complex was studied, whereas Fearon and coworkers used a C3b-dependent assay. These differences can probably be resolved by a careful comparison of preparations and assay methods, as well as by the use of radiolabeled DFP in order to study the specific uptake and incorporation of this compound into the active site of C3PAse.

The C3b-dependent Activation of C3PA and C3

Soon after the activation of C3PA in serum by a proteolytic process had been established, it was observed that treatment of serum with .015 M hydrazine hydrate destroyed the C3PA cleaving ability that is generated in fresh serum upon incubation with inulin. The hydrazine-sensitive factor (HSF) was subsequently purified and was identified as C3, which had long been known to lose its hemolytic activity by treatment with hydrazine. Later, C3 could also be identified with factor A of the properdin system (Goodkofsky, Steward and Lepow 1973). An activated form of this factor (HSFa) was found to mediate cleavage of C3PA in hydrazine-treated and in untreated serum in the absence of one of the typical activating substances. It was identified as C3b. Incubation of isolated C3PA and C3b did not lead to C3PA conversion. An additional protein, C3PA convertase, found to be essential for this reaction, was detected in the α-globulin fraction of serum separated by Pevikon block electrophoresis. C3PA activation was inhibited in the presence of EDTA and could be restored by the addition of 10^{-4}–10^{-3} M Mg^{++}. Addition of isolated native C3 to an incubation mixture containing C3b, C3PAse and C3PA led to the cleavage not only of C3PA, but of C3 as well. The finding that C3b, a product of C3, can switch on the enzyme system that generates C3a and C3b from C3 established the unexpected existence of a positive feedback mechanism within the C3 activator system. Based on these results, a concept was proposed in which C3b acts as an activator or modulator of C3PAse, enabling this enzyme to express its C3PA cleaving activity (Müller-Eberhard and Götze 1972). C3b inactivator, an enzyme known to degrade C3b and destroy its biological activity (Lachmann and Müller-Eberhard 1968; Ruddy and Austen 1971), serves as an important regulator of this part of the properdin system. Pretreatment of C3b with C3bINA reduced its capacity to initiate cleavage of C3PA by C3PAse (Götze and Müller-Eberhard 1973). Removal of C3bINA from serum by immune absorption leads to the activation of C3PA and C3 upon incubation of such serum, presumably through accumulation of C3b (Nicol and Lachmann 1973). Alper, Rosen and Lachmann (1972) postulated that C3bINA acts directly on C3PAse. However, they did not use purified C3PAse in their experiments. Instead, they used the serum of a patient genetically deficient in C3bINA as a source of C3PAse. Probably because of this deficiency, C3b accumulates in the circulation of this patient and is present in large amounts of his serum, which therefore contains C3PA cleaving activity. Thus the experiments of Alper and coworkers do not allow us to distinguish between an action of C3bINA on C3b or on C3PAse. Drawing an analogy from the mode of action of CVF, Vogt et al. (1974) proposed a somewhat different concept of the role of C3b. They envisioned C3PA as forming a loose complex with C3b which leads to a conformational change in C3PA sufficient to expose that bond in the molecule which is susceptible to cleavage by C3PAse. It is conceivable that a complex composed of C3b and C3A is formed as a result of the C3b-dependent activation of C3PA, although an affinity of C3A for C3b has so far not been demonstrated. It is therefore unclear at the present time in which way C3b participates in the activation of native C3 by C3A. It is possible that C3b is required for the stabilization of a labile site on C3A which is essential for the enzymatic attack of C3 by this enzyme. Alternatively, C3b might provide the binding site for native C3 in a C3b-C3A complex. These questions could be resolved if alternative methods or substances other than C3b or CVF were available for the stabilization of the C3A enzyme.

The Properdin-dependent Activation of C3PA and C3

After the discovery of the C3b-dependent feedback mechanism of the activation of C3PA and C3, the question arose as to the manner in which activating substances such as inulin initiate the reaction sequence of the properdin system. Two possibilities may be considered: (1) that an enzyme becomes activated and cleaves C3b from native C3, thereby setting off the feedback mechanism, or (2) that a protein other than C3b is involved in the initial activation of C3PA by C3PAse. Some insight into these questions came from experiments in which an antiserum to purified properdin and isolated P was used. Serum was depleted of P by an immune absorption method. P-depleted serum did not support activation of C3PA or C3 upon incubation with inulin. However, addition of C3b or CVF to such serum still resulted in the activation of C3PA and C3. Activation of these components was also seen when P-depleted serum was incubated with highly purified, isolated P in the absence of any activating substance (Fig. 6). These observations allowed the following conclusions: (1) properdin is an essential early acting factor of the alternative pathway of complement activation, (2) the purification method yielded properdin in an activated state ($\overline{P}$), and (3) properdin exists in fresh serum as a precursor, from which it becomes activated when inulin is added. A requirement for C3, C3PA and C3PAse for the expression of activity of $\overline{P}$ could be demonstrated by using sera specifically depleted of any one of these components. In addition, experiments in which only isolated components were used showed that the conversion of C3PA depended on $\overline{P}$, native C3 and C3PAse, and that conversion of native C3 depended on $\overline{P}$, C3PA and C3PAse (Fig. 4). These results thus clearly demonstrated that $\overline{P}$ requires native C3 in addition to C3PAse in order to initiate C3PA activation, which, in turn, would

C3   C3PA

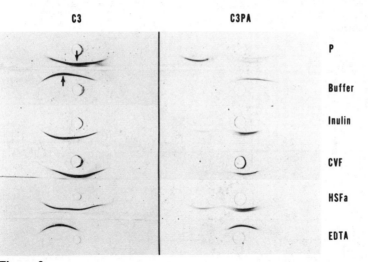

P

Buffer

Inulin

CVF

HSFa

EDTA

Figure 6
Conversion (activation) of C3 and C3PA in P-depleted serum. Fifty
μl of P-depleted serum was incubated for 30 minutes 37°C with 1 μg
of isolated $\overline{P}$, 250 μg of inulin, 5 μg of CVF or 2 μg of HSFa (C3b).
Controls received buffer or EDTA (2×10^{-3} M final concentration).
Total reaction volumes were 65 μl. After incubation, aliquots of the
mixtures were analyzed for C3 conversion (*left*) or for C3PA conversion (*right*) using an antiserum which detects only the C3A fragment of C3PA. Immunoelectrophoresis was performed in 1% ion
agar, and slides were stained with amido black. The anodes were at
the right. (Reprinted, with permission, from Götze and Müller-Eberhard 1974.)

catalyze the activation of C3. It was therefore proposed that $\overline{P}$ confers C3b-
like properties upon native C3 (conceivably through formation of a complex)
thereby allowing $\overline{C3PAse}$ to activate C3PA to C3A. This enzyme would then
generate C3b from native C3, setting the C3b-dependent feedback mechanism
in motion. Alternatively, it is possible that P is an enzyme with a highly restricted activity for C3, and that the molar ratio of $\overline{P}$ to C3b generated approaches a value of unity. A similar explanation was considered by Valet and
Cooper (1974) for their finding that equimolar amounts of $\overline{C1r}$ are required
for full activation of C1s. Both of these possible modes of action of $\overline{P}$ would
be compatible with the finding that, in the absence of C3PA and $\overline{C3PAse}$,
cleavage of isolated C3 by small amounts of P could not be detected by
immunoelectrophoretic analysis.

The present, tentative concept of the reaction mechanisms of the properdin
system is summarized in Figure 7. Upon contact with activating substances,
an initiating factor (IF) is converted to its activated form ($\overline{IF}$), which may
be identical to the C3 nephritic factor. $\overline{IF}$, possibly together with other serum
constituents (X), converts precursor P to an active form ($\overline{P}$). $\overline{P}$ is envisioned
as conferring properties upon native C3 which allow $\overline{C3PAse}$ to act on C3PA,
resulting in the generation of C3A. This enzyme cleaves C3 into C3a and
C3b, thereby setting in motion the positive feedback mechanism which is inde-

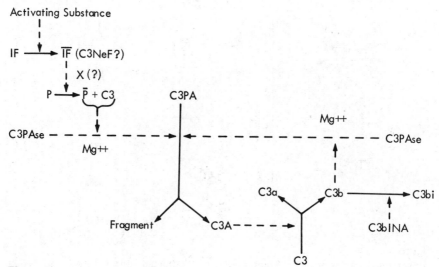

Figure 7
Tentative concept of the molecular mechanisms of the properdin system.

pendent of all earlier reacting components. C3b inactivator (C3bINA) functions as an important control enzyme during the C3b-dependent phase of the reaction sequence. By degrading C3b to C3c and C3d (C3bi in the scheme), it interferes with the continuing activation of C3PA and C3.

The mode of activation of C5 to C5b by properdin system enzymes has not yet been completely clarified, but it does require C3A and C3b and, in addition, at least one other serum factor (Brade et al. 1973; Müller-Eberhard 1975). With the generation of C5b, the $\overline{C5b67}$ and $\overline{C5b}$-9 complexes assemble and exhibit their respective chemotactic and cytolytic activities.

Cobra Venom Factor-dependent Reactions and Assays

The observation that hemolytic complement is consumed by the addition of cobra venom to serum was made more than 70 years ago (Flexner and Noguchi 1903). The active principle of the venom, cobra venom factor (CVF), was used to generate anaphylatoxin (Vogt and Schmidt 1964), to inactivate C3 in vitro and in vivo (Klein and Wellensiek 1965; Nelson 1966), and to passively lyse unsensitized erythrocytes in a complement-dependent reaction (Pickering et al. 1969; Götze and Müller-Eberhard, 1971). CVF is a glycoprotein with a molecular weight of 140,000 daltons (Müller-Eberhard and Fjellström 1971). When added to serum, CVF interacts with C3PA to form a complex which will activate the complement components C3-C9. It is now clear that a number of different interactions of CVF with C3PA have to be distinguished. As reported earlier (Götze and Müller-Eberhard 1971), CVF and C3PA can interact to form a complex. Cooper (1973) found that this interaction results in only feeble C3 activating activity and is reversible in the absence of $\overline{C3PAse}$. A stable, enzymatically very active complex with a molecular weight of 220,000 daltons is formed only in the presence of

$\overline{C3PAse}$ and Mg^{++}. The requirement of $\overline{C3PAse}$ for the development of a highly active CVF-dependent C3 cleaving enzyme was established by Hunsicker, Ruddy and Austen (1973), who failed, however, to detect the formation of a CVF-C3PA complex, and Cooper, who confirmed the formation of a stable, bimolecular CVF-C3PA complex and obtained some indication that, similar to the C3b-dependent activation of C3PA, the CVF-dependent activation also involved cleavage of C3PA to C3A (Cooper 1973). The reaction mechanism was further clarified by Vogt et al. (1974), who showed by immunochemical analysis using monospecific antisera to CVF and C3PA and its fragments that the active complex is composed of CVF and C3A. They also demonstrated the formation of the CVF-C3A complex after activation of C3PA by trypsin in the presence of Mg^{++}. In the presence of EDTA, C3PA was cleaved by trypsin in typical fashion, but a CVF-C3A complex was not generated, suggesting a role for Mg^{++} in the binding of C3PA or C3A to CVF. Once the CVF-C3A complex has been formed, it is able to activate C3 in the presence of EDTA. This separation of the CVF-dependent reaction into a first Mg^{++}-requiring part during which the CVF-C3A complex is formed and a second one during which a stable CVF-C3A complex proceeds to activate C3-C9 in the presence of EDTA has allowed the elaboration of very sensitive assays for C3PA and $\overline{C3PAse}$. Purified CVF is incubated with an excess of isolated C3PA or $\overline{C3PAse}$ and dilutions of the sample containing unknown amounts of $\overline{C3PAse}$ or C3PA. After incubation for 30 minutes at 37°C in the presence of 10^{-3} M Mg^{++} to allow for the formation of the CVF-C3A complex, whole serum diluted in EDTA buffer and human erythrocytes (E) treated with reduced glutathione (GSH) are added and the incubation is continued for another 60 minutes. The EDTA-serum prevents any further utilization of C3PA or $\overline{C3PAse}$, but does provide C3-C9 for the lytic reaction. GSH-treated erythrocytes (Götze and Müller-Eberhard 1972; Stålenheim et al. 1973) are exquisitely sensitive to passive lysis by complement. The degree of the ensuing lysis is proportional to the amount of CVF-C3A generated during the first phase of this titration method. Using this assay, less than 100 ng of C3PA and less than 50 pg of $\overline{C3PAse}$ can be detected.

CONSEQUENCES OF THE ACTION OF THE PROPERDIN SYSTEM

As stated in the introduction, activation of C3-C9 by the alternative pathway leads to the generation of the same activities, fragments and complexes as those resulting from activation of the $\overline{C42}$-dependent pathway. The anaphylatoxins C3a and C5a are formed and the chemotactic C5b67 complex is assembled, as is the potentially cytolytic $\overline{C5b-9}$ complex (Kolb and Müller-Eberhard 1973). Erythrocytes from patients with paroxysmal nocturnal hemoglobinuria and GSH-treated erythrocytes can be lysed (Götze and Müller-Eberhard 1972), and erythrocytes and Burkitt's lymphoma cells may be destroyed after binding of C3b to membrane receptors, subsequent activation of C3PA and C3, and ensuing assembly of the $\overline{C5b-9}$ complex at the cell surface (Fearon, Austen and Ruddy 1973; Theofilopoulos, Bokisch and Dixon 1974). In addition, phagocytosis of bacteria and particles by granulocytes may be promoted through the binding of C3b to their surface (opsonization), and susceptible bacteria may be killed in the absence of immune antibodies. Whether the enzymes of the properdin system have biological functions other than those necessary for the activation of C3 and C5 is uncertain, although this seems improbable in view of their high specificities and complex reaction mechanisms.

CONCLUSIONS

Although our knowledge of the properdin system is still far from complete, we are now beginning to understand the molecular interactions of some of its constituents. As in the classical, $C\overline{4,2}$-dependent activation system of complement, the fusion of proteins, after activation by limited proteolysis, appears to play a key role in the properdin system as well. Whereas enzyme activity toward small substrates can be demonstrated in single isolated components, the natural, macromolecular substrates are attacked efficiently only after the enzyme has entered into a complex with a second protein. Whether the second protein (e.g., C3b, CVF) helps to stabilize a conformationally labile site on the newly activated enzyme or whether it provides a binding site for the macromolecular substrate or has several such functions is not yet known. In addition, by complexing with the zymogen, as is conceivably the case for the activation of C3PA, the second protein may induce a conformational change in the proenzyme, leading to the exposure of a bond, previously inaccessible, which can be cleaved by the activating enzyme (C3PA convertase).

One of the prominent features of the system is its dependence on a positive feedback mechanism through the requirement of C3b for the activation of C3PA to C3A and the efficient cleavage of C3 by C3A. Thus a product of the reaction is not only indispensable as a cofactor for the activation of the enzyme which drives the reaction, but is also vital for the effective functioning of this enzyme. These functions of C3b are efficiently controlled by the enzyme C3bINA, which cleaves C3b into two fragments, C3c and C3d, which are both inactive.

The biological importance of the C3b-dependent feedback mechanism is illustrated by the fact that C3b, regardless of its mode of formation, can switch on the system and recruit the biological activities associated with the activation of C3-C9, including those which result in membrane damage and cytolysis. The fact that besides the $C\overline{4, 2}$ enzyme, trypsin, thrombin, plasmin and lysosomal proteases have been reported to cleave C3 in typical fashion (Bokisch, Müller-Eberhard and Cochrane 1969; Hill and Ward 1971; Lepow 1971) emphasizes that activating polysaccharides and early acting components of the C3 activator system may not be required for the expression of its biological potential in humoral and cell-mediated processes.

Acknowledgments

I am grateful for the competent technical assistance of Ms. M. Yelvington and Ms. L. Wood. This work was performed during the tenure of an Established Investigatorship of the American Heart Association and supported by United States Public Health Service Program Project Grant AI-07007. This is publication number 885, Department of Molecular Immunology, Scripps Clinic and Research Foundation, La Jolla, California.

REFERENCES

Alper, C. A., F. S. Rosen and P. J. Lachmann. 1972. Inactivator of the third component of complement as an inhibitor in the properdin pathway. *Proc. Nat. Acad. Sci.* **69**:2910.

Blum, L., L. Pillemer and I. H. Lepow. 1959. The properdin system and immunity. XIII. Assay and properties of a heat-labile serum factor (factor B) in the properdin system. *Z. Immun. Forsch.* **118**:349.

Boenisch, T. and C. A. Alper. 1970a. Isolation and properties of a glycine-rich β-glycoprotein of human serum. *Biochim. Biophys. Acta* **221**:529.

————. 1970b. Isolation and properties of a glycine-rich γ-glycoprotein of human serum. *Biochim. Biophys. Acta* **214**:135.

Bokisch, V. A., H. J. Müller-Eberhard and C. G. Cochrane. 1969. Isolation of a fragment (C3a) of the third component of human complement containing anaphylatoxin and chemotactic activity and description of an anaphylatoxin inactivator of human serum. *J. Exp. Med.* **129**:1109.

Brade, V., G. D. Lee, A. Nicholson, H. S. Shin and M. M. Mayer. 1973. The reaction of zymosan with the properdin system in normal and C4-deficient guinea pig serum: Demonstration of C3- and C5-cleaving multi-unit enzymes, both containing factor B, and acceleration of their formation by the classical complement pathway. *J. Immunol.* **111**:1389.

Chase, T., Jr. and E. Shaw. 1967. *p*-Nitrophenyl-*p'*-guanidinobenzoate HCl: New active site titrant for trypsin. *Biochem. Biophys. Res. Comm.* **29**:508.

Cohen, P., T. Duewer and E. Fischer. 1972. Measurement of molecular weights by electrophoresis on SDS-acrylamide gel. In *Methods in Enzymology* (ed. C. H. W. Hirs and S. N. Timasheff), vol. 26, p. 3. Academic Press, New York.

Cooper, N. R. 1971. Enzymes of the complement system. In *Progress in Immunology* (ed. B. Amos), p. 567. Academic Press, New York.

————. 1973. Formation and function of a complex of the C3 proactivator with a protein from cobra venom. *J. Exp. Med.* **137**:451.

Fearon, D. T., K. F. Austen and S. Ruddy. 1973. Formation of a hemolytically active cellular intermediate by the interaction between properdin factor B and D and the activated third component of complement. *J. Exp. Med.* **138**:1305.

————. 1974a. Properdin factor D. II. Activation to $\bar{\text{D}}$ by properdin. *J. Exp. Med.* **140**:426.

————. 1974b. Properdin factor D: Characterization of its active site and isolation of the precursor form. *J. Exp. Med.* **139**:355.

Flexner, S. and H. Noguchi. 1903. Snake venom in relation to hemolysis, bacteriolysis and toxicity. *J. Exp. Med.* **6**:277.

Frank, M. M., J. May, T. Gaither and L. Ellman. 1971. *In vitro* studies of complement function in sera of C4 deficient guinea pigs. *J. Exp. Med.* **134**:176.

Gewurz, H., H. S. Shin and S. E. Mergenhagen. 1968. Interactions of the complement system with endotoxin lipopolysaccharides: Consumption of each of the six terminal complement components. *J. Exp. Med.* **128**:1049.

Goodkofsky, I. and I. H. Lepow. 1971. Functional relationship of factor B in the properdin system to C3 proactivator of human serum. *J. Immunol.* **107**:1200.

Goodkofsky, I., A. H. Steward and I. H. Lepow. 1973. Relationship of C3 and factor A of the properdin system. *J. Immunol.* **111**:287.

Götze, O. and H. J. Müller-Eberhard. 1971. The C3-activator system: An alternate pathway of complement activation. *J. Exp. Med.* **134**:90s.

————. 1972. Paroxysmal nocturnal hemoglobinuria: Hemolysis initiated by the C3 activator system. *New Eng. J. Med.* **286**:180.

————. 1973. The C3 activator system of human complement: Molecular mechanisms. In *"Non-specific" Factors Influencing Host Resistance* (ed. W. Braun and J. Ungar), p. 332. Karger, Basel.

————. 1974. The role of properdin in the alternate pathway of complement activation. *J. Exp. Med.* **139**:44.

Hadding, U., M. Dierich, W. Konig, M. Limbert, H. U. Schorlemmer and D. Bitter-Suermann. 1973. Ability of the T cell-replacing polyanion dextran sulfate to trigger the alternate pathway of complement activation. *Eur. J. Immunol.* **8**:527.

Haupt, H. and K. Heide. 1965. Isolierung und Eigenschaften eines β_2-Glykoproteins aus Humanserum. *Clin. Chim. Acta* **12**:419.

Hill, J. H. and P. A. Ward. 1971. The phlogistic role of C3 leukotactic fragments in myocardial infarcts of rats. *J. Exp. Med.* **133**:885.

Hirs, C. H. W. 1967. Performic acid oxidation. In *Methods in Enzymology* (ed. C. H. W. Hirs), vol. 9, p. 197. Academic Press, New York.

Hunsicker, L. G., S. Ruddy and K. F. Austen. 1973. Alternate complement pathway: Factors involved in the cobra venom factor (CoVF) activation of the third component of complement (C3). *J. Immunol.* **110**:128.

Klein, P. G. and H. J. Wellensiek. 1965. Complement: Hemolytic function and chemical properties. *Int. Rev. Exp. Path.* **4**:245.

Kolb, W. P. and H. J. Müller-Eberhard. 1973. The membrane attack mechanism of complement. Verification of a stable C5-9 complex in free solution. *J. Exp. Med.* **138**:438.

König, W., D. Bitter-Suermann, M. Dierich, M. Limbert, H. U. Schorlemmer and U. Hadding. 1974. DNP-antigens activate the alternate pathway of the complement system. *J. Immunol.* **113**:501.

Lachmann, P. J. and H. J. Müller-Eberhard. 1968. The demonstration in human serum of "conglutinogen-activating factor" and its effect on the third component of complement. *J. Immunol.* **100**:691.

Lepow, I. H. 1961. The properdin system: A review of current concepts. In *Immunochemical Approaches to Problems in Microbiology* (ed. M. Heidelberger and O. J. Plescia), p. 280. Rutgers University, New Brunswick, New Jersey.

————. 1971. Biologically active fragments of complement. In *Progress in Immunology* (ed. B. Amos), p. 579. Academic Press, New York.

Minta, J. O. 1974. Inactivation of zymosan-purified properdin in human serum. *Fed. Proc.* **33**:776.

Minta, J. O. and I. H. Lepow. 1974. Studies on the subunit structure of human properdin. *Immunochemistry* **11**:361.

Mosesson, M. W., J. S. Finlayson, R. A. Umfleet and D. Galanakis. 1972. Human fibrinogen heterogeneities. I. Structural and related studies of plasma fibrinogens which are high solubility catabolic intermediates. *J. Biol. Chem.* **247**:5210.

Müller-Eberhard, H. J. 1967. Mechanism of inactivation of the third component of human complement (C′3) by cobra venom. *Fed. Proc.* **26**:744.

————. 1975. C5 convertase of the alternative complement pathway. *J. Immunol.* (in press).

Müller-Eberhard, H. J. and K. E. Fjellström. 1971. Isolation of the anticomplementary protein from cobra venom and its mode of action on C3. *J. Immunol.* **107**:1666.

Müller-Eberhard, H. J. and O. Götze. 1972. C3 proactivator convertase and its mode of action. *J. Exp. Med.* **135**:1003.

Nicol, P. A. E. and P. J. Lachmann. 1973. The alternate pathway of complement activation. The role of C3 and its inactivator (KAF). *Immunology* **24**:259.

Nelson, R. A., Jr. 1966. A new concept of immunosuppression in hypersenitivity reactions and in transplantation immunity. *Survey Ophthalmol.* **11**:498.

Pensky, J., L. Wurz, L. Pillemer and I. H. Lepow. 1959. The properdin system and immunity. XII. Assay, properties and partial purification of a hydrazine-sensitive serum factor (factor A) in the properdin system. *Z. Immun. Forsch.* **118**:329.

Pensky, J., C. F. Hinz, Jr., E. W. Todd, R. J. Wedgwood, J. T. Boyer and I. H. Lepow. 1968. Properties of highly purified human properdin. *J. Immunol.* **100**:142.

Pickering, R. J., M. R. Wolfson, R. A. Good and H. Gewurz. 1969. Passive hemolysis by serum and cobra venom factor: A new mechanism inducing membrane damage by complement. *Proc. Nat. Acad. Sci.* **62**:521.

Pillemer, L., L. Blum, I. H. Lepow, O. A. Ross, E. W. Todd and A. C. Wardlaw. 1954. The properdin system and immunity. I. Demonstration and isolation of a new serum protein, properdin, and its role in immune phenomena. *Science* **120**:279.

Ruddy, S. and K. F. Austen. 1971. C3b inactivator of man. II. Fragments produced by C3b inactivator cleavage of cell-bound or fluid phase C3b. *J. Immunol.* **107**:742.

Ruley, E. J., J. Forristal, N. C. Davis, C. Andres and C. D. West. 1973. Hypocomplementemia of membranoproliferative nephritis. Dependence of the nephritic factor reaction on properdin factor B. *J. Clin. Invest.* **52**:896.

Sandberg, A. L., O. Götze, H. J. Müller-Eberhard and A. G. Osler. 1971. Complement utilization by guinea pig γ1 and γ2 immunoglobulins through the C3 activator system. *J. Immunol.* **107**:920.

Sandberg, A. L., A. G. Osler, H. S. Shin and B. Oliveira. 1970. The biologic activities of guinea pig antibodies. II. Modes of complement interaction with γ1 and γ2 immunoglobulins. *J. Immunol.* **104**:329.

Spitzer, R. E., E. H. Vallota, J. Forristal, E. Sudora, A. Stitzel, N. C. Davis and C. D. West. 1969. Serum C′3 lytic system in patients with glomerulonephritis. *Science* **164**:436.

Stålenheim, G., O. Götze, N. R. Cooper, J. Sjöquist and H. J. Müller-Eberhard. 1973. Consumption of human complement components by complexes of IgG with protein A of staphylococcus aureus. *Immunochemistry* **10**:501.

Tamura, H. and N. Ki. 1972. A new buffer system for disc electrophoresis suitable for slightly basic proteins. *J. Biochem.* **71**:543.

Theofilopoulos, A. N., V. A. Bokisch and F. J. Dixon. 1974. Receptor for soluble C3 and C3b on human lymphoblastoid (Raji) cells: Properties and biological significance. *J. Exp. Med.* **139**:696.

Valet, G. and N. R. Cooper. 1974. Isolation and characterization of the proenzyme form of the C1s subunit of the first complement component. *J. Immunol.* **112**:339.

Vallota, E. H., J. Forristal, R. E. Spitzer, N. C. Davis and C. D. West. 1970. Characteristics of a non–complement-dependent C3-reactive complex formed from factors in nephritic and normal serum. *J. Exp. Med.* **131**:1306.

Vallota, E. H., O. Götze, H. L. Speigelberg, J. Forristal, C. D. West and H. J. Müller-Eberhard. 1974. A serum factor in chronic hypocomplementemic nephritis distinct from immunoglobulins and activating the alternate pathway of complement. *J. Exp. Med.* **139:**1249.

Vogt, W. and G. Schmidt. 1964. Abtrennung des anaphylatoxinbildenden Prinzips aus Cobragift von anderen Giftkomponenten. *Experientia* **20:**207.

Vogt, W., L. Dieminger, R. Lynen and G. Schmidt. 1974. Formation and composition of the complex with cobra venom factor that cleaves the third component of complement. *Hoppe-Seyler's Z. Physiol. Chem.* **355:**171.

Weber, K. and M. Osborn. 1969. The reliability of molecular weight determinations by dodecyl sulfate-polyacrylamide gel electrophoresis. *J. Biol. Chem.* **244:**4406.

Wu, Y. V. and H. A. Scherega. 1962. Studies on soybean trypsin inhibitor. I. Physicochemical properties. *Biochemistry* **1:**698.

Serum Anaphylatoxins: Formation, Characterization and Control

Tony E. Hugli

Department of Molecular Immunology, Scripps Clinic and Research Foundation
La Jolla, California 92037

The term anaphylatoxin was selected by Friedberger (1910) to describe an agent produced in the serum of experimental animals that was capable of inducing shock. Serum taken from a guinea pig, exposed to an immune precipitate, and then injected into the donor animal promoted a syndrome similar to systemic anaphylaxis. Eventually it was discovered that other substances, including antigen-antibody complexes, inulin, dextran, yeast cells, agar and an active component of cobra venom (CVF) (Cochrane, Müller-Eberhard and Aiken 1970) could promote anaphylatoxin formation in serum. At first, anaphylatoxin was an ill-defined entity known only to require multiple blood factors for its formation. Now it is well established that activation of complement components C3[1] and C5 produces anaphylatoxins (Cochrane and Müller-Eberhard 1968). These protein products of the complement activation sequence have therefore been termed C3a and C5a anaphylatoxins. The activities of both fragments in vivo and in vitro presently indicate a biological importance for them that was not fully appreciated in the past.

A number of the biological responses to anaphylatoxins require or result from an induced histamine release from mast cells. A primary example of this effect is the severe cutaneous response in human or animal skin after injection of anaphylatoxin. Wheal and erythema reactions were visible on human skin after application of nanogram quantities of C3a, and biopsy revealed degranulation of the local mast cell population accompanied by free metachromatic granules (Wuepper et al. 1972). These results clearly indicated the potential involvement of anaphylatoxins in the local inflammatory response. Direct evidence for the influence of anaphylatoxin on mast cell histamine release was supplied by demonstration that both radiolabeled C3a and C5a were bound to isolated rat mast cells (Johnson and Müller-Eberhard 1975). Maximal

---

[1] Nomenclature for the complement proteins conforms with the recommendations of the World Health Organization Committee on Complement of 1968. A bar over the number of a component designates an activated state not expressed by the native protein molecule.

binding of each anaphylatoxin resulted in the uptake of 10–12 pmoles of C3a and C5a per 10^6 mast cells. Histamine was released from mast cells by both anaphylatoxins, and their respective effects were additive. In neither case did release exceed 30% of the total store of histamine. The relative efficiency of both C3a and C5a in releasing histamine from mast cells exceeded that of bradykinin by more than tenfold on a molar basis. Interaction of human C3a molecules with rat mast cells has also been demonstrated visually by indirect immunofluorescence (Laan et al. 1974). The extent of C3a binding was shown to correlate with the extent of mast cell degranulation. Low levels of C3a were detected on mast cells which were permitted to release histamine and undergo degranulation. However when histamine release was inhibited by agents such as EDTA and disodium cromoglycate, degranulation could not occur, and significantly larger amounts of C3a were bound as judged by an increased intensity of fluorescence. Obviously a reappraisal is needed if we wish to differentiate radiolabeled anaphylatoxin bound to mast cells from that attached to the granules.

Responses to anaphylatoxins which occur independently of histamine include chemotactic activity and contraction of certain isolated smooth muscle preparations (Vogt, Zeman and Garbe 1969). Chemotaxis was demonstrated for C5a generated from isolated human C5 (Ward 1969) and from activated guinea pig serum, where the active agent was presumed to be the C5a (Shin et al. 1968). Unfortunately, the literature concerning the chemotactic activity of C3a is contradictory, and no definitive study has yet appeared which conclusively resolves this issue. In contrast, the action of anaphylatoxins on smooth muscle tissue is well documented, mostly by assays that measure contractions of guinea pig ileum. Sections of ileum resting in an organ bath are challenged with the test substances. Characteristically this muscle tissue is desensitized to repeated applications of either anaphylatoxin after an initial contraction. Such desensitization or tachyphylaxis has been shown to be independent of species differences since human, rat, guinea pig and porcine C3a are all cross-tachyphylactic. A similar cross-tachyphylaxis exists between C5a molecules obtained from a variety of species. However, C3a and C5a are not tachyphylactic toward one another, precluding interaction at identical receptor sites on the smooth muscle tissue. Figure 1 shows results from assays performed to demonstrate the lability of the smooth muscle receptors to proteases. Although these are preliminary results, they do indicate that the effect of human C3a was largely abolished after just 10 minutes of proteolytic digestion. No gross diminution in the ileum's responsiveness to histamine or bradykinin was detected. Thus it seems fair to conclude that surface proteins are intimately involved in the organization of cellular receptors to this anaphylatoxin.

In previous studies, nanogram amounts of C3a and C5a significantly enhanced vascular permeability, which implied potent effects on vascular smooth muscle (Bodammer and Vogt 1970). Recently, the omentum of the living rabbit was used as a model to assess the local effects of anaphylatoxin on microcirculation (Mahler et al. 1974). Indeed, topical application of C3a to the omentum preparation produced measurable pressure and diameter changes of the arterioles, indicating vasoconstriction (Fig. 2). This effect of C3a on arterioles mimicked the response to norepinephrine. In addition, C3a, like

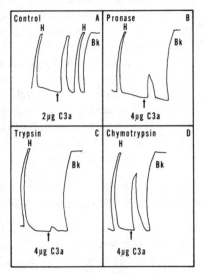

Figure 1
Guinea pig ileum contractions in response to various stimuli. (*A*) Normal contractions of the ileum strip to 0.5 μg of histamine (H), 2 μg of human C3a and 1 μg of bradykinin (Bk). (*B, C, D*) Effects on ileum from pronase, tryptic or chymotryptic digestion in regard to the contractile response to C3a. Each strip of ileum was incubated with 20 μg/ml of the respective protease for 10 minutes, pH 7.5 at 37°C. The muscle strips were washed free of protease before challenge with the agents indicated.

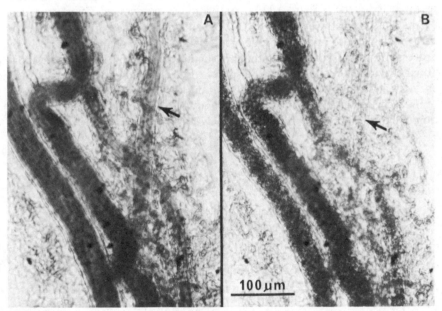

Figure 2
Vasoconstriction of a feeding arteriole in rabbit omentum induced by human C3a. The omentum from an anesthetized rabbit was cradled in a glass disc and irrigated continuously with buffered Ringers solution at pH 7.4. (*A*) An area of vascular bed before treatment. (*B*) A contracting arteriole just 10 seconds after topical application of 2 nmoles of human C3a. Arrows indicate the arteriole which responded to application of the anaphylatoxin.

norepinephrine, was blocked by pentolamine, an alpha adrenergic blocking agent. Therefore, these results suggested that the C3a activity on vascular smooth muscle involved neuromotor mechanisms possibly mediated by nor-epinephrine. Vasoconstriction, unlike contractions of ileum smooth muscle, could be produced repeatedly by additional applications of C3a.

It is now clear that multiple actions and various receptors are involved in the total biological activity of anaphylatoxins. Our understanding of these responses will come with complete clarification of the nature of anaphylatoxin molecules themselves, as well as a molecular description of the receptors and their relationship to the target cells. The remainder of this presentation will deal with one aspect of this problem, namely, a deeper look into the makeup of anaphylatoxin molecules.

Anaphylatoxin Formation in Serum

Modes of generation for C3a and C5a in blood or in serum are outlined in Figure 3. The parent C3 and C5 molecules are composed of two nonidentical polypeptide chains joined by disulfide bridges (Nilsson, Mandle and McConnel-Mapes 1975). There are two separate routes for complement activation in serum, the "classical" and the "alternative," and each involves selective enzymatic attack. The particular enzymes involved in C3 activation are $\overline{C4,2}$ (C3 convertase), intrinsic in the classical route, and the multicomponent C3 activator system, an integral portion of the alternative complement activation pathway (Müller-Eberhard 1972). Activation of component C5 is produced by the classical pathway enzyme $\overline{C4,2,3b}$ and by enzymes of the alternative pathway which are not entirely defined. Details of the alternative activation process are discussed more thoroughly in this volume by Dr. Götze. Additional enzymes in blood, such as plasmin and thrombin, are potential cleavage enzymes and, at least in the case of C3, produce anaphylatoxin.

Bokisch, Müller-Eberhard and Cochrane (1969) found that limited digestion of either C3 or C5 with trypsin produces the respective C3a and C5a mol-

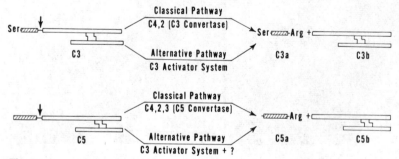

Figure 3
Diagrammatic representation of the two pathways for activating C3 and C5 in serum. Arrows indicate sites on the C3 and C5 molecule where enzymatic attack causes a release of the anaphylatoxins. The parent C3 and C5 molecules are composed of nonidentical polypeptide chains covalently linked by disulfide bridges.

ecules. On the basis of their work and that of others, it can be concluded that the sites of enzymatic attack on the C3 and C5 molecules involve particularly sensitive peptide bonds which can be selectively hydrolyzed by enzymes having a specificity similar to that of trypsin. It has been established that C3a molecules obtained from isolated C3 after activation by either the serum $\overline{C4,2}$ enzyme or by trypsin were essentially identical in function and amino acid composition (Budzko, Bokisch and Müller-Eberhard 1971). More recently, a similar comparison was made between C3a generated by the classical and the alternative pathway enzymes, and these products also appeared identical (Hugli, Vallota and Müller-Eberhard 1975). In addition, it was established that a functionally essential arginine residue occupied the carboxyl-terminal position of human C3a produced by all of these means. This evidence clearly suggests that a single site of cleavage exists on human C3 for attack by the C3-activating enzymes, a site which involves the carboxyl terminus of the C3a molecules. Furthermore, porcine C3a isolated from serum activated by enzymes of the alternative pathway contained an essential carboxyl-terminal arginine residue. It can therefore be more generally stated that a similar site exists on the C3 molecule from porcine and probably other species as well. Similarities have also been shown between the C3a and C5a anaphylatoxins at the molecular level. Since an arginine residue was shown to occupy the carboxyl-terminal position of porcine C5a and was presumed to be essential, as was proven to be the case for C3a molecules, an absolute dependence on a basic residue at the carboxyl-terminal position of all anaphylatoxins is implied.

Similarities between the modes of formation and various properties of the serum anaphylatoxins prompted further comparisons concerning their control mechanisms in vivo. Removal of the carboxyl-terminal arginine residue C3a by an indigenous serum carboxypeptidase clearly constitutes the major systemic defense mechanism against appreciable accumulations of active C3a. The product formed by carboxypeptidase action on C3a ($C3a_{des\ Arg}$) will be referred to hereafter as $C3a_i$. This serum carboxypeptidase has been characterized and resembles pancreatic carboxypeptidase B in its enzymatic specificity (Bokisch and Müller-Eberhard 1970). The effectiveness of normal levels of the serum carboxypeptidase in human blood is illustrated by its capacity to inactivate in 3–6 seconds more C3a than blood would contain if all the circulating C3 was instantaneously converted. This same carboxypeptidase controls biologically active products of the kinin system. Although the serum carboxypeptidase is extremely efficient in controlling the C3a levels in serum, it was found to be less effective against C5a. C5a was only partially inactivated in some species; however in human serum, the carboxypeptidase seemed almost totally effective. Details of these differences are discussed later in this presentation.

Isolation and Characterization of C3a Anaphylatoxin

Until recently, C3a anaphylatoxin had not been isolated directly from activated serum. However when special precautions were taken to inhibit the serum carboxypeptidase during the course of complement activation, active C3a was demonstrated in the treated serum (Vallota and Müller-Eberhard 1973). A lysine analog, ε-aminocaproic acid (EACA), a well-known competitive inhibitor of pancreatic carboxypeptidase B (Folk 1956), was found to be a highly effective inhibitor of the serum carboxypeptidase. In order to produce anaphylatoxin in serum, we needed an inhibitor of serum carboxypeptidase which did not also inhibit complement activation (alternative pathway), and EACA met this requirement.

C3a anaphylatoxin was generated in serum containing 1 M EACA and was isolated by a four-step procedure. The first step involved gel filtration of the activated serum on a Sephadex G-100 column. Second, fractions containing C3a were adsorbed in batches to carboxymethyl (Cm) Sephadex in 0.15 M sodium acetate at pH 3.7. The Cm Sephadex was washed with 1 M NaCl in the sodium acetate buffer to remove weakly bound proteins and was transferred to a glass column for elution with 0.1 M HCl. Third, the acidic effluent containing C3a was collected and lyophilized. Lyophilization from HCl aggregated and denatured a majority of the contaminating proteins exclusive of the C3a molecule. This rather harsh treatment was responsible for the improved separation between contaminating proteins and the C3a molecule when gel filtration through Sephadex G-100 was repeated. Figure 4 A and B

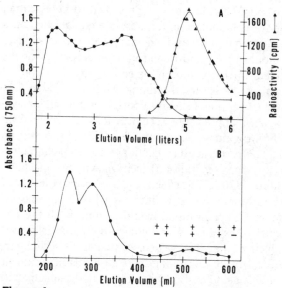

Figure 4

(*A*) Gel filtration pattern of inulin-activated serum eluted from Sephadex G-100. Protein was determined by the colorimetric Folin procedure. Radiolabeled C3a (^{125}I) was introduced into the serum to facilitate monitoring of anaphylatoxin in the eluate.

(*B*) Sephadex G-100 gel filtration of lyophilized C3a-containing fractions obtained from Cm Sephadex eluted with 0.1 M HCl. The C3a was detected by bioassay using guinea pig ileum. Full contractions ($\ddagger$) and partial contractions ($\pm$) of ileum were indicated for 50-μl aliquots of the effluent.

illustrate the elution profiles of whole activated serum and of lyophilized C3a fractions obtained from Cm Sephadex, respectively. ^{125}I-labeled C3a was usually added to the activated serum to aid in monitoring the C3a. A fourth and final step consisted of passing the C3a recovered from the second Sephadex G-100 column over a column of DEAE-Sephadex. Material obtained after the DEAE-Sephadex chromatography step was judged homogeneous by electrophoresis at pH 4.6 and 8.6. The purification scheme described above was originally designed for the isolation of human C3a. However, C3a from various other species were quite similar to human C3a in size and net charge, suggesting that the same scheme should apply to them. Indeed, porcine C3a was isolated by using the procedure described above, and undoubtedly, C3a from numerous other species can be derived similarly. Yields of between 30 and 40% of the total C3a in serum, assuming complete C3 activation, were obtained consistently.

Several unusual properties permit such facile purification of the C3a anaphylatoxin. These properties include stability, a relatively small size and an extremely cationic nature. In fact, the C3a molecule is more cationic than either group IV histones or hen egg lysozyme, both considered exceptionally cationic proteins. Estimates of the molecular weight of C3a isolated from serum disagree somewhat with earlier determinations for C3a derived from purified C3 (Budzko, Bokisch and Müller-Eberhard 1971). A molecular weight of approximately 9000 daltons was obtained for both human and porcine C3a by gel filtration (Fig. 5). The C3a was eluted from Sephadex G-50 with 0.15 M sodium acetate at pH 3.7. Artificially low molecular weight estimates were obtained when elution was performed at neutral pH conditions, presumably due to an interaction of the C3a with the column walls or the Sephadex gel itself. The purity of human and porcine C3a following electrophoresis in polyacrylamide gels at pH 4.6 is indicated in the inset of Figure 5. A difference of approximately one net charge was estimated to exist between human and porcine

Figure 5

This diagram shows the log molecular weight vs. elution volume from a Sephadex G-50 column plotted for a number of selected proteins. The elution volume of human and porcine C3a indicated a molecular weight of approximately 9000. The inset shows polyacrylamide gels containing purified human and porcine C3a. Electrophoresis was performed at pH 4.6, and the upper tip of a wire insert designates the buffer front.

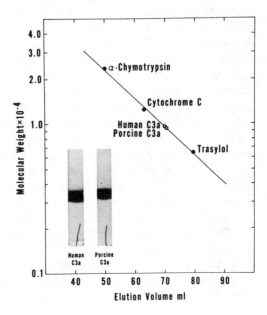

Table 1

Amino Acid Composition of Human and Porcine C3a Anaphylatoxins

Amino acid	Human C3a (residues/mole protein[a])	Porcine C3a (residues/mole protein)
Lysine	7.0 (7)	7.0 (7)
Histidine	2.0 (2)	2.2 (2)
Arginine	11.0 (11)	8.0 (8)
Aspartic acid	5.0 (5)	5.6 (6)
Threonine	3.0 (3)	0.2 (0)
Serine	4.0 (4)	3.8 (4)
Glutamic acid	8.5 (9)	10.9 (11)
Proline	1.9 (2)	2.9 (3)
Glycine	3.7 (4)	4.4 (4)
Alanine	3.9 (4)	5.6 (6)
Half-cystine	5.6 (6)	5.6 (6)
Valine	3.2 (3)	2.1 (2)
Methionine	2.8 (3)	2.8 (3)
Isoleucine	2.0 (2)	1.9 (2)
Leucine	6.7 (7)	6.9 (7)
Tyrosine	1.7 (2)	1.7 (2)
Phenylalanine	3.1 (3)	2.9 (3)
Tryptophan	0.0	ND
Total residues	77	76
Molecular weight	9083	8825

[a] Based on a molecular weight of 9000 daltons.

C3a based on polyacrylamide sheet electrophoresis at pH 8.6, and human C3a was the more cationic of the two anaphylatoxins.

Amino acid compositions for human and porcine C3a were based on the estimated molecular weight of 9000 and are presented in Table 1. Although neither molecule contains tryptophan or carbohydrate, the porcine molecule is also devoid of threonine. Several other differences are apparent; for instance, porcine C3a contains fewer valine and arginine residues, but a greater number of alanine, aspartic acid and glutamic acid residues. The cationic behavior of both C3a molecules requires the numerous acidic residues to occur largely as amides in the intact molecule.

Human C3a anaphylatoxin has been selected for primary structural analysis. Preliminary sequence analysis of the amino terminus of intact C3a by automatic sequencing techniques indicated that two of the three methionine residues occurred at positions 9 and 21 and one of the two tyrosine residues occurred at position 15. Since the second tyrosine residue in C3a was believed to occur in a large fragment at the carboxyl portion of the molecule, C3a was radiolabeled with [125]I (McConahey and Dixon 1966), and cyanogen bromide (CNBr) cleavage was performed. Elution patterns of the radioactive peptides derived from the CNBr cleavage of ^{125}I-labeled C3a are illustrated in Figure 6. Only the large peptides carried radiolabel and thus were detected by this procedure; however, the smaller peptides were readily recovered by collecting the effluent beyond peak IV. Peaks II and IV contained peptides representing

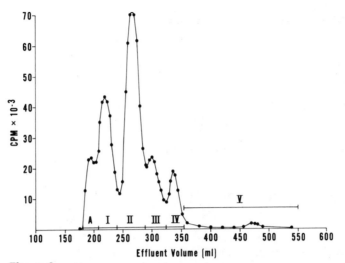

Figure 6
Elution pattern of the radiolabeled peptides formed by cyanogen bromide cleavage of ^{125}I-labeled human C3a. Several overcuts occurred at the three methionine positions, which accounts for the complex elution profile. Pool II contained the carboxyl-terminal fragment of C3a, representing residues 33–77. Pool IV contained a peptide representing residues 10–27 of the primary structure of C3a. Pool V contained two small unlabeled peptides shown to be residues 1–9 and 28–32.

sequence regions later identified as 10–27 and 33–77, respectively, in the primary structure of C3a. Combining data obtained by automatic sequencing from the amino terminus of intact C3a with automatic sequencing results obtained from CNBr fragments and sequences of various peptides isolated from tryptic and chymotryptic digestions of C3a, a complete primary structure was obtained. Figure 7 contains the complete amino acid sequence for human C3a. Features which appeared characteristic of this molecule were the two Cys-Cys sequences at positions 22,23 and 56,57. There was also an unusually high proportion of basic residues near the functionally essential arginine residue at the carboxyl terminus of the molecule. Comparison of human C3a with the partial sequence of porcine C3a was possible for 20 residues at the amino-terminal end, and replacements were found at positions 5,12,14 and 16 where methionine, leucine, glutamic acid and alanine occupy these respective positions in the porcine molecule. The carboxyl-terminal sequences remained unchanged for human and porcine C3a through three residues, as judged by carboxypeptidase analysis. Therefore, a minimum of 20% difference exists between the total primary sequences of human and porcine C3a; however, no differences could be detected between any of the biological activities examined for these two anaphylatoxins (Hugli, Vallota and Müller-Eberhard 1975). Interestingly, the most pronounced difference between human and porcine C3a could be demonstrated immunologically using antibody specific to each of the two small molecules (Fig. 8). Immunological cross-reactivity between human

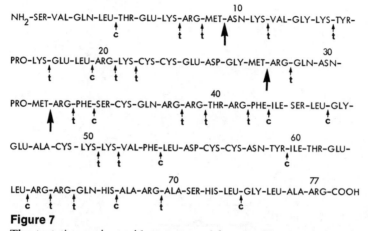

Figure 7

The tentative amino acid sequence of human C3a anaphylatoxin. Heavy arrows indicate the cyanogen bromide cleavage sites (↑). Major tryptic (t) and chymotryptic (c) cleavage sites are indicated. Automatic sequence analysis of intact C3a established the arrangement of 26 amino acid residues from the amino-terminal end. Amino acid sequence determinations of cyanogen bromide fragments and tryptic and chymotryptic peptides helped to establish the primary structure of human C3a.

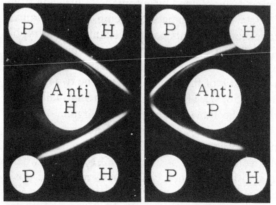

Figure 8

Double immunodiffusion in gel analyses of human and porcine C3a anaphylatoxins. Rabbit anti-human C3a (anti-H) was added to the center well on the left and rabbit anti-porcine C3a was added to the center well on the right. One μl of human C3a (1.2 mg/ml) and 1 μl of porcine C3a (1.1 mg/ml) was applied to the outer wells according to the labels H and P, respectively.

and porcine C3a could not be demonstrated. These results indicate that the amino acid substitutions in the primary structure of these two C3a molecules have involved nonfunctional and presumably many surface-oriented side chains since all antigenic sites on these two molecules were affected, as judged by the total absence of cross-reactivity.

Conformation of the C3a Anaphylatoxin

Circular dichroism (CD) spectra of C3a indicated large negative contributions centered at 208 and 222 nm (Hugli, Morgan and Müller-Eberhard 1975). The α-helical content of the C3a molecule was estimated to be between 40 and 45%, which represents more than four times the percentage of helix estimated in the intact C3 molecule. Figure 9 illustrates the large difference in mean residue ellipticity between intact human C3 and C3a fragments from both human and porcine sources. As is readily apparent, CD spectra obtained from human and porcine C3a were nearly indistinguishable. It should be emphasized here that the secondary structure of the C3a molecule accommodates three disulfide bridges. Coupling this fact with an estimated helical content of more than 40% suggests a highly compact molecular conformation for C3a.

Efforts to establish the relationship between anaphylatoxin structure and function were aided by further CD studies. Reduced and carboxymethylated C3a was biologically inactive, and CD spectra indicated a marked decrease in the ellipticity between 200 and 230 nm (Fig. 10, curve C). Addition of the reducing agent mercaptoethanol to native C3a caused the CD spectra to

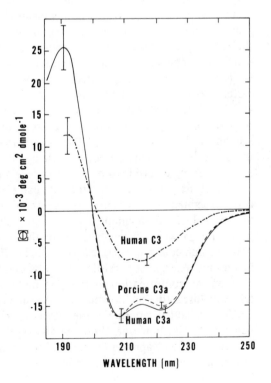

Figure 9
Circular dichroism spectra for human and porcine C3a as compared to human C3. Double minima at 208 and 222 nm were observed with both anaphylatoxins. A negative mean residue ellipticity of 17,000 was estimated for the C3a molecules at 208 nm compared to less than 8000 for intact C3. All spectra were obtained at 27°C in 0.02 M sodium phosphate buffer at pH 7.2.

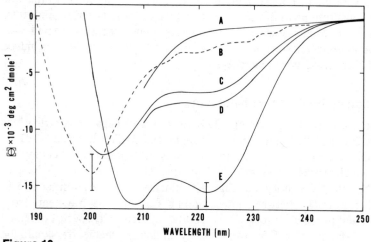

Figure 10

Circular dichroism spectra for human C3a under various conditions.
Curve A was obtained for C3a containing 0.05 M mercaptoethanol and
6 M guanidinium chloride; curve B, C3a heated at 100°C for 3 hours;
curve C, reduced and carboxymethylated C3a; curve D, C3a reduced
by 0.05 M mercaptoethanol; curve E, same treatment as curve A followed by dialysis against 0.02 M phosphate at pH 7.2.

assume a shape similar to that of reduced and carboxymethylated C3a (curve
D); however, no loss of function was detected by the ileum assay. Guanidinium chloride (GuCl) at 6 M significantly reduced ellipticity, but like the
reducing agent, did not seem to impair the function. When both GuCl and
the reducing agent were simultaneously added to C3a, CD spectra indicated
nearly total conversion to a random structure (curve A), and approximately
90% activity loss was recorded. The secondary structure of C3a was fully
restored once the salt and mercaptoethanol had been removed by dialysis,
and C3a activity also returned to normal (curve E). Furthermore, enzymatic
removal of the carboxyl-terminal arginine of human C3a to form C3a$_i$ did
not affect the CD spectrum. The assay used, in which C3a function was
measured by its ability to contract ileum, had definite limitations since C3a
activity can be determined only at low levels of reducing agent or GuCl. Refolding of the perturbed C3a molecule can proceed as the assay is performed,
and consequently, losses in activity are observed only when restoration of
the original C3a conformation takes longer than the almost instantaneous contraction of smooth muscle produced by the application of anaphylatoxin. Since
only an immediate and concerted C3a binding leads to contraction of the
ileum muscle strip, slow refolding can be detected. Thus a functional dependence on the conformation of the C3a molecule is demonstrable only by extensively disrupting the polypeptide. Presumably when the reducing agent and
GuCl were added separately, the effects on conformation also influenced the
activity of C3a; however, these agents were diluted by introduction into the
test chamber, and apparently reversal to a native conformation and recovery
of activity were too rapid to be monitored.

CD results demonstrated that the C3a molecule is remarkably stable. Only minor differences were observed between CD spectra obtained at pH extremes from 1 to 10. The higher the pH, the greater was the ellipticity, and this behavior correlated with the conformational sensitivity to pH recorded for the cationic model peptide polylysine (Greenfield and Fasman 1969). The biological activity of C3a was similarly unaffected after exposure to pH conditions between 1 and 10. Irreversible denaturation of C3a resulted only from prolonged heat treatment at 100°C; three hours at this temperature were required to completely inactivate C3a at neutral pH. The structural effects imparted to C3a by thermal denaturation are portrayed in Figure 10 as curve B. Excluding thermal denaturation, the perturbed C3a molecule demonstrates a remarkable capacity for reassuming a characteristic conformation in solution.

Some Properties of the C5a Anaphylatoxin

C5a is the so-called "classical" anaphylatoxin. It was residual C5a activity resulting from complement activation in certain animal sera that originally led to the discovery of anaphylatoxins. In human serum, however, the indigenous carboxypeptidase is almost totally effective for inactivating the generated C5a. Complement-activated porcine, guinea pig and rat sera all exhibit residual C5a activity which resists further inactivation by their respective serum carboxypeptidases. Paradoxically, it is generally agreed that the levels of serum carboxypeptidase in these animals are higher than in man. Why then does this inconsistency occur, and how does the formation of C5a in sera from various species differ?

Attempts to answer these questions require that we be able to isolate sufficient quantities of C5a for characterization. As is true of many interesting biological products, only minute quantities of C5a can be produced in serum. Maximal amounts of C5a produced from the total enzymatic conversion of the C5 would average only 3–5 mg/liter in human and other mammalian serum. This amount of C5a is seldom available for isolation, since C5 is usually not quantitatively converted.

Porcine blood has become a popular source for economically obtaining C5a in bulk. Lieflander et al. (1972) characterized porcine C5a as a molecule of approximately 9000 molecular weight containing the carboxyl-terminal sequence lysyl-glycyl-leucine. Their studies also indicated that isoelectric focusing was a desirable final purification step for obtaining homogeneous preparations of this trace product. For comparison, porcine C5a was isolated by a similar procedure from serum activated in the presence of EACA (hereafter EACA-C5a) (Vallota and Müller-Eberhard 1973); amino acid compositions for EACA-C5a are given in Table 2. Values under I are from C5a after isoelectric focusing in a 1% ampholine mixture over a pH range of 9–11. The active EACA-C5a focused as a single band at pH 9.5, in agreement with results obtained by Lieflander et al. (1972). II contains values for EACA-C5a recovered in a single band after polyacrylamide gel electrophoresis at pH 4.6. Values of the individual EACA-C5a preparations agree remarkably well. The amino acid compositions reported here were also generally like those of C5a isolated from serum without EACA added (Lieflander

Table 2

Amino Acid Composition of Porcine C5a Anaphylatoxin

	I		*II*	
	moles/ 100 moles	*residues*[a] *(mole/mole C5a)*	*moles/ 100 moles*	*residues (mole/mole C5a)*
Lysine	13.14	16.88 (17)	12.94	16.52 (17)
Histidine	1.79	2.29 (2)	2.32	2.95 (3)
Arginine	7.95	10.17 (10)	8.52	10.82 (11)
Aspartic acid	9.17	11.73 (12)	8.84	11.23 (11)
Threonine	1.82	2.28 (2)	1.68	2.13 (2)
Serine	2.37	3.03 (3)	2.39	3.03 (3)
Glutamic acid	14.55	18.61 (19)	13.74	17.51 (18)
Proline	3.72	4.75 (5)	4.26	5.41 (5)
Glycine	4.74	6.07 (6)	4.26	5.41 (5)
Alanine	10.26	13.12 (13)	10.58	13.44 (13)
Half-cystine[b]	6.47	8.10 (8)	6.37	8.03 (8)
Valine	3.85	4.92 (5)	3.58	4.51 (5)
Methionine	2.37	3.03 (3)	2.52	3.20 (3)
Isoleucine	5.89	7.52 (8)	5.93	7.54 (8)
Leucine	5.13	6.56 (7)	5.29	6.72 (7)
Tyrosine	5.45	6.97 (7)	5.61	7.13 (7)
Phenylalanine	1.47	1.89 (2)	1.68	2.08 (2)
Total residues		129		128
Molecular weight		14,906		14,898

Values under I were obtained from C5a purified by isoelectric focusing. II contains results for C5a purified by polyacrylamide gel electrophoresis at pH 4.6. A gel slice containing 50 μg of C5a was hydrolyzed according to the method of Houston (1971).

[a] Residues per molecule of C5a were based on an assumed MW of 15,000 daltons as estimated by gel filtration studies.

[b] Half-cystine residues were determined as cysteic acid.

et al. 1972). However, their assumed molecular weight of 9000 was considerably lower than the 15,000 daltons assumed here based on gel filtration, sedimentation studies and direct comparisons with the C3a molecule.

The EACA-C5a molecule contained an arginine residue at the carboxyl terminus. Surprisingly, carboxypeptidases from pancreas, yeast or serum were unable to remove the carboxyl-terminal arginine residue of EACA-C5a unless the C5a molecule had been previously denatured by heating at 100°C. This stability of the isolated porcine EACA-C5a suggests the mechanism by which this molecule becomes resistant to inactivation by the carboxypeptidase in serum and undoubtedly applies to C5a obtained from various other species. If one postulates that a conformational rearrangement occurs after cleavage of C5a from the parent C5 molecule, then this structural change could render the carboxyl terminus of C5a inaccessible for carboxypeptidase attack.

Evidence that the C5a, purified previously in the absence of EACA, contains predominately inactivated material is supplied by comparing its activities with those of EACA-C5a. For instance, a full contraction of guinea pig ileum was induced with as little as 3×10^{-10} M EACA-C5a, whereas 7×10^{-9} M

C5a from the untreated serum was required (Vogt 1968). These results would suggest that as much as 90% of the C5a isolated from untreated serum exists in the inactive C5a$_{des\ Arg}$ form. If this interpretation is correct, then the carboxyl-terminal arginine residue of porcine C5a is indeed functionally essential. Additional support for this hypothesis was provided by the fact that after the arginine residue was removed from heat-denatured EACA-C5a by carboxypeptidase B, further digestion with yeast carboxypeptidase released the residues leucine and glycine, the carboxyl-terminal and penultimate residues reported for C5a obtained from EACA free serum.

It was shown by CD analysis that the structure of native porcine C5a was highly ordered (Morgan, Vallota and Müller-Eberhard 1974). The content of α-helical structure was estimated at nearly 40%, which approaches the amount of α-structure in the C3a anaphylatoxins. Presumably the rigidity provided by the numerous disulfide linkages and the regularities of secondary structure could account for a static configuration about the carboxyl terminus which renders the arginyl residue inaccessible to enzymatic attack. The scheme outlined in Figure 11 might therefore explain the persistence of C5a anaphylatoxin activity in porcine and other animal sera. Since human C5a apparently assumes a final conformation which is less resistant to attack by serum carboxypeptidase than other C5a molecules, the control mechanism in man appears more efficient than in other animal species. This may, in turn, explain the need for elevated levels of carboxypeptidase in porcine and guinea pig sera. Adequate control of the C5a generated in the sera of these animals would require a greater turnover of the carboxyl-terminal arginine at the time of C5a release from C5 while the transitional form of the C5a molecule prevails. Only a more extensive structural analysis will provide the data necessary to fully resolve the structure-function relationships of the C5a anaphylatoxin molecule.

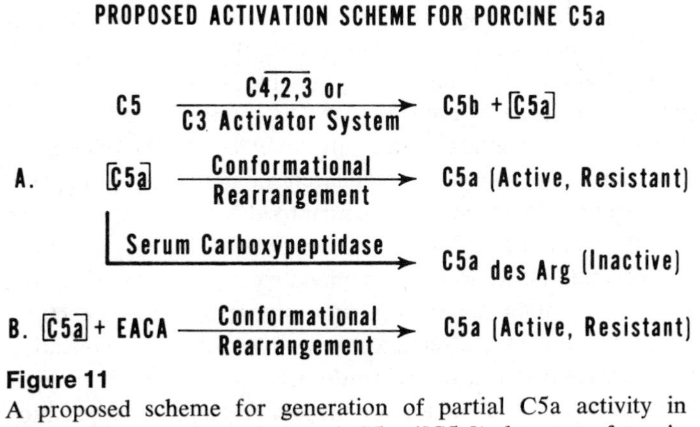

PROPOSED ACTIVATION SCHEME FOR PORCINE C5a

$$C5 \xrightarrow{\overline{C4,2,3}\ \text{or}} C5b + [\overline{C5a}]$$
$$\text{C3 Activator System}$$

A. $[\overline{C5a}] \xrightarrow[\text{Rearrangement}]{\text{Conformational}}$ C5a (Active, Resistant)

$\xrightarrow{\text{Serum Carboxypeptidase}}$ C5a $_{des\ Arg}$ (Inactive)

B. $[\overline{C5a}]$ + EACA $\xrightarrow[\text{Rearrangement}]{\text{Conformational}}$ C5a (Active, Resistant)

Figure 11
A proposed scheme for generation of partial C5a activity in serum. The transitory form of C5a ([C5a]) has two fates in serum. When conformational rearrangement occurs before the carboxyl-terminal arginine is released, a stable C5a is formed which resists attack by the serum carboxypeptidase. Otherwise, the inactive C5a$_{des\ Arg}$ is formed. Addition of EACA inhibits the carboxypeptidase, permitting the conformational rearrangement to occur.

DISCUSSION

A number of conclusions can be drawn from the current data obtained through studies with the anaphylatoxins. Firm evidence now exists that both C3a and C5a molecules are generated in human and other mammalian sera. Significant chemical differences were demonstrated between the C3a molecules obtained from human and animal sources; however, the respective biological activities were essentially identical. The human C3a molecule generated by either the classical or the alternative complement activation pathway were shown to be chemically indistinguishable. By analogy, the C5a generated in serum by either of the indigenous enzymatic routes will undoubtedly prove to be identical molecular entities as well. Recent chemical characterizations of C3a and C5a produced in serum have confirmed previous speculation that the activating enzymes for C3 and C5 exhibit trypsinlike specificity. These various activating enzymes then selectively attack a single peptide linkage in the C3 and C5 molecules to produce the anaphylatoxins.

Although several explicit serum enzymes are known to be involved in the activation of both C3 and C5, a single mechanism is responsible for inactivating the respective anaphylatoxins that are released in the serum. The mechanism involves the enzymatic removal of an essential carboxyl-terminal residue. To date, only the basic amino acid arginine has occupied the carboxyl-terminal position of the C3a and C5a molecules studied, and serum carboxypeptidase is presumed to be the only controlling enzyme. Potent and varied activities of the anaphylatoxins dramatically illustrate the necessity for an efficient in vivo control mechanism. Serum levels of the indigenous carboxypeptidase may realistically govern an animal's chances for survival under conditions resulting in complement activation.

Direct comparisons between the two anaphylatoxins at the molecular level indicated a high degree of similarity. Both C3a and C5a obtained from several animal sources were highly cationic molecules and occurred as single polypeptide chains. The active forms of these molecules all contained an essential arginine amino acid residue at the carboxyl-terminal position. Anaphylatoxins are relatively small molecules: C3a was estimated at 9000 daltons and C5a is slightly larger, at 15,000 daltons. Definite and reversible secondary conformational arrangements are assumed by the C3a and C5a molecules, and in each case, a number of intrachain disulfide bridges are involved. Carboxymethylation of all sulfhydryl groups in reduced C3a rendered the molecule biologically inactive. C5a was reversibly inactivated by mercaptoethanol alone, exclusive of alkylation (Morgan, Vallota and Müller-Eberhard 1974). Thus, disulfides are an important structural feature in both anaphylatoxin molecules. C3a and C5a were each shown to contain rather large amounts of α-helical structure, which presumably contributes to the remarkable stability of these molecules. Stability and rapid reversibility to a native conformation following perturbation were somewhat unexpected behavior for protein fragments representing no more than 5–7% of the parent molecular structure. Hence, the conformations of both anaphylatoxin molecules appear to be intimately associated with biological activity. Finally, it has been shown that C3a and C5a differ primarily in potency and not in function. On a molar basis, C5a is the most potent of the two anaphylatoxins, but is less abundant, a factor which tends to equalize the potential activities of C3a and C5a in serum. Ultimately, the chemical and physical properties of C3a and C5a will provide sufficient information to ascertain the exact nature of the biological responses of certain tissues and cells to anaphylatoxins.

Acknowledgments

I wish to thank Ms. Yvonne David and Mr. Peter Sigrist for excellent technical assistance. This work was supported by United States Public Health Service Program Project Grant HL-16411, American Heart Association Grant 68-666 and an Established Investigatorship from the American Heart Association (72-175). This is publication number 882, Department of Molecular Immunology, Scripps Clinic and Research Foundation, La Jolla, California.

REFERENCES

Bodammer, G. and W. Vogt. 1970. Beeinflussung der Capillarpermeabilität in der Meerschweinchenbaut durch Anaphylatoxin (AT). *Arch. Exp. Pathol. Pharmakol.* **266**:255.

Bokisch, V. A. and H. J. Müller-Eberhard. 1970. Anaphylatoxin inactivator of human plasma: Its isolation and characterization as a carboxypeptidase. *J. Clin. Invest.* **49**:2427.

Bokisch, V. A., H. J. Müller-Eberhard and C. G. Cochrane. 1969. Isolation of a fragment (C3a) of the third component of human complement containing anaphylatoxin and chemotactic activity and description of an anaphylatoxin inactivator of human serum. *J. Exp. Med.* **129**:1109.

Budzko, D. B., V. A. Bokisch and H. J. Müller-Eberhard. 1971. A fragment of the third component of human complement with anaphylatoxin activity. *Biochemistry* **10**:1166.

Cochrane, C. G. and H. J. Müller-Eberhard. 1968. The derivation of two distinct anaphylatoxin activities from the third and fifth components of human complement. *J. Exp. Med.* **127**:371.

Cochrane, C. G., H. J. Müller-Eberhard and B. S. Aikin. 1970. Depletion of plasma complement in vivo by a protein of cobra venom: Its effect on various immunologic reactions. *J. Immunol.* **105**:55.

Folk, J. E. 1956. A new pancreatic carboxypeptidase. *J. Amer. Chem. Soc.* **78**:3541.

Friedberger, E. 1910. Weitere Untersuchungen über Eiweiss-anaphylaxic. IV. Mitteilung. *Z. Immunitatsforsch.* **4**:636.

Greenfield, N. and G. D. Fasman. 1969. Computed circular dichroism spectra for the evaluation of protein conformations. *Biochemistry* **8**:4108.

Houston, L. L. 1971. Amino acid analysis of stained bands from polyacrylamide gels. *Anal. Biochem.* **44**:81.

Hugli, T. E., W. T. Morgan and H. J. Müller-Eberhard. 1975. Circular dichroism of C3a anaphylatoxin: Effects of pH, heat, guanidinium chloride and mercaptoethanol on conformation and function. *J. Biol. Chem.* (in press).

Hugli, T. E., E. H. Vallota and H. J. Müller-Eberhard. 1975. Purification and partial characterization of human and porcine C3a anaphylatoxin. *J. Biol. Chem.* (in press).

Johnson, A. R. and H. J. Müller-Eberhard. 1975. Release of histamine from rat mast cells by the complement peptides C3a and C5a. *J. Immunol.* (in press).

Laan, B., J. L. Molenaar, T. M. Feltkamp-Vroom and K. W. Pondman. 1974. Interaction of human anaphylatoxin C3a with rat mast cells demonstrated by immunofluorescence. *Eur. J. Immunol.* **4**:393.

Lieflander, M., D. Dielenberg, G. Schmidt and W. Vogt. 1972. Structural elements of anaphylatoxin obtained by contact activation of hog serum. *Hoppe-Seyler's Z. Physiol. Chem.* **353**:385.

Mahler, F., M. Intaglietta, T. E. Hugli and A. R. Johnson. 1975. Influences of C3a anaphylatoxin compared to other vasoactive agents on the microcirculation of rabbit omentum. *Microvasc. Res.* (in press).

McConahey, P. J. and F. J. Dixon. 1966. A method of trace iodination of proteins for immunologic studies. *Int. Arch. Allergy* **29**:185.

Morgan, W. T., E. H. Vallota and H. J. Müller-Eberhard. 1974. Circular dichroism of C5a anaphylatoxin of porcine complement. *Biochem. Biophys. Res. Comm.* **57**:572.

Müller-Eberhard, H. J. 1972. The molecular basis of the biological activities of complement. In *The Harvey Lectures,* series 66, pp. 75–104. Academic Press, New York.

Nilsson, U. R., R. Mandle and T. McConnel-Mapes. 1975. Human C3 and C5: Subunit structure and modification by trypsin and $\overline{C4,2}$ and $\overline{C4,2,3}$. *J. Immunol.* **114**:815.

Shin, H. S., R. Snyderman, E. Friedman, A. Mellors and M. M. Mayer. 1968. Chemotactic and anaphylatoxic fragment cleaved from the fifth component of guinea pig complement. *Science* **162**:361.

Valotta, E. H. and H. J. Müller-Eberhard. 1973. Formation of C3a and C5a anaphylatoxins in whole human serum after inhibition of the anaphylatoxin inactivator. *J. Exp. Med.* **137**:1109.

Vogt, W. 1968. Preparation and some properties of anaphylatoxin from hog serum. *Biochem. Pharmacol.* **17**:727.

Vogt, W., N. Zeman and G. Garbe. 1969. Hislaminunabhängige Wirkunger von Anaphylatoxin auf Glatte Muskulatur Isolierter Organe. *Arch. Exp. Pathol. Pharmakol.* **262**:399.

Ward, P. A. 1969. The heterogeneity of chemotactic factors for neutrophils generated from the complement system. In *Cellular and Humoral Mechanisms in Anaphylaxis and Allergy* (ed. H. Z. Movat), pp. 279–288. Karger, Basel.

Wuepper, K. D., V. A. Bokisch, H. J. Müller-Eberhard and R. B. Stoughton. 1972. Cutaneous responses in human C3 anaphylatoxin in man. *Clin. Exp. Immunol.* **11**:13.

On the Generation of Intermediate Plasminogen and Its Significance for Activation

Per Wallén and Björn Wiman

Department of Physiological Chemistry
University of Umeå, Umeå, Sweden

Two types of human plasminogen dissimilar in regard to the structure of the amino-terminal part have been extensively studied during the past decade. One of them has only glutamic acid in the amino-terminal position (Bergström and Wallén 1963; Wallén and Wiman 1970), and the other mainly lysine (Robbins et al. 1967). In the following presentation they will be referred to as Glu-plasminogen and Lys-plasminogen, respectively[1]. By zymographic analysis after starch gel electrophoresis, it has been shown that plasminogen, which has glutamic acid in the amino-terminal position, has an electrophoretic mobility indistinguishable from that of plasminogen in fresh plasma (Wallén and Wiman 1970), whereas plasminogen types, which have lysine or valine at the amino terminus, have, according to their electrophoretic behavior, significantly higher isoelectric points than plasminogen of plasma. It has been suggested that the latter plasminogen preparations have been proteolytically modified during the purification procedure. It was recently shown that limited proteolysis by plasmin converts Glu-plasminogen into a plasminogen form indistinguishable from Lys-plasminogen (Claeys, Molla and Verstraete 1973b).

Robbins et al. (1967) discovered that plasminogen consists of a single polypeptide chain, which, on activation, forms plasmin composed of two polypeptide chains connected by a disulfide bridge. From studies on Lys-plasminogen, it was originally suggested that the activation of plasminogen is achieved by cleavage of a single peptide bond, and that no peptide material is primarily released from the proenzyme (Robbins et al. 1967). However when Glu-plasminogen is activated with urokinase, there is a rapid release of

[1] Several authors have lately included the name of the amino-terminal amino acid in their designation of the plasminogen type. In order to avoid confusion, we will use this nomenclature in the present report, although in earlier publications we used the designations plasminogen A for Glu-plasminogen and plasminogen B for Lys-plasminogen. It should be stressed, however, that plasminogen forms designated Lys-plasminogen in general are heterogeneous in regard to the amino-terminal structure. In addition to lysine, valine and often methionine occur in varying amounts.

peptide material from the amino-terminal part of the proenzyme, which seems to precede the final cleavage into the two-chain structure of plasmin (Rickli and Otavsky 1973; Wiman and Wallén 1973). The modified plasminogen generated by the activator-induced proteolysis in the amino-terminal part of Glu-plasminogen has physicochemical properties very close to those of Lys-plasminogen. It is suggested that the activation of plasminogen is a two-step reaction in which the first step, the generation of the intermediate plasmino-gen, is a prerequisite for the second reaction, the generation of plasmin by cleavage of an internal peptide bond.

This report intends to show the relationship between Glu-plasminogen and Lys-plasminogen and to demonstrate the molecular changes occurring in the amino-terminal part of Glu-plasminogen during activation. The significance of these molecular events for the overall activation process will be discussed.

MATERIALS AND METHODS

Preparation of Plasminogen

Plasminogen with high specific activity (about 30 CTA U/mg protein) with either glutamic acid or lysine/valine as amino-terminal amino acid were pre-pared as described earlier (Wallén and Wiman 1972; Wiman and Wallén 1973).

Activation Procedures

Activation of plasminogen was performed with urokinase (Leo Pharmaceuti-cal, Denmark) in two different ways. When the purpose was to study the structural changes in the plasminogen during activation, activation was per-formed by perfusion of a plasminogen solution through a small column (bed volume about 0.5 ml) packed with insolubilized urokinase. The details of the procedure are described elsewhere (Wiman and Wallén 1973). The degree of activation was controlled by changing the perfusion rate. The effluent was collected in a solution saturated with p-nitrophenyl-p-guanidino benzoate. The rapid perfusion rate (1–2 ml/min) and the effective inhibition of plasmin minimize the autoproteolytic degradation.

The following procedure was used for comparative studies on the activation rates of Glu-plasminogen and Lys-plasminogen. Solutions of plasminogen and urokinase in 0.04 M sodium phosphate, 0.1 M sodium chloride pH 7.5, con-taining 25% glycerol were incubated at 25°C. The plasminogen concentration was 0.5 mg/ml in the final reaction mixture in all experiments, and the uro-kinase concentration varied between 250 and 1250 Plough units/ml. Sub-stances, the influence of which were to be studied, were dissolved in the plasminogen solution prior to the initiation of the reaction with urokinase. The generated plasmin was determined in samples removed at different times (10 min to 24 hr) by measuring the esterolytic activity of p-tosyl-L-arginine methyl ester as described previously (Wiman and Wallén 1973). The estero-lytic activity was converted to CTA units (Committee on Thrombolytic Agents) by comparison with a standard (ampoule 71/268, kindly provided by the Division of Biological Standards, MRC, London).

Various Methods

The different analytical techniques used in this investigation have been described in detail in earlier reports: amino-terminal amino acid and sequence determinations (Wallén and Wiman 1972; Wiman 1973); amino acid analysis, cleavage with cyanogen bromide, peptide mapping and reversible blocking of amino groups (Wiman 1973); insolubilization of proteins and SDS-polyacrylamide gel electrophoresis (Wiman and Wallén 1973); ultracentrifugation methods and spectropolarimetry (Sjöholm, Wiman and Wallén 1973).

RESULTS AND DISCUSSION

Molecular Changes in Amino-terminal Part of Plasminogen during Activation

When using the column technique described above, the solution of zymogen is exposed to the activator for a very short time (0.2–1 min). This considerably reduces the risk for autodigestion by plasmin. Furthermore, the technique provides an easy way to control the degree of activation by changing the perfusion rate, which is of value for studies of intermediate compounds formed during activation (Wiman and Wallén 1973). SDS-polyacrylamide gel electrophoresis on reduced samples of activation mixtures showed five components with the molecular weights 92,000, 86,000, 63,000, 25,000 and 7000. The 92,000-dalton component corresponds to unmodified plasminogen, and the 63,000- and 25,000-dalton fragments correspond to the heavy and light chains described by Robbins et al. (1967). Amino-terminal analysis of the 86,000- and 63,000-dalton fragments showed that both are heterogeneous; thus methionine, lysine and valine were found in substantial amounts, indicating a relationship between these fragments and Lys-plasminogen. In general, methionine was dominating when the fragments were obtained from mixtures with a low degree of activation, whereas there was a gradual increase of valine and lysine on increasing the degree of activation, i.e., the plasmin concentration (Wiman and Wallén 1973).

The structural relationship between plasminogen and the different components demonstrated by SDS-polyacrylamide gel electrophoresis has been shown by structural analyses of components isolated from the activation mixtures. By gel filtration of the activation mixture (before reduction) on Sephadex G-75, the 7000-dalton fragment could be separated from the other components. The salt fraction in this chromatogram contained a pentapeptide, which could be purified by gel filtration on Sephadex G-10. The protein fraction could, after reduction and S-carboxymethylation, be separated into the components with molecular weights of 86,000, 63,000 and 25,000 by gel filtration on Sephadex G-150 in a buffer containing 6 M urea.

The 7000-dalton fragment and the pentapeptide could both be deduced from the amino-terminal part of the molecule. The 7000-dalton fragment is, in fact, the amino-terminal part of Glu-plasminogen (Wiman and Wallén 1973) and together with the pentapeptide, constitutes 68 amino acids in sequence (Wiman 1973).

Structural studies on intact plasminogen, amino-terminal cyanogen bromide fragments and the fragments obtained on digestion of native and maleinylated plasminogen with urokinase have enabled us to define the sequence of 81 amino acids in the amino-terminal part of Glu-plasminogen (Wallén and Wiman 1970; Wiman 1972, 1973; Wiman and Wallén, unpubl.). The sequence and the peptide bonds split by urokinase and plasmin are shown in Figure 1. The sequence of the five carboxyl-terminal amino acids -Lys-Lys-Val-Tyr-Leu- is of special interest since they seem to provide the connection between the Glu-plasminogen and Lys-plasminogen. The four amino acids in the amino-terminal part of Lys-plasminogen are in sequence Lys-Val-Tyr-Leu- (Robbins et al. 1972; Walther et al. 1974). The peptide bonds Lys-63/Ser-64 and Arg-68/Met-69 are specifically split by urokinase, which leads to the release of two peptides subsequently called the activation peptides AP I and AP II and to the generation of an intermediate plasminogen form. The fact that urokinase cleaves only one bond, Arg-68/Met-69, in maleinylated plasminogen indicates strongly that this bond is a specific site for proteolysis by urokinase, although even plasmin may cleave this bond. The intermediate form initially contains methionine as amino-terminal amino acid, but is easily degraded by plasmin. The release of a nonapeptide gives rise to the Lys-plasminogen, which seems to be rather stable. A model for the molecular events occurring during activation of Glu-plasmniogen consistent with the present results is shown in Figure 2, which describes a reaction proceeding in two steps. In the first step, the two activation peptides are released from Glu-plasminogen, forming an intermediate plasminogen. In the second step, the intermediate plasminogen is transformed to plasmin by cleavage of a single -Arg-Val- bond situated about 600 residues from the amino terminus of Glu-plasminogen. The fact that a heavy chain with glutamic acid as amino-terminal amino acid and a molecular weight of about 70,000 has never been found even in trace amounts is in favor of a sequential rather than a random cleavage at the two regions sensitive for urokinase. It could be argued, however, that such a fragment may be extremely sensitive to degradation and thus escape detection. Since there is at present no way to block the release of the activation peptides specifically, the concept of a sequential reaction has to rest largely on indirect evidence, i.e., conformational changes and differences in biological properties between Glu-plasminogen and the intermediate forms.

Conformational Changes of Plasminogen during Activation

By gel filtration studies it was found that Lys-plasminogen, in spite of its lower molecular weight, had a smaller elution volume than Glu-plasminogen (Wallén and Wiman 1972). It was therefore suggested that the conversion of Glu-plasminogen to Lys-plasminogen was associated with an increase in Stoke's radius. A similar increase in Stoke's radius was observed for Glu-plasminogen when 6-aminohexanoic acid was present during the experiment. More detailed studies on the hydrodynamic properties of Glu-plasminogen, Lys-plasminogen and plasmin inactivated by diisopropylphosphofluoridate (DIP-plasmin) were therefore performed in order to investigate the conformational changes occurring during activation of plasminogen (Sjöholm, Wiman

1
Glu -Pro-Leu-Asp-Asp-Tyr-Val -Asn-Thr-Gln-Gly -Ala -Ser-Leu-Phe-Ser-Val -Thr-Lys-Lys-

20

30
Gln -Leu-Gly -Ala -Gly -Ser-Ile -Glu -Glu -Cys-Ala -Gln -Ala -Lys-Cys-Glu -Glu -Asp-Glu -Glu -

40

50
Phe-Thr-Cys -Arg-Ala -Phe-Gln -Tyr-His -Ser-Lys-Glu -Gln-Glu -Cys-Val -Ile -Met-Ala -Glu -

60

UK
70
Asn-Arg-Lys-Ser-Ser-Ile -Ile - Arg-Met-Ser-Asp-Val -Val -Leu-Phe-Glu -Lys-Lys-Val -Tyr -

80

UK

Pli   Pli

Leu-

Figure 1

The primary structure of the amino-terminal part of Glu-plasminogen. The peptide bonds split by urokinase (UK) and plasmin (Pli) are indicated with arrows. The part of the sequence (residues 45–51) which participates in the specific noncovalent interaction is underlined. (*Note added in proof:* Rechecking of the sequence in Fig. 1 has shown that residue 70 is Arg and not Ser.)

295

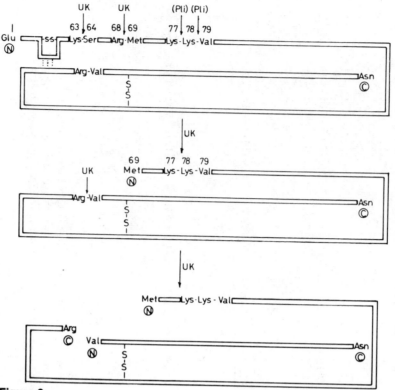

Figure 2

Schematic representation of the two-step model for activation of plasminogen by urokinase. The bonds sensitive for urokinase (UK) and plasmin (Pli) are indicated. The circumscribed N and C designate the amino and carboxyl termini, respectively. The noncovalent interaction is indicated by the dotted lines.

and Wallén 1973). The transformation of Glu-plasminogen to Lys-plasminogen is, according to these studies, associated with the following changes: The sedimentation constant ($s^{\circ}_{20,w}$) decreases from 5.10 to 4.80; Stoke's radius increases from 4.3 to 4.5; and the frictional ratio (f/f_o) increases from 1.50 to 1.55. The findings confirm the earlier postulated changes in Stoke's radius and in conformation.

The conformational changes have also been investigated by spectropolarimetric studies on Glu-plasminogen, Lys-plasminogen and DIP-plasmin (Sjöholm, Wiman and Wallén 1973). The circular dichroism spectra in the far-ultraviolet region are almost identical for all these substances and indicate, furthermore, that they contain about 80% random structure and about 20% β-structure, while no significant amounts of α-helix are detected. However, as demonstrated by Figure 3, which shows the circular dichroism spectra in the near-ultraviolet region (250–308 nm), significant differences are found. These spectra, which reflect conformational changes around certain chromophores (e.g., tryptophan, tyrosine, phenylalanine and disulfide bridges), show a difference between Glu-plasminogen and DIP-plasmin over the whole range. The

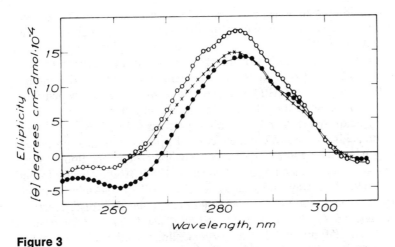

Figure 3
Circular dichroism spectra in the near-ultraviolet region of Glu-plasminogen (•——•), Lys-plasminogen (x——x) and DIP-plasmin (o——o). (Reprinted, with permission, from Sjöholm et al. 1973.)

spectrum of Lys-plasminogen, on the other hand, is similar to that of Glu-plasminogen beyond 280 nm and similar to that of DIP-plasmin below 265 nm. This difference is still better demonstrated by Figure 4, which shows the differential spectra. These results also indicate conformational changes and are suggestive of a stepwise change in conformation during activation of Glu-plasminogen to plasmin. Evidence that the conformational differences ob-

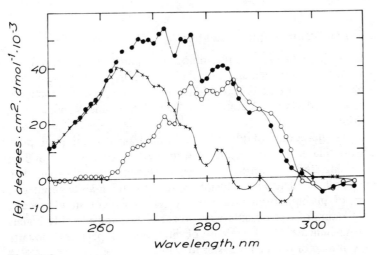

Figure 4
Differential circular dichroism spectra in the near-ultraviolet region obtained from the spectra shown in Figure 3: (•——•) DIP-plasmin minus Glu-plasminogen; (x——x) Lys-plasminogen minus Glu-plasminogen; (o——o) DIP-plasmin minus Lys-plasminogen. (Reprinted, with permission, from Sjöholm et al. 1973.)

served between Glu-plasminogen and Lys-plasminogen are due to the removal of the activation peptides is provided by the fact that the addition of the activation peptide (API), in excess, to a solution of Lys-plasminogen results in an almost complete restoration of the circular dichroism spectrum of Glu-plasminogen. This suggests a noncovalent interaction between the amino-terminal activation peptide (API) and the rest of the molecule, which is of importance for the conformational state of the plasminogen molecule.

The amino acids lysine and 6-aminohexanoic acid (6-AHA) and some other ω-amino acids have a profound effect on the physicochemical properties of plasminogen, seemingly due to a specific stoichiometric interaction between these amino acids and plasminogen (Alkjaersig 1964; Abiko, Iwamoto and Tomikawa 1969; Brockway and Castellino 1971, 1972). 6-Aminohexanoic acid in 0.01 M concentration effects markedly the circular dichroism spectrum of Glu-plasminogen (Sjöholm, Wiman and Wallén 1973). An increase in ellipticity between 250 nm and 305 nm occurs, similar to that observed when Glu-plasminogen is activated to plasmin. Lysine has a similar effect, although a higher concentration is required, in agreement with the results of Brockway and Castellino (1972), showing that the association constant for the complex with plasminogen is higher for 6-aminohexanoic acid than for lysine. The change in conformation caused by 6-aminohexanoic acid and lysine is indicative of a dissociation of the noncovalent interaction between the activation peptide and the rest of the molecule, thus imitating the cleavage occurring in the first step of the activation process.

It is evident from the studies referred to above that the physicochemical properties of plasminogen are markedly influenced by the release of the amino-terminal activation peptides, and it is suggestive to suppose that the conformational changes induced are necessary for the generation of active plasmin.

Kinetic Studies on Plasminogen Activation

Recent studies on the activation of plasminogen demonstrate a marked difference between Glu-plasminogen and Lys-plasminogen in regard to activation rate, indicating that the two-step activation process may have a functional significance (Claeys, Molla and Verstraete 1973a; Wallén and Wiman 1973; Claeys and Vermylen 1974).

Figure 5 shows the course of plasmin generation on activation of plasminogen with urokinase. The experiments were performed at 25°C and in solutions containing 25% glycerol in order to stabilize the plasmin. The concentrations of plasminogen and activator were the same in all experiments. The generation of plasmin from Glu-plasminogen is a very slow process, whereas the activation of Lys-plasminogen proceeds at a considerably higher rate. These results indicate that the release of the activation peptides converts the plasminogen into a form which is easier to activate.

Referring to the two-step model for activation of plasminogen (Fig. 2), this behavior of the two types of plasminogen indicate that the first step is rate-limiting. Since Glu-plasminogen is easily degraded to Lys-plasminogen by plasmin (Claeys, Molla and Verstraete 1973b), a continuous increase in the rate of activation should be expected. Such a course has also been observed

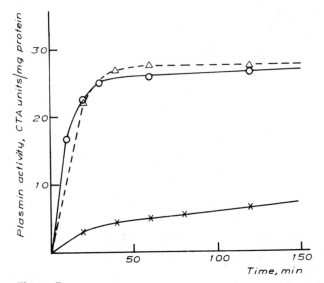

Figure 5
Generation of plasmin during activation of Glu-plasminogen (x——x) and Lys-plasminogen (o——o) with urokinase.
The incubation was performed at 25°C. The reaction mixtures contained 0.5 mg plasminogen, 1250 Ploug units
urokinase and 0.25 ml glycerol per ml. The effect of lysine
(0.01 M) on the activation of Glu-plasminogen is also
shown (△– – –△).

in activation systems free from glycerol (Claeys and Vermylen 1974). The
almost linear course shown in Figure 5 is probably due to the protective effect
of glycerol against plasmic digestion.

As discussed in the previous section, the ω-amino acids, lysine and 6-
aminohexanoic acid induce conformational changes in Glu-plasminogen very
similar to those observed on limited proteolysis in the amino-terminal part of
the molecule. These amino acids also have striking effects on the activation
of Glu-plasminogen, which is in accordance with their effect on the conformation. As shown in Figure 5, 10^{-2} M lysine has a pronounced stimulating effect
on the activation of Glu-plasminogen. In fact, the behavior of Glu-plasminogen becomes indistinguishable from that of Lys-plasminogen. As shown in
Figure 6, 6-aminohexanoic acid has a stimulative effect on the activation of
Glu-plasminogen in concentrations below about 2.5×10^{-3} M, whereas increasing inhibition is observed on increasing the concentration over this value.
The effect of 6-aminohexanoic acid on Lys-plasminogen is different (Fig. 6).
No stimulative effect is observed in any concentration range; instead, an increasing inhibition is observed over the whole range as the concentration of
6-aminohexanoic acid is raised. Similar effects of 6-aminohexanoic acid have
also been reported by Claeys and Vermylen (1974). These results indicate
that the essential event in the first step of the activation process is a dissociation of noncovalent bonds connecting the amino-terminal part of plasminogen with the rest of the molecule. In accordance with this theory, an inhibitory

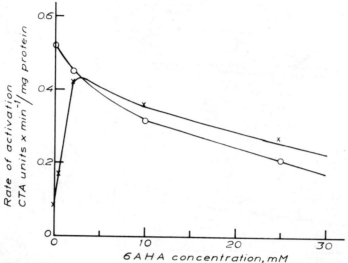

Figure 6
Effect of the concentration of 6-aminohexanoic acid (6-AHA)
on the activation rates of Glu-plasminogen (x——x) and Lys-
plasminogen (o——o). The composition of the reaction mixtures
was the same as given in the legend to Figure 5.

effect on the activation of Lys-plasminogen has been observed when an excess
of the amino-terminal activation peptide (API) is added to the activation
mixture (Wallén, unpubl.).

The effect of urea on the activation of plasminogen is demonstrated in
Figure 7. The activation of Lys-plasminogen is strongly inhibited, even at
rather low urea concentrations, showing an effect at the second step of the
activation process. Nevertheless, the presence of urea has a small but significant stimulative effect on the activation of Glu-plasminogen. This may be due
to dissociation of the noncovalent interaction discussed above.

Affinity Chromatography of Amino-terminal Activation Peptide

The concept that a specific noncovalent interaction participates in the attachment of the activation peptides to plasminogen prompted us to study this
interaction by affinity chromatography in order to localize and define one of
the specific sites involved in this interaction (Wiman and Wallén, unpubl.).
The amino-terminal activation peptide (API) was chosen for this purpose. It
was isolated by previously published methods (Wiman and Wallén 1973).
Lys-plasminogen was covalently coupled to Sepharose using the bromocyan
technique of Axén, Porath and Ernback (1967).

About 10 mg plasminogen was bound per gram wet Sepharose derivative.
At maximal activation with urokinase, approximately 30% of the insolubilized proenzyme was converted to plasmin, as judged from the esterolytic
activity using tosyl-arginine-methyl ester as substrate. A solution of the activation peptide API was perfused through a column, packed with 10 g of insolu-

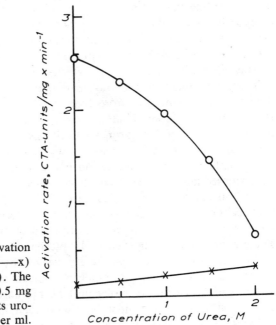

Figure 7
Influence of urea on the activation rates of Glu-plasminogen (x——x) and Lys-plasminogen (o——o). The reaction mixtures contained 0.5 mg plasminogen, 1000 Ploug units urokinase and 0.25 ml glycerol per ml.

bilized plasminogen (containing about 30 mg [0.3 μmole] activable plasminogen), and equilibrated with 0.02 M sodium phosphate buffer pH 7.5. About 0.27 μmole of the activation peptide was adsorbed to the Sepharose derivative and could be eluted with the equilibration buffer containing 6-aminohexanoic acid to a concentration of 0.005 M.

The fact that 6-aminohexanoic acid in this small concentration dissociates the complex between the plasminogen and the peptide indicated that the specific interaction we were looking for resides in the activation peptide API. The same chromatographic procedure was repeated on a tryptic digest of the peptide. It is known from earlier studies on the structure of the amino-terminal activation peptide API that five fragments, called Try-1–Try-5, are formed by tryptic cleavage. Their complete primary structure is known, and they can be distinctly separated by peptide mapping (Wiman 1973). The fraction passing the column unadsorbed contained all tryptic fragments in good yield except one, Try-4, which occurred only in trace amounts. The fraction eluted with buffer containing 6-aminohexanoic acid contained only one peptide, which according to peptide mapping and amino acid analysis was identical with the fragment Try-4. This fragment is a heptapeptide with the sequence Ala-Phe-Gln-Tyr-His-Ser-Lys, which in the activation peptide as well as in the intact plasminogen molecule occupies the positions numbered 45–51.

CONCLUSIONS

Investigations performed at several laboratories during the last few years have shown that activation of human plasminogen is associated with proteolysis at two different sites. One site is situated in the amino-terminal part of the polypeptide chain of plasminogen and consists of a few susceptible peptide bonds. Cleavage in this part of the molecule results in the release of peptides and the generation of a modified plasminogen, Lys-plasminogen. The other site is a single Arg-Val bond situated in the carboxyl-terminal half of the polypeptide chain, the cleavage of which generates active plasmin.

Structural studies on the fragments obtained on rapid activation of plasminogen indicate that the proteolytic attack at these sites occurs in sequence rather than at random. Thus a two-step mechanism has been suggested for the activation of plasminogen by urokinase. The elucidation of the amino acid sequence of 81 amino acids in the amino-terminal part of plasminogen has imparted valuable information concerning the molecular events during the proposed initial phase of activation, since the proteolytic reactions during this phase are confined to this part of the molecule. By cleavages of the peptide bonds Lys-63/Ser-64 and Arg-68/Met-69, two activation peptides are released, which results in the formation of an intermediate plasminogen. These bonds may also be cleaved by plasmin, but at least one of them, Arg-68/Met-69, seems to be specifically cleaved by urokinase, as judged from studies on maleinylated plasminogen, indicating a specific function of the activator in this part of the molecule. This proteolytic reaction seems to induce a dissociation of noncovalent bonds between the amino-terminal part (at residues 45–51) and a site somewhere else in the plasminogen molecule. A conformational change occurs which makes the second site, a single Arg-Val bond, accessible for attack by the activator. The cleavage of this Arg-Val bond is also associated with conformational changes leading to the formation of active plasmin.

Comparative kinetic studies on the activation of the different plasminogen forms indicate that the first step in the activation process is much slower than the second step. Lysine and 6-aminohexanoic acid, which induce conformational changes in plasminogen similar to those observed during activation, enhance the activation considerably. These amino acids, in low concentrations, have the capability of dissociating the noncovalent interaction discussed above. These studies therefore indicate that the first step is rate-limiting, suggesting a regulatory function.

Acknowledgments

This investigation has been financially supported by the Swedish Medical Research Council (project no. 13X3906) and the Magnus Bergvalls Foundation. Skillful technical assistance by Mr. Nils Bergsdorf, Mr. Ove Schedin and Mrs. Rut White is gratefully acknowledged.

REFERENCES

Abiko, Y., M. Iwamoto and M. Tomikawa. 1969. Plasminogen-plasmin system. V. A stoichiometric equilibrium complex of plasminogen and a synthetic inhibitor. *Biochim. Biophys. Acta* **185**:424.

Alkjaersig, N. 1964. The purification and properties of human plasminogen. *Biochem. J.* **93**:171.

Axén, R., J. Porath and S. Ernback. 1967. Chemical coupling of peptides and proteins to polysaccharides by means of cyanogen halides. *Nature* **214**:1302.

Bergström, K. and P. Wallén. 1963. Purification and properties of plasminogen. In *Proceedings of the 9th Congress of the European Society of Haematology* (Lisbon), p. 1325. S. Karger, Basel.

Brockway, W. J. and F. J. Castellino. 1971. The mechanism of the inhibition of plasmin activity by ε-aminocaproic acid. *J. Biol. Chem.* **246**:4641.

————. 1972. Measurement of the binding of antifibrinolytic amino acids to various plasminogens. *Arch. Biochem. Biophys.* **151**:194.

Claeys, H. and J. Vermylen. 1974. Physico-chemical and proenzyme properties of NH₂-terminal glutamic acid and NH₂-terminal lysine human plasminogen. Influence of 6-aminohexanoic acid. *Biochim. Biophys. Acta* **342**:351.

Claeys, H., A. Molla and M. Verstraete. 1973a. Digestion of human plasminogen by human plasmin. In *Proceedings of the 4th International Congress on Thrombosis and Haemostasis* (Vienna), p. 187. G. Gistel & Cie, Vienna.

————. 1973b. Conversion of NH₂-terminal glutamic acid to NH₂-terminal lysine human plasminogen by plasmin. *Thromb. Res.* **3**:515.

Rickli, E. E. and W. I. Otavsky. 1973. Release of an N-terminal peptide from human plasminogen during activation with urokinase. *Biochim. Biophys. Acta* **295**:381.

Robbins, K. C., P. Bernabe, L. Arzadon and L. Summaria. 1972. The primary structure of human plasminogen. I. The NH₂-terminal sequences of human plasminogen and the S-carboxymethyl heavy (A) and light (B) chain derivatives of plasmin. *J. Biol. Chem.* **247**:6757.

Robbins, K. C., L. Summaria, B. Hsieh and R. J. Shah. 1967. The peptide chains of human plasmin. Mechanism of activation of human plasminogen to plasmin. *J. Biol. Chem.* **242**:2333.

Sjöholm, I., B. Wiman and P. Wallén. 1973. Studies on the conformational changes of plasminogen induced during activation to plasmin and by 6-aminohexanoic acid. *Eur. J. Biochem.* **39**:471.

Wallén, P. and B. Wiman. 1970. Characterization of human plasminogen. I. On the relationship between different molecular forms of plasminogen demonstrated in plasma and found in purified preparations. *Biochim. Biophys. Acta* **221**:20.

————. 1972. Characterization of human plasminogen. II. Separation and partial characterization of different molecular forms of human plasminogen. *Biochim. Biophys. Acta* **257**:122.

————. 1973. On the formation and properties of an intermediate form of human plasminogen generated during activation with urokinase. *Proceedings of the 4th International Congress on Thrombosis and Haemostasis* (Vienna), p. 184. G. Gistel & Cie, Vienna.

Walther, P. J., H. M. Steinman, R. L. Hill and P. A. McKee. 1974. Activation of human plasminogen by urokinase. Partial characterization of a pre-activation peptide. *J. Biol. Chem.* **249**:1173.

Wiman, B. 1972. On the structure of the N-terminal fragment of human plasminogen obtained after cleavage with cyanogen bromide. *Thromb. Res.* **1**:89.

————. 1973. Primary structure of peptides released during activation of human plasminogen by urokinase. *Eur. J. Biochem.* **39**:1.

Wiman, B. and P. Wallén. 1973. Activation of human plasminogen by an insoluble derivative of urokinase. Structural changes of plasminogen in the course of activation to plasmin and demonstration of a possible intermediate compound. *Eur. J. Biochem.* **36**:25.

Activation of Plasminogen

Kenneth C. Robbins and Louis Summaria

Michael Reese Research Foundation, Chicago, Illinois 60616;
Department of Medicine, Pritzker School of Medicine
University of Chicago, Chicago, Illinois 60637

Grant H. Barlow

Department of Molecular Biology, Abbott Laboratories
North Chicago, Illinois 60637

One of the critical events that occurs during the activation of human plasminogen to plasmin by catalytic amounts of either urokinase or streptokinase involves the cleavage of an arginyl-valine peptide bond in the carboxyl-terminal portion of the zymogen to give a two-chain molecule (Robbins et al. 1967, 1973). During the activation of Glu-plasminogen, a second major bond cleavage occurs in the amino-terminal portion of the zymogen, which involves the release of a peptide or peptides (Robbins et al. 1973). There is considerable disagreement as to where the peptide bond(s) is cleaved in the amino-terminal segment (Robbins et al. 1973; Wiman and Wallén 1973) and whether the cleavage of the amino-terminal peptide bond is essential for the formation of the enzyme. Also, the question of a preactivation peptide released from the amino-terminal segment of the zymogen (Walther et al. 1974) has not been resolved. We have found that a single peptide bond is cleaved by urokinase, namely, the arginyl-valine bond in the carboxyl-terminal portion of the molecule, when the activation of Glu-plasminogen is carried out in the presence of the plasmin inhibitor Trasylol (Summaria et al. 1975). It appears that the cleavage of peptide bonds at the amino-terminal portion of the zymogen is probably due to plasmin itself.

In order to study the mechanism of activation of human plasminogen to plasmin, the question of the relationship between the Glu- and Lys-plasminogen forms needed to be resolved (Summaria et al. 1973a; Wiman and Wallén 1973; Robbins et al. 1975). The activation of the Glu- and Lys-plasminogen forms by urokinase resulted in the same Lys-plasmin (Robbins et al. 1975), indicating similar mechanisms of activation. The Glu-plasminogen forms appear to be single polypeptide chains, whereas the Lys-plasminogen forms probably arise from the Glu-forms due to cleavage by either plasmin or other plasma serine proteases or perhaps a contaminating plasma plasminogen activator (Robbins et al. 1975). The molecular weights of both the Glu-plasminogen and the Lys-plasminogen forms were found to be similar, with perhaps the Glu- forms being larger in size by 1000–2000 daltons (Robbins et al. 1975). This possible difference in molecular weight between the Glu- and Lys- forms may be explained by the loss of a small peptide from the amino-terminal end of Glu-plasminogen. Also, additional plasmic clips probably occur at X⤓Lys-Val-Tyr-Leu-, the amino-terminal sequence of the heavy (A) chain, and at Arg or Lys⤓X peptide bonds amino-terminal to this sequence without the release of any peptides. The Glu- and Lys-plasminogen isoelectric forms differed markedly in their isoelectric point ranges, pI's 6.2–6.6 and 7.2–8.3, respectively (Summaria et al. 1972; Summaria et al. 1973a). Their sedimentation coefficients, partial specific volumes and calculated frictional coefficients were found to be different, indicating molecules with different conformations (Robbins et al. 1975).

Comparison of Human Glu- and Lys-plasminogen Forms

Comparative physical and chemical data on the human Glu- and Lys-plasminogen forms are summarized in Table 1. The molecular weights of Glu-plasminogen and Lys-plasminogen were determined by sedimentation equilibrium methods to be 83,300 and 82,400, respectively. The sedimentation coefficients, the partial specific volumes and the calculated frictional coefficients were different, indicating molecules with different conformations. The isoelectric points of the Glu- forms ranged between pI 6.2 and 6.6 (4 major forms), whereas the isoelectric points of the Lys- forms ranged between pI 7.2 and 8.3 (5 major forms). Although the amino acid compositions of the Glu- and Lys- forms were similar, several amino acid residues (glutamic acid, alanine, isoleucine, phenylalanine and lysine) were found to be significantly higher in the Glu-plasminogen forms.

Activation of Human Plasminogen by Urokinase

Both the Glu- and Lys-plasminogen forms can be activated with catalytic amounts of urokinase (plasminogen:urokinase molar ratio of 1000:1) to give Lys-plasmin (Robbins et al. 1973). In the same activation system, the Lys- forms are activated more rapidly than the Glu- forms, but both forms

Table 1
Human Plasminogen

	Glu- forms	Lys- forms
Source	serum, Fr. III	Fr. III$_{2,3}$
Amino-terminal residue	Glu	Lys
Carboxyl-terminal residue	Asn	Asn
Isoelectric point (pI)	6.2–6.6(4 forms)	7.2–8.3(5 forms)
Sedimentation coefficient ($s^{\circ}_{20,w}$)	5.0S	4.4S
Partial specific volume ($\bar{v}$)	0.709	0.714
Molecular weight	83,300	82,400
Frictional coefficient (f/f_o)	1.45	1.63
Amino acid composition	— similar	—

Data from Robbins et al. (1967, 1973, 1975) and Summaria et al. (1973a).

Table 2
Human Plasmin and Derived Chains

Property	Plasmin	Heavy (A) chain	Light (B) chain
Amino-terminal residue	Lys, Val	Lys	Val
Carboxyl-terminal residue	Arg, Asn	Arg	Asn
Isoelectric point (pI)	7.4–8.5(5 forms)	4.9(1 form)	5.8–6.0(3 forms)
Sedimentation coefficient ($s^\circ_{20,w}$)	4.3S	2.8S	1.4S
Molecular weight	76,500	48,000	25,700
Active site (serine, histidine)	+	–	+

Data from Robbins et al. (1967, 1972, 1973), Summaria et al. (1967) and Summaria, Robbins and Barlow (1971b).

give the same Lys-plasmin with the same physical and chemical properties (Robbins et al. 1975). The molecular weight of Lys-plasmin was determined by sedimentation equilibrium methods to be 76,500, and its sedimentation co-efficient was determined to be 4.3S (Table 2). The physical and chemical properties of the plasmin-derived heavy (A) and light (B) chains are summarized in Table 2.

Plasminogens prepared from human, cat, rabbit and bovine plasmas (Summaria et al. 1973b) showed homologous amino-terminal sequences (Table 3). Plasmins can be prepared from these mammalian plasminogens, and in addition from dog plasminogen, by activation with urokinase (Summaria et al. 1973b). The isolated plasmin-derived heavy (A) and light (B) chains from these enzymes showed homologous, and nearly identical, amino-terminal sequences (Table 3).

Activation of Human Plasminogen by Urokinase in the Presence of Trasylol

Human plasminogen (Glu- and Lys- forms) can be completely activated by catalytic amounts of urokinase in the presence of Trasylol, a plasmin inhibitor (Summaria et al. 1975). The plasmin-Trasylol complexes were enzymatically active only after dissociation in either 6 M guanidine-HCl or 8 M urea. The amino-terminal sequence of the isolated plasmin-derived heavy (A) chain from Glu-plasminogen was identical to that found in the zymogen; the amino-terminal sequence of the light (B) chain was the same as that found in plasmin prepared in activation mixtures not containing Trasylol. The same results were obtained with the activation of cat, dog and rabbit plasminogens in the presence of the inhibitor. All of the isolated mammalian heavy (A) chains had either the Glu- or Asp- (or blocked) amino-terminal sequences; all of the isolated mammalian light (B) chains had the Val- amino-terminal sequences. The amino-terminal sequence of human Lys-plasminogen was the same as the sequence found for the Lys- heavy (A) chain.

Activation of Human Plasminogen by Streptokinase

Human plasminogen (Glu- and Lys- forms) can be activated by both catalytic amounts of streptokinase, and in the equimolar plasminogen-streptokinase

Table 3

Amino-terminal 5 Residue Sequences

Species	Plasminogen	Plasmin heavy (A) chain	Plasmin light (B) chain
Human	Glu-Pro-Leu-Asp-Asp-	Lys-Val-Tyr-Leu-Leu-	Val-Val-Gly-Gly-Cys-
Cat	Asp-Pro-Leu-Asp-Asp-	Lys-Ile-Tyr-Leu-Val-	Val-Val-Gly-Gly-Cys-
Dog	blocked	Arg-Ile-Tyr-Leu-Gly-	Val-Val-Gly-Gly-Cys-
Rabbit	Glu-Pro-Leu-Asp-Asp-	Lys-Val-Tyr-Leu-Gly-	Val-Val-Gly-Gly-Cys-
Bovine	Asp-Leu-Leu-Asp-Asp-	Lys-Ile-Tyr-Leu-Val-	Ile-Val-Gly-Gly-Cys-

Data from Robbins et al. (1972, 1973).

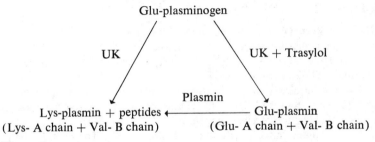

Figure 1
Mechanism of activation of human plasminogen by urokinase.

complex, to give plasmin (Robbins et al. 1967) and the plasmin-streptokinase complex (Summaria, Robbins and Barlow 1971a; Summaria et al. 1974). Although the plasmin-derived light (B) chains appear to be identical, the heavy (A) chains were found to be different. In the equimolar complex prepared from Glu-plasminogen, the heavy (A) chain was found to be larger than the normally produced Lys- heavy (A) chain and similiar in size to the Glu- heavy (A) chain found in the plasmin-Trasylol complex. Similar results were obtained with the equimolar rabbit plasminogen-streptokinase complex, where the heavy (A) chain was found to be similar to the Glu- heavy (A) chain found in the rabbit plasmin-Trasylol complex.

Mechanism of Activation of Human Plasminogen

The mechanism of activation of human Glu-plasminogen to plasmin by urokinase involves two specific bond cleavages: (1) an arginyl-valine peptide bond in the carboxyl-terminal portion of the zymogen is cleaved by urokinase to give a two-chain plasmin molecule containing a Glu- heavy (A) chain and a Val- light (B) chain; and (2) an X-lysine peptide bond in the amino-terminal portion of the Glu- heavy (A) chain is cleaved by plasmin to give a Lys- heavy (A) chain and some peptides (see schematic of activation mechanism in Fig. 1). In the presence of the plasmin inhibitor Trasylol, only one peptide bond is cleaved to give the two-chain molecule with a Glu- heavy (A) chain. Urokinase will not cleave any peptide bonds in the amino-terminal portion of plasminogen when Trasylol is present in the activation mixture. Therefore, the formation of Lys-plasmin from Glu-plasmin is probably due to plasmic cleavages in the amino-terminal portion of the Glu- heavy (A) chain. The mechanism of activation of human plasminogen by streptokinase is similar, and in the equimolar human plasminogen-streptokinase complex, the streptokinase probably blocks the plasmin in the complex from cleaving peptide bonds in the amino-terminal end of the Glu- heavy (A) chain. This proposed mechanism of activation of plasminogen is probably a general mechanism for mammalian plasminogen activation by all activators.

Acknowledgment

This investigation was supported in part by United States Public Health Service Research Grant HL04366 from the National Heart and Lung Institute.

REFERENCES

Robbins, K. C., P. Bernabe, L. Arzadon and L. Summaria. 1972. The primary structure of human plasminogen. I. The amino-terminal sequences of human plasminogen and the S-carboxymethyl heavy (A) and light (B) chain derivatives of plasmin. *J. Biol. Chem.* **247**:6757.

———. 1973. NH_2-Terminal sequences of animal plasminogens and plasmin S-carboxymethyl heavy (A) and light (B) chain derivatives: A re-evaluation of the mechanism of activation of plasminogen. *J. Biol. Chem.* **248**:7242.

Robbins, K. C., L. Summaria, B. Hsieh and R. J. Shah. 1967. The peptide chains of human plasmin: Mechanisms of activation of human plasminogen to plasmin. *J. Biol. Chem.* **242**:2333.

Robbins, K. C., I. G. Boreisha, L. Arzadon, L. Summaria and G. H. Barlow. 1975. Physical and chemical properties of the NH_2-terminal glutamic acid and lysine forms of human plasminogen and their derived plasmins with an NH_2-terminal lysine heavy (A) chain: Comparative data. *J. Biol. Chem.* (in press).

Summaria, L., K. C. Robbins and G. H. Barlow. 1971a. Dissociation of the equimolar human plasmin-streptokinase complex. Partial characterization of the isolated plasmin and streptokinase moieties. *J. Biol. Chem.* **246**:2136.

———. 1971b. Isolation and characterization of the S-carboxymethyl heavy chain derivative of human plasmin. *J. Biol. Chem.* **246**:2143.

Summaria, L., L. Arzadon, P. Bernabe and K. C. Robbins. 1972. Studies on the isolation of the multiple molecular forms of human plasminogen and plasmin by isoelectric focusing methods. *J. Biol. Chem.* **247**:4691.

———. 1973b. Isolation, characterization and comparison of the isolated S-carboxymethyl heavy (A) and light (B) chain derivatives of cat, dog, rabbit and bovine plasmins. *J. Biol. Chem.* **248**:6522.

———. 1974. The interaction of streptokinase with human, cat, dog and rabbit plasminogens. The fragmentation of streptokinase in the equimolar plasminogen-streptokinase complexes. *J. Biol. Chem.* **249**:4760.

———. 1975. The activation of plasminogen to plasmin by urokinase in the presence of the plasmin inhibitor Trasylol: The preparation of plasmin with the same NH_2-terminal heavy (A) chain sequence as the parent zymogen. *J. Biol. Chem.* (in press).

Summaria, L., L. Arzadon, P. Bernabe, K. C. Robbins and G. H. Barlow. 1973a. Characterization of the NH_2-terminal glutamic acid and NH_2-terminal lysine forms of human plasminogen isolated by affinity chromatography and isoelectric focusing methods. *J. Biol. Chem.* **248**:2984.

Summaria, L., B. Hsieh, W. R. Groskopf, K. C. Robbins and G. Barlow. 1967. The isolation and characterization of the S-carboxymethyl β (light) chain derivative of human plasmin. The localization of the active site on the β (light) chain. *J. Biol. Chem.* **242**:5046.

Walther, P. J., H. M. Steinman, R. L. Hill and P. A. McKee. 1974. Activation of human plasminogen by urokinase. Partial characterization of a pre-activation peptide. *J. Biol. Chem.* **249**:1173.

Wiman, B. and P. Wallén. 1973. Activation of human plasminogen by an insoluble derivative of urokinase. Structural changes of plasminogen in the course of activation to plasmin and demonstration of a possible intermediate compound. *Eur. J. Biochem.* **36**:25.

A Comparison of Mechanisms of Activation of Rabbit Plasminogen Isozymes by Urokinase

James M. Sodetz and Francis J. Castellino

Department of Chemistry, Program in Biochemistry and Biophysics
The University of Notre Dame, Notre Dame, Indiana 46556

Plasminogen is the inactive plasma protein precursor of the proteolytic enzyme plasmin. Although plasminogen is a single-chain molecule highly linked by disulfide bonds, plasmin is a two-chain enzyme consisting of a single inter-chain disulfide bond (Robbins et al. 1967; Summaria, Hsieh and Robbins 1967; Sodetz, Brockway and Castellino 1972). The enzyme plasmin contains a heavy chain derived from the amino terminus of the original plasminogen and a light chain derived from the carboxyl terminus of plasminogen (Robbins et al. 1967). The activation of plasminogen to plasmin is accomplished by proteases such as urokinase and diverse tissue activators and involves at least two proteolytic cleavages in plasminogen (Robbins et al. 1973). One bond cleaved results in the evolution of a small peptide of molecular weight 6500–8000 from the amino terminus of plasminogen, and a second bond cleaved in the interior of the molecule results in the characteristic two-chain plasmin molecule. It was proposed, using data derived from the activation of human plasminogen in the presence of proteolytic inhibitors such as p-nitro-phenyl-guanidinobenzoate (NphBzoGdn) and tosyl-L-lysine-chloromethyl ketone (TLCK), that the small peptide was cleaved from plasminogen by urokinase prior to the urokinase-induced cleavage of the internal bond (Wiman and Wallén 1973; Wiman 1973; Walther et al. 1974; Rickli and Otavsky 1973).

Our laboratory has shown that two highly purified forms of plasminogen can be obtained from several species by gradient elution from Sepharose 4B-L-lysine solid supports (Brockway and Castellino 1972). Studies on rabbit plasminogen have shown these two forms to have similar molecular weights and identical amino-terminal sequences (Sodetz, Brockway and Castellino 1972); Castellino et al. 1973). Further, these two forms are independently synthesized and metabolized and are not interconverted (Siefring and Castellino 1974). However, these forms differ in their carbohydrate compositions (Castellino et al. 1973) and consist of subforms which differ in their charge characteristics (Sodetz, Brockway and Castellino 1972).

In light of these structural differences in the circulating plasminogen, we felt it necessary to compare the mechanism of activation of the two rabbit plasminogen affinity chromatography forms to each other and to the mechanism proposed for human plasminogen in order to determine whether a universal mechanism of plasminogen activation exists. This report summarizes our findings on this important topic.

MATERIALS AND METHODS

Proteins

Rabbit plasminogens were prepared from fresh plasma as described previously (Sodetz, Brockway and Castellino 1972). Human plasminogen was prepared from outdated plasma and consisted of a mixture of the two affinity chromatography forms. Urokinase was purchased from Calbiochem. Bovine pancreatic trypsin inhibitor (BTI) was prepared as described by Kassell (1970) using Sepharose 4B-trypsin matrices. The functional purity of this inhibitor was determined by titration with trypsin. The solid-state lactoperoxidase procedure of David (1972) was used to prepare [^{125}I]BTI. Urokinase-free plasmin was prepared utilizing insolubilized urokinase to activate plasminogen. The methods are described elsewhere (Sodetz and Castellino 1975).

Plasminogen Activation in Presence and Absence of BTI

Stock solutions of rabbit or human plasminogen (9.8 mg/ml), BTI (3.7 mg/ml) and urokinase (4420 Plough U/ml) were prepared in 0.05 M Tris · HCl–0.1 M L-lysine pH 8.0. Incubations were performed at 22°C and consisted of mixing 0.020 ml of plasminogen, 0.005 ml buffer and 0.010 ml BTI followed by 0.010 ml urokinase. At appropriate times, aliquots were quenched in SDS-mercaptoethanol 6 M urea buffer and subjected to SDS gel electrophoresis on 5% polyacrylamide gels (Weber and Osborn 1969). Similar experiments consisted of substituting buffer for BTI. Here, the stock urokinase solution used was 2800 Plough units/ml.

Purification and Characterization of Component Polypeptide Chains of Plasmin

Plasminogen b^1 was prepared by digestion of plasminogen a with low levels of urokinase-free plasmin. The component chains of plasmin were isolated by gel filtration following reduction and carboxymethylation as described elsewhere (Sodetz et al. 1974; Sodetz and Castellino 1975). Methods used for amino acid compositions (Sodetz, Brockway and Castellino 1972) and aminoterminal amino acid sequences (Castellino et al. 1973) are as previously described. Sedimentation equilibrium molecular weights were determined in 0.1 M Tris · HCl pH 7.8 for plasminogens and 6 M guanidine hydrochloride for plasmin component chains (Sodetz and Castellino 1974).

Binding of [^{125}I]BTI

Binding of BTI to plasmin a was performed by activating plasminogen a in the presence of [^{125}I]BTI as described above. The complex was isolated by

---

[1] Plasminogen a and plasminogen b refer to plasminogens with amino-terminal glutamic acid or methionine, respectively; Ha and Hb refer to plasmin heavy chains containing an amino-terminal glutamic acid or methionine, respectively; plasmin a and plasmin b refer to plasmins containing either Ha or Hb heavy chain; L refers to plasmin light chain and P refers to the amino-terminal peptide of plasminogen a.

Sepharose 4B-L-lysine affinity chromatography at 4°C using 0.1 M phosphate–0.1 M 6-aminohexanoic acid pH 8.0 as the eluent. The specific radioactivity of the complex was then measured. Binding to plasmin *b* was performed by adding [125I]BTI after activation of plasminogen by urokinase in the absence of this inhibitor. Diisopropylfluorophosphate (DFP)-treated plasmin *b* and plasminogen *a* were incubated with [125I]BTI in the absence of urokinase and binding was measured as above.

Digestion of Plasmin a *with Plasmin and Urokinase*

A stock solution of plasmin *a* · BTI complex isolated as described above was prepared at 1.35 mg/ml in 0.1 M phosphate–0.001 M 6-aminohexanoic acid pH 8.0. A quantity of 0.090 ml of the complex stock solution was mixed with 0.010 ml of 0.05 M Tris · HCl–0.5 M L-lysine pH 8.0 and incubated at 22°C with either 0.005 ml of urokinase-free plasmin (1.97 mg active plasmin/ml) or 0.010 ml of urokinase (11,050 Plough U/ml), both in 0.05 M Tris · HCl–0.05 M L-lysine pH 8.0. Parallel experiments in the presence and absence of BTI and substituting plasminogen *a* for the plasmin *a* · BTI complex at the same initial concentration were performed in order to determine the effectiveness of the urokinase levels used.

RESULTS

The electrophoretic properties of the two rabbit plasminogen *a* affinity chromatography forms in various polyacrylamide gel systems are presented in Figure 1. Both forms exhibit identical mobilities on SDS gels, but differ in their electrophoretic properties on analytical gel systems. In addition, each

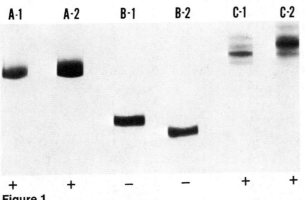

Figure 1
Characterization of the purified affinity chromatography forms of rabbit plasminogen *a* on various polyacrylamide gel systems. Affinity form one and affinity form two plasminogen are designated by 1 and 2, respectively. A's refer to SDS-mercaptoethanol gels; B's refer to pH 4.3 analytical gels; C's refer to pH 9.5 analytical gels. The polarity of the gel is shown at the bottom. Analytical gels were prepared as described earlier (Sodetz, Brockway and Castellino 1972).

affinity chromatography form contains the same amino-terminal amino acid sequence of NH_2-Glu-Pro-Leu-Asp-Asp, as reported earlier (Castellino et al. 1973). Upon activation with urokinase, as described in Methods, both plasminogen a forms are converted to the two-chain plasmin molecule (plasmin b). Plasmin obtained in this manner contains 75–80% active sites as determined by titration with NphBzoGdn according to Chase and Shaw (1970). Gel filtration of the reduced and carboxymethylated plasmin b results in highly purified component plasmin peptide chains (Sodetz et al. 1974). Complete amino acid compositions of the heavy (Hb), light (L) and peptide (P) chains derived from each affinity form of plasminogen a are shown in Table 1. The amino-terminal amino acid sequences of the purified Hb and L chains of each plasmin b are shown in Table 2. Comparison of the component chain sequences with that of plasminogen a indicates that peptide material is released from the amino terminus upon activation of each plasminogen a affinity chromatography form. Activation of human plasminogen a by urokinase, as described in Methods with subsequent purification, produces a heavy chain of molecular weight 62,000, containing an amino-terminal amino acid sequence of NH_2-Lys-Val-Tyr-Leu-X-Glu. This is similar to the sequence reported earlier by Robbins et al. (1972).

Table 1

Amino Acid Compositions of the Component Chains of Plasmin Derived from Rabbit Plasminogen a

Amino acid	Hb		L		P	
	1[a]	2	1	2	1	2
Asp	60	61	21	19	6	6
Thr[b]	39	39	14	14	5	4
Ser[b]	41	37	17	16	6	5
Glu	54	53	22	20	7	8
Pro	59	60	12	13	6	4
Gly	35	37	22	22	10	6
Ala	23	24	16	16	4	5
Val[c]	11	13	18	19	3	5
Met	8	8	1	2	1	1
Ileu[c]	11	12	10	10	2	3
Leu[c]	18	19	20	19	4	5
Tyr	26	27	8	8	2	2
Phe	9	11	4	5	2	2
His	13	13	5	4	2	2
Lys	32	31	10	10	3	4
Arg	34	33	12	12	3	3
Trp[d]	18	18	7	7	0	0
Cys[d]	36	35	9	10	2	2

[a] 1 and 2 represent affinity chromatography form 1 and form 2, respectively. Values are expressed as number of residues/molecule ($\pm 5\%$).

[b] Values obtained by extrapolation to zero hydrolysis time.

[c] Values taken from 72-hour hydrolysis time.

[d] Determined as described in Methods.

Table 2

Amino-terminal Amino Acid Sequences of the Component Chains of
Plasmin Derived from Rabbit Plasminogen *a*

		Amino acid derived from		
Sequence number	plasminogen a[a]	Hb[a]	L[a]	P[b]
1	Glu	Met	Val	Glu
2	Pro	Tyr	Val	Pro
3	Leu	Leu	Gly	Leu
4	Asp	X	Gly	Asp
5	Asp	Glu	X	Asp
6	Tyr	X	Val	Tyr
7	Val	X	Ala	Val
8	Asn	Ileu	X	X

[a] Values reported were found to be the same for both affinity chromatography forms.
[b] Data from Walther et al. (1974).

Table 3 contains the molecular weights of each plasminogen form, as well as the purified component chains of their respective plasmins. Although the molecular weight difference between the two forms of rabbit plasminogen *a* is within experimental error, we have observed affinity chromatography form 1[2] to be consistently higher than the corresponding form 2. Similar behavior is observed between their respective H*b* chains.

Figure 2 contains an SDS gel profile of the time course of urokinase

Table 3

Molecular Weight Values for Rabbit Plasminogens and
Component Plasmin Chains

	Molecular weight*	
Polypeptide	1	2
Plasminogen *a*	90,000–94,000[a]	88,000–92,000[a]
Plasminogen *b*	82,000–86,000[a]	81,000–85,000[a]
Plasmin *a* heavy chain (H*a*)[c]	66,000–69,000[a,b]	65,000–68,000[a,b]
Plasmin *b* heavy chain (H*b*)[c]	60,000–63,000[a,b]	58,000–61,000[a,b]
Light chain (L)[c]	24,000–26,000[b]	24,000–26,000[b]
Peptide (P)[c]	6000–8000[b]	6000–8000[b]
Human plasmin *b* heavy chain (Hb)[c]	60,000–63,000[a]	

* 1 and 2 refer to affinity chromatography forms 1 and 2, respectively.
[a] Determined by sedimentation equilibrium.
[b] Determined by SDS gel electrophoresis using appropriate standards.
[c] The amount of acetate incorporated as a result of carboxymethylation has been subtracted.

[2] Affinity chromatography form 1 refers to the first plasminogen eluted from the Sepharose 4B-L-lysine affinity chromatography columns. Correspondingly, plasminogen form 2 refers to the second plasminogen eluted from these columns (Brockway and Castellino 1972).

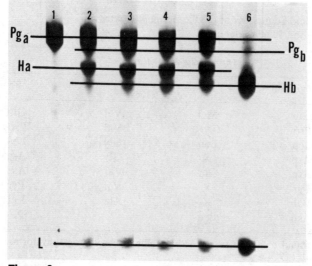

Figure 2
The urokinase activation of affinity chromatography form
1 rabbit plasminogen *a* in the absence of BTI as described
in Methods. The inserts represent migration according to
the molecular weight of plasminogen *a* (Pg *a*), plasminogen *b* (Pg *b*), plasmin *a* heavy chain (H*a*), plasmin *b*
heavy chain (H*b*) and plasmin light chain (L). All gels
are reduced 6 M urea-SDS gels containing 5% polyacrylamide. Gel 1, plasminogen *a*; gel 2, activation for 2
minutes; gel 3, 3 minutes; gel 4, 4 minutes; gel 5, 5
minutes; gel 6, 30 minutes.

activation of affinity form 1 plasminogen. The gel shown at 30 minutes of
activation represents the stable, highly active plasmin *b* molecule containing
the H*b* heavy chain and L light chains characterized above. At early times of
activation, however, a higher molecular weight heavy chain (H*a*) is observed
as a transient component. With time, the H*b* heavy chain appears, as well as
an altered form of plasminogen (plasminogen *b*). After plasminogen depletion, the resulting plasmin *b* contains solely the H*b* chain and characteristic
light (L) chain. Figure 3 shows that affinity form 2 rabbit plasminogen activates in an identical manner.

The plasminogen *b* observed in Figures 2 and 3 was prepared in highly
purified form by gel filtration, as described in Methods. Amino-terminal amino
acid sequence analysis gave NH_2-Met-Tyr-Leu-X-Glu. Activation of plasminogen *b* with urokinase results in only plasmin *b* formation, with no transient
plasmin *a* observable.

In contrast to the proposed human mechanism, the above results strongly
suggested that amino-terminal peptide release need not occur prior to
urokinase cleavage of the internal peptide bond in rabbit plasminogen *a*.
These results were fortified as a result of the investigation of the activation of
rabbit plasminogen *a* in the presence of a 2.5 molar excess of the trypsin and
plasmin inhibitor BTI. This level of BTI instantly and completely inhibits

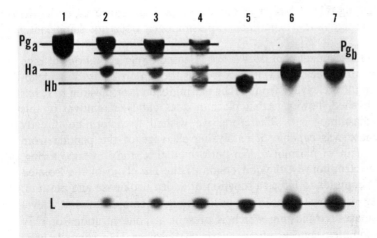

Figure 3
The urokinase activation of affinity chromatography form 2 rabbit plasminogen *a* in the absence and presence of BTI. See Methods for details. Inserts and gel solutions are as described in the legend to Figure 2. Gels 2–5 are activations in the absence of BTI. Gel 1, plasminogen *a;* gel 2, activation for 2 minutes; gel 3, 3 minutes; gel 4, 5 minutes; gel 5, 30 minutes. Gels 6 and 7 are activation in the presence of BTI for 60 minutes, but gel 7 is at a level of urokinase sixfold greater than gel 6. In the presence of BTI, activation was complete before 60 minutes; however, gels at 60 minutes are shown to illustrate H*a* heavy chain stability at this time period.

plasmin. However since urokinase is also slightly inhibited by this agent, higher concentrations of urokinase were employed for the activation in the presence of BTI. Results of these experiments are shown in Figure 3. These results indicate that under conditions where the generated plasmin is completely inhibited by BTI, only plasmin containing H*a* heavy chain is observed. Purification and subsequent amino-terminal amino acid sequence determination of the H*a* chain resulted in a sequence NH$_2$-Glu-Pro-Leu-Asp-Asp. This sequence is identical to the original plasminogen *a* (Sodetz et al. 1974). Further, the molecular weight of the H*a* chain (Table 3) was found to be 6000–8000 daltons higher than the H*b* heavy chain. It can also be observed from Figure 3 that increasing urokinase to a level sixfold greater than used in Methods again results only in plasmin *a* formation. Under these conditions, we have found the H*a* chain to be stable for at least 3 hours.

Since activation in the absence of BTI favors rapid conversion of plasmin *a* to plasmin *b,* we have determined in the only feasible manner that plasmin *a* is an active species. This was done by measuring the extent of incorporation of [^{125}I]BTI into plasmin *a,* as described in Methods. The stoichiometry of [^{125}I]BTI to plasmin *a* (moles/mole) was found to be 0.80. Approximately the same stoichiometry was found for binding of BTI to plasmin *b.* On the other hand, neither plasminogen *a* nor DFP-plasmin *b* significantly bound [^{125}I]BTI under these conditions. We did not investigate whether plasminogen is capable of binding BTI at prolonged incubation periods. However, under the conditions of the activation experiments, plasminogen does not significantly bind BTI.

Since the amino-terminal peptide was not released under conditions of high urokinase in the presence of BTI, it appeared as though urokinase was incapable of catalyzing this cleavage from either plasminogen a or plasmin a. It has been well established, however, that plasmin does catalyze removal of this peptide from plasminogen a. Therefore, we wished to determine directly whether urokinase was capable of catalyzing cleavage of the peptide from the Ha heavy chain of plasmin a. We performed this study by monitoring, with SDS gel electrophoresis, the conversion of the Ha chain of the isolated plasmin $a \cdot$ BTI complex to Hb as a function of added urokinase and plasmin. The results (Fig. 4) show that at low concentrations of complex and at a level of added urokinase-free plasmin which is present at concentrations of only 7–8% of the complex concentration, nearly 50% of the Ha chain is converted to the Hb chain in 30 minutes. At 1 hour of incubation, total conversion of Ha to Hb has occurred. On the other hand, at extremely high levels of

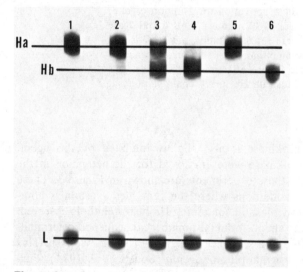

Figure 4
Comparison of the effectiveness of plasmin and urokinase as catalysts for release of the amino-terminal peptide from the *Ha* chain of the rabbit plasmin $a \cdot$ BTI complex. Gel 1, isolated plasmin $a \cdot$ BTI complex at zero time of incubation; gel 2, complex plus added urokinase-free plasmin at 1 minute of incubation (1 plasmin:12 complex); gel 3, as in gel 2 after 1 hour; gel 4, as in gel 2 after 3 hours; gel 5, plasmin $a \cdot$ BTI complex shown in gel 1 treated with urokinase ($\sim$1100 Plough U/ml) for 3 hours; gel 6, plasminogen a at the same initial concentration as the complex in gel 5 treated with urokinase ($\sim$1100 Plough U/ml) for 15 minutes. Inserts are as in Figure 2.

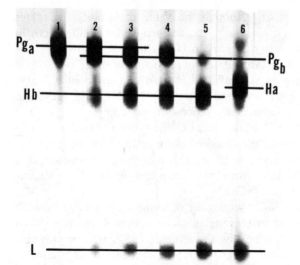

Figure 5
The urokinase activation of human plasminogen *a*
in the absence and presence of BTI as described in
Methods. Gels 2–5 are activations in the absence
of BTI. Gel 1, human plasminogen *a;* gel 2 activation for 1 minute; gel 3, 2 minutes; gel 4, 3 minutes;
gel 5, 30 minutes; gel 6, activation in the presence
of BTI. Although activation was completed earlier,
gel 6 represents 60 minutes of activation. Inserts
and gel solutions are as in Figure 2.

added urokinase, the H*a* heavy chain remains intact for at least 3 hours. The
effectiveness of this urokinase level is also dramatically shown in Figure 4,
where activation of plasminogen *a,* at the same initial concentration as the
plasmin *a* · BTI complex, is complete in only 15 minutes. Similarly, this level
of urokinase activated plasminogen *a* completely to plasmin *a* in the presence
of BTI in 30 minutes. The levels of added plasmin used in these experiments
were purposely kept low to amplify the sensitivity of the H*a* chain to plasminolysis.

Finally, the urokinase activation of human plasminogen in the absence and
presence of BTI is shown in Figure 5. In contrast to the rabbit system, no
transient H*a* heavy chain is observable without added BTI. Furthermore,
conversion of plasminogen *a* to plasminogen *b* occurs much more rapidly in
the human system compared to the rabbit system. However, activation in the
presence of BTI results in only plasmin *a* containing the stable H*a* heavy
chain, as was found for the rabbit system.

DISCUSSION

Our studies on the structural characteristics of plasmin derived from each
affinity chromatography form of rabbit plasminogen indicate that the two
forms are very similar in a number of respects. Molecular weights of the two

plasminogen *a* forms are similar, although the value for affinity form 1 is consistently slightly higher. This is likely due to carbohydrate differences in the two plasminogen forms reported earlier (Castellino et al. 1973). The amino acid compositions were previously found to be similar in the two forms, although as shown in Figure 1, each affinity form is composed of different isoelectric forms (Sodetz, Brockway and Castellino 1972). Data presented in Table 1 show that amino acid compositions of the component plasmin chains in affinity form 1 plasmin are identical to the corresponding chains in affinity form 2. Also, the sum of the individual chain amino acid compositions agrees well with that of plasminogen *a*. Examination of this data indicates that component chains of each affinity form of plasmin are derived from similar regions of the initial plasminogen molecule.

The amino-terminal amino acid sequences of component chains in each plasmin *b* form (Table 2) are interesting in two respects. First, these data represent additional evidence supporting the structural similarity in plasmins derived from each plasminogen *a* form. The H*b* heavy chain amino-terminal sequence was found to be identical for each plasmin *b* form. Similarly, each plasmin *b* had light chain (L) sequences which were identical to each other and to that of human plasmin light chain (Robbins et al. 1972). Second, urokinase activation of each rabbit plasminogen *a* occurs by at least two proteolytic cleavages. As with human plasminogen *a* activation, one cleavage involves the ultimate release of peptide material which is derived from the amino terminus of plasminogen *a*. The other cleavage involves an internal peptide bond and leads to generation of plasmin heavy and light chains. Although the rabbit plasmin *b* heavy chain amino-terminal sequence is slightly different than that of human plasmin *b,* we feel sufficient similarity exists at this time to term them homologous. Presently, we can offer no explanation for the small difference observed between our rabbit plasmin *b* heavy chain sequence and that reported by Robbins et al. (1973).

Molecular weights of the component chains of each plasmin *b* affinity form are very similar. This can be seen from the data presented in Table 3. Although the difference between rabbit H*b* chains is probably within experimental error, the H*b* heavy chain from affinity form 1 plasminogen is consistently higher by 1000–3000 daltons than the H*b* chain of form 2. This suggests that the carbohydrate difference may reside primarily in the heavy chain region of plasminogen.

In light of the above data and those presented in previous studies, it appears as though at this time the only detectable structural difference between affinity chromatography-resolved forms of rabbit plasminogen and plasmin is in carbohydrate content. This difference, along with possible small differences in their primary structures, may be of importance in determining their physiological significance.

As shown in Figures 2 and 3, urokinase activation of each rabbit plasminogen to plasmin *b* occurs through the same transient species, plasmin *a*. Isolation and analysis of the transient components has led us to the following conclusions regarding the activation mechanism: (1) The initial step in urokinase activation of rabbit plasminogen *a* is cleavage of the internal bond to produce the transient plasmin *a*. In the presence of BTI, the H*a* heavy chain of plasmin *a* can be exclusively isolated. Furthermore, in the absence of BTI, the amino-terminal peptide can be removed from plasminogen *a*, producing plasminogen *b*, but this plasminogen is not an obligatory first step of activation. This peptide is also readily cleaved from the H*a* chain of plasmin *a*. (2) Plasmin generated during the course of activation is primarily responsible for catalyzing amino-terminal peptide release from rabbit plasminogen *a* and the H*a* chain of plasmin *a*.

Regarding the first point, it can be seen in Figures 2 and 3 that at early times of activation nearly all of the initial heavy chain formed is H*a*. As the plasmin *a* level increases with time, the amino-terminal peptide is removed from plasminogen *a*, producing plasminogen *b*. This is activated by urokinase directly to plasmin *b*. Concurrently, the amino-terminal peptide is cleaved from the H*a* chain of the initial plasmin *a*, resulting in an accumulation of the final product, plasmin *b*. These experiments alone provide strong evidence against the concept of a preactivation peptide for activation of the rabbit plasminogen system. Clearly if peptide release from rabbit plasminogen *a* was a requisite first step in the activation, H*a* heavy chain would never be observed.

The H*a* chain of plasmin *a* was found to be a very transient species under conditions where the plasmin generated during the course of activation is not inhibited. Therefore, activations were carried out in the presence of excess BTI. These results are also shown in Figure 3. It can be seen that BTI is extremely effective in trapping the initial plasmin *a* formed. In fact, we have increased activation times to 3 hours and used a level of urokinase sixfold greater than described in Methods and have still observed only plasmin *a*. Possible alterations in the activation mechanism in the presence of BTI, other than plasmin inhibition, were investigated by [^{125}I]BTI binding experiments. As described in Results, BTI did not form a stable complex with either plasminogen *a* or DFP-treated plasmin *b* at time periods significant in terms of our proposed mechanism. However, BTI does stoichiometrically bind both plasmin *a* and plasmin *b*. These experiments indicate that a plasmin active site is required for BTI binding and thus provide convincing evidence that plasmin *a* is an active molecule. Therefore, we feel that since only plasmin *a* is observed at excessive urokinase levels and since BTI is not binding to plasminogen *a*, the only alteration in the activation system caused by BTI is plasmin *a* inhibition.

Regarding the second conclusion described above, we feel that urokinase cannot effectively remove the amino-terminal peptide from either rabbit plasminogen *a* or the H*a* heavy chains of rabbit plasmin *a*. Indirect evidence for this is obtained from activations in the presence of BTI, where at high urokinase levels and long incubation times, neither plasminogen *b* nor plasmin *b* are ever observed. Direct evidence showing that plasmin, rather than urokinase, catalyzes this step is presented in Figure 4. The conditions used for this experiment, i.e., low levels of added plasmin, were chosen purposely to emphasize the lability of the H*a* heavy chain to plasminolysis. At added levels of urokinase-free plasmin, which are not readily detectable by electrophoretic methods, efficient removal of the peptide from the H*a* chain occurs. During the course of a usual activation, plasmin levels are continually increasing, and it could be expected that peptide release will occur much more readily than under the conditions chosen for the experiments of Figure 4.

Figure 4 also shows that at extremely high levels of added urokinase, which can fully activate plasminogen *a* to plasmin *b* in 15 minutes, the H*a* heavy chain of plasmin *a* remains intact. Thus we feel these experiments conclusively demonstrate that amino-terminal peptide release, at least for the rabbit system, is catalyzed by plasmin generated during the course of activation, rather than added urokinase.

The mechanism which we wish to propose for the urokinase conversion of rabbit plasminogen to plasmin is shown in Figure 6. The initial step in the reaction is a urokinase-catalyzed cleavage of the internal bond of plasminogen *a*, producing the two-chain plasmin *a* molecule. Plasmin *a* can then autocatalytically remove the amino-terminal peptide from the H*a* heavy chain of plasmin *a*, producing plasmin *b*. Either plasmin *a* or plasmin *b*, or both, can catalyze removal of the peptide from the remaining plasminogen *a*, yielding plasminogen *b*. In addition, urokinase can convert plasminogen *b*, without loss of peptide material, to the final product plasmin *b*. Thus there are several pathways to plasmin *b* formation. Factors such as plasminogen and urokinase concentrations are probably primarily responsible for the pathway(s) which the bulk of plasminogen *a* will follow during the course of activation.

Our activation mechanism determined from studies on the rabbit system only parallels that forwarded for the human system in that at least two proteolytic cleavages in plasminogen *a* occur during activation to plasmin. However, the systems differ in two very important respects: First, we have shown that amino-terminal peptide release from plasminogen *a* is not a requisite first step in the activation mechanism. Thus plasminogen *b* is not an

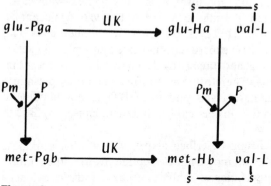

Figure 6
The mechanism of activation of plasminogen *a* by urokinase. Abbreviations used are: Glu-Pg *a*, plasminogen with an amino-terminal glutamic acid; Met-Pg *b*, plasminogen with an amino-terminal methiomine; Glu-H*a*-(S-S)-Val-L, plasmin *a* which contains the H*a* chain disulfide linked to the valine amino-terminal of light chain L; Met-H*b*-(S-S)-Val-L, plasmin *b* containing the H*b*-chain disulfide linked to the valine amino-terminal of light chain L; Pm, plasmin; UK, urokinase; P, peptide derived from the amino terminus of plasminogen *a*.

obligatory intermediate in the rabbit plasminogen activation. Second, plasmin efficiently catalyzes release of this peptide, whereas urokinase does not. The results presented in Figure 5 for the urokinase activation of human plasminogen *a* show, in contrast to the rabbit system, that no plasmin *a* is observed in the absence of BTI. Furthermore, the conversion of plasminogen *a* to plasminogen *b* occurs much more rapidly than observed for rabbit plasminogen. However, activation in the presence of BTI results in only plasmin *a* formation, as observed in the rabbit system. Thus since only human plasmin *a* is observed under conditions of plasmin inhibition and since the significance of transient plasmin *a* in the mechanism has been thoroughly defined in the rabbit system, we feel a similar mechanism may apply for human plasminogen. However, we have not rigorously proven that the mechanisms of rabbit and human plasminogen are identical. Investigations into this area are currently underway.

Finally, at the levels of lysine used in these experiments, plasminogen is known to be conformationally altered (Brockway and Castellino 1972). However, the same plasmin *a* species were observed without lysine both in the presence and absence of BTI. Furthermore, the rate of cleavage of the amino-terminal peptide and the rate of cleavage of the internal peptide bond in plasminogen are much slower in the absence of L-lysine.

Acknowledgments

This work was supported by Grants HL-13423 and HL-15747 from the National Heart and Lung Institute, National Institutes of Health and a grant-in-aid from the American and Indiana Heart Associations. F. J. C. is the recipient of a Career Development Award (1 K04 HL-70717) from the National Institutes of Health and a Teacher-Scholar Grant from the Camille and Henry Dreyfus Foundation.

REFERENCES

Brockway, W. J. and F. J. Castellino. 1972. Measurement of the binding of anti-fibrinolytic amino acids to various plasminogens. *Arch. Biochem. Biophys.* **151**:194.

Castellino, F. J., G. E. Siefring, Jr., J. M. Sodetz and R. K. Bretthauer. 1973. Amino terminal amino acid sequences and carbohydrate of the two major forms of rabbit plasminogen. *Biochem. Biophys. Res. Comm.* **53**:845.

Chase, T. and E. Shaw. 1970. Titration of trypsin, plasmin, and thrombin with *p*-nitrophenyl *p*′-guanidinobenzoate HCl. In *Methods in Enzymology* (ed. G. E. Perlman and L. Lorand), vol. 19, p. 20. Academic Press, New York.

David, G. S. 1972. Solid state lactoperoxidase: A highly stable enzyme for simple, gentle iodination of proteins. *Biochem. Biophys. Res. Comm.* **48**:464.

Kassel, B. 1970. Bovine trypsin-kallikrein inhibitor. In *Methods in Enzymology* (ed. G. E. Perlman and L. Lorand), vol. 19, p. 844. Academic Press, New York.

Rickli, E. E. and W. I. Otavsky. 1973. Release of an N-terminal peptide from human plasminogen during activation with urokinase. *Biochim. Biophys. Acta* **295**:381.

Robbins, K. C., P. Bernabe, L. Arzadan and L. Summaria. 1972. The primary structure of human plasminogen. *J. Biol. Chem.* **247**:6757.

————. 1973. NH$_2$-Terminal sequences of mammalian plasminogens and plasmin S-carboxymethyl heavy (A) and light (B) chain derivatives. *J. Biol. Chem.* **248**:7242.

Robbins, K. C., L. Summaria, B. Hsieh and R. J. Shah. 1967. The peptide chains of human plasmin. *J. Biol. Chem.* **242**:2333.

Siefring, G. E., Jr. and F. J. Castellino. 1974. Metabolic turnover studies on the two major forms of rat and rabbit plasminogen. *J. Biol. Chem.* **249**:1434.

Sodetz, J. M. and F. J. Castellino. 1975. The mechanism of activation of rabbit plasminogen by urokinase. *J. Biol. Chem.* (in press).

Sodetz, J. M., W. J. Brockway and F. J. Castellino. 1972. Multiplicity of rabbit plasminogen. Physical characterization. *Biochemistry* **11**:4451.

Sodetz, J. M., W. J. Brockway, K. G. Mann and F. J. Castellino. 1974. The mechanism of activation of rabbit plasminogen. Lack of a pre-activation peptide. *Biochem. Biophys. Res. Comm.* **60**:729.

Summaria, L., B. Hsieh and K. C. Robbins. 1967. The specific mechanism of activation of human plasminogen to plasmin. *J. Biol. Chem.* **242**:4279.

Walther, P. J., H. M. Steinman, R. L. Hill and P. A. McKee. 1974. Activation of human plasminogen by urokinase. *J. Biol. Chem.* **249**:1173.

Weber, K. and M. Osborn. 1969. The reliability of molecular weight determinations by dodecyl sulfate-polyacrylamide gel electrophoresis. *J. Biol. Chem.* **244**:4406.

Wiman, B. 1973. Primary structure of peptides released during activation of human plasminogen by urokinase. *Eur. J. Biochem.* **39**:1.

Wiman, B. and P. Wallén. 1973. Activation of human plasminogen by an insoluble derivative of urokinase. *Eur. J. Biochem.* **36**:25.

Production of Plasminogen Acti by Tissue Culture Techniques

Grant H. Barlow, Annemarie Rueter and Ilse Tribby

Experimental Biology Division, Abbott Laboratories
North Chicago, Illinois 60604

Urokinase, the plasminogen activator, has been used clinically with success. The major problem involves its availability and cost. Calculations have shown that isolation of the enzyme from human urine requires at least 1500 liters per dose, at a cost exceeding a thousand dollars. The discovery by Bernik and Kwaan (1967, 1969) that human embryonic kidney cells produce activator in culture and the development of the Mass Tissue Culture Propagator (MTCP) by Schleicher and Weiss (1968; Weiss and Schleicher 1968) led us to attempt to produce urokinase on a large scale from human embryo kidney cells by the tissue culture method. This paper will report on studies performed in an attempt to understand the synthesis in order to achieve the maximum production of urokinase.

METHODS

Human embryonic kidney cells (Flow Laboratories, Rockville, Md.) were grown to confluency in Eagle's medium (E-199) containing 10% fetal calf serum (growth medium) at 37°C in 5% CO_2 in air. Cells were grown either in plastic tissue culture flasks (Falcon), each with 40 ml of medium and a 75-cm^2 growth surface, in a humidified incubator or in mass tissue culture propagators containing 16 liters of medium. Confluency was attained in 7–10 days. The cells were then maintained in a 0.5% lactalbumin hydrolysate (Difco) medium (LH-production medium) for 5–6 weeks under the same conditions.

Fibrinolytic activity was measured by the fibrin plate method as described by Brakman (1967). Activity is expressed as Committee on Thrombolytic Agents (CTA) units, as described by Johnson, Kline and Alkjaersig (1969).

The cell culturing process takes approximately 30 days (Fig. 1) at 37°C to achieve maximum activator production. To answer the question as to

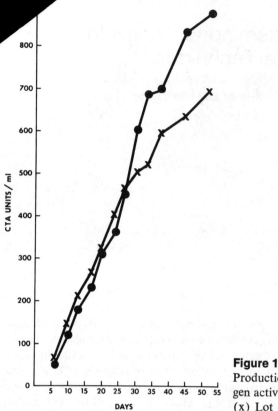

Figure 1

Production of tissue culture plasminogen activator with embryo kidney cells.
(x) Lot 19601; (•) Lot 549.

the stability of the enzyme under these conditions, the following experiment
was performed. From a series of flasks, on production, different percentages
of the media were removed periodically, assayed, and an additive record of
total units produced was made. The media removed were always replaced
with fresh media. The data show that all samples produced the same amount
of activator. This would infer that the enzyme is stable during the production
time.

In the large-scale production of enzyme, one needs large quantities of
cells. If one calculates, based on yield of enzymes/propagator, the cells
needed per propagator and the number of propagators required for the overall
quantity of enzyme desired, the number of fetal kidneys becomes unattainable.
Therefore on a production basis, one cannot use primary cells, but needs to
subculture the cells. Table 1 shows the results obtained with a cell lot during
subculturing as to its ability to produce the enzyme at various subcultivation
levels. The data also show the karyological classification of the first and last
subcultures. At each subculture, the enzyme is produced without any indication of the cells suffering from abnormal karyology. Since each of these subcultures represents three to four cell generations, this constitutes approximately
30 generations. Therefore, all production is made with subcultured cells which
fulfill the requirements of a normal diploid cell karyogram.

Table 1
Cell Passage Evaluation

Growth stage	Urokinase production[a]	Karyology			
		% diploidy	% aneuploidy	% polyploidy	% gaps
Primary	1050/33				
Subculture-1	590/33				
Subculture-2	1350/35	90.0	8.5	1.0	0.5
Subculture-3	1180/35				
Subculture-4	770/37				
Subculture-5	545/40				
Subculture-6	306/34				
Subculture-7	679/39				
Subculture-8	272/32				
Subculture-9	362/33	91.0	6.4	1.1	1.5
Subculture-10	—				

[a] CTA U/ml per days on production.

An interesting experiment is shown in Figure 2. The same number of cells were planted in flasks and allowed to reach confluency. The media were then replaced with different amounts of fresh media varying from 15–100 ml. It can be seen that the amount of urokinase produced is proportional to the volume of liquid covering the cells. To verify that this result was not just the effect of increased nutrients, the experiment was repeated using a constant amount of media (15 ml) and using isotonic NaCl for the rest of the volume; the results were identical in both cases.

To test the possibility that a feedback mechanism is involved in the control

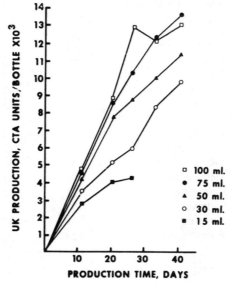

Figure 2
Total activator units produced as a function of the volume in a tissue culture flask of embryo kidney cells.

Table 2

Urokinase-priming Experiment

Medium	UK production, day 34 (CTA U/ml)	Corrected UK production (CTA U/ml)
Control	276	276
Growth + production with UK	576	300
Growth with UK + production	295	295
Growth with UK + production with UK	564	288

of urokinase production, an experiment was performed in which purified urokinase was added to either the growth media, the production media, or to both. The level of added urokinase was equal to what the cells were expected to yield. The results are shown in Table 2. When corrections are made for the added urokinase, all combinations yield the same amount of activator.

To investigate the possible role of amino acids in the production of the enzyme, amino acid analyses were performed at various times during the production cycle. As can be seen in Table 3, there is a rapid decrease in the availability of aspartic acid and glutamic acid from the media. However, addi-

Table 3

Amino Acid Analyses during Urokinase Production

Amino acids	Day, UK potency (μ/ml)				
	0, 0	14, 75	21, 75	28, 95	35, 109
Lysine	318[a]	350[a]	356[a]	327[a]	362[a]
Histidine	68	65	62	37	+
Ammonia (μm/ml)	1.3	4.3	4.5	4.4	6.0
Arginine	117	104	98	87	91
Tryptophan	+	+	+	+	+
Aspartic acid	5.5	—	—	—	—
Threonine	93	145	147	147	145
Serine	279	241	212	206	187
Glutamic acid	142	47	+	—	—
Proline	+	126	110	113	117
Glycine	*	*	*	*	*
Alanine	169	287	323	337	341
Half-cystine	18	12	12	10	14
Valine	168	194	204	185	178
Methionine	101	86	92	77	77
Isoleucine	154	165	162	149	144
Leucine	457	414	401	380	369
Tyrosine	139	148	145	145	137
Phenylalanine	128	138	132	129	125

[a] γ/ml.
* Too high to measure.

Table 4
Enhancement of Urokinase Production by Plasminogen

| | | Urokinase production, day 35 (CTA U/ml) | | |
| | | plasminogen added in LH media (Casein U/ml) | | |
Cell lot	control	0.1	0.01	0.001
816–2	981	1659	1485	1101
642–3	1143	1664	1554	1202
599–2	576	926	830	—

tion of extra amounts of these two amino acids did not increase the production of activator. Since lactalbumin hydrolysate contains peptides as well as free amino acids, it is possible that the cells utilize the peptide source, and when this is used, the amount of these two amino acids is adequate.

The role of amino acids and peptides is also seen in experiments which show the effect of added plasminogen to the media. Table 4 shows that the addition of small amounts of plasminogen to the production media brings about a significant increase in the amount of enzyme produced. This has been further explored by pretreating each component of the production media separately prior to incorporation into the medium. When this is done, it is found that pretreatment of the lactalbumin hydrolysate with the protease has the same effect as when the entire system is treated. This indicates that the reduction of the hydrolysate to smaller peptides and more free amino acids increases the productivity. To further substantiate these findings, the same phenomenon is seen with the addition of other proteases.

Still another approach has been taken to increase the yield of activator. Bernik and Kwaan (1967, 1969) have shown, and we have confirmed, that when the kidney culture is overlayed with fibrin, it appears that discrete cells ($\sim$ 5–10%) are producing the enzyme. Culturing of these specific cells should then lead to a large increase in the quantity of enzyme. Density gradient centrifugation of the primary cells in a serum albumin gradient, as described by Shortman, Williams and Adams (1972), has given at least ten bands. When these ten bands are cultured and placed on production media, several of the bands produce enzyme (Table 5).

Similar centrifugation experiments with subcultured cells from confluent monolayers in the G_1 stage gave only one band, indicating that the separations are based on cell density variations with different stages of the cell cycle.

DISCUSSION

The potential for producing large quantities of biological products from cell culture is in its infancy. Many breakthroughs will be needed before the methodolgy becomes routine. Our first major breakthrough was the finding that subcultured cells would continue to produce enzyme at a rate at least

Table 5

Density Gradient Centrifugation: Embryo Kidney Cells

Band	Outgrowth time (days)	UK production, day 28 (CTA U/ml)
1	20%, day 20	—
2	5%, day 20	—
3	40%, day 20	—
4	contamination	—
5	8	0
6	8	0
7	8	626
8	5	245
9	5	267
10	8	450
Control	8	480

comparable to the primary culture. This reduces the number of primary cells required by at least a factor of one hundred, and with further advances, conceivably even more. It should be pointed out again that karyology studies on the subcultured cells show that there have been no major morphological chromosomal changes during the cell passages.

Two important points should be made about the enzyme. First, this enzyme is shown to be stable during the long period of cell maintenance in culture. The replacement experiment shows this in a reasonable fashion, and it is confirmed in the priming experiment, where the control sample lost no activity. Second is the early finding by Barlow and Lazer (1972) that the enzyme produced in culture is identical to the enzyme produced in urine. This allows for interchange of results on the two enzymes.

The experiment showing the amount of urokinase produced being proportional to the volume of solution is an intriguing one. The observation that it is not nutritional poses several other possibilities. The first is that it may involve a feedback mechanism in which a constant concentration would be achieved. This would account for more total production with the larger volumes and, at the same time, explain the constant concentration per milliliter. The priming experiment aimed at substantiating this theory rules out the possibility that urokinase itself controls the feedback. However, it does not rule out the possibility that some intermediate in the synthesis could be the controlling factor. Another explanation would be a difference in oxygen tension on the cells brought about by the differences in volume above the cells. Such a possibility is being explored.

The results with the enzymes indicate that small peptides or the release of free amino acids into the medium may play an important role in the synthesis of the activator. Each of the proteolytic enzymes used must have an effect on the peptide distribution of the lactalbumin hydrolysate to reduce the size of some peptides and also to make available more free amino acids. These findings should be distinguished from the results of Bernik (1973), who added proteases to cell-free harvest and demonstrated an increase in activity. This result has indicated the possible presence of a proactivator which has not been fully activated during the culture period. It is conceivable that some of the increase seen here is of the same type, but this cannot explain the entire increase.

While all of the methods to increase yield are important, the selection of a cell that produces the enzyme with the elimination of all nonproducers would have a very dramatic effect on the yield of enzyme. If the observation reported by Bernik (1973) that only 5% of the cells are producers is correct, then the culturing of only these cells should lead to a 20-fold increase in enzyme yield. The results of the gradient centrifugation experiments indicate that the bands may reflect only the various stages of the cell cycle that are present, and when the cells are synchronized, this resolution disappears. This is similar to the behavior described by Dicke, vanHooft and Bekkum (1968).

These experiments confirm the feasibility of the tissue culture method being used to produce enzymes, which should make urokinase available to the medical profession in the amounts necessary for use in clinical situations.

Acknowledgment

The research upon which this publication is based was performed pursuant to Contract No. NIH-71-2405 with the National Institutes of Health, Department of Health, Education and Welfare.

REFERENCES

Barlow, G. H. and L. Lazer. 1972. Characterization of the plasminogen activator from human embryo kidney cells. Comparison with urokinase. *Thromb. Res.* **1**:201.

Bernik, M. B. 1973. Increased plasminogen activator (urokinase) in tissue culture after fibrin deposition. *J. Clin. Invest.* **52**:823.

Bernik, M. B. and H. C. Kwaan. 1967. Origin of fibrinolytic activity in cultures of the human kidney. *J. Lab. Clin. Med.* **70**:650.

————. 1969. Plasminogen activator activity in cultures from human tissues. An immunological and histochemical study. *J. Clin. Invest.* **48**:1740.

Brakman, K., ed. 1967. *Fibrinolysis: A Standardized Fibrin Plate Method and a Fibrinolytic Assay of Plasminogen.* Scheltema and Holkema, Amsterdam.

Dicke, K. A., J. I. M. vanHooft and D. W. Bekkum. 1968. The selective elimination of immunologically competent cells from bone marrow and lymphocyte cell mixtures. II. Mouse spleen cell fractionation on a discontinuous albumin gradient. *Transplantation* **6**:562.

Johnson, A. J., D. L. Kline and N. Alkjaersig. 1969. Assay methods and standard preparations for plasmin, plasminogen and urokinase in purified systems. *Thromb. Diath. Haemorrh.* **21**:259.

Schleicher, J. B. and R. E. Weiss. 1968. Application of a multiple surface tissue culture propagator for the production of cell monolayers, virus and biochemicals. *Biotech. Bioeng.* **10**:617.

Shortman, K., N. Williams and P. Adams. 1972. The separation of different cell classes from lymphoid organs. V. Simple procedure for the removal of cell debris, damaged cells and erythroid cells. *J. Immunol. Meth.* **1**:273.

Weiss, R. E. and J. B. Schleicher. 1968. A multisurface tissue propagator for the mass-scale growth of cell monolayers. *Biotech. Bioeng.* **10**:601.

Plasminogen Activator: Secretion by Neoplastic Cells and Macrophages

E. Reich

The Rockefeller University, New York, New York 10021

Since the subject of tumor-associated fibrinolysis was summarized for the first volume in this series (Reich 1974), knowledge of plasminogen activator has been extended considerably. The finding of increased activator production associated with neoplasia (Unkeless et al. 1973; Ossowski et al., 1973a,b) has been confirmed in reports from several laboratories (Christman and Acs 1974; Goldberg 1974; Laug, Jones and Benedict 1975). However, the sum of information remains limited, and we do not know yet how plasminogen activator relates to neoplasia. Fortunately, this state is likely to be only temporary, since improvements in assay procedures and experimental materials should make it possible to delineate the biological parameters of plasminogen activation with some precision in the near future.

During the past year, we have continued to explore plasminogen activation in cell cultures, particularly in relation to oncogenic transformation and to macrophage function, and the current status of this work is summarized below. Other contributions that deal with particular aspects of plasminogen activator appear elsewhere in this volume (Rifkin, Beal and Reich; Danø and Reich; Ossowski, Quigley and Reich).

Assays for Plasminogen Activator Formation by Single Cells and Cell Clones

Almost all studies of plasminogen activation or fibrinolysis in culture have been based on procedures that measure the average behavior of relatively large numbers of cells. Such experiments cannot provide any insights about the nature, distribution or timing of events at the level of individual cells, nor about the range of variation between clones. In view of the obvious importance of this kind of information, we have developed suitable assays for the purpose, in collaboration with the hematology group at the Children's Hospital of Los Angeles (Jones et al. 1975). The methods are based on the use of overlays of fibrin-agar gels and, depending on the choice of experi-

mental conditions, they can discriminate clearly between normal and transformed cells after periods of incubation in the range of 2–24 hours. Compared with Todd's original "fibrin-slide" technique (Todd 1958), this method
has the advantage of being applicable to living cells; activator-positive cells
or clones can be located and thereafter isolated and subcultured. This procedure has already enabled us to demonstrate the control of plasminogen
activator synthesis in normal macrophages by physiological concentrations of
hormones applied in culture. In addition, foci of cells transformed by sarcoma
viruses can be identified and counted 24 hours after infection, when they consist of 1–2 cells, rather than after 1–2 weeks. The method also is applicable
to studies of chemical carcinogenesis in vitro and to the analysis of factors
that control plasminogen activator synthesis and secretion. In comparison
with the casein-agar overlay method reported by Goldberg (1974), the use of
fibrin presents a number of advantages in terms of versatility, sensitivity and
signal-to-background ratio.

Secretion of Serine Enzymes by Transformed Cells

Our first studies of macrophage conditioned medium (Unkeless, Gordon and
Reich 1974) showed that the onset of plasminogen activator synthesis was
coordinated with marked changes in the pattern of serine enzyme secretion.
Like plasminogen activator itself, most of the serine enzymes secreted by
stimulated macrophages are proteases with trypsinlike specificity, whereas
the enzymes secreted by unstimulated macrophages are not (J. C. Unkeless,
A. Piperno and K. Danø, unpubl.). These results prompted more detailed comparisons of serine enzymes in conditioned media from normal and transformed cultures. It has now been found that following transformation by
mouse sarcoma virus, at least five new and distinct serine enzymes accumulate
in mouse fibroblast culture fluids, four of which are proteases. Two of these
are capable of activating plasminogen, although the major activator (MW
$\sim$ 50,000) is quantitatively the most significant enzyme: it accounts for at
least 90% of the total detectable serine enzyme secretion by MSV-transformed cells, and its specific catalytic activity is about tenfold higher than
that of the minor activator (MW $\sim$ 25,000) (K. Danø and E. Reich, unpubl.).
The function of the remaining and so far uncharacterized serine enzymes is
unknown. However since they are apparently produced at rates even lower
than that of plasminogen activator, their total catalytic activity is certainly
low, and they may be activators for other zymogens or precursors of enzyme
cascades.

Inhibitors of Plasminogen Activator and Other Serine Enzymes

We have conducted a modest screening program to define the inhibitor spectrum of plasminogen activator and, to some extent, of the accompanying
serine enzymes found in conditioned media. Some details of this work are
presented elsewhere in this volume (Danø and Reich). An additional convenient assay for potential inhibitors is based on incorporation of [³H]diiso-
propylfluorophosphate into the single serine residue at the active site of
enzymes; the ability of a substance to reduce [³H]DFP incorporation becomes

a semiquantitative measure of its potential as an enzyme inhibitor. The assay is useful as a screening approach since it can be used with impure enzyme mixtures, the incorporated [³H]DFP being measured by autoradiography following SDS-polyacrylamide gel electrophoresis. Tests of this kind have given the interesting result that plasminogen activators, and the associated serine enzymes secreted by macrophages and transformed cells, are apparently unaffected by any of a large number of macromolecular inhibitors, most (or all) of which inhibit other enzymes with tryptic specificity. These secreted enzymes may therefore have certain unique properties in common, as is the case for other categories of proteases.

Correlation of Fibrinolysis and Parameters of Transformation

During the past year, the correlations between enhanced plasminogen activator synthesis and parameters of neoplasia and transformation have continued to be analyzed in several laboratories, including our own. Several reports provided additional documentation of the morphological changes that follow plasminogen activation in culture (G. Acs et al., pers. comm.; Christman et al., this volume; Laug, Jones and Benedict 1975). However, not all morphological changes are the result of plasminogen activation. In some cell systems, morphological changes may be plasminogen independent; in others, such as macrophages, cell morphology may be insensitive to the action of plasmin.

A particularly interesting correlation between enhanced activator production and tumorigenicity has been observed (Christman et al., this volume) in the parallel suppression of both of these phenotypic properties following treatment of mouse melanoma cells with 5-bromo-2'-deoxyuridine.

A recently completed series of experiments has shown that neither plasminogen nor enhanced activation is required to maintain maximal rates or extent of cell growth in culture. The growth of early monolayer cell cultures prepared from chicken, hamster or mouse embryos is unaffected by removal of plasminogen or by incorporation of plasmin inhibitors into the medium; the same is true for the respective transformed derivatives (L. Ossowski, J. P. Quigley and E. Reich, unpubl.).

On the other hand, detailed studies (Ossowski, Quigley and Reich, this volume) have reinforced earlier observations (Ossowski et al. 1973a) indicating that plasminogen activation is a necessary element in cell migration. If the model systems in culture are in fact measuring a cellular function required for migration in animals, this role of plasminogen might be an inviting target for control of tumor invasiveness and metastasis in vivo.

Another suggestive finding comes from a follow-up of two previous independent indications (Ossowski et al. 1973a,b; Pollack et al. 1974) that loss of anchorage dependence was well correlated with elevated production of plasminogen activator. It has now been found that these correlations include the ability of cloned transformants to grow in nude mice. The mechanisms that link the diverse phenotypes of activator production, anchorage independence and growth in immunodeficient mice remain obscure.

Interesting preliminary results have developed from the study of intracellular plasminogen activator in chick embryo cultures infected with Rous sarcoma virus mutants that are temperature sensitive for transformation (Rifkin, Beal and Reich, this volume). These experiments show both the tight coupling between the expression of the transforming function and activator synthesis and the rapid response of the latter. The magnitude of the changes in levels of activity makes this enzyme an excellent object for studying factors that regulate genetic transcription and translation in higher forms.

Cell Lines and Transplantable Tumors

The significance and assessment of correlations between any cellular function and the parameters of oncogenic change depend on many factors, not least of which are the choice of experimental material and the definition of "transformation." The large number of controversies generated by the questions attest to the necessarily vague and subjective quality of many measurements, as well as to the uncertain biological reference points used for their evaluation. In this regard, the widespread use of permanent cell lines and transplantable tumors for studying oncogenic changes presents many difficulties. Cell lines are usually isolated after stringent selection, often under conditions in which many of the environmental variables are uncontrolled. As a rule, only a trivial fraction of the progeny arising from the original inoculum survive. They are often then subjected to repeated cloning, thereby possibly accentuating genotypic and phenotypic differences in comparison with the parent embryonic cells. Moreover, the number of replication cycles in culture, with attendant error frequency and under conditions that encourage variation, greatly exceeds the total genetic activity of any somatic cell in vivo during the lifetime of a mammal. It is well known that the karyotype of cell lines is almost universally abnormal, and the definition of normal or untransformed cell behavior is often arbitrary. Transformed cells, particularly those obtained from cell lines after infection with DNA viruses or treatment with chemical carcinogens, are frequently isolated after long periods of incubation and further recloning, and it is therefore not surprising that experimental results are duplicated only with difficulty. Thus all studies of complex phenotypes, such as oncogenic transformation, that are based on the use of permanent cell lines are difficult to interpret and to relate to neoplasia.

The provocative experiments of Pollack and his associates (Pollack et al. 1974) have provided fresh insight into the association of physiological or phenotypic parameters that are tested in different assays of transformation. They have isolated clones after infection with SV40 virus that behave as transformed in only one or a few of the assays used to define transformation. They have also found that individual clones revertant for one or more of the usual transformation parameters can be isolated from "fully transformed" clones that are positive for all parameters. Their work therefore shows clearly that most of the phenotypic properties correlated with transformation are independent and can be expressed independently in cells presumed to be transformed. Since plasminogen activator is but one of the phenotypic changes associated with oncogenic change, its expression in cloned derivatives of cell lines, transformed or untransformed, can be expected to vary, just as do all of the other cellular parameters under similar conditions. This is in fact the case (L. Ossowski and S. Strickland, unpubl.; Mott et al. 1974). Thus many cell lines of supposedly normal or untransformed origin show enhanced fibrinolysis in culture, whereas some cloned derivatives of nominally transformed cells do not. An interesting case is the so-called "normal" 3T3 cell line, which produces high levels of plasminogen activator: 3T3 is now known to be highly tumorigenic (Boone 1974), and it gives rise to individual subclones whose content of plasminogen activator cover a wide range (L. Ossowski and D. Rifkin, unpubl.). On the other hand, no such variation is observed when primary or early passage cultures are compared with their transformed derivatives, especially those transformed by RNA tumor viruses. For example, several thousand RSV foci have been examined to date and without a single exception, all produced high levels of plasminogen activator.

Another unusual case recently analyzed (D. Rifkin and R. Pollack, in prep.) was a transplanted mouse tumor derived from 3T3 infected with the Kirsten derivative of mouse sarcoma virus. Both the transplanted tumor and cultures derived from it produced low levels of activator. The transforming virus was then rescued from the tumor cells and used to reinfect primary mouse embryo cultures; these newly transformed cells produced the usual high levels of activator. All of these data suggest that the results of phenotypic studies obtained with the use of established cell lines need to be interpreted with caution.

Macrophage Plasminogen Activator

It is well established that many normal tissues contain significant amounts of plasminogen activator, although the levels are, in general, appreciably lower than those found in tumors or in cells transformed by RNA tumor viruses. We do not know the biological function of plasminogen activators, nor how or whether the synthesis or activity of these enzymes is connected with growth of normal or diseased tissue. However, the presence of plasminogen activators in normal adult animals implies some necessary and normal function, and it seemed desirable to explore for a possible involvement of plasminogen activation in normal tissue growth, such as that occurring during cell proliferation in the lymphoid and hemopoietic systems. The macrophage suggested itself as attractive experimental material for a variety of reasons:

(1) Macrophages are known to be involved in the activation of lymphocytes by lectins and antigens, and they are also a rich source of factors regulating granulocyte production and differentiation.

(2) Macrophages are generally associated with chronic inflammatory states, particularly those in which tissue destruction is occurring, and it appeared a reasonable inference that the high levels of circulating plasminogen, which represent a huge potential source of proteolytic activity, might be contributing to such destructive processes.

(3) Macrophages are prominently involved in the reaction to indigestible materials, such as asbestos, mineral oil and other substances that are potent foreign body carcinogens. The work of Potter and his associates is particularly suggestive in this regard (Potter and MacCardle 1964; Potter and Walters 1973). They have shown that both the production and the transplantability of myelomas in Balb/c mice are associated with the formation of adjacent granulomas consisting of macrophages that have ingested mineral oil. Of further

interest is the report showing that such myelomas do not occur in germ-free animals unless their intestinal tract is first repopulated with bacteria. This indicates that immune stimulation is somehow required for the neoplastic process, and many studies attest to the involvement of macrophages in the early response to nonspecific stimuli, such as gram-negative endotoxins.

With the preceding considerations in mind, we have studied peritoneal macrophages obtained from mice following exposure to a number of different treatments and have so far established the following:

(1) Cultures of peritoneal mouse macrophages from suitably stimulated animals produce and secrete a plasminogen activator in vitro similar to the enzyme secreted by transformed mouse fibroblasts (Unkeless, Gordon and Reich 1974); peritoneal macrophages from unstimulated animals do not produce this enzyme.

(2) Intraperitoneally injected materials that give rise to stimulated macrophages include thioglycollate, mineral oil, BCG, *Corynebacterium parvum*, antigen-antibody complexes and asbestos (Gordon, Unkeless and Cohn, 1974; J.-D. Vassalli, J. Hamilton and E. Reich, unpubl.).

(3) The synthesis of plasminogen activator is not a solitary event following macrophage stimulation. Comparisons of conditioned media from cultures of ordinary and stimulated macrophages have shown that there is a marked overall change in the synthetic program, at least as it applies to secreted proteins. The plasminogen activator is but one of a large number of new serine enzymes, mostly proteases, that appear following stimulation. In addition, a number of other differences between the two kinds of macrophages have been found among proteins of unknown function that appear in conditioned media (Unkeless, Gordon and Reich 1974).

(4) The formation of plasminogen activator by stimulated macrophages may take place in a two-step process, at least under some conditions (Gordon, Unkeless and Cohn 1974). Thus peritoneal macrophages obtained from endotoxin-treated animals appear to be "primed" for activator production, in the sense that they synthesize little enzyme when cultured in the ordinary way. However, high levels of enzyme are secreted for long periods after such cultures are permitted to phagocytize particulate materials, such as suspensions of latex beads. Under these conditions, enzyme secretion may continue at maximal levels for the lifetime of the culture (Gordon, Unkeless and Cohn 1974). Although we do not know what are the consequences of plasminogen activator secretion for long periods, it appears possible that this aspect of macrophage function may be significant for foreign body carcinogenesis: if their behavior in culture parallels that in vivo, some indigestible foreign materials can be assumed to stimulate the secretion of numerous macrophage proteins, including plasminogen activator, for the lifetime of the animal. Persistent changes in local microenvironments, involving small numbers of cells, could be produced in this way.

More recently, we have observed (J.-D. Vassalli, J. Hamilton and E. Reich, unpubl.) that very low concentrations of concanavalin A can also stimulate activator production by endotoxin-primed macrophages, indicating that the stimulus which triggers this process need not be particulate.

(5) Of some interest is the recent finding that physiological levels of anti-inflammatory steroids reversibly suppress plasminogen activator secretion by

stimulated macrophages (J.-D. Vassalli, J. Hamilton and E. Reich, unpubl.). The effective concentration ranges are: dexamethasone, $10^{-9}-10^{-8}$ M; prednisolone and hydrocortisone, $10^{-8}-10^{-7}$ M. The obvious potential relevance of these findings for chronic inflammation, foreign body carcinogenesis and macrophage-lymphocyte interaction remain to be explored. As might be expected, the effect of the steroids on macrophages is not limited to plasminogen activator, since there appear to be coordinated changes affecting the secretion of several proteins.

What is the function of plasminogen activator in the body, and why do widely different types, such as vascular endothelium, ovarian cells, stimulated macrophages and neoplastic cells, produce high levels of enzyme? Although we cannot answer these questions at present, some speculation may be warranted. The high concentrations of circulating plasminogen and of plasmin inhibitors suggest that plasminogen activation is likely to be involved in processes other than control of blood coagulation. Plasminogen is a reservoir of potential protease activity that can, in principle, be tapped for any cellular activity requiring local, short-term proteolysis. If we consider the conditions that exist in rapidly growing solid tissues, such as liver, it is apparent that continuing cell division and growth depend on expansion of solid organs. When cells divide, two progeny cells would have to be compressed into the space previously occupied by the single parent cell, unless additional space were created. We may visualize that this space could be created by local, controlled proteolysis of the surrounding connective tissue, following which the newly replicated cells could migrate into the cleared area, attach, and repair the defects in the intercellular structures. If this were so, plasminogen activation, and the possible formation of other proteases, such as collagenase, might be required only for short periods after cells were committed to replication and division. In that case, plasminogen activator synthesis might occur transiently during a limited portion of the cell cycle.

This hypothetical view of plasminogen activation is consistent with a number of observations:

(1) Cultures of normal chick embryo fibroblasts produce small amounts of plasminogen activator. These levels are much lower than those produced by transformed cells, but they are real, and the normal cell activator appears to be the same as that secreted by transformed cells. Unfortunately, the available assay procedures are not yet sufficiently precise to permit the cell cycle analysis of activator production in normal cultures.

These considerations suggest a primarily extracellular role for plasminogen activator and might lend particular significance to two facts: (a) that activator secretion is seemingly coordinated with that of other serine enzymes, and (b) that most, and perhaps all, of the intracellular activator is sedimentible and tightly associated with the membraneous granule fraction. This behavior corresponds to that of zymogen or secretion granules; their contents are membrane-bound and destined ultimately to be released and to function outside of the cells in which they are made.

(2) As already mentioned, certain normal tissues produce high levels of plasminogen activator. Recent observations suggest that this enzyme, by generating active plasmin, may, in some instances, be responsible for connective tissue degradation. One example of this is the rupture of the mature, pre-

ovulatory follicle which occurs by digestion of the surrounding connective tissue theca. In the ovary, the formation of plasminogen activator is transient and is under strict hormonal control (W. Beers and S. Strickland, unpubl.). In contrast, the synthesis of activator in neoplastic cells appears to be continuous and uncontrolled, and the surrounding connective tissue would thereby be exposed to persistent degradation throughout the life of the cell; this could foster cell migration, connective tissue turnover and metastasis, all independently of cell division or growth. We are currently attempting to test these possibilities.

Acknowledgments

The studies in our laboratory summarized here were performed in collaboration with Keld Danø, John Hamilton, David Loskutoff, Lilliana Ossowski, Angela Piperno, James Quigley, Daniel Rifkin, Susannah T. Rohrlich, Sidney Strickland, Annette Tobia, Jay Unkeless, Jean-Dominique Vassalli and Elaine Wilson and were supported by grants from the American Cancer Society, the National Institutes of Health and the Fred P. Goldhirsch Foundation.

REFERENCES

Boone, C. W. 1974. Hemangioendothelionas produced by the subcutaneous inoculation of Balb/3T3 cells attached to glass beads. In *Abstracts of 14th Meeting of the American Society for Cell Biology,* vol. 63, p. 32A. San Diego, California.

Christman, J. and G. Acs. 1974. Purification and characterization of a cellular fibrinolytic factor associated with oncogenic transformation; the plasminogen activator from SV-40 transformed hamster cells. *Biochim. Biophys. Acta* **340**:339.

Goldberg, A. 1974. Increased protease levels in transformed cells: A casein overlay assay for the detection of plasminogen activator production. *Cell* **2**:95.

Gordon, S., J. C. Unkeless and Z. Cohn. 1974. Induction of macrophage plasminogen activator by endotoxin stimulation and phagocytosis. Evidence for a two stage process. *J. Exp. Med.* **140**:995.

Jones, P., S. Strickland, W. Benedict and E. Reich. 1975. Fibrin overlay methods for identification of transformation in single cells and colonies. *Cell* (in press).

Laug, W., P. Jones and W. Benedict. 1975. Studies on the relationship between fibrinolysis of cultured cells and malignancy. *J. Nat. Cancer Inst.* **54**:173.

Mott, D. M., P. H. Fabisch, B. P. Sani and S. Sorof. 1974. Lack of correlation between fibrinolysis and the transformed state of cultured mammalian cells. *Biochem. Biophys. Res. Comm.* **61**:621.

Ossowski, L., J. P. Quigley, G. M. Kellerman and E. Reich. 1973a. Fibrinolysis associated with oncogenic transformation. Requirement of plasminogen for correlated changes in cellular morphology, colony formation in cellular morphology, colony formation in agar and cell migration. *J. Exp. Med.* **138**:1056.

Ossowski, L., J. C. Unkeless, A. Tobia, J. P. Quigley, D. B. Rifkin and E. Reich. 1973b. An enzymatic function associated with transformation of fibroblasts by oncogenic viruses. II. Mammalian fibroblast cultures transformed by DNA and RNA tumor viruses. *J. Exp. Med.* **137**:112.

Pollack, R., R. Risser, S. Conlon and D. Rifkin. 1974. Plasminogen activator production accompanies loss of anchorage regulation in transformation of primary rat embryo cells by simian virus 40. *Proc. Nat. Acad. Sci.* **71:**4792.

Potter, M. and R. C. MacCardle. 1964. Histology of developing plasma cell neoplasia induced by mineral oil in Balb/c mice. *J. Nat. Cancer. Inst.* **33:**497.

Potter, M. and L. Walters. 1973. Effect of intraperitoneal pristane on established immunity to the Adj-TC-5 plasmacytoma. *J. Nat. Cancer Inst.* **51:**875.

Reich, E. 1974. Tumor-associated fibrinolysis. In *Control of Proliferation in Animal Cells* (ed. B. Clarkson and R. Baserga), pp. 351–355. Cold Spring Harbor Laboratory, Cold Spring Harbor, New York.

Todd, A. S. 1958. Fibrinolysis autographs. *Nature* **181:**495.

Unkeless, J. C., S. Gordon and E. Reich. 1974. Secretion of plasminogen activator by stimulated macrophages. *J. Exp. Med.* **139:**834.

Unkeless, J. C., A. Tobia, L. Ossowski, J. P. Quigley, D. B. Rifkin and E. Reich. 1973. An enzymatic function associated with transformation of fibroblasts by oncogenic viruses. I. Chick embryo fibroblasts transformed by avian RNA tumor viruses. *J. Exp. Med.* **137:**85.

Cell-induced Fibrinolysis:
A Fundamental Process

Tage Astrup

The James F. Mitchell Foundation Institute for Medical Research
Washington, D.C. 20015

The purpose of this presentation is to provide the reader with a brief exposé of the developments in the field of cellular fibrinolysis. Before doing this, I would like to say that I have followed with great interest the recent work by Dr. Reich and his associates, which has opened up a wide field of research in cellular fibrinolysis and its possible relation to the cancer problem, a lead which is now being followed by several investigators in various laboratories. I would also like to commend Drs. Reich, Rifkin and Shaw for arranging the conference on proteases and biological control which resulted in this book. Having been active in this field for slightly more than a generation, I can convey the information that this was the first conference in which fibrinolysis was placed in its true perspective as well as in a frame of prospective developments, namely, as a fundamental process of great importance in cell physiology and cell biology. These were the concepts that guided the work of the late Dr. Albert Fischer, and the conference, therefore, was also an indication of a revival and confirmation of his basic ideas. It might be of interest to mention, for the benefit of younger investigators, that Albert Fischer worked with Alexis Carrel at the Rockefeller Institute in the early nineteen twenties. Later he moved to the Kaiser Wilhelm Institute in Berlin to set up a department for tissue culture, in which Otto Warburg had shown great interest. In 1932, he returned to Copenhagen to his own new institute, The Biological Institute of the Carlsberg Foundation, the establishment of which was supported by the Rockefeller Foundation.

In recent work describing an association of fibrinolysis with malignant transformed cells, reference is usually made to an article by Fischer published in 1925. This article reviews experiments dealing with the propagation of normal epithelial cells as well as of tumor cells (Rous chicken sarcoma, Flexner-Jobling rat carcinoma) in tissue cultures. While normal epithelium cultivated in vitro produces some liquefaction of the chicken plasma clot which serves as the substrate, this phenomenon is enhanced in cultures of malignant tissues to such a degree that continuation of the culture is not possible. To

ensure propagation, Fischer introduced a small piece of muscle tissue (living
or dead) in the cultivation of the Rous chicken sarcoma. In 1927, he used
the same technique in the cultivation of the Ehrlich mouse carcinoma (Fischer
1927a). He also found that propagation of the mouse carcinoma succeeded
when a fully heterologous substrate (chicken plasma with chicken embryo
extract and rate plasma) was used (Fischer 1927b). The problem of the liquefaction of the plasma clot and its possible association with malignant cells
was treated in detail in Fischer's handbook on tissue cultivation published in
1930. The study of the liquefaction of the fibrin clot in tissue culture continued to be a major endeavor of Fischer's group. Santesson, working with
Fischer, as well as Carrel, extensively reviewed the field in 1935 in his
monograph on epithelial mouse tumor cells. Interestingly, Santesson developed
a technique based upon the use of slices of heat-denatured egg albumin for
the study of the proteolytic activity causing the liquefaction.

The Tissue Plasminogen Activator

When, in 1937, Fischer asked me to join his group, it was in an effort to
intensify studies dealing with the interactions between the cultured cells and
the substrate—in part the nutritive aspects and in part the problem of substrate liquefaction. Here I shall not dwell on the first part, but will concentrate
on the second part, which became my main responsibility and in which
several associates and visiting investigators participated during subsequent
years. A solid substrate of clotted plasma was then the routine substrate for
the cultivation of tissues and it had been found superior to several other
supporting media. Efforts were first made to elucidate some pertinent problems related to the mechanism of blood coagulation, in particular, its induction by tissue extracts. The knowledge of blood coagulation was at that time in
a state of utter confusion. The significance of a solid fibrin substrate was substantiated much later by Beck et al. (1961, 1962), who found that the
growth of fibroblasts was impaired when the clots were prepared in the absence of the fibrin-stablizing factor (blood coagulation factor XIII). In the
middle forties, we were ready to begin an attack on the problem of plasma
clot liquefaction during cultivation of tissues. This led to the discovery of a
cellular agent which activated a humoral protease precursor (plasminogen)
(Astrup and Permin 1947). The method used in the study of this plasminogen
activator was directly derived from tissue culture techniques and was later
standardized as the fibrin plate assay. This was also the first unambiguous
demonstration of the existence of a physiological activator of the humoral
fibrinolytic system. A review of this phase of development was presented in
1956 (Astrup 1956a).

The tissue plasminogen activator was found to be strongly bound to structural proteins, from which it could not be extracted with solutions commonly
used for such purposes. Saline extracts contained only a small fraction of the
total activity. Only after we had found that strong solutions of potassium
thiocyanate could extract the activator did it become possible to devise a
quantitative assay of the tissue plasminogen activator (Astrup and Albrechtsen 1957). With this assay, several systematic studies were undertaken to
determine the distribution of the tissue activator in various tissues and animal

species. This phase was reviewed in 1966 (Astrup 1966). The most detailed of these studies was later done in the rabbit (Glas and Astrup 1970).

Fibrinolysis and Tissue Repair

A guiding concept in our studies was the involvement of fibrinolysis in the regulation of tissue repair. This led to an hypothesis of the pathogenesis of arterial disease (Astrup 1956b) which served as a basis for many subsequent studies. This hypothesis assumes that injury to the endothelium and intima is an etiological factor in arterial disease, the pattern of the ensuing repair process depending upon conditions in the vessel wall (cellular factors) and in the circulating blood (humoral factors). The concept provides the biochemical and physiological basis of the so-called thrombogenic theory of the pathogenesis of arteriosclerosis proposed by Duguid (1946). It extends the basis of this theory so that it is now possible to understand why hemophilic patients also may be subject to arterial disease. The history of this development was presented in 1967 (Astrup 1967).

Studies of the distribution of plasminogen activator in the layers of the vessel wall and in samples of liver tissue showed high concentrations in certain types of vascularized connective tissue (arterial adventitia) and in repair tissue (cirrhotic livers) (Astrup et al. 1959; Astrup et al. 1960), but an identification of the cellular carrier of the plasminogen activator was, of course, not possible by experiments based on the extraction method. The tool needed for this task was provided when Todd (1959) designed a histochemical fibrin slide technique for demonstration of the tissue plasminogen activator. Frozen sections were covered with a fibrin film and incubated at 37°C followed by fixation and staining with hematoxylin. Clear zones in the blue-stained fibrin identified locations of fibrinolytic agents. With this technique, Todd could identify the vascular endothelial cells as a source of plasminogen activator in tissues. We have used modifications of this technique in systematic studies of processes of tissue repair and of the changes in the vascularization associated with development and involution of organs and tissues. Thus high fibrinolytic activity is brought into an area of tissue repair by the budding capillaries. Following scar formation, the vessels undergo atrophy and the fibrinolytic activity disappears (Kwaan and Astrup 1964). In a study of the developing rat eye, the degree of vascularization could be related to stages of development, differentiation and function of several structures (lens, retina, choroid). Completion of development was followed by involution of the supplying vessels accompanied by loss of fibrinolytic activity (Pandolfi 1967). A review on fibrinolysis in tissue culture and tissue repair presented some of these results in a broader context (Astrup 1968). The occlusion and involution of the ductus arteriosus were seen to be followed by the production of an area with numerous fibrinolytically active vessels, thus representing a physiological repair process (Glas-Greenwalt, Strand and Astrup 1972). More intense injury associated with an inflammatory reaction and infiltration by leukocytes produced a decrease in fibrinolytic activity in the affected tissues (Tympanidis and Astrup 1972). Injected intradermally, the tissue plasminogen activator disappears rapidly (Smokovitis, Kok and Astrup 1975). This is in contrast to the cell-bound activator associated with vascular endothelial cells. In frozen sections, the active cells are often found randomly distributed along the endothelium separated by many inactive cells, and active capillaries are observed surrounded by inactive vessels. These findings pose a problem in cell physiology. Does it mean that endothelial cells function as small glands producing and releasing plasminogen activator at certain periods? Or is the production and release associated with certain phases of the life cycle of the active cell? A possible answer to this puzzling question came from a different angle of approach.

Plasminogen Activator in Epithelial Cells

The association of fibrinolytic activity with vascular endothelial cells has been confirmed by several investigators. However, there are also fibrinolytically active epithelial cells. This observation was first made in a study of the eye (Pandolfi and Astrup 1967). It was found that many corneal epithelial cells of several animal species were highly fibrinolytic. Collected as frozen imprints ("häutschen"), the active cells were observed surrounded by cells of weak activity or even by inactive cells (Fig. 1). Clearly, all epithelial cells were not

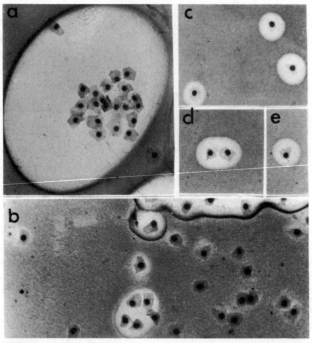

Figure 1
Epithelial cells from cornea of Rhesus monkey. Fibrin slide technique; incubation time, 15 minutes; hematoxylin stain.
(*a*) Large area of lysis produced by group of cells. (*b*) Lysis related to single cells or groups of cells. Some weakly active or inactive cells are seen. (*c, d, e*) Examples of lysis related to single cells. (Reprinted, with permission, from Pandolfi and Astrup 1967.)

active at the same moment. Systematic studies were then undertaken in order to determine whether surface epithelial cells (squamous epithelial cells) generally speaking could be considered to be fibrinolytically active. The second such study (Henrichsen and Astrup 1967) dealt with the epithelium of the rat vagina and immediately led to some important findings. The young desquamated epithelial cells in smears obtained in early estrus were usually found highly active, whereas the anuclear, cornified cells in late estrus were inactive (Fig. 2). A similar pattern was observed in human vaginal smears, although the fact that cornified vaginal cells of man retain their nuclei made the association less clear (Tympanidis, King and Astrup 1968). However, the findings led to several significant conclusions, some of which have been briefly mentioned in a previous review (Astrup 1968).

It was now obvious that cellular fibrinolytic activity was not solely related to vascular endothelial cells. Epithelial cells lining surfaces (squamous epithelium, but not glandular epithelium) also contained plasminogen activator. The presence of plasminogen activator seemed to be related to cells at a certain phase of life since fully degenerated cells always were inactive. The activity was probably related to phases of proliferation and maturation since the activity of the vaginal epithelial cells depended upon the estrus cycle. It could reasonably be assumed that the activity in the endothelial cells would follow a pattern of production and release similar to that observed in surface epithelial cells. Hence, the epithelial cells provided us with a model of cellular fibrinolytic activity which was easier to observe and study than the endothelial cells and which could be influenced by hormones through their effect on proliferation and maturation. In this manner, it was possible to study in detail the biology of cellular fibrinolysis in an effort to disclose mechanisms of production and release.

First, a number of studies were undertaken to verify the finding that surface epithelial cells may possess fibrinolytic activity. Among the tissues thus studied were the epithelium of the oviduct of the rat (Tympanidis and Astrup 1968a) and the bladder epithelium (Tympanidis and Astrup 1971). Sections of the endometrium and the vagina of the rat were studied under the influence of estrogen or progesterone (Tympanidis and Astrup 1969). Some of these studies were complicated by the presence of high vascular fibrinolytic activity in the deeper layers, so that active epithelial cells could be identified only when highly active. Even the surface epithelium of the gastrointestinal tract of the rat was found to contain plasminogen activator, despite the complications caused in some areas by the content of pancreatic enzymes (Myhre-Jensen, Baetz and Astrup 1973). A study of the squamous epithelium of the oral cavity of the rat indicated that the plasminogen activator was absent in young proliferating cells, whereas the intermediate cells were highly active and the fully cornified cells were inactive as usual (Myhre-Jensen and Astrup 1971). There then followed a study of smears from the buccal mucosa of man in which a staining after Papanicolaou was introduced in order to try to obtain a better correlation between fibrinolytic cells and the maturation stage (Wünschmann-Henderson and Astrup 1972). Desquamated, blue cells showed a high frequency of fibrinolytic activity. Pink-stained, nucleated cells were sometimes weakly active, but mostly inactive, and anucleate (pink or orange) cells were always inactive (Fig. 3). Particularly high activity was associated

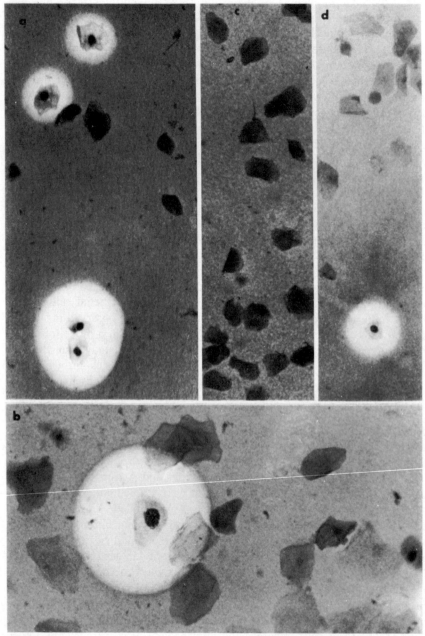

Figure 2

Epithelial cells in vaginal smears from rats. Fibrin slide technique; incubation time on plasminogen-rich fibrin, 10 minutes; hematoxylin stain.

(a, b) Early estrus. Areas of lysis related to nucleated cells. Two degenerating nucleated cells show weak activity. Some inactive nucleated cells and many inactive anuclear cells. (c) Late estrus. Inactive cornified cells. (d) Smears from early estrus on plasminogen-free fibrin incubated for 120 minutes. Weak protease activity around a nucleated cell. Other nucleated or cornified cells are inactive. (Reprinted, with permission, from Henrichsen and Astrup 1967.)

Figure 3
Epithelial cells in smears from human buccal mucosa. Fibrin slide technique; incubation period, 120 minutes; Papanicolaou stain.

(*Center*) Intense lysis around two blue cells; (*top right*) marked lysis around a blue cell adjacent to a pink cell; (*lower center*) weak lysis (barely discernible) around a pink cell. Remaining cells are pink and inactive (Wünschmann-Henderson and Astrup 1972).

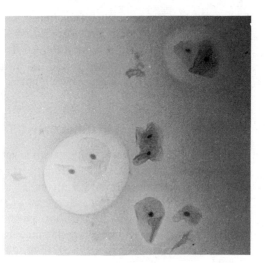

with small blue cells of young age obtained in smears from smokers, presumably indicating a premature desquamation. The activity of cells from buccal smears was low compared to that of many vascular endothelial cells or of surface epithelial cells from the endometrium or vagina, and an incubation period of 120 minutes was required. In a recent study (Wünschmann-Henderson and Astrup 1974), frozen sections of rat embryo tongue kept in organ culture for periods of up to 6 days showed increasing intensity of fibrinolytic activity related to the dorsal epithelium, and the potassium thiocyanide extracts of the explants showed increases in the concentration of plasminogen activator of several hundred times. The supernatant also contained plasminogen activator produced and released by the cells. When the explants were cultivated in the presence of hydrocortisone, the epithelial activity was absent in frozen sections studied on fibrin slides, there was little activity extractable with potassium thiocyanate, and the supernatant was inactive. The vascular endothelial activity was less influenced. Since the production and release of some typical lysosomal enzymes (β-glucuronidase, acid protease) was little influenced by the presence of hydrocortisone, we take this result to indicate that the plasminogen activator is not of lysosomal origin, or that its binding differs from that of regular lysosomal enzymes—at least insofar as concerns the epithelial activity. We have also found that cycloheximide prevents the production of fibrinolytic activity in rat tongue organ cultures (Wünschmann-Henderson and Astrup, in prep.). This result agrees well with the blocking by cycloheximide of plasminogen activator production in Rous sarcoma virus-infected cultures of chicken fibroblasts observed by Dr. Reich's group (Unkeless et al. 1973). The plasminogen activator produced by the cultivation of explants from kidney (Barnett and Baron 1959; Painter and Charles 1962; Bernik and Kwaan 1967), and later studied in more detail by several investigators and produced in larger quantities by Dr. Grant Barlow (this volume), is probably derived from the epithelium lining the urinary conducting tract of the kidney, which possess high fibrinolytic activity in contrast to the low activity of the renal cortex (Epstein, Beller and Douglas 1968; Myhre-

Jensen 1971). The detection of urokinase (the urinary plasminogen activator) by immunochemical methods (Bernik and Kwaan 1969) in several other tissues may reflect common immunological determinants and not an identity of the tissue plasminogen activator and the urinary activator. The activator may be present in the cells as a precursor (Bernik 1973).

Plasminogen Activator in Other Cells

A search for fibrinolytic activity in other surface-protecting cell linings using the histochemical fibrin slide technique revealed that weak activity is associated with serosal cells and with cells lining the synovial membranes of some species (Myhre-Jensen, Bergmann-Larsen and Astrup 1969; Pugatch et al. 1970). It might be added that sperm cells also contain a plasminogen activator, which may be of significance in the mechanism of fertilization (Tympanidis and Astrup, 1968b) (Fig. 4). The plasminogen activator observed in neutrophil leukocytes by several investigators is present in low concentration in comparison with that found in most vascular endothelial cells and in some epithelial cells requiring incubation for 120 minutes. We observed that exposure to endotoxin increased the lytic activity of the leukocytes without affecting their viability, as determined by the ability to exclude trypan blue (Wünschmann-Henderson, Horwitz and Astrup 1972).

So far, cultures of several different cell lines have not in our hands produced fibrinolytically active cells as determined by the histochemical fibrin slide technique. Neither have we observed more than traces of activity in samples of Rous sarcoma assayed by the extraction method. This difference from the results reported by Dr. Reich's group could in part be due to differences in sensitivity of the respective methods and techniques, including

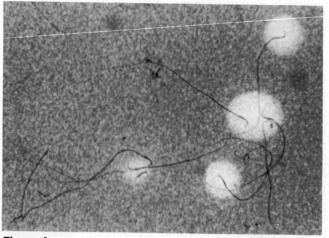

Figure 4
Smear of rat sperm cells prepared from washed semen. Fibrin slide technique; incubation period: 40 minutes. Areas of lysis related to heads of sperm cells. (Reprinted, with permission, from Tympanidis and Astrup 1968.)

the methods of cultivation. Thus, the fibrin slide technique is aimed at identifying and localizing the individual, fibrinolytically active cell, while methods such as those used by Reich's group (Unkeless et al. 1973) and others (Santesson 1935; Åstedt, Pandolfi and Nilsson 1971) determine the breakdown of a substrate by fibrinolytic agents (plasminogen activator, plasmin) released or produced in the culture fluid by all the cells in the culture. The two techniques are, therefore, supplementary.

Cellular Production and Release of Plasminogen Activator

With reference to Figure 5, we believe that our studies have substantiated the following sequence of events:

Proliferating cells, such as are present in the basal cell layers or in rapidly growing tissue cultures, probably have little or no plasminogen activator. This is in agreement with the common rule in cell biology that multiplication excludes differentiation. Later, when certain cells begin to mature and undergo differentiation (in the cases studied here into protective covers for internal and external surfaces), a production of plasminogen activator occurs, probably reaching its maximum just before degeneration begins. During degeneration, the plasminogen activator is released into the surroundings. Fully degenerated cells are inactive. In cell cultures or organ cultures, the plasminogen activator then appears in the culture fluid. In the organism, the endothelial cells release their activity into the bloodstream. The epithelial cells undergo desquamation at epithelial surfaces, possibly assisted by the fibrinolytic enzymes, and the plasminogen activator disappears in the surrounding fluids. Hence, there is a rate of production as well as a rate of release, in addition to the maximal potential activity which can be developed. Whether in a frozen section or in a cell smear many or few cells will show fibrinolytic activity depends upon the number of cells which are in the mature, predegeneration stage. In the case of an epithelial smear, the youngest cells would then be

Potentially fibrinolytically active normal cells:

Vascular endothelial cells,
Surface epithelial cells,
Serosal cells,
Synovial cells,

↓

Phase 1.   Proliferation (inactive)

↓

Phase 2.   Differentiation and maturation (production of activator)

↓

Phase 3.   Desquamation and degeneration (release of activator)

↓

Phase 4:   Cornification (inactive)

Figure 5
Development of cellular fibrinolytic activity.

most active, assuming that desquamation is an indication of maturity and imminent degeneration. The very young proliferating cells are probably not present in carefully obtained smears. If the phase of maturity and production of plasminogen activator covers a prolonged period, more of the cells present will be in this phase, more fibrinolytically active cells will be seen, and more cells will possess high activity. The small, highly active, blue cells in our buccal smear study might represent such a state. The partially blue, partially pink cells and the exclusively pink cells in the buccal smears then represent phases of progressive degeneration with loss of plasminogen activator. If the periods of production and retainment of plasminogen activator are of short duration, then few fibrinolytically active cells will be present in a particular section or smear. Examples exist of a similar mode of production of an active enzyme in cells undergoing degeneration. While alkaline phosphatase is absent in normal fibroblasts, it appears when the cells undergo degeneration (Henrichsen 1956a,b,c). It is of interest, therefore, that we have found good correlation between the presence of plasminogen activator and alkaline phosphatase in vascular endothelial cells at sites of endothelial damage and repair (Glas-Greenwalt, Strand and Astrup, in prep.). In principle, a production and release of active compounds during cellular maturation and degeneration could be looked upon as patterned after the function of the cells of holocrine glands.

In considering the possible induction of an ability to exhibit fibrinolytic activity when normal cells are transformed into neoplastic cells, it should be remembered that a plasminogen activator is present in several types of normal cells. Whether its preferable presence in malignant cells, as indicated in the early literature and now verified in the articles published by Dr. Reich and his associates (Unkeless et al. 1973; see also other papers this volume), is caused by a specific production of the activator in consequence of the transformation, or whether it is a result of changes in the normal rates of production and release of the activator, remains to be seen. The new observations published by Dr. Reich (this volume) will certainly accelerate research in this important field of cell biology. We may look forward to many important findings in the next few years, findings which will elucidate mechanisms of maturation and degeneration of normal cells as well as deal with the transformation of such cells into malignant cells and the properties of the transformed cells. These observations will also add to our understanding of the physiological role of the plasminogen activator and the fibrinolytic system in the living organism, in particular, their role in the desquamation of protective cell layers, in the tissue repair processes aimed at maintaining the integrity of the organism, and in the growth and invasiveness of transformed cells. In conclusion, I want again to emphasize that finally with this conference, the fibrinolytic system has been placed in its true biological context.

Acknowledgment

Supported by USPHS Grant HL-05020 from the National Institutes of Health, National Heart and Lung Institute.

REFERENCES

Åstedt, B., M. Pandolfi and I. M. Nilsson. 1971. Quantitation of fibrinolytic agents released in tissue culture. *Experientia* **27**:358.

Astrup, T. 1956a. Fibrinolysis in the organism. *Blood* **11**:781.

———. 1956b. The biological significance of fibrinolysis. *Lancet* **2**:565.

———. 1966. Tissue activators of plasminogen. *Fed. Proc.* **25**:42.

———. 1967. Blood coagulation, fibrinolysis and the development of the thrombogenic theory of arteriosclerosis. In *Transactions of Conference (June 15–17)*: *Le Role de la Paroi Arterielle dans l'Atherogenese*, p. 535. Centre National de la Recherche Scientifique, Paris, France.

———. 1968. Blood coagulation and fibrinolysis in tissue culture and tissue repair. *Biochem. Pharmacol.* (Suppl.) **17**:241.

Astrup, T. and O. K. Albrechtsen. 1957. Estimation of the plasminogen activator and the trypsin inhibitor in animal and human tissues. *Scand. J. Clin. Lab. Invest.* **9**:233.

Astrup, T. and P. Permin. 1947. Fibrinolysis in the animal organism. *Nature* **159**:681.

Astrup, T., O. K. Albrechtsen, M. Claassen and J. Rasmussen. 1959. Thromboplastic and fibrinolytic activity of the human aorta. *Circ. Res.* **7**:969.

Astrup, T., J. Rasmussen, A. Amery and H. E. Poulsen. 1960. Fibrinolytic activity in cirrhotic liver. *Nature* **185**:619.

Barnett, E. V. and S. Baron. 1959. An activator of plasminogen produced by cell culture. *Proc. Soc. Exp. Biol. Med.* **102**:308.

Beck, E., F. Duckert and M. Ernst. 1961. The influence of fibrin stabilizing factor on the growth of fibroblasts in vitro and wound healing. *Thromb. Diath. Haemorrh.* **6**:485.

Beck, E., F. Duckert, A. Vogel and M. Ernst. 1962. The influence of the fibrin-stabilizing factor on the function and morphology of fibroblasts in vitro. *Z. Zellforsch.* **57**:327.

Bernik, M. B. 1973. Increased plasminogen activator (urokinase) in tissue culture after fibrin deposition. *J. Clin. Invest.* **52**:823.

Bernik, M. B. and H. C. Kwaan. 1967. Origin of fibrinolytic activity in cultures of the human kidney. *J. Lab. Clin. Med.* **70**:650.

———. 1969. Plasminogen activator activity in cultures from human tissues. An immunological and histochemical study. *J. Clin. Invest.* **48**:1740.

Duguid, J. B. 1946. Thrombosis as a factor in the pathogenesis of coronary atheriosclerosis. *J. Path. Bact.* **58**:207.

Epstein, M. D., F. K. Beller and G. W. Douglas. 1968. Kidney tissue activator of fibrinolysis in relation to pregnancy. *Obstet. Gynecol.* **32**:494.

Fischer, A. 1925. Beitrag zur Biologie der Gewebezellen. Eine vergleichend-biologische Studie der normalen und malignen Gewebezellen in vitro. *Arch. Entwicklungs.* **104**:210.

———. 1927a. Dauerzüchtung reiner Stämme von Carcinomzellen in vitro. *Z. Krebsforsch.* **25**:89.

———. 1927b. Carcinomzellen und heterologe Systeme in vitro. *Z. Krebsforsch.* **25**:482.

———. 1930. *Gewebezüchtung. Handbuch der Biologie der Gewebezellen in vitro*. Rudolph Müller and Steinicke, Munich.

Glas, P. and T. Astrup. 1970. Thromboplastin and plasminogen activator in tissues of the rabbit. *Amer. J. Physiol.* **219**:1140.

Glas-Greenwalt, P., C. Strand and T. Astrup. 1972. Fibrinolytic activity in the closed ductus arteriosus. *Experientia* **28**:448.

Henrichsen, E. 1956a. Alkaline phosphatase in osteoblasts and fibroblasts cultivated in vitro. *Exp. Cell Res.* **11**:115.

―――. 1956b. Alkaline phosphatase and the calcification in tissue cultures. *Exp. Cell Res.* **11**:403.

―――. 1956c. Alkaline phosphatase and calcification in tuberculous lymph nodes. *Exp. Cell Res.* **11**:511.

Henrichsen, J. and T. Astrup. 1967. Fibrinolytically active rat vaginal epithelial cells. *J. Path. Bact.* **93**:706.

Kwaan, H. C. and T. Astrup. 1964. Fibrinolytic activity of reparative connective tissue. *J. Path. Bact.* **87**:409.

Myhre-Jensen, O. 1971. Localization of fibrinolytic activity in the kidney and urinary tract of rats and rabbits. *Lab. Invest.* **25**:403.

Myhre-Jensen, O. and T. Astrup. 1971. Fibrinolytic activity of squamous epithelium of the oral cavity and oesophagus of rat, guinea pig and rabbit. *Arch. Oral Biol.* **16**:1099.

Myhre-Jensen, O., B. A. Baetz and T. Astrup. 1973. Fibrinolytic activity in rat gastrointestinal mucosa. *Arch. Path.* **95**:195.

Myhre-Jensen, O., S. Bergmann-Larsen and T. Astrup. 1969. Fibrinolytic activity in serosal and synovial membranes—rats, guinea pigs and rabbits. *Arch. Path.* **88**:623.

Painter, R. H. and A. F. Charles. 1962. Characterization of a soluble plasminogen activator from kidney cell cultures. *Amer. J. Physiol.* **202**:1125.

Pandolfi, M. 1967. Localization of fibrinolytic activity in the developing rat eye. *Arch. Ophthalmol.* **78**:512.

Pandolfi, M. and T. Astrup. 1967. A histochemical study of the fibrinolytic activity. Cornea, conjuctiva and lacrimal gland. *Arch. Ophthalmol.* **77**:258.

Pugatch, M. J., E. A. Foster, D. E. Macfarlane and J. C. F. Poole. 1970. The extraction and separation of activators and inhibitors of fibrinolysis from bovine endothelium and mesothelium. *Brit. J. Haematol.* **18**:669.

Santesson, L. 1935. In *Characteristics of Epithelial Mouse Tumour Cells in Vitro and Tumour Structures In Vivo. A Comparative Study*. Levin and Munksgaard, Copenhagen.

Smokovitis, A., P. Kok and T. Astrup. 1975. Tissue repair in rats in presence of locally applied tissue plasminogen activator. *Exp. Mol. Path.* **22**:109.

Todd, A. S. 1959. The histological localisation of fibrinolysin activator. *J. Path. Bact.* **78**:281.

Tympanidis, K. and T. Astrup. 1968a. Localization of fibrinolytic activity in the rat oviduct. *Obstet. Gynecol.* **31**:727.

―――. 1968b. Fibrinolytic activity of rat, rabbit and human sperm cells. *Proc. Soc. Exp. Biol. Med.* **129**:179.

―――. 1969. Hormonal influence on fibrinolytic activity of uterus and vagina in the juvenile rat. *Acta Endocrinol.* **60**:69.

―――. 1971. Fibrinolytic activity of epithelium of bladder of rat and man. *J. Urol.* **105**:214.

―――. 1972. Fibrinolytic activity in injured rat skin. *Exp. Mol. Path.* **16**:101.

Tympanidis, K., A. E. King and T. Astrup. 1968. Fibrinolytically active cells in human vaginal smears. *Amer. J. Obstet. Gynecol.* **100**:185.

Unkeless, J. C., A. Tobia, L. Ossowski, J. P. Quigley, D. B. Rifkin and E. Reich. 1973. An enzymatic function associated with transformation of fibroblasts by oncogenic viruses. I. Chick embryo fibroblast cultures transformed by avian RNA tumor viruses. *J. Exp. Med.* **137**:85.

Wünschmann-Henderson, B. and T. Astrup. 1972. Relation of fibrinolytic activity in human oral epithelial cells to cellular maturation: The influence of smoking. *J. Path.* **108:**293.

———. 1974. Inhibition by hydrocortisone of plasminogen activator production in rat tongue organ cultures. *Lab. Invest.* **30:**427.

Wünschmann-Henderson, B., D. L. Horwitz and T. Astrup. 1972. Release of plasminogen activator from viable leukocytes of man, baboon, dog and rabbit. *Proc. Soc. Exp. Biol. Med.* **141:**634.

Inhibitors of Plasminogen Activation

Keld Danø and E. Reich

Department of Chemical Biology, The Rockefeller University,
New York, New York 10021

Although plasminogen activators have been known for some time and one of these enzymes has even been crystallized (Lesuk, Termineille and Traver 1965), knowledge of their catalytic properties remains quite limited. Moreover, the literature is controversial with respect to inhibitor effects on proteases of the urokinase type (Alkjaersig, Fletcher and Sherry 1959; Ambrus et al. 1968; Okamoto et al. 1968). In view of the enhanced production of plasminogen activator associated with transformation, neoplasia and macrophage stimulation (Unkeless et al. 1973; Ossowski et al. 1973; Unkeless, Gordon and Reich 1974), it appeared desirable to initiate further studies of this class of enzymatic activities. In particular, the elucidation of the biological role of plasminogen activation would be greatly accelerated if effective, and maximally specific, inhibitors could be identified. With this in mind, we have undertaken a modest screening program to define a spectrum of inhibitory compounds that might provide useful tools for analyzing events connected with the plasminogen activation and plasmin action. The results show that although both plasminogen activator and plasmin are trypsinlike enzymes, there are significant differences in their susceptibility to protease inhibitors. These differences are most pronounced in the case of macromolecular trypsin inhibitors, all of which inhibit plasmin without affecting plasminogen activation.

MATERIALS AND METHODS

Cell Cultures and Serum-free Conditioned Medium

Secondary cultures of mouse embryo fibroblasts were infected with mouse sarcoma virus (MSV), and serum-free conditioned medium was prepared as described by Unkeless et al. (1973). Conditioned medium was concentrated tenfold by filtration in dialysis tubes against Sephadex, dialyzed against 0.1 M Tris-sulfate pH 7.4 for 48 hours, and used as the source for plasminogen

357

activator (cell factor: CF). These preparations contained 2000–3000 units of plasminogen activator per ml (see below).

Plasminogen

Human plasminogen was prepared from outdated human plasma according to the method of Deutsch and Mertz (1970) by two cycles of affinity chromatography on columns of Sepharose 4B substituted with L-lysine. The two species of human plasminogen which appeared on SDS-polyacrylamide gels (see below) were separated by a method similar to that described by Sodetz, Brockway and Castellino (1972) for rabbit plasminogen. The plasminogens were eluted from an L-lysine-substituted Sepharose column with a linear gradient of ϵ-aminocaproic acid (ϵ-ACA, 0–20 mM) with forms a and b eluting at 8 and 12 mM ϵ-ACA, respectively. The plasminogen was labeled with ^{125}I according to the procedure of Helmkamp et al. (1960); a fourfold molar excess of ICl relative to plasminogen was used.

Assays

The assay for plasminogen activation was based on the conversion of the single chain of ^{125}I-plasminogen (specific radioactivity 2–10 mCi/μmole) to the two chains of plasmin. To measure activation, incubations were performed as described in the legend to Figure 3. At the desired times, samples were taken and analyzed by SDS-β-mercaptoethanol-polyacrylamide gel electrophoresis as in Figure 1. The zones corresponding to plasminogen and to the individual chains of plasmin were excised from the gel and their radioactivity determined in a Packard gamma spectrometer.

Another assay for plasminogen activator and plasmin was performed in plastic petri dishes coated with ^{125}I-fibrin, as described by Unkeless et al. (1974). The assay contained 2 μg of human plasminogen in a final volmue of 1 ml 0.1 M Tris-HCl pH 8.1 and was incubated for 1 hour at 37°C. One unit of plasminogen activator was defined as the amount that in 1 hour released 5% of the radioactivity from freshly activated plates. In experiments involving inhibitors, the sensitivity of the plates to fibrinolysis was decreased by storage at 4°C for 24 hours before use. This storage period results in some degree of fibrin cross-linking and slows the rate of fibrinolysis by plasmin.

SDS-β-mercaptoethanol-polyacrylamide gel electrophoresis was performed in a stacking system (Laemmli 1970) using slab gels (0.12 x 9 x 15 cm) of 11% acrylamide overlayered with gels of 4% acrylamide. The gels were run 16–20 hours at 6 mA and 15–25 V. Buffers were as described by Laemmli (1970). Protein was stained with 1% Coomassie blue in 50% (w/v) trichloroacetic acid, and the stained gels were destained in 7% acetic acid.

Bovine pancreatic trypsin inhibitor [BPTI, Kunitz inhibitor, Trasylol (R)], and Germanine (Suramine) were gifts from Bayer AG, Wuppertal, Germany, and nitrophenyl-p-guanidinobenzoate (NPGB) was a gift from Dr. E. Shaw. Trypsin inhibitors prepared from leeches, potato, sea anemone, bovine colostrum, dog submandibularis gland and boar sperm plasma were all gifts from Dr. Hans Fritz. Urokinase from human urine was obtained from Leo Pharmaceutical Industries, Ballerup, Denmark. All other reagents were of the best commercial grade available.

RESULTS AND DISCUSSION

The fibrinolytic activity associated with neoplasia is a two-stage enzymatic process: In the first stage, plasminogen activator, a serine protease released by transformed cells, activates plasminogen by limited proteolysis. The resulting plasmin activity then degrades fibrin or other substrates in the second stage of the reaction (Unkeless et al. 1974). A proper characterization of inhibitors therefore demands the use of assays that permit the effects on each of the two stages to be evaluated individually. Accordingly, we have used two separate assays to define the effect of each inhibitor. In view of its sensitivity, rapidity and convenience, the fibrinolysis assay, which measures the overall reaction by the action of plasmin on an ^{125}I-fibrin substrate, is a useful general screening procedure. Any compound tested at high concentrations that fails to inhibit fibrinolysis under standard conditions of both activator and plasminogen obviously does not inhibit either stage of the reaction. On the other hand, if a compound inhibits the overall reaction, its effect on plasminogen activation can be assessed in a separate and more specific test.

The basis for directly assaying plasminogen activation is shown in Figure 1, where the effect of urokinase on human plasminogen can be seen. As was first shown by Robbins et al. (1967) and Summaria, Hsieh and Robbins (1967), urokinase activates plasminogen by limited proteolysis, in the course of which the single polypeptide chain of plasminogen is converted to the two chains of plasmin. Human plasminogen contains two major components that can be distinguished on SDS-β-mercaptoethanol-polyacrylamide gel electrophoresis (Fig. 1, A), and these are separable by affinity chromatography. When the mixture was incubated with plasminogen activator from transformed mouse cells or with human urokinase, the resulting plasmin appeared to contain two different heavy chains, but only a single species of light chain (Fig. 1, B). When each of the separated plasminogen species (Fig. 1, C and E) was incubated with plasminogen activator, the respective plasmins contained only one species of plasmin heavy chain (Fig. 1, D and F). This shows that the single chain of each of the plasminogen molecules was converted into two chains of plasmin. Both plasmin molecules yielded apparently identical light chains, but the heavy chains reflected the difference in mobility present in the corresponding plasminogens.

To monitor the activation, plasminogen (the crude mixture of *a* and *b*) was first labeled with ^{125}I. The extent of activation was then determined by SDS-β-mercaptoethanol-polyacrylamide gel electrophoresis of aliquots of the reaction. Excision of the zones corresponding to plasminogen and to the two chains of plasmin provided a measure of activation and allowed the recovery of the initial radioactivity to be estimated. Experiments of this kind immediately revealed that extensive proteolysis occurred when pure plasminogen was activated by transformed mouse cell activator or by urokinase. This was evident from the fact that only a minor fraction of the radioactivity in the original ^{125}I-plasminogen could be accounted for by total plasminogen plus plasmin in the gel. Since reactions containing activator alone were stable for relatively long incubation periods, it appeared likely that proteolysis was due to plasmin, and a search was begun for compounds that could block the action of plasmin without affecting plasminogen activation. As seen in Figure

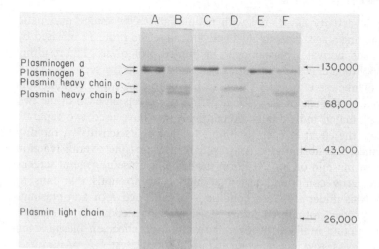

Figure 1

SDS-β-mercaptoethanol-polyacrylamide gel electrophoresis of
(A) human plasminogen, (B) human plasminogen after incubation with urokinase, (C) purified human plasminogen a, (D)
purified human plasminogen a incubated with urokinase, (E)
purified human plasminogen b, and (F) purified plasminogen b
incubated with urokinase. Four micrograms of plasminogen was
applied to each slot. Incubation of plasminogen (160 μg/ml)
with urokinase (100 Ploug units/ml) was for 1 hour at 37°C in
0.1 M Tris-HCl pH 8.1, in presence of BPTI (600 kallikrein
inhibitor units/ml). The reaction was stopped by addition of an
equal volume of buffer (20% glycerol, 6% sodium dodecyl
sulfate, 10% β-mercaptoethanol, 0.125 M Tris-HCl pH 6.8).
After boiling for 2 minutes, samples of 50 μl were applied to a
slab gel of 11% polyacrylamide, and electrophoresis was performed as described under Materials and Methods. When plasminogen activator from transformed mouse cells was used instead of urokinase, similar results were obtained.

2, BPTI, a known plasmin inhibitor, had the desired properties. When increasing concentrations of BPTI were added to reactions in which ^{125}I-plas-
minogen was being activated by transformed mouse cell activator, the overall
recovery of radioactivity increased rapidly; at sufficiently high concentrations
essentially all of the original radioactivity was accounted for in the form of
plasminogen and plasmin. It is significant that the recovery of both plasminogen and plasmin increased in parallel with that of total radioactivity, showing that in the absence of inhibitor, both of these were degraded by any active
plasmin in the reaction. Independent control experiments (results not presented) showed that plasminogen activator was also rapidly destroyed by
plasmin; the presence of BPTI, therefore, prolonged the effective lifetime of
activator molecules and greatly increased the extent of the activation reaction.
In view of these results, all subsequent assays for plasminogen activation were
performed in the presence of high levels of BPTI, no concentration of which
has ever been observed to inhibit the activator.

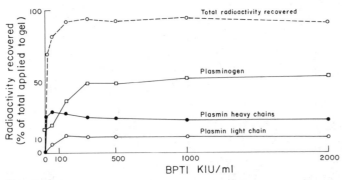

Figure 2

Effect of BPTI on conversion of ^{125}I-plasminogen to plasmin. Samples were incubated for 2 hours at 37°C in 0.1 M Tris-HCl pH 8.1. The incubation mixture contained ^{125}I-plasminogen, 164 μg/ml; plasminogen activator from mouse cells, 1000 units/ml; and BPTI as indicated. Twenty-five μl of each sample was analyzed by gel electrophoresis as described in Figure 1. The bands containing plasminogen and the two chains of plasmin were excised, and the radioactivity was counted in a gamma counter. By "total radioactivity recovered" is meant the radioactivity recovered from all of the slices in each channel expressed as a percentage of that applied for electrophoresis and includes the radioactivity recovered in regions corresponding to plasminogen and plasmin. At low concentrations of inhibitor there is extensive proteolysis; the small peptide fragments emerge from the gel at the electrophoretic front and are not recovered.

The failure of BPTI to inhibit the plasminogen activator has also been proved independently (results not presented) by labeling with [3H]diisopropylfluorophosphate (DFP): BPTI did not reduce the incorporation of [3H]DFP into the plasminogen activator from mouse cells.

Figure 3 shows the activation of ^{125}I-plasminogen (2 μg/ml) as a function of activator concentration and of time. These results form the basis for the standard conditions selected for assaying inhibitors of the activation reaction (50 units plasminogen activator/ml; 1 hr incubation). The SDS-β-mercaptoethanol-polyacrylamide gel electrophoresis radioactivity profile of ^{125}I-plasminogen before and after incubation under these conditions is given in Figure 4. The ratio of radioactivity in the two plasmin chains reflects exactly their molecular weight proportion.

We have tested a series of macromolecular protease inhibitors of plant and animal origin for their effects on the two stages of the fibrinolytic reaction, and the results obtained with soybean and lima bean trypsin inhibitors are qualitatively representative of the entire group. As seen in Figure 5, both of these small proteins block fibrinolysis, but neither inhibits plasminogen activation. Their effects were thus qualitatively the same as those of BPTI, and they acted as plasmin inhibitors, although their potency on a molar basis was appreciably less than that of BPTI. Similar results were observed with macromolecular inhibitors prepared from leeches, potato, sea anemone,

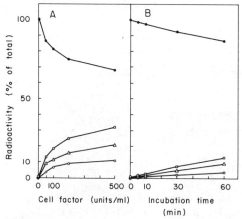

Figure 3

Plasminogen activation as a function of activator concentration and time. (*A*) ^{125}I-plasminogen, 2 μg/ml; BPTI, 300 kallikrein inhibitor units/ml; and plasminogen activator (cell factor) as indicated were incubated for 1 hour in 0.1 M Tris-HCl pH 8.1. The reaction was stopped by addition of an equal volume of buffer (as described in the legend to Fig. 1) and unlabeled, partially activated plasminogen, 100 μg/ml, which provided marker regions for subsequent excision from the gel. Gel electrophoresis, followed by excision of bands and assay of radioactivity, was performed as described in the legend to Figure 2. The background radioactivity throughout the gel was determined by slicing a control channel in which untreated ^{125}I-plasminogen alone had been subjected to electrophoresis, and the resultant values were subtracted from the corresponding regions in the experimental channels. (*B*) Incubation and analyses as in *A*, except that aliquots were mixed with the buffer at the indicated time points. The plasminogen activator concentration during the incubation was 50 units/ml. (●——●) Plasminogen; (○——○) plasmin light chain; (△——△) plasmin heavy chain; (□——□) plasmin total.

bovine colostrum, dog submandibularis gland and boar sperm plasma, all of which inhibited plasmin, but not the plasminogen activator.

We have also screened representatives of numerous classes of small molecules, drugs and electrolytes as potential inhibitors of fibrinolysis. As in the preceding cases, all agents were first tested in the overall fibrinolytic reaction, and those showing activity were examined in the plasminogen activation assay in the presence of BPTI. From the agents tested, those showing a significant degree of inhibition of plasminogen activation are listed in Table 1; almost all

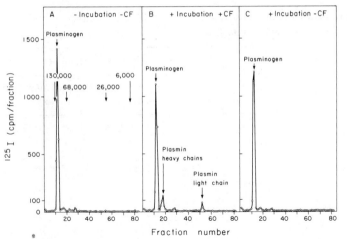

Figure 4

Radioactivity profile of ^{125}I-plasminogen before and after activation with activator from transformed mouse cells. Human ^{125}I-plasminogen, 2 µg/ml; plasminogen activator from transformed cells, 50 units/ml; and BPTI, 300 kallikrein inhibitor units/ml, were incubated for 1 hour in 0.1 M Tris-HCl pH 8.1. The reaction was stopped and electrophoresis was performed as described in the legend to Figure 3. The gel was sectioned longitudinally, and each strip was cut into 1.1-mm slices which were assayed for radioactivity. (*A*) Control, plasminogen activator and incubation omitted; (*B*) incubated with plasminogen activator; (*C*) incubated without plasminogen activator.

Figure 5

Effect of soybean and lima bean trypsin inhibitors on fibrinolysis assay and on plasminogen activation. Both assays were incubated for 1 hour at 37°C. The incubation mixture contained 0.1 M Tris-HCl pH 8.1; human plasminogen, 2 µg/ml; and plasminogen activator from transformed mouse cells, 50 units/ml. The assays for plasminogen activation contained ^{125}I-plasminogen and BPTI, 300 kallikrein inhibitor units/ml. The activation of plasminogen was determined by SDS-β-mercaptoethanol-polyacrylamide gel electrophoresis as described in the legend to Figure 3. Twelve percent of the plasminogen was activated in the control experiments.

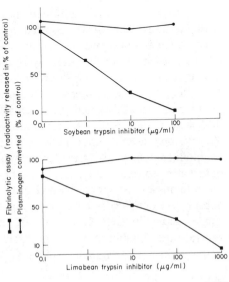

Table 1

Inhibitors of Fibrinolysis and of Activation of Human Plasminogen
by Activator from Transformed Mouse Cells

Inhibitor	Concentration (μM) inhibiting 50%	
	plasminogen activation	overall fibrinolytic assay
NPGB	0.006	0.005
Trypan blue	6	7
$ZnCl_2$	10	4
YCl_3	31	7
$YbCl_3$	31	23
Coomassie blue	32	3
Germanine	70	26
$CdCl_2$	140	70
Diisopropylfluorophosphate	160	140
N-Actyl-L-arginine methyl ester	170	140
$FeCl_3$	230	36
L-Arginine methyl ester	290	260
Benzamidine	330	300
ε-ACA	350	280
$HgCl_2$	450	300
L-Lysine	9000	2400
L-Arginine	14,000	7600
NaCl	34,000	30,000

Experimental conditions were as described in the legend to Figure 4. In both assays, the concentration of plasminogen was 2 μg/ml (22 nM) and that of plasminogen activator was estimated to be about 10 pM. For each inhibitor, the concentration inhibiting 50% was obtained from a log dose-response curve with tenfold increments in concentrations as in Figure 4. (The original observation of inhibitory effect for some of these compounds was made by J. Unkeless, A. Piperno and S. Strickland.)

of the active compounds inhibited the overall fibrinolytic reaction at lower concentrations than those required to inhibit plasminogen activation. The most potent inhibitor was the active-site acylating reagent NPGB, which inhibited both stages of the fibrinolysis assay (results with plasmin not shown) just as it inhibits all other known trypsinlike enzymes. It is of interest that plasminogen activation was sensitive to low concentrations of trypan blue and to other structurally related dyes such as Coomassie blue. The activation reaction was also sensitive to micromolar concentrations of Zn^{++}, Y^{+++}, Yb^{+++} and several other metal ions and to rather higher concentrations of compounds related to lysine and arginine (Fig. 6). We have applied this observation to purification of the plasminogen activator by constructing affinity columns of L-arginine methyl ester-substituted Sepharose, which, under appropriate conditions, have given a purification of well over 1000-fold in a single step (Unkeless et al. 1974). Such columns have been made with several other inhibitors of plasminogen activators and they are also effective in the purification of human urokinase.

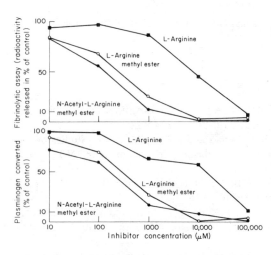

Figure 6
Effect of L-arginine and some arginine derivatives on fibrinolysis and plasminogen activation. The experimental conditions were as described in the legend to Figure 5.

At the present stage of the investigation, no particularly useful or specific inhibitor of plasminogen activator has been identified. Several of the compounds that interfered with plasminogen activation also are known to inhibit other enzymes, including plasmin, with tryptic specificity. It seems reasonable to expect that peptides of appropriate sequence can be constructed to act as specific inhibitors of plasminogen activators of the urokinase type. Knowledge of the amino acid sequence surrounding the activation site in plasminogen should provide a useful basis for the design of such potential inhibitors.

It is noteworthy that none of the macromolecular inhibitors tested so far blocked plasminogen activators of the urokinase type. Thus although these enzymes clearly resemble trypsin in their requirement for substrates containing arginine or lysine residues, the active-site regions must be fundamentally different from those of the large numbers of trypsinlike enzymes that are known to form catalytically inactive complexes with many of these macromolecular inhibitors.

SUMMARY

A direct assay for inhibition of plasminogen activation has been designed. The two-stage fibrinolysis reaction was used to screen a variety of different classes of potential inhibitors. Those showing activity were assayed in the plasminogen activation assay, and several inhibitors of this reaction were found. The overall fibrinolytic assay was more sensitive to a number of the agents than was the activation of plasminogen. Fibrinolysis was blocked by low concentrations of many macromolecular inhibitors of plant and animal origin; all of these inhibited the action of plasmin, but not the activation of plasminogen by enzymes of the urokinase type.

Acknowledgments

Supported in part by a grant from the Goldhirsch Foundation. K. D. holds a fellowship from the University of Copenhagen and the Danish Medical Research Council. We thank Inge Strarup and Myung Chun for excellent technical assistance.

REFERENCES

Alkjaersig, N., A. P. Fletcher and S. Sherry. 1959. ε-Aminocaproic acid: An inhibitor of plasminogen activation. *J. Biol. Chem.* **234**:832.

Ambrus, C. M., J. L. Ambrus, H. B. Lassman and I. B. Mink. 1968. Studies of the mechanism of action of inhibitors of the fibrinolysin system. *Ann. N.Y. Acad. Sci.* **146**:430.

Deutsch, D. G. and E. T. Mertz. 1970. Plasminogen: Purification from human plasma by affinity chromatography. *Science* **170**:1095.

Helmkamp, R. W., R. L. Goodland, W. F. Bale, I. L. Spar and L. E. Mutschler. 1960. High specific activity iodination of γ-globulin with iodine-131 monochloride. *Cancer Res.* **20**:1495.

Laemmli, U. K. 1970. Cleavage of structural proteins during the assembly of the head of bacteriophage T4. *Nature* **227**:680.

Lesuk, A., L. Termineille and J. H. Traver. 1965. Crystalline human urokinase: Some properties. *Science* **147**:880.

Okamoto, S., S. Oshiba, H. Mihara and U. Okamoto. 1968. Synthetic inhibitors of fibrinolysis: In vitro and in vivo mode of action. *Ann. N.Y. Acad. Sci.* **146**:414.

Ossowski, L., J. C. Unkeless, A. Tobia, J. P. Quigley, D. B. Rifkin and E. Reich. 1973. An enzymatic function associated with transformation of fibroblasts by oncogenic viruses. II. Mammalian fibroblast cultures transformed by DNA and RNA tumor viruses. *J. Exp. Med.* **137**:112.

Robbins, K. C., L. Summaria, B. Hsieh, and R. J. Shah. 1967. The peptide chains of human plasmin. Mechanism of activation of human plasminogen to plasmin. *J. Biol. Chem.* **242**:2333.

Sodetz, J. M., W. J. Brockway and F. J. Castellino. 1972. Multiplicity of rabbit plasminogen. Physical characterization. *Biochemistry* **11**:4451.

Summaria, L., B. Hsieh and K. C. Robbins. 1967. The specific mechanism of activation of human plasminogen to plasmin. *J. Biol. Chem.* **242**:4279.

Unkeless, J. C., S. Gordon and E. Reich. 1974. Secretion of plasminogen activator by stimulated macrophages. *J. Exp. Med.* **139**:834.

Unkeless, J. C., K. Danø, G. M. Kellerman and E. Reich. 1974. Fibrinolysis associated with oncogenic transformation: Partial purification and characterization of the cell factor, a plasminogen activator. *J. Biol. Chem.* **249**:4295.

Unkeless, J. C., A. Tobia, L. Ossowski, J. P. Quigley, D. B. Rifkin and E. Reich. 1973. An enzymatic function associated with transformation of fibroblasts by oncogenic viruses. I. Chick embryo fibroblast cultures transformed by avian RNA tumor viruses. *J. Exp. Med.* **137**:85.

Proteinase Inhibitors of Human Plasma— Their Properties and Control Functions

Norbert Heimburger

Behringwerke AG, Marburg/Lahn, West Germany

Proteinase inhibitors (PI's) represent a group of plasma proteins which are characterized by their ability to block the catalytic site of proteolytic enzymes. Since proteases are mediators of the most important biological processes, it is the function of the PI's to control the enzymes involved, i.e., to limit their action to the biological requirements. This is true for blood coagulation, fibrinolysis, inflammation and immunodefense; the latter involve activation of the kallikrein and complement systems. The regulation must occur with high precision since any disturbance of the balanced system may be life-threatening.

Mechanism Controlling Proteinases

The mechanism controlling proteolysis can be summarized as follows:

1. Human plasma proteinases are only present in the form of inactive precursors.
2. The activators are frequently separated from the proteinases by a cell barrier.
3. Liberation of the activators and activation of the proteinases occur only when biologically required.
4. Activation happens locally and is controlled by inhibitors to both activators and proteinases.
5. PI's are found not only in plasma, but also in organs and tissues sensitive to proteolysis, i.e., the lung and mucous membranes of the nasal, respiratory and gastrointestinal tracts.

It is obviously a rule that proteinases become active only locally in order to prevent a systemic proteolysis which requires a high inhibitor capacity. Actual human plasma is extremely rich in PI's.

Evidence of Various Inhibitors of Human Plasma

Combining starch gel electrophoresis and fibrin-containing agar plates in a sandwich technique, human plasma can be shown to contain at least six PI's (Schwick, Heimburger and Haupt 1966). This technique is demonstrated in Figure 1. To prove the presence of PI's, plasma is separated electrophoreti-

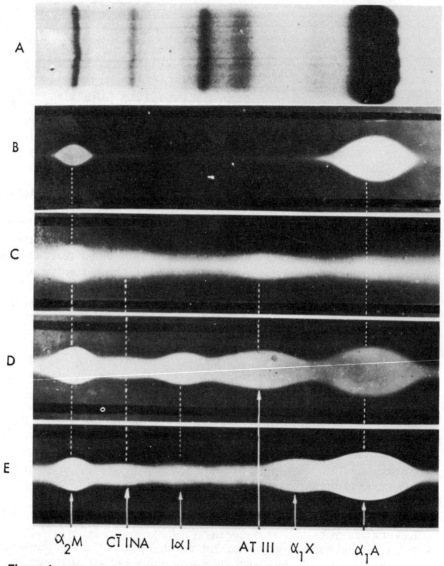

Figure 1

Identification of six proteinase inhibitors in human plasma. Starch gel electrophoresis used in combination with fibrin agar plates according to the sandwich technique. (*A*) Electrophoretogram of plasma stained with amido black; (*B, C, D, E*) fibrin plates after being covered with and lysed by elastase, plasmin, trypsin and chymotrypsin, respectively.

cally in starch gel. The gel is then cut into two discs; the upper one is stained with amido black, and the lower one is used to cover agar plates containing heat-inactivated fibrin. As soon as the proteins have entered the fibrin-agar film, troughs parallel to the migration direction are cut and filled with enzyme solutions. The substrate plates are then incubated at 37°C for about 20 hours. During this time, the proteases enter the gel, and diffusion is evidenced by lysis of the fibrin. Fibrin remains free from attack only in the electrophoretic positions of the inhibitors. By this technique, it can be demonstrated that plasma contains a total of six inhibitors against elastase, plasmin, trypsin and chymotrypsin, five of which have proved to be polyvalent. The PI's were named according to their electrophoretic mobility and their first recognized or most important physiological function, as far as was known at the time of isolation.

Table 1 is a compilation of the PI's together with their useful abbreviations and the normal concentration ranges in which they are found in human plasma. The inhibitors below the line have not yet been isolated in pure and active form. However, they have been traced by their function and can be shown to be not identical to one of the known PI's. Consequently, it might be suggested that in addition to α_2-macroglobulin (α_2M), there is a further inhibitor of thiol proteinases in plasma. There is also evidence for further

Table 1

Proteinase Inhibitors from Human Plasma

Name	Abbreviation	Normal range (mg/100 ml)
α_1-Antitrypsin	α_1A	200–400
α_1-Antichymotrypsin	α_1X	30–60
Inter-α-trypsin inhibitor	IαI	20–70
Antithrombin III	AT III	17–30
C$\overline{1}$ Inactivator	C$\overline{1}$ INA	15–35
α_2-Macroglobulin	α_2M	150–350♂ 175–420♀

Thiol proteinase inhibitor[a]

Inhibitors of clotting factors:
    factor II$_a$, thrombin[b]

Factor XI$_a$, plasmathromboplastin
    antecedent[c]

Inhibitors of plasminogen
    activation[d]

[a] Sasaki (pers. comm.).
[b] A high molecular conjugated protein (Miller-Andersson, Andersson and Borg 1973).
[c] Amir, Ratnoff and Pensky (1972).
[d] A low molecular weight (75,000) and a high molecular weight protein (Hedner 1973).

inhibitors of clotting factors—one involves the inhibition of factor XI_a by a protein not characterized up to now and another, the inhibition of thrombin by a conjugated protein which might be identical to the β-lipoprotein. Finally, the presence of two inhibitors of plasminogen activation has been proved. They differ according to their molecular weights. The low molecular one has already been partially characterized and is an α_2-globulin with a molecular weight of about 75,000 (Hedner 1973).

Historical Review

During the last 20 years, all six PI's of human plasma have been isolated and chemically and biologically characterized. Many scientists have contributed to this work. The earliest observation concerning the antitryptic activity of serum dates back to the end of the last century (Fermi and Pernossi 1894; Camus and Gley 1897; Hahn 1897), and the first attempts to concentrate and isolate the PI's were reported by Landsteiner in 1900. It was not until 1938, however, that Schmitz was actually successful. He applied the methods used by Northrop and Kunitz (1936) in the isolation of the pancreatic inhibitors to the fractionation of the plasma inhibitors. Over a period of years, Despainsmith and Lindsley (1939), Shulman (1952) and Jacobsson (1953, 1955) developed the preparative electrophoresis for the isolation of PI's. The first successful isolation of α_1-antitrypsin (α_1A) was reported by Moll, Sunden and Brown (1958) and one year later by Bundy and Mehl (1959). Three years later (Schultze, Heide and Haupt 1962a), it turned out that the α_1A is identical with the α_1-3,5-glycoprotein already isolated in 1955 by Schultze et al. In 1954–55, α_2M was discovered independently by Brown et al. (1954) and Schultze et al. (1955). In 1962, α_1X-glycoprotein was described (Schultze, Heide and Haupt 1962b), which later proved to be identical to α_1-antichymotrypsin (α_1X) (Heimburger and Haupt 1965).

Antithrombin III (AT III) had been traced by many groups (Loeb 1956; Markwardt and Walsmann 1959; Hensen and Loeliger 1963) before it was isolated by Abildgaard (1967) and ourselves (Heimburger 1967). The inter-α-trypsin inhibitor (IαI) was described in 1965 by our laboratory (Heide, Heimburger and Haupt), although it had already been known before as protein π (Steinbuch 1961).

The inactivator efficiency of human serum for C1 esterase was observed by Ratnoff and Lepow in 1957.

Characterization of the Proteinase Inhibitors

During the last 15 years, we have been successful in isolating the main PI's of human plasma by means of conventional methods (Heimburger, Haupt and Schwick 1971). In the last three years, more modern methods utilizing the principle of affinity chromatography have been introduced. It has been used for both the binding of inhibitors and the elimination of unspecific proteins, for instance albumin when isolating α_1A. For this purpose, the specificity of albumin to combine with its homologous antibody (Myerowitz, Handzel and Robbins 1972) and to bind tightly to anionic aromatic dyes (Travis and

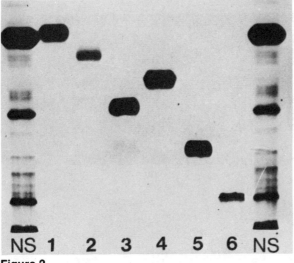

Figure 2
Characterization of the isolated inhibitors in polyacrylamide gel electrophoresis at pH 8.6. NS = normal serum; (1) α_1-antitrypsin; (2) α_1-antichymotrypsin; (3) inter-α-trypsin inhibitor; (4) antithrombin III; (5) C$\overline{1}$ inactivator; (6) α_2-macroglobulin.

Pannell 1973) has been used. For the isolation of α_1A and AT III, the affinity of the inhibitors to concanavalin (Liener, Garrison and Pravda 1973), heparin and dextran sulfate, respectively (Miller-Andersson, Andersson and Borg 1973), has been applied successfully.

Figure 2 shows the characterization of the main plasmatic inhibitors by means of polyacrylamide gel electrophoresis. α_1A and α_1X migrate in front of the plasma proteins; α_2M and the C$\overline{1}$ inactivator (C$\overline{1}$ INA) remain near the starting point, and the IαI and AT III are found in between.

In Figure 3 the plasmatic PI's are characterized by means of immuno-electrophoresis and specific antisera. To indicate the electrophoretic positions of the PI's, human serum was separated on a plate containing heat-denatured fibrin which was developed thereafter by trypsin and an antiserum to total human serum (plate A). It is evident that all plasmatic PI's migrate together with the α_1- and α_2-globulins. Plates B–G show electrophoretograms of human sera after developing with specific antisera to the six inhibitors. By means of the double diffusion technique of Ouchterlony, we were able to demonstrate that the six PI's are individual proteins, not related to each other.

All PI's are glycoproteins; the most relevant data (mean concentrations found in serum, molecular weights, peptide and carbohydrate contents) are compiled in Table 2. Based on concentrations, α_1A and α_2M are predominant, whereas the other inhibitors are only present in trace amounts in human sera. The concentration of all plasma inhibitors amounts to about 700 mg/100ml, which means that 10% of the plasma proteins are carriers of inhibitor func-

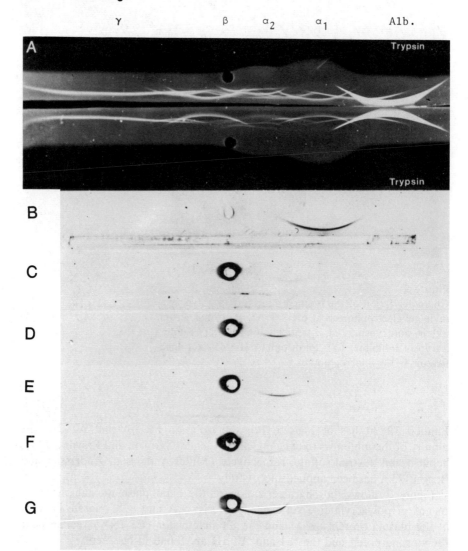

Figure 3

Identification of six proteinase inhibitors in human serum by immunoelectrophoresis using monospecific antisera. On plate A, human serum was electrophoretically separated on an agarose plate containing heat-denatured fibrin which was developed thereafter by trypsin and an antiserum to the total human serum proteins (A-HS). Agarose plates B to G were developed by use of specific antisera from rabbits to the individual inhibitors: A-α_1A, A-α_1X, A-IαI, A-AT III, A-C$\overline{1}$ INA, A-α_2M.

tions. The molecular weights vary from 54,000 for α_1A to 725,000 for α_2M (Jones, Creeth and Kekwick 1972). Even the carbohydrate content differs from one inhibitor to another; especially rich in carbohydrates are α_1A and C$\overline{1}$ INA, containing 25 and 35%, respectively.

Table 3 reflects the individual carbohydrate composition of the inhibitors. Up to now there is no indication that the carbohydrate moiety is necessary for the inhibitor function. By treating with neuraminidase, for instance, the

Table 2

Concentration and Characterization of Inhibitors in Human Sera

Inhibitor	Conc. mean ± SD (mg/100 ml)	MW	Peptide content %	Carbo-hydrate content %
α_1-Antitrypsin	290.0 ± 45.0	54,000	86	12.2
α_1-Antichymotrypsin	48.7 ± 6.5	69,000	73	24.6
Inter-α-trypsin inhibitor	50.0	160,000	90	8.4
Antithrombin III	23.5 ± 2.0	65,000	85	13.4
C$\overline{1}$ Inactivator	23.5 ± 3.0	104,000	65	34.7
α_2-Macroglobulin	260.0 ± 70.0	725,000	92	7.7
	695.7			

Data from Heimburger, Haupt and Schwick (1971).

electrophoretic mobility, but not the biological function, is affected. The carbohydrate moiety, however, might be of importance for the active transport of the inhibitors across the cell membrane. This has been discussed by Bell and Carrell (1973) in connection with the reduced cell secretion of α_1A in cases of deficiency.

Mechanism of Action of Proteinase Inhibitors

The interaction between proteinases and inhibitors is characterized by the formation of an enzymatically inactive complex having an electrophoretic mobility different from free inhibitors, as can be shown by means of immuno- as well as crossed electrophoresis. The exact mechanism of action of the plasmatic inhibitors is still unknown. However, interactions analogous to those found for the low molecular inhibitors are suggested: a proteolytic attack of the inhibitor by the enzyme, hydrolysis of a sensitive peptide bond in the so-called reactive center, followed by acylation of the catalytic site of the enzyme (Laskowski and Sealock 1971).

This mechanism of action is supported by results obtained by masking of the amino acid residues in the active and reactive sites, respectively, of the proteinases and inhibitors (Heimburger, Haupt and Schwick 1971; Cohen 1973). By means of this principle, it can be demonstrated that all plasmatic trypsin inhibitors contain a positively charged amino acid in the reactive site. Recent data of Cohen (1973) indicate that trypsin and chymotrypsin attack α_1A at two different sites.

The velocity at which complexes are formed by proteinases and inhibitors allows the differentiation between immediate and progressive inhibitors. Both types of inhibitors are observed in human plasma. For instance, trypsin and chymotrypsin are blocked by α_1A in an immediate manner, whereas acrosin, plasmin and kallikrein are blocked in a progressive manner. Considering the kinetics of reaction, the main thrombin inhibitor AT III merits special interest; it combines with thrombin in a strong time- and temperature-dependent reaction. The coagulation time of thrombin in a mixture with AT III increases

Table 3
Percent of Carbohydrate Residues in Proteinase Inhibitors

	α_1-Anti-trypsin	α_1-Anti-chymotrypsin	Inter-α-trypsin inhibitor	Antithrombin III	$C\overline{1}$ inactivator	α_2-Macro-globulin
Hexoses	5.0	9.9	3.1	6.2	10.8	3.2
Fucose	0.2	0.7	0.1		0.4	0.1
Acetylhexosamines	3.6	7.4	3.2	4.1	9.2	2.7
Sialic acid	3.4	6.6	2.0	3.1	14.3	1.7
	12.2	24.6	8.4	13.4	34.7	7.7

Data from Heimburger, Haupt and Schwick (1971).

as a function of the incubation time, which means that thrombin is slowly neutralized. By the addition of heparin, however, the inactivation ratio of thrombin by AT III is increased enormously. This effect, which is widely used therapeutically, can be explained by an interaction of AT III with heparin. Lysyl residues of AT III function as binding sites, as has been assessed recently by Rosenberg and Damus (1973). The complex can be dissociated again by the heparin antidote protamine. To explain the action mechanism of heparin, the following hypothesis has been put forth by Rosenberg and Damus (1973): Due to sterical hindrance, the reactive site of AT III, characterized by an arginine residue(s), is only accessible to thrombin with difficulty; by binding of heparin, however, a conformational change is triggered which opens the reactive site to thrombin.

Considering the action mechanism, α_2M differs from the other inhibitors: proteinases bound to α_2M are still highly active against low molecular substrates, whereas they scarcely attack proteins (Mehl, O'Connel and Degroot 1964). Consequently, the conclusion can be drawn that the catalytic sites of the enzymes bound still do work, only that high molecular substrates are excluded by steric hindrance. This special type of inhibition is explained by the "trap" mechanism put forth by Barrett and Starkey (1973). In my opinion, the "trap" theory gives a reasonable explanation for nearly all data collected with α_2M. The "trap" mechanism requires the attack of α_2M by a proteolytic enzyme. This results in a conformational change entrapping the proteinase, which now can neither escape nor be displaced. This might also explain the apparent noncompetitive character of the inhibitor (Barrett and Starkey 1973). A necessity for a functionally active α_2M is a native quaternary structure (Steinbuch and Reuge 1970).

The proteolytic attack and the resulting conformational change have been confirmed by means of different methods. The inhibition mechanism of α_2M represents a more general defense system because it is directed against many proteinases independent of their enzyme class and origin. Considering the principle, it closely resembles the action of antibodies (Heimburger 1974a). It is not so specific, but yet very effective and simple: in contrast to antibodies which are synthesized fitting to each individual antigen, proteinases are sterically caught at the moment when they attack the α_2-macroglobulin.

Biological Functions of Proteinase Inhibitors

In Table 4 the PI's are compiled according to their preferential interaction with proteinases from human plasma and from endogenous origin. "Plus" means that there is strong inhibition on a stoichiometric level, "weak" indicates an unspecific inhibition, and "minus" indicates no inhibition. In some cases, the molar combining ratios of proteinases and inhibitors are still not yet definitively determined. In some inhibitors, discrepancies exist concerning their binding ratios. This applies to α_1A (Johnson, Pannell and Travis 1974) and α_2M (Mehl, O'Connel and Degroot 1964; Barrett and Starkey 1973) and might have two reasons: (1) the denaturation sensitivity of the plasma inhibitors, which complicates their isolation in an absolute active and pure form, and (2) the relatively low specific activity of enzyme preparations used for a long period, which only became evident by the introduction of the active-site titration.

Table 4

Spectrum of Action of the Plasmatic Proteinase Inhibitors

Proteinase name	function	Inhibitor $\alpha_1 A$	$\alpha_1 X$	$I\alpha l$	AT III	$C\bar{1}$ INA	$\alpha_2 M$
Thrombin factor II$_a$	coagulation	−	−	−	+	−	+
Clotting factor X$_a$		−	−	−	+	−	−
Clotting factor XI$_a$		−	−	−	?	+	−
Clotting factor XII$_a$		−	−	−	−	+	−
Plasmin	fibrinolysis	+	−	−	weak	+	+
C1r	complement	−	−	?	−	+	−
C1s		−	−	?	−	+	−
Prekallikrein activator	kallikrein	+	−	−	−	+	+
Plasma kallikrein		+	−	−	−	+	+
PF/Dil	permeability	−	−	?	+	+	?
Acrosin	fertilization	+	−	+	+	−	+
Trypsin	pancreatic hydrolysis	+	−	+	+	weak	+
Chymotrypsin		+	+	weak	−	weak	+
Elastase	phagocytosis	+	−	−	−	−	+
Collagenase		+	−	−	−	−	+
Cathepsin D		−	−	−	?	?	+
Papain	metabolism	−	−	?	?	?	+
Bromelain		−	−	?	?	?	+
Ficin		−	−	?	?	?	+

(+) Strong, probably stoichiometric inhibition; (?) not determined; (−) no inhibition.

From Table 4 it is evident that inhibitors against all proteinases that catalyze important biological processes are present in human plasma: coagulation, fibrinolysis, immunodefense via complement, kinin liberation, enhanced cell permeability by plasma factors, fertility, pancreatic hydrolysis and infection defense on the level of proteolytic enzymes of leukocytes. Proteinases not only of animal origin, but also those from bacteria and plants are inactivated by the plasma inhibitors. This is especially true of α_2M. Up to now, it is the only plasma inhibitor shown to combine with thiol proteinases, as has been demonstrated by Barrett and Starkey (1973) and confirmed by Sasaki et al. (1974).

With the exception of α_1X, all PI's have a broad spectrum of action. However, conclusions regarding their actual physiological function cannot be drawn. For this purpose, the names are also more confusing than helpful because they were chosen according to the functions primarily observed. This was, as already pointed out in the historical review, the antitryptic activity of serum. Due to the fact that the thrombin inhibitor capacity had already been detected by Morawitz (1905), the AT III is the only inhibitor termed functionally correct. There are two improved ways to determine the actual function of the various PI's. The first is to ascertain the molar concentration in plasma and the relative affinity of the inhibitors for different proteolytic enzymes (Ohlsson 1974); in addition, the partition of the inhibitors in the extravascular space should be determined as suggested by Steinbuch and Audran (1974). The second is to discover diseases associated with an isolated deficiency of defined inhibitors. Of course this is only by chance. Based on these results, the various inhibitors were arranged in Table 5 according to their most important physiological functions.

Inhibitors of the Proenzyme Activation

There is no doubt that the $C\overline{1}$ INA controls the first step in the activation of all four proteolytic enzyme systems of the blood: coagulation, fibrinolysis, kallikrein and complement. This results from the inhibition of factor XI_a (PTA, plasmin thromboplastin antecedent), factor XII_a (Hageman factor; Forbes, Pensky and Ratnoff 1970) and its active degradation product which has been revealed to be identical to the prekallikrein activator (Kaplan and Austen 1971).

Furthermore, C1r and C1s, the subcomponents of the first complement factor, are also blocked (Ratnoff et al. 1969). According to the inhibitor activity first observed, the inhibitor was named $C\overline{1}$ INA by Levy and Lepow (1959). The discovery of the $C\overline{1}$ INA became a subject of clinical interest in 1962. At that time, a $C\overline{1}$ INA deficiency was diagnosed by Landermann et al. (1962) in patients suffering from hereditary angioneurotic edema; this was later confirmed by Donaldson and Evans (1963). Since then, some variants of the disease have been observed; one form is characterized by both low antigen and inhibitor levels and another by normal antigen values but a functional deficient protein. In a special variant described by Laurell and Martensson (1971), the $C\overline{1}$ INA is complexed to albumin, resulting in an inactivation. However, the pathogenical mechanism is still a subject of speculation, in part due to the central position of the Hageman factor in the rela-

Table 5
The Function of the Most Important Inhibitors

Inhibitor	Function	Pathology
$\overline{\text{CI}}$ INA	Activation control of blood coagulation, fibrinolysis, kallikrein and complement	Hereditary angioneurotic edema: swelling of skin and mucous membranes due to a decreased or functionally inactive inhibitor
AT III α_2M	Local limitation of coagulation	Hereditary AT III deficiency; thrombo-embolic disorders. Consumption of AT III: hyper-coagulability
α_2M $\overline{\text{CI}}$ INA α_1A	Limitation of fibrinolysis to the pathological substrates	Consumption of α_2M during fibrinolytic therapy
α_2M α_1A IαI	Control of infections and inflammations on the level of proteinases involved, their neutralization and clearance Prevention of autodigestive processes and side reactions to the coagulation system	Hereditary α_1A deficiency: pulmonary emphysema Local consumption of inhibitors during acute inflammations, infections of the nasal mucous membrane, rheumatoid arthritis in the synovial fluid, in burned patients, leukemia and endotoxin shock

tionship between blood coagulation, fibrinolysis, kallikrein and complement.

The inhibitors of plasminogen activation belong to the group of inhibitors controlling enzyme activation. This is in good agreement with observations that low fibrinolytic activity is found in the presence, and high activity in the absence, of these inhibitors (Hedner, Nilsson and Jacobsen 1970).

Thrombin Inhibitors

Blood coagulation is controlled by three PI's: AT III, α_2M and a high molecular conjugated plasma protein. AT III obviously is the main thrombin inhibitor: after elimination of AT III from plasma by adsorption to a resin containing covalently bound heparin, only a weak thrombin inhibitor capacity remains. Also, factor X_a, the prothrombin activator, is blocked by AT III (Biggs et al. 1970). In contrast to α_2M, the inactivation ratio of thrombin by AT III is catalytically accelerated by heparin. The plasma thrombin inhibitor capacity is estimated to be 4–5-fold of the thrombin potential.

The role of the high molecular conjugated thrombin inhibitor is not yet completely clear. However, it might be of interest to consider the fact that it neutralizes thrombin immediately and is inactivated by heparin (Miller-Andersson, Andersson and Borg 1973).

Persons completely deficient in AT III have not yet been observed. It can be postulated that they would not be able to live; in a family with an inherited AT III reduction to 50% of the normal value, Egeberg (1965) discovered a high rate of thrombo-embolic disorders. The AT III assay has been revealed to be of clinical value (for references, see Heimburger 1974a,b). Reductions of the normal value may signal an hypercoagulability. This has been observed during estrogen therapy, after operations, and during disseminated intravascular coagulation (DIC). Obviously it is the physiological role of the thrombin inhibitors to prevent an overshoot of a systemic process, as can be seen in the case of DIC.

Plasmin and Kallikrein Inhibitors

There are three inhibitors in human plasma which neutralize kallikrein as well as plasmin: $\alpha_1 A$, $\alpha_2 M$ and $C\bar{1}$ INA. Concerning the kinetics of inhibition, $\alpha_2 M$ and $C\bar{1}$ INA are inhibitors of the immediate type and $\alpha_1 A$ is of the progressive type. This had already been recognized by Norman (1958) when he localized the inhibitors of plasmin in the α_1- and α_2-globulin zone. Some years later, we were able to identify the inhibitors as $\alpha_1 A$ and $\alpha_2 M$, respectively (Schultze et al. 1963). Considering both the relative affinity and the concentration in plasma, $\alpha_2 M$ is the main plasmin inhibitor. This also can be concluded from assays performed during a fibrinolytic therapy. As $\alpha_2 M$ drops together with the plasminogen, the other PI's are not significantly influenced (Arnesen and Fagerhol 1972). The plasma potential of $\alpha_2 M$ is at least equal to that of plasminogin; in fact, it is probably even twice as much. The half-life of an $\alpha_2 M$ plasmin complex is relatively short, as has been assessed by Blatrix et al. (1973). It is the function of $\alpha_2 M$ to limit the action of plasmin to the pathological substrate, namely, the fibrin occlusion, because in the presence of free plasmin, autodigestive destruction of the endothelial surface will occur. On intravenous injection of plasmin into the ears of rabbits, the following intermittence of the flow, flash, swelling and inflammation associated with a rise of temperature are observed. Application of an equal amount of plasmin in a mixture with a neutralizing dosage of inhibitor is well tolerated (Heimburger, unpubl.). While looking for enzymes dissolving pulmonary hyaline membranes, we observed a proteolysis of lung parenchyms when incubating lung tissues with plasmin (Harms et al. 1975).

Elastase Collagenase Inhibitors and Others

The main function of $\alpha_2 M$ as well as $\alpha_1 A$ is to protect against proteinases involved in infections and inflammations. This has been assessed by in vitro and animal experiments and by observations in patients.

$\alpha_1 A$ and $\alpha_2 M$ can mainly neutralize proteinases found in various infectious organisms, for instance, the subtilopeptidases A and B from *Bacillus subtilis* (Wicher and Dolovich 1973), the proteinases from *Proteus vulgaris* (Kueppers and Bearn 1966) and *Pseudomonas aerugiosa* (Hochstrasser, Theopold and Brandl 1973) and keratinase from *Trichophyton mentagrophytes* (Yu, Grappel and Blank 1972).

α_2M has a stronger affinity to trypsin than to all other PI's. It has been determined by Ganrot (1967) and confirmed and extended to other proteinases, such as elastase and collagenase of granulocytes, by Ohlsson (1974). The latter found a tenfold higher affinity in α_2M than in α_1A for trypsin and collagenase, but they did not differ in their affinity for chymotrypsin and elastase (Ohlsson 1971a). Trypsin infusions in dogs affected α_1A only slightly, but caused a rapid drop of α_2M. Irreversible shock has been found associated with consumption of α_2M (Ohlsson 1971b), but neither the absence nor a considerable reduction of α_2M from circulation have yet been observed. Obviously it is resynthesized very quickly after consumption, as is the case for the acute phase proteins, for instance, α_1A.

α_1A is inherited in a number of genetic variants comprised under the PI system (Fagerhol and Laurell 1967), including forms characterized by a deficient synthesis of α_1A. A heavy deficiency, however, represents a high risk for pulmonary emphysema, as has been shown by Eriksson (1964) and Ganrot, Laurell and Eriksson (1967b). The pathogenetic background of this disease is seen in an autodigestive destruction of lung tissue by proteinases from leukocytes (Mittman 1972).

Consumption of inhibitors in acute inflammations associated with an accumulation of granulocytes has been reported by Ohlsson (1974). In haemorrhagic pancreatitis and appendix peritonitis, he observed complexes of α_1A and α_2M containing elastase and collagenase, which he identified by means of specific antisera.

Mucous membranes are protected from proteolytic enzymes by high inhibitor capacities. This is also true of the nasal mucous membrane. After infections, however, Reichert, Hochstrasser and Werle (1971) observed a rapid fall of the inhibitors and a rise of free enzymes. This could be explained by a consumption of the inhibitors which were found complexed with proteinases.

Functionally inactive α_2M was seen in the synovial fluid during rheumatoid arthritis (Shtacher, Maayan and Feinstein 1973). In burned patients, the metabolism of α_2M was found significantly increased, thereby compensating for the continuous loss from the burned area (Farrow and Baar 1973).

Finally, it should be mentioned that increased α_2M has been found in diabetes mellitus and in several conditions associated with growth; e.g., in pregnancy, in childhood (Ganrot and Scherstén 1967; Ganrot and Bjerre 1970), in regeneration and tumor growth (Heimburger, unpubl.). In cell cultures, it substitutes for serum (Marr, Owen and Wilson 1962; Healy and Parker 1966; Landureau and Steinbuch 1970).

By neutralization of the proteolytic enzymes involved in the defense mechanism, the PI's play an important role in the dynamic equilibrium between the various enzyme systems of blood. From the fact that inflammation areas are bordered by fibrin deposits, it can be concluded that during the inflammation process substances are liberated which catalyze an intravascular coagulation (Fig. 4). These substances might be identical with proteinases, considering the amount released from leukocytes and lysosoma. When liberated in concentrations higher than the inhibitor level in the vicinity, they will cause endothelial lesions (Janoff and Zelios 1968) by the unmasking and

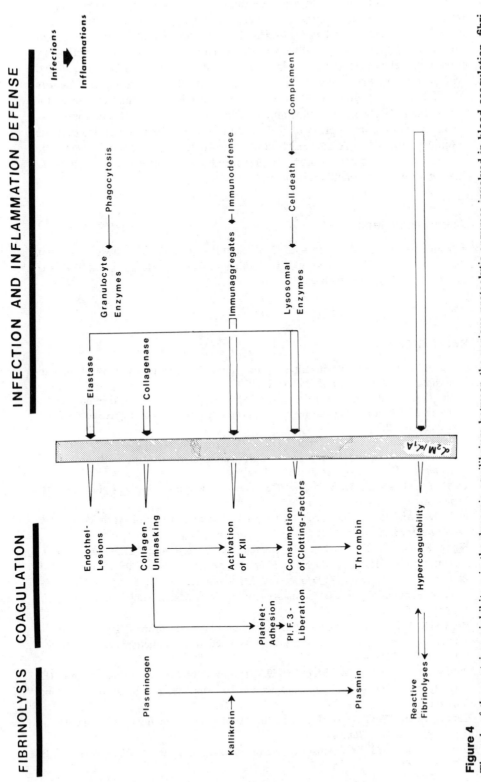

Figure 4
The role of the proteinase inhibitors in the dynamic equilibrium between the various proteolytic enzymes involved in blood coagulation, fibrinolysis, infections and inflammations.

breakdown of collagen. Reactions like these are catalyzed by collagenase, for instance. Endothelial lesions, however, can give rise to an intravascular coagulation via platelet aggregation and liberation of platelet factor 3. Another, still hypothetical, pathway is the inactivation of clotting factors by proteolytic enzymes; the so-called "consumption coagulopathy" associated with severe bleeding diseases. Consumption of fibrinogen and clotting factors V, VIII, XII and XIII has already been diagnosed in leukemia and in endotoxin shock (Egbring, Schmidt and Havemann 1973a, b). It was found associated with a consumption of α_1A. It can be concluded, therefore, that the inhibitors, mainly α_1A and α_2M, contribute to the equilibrium between the various enzyme systems of blood by preventing an uncontrolled activation and inactivation of the coagulation mechanism.

Acknowledgments

I would like to thank Professors H. E. Schultze and H. G. Schwick for their encouragement of this work, and my colleagues Dr. K. Heide, Dr. Th. Kranz and Mr. H. Haupt for their assistance.

REFERENCES

Abildgaard, U. 1967. Purification of two progressive antithrombins of human plasma. *Scand. J. Clin. Lab. Invest.* **19**:190.

Amir, J., O. D. Ratnoff and J. Pensky. 1972. Partial purification and some properties of a plasma inhibitor of activated plasma thromboplastin antecedent. *J. Lab. Clin. Med.* **80**:786.

Arnesen, H. and M. N. Fagerhol. 1972. α_2-Macroglobulin, α_1-antitrypsin and antithrombin III in plasma and serum during fibrinolytic therapy with urokinase. *Scand. J. Clin. Lab. Invest.* **29**:259.

Barrett, A. J. and P. M. Starkey. 1973. The interaction of α_2-macroglobulin with proteinases. *Biochem. J.* **133**:709.

Bell, O. F. and R. W. Carrell. 1973. Basis of the defect in α_1-antitrypsin deficiency. *Nature* **243**:410.

Biggs, R., K. W. E. Denson, N. Ackman, R. Borrett and M. Haddon. 1970. Antithrombin III, antifactor Xa and heparin. *Brit. J. Haematol.* **19**:283.

Blatrix, C., P. Amouch, J. Drouet and M. Steinbuch. 1973. Study on the plasmatic elimination of the α_2-macroglobulin/proteinase complexes. *Path. Biol.* (Suppl.) **21**:11.

Brown, R. K., W. H. Baker, A. Peterkofsky and D. L. Kauffman. 1954. Crystallization and properties of a glycoprotein isolated from human plasma. *J. Amer. Chem. Soc.* **76**:4244.

Bundy, H. F. and J. W. Mehl. 1959. Trypsin inhibitors of human serum. II. Isolation of the α_1-inhibitor and its partial characterization. *J. Biol. Chem.* **234**:1124.

Camus, L. and E. Gley. 1897. Action du serum sanguin sur quelques ferments digestifs. *C. R. Soc. Biol.* **49**:8.

Cohen, A. B. 1973. Mechanism of action of α_1-antitrypsin. *J. Biol. Chem.* **248**: 7055.

Despainsmith, L. and C. H. Lindsley. 1939. Inhibition of proteinases of certain Chlostridia by serum. *J. Bact.* **38**:221.

Donaldson, V. H. and R. R. Evans. 1963. A biochemical abnormality in hereditary angioneurotic edema: Absence of serum inhibitor C1-esterase. *Amer. J. Med.* **35**:37.

Egbring, R., W. Schmidt and K. Havemann. 1973a. Possible destruction of clotting factors (factor I and XIII) by leucocyte proteases in acute leukemia. *4th International Symposium on Thrombosis and Haemostasis,* Abs. 30.

————. 1973b. Untersuchungen zur Äthiologie des Faktor XIII-, Faktor V- und Fibrinogenmangels bei akuten Leukämien. *Verh. Dtsch. Ges. Inn. Med.* **79**: 1351.

Egeberg, O. 1965. Inherited antithrombin deficiency causing thrombophilia. *Thromb. Diath. Haemorrh.* **13**:516.

Eriksson, S. 1964. Pulmonary emphysema and α_1-antitrypsin deficiency. *Acta Med. Scand.* **175**:197.

Fagerhol, M. K. and C. B. Laurell. 1967. The polymorphism of "prealbumins" and α_1-antitrypsin in human serum. *Clin. Chim. Acta* **16**:199.

Farrow, S. P. and S. Baar. 1973. The metabolism of α_2-macroglobulin in mildly burned patients. *Clin. Chim. Acta* **46**:39.

Fermi, C. and L. Pernossi. 1894. Über die Enzyme—Vergleichende Studie. *Zschr. Hyg.* **18**:83.

Forbes, C. D., J. Pensky and O. D. Ratnoff. 1970. Inhibition of activated Hageman factor and activated plasma thromboplastin antecedent by purified serum C1 inactivator. *J. Lab. Clin. Med.* **76**:809.

Ganrot, P. O. 1967. Partition of trypsin between α_2-macroglobulin and the other trypsin inhibitors of serum. *Arkiv Kemi* **26**:577.

Ganrot, P. O. and B. Bjerre. 1970. α_1-Antitrypsin and α_2-macroglobulin concentration in serum during pregnancy. *Acta Obstet. Gynecol. Scand.* **46**:126.

Ganrot, P. O. and B. Scherstén. 1967. Serum α_2-macroglobulin. Concentration and its variation with age and sex. *Clin. Chim. Acta* **15**:113.

Ganrot, P. O., C. B. Laurell and S. Eriksson. 1967. Obstructive lung disease and trypsin inhibitors in α_1-antitrypsin deficiency. *Scand. J. Clin. Lab. Invest.* **19**:205.

Hahn, M. 1897. Zur Kenntnis der Wirkungen des extravasculären Blutes. *Berl. Klin. Wschr.* **34**:499.

Harms, D., G. Buss, N. Heimburger and G. R. Pape. 1975. Experimentelle Untersuchungen zur Aufhebung von pulmonaren hyalinen Membranen. *K. Wochenscrift* (in press).

Healy, G. M. and R. C. Parker. 1966. Cultivation of mammalian cells in defined media with protein and nonprotein supplements. *J. Cell Biol.* **30**:539.

Hedner, U. 1973. Studies on an inhibitor of plasminogen activation in human serum. *Thromb. Diath. Haemorrh.* **30**:414.

Hedner, U., I. M. Nilsson and C. D. Jacobsen. 1970. Demonstration of low content of fibrinolytic inhibitors in individuals with high fibrinolytic capacity. *Scand. J. Clin. Lab. Invest.* **25**:329.

Heide, K., N. Heimburger and H. Haupt. 1965. An inter-alpha trypsin inhibitor of human serum. *Clin. Chim. Acta* **11**:82.

Heimburger, N. 1967. On the proteinase inhibitors of human plasma with especial reference to antithrombin. In *1st International Symposium on Tissue Factors in the Haemostasis of the Coagulation-Fibrinolysis System,* vol. 1, p. 353. Unione Chimica Medicamenti, Firenze, Italy.

————. 1974a. Biochemistry of proteinase inhibitors from human plasma: A review of recent developments. In *Bayer Symposium V: Proteinase Inhibitors* (ed. H. Fritz et al.), p. 14. Springer-Verlag, New York.

————. 1974b. Hemmstoffe der Blutgerinnung—ihr diagnostischer Wert bei Hypercoagulabilität. In *Verhandlungen der Deutschen Arbeitsgemeinschaft für Blutgerinnungsfestsetzung* (ed. E. F. Lüscher), p. 18. Tagung, Bern.

Heimburger, N. and H. Haupt. 1965. Characterisierung von α_1X-Glykoprotein als Chymotrypsin-Inhibitor des Humanplasmas. *Clin. Chim. Acta* **12**:116.

Heimburger, N., H. Haupt and H. G. Schwick. 1971. Proteinase inhibitors of human plasma. In *Proceedings of the 1st International Research Conference on Proteinases Inhibitors* (ed. H. Fritz and H. Tschesche), p. 1. Walter de Gruyter, New York.

Hensen, A. and E. A. Loeliger. 1963. Antithrombin. III. Its metabolism and its function in blood coagulation. *Thromb. Diath. Haemorrh.* (Suppl. 10) **9**:1.

Hochstrasser, K., M. Theopold and O. Brandl. 1973. Zur Hemmbarkeit der Proteinasen aus Pseudomonas aeruginosa durch α_2-Macroglobulin. *Hoppe-Seyler's Z. Physiol. Chem.* **354**:1013.

Jacobsson, K. 1953. Electrophoretic demonstration of two trypsin inhibitors in human serum. *Scand. J. Clin. Lab. Invest.* **5**:97.

————. 1955. Studies on fibrinogen. II. Studies on the trypsin and plasmin inhibitors in human blood serum. *Scand. J. Clin. Lab. Invest.* (Suppl. 14) **7**:66.

Janoff, A. and J. D. Zelios. 1968. Vascular injury and lysis of basement membrane in vitro by neutral proteases of human leucocytes. *Science* **161**:702.

Johnson, D. A., R. N. Pannell and J. Travis. 1974. The molecular stoichiometry of trypsin inhibition by human α_1-proteinase inhibitor. *Biochem. Biophys. Res. Comm.* **57**:584.

Jones, J. M., J. M. Creeth and R. A. Kekwick. 1972. Thiol reduction of human α_2-macroglobulin. The subunit structure. *Biochem. J.* **127**:187.

Kaplan, A. P. and K. F. Austen. 1971. A prealbumin activator of prekallikrein. II. Derivation of activators of prekallikrein from active Hageman factor by digestion with plasmin. *J. Exp. Med.* **133**:696.

Kueppers, F. and A. G. Bearn. 1966. A possible experimental approach to the association of hereditary α_1A-deficiency and pulmonary emphysema. *Proc. Soc. Exp. Med.* **121**:1207.

Landermann, N. S., M. E. Webster, E. L. Becker and H. E. Ratcliffe. 1962. Hereditary angioneurotic edema. Deficiency of inhibitor from serum-globulin-permeability factor and/or plasma kallikrein. *J. Allergy* **33**:330.

Landsteiner, K. 1900. Zur Kenntnis der antifermentativen, lytischen und agglutinierenden Wirkung des Blutserums und der Lymphe. *Z. Bakteriol.* **27**:357.

Landureau, J. C. and M. Steinbuch. 1970. In vitro, cell protective effects by certain antiproteases of human serum. *Z. Naturforsch.* **25b**:231.

Laskowski, M., Jr. and R. W. Sealock. 1971. Protein proteinase inhibitors—Molecular aspects. In *The Enzymes* (ed. P. D. Boyer), p. 375. Academic Press, New York.

Laurell, A. B. and U. Martensson. 1971. C$\overline{1}$ inactivator protein complexed with albumin in plasma from a patient with angioneurotic edema. *Eur. J. Immunol.* **1**:146.

Levy, L. R. and I. H. Lepow. 1959. Assay and properties of serum inhibitor of C1-esterase. *Proc. Soc. Exp. Biol. Med.* **101**:608.

Liener, I. E., O. R. Garrison and Z. Pravda. 1973. The purification of human serum α_1-antitrypsin by affinity chromatography on concanavalin A. *Biochem. Biophys. Res. Comm.* **51**:436.

Loeb, J. 1956. Le cofacteur plasmatique de l'heparine et les rapports avec l'antithrombine. *Arch. Sci. Physiol.* **10**:129.

Markwardt, F. and P. Walsmann. 1959. Untersuchungen über den Mechanismus der Antithrombinwirkung des Heparins. *Hoppe-Seyler's Z. Physiol. Chem.* **317:**64.

Marr, A. G. M., J. A. Owen and G. S. Wilson. 1962. The composition of a gly-coprotein fraction of foetal calf serum possessing tissue culture growth activity. *Biochim. Biophys. Acta* **63:**276.

Mehl, J. W., W. O'Connel and J. Degroot. 1964. Macroglobulin from human plasma which forms an enzymatically active compound with trypsin. *Science* **145:**821.

Mittman, Ch., ed. 1972. *Pulmonary Emphysema and Proteolysis.* Academic Press, New York.

Miller-Andersson, M., L. O. Andersson and H. Borg. 1973. Isolation and characterization of a new immediate antithrombin from human plasma. *4th International Symposium on Thrombosis and Haemostasis,* Abs. 181.

Moll, F. C., S. F. Sunden and J. R. Brown. 1958. Partial purification of the serum trypsin inhibitor. *J. Biol. Chem.* **233:**121.

Morawitz, P. 1905. Die Chemie der Blutgerinnung. *E. Physiol.* **4:**307.

Myerowitz, R. L., Z. T. Handzel and J. B. Robbins. 1972. Human serum α_1-antitrypsin: Isolation and demonstration of electrophoretic and immunologic heterogeneity. *Clin. Chim. Acta* **39:**307.

Norman, P. S. 1958. II. Inhibition of plasmin by serum or plasma. *J. Exp. Med.* **108:**53.

Northrop, J. H. and M. Kunitz. 1936. Isolation from beefs pancreas of crystalline trypsinogen, trypsin, α_1-trypsin inhibitor, and an inhibitor-trypsin compound. *J. Gen. Physiol.* **19:**991.

Ohlsson, K. 1971a. Interactions between bovine α-chymotrypsin and the proteinase inhibitors of human and dog serum in vitro. *Scand. J. Clin. Lab. Invest.* **28:**5.

―――. 1971b. Interactions in vitro and in vivo between dog trypsin and dog plasma protease inhibitors. *Scand. J. Clin. Lab. Invest.* **28:**219.

―――. 1974. Interaction between endogenous proteases and plasma protease inhibitors. In *Bayer Symposium V: Proteinase Inhibitors* (ed. H. Fritz et al.), p. 96. Springer-Verlag, New York.

Ratnoff, O. D. and I. H. Lepow. 1957. Some properties of an esterase derived from preparations of the first component of complement. *J. Exp. Med.* **106:**327.

Ratnoff, O. D., J. Pensky, D. Ogston and G. B. Naff. 1969. The inhibition of plasmin, plasmakallikrein, plasma-permeability factor and the C1 r subcomponent of the first component of complement by serum C1 esterase inhibitor. *J. Exp. Med.* **129:**315.

Reichert, R., K. Hochstrasser and E. Werle. 1971. Der Proteinaseinhibitorspiegel im menschlichen Nasensekret unter physiologischen und pathophysiologischen Bedingungen. *K. Wochenschrift* **49:**1234.

Rosenberg, R. D. and P. S. Damus. 1973. The purification and mechanism of action of human antithrombin-heparin cofactor. *J. Biol. Chem.* **248:**6490.

Sasaki, M., H. Yamamoto, H. Yamamoto and I. Shigeki. 1974. Interaction of human serum proteinase inhibitors with proteolytic enzymes of animal, plant and bacterial origin. *J. Biochem.* **75:**171.

Schmitz, A. 1938. Über den Trypsin-Inhibitor des Blutes. Dritte Mittlg. zur Kenntnis des Plasmatrypsinsystems. *Hoppe-Seyler's Z. Physiol. Chem.* **255:**234.

Schultze, H. E., K. Heide and H. Haupt. 1962a. α_1-Antitrypsin aus humanserum. *K. Wochenschrift* **40:**427.

―――. 1962b. Über ein noch nicht beschriebenes α_1-Glycoprotein des menschlichen Serums. *Naturwissenschaften* **49:**133.

Schultze, H. E., I. Göllner, K. Heide, M. Schönenberger and G. Schwick. 1955. Zur Kenntnis der α-Globuline des menschlichen Normalserums. *Z. Naturforsch.* **10b:**463.

Schultze, H. E., N. Heimburger, K. Heide, H. Haupt, K. Störiko and H. G. Schwick. 1963. Preparation and characterization of α_1-trypsin inhibitor and α_2-plasmin inhibitor of human serum. In *Proceedings of the 9th Congress of the European Society of Haematology* (Lisbon), p. 1315. S. Karger, Basel.

Schwick, H. G., N. Heimburger and H. Haupt. 1966. Antiproteinasen des Humanserums. *Z. Ges. Exp. Med.* **21:**1.

Shtacher, G., R. Maayan and G. Feinstein. 1973. Proteinase inhibitors in human synovial fluid. *Biochim. Biophys. Acta* **303:**138.

Shulman, N. R. 1952. Studies on the inhibition of proteolytic enzymes by serum. II. Demonstration that separate proteolytic inhibitors exist in serum; their distinctive properties and the specificity of their action. *J. Exp. Med.* **95:**539.

Steinbuch, M. 1961. Isolation of an α_2-globulin from human plasma. *Nature* **192:**1196.

Steinbuch, M. and R. Audran. 1974. Biology and pathology of plasma proteinase inhibitors. In *Bayer Symposium V: Proteinase Inhibitors* (ed. H. Fritz et al.), p. 78. Springer-Verlag, New York.

Steinbuch, M. and C. Reuge. 1970. Variation of the activity of α_2-macroglobulin as progressive antithrombin after molecular modification. *Clin. Enzymol.* **2:**61.

Travis, J. and R. Pannell. 1973. Selective removal of albumin from plasma by affinity chromatography. *Clin. Chim. Acta* **49:**49.

Wicher, V. and J. Dolovich. 1973. Effects of human serum inhibitors on immunologic properties of B-subtilis alkaline proteinase. *Immunochemistry* **10:**239.

Yu, R. L., S. F. Grappel and F. Blank. 1972. Inhibition of keratinases by α_2-macroglobulin. *Experientia* **28:**866.

Studies on the Structure and Function of α_2-Macroglobulin and $\overline{\text{C1}}$ Inactivator

Peter C. Harpel

Department of Medicine, Division of Hematology
The New York Hospital-Cornell Medical Center, New York 10021

Michael W. Mosesson

Department of Medicine, Hemostasis Section, Downstate Medical Center
State University of New York, Brooklyn, New York 11203

Neil R. Cooper

Department of Molecular Immunology, Scripps Clinic and Research Foundation
La Jolla, California 92037

The potential pathophysiologic importance of the human plasma proteins which serve to inhibit proteolytic enzymes is suggested by their comprising approximately 10% of the total plasma protein concentration (Heimburger, Haupt and Schwick 1971). Inherited deficiencies of three inhibitory proteins are associated with disease and therefore give clues as to the function of these particular circulating inhibitors. Thus α_1-antitrypsin deficiency predisposes to emphysema and hepatic cirrhosis (Mittman 1972), $\overline{\text{C1}}$ inactivator deficiency underlies hereditary angioneurotic edema (Donaldson and Evans 1963), and antithrombin III deficiency is associated with thrombo-embolic disease (Marciniak, Farley and DeSimone 1974).

In this communication, we will review our studies concerning the structure and function of the plasma proteolytic enzyme inhibitors α_2-macroglobulin and $\overline{\text{C1}}$ inactivator. The inhibitory spectrum of these two proteins (Fig. 1) indicates that they interact with a variety of proteases which participate in inflammatory and hemostatic reactions. The α_2-macroglobulin forms a high molecular weight complex with enzymes of the kinin-forming, coagulation and fibrinolytic pathways, thereby modifying their activity. These include kallikrein (Harpel 1970a, 1971), thrombin (Lanchantin et al. 1966) and plasmin (Schultze et al. 1963; Steinbuch, Quentin and Pejaudier 1965; Ganrot 1967), as well as other proteases, including trypsin, chymotrypsin, elastase and others which have been recently summarized (Barrett and Starkey 1973). $\overline{\text{C1}}$ inactivator inhibits enzymes in the complement, kinin-forming, coagulation and fibrinolytic systems. These include $\overline{\text{C1}}$ and its subcomponents $\overline{\text{C1s}}$ and $\overline{\text{C1r}}$ (Levy and Lepow 1959; Pensky, Levy and Lepow 1961; Lepow and Leon 1962; Gigli, Ruddy and Austen 1968; Ratnoff et al. 1969), kallikrein, PF/DiI (Landerman et al. 1962; Kagen and Becker 1963; Gigli et al. 1970), plasmin (Ratnoff et al. 1969; Schreiber, Kaplan and Austen 1973), Hageman factor (Forbes, Pensky and Ratnoff 1970) and its active fragments (Schreiber et al. 1973), and coagulation factor XI (Forbes, Pensky and Ratnoff 1970; Harpel 1971). Trypsin, however, is poorly inhibited by $\overline{\text{C1}}$ inactivator (Pensky, Levy and Lepow 1961; Haupt et al. 1970).

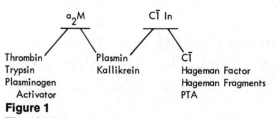

Figure 1
The physiologically important proteases that are
inhibited by human plasma α_2-macroglobulin and
by $C\overline{1}$ inactivator are indicated.

Although the inhibitory spectrum of these human plasma inhibitors is relatively well defined, the nature of the molecular interactions that occur when
these inhibitors react with the enzymes they inhibit has remained obscure.
Our studies examining some of these reactions have suggested that a variety
of events occur depending upon the inhibitor and the enzyme investigated.
The α_2-macroglobulin has been found to be proteolytically cleaved by the
enzymes which it inhibits and the altered inhibitor molecule is incorporated
into a complex with the enzyme (Harpel 1973). Furthermore, the α_2-macro-
globulin–plasmin complex retained fibrinogenolytic activity (Harpel and
Mosesson 1973), demonstrating the catalytic potential of the complex. In
investigations presented in greater detail elsewhere (Harpel and Cooper
1975), the interactions between $C\overline{1}$ inactivator and $C\overline{1}$s, plasmin, plasma
kallikrein and trypsin have been studied. Incubation of the inhibitor with
either $C\overline{1}$s or kallikrein led to the formation of a 1:1 molar complex with
associated loss of enzymatic activity. The enzyme-inhibitor complexes resisted
the denaturing agents sodium dodecyl sulfate (SDS) and urea. Similar complex formation with functional inactivation occurred on adding plasmin to $C\overline{1}$
inactivator; however, plasmin cleaved the inhibitor and produced several characteristic derivatives, at least one of which retained the ability to form a stable
complex with plasmin. The light chain of both $C\overline{1}$s and plasmin was found
to provide the binding site(s) for the inhibitor. Trypsin failed to form a complex with $C\overline{1}$ inactivator and degraded the inhibitor in a sequential and
limited manner, producing nonfunctional derivatives. These studies have
provided new information regarding the molecular interactions involved in
enzyme regulation by α_2-macroglobulin and by $C\overline{1}$ inactivator.

METHODS

Purification of Human Plasma Proteins

The following proteins were isolated from plasma by methods which have
been described previously: α_2-macroglobulin (Harpel 1973), fibrinogen
(Mosesson and Finlayson 1963), plasminogen (Deutsch and Mertz 1970),
$C\overline{1}$s (Valet and Cooper 1974; Harpel and Cooper 1975) and plasma kallikrein (Harpel 1972). Plasminogen was freshly activated for each experiment
in 25% glycerol with urokinase, 20 units per 100 μg plasminogen, at 25°C
for 18 hours. The plasminogen was completely activated by this procedure

since SDS-acrylamide gel electrophoretic analysis of reduced samples of the incubation mixture demonstrated conversion of the plasminogen band to the heavy and light chain of plasmin.

Human plasma C$\overline{1}$ inactivator was purified as detailed elsewhere (Harpel and Cooper 1975) by methods designed to minimize activation of plasma proteases. This included fractional precipitation with polyethylene glycol, DEAE-cellulose and gel filtration (Bio-Gel A-5m, Bio Rad Laboratories) column chromatography, preparative Pevikon block electrophoresis, and concanavalin A-Sepharose (Pharmacia Fine Chemicals) affinity chromatography. The final product formed a single precipitin arc in the α_2-region with rabbit antihuman serum antibody by immunoelectrophoresis. The specific activity of the C$\overline{1}$ inactivator, measured using acetyl-L-lysine methyl ester as a substrate for C$\overline{1}$s (Harpel 1970b), was 217 inhibitor units/mg C$\overline{1}$ inactivator.

SDS-acrylamide gel electrophoretic analysis of the inhibitor preparations and of the interaction between the inhibitors and their proteases was performed by the method of Weber and Osborn (1969). The methods for preparation of the samples for electrophoresis and the proteins used for molecular weight markers were as described previously (Harpel and Mosesson 1973).

RESULTS

Interaction of α_2-Macroglobulin with Various Enzymes

The reaction of α_2-macroglobulin with trypsin, thrombin, plasmin and plasma kallikrein was assessed by polyacrylamide gel electrophoresis of reduced samples in the presence of sodium dodecyl sulfate (Fig. 2) (Harpel 1973). After reduction of α_2-macroglobulin which had not been reacted with an enzyme, one protein band with an apparent molecular weight of 185,000 daltons was observed (Fig. 2, 1st gel). Following reaction with the enzymes, a derivative chain (molecular weight 85,000 daltons) was identified. The electrophoretic mobility of the derivative chain resulting from the action of each enzyme was identical, suggesting that these proteases attacked a similar region of the α_2-macroglobulin molecule. The derivative subunit chains resulting from the α_2-macroglobulin–protease interaction were identified only following disulfide bond cleavage, demonstrating that these products were linked to the parent molecule by disulfide bonds.

The native structure of α_2-macroglobulin was required to limit the hydrolytic attack to a specific region of the molecule. After acidification of the α_2-macroglobulin molecule, its structure as assessed by SDS-polyacrylamide gel electrophoresis was unchanged; however, it failed to inhibit trypsin and was hydrolyzed by this enzyme into multiple lower molecular weight products (Harpel 1973).

These results suggest the model illustrated in Figure 3 for the subunit structure of α_2-macroglobulin. Prior studies have shown that the α_2-macroglobulin dissociates into half-molecules under denaturing conditions, suggesting that it has a noncovalently linked dimeric structure (Jones, Creeth and Kekwick 1972). Recent estimates of the molecular weight of α_2-macroglobulin are on the order of 650,000 (Saunders et al. 1971) and 725,000 daltons

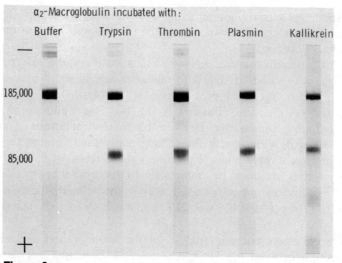

Figure 2
Acrylamide gel (5%) electrophoresis in the presence of SDS and following reduction with dithiothreitol of α_2-macroglobulin incubated with bovine trypsin, human thrombin, plasmin or plasma kallikrein. These enzymes were incubated with α_2-macroglobulin (165 μg/system) in the following approximate molar ratios: trypsin, 1 (5.4 μg); thrombin, 0.5 (12 NIH units); plasmin, 0.4 (0.15 caseinolytic units); plasma kallikrein, 0.7 (0.14 TAMe esterase units). The reaction was stopped after 1 hr at 37°C and the mixture reduced by the addition of an SDS-urea-DTT solution. (Reprinted, with permission, from Harpel 1973.)

(Jones, Creeth and Kekwick 1972), indicating that the molecule must consist of four chains, since each subunit chain formed after disulfide bond cleavage has a molecular weight of 185,000. Each dimer, then, consists of two subunit chains linked by disulfide bridges. These subunit chains provide the cleavage site(s) for the proteolytic enzymes studied and are hydrolyzed at or near their center, yielding a maximum of eight derivative products per α_2-macroglobulin molecule.

α_2-Macroglobulin–Enzyme Complexes Degrade Fibrinogen

The following studies were designed to examine the possibility that in addition to its function as a plasmin inhibitor, α_2-macroglobulin might participate in fibrinogen catabolism by forming a biologically active complex with plasmin (Harpel and Mosesson 1973). α_2-Macroglobulin was purified from plasma whose plasminogen was activated to plasmin by preincubation with several different concentrations of urokinase. These α_2-macroglobulin preparations possessed esterolytic activity against the substrate tosyl-L-arginine methyl ester HCl (TAMe) which was commensurate with the concentration of urokinase added to the plasma and which was greatly increased over the α_2-macroglobulin purified from nonactivated plasma (Table 1). Although

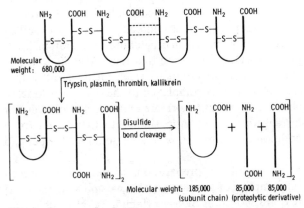

Molecular weight: 680,000

Trypsin, plasmin, thrombin, kallikrein

Disulfide bond cleavage

Molecular weight: 185,000   85,000   85,000
(subunit chain) (proteolytic derivative)

Figure 3

A schematic model of a proposed subunit structure of human plasma α_2-macroglobulin showing the alterations induced in the molecule by interaction with the proteases it inhibits. The placement of the amino- and C-terminal ends of the subunit chains is arbitrary, and the minimum number of inter- and intrachain disulfide bridges is indicated. The half-molecules of α_2-macroglobulin are held together by noncovalent bonds (dashed lines). In the illustration, one-half of the precursor subunit chains have been cleaved. (Reprinted, with permission, from Harpel 1973.)

Table 1

The Esterolytic Activity of α_2-Macroglobulin
Prepared from Plasma Incubated with Urokinase

Incubation, plasma +	TAMe esterase activity[a]
Buffer	0.5
100 U UK/ml plasma	7.3
500 U UK/ml plasma	20.0

Data from Harpel and Mosesson (1973).

[a] The TAMe esterase activity of the α_2-macroglobulin preparations was measured in the presence and absence of pancreatic or soybean trypsin inhibitor (100 μg/ml, final conc.). These inhibitors failed to change the activity shown in the table. Plasmin hydrolyzed 28.2 μmoles TAMe/caseinolytic U/hr; this activity was completely inhibited by either pancreatic or soybean trypsin inhibitor. Activity is expressed as μmoles hydrolyzed/mg α_2-macroglobulin/hr.

both pancreatic and soybean trypsin inhibitor completely inhibited the TAMe esterase activity of plasmin, these inhibitors did not inhibit the esterolytic activity of the α_2-macroglobulin preparations studied. Thus the observed activity appeared to be a property of the α_2-macroglobulin preparation itself, and not of a dissociating enzyme. Documentation that the TAMe-hydrolyzing enzyme was plasmin was provided by studies which demonstrated that immunologically identifiable plasmin was complexed to the α_2-macroglobulin. Further, the α_2-macroglobulin from urokinase-treated plasma hydrolyzed four different amino acid ester substrates in a ratio similar to that found for a complex formed by the addition of plasmin to α_2-macroglobulin (Harpel and Mosesson 1973).

The ability of these α_2-macroglobulin preparations to degrade fibrinogen was examined by sequential sampling of a mixture of the two and analysis of the resulting subunit structure of fibrinogen by SDS-acrylamide gel electrophoresis as detailed previously (Harpel and Mosesson 1973). All α_2-macroglobulin preparations induced progressive changes in the subunit chains of fibrinogen, as was quantitated by densitometric scans of the gels in which the depletion of the A α- or B β-chains was compared with the intact γ-chains. As indicated in Table 2, the first, and ultimately most extensive, change was depletion of the A α-chain position. The B β-chain was also depleted, but at a slower rate. This sequence of degradation was similar to that observed in plasmin-catalyzed digests, in that A α-chain hydrolysis preceded that of B β-chain (Mosesson et al. 1972), and provided further support for the identification of the enzyme bound to α_2-macroglobulin as plasmin. The α_2-macroglobulin obtained from urokinase-treated plasma degraded fibrinogen far more extensively than did the α_2-macroglobulin from nonactivated plasma. Fibrinogen incubated with urokinase alone showed no change in its structure, indicating that the alterations which occurred were due to the proteolytic activity associated with the α_2-macroglobulin. Estimates of the fibrinogenolytic activity associated with the α_2-macroglobulin suggested that in comparison with the amount of plasmin bound to the inhibitor as determined by its TAMe esterase activity, the proteolytic activity of the bound plasmin had been reduced to less than 0.1% of that of an equivalent amount of the free enzyme.

Table 2

Degradation of Fibrinogen Chains by α_2-Macroglobulin Preparations

	A α-/γ-chain (%)[a]			B β-/γ-chain (%)		
Time of incubation:	6 hr	24 hr	48 hr	6 hr	24 hr	48 hr
α_2M (normal plasma)	82	50	43	100	92	79
α_2M (UK-treated plasma)	65	25	12	88	69	65

Data from Harpel and Mosesson (1973).

[a] The percentage of fibrinogen subunit chains remaining following varying periods of incubation with two α_2-macroglobulin preparations was determined by densitometric scans of SDS-acrylamide gels. The α_2-macroglobulins were purified from normal plasma or from plasma preincubated with urokinase (500 U/ml plasma). The ratio of the area under the A α- and B β-chain positions, respectively, to that of the γ-chain position is expressed as percent of the ratio found at time 0.

Two approaches were used to demonstrate that the fibrinogenolytic activity observed was due to an α_2-macroglobulin–enzyme complex and not to free enzyme. First, when pancreatic trypsin inhibitor was added to fibrinogen, it completely inhibited the proteolytic activity of free plasmin, but did not prevent the degradation of fibrinogen by the α_2-macroglobulin preparations. Second, immunoprecipitation of α_2-macroglobulin by specific antiserum removed both the α_2-macroglobulin and all proteolytic activity (Harpel and Mosesson 1973).

Identification of Two C$\overline{1}$ Inactivator Molecules

The following studies, presented in detail elsewhere (Harpel and Cooper 1975), were performed to further define the structure of C$\overline{1}$ inactivator. SDS-polyacrylamide gel electrophoresis of purified C$\overline{1}$ inactivator preparations demonstrated two bands, designated I and II, with apparent molecular weights of 105,000 and 96,000 daltons, respectively (Fig. 4, gel b). Band I was usually the major component, and the proportion of band I and band II in the final preparation was not affected by the use of proteolytic enzyme inhibitors during the purification procedures. The immunologic identity of bands I and II, with respect to one another, was established by immunodiffusion of rabbit antiserum to C$\overline{1}$ inactivator against a longitudinal slice of the SDS-acrylamide gel containing the separated C$\overline{1}$ inactivator preparation. Two precipitin arcs were identified which joined without spur formation in positions corresponding to bands I and II of the C$\overline{1}$ inactivator. Furthermore, following crossed electrophoresis (Laurell 1965) using the SDS-acrylamide gel containing the inhibitor, a larger and a smaller peak of precipitation, corresponding to bands I and II, respectively, were identified (Harpel and Cooper 1975). These results have documented the existence of two immunologically identical forms of C$\overline{1}$ inactivator, differing by 9000 daltons in molecular weight; however, it remains unclear whether the lower molecular weight C$\overline{1}$ inactivator-related protein represented a circulating form of the inhibitor or a derivative produced during purification.

Interaction of C1 Inactivator with Various Enzymes

The C$\overline{1}$ Inactivator–C$\overline{1}$s Reaction

As indicated in the studies to be described, C$\overline{1}$ inactivator formed a stoichiometric complex with purified C$\overline{1}$s, and the light chain of the enzyme, containing the active site (Barkas, Scott and Fothergill 1973), provided the binding site(s) (Harpel and Cooper 1975). Electrophoretic analysis in SDS-acrylamide gels of equimolar concentrations of C$\overline{1}$ inactivator and C$\overline{1}$s demonstrated two new slower moving bands (complex I and II) (Fig. 4, gel c). As compared to the pattern produced by C$\overline{1}$s and C$\overline{1}$ inactivator alone (Fig. 4, gels a and b), the concentration of material in the position of the two inhibitor bands, as well as in the concentration of the C$\overline{1}$s band, was decreased. The molecular weights of the new bands were approximately 180,000 and 170,000 daltons, respectively, consistent with the formation of a 1:1 molar complex between the inhibitor bands I and II and C$\overline{1}$s (72,000 daltons). These results

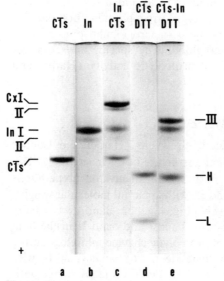

Figure 4

SDS-acrylamide gel (5%) electrophoretic analysis of the interaction between $C\overline{1}$ inactivator and $C\overline{1}$s. $C\overline{1}$s (0.9 mg/ml) was incubated 10 minutes at 37°C with 0.1 M sodium phosphate buffer pH 7.6, or with $C\overline{1}$ inactivator (1.5 mg/ml). The samples were then incubated in an equal volume of a solution containing 10 M urea, 2% SDS either with or without 14 mM dithiothreitol for 45 minutes at 37°C and applied to the gels as follows: gel a, $C\overline{1}$s; gel b, $C\overline{1}$ inactivator; gel c, $C\overline{1}$ inactivator-$C\overline{1}$s mixture; gel d, $C\overline{1}$s reduced; gel e, $C\overline{1}$ inactivator-$C\overline{1}$s mixture reduced, "CxI" and "II" indicate the positions occupied by the complex formed between $C\overline{1}$s and $C\overline{1}$ inactivator bands I and II. "In I" and "II" are designations for the major and minor $C\overline{1}$ inactivator bands. The complex formed by the light chain of $C\overline{1}$s and $C\overline{1}$ inactivator band I observed following reduction (gel e) is designated "III." The heavy (H) and light (L) chains of $C\overline{1}$s are indicated. The direction of electrophoresis is toward the anode indicated by "+." (Reprinted, with permission, from Harpel and Cooper 1975.)

394

have provided further support for the identity of band II as a lower molecular weight form of C$\overline{\text{I}}$ inactivator since it formed a complex with C$\overline{\text{Is}}$.

Disulfide bond cleavage altered the electrophoretic pattern of the inhibitor C$\overline{\text{Is}}$ mixture as compared to the unreduced sample. The high molecular weight complex bands I and II were replaced by a new band (complex band III) with an apparent molecular weight of 125,000 daltons (Fig. 4, gel e). Although no C$\overline{\text{Is}}$ light chain could be identified in the reduced mixture, the heavy chain reappeared in a concentration similar to that of the reduced C$\overline{\text{Is}}$ preparation (Fig. 4, gel d). Thus the new band in the reduced sample represented formation of a complex between the C$\overline{\text{Is}}$ light chain and the major component of the C$\overline{\text{I}}$ inactivator preparation.

The C$\overline{\text{I}}$ Inactivator–Plasmin Reaction

Analysis of the interaction between plasmin and C$\overline{\text{I}}$ inactivator showed several features which differed from those observed with C$\overline{\text{Is}}$. Two higher molecular weight bands (180,000 and 170,000 daltons, respectively) were formed when C$\overline{\text{I}}$ inactivator was incubated with plasmin (Fig. 5, gel c); however, in contrast to the predominance of the C$\overline{\text{I}}$ inactivator-C$\overline{\text{Is}}$ band I complex over the band II complex (Fig. 5, gel b), in the plasmin-inhibitor reaction, a higher concentration of the band II complex was apparent (Fig. 5, gels c, e, g, i). These results suggested that plasmin had degraded the inhibitor molecule, thereby producing a derivative similar in molecular weight to the band II of the purified C$\overline{\text{I}}$ inactivator preparations used, which also appeared to form a complex with plasmin. Furthermore, the residual C$\overline{\text{I}}$ inactivator not incorporated into the complex was converted to a pattern in which material occupying the band II position represented a major proportion of the C$\overline{\text{I}}$ inactivator preparation. This was especially apparent when the effects of increasing molar ratios of plasmin relative to C$\overline{\text{I}}$ inactivator were examined by SDS-acrylamide gel electrophoresis (Fig. 5, gels c, e, g, i). Even at plasmin concentrations in excess of the established 1:1 molar binding ratio between the enzyme and the inhibitor, a relatively large amount of plasmin-induced C$\overline{\text{I}}$ inactivator band II was apparent (Fig. 5, gels g, i). This observation indicated that this derivative was not capable of forming a complex with plasmin under the experimental conditions described. It can be concluded, therefore, that plasmin had cleaved the C$\overline{\text{I}}$ inactivator molecule, producing two derivatives with similar molecular weights, one of which was functionally active in forming a complex with plasmin, whereas the other was inactive.

Experiments examining the effect of disulfide bond cleavage on the SDS-acrylamide gel electrophoretic pattern of the C$\overline{\text{I}}$ inactivator-plasmin mixture demonstrated that analogous to the inhibitor-C$\overline{\text{Is}}$ reaction, the light chain of the enzyme containing the active site of plasmin (Summaria et al. 1967) provided the binding site(s) for the inhibitor (Harpel and Cooper 1975).

The C$\overline{\text{I}}$ Inactivator, Plasmin and C$\overline{\text{Is}}$ Reactions

Previous studies have established that plasmin renders C$\overline{\text{I}}$ inactivator incapable of inhibiting the esterolytic activity of C$\overline{\text{Is}}$ (Harpel 1970c). In order to further explore the mechanisms underlying this inactivation of the inhibitor, the residual C$\overline{\text{Is}}$-binding and -inhibiting activity of C$\overline{\text{I}}$ inactivator in the

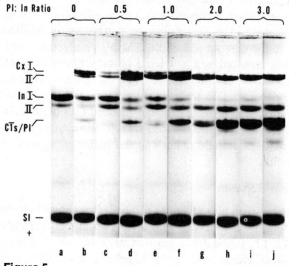

Figure 5

SDS-acrylamide gel electrophoretic analysis of the reaction between increasing amounts of plasmin (Pl)
and C$\overline{I}$ inactivator (data from Harpel and Cooper
1975). Equal volumes of increasing concentrations of
plasmin (0.55, 1.1 and 3.1 mg/ml) were incubated
with C$\overline{I}$ inactivator (1.5 mg/ml) for 30 minutes at
37°C. Following the addition of an equal volume of soybean trypsin inhibitor (SI, 2 mg/ml) to inhibit residual
plasmin, either buffer (gels c, e, g, i) or C$\overline{I}$s (1.0 mg/
ml; gels d, f, h, j) was added and incubation continued
for an additional 10 minutes. The following samples
were included for comparative purposes: gel a, C$\overline{I}$ inactivator plus soybean trypsin inhibitor; gel b, C$\overline{I}$-in-
activator-C$\overline{I}$s incubation mixture (10 min, 37°C), following which soybean trypsin inhibitor was added. All
mixtures were analyzed in the absence of reduction.
The molar ratio of plasmin to C$\overline{I}$ inactivator used in
each sample is indicated. "C$\overline{I}$s/Pl" designates the
position occupied by both C$\overline{I}$s and plasmin.

plasmin-C$\overline{I}$ inactivator incubation mixture was evaluated by SDS-acrylamide
gel electrophoresis (Harpel and Cooper 1975). Soybean trypsin inhibitor was
added to the incubation mixture to neutralize residual proteolytic activity,
following which concentrations of C$\overline{I}$s equimolar to the C$\overline{I}$ inactivator were
added. The addition of C$\overline{I}$s induced a loss of material from C$\overline{I}$ inactivator
bands I and II of the samples of C$\overline{I}$ inactivator previously treated with 0.5
and 1.0 molar ratios of plasmin, and there was an increase in bands I and II
of the complex (Fig. 5, gels d, f). At higher plasmin concentrations of 2 and
3 molar, no loss of protein in the C$\overline{I}$ inactivator band II position was apparent
after C$\overline{I}$s had been added, nor was there an increase in band II position of the
complex as compared to the samples without C$\overline{I}$s (Fig. 5, gels h, j). Thus

the proteolytic derivative generated by an amount of plasmin in excess of its binding ratio and which did not bind plasmin also failed to form a complex with C$\bar{\text{1}}$s. The ability of C$\bar{\text{1}}$ inactivator in the plasmin-inhibitor incubation mixture to inhibit C$\bar{\text{1}}$s, as tested in a C4 inactivating assay was studied. The plasmin-treated inhibitor's loss of ability to complex with C$\bar{\text{1}}$s, as documented by SDS-acrylamide gel electrophoresis, was also paralleled by the loss of functional C$\bar{\text{1}}$ inhibiting activity. Therefore, plasmin inactivated C$\bar{\text{1}}$ inhibitor by forming a complex with it and presumably blocking C$\bar{\text{1}}$s binding sites, as well as by producing a derivative which could not bind to C$\bar{\text{1}}$s.

The C$\bar{\text{1}}$ Inactivator, Trypsin and C$\bar{\text{1}}$s Reactions

Trypsin failed to form a complex with the inhibitor and degraded it in a limited, sequential fashion, thereby destroying its C$\bar{\text{1}}$s binding and inhibitory properties (Fig. 6) (Harpel and Cooper 1975). Increasing concentrations of trypsin (quantitated by the method of Chase and Shaw 1970) caused a progressive depletion of C$\bar{\text{1}}$ inactivator band I with a proportional increase in a band occupying the band II position (Fig. 6, gels c, e, g). At higher molar ratios of trypsin, and following depletion of band I material, an additional band with an approximate molecular weight of 86,000 daltons was observed (band III, Fig. 6, gel g).

Trypsin destroyed the C$\bar{\text{1}}$s binding activity of C$\bar{\text{1}}$ inactivator (Fig. 6, gels d, f, h). Complex formation with C$\bar{\text{1}}$s was partially inhibited by pretreatment of the inhibitor with molar ratios of trypsin as low as 0.025 and was totally inhibited at a molar ratio of 0.1. Neither of the trypsin-produced derivative bands II or III possessed C$\bar{\text{1}}$s binding activity. Other studies demonstrated that the function of C$\bar{\text{1}}$ inhibitor in C$\bar{\text{1}}$s inactivation was lost following incubation with trypsin, in parallel with the loss of binding activity.

The C$\bar{\text{1}}$ Inactivator–Plasma Kallikrein Reaction

C$\bar{\text{1}}$ inactivator was labeled with ^{125}I by the method of McConahey and Dixon (1966) and was found to have an electrophoretic mobility on SDS-acrylamide gels identical to that of unlabeled inhibitor as determined by counting gel segments. The radiolabeled inhibitor proved to be functionally competent since it formed a complex with C$\bar{\text{1}}$s, as indicated by a peak of radioactivity with a slower mobility than that displayed by the C$\bar{\text{1}}$ inactivator alone (Fig. 7a, b). In an analogous manner, the inhibitor protein also was found to form a 1:1 molar complex with a partially purified preparation of plasma kallikrein (Fig. 7c). As was the case for the plasmin and C$\bar{\text{1}}$s-inhibitor reaction, the complex formed between C$\bar{\text{1}}$ inactivator and kallikrein was resistant to the denaturing agents SDS and urea.

DISCUSSION

These studies have further defined some of the structural and functional features of α_2-macroglobulin and C$\bar{\text{1}}$ inactivator plasma proteins which react with several different enzyme systems participating in the inflammatory and hemostatic response. As outlined in Figure 8, the activation of Hageman

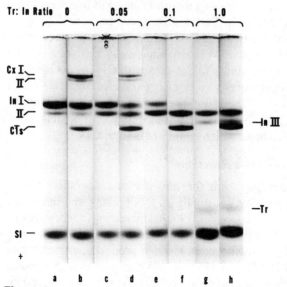

Figure 6

SDS-acrylamide gel (5%) electrophoretic analysis of
the reaction between increasing amounts of trypsin
(Tr) and C$\overline{\text{1}}$ inactivator (data from Harpel and
Cooper 1975). Equal volumes of increasing molar
ratios of bovine trypsin (16, 32, and 320 μg/ml)
were incubated with C$\overline{\text{1}}$ inactivator (1.5 mg/ml) for
10 minutes at 37°C. Following the addition of an
equal volume of soybean trypsin inhibitor (2.0 mg/
ml), either buffer (gels c, e, g) or C$\overline{\text{1}}$s (1.4 mg/ml;
gels d, f, h) was added and incubation continued for
an additional 10 minutes. The following samples were
also included: gel a, C$\overline{\text{1}}$ inactivator plus soybean trypsin inhibitor; gel b, C$\overline{\text{1}}$ inactivator plus C1s and soybean trypsin inhibitor. The mixtures were analyzed in
the absence of reduction. The molar ratio of trypsin to
C$\overline{\text{1}}$ inactivator used in each sample is indicated. "In
III" designates the new C$\overline{\text{1}}$ inactivator proteolytic
derivative which appeared in the C$\overline{\text{1}}$ inactivator-
trypsin incubation mixture following depletion of
the inhibitor band I (gels g, h). "Tr" indicates the
position of the trypsin protein band, and "SI," that
of the soybean trypsin inhibitor.

factor leads to the generation of thrombin, plasmin and kallikrein (Cochrane,
Revak and Wuepper 1973). These enzymes form complexes with the α_2-
macroglobulin, suggesting that the α_2-macroglobulin may modulate the activity of these proteases. Support for this concept has been obtained by McConnell (1972) and by Dyce et al. (1967), who have identified an α_2-macro-
globulin fraction possessing kallikreinlike activity, and by Rinderknecht and
Geokas (1973), who have demonstrated that an α_2-macroglobulin fraction
from serum possesses clotting activity.

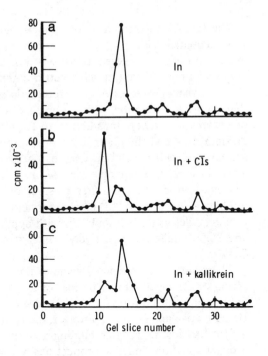

Figure 7
SDS-acrylamide gel (5%) electrophoretic analysis of the reaction between ¹²⁵I-labeled C$\overline{1}$ inactivator, C$\overline{1}$s and plasma kallikrein. Radiolabeled C$\overline{1}$ inactivator was incubated with buffer, C$\overline{1}$s or plasma kallikrein (prepared as detailed previously, Harpel 1972) for 10 minutes at 37°C and the mixtures analyzed in the absence of reduction following the addition of an SDS-urea solution. After electrophoresis, the gels were sliced into 2.0-mm segments and the radioactivity counted. The direction of electrophoresis is from left to right.

Our findings that the α_2-macroglobulin exhibits both esterolytic and fibrinogenolytic activities imply that these activities are present in the circulation. In addition, the binding of plasmin to α_2-macroglobulin occurs in whole plasma, as demonstrated by the enhancement in plasminlike activity associated with α_2-macroglobulin preparations isolated from plasma treated with plasminogen activators. Thus plasmin binds to the α_2-macroglobulin in the presence of other known plasma proteolytic enzyme inhibitors. Further, the inability of pancreatic trypsin inhibitor to limit the fibrinogenolytic activity of the α_2-macroglobulin–plasmin complex supports the concept that the α_2-macroglobulin may both preserve and protect a portion of the biologic activity of this enzyme in the presence of circulating inhibitors. The α_2-macroglobulin, therefore, has the potential for mediating the physiologic fibrinogenolytic catabolic pathway that accounts for a major proportion of fibrinogen turnover (Mosesson et al. 1972; Sherman, Fletcher and Sherry 1969). It can be concluded that the function of α_2-macroglobulin as an enzyme-binding protein may provide a mechanism by which the activity of proteases which participate in hemostatic and inflammatory reactions can be expressed in vivo.

Figure 8
A schema for the proposed function of α_2-macroglobulin in providing a mechanism for the preservation of the biologic activity of Hageman factor pathway-dependent proteases.

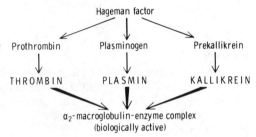

The several different reactions which were observed when C$\bar{1}$ inactivator was incubated with C$\bar{1}$s, plasmin or trypsin are summarized in schematic outline in Figure 9. These interactions were characterized by the formation of SDS and urea-resistant stoichiometric complexes between the inhibitor and C$\bar{1}$s or plasmin. Studies of the proteolytic activity of these enzymes showed that complex formation with C$\bar{1}$ inactivator was associated with inhibition of enzymatic activity. In addition to forming a complex with the inhibitor, plasmin degraded the C$\bar{1}$ inactivator, producing at least two derivatives with similar molecular weights. One of these derivatives, designated IIa, was active in forming a complex with plasmin, whereas the other, designated IIi, was inactive in binding either plasmin or C$\bar{1}$s. Trypsin destroyed the C$\bar{1}$s inhibitory function of C$\bar{1}$ inactivator and yielded two identifiable cleavage products, designated IIi and IIIi, with apparent molecular weights of 96,000 and 86,000 daltons, respectively, which were inactive in binding or inhibiting C$\bar{1}$s.

These studies have defined some of the structural features required for the action of C$\bar{1}$ inactivator. The loss of a 9000-dalton portion of the molecule did not prevent the formation of a stable complex between the inhibitor band IIa derivative and plasmin; however, the IIi derivative with a similar molecular weight had lost its enzyme-binding activity. The detailed mechanism for the production of this inactive product has not as yet been clarified. One alternative is that there are two regions in the C$\bar{1}$ inactivator molecule susceptible to hydrolytic attack by trypsin and plasmin. Proteolysis of one region would yield an active derivative, whereas cleavage of two regions would produce an inactive product. It is reasonable to assume that the additional cleavage would

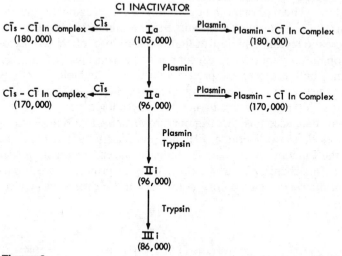

Figure 9

A schematic representation of the interaction between C$\bar{1}$ inactivator, C$\bar{1}$s, plasmin and trypsin. The explanation for the nomenclature used is found in the text. The apparent molecular weights of the complexes and of the C$\bar{1}$ inactivator derivatives are indicated in parentheses.

result in the release of a small peptide, thereby producing a molecular weight change in the parent chain too small to be identified by SDS-acrylamide gel electrophoresis. Although the cleaved peptide may have contained a critical binding site for the enzyme, it is more likely that the additional cleavage changed the conformation of the molecule, thereby destroying its function. Data supporting this concept (not presented in this communication) have demonstrated that acid treatment of $\overline{\text{C1}}$ inactivator, a procedure which alters its native conformation but not its primary structure, destroyed its enzyme-binding and -inactivating activity (Harpel and Cooper 1975).

The importance of peptide bond cleavage in contributing to complex formation between proteases and their naturally occurring inhibitors remains a controversial question. Laskowski (1972) has concluded from his studies of the reaction between trypsin and plant-derived trypsin inhibitors that hydrolysis of inhibitors by proteases is a requirement for complex formation and enzyme inhibition. In this reaction it is postulated that cleavage of the inhibitor leads to the formation of an ester bond between the active-site serine of trypsin and the arginine or lysine reactive site of the inhibitor (Laskowski 1972). In contrast, Feeney (1972) has contended that the strength of binding between enzyme and its inhibitor is secondary to the summation of many hydrophobic and electrostatic interactions. According to this concept, hydrolysis of the inhibitor is a coincidental event. Ako, Foster and Ryan (1974) have provided data supporting this theory by demonstrating that an enzymatically inactive trypsin derivative can form a complex with soybean trypsin inhibitor with an energy of inhibitor binding identical to that found for trypsin.

In our studies with α_2-macroglobulin, the formation of a complex with the enzymes trypsin, thrombin, plasmin and plasma kallikrein was accompanied by limited proteolysis of a region located in a disulfide loop; however, the active site of the enzyme in the complex retained the capacity to hydrolyze synthetic amino acid esters and in the case of the α_2-macroglobulin-plasmin complex, retained fibrinogenolytic activity. Therefore, for the α_2-macroglobulin–enzyme reaction, complex formation, peptide bond cleavage and active-site inhibition appear to be dissociated events. In the case of $\overline{\text{C1}}$ inactivator, hydrolysis of the single polypeptide chain of the inhibitor by $\overline{\text{C1}}$s was not documented when evaluated following disulfide bond cleavage by SDS-acrylamide gel electrophoresis. Furthermore, plasmin was capable of forming a high molecular weight complex with both the intact inhibitor chain (complex I) as well as with a degraded inhibitor molecule (complex II), thereby suggesting that a proteolytic event may not be an essential feature of the interaction between $\overline{\text{C1}}$ inactivator and the proteases which it inhibits.

Acknowledgments

These studies were performed with the skilled technical assistance of Mr. Tsun-San Chang and Miss Linda Nyari. Drs. Harpel and Cooper are recipients of U.S. Public Health Service Research Career Development Awards, 1-K04-HE50, 285 and 5-K4-A1-33, 630. These studies were also supported by U.S. Public Health Service Grants HL-16672, HL-14810, A107007 and HL-11409. Part of this study was performed by Dr. Harpel while a Visiting Investigator at the Scripps Clinic and Research Foundation.

REFERENCES

Ako, H., R. J. Foster and C. A. Ryan. 1974. Mechanism of action of naturally occurring proteinase inhibitors. Studies with anhydrotrypsin and anhydrochymotrypsin purified by affinity chromatography. *Biochemistry* **13**:132.

Barkas, T., G. K. Scott and J. E. Fothergill. 1973. Purification, characterization, and active site studies on human serum complement subcomponent C1s. *Biochem. Soc. Trans.* **1**:1219.

Barrett, A. J. and P. M. Starkey. 1973. The interaction of α_2-macroglobulin with proteinases. *Biochem. J.* **133**:709.

Chase, T., Jr. and E. Shaw. 1970. Titration of trypsin, plasmin, and thrombin with *p*-nitrophenyl *p'*-guanidinobenzoate HCl. In *Methods in Enzymology* (ed. G. E. Perlmann and L. Lorand), vol. 19, p. 20. Academic Press, New York.

Cochrane, C. G., S. D. Revak and K. D. Wuepper. 1973. Activation of Hageman factor in solid and fluid phases. A critical role of kallikrein. *J. Exp. Med.* **138**:1564.

Deutsch. D. G. and E. T. Mertz. 1970. Plasminogen: Purification from human plasma by affinity chromatography. *Science* **170**:1095.

Donaldson, V. H. and R. R. Evans. 1963. Biochemical abnormality in hereditary angioneurotic edema: Absence of serum inhibitor of C'1 esterase. *Amer. J. Med.* **35**:37.

Dyce, B., T. Wong, N. F. Adham, J. Mehl and B. J. Haverback. 1967. Human plasma kallikrein esterase associated with the α_2-M binding protein. *Clin. Res.* **15**:101.

Feeney, R. E. 1972. The non-bond splitting mechanism of action of inhibition of proteolytic enzymes—The conservative interpretation. In *Proceedings of the 1st International Research Conference on Protease Inhibitors* (ed. H. Fritz and H. Tschesche), p. 162. Walter de Gruyter, New York.

Forbes, C. D., J. Pensky and O. D. Ratnoff. 1970. Inhibition of activated Hageman factor and activated plasma thromboplastin antecedent by purified serum C$\overline{1}$ inactivator. *J. Lab. Clin. Med.* **76**:809.

Ganrot, P. O. 1967. Inhibition of plasmin activity by α_2-macroglobulin. *Clin. Chim. Acta* **16**:328.

Gigli, I., S. Ruddy and K. F. Austen. 1968. The stoichiometric measurement of the serum inhibitor of the first component of complement by the inhibition of immune hemolysis. *J. Immunol.* **100**:1154.

Gigli, I., J. W. Mason, R. W. Colman and K. F. Austen. 1970. Interaction of plasma kallikrein with the C$\overline{1}$ inhibitor. *J. Immunol.* **104**:574.

Harpel, P. C. 1970a. Human plasma alpha 2-macroglobulin. An inhibitor of plasma kallikrein. *J. Exp. Med.* **132**:329.

———. 1970b. A sensitive, colorimetric method for the measurement of serum C$\overline{1}$ inactivator using the substrate N-α-acetyl-L-lysine methyl ester. *J. Immunol.* **104**:1024.

———. 1970c. C$\overline{1}$ inactivator inhibition by plasmin. *J. Clin. Invest.* **49**:568.

———. 1971. Separation of plasma thromboplastin antecedent from kallikrein by the plasma α_2-macroglobulin, kallikrein inhibitor. *J. Clin. Invest.* **50**:2084.

———. 1972. Studies on the interaction between collagen and a plasma kallikrein-like activity. Evidence for a surface-active enzyme system. *J. Clin. Invest.* **51**:1813.

———. 1973. Studies on human plasma α_2-macroglobulin-enzyme interactions. Evidence for proteolytic modification of the subunit chain structure. *J. Exp. Med.* **138**:508.

Harpel, P. C. and N. R. Cooper. 1975. Studies on human plasma C$\overline{1}$ inactivator-enzyme interactions. I. Mechanisms of interaction with C$\overline{1}$s, plasmin and trypsin. *J. Clin. Invest.* **55:**593.

Harpel, P. C. and M. W. Mosesson. 1973. Degradation of human fibrinogen by plasma α_2-macroglobulin-enzyme complexes. *J. Clin. Invest.* **52:**2175.

Haupt, H., N. Heimburger, T. Kranz and H. G. Schwick. 1970. Ein Beitrag zur isoleirung und charakterisierung des C$\overline{1}$-Inaktivators aus human Plasma. *Eur. J. Biochem.* **17:**254.

Heimburger, N., H. Haupt and H. G. Schwick. 1971. Proteinase inhibitors of human plasma. In *Proceedings of the 1st International Research Conference on Proteinase Inhibitors* (ed. H. Fritz and H. Tschesche), p. 1. Walter de Gruyter, New York.

Jones, J. M., J. M. Creeth and R. A. Kekwick. 1972. Thiol reduction of human α_2-macroglobulin. The subunit structure. *Biochem. J.* **127:**187.

Kagen, L. J. and E. L. Becker. 1963. Inhibition of permeability globulins by C′1-esterase inhibitor. *Fed. Proc.* **20:**613.

Lanchantin, G. F., M. L. Plesset, J. A. Friedmann and D. W. Hart. 1966. Dissociation of esterolytic and clotting activities of thrombin by trypsin-binding macroglobulin. *Proc. Soc. Exp. Biol. Med.* **121:**444.

Landerman, N. S., M. E. Webster, E. L. Becker and H. E. Ratcliffe. 1962. Hereditary angioneurotic edema. II. Deficiency of inhibitor for serum globulin permeability factor and/or plasma kallikrein. *J. Allergy* **33:**330.

Laskowski, M., Jr. 1972. Interaction of proteinases with protein proteinase inhibitors. In *Pulmonary Emphysema and Proteolysis* (ed. C. Mittman), p. 311. Academic Press, New York.

Laurell, C.-B. 1965. Antigen-antibody crossed electrophoresis. *Anal. Biochem.* **10:**358.

Lepow, I. H. and M. A. Leon. 1962. Interaction of a serum inhibitor of C′1-esterase with intermediate complexes of the immune haemolytic system. I. Specificity of inhibition of C′1 activity associated with intermediate complexes. *Immunology* **5:**222.

Levy, L. R. and I. H. Lepow. 1959. Assay and properties of serum inhibitor of C′1-esterase. *Proc. Soc. Exp. Biol. Med.* **101:**608.

Marciniak, E., C. H. Farley and P. A. DeSimone. 1974. Familial thrombosis due to antithrombin III deficiency. *Blood* **43:**219.

McConahey, P. J. and F. J. Dixon. 1966. A method of trace iodination of proteins for immunologic studies. *Int. Arch. Allergy Appl. Immunol.* **29:**185.

McConnell, D. J. 1972. Inhibitors of kallikrein in human plasma. *J. Clin. Invest.* **51:**1611.

Mittman, C., ed. 1972. *Pulmonary Emphysema and Proteolysis,* p. 3. Academic Press, New York.

Mosesson, M. W. and J. S. Finlayson. 1963. Subfraction of human fibrinogen. Preparation and analysis. *J. Lab. Clin. Med.* **62:**663.

Mosesson, M. W., J. S. Finlayson, R. A. Umfleet and D. Galanakis. 1972. Human fibrinogen heterogeneities. I. Structural and related studies of plasma fibrinogens which are high solubility catabolic intermediates. *J. Biol. Chem.* **247:**5210.

Pensky, J., L. R. Levy and I. H. Lepow. 1961. Partial purification of serum inhibitor of C′1-esterase. *J. Biol. Chem.* **236:**1674.

Ratnoff, O. D., J. Pensky, D. Ogston and G. B. Naff. 1969. The inhibition of plasmin, plasma kallikrein, plasma permeability factor, and the C1r subcomponent of the first component of complement by serum C′1 esterase inhibitor. *J. Exp. Med.* **129:**315.

Rinderknecht, H. and M. C. Geokas. 1973. On the physiological role of α_2-macroglobulin. *Biochim. Biophys. Acta* **295:**233.

Saunders, R., B. J. Dyce, W. E. Vannier and B. J. Haverback. 1971. The separation of alpha-2 macroglobulin into five components with differing electrophoretic and enzyme-binding properties. *J. Clin. Invest.* **50:**2376.

Schreiber, A. D., A. P. Kaplan and K. F. Austen. 1973. Inhibition by C$\overline{1}$ INH of Hageman factor fragment activation of coagulation, fibrinolysis, and kinin generation. *J. Clin. Invest.* **52:**1402.

Schultze, H. E., N. Heimburger, K. Heide, H. Haupt, K. Störiko and H. G. Schwick. 1963. Preparation and characterization of α_1-trypsin inhibitor and α_2-plasmin inhibitor of human serum. In *Proceedings 9th Congress European Society Haematology (Lisbon)*, p. 1315. S. Karger, Basel.

Sherman, L. A., A. P. Fletcher and S. Sherry. 1969. In vivo transformation between fibrinogens of varying ethanol solubilities: A pathway of fibrinogen catabolism. *J. Lab. Clin. Med.* **73:**574.

Steinbuch, M., M. Quentin and L. Pejaudier. 1965. Technique d'isolement et étude de l'α_2-macroglobuline. In *Protides of the Biological Fluids* (ed. H. Peeters), p. 375. Elsevier, Amsterdam.

Summaria, L., B. Hsieh, W. R. Groskopf, K. C. Robbins and G. H. Barlow. 1967. The isolation and characterization of the S-carboxylmethyl β (light) chain derivative of human plasmin. *J. Biol. Chem.* **242:**5046.

Valet, G. and N. R. Cooper. 1974. Isolation and characterization of the proenzyme form of the C1s subunit of the first complement component. *J. Immunol.* **112:**339.

Weber, K. and M. Osborn. 1969. The reliability of molecular weight determinations by dodecyl sulfate-polyacrylamide gel electrophoresis. *J. Biol. Chem.* **244:**4406.

Function and Chemical Composition of α_1-Antitrypsin

Jan-Olof Jeppsson and Carl-Bertil Laurell
Department of Clinical Chemistry, University of Lund
Malmö General Hospital, Malmö, Sweden

The plasma proteins are often named according to their assumed biologic function, reflecting the findings made by those workers who detected or identified these proteins. There is a wild flora of names on the protease inhibitors.

Jacobsson (1955) first localized the major trypsin inhibitor to the electrophoretic α_1-zone, and Bundy and Mehl (1959) isolated the α_1-trypsin inhibitor. Schultze et al. (1955) isolated a 3.5S α_1-glycoprotein and later coined the name α_1-antitrypsin (α_1-AT) after having observed its capacity to link serine proteases (Schultze, Heide and Haupt 1962). Norman (1958) introduced α_1-antiplasmin, and Rimon, Yeheskel and Benyamin (1966), slow or progressive antiplasmin, but they have never proved that it forms complexes with plasmin in vivo. Human α_1-protease inhibitor is a recent proposal (Johnson, Pannell and Travis 1974), fitting the abbreviation Pi used in the system for classification of genetic α_1-AT variants. It is also appropriate for a recently documented essential function of α_1-AT, namely, the inactivation of granulocytal enzymes, primarily elastase (Janoff 1972; Ohlsson and Olsson 1973, 1974).

Genetics and Biologic Variation

During systematic studies of serum proteins, Laurell and Eriksson (1963) observed for the first time that the main component α_1-AT in the α_1-globulin band on paper electrophoresis was lacking in some human plasma. Most of these patients had pulmonary emphysema. α_1-Antitrypsin has since been shown to have a number of genetic variants, called Pi types for protease inhibitor (Fagerhol and Laurell 1967).

Conventionally, the most common allele with a product of an intermediate mobility in electrophoresis has been designated M; faster alleles, F; slower ones, S; and much slower ones, Z. The most common phenotype in all population studies is Pi MM, and so far 23 Pi alleles have been recognized

(Fagerhol, pers. comm.). A two-dimensional electrophoretic system with high resolution and high specificity has been developed by Fagerhol and Laurell (1967). Acid discontinuous starch gel (pH 4.95) was used in the first dimension and crossed immunoelectrophoresis with rabbit anti-α_1-AT in the second dimension. A system with acid agarose electrophoresis (Laurell and Persson 1973) or electrofocusing in polyacrylamide (Lebas, Hayeme and Martin 1974) instead of starch gel electrophoresis in the first dimension has been proposed. Electrofocusing on thin-layer polyacrylamide shows a microheterogeneity with isoionic points between pH 4.5–4.9. However, the primary system as improved by Fagerhol (1968) remains the basic system for classification of genetic Pi variants. In a few variants, the charge is more apparent on conventional agarose gel at pH 8.6 than at pH 5. The charge of α_1-AT from two phenotypes may be the same, but the plasma concentration may be subnormal for one of them. When describing new genetic variants, it is therefore considered necessary to analyze α_1-AT (1) with starch gel electrophoresis followed by (2) crossed immunoelectrophoresis with (3) agarose gel electrophoresis pH 8.6 and (4) to quantitate α_1-AT immunochemically and (5) measure its specific trypsin inhibitory activity. A Pi committee conducted by Dr. M. Fagerhol (Blodbank og Immunhematologiskt laboratorium, Ullevål sykehus, Oslo, Norway) acts as a reference center for analyses of Pi variants.

A few alleles give phenotypes with reduced α_1-AT concentration in plasma. These different Pi alleles give the following relative concentration of α_1-AT compared to Pi^M 1.0: Pi$^-$, 0 (Talamo et al. 1973); Pi^Z, 0.15 (Laurell 1972); Pi^P, 0.25 (Fagerhol and Hauge 1968); Pi^S, 0.60 (Fagerhol 1968). The normal mean plasma level for homozygotes ZZ and SS are 15 and 60% and for heterozygotes MZ and MS, 57.5 and 80%, respectively, of the average adult normal phenotype MM. In a Swedish population (Eriksson 1965), the frequency of Pi MZ heterozygotes is around 4% and of Pi ZZ homozygotes, 0.07%.

The normal plasma concentration is estimated immunochemically at 2.0–2.2 g/liter (Kueppers 1967; Ganrot 1972). In most acute inflammatory conditions caused by infections or necrosis, the α_1-AT increases like other acute phase reactants. The doubling of its plasma level during pregnancy and an appreciable increase (50–100%) during the use of estrogens for contraceptive purposes may be attributed to a synthesis effect of estrogens (Fagerhol and Laurell 1970).

Purification and Physiochemical Characteristics

α_1-Antitrypsin had been fractionated earlier by use of organic solvents and pH values as low as 4.5 (Schultze et al. 1955; Bundy and Mehl 1959). Later it was shown that α_1-AT loses its trypsin inhibitory capacity at pH below 5 (Schultze, Heide and Haupt 1962). In the last two years several methods have been published using modern equipment for protein fractionation (Crawford 1973; Chan, Luby and Wu 1973; Kress and Laskowski 1974). The main impurity, albumin, which has very similar solubility, can be eliminated either by running affinity chromatography on Sepharose coupled to antibodies against albumin (Myerowitz, Handzel and Robbins 1972) or Sepharose 4B coupled with blue dextran (Travis and Pannell 1973). Other impurities can be eliminated by affinity chromatography on Sepharose 4B-coupled antibodies raised to proteins with chromatographic properties similar to α_1-AT (Jeppsson and Laurell 1974).

Figure 1 shows two alternatives for preparation of α_1-AT. The left section, showing DEAE chromatography after ammonium sulfate precipitation and affinity chromatography for removing impurities, gives a native α_1-AT with preserved microheterogeneity and 70–80% yield. The right alternative concerns affinity chromatography on a Sepharose 4B column conjugated with rabbit IgG-anti-α_1-AT after initial ammonium sulfate precipitation. The elution is affected by 3 M NaSCN. The second alternative gives pure material, but the immunoreactivity to its corresponding antibodies is diminished, which suggests a conformational change when the antigen-antibody complexes are dissociated by the strong thiocyanate solution.

α_1-Antitrypsin is a single polypeptide chain with a molecular weight of approximately 55,000 daltons. It contains 12% carbohydrates, including 6–7 residues of sialic acid. The N-terminal amino acid is blocked, and the half-cystines content has been reported to be one (Crawford 1973) or two residues (Heimburger, Haupt and Schwick 1971; Jeppsson and Laurell 1974) per mole α_1-AT. The absorbance $A^{1\%}_{280\ nm}$ has been reported as 5.3 (Schönenberger 1955) and 4.36 (Crawford 1973).

Characterization by the fingerprint technique revealed a very high resistance of the arginine and lysine bonds to trypsin even after denaturation and aminoethylation of α_1-AT. Thermolysin digestion, however, gave too many spots on the peptide map.

Cyanogen bromide (CNBr) degradation was performed after reduction and carboxymethylation. Eight to nine fragments can be identified and isolated. Some of them have 3–20 amino acid residues. Those which are stainable by Coomassie blue are identified by electrofocusing on thin-layer polyacrylamide gel. The fragments are only poorly soluble in ordinary buffers, so electrofocusing must be done in freshly prepared 6 M urea solution.

Figure 2 compares CNBr fragments of isolated α_1-AT from phenotypes FF, MM and SZ. There is no difference in the larger CNBr fragments between phenotype MM and FF, which suggests that the difference lies in the smaller

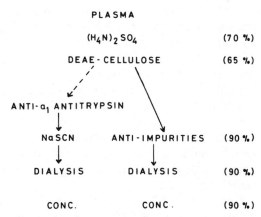

Figure 1
Fractionation schedule for α_1-AT. The yields in the individual steps are given to the right.

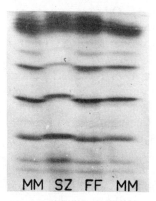

MM SZ FF MM

Figure 2

Thin-layer electrofocusing at pH 3–10 in 6 M urea of cyanogen bromide fragments from α_1-AT Pi MM, FF and SZ.

peptides not stainable by Coomassie blue. Comparison of MM and SZ showed a clear difference in charge between the fragments. Recently the same comparison was made between MM and ZZ, and ZZ showed different carbohydrate-containing fragments. Preliminary carbohydrate analysis showed reduced amounts of sialic acid in ZZ as compared to MM.

Liver Involvement in α_1-AT Deficiency

The liver is probably the major source of α_1-AT, which is synthesized in in vitro cultures of hepatocytes. Further evidence for synthesis of α_1-AT in the hepatocytes has been obtained from clinical pathology. Individuals with α_1-AT deficiency, homozygous for Pi^{ZZ}, are predisposed to emphysema and liver disease with cirrhosis as a late phenomena.

Sharp (1971) detected large amounts of inclusion bodies in the rough endoplasmic reticulum of the hepatocytes from Pi ZZ babies with neonatal hepatitis. These globules were PAS (periodic acid Schiff)-positive, which meant that they contained carbohydrates. The material appears as round or oval globules ranging in size from 1 to 40 μm. The number varies from one hepatocyte to another. They are not seen in Kupfer cells. By using immuno-fluorescence, Sharp (1971) showed the existence of material antigenically similar to α_1-AT. Electron microscopy studies (Sharp 1971) have shown amorphous material within the lumen of the endoplasmic reticulum, which was dilated. Others (Lieberman, Mittman and Gordon 1972) have confirmed this only in the rough or both in rough and smooth endoplasmic reticulum, but never in the Golgi apparatus. Homozygotes and heterozygotes for the Pi^Z allele regularly have more or fewer of these inclusion bodies in their periportal hepatocytes (Berg and Eriksson 1972; Gordon et al. 1972). They have not been observed in subjects with any other Pi allele.

The inclusion bodies were recently isolated and proven to consist of aggregated α_1-AT (Eriksson and Larsson 1974). The material reacted with antibodies raised against plasma-α_1-AT, and antibodies raised from aggregated material reacted with several different α_1-AT variants. The molecular weight as determined by SDS-polyacrylamide electrophoresis was approximately the same.

Comparison of CNBr fragments from isolated variants MM, ZZ and isolated ZZ protein from liver (PAS-positive inclusion bodies) showed a clear difference in charge on electrofocusing in 6 M urea. There were also fewer carbohydrate-containing fragments in ZZ materials than in MM (Fig. 3). This may be interpreted as a reduced amount of carbohydrate chains.

Independent studies from two other laboratories (Bell and Carrel 1973; Cox 1973) show that α_1-AT of Pi type ZZ may lack some of the sialic acid compared with that of Pi type MM. Cox (1975) incubated sera of different Pi types with neuraminidase, and the reaction products were studied with two types of electrophoresis. The results suggested that Z type had less sialic acid than types M, F and S. It was thus proposed that the Z protein lacks a carbohydrate chain with two terminal sialic residues.

It is known that some asialoglycoproteins or glycoproteins with reduced sialic acid content are eliminated from the circulation (Morrell et al. 1971). They are removed by their attachment to the sialic acid-containing plasma membrane of the hepatocyte provided that the next galactose residues are intact (Pricer and Ashwell 1971). Electron microscopic studies showed the presence of aggregated α_1-AT in rough or smooth endoplasmic reticulum (RER or SER). A few carbohydrate molecules must have been added to the polypeptide on the way through RER or SER since histological PAS staining demonstrated carbohydrates. Since sialyltransferases are located to the Golgi apparatus (Schachter et al. 1970), the chance of adding terminal sialic acid must be small if the electron microscopic findings are correctly interpreted. Thus the question is still open as to why incomplete α_1-AT of ZZ type is precipitated in the liver cells.

Interaction with Endopeptidases and Other Proteins

The pancreatic enzymes, trypsin, chymotrypsin and elastase (Kaplan, Kuhn and Pierce 1973), the granulocytal enzymes, elastase (Ohlsson 1971a; Janoff 1972; Lieberman and Kaneshiro 1972; Ohlsson and Olsson 1974) and collagenase (Ohlsson and Olsson 1973), and the skin and synovial collagenases (Tokoro, Eisen and Jeffrey 1972; Harris, DiBona and Krane 1969) are firmly bound and inactivated by α_1-AT in an immediate reaction with a molar ratio 1:1 as first shown for trypsin by Bundy and Mehl (1959). Both the proteolytic and esterolytic activity of these enzymes are inhibited. The mechanism behind the blocking of the active site is largely unknown. Modification of the lysyl residues of α_1-AT with maleic anhydride results in faster

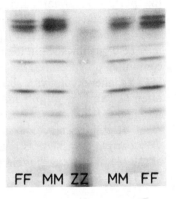

Figure 3
Thin-layer electrofocusing at pH 3–10 in 6 M urea of cyanogen bromide fragments from α_1-AT Pi MM, FF and ZZ from liver. The carbohydrate-containing fragments are at the top.

FF MM ZZ MM FF

electrophoretic migration and loss of antitrypsin activity (Heimburger, Haupt and Schwick 1971). Phenylglyoxalhydrate blocks the action of α_1-AT on trypsin, giving support to a modified arginine residue (Cohen 1973).

The α_1-AT reaction with the proteases implies formation of complexes with covered active sites of the enzymes. The complexes are stable on electrophoresis (pH 9–5) and have an electrophoretic mobility intermediate between that of the reactants.

The stability of the complexes on chromatographic and electrophoretic separation are indications of the firmness of the linkage, but the reaction is reversible since a slow transfer of [^{125}I]trypsin has been observed from α_1-AT to α_2-macroglobulin in vitro and in vivo (Ohlsson 1971b). All microheterogeneous α_1-AT fractions appear to have very similar antiprotease activity as judged from the pattern of free and complexed α_1-AT observed on crossed immunoelectrophoretic analyses of plasma at pH 5.1 after stepwise addition of trypsin.

Another group of mammalian proteolytic enzymes, plasmin (Rimon, Yeheskel and Benyamin 1966; Heimburger, Haupt and Schwick 1971), kallikrein (Fritz et al. 1969) and the spermal acrosine (Fritz et al. 1972), are slowly inactivated by α_1-AT in a reaction running for hours in vitro. The reaction products have not been defined. Inhibition of bacterial enzymes such as Aspergillus protease (Bergkvist 1963) and subtilopeptidase A (Wicher and Dolowich 1973) have been observed.

Complexes between enzymes and α_1-AT are normally not detectable in plasma. Ohlsson (1971a, 1974) has clearly shown that α_1-AT complexes appear in the exudate from the pancreas in acute pancreatitis and in the ascitic fluid, but that they disappear during the draining of the fluids through the lymphatic. Granulocytal protease-α_1-AT complexes have been detected in ascitic fluid in acute peritoneitis (Ohlsson 1974), in synovial fluid in arthritis (Shtacher, Maayan and Feinstein 1973) and in cerebrospinal fluid during leukocytosis (Hochstrasser et al. 1972). No plasmin-α_1-AT complexes are recognized in plasma even after complete plasminogen activation with streptokinase (Niléhn and Ganrot 1967; Arnesen and Fagerhol 1972) or urokinase.

Normally about one percent of the α_1-AT of plasma occurs as an IgA complex. The α_1-AT is joined through a disulfide link probably with the heavy IgA chain or through J chains. The complex is sensitive for mild reduction and is cleaved on ingestion of penicillamine for some days. Less than one percent is also regularly linked to fibrinogen by a more stable bond (Laurell and Thulin 1975). In myeloma with Bence-Jones proteinemia of κ type, the plasma invariably contains monomeric κ-α_1-AT complexes 1:1 with covalent bonds. They are formed by disulfide interchange not affecting the trypsin binding capacity of α_1-AT (Laurell and Thulin 1974). This principle may be of significance in the inactivation and transport of thiol- and disulfide-containing polypeptides.

Pi ZZ phenotypes with about 15% of the average α_1-AT content in plasma often and early develop panacinar emphysema in the lungs. This phenomena has focused interest on possible relations between proteolysis and development of emphysema (Eriksson 1965). The perfusion of the lower lobes is very early diminished, as is the elastic recoil of these lung parts (Kanner et al. 1973). The latter finding suggests lesions of the elastic fibers. Leukocytes

. are normally abundant, sequestered in these parts of the lung. Too slow inactivation of proteolytic enzymes released by granulocytes may cause premature aging of the fibrous network of the lung. Emphysema has also been induced experimentally by pancreatic elastase (Kaplan, Kuhn and Pierce 1973). Any factor tending to promote hydrolysis of lung elastin will increase the risk of wasting of the lung tissue.

Recent analysis of granulocytal proteases has shown that elastase and collagenase constitute about 5% of the dry weight of the cells (Ohlsson and Olsson 1974). In all likelihood, larger amounts of these enzymes are released from the granulocytes into the extracellular space than are proteases from any other cells. α_1-Antitrypsin is the inhibitor available in the largest amount to govern the effect of the enzymes released. Therefore, a major biologic function attributed to α_1-AT may be the inactivation of proteases released mainly intercellularly during the activity of granulocytes. This mechanism for inactivation of proteases in the intercellular fluid, although reasonable, is difficult to prove experimentally since the complexes are probably eliminated through the reticuloendothelial system and macrophages before arrival and dilution in the blood. An interaction between α_1-AT and α_2-M seems also to result in transfer of the enzyme to α_2-M and elimination.

SUMMARY

α_1-Antitrypsin is one of the main protease inhibitors in plasma. It is a single chain with blocked N terminal, molecular weight 50–55,000 daltons, and 12% carbohydrates. A microheterogeneity is obtained by prolonged acid electrophoresis (pH 4.95) or electrofocusing.

An adult form of pulmonary emphysema and juvenile cirrhosis is associated with inherited variants with low levels of α_1-AT. Individuals homozygous for Pi ZZ possess only 10–20% the amount of α_1-AT compared with subjects homozygous for the normal Pi MM. The frequency of Pi MZ is around 4% and Pi ZZ, 0.07%. Selective low values are compatible with genetic variants with Pi alleles null, Z or S. At least one allele—the Z allele—causes synthesis of a modified α_1-AT with abnormal retention in the endoplasmic reticulum in the liver.

Comparison of CNBr fragments of isolated α_1-AT from phenotypes MM, SZ, ZZ and isolated ZZ protein from liver (PAS-positive inclusion bodies) showed a clear difference in charge and carbohydrate-containing fractions.

The pancreatic enzymes, trypsin, chymotrypsin and elastase, and the granulocytal enzymes, elastase and collagenase, are firmly bound and inactivated by α_1-AT in an immediate reaction with a molar ratio 1:1. The major biologic function is presumably the inactivation of proteases released during the activity of granulocytes.

REFERENCES

Arnesen, H. and M. K. Fagerhol. 1972. α_2-Macroglobulin, α_1-antitrypsin and antithrombin III in plasma and serum during fibrinolytic therapy with urokinase. *Scand. J. Clin. Lab. Invest.* **29**:259.

Bell, O. F. and R. W. Carrell. 1973. Basis of the defect in α_1-1-antitrypsin deficiency. *Nature* **243**:410.

Berg, N. O. and S. Eriksson. 1972. Liver disease in adults with alpha$_1$-antitrypsin deficiency. *New Eng. J. Med.* **287**:1264.

Bergkvist, R. 1963. The proteolytic enzymes of *Aspergillus oryzae*. *Acta Chem. Scand.* **17**:2239.

Bundy, H. F. and J. W. Mehl. 1959. Trypsin inhibitors of human serum. *J. Biol. Chem.* **234**:1124.

Chan, S. K., J. Luby and Y. C. Wu. 1973. Purification and chemical compositions of human α_1-antitrypsin of the MM type. *FEBS Letters* **35**:79.

Cohen, A. B. 1973. Mechanism of action of α-1-antitrypsin. *J. Biol. Chem.* **248**: 7055.

Cox, D. 1973. Defect in alpha$_1$-antitrypsin deficiency. *Lancet* **2**:844.

———. 1975. The effect of neuraminidase on genetic variants of alpha$_1$-antitrypsin. *Amer. J. Human Genet.* **27**:165.

Crawford, I. P. 1973. Purification and properties of normal human α_1-antitrypsin. *Arch. Biochem. Biophys.* **156**:215.

Eriksson, S. 1965. Studies in α_1-antitrypsin deficiency. *Acta Med. Scand.* (Suppl.) **177**:432.

Eriksson, S. and Ch. Larsson. 1974. Purification and partial characterization of PAS-positive inclusion bodies from the liver in alpha$_1$-antitrypsin deficiency. *New Eng. J. Med.* **292**:176.

Fagerhol, M. K. 1968. The Pi-system. Genetic variants of serum α_1-antitrypsin. *Ser. Haematol.* **1**:153.

Fagerhol, M. K. and H. E. Hauge. 1968. The Pi phenotype MP. Discovery of a ninth allele belonging to the system of inherited variants of serum α_1-antitrypsin. *Vox Sang.* **15**:396.

Fagerhol, M. K. and C.-B. Laurell. 1967. The polymorphism of "prealbumins" and α_1-antitrypsin in human sera. *Clin. Chim. Acta* **16**:199.

———. 1970. The Pi system—Inherited variants of serum α_1-antitrypsin. In *Progress in Medical Genetics* (ed. A. Steinberg and A. Bearn), p. 96. Grune & Stratton, London.

Fritz, H., B. Brey, A. Schmal and E. Werle. 1969. Zur Identität des Progressiv-Antikallikreins mit α_1-Antitrypsin aus Humanserum. *Hoppe-Seyler's Z. Physiol. Chem.* **350**:1551.

Fritz, H., N. Heimburger, M. Meier, M. Arnhold, L. J. D. Zaneveld and G. F. B. Schumacher. 1972. Humanakrosin: Zur Kinetek der Hemmung durch Humanseruminhibitoren. *Hoppe-Seyler's Z. Physiol. Chem.* **353**:1953.

Ganrot, P.-O. 1972. Variation of the concentrations of some plasma proteins in normal adults, in pregnant women and in newborns. *Scand. J. Clin. Lab. Invest.* (Suppl. 124) **29**:83.

Gordon, H. W., J. Dixon, J. C. Rogers, C. Mittman and J. Lieberman. 1972. Alpha$_1$-antitrypsin (α_1-AT) accumulation in livers of emphysematous patients with α_1-AT deficiency. *Human Pathol.* **3**:361.

Harris, E. D., D. R. DiBona and S. M. Krane. 1969. Collagenases in human synovial fluid. *J. Clin. Invest.* **48**:2104.

Heimburger, N., H. Haupt and H. G. Schwick. 1971. Proteinase inhibitors of human plasma. In *Proceedings of the International Research Conference on Proteinase Inhibitors* (ed. H. Fritz and H. Tscheche), p. 1. Walter de Gruyter, New York.

Hochstrasser, K., R. Schuster, R. Reichert and N. Heimburger. 1972. Nachweis und quanitative Bestimmung von Komplexen zwischen Leukocytenproteasen und α_1-Antitrypsin in Körpersekreten und Körperflüssigkeiten. *Hoppe-Seyler's Z. Physiol. Chem.* **353:**1120.

Jacobsson, K. 1955. Studies on the trypsin and plasmin inhibitors in human blood serum. *Scand. J. Clin. Lab. Invest.* (Suppl.) **14:**55.

Janoff, A. 1972. Inhibition of human granulocyte elastase by serum α_1-antitrypsin. *Amer. Rev. Respir. Dis.* **105:**121.

Jeppsson, J.-O. and C.-B. Laurell. 1974. Isolation and fragmentation of α_1-antitrypsin. In *Bayer Symposium II: Proteinase Inhibitor II* (ed. H. Fritz), p. 47. Springer-Verlag, New York.

Johnson, D., R. Pannell and J. Travis. 1974. The molecular stoichiometry of trypsin inhibition by human alpha-1-proteinase inhibitor. *Biochem. Biophys. Res. Comm.* **57:**584.

Kanner, R. E., M. R. Klauber, S. Watanabe, A. D. Renzetti and A. Bigler. 1973. Pathologic patterns of chronic obstructive pulmonary disease in patients with normal and deficient levels of alpha$_1$-antitrypsin. *Amer. J. Med.* **54:**706.

Kaplan, P. D., C. Kuhn and J. A. Pierce. 1973. The induction of emphysema with elastase. I. The evolution of the lesion and the influence of serum. *J. Lab. Clin. Med.* **82:**349.

Kress, L. F. and M. Laskowski, Sr. 1974. Purification, properties and composition of alpha-1-trypsin inhibitor from human plasma. In *Bayer Symposium V: Proteinase Inhibitors* (ed. H. Fritz et al.), p. 23. Springer-Verlag, New York.

Kueppers, F. 1967. Immunologic assay of alpha$_1$-antitrypsin in deficient subjects and their families. *Humangenetik* **5:**54.

Laurell, C.-B. 1972. Variation of the alpha$_1$-antitrypsin level of plasma. In *Pulmonary Emphysema and Proteolysis* (ed. Mittman), p. 161. Academic Press, New York.

Laurell, C.-B. and S. Eriksson. 1963. The electrophoretic α_1-globulin pattern of serum in α_1-antitrypsin deficiency. *Scand. J. Clin. Lab. Invest.* **15:**132.

Laurell, C.-B. and U. Persson. 1973. Analysis of plasma α_1-antitrypsin variants and their microheterogeneity. *Biochim. Biophys. Acta* **310:**500.

Laurell, C.-B. and E. Thulin. 1974. Complexes in plasma between light chain κ immunoglobulins and alpha$_1$-antitrypsin, respectively, prealbumin. *Immunochemistry* **11:**703.

————. 1975. Bence Jones proteins bound to α_1-antitrypsin, prealbumin and albumin through sulfhydryl-disulphide interchange. *J. Immunol.* (in press).

Lebas, J., A. Hayeme and J. P. Martin. 1974. Étude des variants génétiques de 1-α_1-antitrypsine en immunoélectrofocalisation bidimensionelle. *C. R. Acad. Sci.* **278:**2359.

Lieberman, J. and W. Kaneshiro. 1972. Inhibition of leukocytic elastase from purulent sputum by alpha$_1$-antitrypsin. *J. Lab. Clin. Med.* **80:**88.

Lieberman, J., C. Mittman and H. W. Gordon. 1972. Alpha$_1$-antitrypsin in the livers of patients with emphysema. *Science* **175:**63.

Morell, A. G., G. Gregoriados, H. Scheinberg, J. Hickman and G. Ashwell. 1971. The role of sialic acid in determining the survival of glycoproteins in the circulation. *J. Biol. Chem.* **246:**1461.

Myerowitz, R. L., Z. T. Handzel and J. B. Robbins. 1972. Human serum α_1-antitrypsin: Isolation and demonstration of electrophoretic and immunologic heterogeneity. *Clin. Chim. Acta* **39:**307.

Niléhn, J.-E. and P.-O. Ganrot. 1967. Plasmin, plasmin inhibitors and degradation products of fibrinogen in human serum during and after intravenous infusion of streptokinase. *Scand. J. Clin. Lab. Invest.* **20:**113.

Norman, P. S. 1958. Studies of the plasmin system. III. Physical properties of the two plasmin inhibitors in plasma. *J. Exp. Med.* **108:**31.

Ohlsson, K. 1971a. Interaction between human or dog leucocyte proteases and plasma inhibitors. *Scand. J. Clin. Lab. Invest.* **28**:225.

————. 1971b. Interactions in vitro and in vivo between dog trypsin and dog plasma protease inhibitors. *Scand. J. Clin. Lab. Invest.* **28**:219.

————. 1974. Interaction between endogenous proteases and plasma protease inhibitors in vitro and in vivo. In *Bayer Symposium II: Proteinase Inhibitors II* (ed. H. Fritz and H. Tschesche), p. 1. Springer-Verlag, New York.

Ohlsson, K. and I. Olsson. 1973. The neutral protease of human granulocytes. Isolation and partial characterization of two granulocyte collagenases. *Eur. J. Biochem.* **36**:473.

————. 1974. The neutral proteases of human granulocytes. Isolation and partial characterization of granulocyte elastases. *Eur. J. Biochem.* **42**:519.

Pricer, W. E., Jr. and G. Ashwell. 1971. The binding of desialylated glycoproteins by plasma membranes of rat liver. *J. Biol. Chem.* **246**:4825.

Rimon, A., S. Yeheskel and S. Benyamin. 1966. The plasmin inhibitor of human plasma. IV. Its action on plasmin, trypsin, chymotrypsin and thrombin. *J. Biol. Chem.* **241**:5102.

Schachter, H., I. Jabbal, R. L. Hudgin, L. Pinteric, E. J. McGuire and S. Roseman. 1970. Intracellular localisation of liver sugar nucleotide glycoprotein glyco-syltransferases in a Golgi-rich fraction. *J. Biol. Chem.* **245**:1090.

Schönenberger, M. 1955. Streulichtmessungen an plasmaproteinen. *Z. Naturforsch.* **10b**:474.

Schultze, H. E., K. Heide and H. Haupt. 1962. α_1-Antitrypsin aus Humanserum. *K. Wachenschrift* **40**:427.

Schultze, H. E., I. Göllner, K. Heide, M. Schönenberger and G. Schwick. 1955. Über das α_2-Makroglobulin. *Z. Naturforsch.* **10b**:463.

Sharp, H. L. 1971. Alpha-1-antitrypsin deficiency. *Hosp. Pract.* **6**:83.

Shtacher, G., R. Maayan and G. Feinstein. 1973. Proteinase inhibitors in human synovial fluid. *Biochim. Biophys. Acta* **303**:138.

Talamo, R. C., C. E. Langley, C. E. Reed and S. Makino. 1973. Alpha-1-anti-trypsin deficiency: A variant with no detectable alpha-1-antitrypsin. *Science* **181**:70.

Tokoro, Y., A. Z. Eisen and J. J. Jeffrey. 1972. Characterization of a collagenase from rat skin. *Biochim. Biophys. Acta* **258**:289.

Travis, J. and R. Pannell. 1973. Selective removal of albumin from plasma by affinity chromatography. *Clin. Chim. Acta* **49**:49.

Wicher, V. and J. Dolowich. 1973. Effects of human serum inhibitors on im-munologic properties of *B. subtilis* alkaline proteinase. *Immunochemistry* **10**:239.

Serum Sialyltransferase Activity in α_1-Antitrypsin Deficiency and Hepatic Cirrhosis

**Mark S. Kuhlenschmidt, Carol J. Coffee,
Stephen P. Peters and Robert H. Glew**

Department of Biochemistry, School of Medicine, University of Pittsburgh
Pittsburgh, Pennsylvania 15261

Harvey L. Sharp

Department of Pediatrics, School of Medicine, University of Minnesota
Minneapolis, Minnesota 55455

The genetic deficiency of the serum glycoprotein α_1-antitrypsin (α_1-AT) in children is associated with a severe form of hepatic cirrhosis (Sharp 1971; Williams and Fajardo 1974; Cooper and Gupta 1974; Kueppers and Black 1974). These patients display, in addition to a pronounced reduction in the concentration of serum α_1-AT, the accumulation of α_1-AT cross-reacting material in smooth, membrane-delimited, globular deposits in their liver cells. The metabolic basis for the deficiency of serum α_1-AT and the accumulation of cross-reacting material in liver has yet to be determined.

Using a liver biopsy, we recently demonstrated that the carbohydrate-rich material which had accumulated in the liver of a patient with this disease failed to take up reagents which would ordinarily stain sialic acid-containing macromolecules (Kuhlenschmidt et al. 1974). These results suggested that the α_1-AT cross-reacting material might represent the asialo derivative of α_1-AT. Prompted by these observations, we were anxious to examine the capability of serum from patients with α_1-AT deficiency and hepatic cirrhosis to transfer sialic acid from cytidine-5'-monophosphate-N-acetyl neuraminic acid (CMP-NANA) to the asialo derivatives of various serum glycoproteins.

A prerequisite for these studies on sialyltransferase activity was the purification of relatively large amounts of serum α_1-AT, which was required as the asialo derivative, as an acceptor of radioactively labeled sialic acid in the transferase assay. The present report describes a relatively rapid, high-yield procedure for the purification of normal (Pi^{MM}) α_1-AT from human plasma using conventional column chromatographic methods. The last step in the purification results in the resolution of three major isomeric forms of α_1-AT, which seem to differ only in terms of sialic acid content or distribution. We have used a neuraminidase-treated preparation of α_1-AT as well as asialofetuin and asialoceruloplasmin to demonstrate that serum from six patients with α_1-AT deficiency and hepatic cirrhosis is deficient in sialyltransferase activity. In an effort to evaluate the possible causal relationship between sialyltransferase activity and the disease state involving α_1-AT deficiency and

415

hepatic cirrhosis, we conducted similar studies using serum from asymptomatic heterozygous and homozygous relatives. The results presented here indicate that the decrease in serum sialyltransferase activity in these patients is probably a consequence of extensive liver damage involving hepatic accumulation of α_1-AT rather than the cause of such disease.

EXPERIMENTAL PROCEDURES

Purification of α_1-Antitrypsin

One unit of fresh human plasma (approximately 300 ml) was purified through the QAE-Sephadex step described by Crawford (1973), except that the gradient employed was 0.1–0.3 M in NaCl. Fractions containing α_1-AT activity were pooled, dialyzed against 50 mM Tris pH 8.6, and layered onto the same QAE-Sephadex column described above, but which had been reequilibrated with 50 mM Tris pH 8.6 containing 0.1 M NaCl. α_1-Antitrypsin was eluted with 50 mM Tris buffer pH 8.6 containing 0.2 M NaCl. Fractions containing α_1-AT activity were pooled and dialyzed against 10 mM Tris buffer pH 7.6.

The dialysate from the second QAE-Sephadex column was concentrated to 2 ml using a pressure dialysis cell (Amicon) and chromatographed on Sephadex G-150 according to Crawford (1973).

The α_1-AT-containing fractions from the previous gel filtration column were pooled and layered onto a second DEAE-cellulose column (2 x 87 cm) equilibrated with 10 mM sodium phosphate pH 7.0. Resolution of the α_1-AT into three activity peaks was accomplished by equilibrium chromatography using a step gradient employing the same phosphate buffer, but containing different concentrations of NaCl (Fig. 1). The first fraction containing α_1-AT activity, pool A, was eluted with 0.08 M NaCl. After 3–4 column volumes, the NaCl concentration was increased to 0.1 M and fractions B and C containing α_1-AT activity were eluted. The NaCl concentration was then increased to 0.12 M, and finally to 0.2 M, which eluted additional protein peaks, D and E, neither of which possessed measurable α_1-AT activity. When pools A and C were rechromatographed individually on DEAE-cellulose under similar conditions of equilibrium elution, these α_1-AT fractions maintained their relative elution positions. The purification scheme is summarized in Table 1.

Enzyme Assays

Serum, obtained by centrifugation (500g, 10 min) one hour after drawing venous blood samples, was frozen immediately ($-70°$C). In the case of specimens requiring transportation between Minneapolis and Pittsburgh, serum samples were sent by air transport packed in dry ice; control sera drawn at the same time were also included in such shipments.

Sialyltransferase activity was measured essentially as described by Kuhlenschmidt et al. (1974), using the asialo derivatives of α_1-AT, ceruloplasmin and fetuin, each of which was prepared by treatment of the native proteins with neuraminidase as described by Carlson et al. (1973). Commercial Cl. perfringens neuraminidase (Worthington Biochemical) was used after further

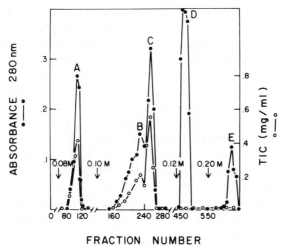

Figure 1

Equilibrium chromatography of α_1-antitrypsin on DEAE-cellulose. α_1-Antitrypsin from the Sephadex G-150 step was layered over a column of DEAE-cellulose and eluted discontinuously with NaCl as described in Experimental Procedures. (●——●) Absorbance at 280 nm; (○——○) α_1-antitrypsin as TIC U/ml.

Table 1

Purification of α_1-Antitrypsin

Step	α_1-AT activity[a] (total TIC units)	Protein[b] (mg)	Specific activity (TIC U/mg)
Crude plasma	302	22,208	0.014
50–75% Ammonium sulfate	295	9516	0.031
1st DEAE-cellulose	279	3583	0.078
1st QAE-Sephadex	270	2617	0.103
2nd QAE-Sephadex	250	542	0.461
Sephadex G-150	183	329	0.556
2nd DEAE-cellulose			
pool A	32.6	55.8	0.580
pool B	68.5	99.6	0.687
pool C	25.9	40.6	0.638
	127 (42%)	196 mg	

[a] α_1-Antitrypsin was assayed according to Eriksson (1965).
[b] Protein was estimated by the procedure of Lowry et al. (1951) using bovine serum albumin as standard.

purification by an affinity column procedure described by Cuatrecasas and Illiano (1971). Neuraminidase was inactivated by heat (60°C, 1 hr) as described by Carlson et al. (1973), and potential sialic acid acceptor sites were estimated according to the thiobarbituric acid method of Warren (1959), using neuramine lactose (Sigma Biochemical) as a standard. Unless otherwise indicated, each incubation contained the following components in a final volume of 0.080 ml: 0.088 mg asialo-α_1-AT; 0.178 mg asialofetuin (or 0.075 mg asialoceruloplasmin); 8.0 mmoles sodium acetate (pH 6.5); 0.80 mg Triton X-100; 0.205 μmoles CMP-NANA, 110,000 cpm ([4,5,6,7,8,9-^{14}C]-NANA); and 2–8 μl serum. Incubations were conducted at 37°C for 60 minutes and analyzed for extent of N-[^{14}C]-acetyl neuraminic acid incorporation into acid-insoluble product using trichloroacetic acid-phosphotunstic acid precipitation as described elsewhere (Kuhlenschmidt et al. 1974). Under these conditions, the incorporation of radioactivity into acid-insoluble product was linear with time and enzyme for serum from both homozygous (Pi^{zz}) and normal individuals. The galactosyltransferase assay measured the transfer of [^{14}C]galactose from UDP (uridine diphosphate)-[^{14}C]galactose (U-[^{14}C]galactose) to ovalbumin and is essentially as described elsewhere (Kuhlenschmidt et al. 1974), except that each incubation contained Triton X-100 at a final concentration of 1%.

RESULTS

Comments on the Purification

The purification of human serum α_1-AT which is summarized in Table 1 involves methods employed by other laboratories (Crawford 1973; Kress and Laskowski 1973). The specific activity of the α_1-AT preparation after the Sephadex G-150 step was comparable to that obtained by these other investigators and yielded a single protein component when subjected to disc gel electrophoresis in the absence and presence of SDS. However when replicate SDS gels were stained for carbohydrate using the periodic acid-Schiff's (PAS) reagent (Zacharius et al. 1969), two PAS-positive bands developed; the major band was coincident with the protein band that coelectrophoresed with the heavy subunit of human immunoglobulin (50,000 daltons) and represents α_1-AT; the second PAS-positive band (42,000 daltons) did not take up detectable Coomassie blue stain. In an effort to eliminate this carbohydrate-rich material, we subjected the α_1-AT preparation to equilibrium chromatography on DEAE-cellulose (Fig. 1). This second DEAE-cellulose chromatography step (1) resolved α_1-AT into three fractions with trypsin-neutralizing activity, and (2) resolved two additional protein components (Fig. 1, pools D and E) which did not possess measurable α_1-AT activity. Upon SDS-disc gel electrophoresis, pool D coelectrophoresed with the PAS-positive, Coomassie blue-negative component present in the α_1-AT preparation after the Sephadex G-150 step.

The overall purification of normal α_1-AT requires approximately 10 days and represents a relatively high yield (40%) procedure. It is the only procedure to our knowledge which permits the resolution of isomeric forms of α_1-AT on a preparative scale. The procedure can be scaled-up at least ten-fold without difficulty; we have purified 800 mg of α_1-AT from 2 liters of human plasma in approximately three weeks. The final preparation is stable to freezing ($-20°$C) for at least four months and can be maintained indefinitely in saturated ammonium sulfate solution at $4°$C, provided β-mercaptoethanol (ME) is present. A desirable feature of the purification procedure just described is that it can also be used to purify the Pi^{ZZ} variant of α_1-antitrypsin.

Properties of α_1-Antitrypsin

The molecular weights of α_1-AT pools A, B and C were determined in the analytical ultracentrifuge according to Yphantis (1964) and found to be 46,200, 46,000 and 46,000 daltons, respectively. The sedimentation coefficients of α_1-AT preparations A, B and C are 3.33, 3.25 and 3.33, respectively. All three preparations of α_1-AT bind 1.0 ± 0.07 moles of trypsin per mole of α_1-AT when titrated with active-site-titrated trypsin (0.91 moles of active site per mole) according to Chase and Shaw (1967).

Composition of α_1-Antitrypsin

The amino acid compositions of the three α_1-AT fractions from the equilibrium DEAE-cellulose column are shown in Table 2. The compositions of the three isomers of α_1-AT are nearly identical; the small differences recorded in Table 2 for fractions A, B and C are considered to be within the limits of experimental error. The amino acid data themselves show no unusual features, with the possible exception of aspartic and glutamic acids, which together comprise 28% of the total composition. However since amide analyses were not performed, it is uncertain how much of this arises from asparagine and glutamine. The amino acid compositions of α_1-AT fractions A, B and C are in reasonable agreement with those reported by Crawford (1973) for a preparation which corresponds to our preparation prior to resolution of isomeric forms. The most notable difference between the compositions reported here and elsewhere is in the half-cystine content. After performic acid oxidation, we consistently find 2 residues of cysteic acid per mole of α_1-AT, whereas Crawford reports 1 residue of half-cystine per 50,000 daltons.

The carbohydrate composition of each preparation of α_1-AT from the equilibrium DEAE-cellulose column is also summarized in Table 2. The carbohydrate contents of fractions A, B and C are 13.6, 13.4 and 13.8%, respectively, and each contains sialic acid, glucosamine and the neutral monosaccharides, galactose and mannose. Differences in the number of residues of any one sugar between the various fractions A, B and C were not remarkable and were within the accuracy of the colorimetric methods employed in their estimation. Each preparation of α_1-AT contained 7–8 residues of sialic acid (per 46,000 daltons). Differences in the electrophoretic mobility of the three α_1-AT preparations on starch and polyacrylamide gels may be due to slight differences in their content of sialic acid residues, which we are unable to distinguish, or to possible differences in the distribution of the same number of sialic acid residues in the three preparations. Nevertheless, the existence of resolvable isomers of α_1-AT is probably due to sialic acid substituents since this apparent microheterogeneity can be eliminated completely by removal (98%) of sialic acid from each preparation with neuraminidase.

Table 2

Composition of α_1-Antitrypsin Residues per 46,000 g

	A	B	C
Lysine	34.0 (34)	33.5 (34)	33.7 (34)
Histidine	12.3 (12)	12.2 (12)	12.5 (12)
Arginine	6.7 (7)	6.8 (7)	6.8 (7)
Aspartic acid	46.8 (47)	44.6 (45)	43.5 (44)
Threonine	26.6 (27)	27.4 (27)	26.9 (27)
Serine	17.9 (18)	19.1 (19)	17.8 (18)
Glutamic acid	55.4 (55)	56.9 (57)	56.8 (57)
Proline	15.9 (16)	15.9 (16)	15.9 (16)
Glycine	21.7 (22)	21.7 (22)	20.9 (21)
Alanine	23.4 (23)	24.3 (24)	22.5 (23)
Half-cystine[a]	1.8 (2)	1.7 (2)	1.8 (2)
Valine	19.2 (19)	19.2 (19)	20.1 (20)
Methionine	6.9 (7)	6.8 (7)	6.4 (6)
Isoleucine	18.4 (18)	18.5 (18)	18.4 (18)
Leucine	36.5 (36)	36.3 (36)	35.8 (36)
Tyrosine	5.1 (5)	5.3 (5)	5.1 (5)
Phenylalanine	21.9 (22)	21.8 (22)	22.4 (22)
Tryptophan[b]	2.6 (3)	2.7 (3)	2.6 (3)
Glucosamine	8.9 (9)	7.3 (7)	9.6 (10)
Sialic acid[c]	7.2 (7)	7.1 (7)	7.9 (8)
Neutral sugars[d]	12.0 (12)	12.6 (13)	11.9 (12)

Amino acid analysis was performed according to Moore and Stein (1963).

[a] Determined as cysteic acid (Moore 1963).

[b] Determined spectrophotometrically (Bencze and Schmidt 1957).

[c] Determined by the thiobarbituric acid method (Warren 1959).

[d] Determined from the reducing sugar method of Park and Johnson (1949) and the anthrone procedure of Seifter et al. (1950).

When neuraminidase-treated fractions A, B and C were subjected to slab gel electrophoresis at pH 8.9, differences in their electrophoretic mobility were eliminated and all of the asialo derivatives coelectrophoresed. Similar results were obtained when these same asialo-α_1-AT preparations were subjected to the crossed antigen-antibody immunoelectrophoresis procedure of Fagerhol and Laurell (1967); of the native α_1-AT preparations, pool C was the most firmly bound to the equilibrium DEAE-cellulose column and had the greatest anodal mobility. After neuraminidase treatment, all three asialo-α_1-AT preparations were indistinguishable when subjected to this same analytical procedure. Serum from a homozygous α_1-AT-deficient individual of the Pi[zz] phenotype migrated to a position intermediate between native α_1-AT and asialo-α_1-AT (which had a cathodal mobility following neuraminidase treatment). This observation suggests a sialic acid content of 2–4 residues per molecule of ZZ α_1-AT, assuming that the only difference between normal MM and ZZ phenotypes is in their sialic acid contents. Extensive neuraminidase treatment did not affect the trypsin inhibitory capacity (TIC) of purified α_1-AT.

Sialic acid-dependent microheterogeneity has also been suggested by results of Bell and Carrell (1973) and Cox (1973). In each of these studies it was shown that limited neuraminidase digestion of the Pi[MM] serum produced an electrophoretic pattern characteristic of the Pi[ZZ] phenotype. Most extensive treatment with neuraminidase produced an α_1-AT preparation that continued to display charge heterogeneity; however, migration was towards the cathode rather than the anode (Cox 1973). Bell and Carrell (1973) have suggested that differences in the behavior of a α_1-AT in serum from individuals with α_1-AT deficiency and from controls on crossed antigen-antibody electrophoresis can be explained by a deficiency of sialic acid in the former. Our results support those mentioned above and in addition show that extensive digestion of α_1-AT with neuraminidase completely eliminates the microheterogeneity, resulting in a single band on crossed antigen-antibody electrophoresis.

Serum Glycosyltransferase Activities

The core population of the present study involves six patients (Table 3, patients I-VI) who share the following common properties: (1) homozygous with respect to serum α_1-AT levels (TIC <0.20), (2) α_1-AT, type Pi[ZZ], (3) presence of diastase-resistant, PAS-positive, amorphous deposits contained in cytoplasmic membranes of parenchymal cells confirmed by light and electron microscopy, (4) extensive liver disease. Serum indices of current liver disease, including serum glutamate-pyruvate transaminase (SGPT), alkaline phosphatase and γ-glutamyl transpeptidase (GGTP), are substantially elevated in each of these patients (Table 3). Circulating levels of ceruloplasmin in these patients with α_1-AT deficiency and hepatic cirrhosis are essentially normal, with the exception of patient VI whose serum ceruloplasmin is slightly reduced (72% of the mean for controls). Also included in Table 3 are serum parameters for relatives of several of these patients. Two asymptomatic, homozygous siblings (IIIa and IVa) with residual α_1-AT of the Pi[ZZ] type exhibit elevated levels of SGPT and alkaline phosphatase, whereas their serum levels of GGTP and ceruloplasmin were essentially normal. Serum from one other homozygous sibling (IIIb) was normal in terms of the levels of SGPT, GGTP, alkaline phosphatase and ceruloplasmin. These same serum parameters in the heterozygous parents of patient III were not significantly different from control values (IIIc and IIId).

Individuals I–VI are all pediatric patients. Therefore, control values for the various glycosyltransferase assays were established separately for serum from adults and children since the activities of children are consistently higher than those of adults. The sialyltransferase activity in serum from patients I–VI was consistently and significantly less than that from the control group; the ability of serum from these patients to catalyze the transfer of sialic acid from CMP-NANA to various asialoglycoprotein acceptors, including asialofetuin, asialoceruloplasmin and asialo-α_1-AT, was 38, 56 and 48%, respectively, of the mean values obtained for serum specimens from the pediatric control group (Table 4). On the other hand, in each patient the activity of another glycosyltransferase, galactosyltransferase (which transfers galactose from UDP-galactose to ovalbumin), was within the normal range of values

Table 3
Serum Parameters in Controls and Individuals with α_1-Antitrypsin Deficiency

Serum source	Disease description	α_1-AT TIC	Pi type	SGPT[a]	GGTP[b] ($\times 10^3$)	Alkaline[c] phosphatase	Ceruloplasmin (mg/100 ml)
Patient I	α_1-AT	0.169	ZZ	36	54	117	24
II	deficiency	0.124	ZZ	24	56	18	24
III	and hepatic	0.093	ZZ	34	86	52	24
IV	cirrhosis	0.158	ZZ	47	76	194	31
V		0.173	ZZ	37	61	131	29
VI		0.110	ZZ	39	36	140	17
Relative IIIa	asymptomatic	0.079	ZZ	43	11	70	21
IIIb	α_1-AT	0.15	ZZ	9.0	11	14	34
IIIc	deficiency	0.38	MZ	9.0	12	26	21
IIId		0.55	MZ	9.0	12	4.0	25
IVa		0.273	ZZ	25	14	56	33
Controls[d]							
range	—	0.814–1.05	MM	5.0–9.0	3.8–10.0	10.8–16.2	19.8–29.0
mean	—	0.926		6.7	6.3	14.1	24.0

All assays were performed as described in Experimental Procedures and duplicated with less than 8% variation. SGPT was assayed according to the procedure of Henry et al. (1960). Ceruloplasmin was quantitated using the phenylenediamine oxidase method of King (1967). Alkaline phosphatase was assayed as described by Glew and Heath (1971). GGTP was assayed by the procedure of Szasz (1969).

[a] Serum glutamate-pyruvate transaminase, measured in I.U., i.e., that amount of enzyme producing 1 μmole product/min/ml serum.
[b] γ-Glutamyl transpeptidase, measured in I.U.
[c] Measured in I.U.
[d] Controls were averaged from 5 adults and 2 children.

Table 4

Glycosyltransferase Activity in Serum from Various Controls
and Individuals with α_1-Antitrypsin Deficiency

| | Glycosyltransferase activity (μmoles/hr/ml serum [$\times 10^{-1}$]) | | | |
| | sialytransferase | | | galactosyltransferase |
Acceptor	asialofetuin	asialoceruloplasmin	asialo-α_1-AT	ovalbumin
Patient I	2.77 (46)[a]	1.16 (62)	0.70 (51)	0.396 (99)
II	2.39 (39)	0.96 (51)	0.64 (46)	0.334 (84)
III	1.75 (29)	1.02 (54)	0.64 (46)	0.472 (118)
IV	2.14 (35)	1.08 (57)	0.74 (54)	0.404 (101)
V	2.12 (35)	1.05 (56)	0.69 (50)	0.414 (104)
VI	2.63 (43)	1.05 (56)	0.61 (44)	0.410 (102)
Mean (I–VI)	2.30 (38)	1.05 (56)	0.67 (48)	0.404 (101)
Relative IIIa	3.54 (58)[a]	1.80 (96)[a]	1.51 (109)[a]	0.294 (74)[a]
IIIb	5.00 (120)[b]	1.31 (81)[b]	1.22 (106)[b]	0.242 (82)[b]
IIIc	3.89 (93)[b]	1.54 (96)[b]	1.21 (105)[b]	0.313 (106)[b]
IIId	5.86 (140)[b]	1.48 (92)[b]	1.26 (110)[b]	0.242 (82)[b]
IVa	4.53 (75)[a]	2.00 (106)[a]	1.37 (99)[a]	0.300 (75)[a]
Other[c] VII	6.13 (147)[b]	2.78 (173)[b]	1.96 (170)[b]	0.352 (119)[b]
Adult controls[d]				
range	3.92–4.53	1.50–1.72	0.78–1.17	0.249–.322
mean	4.17	1.61	1.15	0.296
Pediatric controls[e]				
range	4.04–6.64	1.68–2.09	1.24–1.66	0.259–.588
mean	6.06	1.88	1.38	0.400

[a] Percentage of pediatric control group, mean value. All numbers in parenthesis for patients are percentages of pediatric control group mean value.
[b] Percentage of adult control group, mean value.
[c] This patient is a 47-year-old male with advanced emphysema; he is homozygous for α_1-AT deficiency (Pi^{ZZ}) and has a serum TIC of 0.205.
[d] The controls consisted of 10 adults, ages 24–47.
[e] The controls consisted of 5 children, ages 6–11.

obtained for the pediatric control groups. Thus the serum sialyltransferase deficiency which we reported earlier in one patient (III) with α_1-AT deficiency and hepatic cirrhosis is also observed in these five additional patients with the same disease. Furthermore, of the serum parameters that we have measured, the sialyltransferase enzyme appears to be the only index that we have found to be consistently and significantly deficient. In addition, of the various serum specimens that we have studied from patients exhibiting a range of liver diseases, in no other instance have we observed a significant deficiency in sialyltransferase activity (Table 5). Included in Table 5 are the results of serum sialyltransferase assays in serum from individuals with Wilson's disease, cystic fibrosis, hepatitis and congenital polycystic liver disease. In general, the sialyltransferase activity in the serum of these individuals with liver disease was 25–50% greater than that in controls. Therefore if the serum specimens

Table 5

Sialyltransferase Activities in Serum from Individuals with
Various Forms of Liver Disease

Patient description	Sialyltransferase activity ($\mu moles\ NANA\ transferred/hr/ml\ serum\ [\times\ 10^{-1}]$)		
	asialofetuin	asialoceruloplasmin	asialo-α_1-AT
Controls[a]			
range	3.98–4.75	1.09–1.91	2.40–3.94
mean	4.75	1.61	3.28[b]
Wilson's disease-1	6.44	2.76	5.55
Wilson's disease-2	6.39	1.63	4.07
Wilson's disease-3	5.66	2.19	4.07
Cystic fibrosis-1	5.49	2.69	5.41
Cystic fibrosis-2	5.82	2.58	5.68
Cystic fibrosis, Pi[MM], with portacaval shunt	4.18	2.08	3.87
Cystic fibrosis, Pi[MZ], with portacaval shunt	4.75	1.92	4.30
Hepatitis, Australian antigen-positive-1	3.62	2.28	3.20
Hepatitis, Australian antigen-positive-2	7.72	3.70	7.06
Asymptomatic, Australian antigen-positive-1	4.26	2.06	3.61
Asymptomatic, Australian antigen-positive-2	6.05	2.27	3.74
Congenital polycystic liver disease	5.58	2.40	4.56
Hepatic cirrhosis Pi[MS]	3.66	2.24	4.56
Mean for liver disease[c]	5.39	2.41	4.76

[a] The controls consisted of 3 adults whose sialyltransferase activity is characteristic of our control population.

[b] The sialyltransferase activity using asialo-α_1-AT as acceptor is considerably higher here than in Table 4 because different preparations of asialo-α_1-AT were used in each of these experiments.

[c] These values are calculated having excluded the data obtained for the asymptomatic Australian antigen-positive individuals.

from individuals with liver disease are considered as reference standards for patients I–VI, these patients with α_1-AT deficiency and hepatic cirrhosis are characterized by serum that is approximately 70% deficient in sialyltransferase activity.

Also included in Table 4 are the results of sialyltransferase determinations on serum from homozygous and heterozygous relatives of patients III and IV.

When compared to adult controls, both parents of patient III (IIIc and IIId) possess essentially normal sialyltransferase activities when assayed using the three asialoglycoprotein acceptors. This result is in contrast to results previously reported (Kuhlenschmidt et al. 1974) in which these two individuals showed a 57–72% deficiency of sialyltransferase when asialo-α_1-AT was used as the sialic acid acceptor. These conflicting results could be due to the fact that in our previous study asialoglycoprotein acceptors were prepared by acid hydrolysis, whereas in the present study asialoglycoprotein acceptors were prepared by neuraminidase treatment.

In general, asymptomatic individuals, homozygous for α_1-AT deficiency (IIIa, IIIb, IVa), did not display the marked serum sialyltransferase deficiency characteristic of their siblings (Table 4). However, in several instances, depending upon the sialic acid acceptor used in the assay, reduced levels of sialyltransferase activity were noted. For example, serum specimens IIIa and IVa contained only 58 and 75% as much sialyltransferase activity, respectively, as pediatric control serum when assayed using asialofetuin as the acceptor. Similarly, serum from IIIb contained moderately decreased sialyltransferase activity when assayed using asialoceruloplasmin as acceptor.

We have also evaluated the serum sialyltransferase activity in serum from an emphysematous adult who is homozygous (Pi^{ZZ}) for α_1-AT deficiency (Table 4, patient VII). Light microscopic examination of a liver biopsy from this patient revealed the presence of amorphous deposits characteristic of the PAS-positive material observed in the pediatric patients (I–VI) with hepatic cirrhosis. The serum sialyltransferase activity in this individual was consistently elevated in comparison to adult controls when assayed with all three asialoglycoprotein acceptors. Thus the pronounced serum sialyltransferase deficiency is observed only in those homozygous α_1-AT-deficient patients with extensive liver disease.

DISCUSSION

The most important observation of the present report is the consistent finding that all six pediatric patients with homozygous α_1-AT (Pi^{ZZ}) deficiency and hepatic cirrhosis exhibited significantly decreased serum sialyltransferase activity. This statement is supported by the results presented in Table 4 in which the serum sialyltransferase activity of these individuals averaged 38, 56 and 48% of the control pediatric mean when assayed using asialofetuin, asialoceruloplasmin and asialo-α_1-AT, respectively, as the sialic acid acceptor. However, this decreased level of sialyltransferase activity is not associated with a deficiency in UDP-galactose:ovalbumin galactosyltransferase, a closely related enzyme of the glycosyltransferase family (Roseman 1970). In patients I–VI, the galactosyltransferase activities in serum were within the range of those of the control group. The low level of sialyltransferase activity in the serum of these patients is unique in that all of the following serum parameters in these same cirrhotic patients were either elevated or within the control range: ceruloplasmin, alkaline phosphatase, SGPT, and GGPT (Table 3), and β-glucosidase, β-galactosidase and acid phosphatase (R. H. Glew, unpubl.). The decreased sialyltransferase activity in α_1-AT-deficient individuals with hepatic cirrhosis is probably not due to their general state of liver dysfunction, since serum from 11 patients with a variety of liver diseases unrelated to α_1-AT deficiency did not exhibit a sialyltransferase deficiency (Table 5). In fact, serum from patients with cystic fibrosis, Wilson's disease, cirrhosis and hepatitis often display significantly elevated levels of sialyltransferase activity.

The results of glycosyltransferase measurements on serum from asymptomatic homozygous Pi^{zz} siblings of cirrhotic patients indicate, in general, that a marked sialyltransferase deficiency does not exist in their serum. Nevertheless, serum from one healthy, homozygous α_1-AT-deficient sibling (IIIa) was consistently and significantly reduced in sialyltransferase content when compared with the pediatric control group using asialofetuin as the sialic acid acceptor. It will be of interest to follow the level of serum sialyltransferase activity in this individual since the appearance of liver dysfunction may be paralleled by a decrease in serum sialyltransferase activity.

The heterozygous Pi^{MZ} mother (IIId) and father (IIIc) of patient III possessed normal serum sialyltransferase activity. In addition, an α_1-AT Pi^{zz} adult with emphysema, but without apparent liver dysfunction, displayed elevated serum sialyltransferase values. These results suggest that decreased serum sialyltransferase activity is secondary to the extensive liver damage related to the α_1-AT deficiency and accumulation of PAS-positive material in liver. That is, only when homozygous α_1-AT-deficient individuals reach a state of severe liver dysfunction does the sialyltransferase deficiency occur. It may be that the sialyltransferase of serum is a secreted protein and that the continued production and secretion of sialyltransferase requires a healthy liver with a competent secretory apparatus.

Practical considerations arising from this study pertain to the application of the serum sialyltransferase assay to the management of patients with α_1-AT deficiency and hepatic cirrhosis. Invariably, these patients develop portal hypertension and bleeding from esophageal varices which is often treated by a portacaval anastamosis. However, the problem often arises as to when this surgical procedure should be carried out. Frequently the procedure is initiated only after extensive liver damage has occurred and the patient is in a debilitated state. Clearly, it would be of value to have available objective criteria which would indicate the most opportune time for surgical action. It may be that the measurement of serum sialyltransferase activity in these patients will provide such a readily accessible, quantitative index.

Acknowledgments

We are indebted to Dr. Robert D. Gray for the gift of Cohn fraction IV from which ceruloplasmin was isolated; Dr. Joan Rodnan for the generous gift of serum from patients with cystic fibrosis; Dr. Paul Gaffney for serum from patients with Wilson's disease; Dr. Ned Maxwell and the Blood Bank at Presbyterian University Hospital for the serum from which α_1-AT was purified. We gratefully acknowledge Dr. Richard Iammarino for conducting crossed antigen-antibody electrophoresis of α_1-AT preparations and Dr. Jeri Tuma for the preparation of active-site-titrated trypsin. We thank Ingrid Kuo and Cynthia Solano for their technical assistance.

M. S. K. is a predoctoral trainee under United States Public Health Grant 5 TO1 GM00149. This work was supported by grants from the United States Public Health Service (AM-17465) and the National Foundation, March of Dimes.

REFERENCES

Bell, O. F. and R. W. Carrell. 1973. Basis of the defect in alpha-1-antitrypsin deficiency. *Nature* **243**:410.

Bencze, W. L. and K. Schmidt. 1957. Determination of tyrosine and tryptophan in proteins. *Anal. Chem.* **29**:1193.

Carlson, D. M., E. J. McGuire, G. W. Jourdian and S. Roseman. 1973. Isolation of a mucin sialyltransferase from sheep submaxillary gland. *J. Biol. Chem.* **248**:5763.

Chase, T., Jr. and E. Shaw. 1967. *p*-Nitrophenyl-*p'*-guanidinobenzoate HCl: A new active site titrant for trypsin. *Biochem. Biophys. Res. Comm.* **29**:508.

Cooper, H. S. and P. K. Gupta. 1974. Hepatic cytopathology in alpha-1-antitrypsin deficiency. *Amer. J. Clin. Pathol.* **62**:118.

Cox, D. W. 1973. Defect in alpha-1-antitrypsin deficiency. *Lancet* **2**:844.

Crawford, I. P. 1973. Purification and properties of normal human alpha-1-antitrypsin. *Arch. Biochem. Biophys.* **156**:215.

Cuatrecasas, P. and G. Illiano. 1971. Purification of neuraminidase from Vibrio cholera, *Clostridium perfringens,* and influenza virus by affinity chromatography. *Biochem. Biophys. Res. Comm.* **44**:178.

Eriksson, S. 1965. Determination of serum trypsin inhibitor capacity. *Acta Med. Scand.* (Suppl. 175) **177**:6.

Fagerhol, M. K. and C. B. Laurell. 1967. The polymorphism of "prealbumins" and α-1-antitrypsin in human sera. *Clin. Chim. Acta* **16**:199.

Glew, R. H. and E. C. Heath. 1971. Studies on the extracellular alkaline phosphatase of *Micrococcus sodonensis.* I. Isolation and characterization. *J. Biol. Chem.* **246**:1556.

Henry, R. J., N. Chaimori, O. S. Golub and S. Bergman. 1960. Revised spectrophotometric methods for the determination of glutamic-oxalacetic transaminase, glutamic-pyruvic transaminase, and lactic dehydrogenase. *Amer. J. Clin. Pathol.* **34**:381.

King, J. 1967. Ceruloplasmin. In *Practical Clinical Enzymology,* p. 108. D. Van Nostrand, New York.

Kress, L. F. and M. Laskowski, Sr. 1973. Large scale purification of alpha-1-antitrypsin inhibitor from human plasma. *Prep. Biochem.* **3**:541.

Kueppers, F. and L. F. Black. 1974. State of the art—Alpha-1-antitrypsin and its deficiency. *Amer. Rev. Resp. Dis.* **110**:176.

Kuhlenschmidt, M. S., E. J. Yunis, R. M. Iammarino, S. J. Turco, S. P. Peters and R. H. Glew. 1974. Sialyltransferase and alpha-1-antitrypsin deficiency. *Lab. Invest.* **31**:413.

Lowry, O. H., N. J. Rosebrough, A. L. Farr and R. J. Randall. 1951. Protein measurements with the folin phenol reagent. *J. Biol. Chem.* **193**:265.

Moore, S. 1963. On the determination of cystine as cysteic acid. *J. Biol. Chem.* **238**:235.

Moore, S. and W. H. Stein. 1963. Chromatographic determination of amino acids by the use of automatic recording equipment. In *Methods in Enzymology* (ed. S. P. Colowick and N. O. Kaplan), vol. 6, p. 819. Academic Press, New York.

Park, J. T. and M. J. Johnson. 1949. A submicro determination of glucose. *J. Biol. Chem.* **181**:149.

Roseman, S. 1970. The synthesis of complex carbohydrates by multiglycosyl-transferase systems and their potential function in intercellular adhesion. *Chem. Phys. Lipids* **5**:270.

Seifter, S., S. Dayton, B. Novic and E. Muntwyler. 1950. The estimation of glycogen with the anthrone reagent. *Arch. Biochem.* **25**:191.

Sharp, H. 1971. Alpha-1-antitrypsin deficiency. *Hosp. Pract.* **6**:83.

Szasz, G. 1969. A kinetic photometric method for serum γ-glutamyl transpeptidase. *Clin. Chem.* **15**:124.

Warren, L. 1959. The thiobarbituric acid assay of sialic acids. *J. Biol. Chem.* **234**:1971.

Williams, W. D. and L. F. Fajardo. 1974. Alpha-1-antitrypsin deficiency: A hereditary enigma. *Amer. J. Clin. Pathol.* **61**:311.

Yphantis, D. A. 1964. Equilibrium ultracentrifugation of dilute solutions. *Biochemistry* **3**:297.

Zacharius, R. M., T. E. Zell, J. H. Morrison and J. J. Woodlock. 1969. Glycoprotein staining following electrophoresis on acrylamide gels. *Anal. Biochem.* **30**:148.

Structures and Activities of Protease Inhibitors of Microbial Origin

Takaaki Aoyagi and Hamao Umezawa

Institute of Microbial Chemistry, Tokyo, Japan

Specific enzyme inhibitors can be considered as useful tools in analyzing the complicated network of enzymatic reactions in living organisms and even in understanding disease processes. Numerous compounds with antimicrobial or antitumor activities have been discovered in cultured broths of microorganisms, and research on antibiotics has demonstrated an unexpectedly enormous ability of microorganisms to produce compounds of various structures. In 1965, Umezawa (1972) and his collaborators initiated the screening of enzyme inhibitors in microbial culture filtrates. We started with the screening of protease inhibitors in the culture filtrates of actinomycetes, which were known to produce strong proteases, and succeeded in finding leupeptins which inhibit trypsin, plasmin, papain and cathepsin B; antipain which inhibits papain, trypsin, cathepsin A and B and plasmin (weak); chymostatins which inhibit chymotrypsin and cathepsins (weak); and elastatinal which inhibits elastase. Furthermore, we discovered pepstains, pepstanones and hydroxy-pepstatins which inhibit acid proteases, cathepsin D and renin; and phosphoramidon which inhibits thermolysin and collagenase (weak). In this paper, we report on these protease inhibitors.

EXPERIMENTAL PROCEDURES

Screening Protease Inhibitors in Actinomycetes Culture Filtrates

Strains of actinomycetes were isolated from soils collected from various parts of Japan and cultured on a rotary or reciprocating shaker at 27°C. After 3–6 days' culture, the broths were filtered and 10–50-μl aliquots of the filtrates, which had been heated at 100°C for 5 minutes, were added to reaction mixtures.

The inhibitory activity against various proteases was determined as previously described: inhibition of trypsin, plasmin, papain, kallikrein, thrombin, thrombokinase and chymotrypsins (Aoyagi et al. 1969a,b); inhibition of elastase (Umezawa et al. 1973a); inhibition of pepsin, gastricsin and acid proteases (Aoyagi et al. 1971); inhibition of cathepsin A, B, D and renin (Aoyagi et al. 1972; Ikezawa et al. 1971); inhibition of thermolysin (Suda et al. 1973).

Extraction and Purification of Protease Inhibitors

Leupeptins, antipain, chymostatins, elastatinal, pepstatins, pepstanones, hydroxypepstatins and phosphoramidon were extracted from the cultured broths by ion exchange resin, carbon adsorption or solvent extraction processes and purified by resin or Sephadex chromatography. These processes have been described in detail elsewhere: leupeptins (Kondo et al. 1969); antipain (Suda et al. 1972; Umezawa et al. 1972a); chymostatin (Umezawa et al. 1970a); elastatinal (Umezawa et al. 1973a); pepstatins (Umezawa et al. 1970b; Aoyagi et al. 1971); pepstanones (Miyano et al. 1972); hydroxypepstatins (Umezawa et al. 1973b); phosphoramidon (Suda et al. 1973; Umezawa et al. 1972b).

Determination of Inhibitor Concentration from Percent Inhibition of an Enzyme Reaction

The product of an enzyme reaction can be determined either by spectroscopic analysis or by measuring radioactivity after separation from the residual substrate. The quantity of inhibitor is obtained by measuring the product in the reaction mixture without inhibitor, a, and the product in the mixture with inhibitor, b, and calculating the percent inhibition by $(a - b) / a \times 100$. The probit of the percent inhibition was plotted on the ordinate, and the logarithm of the value corresponding to the concentration of the inhibitor was plotted on the abscissa. From this graph, a 50% inhibition concentration (ID_{50}) was obtained and this value was employed to express the activity.

RESULTS

Inhibitors of Serine and Thiol Proteases

Leupeptins

We were interested in inhibitors of serine and thiol proteases, such as plasmin, trypsin, kallikrein, chymotrypsin and cathepsins, because these enzymes are involved in blood coagulation, fibrinolysis and kinin formation (Fig. 1). An inhibitor of these enzymes would thus be useful in studies on related diseases.

The screening of actinomycetes culture filtrates for activity to inhibit plasmin or trypsin led to the discovery of leupeptin (Aoyagi et al. 1969a, b). Leupeptin is the name of a group of streptomyces products inhibiting proteases and includes acetyl- or propionyl-L-leucyl-L-leucyl-argininal and their analogs which contain L-isoleucine or L-valine instead of L-leucine (Kondo et al. 1969). The signal of the proton on aldehyde carbon in n.m.r. indicates that leupeptin is present both in the hydrated form and in the cyclic form, as shown in Table 1. These two forms may be responsible for the possible appearance of two spots on thin-layer chromatography (Maeda et al. 1971). α-N-Acetylargininal also shows two spots.

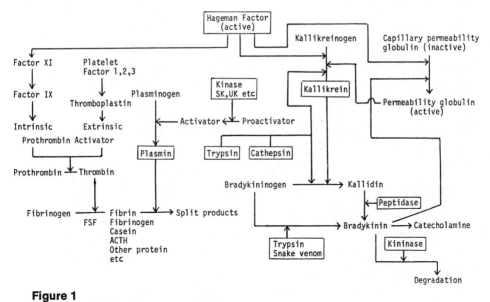

Figure 1
Mechanism of blood coagulation, fibrinolysis and kinin formation.

Leupeptin strongly inhibits plasmin, trypsin, papain and cathepsin B, but it has little or no activity in inhibiting chymotrypsin, elastase, pepsin, cathepsin A and D and thermolysin. It is interesting to note that the lysosomal enzyme cathepsin B is also inhibited by leupeptins. Hydrolysis of leupeptin acid, which is obtained by permanganate oxidation of leupeptin, gives racemic arginine. Studies on the synthetic leupeptin indicate that the L-argininal form is active and the D-argininal form is not active (Kawamura et al. 1969; Shimizu et al. 1972). Leupeptin acid, leupeptinol and the dibutyl acetal of leupeptin are not active, and therefore the aldehyde of L-argininal moiety can be considered as essential for the antiprotease activity. Leupeptin analogs have been synthesized, and their activities are shown in Table 2. α-N-Acetylargininal is not active, but acetyl-L-leucyl-argininal is active. Acetyl-L-isoleucyl-L-leucyl-argininal and acetyl-L-valyl-L-leucyl-argininal were found to be more active than natural leupeptins in inhibiting plasmin. As is well known, plasmin, trypsin and papain cleave a peptide bond of the carboxyl side of basic amino acid moieties, and the structure of leupeptin gives us important information on other inhibitors. It can be predicted that a peptide containing L-α-aminoaldehyde would exhibit inhibition of an enzyme which cleaves the peptide bond of the carboxyl side of the corresponding L-amino acid moiety. As shown later, this idea is also supported by the structures of chymostatin and elastatinal which inhibit chymotrypsin or elastase. Inhibition of cathepsin B by leupeptin suggests the trypsinlike activity of this enzyme (Ikezawa et al. 1971).

Antipain

In testing the ability of the culture filtrates of actinomycetes to inhibit papain, we discovered another inhibitor named antipain (Suda et al. 1972). S. Umezawa and associates (pers. comm.) found a new compound among

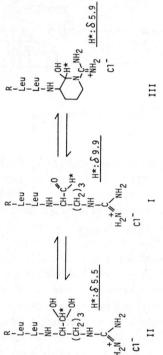

H*: δ 5.5   II

H*: δ 9.9   I

H*: δ 5.9   III

Table 1

Activities and Structure of Leupeptins

Inhibitor	ID_{50} ($\mu g/ml$)					cathepsin		
	plasmin	trypsin	papain	chymotrypsin	pepsin	A	B	D
Leupeptin[a]	8	2	0.5	>500	>500	1680	0.44	109
Leupeptinol	>500	>500	>500	>500	>500	—	—	>500
Leupeptin acid	>500	>500	>500	>500	>500	—	—	>500
Dibutylacetal of leupeptin	>500	>500	133	>500	>500	—	—	>500

[a] Mixture of propionyl-L-Leu-L-Leu-argininal and acetyl-L-Leu-L-Leu-argininal in 3:1.

Table 2

Inhibition of Proteases by Leupeptin and Its Analogs

Compound	ID_{50} ($\mu g/ml$)		
	plasmin[a]	papain[b]	thrombokinase
α-N-Ac-Argal	>250	>100	—
Ac-L-Leu-Argal	8	0.3	38
Pr-L-Leu-Argal	8	—	—
Ac-L-Ile-Argal	33	—	—
Ac-L-Ile-L-Leu-Argal	2	2.0	11
Ac-L-Val-L-Leu-Argal	2	1.7	9
Ac-L-Ile-L-Ile-Argal	35	6.5	—
Ac-L-Ile-L-Val-Argal	314	15.5	—
Ac-L-Leu-L-Leu-Argal[c]	9	0.8	14
Pr-L-Leu-L-Leu-Argal[c]	8	0.5	14
Ac-L-Leu-L-Leu-L-Argal	4.5	0.18	6.7
Ac-L-Leu-L-Leu-D-Argal	70.0	1.9	103

Ac, acetyl; Pr, propionyl; Leu, leucyl; Ile, isoleucyl; Val, valyl; Argal, argininal.

[a] Fibrinogenolysis by plasmin.

[b] Caseinolysis by papain.

[c] Separated from natural leupeptin. Natural leupeptin is a mixture of Pr-L-Leu-L-Leu-Argal and Ac-L-Leu-L-Leu-Argal in a ratio of 3:1.

Sakaguchi-positive metabolites of actinomycetes. Antipain was identical with this guanidine peptide. S. Umezawa et al. (1972a) proposed the structure [1-carboxy-2-phenyl-ethyl]carbamoyl-L-arginyl-L-valyl-argininal, as shown in Table 3. Antipain is related to leupeptin in that they both contain argininal at the terminal carbon. Antipain and leupeptin both inhibit trypsin, papain and cathepsin B. However, the former is weaker in inhibiting plasmin and inhibits cathepsin A more than the latter (Suda et al. 1972; Ikezawa et al. 1972). This suggests that a structural part other than L-argininal moiety may be involved in interaction with cathepsin A.

When the mechanism of inhibition of trypsin by leupeptin and antipain was investigated, the following unusual observations were obtained. Leupeptin shows competitive inhibition of the hydrolysis of TAMe (N-p-toluene sulfonyl-L-arginine methyl ester) and BAEe (α-N-benzoyl-L-arginine ethyl ester) by crystalline trypsin. Its dissociation constant (K_i) is 3.4×10^{-7} M with TAMe and 1.3×10^{-7} M with BAEe. But the inhibitory mechanism of leupeptin is noncompetitive to the hydrolysis of BAPA (α-N-benzoyl-L-arginine-p-nitroanilide) with almost the same dissociation constant. Leupeptin inhibits competitively substrate activation phenomena of trypsin.

In the presence of antipain, a few minutes were required before the steady-state reaction was established. This can be understood by considering that binding between trypsin and antipain proceeds slowly. A Lineweaver-Burk plot under the steady-state hydrolysis of BAPA by trypsin in the presence of antipain gives a concave curve (Fig. 2). The same phenomena shown in the effect of antipain on trypsin occur in that of leupeptin on trypsin previously

Structure of Antipain

$$\underset{\begin{array}{c}\text{(S)}\\ \text{HOOC-CH-NH-CO-NH-CH-CO-NH-CH-CO-NH-CH-CH-CHO}\end{array}}{\quad}$$

HOOC-CH-NH-CO-NH-CH-CO-NH-CH-CO-NH-CH-CHO
  (S)   (S)   (S)
  | | | |
  CH$_2$ (CH$_2$)$_3$ CH-CH$_3$ (CH$_2$)$_3$
  | | | |
 (ring)   NH   CH$_3$ NH
                     | |
                     C=NH   C=NH
                     | |
                     NH$_2$ NH$_2$

[1-carboxy-2-phenyl-ethyl]carbamoyl-L-arginyl-L-arginyl-L-valyl-argininal

Table 3

Effect of Antipain against Proteases

Inhibitor	ID_{50} ($\mu g/ml$)				cathepsin		
	plasmin	trypsin	papain	chymotrypsin	pepsin	A	B	D
Antipain	93	0.26	0.16	>250	>250	1.19	0.59	>125

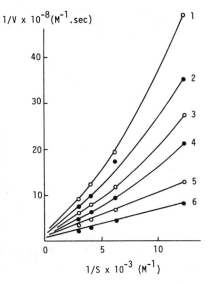

$1/V \times 10^{-8} (M^{-1}.sec)$

$1/S \times 10^{-3} (M^{-1})$

Figure 2
Effect of antipain on trypsin-catalyzed hydrolysis of BAPA. BAPA was used at a concentration of 8×10^{-5} M to 3.2×10^{-4} M. The concentration of trypsin was 0.2 μg/ml. Antipain was used at a concentration of (1) 6.61×10^{-8} M, (2) 5.29×10^{-8} M, (3) 3.97×10^{-8} M, (4) 2.64×10^{-8} M, (5) 1.32×10^{-8} M, and (6) none. The reaction was done at 23°C in 0.02 M Tris-HCl buffer pH 7.8, containing 0.02 M CaCl$_2$ and 0.32% DMSO.

modified by 3-diazoquinolin (kind gift of Dr. Ishii); i.e., slow binding between leupeptin and modified trypsin and a concave of a Lineweaver-Burk plot. The complicated observations for the effect of leupeptin on trypsin and the suggestion that leupeptin did not bind to anhydrotrypsin in which the active serine residue was converted into a dehydroalanine residue were personally communicated by Dr. Ishii (Faculty of Pharmaceutical Sciences, University of Hokkaido). The above-mentioned observations cannot be explained by any simple competitive or noncompetitive mechanism. The idea that these inhibitors act as transition state analogs, though attractive, cannot explain all the kinetic data.

Chymostatin

Our screening studies, utilizing the determination of antichymotrypsin activity, resulted in the isolation of an inhibitor, named chymostatin, from seven strains of actinomycetes (Umezawa et al. 1970a). Tatsuta et al. (1973) determined the structure of chymostatin as N-[((S)-1-carboxy-2-phenylethyl)carbamoyl]-α-[2-iminohexahydro-4(S)-pyrimidyl]-L-glycyl-L-leucyl-phenylalaninal (Table 4). Chymostatin was crystallized, but amino acid analysis suggested that it was a mixture of peptides. It is interesting to note that the terminal moiety in chymostatin is phenylalaninal. Chymotrypsin cleaves the carboxyl side of phenylalanine in peptide or protein. Therefore, it is possible that the phenylalaninal moiety in chymostatin plays an essential role in inhibiting this enzyme. Ito et al. (1972) synthesized acetyl-L-leucyl-L-leucyl-phenylalaninal which exhibited about 1/100 the activity of chymostatin in inhibiting chymotrypsin. This suggests that the N-[((S)-1-carboxy-2-phenylethyl)carbamoyl]-α-[2-iminohexahydro-4(S)-pyrimidyl]-L-glycyl moiety is also involved in increasing the inhibitory activity. As shown in Table 4, chymostatin is a strong inhibitor of chymotrypsin, a weak inhibitor of papain, cathepsin A, B and D, but does not inhibit trypsin.

Structure of Chymostatin

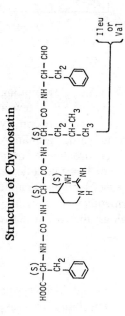

N-[((S)-1-carboxy-2-phenylethyl)carbamoyl]-*α*-[2-iminohexahydro-4(S)-pyrimidyl]-L-glycyl-L-leucyl-phenylalaninal

Table 4

Effect of Chymostatin against Proteases

| Inhibitor | ID_{50} ($\mu g/ml$) | | | | | cathepsin | |
	plasmin	trypsin	papain	chymotrypsin	pepsin	A	B	D
Chymostatin	>250	>250	7.5	0.15	>250	62.5	2.6	49.0

Elastatinal

Actinomycetes also produces an inhibitor of elastase, which was isolated and named elastatinal (Umezawa et al. 1973a). We elucidated the structure *N*-(1-carboxy-isopentyl) carbamoyl-α-(2-iminohexahydro-4-pyrimidyl) glycyl-glutaminyl-alaninal shown in Table 5. As expected, the C-terminal moiety of elastatinal was alaninal. Before our discovery of elastatinal, Thompson (1973) synthesized a peptide containing alaninal which inhibited elastase. Elastatinal is soluble in water. The elastatinal molecule contains an unusual peptide, *N*-(1-carboxy-isopentyl)carbamoyl-α-(2-iminohexahydro-4-pyrimidyl)glycyl, which is thought to be more resistant to proteases in vivo than those containing ordinary L-amino acids. Elastatinal shows that strong inhibition against elastase and elastatinal is competitive with the substrate. Its dissociation constant (K_i) is 2.4×10^{-7} M with acetyl-alanyl-alanyl-alanine *p*-nitroanilide and 2.1×10^{-7} M with acetyl-alanyl-alanyl-alanine methyl ester. But it does not inhibit the other serine or thiol enzymes. Two derivatives which contain carboxyl or carbinol instead of the aldehyde group showed 1/200 the activity of elastatinal.

So far, we have discussed leupeptin, antipain, chymostatin and elastatinal. As summarized in Table 6, these are specific for serine or thiol enzymes. Structure-function relationships of these inhibitors are shown in Table 7. Leupeptin and antipain have the argininal residue at their terminal carbon. They inhibit the common enzymes trypsin and papain which cleave the carboxyl side of basic amino acids such as arginine or lysine.

Chymotrypsin cleaves the carboxyl side of aromatic amino acid. Against this enzyme, we have chymostatin, which has phenylalaninal at the terminal carbon.

Structure of Elastatinal

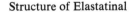
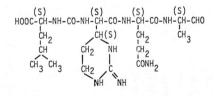

N-(1-carboxy-isopentyl)carbamoyl-α-(2-iminohexahydro-4-pyrimidyl) glycyl-glutaminyl-alaninal

Table 5

Inhibitory Activity of Elastatinal and Its Derivatives against Proteases

Inhibitor	ID_{50} ($\mu g/ml$)					
	trypsin	papain	chymotrypsin	pepsin	thermolysin	elastase
Elastatinal	>250	>250	>250	>250	>250	0.29
Elastatinol	>250	>250	>250	>250	>250	64
Elastatinic acid	>250	>250	>250	>250	>250	98

Table 6

Inhibitory Effects of Leupeptin, Antipain, Chymostatin and Elastatinal on Various Proteases

Enzyme	Substrate	ID_{50} ($\mu g/ml$)			
		leupeptin	antipain	chymostatin	elastatinal
Thrombokinase	plasma	15	20	>250	>250
Thrombin	TAMe[a]	10,000	>250	>250	>250
Plasmin	fibrinogen	8	93	>250	>250
Trypsin	casein	2	0.26	>250	>250
Papain	casein	0.5	0.16	7.5	>250
Kallikrein	BAEe[b]	75		>250	>250
α-Chymotrypsin	casein	>500	>250	0.15	>250
β,γ,δ-Chymotrypsin	casein	>500	>250	0.15	>250
Elastase	elastin-congo red	>500	>250	>250	1.8
Pepsin	casein	>500	>250	>250	>250
	hemoglobin	>500	>250	>250	>250
Proctase A	casein	>250	>250	>250	>250
Proctase B	casein	>250	>250	>250	>250
Cathepsin A	Cb-Glut-Tyr[c]	1680	190	26.5	—
Cathepsin B	BAA[d]	0.44	1.19	62.5	—
Cathepsin D	hemoglobin	109	0.595	2.6	>250
Renin	—[e]	>250	>125	49.0	>250
Thermolysin	casein	>250	>250	>250	>250

[a] α-N-(p-toluene sulfonyl)-L-arginine methyl ester HCl.
[b] α-N-benzoyl-L-arginine ethyl ester HCl.
[c] Carbobenzoxy-L-glutamyl-L-tyrosine.
[d] α-N-benzoyl-L-arginine amide HCl.
[e] His-Pro-Phe-His-Leu-Leu-([³H]Val)-Try-Ser.

Table 7
Structure of Enzyme Inhibitors against Serine and Thiol Enzymes

Inhibitor		Enzyme
name	structure	
Leupeptin	R_1-Leu-Leu-argininal	trypsin, plasmin, papain, cathepsin B
Antipain	R_2-Arg-Leu-argininal	papain, trypsin, cathepsin A and B
Chymostatin	R_2-X-Leu-phenylalaninal	chymotrypsin, papain
Elastatinal	R_3-X-Gln-alaninal	elastase

R_1, acetyl or propionyl; R_2, ((S)-1-carboxy-2-phenylethyl)carbamoyl; R_3, (1-carboxy-isopentyl)carbamoyl; X, α-(2-iminohexahydro-4-pyrimidyl)glycyl.

Elastase cleaves the carboxyl side of alanine. Against this enzyme, we have elastatinal, which has alaninal at the terminal carbon. The data are very useful in understanding the relationship between the catalytic site of the enzyme and the essential structure of its inhibitor.

Effect of Pepstatin, Pepstanone and Hydroxypepstatin against Acid Proteases and Renin

As discussed previously, a specific enzyme inhibitor is useful for the analysis of disease processes and may even be effective in the treatment of certain kinds of diseases. In this respect, among protease inhibitors a specific pepsin inhibitor is most valuable, because no compound exhibiting such activity has been available before. By testing for antipepsin activity of actinomycetes culture filtrates, we found pepstatin (Umezawa et al. 1970b). As shown in Table 8, pepstatin specifically inhibits acid proteases except for proctase type A, cathepsin D and renin. It is interesting that pepstatin inhibits the activity of renin, that is, producing angiotensin I from angiotensinogen. However, pepstatin was not inhibitory against other proteases (Aoyagi et al. 1971, 1972).

Extraction of culture filtrate with butanol, concentration under reduced pressure, and decolorization with carbon yielded pepstatin crystals. The structure of this pepstatin was determined by isolating iso-valeric acid, L-valine (2), L-alanine (1) and 4-amino-3-hydroxy-6-methylheptanoic acid (2) from the hydrolysate in the molar ratio shown in parentheses and by the high-resolution mass spectroscopy of permethylated pepstatin. Thus iso-valeryl-L-valyl-L-valyl-(3S,4S)-4-amino-3-hydroxy-6-methylheptanoyl-L-alanyl-(3S,4S)-4-amino-3-hydroxy-6-methylheptanoic acid is proposed as the structure of pepstatin (Morishima et al. 1970).

Pepstatin-producing strains of *Streptomyces testaceus* produced several kinds of pepstatin which differ from one another in the fatty acid moiety (C_2-C_{20}) or in the amino acid moiety. Six of these were obtained in amounts sufficient for us to study their activities in inhibiting pepsin, cathepsin D and renin. All of them showed almost the same activity (Table 9) against pepsin and cathepsin D as the original pepstatin, which was described above and designated pepstatin A. However, the activity of the pepstatins against renin

Structure of Pepstatin A

```
                               CH3
                CH3          CH-CH3
   CH3 CH3    CH-CH3        CH2OH              CH3        CH2OH
   CH-CH3     CH-CH3        CH2                CH-CH3     CH2
   CH2        CH2
CO-NH-CH-CO-NH-CH-CO-NH-CH-CH-CH2-CH-CO-NH-CH-CO-NH-CH-CH-CH2-CH-CH-CH2-COOH
   (S)          (S) (S)              (S)(S)              (S)          (S)(S)
```

IVA-L-Val-L-Val-AHMHA-L-Ala-AHMHA

Table 8
Inhibitory Activity of Pepstatin A on Various Proteases

Enzyme	Substrate	ID_{50} (M)	Enzyme	Substrate	ID_{50} (M)
Pepsin	casein	1.5×10^{-8}	trypsin	casein	$>3.6 \times 10^{-4}$
	hemoglobin	4.5×10^{-9}	plasmin	fibrinogen	$>3.6 \times 10^{-4}$
Proctase A	casein	$>3.6 \times 10^{-4}$	papain	casein	$>3.6 \times 10^{-4}$
Proctase B	casein	1.3×10^{-8}	chymotrypsin	casein	$>3.6 \times 10^{-4}$
Cathepsin D	hemoglobin	1.5×10^{-8}	elastase	elastin-congo red	$>3.6 \times 10^{-4}$
Renin	*	6.6×10^{-6}	thermolysin	casein	$>3.6 \times 10^{-4}$

IVA, isovaleric acid; AHMHA, (3S,4S)-4-amino-3-hydroxy-6-methylheptanoic acid.
* His-Pro-Phe-His-Leu-Leu-([3H]Val)-Tyr-Ser.

Table 9
Inhibitory Activity of Pepstatin Analogs on Pepsin, Cathepsin D and Renin

$$R_1-NH-CH-CO-NH-CH-CO-NH-CH-CH_2-CO-NH-CH-CH_2-CO-NH-CH-CH-C-CH_2-R_4$$

(with substituents: CH_3, $CH-CH_3$ at (S) positions; CH_3, $CH-CH_3$, CH_2OH at (S)(S); R_2 at (S); CH_3, $CH-CH_3$, CH_2, R_3 at (S))

Pepstatin	R_1	R_2	R_3	R_4	ID$_{50}$ pepsin ($\times 10^{-8}$ M)	cathepsin D ($\times 10^{-9}$ M)	renin ($\times 10^{-6}$ M)
Ac	acetyl	$-CH_3$	$-H$ / $-OH$	$-COOH$	1.5	9.3	24.9
Pr	propionyl	$-CH_3$	$-H$ / $-OH$	$-COOH$	1.5	9.1	15.2
Bu	n-butyl	$-CH_3$	$-H$ / $-OH$	$-COOH$	1.4	8.9	9.7
A	iso-valeryl	$-CH_3$	$-H$ / $-OH$	$-COOH$	1.4	8.8	6.6
B	n-caproyl	$-CH_3$	$-H$ / $-OH$	$-COOH$	1.4	9.3	4.3
F	anteiso-heptanoyl	$-CH_3$	$-H$ / $-OH$	$-COOH$	1.4	9.1	2.5
G	n-capryl	$-CH_3$	$-H$ / $-OH$	$-COOH$	1.4	8.9	1.7
Pepstanone A	iso-valeryl	$-CH_3$	$=O$ / $-H$	$-H$	2.0	17	39
Hydroxypepstatin A	iso-valeryl	$-CH_2OH$	$-H$ / $-OH$	$-COOH$	1.7	11.4	20

was dependent on the number of carbon atoms in the fatty acid moiety. The activity against renin increased with increasing number of carbon atoms in the acyl group, that is, with increase of their hydrophobic properties (Aoyagi et al. 1973). After we reported the isolation and the chemistry of pepstatin, Murao and Satoi (1970) reported the isolation of a pepstatin containing an acetyl group. The streptomyces strains producing various pepstatins also produced pepstanones which differ from pepstatins in the C-terminal moieties (Table 9). We also isolated hydroxypepstatin in which serine replaces alanine. Pepstanones are considered to be derived from pepstatins by oxidative decarboxylation of the latter (Miyano et al. 1972). Another strain, which belonged to *Streptomyces parvisporogenes,* produced the other group of pepstatins, which were different from one another in their fatty acid moieties (Aoyagi et al. 1973). As in the case of pepstatins, we named the pepstanone-containing iso-valeryl group pepstanone A and the hydroxypepstatin-containing iso-valeryl group hydroxypepstatin A. Pepstanone A and hydroxypepstatin A showed almost equal activity against pepsin and cathepsin D as pepstatin A, but less activity against renin. Gas chromatography of an hydrolysate of pepstanone and hydroxypepstatin suggested the presence of analogs which are different in the fatty acid moiety.

Pepstanone, which does not contain the terminal carboxyl group, showed the same activity as pepstatin. As shown in Table 10, the derivatives in which the carboxyl group of pepstatin was converted to ester group or reduced to alcohol or aldehyde showed the same activity against pepsin as the parent compound. Thus it is certain that the carboxyl group is not involved in the binding of

Structure of Pepstatin A

IVA-L-Val-L-Val-AHMHA-L-Ala-AHMHA

Table 10

Inhibitory Activity of Pepstatin A and Its Derivatives against Pepsin

Inhibitor	ID_{50} ($\mu g/ml$)	Relative activity
Pepstatin A	0.01	100
methyl ester	0.008	125
ethyl ester	0.01	100
p-bromophenacyl ester	0.01	100
Na, Mg, Ca	0.01	100
Pepstatin-CHO	0.012	84
Pepstatin-CH$_2$OH	0.012	84
Dehydropepstatin	0.015	67
Dehydroacetyl pepstatin	1.31	0.76

pepstatin to pepsin. One of two hydroxyl groups in pepstatin seems to be essential in the activity. Dehydropepstatin, in which the terminal amino acid has been dehydrated, was slightly less active than pepstatin. On the other hand, dehydroacetyl pepstatin, in which the hydroxyl group of dehydropepstatin was acetylated, was much less active. These results suggest that not only the hydrophobic binding to the enzyme, but also hydrogen bonding through one of the hydroxyl groups may be necessary for pepstatin to inhibit the activity of pepsin (Aoyagi et al. 1971). Derivatives in which the terminal carboxyl group links to hexosamine or hexose were 10 times more soluble in water than pepstatin and showed about one-fifth the activity of the latter.

Pepstatin analogs (shown in Table 11) were synthesized and their activities were compared with pepstatin. Neither (3S, 4S)-4-amino-3-hydroxy-6-methylheptanoic acid (AHMHA) nor its N-acyl derivatives nor partial structures of pepstatin not containing AHMHA were active. However if N-acyl amino acid was bound to the amino group of AHMHA, the inhibitory activity against pepsin and cathepsin D appeared. Iso-valeryl-L-Val-AHMHA-L-Ala-AHMHA, which lacks one L-valine, showed the same activity as pepstatin, but it showed no activity against renin. It seems that the lipophilic property is more important for activity against renin than against pepsin (Aoyagi et al. 1972).

As described above, pepstatins are strongly active in inhibiting acid proteases. The activity can be explained by the strong binding of pepstatin A to active pepsin. Sephadex G-50 column chromatography of a mixture of ^{3}H–labeled pepstatin A and active pepsin indicated the presence of an equimolar complex. As shown in Figure 3, based on this strong equimolar binding, pepstatin can be used for titration of the amount of active pepsin. In the presence of pepstatin in a reaction mixture containing pepsin and its substrate (N-acetyl-L-phenylalanyl-L-diiodotyrosine, APDT), the reaction occurred

Table 11

Biological Activity of Pepstatin and Its Partial Peptides

Compounds	ID_{50} ($\mu g/ml$)		
	pepsin	cathepsin D	renin
AHMHA	>250	>250	>250
IVA-Val-Val	>250	>250	>250
Ala-, Leu-, or Val-AHMHA	>250	>250	>250
Ac-Ala-AHMHA	26	280	>250
Ac-Leu-AHMHA	9.3	8.5	>250
Ac-Val-AHMHA	2.4	4.4	>250
Val-Val-AHMHA	18	3.2	>250
Ac-Val-Val-AHMHA	6	200	>250
Val-AHMHA-Val-AHMHA	10	6.5	>250
Ac-Val-AHMHA-Val-AHMHA	0.031	0.42	>250
IVA-Val-AHMHA-Val-AHMHA	0.01	0.05	>250
IVA-Val-Val-AHMHA-Ala-AHMHA	0.01	0.005	4.5

AHMHA, (3S,4S)-4-amino-3-hydroxy-6-methylheptanoic acid; Ac, acetyl; IVA, iso-valeryl.

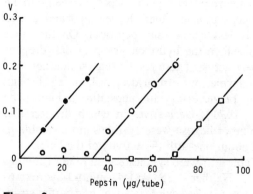

Figure 3

Plots of initial velocity against pepsin concentration at different concentrations of pepstatin. (•——•) A tube contains 0.17 ml of 2 mM acetyl-L-phenyl-L-diiodotyrosine in 4 mM caustic soda, 3.03 ml of 0.1 N HCl, 0.2 ml of 0.001 N HCl containing varied amounts of pepsin. (o——o), (□——□) Same as above, but with 0.5 μg of pepstatin per tube or with 1.0 μg of pepstatin per tube, respectively. (V) Initial velocity of hydrolysis: after 10 min at 37°C, hydrolysis was determined by the ninhydrin method, reading the absorbancy at 570 nm.

when the amount of pepsin exceeded the equimolar amount of pepstatin, and the slope of the reaction rate was the same as that in the absence of pepstatin. The data in Figure 3 indicate that the purity of this pepsin preparation was 77% (Kunimoto et al. 1972).

The equimolar binding of pepstatin to pepsin can also be shown by the ultraviolet difference spectra of pepsin-pepstatin complex. As shown in Figure 4, an increase in absorption at 279, 287 and 293 nm is linear and stoichiometric with pepstatin concentration. Normality titration of pepsin can be done more easily by the difference spectral technique, as shown in Figure 4b. In this case, the ratio of normality/molarity was 0.78, a value almost equal to the result of 0.77 obtained by the activity-measuring method. However, the optical rotatory dispersion spectrum in the far-ultraviolet region of pepsin was not altered by the addition of equimolar pepstatin. Ultraviolet difference spectra suggest that at least around the tyrosine and tryptophan environment in the pepsin molecule certain changes are caused by pepstatin.

Because of the strong binding of pepstatin to pepsin, the kinetic analysis has many limitations. Therefore we studied the kinetics of inhibition of pepsin by a 10,000 times less active derivative, diacetylpepstatin. Diacetylpepstatin was a competitive inhibitor, as shown in a Lineweaver-Burk plot and a Dixon plot using APDT as the substrate. Fruton (Medzihradszky et al. 1970) recommended the synthetic substrate shown in Figure 5 as a suitable substrate for the kinetic study because K_{cat} is very large and experiments can be done at very low enzyme concentrations and in a short reaction time. The inhibition curve of pepstatin for peptic hydrolysis of this substrate is shown in Figure 5.

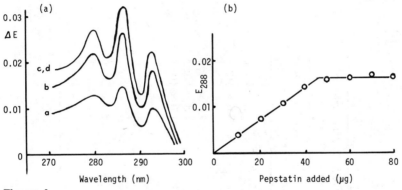

Figure 4

Ultraviolet difference spectra of pepsin-pepstatin complex vs. pepstatin.
(*a*) Difference spectrum. Pepsin was dissolved at 1.2 mg/ml in 0.04 M
formate buffer pH 4.0. Pepstatin concentrations were (a) 6.66 μg/ml,
(b) 13.3 μg/ml, (c) 20.0 μg/ml and (d) 26.6 μg/ml. The difference ultraviolet adsorption spectra show an increase in absorption at 279, 287 and
293 nm. (*b*) The increase in absorptivity at 288 nm against added pepstatin.
Pepstatin was added to 3 ml of 1 mg/ml pepsin solution. The increase at
288 nm is linear and stoichiometric with pepstatin concentration.

The inhibitor concentration (I) at which the tangent to this curve at I_t (total
inhibitor) $= 0$ cut the abscissa will be given by the equation obtained by
Morrison (1969), $I = E_t + \{K_m + S/(K_m/K_i)\}$, if pepstatin is a competitive
inhibitor. From this equation, K_i is calculated as 9.7×10^{-11} M.

In order to know the binding site of pepstatin, we studied the effect of
pepstatin on chemically modified pepsins. As shown in Table 12, the order of

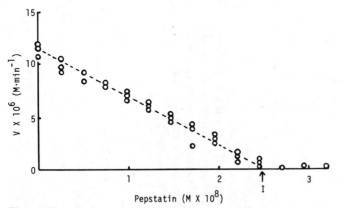

Figure 5

Effect of pepstatin concentration on the hydrolysis of Phe-Gly-
His-Phe(NO$_2$)-Phe-Ala-PheOMe by pepsin. Incubation mixture was 3 ml 0.04 M formate buffer pH 4.0, containing $1.67 \times
10^{-4}$ M substrate, 2.43×10^{-8} N pepsin and various concen-
trations of pepstatin. K_m of this substrate was 4×10^{-5} M
from Lineweaver-Burk plot.

Table 12

Enzyme Activity and Pepstatin-binding Activity of Modified Pepsins and Pepsinogen

Enzyme	Enzyme activity		Binding of pepstatin determined by gel filtration	
	hemoglobin	APDT	column method	batch method (K_i) (M)
Pepsin	100	100	100	8.9×10^{-6}
EPNP-pepsin	16		20	3.0×10^{-6}
Biacetyl-pepsin	29		43	
p-Bromophenacyl bromide and				
α-diazo-p-bromoacetophenone-pepsin	0.41		10	2.1×10^{-6}
α-Diazo-p-bromoacetophenone-pepsin	0.74		64	1.8×10^{-6}
p-Bromophenacyl bromide-pepsin	25		81	1.0×10^{-6}
Acetylimidazole-pepsin	103	196	101	
	81	194	93	
	74	331	88	
	46	311	80	
Heat-denatured pepsin	0			1.0×10^{-5}
Pepsin (pH 7.5)	0			4.6×10^{-5}
Pepsinogen (pH 7.5)	0			4.1×10^{-5}

weakening the binding by modification was 1,2-epoxy-3-(p-nitrophenoxy)propane (EPNP), biacetyl, α-diazo-p-bromoacetophenone and p-bromophenacyl bromide except with the double modification by p-bromophenacyl bromide and diazo-p-bromoacetophenone. All the modification except diazo-acetylnorleucine methyl ester (DAN) increased the dissociation constant of the E-I complex from 10^{-10} M to the order of 10^{-6} M. We also tested the binding of pepstatin to denatured pepsin, pepsinogen and serum albumin. These K_i values were all of the order of 10^{-5} M. This was considered to be due to the lypophilic binding of pepstatin with denatured enzyme protein. These studies on chemically modified pepsins and denatured pepsin suggest that pepstatin interacts with pepsin through both its binding to the active site and the lypophilic binding to the lypophilic area around the active site (Kunimoto et al. 1974).

Effect of Phosphoramidon against Metalloendopeptidase

We were also interested in inhibitors of metalloendopeptidase and found phosphoramidon in our screening studies to be a thermolysin inhibitor (Suda et al. 1973). The structure which was proposed by S. Umezawa and associates (1972b) is very interesting (Table 13). Phosphoramidon inhibited the metalloendopeptidases specifically and showed a competitive type of inhibition. Its K_i is 2.8×10^{-8} M with carbobenzoxyglycyl-L-leucine amide (Z-Gly-Leu NH$_2$). The structure of phosphoramidon suggests that its phosphate group should bind to the active site of thermolysin, but whether zinc ion is involved in binding with this inhibitor is not yet known.

Several analogs were prepared and compared with phosphoramidon. As shown in Table 14, leucyl-leucine and rhamnosyl-phosphoryl-leucine methyl ester were not active. Leucyl-tryptophan is slightly active and rhamnosyl-phosphoryl-leucyl-histidine (kind gift of Dr. S. Umezawa) is less active than phosphoramidon. Phosphoryl-leucyl-tryptophan is found to be more active than phosphoramidon and showed a competitive type of inhibition. Its K_i is 2.0×10^{-9} M with Z-Gly-Leu NH$_2$. As discussed above, phosphoryl-leucyl-tryptophan moiety has been proven to be the active group in phosphoramidon and the rhamnose moiety is not essential to the action. Besides phosphoro-L-leucine moiety, another amino acid residue such as tryptophane or histidine may be necessary for the activity against thermolysin, because the methyl ester of N-(α-L-rhamnopyranosyloxyhydroxyphosphinyl)-L-leucine showed no activity. The binding of phosphoramidon or N-phosphoroleucyl-tryptophan to thermolysin was studied with the ultraviolet difference spectra (Fig. 6). The spectral changes were increased with the addition of these inhibitors up to amounts equimolar to the enzyme.

The binding properties of phosphoramidon and its analogs to thermolysin were studied with the ultraviolet difference spectra. Phosphoramidon and its analogs showed difference spectra of the red shift of tryptophan (Table 15), and spectrophotometric titration studies (Fig. 6) showed that the spectral changes were increased with the addition of these inhibitors up to their equimolar amounts of the enzyme. And the order of apparent K_d values showed good agreement with the order of inhibitory activity. As rhamnosyl-phosphoryl-leucyl-histidine also showed the difference spectrum of tryptophan,

Structure of Phosphoramidon

N-(α-L-rhamnopyranosyloxyhydroxyphosphinyl)-L-leucyl-L-tryptophan

Table 13
Inhibitory Activity of Phosphoramidon

Inhibitor	ID_{50} ($\mu g/ml$)					
	trypsin	papain	α-chymotrypsin	pepsin	thermolysin	collagenase
Phosphoramidon	>250	>250	>250	>250	0.4	33

Table 14
Inhibition of Thermolysin by
Phosphoramidon and Its Analogs

Compound	ID_{50} ($\mu g/ml$)
Leu-Leu	>500
Leu-Trp	190
P-Leu-Trp	0.018
Rha-P-Leu-Trp	0.36
Rha-P-Leu-His	12.5
Rha-P-Leu-OMe	>500

Rha, rhamnose; Leu, leucine; Trp, tryptophan; His, histidine. All amino acids used had the L-configuration.

the tryptophanyl group which showed the difference spectrum may belong to the thermolysin. However, the optical rotatory dispersion spectrum in the far-ultraviolet region of thermolysin was not altered by the addition of equimolar phosphoramidon. The structure of phosphoramidon and its analogs with strong activity suggests that an N-phosphoro peptide might be a potential

Figure 6
Spectrophotometric titration of thermolysin with phosphoramidon and P-Leu-Trp. (A) The change in absorbancy at 295 nm on addition of increasing amounts of phosphoramidon to 3.0×10^{-5} M of thermolysin. (B) The change in absorbancy at 295 nm on addition of increasing amounts of P-Leu-Trp to 3.1×10^{-5} M of thermolysin. The buffer solution consisted of 0.1 M Tris-HCl pH 7.0, 10 mM $CaCl_2$ and 60 mM NaCl. The difference spectrum was recorded with the dual cell having 0.44-cm light-path compartments for each light beam at 10-fold magnification with Hitachi recording spectrophotometer EPS 3T. The reference consisted of the enzyme and inhibitor separately, whereas the sample consisted of the enzyme-inhibitor mixture in one compartment and the buffer solution in the other.

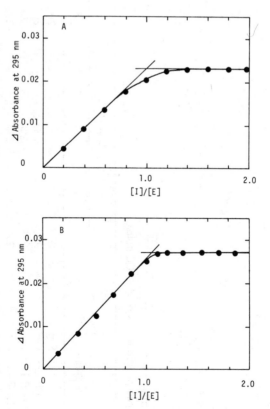

Table 15
Spectral Properties of E · I Complex

Inhibitor	ΔA_{max} (nm)			$\Delta\varepsilon$ 294 ~ 296 nm ($M^{-1} cm^{-1}$)	Inhibitor binding [I]/[E]	Apparent K_d[a] (M)
	1	2	3			
Leu-Trp	277	286	294	1200	1:1	3.7×10^{-5}
P-Leu-Trp	277	285.5	295	2040	1:1	5.1×10^{-8}
Rha-P-Leu-Trp	279	286	294.5	1640	1:1	1.9×10^{-7}
Rha-P-Leu-His	278	286	296	197	1:1	5.6×10^{-7b}

[a] K_d values are calculated by the following equation:

$$K_d = \frac{(E_0 - EI)(I_0 - EI)}{(EI)} = \frac{(E_0 - \Delta A/\Delta\varepsilon)(I_0 - \Delta A/\Delta\varepsilon)}{\Delta A/\Delta\varepsilon}$$

E_0 = initial molar concentration of thermolysin; I_0 = initial molar concentration of inhibitor.
[b] K_d values are calculated by Δ absorbance at 290 nm.

inhibitor of a metalloendopeptidase which cleaves the amino side of an amino acid moiety in the substrates.

As is well known, thermolysin cleaves the amino acid side of hydrophobic amino acid moieties such as leucine, valine, etc., in peptides or proteins.

DISCUSSION

Protease inhibitors isolated from plant and animal tissues are generally of a macromolecular nature. However, as described above, we have found inhibitors with small molecular weights in the culture filtrates of actinomycetes, which usually produce strong protease. Inhibitors could not be found among fungi and cultivated mushrooms. Though not yet proved, protease inhibitors of small molecular nature may be utilized by actinomycetes for their growth. We studied biosyntheses of leupeptins and pepstatin A. All radioactive amino acid constituents of leupeptin added to media were incorporated into the inhibitor, suggesting that the argininal moiety of leupeptin is derived from arginine. It is possible that the biosynthesis of peptide chains of inhibitors may be carried out on a multienzyme system, as proposed by Lipmann (1971) for bacterial peptide antibiotics. The biosynthesis of a new amino acid in the pepstatin molecule is interesting. ^{14}C-leucine was incorporated into from C_3 to C_8 carbon atoms and ^{14}C-malonic acid was incorporated into C-terminal two-carbon atoms of (3S , 4S)-4-amino-3-hydroxy-6-methylheptanoic acid (Morishima et al. 1974). This suggests that iso-valeric acid, two L-valines, L-leucine, malonic acid, L-alanine, L-leucine and malonic acid are sequentially linked by a multienzyme system to produce pepstatin A.

Actinomycetes can be considered to have many multienzyme systems to produce peptides of various structures, and we might happen to find inhibitors among them. When compared with synthetic inhibitors, inhibitors which are produced by protease-producing microorganisms can be considered to be more resistant to proteolytic enzymes. Leupeptins or pepstatins injected into animals were excreted in urine without any hydrolysis. This suggests that inhibitors of microbial origin would also exhibit their action in vivo. Leupeptin ointment, when applied immediately to burns, suppressed pain and blister formation, which may be due to inhibition of kinin formation. Leupeptin was found by Hozumi et al. (1972) to inhibit tumorigenesis in rat skin induced by 7,12-dimethylbenzanthracene (DMBA) and croton oil. Administration of leupeptin with feed also inhibited tumorigenesis in rat colon induced by azoxmethane (Yamamoto et al. 1974). More surprisingly, Sugimura reported that in Balb/c 3T3 cells in tissue culture, leupeptin inhibits malignant transformation induced by DMBA. Furthermore, leupeptin inhibits the PHA-stimulated DNA synthesis in guinea pig blood lymphocytes (Saito et al. 1972, 1973). Leupeptin is well absorbed when given orally. Injection or oral administration of leupeptin or chymostatin showed an anti-inflammatory effect against carrageenin edema. Pepstatin is practically unadsorbed on oral administration, but is excreted in feces. Pepstatin was shown to be very effective in suppressing ulcers that occur in the stomach of the pylorus-ligated rat (Aoyagi et al. 1971). There are so many fields for application of protease inhibitors that more detailed study on each inhibitor and its derivatives is obviously needed.

As already described, inhibitors which we obtained as specific for each group of proteases seem to be useful in identifying uncertain enzymes, especially in typing of cathepsins and elastase. As reported by Kato et al. (1972), leupeptin, which inhibits cathepsin B, antipain, which inhibits cathepsin A and B, and pepstatin, which inhibits cathepsin D, were utilized for the identification of cathepsin-type enzymes in macrophages. Recently, Urata and Aoki (pers. comm.) tried to identify an enzyme in bone marrow which inactivates a δ-aminolevulinate synthetase involved in heme biosynthesis. This enzyme was inhibited only by elastatinal. We found that the substrate specificity of this enzyme was similar to elastase. Thus this enzyme was confirmed to be an elastaselike enzyme.

We are continuing to search for the protease inhibitors in actinomycetes culture filtrates. These organisms seem to be producing more useful inhibitors. We will be able to report on inhibitors of aminopeptidases A and B, carboxypeptidases A and B, etc., and on their structures in the near future.

SUMMARY

We have found protease inhibitors of small molecular nature in actinomycetes culture filtrates, and the structures of these inhibitors were determined as follows: leupeptins which inhibit plasmin, trypsin, kallikrein, papain, cathepsin B, propionyl or acetyl-L-leucyl-L-leucyl-arginal and their analogs contain L-valine or L-isoleucine instead of L-leucine; antipain which inhibits trypsin, papain, cathepsin A and B, [1-carboxy-2-phenyl-ethyl]carbamoyl-L-valyl-arginal; chymostatin which inhibits chymotrypsin, papain (weakly), N-[((S)-1-carboxy-2-phenylethyl)carbamoyl]-α-[2-iminohexahydro-4(S)-pyrimidyl]-L-glycyl-L-leucyl-phenylalaninal and their analogs contain L-valine or L-isoleucine instead of L-leucine; elastatinal which inhibits elastase, N-(1-carboxy-isopentyl)carbamoyl-α-(2-iminohexahydro-4-pyrimidyl)-glycyl-glutaminyl-alaninal; pepstatins, pepstanones and hydroxypepstatins which inhibit acid proteases, cathepsin D and renin; structure of pepstatin A is iso-valeryl-L-valyl-L-valyl-(3S,4S)-4-amino-3-hydroxy-6-methylheptanoyl-L-alanyl-(3S,-4S)-4-amino-3-hydroxy-6-methylheptanoic acid; phosphoramidon which inhibits thermolysin and collagenase, N-(α-L-rhamnopyranosyloxyhydroxyphosphinyl)-L-leucyl-L-tryptophan.

The structures of leupeptin, antipain, chymostatin and elastatinal indicate that an N-acyl peptide containing the terminal aldehyde instead of carboxyl is inhibitory against the enzymes which cleave the carboxyl side of that amino acid in peptides. The structure of phosphoramidon suggests an inhibitor structure of an enzyme which cleaves the amino side of an amino acid residue in peptides. We have also carried out more detailed studies on the structure-activity relationship in leupeptins, pepstatins and phosphoramidon.

The data of kinetic studies on leupeptin, elastatinal, pepstatin and phosphoramidon were described, and the medical utilization of the inhibitors which do not undergo metabolism in vivo were discussed. We have also described the use of leupeptin, antipain, pepstatin and elastatinal for identification of cathepsins A, B, D and elastase.

REFERENCES

Aoyagi, T., S. Kunimoto, H. Morishima, T. Takeuchi and H. Umezawa. 1971. Effect of pepstatin on acid protease. *J. Antibiotics* **24:**687.

Aoyagi, T., H. Morishima, R. Nishizawa, S. Kunimoto, T. Takeuchi and H. Umezawa. 1972. Biological activity of pepstatins, pepstanone A and partial peptides on pepsin, cathepsin D and renin. *J. Antibiotics* **25:**689.

Aoyagi, T., Y. Yagisawa, M. Kumagai, M. Hamada, H. Morishima, T. Takeuchi and H. Umezawa. 1973. New pepstatins, pepstatins BU, PR and AC produced by streptomyces. *J. Antibiotics* **26:**539.

Aoyagi, T., T. Takeuchi, A. Matsuzaki, K. Kawamura, S. Kondo, M. Hamada, K. Maeda and H. Umezawa. 1969a. Leupeptins, new proteinase inhibitors from actinomycetes. *J. Antibiotics* **22:**283.

Aoyagi, T., S. Miyata, M. Nanbo, F. Kojima, M. Matsuzaki, M. Ishizuka, T. Takeuchi and H. Umezawa. 1969b. Biological activities of leupeptin. *J. Antibiotics* **22:**558.

Hozumi, M., M. Ogawa, T. Sugimura, T. Takeuchi and H. Umezawa. 1972. Inhibition of tumorigenesis in mouse skin by leupeptin, a protease inhibitor from actinomycetes. *Cancer Res.* **32:**1725.

Ikezawa, H., T. Aoyagi, T. Takeuchi and H. Umezawa. 1971. Effect of protease inhibitors of actinomycetes on lysosomal peptide hydrolases from swine liver. *J. Antibiotics* **24:**488.

Ikezawa, H., K. Yamada, T. Aoyagi, T. Takeuchi and H. Umezawa. 1972. Effect of antipain on lysosomal peptide-hydrolases from swine liver. *J. Antibiotics* **25:**738.

Ito, A., K. Tokawa and B. Shimizu. 1972. Peptide aldehydes inhibiting chymotrypsin. *Biochem. Biophys. Res. Comm.* **49:**343.

Kato, T., K. Kojima and T. Murachi. 1972. Proteases of macrophages in rat peritoneal exudate, with special reference to the effects of actinomycete protease inhibitors. *Biochim. Biophys. Acta* **289:**187.

Kawamura, K., S. Kondo, K. Maeda and H. Umezawa. 1969. Structures and syntheses of leupeptins Pr-LL and Ac-LL. *Chem. Pharm. Bull.* **17:**1902.

Kondo, S., K. Kawamura, J. Iwanaga, M. Hamada, T. Aoyagi, K. Maeda, T. Takeuchi and H. Umezawa. 1969. Isolation and characterization of leupeptins produced by actinomycetes. *Chem. Pharm. Bull.* **17:**1896.

Kunimoto, S., T. Aoyagi, H. Morishima, T. Takeuchi and H. Umezawa. 1972. Mechanism of inhibition of pepsin by pepstatin. *J. Antibiotics* **25:**251.

Kunimoto, S., T. Aoyagi, R. Nishizawa, T. Komai, T. Takeuchi and H. Umezawa. 1974. Mechanism of inhibition of pepsin by pepstatin. II. *J. Antibiotics* **27:**413.

Lipmann, F. 1971. Attempts to map a process evolution of peptide biosynthesis. *Science* **173:**875.

Maeda, K., K. Kawamura, S. Kondo, T. Aoyagi, T. Takeuchi and H. Umezawa. 1971. The structure and activity of leupeptins and related analogs. *J. Antibiotics* **24:**402.

Medzihradszky, K., I. M. Voynick, H. Medzihradszky-Schweiger and J. S. Fruton. 1970. Effect of secondary enzyme-substrate interactions on the cleavage of synthetic peptides by pepsin. *Biochemistry* **9:**1154.

Miyano, T., M. Tomiyasu, H. Iizuka, S. Tomisaka, T. Takita, T. Aoyagi and H. Umezawa. 1972. New pepstatins, pepstatins B and C, and pepstanone A, produced by streptomyces. *J. Antibiotics* **25:**489.

Morishima, H., T. Takita, T. Aoyagi, T. Takeuchi and H. Umezawa. 1970. The structure of pepstatin. *J. Antibiotics* **23:**263.

Morishima, H., T. Sawa, T. Takita, T. Aoyagi, T. Takeuchi and H. Umezawa. 1974. Biosynthetic studies on pepstatin. Biosynthesis of (3S,4S)-4-amino-3-hydroxy-6-methylheptanoic acid moiety. *J. Antibiotics* **27**:267.

Morrison, J. F. 1969. Kinetics of the reversible inhibition of enzyme-catalyzed reactions by tight-binding inhibitors. *Biochim. Biophys. Acta* **185**:269.

Murao, S. and S. Satoi. 1970. New pepsin inhibitors (S-P1) from streptomyces EF-44-201. *Agr. Biol. Chem.* **34**:1265.

Saito, M., T. Hagiwara, T. Aoyagi and Y. Nagai. 1972. Leupeptin, a protease inhibitor, inhibits the PHA-stimulated DNA synthesis in guinea pig blood lymphocytes. *Japan. J. Exp. Med.* **42**:509.

Saito, M., T. Yoshizawa, T. Aoyagi and Y. Nagai. 1973. Involvement of proteolytic activity in early events in lymphocyte transformation by phytohemagglutinin. *Biochem. Biophys. Res. Comm.* **52**:569.

Shimizu, B., A. Saito, A. Ito, K. Tokawa, K. Maeda and H. Umezawa. 1972. Synthetic studies on leupeptins and their analogs. *J. Antibiotics* **25**:515.

Suda, H., T. Aoyagi, T. Takeuchi and H. Umezawa. 1973. A thermolysin inhibitor protease by actinomycetes: Phosphoramidon. *J. Antibiotics* **26**:621.

Suda, H., T. Aoyagi, M. Hamada, T. Takeuchi and H. Umezawa. 1972. Antipain, a new protease inhibitor isolated from actinomycetes. *J. Antibiotics* **25**:263.

Tatsuta, K., N. Mikami, K. Fujimoto, S. Umezawa, H. Umezawa and T. Aoyagi. 1973. The structure of chymostatin, a chymotrypsin inhibitor. *J. Antibiotics* **26**:625.

Thompson, R. C. 1973. Use of peptide aldehydes to generate transition state analogs of elastase. *Biochemistry* **12**:47.

Umezawa, H. 1972. *Enzyme Inhibitors of Microbial Origin.* University of Tokyo Press, Tokyo.

Umezawa, H., T. Aoyagi, H. Morishima, M. Matsuzaki, M. Hamada and T. Takeuchi. 1970b. Pepstatin, a new pepsin inhibitor produced by actinomycetes. *J. Antibiotics* **23**:259.

Umezawa, H., T. Aoyagi, A. Okura, H. Morishima, T. Takeuchi and Y. Okami. 1973a. Elastatinal, a new elastase inhibitor produced by actinomycetes. *J. Antibiotics* **26**:787.

Umezawa, H., Aoyagi, H. Morishima, S. Kunimoto, M. Matsuzaki, M. Hamada and T. Takeuchi. 1970a. Chymostatin, a new chymotrypsin inhibitor produced by actinomycetes. *J. Antibiotics* **23**:425.

Umezawa, H., T. Miyano, T. Murakami, T. Takita, T. Aoyagi, T. Takeuchi, H. Naganawa and H. Morishima. 1973b. Hydroxypepstatin, a new pepstatin produced by streptomyces. *J. Antibiotics* **26**:615.

Umezawa, S., K. Tatsuta, K. Fujimoto, T. Tsuchiya, H. Umezawa and H. Naganawa. 1972a. Structure of antipain, a new Sakaguchi-positive product of streptomyces. *J. Antibiotics* **25**:267.

Umezawa, S., K. Tatsuta, O. Izawa, T. Tsuchiya and H. Umezawa. 1972b. A new microbial metabolite phosphoramidon. *Tetrah. Letters* **1**:97.

Yamamoto, R. S., H. Umezawa, T. Takeuchi, T. Matsuhima, K. Hara and T. Sugimura. 1974. Effect of leupeptin on colon carcinogenesis in rats with azoxymethane. *Proc. Amer. Assn. Cancer Res.* **15**:38.

Synthetic Protease Inhibitors
Acting by Affinity Labeling

Elliott Shaw

Biology Department, Brookhaven National Laboratory
Upton, New York 11973

Affinity labeling was introduced as a new approach to the chemical modification of proteins in order to identify functional groups (Schoellmann and Shaw 1963). Studies on proteolytic enzymes have been an important part of the development of this type of protein chemistry, which has by now been applied to a great variety of enzymes (Shaw 1970). Such studies are generally carried out on purified enzymes in which the site of covalent binding can be determined. In the examples that best demonstrate the selectivity possible with affinity labeling, only a single covalent modification occurs per target molecule. Since this event takes place at the active center, enzymatic activity is abolished. I shall be discussing chloromethyl ketones derived from small molecules which satisfy the primary specificity requirements of the target proteases such as TPCK and TLCK (Table 1), which inactivate chymotrypsin and trypsin, respectively. It has been shown that satisfying only the primary specificity of elastase, namely with Ala at P_1[1], is not sufficient to obtain covalent modification of elastase (Powers and Tuhy 1973), but that additional amino acid residues are required (X and Y in Table 1). The affinity labels for chymotrypsin and trypsin acquire increased effectiveness in peptide form. In the early work with TPCK and TLCK, the striking observation was made that each of these reagents inactivated only that serine proteinase whose specificity was satisfied in the reagent structure, although in each case the modification was the same, i.e., alkylation of the active-center histidine (Schoellmann and Shaw 1963; Shaw, Mares-Guia and Cohen 1965). This result indicated that in contrast to DFP (diisopropylfluorophosphate), which inactivates serine proteinases as a class, substrate-derived chloromethyl ketones are capable of more selective inhibition.

More recently, the use of these protease inhibitors in animals or with intact cells has become commonplace, and in such complex biochemical environ-

---

[1] The notation P_1, P_1', etc., is that introduced by Schechter and Berger (1967) to designate the location of amino acid residues of a protein substrate relative to the site of hydrolysis. An example of its use may be found in the notes to Table 3.

Table 1

Substrate-derived Chloromethyl Ketones That
Inactivate a Particular Class of Serine Proteases

P_1	Serine protease inactivated
Tos-Phe-CH$_2$Cl (TPCK)[a]	chymotrypsin
Cbz-Phe-CH$_2$Cl[b]	
Y-X-Phe-CH$_2$Cl[c, d]	
	elastase
Tos-Ala-CH$_2$Cl[e]	(no)
Y-X-Ala-CH$_2$Cl[f, g]	yes
Tos-Lys-CH$_2$Cl (TLCK)[h]	trypsin
Y-X-Lys-CH$_2$Cl[i]	

[a] Schoellmann and Shaw (1963); [b] Shaw and Ruscica (1971); [c] Morihara and Oka (1970); [d] Segal et al. (1971); [e] Powers and Tuhy (1973); [f] Thompson and Blout (1974); [g] Thomson and Denniss (1973); [h] Shaw et al. (1971); [i] Coggins, Kray and Shaw (1974).

ments, the possibility of side reaction is great, particularly at -SH groups. The effects which are produced in these experiments are difficult to interpret safely because of possible side reactions. However in view of the great number of physiologically important serine proteases continually being discovered, there is considerable incentive to improve this type of inhibitor to help in understanding the function of a given enzyme in its complex physiological environment. There appear to be two problems with which to contend: the need to reduce side reactions, and the need to discriminate among enzymes of closely related specificity. There are so many important trypsinlike enzymes, for example, that it would seem to be very difficult to be able to inactivate a single one of them in the presence of the others. However, I believe we are making progress in dealing with these goals.

Basis for Effectiveness of Affinity-labeling Agents

In connection with the problem of side reactions evoked by chloromethyl ketones, it is worthwhile considering the basis for the effectiveness of affinity-labeling reagents. The substratelike reagent (I) becomes complexed to the active center. Within the complex, the reacting groups must be in the appropriate juxtaposition for a chemical reaction to take place:

$$E + I \rightleftharpoons E \cdot I \rightarrow E - I'.$$
COMPLEX

Both of these features offer an opportunity to diminish side reactions with other cellular constituents. Thus, increasing the affinity of the reagent for the target active center would permit its use at lower concentrations. In addition

Table 2

Comparison of Two Rates of Alkylation
by ZPCK

	Rate[a]
Cbz-Phe-CH₂Cl + His-57 in chymotrypsin	69
Cbz-Phe-CH₂Cl + Ac-His	4.5×10^{-5}

[a] pH 7, k_{2nd} ($\text{M}^{-1} \cdot \text{sec}^{-1}$).

to decreasing side reactions at sites not related to the specificity satisfied by the reagent structure, the possibility of effectiveness at greater dilutions might also provide a means for achieving selectivity among enzymes of closely related specificity (see below).

The inactivation of chymotrypsin by benzyloxycarbonyl-phenylalanyl chloromethyl ketone (ZPCK) demonstrates the rate enhancement possible by affinity labeling. ZPCK (or Cbz-Phe-CH₂Cl) alkylates His-57 in chymotrypsin very rapidly. Compared with acetyl histidine as a model, the reaction with chymotrypsin proceeds at a rate at least 10^6 greater (Table 2) (Shaw and Ruscica 1971). This rate enhancement results from a proximity effect within the enzyme-complexed reagent and accounts for the selectivity of ZPCK for His-57 in the presence of other nucleophilic side chains of chymotrypsin.

The effectiveness of chloromethyl ketone substrate derivatives for serine proteinases is somewhat surprising in view of the fact that they contain an extra methylene group that might be expected to provide unfavorable geometry with respect to amino acid side chains at the active center. However, it was suggested by Bartlett (see discussion in Streitwieser 1962) that nucleophilic displacement in α-halo-ketones may be facilitated by energy-lowering contributions to the transition state of an intermediate resulting from attack at the carbonyl (Fig. 1). This possibility, in the case of serine proteinases, would account for the heightened reactivity of substrate-derived halomethyl ketones, since the geometry within the complex required for this intermediate is similar to that for normal enzymatic hydrolysis. Therefore in complexing with the enzyme, a reagent can occupy the same binding sites of the enzyme active center as the substrate, a feature of importance in attempts to use the extended binding region of protease active centers to achieve selectivity among closely related proteolytic enzymes whose specificity determinants may involve subsites somewhat distant from the hydrolytic center.

Not only in the reagent-protease complex is the geometry favorable for alkylation in spite of the extra methylene group, but, in addition, the product of alkylation appears to have the structure of a normal intermediate of the enzymatic process. A three-dimensional structural analysis of the product of alkylation of chymotrypsin by Ala-Ala-Ala-Phe-CH₂Cl (Segal et al. 1971) has shown that the phenylalanine side chain is located in the aromatic binding pocket deduced to be the specificity-determining part of the chymotrypsin active center. Moreover, the next two residues are hydrogen bonded to peptide bonds in the enzymatic chain (Fig. 2). This has been interpreted as represen-

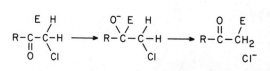

Figure 1
Suggested role of carbonyl group in facilitating nucleophilic displacement reactions of α-chloroketone.

tative of normal substrate binding since, in the model of the structure, an ester bond can be formed between the carboxyl group of the phenylalanine residue and the active center serine without disturbing this interaction. This would be the structure of the normal intermediate of catalysis, an acyl enzyme. Thus the geometry for chloromethyl ketone peptide derivatives appears to correspond to the geometry for normal substrate interaction with chymotrypsin in both the initial complex and after alkylation.

Modulation of the Reactivity of Alkylating Agents by Alterations in Departing Group

The above discussion has dealt with the commonly studied chloromethyl ketones of protease substrates in which chloride is the departing group. It was expected that other departing groups would provide affinity-labeling reagents whose reactivity would correlate with the ease of displacement of the

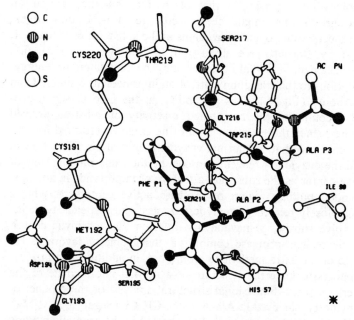

Figure 2
Atomic model of chymotrypsin Aγ inhibited by Ala-Ala-Ala-Phe-CH$_2$Cl (dark bonds) showing H-bonded interactions with enzyme whose structure is only partially shown (Segal et al. 1971).

departing anion. This appears to be the case judging from initial results of a study of the susceptibility of chymotrypsin to Cbz-Phe-Ch₂X, where X is a potential ion of varied displaceability from a series including members both more and less displaceable than chloride (Larsen and Shaw, unpubl.). It seems attractive to choose groups less reactive than chloro in order to diminish side reactions. For example, fluoro is of interest since fluoroacetate does not alkylate SH groups in contrast to the other haloacetates (Webb 1966). However, it can be expected that some reactivity in an inert group of this type can be elicited towards an active side chain within an enzyme complex in view of the millionfold increase in the rate of chloride displacement provided by the action of chymotrypsin on Cbz-Phe-CH₂Cl cited above. A reagent containing a relatively poor departing group may act on the target enzyme slowly; however, this may be compensated for by increased specificity. Studies of this type are continuing. An additional possibility is the utilization of departing group subsites on the enzyme surface for greater selectivity of action.

Selective Inhibition of Trypsinlike Enzymes

It was pointed out above that one of the challenges to affinity labeling of proteases in vivo was the need for agents capable of discriminating among enzymes of closely related specificity. The trypsinlike enzymes represent an important class of protein-processing hydrolases, as many papers in this volume testify. In contrast to the digestive enzymes, these plasma and tissue proteases cleave at only one or two sites in physiological substrates. For some, the site of cleavage is at an arginine residue, for others, it is at a lysine residue. However on simple substrates, this degree of specificity is generally not shown. One may conclude that the physiological specificity depends on some larger structural features than a single basic side chain. At the same time, the active centers of physiologically important trypsinlike enzymes appear to have lost affinity for small substrates such as simple amino acid derivatives. This probably accounts for the observed insensitivity to TLCK of plasma kallikrein and factor X_a, for example. However, peptide derivatives of lysyl chloromethyl ketone readily inactivate human plasma kallikrein (Coggins, Kray and Shaw 1974). This finding is similar to observations with elastase, which is susceptible to peptide chloromethyl ketones terminating in Ala-CH₂Cl, but not to Tos-Ala-CH₂Cl (Table 1). An inactivation of plasma kallikrein by Ala-Phe-Lys-CH₂Cl is shown in Figure 3. This reagent was synthesized to provide the phenylalanine residue penultimate to the site of cleavage in kininogen when kallikrein acts to form bradykinin as well as a third amino acid to permit the hydrogen-bonding interactions that are probably a general feature of protease-protein interactions. It is satisfying to note that the inactivation of plasma kallikrein proceeds with micromolar concentrations of this inhibitor. On the other hand, thrombin is insensitive to Ala-Phe-Lys-CH₂Cl unless elevated concentrations are used (Fig. 3). This is not due to a strict requirement of thrombin for arginine at P_1, but rather to a limited ability to accommodate to large side chains at P_2, since thrombin is susceptible to tripeptides with a C-terminal lysine chloromethyl ketone containing smaller side chains at P_2 (Shaw, Latham and Ruscica, unpubl.). Thus this type of affinity label is able to discriminate among trypsinlike enzymes, although it is

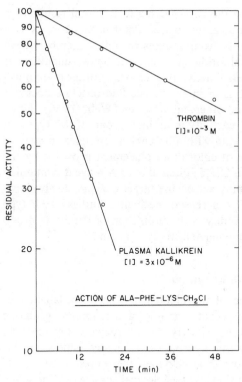

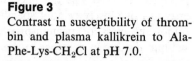

Figure 3

Contrast in susceptibility of thrombin and plasma kallikrein to Ala-Phe-Lys-CH$_2$Cl at pH 7.0.

not clear yet how far we shall be able to extend this. For example, Ala-Phe-Lys-CH$_2$Cl rapidly inactivates plasmin as well as kallikrein. However, additional examples of selectivity have been found. Initial observations with a plasminogen activator of tissue origin demonstrate this possibility. We have been examining a number of activators from normal and transformed cells in collaboration with Dr. Allan Goldberg. Because of the minute amount of activator typically available for study, we have developed a sensitive fluorescent assay based on the methylumbelliferyl ester of Z-lysine which permits the quantitation of plasmin in the 1–5 nM range. When evaluating the effect of a chloromethyl ketone on a plasminogen activator, we have to distinguish between an inhibition of the activator and of the resultant plasmin. For this purpose, after exposure of the activator to a chloromethyl ketone, the reaction mixture is maintained at pH 9 for 30 minutes to destroy the inhibitor. This treatment does not affect the activator, whose ability to form plasmin then can be subsequently assessed. Using these procedures, it was observed that the plasminogen activator from cultured bovine kidney cells (MDBK) was not inhibited by a number of peptides terminating in lysine chloromethyl ketone: for example, Ala-Phe-Lys-CH$_2$Cl, even at 10^{-3} M (pH 7.4, 30 min), had no measurable effect on this activator, although it is quite effective in inactivating plasmin (see Fig. 4, L-form of reagent). Both of these are trypsinlike enzymes. In contrast, the isomer Phe-Ala-Lys-CH$_2$Cl does inhibit the plasminogen activator: at 10^{-4} M, the activity dropped 35% in 30 minutes or completely with higher concentrations in the same time interval (Coleman et al., unpubl.). Thus these preliminary results show that the sequence in the peptide chloromethyl ketones offers a means of discriminating among trypsinlike

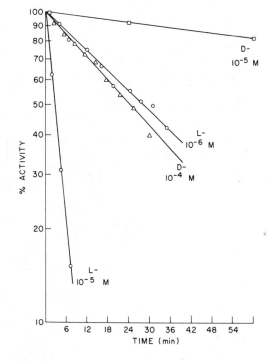

Figure 4

Stereospecificity in the inactivation of plasmin by isomers of Ala-Phe-Lys-CH$_2$Cl at pH 7. Configuration of lysine residue indicated.

enzymes. To extend this work, one must consider the basis for the specificity of the enzymes themselves, since these inhibitors utilize substrate binding sites. If specificity resides in only a limited amino acid sequence, then inhibition by affinity labeling with peptide chloromethyl ketones may be capable of considerably greater selectivity than has thus far been observed with a few tripeptide derivatives. In view of the large size of the physiological substrates of these specialized proteases, it may seem improbable that specificity be determined by a limited sequence alone. However, very little information is available on this subject. The studies of Scheraga and his colleagues (Hageman and Scheraga 1974) with thrombin are of interest. Some idea of the specificity of thrombin (or any protease) can be obtained by determining the site of cleavages in various protein substrates, but kinetic studies provide the most significant data. Specificity can be expressed in the second-order rate constant, k_{cat}/K_m, which combines turnover and affinity constants (Bender and Kezdy 1965). The initial cleavage of fibrinogen by thrombin results in the rapid liberation of fibrinopeptides. In the A-α-chain this occurs at Arg-16, and the process has been kinetically analyzed (Table 3, first entry). A number of small peptides containing the sequence of the immediate region of this site are cleaved by thrombin at a much lower rate (cf. the heptapeptide, third entry in Table 3), due to both diminished binding and turnover. However, a somewhat larger fragment of fibrinogen produced by cyanogen bromide cleavage and reduction and carboxymethylation contains the structural elements sufficient to explain the specificity of the action of thrombin on fibrinogen (Table 3) (Hageman and Scheraga 1974). In view of the chemical treatment, loss of secondary and tertiary structure in the fragment appears likely. At this stage of the study, it would appear that sequence alone accounts for specificity in

Table 3
Action of Thrombin

Substrate	K_m (M)	k_{cat}	k_{cat}/K_m $[(NIH\ U/liter)sec]^{-1}$
Fibrinogen	1.2×10^{-5}	6.3×10^{-10}	50×10^{-6}
Fibrinogen fragment (A-α-CNBr-RCM)[a]	4.7×10^{-5}	4.8×10^{-10}	10×10^{-6}
Gly-Val-Arg-Gly-Pro-Arg-Leu-NH$_2$	1.8×10^{-3}	7.5×10^{-12}	4×10^{-9}

Data from Hageman and Scheraga (1974).
[a] Cyanogen bromide cleavage product of reduced and carboxymethylated A-α-chain of fibrinogen with partial structure shown below. Arrow indicates initial site of thrombin action:

Ala$_1$-//-Gly$_{12}$-Gly$_{13}$-Gly$_{14}$-Val$_{15}$-Arg$_{16}$-Gly$_{17}$-Pro$_{18}$-Arg$_{19}$-Val$_{20}$-//-HSER$_{51}$.

P_5 P_4 P_3 P_2 P_1 P_1' P_2' P_3'

this case, and future work may define precisely the essential substrate features. Presumably this information would provide the basis for designing a most effective affinity-labeling reagent for thrombin. The synthesis might be difficult, but probably within the reach of current methods. Similar considerations apply to obtain a specific inhibitor for other trypsinlike enzymes.

Stereospecificity of Affinity Labeling

Serine proteinases are stereospecific, hydrolyzing only derivatives containing the L-form of residues bound in the enzymatic primary specificity site. As expected, the D-forms of TPCK (Stevenson and Smillie 1965), ZPCK (Shaw and Ruscica 1971) and TLCK (Shaw and Glover 1970) are without effect on their target enzymes. In examining the effect of protease inhibitors in complex systems such as cultured cells, therefore, the stereospecificity of affinity-labeling reagents should provide a useful control, since nonspecific effects due to alkylation would not be expected to be stereospecific.

Observations on the stereospecificity of protease inhibitors have been extended to include the peptides of Lys-CH$_2$Cl. Thus the difference in effectiveness of isomers of Ala-Phe-Lys-CH$_2$Cl in which the lysine residue is either of the L or D configuration is about 100:1 in the case of plasmin (Fig. 4). This requirement for the natural configuration to obtain an effective reagent exists only at the primary specificity site according to present knowledge. At other positions in the peptide affinity label, the D-isomer of a residue may actually provide a reagent that is more effective than the corresponding form containing an L-isomer at the same position (Fig. 5).

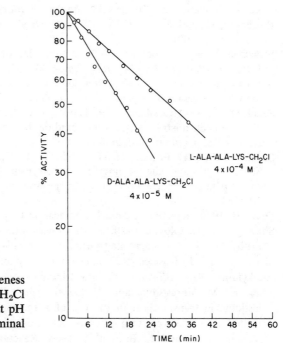

Figure 5
Comparison of the effectiveness of isomers of Ala-Ala-Lys-CH$_2$Cl as inactivators of thrombin at pH 7. Configuration of N-terminal alanine indicated.

The observations cited above provide a basis for optimism that reagents which inactivate proteases by affinity alkylation can be produced which are highly effective and capable of discrimination among closely related proteases to a degree that will render them useful in vivo. While this discussion has been restricted to chloromethyl ketone derivatives, other chemical types of affinity labels for protease inhibition are also proving to be of value (Wong and Shaw 1974).

Acknowledgments

Research was carried out at Brookhaven National Laboratory under the auspices of the U.S. Atomic Energy Commission and U.S. Public Health Service through Grant No. GM17849.

REFERENCES

Bender, M. L. and F. J. Kezdy. 1965. Mechanism of action of proteolytic enzymes. *Annu. Rev. Biochem.* **34**:49.

Coggins, J. R., W. Kray and E. Shaw. 1974. Affinity-labeling of proteases with tryptic specificity by peptides with C-terminal lysine chloromethyl ketone. *Biochem. J.* **138**:479.

Hageman, T. C. and H. A. Scheraga. 1974. Mechanism of action of thrombin on fibrinogen. V. Reaction of the N-terminal CNBr fragment from the A_α chain of human fibrinogen with bovine thrombin. *Arch. Biochem. Biophys.* **164**:707.

Morihara, K. and T. Oka. 1970. Subtilisin BPN': Inactivation by chloromethyl ketone derivatives of peptide substrates. *Arch. Biochem. Biophys.* **138**:526.

Powers, J. C. and P. M. Tuhy. 1973. Active-site specific inhibitors of elastase. *Biochemistry* **12**:4767.

Schechter, I. and A. Berger. 1967. On the size of the active site in proteases. I. Papain. *Biochem. Biophys. Res. Comm.* **27**:157.

Schoellmann, G. and E. Shaw. 1963. Direct evidence for the presence of histidine in the active center of chymotrypsin. *Biochemistry* **2**:252.

Segal, D. M., G. H. Cohen, D. R. Davies, J. C. Powers and P. E. Wilcox. 1971. The stereochemistry of substrate binding to chymotrypsin A_γ. *Cold Spring Harbor Symp. Quant. Biol.* **36**:85.

Segal, D. M., J. C. Powers, G. H. Cohen, D. R. Davies and P. E. Wilcox. 1971. Substrate binding site in bovine chymotrypsin A_α. A crystallographic study using peptide chloromethyl ketones as site-specific inhibitors. *Biochemistry* **10**:3728.

Shaw, E. 1970. Selective chemical modification of proteins. *Physiol. Rev.* **50**:244.

Shaw, E. and G. Glover. 1970. Further observations on substrate-derived chloromethyl ketones that inactivate trypsin. *Arch. Biochem. Biophys.* **139**:298.

Shaw, E. and J. Ruscica. 1971. The reactivity of His-57 in chymotrypsin to alkylation. *Arch. Biochem. Biophys.* **145**:484.

Shaw, E., M. Mares-Guia and W. Cohen. 1965. Evidence for an active center histidine in trypsin through the use of a specific reagent, TLCK, the chloromethyl ketone derived from N-tosyl-L-lysine. *Biochemistry* **4**:2219.

Stevenson, K. J. and L. B. Smillie. 1965. Reaction of phenoxymethyl chloromethyl ketone with nitrogen 3 of histidine-57 of chymotrypsin. *J. Mol. Biol.* **12:**937.

Streitwieser, A., Jr. 1962. *Solvolytic Displacement Reactions,* p. 28. McGraw-Hill, New York.

Thompson, R. C. and E. R. Blout. 1974. Peptide chloromethyl ketones as irreversible inhibitors of elastase. *Biochemistry* **12:**44.

Thomson, A. and I. S. Denniss. 1973. The reaction of active-site inhibitors with elastase using a new assay substrate. *Eur. J. Biochem.* **38:**1.

Webb, J. L. 1966. Iodoacetate and iodoacetamide. In *Enzyme and Metabolic Inhibitors,* vol. 3, p. 11. Academic Press, New York.

Wong, S. C. and E. Shaw. 1974. Differences in active center reactivity of trypsin homologs. Specific inactivation of thrombin by nitrophenyl-*p*-amidinophenyl-methanesulfonate. *Arch. Biochem. Biophys.* **161:**536.

Lysosomal and Related Proteinases

Alan J. Barrett

Tissue Physiology Department, Strangeways Research Laboratory
Cambridge, England

Enzymes which hydrolyze peptide bonds form a major group among the enzymes of lysosomes and related organelles. Before the properties of the individual enzymes are considered, it will be desirable to remember something of the general classification of these enzymes.[1]

Classification and Nomenclature

The peptide hydrolases (or peptidases) are divided into the *endopeptidases*, which cleave bonds away from the ends of polypeptide chains (but often near the ends, also), and the *exopeptidases*, which cleave bonds only near the ends. The exopeptidases can be named semi-systematically, i.e., partially on the basis of their specificity (see section on cathepsins A, B2 and C), but the endopeptidases are more difficult. No satisfactory scheme has been produced for the classification and naming of endopeptidases on the basis of their specificity. Instead, a useful classification system has been based on the recognition of four distinct classes of catalytic mechanisms among the endopeptidases (Hartley 1960), and it has been proposed that the *tissue proteinases* be classified as *carboxyl, thiol, serine or metallo*-proteinases (see Appendix following this article).

Trivial names have to be used for the endopeptidases, and in the opinion of this author, those who discover a new enzyme of this type will do well to break away from the usual conventions in the naming of enzymes and propose something memorable and truly trivial, in the sense of not being seriously descriptive. The value of a name such as *trypsin* or *pepsin* has been that it provides a label for a particular enzyme, whereas attempts to convey information (on the distribution or function of an enzyme, for example) through a name that is normally conferred when only sketchy information is available are more often confusing than helpful.

[1] The nomenclature described here is based on the Recommendations of the IUPAC-IUB Commission (Florkin and Stotz 1972), but does not follow it in all particulars.

467

The enzyme originally named "kathepsin" (Willstätter and Bamann 1928) is that which we now call cathepsin D, but the word cathepsin has become recognized as a generic term for the enzymes which have also been loosely referred to as "tissue proteinases," those proteinases which occur in cells and tissues of higher animals, but excluding those of the digestive tract and blood plasma. Although cathepsins A–E are a very heterogeneous group of peptide hydrolases, all with pH optima in the acid range, it has recently been proposed that the term "cathepsin" be retained for further use and be applied to cell and tissue proteinases regardless of their pH optima (see Appendix).

The best known of the tissue proteinases are located in cells either in the lysosomes proper or in granules such as those of the neutrophil granulocyte, which have important functions in common with the lysosomes of other cells, but are far from typical lysosomes. Since the properties of the lysosomal enzymes have been reviewed (Barrett 1972a), the emphasis here will be on recent advances. It is convenient that four of the better known tissue proteinases may be taken as examples to illustrate the main characteristics of endopeptidases within the four catalytic groups. Inevitably, some others fall into a fifth group of proteinases of unknown catalytic mechanism.

Cathepsin D, a Carboxyl Proteinase

To the biochemist, cathepsin D (EN 3.4.23.5) is the most easily detected proteinase in most tissues, and subcellular fractionation studies have clearly shown it to be lysosomal.

Cathepsin D shows only modest activity against substrates of low molecular weight (Sampath-Kumar and Fruton 1974), and the usual test substrates are proteins. With hemoglobin, the pH optimum is 3.0–3.5, but maximal activity is seen at higher pH values with most other proteins. The peak of activity against cartilage proteoglycan is at pH 5, and appreciable activity persists at pH 6, but cathepsin D appears to be totally inactive at pH 7 or above. It has been discovered that cathepsin D from bovine (Woessner 1973) and human (Fig. 1) tissues tends to be associated with a distinct neutral proteinase, until the latest stages of purification.

In our laboratory, cathepsin D has been isolated by a procedure which also yields cathepsin B1 (Barrett 1970, 1973); the method involves autolysis, fractional precipitation with acetone, chromatography on CM-cellulose, on an organomercurial adsorbent and DEAE-cellulose, and isoelectric focusing. In the final stage, the enzyme is resolved into three major forms with slightly differing isoelectric points. The three forms were immunologically identical (Dingle, Barrett and Weston 1971). The amino acid composition of one of the forms (Table 1) has been compared with those of two forms of chicken cathepsin D (Barrett 1971). The molecular weight of cathepsin D from most sources appears closely similar to that of ovalbumin (44,600, as reported by Castellino and Barker 1968) in gel chromatography with Sephadex G-100. A value of 42,000 has been obtained for human cathepsin D by G. C. Knight (unpubl.) by titration with pepstatin and dry weight determination. When cathepsin D is subjected to electrophoresis in the presence of sodium dodecyl sulfate, with or without reduction, bands of approximately 28,000 and 14,000 are commonly obtained. Some forms of the bovine spleen enzyme retained the

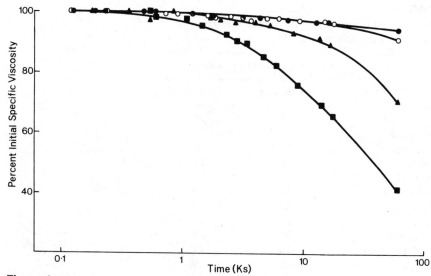

Figure 1

Disappearance of neutral proteinase activity during the purification of cathepsin D from human liver. The purification procedure was that of Barrett (1973) in which the last two steps are chromatography on DEAE-cellulose and preparative isoelectric focusing. Neutral proteinase activity was detected as the loss of the viscosity in a solution (5 mg/ml) of bovine nasal proteoglycan in 200 mM NaCl, 50 mM sodium potassium phosphate, pH 7.0, at 37°C. The samples were cathepsin D before DEAE-cellulose (■), cathepsin D after DEAE-cellulose (▲), cathepsin D after isoelectric focusing (●) and boiled enzyme (○). The cathepsin D samples each produced a final activity of 50 units (at pH 3.5) per ml.

It can be seen that neutral proteinase activity was reduced by the ion exchange chromatography and eliminated by isoelectric focusing.

1972). Ultracentrifuge experiments suggest that there may be some dissociation of cathepsin D even under quite mild conditions (Smith et al. 1969; Ferguson et al. 1973).

Specific antisera have been raised against cathepsin D, and the characteristics of immunoinhibition of the enzyme have been examined (Dingle, Barrett and Weston 1971). It was concluded that the antisera should be useful tools with which to investigate the physiological functions of cathepsin D. It was indeed possible to show that much of the intracellular digestion of hemoglobin endocytosed by rabbit alveolar macrophages is mediated by cathepsin D (Dingle et al. 1973).

The usefulness of antisera against cathepsin D in the localization of the enzyme in cells and tissues has been fully exploited by Poole and coworkers. The localization of the enzyme in discrete, lysosomelike organelles in a variety of cells was demonstrated microscopically for the first time (Poole, Dingle and Barrett 1972). It was also possible to show the secretion of cathepsin D from certain cells within a tissue by a new technique in which the living tissue was exposed to the antiserum. The antibodies did not enter the living cells to a significant extent, but did form extracellular precipitates around cells

Table 1

Amino Acid Compositions of Human Cathepsin D
(γ-form) and Cathepsin B1

	Residues/mole	
	cathepsin D	*cathepsin B1*
Lysine	25	10
Histidine	5	8
Arginine	10	8
Cysteine	8	12
Aspartic	32	23
Methionine	14	4
Threonine	20	10
Seine	26	17
Glutamic	36	23
Proline	23	17
Glycine	40	29
Alanine	20	13
Valine	31	15
Isoleucine	22	13
Leucine	35	10
Tyrosine	21	13
Phenylalanine	15	7
Tryptophan	4	7
Glucosamine	7	0

The compositions are calculated on the basis of molecular weights of 42,000 (cathepsin D) and 25,000 (cathepsin B1). The figure for tryptophan is derived from a direct spectrophotometric analysis made by Dr. C. G. Knight.

that secreted cathepsin D, and these were subsequently labeled with a second antibody (Poole, Hembry and Dingle 1973).

Cathepsin D is inactivated by some of the diazonium compounds which are also effective against pepsin, always in the presence of cupric ions; a typical compound of this type is diazoacetyl-norleucine methyl ester. Such compounds could not be expected to give specific inhibition in biological systems, however, so that the discovery of pepstatin by Umezawa and coworkers (see Aoyagi and Umezawa, this volume) was a most exciting development. The inhibition of cathepsin D from rabbit and man by pepstatin was reported by Barrett and Dingle (1972). Calf renin and rabbit cathepsin E were other sensitive carboxyl proteinases, whereas trypsin, chymotrypsin, papain and cathepsin B1 were unaffected. Renin is also known to be inhibited by pepstatin, although it is less sensitive than the other enzymes (Overturf, Leonard and Kirkendall 1974); renin probably is a lysosomal carboxyl proteinase (Onoyama et al. 1973). It has recently been shown that the binding of pepstatin by human cathepsin D is a very pH-dependent process, for K_{ass} falls from about 2×10^9 M to 6×10^5 M in the range pH 3.5–6.4 (C. G. Knight and A. J. Barrett, unpubl.). Clearly, however, the inhibitor has a very high affinity for cathepsin D over the whole pH range of activity, and this fact, together with the very low toxicity of the inhibitor, suggested that it might produce dramatic effects in various biological systems in which cathepsin D was thought to be acting. In our laboratory, and as far as I know in other laboratories, few such effects have been observed. One obvious explanation is that cathepsin D is not essential for such processes as the resorption of the extracellular matrix of living cartilage, contrary to what had been thought previously. However, several other possible explanations have yet to be definitely ruled out. It is known that pepstatin is excreted rapidly from whole animals, and preliminary results in our laboratory have indicated that it penetrates tissues and cells only poorly. Further work is being done to determine whether these considerations should affect the interpretation of results obtained with pepstatin.

When the specificity of cathepsin D is examined by use of the oxidized B chain of insulin as substrate, there is a large degree of cleavage of all of the peptide bonds in the distinctly hydrophobic sequences between Glu-13–Leu-17 and Phe-24–Tyr-26 (reviewed by Barrett 1972a). Fruton and coworkers have recently compared the specificity of action on synthetic peptides of low molecular weight with those of pepsin and a fungal proteinase (Ferguson et al. 1973; Sampath-Kumar and Fruton 1974). It was found that cathepsin D had a much greater requirement for a high molecular weight of substrate than does pepsin. Cathepsin E has been studied only with rabbit bone marrow as source (Lapresle 1971); it certainly is a carboxyl proteinase distinct from cathepsin D and probably occurs in the granules of rabbit neutrophil granulocytes or platelets, or both.

Cathepsin B1 and the Thiol Proteinases

The thiol proteinases include papain, ficin and bromelain, from the plant kingdom, as well as a few bacterial enzymes. Cathepsin B1 (EN 3.4.22.1) is the only well-characterized thiol proteinase from the tissues of higher animals.

"Cathepsin B" was the name originally given to the "enzyme" from bovine spleen that deamidated benzoylarginine amide. Recently, however, it has been shown that this activity is due to two quite distinct enzymes, and they have been named cathepsins B1 and B2 (Otto 1971; Barrett and Dingle 1971). Cathepsin B1 may be described as a thiol-dependent lysosomal proteinase, which is also active against several synthetic substrates and has a molecular weight of about 25,000.

Cathepsin B1 has been purified from human liver by a procedure that also yields cathepsin D (Barrett 1973). The two enzymes are separated by the use of an organomercurial-Sepharose affinity column, which binds cathepsin B1 while cathepsin D passes straight through. Like cathepsin D, cathepsin B1 is obtained finally as a number of forms differing slightly in isoelectric point, but it is not clear whether these occur in the tissue or are artifacts of the purification procedure.

A new color reaction has been used to assay cathepsin B1 in the test tube (Barrett 1972b) and also to stain electrophoresis and isoelectric focusing gels for activity (Barrett 1973). In this procedure, the synthetic substrate Bz-

D,L-Arg 2-naphthylamide[2] is used, and the 2-naphthylamine released is converted to a red product by coupling with a diazonium salt, Fast Garnet GBC. In principle, this is a well-known reaction, but the thiol compound required for full activity of cathepsin B1 interferes in the coupling reaction and has to be blocked with p-chloromercuribenzoate.

The amino acid composition of human cathepsin B1 is given in Table 1. The analysis showed no hexosamine, which presumably means that cathepsin B1 is not a glycoprotein. Cathepsin D, and the several other lysosomal enzymes that had been adequately characterized, had been found to contain carbohydrate, and it had been thought that the presence of carbohydrate (and possibly disulfide bonds) might be a general characteristic of lysosomal enzymes.

The specificity of hydrolysis of insulin B chain by bovine spleen cathepsin B1 has been reported by Otto (1971) and Keilová (1971). Neither preparation of the enzyme had been shown to be homogeneous, and it is perhaps not surprising that the two authors did not fully agree, but two points emerge clearly: the specificity is broader than that of cathepsin D and quite unlike that of trypsin; also, many of the bonds split are susceptible to papain.

There has been particular interest in the activity of cathepsin B1 in producing limited cleavage of other enzyme molecules, thus altering their properties. The enzyme was first discovered by Otto because of its property of inactivating glucokinase (EN 2.7.1.2), and it was found also to destroy the activity of fructose-bisphosphate aldolase (EN 4.1.2.13) and pyruvate kinase (EN 2.7.1.40). Tyrosine aminotransferase (EN 2.6.1.5) is also inactivated by an enzyme with the properties of cathepsin B1 (Auricchio, Mollica and Liguori 1972), and hexosediphosphatase (EN 3.1.3.11) has its pH optimum shifted to a much higher value (Nakashima and Ogino 1974).

Cathepsin B1 shares with papain two aspects of specificity that make it particularly interesting in the context of the catabolism of the extracellular macromolecules of connective tissues. When acting on cartilage proteoglycan, it is able to cleave the relatively resistant, short peptide between pairs of chondroitin sulfate chains (Mathews 1971), liberating single polysaccharide chains (Morrison et al. 1973). Cathepsin B1 also degrades native collagen at acid pH and may well play a part in the intracellular catabolism of this component (Burleigh, Barrett and Lazarus 1974).

Human cathepsin B1 has proved to be an unsatisfactory immunogen in our hands, in that antibodies were raised only against a denatured form of the protein (Barrett 1973). Immunohistochemical localization of cathepsin B1 has been claimed (Sylvén and Snellman 1974), but without evidence of the specificity of the antiserum.

There are a large number of inactivators of the thiol proteinases, because of the high reactivity of the essential thiol group, but specific inhibition of particular thiol proteinases is difficult. Cathepsin B1 is powerfully inhibited by the chloromethyl ketones of tosyl-lysine, tosyl-phenylalanine and acetyl-tetra-alanine, which are selective for certain serine proteinases. Dr. Zena Werb, in our laboratory, has recently cultured rabbit synovial fibroblasts in the presence of 100 μM Tos-Lys-CH$_2$Cl; during two days the inhibitor caused no fall in

[2] Standard abbreviations are used for the peptides and peptide derivatives (*J. Biol. Chem.* **247**:977–983).

growth rate or acid phosphatase content of the cells, but the activity of cathepsin B1 in the cells fell to about 20% of the control value during 24 hours. The fall was reversible after 24 hours, in that cells transferred at this time to control medium rapidly regained the normal activity.

Ikezawa et al. (1971, 1972) reported the inhibition of the amidase activity of cathepsin B by two naturally occurring peptide aldehyde derivatives, leupeptin and antipain. Barrett (1973) showed that leupeptin-Pr strongly inhibits human cathepsin B1.

The papain inhibitor that has been found in chicken egg white (Sen and Whitaker 1973) is also a powerful inhibitor of cathepsin B1 (Keilová and Tomášek 1974). This inhibitor is a heat-stable protein of 12,700 molecular weight.

Starkey and Barrett (1973) showed that the major inhibitor of cathepsin B1 in human serum is α_2-macroglobulin. The study of the characteristics of interaction between the enzyme and the macroglobulin led to the realization that α_2-macroglobulin has the remarkable capacity to bind almost any active endopeptidase and yet be specific for this functional group of enzymes. A molecular mechanism was proposed to account for the activity of the macroglobulin, and the probable physiological importance of this protein was discussed (Barrett and Starkey 1973).

Granulocyte Elastase, a Serine Proteinase

The elastase of neutrophil granulocytes (EN 3.4.2.1-) is a serine proteinase that may be important in several physiological processes. Recent work in our laboratory (P. M. Starkey and A. J. Barrett, unpubl.) has been directed toward the characterization of the serine proteinases of human spleen. The result has been the purification and characterization of two proteinases which together account for all of the activity of homogenates of this tissue against azo-casein in the neutral–alkaline pH range. One of the enzymes has proved to be the granulocyte elastase, the other we have named "cathepsin G." Each of the enzymes has a marked tendency to be adsorbed by surfaces and insoluble material, except in the presence of neutral detergents or high concentrations of salt. At several stages of the purification, this anomalous behavior has been turned to advantage in facilitating the separation of the enzymes from inactive proteins of similar electrostatic charge and molecular weight. Elastase has been purified from normal white cells (Janoff 1973) and cells from leukemic patients (Ohlsson and Olsson 1974), but has not been obtained previously in appreciable amounts from a solid tissue.

Both of the neutral proteinases of the spleen are active against azo-casein, a convenient test substrate (Charney and Tomarelli 1947), and this was used for assays during the early part of the purification. The enzymes also share activity against one of the histochemists' esterase substrates, chloroacetyl-naphthol AS-D. Assays more specific for the elastase were made with elastin or Z-Ala-2-naphthol ester. The pH-optimum for activity against elastin and Z-Ala-2-naphthol ester is pH 8.5–8.7.

Granulocyte elastase has a remarkably broad specificity for protein substrates. The digestion of elastin itself is conveniently demonstrated by the use of an elastin-agarose plate assay (Schumacher and Schill 1972) or of congo red-elastin (Naughton and Sanger 1961). The likelihood that the elastase degrades the proteoglycan complex of cartilage was suggested by the results of Janoff and Blondin (1970) with a partially purified fraction. This has recently been fully confirmed by us with the highly purified enzyme. Furthermore, preliminary results (M. C. Burleigh, P. M. Starkey and A. J. Barrett, unpubl.) strongly suggest that the enzyme is capable of digesting insoluble collagen from both tendon and cartilage. This wide range of activities makes elastase unique in this capacity to break down connective tissue structures.

The molecular weight of the elastase, determined by SDS gel electrophoresis, is about 28,000, and it is a markedly basic protein. Like most serine proteinases, it is inactivated by diisopropylflurophosphate. Powers and Tuhy (1973) have described a series of peptide chloromethyl ketone inhibitors for pancreatic elastase which are also active against the granulocyte enzyme. These have been used in several laboratories, including our own, to distinguish the activity of the elastase from that of other proteinases. In performing such experiments, however, it is important to remember that although the difference in rate of inactivation of elastase as compared with some other serine proteinases may be large, the reaction, given time, will proceed to the stage of virtually complete inactivation of other enzymes, too. Also, of course, all of the chloromethyl ketones are highly active against the thiol proteinases. Another inhibitor of the thiol proteinases that is strongly inhibitory for elastase is gold thiomalate.

Of the naturally occurring inhibitors, soybean trypsin inhibitor is particularly effective. Ohlsson (this volume) has made a full study of the reactivity of α_1-trypsin inhibitor and α_2-macroglobulin with granulocyte elastase.

The second of the spleen proteinases, cathepsin G, may well be the "chymotrypsinlike" enzyme detected in the granules of human neutrophils (Rindler and Braunsteiner 1973; Rindler et al. 1973). We have found cathepsin G to be active against azo-casein and chloroacetyl-naphthol AS-D, like the elastase, but it is distinguishable by its activity against Bz-D,L-Phe-2-naphthol ester. The pH optimum is near pH 7.5. The enzyme is not readily inactivated by diisopropylfluorophosphate or Tos-Phe-CH$_2$Cl, but is sensitive to Z-Phe-CH$_2$BR (kindly provided by Dr. Elliott Shaw). Like the granulocyte elastase, cathepsin G is a basic protein of about 28,000 molecular weight; it is not active against elastin, but does degrade cartilage proteoglycan and native collagen from tendon and cartilage.

Among the other known serine proteinases of lysosomes or related organelles are the acrosomal proteinases of the sperm and the chymotrypsinlike enzyme, or "chymase," of mast cell granules (probably distinct from cathepsin G) (Kawiak et al. 1971; Vensel, Komender and Barnard 1971).

Collagenase, a Metallo-proteinase

By far the best known metallo-proteinase of higher animals is collagenase, and indeed, the literature on this enzyme is so extensive that the reader must be referred to reviews such as those of Seifter and Harper (1970) and Lazarus (1973) for full information. The collagenase that is found in human neutrophil leukocytes (Lazarus et al. 1968, 1972) certainly is localized in the granules of these cells, probably in the azurophil granules. For other cell types, the localization of collagenase is more uncertain; the enzyme is seldom detectable in homogenates, but is secreted from fragments of many different tissues when they are cultured in the absence of serum (which contains the inhibitory α_2-macroglobulin). Recently in our laboratory, Werb and Burleigh (1974) have studied the secretion of a specific collagenase from rabbit fibroblasts in monolayer culture. The intracellular localization of the enzyme before secretion has not yet been established, but the mechanics of secretion would seem to require that it be present in membrane-limited organelles.

Collagenase itself is a particularly specific enzyme, and the same may be true for a metal-dependent lens proteinase that has been studied recently (A. M. J. Blow, R. van Heyningen and A. J. Barrett, unpubl.). A zinc-containing endopeptidase from rabbit kidney brush border has been the subject of a detailed study by Kerr and Kenny (1974a,b), and Harris and Krane (1972) have described a metal-dependent "gelatinase" from human rheumatoid synovial fluid.

Inhibitors that may be useful in the identification of metallo-proteinase activity are mainly chelating agents, since enzymatic activity usually seems to be dependent on the availability of calcium and zinc ions. 1,10-Phenanthroline, EDTA and thiol compounds have been found effective. Penicillamine would be a possible choice of inhibitor for use in a living tissue system. Umezawa and his colleagues (Suda et al. 1973) have described a natural inhibitor of thermolysin, called phosphoramidon, which may also be effective against collagenase and other metallo-proteinases.

The inhibition of collagenase and other metallo-proteinases by α_2-macroglobulin has been described by Werb et al. (1974). Since the mechanism of interaction of the enzymes with the macroglobulin requires proteolytic cleavage of the inhibitor, the interaction of so specific an endopeptidase as collagenase is a remarkable illustration of the broad specificity of α_2-macroglobulin for the control of proteolytic activity.

Proteinases of Unknown Catalytic Mechanism

A sub-subgroup of the endopeptidases (EN 3.4.99) is reserved for enzymes of unknown mechanism. Work in our laboratory (J. T. Dingle, A. J. Barrett and A. M. J. Blow, unpubl.) has led to the detection of a proteinase in rabbit and human cartilage that degrades cartilage proteoglycan maximally at pH 4.5. This enzyme, which has tentatively been named "cathepsin F," is resistant to pepstatin and most of the other inhibitors that have been tried and therefore has not been assigned to a catalytic group. Cathepsin F, however, is inhibited by α_2-macroglobulin. The intracellular localization of the enzyme has not been established.

Cathepsins A, B2, C and Other Exopeptidases

It would not be appropriate to deal with the exopeptidases at all fully here, but for clarity, those which have been named "cathepsin" should be mentioned, and some others may be alluded to in order to cover the bare bones of the system of classification. As is mentioned in the notes contained in the Appendix, it will probably be desirable to abandon the use of the word "cathepsin" for those tissue proteinases which prove to be exopeptidases, since these enzymes, unlike the proteinases, can be named at least semi-systematically. Five classes of exopeptidase are recognized; these comprise the aminopeptidases and carboxypeptidases, which cleave single amino acid residues from N and C termini, respectively; the dipeptidylpeptide hydrolases and peptidyldipeptide hydrolases, which cleave dipeptides from N and C termini; and dipeptidases, which split dipeptides.

The activity of the exopeptidases is such that it is difficult to detect partially degraded protein molecules in cells; it seems that once the endopeptidases have opened the way for the exopeptidases, extensive digestion takes place rapidly.

Iodice (1967) reported that bovine spleen *cathepsin A* exerted a carboxypeptidase action on glucagon, the release of residues from the C terminus being slowed by the appearance of basic amino acids. Further characterization of the enzyme from rat liver has been reported by Taylor and Tappel (1974a, b) who distinguished two forms of the enzyme, cathepsins A1 and A2, on the basis of gel chromatographic behavior. Cathepsin A1 was most active in the liberation of hydrophobic amino acid residues and was activated by halide ions. A detailed characterization of cathepsin A from pig kidney, again with recognition of two molecular forms, has been made by Doi (1974; Doi et al. 1974). Reports of endopeptidase activity in cathepsin A preparations (Lichtenstein and Fruton 1960; Logunov and Orekhovich 1972) may be due to trace contamination of the enzyme samples.

Ninjoor, Taylor and Tappel (1974) have also reported that *cathepsin B2* of rat liver appears to show carboxypeptidase activity. In the absence of definite evidence for endopeptidase activity of cathepsin B2, it seems appropriate to consider whether the enzyme could be a carboxypeptidase, in spite of its amidase action against Bz-Arg-NH$_2$.

There are several other lysosomal carboxypeptidases, some of which have been called catheptic carboxypeptidases, and information on these is provided by Taylor and Tappel (1974a).

Cathepsin C, also called dipeptidyl aminopeptidase I, is now properly referred to as dipeptidylpeptidase I (EN 3.4.14.1); it has been the subject of much study, particularly since it was shown to have considerable promise in the sequencing of proteins (McDonald, Callahan and Ellis 1971, 1972).

The most convenient test substrate for dipeptidylpeptidase I is Gly-Phe-2-naphthylamide, the release of naphthylamine being followed by direct fluorometry (McDonald, Zeitman and Ellis 1970) or the coupling reaction originally developed for cathepsin B1 (Barrett 1972a). The latter method has been used to show the presence of cathepsin C in the granules of human leukocytes (Davies, Allison and Hylton 1974).

Dipeptidylpeptidase I has a broad specificity for the removal of dipeptides from the N terminus of polypeptides, but is blocked by the appearance of terminal arginine or lysine or of proline in the penultimate position (McDonald, Callahan and Ellis 1971). The enzyme was shown to cleave 13 bonds in the B chain of insulin, more than trypsin, chymotrypsin and pepsin together!

Dipeptidylpeptidase I is a protein of 200,000 molecular weight, containing eight polypeptide chains. The active center has an essential thiol group, so that the enzyme is readily inactivated by thiol-blocking reagents; the enzyme also has a requirement for a halide ion, for activity.

Another lysosomal dipeptidylpeptidase (dipeptidylpeptidase II) has been described by McDonald and coworkers under the name of dipeptidylaminopeptidase II (reviewed by McDonald, Callahan and Ellis 1971). The specificity of the enzyme includes polypeptides with N-terminal lysine and arginine, which are resistant to dipeptidylpeptidase I.

CONCLUSIONS

We are beginning to acquire some detailed knowledge of the identity and properties of the lysosomal and related proteinases. The justification for the work which has led to this has been the belief that the enzymes have important functions in physiology and pathology, and that knowledge of their properties would help us towards an understanding of these functions. Ultimately, it might be possible to devise ways of correcting malfunctions.

The stage in which we should elucidate the functions of the enzymes is, as yet, hardly started. Real progress seems to be awaiting the application of techniques which are, as yet, not fully developed. It seems to me that two groups of such techniques stem rather directly from our present work on the purification and characterization of the individual enzymes. In the first place, the availability of a pure enzyme can lead to reliable information on its specificity, and thus, hopefully, to the design of truly specific inhibitors. Such inhibitors may be used to set up assays (e.g., by titration) to identify the activity of the enzyme and perhaps (in labeled form) to localize the enzyme in cells and tissues. Second, specific antisera can be raised against the enzyme, and these have proven usefulness in the realms of immunoassay, specific inhibition, localization and studies on biosynthesis and translocation. Undoubtedly, there are other profitable approaches to the study of the function of the tissue proteinases, but those outlined above are those which are receiving particular attention in our laboratory.

REFERENCES

Auricchio, F., L. Mollica and A. Liguori. 1972. Inactivation of tyrosine aminotransferase in neutral homogenates and rat liver slices. *Biochem. J.* **129:**1131.

Barrett, A. J. 1970. Cathepsin D. Purification of isoenzymes from human and chicken liver. *Biochem. J.* **117:**601.

————. 1971. Purification and properties of cathepsin D from liver of chicken, rabbit and man. In *Tissue Proteinases* (ed. A. J. Barrett and J. T. Dingle), p. 109. North-Holland, Amsterdam.

————. 1972a. Lysosomal enzymes. In *Lysosomes: A Laboratory Handbook* (ed. J. T. Dingle), p. 46. North-Holland, Amsterdam.

————. 1972b. A new assay for cathepsin B1 and other thiol proteinases. *Anal. Biochem.* **47:**280.

————. 1973. Human cathepsin B1. Purification and some properties of the enzyme. *Biochem. J.* **131:**809.

Barrett, A. J. and J. T. Dingle. 1971. Terminology of the tissue proteinases. In *Tissue Proteinases* (ed. A. J. Barrett and J. T. Dingle), p. ix. North-Holland, Amsterdam.

————. 1972. Inhibition of tissue acid proteinases by pepstatin. *Biochem. J.* **127:**439.

Barrett, A. J. and P. M. Starkey. 1973. The interaction of α_2-macroglobulin with proteinases. Characteristics and specificity of the reaction, and a hypothesis concerning its molecular mechanism. *Biochem. J.* **133**:709.

Burleigh, M. C., A. J. Barrett and G. S. Lazarus. 1974. Cathepsin B1. A lysosomal enzyme that degrades native collagen. *Biochem. J.* **137**:387.

Castellino, F. J. and R. Barker. 1968. Examination of the dissociation of multichain proteins in guanidine hydrochloride by membrane osmometry. *Biochemistry* **7**:2207.

Charney, J. and R. M. Tomarelli. 1947. A colorimetric method for the determination of the proteolytic activity of duodenal juice. *J. Biol. Chem.* **171**:501.

Davies, P., A. C. Allison and W. J. Hylton. 1974. The identification, properties and subcellular distribution of cathepsins B1 and C (dipeptidyl aminopeptidase I) in human peripheral-blood leucocytes. *Biochem. Soc. Trans.* **2**:432.

Dingle, J. T., A. J. Barrett and P. D. Weston. 1971. Cathepsin D. Immunoinhibition by specific antisera. *Biochem. J.* **123**:1.

Dingle, J. T., A. R. Poole, G. S. Lazarus and A. J. Barrett. 1973. Immunoinhibition of intracellular protein digestion in macrophages. *J. Exp. Med.* **137**: 1124.

Doi, E. 1974. Stabilization of pig kidney cathepsin A by sucrose and chloride ion, and inhibition of the enzymic activity by diisopropylfluorophosphate and sulfhydryl reagents. *J. Biochem.* **75**:881.

Doi, E., Y. Kawamura, T. Matoba and T. Hata. 1974. Cathepsin A of two different molecular sizes in pig kidney. *J. Biochem.* **75**:889.

Ferguson, J. B., J. R. Andrews, I. M. Voynick and J. S. Fruton. 1973. The specificity of cathepsin D. *J. Biol. Chem.* **248**:6701.

Florkin, M. and E. H. Stotz, eds. 1972. *Comprehensive Biochemistry,* vol. 13, 3rd ed. Elsevier, Amsterdam.

Harris, E. D. and S. M. Krane. 1972. An endopeptidase from rheumatoid synovial tissue culture. *Biochim. Biophys. Acta* **258**:566.

Hartley, B. S. 1960. Proteolytic enzymes. *Annu. Rev. Biochem.* **29**:45.

Ikezawa, H., T. Aoyagi, T. Takeuchi and H. Umezawa. 1971. Effect of protease inhibitors of actinomycetes on lysosomal peptide-hydrolases of swine liver. *J. Antibiotics* **24**:488.

Ikezawa, H., K. Yamada, T. Aoyagi, T. Takeuchi and H. Umezawa. 1972. Effect of antipain on lysosomal peptide hydrolases from swine liver. *J. Antibiotics* **25**: 738.

Iodice, A. A. 1967. The carboxypeptidase nature of cathepsin A. *Arch. Biochem. Biophys.* **121**:241.

Janoff, A. 1973. Purification of human granulocyte elastase by affinity chromatography. *Lab. Invest.* **29**:458.

Janoff, A. and J. Blondin. 1970. Depletion of cartilage matrix by a neutral protease fraction of human leucocyte lysosomes. *Proc. Soc. Exp. Biol. Med.* **135**: 302.

Kawiak, J., W. H. Vensel, J. Komender and E. A. Barnard. 1971. Non-pancreatic proteases of the chymotrypsin family. 1. A chymotrypsin-like protease from rat mast cells. *Biochim. Biophys. Acta* **235**:172.

Keilová, H. 1971. On the specificity and inhibition of cathepsins D and B. In *Tissue Proteinases* (ed. A. J. Barrett and J. T. Dingle), p. 45. North-Holland, Amsterdam.

Keilová, H. and V. Tomášek. 1974. Effect of papain inhibitor from chicken egg white on cathepsin B1. *Biochim. Biophys. Acta* **334**:179.

Kerr, M. A. and A. J. Kenny. 1974a. The purification and specificity of a neutral endopeptidase from rabbit kidney brush border. *Biochem. J.* **137**:477.

————. 1974b. The molecular weight and properties of a neutral metallo-endo-peptidase from rabbit kidney brush border. *Biochem. J.* **137**:489.

Lapresle, C. 1971. Rabbit cathepsins D and E. In *Tissue Proteinases* (ed. A. J. Barrett and J. T. Dingle), p. 135. North-Holland, Amsterdam.

Lazarus, G. S. 1973. Studies on the degradation of collagen by collagenases. In *Lysosomes in Biology and Pathology* (ed. J. T. Dingle), vol. 3, p. 338. North-Holland, Amsterdam.

Lazarus, G. S., J. R. Daniels, R. S. Brown, H. A. Bladen and H. M. Fullmer. 1968. Degradation of collagen by a human granulocyte collagenolytic system. *J. Clin. Invest.* **47**:2622.

Lazarus, G. S., J. R. Daniels, J. Lian and M. C. Burleigh. 1972. Role of granulocyte collagenase in collagen degradation. *Amer. J. Pathol.* **68**:565.

Lichtenstein, N. and J. S. Fruton. 1960. Studies on beef spleen cathepsin A. *Proc. Nat. Acad. Sci.* **46**:787.

Logunov, A. I. and V. N. Orekhovich. 1972. Isolation and some properties of cathepsin A from bovine spleen. *Biochem. Biophys. Res. Comm.* **46**:1161.

Mathews, M. B. 1971. Comparative biochemistry of chondroitin-sulphate proteins of cartilage and notochord. *Biochem. J.* **125**:37.

McDonald, J. K., P. X. Callahan and S. Ellis. 1971. Polypeptide degradation by dipeptidyl aminopeptidase I (cathepsin C) and related peptidases. In *Tissue Proteinases* (ed. A. J. Barrett and J. T. Dingle), p. 69. North-Holland, Amsterdam.

————. 1972. Preparation and specificity of dipeptidyl aminopeptidase I. In *Methods in Enzymology* (ed. C. H. W. Hirs and S. N. Timasheff), vol. 25, p. 272. Academic Press, New York.

McDonald, J. K., B. B. Zietman and S. Ellis. 1970. Leucine naphthylamide: An inappropriate substrate for the histochemical detection of cathepsins B and B1. *Nature* **225**:1048.

Morrison, R. I. G., A. J. Barrett, J. T. Dingle and D. Prior. 1973. Cathepsins B1 and D. Action on human cartilage proteoglycans. *Biochim. Biophys. Acta* **302**:411.

Nakashima, K. and K. Ogino. 1974. Regulation of rabbit liver fructose-1, 6-diphosphatase. II. Modification by lysosomal cathepsin B1 from the same cell. *J. Biochem.* **75**:355.

Naughton, M. A. and F. Sanger. 1961. Purification and specificity of pancreatic elastase. *Biochem. J.* **78**:156.

Ninjoor, V., S. L. Taylor and A. L. Tappel. 1974. Purification and characterization of rat liver lysosomal cathepsin B2. *Biochim. Biophys. Acta* **370**:308.

Ohlsson, K. and I. Olsson. 1974. The neutral proteases of human granulocytes. Isolation and partial characterization of granulocyte elastases. *Eur. J. Biochem.* **42**:519.

Onoyama, K., M. Hara, K. Tanaka and T. Omae. 1973. Renin content in subcellular fractions of normal rat kidney extract. *Japan. Heart J.* **14**:440.

Otto, K. 1971. Cathepsins B1 and B2. In *Tissue Proteinases* (ed. A. J. Barrett and J. T. Dingle), p. 1. North-Holland, Amsterdam.

Overturf, M., M. Leonard and W. M. Kirkendall. 1974. Purification of human renin and inhibition of its activity by pepstatin. *Biochem. Pharmacol.* **23**:671.

Poole, A. R., J. T. Dingle and A. J. Barrett. 1972. The immunocytochemical demonstration of cathepsin D. *J. Histochem. Cytochem.* **20**:261.

Poole, A. R., R. M. Hembry and J. T. Dingle. 1973. Extracellular localization of cathepsin D in ossifying cartilage. *Calc. Tis. Res.* **12**:313.

Powers, J. C. and P. M. Tuhy. 1973. Active-site specific inhibitors of elastase. *Biochemistry* **12**:4767.

Rindler, R. and H. Braunsteiner. 1973. Soluble proteins from human leukocyte granules. Esterase activity of cationic proteins. *Blut* **27:**26.

Rindler, R., H. Hörtnagl, F. Schmalzl and H. Braunsteiner. 1973. Hydrolysis of a chymotrypsin substrate and of naphthol AS-D cloroacetate by human leukocyte granules. *Blut* **26:**239.

Sampath-Kumar, P. S. and J. S. Fruton. 1974. Studies on the extended active sites of acid proteinases. *Proc. Nat. Acad. Sci.* **71:**1070.

Sapolsky, A. I. and J. F. Woessner. 1972. Multiple forms of cathepsin D from bovine uterus. *J. Biol. Chem.* **247:**2069.

Schumacher, G. F. B. and W. B. Schill. 1972. Radial diffusion in gel for micro determination of enzymes. II. Plasminogen activator, elastase and nonspecific proteases. *Anal. Biochem.* **48:**9.

Seifter, S. and E. Harper. 1970. Collagenases. In *Methods in Enzymology* (ed. G. E. Perlmann and L. Lorand), vol. 19, p. 613. Academic Press, New York.

Sen, L. C. and J. R. Whitaker. 1973. Some properties of a ficin-papain inhibitor from avian egg white. *Arch. Biochem. Biophys.* **158:**623.

Smith, G. D., M. A. Murray, L. W. Nichol and V. M. Trikojus. 1969. Thyroid acid proteinase. Properties and inactivation by diazoacetyl-norleucine methyl ester. *Biochim. Biophys. Acta* **171:**288.

Starkey, P. M. and A. J. Barrett. 1973. Human cathepsin B1. Inhibition by α_2-macroglobulin and other serum proteins. *Biochem. J.* **131:**823.

Suda, H., T. Aoyagi, T. Takeuchi and H. Umezawa. 1973. A thermolysin inhibitor produced by actinomycetes: Phosphoramidon. *J. Antibiotics* **26:**621.

Sylvén, B. and O. Snellman. 1974. The immunofluorescent demonstration of cathepsin B1 in tissue sections. *Histochemie* **38:**35.

Taylor, S. L. and A. L. Tappel. 1974a. Identification and separation of lysosomal carboxypeptidases. *Biochim. Biophys. Acta* **341:**99.

———. 1974b. Characterization of rat liver lysosomal cathepsin A1. *Biochim. Biophys. Acta* **341:**112.

Vensel, W. H., J. Komender and E. A. Barnard. 1971. Nonpancreatic proteases of the chymotrypsin family. II. Two proteases from a mouse mast cell tumour. *Biochim. Biophys. Acta* **250:**395.

Werb, Z. and M. C. Burleigh. 1974. A specific collagenase from rabbit fibroblasts in monolayer culture. *Biochem. J.* **137:**373.

Werb, Z., M. C. Burleigh, A. J. Barrett and P. M. Starkey. 1974. The interaction of α_2-macroglobulin with proteinases: Binding and inhibition of mammalian collagenases and other metal proteinases. *Biochem. J.* **139:**359.

Willstätter, R. and E. Bamann. 1928. Uber die proteasen der Magenschleimhaut. Erste Abhandlung über die ewzyme der lukocyten. *Hoppe-Seyler's. Z. Physiol. Chem.* **180:**127.

Woessner, J. F. 1973. Cartilage cathepsin D and its action on matrix components. *Fed. Proc.* **32:**1485.

APPENDIX

Some Remarks Concerning the Naming and Description of Tissue Proteinases

The following remarks represent a summary of the conclusions of discussions on the difficult, but important, subject of nomenclature of tissue proteinases that occurred during the Friedrichroda meeting of May, 1973.[1]

The Naming of a New Tissue Proteinase

The word "cathepsin" is generally understood to apply to enzymes of animal tissues that hydrolyze the peptide bonds of proteins or polypeptides. In spite of some confusion surrounding the term in the earlier literature, we would recommend that "cathepsin" continue to be used in this context. Trivial names for individual enzymes can be constructed on the alphabetical system already established for cathepsins A–E, perhaps with numbered subgroups, as have been used for cathepsins B1 and B2. It is anticipated that enzymes will often have to be named in a preliminary way before detailed knowledge is available concerning their specificity. When such detailed information becomes available, it will be natural to retain the name "cathepsin" for an endopeptidase, but consideration should be given to adopting a more systematic name for an exopeptidase.

We do not feel that there is any need to restrict the use of the term "cathepsin" to tissue proteinases with acidic pH optima. We would wish to discourage the naming of cathepsins in terms of particular protein substrates, unless a very high degree of specificity for one protein has been proven.

Characterization of Tissue Proteinases

We have been very impressed by the value of the classification system proposed by Hartley (1960) for proteolytic enzymes, as applied to tissue proteinases. Experiments with the selective inhibitors of the various catalytic classes of proteinases often allow one to assign an enzyme to a class. Briefly, the carboxyl proteinases, previously called acid proteinases, are inactivated by various diazomethylketo compounds in the presence of Cu^{++} and, at least in many cases, are powerfully inhibited by pepstatin; they are resistant to the characteristic inhibitors of the other groups of proteinases. The thiol proteinases are readily inactivated by such compounds as *p*-chloromercuribenzoate, iodoacetate and *N*-ethylmaleimide. The metal proteinases are inactivated by EDTA, 1, 10-phenanthroline and usually by thiol compounds. The most general reagent for the serine proteinases is diisopropylphosphorofluoridate, but phenylmethane sulfonyl fluoride and soybean trypsin inhibitor also are often useful.

The assignment of a newly discovered proteinase to one of these groups, or the clearly stated conclusion that it seems to belong to some new class, allows it to be placed provisionally in one of the EN 1972 sub-subclasses

[1] These remarks are produced by permission of Professor H. Hanson, having been compiled at his suggestion during the International Meeting on Intracellular Protein Catabolism at Freidrichroda, D.D.R.

3.4.21–24 or 99 and is usually of greater value than the determination of pH optimum, for example.

Criteria by Which an Enzyme May Be Considered Different from One Previously Described

It is obvious that differences in catalytic activity are useful in the identification of enzymes, but the value of physicochemical properties, also, is illustrated by the fact that the clearest distinction between cathepsin D and cathepsin E is still on the basis of molecular weight and charge.

If two proteinases are found to be similarly affected by a range of inhibitors, and not to differ widely in pH optimum, it is often necessary to examine their specificity in some detail. The value of very low molecular weight substrates now seems more limited than at one time. Important results have been obtained with penta- and hexapeptides, and the polypeptide of preference is undoubtedly the oxidized B chain of insulin.

By use of gel filtration and isoelectric focusing, it is usually possible to estimate the molecular weight and isoelectric point of an enzyme without having purified it. Large differences in one or both of these parameters may distinguish a new enzyme from one already known, whereas small differences, especially in isoelectric point, may simply indicate species differences or multiple forms of a single enzyme.

The Role of the Commission on Biochemical Nomenclature

The suggestions made in the first section above refer to the selection of trivial names for tissue proteinases. Such names may often be adopted at a stage when little detailed information about the enzyme has been obtained. When fuller information on purification and properties has been obtained, and presumably published, the data might be sent to a member of the Expert Committee on Enzymes of the IUPAC/IUB Commission on Biochemical Nomenclature; the Commission will then decide whether to allocate a number and perhaps a systematic name to the new enzyme.

Official publications giving valuable guidance on nomenclature for enzymes, substrates and inhibitors include:

1. The nomenclature of multiple forms of enzymes. Recommendations (1971). *Biochim. Biophys. Acta* (1972) **258**:1–3.
2. Symbols for amino acid derivatives and peptides. Recommendations (1971). *Biochim. Biophys. Acta* (1972) **263**:205–212.
3. Enzyme nomenclature. In *Comprehensive Biochemistry* (ed. M. Florkin and E. H. Stotz), vol. 13, 3rd ed. (1973). Elsevier, Amsterdam.

May, 1973

A. J. Barrett
H. Keilová
H. Kirschke
K. Otto
G. Siebert
V. M. Stepanov
V. Turk

The Role of Proteases
in Macrophage Physiology

Zanvil A. Cohn

Department of Cellular Physiology and Immunology
The Rockefeller University, New York, New York 10021

Macrophages represent the most mature element of a system of cells known as the "mononuclear phagocytes." Along with the neutrophilic leukocytes, they constitute the most important endocytic cells of all multicellular animals. Arising from bone marrow precursors, they enter the circulation as *monocytes* and are widely distributed throughout the body. After emigrating between the endothelial cells of postcapillary venules, monocytes reach the extravascular space and take on the properties of the larger and more active *macrophages* or tissue histiocytes. Large numbers are found on the walls of liver and spleen sinusoids, as well as being present in other organs and all areas of connective tissue. In these loci, they are involved in a large number of physiological and pathological processes, mainly focused on their capacity to recognize and interiorize both soluble and particulate environmental molecules and to degrade them within their vacuolar system (Steinman and Cohn 1974; Gordon and Cohn 1973).

In this report, I will review our knowledge of the role of lysosomal hydrolases, with particular reference to proteases, in the degradation of exogenous and endogenous substrates. Although these events are occurring continually in the environs of the vertebrate host, I will restrict my remarks to tissue culture situations in which the mouse peritoneal macrophage expresses these functions under a more controlled environment.

Endocytosis and the Uptake of Exogenous Substrates

Macrophages interiorize a wide spectrum of soluble and particulate molecules ranging from simple polypeptides to intact microbial and mammalian cells. The rate of uptake and therefore the rate at which substrate flows into the vacuolar system varies considerably. Soluble proteins such as horseradish peroxidase (HRP) (Steinman and Cohn 1972a), which do not bind appreciably to the plasma membrane, are interiorized at a rate proportional to their extracellular concentration. Cultivated macrophages take up $0.0025\%/10^6$ cells/hour over a wide range of concentrations, and this material is sequestered in endocytic vesicles. Although a seemingly small amount, it should be realized that when macrophages are cultivated in 50% calf serum, they are exposed to ~ 3.0 gram percent of protein (primarily albumin) and at a minimum would be expected to interiorize 3% of their total protein per hour. Comparable values have been obtained for human serum albumin and hemoglobin (Ehrenreich and Cohn 1967, 1968). The rate of peroxidase uptake by mammalian fibroblasts is approximately 10–20% that of the macrophage on a cell protein basis (R. Steinman, J. Silver and Z. A. Cohn, unpubl.), but may be increased 2–3-fold under conditions of confluent growth. Nonadsorptive pinocytosis of this nature is probably reflected by many if not all cells in culture.

The rate at which soluble and particulate substances are pinocytized or phagocytized may be markedly increased under conditions in which an initial binding to the plasmalemma takes place. For example, equivalence complexes of HRP–anti-HRP, which bind to the macrophage surface by means of the Fc receptor, are interiorized at a rate proportional to the external load. In this instance, the rate of uptake is 4000-fold greater than with soluble HRP. Fibroblasts which lack the Fc receptor exclude such immune complexes. Finally, cationic proteins and polypeptides which bind to the plasma membrane of macrophages (Seljelid, Silverstein and Cohn 1973) are interiorized at a rate which may be 100-fold greater than with nonbinding proteins. In addition, once bound to the membrane, they may interact with macromolecular anionic moieties, and the resulting complex can subsequently be taken up.

It is apparent from the above that large quantities of protein substrates may flow through the vacuolar system of macrophages. Under in vivo conditions or following the ingestion of intact cells, the macrophage is presented simultaneously with a complex array of degradable, and on occasion nondegradable, substrates. In addition, because of the long life of the cell and the ability to maintain endocytic activity, the turnover of substrates by individual cells may be appreciable in terms of body economy.

Endogenous Substrates

There are at least two mechanisms by which endogenous substrates of macrophage origin are exposed to acid hydrolases and subsequently degraded. The first relates to the process of autophagy, in which organelles or soluble components of the cytosol may be enclosed within the vacuolar apparatus. Evidence with macrophage cultures suggests that this occurs by the invagination of the limiting membrane of endocytic vacuoles or secondary lysosomes. This is followed by membrane fusion and the formation of a multivesicular body (Cohn 1970). Certain drugs, such as chloroquine, stimulate this process, yielding large autophagic vacuoles containing recognizable cytoplasmic organelles (Fedorko, Hirsch and Cohn 1968a,b). Under physiological conditions, an analogous situation at a "micro" level leads to the usual multivesicular body (Hirsch, Fedorko and Cohn 1968). Shortly after the formation of such structures, the intravacuolar membranes disappear at the ultrastructural level and are presumably degraded. Technical difficulties in finding appropriately labeled markers and models have so far inhibited a more quantitative analysis of the importance of autophagy in the turnover of cytoplasmic constituents. It nevertheless seems reasonable that endogenous components are degraded as readily as exogenous substrates.

The second form of endogenous substrate represents the external surface of plasma membrane, interiorized by endocytosis, and forming the inner surface of endocytic vacuole. Evidence for the modification of externally disposed membrane polypeptides following endocytosis will be covered in a subsequent section.

The Nature and Control of Lysosomal Proteases

The acid proteases and peptidases of lysosomes have been reviewed by Barrett (Barrett and Dingle 1971), and macrophages are particularly rich in these enzymes (Cohn 1968). Within a series of phagocytic cells of the rabbit, the relative specific activity of the major protease, cathepsin D, is 60-fold higher in induced alveolar macrophages than in the polymorphonuclear leukocyte. This enzyme is found in a class of particles consistent with dense lysosomes and in the company of a variety of other acid hydrolases (Cohn and Wiener 1963). None of the macrophage enzymes have been obtained in a purified state, although considerable enrichment of an acid protease has been obtained by Dannenberg (Dannenberg and Smith 1955). More recent evidence indicates that the cathepsin D of macrophages is strongly inhibited by the pentapeptide "pepstatin" (McAdoo et al. 1973).

The levels of macrophage acid proteinases, as well as other hydrolases, fluctuate widely with the physiological activity of the cell. These changes are evident both in vivo (Cohn and Wiener 1963; Dannenberg and Bennett 1964) and in the confines of a tissue culture environment (Cohn and Benson 1965a, b). Under conditions in which the cell is imbibing large amounts of medium by means of pinocytosis, the resting level of cathepsin D may increase 5–6-fold. This accumulation depends upon continuing protein synthesis, although the rate of degradation of this enzyme within macrophages has not been quantitatively evaluated. The process is reversible, in that reducing the rate of pinocytosis leads to a reduction in cathepsin and other intracellular hydrolases (Cohn and Benson 1965c). Several such cycles can be accomplished in culture, and the phagocytosis of particulates leads to similar events. The factors regulating the level of proteinase are still unclear, but seem to be linked to the uptake of substrate by endocytosis (Axline and Cohn 1970). Digestible substrates such as simple proteins and L-amino acid homopolymers lead to increased levels of many lysosomal enzymes, including those not involved in the degradation of the substrate. In contrast, nondegradable amino acid homopolymers prepared with D-amino acids and nondigestible particles such as polystyrene latex do not enhance the levels of acid hydrolases. These data suggest that the control of lysosomal enzyme accumulation occurs at a post-digestion stage and appears unrelated to the interiorization of membrane per se.

Intralysosomal Digestion

Ultrastructural and cytochemical evidence strongly suggests that all macromolecular exogenous substrates remain within the membrane systems of the vacuolar apparatus. The fusion of the endocytic vacuole with either primary lysosomes of Golgi origin (Cohn, Fedorko and Hirsch 1966; Cohn and Fedorko 1969) or preexisting secondary lysosomes leads to the mixing of hydrolases with substrates and initiates the process of intracellular digestion.

The rate and extent of intralysosomal hydrolysis of proteins has been studied with viable macrophages. These experiments employ homogeneous populations of cells pulsed with various proteins, washed, and then cultivated in their absence. Such proteins have either been radiolabeled or their enzymatic activity followed or both parameters have been employed with enzymes such as horseradish peroxidase. Following a pulse, it is assumed that the majority of the exogenous molecule has rapidly entered the lysosomal compartment. Prior publications have commented on the fate of albumin (Ehrenreich and Cohn 1967), hemoglobin (Ehrenreich and Cohn 1969) and peroxidase (Steinman and Cohn 1972a,b). When care is taken to remove all the macromolecular tracer from the cell surface and culture dish, it is found that proteins are degraded intracellularly by means of a temperature-sensitive, exponential process. Inhibitors which block glycolysis and oxidative metabolism have had no apparent influence on the rate of digestion. The rate at which proteins are degraded to acid-soluble components varies with the substrate. Figure 1 illustrates some representative data on this point. The intracellular half-lives may vary from 5 $\geq$ 40 hours. At this point in time, the enzymatic mechanisms underlying these differences are not known.

In each instance thus far studied, the only acid-soluble product which accumulates in the external medium is the appropriate radiolabeled amino acid, i.e., moniodotyrosine or [^{3}H]leucine. Polypeptides composed of the unnatural

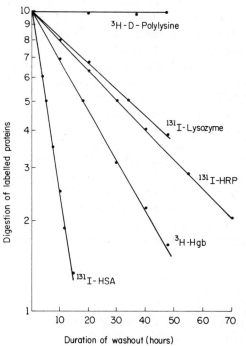

Figure 1

The digestion of radiolabeled polypeptides within the lysosomes of cultivated mouse macrophages.

D-isomers are not appreciably degraded. In addition, the examination of the state of intracellular label, during the washout period, has revealed only small amounts (<10%) of intermediate products. In the case of peroxidase, the half-life of intracellular enzymatic activity is as short as 7–8 hours, whereas more complete hydrolysis to amino acids requires some 30 hours. The digestion products of iodinated proteins, i.e., amino acids, are not appreciably influenced by the presence of serum peptidases, although a certain degree of deiodination does occur.

More complex protein substrates contained in the bacterial (Cohn 1963) and viral particles (Silverstein and Dales 1968) are degraded in a similar fashion.

Lysosomal Permeability and the Escape of Digestion Products

One of the questions left unanswered by the prior experiments is related to the size of the peptides capable of crossing the lysosomal membrane and entering the cytosol. Although amino acids accounted for the vast majority of acid-soluble digestion products, it seemed possible that nonlysosomal, cytosol-localized peptidases might play a role. To examine this question in more detail, an examination of the permeability of lysosomes in living macrophages was developed (Cohn and Ehrenreich 1969; Ehrenreich and Cohn 1969). The results of this method are in general agreement with studies on human fibroblasts (Shulman and Bradley 1970) and other methods employing isolated lysosomes (Coffey and de Duve 1968; Lloyd 1971; Goldman and Naider 1974).

The results of such experiments performed with D-peptides are summarized in Figure 2 and are based upon a morphological method in which the storage of a nonpermeable peptide within the lysosome leads to swelling of the organelle. In general, the lysosomal membrane is impermeable to all the tripeptides examined, whereas most dipeptides and amino acids are able to permeate through the membrane. Natural peptides containing L-amino acids are rapidly hydrolyzed and cannot be screened by this method. The highly charged (D-Glu)$_2$ is not permeable. Although permeability of peptides probably reflects parameters of the molecule other than molecular weight (hydrophilicity, charge, etc.), most peptides with a size greater than 240 daltons do not easily penetrate the membrane. Similar studies have been conducted with carbohydrates (Cohn and Ehrenreich 1969; Lloyd 1969).

In concert, these data suggest that only relatively small products of intralysosomal digestion can escape the confines of this organelle and interact with the cytosol. The mechanism(s) of transport has not been elucidated nor have the effects of dipeptides on the protein synthetic apparatus been examined.

Role of Concanavalin A in Modifying the Vacuolar Apparatus

One of the basic attributes of the vacuolar system of macrophages and other cells is the ability of various membrane-bound organelles to mix their contents by means of membrane fusions. This primarily involves membrane

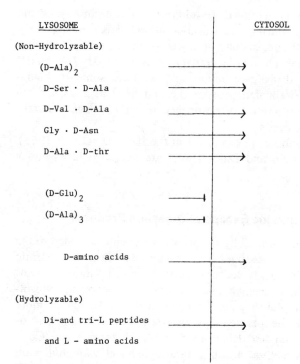

Figure 2
The permeability of the secondary lysosomes of
mouse peritoneal macrophages to a selected group of
peptides and amino acids.

derived from the plasmalemma and Golgi apparatus, and the underlying
mechanisms have remained elusive. Recent studies by Dr. Paul Edelson of
this laboratory have uncovered a pronounced effect of the plant lectin concanavalin A (ConA) on inhibiting the fusion of endocytic vacuoles with
lysosomes and the subsequent digestion of exogenous proteins (Edelson and
Cohn 1974a,b).

The morphological and cytochemical effects of ConA are schematized in
Figure 3. The initial interaction of ConA on the macrophage membrane is its
binding presumably to exteriorly disposed glycoproteins. This is followed by
a stimulation of pinocytic activity, which represents a 3–4-fold increase over
resting levels. Subsequently, large phase and electron lucent vacuoles accumulate in the cytoplasm and persist in this form for long periods of time. The
inner surface of these vacuoles is lined with ConA, and the membrane of the
vacuoles is derived from the plasma membrane by endocytosis. In contrast to
the control situation diagrammed on the left side of Figure 3, there is a
failure of fusion between the ConA vacuoles and both preexisting secondary
lysosomes and primary lysosomes. This is indicated by the failure of acid
phosphatase (an electron microscopic lysosomal marker) to enter the vacuole
and in the transfer of a number of enzymatic and electron dense markers
from preloaded secondary lysosomes.

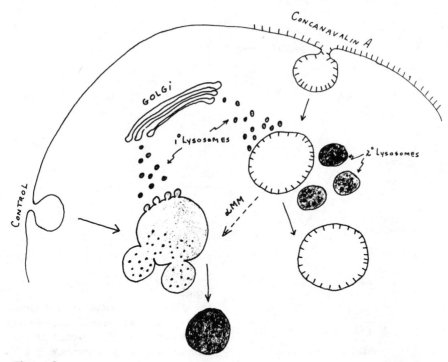

Figure 3
Schematic representation of the vacuolar system of macrophages. (*Left*) The fusion of primary (1°) and secondary (2°) lysosomes with endocytic vacuoles under control conditions. (*Right*) The formation of a ConA-coated endocytic vacuole and the lack of fusion with lysosomes. The addition of α-methyl mannoside promptly yields fusion with lysosomes.

The addition of 100 mM mannose to ConA-treated macrophages leads to the reversal of the above structural defects. This includes the disappearance of the large vacuoles, the dissociation of ConA from the membrane, and the fusion of secondary lysosomes with the vacuoles. These changes do not occur following the addition of galactose to the medium.

As might be expected, ConA treatment of macrophages also leads to the inhibition of the digestion of exogenous proteins. An example employing radioiodinated albumin is given in Figure 4. The uptake of albumin by control cells is followed by rapid digestion and an intracellular half-life of this molecule of approximately 5 hours. In contrast, albumin digestion in ConA-treated cells is slowed, and the half-life is increased to more than 15 hours. The addition of α-MM to ConA-treated, albumin-containing cells leads to a prompt increase in albumin digestion to control rates. Galactose has no influence on the ConA-treated cell nor does α-MM modify the control state.

It therefore appears that ConA bound to the internal surface of the endocytic vesicle modifies the fusibility of the organelle with lysosomes. The mechanism is currently under investigation with ConA and other lectins, and possibilities include the modification of transmembraneous proteins and the fluidity of the membrane.

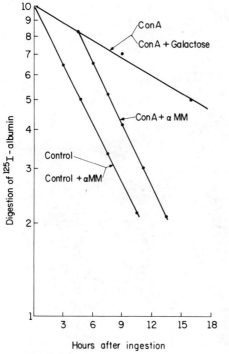

Figure 4
The digestion of radioiodinated albumin by control and ConA-treated macrophages and the reversal of the effects with α-methyl mannoside.

Modification of the Plasma Membrane following Endocytosis

Observations over the past few years have suggested that components of the plasma membrane may be altered in macrophages following the ingestion of large numbers of inert polystyrene latex particles (Werb and Cohn 1972). During the course of this work, the activity of a plasma membrane enzyme, 5′-nucleotidase, was followed temporally in isolated phagolysosomes. The vast majority of this enzyme is present on the plasma membrane of the macrophage and shortly after endocytosis can be found in the phagolysosomal membrane. Thereafter, the activity of this enzyme in the phagolysosomal fraction declines with a half-life of ~3 hours, and by 5–6 hours, activity is not measurable in this compartment (Fig. 5). One interpretation of this data is that nucleotidase, an "ectoenzyme," is being inactivated in the presence of the lysosomal hydrolases following the conversion of the endocytic vesicle to a phagolysosome.

More recently, similar experiments have been conducted by Dr. Ann Hubbard in mouse L-cell fibroblasts (Hubbard and Cohn 1975a,b). Employing another plasma membrane marker, alkaline phosphodiesterase I, a similar fate of this enzyme occurred (Fig. 4). Since we were assaying enzyme activity alone, many interpretations of these data were possible.

Other studies with L cells suggested, however, that extensive degradation of plasma membrane polypeptides takes place following endocytosis. This analysis by Dr. Hubbard made use of the technique of surface iodination (Hubbard and Cohn 1972), employing ^{125}I, lactoperoxidase and glucose oxidase and

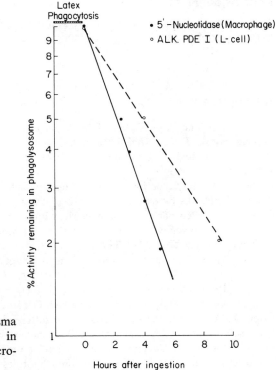

Figure 5
The inactivation of plasma membrane marker enzymes in the phagolysosomes of macrophages and L cells.

glucose as a peroxide-generating system. Using L cells grown in suspension, it was found that these fibroblasts could be effectively labeled without loss of viability. Analysis of the incorporated label showed it to be present in approximately 15 endogenous polypeptides exteriorly disposed on the plasma membrane.

Following ingestion of polystyrene latex, as much as 30% of the plasma membrane could be interiorized surrounding the particle and isolated as phagolysosomes. The interiorized membrane initially had all of the components of the original plasma membrane. With the passage of time, however, >70% of the acid-precipitable labeled proteins were lost with a half-life of 1–2 hours. This loss of label could be correlated with the appearance of acid-soluble radioactivity in the medium as monoiodotyrosine. SDS slab gels revealed that the majority of the labeled membrane polypeptides were cleaved and indicated that extensive proteolysis had occurred following membrane interiorization. It was of interest, however, that the slowly degraded label in the phagolysosomal compartment could be accounted for by long-lived polypeptides of 85,000 and 8–15,000 daltons. In these studies, fusion with preexisting lysosomes had taken place, and these organelles may be supplying the proteases that are attacking the inner membrane surface of the phagolysosome. At the present time, we do not have information concerning the cytosol face of the membrane, nor whether such modifications influence the transport properties of the phagolysosome.

Acknowledgment

This work was supported in part by National Institutes of Health Grants AI 07012 and AI 01831.

REFERENCES

Axline, S. G. and Z. A. Cohn. 1970. *In vitro* induction of lysosomal enzymes by phagocytosis. *J. Exp. Med.* **131:**1239.

Barrett, A. J. and J. T. Dingle, eds. 1971. *Tissue Proteinases.* North-Holland, London.

Coffey, J. W. and C. de Duve. 1968. Digestive activity of lysosomes. I. The digestion of proteins by extracts of rat liver lysosomes. *J. Biol. Chem.* **243:**3255.

Cohn, Z. A. 1963. The fate of bacteria within phagocytic cells. I. The degradation of isotopically labeled bacteria by polymorphonuclear leucocytes and macrophages. *J. Exp. Med.* **117:**27.

———. 1968. The differentiation of macrophages. In *Differentiation and Immunology* (ed. K. B. Warren), vol. 7, p. 101. Academic Press, New York.

———. 1970. Lysosomes in mononuclear phagocytes. In *Mononuclear Phagocytes* (ed. R. van Furth), p. 50. Blackwell, Oxford.

Cohn, Z. A. and B. Benson. 1965a. The differentiation of mononuclear phagocytes. Morphology, cytochemistry and biochemistry. *J. Exp. Med.* **121:**153.

———. 1965b. The *in vitro* differentiation of mononuclear phagocytes. The influence of serum on granule formation, hydrolase production and pinocytosis. *J. Exp. Med.* **121:**835.

———. 1965c. The *in vitro* differentiation of mononuclear phagocytes. III. Reversibility of granule and hydrolytic enzyme formation and the turnover of granule constituents. *J. Exp. Med.* **122:**455.

Cohn, Z. A. and B. A. Ehrenreich. 1969. The uptake, storage and intracellular hydrolysis of carbohydrates by macrophages. *J. Exp. Med.* **129:**201.

Cohn, Z. A. and M. E. Fedorko. 1969. The formation and fate of lysosomes. In *Lysosomes in Biology and Pathology* (ed. J. T. Dingle and H. B. Fell), vol. 1, p. 43. North-Holland, London.

Cohn, Z. A. and E. Wiener. 1963. The particulate hydrolases of macrophages. I. Comparative enzymology, isolation and properties. *J. Exp. Med.* **118:**991.

Cohn, Z. A., M. E. Fedorko and J. G. Hirsch. 1966. The formation of macrophage lysosomes. *J. Exp. Med.* **123:**757.

Dannenberg, A. M., Jr. and W. E. Bennet. 1964. Hydrolytic enzymes of rabbit mononuclear exudate cells. *J. Cell Biol.* **21:**1.

Dannenberg, A. M., Jr. and E. L. Smith. 1955. Proteolytic enzymes of lung. *J. Biol. Chem.* **215:**45.

Edelson, P. J. and Z. A. Cohn. 1974a. Effects of concanavalin A on mouse peritoneal macrophages. I. Stimulation of endocytic activity and inhibition of phagolysosome formation. *J. Exp. Med.* **140:**1364.

———. 1974b. Effects of concanavalin A on mouse peritoneal macrophages. II. Metabolism of endocytised proteins, and reversibility of the effects of mannose. *J. Exp. Med.* **140:**1387.

Ehrenreich, B. A. and Z. A. Cohn. 1967. Fate of hemoglobin pinocytized by macrophages *in vitro. J. Cell Biol.* **38:**244.

———. 1968. The uptake and digestion of iodinated human serum albumin by macrophages *in vitro. J. Exp. Med.* **126:**941.

————. 1969. The fate of peptides pinocytosed by macrophages *in vitro*. *J. Exp. Med.* **129:**227.

Fedorko, M. E., J. G. Hirsch and Z. A. Cohn. 1968a. Autophagic vacuoles produced *in vitro*. I. Studies on cultured macrophages exposed to chloroquine. *J. Cell Biol.* **38:**392.

————. 1968b. Studies on the mechanism of formation of autophagic vacuoles produced by chloroquine. *J. Cell Biol.* **38:**392.

Goldman, R. and F. Naider. 1974. Permeation and stereospecificity of hydrolysis of peptide esters within intact lysosomes *in vitro*. *Biochim. Biophys. Acta* **338:** 224.

Gordon, S. and Z. A. Cohn. 1973. The macrophage. *Int. Rev. Cytol.* **36:**171.

Hirsch, J. G., M. E. Fedorko and Z. A. Cohn. 1968. Vesicle fusion and formation at the surface of pinocytic vesicles in macrophages. *J. Cell Biol.* **38:**629.

Hubbard, A. L. and Z. A. Cohn. 1972. The enzymatic iodination of the red cell membrane. *J. Cell Biol.* **55:**390.

————. 1975a. Externally-disposed plasma membrane proteins. I. Enzymatic iodination of mouse L cells. *J. Cell Biol.* **64:**438.

————. 1975b. Externally disposed plasma membrane proteins. II. Metabolic fate of iodinated polypeptides of mouse L cells. *J. Cell Biol.* **64:**461.

Lloyd, J. B. 1969. Studies on the permeability of rat liver lysosomes to carbohydrates. *Biochem. J.* **115:**703.

————. 1971. A study of permeability of lysosomes to amino acids and small peptides. *Biochem. J.* **121:**245.

McAdoo, M. H., A. M. Dannenberg, C. J. Hayes, S. P. James and J. H. Sanner. 1973. Inhibition of the cathepsin D-type proteinase of macrophages by pepstatin, a specific pepsin inhibitor, and other substances. *Infect. Immunol.* **7:**655.

Seljelid, R., S. Silverstein and Z. A. Cohn. 1973. The effect of poly-L-lysine on the uptake of reovirus double-stranded RNA in macrophages. *J. Cell Biol.* **57:** 484.

Shulman, J. D. and K. H. Bradley. 1970. The metabolism of amino acids, peptides and disulfides in lysosomes of fibroblasts cultured from normal individuals and those with cystinosis. *J. Exp. Med.* **132:**1090.

Silverstein, S. C. and S. Dales. 1968. The penetration of reovirus RNA and initiation of its genetic function in L-strain fibroblasts. *J. Cell Biol.* **36:**197.

Steinman, R. and Z. A. Cohn. 1972a. The interaction of soluble horseradish peroxidase with mouse peritoneal macrophages *in vitro*. *J. Cell Biol.* **55:**186.

————. 1972b. The interaction of particulate horseradish peroxidase (HRP)-anti-HRP immune complexes with mouse peritoneal macrophages *in vitro*. *J. Cell Biol.* **55:**616.

————. 1974. The metabolism and physiology of the mononuclear phagocytes. In *The Inflammatory Process* (ed. B. W. Zweifach, L. Grant and R. T. McClusky), vol. I, p. 450. Academic Press, New York.

Werb, Z. and Z. A. Cohn. 1972. Plasma membrane synthesis in the macrophage following phagocytosis of polystyrene latex particles. *J. Biol. Chem.* **247:**2439.

Secretion and Regulation of Plasminogen Activator in Macrophages

Jay C. Unkeless* and Saimon Gordon

The Rockefeller University, New York, New York 10021

The macrophage plays an important role in host defense against infection, in the course of which it may become activated, display increased metabolic and phagocytic activity and gain an enhanced capacity to kill microorganisms and possibly tumor cells (Nelson 1972; Karnovsky et al. 1970). However, the biochemical changes in "activated" macrophages compared to normal peritoneal macrophages are unknown.

Following transformation by oncogenic viruses or chemicals, avian and mammalian fibroblasts release an enzyme that functions as a plasminogen activator (Unkeless et al. 1973; Ossowski et al. 1973) and is a diisopropyl-fluorophosphate (DFP)-sensitive protease most active at neutral to slightly alkaline pH (Unkeless, Gordon and Reich 1974). Macrophages are a rich source of acid hydrolases (Cohn and Benson 1965), which are thought to play a role in intracellular digestion, but neutral proteases from macrophages, capable of acting extracellularly at neutral pH, have not been well characterized. We have found that the ability of macrophages to solubilize a thin film of radioactive fibrin in the presence of plasminogen can be stimulated at least 100-fold following injection of an intraperitoneal irritant, thioglycollate medium (Unkeless, Gordon and Reich 1974). The enzyme responsible for the fibrinolytic activity of thioglycollate-stimulated macrophages is a plasminogen activator similar to that released by mouse fibroblasts transformed by the mouse sarcoma virus.

We have also shown that macrophages harvested from mice after intraperitoneal injection of endotoxin contain and release little plasminogen activator, but such stimulated cells can be triggered to secrete high levels of the enzyme by subsequent phagocytosis of latex or bacteria (Gordon, Unkeless and Cohn 1974). Fibrinolytic activity and secretion of the plasminogen activator were prolonged after ingestion of a nondigestible particle such as latex, but were

* Present address: Center for Cancer Research, Department of Biology, Massachusetts Institute of Technology, Cambridge, Massachusetts 02139.

transient following ingestion of easily digested particles such as *Micrococcus lysodeikticus.*

EXPERIMENTAL PROCEDURES

Reagents were obtained as follows: Brewers thioglycollate medium (Difco Laboratories, Detroit, Mich.); lactalbumin hydrolysate (Nutritional Biochemicals, Cleveland, Ohio); endotoxin and lipid A, provided by Dr. O. Westphal (Max Planck Institute for Immunobiology, Freiburg, Germany). The endotoxin, batch AE 3990-S$_3$, was purified from *Salmonella abortus equi* and was a "high molecular weight" preparation; bacillus Calmette-Guérin (BCG) (Aronson), lot 581, a water-washed, lyophylized preparation, was suspended in sterile saline, 1 mg/ml, sonicated, and stored at $-20°$C; polystyrene latex particles, 1.01 μ (Dow Diagnostics, Indianapolis); *M. lysodeikticus,* spray dried, and soybean trypsin inhibitor (STI), fraction VI (Miles Chemical, Kankakee, Ill.).

Cell Cultures

Female NCS (Rockefeller) strain mice weighing 25–30 g were used. Macrophages were harvested and cultured by standard procedures (Gordon, Todd and Cohn 1974) from control mice or mice injected intraperitoneally 4 days previously with 0.75 ml thioglycollate medium or 30 μg endotoxin. The cell yields and conditions for cultivation have been described previously (Gordon, Unkeless and Cohn 1974). Human monocytes and polymorphonuclear leukocytes (PMN) were purified as described (Gordon, Todd and Cohn 1974); the monocyte populations were more than 98% pure and contained 0–2% PMN.

Fibrinolytic Assays

Cells were plated in sterile, 35-mm, ^{125}I-labeled fibrin-coated dishes containing 10 μg fibrin/cm^2 prepared as described previously (Quigley, Ossowski and Reich 1974) and cultivated in Dulbecco's medium supplemented with 5–10% heat-inactivated fetal calf serum (FCS), to which was added 60 μg/ml STI to block fibrinolysis (Gordon, Unkeless and Cohn 1974). Fibrinolytic assays were initiated by washing the cells 3 times with Hank's balanced salt solution (HBSS), followed by incubation in Dulbecco's medium supplemented with 5% acid-treated fetal calf serum (AT-FCS). The use of AT-FCS (prepared as described in Unkeless, Gordon and Reich 1974) amplified detection of fibrinolytic activity since most of the serum inhibitors of plasmin are destroyed by such treatment (Loskutoff and Reich, in prep.). Conditioned media were prepared by incubating cells for up to 48 hours in serum-free medium consisting of Eagle's minimal medium (MEM) or Dulbecco's supplemented with 0.02–0.05% (w/v) lactalbumin hydrolysate. After such treatment, the cells were viable and continued to secrete lysozyme normally. Cell lysates were made in Triton X-100, 0.2% w/v. Lysates and conditioned media were assayed on ^{125}I-labeled fibrin plates as described (Unkeless, Gordon and Reich 1974) using 8 μg/ml of human plasminogen prepared from outdated plasma by affinity chromatography on lysine-Sepharose 4B

(Deutsch and Mertz 1970). One unit of activity released 10% of the total radioactivity on the dish in 4 hours.

Polyacrylamide Gel Electrophoresis

Samples were prepared for electrophoresis by dialysis against 0.1–0.2% sodium dodecyl sulfate (SDS) followed by lyophylization of 1–2 ml of dialysate. Such samples were then dissolved in 25–50 μl of sample buffer consisting of 3% ethylene glycol and 0.0625 M Tris-HCl pH 6.8. Samples were not reduced or boiled. Plasminogen activator was not inactivated by this procedure. When necessary, after addition of 1/10 volume of 1 M sodium sulfate, conditioned medium was first concentrated to 1–2 ml by placing the solution in a dialysis bag and packing it in dry Sephadex. SDS-polyacrylamide slab gel electrophoresis was performed according to Maizel (1971), and molecular standards used were bovine serum albumin, pepsin, chymotrypsinogen and lysozyme.

[^{3}H]DFP Labeling

Conditioned medium from 2×10^7 thioglycollate-stimulated or 7×10^7 control macrophages incubated in serum-free medium for 48 hours was concentrated eightfold to 2 ml as described. After addition of 1/10 volume 1 M Tris-sulfate pH 7.4, 0.1 ml [^{3}H]DFP (0.21 mg/ml, 0.9 Ci/mmole) was added in the presence or absence of p-nitrophenyl-p-guanidinobenzoate (NPGB) (8×10^{-5} M), and the mixture incubated 20 hours at room temperature. The plasminogen activator was inhibited 94% by this procedure. The solution was then dialyzed extensively against 0.1% SDS and prepared for electrophoresis as described. After electrophoresis, the gel was stained and destained, and the slices were swollen and counted in 4 M ammonium hydroxide:Protosol: Liquifluor 1:5:50 (Unkeless, Gordon and Reich 1974).

Phagocytosis

Particles for phagocytosis experiments were prepared as follows: Latex particles and formaldehyde-treated sheep erythrocytes (SRBC) were washed seven times, resuspended in Dulbecco's medium at a concentration of 5×10^9 particles/ml, irradiated with ultraviolet light, and stored at 4°C until used. *M. lysodeikticus* was suspended in Dulbecco's medium, 2.5 mg/ml, autoclaved, and washed three times before use. Dr. R. Winchester (The Rockefeller University) provided heat-aggregated human γ-globulin, which consisted of soluble 30–150S aggregates. The aggregates were centrifuged for 1 hour at 100,000g and resuspended before use. Dr. R. Steinman (The Rockefeller University) provided immune complexes of horseradish peroxidase (HRP)–rabbit anti-HRP (Gordon, Unkeless and Cohn 1974).

Phagocytosis experiments were performed using cells in culture 24 hours. STI was included in the medium containing 5% FCS during phagocytosis in all experiments on radioactive plates. Phagocytosis was observed by phase contrast microscopy and stopped when >95% of the cells had ingested a large number of particles as follows: latex, 20–50 particles; SRBC, 5–10 cells; and *M. lysodeikticus* more than 50 bacteria per cell. After phagocytosis, the cells were washed and shifted to medium containing either 10% FCS and 60 μg STI/ml or 5% AT-FCS as desired. Cultures on nonradioactive plates were treated similarly, except STI was omitted from the medium, and cells were washed and transferred to serum-free medium with 0.05% lactalbumin hydrolysate.

Other Procedures

Lysozyme, N-acetyl-β-glucosaminidase and cathepsin D were assayed as described elsewhere (Gordon, Todd and Cohn 1974). Protein was assayed by the method of Lowry et al. (1951).

RESULTS

Fibrinolysis by Macrophages, PMN and Monocytes

Fibrinolytic activity of macrophages was measured after plating on ^{125}I-labeled fibrin-coated dishes. As seen in Figure 1, thioglycollate-stimulated macrophages showed rapid fibrinolysis; the fibrinolytic activity of the unstimulated cells was very low and barely exceeded the background level in cell-free control dishes. The rate of fibrinolysis was linear with time and in other experiments, was also proportional to the number of stimulated macrophages, in the range of $1\text{--}6 \times 10^5$ macrophages per 35-mm dish. The fibrinolytic activity per cell of thioglycollate-stimulated cells was at least 100-fold higher than that

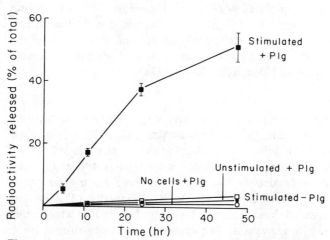

Figure 1
Fibrinolysis by thioglycollate-stimulated and unstimulated macrophages. Macrophages, 1.5×10^6 stimulated or 4×10^6 unstimulated, were plated on 35-mm ^{125}I-labeled fibrin-coated dishes for 24 hours, washed, and placed in 3 ml of Eagle's minimal medium containing 5% human serum or 5% plasminogen-depleted human serum. Samples of 0.3 ml were removed as indicated: ($\blacksquare$——$\blacksquare$) stimulated cells, human serum; ($\square$——$\square$) stimulated cells, plasminogen-depleted serum; ($\bullet$——$\bullet$) unstimulated cells, human serum; ($\circ$——$\circ$) control dishes without cells, human serum. (Reprinted, with permission, from Unkeless, Gordon and Reich 1974.)

of control macrophages. Virtually all the fibrinolytic activity of thioglycollate-stimulated macrophages was plasminogen dependent, since it was not observed when serum was depleted of plasminogen by passage over a lysine-Sepharose column (Fig. 1). The difference in rates of fibrinolysis between thioglycollate-stimulated and unstimulated macrophages remained unaltered during at least 4 days of cultivation.

Since macrophages harvested after intraperitoneal injection of an irritant such as thioglycollate medium could contain macrophage precursors as well as contaminant PMN, fibrinolytic activity of such cell types was of interest. Human monocytes and PMN were used because they can be readily purified in sufficient numbers. Purified PMN plated on ^{125}I-labeled fibrin-coated plates in AT-FCS show active fibrinolysis (Fig. 2A); freshly plated monocytes show 18% the activity of PMN (Fig. 2B). However, PMN die after 6–12 hours of cultivation in vitro, and no fibrinolysis was detected when PMN were preincubated in 5% FCS and STI for one day before assay. In order to avoid contamination by PMN, macrophage cultures were routinely cultivated in STI for one day before assay in AT-FCS. In contrast to the PMN, monocytes were still active after 3 days in culture.

Fibrinolysis by Conditioned Medium and Cell Lysates

We then looked for secretion of the plasminogen activator from macrophages into serum-free medium. Both conditioned medium and cell lysates from thioglycollate-stimulated macrophages showed plasminogen-dependent fibrinolysis, whereas unstimulated macrophages did not (Table 1). The unstimulated macrophages had some fibrinolytic activity in the absence of plasminogen,

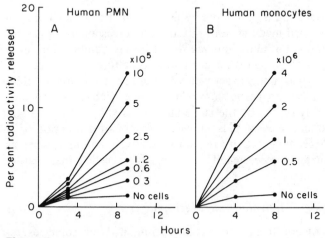

Figure 2

Fibrinolysis by purified human PMN (*A*) and monocytes (*B*). Serial dilutions of human PMN were assayed immediately; human monocytes were assayed after one day in culture. (Reprinted, with permission, from Gordon, Unkeless and Cohn 1974.)

Table 1

Fibrinolysis by Thioglycollate-stimulated and
Unstimulated Macrophages

	Fibrinolysis[a] (U/dish)			
	stimulated plasminogen		unstimulated plasminogen	
	present	absent	present	absent
Cell lysates				
24 hr[b]	18	1.0	23	23
72 hr	55	0.40	12	7
Conditioned medium				
24–72 hr	800	26	3.4	2.8

Data from Unkeless, Gordon and Reich (1974).

[a] Background values (1.1–1.4% of total radioactivity) were subtracted; 25–50-μl samples were assayed; all assays were performed in duplicate and the average is reported.

[b] Cells were cultured for 24 hours in Dulbecco's medium containing 10% FCS and incubated 2 days in Dulbecco's medium containing 0.05% lactalbumin hydrolysate.

presumably due to attack of the fibrin by other proteases (Table 1). The specific activity of the plasminogen activator secreted into the medium was 60-fold higher than in cell lysates, suggesting that the enzyme is actively secreted by the cells. Lysozyme, another macrophage product, showed a comparable increase in specific activity in the medium compared to cell lysates, but was released from both stimulated and unstimulated cells in equal amounts (Table 2).

To test for the production of inhibitors by unstimulated macrophages, cell lysates and/or conditioned medium were mixed with corresponding fractions from stimulated cultures; no inhibition was found in any combination, even when a fourfold excess of protein from unstimulated cultures was used.

The time course of plasminogen activator formation in cultures of thioglycollate-stimulated macrophages is shown in Figure 3 (left); for comparison, the formation of lysozyme by the same cultures is shown on the right. The cells synthesized and released plasminogen activator over a period of at least 3 days. The intracellular activity remained relatively constant, whereas the activity in the medium rose steadily and greatly exceeded that present initially in the cells.

Properties of Plasminogen Activator from Macrophages

The plasminogen activator from thioglycollate-stimulated macrophages resembles that secreted by chicken and mouse fibroblasts transformed by sarcoma viruses. The enzyme was not inhibited by incubation for 24 hours at room temperature with iodoacetamide (10 mM), iodoacetate (10 mM), N-ethylmaleimide (10 mM), or p-chloromercuribenzoate (0.1 mM), but was inhibited by treatment with dithiothreitol (10 mM). Diisopropylfluorophosphate

Table 2
Specific Activities of Plasminogen Activator and Lysozyme

	Total macrophage no.	Plasminogen-dependent fibrinolysis (U/dish)	Total protein (mg)	Plasminogen activator specific activity (U/mg)	Lysozyme (μg/dish)	Lysozyme specific activity (μg lysozyme/mg protein)
Stimulated macrophages	4×10^6					
Cell lysate (24 hr)		17	0.60	28	0.43	0.70
Cell lysate (72 hr)		55	0.75	74	1.3	1.7
Conditioned medium (24–72 hr)		770	0.17	4500	11	62
Unstimulated macrophages	12×10^6					
Cell lysate (24 hr)		0	0.70	0	0.40	0.57
Cell lysate (72 hr)		5.0	0.40	12.0	0.85	2.1
Conditioned medium (24–72 hr)		0.60	0.22	3.0	11	50

Data from Unkeless, Gordon and Reich (1974).

501

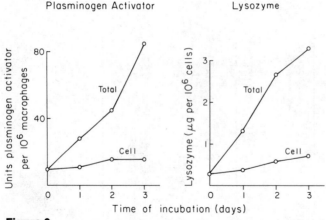

Plasminogen Activator   Lysozyme

Figure 3

Production and secretion of plasminogen activator and lysozyme. Thioglycollate-stimulated macrophages (4×10^6) on
60-mm dishes after 24 hours in culture were washed and incubated in Eagle's minimal medium containing 0.02% lactalbumin hydrolysate. Cell lysates and conditioned media were
prepared daily from replicate dishes. (Reprinted, with permission, from Unkeless, Gordon and Reich 1974.)

at a concentration of 0.25 mM in 0.1 M Tris-sulfate pH 7.4 inhibited the plasminogen activator irreversibly, with a half-time of 10 minutes at room temperature. p-Nitrophenyl-p-guanidinobenzoate, a compound known to acylate
the active serine of a number of trypsinlike proteases (Fasco and Fenton
1973), inhibited the plasminogen activator 50% at a concentration of
10^{-8} M. Inhibition by NPGB was reversible upon prolonged dialysis, consistent with the expected deacylation of the guanidinobenzoate ester. The
enzyme was stable in 2% SDS solutions at neutral pH, and SDS did not affect
the fibrin plate assay provided the final concentration of SDS was less than
0.006%.

SDS-Polyacrylamide Gel Electrophoresis

The stability of the plasminogen activator in SDS enabled us to identify the
molecular species present by SDS-polyacrylamide gel electrophoresis. Conditioned serum-free media from thioglycollate-stimulated macrophages,
SV40-transformed 3T3 cells (strain SV101), and mouse sarcoma virus-transformed mouse fibroblasts were electrophoresed, after which separate lanes
from the slab were sliced and incubated on radioactive fibrin dishes in the
presence of plasminogen. A major symmetrical peak of activity was observed
in all three samples at an apparent molecular weight of 48,000 (Fig. 4).
There was also a minor plasminogen activator at 28,000 and a small amount
of fibrinolysis associated with heterogeneous material larger than the main
active peak. The SV101 culture also produced an activator of 83,000 molecular weight. Control experiments established the lack of fibrinolysis in the
absence of plasminogen.

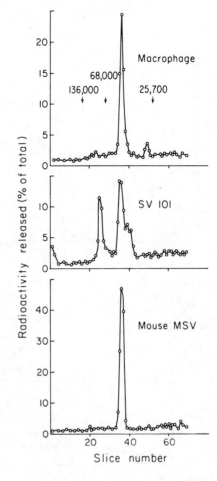

Figure 4
Comparison of plasminogen activators from stimulated macrophages, SV40-transformed 3T3 cells (strain SV101) and mouse sarcoma virus-transformed mouse embryo fibroblasts. Conditioned medium (2–4 ml) was concentrated, electrophoresed in a 10–15% polyacrylamide gradient SDS slab gel and assayed as described in the text. (Reprinted, with permission, from Unkeless, Gordon and Reich 1974.)

SDS-Polyacrylamide Gel Electrophoresis of [³H]DFP-labeled Conditioned Medium

Encouraged by the striking difference in plasminogen activator activity of thioglycollate-stimulated and unstimulated macrophages, we attempted to extend the comparison to include [³H]DFP-labeled products in the respective conditioned media. Accordingly, concentrated conditioned medium was incubated with [³H]DFP at neutral pH; the radioactivity profile in the gel should therefore give an indication of the spectrum of serine hydrolases present. As seen in Figure 5A, conditioned medium from thioglycollate-stimulated macrophages again contained the 48,000 and 28,000 molecular weight species of plasminogen activator. Figure 5B shows the radioactivity profile of [³H]DFP-labeled proteins. There were two DFP-labeled peaks in the medium from thioglycollate-stimulated macrophages which corresponded exactly with the plasminogen activator activity in Figure 5A; both of these were absent from unstimulated macrophage conditioned medium. The ratio of plasminogen activator activity to [³H]DFP was fivefold greater in the 48,000 molecular weight peak. Using previous studies of the counting efficiency of ³H in such gels (Unkeless, Gordon and Reich 1974), the amount of [³H]DFP in the

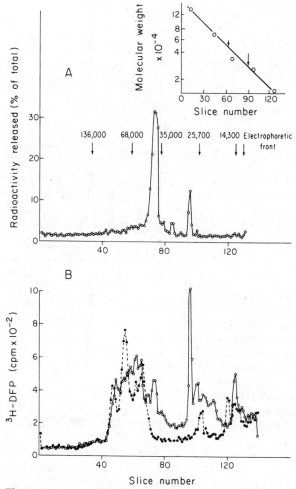

Figure 5

Correlation of plasminogen activator and [³H]DFP labeling in conditioned media from stimulated and unstimulated macrophages. (*A*) Plasminogen activator activity.
(*B*) [³H]DFP labeling of thioglycollate-stimulated macrophage conditioned medium (o——o); unstimulated
macrophage conditioned medium (•——•). (Reprinted,
with permission, from Unkeless, Gordon and Reich
1974.)

48,000 molecular weight peak was calculated to be 3 pmoles. Assuming 1
mole of DFP was found per mole of enzyme, this represents 0.15 μg of
plasminogen activator released by 2×10^7 cells in 48 hours. For comparison,
this culture produced 40 μg of lysozyme during the same period.

In an attempt to make a preliminary characterization of the serine enzymes
secreted by macrophages, conditioned media were labeled with [³H]DFP in
the presence of NPGB. Since NPGB acylates the same active site as DFP,

the reaction with [³H]DFP should be reduced; the extent of the reduction of labeling should reflect the affinity of the enzyme for NPGB relative to DFP and the rate of deacylation of the guanidinobenzoate ester formed. As shown by the gel profiles in Figure 6A, the labeling of the two plasminogen activator species in medium from thioglycollate-stimulated macrophages was strongly inhibited by pretreatment with NPGB, consistent with the sensitivity of the activator to NPGB. The labeling of a number of other species in the high molecular weight region of the gel was also reduced. In contrast, the labeling of all the peaks in the unstimulated macrophage conditioned medium was resistant to NPGB treatment (Fig. 6B). This would suggest that serine enzymes released by the two kinds of macrophages differ in specificity and electrophoretic mobility.

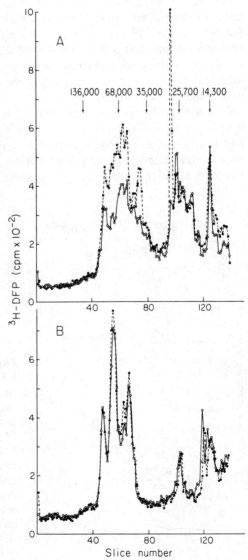

Figure 6

Inhibition by NPGB of [³H]DFP labeling of proteins in conditioned medium from thioglycollate-stimulated and unstimulated macrophages. (*A*) Stimulated macrophage conditioned medium without NPGB (•– – –•) and with 8×10^{-5} M NPGB (o——o). (*B*) Unstimulated macrophage conditioned medium without NPGB (•– – –•) and with 8×10^{-5} M NPGB (o——o). (Reprinted, with permission, from Unkeless, Gordon and Reich 1974.)

Induction of Fibrinolytic Activity

Since thioglycollate-stimulated macrophages contain inclusions of undegradable constituents, probably derived from phagocytic activity in the peritoneal cavity (Hirsch and Fedorko 1970), the hypothesis that phagocytosis might also play a role in induction of fibrinolytic activity was tested by feeding latex particles to cultures in vitro, obtained 4 days after injection of a number of different intraperitoneal irritants. The results of such an experiment are shown in Figure 7. Thioglycollate-stimulated cells were most active; endotoxin, mineral oil and BCG provided an intermediate stimulus, whereas FCS and saline-treated cells approached normal cells in activity. Fibrinolytic activity of the macrophages varied widely, but in all cases, except thioglycollate-stimulated macrophages, phagocytosis of latex enhanced fibrinolytic activity 3–6-fold.

The cells tested for fibrinolysis in Figure 7 were also tested for secretion of the plasminogen activator. Thioglycollate-stimulated cells and the same cells fed latex secreted 130 and 163 units/mg cell protein respectively in 48 hours. Endotoxin-stimulated cells secreted no detectable activator; after ingestion of latex they secreted 41 units/mg cell protein, 25% that of the thioglycollate cells. No other cells tested secreted detectable plasminogen activator, even after phagocytosis of latex.

The relationship between endotoxin dose and fibrinolysis was studied by injecting groups of animals intraperitoneally with graded doses of endotoxin and testing the macrophages harvested 4 days later for fibrinolytic activity after feeding them latex for 1 hour. Fibrinolysis was measured one day after phagocytosis. Maximum fibrinolysis was seen with a dose of 30 μg (Fig. 8). The

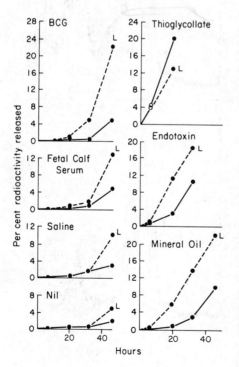

Figure 7

Effect of different intraperitoneal irritants and phagocytosis of latex on macrophage fibrinolytic activity. Injected materials were: saline, 1 ml; FCS, 1 ml; BCG, 100 μg; mineral oil, 1 ml; endotoxin, 30 μg; thioglycollate medium, 0.75 ml. Thioglycollate-stimulated macrophages were plated at a density of 6×10^5 per 35-mm dish, others at 4–6×10^6. (●——●) No latex; (●– – –●) fed latex. (Reprinted, with permission, from Gordon, Unkeless and Cohn 1974.)

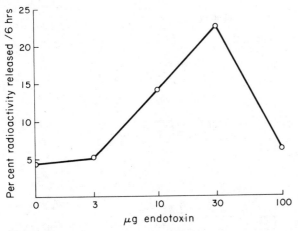

Figure 8
The relation between intraperitoneal dose of endotoxin
and macrophage fibrinolysis. Macrophages (4×10^6)
were cultured on ^{125}I-labeled fibrin plates for 24 hours
in the presence of STI. After latex phagocytosis, cells
were cultured for one day before assay in AT-FCS.
(Reprinted, with permission, from Gordon, Unkeless
and Cohn 1974.)

morphologic appearance was correlated with subsequent fibrinolysis; the most
active macrophages phagocytosed the latex more rapidly, showed more membrane ruffling, and contained prominent cytoplasmic granules. A similar experiment was performed with 0–200 μg of a 1:1 lipid A:bovine serum
albumin complex, since lipid A contains the moiety responsible for some of
the biological activities of endotoxin (Lüderitz et al. 1973). However, 200 μg
of the complex was required to achieve the same level of stimulation as 30 μg
of endotoxin.

The influence of phagocytosis on secretion of plasminogen activator from
endotoxin-stimulated cells was examined in more detail by measuring intracellular enzyme levels in Triton X-100 lysates and secretions from the
same cells. In vivo endotoxin stimulation followed by in vitro phagocytosis
of latex resulted in a striking increase of plasminogen activator in cell lysates
(Fig. 9A), as well as in conditioned medium (Fig. 9B). This increase was
observed one day after phagocytosis of latex and further increased fourfold
in the lysate and the medium during the next day. Such cells secreted 60–80%
of their total content of plasminogen activator produced per day. In contrast,
endotoxin-stimulated cells secreted little or no activator.

Measurement of secretion gives results somewhat different than would be
expected from fibrinolytic assays of living cells—endotoxin-stimulated cells
have 30% the fibrinolytic activity of the same cells fed latex (Fig. 7). This
discrepancy may be due to the intimate contact of the cell with the fibrin substratum, resulting in significant fibrinolysis by an amount of secreted plasminogen activator too dilute to assay.

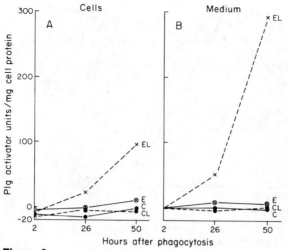

Figure 9

The effect of endotoxin stimulation and latex phagocytosis on plasminogen activator in cell lysates (*A*)
and conditioned media (*B*). Macrophages, 4×10^6
in 35-mm dishes, were cultivated for one day, fed
latex, and placed in Dulbecco's medium with 0.05%
lactalbumin hydrolysate either immediately or 26 hours
after phagocytosis. Plasminogen-dependent fibrinolysis was defined as the difference in radioactivity released in the presence and absence of plasminogen. The
results show secretion between 2–26 and 26–50 hours,
respectively. (●——●) Unstimulated macrophages;
(●– – –●) unstimulated macrophages fed latex; (o——o)
endotoxin-stimulated macrophages; (x– – –x) endotoxin-stimulated cells fed latex. (Reprinted, with permission, from Gordon, Unkeless and Cohn 1974.)

Selectivity of Production and Release of Macrophage Enzymes

Evidence for the selective induction of plasminogen activator was obtained by
comparing the effect of endotoxin, latex and thioglycollate treatment on plasminogen activator, lysozyme and two intracellular acid hydrolases, cathepsin
D and N-acetyl-β-D-glucosaminidase. Table 3 shows that stimulation by
thioglycollate and endotoxin followed by latex phagocytosis selectively induced production and secretion of the plasminogen activator. Unstimulated
macrophages displayed little fibrinolytic activity which was plasminogen dependent. Latex or endotoxin alone stimulated a small fraction, 2 and 7%,
respectively, of thioglycollate cell activity compared with 31% after combined
treatment; 83–100% of the induced enzyme activity was secreted into the
medium. In contrast, all groups of macrophages produced comparable levels
of lysozyme, predominantly as secretion, and of cathepsin D and N-acetyl-β-
D-glucosaminidase, which remained largely intracellular.

Table 3

Effect of Stimulation and Phagocytosis on Production and Release of Enzymes by Macrophages

Macrophage	Cell protein/ 1×10^7 macrophages	Plasminogen activator		Lysozyme		Cathepsin D		N-acetyl-β-D-glucosaminidase	
		U^a/mg cell protein	$\%^b$ secreted	μg/mg cell protein	% secreted	U^c/mg cell protein	% secreted	μmole/ min/mg cell protein	% secreted
Unstimulated	336	−3.6	—	56	95	0.38	20	0.15	8
Unstimulated fed latex	738	15	—	31	92	0.44	11	0.14	6
Endotoxin	246	55	100	80	98	0.39	20	0.095	16
Endotoxin fed latex	594	250	87	39	95	0.43	11	0.13	5
Thioglycollate	864	800	83	37	94	0.52	25	0.19	29

Macrophages (2×10^7) were cultivated 24 hours in Dulbecco's medium containing 10% FCS, fed latex, and placed in medium containing 0.05% lactalbumin hydrolysate for 48 hours. (Data from Gordon, Unkeless and Cohn 1974.)

a Cell lysate + medium.

b Medium/cell lysate + medium × 100.

c Units in chromogenic equivalents of 1 mg/ml egg lysozyme released/min.

Effect of Different Phagocytosed Particles

In view of the phagocytosis of latex triggering the intracellular accumulation and release of plasminogen activator, further experiments were undertaken to study the effect of other types of phagocytic loads on fibrinolytic activity by endotoxin-stimulated macrophages. Table 4 illustrates that a variety of particles, including *M. lysodeikticus,* heat-aggregated γ-globulin and immune complexes of HRP–anti-HRP, could stimulate fibrinolysis as effectively as latex. The degree of stimulation of fibrinolysis was proportional to particle dose.

Comparable experiments with unstimulated macrophages were carried out using formaldehyde-treated sheep red blood cells, which were digested within 24–36 hours after phagocytosis and latex. Ingestion of latex increased the fibrinolytic activity of control macrophages threefold, and activity continued to increase on consecutive days. However, fibrinolytic activity after ingestion of red blood cells was maximal 2 days after phagocytosis and decreased slightly thereafter (Fig. 10). This result suggested that the duration of fibrinolysis might be related to the persistence of the phagocytosed particle within the cell.

To test this hypothesis, endotoxin-stimulated macrophages on ^{125}I-labeled fibrin plates were fed either latex or *M. lysodeikticus,* which was chosen because macrophages digest it readily. The results of such an experiment (Fig. 11) show that fibrinolysis induced by latex increased and persisted, whereas the activity of the cultures fed *M. lysodeikticus* diminished within a day and approached that of unfed cells after further cultivation. Similar experiments on nonradioactive dishes confirmed that plasminogen activator was secreted

Table 4

Phagocytosis and Fibrinolysis by Endotoxin-stimulated Macrophages

Substance phagocytosed	Load	Radioactivity released (%/12 hr)	% Control
Nil	—	5.3	100
Latex	5×10^9	21	400
	1.5×10^8	17	330
	0.5×10^8	12	250
M. lysodeikticus	0.5 mg	16	310
	0.17 mg	11	210
	0.06 mg	5.6	130
Heat-aggregated γ-globulin	150 μg	34	650
	15 μg	18	350
HRP–anti-HRP complexes	30 μg	28	540
	3 μg	12	240

Macrophages (3×10^6) were cultivated for 24 hours on ^{125}I-labeled fibrin plates in the presence of STI, washed, and fed particulates for 60 minutes in 2.5 ml of medium containing STI. The cultures were then washed and placed in AT-FCS for assay. (Data from Gordon, Unkeless and Cohn 1974.)

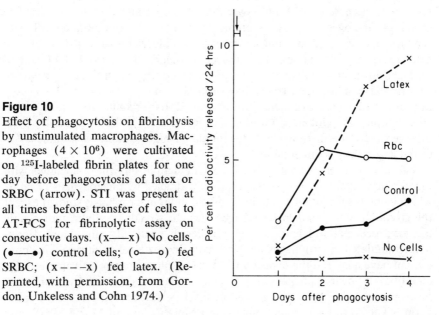

Figure 10
Effect of phagocytosis on fibrinolysis by unstimulated macrophages. Macrophages (4×10^6) were cultivated on ^{125}I-labeled fibrin plates for one day before phagocytosis of latex or SRBC (arrow). STI was present at all times before transfer of cells to AT-FCS for fibrinolytic assay on consecutive days. (x——x) No cells, (●——●) control cells; (o——o) fed SRBC; (x – – –x) fed latex. (Reprinted, with permission, from Gordon, Unkeless and Cohn 1974.)

only transiently by endotoxin-stimulated macrophages after ingestion of *M. lysodeikticus;* secretion by such cells fed latex continued to increase on subsequent days (Fig. 9B).

DISCUSSION

The data presented show clearly that there are substantial differences between thioglycollate-stimulated cells and unstimulated macrophages with respect to plasminogen activator formation. Plasminogen-dependent fibrinolysis by unstimulated cells, cell lysates or conditioned media barely exceeded background

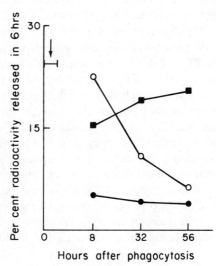

Figure 11
Fibrinolysis by endotoxin-stimulated macrophages after phagocytosis of latex or *M. lysodeikticus*. Macrophages (5×10^6) were cultured on ^{125}I-labeled fibrin dishes for 24 hours. After phagocytosis (arrow), cultures were maintained in Dulbecco's medium containing 10% FCS and 60 μg/ml STI until assay in AT-FCS at 2, 26 and 50 hours. (●——●) Endotoxin-stimulated macrophages; (o——o) fed *M. lysodeikticus;* (■——■) fed latex. (Reprinted, with permission, from Gordon, Unkeless and Cohn 1974.)

levels and was at least 100-fold lower than thioglycollate-stimulated cells. The major plasminogen activator secreted by thioglycollate-stimulated cells resembles that secreted by mouse embryo fibroblasts transformed by mouse sarcoma virus; it has an apparent molecular weight of 48,000, is NPGB- and DFP-sensitive, and activates plasminogen efficiently by proteolytic cleavage to form plasmin, a potent trypsinlike protease. The precise intracellular location of the enzyme is not clear, but, as with Rous sarcoma virus-transformed chick fibroblasts (Unkeless, Gordon and Reich 1974), the intracellular plasminogen activator in thioglycollate-stimulated macrophages is granule associated and readily sedimentable.

The plasminogen activator is thus an inductible enzyme associated with macrophage stimulation. It is also markedly affected by phagocytosis of latex, which increases fibrinolytic activity 3–5-fold in all macrophages studied except thioglycollate-stimulated macrophages, which already contain indigestible phagocytosed material and can be considered maximally stimulated. Endotoxin-stimulated macrophages have a relatively low level of fibrinolytic activity, but have acquired the capacity of secreting plasminogen activator in large quantities following phagocytosis of latex. The mechanism by which the macrophage is thus "primed" by endotoxin may well be complex, since endotoxin is known to have potent effects on PMN (Cohn and Morse 1960) and lymphocytes (Andersson et al. 1973) as well as activating humoral factors such as complement (Galanos et al. 1971), all of which could affect macrophage function indirectly. Attempts to "prime" populations of cultured unstimulated macrophages in vitro have thus far been unsuccessful.

Although the function of the plasminogen activator may not be limited to the activation of plasminogen, the formation of plasmin may have important physiological sequelae. The high level of circulating plasminogen may act as a reservoir of potential proteolytic activity which could be recruited by cells releasing the activator. The interactions of plasmin with enzymatic cascades involved in complement activation (Ratnoff and Naff 1967) and kinin formation (Kaplan and Austen 1971) provide several pathways for generating peptides that are pharmacologically active, particularly in their effects on vascular permeability. In addition, plasmin may affect the properties of other cells in the immediate environment, as has been demonstrated in tissue culture (Ossowski et al. 1973; Ossowski, Quigley and Reich 1974).

Continued production and secretion of plasminogen activator is associated with persistence of phagocytosed particles; readily digested particles such as SRBC and M. lysodeikticus result in only a transient increase in fibrinolytic activity. This is in marked contrast to the behavior of acid hydrolases, which are induced solely by digestible particles (Axline and Cohn 1970). The long-term accumulation and secretion of the plasminogen activator provides some evidence for synthesis of the enzyme, rather than conversion of a precursor as a result of phagocytosis. Lysozyme is an example of another class of secreted products from macrophages. Lysozyme secretion is unaffected by intraperitoneal irritants or by phagocytosis. These differences in enzyme regulation suggest different functions for each class of enzymes.

The concept of activation of macrophages was developed to explain the enhancement of bactericidal activity observed after infection with BCG and other agents (Mackaness 1970). Endotoxin treatment has been reported to enhance phagocytosis and killing of bacteria by macrophages (Shilo 1959) and promote their cytocidal and cytostatic capacity for tumor cells (Alexander and Evans 1971; Lohmann-Matthes and Fischer 1973). Our studies show that plasminogen activator accompanies these parameters of macrophage activation and suggest that it may play a role in the effector functions of the activated macrophage (Bast et al. 1974). The requirement of phagocytosis for secretion of the plasminogen activator and the fall in fibrinolytic activity and secretion after digestion of the particles may serve to regulate closely the activation of plasminogen to plasmin by macrophages.

Acknowledgments

We thank Professors E. Reich and Z. Cohn for their stimulating interest in this work. J. C. U. is a Jane Coffin Childs Fellow. S. G. is a Scholar of the Leukemia Society of America, supported in part by a grant from The Rockefeller Foundation.

REFERENCES

Alexander, P. and R. Evans. 1971. Endotoxin and double stranded RNA render macrophages cytotoxic. *Nature New Biol.* **232:**76.

Andersson, J., F. Melchers, C. Galanos and O. Lüderitz. 1973. The mitogenic effect of lipopolysaccharide on bone marrow-derived mouse lymphocytes. Lipid A as the mitogenic part of the molecule. *J. Exp. Med.* **137:**943.

Axline, S. G. and Z. Cohn. 1970. *In vitro* induction of lysosomal enzymes by phagocytosis. *J. Exp. Med.* **131:**1239.

Bast, R. C., Jr., R. D. Cleveland, B. H. Littman, B. Zbar and H. J. Rapp. 1974. Acquired cellular immunity: Extracellular killing of *Listeria monocytogenes* by a product of immunologically activated macrophages. *Cell Immunol.* **10:** 248.

Cohn, Z. and B. Benson. 1965. The *in vitro* differentiation of mononuclear phagocytes. II. The influence of serum on granule formation, hydrolase production, and pinocytosis. *J. Exp. Med.* **121:**835.

Cohn, Z. and S. Morse. 1960. Functional and metabolic properties of polymorphonuclear leukocytes. II. The influence of a lipopolysaccharide endotoxin. *J. Exp. Med.* **111:**689.

Deutsch, D. G. and E. T. Mertz. 1970. Plasminogen: Purification from human plasma by affinity chromatography. *Science* **170:**1095.

Fasco, M. J. and J. W. Fenton II. 1973. Specificity of thrombin. I. Esterolytic properties of thrombin, plasmin, trypsin, and chymotrypsin with N-β-substituted guanidino derivatives of *p*-nitrophenyl-*p'*-guanidinobenzoate. *Arch. Biochem. Biophys.* **159:**802.

Galanos, C., E. T. Rietschel, O. Lüderitz and O. Westphal. 1971. Interaction of lipopolysaccharides and lipid A with complement. *Eur. J. Biochem.* **19:**143.

Gordon, S., J. Todd and Z. A. Cohn. 1974. *In vitro* synthesis and secretion of lysozyme by mononuclear phagocytes. *J. Exp. Med.* **139:**1228.

Gordon, S., J. C. Unkeless and Z. A. Cohn. 1974. Induction of macrophage plasminogen activator by endotoxin stimulation and phagocytosis. *J. Exp. Med.* **140:**995.

Hirsch, J. G. and M. E. Fedorko. 1970. Morphology of mouse mononuclear phagocytes. In *Mononuclear Phagocytes* (ed. R. van Furth), p. 7. Blackwell Scientific Publications, Oxford.

Kaplan, A. P. and K. F. Austen. 1971. A prealbumin activator of prekallikrein. II. Derivation of activators of prekallikrein from active Hageman factor by digestion with plasmin. *J. Exp. Med.* **133**:696.

Karnovsky, M. L., S. Simmons, E. A. Glass, A. W. Shafer and P. D'arcy Hart. 1970. Metabolism of macrophages. In *Mononuclear Phagocytes* (ed. R. van Furth), p. 103. Blackwell Scientific Publications, Oxford.

Lohman-Matthes, M. L. and H. Fischer. 1973. T-cell cytotoxicity and amplification of the cytotoxic reaction by macrophages. *Transplant. Rev.* **17**:150.

Lowry, O. H., N. J. Rosebrough, A. L. Farr and R. J. Randall. 1951. Protein measurement with the folin-phenol reagent. *J. Biol. Chem.* **193**:265.

Lüderitz, O., C. Galanos, V. Lehman, M. Nurminen, E. T. Rietschel, G. Rosenfelder, M. Simon and O. Westphal. 1973. Lipid A: Chemical structure and biological activity. *J. Infect. Dis.* (Suppl.) **128**:17.

Mackaness, G. B. 1970. Cellular immunity. In *Mononuclear Phagocytes* (ed. R. van Furth), p. 461. Blackwell Scientific Publications, Oxford.

Maizel, J. 1971. Polyacrylamide gel electrophoresis of viral proteins. In *Methods in Virology* (ed. K. Maramorosch and H. Koprowski), vol. 5, p. 179. Academic Press, New York.

Nelson, D. S. 1972. Macrophages as effectors of cell mediated immunity. *Crit. Rev. Microbiol.* **1**:353.

Ossowski, L., J. P. Quigley and E. Reich. 1974. Fibrinolysis associated with oncogenic transformation. Morphological correlates. *J. Biol. Chem.* **249**:4312.

Ossowski, L., J. P. Quigley, G. M. Kellerman and E. Reich. 1973. Fibrinolysis associated with oncogenic transformation. Requirement of plasminogen for correlated changes in cellular morphology, colony formation in agar, and cell migration. *J. Exp. Med.* **138**:1056.

Ossowski, L., J. C. Unkeless, A. Tobia, J. P. Quigley, D. B. Rifkin and E. Reich. 1973. An enzymatic function associated with transformation of fibroblasts by oncogenic viruses. II. Mammalian fibroblast cultures transformed by DNA and RNA tumor viruses. *J. Exp. Med.* **137**:112.

Quigley, J. P., L. Ossowski and E. Reich. 1974. Plasminogen: The serum proenzyme activated by factors from cells transformed by oncogenic viruses. *J. Biol. Chem.* **249**:4306.

Ratnoff, O. D. and P. B. Naff. 1967. The conversion of C′1s to C′1 esterase by plasmin and trypsin. *J. Exp. Med.* **125**:337.

Shilo, M. 1959. Nonspecific resistance to infection. *Annu. Rev. Microbiol.* **13**:225.

Unkeless, J. C., S. Gordon and E. Reich. 1974. Secretion of plasminogen activator by stimulated macrophages. *J. Exp. Med.* **139**:834.

Unkeless, J. C., K. Danø, G. Kellerman and E. Reich. 1973. Fibrinolysis associated with oncogenic transformation: Partial purification and characterization of the cell factor—A plasminogen activator. *J. Biol. Chem.* **249**:4295.

Unkeless, J. C., A. Tobia, L. Ossowski, J. P. Quigley, D. B. Rifkin and E. Reich. 1973. An enzymatic function associated with transformation of fibroblasts by oncogenic viruses. I. Chick embryo fibroblast cultures transformed by avian RNA tumor viruses. *J. Exp. Med.* **137**:85.

Properties of Protein Turnover in Animal Cells and a Possible Role for Turnover in "Quality" Control of Proteins

Robert T. Schimke and Mathews O. Bradley*

Department of Biological Sciences, Stanford University
Stanford, California 94305

Protein turnover is a continual process in essentially all living forms, including both prokaryotes and eukaroytes, and in this paper the general properties of such turnover will be reviewed. (For earlier reviews of turnover, as well as the methodology involved in its measurement, see Schimke 1970; Schimke and Doyle 1970; Dice and Goldberg 1974.) In addition, the possible functions for this seemingly wasteful process of continual degradation of proteins will be discussed in the light of properties of turnover, particularly the role of turnover in changing enzyme complements in slowly growing cells and its possible role in maintaining cells free of proteins containing errors as a result of information transfer from DNA ultimately to proteins.

GENERAL PROPERTIES OF DEGRADATION OF PROTEINS IN ANIMAL TISSUES

Turnover is extensive. Studies of Swick (1958), Buchanan (1961) and Schmike (1964) have shown that essentially all proteins of rat liver undergo continual replacement. Buchanan (1961) estimated that about 70% of rat liver protein was replaced every 4–5 days from a dietary source. In cultured cells, the rate has been estimated at approximately 1–2% per hour (Eagle et al. 1959; Klevecz 1971).

Turnover is largely intracellular. Turnover includes synthesis and secretion of proteins, cell replacement, and intracellular synthesis and degradation. The life-span of hepatic cells is of the order of 160–400 days (Swick 1956; Mac-Donald 1961). Hence the extensive turnover occurring in 4–5 days in liver indicates that protein turnover in this tissue is largely intracellular.

There is a marked heterogeneity of rate of replacement of different proteins

* Present address: National Cancer Institute, National Institutes of Health, Bethesda, Maryland 20014.

Table 1

Half-lives of Specific Enzymes and Subcellular Fractions of Rat Liver

Enzyme	Half-life
Ornithine decarboxylase (soluble)	10 min[a]
δ-Aminolevulinate synthetase (mitochondria)	60 min[b]
Alanine-aminotransferase (soluble)	0.7–1.0 day[c]
Catalase (peroxisomal)	1.4 day[d]
Tyrosine aminotransferase (soluble)	1.5 hr[e]
Tryptophane oxygenase (soluble)	2 hr[f]
Glucokinase (soluble)	1.25 day[g]
Arginase (soluble)	4–5 day[h]
Glutamic-alanine transaminase	2–3 day[i]
Lactate dehydrogenase isozyme-5	16 day[j]
Cytochrome c reductase (endoplasmic reticulum)	60–80 hr[k]
Cytochrome b_5 (endoplasmic reticulum)	100–200 hr[k]
NAD glycohydrolase (endoplasmic reticulum)	16 day[l]
Hydroxymethylglutaryl CoA reductase (endoplasmic reticulum)	2–3 hr[m]
Acetyl CoA carboxylase (soluble)	2 day[n]
Cell fractions	
nuclear	5.1 day[o]
supernatant	5.1 day[o]
endoplasmic reticulum	2.1 day[o]
plasma membrane	2.1 day[o]
ribosomes	5.0 day[o]
mitochondria	4–5 day[c]

[a] Russell and Snyder (1968); [b] Marver et al. (1966); [c] Swick, Rexroth and Stange (1968); [d] Price et al. (1962); [e] Kenney (1967); [f] Schimke, Sweeney and Berlin (1965); [g] Niemeyer (1967); [h] Schimke (1964); [i] Segal and Kim (1963); [j] Fritz et al. (1969); [k] Kuriyama et al. (1969); [l] Bock, Siekevitz and Palade (1971); [m] Higgins, Kawachi and Rudney (1971); [n] Majerus and Kilburn (1969); [o] Arias, Doyle and Schimke (1969).

(*enzymes*). Table 1 gives a representative listing of rates of degradation of various enzymes of rat liver. The variation is remarkable, ranging from 10 minutes for ornithine decarboxylase to 16 days for LDH$_5$ and NAD glycohydrolase. There is no correlation between half-lives and the cell fraction from which the enzyme is isolated. Thus δ-aminolevulinate synthetase and ornithiae aminotransferase, both of which are associated with mitochondria, have half-lives of 60 minutes and 1 day, respectively, whereas total mitochondrial protein has a half-life of 4–5 days. Equally remarkable are the marked differences in half-lives of enzymes associated with the endoplasmic reticulum, ranging from 2 hours for HMG-CoA reductase to 16 days for NAD glycohydrolase.

In addition, the rates at which specific proteins are degraded vary with the physiological state. Schimke, Sweeney and Berlin (1965) showed that administration of tryptophan to animals results in accumulation of trypto-phane oxygenase, an effect that results from continued enzyme synthesis with cessation of the normally occurring rapid degradation (half-life of 2 hr). Schimke (1964) also demonstrated that arginase accumulates in rat liver during starvation as a result of decreased degradation in the presence of continued synthesis. Majerus and Kilburn (1969) have shown that starvation increases the rate of degradation of acetyl CoA carboxylase.

The conclusion to be drawn from such studies is that there is a continual flux in which the total complement of any given enzyme is replaced at different rates, and those rates can be altered by a variety of physiological conditions.

The degradation of an enzyme molecule, once synthesized, is a random process. This conclusion is based on the essentially universal finding that the decay of isotopically labeled enzyme after a single isotope administration follows first-order kinetics. The most likely interpretation of this finding is that once a newly synthesized enzyme molecule enters a pool of like molecules, its chance of being degraded (or otherwise removed from the pool of like molecules) is a random process. Thus there is no evidence for an accumulation of damage, i.e., aging, to explain why a given enzyme molecule is degraded.

There is a correlation between protein size and rate of degradation. Dehlinger and Schimke (1970) first showed that large proteins have greater relative rates of degradation than small proteins, as measured by the double-isotope method of Arias, Doyle and Schimke (1969). Figure 1 shows this correlation for rat liver "soluble" proteins. In this technique, an animal receives a single administration of one form of a labeled amino acid ([^{14}C]-leucine) and at some subsequent time (4 days) a single administration of [^{3}H]-leucine. Proteins with high rates of turnover will have large ^{3}H/^{14}C ratios. This same correlation holds for proteins of organelles, including those associated with chromatin (Dice and Schimke 1973) and membranes (Dehlinger and Schimke 1971), as well as cytoplasmic proteins from a variety of tissues of the rat, of HeLa cells and of pea seedlings (Dice, Dehlinger and Schimke 1973).

Such studies have led us to propose that the correlation of size and rate of degradation is based on the overall greater change of a larger protein being "hit" by a protease, producing an initial rate-limiting peptide bond cleavage, with subsequent unfolding and rapid degradation to amino acids. The unexpected finding that intracellular organelles are not turning over as units also leads us to propose that there is a continual association and dissociation of assembled macromolecular structures, and that degradation occurs in the dissociated state. This suggestion has been supported by studies of Tweto, Dehlinger and Larrabee (1972), showing that the dissimilar subunits of the fatty acid synthetase complex of rat liver are turning over at different rates, with large subunits turning over more rapidly than small subunit components. More recently, J. F. Dice and A. L. Goldberg (pers. comm.), based on their review of the literature of these two parameters, have found a significant correlation between subunit size and the rate of degradation of multimeric proteins; however, the correlation is not nearly as strong when based on the multimeric size of the protein.

There is a correlation between the rate of turnover of proteins in vivo and their susceptibility to proteolysis by exogenous proteases. Figure 2 shows the results of pronase digestion of cytoplasmic liver proteins that have been double-labeled in vivo to differentially labeled proteins with rapid turnover, i.e., high ^{3}H/^{14}C ratios. Pronase preferentially releases to a soluble form radioactivity with a high ^{3}H/^{14}C ratio, indicating that proteins that turn over rapidly in the intact animal are preferentially degraded in vitro. The preferen-

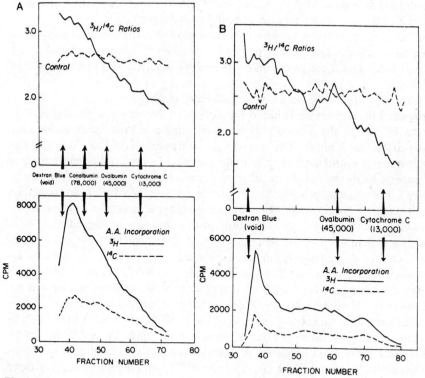

Figure 1

Relative rate of degradation of "soluble" proteins of rat liver as a function of molecular size. Relative rates of degradation were estimated by the double-isotope method of Arias, Doyle and Schimke (1969) in which [14C]leucine is administered to rats four days prior to [3H]leucine administration, with death of animals 4 hours later. The "control" indicates rats receiving both isotope forms of leucine at the same time. High 3H/14C ratios indicate relatively high rates of degradation. Proteins in (A) absence and (B) presence of SDS (to disrupt multimeric proteins) were chromatographed on Sephadex G-200 columns. Details are given in Dehlinger and Schimke (1970). (Reprinted, with permission, from Dehlinger and Schimke 1970.)

tial degradation of these proteins appears to be related to their conformational state, since when the proteins are first unfolded by denaturation in urea and the sulfhydryl groups blocked by aminoethylation, the preferential degradation is not observed (Dice, Dehlinger and Schimke 1973). These same workers have also separated rat liver proteins on the basis of size by chromatography on Sephadex G-100 and have found that both the rate and extent of proteolysis by pronase is dependent on the size of the proteins (Fig. 3).

Abnormal proteins tend to be degraded more rapidly in vivo. The most striking examples of this turnover characteristic are found in instances in which structural mutations in the *lac* repressor (Platt, Miller and Weber 1970) and β-galactosidase (Goldschmidt 1970) result in such rapid degradation that steady-state levels of these proteins are essentially nonexistant. More recently, Goldberg (1972) has shown that incorporation of amino acid analogs into proteins of *E. coli* results in rapid degradation of proteins con-

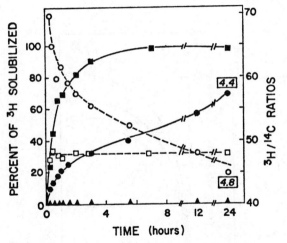

Figure 2

Susceptibility of double-labeled supernatant proteins of rat liver to proteolysis by pronase. A rat weighing 120 g was given 200 μCi of [^{14}C]leucine (uniformly labeled, specific activity 300 mCi/mM) intraperitoneally, followed 5 days later with 400 μCi of L-[^{3}H]leucine (3000 μCi/mM). The animal was killed 6 hours later, and following homogenization in 0.25 M sucrose (2.5:1, sucrose to wet weight liver) and initial centrifugation at 1000 and 10,000g for 30 minutes each, a 100,000g supernatant fraction was obtained. This fraction was freed of amino acids by passage through a column of Sephadex G-25 equilibrated with 0.05 M potassium phosphate pH 7.5. Urea was added to a fraction of the supernatant to the final molarity of 8.0, and pH was adjusted to 9.5 with 3.0 M Tris-OH. After standing at room temperature for 60 minutes, the sulfhydryl groups were blocked by aminoethylation. The denatured protein was dialyzed overnight against a large excess of 2.0 M urea in 0.05 M potassium phosphate pH 7.5. The concentration of both the native and denatured proteins was adjusted to 15 mg/ml, and pronase was added to a final concentration of 80 μg/ml. Two-ml samples were incubated at 37°C with and without pronase, and at the times indicated, 100-μl aliquots were removed and added to 0.05 ml of 10% trichloroacetic acid. The precipitates, after being allowed to sediment at 4°C overnight, were centrifuged, and 0.25-ml samples were removed and extracted 3 times with 2 ml of ethyl ether. Then 0.20-ml samples were counted in a standard dioxane scintillation mixture in the presence of 0.5 ml of NCS solubilizer. Samples of proteins were solubilized in 0.5 ml of NCS solubilizer and counted in the same manner. (▲) No pronase; (●) native proteins, cpm; (○) native proteins, ^{3}H/^{14}C; (■) denatured proteins, cpm; (□) denatured proteins, ^{3}H/^{14}C. (Reprinted, with permission, from Dice, Dehlinger and Schimke 1973.)

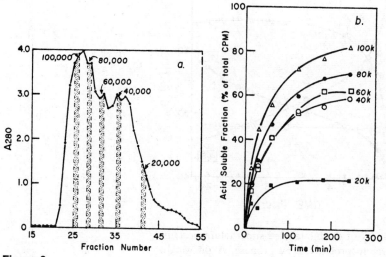

Figure 3

Fractionation of rat liver soluble proteins on Sephadex G-100 in the absence of SDS and in vitro degradation of rat liver soluble proteins of specific molecular size ranges. A rat was given 1 mCi of [³H]leucine and then killed 4 hours later. The soluble proteins were then fractioned on Sephadex G-100. Fractions containing proteins of specific molecular size (shaded areas in *a*) were collected based on the elution volumes of protein markers of known molecular weights (elution positions not shown) and were adjusted by dilution to the same protein concentration (2.0 mg/ml). Proteins of each molecular size range were digested with pronase (100 µg/ml) at 20°C. Samples were removed and acid-soluble radioactivity determined. The acid soluble radioactivity was compared to the total acid-insoluble radioactivity of an equal aliquot before the addition of pronase in order to determine the percentage of the total proteins which had been degraded. The total acid-insoluble radioactivity of 100-µl aliquots before the addition of pronase was between 4000 and 7000 cpm. (*a*) Fractionation of rat liver soluble proteins on Sephadex G-100 in the absence of SDS. (*b*) In vitro degradation of rat liver soluble proteins of approximately (△) 1,000,000; (●) 80,000; (□) 60,000; (○) 40,000; and (■) 20,000 molecular weight. (Reprinted, with permission, from Dice, Dehlinger and Schimke 1973.)

taining amino acid analogs. The same phenomenon exists in animal cells, as shown in the experiment of Figure 4 with cultured WI-38 cells. In these experiments, cells were labeled for 40 hours with [¹⁴C]leucine, or for 30 minutes with [³H]leucine. Labeling for these different time periods will differentially label proteins turning over rapidly (30-min labeling period) or those turning over slowly (40-hr labeling period). In all of the turnover experiments to be described, the cells were subsequently washed ten times with medium containing a large excess of [¹²C]leucine (or [¹²C]arginine, where arginine and canavinine were labeled) to reduce the free labeled amino acid pool. The medium into which the labeled amino acids were subsequently released with time also contained a large pool of ¹²C-amino acid. Details of these methods designed to minimize isotope reutilization are given by Bradley and Schimke

Figure 4

Effect of incorporation of amino acid analogs on protein turnover in WI-38 cells. WI-38 cells (30th passage) were grown in Eagle's basal medium containing 10% calf serum in T-25 flasks. Cells in mid-log phase were labeled with [^{14}C]-leucine (specific activity 208 μCi/mM, 1 μCi per flask) for a total of 40 hours. The medium was then replaced with fresh medium containing either (o) complete medium, (■) one in which 5×10^{-3} M canavinine was substituted for arginine, (△) one in which 5×10^{-3} M azeditine carboxylic acid substituted for proline, or (▲) one in which 5×10^{-3} M p-fluorophenylalanine substituted for phen-

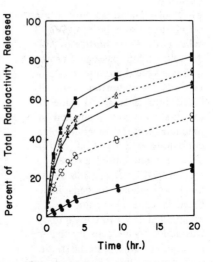

ylalanine. After 30 minutes incubation, 3 μCi of [^{3}H]leucine (specific activity 388 μCi/mM) was added to each flask and incorporation was allowed to take place for 30 minutes. At the end of this time, the flasks were thoroughly washed 10 times with complete medium containing 2 mM [^{12}C]leucine. We have shown (Bradley and Schimke, unpubl.) that this procedure does not alter the turnover characteristics of the cells, but does effectively remove the free intracellular-labeled amino acid pool. Following the washing, 5 ml of complete medium minus phenol red and including [^{12}C]leucine at a concentration of 1×10^{-3} M was added to each flask. At hourly intervals, the medium was removed and replaced with another 5 ml of (prewarmed) medium. One ml of the medium was counted in 10 ml of Instagel (Packard Instruments). We have shown that essentially all the radioactivity released is in the form of free amino acids. At the end of the experiment, the protein in the cells was also counted to obtain total radioactivity present at zero time (i.e., protein plus all radioactivity released). The top four lines in the figure describe the release of short-term ^{3}H label. The bottom line (●) represents the averages of all flasks for the release of the long-term ^{14}C label. All cells were kindly supplied by Dr. Hayflick, Stanford University.

(1975). Figure 4 is a composite of several types of experiments, showing that incorporation of amino acid analogs, canavinine, azeditine carboxylic acid, or p-fluorophenylalanine accelerates the rate of release of isotope in protein to a soluble form. Canavinine is most effective at accelerating degradation. That the analogs do not affect the release of prelabeled protein is shown by the fact that the analogs did not accelerate degrading of the proteins labeled with [^{14}C]leucine prior to analog addition. This indicates that the incorporation of analogs does not accelerate the degradation of cell proteins generally, but rather only those that are newly synthesized in the presence of analogs. We have also demonstrated (Fig. 5) that proteins in which canavinine is incorporated are more rapidly degraded in the test tube by added exogenous protease.

ON THE MECHANISM(S) OF PROTEIN DEGRADATION

As with any biochemical reaction, two components are necessary: the protein molecule to be degraded and the enzymatic apparatus involved in peptide

Figure 5

Effect of canavinine substitution on in vitro proteolysis of WI-38 cells. Four 32-oz bottles of WI-38 cells (28th passage) grown in Eagle's basal medium containing 10% calf serum were divided into 2 lots. At time 0, the first lot was incubated in medium lacking arginine but containing 4×10^{-3} M L-arginine. Fifteen minutes later, 3 μCi of [^{3}H]leucine (388 mCi/mM) was added to cells incubated in canavinine, and 1 μCi of [^{14}C]leucine (208 mCi/mM) was added to the cells incubated in arginine. One hour later, the

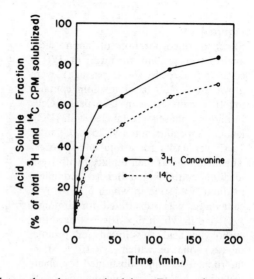

cells were washed, harvested, mixed together, homogenized in a Dounce homogenizer and centrifuged 90 minutes at 100,000g. The supernatant was passed through a Sephadex G-25 column equilibrated with a sodium phosphate buffer, 50 mM, pH 7.5, containing 0.15 M NaCl. The final protein concentration was 1.0 mg/ml. Pronase (Calbiochem B grade) was added at a final concentration of 50 μg/ml, and digestions were undertaken in closed test tubes at 20°C. Aliquots (100 μl) were removed and mixed in 400-μl Beckman plastic microfuge tubes containing 25 μl of 50% trichloroacetic acid and 25 μl of a 10 mg/ml solution of bovine serum albumin. Following centrifugation (15,000g) for 2 minutes, 100 μl of the supernatant was removed and dissolved in 0.7 ml of NCS. Samples of both total precipitated protein and the radioactivity present in a soluble form were counted. Radioactivity released is expressed as a percent of that originally present in the sample. In all studies (this and subsequent), protein radioactivity ranged from 10,000 to 100,000 cpm/mg protein. The NCS overnight digests (37°C) were counted on 10 ml of a toulene-based scintillation fluid employing appropriate double-labeling techniques.

bond hydrolysis. Either factor may be important and rate-limiting for protein degradation.

Protein Molecules as Substrates

Protein molecules exist in various conformational states of differing degrees of detection. Hence one can conceive that a protein molecule would be subject to degradation only when it assumes certain conformations where susceptible peptide bonds are exposed. Heterogeneity of degradation rates would depend on the number or nature of the particularly labile peptide bonds exposed in certain conformational states. Obviously, the conformational state of a protein, including the state of aggregation, will depend on interaction with a variety of ligands, including phospholipids (membranes), cofactors, metabolic intermediates and other proteins. The importance of ligand interaction is suggested by the studies of Litwack and Rosenfield (1973), showing a correlation between in vivo turnover of several pyridoxal-containing enzymes and their affinity for pyridoxal phosphate. Katanuma (1973) has shown that ornithine transaminase is subject to specific proteolysis only in the apo form (without a prosthetic group). The importance of the structure of a protein in determining its rate of degradation is further shown by the studies reviewed herein, and, in particular, by the correlation between in vivo rates of turnover and in vitro rates of proteolysis, and the fact that proteins containing amino acid analogs turn over more rapidly in intact cells and are more rapidly degraded by exogenous proteases.

The model that emerges, then, is one in which protein molecules are individually available to a degradative process which is present at all times. Shifting concentrations of substrates, cofactors, etc., as occur under various metabolic and growth conditions, would lead to a variety of effects on specific enzymes, either to stabilize or labilize them.

Nature and Activity of the Degradative Process

Obviously, the rate of degradation may also be dependent on the activity of the degrading system, as controlled by activation-inhibition, translocation within a cell, or de novo synthesis. Considerations of enzymatic mechanisms are hampered by lack of suitable mutants in the degradative process itself in animal tissues. One obvious candidate for a degradative system is the lysosome, which occurs in virtually all cells (deDuve and Wattiaux 1966). Lysosomes are intracellular organelles that contain acid hydrolases and are currently envisaged to be involved in the autophagy of discrete areas of cytoplasm. It is difficult to conceive that lysosomes are involved in that protein degradation whose properties involve randomness and the heterogeneity of degradation rates of "soluble" proteins or those associated with organelles. Thus some mechanism would be required for the recognition of whether a protein molecule were to be degraded, perhaps involving transport into a lysosome or modification by acetylation, phosphorylation, deamidation or limited cytoplasmic proteolysis. To us, therefore, it seems reasonable to propose that the system of lysosomes is important where cell involution or gross changes in rates of protein degradation occur, such as in starvation and cell death, whereas the degradation that occurs in normal steady-state conditions involves a system(s) not clearly understood at present, perhaps involving lysosomes acting as a "sieve" rather than in an "all-or-none" fashion.

Lysosomes by no means exhaust the possible range of proteases. Most notable among specific proteases are those studied by Katanuma and his group (1973) involving so-called "group-specific" proteases that carry out limited cleavage of pyridoxal-containing enzymes, but only in the apo form. The level of these proteases is also regulated, being elevated in animals on a pyridoxal-deficient diet.

On the Function of Protein Turnover

Certain types of peptide bond cleavage can be readily understood. These include the degradation of N-terminal portions of newly synthesized proteins, a process that can be considered as a terminal step in protein synthesis and folding. Likewise, the extensive degradation of protein that occurs during periods of starvation, when proteins must serve as an energy source or as a source of a limiting amino acid, can be appreciated, although the mechanisms for regulation of these processes are little understood (see Dice and Goldberg 1974). More difficult to explain is the fact that turnover is a continual and extensive process under normal steady-state nutritional conditions in animal tissues, a seemingly wasteful process.

Two functions of protein turnover in animal tissues likely require that turnover be continuous. One is the necessity for removal of enzymatic machinery when that machinery, i.e., metabolic enzymes, are no longer required (see Schimke and Doyle 1970 for examples of shifting enzyme levels in animal tissues). In rapidly growing cells, continual degradation of proteins is not necessary since the cell can outgrow its enzyme complement by dilution of preexisting enzymes. In animal tissues which are not normally growing, degradation of proteins is the only mechanism available for their removal. One possible means of removing enzymes is to have a specific protease that degrades a specific enzyme when no longer required. In theory, then, this would require an enzyme to degrade an enzyme ad infinitum. From an evolutionary standpoint, a mechanism involving unique specific degrading enzymes would require the development of not only the genes for each degradative enzyme, but also the regulatory system to recognize when a given enzyme was to be degraded. It seems highly likely that such a mechanism was too difficult to develop, and hence a system evolved with nonspecific proteases present at all times and with degradation dependent on the properties of the substrate (protein). Coincident with such evolution, then, would be the retention of amino acid substitutions in proteins that either favor stability or lability, depending on whether the enzyme should turn over slowly (and be regulated by activation-inhibition mechanisms) or turn over rapidly, and hence be capable of changing enzyme levels rapidly (see Schimke, Sweeney and Berlin 1965). Such an evolutionary process would then explain the heterogeneity of turnover rates of proteins as being dependent on their nature as substrates.

The other requirement for continual protein turnover relates to the potential errors in information transfer from DNA to RNA to protein, ultimately manifest as functionally and structurally abnormal proteins. Although there are essentially no estimates on error frequency in DNA replication and DNA-dependent RNA synthesis in animal tissues, information is available on error frequency in protein synthesis. Loftfield (1963) has shown that isoleucine, valine and leucine substitute for each other approximately once in every 3000 residues in oviduct synthesis of ovalbumin, and during hemoglobin synthesis in rabbit reticulocytes, valine substitutes for isoleucine approximately once in every 3300 residues. Assuming that such an error frequency is similar for all amino acids, then one can readily calculate that 15% of all proteins in a cell, assuming a mean molecular weight of 50,000 daltons, would contain one amino acid substitution. Some of these errors will allow normal function, others will produce partially functional or nonfunctional proteins.

If proteins with amino acid substitutions were allowed to accumulate in cells, malfunction would likely ensue; this would be of varying degrees of severity depending on whether the abnormalities were in the regulation of cell metabolism only or if they affected the fidelity of protein synthesis or the fidelity of DNA replication. Ultimately, such a process could lead to an increased mutational rate, development of abnormal growth control, or an escalation of protein error frequency of the type proposed by Orgel (1963, 1970) to explain cell aging.

Protein Turnover in Human Fibroblasts as a Function of Age and Viral Transformation

To examine the possibility that errors in proteins accumulate, we have employed the human fibroblast (WI-38) system as studied by Hayflick (1965). These diploid cells undergo approximately 50–60 doublings, following which the growth rate slows and the cells ultimately die. Virally transformed WI-38 cells, in contrast, are capable of unlimited growth (Girardi, Jensen and Koprowski 1965). The system has obvious parallels with aging and has been employed as one model for the aging process. The basic question asked was whether "senescent" cells contained abnormal proteins. If they did, one class of such proteins may be subject to more rapid turnover (by analogy to the studies of Fig. 4 with amino acid analogs) as well as to more rapid degradation by exogenous protease (by analogy to the study of Fig. 5 with canavinine). Figure 6 shows the rate of proteolysis of proteins by pronase from "young" and "senescent" WI-38 cells whose proteins have been labeled for 7 days with [^{3}H]- and [^{14}C]leucine, respectively, and digested in the same reaction mixture. It is evident that the proteins from the "senescent" cells have a fraction of proteins that are degraded more rapidly. Controls include the fact that when either "young" or "old" cells are separately double-labeled with [^{3}H]- and [^{14}C]leucine (separate flasks), combined, and subjected to proteolysis, no difference in release of ^{3}H or ^{14}C is observed. In addition, electrophoresis of proteins from the "young" and "senescent" cells are similar, indicating no gross change in protein makeup of cells of different ages. In addition, proteins from virally transformed cells (SV40) and "young" WI-38 cells show the same rate of exogenous proteolysis and the same protein pattern on acrylamide gels.

In keeping with our previous studies and the evidence of Figure 6, showing a class of more rapidly degradable proteins, we would predict that the rate of protein turnover would be greater in "senescent" cells. This is the case, as shown in Figure 7, in which cells have been labeled with [^{14}C]leucine for 40 hours, followed by a medium replacement containing [^{3}H]leucine for an additional 25 minutes. In all experiments care was taken to insure that cells

Figure 6
In vitro proteolysis of proteins from "young" and "senescent" WI-38 cells by pronase. WI-38 cells were incubated for 7 days with either [^{3}H]leucine (0 passaged doublings before death, pdbd) or with [^{14}C]leucine (33 pdbd) as described in the legend to Figure 5. The cells were harvested and cytoplasmic proteins prepared, incubated with pronase, and radioactivity counted as described in the legend to Figure 5.

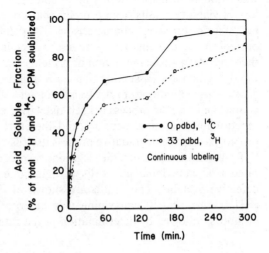

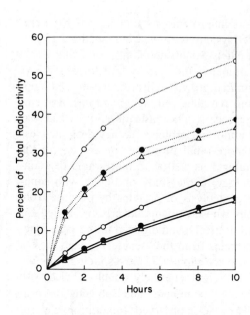

Figure 7
Kinetics of protein degradation in "young," "senescent," and SV40-transformed WI-38 cells. Monolayer cultures were labeled for 40 hours with [^{14}C]leucine starting in mid-log phase of growth. They were subsequently washed 10 times in complete medium containing no label and then incubated for 25 minutes in complete medium containing [^{3}H]leucine. The cells were again washed as described in the legend to Figure 4, and the kinetics of trichloroacetic acid-soluble isotope released was determined as described in the same figure legend. (o) "Senescent" cells (pdbd); (•) "young" cells (33 pdbd); (△) SV40 cells. (····) 25 minutes, [^{3}H]leucine; (——) 40 hours, [^{14}C]leucine.

were in the same growth phase and that comparable radioactivity was incorporated into proteins of cells of each type. In Figure 7 "young" and transformed cells show the same rates of release of radioactivity to the medium, and these rates are similar to those shown in Figure 4. In contrast, the "senescent" cells show elevated release rates both of the rapidly turning over component (^{3}H-labeled) and the slowly turning over component (^{14}C-labeled). These results suggest, therefore, that old cells do contain proteins that are more rapidly degraded, and furthermore, that old cells contain "abnormal" proteins.

The next question, then, is why are the "senescent" cells incapable of sufficiently rapid degradation of "abnormal" proteins to maintain viability. One possibility is that these cells are capable of removing those proteins in which the abnormality results in enhanced degradation in vivo, and that the loss of viability results from proteins that are abnormal in enzymatic function but not in turnover characteristics. Our finding of increased exogenous proteolysis of proteins from "senescent" cells would suggest, however, that there has been an accumulation of proteins that potentially are subject to more rapid degradation (assuming that more rapid exogenous proteolysis indicates greater susceptibility to proteolysis in intact cells). An alternative explanation is that the turnover process is itself defective, either because the proteases are functioning at optimal capacity and therefore incapable of a higher rate of proteolysis, or the degradative process itself is "abnormal."

To determine whether the degradative process was defective, we incorporated [^{14}C]canavinine into cells to determine their capacity to degrade the defective proteins. The results are shown in Figure 8. In these experiments, [^{3}H]arginine or [^{14}C]canavinine was incorporated into separate bottles for 40 hours, followed by determination of the rate of release of the radioactivity.

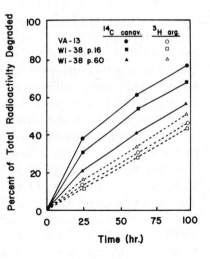

Figure 8

Kinetics of protein degradation in "young," "senescent," and SV40-transformed WI-38 cells labeled with either [^{14}C]canavinine or [^{3}H]arginine. The cells (similar to those employed in Fig. 7) were grown for 40 hours in the presence of either [^{3}H]arginine (····) or [^{14}C]canavinine (——). The cells were washed and then incubated in a complete medium containing 2 mM additional L-[^{12}C]arginine. Release of radioactivity was determined as described in the legend to Figure 4.

The results of the release of [^{3}H]arginine are similar to those shown in Figure 7 with leucine, i.e., "senescent" cells turn over proteins slightly more rapidly than either the "young" or transformed cells. In contrast, the "senescent" cells are less capable of removing the [^{3}H]canavinine-containing proteins than are the "young" cells, indicating an impairment in the degradation process. Most interesting is the finding that the SV40-transformed cells are more capable of removing the abnormal proteins than are the "young" cells. These findings are consistent with our preliminary results indicating that "senescent" cells (6 passage doublings before death) maintained for 24 hours in a medium containing 5×10^{-3} M canavinine with arginine are incapable of recovering growth after replacement with a canavinine-free, arginine-containing medium, whereas the transformed cells recover more rapidly than the "young" cells. Thus these results indicate that "senescent" cells are less capable of degrading abnormal proteins and that the transformed cells are more efficient than "young" cells.

Our results with WI-38 cells support the concept that one function of protein turnover is that of "quality" control of proteins synthesized by a cell. Although repair of DNA is well known, there is no obvious manner in which a protein containing an incorrect amino acid can be repaired without complete degradation. The recent observation of Apte and Zipser (1973) demonstrating covalent linkage of two nonsense fragments of β-galactosidase in *E. coli* to form a functioning enzyme is, however, intriguing in this respect. Our results suggest that abnormal proteins accumulate in WI-38 cells during a period when they become less capable of replication, and that some of these are more rapidly removed than proteins in younger cells. The accumulation of abnormal proteins is consistent with studies of Holliday and Tarrant (1972), who with glucose-6-phosphate dehydrogenase and 6-phosogluconate dehydrogenase demonstrated accumulation of kinetically and physically altered enzyme activity. Our studies do not shed information on the cause of the accumulation of abnormal proteins, but we believe that it can be explained on the basis of error frequency of protein synthesis with the subsequent malfunction of enzymes in a variety of cell processes. Our results also indicate that as the cells senesce, their capacity for degradation of abnormal proteins is reduced. This finding is consistent with demonstrations of increased numbers of lysosomes during aging of WI-38 cells (Cristofalo, Parris and Kritchevsky 1967; Lipetz and Cristofalo 1972; Brock and Hay 1971; Brunk et al. 1973), but our results cannot differentiate at this time between an overwhelming of the degradative process by an excess of proteins capable of degradation, or a degradative mechanism that is itself defective.

One of the most interesting points for speculation is the finding that transformed WI-38 cells appear to be more capable of degrading abnormal proteins than their normal counterparts. This may be consistent with recent findings of increased protease activity in transformed cells. It also provides for speculation that transformed cells can exhibit an infinite life-span because of an enhanced capacity for removing proteins which are otherwise abnormal and would eventually lead to loss of viability and death. Clearly, these speculations require far more experimental inquiry.

Acknowledgment

The work reported here was supported by Research Grant GM 14931 from the National Institutes of Health.

REFERENCES

Apte, B. N. and D. Zipser. 1973. In vivo splicing of protein: One continuous polypeptide from independently functioning operons. *Proc. Nat. Acad. Sci.* **70:**2969.

Arias, I. M., D. Doyle and R. T. Schimke. 1969. Studies on the synthesis and degradation of proteins of the endoplasmic reticulum of rat liver. *J. Biol. Chem.* **244:**3303.

Bock, K. W., P. Siekevitz and G. E. Palade. 1971. Localization and turnover studies of membrane nicotinamide adenine dinucleotide glycohydrolase in rat liver. *J. Biol. Chem.* **246:**188.

Bradley, M. O. and R. T. Schimke. 1975. Protein degradation in normal, transformed, and senescent human fibroblasts. In *Protein Turnover* (ed. R. T. Schimke and N. Katauuma). Academic Press, New York (in press).

Brock, M. A. and E. Hay. 1971. Comparative ultrastructure of chick fibroblasts *in vitro* at early and late stages during their growth span. *J. Ultrastruct. Res.* **36:**291.

Brunk, U., J. L. Ericsson, J. Ponten and B. Westermark. 1973. Residual bodies and "aging" in cultured human glia cells. *Exp. Cell. Res.* **79:**1.

Buchanan, D. L. 1961. Total carbon turnover measured by feeding a uniformly labeled diet. *Arch. Biochem. Biophys.* **94:**500.

Cristofalo, V. J., N. Parris and D. Kritchevsky. 1967. Enzyme activity during the growth and aging of human cells *in vitro*. *J. Cell. Physiol.* **69:**263.

deDuve, C. and R. Wattiaux. 1966. Functions of lysosomes. *Annu. Rev. Physiol.* **28:**435.

Dehlinger, P. J. and R. T. Schimke. 1970. Effect of size on the relative rate of degradation of rat liver soluble proteins. *Biochem. Biophys. Res. Comm.* **40:**1473.

————. 1971. Size distribution of membrane proteins of rat liver and their relative rates of degradation. *J. Biol. Chem.* **246**:2574.

Dice, J. F. and A. L. Goldberg. 1974. Intracellular protein degradation in mammalian and bacterial cells. *Annu. Rev. Biochem.* **43**:835.

Dice, J. F. and R. T. Schimke. 1973. Turnover of chromosomal proteins. *Arch. Biochem. Biophys.* **158**:93.

Dice, J. F., P. J. Dehlinger and R. T. Schimke. 1973. Studies on the correlation between size and relative degradation rate of soluble proteins. *J. Biol. Chem.* **248**:4220.

Eagle, H., K. A. Piez, R. Fleischman and V. I. Oyama. 1959. Protein turnover in mammalian cell cultures. *J. Biol. Chem.* **234**:592.

Fritz, P. J., E. L. White, E. S. Vesell and K. M. Pruitt. 1969. The roles of synthesis and degradation in determining tissue concentrations of lactate dehydrogenase-5. *Proc. Nat. Acad. Sci.* **62**:558.

Girardi, A. J., F. C. Jensen and H. Koprowski. 1965. SV_{40}-induced transformation of human diploid cells: Crisis and recovery. *J. Cell. Comp. Physiol.* **65**:69.

Goldberg, A. L. 1972. Correlation between rates of degradation of bacterial proteins *in vivo* and their sensitivity to proteases. *Proc. Nat. Acad. Sci.* **69**:422.

Goldschmidt, R. 1970. *In vivo* degradation on nonsense fragments in *E. coli. Nature* **228**:1151.

Hayflick, L. 1965. Limited *in vitro* lifetime of human diploid cell strains. *Exp. Cell Res.* **37**:614.

Higgins, M., T. Kawachi and H. Rudney. 1971. The mechanisms of the diurnal variation of hepatic HMG-CoA reductase activity in the rat. *Biochem. Biophys. Res. Comm.* **45**:138.

Holliday, R. and G. M. Tarrant. 1972. Altered enzymes in aging human fibroblasts. *Nature* **238**:26.

Katanuma, N. 1973. Enzyme degradation and its regulation by group-specific proteases in various organs of rats. In *Current Topics in Cellular Regulation* (ed. B. L. Horecker and E. R. Stadtman), vol. 7, p. 175. Academic Press, New York.

Kenney, F. T. 1967. Turnover of rat liver tyrosine transaminase: Stabilization after inhibition of protein synthesis. *Science* **156**:525.

Klevecz, R. 1971. Rapid protein catabolism in mammalian cells is obscured by reutilization of amino acids. *Biochem. Biophys. Res. Comm.* **43**:76.

Kuriyama, T., T. Omura, P. Siekevitz and G. E. Palade. 1969. Effects of phenobarbital on the synthesis and degradation of the protein components of rat liver microsomal membranes. *J. Biol. Chem.* **244**:2017.

Lipetz, J. and V. J. Cristofalo. 1972. Ultrastructural changes accompanying the aging of human diploid cells in culture. *J. Ultrastruct. Res.* **39**:43.

Litwack, G. and S. Rosenfield. 1973. Coenzyme dissociation, a possible determinant of short half-life of inducible enzymes in mammalian liver. *Biochem. Biophys. Res. Comm.* **52**:181.

Loftfield, R. B. 1963. The frequency of errors in protein biosynthesis. *Biochem. J.* **89**:82.

MacDonald, R. A. 1961. "Life-span" of liver cells. *Arch. Int. Med.* **107**:335.

Marjerus, P. W. and E. Kilburn. 1969. Acetyl coenzyme A carboxylase. The roles of synthesis and degradation in regulation of enzyme levels of rat liver. *J. Biol. Chem.* **244**:6254.

Marver, H. S., A. Collins, D. P. Tschudy and M. Rechcigl, Jr. 1966. δ-Aminolevulinic acid synthetase. *J. Biol. Chem.* **241**:4323.

Niemeyer, H. 1967. Regulation of glucose-phosphorylating enzymes. In *National*

Cancer Institute Monograph (ed. M. P. Stulberg), vol. 27, p. 29. NCI, Washington, D.C.

Orgel, L. E. 1963. The maintenance of the accuracy of protein synthesis and its relevance to aging. *Proc. Nat. Acad. Sci.* **49:**517.

————. 1970. Maintenance of the accuracy of protein synthesis and its relevance to aging: A correction. *Proc. Nat. Acad. Sci.* **67:**1476.

Platt, T., J. H. Miller and K. Weber. 1970. *In vivo* degradation of mutant *lac* repressor. *Nature* **228:**1154.

Price, V. E., W. R. Sterling, V. A. Tarantola, R. W. Hartley, Jr. and M. Rechcigl, Jr. 1962. The kinetics of catalase synthesis and destruction *in vivo*. *J. Biol. Chem.* **237:**3468.

Russell, D. and S. H. Snyder. 1968. Amino synthesis in rapidly growing tissues: Ornithine decarboxylase activity in regenerating rat liver, chick embryo, and various tumors. *Proc. Nat. Acad. Sci.* **60:**1420.

Schimke, R. T. 1964. The importance of both synthesis and degradation in the control of arginase levels in rat liver. *J. Biol. Chem.* **239:**3808.

————. 1970. Regulation of protein degradation in mammalian tissues. In *Mammalian Protein Metabolism* (ed. H. N. Munro), vol. 4, p. 177. Academic Press, New York.

Schimke, R. T. and D. Doyle. 1970. Control of enzyme levels in animal tissues. *Annu. Rev. Biochem.* **39:**929.

Schimke, R. T., E. W. Sweeney and C. M. Berlin. 1965. The roles of synthesis and degradation in the control of rat liver tryptophan pyrolase. *J. Biol. Chem.* **240:**322.

Segal, H. L. and Y. S. Kim. 1963. Glucocorticoid stimulation of the biosynthesis of a glutamic-alanine transaminase. *Proc. Nat. Acad. Sci.* **50:**912.

Swick, R. W. 1956. The measurement of nucleic acid turnover in rat liver. *Arch. Biochem. Biophys.* **63:**226.

————. 1958. Measurement of protein turnover in rat liver. *J. Biol. Chem.* **231:**751.

Swick, R. W., A. K. Rexroth and J. L. Stange. 1968. The metabolism of mitochondrial proteins. III. The dynamic state of rat liver mitochondria. *J. Biol. Chem.* **243:**3581.

Tweto, J., P. J. Dehlinger and E. Larrabee. 1972. Relative turnover rates of subunits of rat liver fatty acid synthetase. *Biochem. Biophys. Res. Comm.* **48:**1371.

Proteolytic Mechanisms in the Biosynthesis of Polypeptide Hormones

**Donald F. Steiner, Wolfgang Kemmler,* Howard S. Tager,
Arthur H. Rubenstein, Åke Lernmark,† and Hartmut Zühlke****

Departments of Biochemistry and Medicine, The University of Chicago
Chicago, Illinois 60637

The biosynthesis of proteins and polypeptides by way of larger precursor forms is not in itself a novel concept. Zymogen forms of a number of enzymes have been recognized for many years (Neurath, this volume). These precursors are secreted by a variety of cells into the gut or the circulating plasma, where activation occurs by limited extracellular proteolysis. Many of these systems are discussed in this volume. In recent years, however, it has become increasingly apparent that such proteolytic mechanisms are not limited to zymogen activation or to general protein catabolism (Schimke and Bradley, this volume), but are also important aspects of the biosynthesis of many other biologically active proteins.

With the discovery of proinsulin in 1967 (Steiner and Oyer 1967; Steiner et al. 1967), it was recognized that limited proteolysis also plays an important role in the biosynthesis of insulin. Since then, similar mechanisms have been found to participate in the formation of a variety of other peptide hormones as well as of viral capsule proteins (Jacobson and Baltimore 1968; Korant, this volume), serum albumin (Geller, Judah and Nicholls 1972; Judah, Gamble and Steadman 1973) and even connective tissue structural proteins such as collagen (Bornstein, Davidson and Monson, this volume). A distinctive feature of the process of insulin biosynthesis—and one which sets it apart from classical zymogen activation—is the proteolytic conversion of the precursor to the hormone prior to its storage and secretion from the beta cells of the pancreatic islets (Steiner 1967). The intracellular localization and precise mechanism of this proteolytic process in the beta cell is a problem of considerable interest from a cell biologist's point of view. This system may also prove to be a useful model for the study of the biosynthesis of a number of important

* Present address: Städt. Krankenhaus München-Schwabing, 8000 München 23, Kölner Platz 1, F.R.G.

† Visiting scientist from the Department of Histology, University of Umeå, Umeå, Sweden.

** Visiting WHO Fellow from the Institute for Diabetes "Gerhard Katsch," Karlsburg, G.D.R.

cellular constituents, such as membrane-localized proteins, various intracellular organelles, and perhaps even of intracellular enzymes (Steiner et al.
1972). This review will summarize available information on the structures of
insulin, proinsulin and other secretory products of the beta cell and on the
intracellular localization, specificity and mechanism of the proteolytic cleavages occurring during the biosynthesis of insulin. The probable existence of
similar cleavage processes active in the biosynthesis of a number of other small
polypeptide hormones from larger precursor polypeptides and the biological
significance of these precursor forms will also be considered briefly.

The Structures of Insulin and Proinsulin

Insulin is a small globular protein, the amino acid sequence and three-dimensional structure of which are known (Ryle et al. 1955; Smith 1972a; Blundell
et al. 1972). Perhaps because it contains a high proportion of hydrophobic
residues, insulin readily associates in solution to form isologous dimers and,
under certain conditions, higher polymers (Fig. 1). Hexamer formation,
which occurs in most mammalian insulins, is initiated during crystallization by
the formation of a coordination complex between 6 molecules of insulin (3
dimers) and 2 atoms of zinc, forming a major threefold symmetry axis in the
crystals (Blundell et al. 1972).

Proinsulin, the biosynthetic precursor of insulin, consists of a single
polypeptide chain ranging in length from 78 (dog) to 86 (human, horse, rat)
amino acid residues (Steiner et al. 1973; Chance, Ellis and Bromer 1968).
The variations in length in the mammalian proteins occur only in the connecting polypeptide portion which links the carboxyl terminus of the insulin B
chain to the amino terminus of the insulin A chain. The primary structure of
bovine proinsulin (Nolan et al. 1971) is shown in Figure 2. All the known
mammalian proinsulins have pairs of basic residues at either end of the connecting peptide which link the connecting polypeptide to the insulin chains.
These residues are excised during the conversion of proinsulin to insulin, and
the resulting products are native insulin plus the remainder of the connecting
polypeptide segment lacking amino- or carboxyl-terminal basic residues
(Steiner et al. 1971). This peptide, which has been designated the C peptide,
is retained in the secretion granules along with the insulin and is liberated with
it in essentially equimolar amounts during secretion (Rubenstein et al. 1969).
The biological activity, metabolism and significance of circulating proinsulin
and C peptide in humans and other species has been studied intensively in
several laboratories (for reviews, see Rubenstein, Melani and Steiner 1972;
Horwitz and Rubenstein 1974).

The three-dimensional structure of proinsulin has not yet been determined.
Crystallization has been accomplished by Low and coworkers (Fullerton,
Potter and Low 1970), however, and further progress by means of X-ray
analysis is to be anticipated. Evidence from a variety of preliminary studies
strongly suggests that the conformation of the insulin moiety in proinsulin is
nearly identical to that of insulin itself (Steiner et al. 1972). It is of interest
that the connecting peptide is much larger than would seem to be required to
bridge the short 8-Å gap between the ends of the B and A chains (see Fig. 1).
Although the connecting peptide may be folded over a portion of the surface

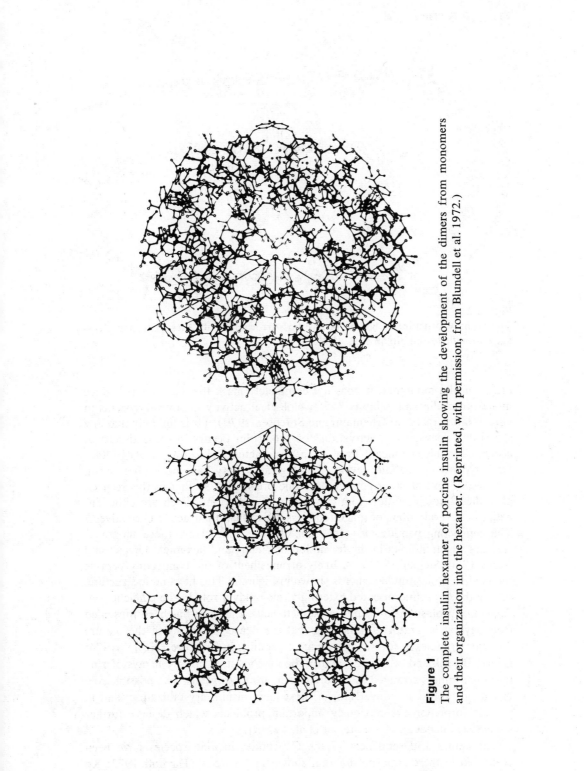

Figure 1
The complete insulin hexamer of porcine insulin showing the development of the dimers from monomers and their organization into the hexamer. (Reprinted, with permission, from Blundell et al. 1972.)

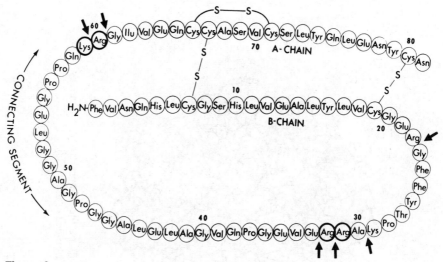

Figure 2
The primary structure of bovine proinsulin. Arrows indicate sites of tryptic cleavage. (Modified from Nolan et al. 1971.)

of the insulin monomer, it does not completely mask the "active site," since intact proinsulin still exhibits 3–5% biological activity in several systems in vitro (Narahara 1972; Gliemann and Sorensen 1970). (It is unlikely that any significant cleavage or "activation" of proinsulin occurs in these tissues to account for this significant level of intrinsic activity [Lazarus et al. 1970b]. Similarly low but definite levels of biological activity have been observed in the case of several other peptide hormone precursors, so that in this respect, also, these proteins differ from classical zymogen forms which are either inactive or several orders of magnitude less active than their active derivatives.) The connecting peptide also does not obscure appreciably those monomer surfaces which interact in the formation of dimers and hexamers (Frank and Veros 1970; Steiner 1973). A likely arrangement of the connecting peptide moiety in a proinsulin hexamer is shown in Figure 3. This hexameric structure, having the C peptide oriented externally, may play a role in the efficient conversion of proinsulin to insulin in the beta cells. Evidence has been presented from several laboratories indicating that the dense inclusions in the mature secretion granules in several species are crystalline in structure (Howell 1974; Lange, Boseck and Ali 1972) and probably consist of orderly arrays of zinc insulin hexamers arranged as in porcine zinc insulin crystals. The possible role of zinc in proinsulin conversion, in subsequent insulin crystallization, and in granule formation are obviously important problems which deserve further study (Falkmer et al. 1973; Steiner et al. 1974a).

The amino acid sequences of the C peptides in nine species have been combined in the composite diagram shown in Figure 4 (Dayhoff 1973; Ko et al. 1971; Markussen and Sundby 1972; Smyth, Markussen and Sundby 1974; Oyer et al. 1971; Peterson et al. 1972; Sundby and Markussen 1970; Tager and Steiner 1972). These peptides show much greater variability in

Figure 3

Hypothetical proinsulin hexamer as viewed along the threefold axis of the hexamer. The central density represents the two zinc atoms which coordinate to the six histidine side chains at position 10 in the B chains. The densities surrounding the zinc atoms and extending toward them represent the six insulin monomers outlined according to the data of Blundell et al. (1972). The connecting polypeptide region is shown in lighter gray around the peripheral portion of the insulin hexamer, and the sites of attachment to the insulin chains are indicated by the arms.

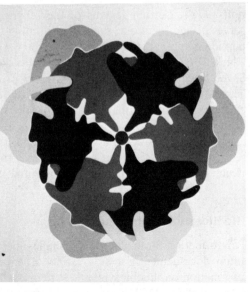

structure than do the corresponding insulins, a finding consistent with the possibility that this region in the proinsulin molecule does not contain an active center for a specific hormonal function. Among known proteins, only the fibrinopeptides have a higher rate of mutation than the proinsulin C peptides. The much lower rate of mutation of insulin, however, is similar to that of many other functional proteins, such as hemoglobin or cytochrome c (Day-

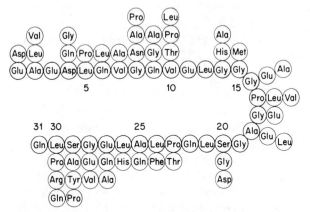

Figure 4

Amino acid sequence of human proinsulin C peptide combined with the known substitutions occurring in eight other mammalian and one avian C peptide shown alongside. Deletions occur in the dog (residues 4–11), pig (residues 18, 19), sheep and ox (residues 22–26) and guinea pig (residues 25, 26). (These sequences do not include the basic residues at either end which link the C peptide to the insulin chains.)

hoff 1972). Certain regions of the relatively large connecting peptide may serve specific functions, such as facilitating the folding of the proinsulin polypeptide chain and the formation of the correct disulfide bonds (Steiner and Clark 1968) or guiding the enzymatic cleavage of proinsulin to insulin. Several acidic residues are consistently present in the connecting peptides. These tend to offset the cationic charges due to the basic residues at the cleavage sites so that the isoelectric pH of proinsulin is nearly the same as that of insulin, i.e., pH 5.1–5.5 (Steiner et al. 1972; Kohnert et al. 1972; Kohnert et al. 1973). It is also possible that translational constraints such as mRNA sequence, size or secondary structure may play a role in dictating the primary structure of some regions within proinsulin.

The Biosynthesis of Insulin

The formation of insulin via the single-chain polypeptide proinsulin has been amply documented in a wide variety of vertebrate species, and much relevant information on this biosynthetic system has been presented in several reviews (Steiner et al. 1972; Steiner et al. 1969; Grant and Coombs 1971; Steiner et al. 1974a). The biosynthesis of the polypeptide chain occurs on the rough endoplasmic reticulum (Sorensen, Steffes and Lindall 1970; Permutt and Kipnis 1972a,b), and the folding of the molecule accompanied by disulfide bond formation probably occurs shortly after synthesis. The newly synthesized proinsulin is then transferred via an energy-dependent process from the cisternae of the rough endoplasmic reticulum to the Golgi apparatus, where cleavage to insulin begins (Steiner et al. 1970). Granules are rapidly formed by enclosure of a portion of the Golgi apparatus contents within membranes derived from the Golgi lamellae. The granules then move into the cytosol where they undergo biochemical maturation associated with further cleavage of the proinsulin to insulin and the ultimate condensation and crystallization of the derived insulin. The important features of this process are summarized schematically in Figure 5.

Conversion of Proinsulin to Insulin

Our hypotheses regarding the mechanism of conversion of proinsulin to insulin are based on several lines of evidence. These include: (1) the known structures of the cleavage products and of a number of intermediate forms, (2) model studies with known proteolytic enzymes, and (3) the detection of converting enzyme activities in whole islet preparations or appropriate subcellular fractions (Steiner et al. 1974b). To date, these approaches have not provided definitive evidence on the origin or subcellular localization of the proteases involved, or on their mechanism and specificity. Although it has been possible to detect activities which produce insulinlike material from proinsulin in extracts of whole pancreas (Yip 1971) or in homogenates of islets of Langerhans (Zühlke et al. 1974; Smith 1972b; Smith and Van Frank 1974; Sorensen, Shank and Lindall 1972), the intracellular origin and mechanism of action of these enzymes are uncertain.

The major types of proteolytic cleavage required for the conversion of pro-

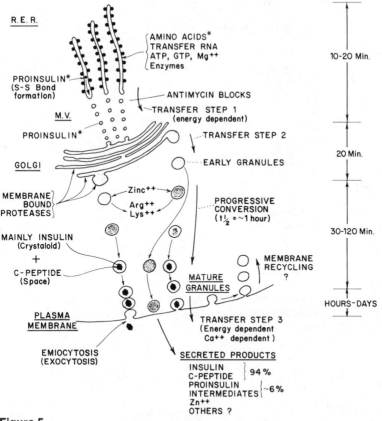

Figure 5

Schematic summary of the insulin biosynthetic mechanism of the pancreatic beta cells. (See text for details regarding this process.) The time scale on the right side of the figure indicates the time required for each of the major stages in biosynthesis. (R.E.R., rough endoplasmic reticulum; M.V., microvesicles.)

insulin to insulin are shown in Figure 6. This scheme envisions the combination of a trypsinlike protease with another having specificity similar to that of carboxypeptidase B. The latter enzyme is necessary to remove the C-terminal basic residues left after tryptic cleavage, giving rise to the important naturally occurring products, the C peptide and native insulin. We have shown that appropriate mixtures of pancreatic trypsin and carboxypeptidase B can quantitatively convert proinsulin to insulin in vitro (Kemmler, Peterson and Steiner 1971). This model system can account for the known major intermediate forms and products that occur naturally in pancreatic extracts (Nolan et al. 1971; Steiner et al. 1971). In some species, such as rats, additional cleavages occur in the C-peptide region of proinsulin which appear to be due to a protease having chymotrypsinlike activity (Tager et al. 1973; Chance 1971). The role of this additional C-peptide cleavage in conversion remains unclear, however, and it probably occurs only in species where the proinsulin C peptide contains sites of high chymotryptic sensitivity (Fig. 7).

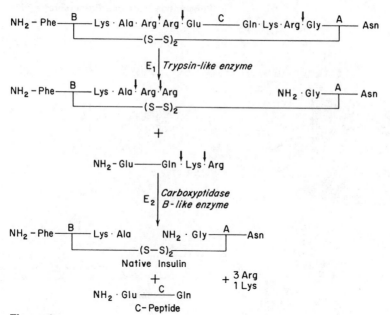

Figure 6
Stages in the cleavage of proinsulin by the combined action of trypsin-like and carboxypeptidase B-like proteases. (See text for further discussion of this model system.)

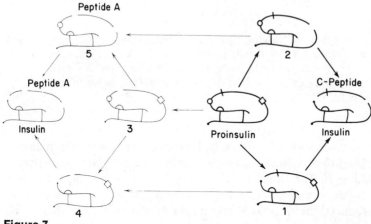

Figure 7
Diagrammatic representation of the pathways for the conversion of proinsulin to insulin in the rat. Structures 1–5 represent possible proinsulinlike intermediates. The major pathway for proinsulin conversion is shown in heavy lines. The sites on either side of the C peptide filled by pairs of basic amino acid residues are indicated by circles and squares. Appropriate cleavage at these sites to yield insulin and C peptide requires both trypsinlike and carboxypeptidase B-like activities. The slash represents a chymotrypsin-sensitive site in the C-peptide region of proinsulin. The NH_2 termini of the A and B chains of insulin are indicated by dots.

These findings suggest that the beta granules may contain a mixture of proteases, many of which are similar to those that occur in the exocrine pancreas. Thus the specific cleavage of precursor forms may be dictated partially by the high sensitivity of certain regions in the substrate molecules to a variety of known proteases, as well as by restricted specificities or special substrate adaptations on the part of the converting proteases.

Structural data on insulins and proinsulins in fish suggest that the conversion of proinsulin to insulin in many of these species may be carried out solely by trypsinlike enzymes (Yamaji, Tada and Trakatellis 1972; Steiner et al. 1974a). A trypsinlike protease has been detected in codfish islet tissue by Grant and coworkers (Grant and Coombs 1971; Grant et al. 1971). However in the hagfish (a primitive cyclostome), the presence of a C-terminal neutral residue (Met) on the insulin B chain, as in the higher forms, suggests the possible additional involvement of a carboxypeptidase (Peterson et al. 1974). Carboxypeptidase-B activity is known to occur in the exocrine pancreas of several species of fish, as well as in echinoderms (Neurath, Walsh and Winter 1968). Chymotrypsinlike cleavages have not yet been encountered in species below mammals.

Thus far we have had only partial success in demonstrating the existence of trypsinlike and carboxypeptidase B-like proteolytic activities in isolated beta cell granule preparations (Kemmler, Steiner and Borg 1973; Steiner et al. 1974b). However, crude granule fractions from rat islets previously labeled with [^{3}H]leucine convert the endogenously labeled proinsulin to insulin during incubation at 37°C and pH 6.3. The initial rates of conversion are comparable to those observed in vivo (Kemmler and Steiner 1970). Externally added labeled bovine or rat proinsulin is not converted, however, indicating that the proteolysis is occurring only within the particulate elements. Disruption of the granules by extremes of pH, various detergents, sonication or repeated freeze-thawing markedly inhibits the conversion (Kemmler and Steiner 1970).

We have examined freeze-thaw lysates of these crude granules for proteolytic enzymes active in the pH range 6.5–8.0, but these lysed preparations do not measurably convert labeled proinsulin to insulin. Low levels of esterase activity have been detected, however, using [*methyl*-^{3}H]TAME (*N*-α-tosyl-L-arginyl methyl ester), a substrate for many of the known trypsinlike enzymes (Kemmler, Steiner and Borg 1973). Similarly, esterase activity could also be detected in the granule fraction prepared from a human islet cell tumor. The tumor enzyme was inhibited by DFP (diisopropylfluorophosphate) and TLCK (*N*-α-tosyl-L-lysyl chloromethane ketone), but not by pancreatic or soybean trypsin inhibitor (Kemmler, Steiner and Borg 1973).

More recently, we have increased the sensitivity and reliability of the [^{3}H]TAME assay of Troll and coworkers (Roffman, Sanocka and Troll 1970) by adding a protective protein to the buffer and carrying out the reaction in small siliconized test tubes to minimize losses of enzyme due to adsorption. With this improved assay procedure, we can detect as little as 5 pg of trypsin (Steiner, unpubl.). In preliminary studies we have detected low levels of esterase activity in islet homogenates and have also observed that the esterase activity is liberated from intact islets during incubation in vitro. The amount of enzymatic activity released is increased by glucose combined with the phosphodiesterase inhibitor IBMX (3-isobutyl-1-methyl xanthine), both stimulators of insulin secretion (Steiner and Lernmark, unpubl.). This activity can be detected at pH 7.5 and is inhibited by both DFP and soybean trypsin inhibitor. This approach may provide a convenient means for further characterizing these proteases under conditions limiting the presence of contaminating cellular proteases.

It has been less difficult to demonstrate the presence of carboxypeptidase B-like activity in crude granule lysates using [^{3}H]arginine-labeled peptides derived from [^{3}H]arginine-labeled proinsulin by treatment with limiting amounts of trypsin, i.e., insulin B chain Arg-Arg-COOH and C peptide-Lys-Arg-COOH (Kemmler, Steiner and Borg 1973; Kemmler et al. 1972). When these substrates were incubated for 5 hours with granule lysates prepared from rat islets, the lysates liberated approximately 80% as much free labeled arginine as did treatment with excess carboxypeptidase B. These results indicate the presence of an exopeptidase in the granule fraction that has cleavage specificity similar to that of carboxypeptidase B.

To study this carboxypeptidase B activity in greater detail, we have recently utilized a fluorometric procedure using hippuryl arginine as the substrate and fluorescamine to detect the release of free arginine (Zühlke and Steiner, unpubl.). With this method, it its possible to detect carboxypeptidase B-like activity (inhibited by EDTA and o-phenanthroline, but not by DFP or iodoacetamide) in homogenates of whole islets as well as in the incubation medium of incubated islets (Zühlke and Steiner, unpubl.). These results thus confirm our earlier findings of exopeptidase activity directed toward C-terminal basic residues in lysed granule fractions (Kemmler, Steiner and Borg 1973) and provide further evidence that a carboxypeptidase B-like enzyme is indeed present in the beta cells.

The lack of sufficient amounts of endopeptidase (tryptic) activity in the lysed granule fraction to convert small amounts of added beef or rat proinsulin (labeled with either ^{3}H-amino acids or ^{131}I) was mentioned earlier, and this may be due to inactivation of the enzyme during disruption of the granules or to dilution of substrate and enzyme after lysis. Indeed, Zühlke and coworkers (Zühlke et al. 1974) have recently shown that disrupted secretion granule fractions convert proinsulin to insulinlike components in a DFP-sensitive reaction when incubated at pH 6.5 with considerably higher concentrations of proinsulin than were used in the experiments cited above. These workers also have described the splitting of a synthetic heptapeptide derived from the amino acid sequence of porcine proinsulin (i.e., residues 28–34 having [^{14}C]arginine at position 31) by islet homogenates and crude secretion granule fractions (Zühlke, Jahr, Schmidt and Kirschke, unpubl.). Free [^{14}C]arginine could be detected, providing evidence for the action of both an endopeptidase and a carboxypeptidase B-like enzyme.

Sun, Lin and Haist (1973) have recently provided evidence that the proinsulin-converting activity of disrupted granule fractions is localized in a "membrane fraction." Unfortunately, this fraction was not characterized very extensively. Such a membrane association of the endopeptidase activity might limit proteolytic activity in vivo and thus afford a means for protecting insulin from further proteolytic degradation in the granules. The possibility that the converting enzymes are related to some of the lysosomal proteases has been considered by several workers (Steiner et al. 1972; Smith 1972b). However, little of the currently available information on these enzymes is consistent with a lysosomal origin (for a more extensive discussion, see Steiner et al. 1972).

The relationship of the proinsulin-transforming enzymes to the exocrine pancreatic proteases poses interesting questions. Some investigators believe that the islet cells and the exocrine pancreatic cells share a common embryological derivation from the endoderm of the gut (Wessels and Evans 1968; Falkmer et al. 1973). Moreover, Rutter and coworkers have shown that trypsin and carboxypeptidase B appear together at a later time during the embryonic development of the pancreas than do many of the other pancreatic hydrolases (Rutter et al. 1968). This behavior suggests that these two enzymes are closely coordinated both developmentally as well as functionally. The recent finding of Frazier, Angeletti and Bradshaw (1972) of striking sequence homologies between proinsulin and the mouse submaxillary gland nerve growth factor also would seem consistent with an endodermal origin of the beta cells (Steiner 1974). Further comparative structural studies of proinsulins from primitive vertebrates will hopefully elucidate the evolutionary origin of this precursor and shed more light on the evolution of the associated cleavage enzymes.

Other Peptide Hormone Precursors

As indicated earlier in this report, a wide variety of zymogen proteins have been known for many years, but their proteolytic activation has been observed only in extracellular spaces. The intracellular cleavage of proinsulin to insulin before secretion is an unusual feature of the insulin biosynthetic mechanism and is one which has now been recognized to occur in the biosynthesis of at least one other peptide hormone, the parathyroid hormone (Cohn et al. 1972; Hamilton et al. 1971; Kemper et al. 1972), and it probably occurs in others as well. The time course of biosynthesis and conversion of proparathyroid hormone in the parathyroid glandular cells is strikingly similar to the observed time course for proinsulin biosynthesis in the beta cells (Cohn et al. 1972). Likewise, as has been observed in about 80% of pancreatic beta cell tumors (Lazarus et al. 1970a; Melani et al. 1970; Sherman et al. 1971), parathyroid adenomas often secrete abnormally large amounts of unprocessed precursor into the circulation, a finding that can be diagnostically useful (Habener et al. 1972).

Highly suggestive evidence also indicates the existence of precursor forms for gastrin (Yalow and Berson 1971; Gregory and Tracy 1972), glucagon (Noe and Bauer 1971, 1973; Tung and Zerega 1971; Tager and Steiner 1973) and MSH (Chretien and Li 1967; Scott et al. 1973). Currently available structural information on these forms is summarized schematically in Figure 8, which shows proinsulin, the proparathyroid hormone (Hamilton et al. 1973; Potts et al. 1973), the large gastrins (Gregory and Tracy 1972; Gregory 1974), a proglucagon fragment (Tager and Steiner 1973), β-MSH (via β- and γ-lipotropins) (Chretien and Li 1967) and α-MSH (via ACTH as its precursor) (Scott et al. 1973). Sachs and coworkers (Sachs et al. 1969) have also suggested that vasopressin may be synthesized via a precursor form, but the nature of this protein has not been established. Size heterogeneity in the circulating forms of still other peptide hormones has been reported, in-

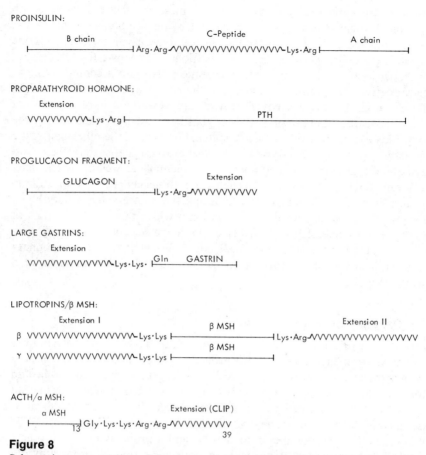

Figure 8

Schematic representation of the structures of several peptide hormone precursor forms. (In the cases of the large gastrin, MSH, and glucagon-containing peptides, definitive biosynthetic evidence of a precursor relationship is lacking at this time.)

cluding ACTH (Yalow and Berson 1972), but no definitive information on the structure or origin of these forms is now available. In view of the marked heterogeneity in circulating ACTH, parathyroid hormone and others, it seems prudent to redefine the hormone in each case as the *major, active* circulating peptide derived from the precursor in the healthy individual.

Inspection of the data (Fig. 8) reveals no consistent pattern of ordering within the precursor peptide chains, i.e., the hormonally active moiety may occur either at the N terminus (ACTH/α-MSH, proglucagon), at the C terminus (proparathyroid hormone, large gastrin), at both termini (proinsulin) or only internally (β-lipotropin/β-MSH). On the other hand, the data shown in Figure 7 strongly suggest the existence of similar proteolytic cleavage mechanisms for all of these putative precursors, i.e., all these peptides contain a pair (or more) of basic residues at the sites where cleavage is required to liberate the hormone. This consistent pattern suggests the existence of intracellular cleavage enzymes having trypsinlike and, in at least some instances, carboxypeptidase B-like specificity. Moreover, the occurrence of the basic residues in pairs suggests the possibility that the intracellular trypsinlike protease in each case may have a special restriction in its specificity (i.e., to require a pair of basic residues) as an adaptation to its role in converting peptide hormone precursors. Isolation of these converting proteases will be necessary to confirm this interesting possibility.

A somewhat more general interpretation of the information summarized in Figure 8 would be that trypsinlike enzymes, with or without accessory carboxypeptidase B-like components, may occur widely in cells and participate in the biosynthesis or turnover of a variety of cellular proteins as well as of secretory products. The biosynthesis of cellular macromolecular complexes, membrane proteins and organelles might be facilitated and coordinated by such precursor processing mechanisms (Steiner et al. 1972). This idea appears to be supported by the finding that proteolytic cleavage of larger precursor polypeptides occurs in the formation of the capsids of the picorna group of animal viruses (Butterworth et al. 1971; Jacobson and Baltimore 1968; Kiehn and Holland 1970). However, the origin, intracellular localization and nature of the proteases involved in these processing events have not yet been fully clarified (Summers et al. 1972; Korant, this volume).

On the other hand, the occurrence of a common mechanism of precursor processing among the diverse endocrine systems mentioned above may be consistent with the notion that the associated glands have all originated from embryologically related cell types, such as endoderm (Adelson 1971) or neural crest cells (Pearse 1974). Such a developmental relationship, however, has not yet been established for most endocrine cells. Of course, it is likely that most of the peptide hormones have been derived in evolution from previously existing gene products which originally served some other role in the cellular economy, perhaps as secreted enzymes such as digestive hydrolases (Steiner et al. 1969; Adelson 1971; Steiner et al. 1973), as enzymes in biosynthetic pathways, or even as cell surface constituents. Proteolytic cleavage of these gene products may have produced fragments having effector activity as well as other desirable metabolic characteristics, such as rapid turnover in the circulation. Thus these fragments may have eventually emerged as the major secreted forms of the hormones. The fact that the "unwanted" regions in these precursors have not been trimmed away at the level of the structural genes suggests the existence of some transcriptional or translational constraints with regard either to gene or mRNA size and structure. On the other hand, the polypeptide hormones may have arisen relatively recently in evolution and due to the preexistence of suitable processing mechanisms within their cells of origin, may have been subject to relatively little selective pressure to reduce the size of the translated unit.

Hormone precursor forms may serve many useful functions either in their cells of origin during biosynthesis, storage and secretion or in the whole organism. Some possible functions of peptide hormone precursors are listed below:

1. Structure-making, i.e., promoting peptide chain folding and disulfide bond formation (proinsulin).
2. Overcoming transcriptional or translational constraints.

3. Regulating hormone formation: (a) peptide chain biosynthesis (rate and localization within cell); (b) vectorial discharge and sequestration; (c) transport within cell; (d) specific cleavage; (e) stabilization and storage (secretion granule formation); (f) feedback control of hormone secretion.
4. Providing hormone-related material having altered biological characteristics: (a) biologic activity (receptor-binding properties); (b) metabolism and degradation (biological half-life); (c) vascular permeability.
5. Providing scientists with indications as to the evolutionary origins of hormones: (a) modification of preexisting gene products (enzymes, structural proteins); (b) derivation by cleavage from regulatory proteins in cell membranes; (c) persistence due to lack of selective pressure against nonfunctional sequences.

Much more structural, molecular biological and physiological information will obviously be necessary before a clear understanding of the significance of these forms can be achieved.

SUMMARY

There is increasing evidence indicating that a number of small peptide hormones are derived in biosynthesis as cleavage products of larger precursor polypeptides. Proinsulin, the first such precursor form to be isolated, consists of a single polypeptide chain: NH_2-B chain-Arg-Arg-C peptide-Lys-Arg-A chain-COOH. The proteolytic conversion of proinsulin to insulin occurs within the beta cells of the islets of Langerhans during the formation and maturation of the secretory granules and results in the liberation of insulin and the free C peptide. The intracellular localization and kinetic behavior of this process was discussed. Attempts to isolate or characterize the converting protease(s) have been only partially successful, but the presently available data, as well as in vitro model studies with known proteases, indicate that both endopeptidase and exopeptidase activities are required. A combination of trypsinlike and carboxypeptidase B-like activities would appear to fulfill the major requirement, but in some species, chymotrypsinlike minor cleavages in the C-peptide region also occur. From the limited structural data available on glucagon, gastrin, parathyroid hormone and melanocyte-stimulating hormone (MSH) precursors, it is noteworthy that paired basic residues occur in every case at cleavage sites. This idiomatic pattern suggests the possible existence in a number of tissues of more specialized trypsinlike enzymes. Alternate theories and further experimental approaches to these problems were considered.

Acknowledgments

The authors are indebted to the many students and colleagues who have made important contributions to various aspects of this work, including E. Margoliash, C. Nolan, F. Melani, H. Sando, J. Clark, P. Oyer, J. D. Peterson and S. Terris. Various aspects of this research have been supported by grants from the USPHS (AM-1314); the Lolly Coustan Memorial Fund; the Novo Research Institute; the Lilly Research Institute; the Kroc Foundation; the UC

Diabetes-Endocrinology Center (AM-17046); and the UCCRC (CA-14599). We thank Miss Roberta Erfurth for her assistance in the preparation of this manuscript.

REFERENCES

Adelson, J. W. 1971. Enterosecretory proteins. *Nature* **229**:321.

Blundell, T., G. Dodson, D. Hodgkin and D. Mercola. 1972. Insulin: The structure in the crystals and its reflection in chemistry and biology. *Adv. Protein Chem.* **26**:279.

Butterworth, B. E., L. Hall, C. M. Stoltzfus and R. R. Rueckert. 1971. Virusspecific proteins synthesized in encephalomyocarditis virus-infected HeLa cells. *Proc. Nat. Acad. Sci.* **68**:3083.

Chance, R. E. 1971. Chemical, physical, biological and immunological studies on porcine proinsulin and related polypeptides. In *Proceedings of the 7th Congress of the International Diabetes Federation* (ed. R. R. Rodriquez and J. J. Vallance-Owen), p. 292. Excerpta Medica, Amsterdam.

Chance, R. E., R. M. Ellis and W. W. Bromer. 1968. Porcine proinsulin: Characterization and amino acid sequence. *Science* **161**:165.

Chretien, M. and C. H. Li. 1967. Isolation, purification and characterization of γ-lipotropic hormone from sheep pituitary glands. *Can. J. Biochem.* **45**:1163.

Cohn, D. V., R. R. MacGregor, L. L. H. Chu, J. R. Kimmel and J. W. Hamilton. 1972. Calcemic fraction-A: Biosynthetic peptide precursor of parathyroid hormones. *Proc. Nat. Acad. Sci.* **69**:1521.

Dayhoff, M. O., ed. 1972. *Atlas of Protein Sequence and Structure*, vol. 5. Biomedical Research Foundation, Bethesda, Maryland.

————. 1973. *Atlas of Protein Sequence and Structure*, vol. 5, suppl. 1. Biomedical Research Foundation, Bethesda, Maryland.

Falkmer, S., S. Emdin, N. Havu, G. Lundgren, M. Marques, Y. Ostberg, D. F. Steiner and N. W. Thomas. 1973. Insulin in invertebrates and cyclostomes. *Amer. Zool.* **13**:625.

Frank, B. H. and A. J. Veros. 1970. Interaction of zinc with proinsulin. *Biochem. Biophys. Res. Comm.* **38**:284.

Frazier, W. A., R. H. Angeletti and R. A. Bradshaw. 1972. Nerve growth factor and insulin. Structural similarities indicate an evolutionary relationship reflected by physiological action. *Science* **176**:482.

Fullerton, W. W., R. Potter and B. W. Low. 1970. Proinsulin: Crystallization and preliminary X-ray diffraction studies. *Proc. Nat. Acad. Sci.* **66**:1213.

Geller, D. M., J. D. Judah and M. R. Nicholls. 1972. Intracellular distribution of serum albumin and its possible precursors in rat liver. *Biochem. J.* **127**:865.

Gliemann, J. and H. H. Sorensen. 1970. Assay of insulin-like activity by the isolated fat cell method. IV. The biological activity of proinsulin. *Diabetologia* **6**:499.

Grant, P. T. and T. L. Coombs. 1971. Proinsulin, a biosynthetic precursor of insulin. In *Essays in Biochemistry* (ed. P. N. Campbell and G. D. Greville), vol. 6, p. 69. Academic Press, London.

Grant, P. T., T. L. Coombs, N. W. Thomas and J. R. Sargent. 1971. The conversion of [14C]proinsulin to insulin in isolated subcellular fractions of fish islet preparations. In *Subcellular Organization and Function in Endocrine Tissues* (ed. H. Heller and K. Lederis), vol. 19, p. 481. Cambridge University Press, Cambridge.

Gregory, R. A. 1974. The gastrointestinal hormones: A review of recent advances. *J. Physiol.* **241**:1.

Gregory, R. A. and H. J. Tracy. 1972. Isolation of two "big gastrins" from Zollinger-Ellison tumour tissue. *Lancet* **2**:797.

Habener, J. F., B. Kemper, J. T. Potts, Jr. and A. Rich. 1972. Proparathyroid hormone: Biosynthesis by human parathyroid adenomas. *Science* **178**:630.

Hamilton, J. W., R. R. MacGregor, L. L. H. Chu and D. V. Cohn. 1971. The isolation and partial purification of a non-parathyroid hormone calcemic fraction from bovine parathyroid glands. *Endocrinology* **89**:1440.

Hamilton, J. W., H. D. Niall, H. T. Keutmann, J. T. Potts, Jr. and D. V. Cohn. 1973. Amino terminal sequence of bovine proparathyroid hormone (calcemic fraction A). *Fed. Proc.* **32**:269.

Horwitz, D. L. and A. H. Rubenstein. 1974. Heterogeneity of circulating insulin and proinsulin in man. *Israel J. Med. Sci.* **10**:1201.

Howell, S. L. 1974. The molecular organization of the β granule of the islets of Langerhans. In *Advances in Cytopharmacology: Cytopharmacology of Secretion* (ed. B. Ceccarelli, J. Meldolesi and F. Clementi), vol. 2, p. 319. Raven Press, New York.

Jacobson, M. F. and D. Baltimore. 1968. Morphogenesis of poliovirus. I. Association of the viral RNA with coat protein. *J. Mol. Biol.* **33**:369.

Judah, J. D., M. Gamble and J. H. Steadman. 1973. Biosynthesis of serum albumin in rat liver. Evidence for the existence of "proalbumin." *Biochem. J.* **134**:1083.

Kemmler, W. and D. F. Steiner. 1970. Conversion of proinsulin to insulin in a subcellular fraction from rat islets. *Biochem. Biophys. Res. Comm.* **41**:1223.

Kemmler, W., J. D. Peterson and D. F. Steiner. 1971. Studies on the conversion of proinsulin to insulin. I. Conversion *in vitro* with trypsin and carboxypeptidase B. *J. Biol. Chem.* **246**:6786.

Kemmler, W., D. F. Steiner and J. Borg. 1973. Studies on the conversion of proinsulin to insulin. III. Studies *in vitro* with a crude secretion granule fraction isolated from islets of Langerhans. *J. Biol. Chem.* **248**:4544.

Kemmler, W., J. D. Peterson, A. H. Rubenstein and D. F. Steiner. 1972. On the biosynthesis, intracellular transport and mechanism of conversion of proinsulin to insulin and C-peptide. *Diabetes* **21**:572.

Kemper, B., J. F. Habener, J. T. Potts, Jr. and A. Rich. 1972. Proparathyroid hormone: Identification of a biosynthetic precursor to parathyroid hormone. *Proc. Nat. Acad. Sci.* **69**:643.

Kiehn, E. D. and J. J. Holland. 1970. Synthesis and cleavage of enterovirus polypeptides in mammalian cells. *J. Virol.* **5**:358.

Ko, S. C., D. G. Smyth, J. Markussen and F. Sundby. 1971. The amino acid sequence of the C-peptide of human proinsulin. *Eur. J. Biochem.* **20**:190.

Kohnert, K.-D., M. Ziegler, H. Zühlke and H. Fiedler. 1972. Isoelectric focusing of proinsulin and intermediate in polyacrylamide gel. *FEBS Letters* **28**:177.

Kohnert, K.-D., H. Zühlke, M. Ziegler and S. Schmidt. 1973. Isolierung und partielle Charakterisierung von Kristall-Insulin sowie Proinsulin-haltiger b-Komponente aus Human-Pancreas. *Acta Biol. Med. Germ.* **31**:515.

Lange, R. H., S. Boseck and S. S. Ali. 1972. Kristallographische Interpretation der Feinstruktur der B-Granula in den Langerhansschen Inseln der Ringelnatter, Natrix n. natrix (L.). *Z. Zellforsch.* **131**:559.

Lazarus, N. R., T. Tanese, R. Gutman and L. Recant. 1970a. Synthesis and release of proinsulin and insulin by human insulinoma tissue. *J. Clin. Endocrinol. Metab.* **30**:273.

Lazarus, N. R., J. C. Penhos, T. Tanese, L. Michaels, R. Gutman and L. Recant.

1970b. Studies on the biological activity of porcine proinsulin. *J. Clin. Invest.* **49:**487.

Markussen, J. and F. Sundby. 1972. Rat proinsulin C-peptides. Amino acid sequences. *Eur. J. Biochem.* **25:**153.

Melani, F., W. G. Ryan, A. H. Rubenstein and D. F. Steiner. 1970. Proinsulin secretion by a pancreatic beta cell adenoma: Proinsulin and C-peptide secretion. *New Eng. J. Med.* **283:**713.

Narahara, H. T. 1972. Biological activity of proinsulin. In *Insulin Action* (ed. I. Fritz), p. 63. Academic Press, New York.

Neurath, H., K. A. Walsh and W. P. Winter. 1968. Evolution of structure and function of proteases. *Science* **158:**1638.

Noe, B. D. and G. E. Bauer. 1971. Evidence for glucagon biosynthesis involving a protein intermediate in islets of the anglerfish (*Lophius americanus*). *Endocrinology* **89:**642.

―――. 1973. Further characterization of a glucagon precursor from anglerfish islet tissue. *Proc. Soc. Exp. Biol. Med.* **142:**210.

Nolan, C., E. Margoliash, J. D. Peterson and D. F. Steiner. 1971. The structure of bovine proinsulin. *J. Biol. Chem.* **246:**2780.

Oyer, P. E., S. Cho, J. D. Peterson and D. F. Steiner. 1971. Studies on human proinsulin. Isolation and amino acid sequence of the human pancreatic C-peptide. *J. Biol. Chem.* **246:**1375.

Pearse, A. G. E. 1975. In *Biological Council Symposium on Peptide Hormones* (ed. J. A. Parsons). Macmillan, London (in press).

Permutt, M. A. and D. M. Kipnis. 1972a. Insulin biosynthesis: Studies of islet polyribosomes. *Proc. Nat. Acad. Sci.* **69:**506.

―――. 1972b. Insulin biosynthesis. I. On the mechanism of glucose stimulation. *J. Biol. Chem.* **247:**1194.

Peterson, J. D., S. Nehrlich, P. E. Oyer and D. F. Steiner. 1972. Determination of the amino acid sequence of the monkey, sheep, and dog proinsulin C-peptides by a semi-micro Edman degradation procedure. *J. Biol. Chem.* **247:**4866.

Peterson, J. D., C. L. Coulter, D. F. Steiner, S. O. Emdin and S. Falkmer. 1974. Hagfish insulin: Structural and crystallographic observations on a primitive protein hormone. *Nature* **251:**239.

Potts, J. T., Jr., H. D. Niall, G. W. Tregear, J. VanRietschoten, J. F. Habener, G. V. Segre and H. T. Keutmann. 1973. Chemical and biologic studies of proparathyroid hormone: Analysis of hormone biosynthesis and metabolism. *Mount Sinai J. Med.* **40:**448.

Roffman, S., U. Sanocka and W. Troll. 1970. Sensitive proteolytic enzyme assay using differential solubilities of radioactive substrates and products in biphasic systems. *Anal. Biochem.* **36:**11.

Rubenstein, A. H., F. Melani and D. F. Steiner. 1972. Circulating proinsulin: Immunology, measurement, and biological activity. In *Handbook of Physiology—Endocrinology I* (ed. D. F. Steiner and N. Freinkel), p. 515. Williams and Wilkins, Baltimore.

Rubenstein, A. H., J. L. Clark, F. Melani and D. F. Steiner. 1969. Secretion of proinsulin C-peptide by pancreatic β cells and its circulation in blood. *Nature* **244:**697.

Rutter, W. J., W. R. Clark, J. D. Kemp, W. S. Bradshaw, T. G. Sanders and W. D. Ball. 1968. Multiphoric regulation in cytodifferentiation. In *Epithelial-Mesenchymal Interaction,* p. 114. Williams and Wilkins, Baltimore.

Ryle, A. P., F. Sanger, L. F. Smith and R. Kitai. 1955. The disulfide bonds of insulin. *Biochem. J.* **60:**541.

Sachs, H., P. Fawcett, Y. Takabatake and R. Portanova. 1969. Biosynthesis and release of vasopressin and neurophysin. *Rec. Prog. Horm. Res.* **25**:447.

Scott, A. P., J. G. Ratcliffe, L. H. Rees, H. P. J. Bennett, P. J. Lowry and C. Mc-Martin. 1973. Pituitary peptide. *Nature New Biol.* **244**:65.

Sherman, B. M., S. Pek, S. S. Fajans, J. C. Floyd, Jr. and J. W. Conn. 1971. Plasma proinsulin in twenty patients with insulin secreting islet cell tumors. *Diabetes* **20**:334.

Smith, L. F. 1972a. Amino acid sequences of insulins. *Diabetes* **21**:457.

Smith, R. E. 1972b. Summary of discussion. *Diabetes* **21**:581.

Smith, R. E. and R. M. Van Frank. 1974. Substructural localization of an enzyme in β cells of rat pancreas with the ability to convert proinsulin to insulin. *Endocrinology* (Suppl. abstr. of Endocrine Soc. Meet., June 1974) **94**:A190.

Smyth, D. G., J. Markussen and F. Sundby. 1974. The amino acid sequence of guinea pig C-peptide. *Nature* **248**:151.

Sorensen, R. L., R. D. Shank and A. W. Lindall. 1972. Effect of pH on conversion of proinsulin to insulin by a subcellular fraction of rat islets. *Proc. Soc. Exp. Biol. Med.* **139**:652.

Sorensen, R. L., M. W. Steffes and A. W. Lindall. 1970. Subcellular localization of proinsulin to insulin conversion in isolated rat islets. *Endocrinology* **86**:88.

Steiner, D. F. 1967. Evidence for a precursor in the biosynthesis of insulin. *Trans. N.Y. Acad. Sci.* (Ser. II) **30**:60.

———. 1973. Cocrystallization of proinsulin and insulin. *Nature* **243**:528.

———. 1975. Peptide hormone precursors: Biosynthesis, processing and significance. In *Biological Council Symposium on Peptide Hormones* (ed. J. A. Parsons). Macmillan, London (in press).

Steiner, D. F. and J. L. Clark. 1968. The spontaneous reoxidation of reduced beef and rat proinsulins. *Proc. Nat. Acad. Sci.* **60**:622.

Steiner, D. F. and P. E. Oyer. 1967. The biosynthesis of insulin and a probable precursor of insulin by a human islet cell adenoma. *Proc. Nat. Acad. Sci.* **57**:473.

Steiner, D. F., D. D. Cunningham, L. Spigelman and B. Aten. 1967. Insulin biosynthesis: Evidence for a precursor. *Science* **157**:697.

Steiner, D. F., W. Kemmler, H. S. Tager and J. D. Peterson. 1974a. Proteolytic processing in the biosynthesis of insulin and other proteins. *Fed. Proc.* **33**:2105.

Steiner, D. F., W. Kemmler, H. S. Tager and A. H. Rubenstein. 1974b. Molecular events taking place during intracellular transport of exportable proteins. The conversion of peptide hormone precursors. In *Advances in Cytopharmacology: Cytopharmacology of Secretion* (ed. B. Ceccarelli, J. Meldolesi and F. Clementi) vol. 2, p. 195. Raven Press, New York.

Steiner, D. F., W. Kemmler, J. L. Clark, P. E. Oyer and A. H. Rubenstein. 1972. The biosynthesis of insulin. In *Handbook of Physiology—Endocrinology I* (ed. D. F. Steiner and N. Freinkel), p. 175. Williams and Wilkins, Baltimore.

Steiner, D. F., S. Cho, P. E. Oyer, S. Terris, J. D. Peterson and A. H. Rubenstein. 1971. Isolation and characterization of proinsulin C-peptide from bovine pancreas. *J. Biol. Chem.* **246**:1365.

Steiner, D. F., J. D. Peterson, H. S. Tager, S. Emdin, Y. Ostberg and S. Falkmer. 1973. Comparative aspects of proinsulin and insulin structure and biosynthesis. *Amer. Zool.* **13**:591.

Steiner, D. F., J. L. Clark, C. Nolan, A. H. Rubenstein, E. Margoliash, B. Aten and P. E. Oyer. 1969. Proinsulin and the biosynthesis of insulin. *Rec. Prog. Horm. Res.* **25**:207.

Steiner, D. F., J. L. Clark, C. Nolan, A. H. Rubenstein, E. Margoliash, F. Melani and P. E. Oyer. 1970. The biosynthesis of insulin and some speculations regarding the pathogenesis of human diabetes. In *The Pathogenesis of Diabetes Mellitus: Proceedings of the 13th Nobel Symposium* (ed. E. Cerasi and R. Luft), p. 123. Almqvist and Wiksell, Stockholm.

Summers, D. F., E. N. Shaw, M. L. Stewart and J. V. Maizel, Jr. 1972. Inhibition of cleavage of large poliovirus-specific precursor proteins in infected HeLa cells by inhibitors of proteolytic enzymes. *J. Virol.* **10**:880.

Sun, A. M., B. J. Lin and R. E. Haist. 1973. Studies on the conversion of proinsulin to insulin in the isolated islets of Langerhans in the rat. *Can. J. Physiol. Pharmacol.* **51**:175.

Sundby, F. and J. Markussen. 1970. Preparative method for the isolation of C-peptides from ox and pork pancreas. *Horm. Metab. Res.* **2**:17.

Tager, H. S. and D. F. Steiner. 1972. Primary structures of the proinsulin connecting peptides of the rat and the horse. *J. Biol. Chem.* **247**:7936.

————. 1973. Isolation of a glucagon-containing peptide: Primary structure of a possible fragment of proglucagon. *Proc. Nat. Acad. Sci.* **70**:2321.

Tager, H. S., S. O. Emdin, J. L. Clark and D. F. Steiner. 1973. Studies on the conversion of proinsulin to insulin. II. Evidence for a chymotrypsin-like cleavage in the connecting peptide region of insulin precursors in the rat. *J. Biol. Chem.* **248**:3476.

Tung, A. K. and F. Zerega. 1971. Biosynthesis of glucagon in isolated pigeon islets. *Biochem. Biophys. Res. Comm.* **45**:387.

Wessells, N. K. and J. Evans. 1968. Ultrastructural studies of early morphogenesis and cytodifferentiation in the embryonic mammalian pancreas. *Develop. Biol.* **17**:413.

Yalow, R. S. and S. A. Berson. 1971. Further studies on the nature of immunoreactive gastrin in human plasma. *Gastroenterology* **60**:203.

————. 1972. Nature of plasma and pituitary ACTH. In *Proceedings of the 4th International Congress of Endocrinology* (ed. R. O. Scow), Series 273, p. 585. Excerpta Medica, Amsterdam.

Yamaji, K., K. Tada and A. C. Trakatellis. 1972. On the biosynthesis of insulin in anglerfish islets. *J. Biol. Chem.* **247**:4080.

Yip, C. C. 1971. A bovine pancreatic enzyme catalyzing the conversion of proinsulin to insulin. *Proc. Nat. Acad. Sci.* **68**:1312.

Zühlke, H., H. Jahr, S. Schmidt, D. Gottschling and B. Wilke. 1974. Catabolism of proinsulin and insulin. Proteolytic activities in Langerhans islets of rat and mice pancreas in vitro. *Acta Biol. Med. Germ.* **33**:407.

Some Properties of a Ca^{++}–activated Protease That May Be Involved in Myofibrillar Protein Turnover

William R. Dayton, Darrel E. Goll, Marvin H. Stromer, William J. Reville, Michael G. Zeece and Richard M. Robson

Muscle Biology Group, Departments of Animal Science
Biochemistry and Biophysics, and Food Technology
Iowa State University, Ames, Iowa 50010

Although it is obvious that net protein accumulation in a cell depends on the difference between rate of general protein synthesis and rate of general protein degradation in that cell, not all intracellular proteins are synthesized or degraded at identical rates (for reviews of half-lives of different proteins, see Schimke 1969, 1970, 1973; Schimke and Doyle 1970). For example, proteins in rat liver have an average half-life of 3.5 days (Schimke 1970), but the half-lives of specific proteins in this organ vary from 11 minutes for ornithine decarboxylase (Russell and Synder 1971) to 19 days for isozyme 5 of lactate dehydrogenase (Fritz et al. 1973). Moreover, recent evidence suggests that both the rate of synthesis and the rate of degradation of individual proteins can vary in response to physiological or environmental stimuli. For example, administration of either hydrocortisone or tryptophan results in an increase in the amount of tryptophan oxygenase in rat liver cells. Schimke, Sweeney and Berlin (1964, 1965) have shown that the increased level of tryptophan oxygenase in response to hydrocortisone results from a four- to fivefold increase in rate of synthesis with no change in rate of degradation. The increase in tryptophan oxygenase in response to tryptophan, however, results from a decreased rate of degradation with no change in rate of synthesis. Similarly, Goldberg (1969) has shown that work-induced hypertrophy of rat skeletal muscle results from both an increased rate of protein synthesis and a decreased rate of protein degradation, but muscle hypertrophy in response to growth hormone results from an increased rate of protein synthesis with little or no change in rate of protein degradation. These examples illustrate the remarkable ability of cells to alter either their rate of protein synthesis or their rate of protein degradation, or both, in response to physiological needs or stimuli. This ability has evoked much interest in the mechanism used by cells to vary synthetic and degradative rates of specific proteins, but although considerable information is now available on the mechanism of protein synthesis and some of the factors that might alter the rate of protein synthesis, very little is known about the mechanism for degradation of specific proteins (Schimke 1970; Swick and Song 1974).

Turnover of Myofibrillar Proteins in Striated Muscle

Because of the highly organized macromolecular nature of the myofibril in striated muscle (Huxley 1972; see also Fig. 1a), turnover of the myofibrillar proteins presents a special problem when devising a mechanism for protein degradation. Myofibrils consist of a long array of sarcomeres (Fig. 1a) linked in series and extending unbroken from one end of the muscle cell to the other (Huxley 1972). This continuous, unbroken series is essential for shortening of the individual sarcomeres to be transmitted to the ends of the muscle cell and results in shortening or contraction of the entire cell. Consequently, even a single break or interruption in the long series of sarcomeres constituting a myofibril will disable that entire myofibril. Yet, numerous studies (Desmond and Harary 1972; Dreyfus, Kruh and Schapira 1960; Funabiki and Cassens 1972; Low and Goldberg 1973; Martonosi and Halpin 1972; McManus and Mueller 1966; Schapira, Dreyfus and Kruh 1962; Swick and Song 1974; Velick 1956; Wikman-Coffelt et al. 1973; Zak, Ratkitzis and Rabinowitz 1971) have shown that myofibrillar proteins turn over, although half-lives of the different myofibrillar proteins are still very uncertain. Because destruction of an entire sarcomere would completely disable the myofibril containing that sarcomere, it would be physiologically advantageous for myofibrillar proteins to turn over by degradation of those proteins on the surface of myofibrils, thereby reducing the diameter of the myofibril without actually severing it. Structural studies on atrophying muscle where myofibrillar degradation is rapid (Bhakthan, Borden and Nair 1970; Engel and Stonnington 1974; Pellegrino and Franzini 1963; Stonnington and Engel 1973; Tomanek and Lund 1973) have shown that a gradual decrease in myofibril diameter characteristically occurs in such muscle, but that missing sarcomeres are not observed. Moreover, Morkin (1970) has reported that newly synthesized myofibrillar proteins are added at the periphery of existing myofibrils; this is exactly where such protein would be expected to be deposited if degradation of myofibrillar protein also occurs at the periphery of myofibrils.

Simply limiting degradation to the periphery of myofibrils, however, does not provide an adequate mechanism for turnover of the myofibrillar proteins. The eight or nine known myofibrillar proteins (Table 1) are assembled into

Table 1

The Known Myofibrillar Proteins and Their Location in
Skeletal Muscle Myofibrils

Protein	Approx. percent by weight in myofibril[a]	Location
Myosin	50–55	thick filaments
Actin	15–20	thin filaments
Tropomyosin	4–6	thin filaments and Z-disk
Troponin	4–6	thin filaments
C-Protein	2–3	thick filaments
α-Actinin	1–3	Z-disk
β-Actinin	1	thin filament
M-Protein(s)	3–5	M-line

[a] The percentage composition figures are for myofibrils from rabbit skeletal muscle.

highly organized filamentous structures (Huxley 1972), and these structures exist at the periphery of the myofibril as well as at its center. Just as it seems physiologically inefficient to turn over entire sarcomeres and thereby disable a complete myofibril, it seems equally inefficient to turn over a single actin or myosin molecule and thereby disable an entire thick or thin filament containing approximately 300 to 400 myosin or actin molecules (Huxley 1972). The mechanism by which individual myofibrillar proteins, assembled into their native filamentous structures, are turned over remains unclear. Although several reports (Funabiki and Cassens 1972; Low and Goldberg 1973) indicate that tropomyosin and troponin turn over more rapidly than myosin and actin, other reports (Rabinowitz 1973; Zak, Ratkitzis and Rabinowitz 1971) suggest that the myofibrillar proteins all turn over with identical half-lives.

In addition to a lack of information on the mechanism for turnover of myofibrillar proteins, almost no information exists on the agent or agents that cause myofibrillar protein degradation. It is generally assumed that the lysosomal system (de Duve and Wattiaux 1966) is involved in gross cellular autophagy, but it is not clear that lysosomal degradation alone is sufficient to explain the variations in rates of degradation among different proteins and the changes in rates of degradation in response to physiological stimuli (Schimke 1970; Schimke and Doyle 1970). Although structures similar to the lysosomal structures seen in liver or kidney cells have never been observed in skeletal muscle cells (Canonico and Bird 1970; Seiden 1973; Stauber and Bird 1974), biochemical studies have established that lysosomal-like particles can be isolated from homogenates of skeletal muscle (Canonico and Bird 1970; Stauber and Bird 1974). Recent evidence indicates that lysosomes in skeletal muscle cells are part of the sarcotubular system in these cells, and that this has prevented their earlier identification in electron micrographs of skeletal muscle (Seiden 1973; Stauber and Bird 1974). Although it is clear that skeletal muscle cells contain very few lysosomes compared with liver or kidney cells (Bird et al. 1969; Canonico and Bird 1970; Iodice, Leong and Weinstock 1966; Iodice et al. 1972; Tappel 1966), the number of lysosomes in muscle cells increases dramatically during periods of rapid protein degradation associated with denervation (Pellegrino and Franzini 1963; Schiaffino and Hanzlikova 1972), muscular dystrophy (Desai et al. 1964; Kar and Pearson 1972; Tappel et al. 1962; Zalkin et al. 1962) and insect metamorphosis (Bhakthan, Borden and Nair 1970; Lockshin and Beaulaton 1974a, b). Because muscle lysosomes contain cathepsins (Canonico and Bird 1970) and because the concentration of catheptic enzymes in muscle also increases during periods of muscular atrophy (Iodice, Leong and Weinstock 1966; Iodice et al. 1972; McLaughlin, Abood and Bosmann 1974), it has generally been assumed that lysosomal cathepsins are responsible for degradation of myofibrillar proteins. Several recent studies have shown, however, that neither myofibrils nor thick and thin filaments are ever seen in lysosomal vacuoles during periods of rapid muscle atrophy (Lockshin and Beaulaton 1974a,b; Schiaffino and Hanzlikova 1972). Indeed, it seems improbable that lysosomes could easily engulf whole segments of myofibrils. Moreover, the activity of catheptic enzymes isolated from muscle has almost always been measured by using hemoglobin or other nonmyofibrillar proteins as substrates and by assaying at pH values of 5.0 or less (Canonico and Bird 1970; Iodice, Leong and

Weinstock 1966; Iodice et al. 1972; Park and Pennington 1967; Tappel 1966). It is very unlikely that extra lysosomal pH in muscle cells ever decreases to the 5.0 or less required for such catheptic activity. Even more importantly, in the few instances where such studies have been done, muscle cathepsins have had no effect on intact myofibrils or thick and thin filaments (Bodwell and Pearson 1964; Friedman, Laufer and Davies 1969; Fukazawa and Yasui 1967; Park and Pennington 1967; Sharp 1963).

Consequently, it seems that myofibrillar proteins must be disassembled to their monomeric form before lysosomes can be involved in their degradation. No agent capable of such selective disassembly has as yet been reported. We have recently discovered a Ca^{++}-activated factor (CAF) that can be isolated from striated muscle and that has the ability to degrade intact myofibrils (Busch et al. 1972). We have now purified this factor and have shown that it is a potent proteolytic enzyme that causes a very selective degradation of myofibrillar proteins (Dayton et al. 1974). This paper describes some of the effects of CAF on purified myofibrillar proteins and proposes a hypothetical mechanism showing how CAF could have a pivotal role in disassembly of the myofibril to initiate turnover of the myofibrillar proteins.

Discovery of CAF

The presence in skeletal muscle cells of a Ca^{++}-activated factor capable of removing Z-disks from intact myofibrils was discovered during electron microscope examination of rabbit psoas strips that had been incubated in Ca^{++}-containing solutions for 9 hours at pH 7.2 and 37°C (Busch et al. 1972; Goll et al. 1970). Although the ultrastructure of such strips was completely normal immediately after death (Fig. 1a), Z-disks in myofibrils in these strips were completely gone after 9 hours of incubation in a 1 mM Ca^{++}-containing solution at pH 7.2 and 37°C (Fig. 1b). The effects of Ca^{++} seem specific because 9 hours of incubation causes little ultrastructurally detectable damage to the A-band (Fig. 1b), and thin filament overlap in the center of the A-band after 9 hours of incubation in Ca^{++} suggests that extensive degradation of thin filaments has not occurred (Fig. 1b). Close examination of the Z-disk area in these strips incubated in 1 mM Ca^{++} confirms that Z-disks are completely gone, leaving a gap in the space formerly occupied by the Z-disk (Fig. 2). Moreover, the abundant thin filaments terminating in the Z-disk region (Fig. 2) also support the suggestion that Z-disk removal is not accompanied by noticeable shortening of the thin filament. Because tropomyosin and troponin are difficult to see in ordinary electron micrographs of thin filaments, it is impossible to determine from these micrographs alone whether the tropomyosin and troponin on the thin filaments are still intact. Some experiments to be discussed later indicate that at least some of the troponin and tropomyosin in these muscle strips has probably been degraded after 9 hours of incubation in a 1 mM Ca^{++}-containing solution.

Incubation of strips from the same rabbit under identical conditions (9 hr at 37°C) and in a solution identical to the solution that caused Z-disk removal but containing 1 mM EGTA (a Ca^{++} chelator) in place of 1 mM Ca^{++} has no detectable effect upon myofibril ultrastructure (not shown here, but see Fig. 9 in Busch et al. 1972). Consequently, Z-disk removal observed after 9 hours of incubation at 37°C in a 1 mM Ca^{++}-containing solution must be due specifically to the Ca^{++} in that solution. Incubation of purified myofibrils with 1 mM Ca^{++} under conditions causing Z-disk removal in intact muscle cells has no effect on the Z-disks in these myofibrils (Busch et al. 1972). Because the sarcoplasm is the principal material removed during myofibril preparation, this finding indicates that Ca^{++} does not act directly to remove Z-disks, but that Ca^{++} probably activates a sarcoplasmic factor that then removes Z-disks. To isolate this factor, muscle was homogenized in 6 volumes (v/w) of 4 mM EDTA pH 7.0 to nullify any susceptibility of this hypothetical factor to autolysis (Busch et al. 1972). After sedimentation of the myofibrillar and connective tissue proteins, which are insoluble in 4 mM EDTA, the sarcoplasmic proteins in the supernatant were subjected to iso-electric precipitation between pH 4.9 and 6.2. The protein precipitated in this pH range caused Z-disk removal when added to purified myofibrils in the presence of 1 mM Ca^{++}. The isoelectrically precipitated protein was dissolved in 30 mM Tris-acetate pH 7.0, 1 mM EDTA, clarified, and then salted out between 0 and 40% ammonium sulfate saturation (Busch et al. 1972). The ammonium sulfate precipitate was dissolved in 1 mM $KHCO_3$, 1 mM EDTA, dialyzed against the same solvent, and then clarified to produce a crude P_{0-40} CAF fraction (Busch et al. 1972). Incubation of myofibrils for 24 hours at 25°C with this crude P_{0-40} CAF fraction in the presence of 10 mM Ca^{++} at pH 7.0 causes complete removal of Z-disks (Fig. 3c). If 10 mM EDTA is substituted for 10 mM Ca^{++} or if the crude P_{0-40} CAF fraction is omitted from the Ca^{++}-containing incubation, Z-disks are completely intact after 24 hours at 25°C and pH 7.0 (Fig. 3a, b). These results show that a Ca^{++}-activated factor capable of removing Z-disks from myofibrils can be prepared from muscle tissue. Preliminary studies suggested that this factor was a protein and probably an enzyme (Busch et al. 1972). Because the crude P_{0-40} CAF fraction prepared by isoelectric precipitation and ammonium sulfate fractionation was still very heterogeneous, it was impossible to determine whether the Ca^{++}-activated removal of Z-disks was due to one or to several enzymes and whether these enzymes were proteases, amylases, lipases or some synergistic combination of these different types. Because these questions could not be answered definitively without a purified enzyme preparation, initial studies on CAF were directed at purification of the factor or factors responsible for Ca^{++}-activated Z-disk removal. These studies were done with porcine skeletal muscle and showed that Ca^{++}-activated Z-disk removal is due to a single proteolytic enzyme (Dayton et al. 1974). In addition to rapidly removing Z-disks, this enzyme rapidly degrades casein to peptides that are soluble in 2% trichloroacetic acid. Porcine skeletal muscle evidently contains very little of this enzyme because only 4 to 10 mg of purified enzyme can be obtained from 10,000 g of muscle fresh weight. The unique specificity of CAF against purified myofibrillar proteins suggests a possible role for this enzyme in metabolic turnover of myofibrils.

Properties of Purified CAF

Purification of CAF requires column chromatography of the crude P_{0-40} CAF fraction through the following sequence of columns: (1) 6% agarose, (2)

Figure 1

Comparison of rabbit psoas muscle sampled immediately after death and again after 9 hours of incubation at 37°C in a saline solution containing 1 mM Ca++ and 1 mM deoxycholate.

(a) Rabbit psoas muscle sampled immediately after death. All the usual bands are present. Note particularly the dense fibrillar Z-disks. 33,000×.

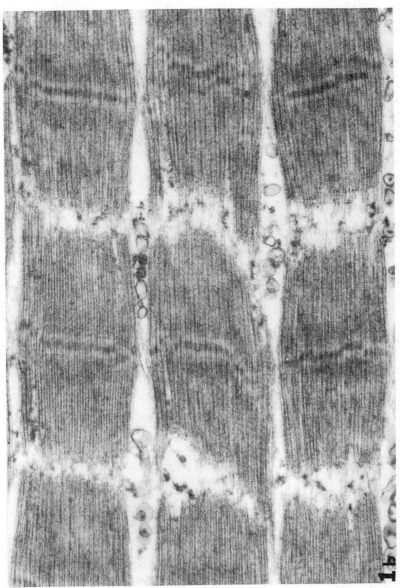

Figure 1 (continued)

(*b*) Rabbit psoas muscle sampled after incubation for 9 hours at 37°C in 80 mM KCl, 60 mM Tris-HCl pH 7.2, 5 mM MgCl₂, 1 mM CaCl₂, 1 mM NaN₃, 1 mM deoxycholate has no Z-disks. Because strips of psoas muscle were held isometrically during incubation, shortening to the extent of double overlap of thin filaments in the center of the A-band results in gaps between sarcomeres. Both thin and thick filaments seem intact after this incubation. 37,600×. (Reprinted, with permission, from Busch et al. 1972).

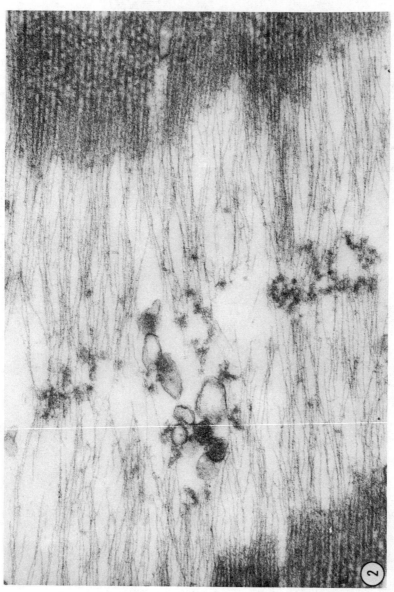

Figure 2

This micrograph is a higher magnification of the I-band from a strip also incubated for 9 hours at 37°C in 80 mм KCl, 60 mм Tris-HCl pH 7.2, 5 mм MgCl₂, 1 mм CaCl₂, 1 mм NaN₃, 1 mм deoxycholate but less extensively shortened than the sample shown in Figure 1b. A few darkly staining pieces of material remain in the gap formerly occupied by the Z-disk, but suspension in a Ca^{++}-containing solution does not seem to shorten thin filaments. 72,100×. (Reprinted, with permission, from Busch et al. 1972.)

DEAE-cellulose, (3) Sephadex G-200, (4) DEAE-cellulose with a very shallow KCl gradient, and (5) Sephadex G-100. Details of this purification procedure and some of the physical properties of purified CAF will be published elsewhere (Dayton et al., in prep.). The similarity in the procedures necessary for purification and in some properties of the purified enzymes makes it very likely that CAF is identical to the kinase activating factor described by Huston and Krebs (1968), although Huston and Krebs did not realize the effect that this enzyme has on myofibrils. Figure 4 shows polyacrylamide gels of purified CAF run both in a nondenaturing buffer at pH 7.5 (left gel) and in the presence of 0.1% SDS (right gel). Most of the protein in purified CAF preparations migrates as a single electrophoretic band in nondenaturing buffers, with a small amount of protein migrating slightly faster than this principal band. In 0.1% SDS, however, purified CAF migrates as two major bands having molecular weights of approximately 80,000 and 30,000 daltons. Densitometry of SDS-polyacrylamide gels indicates that these two protein bands exist in approximately equal molar ratios in purified CAF preparations. Because we have never succeeded during our extensive purification efforts in separating the proteins in these two bands and because CAF activity always coincides exactly with the appearance of these two bands in equal molar ratios, we assume that the proteins in these 80,000- and 30,000-dalton bands are subunits of a parent 110,000-dalton CAF molecule. Elution profiles from calibrated gel permeation columns indicate that the undenatured CAF molecule is not larger than 110,000 daltons. A small amount of material migrating with a molecular weight of 60,000 daltons can be detected in SDS-polyacrylamide gel electrophoresis (Fig. 4). Evidence to be discussed in detail elsewhere (Dayton et al., in prep.) suggests that this 60,000-dalton component probably corresponds to the small amount of rapidly migrating contaminant seen in polyacrylamide gels run in nondenaturing buffer at pH 7.5 (Fig. 4, left gel). This contaminating 60,000-dalton material has no Ca⁺⁺-activated proteolytic activity (Dayton el al., in prep.), and this minor contaminant therefore has no influence on assays of enzymatic activity of purified CAF.

Because CAF is the first proteolytic enzyme with the ability to degrade intact myofibrils that has been purified from skeletal muscle and because purified CAF is maximally active near the in vivo pH in living muscle cells (Dayton et al. 1974), it seemed possible that CAF was involved in metabolic turnover of the myofibrillar proteins. Therefore, studies were done to determine whether CAF was localized in lysosomes and to ascertain the effects of purified CAF on purified myofibrillar proteins. Thorough studies involving differential centrifugation of porcine skeletal muscle homogenates have shown that although catheptic enzyme activities can be located in a lysosomal-like fraction from such homogenates, CAF activity was located in the supernatant fraction from these same homogenates (Reville et al., unpubl.). That catheptic activity could be located in a lysosomal fraction in these homogenates eliminates the possibility that CAF activity was found in the supernatant only because lysosomes in the muscle homogenate had been ruptured during isolation. These results, and the previously described results demonstrating that incubation of muscle strips in a Ca⁺⁺-containing solution causes Z-disk removal before any exogenous enzymes could penetrate the cell mem-

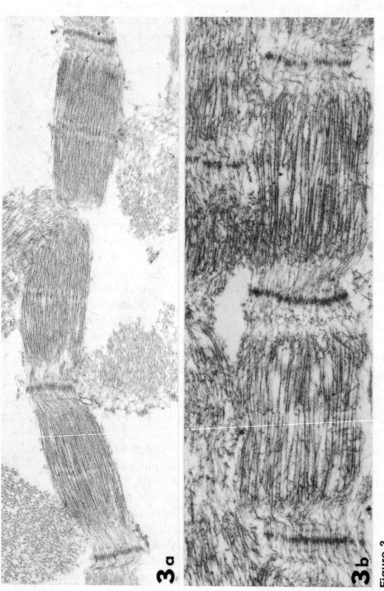

Figure 3

Electron micrographs of sectioned myofibrils after incubation for 24 hours at 25°C with different combinations of crude P_{0-40} CAF fraction, EDTA and Ca^{++}.

(*a*) Myofibrils incubated in 100 mM KCl, 10 mM EDTA, 20 mM Tris-acetate pH 7.0, 1 mM NaN_3, 10 mg myofibrillar protein/1 mg crude P_{0-40} CAF fraction have intact and normal Z-disks. 24,600×.

(*b*) Myofibrils incubated in 100 mM KCl, 10 mM Ca^{++}, 20 mM Tris-acetate pH 7.0, 1 mM NaN_3 with no added crude P_{0-40} CAF fraction also have intact Z-disks. Therefore, Ca^{++} does not affect Z-disks directly. 29,210×.

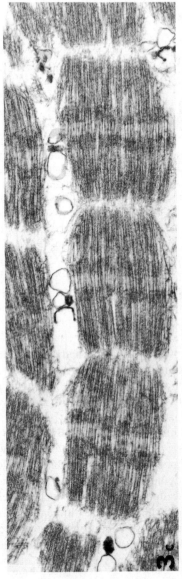

Figure 3 (*continued*)

(*c*) Myofibrils incubated in 100 mM KCl, 10 mM Ca++, 20 mM Tris-acetate pH 7.0, 1 mM NaN₃; 10 mg myofibrillar protein/1 mg crude P₀₋₄₀ CAF fraction have no Z-disks and frequently have double overlap of thin filaments in the center of the A-band. Although M-lines are also gone from this sample, this is not a consistent result, and it is uncertain whether the crude P₀₋₄₀ CAF fraction attacks M-line in the presence of Ca++. 24,600×. (Reprinted, with permission, from Busch et al. 1972.)

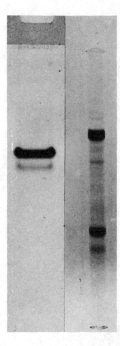

Figure 4
Polyacrylamide gel electrophoresis of purified CAF. Gel
a shows purified CAF when electrophoresed in a 7½ %
polyacrylamide gel in the presence of 70 mM Tris-HCl
pH 7.5, 1 mM EDTA. Gel *b* shows the same sample of
purified CAF when electrophoresed in a 7½ % polyacrylamide gel in the presence of 0.1% SDS, 100 mM
Na-phosphate pH 7.0 (Weber and Osborn 1969). Gels
a and *b* are each loaded with 20 µg protein.

branes in these strips, show that CAF is located in the sarcoplasm of the
muscle cell. This finding is not surprising because CAF has a pH optimum of
7.0 to 8.0 and the pH optimum of most lysosomal enzymes is below 5.0.

Because CAF is located in the sarcoplasm and is not confined to lysosomes,
the muscle cell must have some mechanism for regulating CAF activity to
prevent indiscriminate and continuous degradation of Z-disks. Although the
nature of this regulation is not yet clear, it is possible that Ca^{++} is involved.
Purified CAF requires 1 mM Ca^{++} for maximal activity (Dayton et al., in
prep.), and intracellular free Ca^{++} concentrations are generally 10^{-5} M or
less in living skeletal muscle cells (Jöbsis and O'Conner 1966; Ridgway
and Ashley 1967; Weber 1966). Several reports (Isaacson and Sandow 1968;
Varley and Dhalla 1973; Vihert and Pozdyunina 1969; Young, Young and
Edelman 1959) have indicated that intracellular free Ca^{++} levels increase
in pathological or necrotic states in muscle cells, and this increase in free
Ca^{++} may be sufficient to activate CAF under such conditions. Alternatively,
the Ca^{++} requirement of CAF may be modified in response to physiological
stimuli by phosphorylation of one of the CAF subunits or by some other as
yet unknown mechanism.

Studies on the ability of CAF to degrade native myofibrillar proteins have
involved myosin, actin, C-protein, α-actinin, tropomyosin and troponin.
Myosin was prepared according to the procedure of Seraydarian et al. (1967),
and some preparations of myosin were further purified by chromatography on
DEAE-cellulose as described by Richards et al. (1967). Actin was also prepared according to Seraydarian et al. (1967), but was polymerized in 0.5 M
KCl (Spudich and Watt 1971) instead of in 0.1 M KCl. This modification

greatly reduced tropomyosin and troponin contamination in actin preparations, but despite the 3.3 M KCl treatment to remove α-actinin (Seraydarian et al. 1967), all actin preparations contained a small amount of α-actinin contamination that could be removed only by chromatography on Sephadex G-150 or 8% agarose columns. C-protein was made by DEAE-cellulose chromatography of myosin as described by Offer, Moos and Starr (1973).

α-Actinin was extracted according to Arakawa, Robson and Goll (1970) and was purified on DEAE-cellulose columns (Goll et al. 1972; Robson et al. 1970) followed by a hydroxylapatite column (Suzuki and Goll, unpubl.). Purified tropomyosin and crude troponin were prepared as described by Arakawa, Goll and Temple (1970). The crude troponin was then purified by DEAE-cellulose chromatography (van Eerd and Kawasaki 1973). All protein preparations were from porcine skeletal muscle. The effects of purified CAF on these purified myofibrillar proteins were assayed by incubating one part of purified CAF with 200 parts by weight of the purified myofibrillar protein for 60 minutes at 25°C in the presence of 5 mM Ca^{++}. Because purified CAF completely removes Z-disks from myofibrils within 5 minutes under these conditions, it seemed probable that any effect of CAF on the myofibrillar proteins would be evident after 60 minutes of incubation. The action of CAF was stopped by adding EDTA to a final concentration of 10 mM, and the treated proteins were analyzed by SDS-polyacrylamide gel electrophoresis (Weber and Osborn 1969). Controls were run by incubating CAF and the purified myofibrillar protein for 1 hour in an incubation medium containing 10 mM EDTA instead of 5 mM Ca^{++}. All gels were stained with Coomassie blue (Weber and Osborn 1969). Additional details regarding ionic composition of the incubation medium are given in the figure legends.

Figure 5 shows the effect of purified CAF on a myosin preparation that contained C-protein. CAF has no detectable effect on the mass of either the heavy subunit chains (cf. gels a and b) or the light subunit chains (cf. gels c and d) of myosin. This finding demonstrates a remarkable specificity for CAF because purified myosin is known to be degraded quickly by a wide variety of proteolytic enzymes, including trypsin (Lowey et al. 1969), chymotropsin (Gergely, Gouvea and Karibian 1955), papain (Kominz et al. 1965) and subtilisin (Middlebrook 1959). The C-protein in this myosin preparation is degraded by CAF (Fig. 5, gels a and b), with a peptide chain of slightly lower mass than the peptide chain of the native C-protein ostensibly being one of the products of this degradation (Fig. 5, gel b). That C-protein is degraded during incubation of this myosin preparation with CAF demonstrates that CAF was active during this incubation and that the lack of effect on myosin was not due to inactivation of CAF by the incubation conditions.

The gels in Figure 6 show that CAF has no detectable effect on either purified actin (cf. gels a and b) or purified α-actinin (cf. gels c and d). The very faint protein bands in the 80,000-dalton region in gels a, b and d represent the 80,000-dalton subunit of the CAF added to these protein preparations. Only one part of CAF was added to every 200 parts of myofibrillar protein. Hence, the ability to detect the heavy subunit of CAF in these gels demonstrates the sensitivity of the SDS-polyacrylamide gel assay for proteolysis. Because α-actinin is in the Z-disk (Table 1) and binds to actin in vitro (Goll et al. 1972) and because CAF degrades Z-disks, it was expected that CAF

564   **W. R. Dayton et al.**

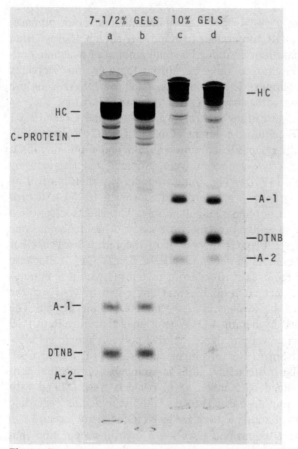

Figure 5

SDS-polyacrylamide gel electrophoresis of porcine skeletal myosin containing C-protein before and after CAF treatment. Gels *a* and *c* show control samples where EDTA was added before CAF. Gels *b* and *d* show myosin and C-protein after 60 minutes of incubation with CAF in 500 mM KCl, 100 mM Tris-acetate pH 7.5, 5 mM $CaCl_2$, 0.1 mM EDTA, 5.0 mg myosin plus C-protein/ml, 0.025 mg purified CAF/ml, 25°C. Both 7½ and 10% polyacrylamide gels were run as indicated. HC = heavy chains of myosin; A-1, DTNB and A-2 = light subunits of myosin (Weeds and Lowey 1971). Each gel is loaded with 25 μg protein.

would degrade either α-actinin or actin, or both. That CAF has no effect on either α-actinin or actin, however, has been confirmed by showing that neither the N- nor C-terminal amino acids of α-actinin and actin are altered during incubation with CAF (Singh and Dayton, unpubl.). This finding precludes the possibility that CAF acts on α-actinin or actin by removing a very small peptide off one end of the peptide chain to leave a fragment with a molecular weight so similar to the parent molecule that they cannot be distinguished on SDS-polyacrylamide gels.

Figure 6
SDS-polyacrylamide gel electrophoresis of porcine skeletal actin and α-actinin before and after CAF treatment. All gels are 7½% polyacrylamide. Gels *a* and *b* = actin; gels *c* and *d* = α-actinin. Gel *a* shows a control sample where EDTA was added before CAF. Gel *c* shows a control sample incubated with Ca++ for 60 minutes. The very faint 90,000-dalton band just under the main α-actinin band is a contaminant, possibly a degradation product of α-actinin produced during the long procedure required to prepare α-actinin. Gels *b* and *d* show samples after 60 minutes of incubation at 25°C with CAF. CAF treatment of both actin and α-actinin was done in 100 mM KCl, 100 mM Tris-acetate pH 7.5, 5 mM CaCl₂, 0.1 mM EDTA, 5.0 mg of actin or α-actinin/ml. CAF to protein ratios of 1:200 by weight. Gels are loaded with 30 μg of protein/gel.

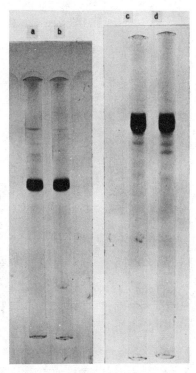

Although CAF has no evident effect on myosin, actin and α-actinin, it does degrade tropomyosin and troponin (Figs. 7 and 8). CAF acts on tropomyosin to break it into a group of fragments having similar molecular weights in the range 13,000–18,000 daltons and into a series of yet smaller pieces (Fig. 7, gels c and d). These 13,000- to 18,000-dalton fragments exist early in the degradation process and therefore are seemingly primary rather than secondary products of hydrolysis (Fig. 7, gels e and f). Porcine skeletal tropomyosin has two subunits of slightly different molecular weight (approx. 34,000 and 36,000), but CAF seems to degrade either subunit equally well (Fig. 7, gels d and f). The molecular weights of the group of fragments produced by CAF suggests that CAF attacks tropomyosin near the center of its subunit peptide chain. Although the porcine skeletal troponin shown in Figure 8 contains a small amount of tropomyosin contamination (see doublet bands just under troponin-T in Fig. 8, gel a) and some 14,000-dalton material that ostensibly originates from proteolytic degradation of troponin-I during preparation (Drabikowski, Drabrowska and Barylko 1973), purity of the preparation is completely adequate to demonstrate that CAF degrades troponin-T and troponin-I, but not troponin-C (Fig. 8). Two kinds of fragments, with molecular weights of approximately 10,000 and 14,000 daltons, seem to be produced by CAF degradation of troponin-T and troponin-I. Experiments done at CAF to troponin ratios of 1:2000 by weight show that troponin-T is degraded before troponin-I (Fig. 8, gels c and d). Although the fragments produced by treatment of troponin at a 1:2000 ratio of CAF to troponin by weight also have molecular weights between 10,000 and 14,000 daltons,

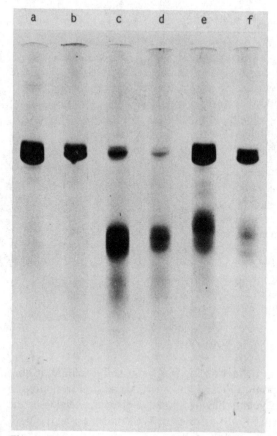

Figure 7

SDS-polyacrylamide gel electrophoresis of porcine skeletal tropomyosin before and after CAF treatment. All gels are 10% polyacrylamide. Gel *a* shows a control sample where EDTA was added before CAF; gel is loaded with 25 µg tropomyosin. Gel *b* shows another control sample, but the gel is loaded with only 7 µg tropomyosin to permit resolution of the two different subunits in porcine skeletal tropomyosin. Gels *c* and *d* show tropomyosin after 60 minutes of incubation with CAF at 25°C and CAF to tropomyosin ratios of 1:200 by weight; gels *c* and *d* are loaded with 25 and 7 µg tropomyosin, respectively. Gels *e* and *f* show tropomyosin after 60 minutes of incubation with CAF at 25°C, but at CAF to tropomyosin ratios of 1:2000 by weight; gels *e* and *f* are loaded with 25 and 7 µg tropomyosin, respectively. CAF treatment of all samples was done in 100 mM KCl, 100 mM Tris-acetate pH 7.5, 5 mM $CaCl_2$, 0.1 mM EDTA, 5.0 mg tropomyosin/ml.

566

Figure 8
SDS-polyacrylamide gel electrophoresis of porcine skeletal troponin before and after CAF treatment. All gels are 10% polyacrylamide. Gels *a* and *c* show control samples where EDTA was added before CAF; 25 μg troponin on both gels *a* and *c*. Gel *b* shows troponin after 60 minutes of incubation with CAF at 25°C, and CAF to troponin ratios of 1:200 by weight. Gel *d* also shows troponin after 60 minutes of incubation with CAF but at CAF to troponin ratios of 1:2000 by weight; 25 μg of troponin loaded. CAF treatment of all samples was done in 100 mM KCl, 100 mM Tris-acetate pH 7.5, 5 mM CaCl$_2$, 0.1 mM EDTA, 5.0 mg troponin/ml. Gels *a* and *b* were run at different times than gels *c* and *d* and hence show slightly greater separation of the troponin subunits.

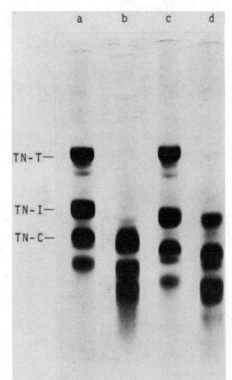

treatment of troponin at a 1:2000 CAF to troponin ratio by weight for 5 minutes has shown that CAF degrades troponin-T to a 30,000-dalton fragment before reducing this fragment to yet smaller pieces. Consequently, the effect of CAF on troponin very much resembles the effects that trypsin has on troponin (Drabikowski, Drabrowska and Barylko 1973). The inability of CAF to degrade myosin, actin or α-actinin, however, clearly distinguishes it from trypsin.

Although the experiments described in the preceding paragraphs define the effects of CAF on purified myofibrillar proteins, it is possible that conformation of these proteins is altered by their assembly into a myofibril and that any such conformational alteration may also change the susceptibility of these proteins to CAF. If CAF has a role in metabolic turnover of myofibrils, it would be important to know the effects of CAF on myofibrillar proteins while they are assembled into myofibrils. Consequently, thoroughly washed myofibrils free from contaminating membranes were prepared from porcine skeletal muscle and were then treated with CAF. Although it is not possible to interpret SDS-polyacrylamide gels of entire myofibrils unequivocally, the results suggest that assembly into myofibrils decreases the rate of degradation of tropomyosin by CAF and also seems to slow the rate of degradation of troponin-T (Fig. 9). On the other hand, there is no evidence in the gels shown in Figure 9 to indicate that assembly into myofibrils makes any of the previously unsusceptible proteins of myosin, actin, α-actinin and troponin-C vulnerable to CAF. These results confirm the conclusions obtained from

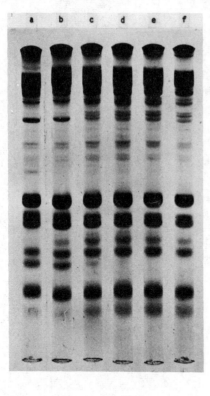

Figure 9

SDS-polyacrylamide gel electrophoresis of porcine skeletal muscle myofibrils before and after CAF treatment. All gels are 7½% polyacrylamide. After incubation with CAF, myofibrils were sedimented at 2000*g* at 25°C, and the sedimented myofibrils were washed three times at 4°C by suspension in 2 ml 100 mM NaCl and sedimentation at 2000*g*. The sedimented myofibrils were dissolved by adding 2 ml 1% SDS and placing in a boiling water bath for 15 minutes. Gel *a* shows a control sample where EDTA was added before CAF. Gels *b, c, d, e* and *f* show myofibrils after 1, 5, 10, 20 and 30 minutes of incubation, respectively, with CAF at 25°C. CAF treatment of all samples was done in 100 mM KCl, 100 mM Tris-acetate pH 7.5, 5 mM $CaCl_2$, 0.1 mM EDTA, 5.0 mg myofibrillar protein/ml. CAF to myofibrillar protein ratios were 1:200 by weight.

electron micrographs of muscles incubated in 1 mM Ca^{++}-containing solutions where it was shown that thick and thin filaments seemed to remain intact even though Z-disks were removed (Figs. 1b and 2). CAF treatment of myofibrils for 1 minute at a CAF to myofibril ratio of 1:200 by weight removes approximately 50% of the α-actinin and degrades some of the C-protein originally in the myofibril (Fig. 9, gel b). After 5 minutes of incubation with CAF, nearly all the α-actinin is gone, degradation of C-protein and troponin-I is nearly complete, and some 30,000-dalton component begins to appear (Fig. 9, gel c). This 30,000-dalton component presumably originates from troponin-T, but because troponin-T comigrates with the heavy subunit of tropomyosin in these gels, destruction of troponin-T can only be observed as a gradual decrease in density of the band corresponding to the heavy subunit of tropomyosin with longer incubation times (cf. gels a and f).

α-Actinin released from myofibrils by CAF comigrates on SDS-polyacrylamide gels with untreated α-actinin. This finding indicates that CAF does not cause extensive degradation of α-actinin, whether α-actinin is in the myofibril or in the purified state. The nature of the three very faint protein bands that appear in the 100,000-dalton region formerly occupied by α-actinin (see Fig. 9, gels c, d, e and f) is not known. Because these bands do not increase in intensity with increasing time of incubation, it is very unlikely that they are degradation products of myosin or the 150,000-dalton M-protein. It is possible that these bands are simply proteins associated with the myofibril, but whose presence is masked by α-actinin before its removal. As suggested by the complete removal of α-actinin, Z-disks in the myofibrils shown in Figure 9 were completely gone within the first 5 minutes of incubation.

Possible Role of CAF in Myofibrillar Protein Turnover

The results discussed in the previous section show that CAF exhibits a remarkable specificity in its ability to degrade myofibrillar proteins. Based on this remarkable specificity, this section will propose a possible role for CAF in metabolic turnover of myofibrils. It should be clearly pointed out, however, that we have not yet initiated studies on the physiological role of CAF and therefore have no direct evidence to indicate that CAF is actually involved in turnover of myofibrillar proteins. CAF activity would be expected to increase during rapid muscular atrophy following denervation or in the muscular dystrophies if CAF has a role in myofibril turnover. Because we have no information on CAF activity in muscle cells in these different physiological states, our proposed role of CAF in turnover of myofibrillar proteins is speculative. The unique effects of CAF on myofibrils and the large number of observations that have been made concerning muscle atrophy, however, suggest a very specific role for CAF in turnover of myofibrillar proteins, and it may be instructive to outline such a proposed role and the indirect evidence that supports it.

First, Kohn (1969) has found that rat skeletal muscle contains a soluble proteolytic component whose activity increases significantly in atrophying muscle. Although Kohn (1969) did not purify this component, all the properties described for the crude component (effect of sulfhydryl reagents, pH optimum, requirement for Ca^{++}) are very similar to the properties we have found for purified CAF. If this soluble proteolytic component is CAF, then Kohn's data (Kohn 1969) suggest that CAF increases in atrophying muscle and support the hypothesis that CAF is involved in myofibril turnover.

Second, a large number of ultrastructural studies on muscle that is atrophying due to denervation or muscular dystrophy have shown that atrophy is accompanied by marked changes in Z-disk structure (Engel 1968). These Z-disk alterations include hypertrophy (Resnick, Engel and Nelson 1968; Santa 1969; Shafiq et al. 1969), streaming (Engel and Stonnington 1974; Gauthier and Dunn 1973; Santa 1969; Stonnington and Engel 1973) and disintegration and disappearance (Johnson 1969; Pellegrino and Franzini 1963; Price, Pease and Pearson 1962; Tomanek and Lund 1973). Based on their thorough ultrastructural study of denervation atrophy in rat muscle, Pellegrino and Franzini (1963) indicated that myofibrillar atrophy occurs in two phases: a degenerative, autolytic phase that accounts for a gross weight loss of 50%, and a simple, continuous phase. The first myofibrillar alteration in the degenerative phase of atrophy is a degradation and gradual disappearance of the Z-disk. The thick and thin filaments then become disordered and eventually disperse, leaving large vacant areas (Fig. 10a,b). The second, continuous phase of atrophy proceeds simultaneously with the degenerative phase, but its effects are most noticeable later in atrophy. In the continuous phase, the diameter of individual myofibrils is reduced without disturbing the three-dimensional arrangement of thin and thick filaments in the remaining myofibrils. This reduction in diameter seems to occur by detachment of filaments from the periphery of myofibrils and subsequent breakdown of the detached filaments in the interfibrillar spaces (Pellegrino and Franzini 1963).

These observations on muscle atrophy and our results on specificity of CAF in degrading the myofibrillar proteins suggest that CAF could be responsible

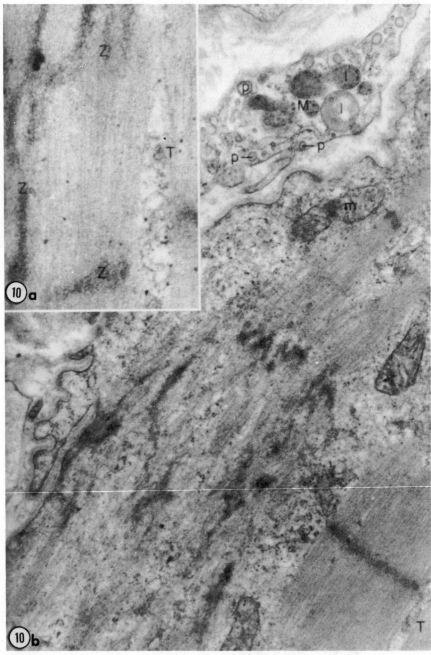

Figure 10
Electron micrographs of rat soleus muscle 1 week after denervation showing the
marked degeneration that occurs in Z-disks during denervation atrophy. (*a*) The beginning of the degradation is illustrated in *b*. The Z-disk is wavy, and Z-disk material
is near the A-band. Approx. 50,000×. (*b*) This micrograph of a peripheral area
of a soleus fiber shows a large area of degeneration in which both Z-disks and the
thick and thin filament area are undergoing disorganization. Approx. 22,000×.
(Reprinted, with permission, from Pellegrino and Franzini 1963.)

570

for the initial steps in turnover of myofibrillar proteins. Careful electron microscope studies indicate that neither intact myofibrils nor thick and thin filaments are digested intralysosomally (Lockshin and Beaulaton 1974a,b; Schiaffino and Hanzlikova 1972), and the observations on denervation atrophy suggest that myofibril turnover is initiated by degradation of the elements that hold the thick and thin filaments in their interdigitating, three-dimensional array. Degradation of these elements then leads to disordering and dispersion of the thick and thin filaments. Three structures are primarily responsible for holding the thick and thin filaments in the proper three-dimensional order: (1) the Z-disk, (2) C-protein filaments around the thick filament (Offer 1973) and (3) the M-line (Huxley 1972). Our results show that CAF destroys two of these three structures. Although some preliminary studies (Reville, unpubl.) indicate that CAF may also affect the M-line, under the incubation conditions we have used thus far, CAF causes only partial M-line degradation, and this partial degradation occurs so slowly that it seems unlikely that M-line is a primary target of CAF.

These observations, therefore, suggest that CAF may initiate myofibril turnover by the mechanism shown schematically in Figure 11. CAF degradation of Z-disks and C-protein (Fig. 11, top) should cause disordering of the three-dimensional array of thick and thin filaments (Fig. 11, middle). Moreover, degradation of tropomyosin and troponin by CAF could initiate disassembly of thin filaments to actin monomers or dimers (Fig. 11, bottom), and degradation of the M-line, possibly by some agent other than CAF, could initiate disassembly of the thick filaments (Fig. 11, bottom). The actin and myosin monomers released by this disassembly might then be degraded by lysosomes or, alternatively, could reassemble with new tropomyosin and M-protein to form new myofibrils. This mechanism is very similar to the mechanism proposed for myofibril degradation by Pellegrino and Franzini (1963) and Schiaffino and Hanzlikova (1972) on the basis of ultrastructural observations, although neither of these two groups of authors knew of the existence of CAF. The slow decrease in myofibril diameter observed by Pellegrino and Franzini (1963) is also explained directly by the proposed mechanism because gradual reduction of myofibril diameters would result from reduced amounts of CAF (or alternatively, CAF whose activity has been attenuated by lack of Ca++ or some other mechanism) slowly degrading Z-disks and C-protein on the periphery of the myofibril and releasing one layer of thick and thin filaments at a time.

Although the mechanism proposed in Figure 11 does not require that myofibrillar proteins be turned over asynchronously, it does indicate that the half-lives for actin and myosin would be longer than the half-lives for troponin and tropomyosin, which are degraded during disassembly of the myofibril, *if* any actin and myosin monomers (Fig. 11, bottom) are reassembled into filaments without being degraded to amino acids. Those studies that have reported asynchronous turnover of myofibrillar proteins (Funabiki and Cassens 1972; Low and Goldberg 1973) have found that troponin and tropomyosin turn over faster than actin and myosin. Moreover, Furakawa and Peter (1972) have found that troponin activity is greatly decreased in muscle that is atrophying due to muscular dystrophy; this also suggests that troponin is particularly labile to those processes that cause myofibrillar degradation.

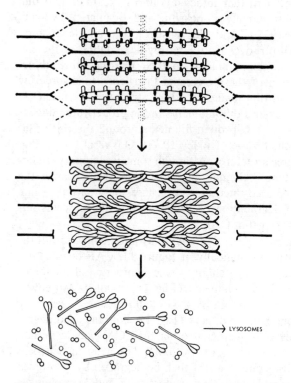

LYSOSOMES

Figure 11

A schematic diagram showing how CAF could act
to initiate a proposed mechanism for metabolic
turnover of myofibrils. A single sarcomere of an
intact myofibril is shown at the top of the diagram
with C-protein bands represented as proposed by
Offer (1973). The schematic is not drawn to
scale, and no attempt has been made to include
all cross-bridges or C-protein bands in each thick
filament. The middle diagram represents the sarcomere after CAF degradation of Z-disks, tropomyosin, troponin and C-protein. Because the proteins binding actin and myosin monomers into
thin and thick filaments have been destroyed by
CAF, the thick filaments are becoming disor-
ganized and the thin filaments are freed from the
myofibril. After dissociation into monomers and
dimers (bottom diagram), actin and myosin molecules must be degraded to amino acids by enzymes other than CAF because CAF has no effect
on purified actin and myosin. As indicated in the
diagram, these other enzymes may be lysosomal
cathepsins.

572

Although the mechanism proposed in Figure 11 for the role of CAF in myofibrillar degradation is in agreement with much of the available evidence concerning degradation of myofibrils, there is as yet no direct evidence to show that CAF is actually involved in myofibril degradation. Future studies on the activity of CAF in atrophying muscle and on the ability of lysosomal cathepsins to degrade myofibrils after CAF treatment may be expected to provide definitive information on the mechanism of myofibril degradation and on the role of CAF in this mechanism.

Acknowledgments

We are grateful to Darlene Markley for drawing Figure 11 and for devoted and expert technical assistance. We thank Mary Arthur for darkroom assistance, and Joan Andersen and Barbara Hallman for their help in preparing this manuscript. We also gratefully acknowledge Allen Christian and John Olson for their help in keeping muscle available for this study. W. J. R. is a Postdoctoral Fellow of the Muscular Dystrophy Association of America.

Journal Paper No. J-8016 of the Iowa Agriculture and Home Economics Experiment Station, Ames. Project Nos. 1795, 1796, and 2025. This research was supported in part by NIH Grants Nos. AM-12654 and HL-15679, American Heart Association Grant No. 71-679, and by grants from the Iowa Heart Association and the Muscular Dystrophy Association of America.

REFERENCES

Arakawa, N., D. E. Goll and J. Temple. 1970. Molecular properties of post-mortem muscle. 9. Effect of temperature and pH on tropomyosin-troponin and purified α-actinin from rabbit muscle. *J. Food Sci.* **35:**712.

Arakawa, N., R. M. Robson and D. E. Goll. 1970. An improved method for the preparation of α-actinin from rabbit striated muscle. *Biochim. Biophys. Acta* **200:**284.

Bhakthan, N. M. G., J. H. Borden and K. K. Nair. 1970. Fine structure of degenerating and regenerating flight muscles in a bark beetle, *Ips. confusus.* I. Degeneration. *J. Cell Sci.* **6:**807.

Bird, J. W. C., T. Berg, A. Minalesi and W. T. Stauber. 1969. Lysosomal enzymes in aquatic species. I. Distribution and aquatic properties of muscle lysosomes of the goldfish. *Comp. Biochem. Physiol.* **30:**457.

Bodwell, C. E. and A. M. Pearson. 1964. The activity of partially purified bovine catheptic enzymes on various natural and synthetic substrates. *J. Food Sci.* **29:** 602.

Busch, W. A., M. H. Stromer, D. E. Goll and A. Suzuki. 1972. Ca^{2+}-specific removal of Z-lines from rabbit skeletal muscle. *J. Cell Biol.* **52:**367.

Canonico, P. G. and J. W. C. Bird. 1970. Lysosomes in skeletal muscle tissue. Zonal centrifugation evidence for multiple cellular sources. *J. Cell Biol.* **45:**321.

Dayton, W. R., D. E. Goll, W. J. Reville, M. G. Zeece, M. H. Stromer and R. M. Robson. 1974. Purification and some properties of a muscle enzyme that degrades myofibrils. *Fed. Proc.* **33:**1580.

de Duve, C. and R. Wattiaux. 1966. Function of lysosomes. *Annu. Rev. Physiol.* **28:**435.

Desai, I. D., C. C. Calvert, M. L. Scott and A. L. Tappel. 1964. Peroxidation and lysosomes in nutritional muscular dystrophy of chicks. *Proc. Soc. Exp. Biol. Med.* **115:**462.

Desmond, W., Jr. and I. Harary. 1972. *In vitro* studies of beating heart cells in culture. Myosin turnover and the effect of serum. *Arch. Biochem. Biophys.* **151:**285.

Drabikowski, W., R. Drabrowska and B. Barylko. 1973. Properties of troponin and its constituents. *Acta Biochim. Polon.* **20:**181.

Dreyfus, J. C., J. Kruh and G. Schapira. 1960. Metabolism of myosin and life time of myofibrils. *Biochem. J.* **75:**574.

Engel, A. G. 1968. Ultrastructural reactions in muscle disease. *Med. Clinics N. Amer.* **52:**909.

Engel, A. G. and H. H. Stonnington. 1974. Morphological effects of denervation of muscle. A quantitative ultrastructural study. *Ann. N.Y. Acad. Sci.* **228:**68.

Friedman, I., A. Laufer and A. M. Davies. 1969. Studies on lysosomes in rat heart cell cultures. II. The effect of exogenous lysosomes. *Experientia* **25:**1092.

Fritz, P. J., E. L. White, K. M. Pruitt and E. S. Vesell. 1973. Lactate dehydrogenase isozymes. Turnover in rat heart, skeletal muscle, and liver. *Biochemistry* **12:**4034.

Fukazawa, T. and T. Yasui. 1967. The change in zigzag configuration of the Z-line of myofibrils. *Biochim. Biophys. Acta* **140:**534.

Funabiki, R. and R. G. Cassens. 1972. Heterogeneous turnover of myofibrillar protein. *Nature New Biol.* **236:**249.

Furakawa, T. and J. B. Peter. 1972. Muscular dystrophy and other myopathies. Troponin activity of natural actomyosin from skeletal muscle. *Arch. Neurol.* **26:**385.

Gauthier, G. F. and R. A. Dunn. 1973. Ultrastructural and cytochemical features of mammalian skeletal muscle fibres following denervation. *J. Cell Sci.* **12:**525.

Gergely, J., M. A. Gouvea and D. Karibian. 1955. Fragmentation of myosin by chymotrypsin. *J. Biol. Chem.* **212:**165.

Goldberg, A. L. 1969. Protein turnover in skeletal muscle. I. Protein catabolism during work-induced hypertrophy and growth induced with growth hormone. *J. Biol. Chem.* **244:**3217.

Goll, D. E., N. Arakawa, M. H. Stromer, W. A. Busch and R. M. Robson. 1970. Chemistry of muscle proteins as a food. In *The Physiology and Biochemistry of Muscle as a Food* (ed. E. J. Briskey, R. G. Cassens and B. B. Marsh), vol. 2, pp. 755–800. University of Wisconsin Press, Madison, Wisconsin.

Goll, D. E., A. Suzuki, J. Temple and G. R. Holmes. 1972. Studies on purified α-actinin. I. Effect of temperature and tropomyosin on the α-actinin/F-actin interaction. *J. Mol. Biol.* **67:**469.

Huston, R. B. and E. G. Krebs. 1968. Activation of skeletal muscle phosphorylase kinase by Ca^{2+}. II. Identification of the kinase activating factor as a proteolytic enzyme. *Biochemistry* **7:**2116.

Huxley, H. E. 1972. Molecular basis of contraction in cross-striated muscles. In *The Structure and Function of Muscle* (ed. G. H. Bourne), vol. 1, 2nd ed., pp. 301–387. Academic Press, New York.

Iodice, A. A., V. Leong and I. M. Weinstock. 1966. Separation of cathepsins A and D of skeletal muscle. *Arch. Biochem. Biophys.* **117:**477.

Iodice, A. A., J. Chin, S. Perker and I. M. Weinstock. 1972. Cathepsins A, B, C, D, and autolysis during development of breast muscle of normal and dystrophic chickens. *Arch. Biochem. Biophys.* **152:**166.

Issacson, A. and A. Sandow. 1968. Zinc content of muscle and tendon in normal and dystrophic mouse and chicken. *Life Sci.* **7:**369.

Jöbsis, F. F. and M. J. O'Conner. 1966. Calcium release and reabsorption in the sartorius muscle of the toad. *Biochem. Biophys. Res. Comm.* **25**:246.

Johnson, A. G. 1969. Alterations of the Z-lines and I-band myofilaments in human skeletal muscle. *Acta Neuropathol.* **12**:218.

Kar, N. C. and C. M. Pearson. 1972. Acid hydrolases in normal and diseased human muscle. *Clin. Chim. Acta* **40**:341.

Kohn, R. R. 1969. A proteolytic system involving myofibrils and a soluble factor from normal and atrophying muscle. *Lab. Invest.* **20**:202.

Kominz, D. R., E. R. Mitchell, T. Nihei and C. M. Kay. 1965. The papain digestion of skeletal myosin A. *Biochemistry* **4**:2373.

Lockshin, R. A. and J. Beaulaton. 1974a. Programmed cell death. Cytochemical evidence for lysosomes during the normal breakdown of the intersegmental muscles. *J. Ultrastruc. Res.* **46**:43.

―――. 1974b. Programmed cell death. Cytochemical appearance of lysosomes when the death of the intersegmental muscles is prevented. *J. Ultrastruc. Res.* **46**:63.

Low, R. B. and A. L. Goldberg. 1973. Nonuniform rates of turnover of myofibrillar proteins in rat diaphragm. *J. Cell Biol.* **56**:590.

Lowey, S., H. S. Slayter, A. G. Weeds and H. Baker. 1969. Substructure of the myosin molecule. I. Subfragments of myosin by enzymic digestion. *J. Mol. Biol.* **42**:1.

Martonosi, A. and R. A. Halpin. 1972. The turnover of proteins and phospholipids in sarcoplasmic reticulum membranes. *Arch. Biochem. Biophys.* **152**:440.

McLaughlin, J., L. G. Abood and H. B. Bosmann. 1974. Early elevations of glycosidase, acid phosphatase, and acid proteolytic enzyme activity in denervated skeletal muscle. *Exp. Neurol.* **42**:541.

McManus, I. R. and H. Mueller. 1966. Metabolism of myosin and the meromyosins in rabbit skeletal muscle. *J. Biol. Chem.* **241**:5967.

Middlebrook, W. R. 1959. Individuality of the meromyosins. *Science* **130**:621.

Morkin, E. 1970. Postnatal muscle fiber assembly: Localization of newly synthesized myofibrillar proteins. *Science* **167**:1499.

Offer, G. 1973. C-protein and the periodicity in the thick filaments of vertebrate skeletal muscle. *Cold Spring Harbor Symp. Quant. Biol.* **37**:87.

Offer, G., C. Moos and R. Starr. 1973. A new protein of the thick filaments of vertebrate skeletal myofibrils. Extraction, purification, and characterization. *J. Mol. Biol.* **74**:653.

Park, D. C. and R. J. Pennington. 1967. Proteinase activity in muscle particles. *Enzymol. Biol. Clin.* **8**:149.

Pellegrino, C. and C. Franzini. 1963. An electron microscope study of denervation atrophy in red and white skeletal muscle fibers. *J. Cell Biol.* **17**:327.

Price, H. M., D. C. Pease and C. M. Pearson. 1962. Selective actin filament and Z-band degeneration induced by plasmocid. *Lab. Invest.* **11**:549.

Rabinowitz, M. 1973. Protein synthesis and turnover in normal and hypertrophied heart. *Amer. J. Cardiol.* **31**:202.

Resnick, J. S., W. K. Engel and P. G. Nelson. 1968. Changes in the Z-disk of skeletal muscle induced by tenotomy. *Neurology* **18**:737.

Richards, E. G., C-S. Chung, D. B. Menzel and H. S. Olcott. 1967. Chromatography of myosin on diethylaminoethyl-Sephadex A-50. *Biochemistry* **6**:528.

Ridgway, E. B. and C. C. Ashley. 1967. Calcium transients in single muscle fibers. *Biochem. Biophys. Res. Comm.* **29**:229.

Robson, R. M., D. E. Goll, N. Arakawa and M. H. Stromer. 1970. Purification and properties of α-actinin from rabbit skeletal muscle. *Biochim. Biophys. Acta* **200**:296.

Russell, D. H. and S. H. Snyder. 1971. Amine synthesis in regenerating rat liver: Extremely rapid turnover of ornithine decarboxylase. *Mol. Pharmacol.* **5**:253.

Santa, T. 1969. Fine structure of the human skeletal muscle in myopathy. *Arch. Neurol.* **20**:479.

Schapira, G., J. C. Dreyfus and J. Kruh. 1962. The interconversion of amino acids after their incorporation into haemoglobin and myosin. *Biochem. J.* **82**:290.

Schiaffino, S. and V. Hanzlikova. 1972. Studies on the effect of denervation in developing muscle. II. The lysosomal system. *J. Ultrastruc. Res.* **39**:1.

Schimke, R. T. 1969. On the roles of synthesis and degradation in regulation of enzyme levels in mammalian tissues. *Curr. Topics Cell. Reg.* **1**:77.

––––––. 1970. Regulation of protein degradation in mammalian tissues. In *Mammalian Protein Metabolism* (ed. H. N. Munro), vol. 4, pp. 177–228. Academic Press, New York.

––––––. 1973. Control of enzyme levels in mammalian tissues. *Adv. Enzymol.* **37**:135.

Schimke, R. T. and D. Doyle. 1970. Control of enzyme levels in animal tissues. *Annu. Rev. Biochem.* **39**:929.

Schimke, R. T., E. W. Sweeney and C. M. Berlin. 1964. An analysis of the kinetics of rat liver tryptophan pyrrolase induction: The significance of both enzyme synthesis and degradation. *Biochem. Biophys. Res. Comm.* **15**:214.

––––––. 1965. The roles of synthesis and degradation in the control of rat liver tryptophan pyrrolase. *J. Biol. Chem.* **240**:322.

Seiden, D. 1973. Effects of colchicine on myofilament arrangement and the lysosomal system in skeletal muscle. *Z. Zellforsch.* **144**:467.

Seraydarian, K., E. J. Briskey and W. F. H. M. Mommaerts. 1967. The modification of actomyosin by α-actinin. I. A survey of experimental conditions. *Biochim. Biophys. Acta* **133**:399.

Shafiq, S. A., M. A. Gorycki, S. A. Asiedu and A. T. Milhorat. 1969. Tenotomy. Effect on the fine structure of the soleus of the rat. *Arch. Neurol.* **20**:625.

Sharp, J. G. 1963. Aseptic autolysis in rabbit and bovine muscle during storage at 37°. *J. Sci. Food Agr.* **14**:468.

Spudich, J. A. and S. Watt. 1971. The regulation of rabbit skeletal muscle contraction. I. Biochemical studies of the interaction of the tropomyosin-troponin complex with actin and the proteolytic fragments of myosin. *J. Biol. Chem.* **246**:4866.

Stauber, W. T. and J. W. C. Bird. 1974. S-ρ Zonal fractionation studies of rat skeletal muscle lysosome-rich fractions. *Biochim. Biophys. Acta* **338**:234.

Stonnington, H. H. and A. G. Engel. 1973. Normal and denervated muscle. A morphometric study of fine structure. *Neurology* **23**:714.

Swick, R. W. and H. Song. 1974. Turnover rates of various muscle proteins. *J. Animal Sci.* **38**:1150.

Tappel, A. L. 1966. Lysosomes: Enzymes and catabolic reactions. In *The Physiology and Biochemistry of Muscle as a Food* (ed. E. J. Briskey, R. G. Cassens and J. C. Trautman), vol. 1, pp. 237–249. University of Wisconsin, Madison, Wisconsin.

Tappel, A. L., H. Zalkin, K. A. Caldwell, I. D. Desai and S. Shibko. 1962. Increased lysosomal enzymes in genetic muscular dystrophy. *Arch. Biochem. Biophys.* **96**:340.

Tomanek, R. J. and D. D. Lund. 1973. Degeneration of different types of skeletal muscle fibres. I. Denervation. *J. Anat.* **116**:395.

van Eerd, J-P. and Y. Kawasaki. 1973. Effect of calcium (II) on the interaction between the subunits of troponin and tropomyosin. *Biochemistry* **12**:4972.

Varley, K. G. and N. S. Dhalla. 1973. Excitation-contraction coupling in heart. XII. Subcellular calcium transport in isoproterenol-induced myocardial necrosis. *Exp. Mol. Pathol.* **19**:94.

Velick, S. F. 1956. The metabolism of myosin, the meromyosins, actin, and tropomyosin in the rabbit. *Biochim. Biophys. Acta* **20**:228.

Vihert, A. M. and N. M. Pozdyunina. 1969. Changes in enzyme activity and electrolite content in the myocardium in experimental myocardial hypertrophy and insufficiency. *Virchows Arch. Abt. A. Path. Anat.* **347**:44.

Weber, A. 1966. Energized calcium transport and relaxing factors. In *Current Topics in Bioenergetics* (ed. D. R. Sanadi), vol. 1, pp. 203–254. Academic Press, New York.

Weber, K. and M. Osborn. 1969. The reliability of molecular weight determinations by dodecyl sulfate-polyacrylamide gel electrophoresis. *J. Biol. Chem.* **244**:4406.

Weeds, A. G. and S. Lowey. 1971. Substructure of the myosin molecule. II. The light chains of myosin. *J. Mol. Biol.* **61**:701.

Wikman-Coffelt, J., R. Zelis, C. Fenner and D. T. Mason. 1973. Studies on the synthesis and degradation of light and heavy chains of cardiac myosin. *J. Biol. Chem.* **248**:5206.

Young, H. L., W. Young and I. S. Edelman. 1959. Electrolyte and lipid composition of skeletal and cardiac muscle in mice with hereditary muscular dystrophy. *Amer. J. Physiol.* **197**:487.

Zak, R., E. Ratkitzis and M. Rabinowitz. 1971. Evidence for simultaneous turnover of four cardiac myofibrillar proteins. *Fed. Proc.* **30**:1147.

Zalkin, H., A. L. Tappel, K. A. Caldwell, S. Shibko, I. D. Desai and T. A. Holliday. 1962. Increased lysosomal enzymes in muscular dystrophy of vitamin E-deficient rabbits. *J. Biol. Chem.* **237**:2678.

The Biosynthetic Conversion of Procollagen to Collagen by Limited Proteolysis

Paul Bornstein, Jeffrey M. Davidson and Janet M. Monson

Departments of Biochemistry and Medicine, University of Washington
Seattle, Washington 98195

The recent discovery and partial characterization of procollagen, a biosynthetic precursor of collagen (for reviews, see Schofield and Prockop 1973; Bornstein 1974a,b; Martin, Byers and Piez 1975), has raised a number of new issues, among them the nature of the proteolytic mechanism by which the precursor is converted to collagen. Initial structural studies of procollagen were based on analyses of dermatosparactic procollagen, isolated from the skin of calves with a hereditary disorder involving conversion of the precursor (Lenaers et al. 1971; Stark et al. 1971; Lapiere, Lenaers and Kohn 1971), and on procollagen obtained by acetic acid extraction of embryonic chick bone (Bornstein et al. 1972; von der Mark and Bornstein 1973). On the basis of these studies, the precursor chains were found to contain additional amino-terminal extensions with molecular weights of approximately 15,000. These sequences differed markedly in amino acid composition from collagen and contained cystine, but not interchain disulfide bonds. Subsequently, interchain disulfide bonds were demonstrated in procollagen identified in the medium of cultured fibroblasts (Dehm et al. 1972; Goldberg, Epstein and Sherr 1972). When care was taken to avoid partial degradation of the precursor during extraction, interchain disulfide bonds were also found in procollagen isolated from chick bone (Fessler et al. 1973; Monson and Bornstein 1973). Furthermore, the precursors isolated from the medium of cultured fibroblasts and from bone by extraction at neutral pH in the presence of a number of inhibitors of proteolytic enzymes were demonstrably higher in molecular weight than procollagen obtained by prolonged extraction at acid pH (Monson and Bornstein 1973).

These structural studies are of direct concern to the investigation of the limited proteolytic conversion of procollagen. Experiments describing an activity (procollagen peptidase) initially thought to be responsible for conversion of procollagen to collagen have been performed either with dermatosparactic procollagen (Lapiere, Lenaers and Kohn 1971; Kohn et al. 1974) or with acid-extracted chick bone procollagen (Bornstein, Ehrlich and Wyke

1972; Bornstein et al. 1973). It is now apparent that the substrates used in these studies are derivatives of procollagen rather than the intact precursor. Thus only one step of the conversion process may have been uncovered. This reservation gains further emphasis from recent reports which suggest that a nonhelical carboxyl-terminal domain may also exist in procollagen (Tanzer et al. 1974; Church, Yaeger and Tanzer 1974).

In this report, we discuss the role of limited proteolysis in the biogenesis of collagen and consider the evidence for, and possible function of, intermediates in the conversion of procollagen to collagen. Several alternate schemes are evaluated in the light of what is known of procollagen structure and of some of the procollagen-derived peptides that have been identified.

METHODS

Preparation of Procollagen and Procollagen Peptidase

Procollagen was extracted from cranial bones of 17-day-old chick embryos with 0.5 M acetic acid (Bellamy and Bornstein 1971) or with 1 M NaCl, 0.05 Tris pH 7.5 in the presence of protease inhibitors (Monson and Bornstein 1973; Monson 1974). Preparations of procollagen peptidase were obtained from calf tendon as described by Kohn et al. (1974) and from chick calvaria (Bornstein, Ehrlich and Wyke 1972).

Acrylamide Gel Electrophoresis

Acrylamide gel electrophoresis was performed in 5% acrylamide containing 0.1% SDS and 0.5 M urea (Goldberg, Epstein and Sherr 1972) modified by substitution of an equimolar amount of diallyl tartardiamide for bis-methylene acrylamide. Samples were electrophoresed on gels (13 × 0.6 cm) at 15 mA per gel for 4 hours. The gels were frozen, cut into 1-mm slices, digested in 0.5 ml of 2% periodic acid, and counted by liquid scintillation spectrometry.

Assays for Procollagen Peptidase

The activity of procollagen peptidase preparations was assayed by one of several means:

(1) *CM-cellulose chromatography*. The enzymatic conversion of acid-extracted procollagen to collagen was monitored by the shift of radioactivity from the position of elution of pro-$\alpha 1$ to that of $\alpha 1$ when chromatographed on CM-cellulose. In a typical assay, approximately 200,000 cpm of [³H] proline-labeled substrate was mixed with partially purified procollagen peptidase and incubated at 20°C for 24 hours. The reaction was terminated by lowering the pH to 2, and the mixture was dialyzed for subsequent chromatography (Bellamy and Bornstein 1971).

(2) *SDS-acrylamide gel electrophoresis*. Reaction mixtures containing enzymatic activities and either neutral- or acid-extracted [³H]proline-labeled procollagen substrates were precipitated with 10% TCA at 0°C, collected by centrifugation, desalted and dehydrated by successive washings with ethanol: ethanol-ether (3:1) and ether, dried in vacuo, and dissolved directly in gel

electrophoresis sample buffer containing 25 mM iodoacetamide. Enzymatic activity was detected by the appearance of lower molecular weight reaction products.

(3) *Immunoprecipitation.* Reactions were terminated either by the addition of 25 mM EDTA or by thermal denaturation at 100°C for 5 minutes. Undegraded [³H]proline-labeled substrate plus enzyme-liberated fragments containing antigenic determinants were precipitated from the reaction mixture by a double-antibody technique utilizing rabbit antibodies specific to the precursor sequences of purified, neutral salt-extracted procollagen and sheep anti-rabbit γ-globulin serum (von der Mark, Click and Bornstein 1973). In control experiments, 95–100% of the radioactivity could be precipitated by these means. After centrifugation, radioactivity remaining in the supernatant was considered to represent enzyme-released collagen.

(4) *Adsorption to DEAE-cellulose.* Reaction mixtures were inactivated with EDTA and mixed with a slurry (10% w/v) of microgranular DEAE-cellulose in 50 mM Tris-HCl pH 7.5 at 4°C. Under these conditions, procollagen binds to the adsorbent, while collagen does not. The unbound radioactivity remaining in the supernatant after centrifugation was used as a measure of conversion of the substrate.

RESULTS

The ability of chick procollagen peptidase to convert acid-extracted procollagen to collagen is shown in Figure 1. The top panel illustrates the distribution of radioactive label in the control, whereas the bottom panel depicts the effect of treating the substrate with the enzyme preparation. It can be seen that treatment with an active enzyme caused the label to shift from a position of elution preceding α1 to the α1 position, suggesting cleavage of procollagen to form collagen (Bellamy and Bornstein 1971).

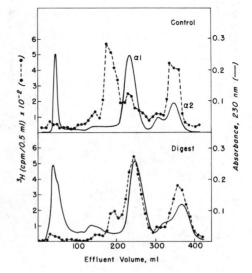

Figure 1
CM-cellulose chromatograms of acetic acid-extracted [³H]proline-labeled procollagen. (*Top*) Control preparation; (*bottom*) treated with procollagen peptidase at 20°C for 24 hours. Chromatography was performed at 40°C in 0.04 M sodium acetate–4 M urea pH 4.8. Elution was achieved with a linear gradient of NaCl from 0–0.1 M over a total of 700 ml. The continuous trace reflects the elution of added carrier rat skin collagen.

In order to determine whether chick procollagen peptidase would convert the larger, disulfide-bonded form of procollagen to collagen, a portion of this substrate was also treated with enzyme and the products examined by CM-cellulose chromatography. A comparison of the control (Fig. 2, top) and the digest (Fig. 2, bottom) after chromatography on CM-cellulose shows that there is no difference between the two radioactivity patterns. The lack of a shift of radioactivity to the $\alpha 1$ region indicates that the disulfide-bonded procollagen was not converted to collagen by this preparation of procollagen peptidase. When the reaction products were examined on SDS gels, the disulfide-bonded procollagen substrate remained intact with no change in apparent molecular weight (data not shown). However in preliminary experiments, when an enzymatic activity was prepared by extracting calvaria with neutral salt solutions of higher ionic strength, a significant conversion of procollagen to discrete lower molecular weight products was observed. A fraction of the converted material still contained interchain disulfide bonds. Further studies will be directed toward the characterization of the activity (or activities) and toward determining the nature of the reaction products.

Analyses of procollagen-converting activity by CM-cellulose chromatog-

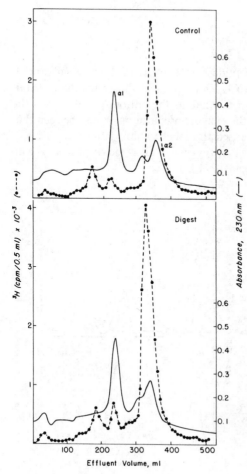

Figure 2

CM-cellulose chromatograms of neutral salt-extracted [³H]proline-labeled procollagen. (*Top*) Control preparation; (*bottom*) treated with procollagen peptidase at 20°C for 24 hours. See legend to Figure 1 for additional details.

raphy or SDS gel electrophoresis are laborious and not suited to assays of the enzymes during their purification. We have therefore developed two procedures, an immunoprecipitation assay and a batch fractionation assay using DEAE-cellulose, which show promise in detecting conversion of procollagen to collagen. The immunoprecipitation assay depends on the presence in the substrate of antigenic determinants unique to the precursor. The effectiveness of this assay was demonstrated by the observation that neither neutral nor acid-extracted procollagen, after brief treatment with chymotrypsin under nondenaturing conditions, was precipitable by antibody to procollagen. Furthermore, when calvarial fractions containing procollagen-converting activity were preincubated with inhibitory concentrations of EDTA (25 mM) or were heat-denatured, immune precipitation equivalent to control values was achieved, indicating inactivation of the enzymatic activities involved.

Similar positive and negative controls have been performed with the DEAE-cellulose batch fractionation assay. In addition, conversion to collagen could be confirmed by recovery of the ^{3}H-labeled protein from the supernatant of the assay mixture. However, both the immune precipitation and DEAE-cellulose assays function most reliably when conversion of the precursor to collagen is complete and, in their present state of development, cannot be used to detect all intermediate steps in the conversion process. Currently, SDS-acrylamide gel electrophoresis remains the most reliable method for detection of such intermediates.

DISCUSSION

The biogenesis of the collagen molecule includes an unusually large number of posttranslational modifications which are required for the normal structure and function of the protein. This sequence includes:

1. peptide bond formation,
2. peptidyl hydroxylation,
3. triple helix formation,
4. interchain disulfide bond formation,
5. glycosylation of peptidyl hydroxylysine,
6. intracellular translocation,
7. exocytosis,
8. conversion of procollagen to collagen by limited proteolysis,
9. molecular packing and cross-link formation in extracellular fibrillogenesis.

Steps 1–3 probably occur as a concerted process in molecular assembly. The position of glycosylation in this scheme has not been established, and peptidyl hydroxylysine may be glycosylated prior to completion of the triple helix.

This complex sequence suggests several opportunities for regulation of synthesis and secretion, in addition to the potential for disordered function in diseases resulting from hereditary defects or from dysfunction on an acquired basis (Bornstein 1974a). For example, control of peptidyl hydroxylation, and more specifically of prolyl hydroxylase activity, has been suggested as an important regulatory step in collagen biosynthesis (Cardinale and Udenfriend 1974). Since underhydroxylated procollagen (protocollagen) is incapable of forming a stable triple helix at 37°C and is therefore inefficiently secreted, inhibition of hydroxylation ultimately leads to an inhibition of synthesis of the protein.

The limited proteolytic conversion of procollagen to collagen provides another step with potential for regulatory control. Since the enzymes involved in conversion are synthesized within cells but function extracellularly (see below), these enzymes may also exist in the form of inactive precursors and may be subject both to cellular control and to the exigencies of the extracellular environment in which the product, collagen, functions. If one or more of the products of the reaction possess catalytic properties, the system may function as another example of amplification by cascade (Neurath, this volume). Finally, peptides released from procollagen by limited proteolysis may also play a feedback regulatory role at the cellular level.

Site of Conversion of Procollagen

Fibroblasts in culture secrete collagen precursors into the culture medium, indicating that in this system, conversion to collagen is not a requirement for secretion (Dehm et al. 1972; Goldberg, Epstein and Sherr 1972). Culture medium procollagen was shown to be capable of conversion to collagen by application of the protein to fibroblast monolayers; presumably under these circumstances conversion occurred extracellularly (Goldberg and Sherr 1973; Taubman and Goldberg 1974). An activity that could convert acid-extracted or dermatosparactic procollagen to collagen was demonstrated in the medium of cultured human fibroblasts, 3T3 and 3T6 cells, but not in the cell layers (Layman and Ross 1973; Kerwar et al. 1973). The absence of a converting activity in cells is of interest and could indicate that the proteolytic enzymes themselves exist as inactive precursors.

An extracellular site for conversion of procollagen is also supported by studies which demonstrate that the inhibition of secretion of procollagen, resulting from the in vitro application of anti-microtubular agents, such as colchicine and vinblastine, leads to an accumulation of procollagen within cells (Ehrlich and Bornstein 1972; Dehm and Prockop 1972; Ehrlich, Ross and Bornstein 1974). In analogous experiments, Pontz, Müller and Meigel (1973) showed that the appearance of procollagen-derived peptides in the medium of cultured bone was inhibited by α,α'-dipyridyl, an iron chelator known to inhibit prolyl hydroxylase and hence collagen secretion (Bornstein 1974a). However, these studies, while consistent with an extracellular site of conversion, do not exclude some intracellular modification during the secretory process.

Since fibroblasts in culture both synthesize and secrete procollagen and convert procollagen to collagen, some means must exist to prevent complete conversion prior to secretion. Conceivably, compartmentalization of secretory pathways within the cell segregates the substrate from the proteolytic activities responsible for its conversion. Alternatively, activation of such proteases may occur extracellularly, possibly by enzymes present in the serum used as a component of culture medium; serum itself is incapable of converting acid-extracted procollagen to collagen (Bornstein, Ehrlich and Wyke 1972). It is of interest that serum is routinely omitted from the cell culture medium harvested for preparation of procollagen to simplify the separation of the protein from bulk serum proteins; such omission may also inadvertently enhance preservation of the precursor.

Procollagen Peptidase

The enzymatic activity capable of converting dermatosparactic procollagen (Lapiere, Lenaers and Kohn 1971; Kohn et al. 1974) or acid-extracted procollagen (Bornstein, Ehrlich and Wyke 1972) to collagen has been termed procollagen peptidase. The activities extracted from normal bovine tissues and embryonic chick bone have many properties in common, including a neutral pH optimum, a divalent metal requirement suggested by inhibition with EDTA, and resistance to low concentrations of sulfhydryl reagents or inhibitors of serine proteases.

On the strength of the finding that this enzymatic activity was demonstrable in normal bovine tissues but not in tissues of animals with dermatosparaxis (Lenaers et al. 1971), Kohn et al. (1974) have purified the activity. The resulting protein tended to aggregate, but contained a subunit of 70,000 molecular weight. Quantitation of enzymatic activity was achieved by electrophoretic monitoring of conversion of measured quantities of dermatosparactic procollagen. The preparations with the highest specific activity were capable of converting dermatosparactic collagen to collagen (within 6 hr) only when added in a quantity comparable, on a molar basis, to that of the substrate. This observation, together with the finding described in this report that preparations of procollagen peptidase obtained from cranial bone (or from calf tendon as described by Kohn et al. [1974]) were inactive against a more intact precursor, suggests that dermatosparactic procollagen does not represent a true intermediate in the conversion of procollagen to collagen. Conceivably, the procollagen intermediate that accumulates in dermatosparaxis is further degraded, possibly in a relatively specific manner, by adventitious extracellular enzymes. Similarly, the procollagen obtained by acid extraction of bone does not appear to represent a physiologic intermediate (Monson and Bornstein 1973) and may arise from the action of lysosomal enzymes released during homogenization of bone. It is known that conversion of procollagen can be carried to completion at neutral pH (Bornstein, Ehrlich and Wyke 1972; Vuust and Piez 1972) and does not require the intervention of an acid protease.

Procollagen-derived Fragments and Possible Mechanisms for the Conversion of Procollagen to Collagen

Initial studies of dermatosparactic procollagen and of the peptides released during conversion of procollagen (Lenaers et al. 1971; Pontz, Müller and Meigel 1973) suggested that conversion occurred as a result of a single cleavage (in each chain) (Fig. 3, scheme A). There is, however, a growing appreciation of the complexity of the procollagen molecule and increasing evidence for intermediates in the conversion process. Veis and coworkers (Veis et al. 1972, 1973; Clark and Veis 1972) have detected proteins in bovine and rat skin with chains higher in molecular weight than collagen

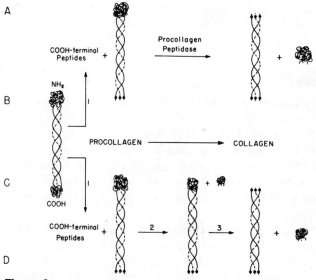

Figure 3
Four possible schemes for the conversion of procollagen to
collagen. Scheme A depicts the action of a previously described enzymatic activity, procollagen peptidase, which
converts procollagen to collagen by means of a single
proteolytic cleavage (in each chain) in a nonhelical region
near the amino-terminal end of the molecule. Although
this activity appears to function on several derivatives of
procollagen, little or no effect is observed if the intact
precursor is used as a substrate. In scheme D, two separate
steps, possibly catalyzed by different enzymes, are involved
in the conversion of the precursor by proteolysis of the
amino-terminal domain of procollagen. Schemes B and C
envision a nonhelical domain near the carboxyl-terminal
end of the molecule which is removed prior to excision of
the amino-terminal region. If a carboxyl-terminal nonhelical
domain exists, it is perhaps more likely, as depicted in
scheme C, that conversion of the precursor involves three
separate enzymatic steps. The number and distribution of
interchain disulfide bonds, indicated by solid bars, is arbitrary. A small dot indicates the presence of a lysyl residue
in a residual nonhelical extension (telopeptide) which is
susceptible to oxidative deamination by lysyl oxidase. These
are sites of subsequent cross-link formation in the collagen
fiber.

α chains. On the basis of labeling studies, such higher molecular weight proteins appear to be intermediates in the biosynthesis of collagen. In more
recent studies, Byers et al. (1974) have provided evidence for similar intermediates, termed pα chains, of type I and type III collagens in rat skin.

The type I procollagens described in the above-mentioned studies lacked
interchain disulfide bonds, since pα chains were demonstrable by acrylamide

gel electrophoresis in the absence of prior reduction. It is possible that the immediate precursors of the isolated proteins were disulfide-bonded and somewhat larger, but were converted, either in vivo or during extraction, to be more stable p-collagen, in a manner similar to that experienced with dermatosparactic procollagen. Preliminary pulse-chase experiments with embryonic chick cranial bone (Davidson and Bornstein 1975) indeed suggest that a lower molecular weight disulfide-bonded procollagen exists. No evidence was found for an intermediate similar to that of dermatosparactic or acid-extracted procollagen.

On the basis of these studies, a conversion mechanism (Fig. 3, scheme D) may be proposed. A similar scheme was advanced by Goldberg and Sherr (1973). In an initial proteolytic step, peptides from the amino-terminal ends of the procollagen molecule are removed. The nature and size of these peptides is not known. Derivatives of the intermediate may represent the procollagen isolated from the skin of dermatosparactic and normal animals. The intermediate appears to be short lived and is converted in a second step to collagen with the release of a disulfide-bonded peptide. The peptide isolated from the medium of cultured bone (Bornstein et al. 1974) may represent such a product.

Scheme D accounts for much of what is currently known of procollagen structure and of the limited proteolytic steps involved in procollagen conversion. It also provides a plausible explanation for the inability of procollagen peptidase (presumably enzyme 3) to act on intact procollagen (the substrate for enzyme 2). Recently, however, evidence has been presented to suggest the existence of a nonhelical domain at the carboxyl terminus of the procollagen molecule (Tanzer et al. 1974; Church, Yaeger and Tanzer 1974). The existence of such a precursor would require additional steps during conversion (Fig. 3, schemes B and C). The number and nature of such postulated steps are totally unknown, and in our current state of knowledge, it would seem best to defer discussion of removal of carboxyl-terminal extensions until more information is available.

SUMMARY

Studies of the conversion of procollagen to collagen have been hampered by the fact that the substrates used to date represent derivatives of the intact precursor and therefore may not be efficiently converted by the activities tested. In addition, the activity which has been studied best, procollagen peptidase, has been shown to be inactive against an intact procollagen which retains interchain disulfide bonds. It is likely that more than one enzymatic step (in each chain of the triple-stranded molecule) is required for limited proteolysis of the nonhelical domain at the amino-terminal end of the precursor; if an extended carboxyl-terminal nonhelical domain exists, additional steps would be required for removal of this region. Several assays have been developed which will assist in the clarification of the conversion process. A scheme involving several steps in the conversion of procollagen to collagen is consistent with preliminary evidence for the existence of intermediates and with limited available data relating to the nature of procollagen-derived peptides. Such a scheme provides for potential control of conversion of procollagen by regulation of the enzymatic activities involved in limited proteolysis. In addition, procollagen-derived fragments may function in a feedback regulatory capacity at an enzymatic or at a cellular level.

Acknowledgment

These studies were supported by National Institutes of Health Grants AM11248 and DE02600.

REFERENCES

Bellamy, G. and P. Bornstein. 1971. Evidence for procollagen, a biosynthetic precursor of collagen. *Proc. Nat. Acad. Sci.* **68:**1138.

Bornstein, P. 1974a. The biosynthesis of collagen. *Annu. Rev. Biochem.* **43:**567.

———. 1974b. The structure and assembly of procollagen—A review. *J. Supramol. Struct.* **2:**108.

Bornstein, P., H. P. Ehrlich and A. W. Wyke. 1972. Procollagen: Conversion of the precursor to collagen by a neutral protease. *Science* **175:**544.

Bornstein, P., K. von der Mark, H. P. Ehrlich and J. M. Monson. 1973. The synthesis and secretion of collagen. *Miami Winter Symp.* **6:**263.

Bornstein, P., K. von der Mark, W. H. Murphy and L. Garnett. 1974. Isolation and characterization of procollagen-derived peptides. *Fed. Proc.* **33:**1595.

Bornstein, P., K. von der Mark, A. W. Wyke, H. P. Ehrlich and J. M. Monson. 1972. Characterization of the proα1 chain of procollagen. *J. Biol. Chem.* **247:** 2808.

Byers, P. H., K. H. McKenney, J. R. Lichtenstein and G. R. Martin. 1974. Preparation of type III procollagen and collagen from rat skin. *Biochemistry* **13:**5243.

Cardinale, G. and S. Udenfriend. 1974. Prolyl hydroxylase. *Adv. Enzymol.* **41:**245.

Church, R. L., J. A. Yaeger and M. L. Tanzer. 1974. Isolation and partial characterization of procollagen fractions produced by a clonal strain of calf dermatosparactic cells. *J. Mol. Biol.* **86:**785.

Clark, C. C. and A. Veis. 1972. High molecular weight α chains in acid soluble collagen and their role in fibrillogenesis. *Biochemistry* **11:**494.

Davidson, J. M. and P. Bornstein. 1975. Evidence for multiple steps in the limited proteolytic conversion of procollagen to collagen. *Fed. Proc.* **34:**562.

Dehm, P. and D. J. Prockop. 1972. Time lag in the secretion of collagen by matrix-free tendon cells and inhibition of the secretory process by colchicine and vinblastine. *Biochim. Biophys. Acta* **264:**375.

Dehm, P., S. A. Jimenez, B. R. Olsen and D. J. Prockop. 1972. A transport form of collagen from embryonic tendon: Electron microscopic demonstration of a NH$_2$-terminal extension and evidence suggesting the presence of cystine in the molecule. *Proc. Nat. Acad. Sci.* **69:**60.

Ehrlich, H. P. and P. Bornstein. 1972. Microtubules in transcellular movement of procollagen. *Nature New Biol.* **238:**257.

Ehrlich, H. P., R. Ross and P. Bornstein. 1974. Effects of antimicrotubular agents on the secretion of collagen. A biochemical and morphological study. *J. Cell Biol.* **62:**390.

Fessler, L. I., R. E. Burgeson, N. P. Morris and J. H. Fessler. 1973. Collagen synthesis: A disulfide-linked collagen precursor in chick bone. *Proc. Nat. Acad. Sci.* **70**:2993.

Goldberg, B. and C. J. Sherr. 1973. Secretion and extracellular processing of procollagen by cultured human fibroblasts. *Proc. Nat. Acad. Sci.* **70**:361.

Goldberg, B., E. H. Epstein, Jr. and C. J. Sherr. 1972. Precursors of collagen secreted by cultured human fibroblasts. *Proc. Nat. Acad. Sci.* **69**:3655.

Kerwar, S. S., G. J. Cardinale, L. D. Kohn, C. L. Spears and F. L. H. Stassen. 1973. Cell-free synthesis of procollagen: L-929 fibroblasts as a cellular model for dermatosparaxis. *Proc. Nat. Acad. Sci.* **70**:1378.

Kohn, L. D., C. Isersky, J. Zupnik, A. Lenaers, G. Lee and C. M. Lapiere. 1974. Calf tendon procollagen peptidase: Its purification and endopeptidase mode of action. *Proc. Nat. Acad. Sci.* **71**:40.

Lapiere, C. M., A. Lenaers and L. D. Kohn. 1971. Procollagen peptidase: An enzyme excising the coordination peptides of procollagen. *Proc. Nat. Acad. Sci.* **68**:3054.

Layman, D. L. and R. Ross. 1973. The production and secretion of procollagen peptidase by human fibroblasts in culture. *Arch. Biochem. Biophys.* **157**:451.

Lenaers, A., M. Ansay, B. V. Nusgens and C. M. Lapiere. 1971. Collagen made of extended α-chains, procollagen, in genetically defective dematosparaxic calves. *Eur. J. Biochem.* **23**:533.

Martin, G. R., P. H. Byers and K. A. Piez. 1975. Procollagen. *Adv. Enzymol.* **42**:167.

Monson, J. M. 1974. Identification of a disulfide-bonded procollagen as a biosynthetic precursor of chick-bone collagen. Ph.D. thesis, University of Washington, Seattle.

Monson, J. M. and P. Bornstein. 1973. Identification of a disulfide-linked procollagen as the biosynthetic precursor of chick-bone collagen. *Proc. Nat. Acad. Sci.* **70**:3521.

Pontz, B. F., P. K. Müller and W. N. Meigel. 1973. A study on the conversion of procollagen. Release and recovery of procollagen peptides in the culture medium. *J. Biol. Chem.* **248**:7558.

Schofield, J. D. and D. J. Prockop. 1973. Procollagen—A precursor form of collagen. *Clin. Orth. Rel. Res.* **97**:175.

Stark, M., A. Lenaers, C. Lapiere and K. Kuhn. 1971. Electronoptical studies of procollagen from the skin of dermatosparaxic calves. *FEBS Letters* **18**:225.

Tanzer, L., R. L. Church, J. A. Yaeger, D. E. Wampler and E. Park. 1974. Procollagen: Intermediate forms containing several types of peptide chains and non-collagen peptide extensions at NH$_2$ and COOH ends. *Proc. Nat. Acad. Sci.* **71**:3009.

Taubman, M. B. and B. Goldberg. 1974. Efficiency of conversion of procollagen to tropocollagen by cultured normal, Ehlers-Danlos and Marfan's fibroblasts. *Fed. Proc.* **33**:617.

Veis, A., J. R. Anesey, J. E. Garvin and M. T. Dimuzio. 1972. High molecular weight collagen: A long-lived intermediate in the biogenesis of collagen fibrils. *Biochem. Biophys. Res. Comm.* **48**:1404.

Veis, A., J. R. Anesey, L. Yuan and S. J. Levy. 1973. Evidence for an amino-terminal extension in high molecular weight collagens from mature bovine skin. *Proc. Nat. Acad. Sci.* **70**:1464.

von der Mark, K. and P. Bornstein. 1973. Characterization of the proα1 chain of procollagen. Isolation of a sequence unique to the precursor chain. *J. Biol. Chem.* **248**:2285.

von der Mark, K., E. M. Click and P. Bornstein. 1973. The immunology of procollagen. I. Development of antibodies to determinants unique to the proα1 chain. *Arch. Biochem. Biophys.* **156:**356.

Vuust, J. and K. A. Piez. 1972. A kinetic study of collagen biosynthesis. *J. Biol. Chem.* **247:**856.

Granulocyte Collagenase and Elastase and Their Interactions with α_1–Antitrypsin and α_2–Macroglobulin

Kjell Ohlsson

Department of Clinical Chemistry and Surgery, University of Lund
Malmö General Hospital, Malmö, Sweden

My interest in human granulocyte proteases originated from the idea that they may be involved in the destruction of the lungs seen in homozygous α_1-antitrypsin deficiency. The discovery of collagenase and elastase activity in human granulocytes by Lazarus et al. (1968) and by Janoff and Scherer (1968) seemed to lend support to this idea. In addition, I also have had a wide interest in the defense mechanism against neutral proteases released into the tissue fluids. The demonstration of the release of lysosomal proteases (Henson 1971; Weissmann et al. 1971) from granulocytes on phagocytosis of particles, such as zymosan and immune complexes, stimulated my interest in granulocyte proteases still more.

Tissue resolution by granulocytal proteases and demarcation of the inflammatory reaction is a daily reality to me as a surgeon. Problems bearing on well-balanced intercellular protease inhibitor systems fall not only in the field of enzyme kinetics, but also of practical medicine.

Thanks to the preparation of pure collagenase and elastase from human granulocytes, as well as of the dominating plasma protease inhibitors, α_1-antitrypsin and α_2-macroglobulin, it was possible to develop an immunochemical approach in the investigation of interactions between the enzymes and the inhibitors.

Purification of Human Granulocyte Collagenase and Elastase

Successful purification was dependent on having a rich source of starting material of granulocytes collected from the peripheral blood of patients with chronic myeloid leukemia. Another advantage of this source was that almost 100% (more than 99%) pure granulocytes were obtained instead of the usual 15% mixture of mononuclears.

The granule fraction isolated by differential centrifugation was originally extracted with 0.2 M sodium acetate buffer pH 4.0. In this method, collagenase (Ohlsson and Olsson 1973) as well as elastase (Ohlsson and Olsson 1974a)

591

was solubilized. However, several extraction media have been tested and it was found that addition of 0.3% CTAB (cetyltrimethyl ammonium bromide) to dilute phosphate buffer pH 7.0 gives a fourfold yield (Ohlsson and Olsson, unpubl.). A drawback is the difficulty in getting rid of all the CTAB.

The proteins of the granule extract can be separated very well by chromatography on Sephadex G-75 (Fig. 1). Myeloperoxidase is identified in the void volume peak and lysozyme in the last protein peak. Collagenase and elastase activities are recovered almost completely separated in peaks 3 and 4, respectively.

The collagenase was purified further by chromatography on Bio-Rex 70 at pH 7.4 utilizing a linear sodium acetate gradient (Fig. 2). This step gave 88% pure enzyme, which was distributed between two components eluted close together.

The impurities of the collagenase preparation thus obtained were separated off by affinity chromatography on a collagen-Sepharose 4B column. On preparative disc electrophoresis, the collagenase recovered separated into two components, a major one and a minor one. Both possessed not only collagenase activity, but also fibrinolytic activity. Reelectrophoresis of the two protein fractions obtained revealed no impurities (see Fig. 4). Both exhibited collagenolytic activity. Rabbits immunized by either of the collagenases produced monospecific antisera and thereby corroborated the purity of the collagenases. The yield of purified collagenase was 38%.

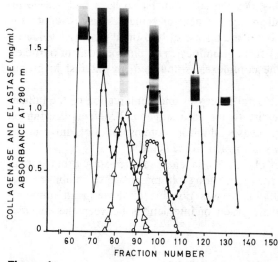

Figure 1

Chromatography on Sephadex G-75 of concentrated granule extract. The distribution in the effluent of absorbance at 280 nm (•——•), collagenase (△——△) and elastase (○——○) was determined by single radial immunodiffusion. The electrophoretic mobility of the protein peaks on polyacrylamide gels is included (anode at top). For details, see Ohlsson and Olsson (1973).

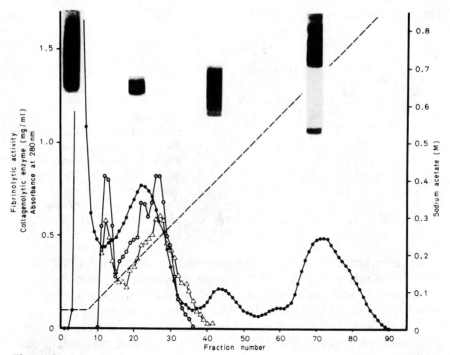

Figure 2

Chromatography on Bio-Rex 70 of the collagenolytic fraction obtained from the Sephadex G-75 column. The fractions were assayed for absorbance at 280 nm (●——●), fibrinolytic activity (△——△) and collagenase (○——○). For further details, see Figure 1 and Ohlsson and Olsson (1973). (Reprinted, with permission, from Ohlsson and Olsson 1973.)

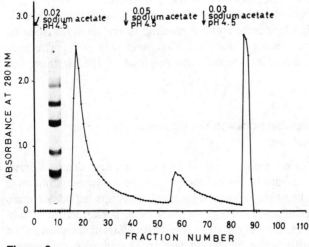

Figure 3

Chromatography on ε-Ahx-Sepharose of the elastase-containing material obtained from the Sephadex G-75 column. All elastase activity was recovered by elution with 0.02 M sodium acetate. The electrophoretic mobilities of the proteins of this fraction are included. For details, see Ohlsson and Olsson (1974a).

593

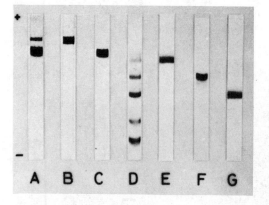

Figure 4
Polyacrylamide gel electrophoresis of (A) collagenase fraction from collagen-Sepharose column, (B-C) the two purified collagenases, (D) elastase fraction from ε-Ahx-Sepharose column, and (E-G) the three elastases.

The granulocyte elastase fraction recovered from the Sephadex G-75 chromatography was purified further by chromatography on an ε-Ahx-Sepharose 4B column, a weak ion adsorbant, which was eluted stepwise with sodium acetate at pH 4.5 (Fig. 3). All elastolytic activity was recovered in a protein peak eluted with 0.02 M sodium acetate. On agarose and polyacrylamide gel electrophoresis (Fig. 4), the elastolytic activity separated into three protein fractions.

On gel electrophoresis, the cationic granule proteins, such as the elastases, interacted with components in commercial agarose to produce pronounced trailing. Moreover, false precipitates tended to appear in immunodiffusion studies. These difficulties were avoided by chromatography of the agarose on QAE-Sephadex at pH 8.1 in dilute Tris-HCl buffer (Johansson and Hjertén 1974). This procedure gave an agarose suitable for the separation of cationic proteins, but unsuitable for a separation of anionic proteins.

Preparative agarose gel electrophoresis of the elastase-containing fraction obtained from the ε-Ahx-Sepharose column resulted in the separate isolation of the three granulocyte elastases (Fig. 4). This yield was 59%. The antiserum obtained from rabbits immunized with any one of the elastases was monospecific.

Immunological Comparison between the Granulocyte Collagenases and Elastases

The results of gel diffusion studies and of crossed immunoelectrophoresis (Figs. 5 and 6) indicated immunological identity of the enzymes within each group. Some experiments on the immunological relationships between human skin collagenase and granulocyte collagenase have been performed in collaboration with Arthur Eisen. On gel diffusion, the two collagenases showed the reaction of partial identity when antiserum against granulocyte collagenase was used. Thus separate immunochemical quantitation of the two collagenases in biological fluids is difficult.

Granulocyte elastase and pancreatic elastase showed no cross precipitation when determined by immunodiffusion with antisera against both enzymes.

Judging from crossed immunoelectrophoresis, granulocyte collagenase and elastase occur as at least two and three fractions, respectively, in the granule

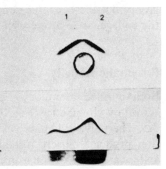

Figure 5
Analysis with immunodiffusion and crossed immunoelectrophoresis of the two collagenases (1, 2). Identical results were obtained with antibodies against either of the two collagenases.

extract. The separation of the enzymes into several components thus appears not to be an artifact due to the purification procedures. Independent evidence of the existence of separate elastase iso-enzymes recently has been produced by Janoff (1973).

Composition and Molecular Weight

Thin-layer chromatography gave a molecular weight of 76,000 for both collagenases. Examination of the denatured and reduced collagenases in the SDS-polyacrylamide electrophoretic system revealed that each of the two enzymes had two components with molecular weights of 42,000 and 33,000, respectively. The $s°_{20,w}$ was 4.5S, in agreement with the results of the thin-layer chromatography. Each of the two native enzymes therefore seemed to be built up of two subunits. Previous results suggest that synovial collagenases (Harris, Di Bona and Krane 1969) exist in polymeric forms.

The molecular weight of elastases as estimated with SDS electrophoresis was about 34,000. These enzymes had an $s°_{20,w}$ of 2.6S.

Daniels, Lian and Lazarus (1969) found a molecular weight of approximately 70,000 for granulocyte collagenase. Rindler-Ludwig, Schmalzl and Braunsteiner (1974) recently reported molecular weights of about 70,000 and 33,000 for the neutral proteases of human granulocytes.

The amino acid compositions differ by only a few residues within each

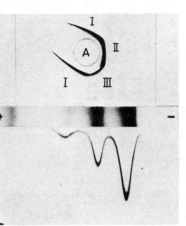

Figure 6
(*Top*) Precipitation patterns obtained on immunodiffusion analysis of the purified granulocyte elastases (I–III). (*Bottom*) Precipitate pattern obtained on crossed immunoelectrophoresis of a partially purified elastase fraction prepared from normal human granulocytes. The IgG fraction from the antisera against each of the elastases (I–III) gave the same pictures.

group, but the elastases contain a larger amount of basic amino acids than the collagenases, fitting well the difference in electrophoretic properties.

Enzymatic Activity

Both purified granulocyte collagenases hydrolyze solubilized collagen, as shown by viscometry (Fig. 7) at 25 °C with an optimal pH between 8.0 and 8.5. The polyacrylamide gel electrophoretic patterns of heat-denatured reaction products closely resemble those of other human collagenases (Eisen, Jeffrey and Gross 1968) on this type of substrate. Furthermore, both enzymes exhibit fibrinolytic activity. As yet, it has not been possible to assign the collagenolytic and fibrinolytic activity to separate enzymes. A broader substrate specificity of granulocyte collagenase than that of the specific peptide bond of the collagen molecule is in accord with the theories concerning the collagenase–α_2-macroglobulin interaction. A peptide bond is cleaved in α_2-macroglobulin as an initial step in the inhibitor-enzyme interaction (for review, see Barrett and Starkey 1973).

The three granulocyte elastases showed closely similar molecular activities on undyed, insoluble elastin as well as on fibrin and low-molecular alanine esters. The optimal pH for the enzyme activity was 8.5.

Indirect evidence suggests that granulocyte collagenase and elastase may be mediators of the breakdown of collagen and elastin in a series of conditions, such as abscess formation, acute arteritis (Janoff 1970) and in familial emphysema (Ohlsson 1971a). After the initial cleavage of the collagen mole-

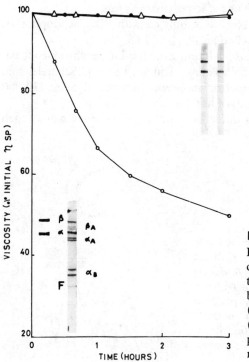

Figure 7

Effect of granulocyte collagenase on the viscosity of a collagen solution before (o——o) and after incubation with human α_1-antitrypsin (•——•) and α_2-macroglobulin (△——△). The disc electrophoretic patterns of control and reaction mixtures are included.

cule by the collagenases, the fragments evidently (Ohlsson and Tegner 1975) constitute substrates also for the elastases, causing hydrolysis into smaller peptides. Recent data published by Janoff and Blondin (1973) suggest that granulocyte elastase is involved in the digestion of some bacterial cell walls within leukocytes. Human leukocyte intravacuolar pH is maintained near neutrality for long periods during phagocytosis.

Formation of Complexes between Granulocyte Collagenase and Elastase and the Plasma Protease Inhibitors

On gel filtration of reaction mixtures of human serum and ^{125}I-labeled granulocyte collagenase corresponding to 1% saturation of the collagenase-inhibiting capacity, 50% was eluted complexed with α_2-macroglobulin and the rest mainly complexed with α_1-antitrypsin. The corresponding figures for granulocyte elastase are about 8% as α_2-macroglobulin and most of the rest as α_1-antitrypsin complexes (Ohlsson and Olsson 1974b). This suggests that α_1-antitrypsin is the major antielastase, whereas α_1-antitrypsin and α_2-macroglobulin seem to be of equal importance as anticollagenase.

Agarose gel electrophoresis was performed on reaction mixtures of serum and ^{125}I-labeled granulocyte collagenase and elastase. Autoradiography of the electrophoretic patterns obtained with reaction mixtures of human serum and human granulocyte collagenase revealed one radioactive zone in the α_2 region. This α_2 zone was shown immunochemically to contain α_1-antitrypsin as well as α_2-macroglobulin complexes with ^{125}I-labeled granulocyte collagenase (Fig. 8). Corresponding analysis with granulocyte elastase showed two radioactive zones, one in the α_2 region and one in the β_2 region. On

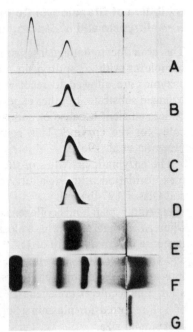

Figure 8
Patterns obtained on crossed immunoelectrophoresis of reaction mixtures of human serum and ^{125}I-labeled granulocyte collagenase with antihuman α_1-antitrypsin (A) and antihuman α_2-macroglobulin (C). (B and D) Autoradiographs of A and C. (E) Autoradiograph of the corresponding agarose gel electrophoretic pattern. The electrophoretic patterns of human serum (F) and granulocyte collagenase (G) are given for reference.

598   K. Ohlsson

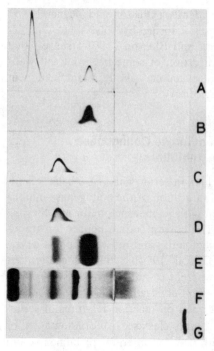

Figure 9
Patterns obtained on crossed immunoelectrophoresis of a reaction mixture of
human serum and ^{125}I-labeled granulocyte elastase using antihuman α_1-anti-
trypsin (A) and α_2-macroglobulin (C).
(B and D) Autoradiographs of A and C.
(E) Autoradiograph of corresponding
agarose gel electrophoretic pattern. The
electrophoretic patterns of human serum
(F) and granulocyte elastase (G) are
given for reference.

crossed immunoelectrophoresis, the α_2 zone was identified as elastase complexes with α_2-macroglobulin. The β_2 zone contained α_1-antitrypsin complexes with elastase (Fig. 9).

Inhibition of Granulocyte Collagenase and Elastase by
α_1-Antitrypsin and α_2-Macroglobulin

The granulocyte collagenases are effectively inactivated by the formation of
complexes with α_1-antitrypsin and α_2-macroglobulin (Figs. 7, 8, 10) if the
enzymes are allowed to react with the inhibitors before the addition of the
collagen substrate. In other experimental models, the inhibitors may appear to
be less efficient because of competition between inhibitor and collagen molecules for the enzyme. This helps to explain previous incompatible results
(Lazarus et al. 1968).

The enzymatic activities of the granulocyte elastases are inhibited by complex formation with α_1-antitrypsin as well as α_2-macroglobulin (Ohlsson
and Olsson 1974b).

α_1-Antitrypsin binds collagenases as well as elastase in a molar ratio of 1:1,
whereas α_2-macroglobulin, at least according to preliminary data, binds collagenase in a molar ratio of 1:1, but elastase in a molar ratio of 1:2 (Ohlsson
and Olsson 1974b). This may be explained by a change in the tertiary structure around the second binding site of α_2-macroglobulin on binding of such
a large molecule as collagenase. This would also be in accord with the ratio
of 1:1 reported for plasmin (Sugihara, Nagasawa and Suzuki 1971), which
is of similar size.

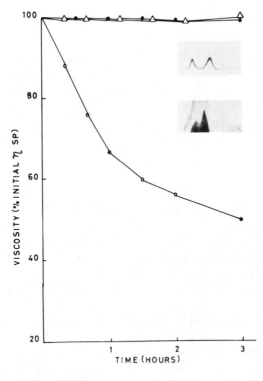

Figure 10
The same as Figure 7, but instead of the disc electrophoretic patterns, the results of the analysis with crossed immunoelectrophoresis of the reaction mixtures containing inhibitors are included: above, α_1-antitrypsin; below, α_2-macroglobulin. (The first step of the α_2-macroglobulin crossed immunoelectrophoresis was isoelectric focusing in polyacrylamide gel.)

Ultrastructural Localization of Collagenase and Elastase in Human Granulocytes

Spitznagel et al. (1974) have demonstrated that about 80% of the neutral protease of the human granulocytes is localized in the slow fraction of the azurophil granules, together with 90% of the myeloperoxidase and about 50% of the lysozyme. In a collaborative study with John Spitznagel and Inge Olsson, it was determined that most of the collagenase and the elastase, as well as the antibacterial cationic proteins, resided in this slow fraction of the azurophil granules.

Release of Collagenase and Elastase from Human Granulocytes

Human leukocytes have been reported to release neutral protease during phagocytosis of antigen/antibody complexes (i.e., Henson 1971; Weissmann et al. 1971). Having specific antisera against collagenase and elastase made it possible to follow the release, if any, of these enzymes during phagocytosis of antigen/antibody complexes also in the presence of autologous serum (Ohlsson and Olsson, unpubl.). Figure 11 shows the results of such an experiment.

Collagenase and elastase were released simultaneously. After 20 minutes, the amounts of enzymes released increased only slightly (Fig. 11). The maximal measurable amount of collagenase released in this experiment was 23 μg/5 $\times$ 10^6 cells. The corresponding figure for elastase was 13 μg. When the medium contained 10% serum, the enzymes were captured by the

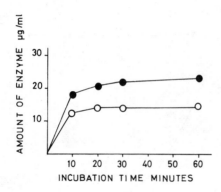

Figure 11

Extracellular release as a function of time of collagenase (o——o) and elastase (•——•) into the medium from granulocytes (5×10^6/ml) on phagocytosis of immune precipitates (600 μg/ml). The incubation medium contained 10% serum.

protease inhibitors α_1-antitrypsin and α_2-macroglobulin and no free collagenase or elastase was demonstrated in the incubation medium. The enzymes are easily identified in the complexes with α_1-antitrypsin on crossed immunoelectrophoresis with the specific antisera against the inhibitor as well as the enzymes. However, enzymes bound by α_2-macroglobulin are less well demonstrated immunochemically as the antibodies to the enzyme usually do not gain access to the enzymes of the complex. This may explain the differences found for the percentage of collagenase and elastase released on phagocytosis, as the affinity of α_2-macroglobulin is higher for collagenase than for elastase. Thus autologous human serum did not affect the release of enzyme, but the collagenase and elastase released were inhibited by α_1-antitrypsin and α_2-macroglobulin.

In vivo, large amounts of granulocyte collagenase and elastase are released into the peritoneal fluid during acute peritonitis and captured by α_1-antitrypsin and α_2-macroglobulin (Ohlsson 1975).

The fate of these complexes is presently being studied in this laboratory, and the key role of α_2-macroglobulin (Ohlsson 1971b) has been confirmed. The results indicate that on infusion or release of granulocyte elastase in vivo, the enzymes are complexed and inhibited by α_1-antitrypsin and α_2-macroglobulin, but mostly eliminated as α_2-macroglobulin complexes after transfer from α_1-antitrypsin to free α_2-macroglobulin.

SUMMARY

Human granulocytes contain, within the slow fraction of the azurophil granule, two collagenases and three elastases. The molecular weight of the collagenases is about 76,000, with an $s^\circ_{20,w}$ value of 4.5S. The corresponding values for the elastases are about 34,000 MW and an $s^\circ_{20,w}$ value of 2.6S.

The results of amino acid analysis and immunochemical work indicate that the chemical differences within each group of enzymes are small. Granulocyte collagenase shows the reaction of partial identity with skin collagenase.

The collagenases cleave collagen molecules in solution into the two specific fragments, but also act on fibrin. The elastases not only hydrolyze native undyed elastin, but also have broad substrate activity. The granulocyte collagenases and elastases are inhibited and complexed by both α_1-antitrypsin and α_2-macroglobulin. On phagocytosis, both enzymes are released from the granulocytes.

Acknowledgments

This investigation was supported by the Swedish Medical Research Council (project no. B74-17X-3910-02A), Torsten and Ragnar Söderbergs Stiftelse, Alfred Österlunds Stiftelse and the Medical Faculty of Lund.

REFERENCES

Barrett, A. J. and P. M. Starkey. 1973. The interaction of α_2-macroglobulin with proteinases. *Biochem. J.* **133:**709.

Daniels, J. R., J. Lian and G. Lazarus. 1969. Polymorphonuclear leukocyte collagenase; isolation and kinetic characterization. *Clin. Res.* **17:**154.

Eisen, A. Z., J. J. Jeffrey and J. Gross. 1968. Human skin collagenase. Isolation and mechanism of attack on the collagen molecule. *Biochim. Biophys. Acta* **151:**637.

Harris, E. D., Jr., D. R. Di Bona and S. M. Krane. 1969. Collagenase in human synovial fluid. *J. Clin. Invest.* **48:**2104.

Henson, P. M. 1971. The immunologic release of constituents from neutrophil leukocytes. I. The role of antibody and complement on nonphagocytosable surfaces of phagocytosable particles. *J. Immunol.* **107:**1535.

Janoff, A. 1970. Mediators of tissue damage in leukocyte lysosomes. X. Further studies on human granulocyte elastase. *Lab. Invest.* **22:**228.

———. Purification of human granulocyte elastase by affinity chromatography. *Lab. Invest.* **29:**458.

Janoff, A. and J. Blondin. 1973. The effect of human granulocyte elastase on bacterial suspensions. *Lab. Invest.* **29:**454.

Janoff. A. and J. Scherer. 1968. Mediators of inflammation in leukocyte lysosomes. IX. Elastinolytic activity in granules of human polymorphonuclear leukocytes. *J. Exp. Med.* **128:**1137.

Johansson, S. and B. G. Hjerten. 1974. Electrophoresis, crossed immunoelectrophoresis and isoelectric focusing in agarose gels and reduced electroendosmotic flow. *Anal. Biochem.* **59:**200.

Lazarus, G. S., J. R. Daniels, R. S. Brown, H. A. Bladen and H. M. Fullmer. 1968. Degradation of collagen by a human granulocyte system. *J. Clin. Invest.* **47:**2622.

Ohlsson, K. 1971a. Neutral leukocyte proteases and elastase inhibited by plasma α_1-antitrypsin. *Scand. J. Clin. Lab. Invest.* **28:**251.

———. 1971b. Elimination of ^{125}I-trypsin-α-macroglobulin complexes from blood by reticuloendothelial cells in dog. *Acta Physiol. Scand.* **81:**269.

———. 1975. Granulocyte collagenase and elastase and the plasma protease inhibitors in diffuse peritonitis following appendicitis. *J. Clin. Invest.* (in press).

Ohlsson, K. and I. Olsson. 1973. The neutral proteases of human granulocytes. Isolation and partial characterization of two granulocyte collagenases. *Eur. J. Biochem.* **36:**473.

———. 1974a. The neutral proteases of human granulocytes. Isolation and partial characterization of granulocyte elastases. *Eur. J. Biochem.* **42:**519.

———. 1974b. Neutral proteases of human granulocytes. III. Interaction between human granulocyte elastase and plasma protease inhibitors. *Scand. J. Clin. Lab. Invest.* **34**:349.

Ohlsson, K. and H. Tegner. 1975. Granulocyte collagenase, elastase and plasma protease inhibitors in purulent sputum. *Eur. J. Clin. Invest.* (in press).

Rindler-Ludwig, R., F. Schmalzl and H. Braunsteiner. 1974. Esterases in human granulocytes. Evidence for their protease nature. *Brit. J. Haematol.* **27**:57.

Spitznagel, J. K., F. G. Dalldorf, M. S. Leffell, J. D. Folds, I. R. H. Welsh, M. H. Cooney and L. E. Martin. 1974. Character of azurophil and specific granules purified from human polymorphonuclear leukocytes. *Lab. Invest.* **30**:774.

Sugihara, H., S. Nagasawa and T. Suzuki. 1971. Studies on α_2-macroglobulin in bovine plasma. III. Its actions on bovine plasma kallikrein, plasmin and thrombin. *J. Biochem.* **70**:649.

Weissmann, G., R. B. Zurièr, P. J. Spieler and I. M. Goldstein. 1971. Mechanisms of lysosomal enzyme release from leukocytes exposed to immuno complexes and other particles. *J. Exp. Med.* (Suppl.) **134**:149s.

Human Neutrophil Elastase: In Vitro Effects on Natural Substrates Suggest Important Physiological and Pathological Actions

Aaron Janoff, Joanne Blondin, Robert A. Sandhaus, Anne Mosser* and Charles Malemud

Department of Pathology, State University of New York at Stony Brook
Stony Brook, New York 11790

The demonstration of a neutral, serine protease with elastin-degrading activity in cytoplasmic granules of human neutrophil leukocytes (Janoff and Scherer 1968; Janoff 1970) which also hydrolyzes specific synthetic substrates of pancreatic elastase (Janoff 1969; Janoff and Basch 1971) has stimulated interest in defining the physiological and pathological functions of this enzyme in man. The isolation of three[1] elastases from human neutrophils (Janoff 1973) and the purification, separation and physicochemical characterization of each of these enzymes (Ohlsson and Olsson 1974) has recently been achieved. A detailed description of the purification and characterization of neutrophil elastases is given by Ohlsson (this volume) along with a discussion of their interactions with the major serum antiproteinases, α_1-antitrypsin and α_2-macroglobulin. The demonstration that α_1-antitrypsin is a major endogenous inhibitor of neutrophil elastase (Ohlsson 1971; Janoff 1972; Ohlsson and Olsson 1974) lends support to the proposed role of the leukocyte enzyme in pathological states associated with deficiencies of that inhibitor (Galdston, Janoff and Davis 1973). The purpose of the present paper will be to review some of our different in vitro observations on the actions of neutrophil elastase upon microbial, connective tissue and cellular targets which suggest possible physiological and pathological roles for these enzymes.

---

* Present address: Biophysics Laboratory, The University of Wisconsin, Madison, Wisconsin 53706.

[1] The presence, in neutrophil granules, of three separate enzymes possessing identical activity on elastase substrates has recently been questioned (Barrett, this volume). Working with a human spleen elastase most likely derived from the neutrophil population of that organ, Barrett has found that rapid addition of diisopropylfluorophosphate to spleen enzyme fractions results in recovery of a single species of elastase. The latter corresponds closely to the major component of the elastase triad characteristically isolated from neutrophil granules. Thus the two minor components of this triad may represent artifacts resulting from limited autodigestion of the enzyme during isolation procedures.

The Basic Experimental Model

In most of the functional studies to be described below, the basic experimental model employed is the same. Leukocyte granules are reacted with a suspected natural target of neutrophil elastase. The effects are measured and then compared to effects obtained in parallel experiments using granules that have been pretreated with a specific and irreversible elastase inhibitor to abolish the activity of their elastase component. In this way, if the neutrophil elastase participates significantly in the overall physiologic or pathologic degradative process brought about by the leukocyte granules, the inhibitor-treated granules will be less effective. This result will be obtained even though other granule enzymes may also be required for degradation to be complete. For example, elastase could participate in granule-mediated digestion in one of three ways: (1) by acting alone to accomplish the digestive action, (2) by acting in sequence or in concert with other granule enzymes, each of the several enzymatic steps being obligatory to the final result, or (3) by acting together with other granule enzymes, any one of which is independently capable of accomplishing the same result. Inhibitor studies would demonstrate actions (1) and (2) of elastase, but would fail to demonstrate (3). (Experiments with purified neutrophil elastase would be required for demonstration of the latter effect. Such an experiment will be described below in our study of the parallel actions of neutrophil elastase and chymotrypsinlike protease, or cathepsin G, upon cartilage matrix proteoglycan.) The elastase inhibitors employed in our experimental model are the peptide chloromethyl ketones: N-acetyl-L-alanyl-L-alanyl-L-prolyl-L-alanine chloromethyl ketone (AAPACK), N-acetyl-tri-alanyl chloromethyl ketone (AAACK) and N-acetyl-tetra-alanyl chloromethyl ketone (AAAACK). In order to rule out nonspecific suppression of thiol-dependent proteases by chloromethyl ketones, a control agent, TLCK (N-α-tosyl-L-lysine chloromethyl ketone) was compared with the elastase inhibitors in these experiments. Unless otherwise stated, the neutrophil granule extracts were preincubated for 1 hour at 22°C in a 1–2 mM concentration of the inhibitor. We are greatly indebted to James Powers and Peter Tuhy of the Georgia Institute of Technology, who synthesized nearly all of the chloromethyl ketones we used and generously provided us with samples of them, thus making possible many of the experiments to be described below. These investigators have recently studied the kinetics of inhibition of purified pancreatic and neutrophil elastases by such synthetic, active-site-directed inhibitors (Tuhy and Powers 1975).

Since most of the experiments to be outlined below have already been described extensively elsewhere, the following brief review will emphasize results without including many details of the methods employed.

Experiments Suggesting Pathological Roles of Human Neutrophil Elastase

Bacterial Wall Lysis

We became interested in the chemical analogies between the cross-links of some bacterial peptidoglycans and mammalian elastin fibers. An example of this is the following:

Micrococcus roseus peptidoglycan cross-link

$$-\text{L-Ala-D-Glu-L-Lys-D-Ala-}$$

$$\Big| \; \epsilon\text{NH}_2$$

$$-\text{L-Ala-L-Ala-L-Ala-L-Thr}$$

and one major type of elastin cross-link

$$-\text{L-Ala-L-Ala-L-Ala-} \quad -\text{L-Ala-L-Ala-L-Ala-}$$

$$\Big|$$

$$- \text{desmosine} -$$

$$\Big| \qquad \text{(a condensation product of four}$$
$$\text{lysaldehyde residues)}.$$

In the foregoing example, repeating L-alanine sequences are found adjacent to a lysine-derivatized cross-link in both instances. In other bacterial peptidoglycans, the cross-linking amino acid is diaminopimelic acid, which resembles lysinenorleucine, another species of elastin cross-linking amino acid:

COOH   COOH

| |

CH—NH$_2$ CH—NH$_2$

| |

(CH$_2$)$_3$ (CH$_2$)$_3$

| |

CH—NH$_2$ CH$_2$

| |

COOH   NH

                    |

                    CH$_2$

                    |

                    (CH$_2$)$_3$

                    |

                    CH—NH$_2$

                    |

                    COOH

diaminopimelic acid   lysinenorleucine

    These structural similarities between bacterial cell wall supporting skeletons and certain regions in elastin fibers led us to examine the activity of neutrophil elastase against bacterial suspensions in vitro (Janoff and Blondin 1973). One of the results we obtained using suspensions of *Micrococcus roseus,* the organism cited above, is shown in Figure 1. None of several enzymes tested decreased the optical density of the bacterial suspension if freshly grown, washed, but otherwise untreated cells were used. However, suspensions of heat-killed organisms were cleared by pancreatic elastase, trypsin and neutrophil granule extracts, although not by chymotrypsin or hen egg white lysozyme. The action of neutrophil granules was prevented by pretreatment with a specific chloromethyl ketone elastase inhibitor, whereas identical pretreatment of granules with TLCK (control for nonspecific effects of chloromethyl

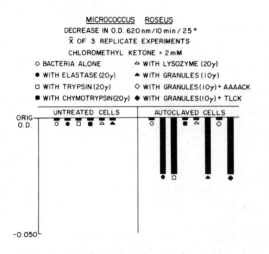

Figure 1

Decreases in optical density of *M. roseus* suspensions treated with various enzymes. Data shown are for phosphate buffer, pH 7.0. Enzymes and their amounts are identified in the figure. Starting optical density values varied (0.200 ± 0.050), but they have been normalized in the figure for purposes of clarity. (Reprinted, with permission, from Janoff and Blondin 1973.)

ketones) had no such effect. The resistance of *M. roseus* to lysozyme, even after autoclaving, and the complete inhibition of granules by the elastase inhibitor argued that the effect of the granules upon *M. roseus* was due to their elastase component. This in vitro observation suggests that elastase, while most likely acting with other granule components, may nevertheless play a crucial role in the lysis of phagocytosed microorganisms of the *M. roseus* type within human neutrophils in vivo. Similar results were also obtained with *Staphylococcus aureus,* an organism with a peptidoglycan structure closely related to that of *M. roseus.*

On the other hand, results obtained with autoclaved bacteria may not be applicable to bacterial digestion in vivo, since microbial killing in the latter case is achieved by the action of cationic proteins or by halide ions acting together with superoxide and other free radicals generated by the H_2O_2-myeloperoxidase system. It should also be noted that peptidoglycan cross-links are not the sole target available for neutrophil elastase acting on bacterial walls. Structural proteins investing the bacterial envelope constitute another potential substrate for the action of granule elastase (and other proteases), and degradation of these elements can also lead to loss of cellular integrity (Inouye and Yee 1972). Additional studies are now in progress to help resolve some of these questions.

Bacterial Protein Digestion

The ability of leukocyte elastase to attack bacterial proteins in general, by virtue of the extraordinarily broad substrate specificity of the elastaselike serine proteases, introduces a second possible role for this enzyme in the physiological functions of human neutrophils. Specifically, the envisioned role is digestion of the bacterial protoplast subsequent to killing and cell wall lysis, whether the latter is mediated by elastase or not. Figure 2 shows this effect of neutrophil elastase on *Escherichia coli* cytoplasmic proteins in vitro. The figure shows release of TCA-soluble peptides from sonicated *E. coli* by neutrophil granules alone and after various treatments. A mutant strain of *E. coli* was grown in the presence of [³H]arginine. Cytoplasmic and cell wall

Figure 2

Release of labeled peptides from sonicated *E. coli*. (•) Sonicate + human PMN granule extract (S + G); (□) S + G pretreated with 0.002 M AAACK; (■) S + G pretreated with 0.002 M AAPACK; (△) S + G pretreated with 0.002 M TLCK; (○) bacterial sonicate alone. Incubation conditions: 34 µg/ml PMN granule protein, pH 7.6, 37°C, final reaction volume = 1.4 ml. Numbers in parentheses = total experiments. Vertical lines represent standard deviation. (Reprinted, with permission, from Janoff and Blondin 1974.)

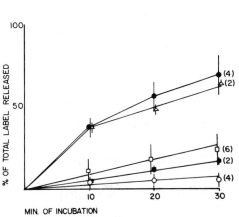

proteins were labeled in a 9:1 ratio, respectively, by this procedure. TCA-soluble peptide release after incubation with neutrophil granules was monitored by counting radioactivity in the TCA supernatant. Release was negligible at pH 3, but occurred at a significant rate at pH 7.6. Release at this pH was inhibited 82 and 94% by pretreatment of the granule extract with 2 mM AAACK or AAPACK, respectively. Preincubation with 2 mM TLCK had no effect. Such results lend support to the suggestion put forward above, namely, that a second function of neutrophil elastase may be participation in the general digestion of protein of phagocytosed microbes.

Experiments Suggesting Physiological Roles of Human Neutrophil Elastase

Lung Elastin Degradation

One natural substrate of elastase is elastin, suggesting potential pathological roles for the neutrophil enzyme in a variety of elastin-rich tissues and organs whose functions depend vitally on their elastic properties. The ability of human neutrophil granule extracts to alter arterial elastic fibers in vitro and in vivo has been reported elsewhere (Janoff and Scherer 1968; Janoff 1970). Lung elastin represents a second possible target for pathological degradation by the neutrophil enzyme and is of special interest in view of growing evidence implicating leukocyte proteases in the pathogenesis of pulmonary emphysema associated with inherited deficiency of α_1-antitrypsin (Mittman 1972). The latter serum antiproteinase is a major inhibitor of leukocyte elastase.

Table 1 shows that human lung elastin fibers are in fact degraded by human neutrophil granule extracts, and that this digestive activity is completely abolished by an elastase inhibitor. The substrate (prepared by V. D. Hospelhorn of the Will Rogers Hospital, Saranac Lake, New York) contained 40% native elastin, 40% denatured collagen and 20% elastin-associated glycoproteins. Release of elastin peptides into digestion supernatants was monitored

Table 1

Solubilization of Elastin Peptides from Human Lung

| Substrate[a] (mg) | Enzyme (μg) | Inhibitor (mM) | Total protein released (mg) | | q-Desmosines recovered[b] (res/1000) |
			Lowry assay	amino acid analysis	
5.0	trypsin (25)	—	0.60	0.80	0
5.0	elastase (pancreas) (25)	—	3.10	2.94	4
5.0	elastase (bacterial)[c] (10)	—	3.30	3.05	4
5.0	PMN granules (500)	—	0.46	0.90	3
5.0	PMN granules (500)	AAPACK (1)	0.01	0.00	0
5.0	elastase (pancreas) (25)	AAPACK (1)	0.00	0.00	0

Experimental conditions: 4 hours, 37°C, 0.05 M Tris + 0.1 M NaCl + 0.001 M CaCl$_2$, pH 7.6; results are the average of two separate experiments.

[a] A fraction of normal, adult human lung containing 40% native elastin, 40% denatured collagen and 20% "elastin-associated" glycoprotein.

[b] Sum of quarter-desmosines and iso-desmosines recovered from enzyme-solubilized peptides. Zero/1000 q-desmosines were recovered when substrate or enzymes were incubated alone.

[c] Thermolysin.

by measuring solubilized desmosine (the cross-link amino acid unique to elastin). As can be seen, pancreatic and bacterial elastases and neutrophil granules solubilized peptides from the substrate which contained desmosine. In contrast, trypsin (which cannot attack undenatured elastin) solubilized only nondesmosine-containing peptides, presumably from the denatured collagen present in the substrate. (Hydroxyproline was recovered in the trypsin digestion supernatants.) AAPACK (1 mM) completely abolished the degradative effects of pancreatic elastase and neutrophil granules. These in vitro results support the suggestion that neutrophil elastase may participate in damage to lung elastin during the development of pulmonary emphysema, at least in those cases in which leukocyte proteases released into the lung are thought to be pathogenic determinants of the disease. A more complete description of these experiments has been reported elsewhere (Janoff et al. 1972). Similar results have also been obtained using whole lung slices as substrate and enzymes derived from purulent sputum (Lieberman and Gawad 1971). Some further support for the hypothetical role of neutrophil elastase in pulmonary emphysema associated with α_1-antitrypsin deficiency has been developed in a recent clinical study (Galdston, Janoff and Davis 1973).

Cartilage Proteoglycan Degradation

The same broad substrate specificity of elastases which attracted our attention to their possible role in neutrophil antibacterial functions has also led us to consider their possible role in neutrophil-mediated damage to connective tissue components other than elastin fibers. One such component is cartilage matrix proteoglycan. In earlier work (Janoff and Blondin 1970), we reported depletion of metachromatic staining substances from rabbit articular cartilage upon treatment with human neutrophil granules at neutral pH, or with granule fractions enriched in elastase activity, or with purified pancreatic elastase alone. More recently, Oronsky, Ignarro and Perper (1973) and Ignarro, Oronsky and Perper (1973) have reported neutral protease activity in human neutrophil granules which is capable of solubilizing radiosulfate-containing material from labeled rabbit ear cartilage. This neutral protease is released from viable leukocytes by exposure of the cells to immune aggregates adsorbed onto cartilage surfaces. In an effort to identify the neutral proteases responsible for cartilage proteoglycan degradation by neutrophil granules, we have recently carried out experiments in which granule extracts were fractionated by preparative polyacrylamide disc gel electrophoresis and the resultant fractions tested against $^{35}SO_4$-labeled rabbit joint cartilage. More detailed results will be reported elsewhere (Malemud and Janoff 1975), but the essential findings are described below.

A typical cationic electrophoretogram of neutrophil granule extract is shown in Figure 3. Gels prepared in this way were cut into zones as diagrammed in the figure, and the contents of each zone and of the control (blank) gel were then incubated with $^{35}SO_4$-labeled rabbit articular cartilage at 37°C for 30 minutes at neutral pH. Table 2 summarizes the results of the incubation and shows that zones 2 and 3 were very active in releasing radiosulfate. Zone 2 is known to contain the three elastase iso-enzymes (see footnote 1) of neutrophil granules (Janoff 1973; Sweetman and Ornstein 1974; Ohlsson and Olsson 1974), whereas zone 3 contains lysozyme (Janoff 1973) and two chymotrypsinlike esterases (Rindler, Schmalzl and Braunsteiner

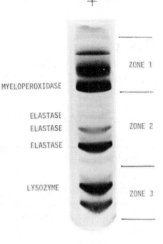

Figure 3
Cationic disc gel electrophoretogram of human neutrophil granule extract (100 μg protein). (*Center*) photograph showing the major protein bands of a gel stained with Amido Schwartz; (*right*) the zones into which such gels were cut for testing proteoglycan degradation by separated protein components; (*left*) three of the major components identified (myeloperoxidase, elastase and lysozyme).

Table 2

$^{35}SO_4$-Articular Cartilage Proteoglycan-degrading Activity of Electrophoretically Separated Human PMN Granule Components

Electro-phoretogram zone	Major identified components	% $^{35}SO_4$ released in 0.5 hr[a]	% $^{35}SO_4$ released by whole granule extract
—	whole granules[b]	79.5	100
1	myeloperoxidase,	9.6	12
2	elastase	74.6	94
3	lysozyme chymotrypsinlike esterases[c]	78.7	99
Control[d]	—	3.6	5

[a] One sample in each determination.

[b] The amount of granule protein used was the same as that placed on the gel in the same experiment (38 μg).

[c] Esterases active against N-acetyl-L-phenylalanine-α-naphthyl ester.

[d] Blank gel processed identically to sample containing gels as above.

1974). Zone 1 released relatively little radiosulfate in comparison to the other zones. At the concentrations employed, the protein components in zones 2 and 3 were equally effective in releasing $^{35}SO_4$ from the substrate and matched the activity of the original starting protein of the whole granule extract.

Addition of a specific elastase inhibitor (AAPACK) to zone 2 completely blocked the release of $^{35}SO_4$ from joint cartilage (see Table 3). By contrast, TLCK was completely ineffective in blocking $^{35}SO_4$ release by neutrophil granules. These data suggest that elastase is one of the neutral proteases contributing to the degradation of cartilage proteoglycan by neutrophils.

The activity of zone 3 was not due to lysozyme since 10 μg of purified human leukocyte lysozyme released no detectable $^{35}SO_4$ from labeled cartilage. On the other hand, 1 μg/ml of pancreatic chymotrypsin released 71% of the available radiosulfate from the substrate in half an hour. Similarly, Ignarro,

Table 3

Effect of a Specific Elastase Inhibitor on Release of Radiosulfate by Zone 2 of PMN Electrophoretogram

Electro-phoretogram zone	% $^{35}SO_4$ released in 0.5 hr	% Inhibition
Control (blank gel)	1	—
Zone 2	94.1	—
Zone 2 + NAcAAPACK[a]	6.3	93

[a] Preincubated with retentate at 1 mM concentration for 1 hour at 22°C.

Oronsky and Perper (1973) previously found that pancreatic chymotrypsin degraded rabbit ear cartilage proteoglycan. Figure 4 shows a cationic zymogram of human neutrophil extract stained with N-acetyl-L-phenylalanine-α-naphthyl ester, a chymotrypsin substrate. It is clear that two bands with mobility similar to that of lysozyme and possessing chymotrypsinlike esterase activity are present in zone 3. Our tentative conclusion is that these latter enzymes are responsible for the proteoglycan-degrading activity of this zone.

The presence in human leukocyte granules of chymotrypsinlike enzymes with electrophoretic mobility similar to that of lysozyme has been reported by Rindler, Schmalzl and Braunsteiner (1974). However, there is some evidence to suggest that these enzymes may not be completely chymotrypsinlike in nature. Barrett (this volume) has isolated a similar enzyme from human spleen which apparently originates in the neutrophils of that organ and which has cartilage proteoglycan-degrading activity at neutral pH. This enzyme also migrates, during electrophoresis, in the region of lysozyme. While the spleen enzyme resembles pancreatic chymotrypsin in certain respects, in other properties, it is quite different from chymotrypsin. Thus it is unaffected by the chymotrypsin inhibitor, N-α-tosyl-L-phenylalanine chloromethyl ketone (Barrett, pers. comm.). For this reason, the enzyme is referred to as cathepsin G rather than as leukocyte chymotrypsin. Until further definition of these enzymes is achieved, we have elected to follow Barrett's terminology in referring to the two cartilage proteoglycan-degrading enzymes isolated from neutrophil electrophoretograms (zone 3).

The in vitro experiments described here demonstrate that at least two kinds of human neutrophil enzymes are potential pathogenetic determinants in inflammatory joint disease; namely, elastase (three iso-enzymes) and cathepsin G (two iso-enzymes). Further studies will be required to define the precise role of these enzymes in particular joint disorders.

Figure 4

Demonstration of chymotrypsinlike esterases (cathepsin G) in zone 3 of a cationic electrophoretogram of neutrophil granules. (*Left*) Cationic electrophoretogram (100 μg) stained with Amido Schwartz. MPO, myeloperoxidase; E, elastase; L, lysozyme. Compare with Figure 3 for identification of major bands. (*Center*) A similar gel stained with N-acetyl-L-phenylalanine-α-naphthyl ester and hexazonium pararosanaline as coupling reagent. Two cotton threads passed through the gel mark the location of the two major esterases active against this substrate. (*Right*) Identical gel as that shown in the center, but after counterstaining with Amido Schwartz. The cotton threads now reveal the position of the two esterases relative to the other components of the pattern and show that their mobility is similar to that of lysozyme. Arrows show the beginning of the separation gels in each case (gels were photographed at different magnifications).

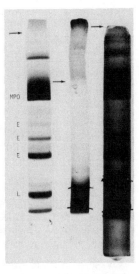

Perturbation of Cell Surface Membranes

In addition to alterations in intercellular substances brought about by neutrophil proteases, of which two examples (elastin fibers and cartilage proteoglycan) have already been cited, these enzymes can also produce alterations of cellular structure, more specifically of the surface membranes of living cells. Again, the elastase component appears to be a significant contributor to this effect.

In a previously reported study (Mosser, Janoff and Blondin 1973), it was shown that brief exposure of cultured 3T3/4 mouse fibroblasts to extracts of human neutrophil granules at neutral pH rendered the cells agglutinable by low concentrations of concanavalin A, a plant lectin. Such enhanced agglutination reactions are indicative of alterations in cell surface architecture. Untreated cells were not agglutinated by the same concentration of the lectin, and methyl-α-D-mannoside, a haptenic sugar specific for concanavalin A binding sites, prevented the agglutination reaction of treated cells. These results are summarized in Table 4. Also shown in the table are the inhibitions of granule-enhanced agglutinability resulting from various treatments of the granule extract (heating, admixture with a chloromethyl ketone elastase inhibitor, reaction with cells in the cold). It can be seen that inhibition occurred under conditions unfavorable to granule-mediated enzymatic hydrolysis (preheating, cold reaction), and that significant inhibition also resulted from neutralization of the elastase activity of the granules by AAPACK.

Further evidence of the role of elastase in granule-mediated enhancement of concanavalin A-induced agglutination was obtained in a fractionation experiment, as described in Tables 5 and 6. A comparison of the data given

Table 4

Activity of PMNg in Promoting Agglutination of 3T3/4 Cells by Con A

Material (μg protein/ml)	Treatment	Con A	Con A + methyl-α-D-mannoside (1 mg/ml)	Con A + N-acetyl-D-glucosamine (1 mg/ml)
			Agglutination[a]	
Buffer		0	0	0
Trypsin, 1.25		+	0	+
PMNg, 50		++	0	++
PMNg, 25		+	0	+
PMNg, 50	56°C for 2 hr	+	0	+
PMNg, 50	100°C for 30 min	0	0	0
PMNg, 50	[b]	±		±
PMNg, 50	react with cells at 0°C	+	0	+

Data from Mosser, Janoff and Blondin (1973). PMNg, polymorphonuclear leukocyte lysosomal extract; Con A, concanavalin A (50 μg/ml).

[a] Scored from 0 (monodisperse suspension of cells) to ++++ (all cells in large clumps as obtained with concentrations of Con A $\geqq$ 200 μg/ml); ±, barely detectable clumping.

[b] N-Acetyl-Ala-Ala-Pro-Ala-Ch$_2$Cl, 1.25×10^{-3} M, preincubated with PMNg for 2 hours at room temperature.

Table 5

Fractionation of Polymorphonuclear Leukocyte Lysosomal Extract
on Sephadex G-75

Fraction	Effluent volume (ml)	Protein[a] (mg)	Esterolytic activity against t-butyloxycarbonyl-L-*Ala*-p-nitrophenol[b]
Polymorphonuclear leukocyte lysosomal extract (starting material)		3.80	0.78
Pool 1	14–20[c]	1.48	0.02
Pool 2	22.5–27	0.59	1.68
Pool 3	30–33	0.17	0

Data from Mosser, Janoff and Blondin (1973).
[a] Total protein recovery including unpooled as well as pooled fractions = 81.5%.
[b] $\Delta 347.5$ nm/30 sec/mg protein.
[c] Void volume, 11.5 ml.

in these two tables shows that the fraction containing the highest elastase activity (pool 2) was also the most active in enhancing cell agglutination. Moreover, activity of this fraction was destroyed by mild heating, markedly diminished by reaction with cells in the cold, and almost completely abolished by the elastase inhibitor. By contrast, other fractions of the granule extract had essentially no activity (pool 3) or weak activity (pool 1), which was relatively unaffected by heating, reaction with cells in the cold or the elastase inhibitor. While it is recognized that pool 2 very likely contained other granule enzymes (including other neutral proteases) which might have contributed to its activity, the pronounced effect of the elastase inhibitor suggests that neutrophil elastase was a major participant in the pool-2 reaction. The weak activity of pool 1, seemingly due to a nonenzymatically mediated reaction, remains unexplained.

Trypsin and certain plant proteases are also capable of inducing enhanced agglutinability of cultured cells by lectins and have even been reported to alter growth behavior leading to transient loss of density-dependent growth control of nontransformed cells (Burger 1970; Sefton and Rubin 1970). Our observation of enhanced agglutination of cultured mouse fibroblasts by concanavalin A following their exposure to leukocyte lysosomal materials in vitro raises the possibility that a similar alteration of cell surfaces occurs in inflamed tissues as a consequence of leukocyte infiltration and degranulation. If enhanced agglutination of animal cells by plant lectins is sometimes indicative of cell surface changes associated with loss of growth control, a similar alteration of cell surfaces in vivo by leukocyte lysosomes might partially explain the tumor-promoting effect of inflammatory agents. However in several experiments, we failed to detect increases in thymidine uptake or cell numbers in cultures of 3T3/4 mouse fibroblasts treated with trypsin or neutrophil granules, despite detectable increases in lectin-induced agglutination of the treated cells (Mosser, unpubl.). Similarly, both Cunningham and Holley (this

Table 6

Activity of Sephadex G-75 Fractions of Polymorphonuclear Leukocyte Lysosomal Extract in Promoting Agglutination of 3T3/4 Cells by Con A

Material (μg protein/ml)	Treatment	Agglutination		
		Con A	Con A + methyl-α-D-mannoside (1 mg/ml)	Con A + N-acetyl-D-glucosamine (1 mg/ml)
Buffer	a	0	0	0
Buffer		0	0	0
Polymorphonuclear leukocyte lysosomal extract, 30		+[b]	0	+
Pool 1, 27.5	56°C for 2 hr	±,+[c]	0	±
Pool 1, 27.5	react with cells at 0°C	+	0	+
Pool 1, 27.5	a	±,+[c]	0	±
Pool 2, 12.5		++	0	++
Pool 2, 6.3		+	0	+
Pool 2, 12.5	56°C for 2 hr	0	0	0
Pool 2, 12.5	react with cells at 0°C	+	0	+
Pool 2, 12.5	a	±	0	±
Pool 3, 23.0		±	0	±

Data from Mosser, Janoff and Blondin (1973). Con A, concanavalin A (50 μg/ml).

a N-Acetyl-Ala-Ala-Pro-Ala-CH$_2$Cl, 2.5 mM, preincubated with test material for 2 hours at room temperature.

b Scoring system is described in Table 4, footnote a.

c In these instances, the + reaction was obtained with Con A (50 μg/ml) in 1 of 4 experiments, and with Con A (100 μg/ml) in 3 of 4 experiments.

volume) have reported that increased agglutination of cells by lectins, follow-ing protease treatment, is not necessarily associated with increased thymidine incorporation or cell division in the majority of cases. For example, chick embryo fibroblasts show parallel loss of density-dependent growth control, whereas mouse fibroblasts do not. Thus while it is an intriguing possibility that neutrophil proteases may stimulate cell division in inflamed tissues, it is by no means established that they do so.

Experiments Demonstrating Endogenous Inhibitors of Human Neutrophil Elastase

Serum Inhibitors

It is well recognized that neutrophil elastase is subject to inhibition by the major antiproteinases of serum, α_1-antitrypsin and α_2-macroglobulin (Ohls-son 1971; Janoff 1972; Ohlsson and Olsson 1974). A broader discussion of the interactions of these agents with the leukocyte elastase is presented by Ohlsson (this volume). It may be of value to note preliminary evidence (Galdston, Janoff and Davis 1973) suggesting that an imbalance between neutrophil elastase and serum α_1-antitrypsin is a determinant in the pathogen-esis of familial emphysema.

Cellular Inhibitors

Of additional interest, neutrophilic leukocytes themselves possess an inhibitor of their own elastase, which appears to be different from the two serum inhib-itors discussed above. Our observations regarding the intracellular inhibitor have been published elsewhere (Janoff and Blondin 1971a,b). The principal findings can be summarized as follows.

The soluble phase of disrupted human neutrophils, freed of granules by centrifugation, inhibits esterolysis of a synthetic elastase substrate by the granule fraction. Hydrolysis of the same substrate by pancreatic elastase is not affected, however, nor is the esterolytic activity of trypsin against benzoyl-D,L-arginine-p-nitroanalide (BAPA) or that of chymotrypsin against glutaryl-L-phenylalanine-p-nitroanalide (GPANA). This selectivity of esterase inhibi-tion is not characteristic of α_1-antitrypsin or α_2-macroglobulin and appears to rule out the two major serum antiproteinases as being responsible for the inhibitory activity within leukocytes. (On the other hand, selectivity of enzyme inhibition was not tested using protein substrates.) Furthermore, the inhibitory cell extracts fail to cross-react with antisera to human α_1-antitrypsin or α_2-macroglobulin, even though cell extract protein has a higher specific inhibitory activity against granule-mediated esterolysis than does serum. There is no evidence favoring DNA, RNA or hexuronic acid-containing substances as being responsible for the inhibition, whereas pronase treatment causes a loss of inhibitor activity approximately equivalent to the degree of overall cell protein digestion resulting from pronase addition. The intracellular inhibitor also suppresses elastin digestion by granule extracts, and it can be demon-strated in rabbit as well as human neutrophils. Inhibitory activity is retained by XM-300 Diaflo membranes and is mainly excluded by Sephadex G-200 but not by Bio Gel A, 1.5 M. Activity is also excluded by carboxymethyl cellulose at pH 6.0. These results suggest a high molecular weight and a net negative charge for the leukocyte inhibitor. Some of the foregoing data are presented in Tables 7 and 8.

In our experiments on the leukocyte inhibitor, granule-free supernatants of disrupted cells were used as the source of inhibitory activity. More recently, other workers (Folds, Welsh and Spitznagel 1972) determined that the nuclei of neutrophils are the source of the intracellular inhibitor. Breakage of nuclei during neutrophil homogenization allows the inhibitor to be released into the soluble phase of the homogenate. Their observations have been confirmed by us using modified homogenization techniques which protect the nuclei against disruption. The biological significance of intranuclear proteinase inhibitors is presently unknown. One possible function of such agents might be preservation of nucleohistones from degradation by lysosomal proteases released during mitosis. In the case of the neutrophil, such a protective mechanism might be required in precursor cells (myeloblasts, promyelocytes) undergoing division in the marrow. Indeed, an inhibitor of histone-degrading neutral protease has been reported in rabbit neutrophils (Davies and Weissmann 1969; Davies et al. 1971).

The role of intracellular inhibitors in the physiological and pathological functions of mature neutrophils is poorly understood. Neither they nor the serum inhibitors are likely to interfere significantly with intracellular digestive functions of neutrophil proteases, as the latter are activated within cytoplasmic vacuoles, largely separated from serum components in the cell's environment, and certainly separated from components of the cell nucleus. Serum inhibitors clearly can affect the *extracellular* action of neutrophil proteases released into connective tissue spaces, but the capacity of serum antiproteinases to complex neutrophil collagenase and elastase may become saturated under certain conditions, allowing free enzymes to damage connective tissue structures (see Ohlsson, this volume). Also, the relative concentrations of serum antiproteinases and neutrophil proteinases are likely to favor the latter when enzymes

Table 7

Inhibition of Elastoproteolytic and Esterolytic Activities of Human Leukocyte Granules by Cytosol

Cytosol protein (μg)	% Inhibition of granules	
	esterolysis[a]	elastolysis[b]
200	50	15
400	70	36
600	70	50

Data from Janoff and Blondin (1971b).

[a] Substrate = t-BOC-1-alanine-p-nitrophenol (0.0002 M); 25 μg of granule protein; volume of reaction mixture = 3.0 ml; pH = 6.5; 25°C; cytosol and enzyme added separately.

[b] Substrate = elastin-orcein (3 mg per ml); 225 μg of granule protein; volume of reaction mixture = 1.5 ml; pH = 8.8; 37°C; cytosol and enzyme premixed.

Table 8
Inhibition of Esterase Activity by Leukocyte Cytosol

	% Inhibition of			
Cytosol protein (μg)	human leukocyte granule protein[a] (25 μg)	pancreatic elastase[a] (10 μg)	trypsin[b] (5 μg)	chymotrypsin[c] (12 μg)
Human cytosol				
50	11	0	0	0
100	25	0	0	0
200	53	0	0	0
300	64			
400	70	0		
500	72			
600	72	0		
1000			0	
	human granules[a] (25 μg)			
Rabbit cytosol				
230	60	—	—	—
460	71	—	—	—

Data from Janoff and Blondin (1971a).

[a] Esterase activity measured with t-BOC-1-alanine-*p*-nitrophenol. Incubation volume, 3.0 ml.

[b] Esterase activity measured with benzoyl-D,L-arginine-*p*-nitroanalide. Incubation volume, 5.0 ml.

[c] Esterase activity measured with glutaryl-L-phenylalanine-*p*-nitroanalide. Incubation volume, 5.0 ml.

are released into a narrow space between the cell and its target (for example, immune complexes deposited on vascular basement membrane or on a cartilage surface in a joint). As a result, essentially unrestricted activity of the proteases may be permitted under certain conditions. The intranuclear inhibitor conceivably complements the action of serum antiproteinases in tissue inflammations characterized by significant necrosis and autolysis of infiltrated leukocytes (for example, pyogenic infections and gouty arthritis). It would probably not interfere significantly with tissue damage mediated by degranulation of intact, viable neutrophils.

The foregoing considerations of the limitations on proteinase inhibitors affecting neutrophil enzymes lead us to view the in vivo physiological and pathological actions of neutrophil proteases as being potentially quite broad. The elastase, especially in view of its wide substrate specificity, may exhibit a remarkable spectrum of effects in health and disease.

CONCLUDING REMARKS

Evidence has been presented which suggests that neutrophil elastase may have a significant number of physiological and pathological actions in man. With respect to physiological functions of neutrophils, the elastase may aid in the lysis of bacterial walls by attacking susceptible linkages in the peptide bridges of peptidoglycans or in other structural wall proteins investing the bacterial envelope. In this capacity, the elastase could operate in complementary fashion to the neutrophil endoglycanohydrolase, lysozyme. Still another physiological role of the elastase may lie in the digestion of bacterial protoplast proteins once cell walls have been removed or broken apart. In this latter function, elastase undoubtedly acts interdependently with other granule enzymes, including other neutral proteases. The in vitro digestion experiment using solubilized *E. coli* protoplasts and chloromethyl ketone inhibitors suggests, however, that the role of elastase in this latter digestive reaction may be a major one.

With respect to pathological actions of neutrophil elastase, the in vitro observations described in the preceding sections suggest several likely possibilities. Whether released from viable cells in the course of their phagocytic functions (Henson 1972; Weissmann, Zurier and Hoffstein 1972) or from damaged and dying leukocytes, the elastase (under those previously discussed conditions in which inhibitors become saturated) may attack elastin fibers or other connective tissue components in a variety of tissues. Degradation of arterial elastin by neutrophils may be involved in the pathogenesis of arteritis; similarly, alterations in lung elastin during the progression of emphysema may be mediated by this enzyme. Attack upon cartilage proteoglycan by elastase could play a part in certain inflammatory joint disorders, especially the arthritides associated with synovial infections, gout and the accumulation of chemotactic immune complexes in joints. Perturbation of cell membranes by neutrophil elastase may contribute to the cocarcinogenic sequela of recurrent inflammation. Finally, production of inflammatory kinin by human leukocytes was recently reported to be due to their elastase component (Movat et al. 1974), although an alternate pathway for kinin generation by these cells has also been reported (Wintroub, Goetzl and Austen 1974). If kinin generation by human neutrophil elastase is confirmed in further studies, it would greatly extend the enzyme's role in generalized tissue injury.

Other neutrophil proteases may also participate in many of these injurious events. A clear example is provided by our observation of the independent solubilization of cartilage proteoglycan by elastase and cathepsin G of human neutrophils, and other examples will undoubtedly be found as experimentation in this area proceeds. One overall goal of future studies, then, should be to delineate the interactions of neutrophil proteases, their tissue targets and their endogenous inhibitors in an effort to further understand the mediation of tissue damage by these cells in man. At the same time, the physiological actions of these enzymes in the antimicrobial functions of leukocytes may also provide a productive area for investigation.

Acknowledgments

This work was supported by USPHS Grants HL 14262 and 1-T01-CA05243 and by a Phebe Cornell Maresi Center Grant of the N.Y. State Arthritis Foundation.

REFERENCES

Burger, M. 1970. Proteolytic enzymes initiating cell division and escape from contact inhibition of growth. *Nature* **227:**170.

Davies, D. T. P. and G. Weissmann. 1969. Hydrolysis of histones at neutral pH by granulocyte lysosomes. *J. Cell Biol.* **43:**29a.

Davies, D. T. P., G. A. Rita, K. Krakauer and G. Weissmann. 1971. Characterization of a neutral protease from lysosomes of rabbit polymorphonuclear leucocytes. *Biochem. J.* **123:**559.

Folds, J. D., I. R. H. Welsh and J. K. Spitznagel. 1972. Localization of proteases in a single granule fraction of human polymorphonuclear leukocytes. *Fed. Proc.* **31:**282.

Galdston, M., A. Janoff and A. L. Davis. 1973. Familial variation of leukocyte lysosomal protease and serum alpha 1 antitrypsin as determinants in chronic obstructive pulmonary disease. *Amer. Rev. Resp. Dis.* **107:**718.

Henson, P. M. 1972. Pathologic mechanisms in neutrophil-mediated injury. *Amer. J. Pathol.* **68:**593.

Ignarro, L., A. Oronsky and R. J. Perper. 1973. Breakdown of noncollagenous chondromucoprotein matrix by leukocyte lysosome granule lysates from guinea pig, rabbit and human. *Clin. Immunol. Immunopathol.* **2:**36.

Inouye, M. and M-L. Yee. 1972. Specific removal of proteins from the envelope of *Escherichia coli* by protease treatments. *J. Bact.* **112:**585.

Janoff, A. 1969. Alanine-*p*-nitrophenyl esterase activity of human leucocyte granules. *Biochem. J.* **114:**157.

————. 1970. Mediators of tissue damage in leukocyte lysosomes. X. Further studies on human granulocyte elastase. *Lab. Invest.* **22:**228.

————. 1972. Inhibition of human granulocyte elastase by serum alpha 1 antitrypsin. *Amer. Rev. Resp. Dis.* **105:**121.

————. 1973. Purification of human granulocyte elastase by affinity chromatography. *Lab. Invest.* **29:**458.

Janoff, A. and R. S. Basch. 1971. Further studies on elastase-like esterase in human leukocyte granules. *Proc. Soc. Exp. Biol. Med.* **136:**1045.

Janoff, A. and J. Blondin. 1970. Depletion of cartilage matrix by a neutral protease fraction of human leukocyte lysosomes. *Proc. Soc. Exp. Biol. Med.* **135:**302.

————. 1971a. Inhibition of the elastase-like esterase in human leukocyte granules by human leukocyte cell sap. *Proc. Soc. Exp. Biol. Med.* **136:**1050.

————. 1971b. Further studies on an esterase inhibitor in human leukocyte cytosol. *Lab. Invest.* **25:**565.

————. 1973. The effect of human granulocyte elastase on bacterial suspensions. *Lab. Invest.* **29:**454.

————. 1974. The role of elastase in the digestion of *E. coli* proteins by human polymorphonuclear leukocytes. I. Experiments *in vitro. Proc. Soc. Exp. Biol. Med.* **145:**1427.

Janoff, A. and J. Scherer. 1968. Mediators of inflammation in leukocyte lysosomes. IX. Elastinolytic activity in granules of human polymorphonuclear leukocytes. *J. Exp. Med.* **128:**1137.

Janoff, A., R. A. Sandhaus, V. D. Hospelhorn and R. Rosenberg. 1972. Digestion of lung proteins by human leukocyte granules *in vitro. Proc. Soc. Exp. Biol. Med.* **140:**516.

Lieberman, J. and M. Gawad. 1971. Inhibitors and activators of leukocytic proteases in purulent sputum. *J. Lab. Clin. Med.* **77:**713.

Malemud, C. J. and A. Janoff. 1975. Identification of neutral proteases in human neutrophil granules which degrade articular cartilage proteoglycan. *Arthritis Rheum.* (in press).

Mittman, C. 1972. Summary of symposium on pulmonary emphysema and proteolysis. *Amer. Rev. Resp. Dis.* **105**:430.

Mosser, A. G., A. Janoff and J. Blondin. 1973. Increased concanavalin A-dependent agglutinability of mouse fibroblasts after treatment with leukocyte lysosomal preparations. *Cancer Res.* **33**:1092.

Movat, H. Z., N. S. Ranadive, D. R. L. Macmorine and S. G. Steinberg. 1974. Liberation of kinin from kininogen by human PMN-leukocyte neutral protease. *Fed. Proc.* **33**:641.

Ohlsson, K. 1971. Interaction between human or dog leucocyte proteases and plasma protease inhibitors. *Scand. J. Clin. Lab. Invest.* **28**:225.

Ohlsson, K. and I. Olsson. 1973. The neutral proteases of human granulocytes. Isolation and partial characterization of two granulocyte collagenases. *Eur. J. Biochem.* **36**:473.

———. 1974. The neutral proteases of human granulocytes. Isolation and partial characterization of granulocyte elastases. *Eur. J. Biochem.* **42**:519.

Oronsky, A., L. Ignarro and R. J. Perper. 1973. Release of cartilage mucopolysaccharide-degrading neutral protease from human leukocytes. *J. Exp. Med.* **138**:461.

Rindler, R., F. Schmalzl and H. Braunsteiner. 1974. Isolierung und Charakterizierung der chymotrypsinähnlichen Protease aus neutrophilen Granulozyten des Menschen. *Schweiz. Med. Wochenschr.* **104**:132.

Sefton, B. M. and H. Rubin. 1970. Release from density-dependent growth inhibition by proteolytic enzymes. *Nature* **227**:843.

Sweetman, F. and L. Ornstein. 1974. Electrophoresis of elastase-like esterases from human neutrophils. *J. Histochem. Cytochem.* **22**:327.

Tuhy, P. M. and J. C. Powers. 1975. Inhibition of human leukocyte elastase by peptide chloromethyl ketones. *Biochem. Biophys. Res. Comm.* (in press).

Weissmann, G., R. B. Zurier and S. Hoffstein. 1972. Leukocyte proteases and the immunologic release of lysosomal enzymes. *Amer. J. Pathol.* **68**:539.

Wintroub, B. U., E. J. Goetzl and K. F. Austen. 1974. A neutrophil-dependent pathway for the generation of a neutral peptide mediator. Partial characterization of components and control by alpha 1 antitrypsin. *J. Exp. Med.* **140**:812.

Regulation of Animal Virus Replication by Protein Cleavage

Bruce D. Korant

Central Research Department, Experimental Station
E. I. du Pont de Nemours and Co., Inc., Wilmington, Delaware 19898

Animal viruses, with few exceptions, form some of their proteins by post-translational cleavage of precursors (reviewed by Summers, Roumiantzeff and Maizel 1971; Sugiyama, Korant and Lonberg-Holm 1972). The types of cleavages range from a minor modification of a structural polypeptide during maturation (virion formation) of myxoviruses, to multistep, sequential cleavages which produce all the viral polypeptides. Examination of Table 1 will provide a brief introduction to the kinds of proteolysis found in the various animal virus groups and the experimental methods used to obtain evidence for cleavage.

First reports describing production of viral proteins in proteolytic reactions were from laboratories studying the small RNA (picorna) viruses (Summers and Maizel 1968; Holland and Kiehn 1968; Jacobson and Baltimore 1968a). These reports were soon confirmed and then extended to other virus groups as the methodology became known and protein cleavages were looked for. It is not surprising that picornaviruses became the prototype for viral proteolysis. These viruses have a relatively simple, defined composition and inhibit rapidly and completely cellular protein synthesis. This permits easy identification of viral specific polypeptides. Also with the picornaviruses, all the viral poly-peptides are derived from a single giant precursor, or polyprotein (Jacobson, Asso and Baltimore 1970), and are readily detected by brief addition of radioactive amino acids to infected cells (pulse) followed by some extended period in excess unlabeled amino acids (chase), during which the proteolytic processing may be examined by gel electrophoresis in sodium dodecyl sulfate (SDS) (Summers, Maizel and Darnell 1965).

By comparison, most of the other mammalian viruses present additional complications if proteolytic reactions in replication are studied. Many of the viruses are ill defined with regard to virion composition, and the most highly purified virions available may still contain host determinants. Even more serious is the inability of some of these viruses to inhibit host protein synthesis, making it necessary to resort to complicated immunologic or other

621

Table 1
Proteolysis in Animal Virus Replication

Virus	Protein cleaved	Function	Analytical method[a]
RNA genome			
Picornaviruses	structural and nonstructural	production of capsid polypeptides; maturation; regulation of RNA synthesis?	pulse-chase (1,2,3) amino acid analogs (4,5,6,7) high temperature (7,8,9) guanidine (10) hypotonic medium (11) serine protease inhibitors (7,12,13,14,15,54) DFP (5) iodoacetamide (16) zinc ion (17,18) in vitro translation (19,20,21) peptide mapping (5,22)
Togaviruses (Sindbis, etc.)	structural; nonstructural?	production of capsid polypeptides	pulse-chase (23,24,25,26) amino acid analogs (24,27) high temperature (28) protease inhibitors (27,29) slowed cleavages in certain hosts (26,27,30) zinc ions (31) in vitro translation (32,33) peptide mapping (25)
Myxoviruses (influenza, etc.)	structural (hemagglutinin)	?	pulse-chase (34,35,36) amino acid analogs (34,37) DFP (37) D-glucosamine (38) peptide mapping (34) differences between host cells (34,35,39)
Paramyxoviruses (Sendai, etc.)	structural (glycoprotein)	infectivity of virions (uncoating)	pulse-chase (40) differences between host cells (41,42) trypsin action in vitro (42,43)
Reovirus	structural	"early"-uncoating; activation of virion RNA polymerase	proteolytic removal of capsid polypeptides from infecting virions (44,45)

Virus	Protein	Function	Evidence (method) [a]
	structural	"late"-maturation	chymotrypsin activation of capsid-associated polymerase in vitro (46,47,48); pulse-chase (49)
RNA tumor	structural	maturation	pulse-chase (50,51); pulse-chase of immunologically related precursors (51); peptide maps (51)
DNA genome Vaccinia (pox)	structural	formation of core; maturation	pulse-chase (52); amino acid analogs (53); inhibition by rifampicin (52)
Adenovirus	structural	maturation	pulse-chase (54,55); peptide-mapping (54)
Herpes simplex	?	?	pulse-chase (56)
Polyoma	structural	?	peptide maps (no direct evidence of precursor-product relationship) (57)

[a] References:
1. Summers and Maizel (1968)
2. Holland and Kiehn (1968)
3. Butterworth (1973)
4. Jacobson and Baltimore (1968a)
5. Jacobson et al. (1970)
6. Kiehn and Holland (1970)
7. Dobos and Martin (1972)
8. Cooper et al. (1970)
9. Garfinkle and Tershak (1971)
10. Jacobson and Baltimore (1968b)
11. Agol et al. (1970)
12. Korant (1972)
13. Summers et al. (1972)
14. Adler and Tershak (1974)
15. Lucas-Lenard (1974)
16. Korant (1973)
17. Korant et al. (1974)
18. Butterworth and Korant (1974)
19. Roumiantzeff et al. (1971)
20. Eggen and Shatkin (1972)
21. Esteban and Kerr (1974)
22. Butterworth et al. (1971)
23. Strauss et al. (1969)
24. Burrell et al. (1970)
25. Schlesinger and Schlesinger (1973)
26. Snyder and Sreevalsan (1974)
27. Morser and Burke (1974)
28. Scheele and Pfefferkorn (1970)
29. Pfefferkorn and Boyle (1972)
30. Snyder and Sreevalsan (1973)
31. Snyder and Sreevalsan (pers. comm.)
32. Cancedda and Schlesinger (1974)
33. Simmons and Strauss (1974)
34. Etchison et al. (1971)
35. Lazarowitz et al. (1971)
36. Skehel (1972)
37. Klenk and Rott (1973)
38. Klenk et al. (1972)
39. Lazarowitz et al. (1973)
40. Samson and Fox (1973)
41. Homma and Ohuchi (1973)
42. Scheid and Choppin (1974)
43. Homma (1971)
44. Chang and Zweerink (1971)
45. Silverstein et al. (1972)
46. Shatkin and Sipe (1968)
47. Smith et al. (1969)
48. Levin et al. (1970)
49. Zweerink and Joklik (1970)
50. Shanmugan et al. (1972)
51. Vogt and Eisenman (1973)
52. Katz and Moss (1970a)
53. Katz and Moss (1970b)
54. Anderson et al. (1973)
55. Ishibashi and Maizel (1974)
56. Honess and Roizman (1973)
57. Friedman (1974)

procedures to determine which intracellular proteins are viral. Finally, the proteins of some of these viruses contain variable amounts of carbohydrate residues (which picornavirus proteins lack altogether), and there are difficulties in determining the molecular size, and therefore the precursor-product relationships, of glycoproteins.

Proteolysis Accompanies Maturation

From examination of Table 1, it is obvious that the most common viral process in which cleavage is observed is maturation. This suggests that proteolysis is a mandatory step in the combination of viral structural proteins with nucleic acid. The cleavage reaction may produce a specific conformation in the affected protein, or a free amino terminus, so that recognition of the homologous viral nucleic acid becomes assured. Another possibility is that the cleavage stabilizes the virion and prohibits other nucleic acids or nucleases from entering. In fact, there is no clear explanation of why proteolysis so often occurs in maturation. Further discussion of this point is included in the section below on regulation. Experiments with poliovirus, which uses its entire genome RNA as the messenger, led to the suggestion that cleavage of proteins was occurring because eukaryotic cells could not translate polycistronic messages (Jacobson and Baltimore 1968a). However, the results with viruses which have monocistronic messages—reovirus (Zweerink and Joklik 1970), myxoviruses (Etchison et al. 1971; Lazarowitz, Compans and Choppin 1971) and the DNA viruses—imply that cleavages in maturation have a role other than just simplifying the translation process.

Inhibition of Cleavage by Altering the Substrate

Attempts to detect cleavages and to characterize the nature of the substrates and the proteases have often relied upon various methods of inhibiting the proteolytic reactions. In general, the approach to blocking cleavage has been either to alter the polypeptide substrate or to inhibit the enzymes involved.

Among the first techniques applied, and very useful still, is the incorporation of amino acid analogs (e.g., p-fluorophenylalanine) into viral polypeptides (see Table 1). The addition of mixtures of several of these analogs has often led to accumulation of modified polypeptides in infected cells and permitted the interpretation that, in the absence of analogs, the large polypeptides are precursors of other viral proteins. A difficulty inherent in using analogs to investigate cleavage is that the analog-containing proteins are often not cleaved at all, or only slowly, and to abnormal end products. However, other reversible inhibitors which alter substrates have become available. These include guanidine, which blocks maturation cleavages with some picornaviruses (Jacobson and Baltimore 1968b), rifampicin with vaccinia virus (Katz and Moss 1970a), and zinc ions with picornaviruses (Korant, Kauer and Butterworth 1974; Butterworth and Korant 1974) and a togavirus (Synder and Sreevalsan, pers. comm.). Reversible inhibition by these chemicals provides more direct evidence of the proteolytic nature of the processing. Zinc ions, in particular, probably reacting nonspecifically with both cystinyl and histidyl residues, can lead to accumulation of very large viral precursors, which are cleaved to the appropriate end products once the inhibitor is washed from the treated cells (Korant and Butterworth, in prep.; Snyder and Sreevalsan, pers. comm.).

Production of altered, stable precursors can be accomplished by high (nonpermissive) temperatures during infection (Cooper, Summers and Maizel 1970; Scheele and Pfefferkorn 1970; Garfinkle and Tershak 1971), and in one example, the inhibition was reversed by transfer of the infected cultures to a slightly lower temperature (Garfinkle and Tershak 1971). There was a similar finding when poliovirus-infected cultures were briefly exposed to low-tonicity medium (Agol et al. 1970).

Chemical Protease Inhibitors

Another approach has been to select some well-known, irreversible inhibitors of proteolytic enzymes, e.g., chloromethyl ketones (Schoellmann and Shaw 1963; Shaw, Mares-Guia and Cohen 1965; Korant 1972; Summers et al. 1972), diisopropylfluorophosphate (DFP) (Jacobson, Asso and Baltimore 1970) and iodoacetamide (Korant 1973; Snyder and Sreevalsan, pers. comm.), add them to infected cells, and observe whether larger viral proteins accumulate. In a number of cases using nonrelated viruses, these inhibitors have had positive results (Table 1). A secondary benefit of these inhibitors is that the polypeptide substrate is stabilized, but not altered, and may be used in vitro to assay proteolysis (see below). Results with DFP and the chloromethyl ketones have indicated that serine proteases may be involved. Moreover, where experiments have used the same or similar virus strains in different cell lines, results with protease inhibitors, particularly phenylalanyl chloromethyl ketone (TPCK) (Korant 1972; cf. Dobos and Martin 1972 and Lucas-Lenard 1974), have implied a host origin for part of the protease activity in the infected cells (see below).

There are some difficulties in the interpretation of effects of chemical inhibitors on cleavages. A large problem is their lack of specificity. For example, DFP can inhibit many enzymes with histidine or serine in the active site, and iodoacetamide reacts with any exposed sulfhydryl groups. Even the more elegant chloromethyl ketones may react with some nonserine proteases and are reported to inhibit protein synthesis in some cell lines (Summers et al. 1972; Chou, Black and Roblin 1974). Furthermore, some inhibitors which do react with serine proteases in vitro (e.g., benzamidines, proflavin, phenyl boronic acid) do not block cleavage in TPCK-sensitive cells (Korant, unpubl.). It is not likely that the chemical inhibitors are blocking proteolysis by directly interfering with protein or nucleic acid synthesis because cycloheximide (Summers, Roumiantzeff and Maizel 1971) and bisbenzylimidazole (Korant, unpubl.) do not affect cleavage, although they are able to drastically block protein and RNA synthesis. Inhibition of initiation of protein synthesis by the antibiotic pactamycin or by hyperionic medium has been utilized in virus systems where proteolysis is well established to decide in what order viral polypeptides are translated. This permits an extrapolation to the sequence of genes in the viral message (Taber, Rekosh and Baltimore 1971; Butterworth and Rueckert 1972; Schlesinger and Schlesinger 1973; Saborio, Pong and Koch 1974).

626   B. D. Korant

Origin of Proteases in Viral Infections

In studying the intimate association between a virus parasite and its host, particularly with the goal of designing inhibitors of processes specific to the virus, it is important to know whether the enzymes involved are virus specified or produced by the host cell. In the animal virus systems, most evidence points to host-coded proteases playing an important role.

This situation is most dramatic with Sendai, a paramyxovirus. Sendai virus produced in chick embryos is infectious, hemolytic and able to cause fusion of cells (Homma 1971; Homma and Ohuchi 1973). However, Sendai grown in mouse L cells, human HeLa cells or bovine kidney cells lacks these activities (Homma and Ohuchi 1973; Scheid and Choppin 1974). The basis for the difference is a proteolytic (trypsinlike) activity, present in chick embryos, but lacking, or not available, to Sendai virus in the other cell lines tested (Homma 1971; Scheid and Choppin 1974). The cellular protease must cleave a glycoprotein in the virions in order to activate them.

A pattern reminiscent of this is seen with certain influenza virus strains. Depending on the cell line chosen, a virion glycoprotein whose function is hemagglutination (HA) may be cleaved or not (Lazarowitz, Compans and Choppin 1971). The situation is further complicated because the cellular activity appears to activate serum plasminogen. The active plasmin then cleaves the flu HA during maturation (Lazarowitz, Goldberg and Choppin 1973). The cell protease was suggested by Lazarowitz, Goldberg and Choppin (1973) to be released from lysosomes following infection. However, unlike Sendai, flu with uncleaved HA is still active. Thus the observed cleavages may be without relevance to the virus, but useful to the biochemist.

An interesting feature of the cleavages that produce the capsid polypeptides of Sindbis virus is that they do not go on at all in aged (7-day-old) chick embryo fibroblast cells, but occur readily in 2–3-day-old chick cells (Snyder and Sreevalsan 1973). Treatment of the cultures with actinomycin D allows aged cells to again cleave the viral proteins. The authors propose that actinomycin releases cellular hydrolases from lysosomes. Also, with the α-group togaviruses, the rates of cleavage of viral proteins differ depending on the cell line chosen (Morser and Burke 1974; Snyder and Sreevalsan 1974). The protease inhibitor TPCK blocks cleavage of Sindbis virus polypeptides in chick cells (Pfefferkorn and Boyle 1972), but not in HeLa cells (T. Sreevalsan, pers. comm.).

The picornaviruses are as autonomous of host functions as any viruses known; nevertheless, poliovirus and cardioviruses apparently utilize host cell proteases for some of the primary cleavages of their polyprotein precursors (Korant 1972; Eggen and Shatkin 1972). Inhibitor studies indicate that these cellular proteases are of the serine type (see Table 1), and that they may be mimicked in vitro by trypsin, or better, chymotrypsin, as assayed by size and antigenicity of the products (Korant 1972). A consequence predicted by the use of cellular enzymes to cleave poliovirus proteins is that the virus may differ depending on the cell line used (see above for myxo- and paramyxoviruses). A study of the serology of poliovirus grown either in HeLa (human) or monkey kidney cells reported differences in the neutralizability of the two pools of virus (Lewenton-Kriss and Mandel 1972). It should be asked why host cells contain serine proteases which are available to cleave viral proteins and whether some of these are similar to those proteases implicated in control of division of uninfected cells (Schnebli and Burger 1972).

Picornaviruses Synthesize or Activate Proteases

Following completion of primary picornavirus cleavages, other proteolytic reactions produce the capsid polypeptides, as well as stable, nonstructural proteins. The origin of the enzymes active in these steps is unclear. Iodoacetamide can block some of these cleavages (Korant 1973), indicating a sulfhydryl protease is utilized. It was found that extracts of several types of uninfected cells are not able to carry out these latter cleavages (Korant 1972; Eggen and Shatkin 1972), and that infected cells have additional protease activities (Flanagan 1966) which are able to process the picornavirus capsid precursors (Kiehn and Holland 1970; Korant 1972, 1973; Esteban and Kerr 1974). These results indicate an accumulation of viral specific enzymes during infection, or an activation of cellular proteases as infection proceeds. A similar situation exists in bacterial cells infected with certain bacterial viruses (Goldstein and Champe 1974; Bachrach and Benchetrit 1974). A report of proteases intimately bound to virions of the picornavirus, myxovirus and rhabdovirus groups (Holland et al. 1972) may facilitate further characterization of the proteases involved in maturation.

The pattern of processing of poliovirus polypeptides in the cytoplasm of infected cells is summarized in Figure 1. Using the nomenclature of Summers and Maizel (1968) and Jacobson, Asso and Baltimore (1970), NCVP refers to noncapsid viral proteins and VP refers to polypeptides which may be found also in empty capsids and virions. The molecular weights given are estimated from coelectrophoresis of known proteins in SDS gels. $NCVP_{1a}$ is the precursor of VP_0 (39,000), VP_1 (35,000) and VP_3 (26,000). $NCVP_2$ is the precursor of $NCVP_4$ (55,000) and $NCVP_7$ (20,000). $NCVP_x$ is stable. Figure 1 is over simplified, in that several other noncapsid and capsid polypeptides which have been identified are not included. The cleavage of polio polyprotein to its primary products probably occurs while the nascent polypeptide is on the polyribosomes (Jacobson, Asso and Baltimore 1970). The remainder of the proteolytic reactions which produce stable end products occur on cytoplasmic membranes (Korant 1972, 1973) and in subviral particles (Jacobson and Baltimore 1968b), that is, well after the viral polypeptides have been completed and released from the polyribosomes.

The specificity of the cleavage reactions is little known. Certainly a native substrate is required for appropriate cleavage reactions to ensue (Korant 1972), and slow or abnormal cleavages may be explained by denaturation of

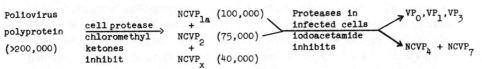

Figure 1
Steps in the cleavage of poliovirus type 2 proteins.

the large polypeptide substrates (Scheele and Pfefferkorn 1970; Cooper, Summers and Maizel 1970; Garfinkle and Tershak 1971; Ginevskaya et al. 1972; Korant 1972; Dobos and Martin 1972). It has also been proposed that some precursors may not be properly cleaved unless they are first glycosylated (Klenk, Scholtissek and Rott 1972). Exact sequences of cleavage sites are not known, but end-group determinations have been performed on two picorna-virus virions. Those results are leucine, isoleucine and threonine for N-terminal polypeptide of foot-and-mouth virus (Laporte 1969), glutamine, leucine and glutamic acid for the C termini of a strain of the same virus (Bachrach, Swaney and Van de Woude 1973), and aspartic acid, glycine and serine for poliovirion N termini (Burrell and Cooper 1973).

Regulation of Viral Functions by Proteolysis

Maturation

As discussed above, maturation of many kinds of viruses includes a processing step in which a precursor is cleaved to yield one or more structural polypep-tides. It is apparent with many of these viruses that no infectious virions are produced which are markedly deficient in these cleavages; for example, in picornaviruses, fewer than 5% of virion polypeptides are uncleaved precur-sors. This implies that virion formation is dependent on the successful com-pletion of this proteolysis, and that maturation may partly be regulated by the activity of the protease(s) involved or is rate-limited by the speed of the processing reactions.

Guanidine, zinc ions and rifampicin are examples of rapid-acting inhibitors of maturation (Jacobson and Baltimore 1968b; Korant, Kauer and Butter-worth 1974; Katz and Moss 1970a). They apparently react with specific regions of the capsid polypeptides (Cooper 1969; Korant and Butterworth, in prep.) and alter them, thereby preventing proteolytic cleavage and com-bination of viral nucleic acid and capsid protein. These compounds do not block maturation because they affect the overall supply or synthesis of viral nucleic acid. In fact, a much superior RNA synthesis inhibitor, bisbenzyl-imidazole (Abbott Labs), does not affect cleavage of picornaviral proteins or formation of virions until long after it has blocked synthesis of viral RNA (F. Yin and B. Korant, unpubl.).

Difficulties are encountered in attempts to follow the process of cleavage of viral polypeptides that occurs concurrently with virion formation. First, in vivo, large pools of capsid precursors complicate the detailed study of those few which enter the actual maturation process. Nevertheless, there is some evidence that nucleic acid is combined not with soluble precursors, but rather into particulate, capsidlike structures (Jacobson and Baltimore 1968b; Katz and Moss 1970b; Ishibashi and Maizel 1974). This permits the study of maturation in infected cells, and some recent progress has been made (Fer-nandez-Tomas and Baltimore 1973). Unfortunately, in vitro, there have been no biochemical demonstrations of a measurable maturation process, which begins with nucleic acid and an empty capsid composed of precursor polypep-tides and ends with a virion. To date, therefore, this important reaction re-mains vague and requires additional study. Pertinent to the subject of this article is the tempting possibility that judicious application of proteolytic enzymes in vitro may facilitate a more complete maturation reaction.

Nucleic Acid Synthesis

A different feature of virus replication in which regulation may be demonstrated is the synthesis of viral nucleic acid (reviewed by Baltimore 1971). However, in very few examples are the specific host or viral polypeptides which make up the replicases known, and there are no clear examples of proteolysis regulating any of the synthetic reactions—initiation, elongation or release of newly completed nucleic acid molecules.

Regulation of the rate of RNA synthesis has been observed with the picornaviruses (reviewed by Baltimore 1969). Most obvious is the onset of fairly rapid decline in the rate of RNA synthesis, particularly in initiation of new single strands, after about 3 hours of infection. This decline has been variously attributed to the breakdown of replicase molecules, increased ribonuclease activity or general metabolic disruption of the host cells.

Since the proteolytic reactions undergone by picornavirus polypeptides are so easily observed, it would seem that any correlation of proteolytic processes and RNA synthesis control could most readily be studied with a virus of this group. The following results are offered to provide evidence that, for poliovirus, the cleavage of a precursor polypeptide serves to regulate synthesis of viral RNA.

EXPERIMENTAL PROCEDURES

Virus Strains and Cell Cultures

Poliovirus type 2, strain P712ch2ab, was used in most experiments. It was designated strain P2. Derived from this strain by a selection protocol described in Results is a second strain, serologically identical, designated strain P2r.

The cell lines used were either HeLa, strain r (Korant et al. 1972), or LLCMK2 rhesus monkey kidney cells (Korant 1972). Cell monolayers were grown in 100-mm plastic petri dishes using McCoy's 5A medium (Grand Island) containing 10% calf serum. Cells were generally used for infection and isotopic labeling when confluent (approx. 1 x 10⁷ cells/dish).

Radioisotope Precursors and Other Chemicals

All isotopes (^{3}H- and ^{14}C-amino acid mixtures and [^{3}H]uridine) were purchased from New England Nuclear Corp.

Phenyl methyl sulfonyl fluoride (PMSF) was purchased from California Biochemical Co. Tosyl-lysine chloromethyl ketone (TLCK) was purchased from Sigma Chemical Co., and iodoacetamide was purchased from Mann Biochemical Co.

Polyacrylamide Gel Electrophoresis in SDS

Electrophoresis in cylindrical gels using phosphate buffer was performed as previously described (Korant et al. 1972). Electrophoresis in discontinuous buffers containing SDS was performed in slab gels, as described by Laemmli

(1970), in an apparatus purchased from the Hoefer Scientific Co., San Francisco.

RESULTS

Instability of the Viral Replicase

For some years, it has been accepted that the replicase of many picornaviruses, that is, the enzyme which takes part in production of new viral RNA in infected cells, is unstable (reviewed by Baltimore 1969). Evidence for this instability is indirect, based on the decline of RNA synthetic activity after an inhibitor of protein synthesis is added to infected cells. The basis of the proposed instability is unknown, partly due to the failure to date in obtaining in vitro RNA synthesis which resembles the in vivo reactions and the lack of secure knowledge of the proteins that function in the enzyme.

An interesting feature of the replicase's instability is that it appears to increase as infection proceeds. In Figure 2 is shown the rate of poliovirus RNA

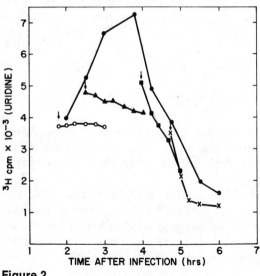

Figure 2
Rate of synthesis of poliovirus RNA in infected HeLa cells. Monolayer cultures were infected with 50 plaque-forming units/cell of poliovirus (strain P2). After 1 hour at 35°C, medium containing 5 μg/ml actinomycin D was added. At the times indicated, 1 μCi/ml [³H]uridine was added for 15 minutes and incorporation ended by addition of cold 10% trichloroacetic acid (TCA) to the cultures. The precipitated cells were dissolved in 0.1 N NaOH and counted by liquid scintillation (●——●). To other cultures, cycloheximide was added (100 μg/ml, final conc.) at the times indicated by the arrows, and labeling for 15 minutes and counting were done as above.

synthesis at various times after infection of HeLa cells. Actinomycin D was included to fully block incorporation of the uridine into host RNA. At the indicated times, 100 μg/ml of cycloheximide was added to some of the cultures in order to abolish protein synthesis. At all times, a decline in the incorporation of [^{3}H]uridine was noted; however, the decline became more rapid as the time after infection increased. The question is: Why does the "instability" of the replicase appear to be increasing as virus multiplication proceeds?

Effects of Cleavage Inhibitors on Polio RNA Synthesis

A possible explanation for the observations of Figure 2 is that instability is due to cleavage of a subunit of the replicase by a proteolytic enzyme(s). The point was made above that several nonstructural polio polypeptides are unstable. If one or more of these had a role in the replicase (Cooper 1969; Caliguiri and Mosser 1971), then activity of the enzyme could be altered by a peptide bond scission. It is established that there are increases of general proteolytic activities as picornavirus infections proceed (Flanagan 1966), and increases as well in the rates of cleavage of viral polypeptide precursors (Kiehn and Holland 1970; Korant 1972, 1973; Esteban and Kerr 1974). Thus, increasing proteolytic activity could be used to explain the ever-decreasing half-life of the replicase enzyme.

In HeLa cells, proteolysis of several polio noncapsid polypeptides is blocked by iodoacetamide (Korant 1973). If one of the polypeptides stabilized by iodoacetamide were a part of the replicase, enzymatic activity might also be stabilized. Stabilization of polio RNA synthesis by iodoacetamide is seen in Figure 3. The inhibitory effect of cycloheximide on viral RNA synthesis may be overcome by a prior (30 min) addition of iodoacetamide. Guanidine, which is an effective inhibitor of polio RNA, but not of protein synthesis, was able to markedly reduce viral RNA synthesis, even though iodoacetamide was present. Therefore, cycloheximide and guanidine do not share a common mechanism in the inhibition of RNA synthesis.

Iodoacetamide treatment accumulates several intermediate-sized polio precursors (Korant 1973), including NCVP$_{1a}$ and NCVP$_2$. TLCK in HeLa cells causes accumulation of even larger viral precursors (Korant 1972; Adler and Tershak 1974). But apparently, these larger proteins are not part of the replicase because TLCK did not subvert cycloheximide action.

A difficulty in interpreting the results of Figure 3 is that iodoacetamide blocks several proteolytic reactions in polio-infected cells. A more specific effect is caused by PMSF, an inhibitor of serine proteases (Fahrney and Gold 1963). In polio-infected HeLa cells, PMSF has little effect; however, in monkey kidney cells, PMSF rather specifically blocks production of NCVP$_4$ and NCVP$_7$, and NCVP$_2$ is accumulated (Fig. 4a). This information may be applied to the interpretation of results in Figure 4b. In infected monkey kidney cells, but not in HeLa cells, PMSF appears to stabilize the viral replicase, as measured by a prolonged high rate of viral RNA synthesis in comparison to infected controls. The results of Figure 4 indicate that stabilization of NCVP$_2$ and stabilization of replicase may be related. The activity of PMSF in monkey kidney cells, but not in HeLa cells, with regard to accumulation of NCVP$_2$ suggests a host origin for the protease involved.

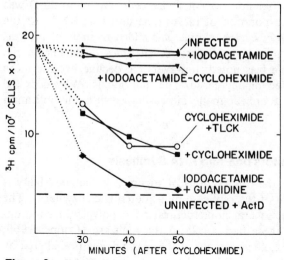

Figure 3
HeLa cell monolayer cultures were infected and incubated as in Figure 2. At 2.5 hours after infection,
some of the cultures received TLCK (10^{-4} M) or
iodoacetamide (5×10^{-4} M). Thirty minutes later,
some of the treated cultures, and additional ones as
well, received cycloheximide (100 μg/ml) or guanidine (2 mM). Thirty minutes later, the cultures were
labeled with [³H]uridine for 5 minutes, at the indicated times, and treated as above for counting.
Uninfected cells, treated with actinomycin D, incorporated about 400 cpm of [³H]uridine (– – –).

Selection of a Poliovirus Variant with a Stable Replicase

If, as is believed, the picornaviruses provide a genetically determined protein
component to the replicase in infected cells, then it might be possible to
manipulate some condition during replication so that a virus mutant with an
altered replicase would be selected. The results given in Figure 2 suggest that
a poliovirus with a stable replicase could be selected using cycloheximide.
Assumptions included in the selection protocol shown in Figure 5 are that
virus maturation will occur in the presence of cycloheximide as long as RNA
synthesis continues and until the pool of coat protein is used up. The neutral
red treatment of one-hour duration was to exclude those parental-type viruses
which, for physiological (nongenetic) reasons, were still maturing one hour
after cycloheximide had been added.

After five such passages, virus titer dropped below 10^5 infectious units per
ml, but subsequently began to rise, and by 12 passages, it was again back to
the level at the beginning of the experiment ($>10^9$ infectious units per ml).
Plaque size of the selected virus and neutralization by antipolio type 2 serum
were very similiar to parental virus.

However, a marked difference between the variant (designated strain P2r)
and its parent (P2) was observed when their respective rates of RNA

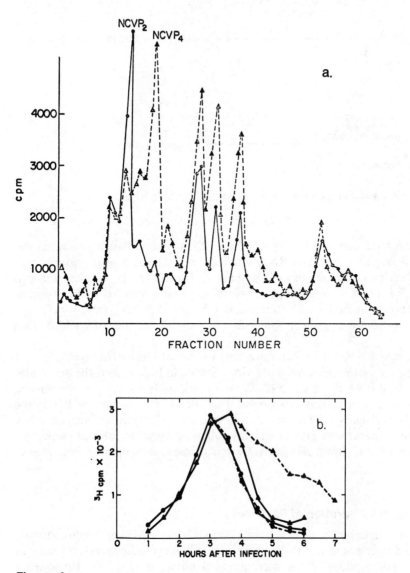

Figure 4

Effect of PMSF on poliovirus (strain P2) protein and RNA synthesis.

(*a*) SDS gel electrophoresis of polio cytoplasmic polypeptides in infected monkey kidney cells. Cells were labeled with radioactive amino
acids (5 µCi/ml) from the fourth to fifth hour postinfection. (▲----▲)
Infected control; (●——●) PMSF (5 × 10⁻⁴ M) added with the label.
Electrophoresis was from left to right.

(*b*) Poliovirus RNA synthesis in HeLa or monkey kidney cells treated
with PMSF. Cells were infected and incubated as in Figure 2. To some
of the cultures, PMSF (5 × 10⁻⁴ M) was added at 3 or 3.5 hours postinfection. Other cultures received no inhibitor. Labeling with [³H]uridine
was for 15 minutes at the times indicated. (●——●) HeLa, (●----●) HeLa
plus PMSF, (▲——▲) monkey kidney, (▲----▲) monkey kidney plus
PMSF.

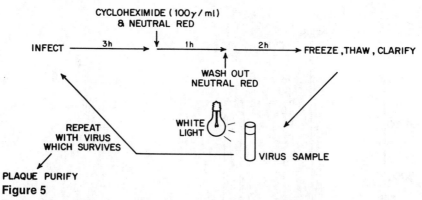

Figure 5
Protocol for selection of a poliovirus variant with a stable RNA replicase.

synthesis were compared (Fig. 6a). The P2r strain, although somewhat delayed in attaining maximal RNA synthesis, maintained a near-maximal rate for at least 60 minutes beyond its parent. The difference between the two strains was not due to an increased synthesis of viral proteins by P2r (circled triangles) compared to P2 (circled dots), nor did P2r virus display any ability to continue protein synthesis after cycloheximide was added (not shown).

The genetic change in P2r leads also to altered viral RNA synthesis at a temperature nonpermissive for P2 virus. Shown in Figure 6b is the accumulation of viral RNA at 35 and 39.5°C. As indicated, 39.5°C is a nonpermissive temperature for P2 virus (Adler and Tershak 1974), but not for P2r in the time interval chosen. The results indicate that P2r has an altered RNA synthesis pattern, with greater stability following cycloheximide at permissive temperature and lower sensitivity to inactivation at nonpermissive temperature.

Basis for the Alteration of P2r Virus

In order to uncover the genetic alteration in P2r virus, SDS gel electrophoresis was used to compare it with its parent. No difference in the sizes of the cytoplasmic polypeptides of the two viruses is observed (Fig. 7). However, a further analysis of the kinds and ratios of viral polypeptides at different times after infection showed that the pair differs in one significant respect—the stability to proteolysis of $NCVP_2$. At early times, up to three hours postinfection, no great difference is observed (Fig. 7a). However by about four hours, there is much less $NCVP_2$ (b, fractions 18–20), relative to its cleavage products $NCVP_4$ (fractions 24–26) and $NCVP_7$ (fractions 59–61), with P2 virus than with P2r. By five hours postinfection, the disparity is even greater (c). There was no other reliable difference between the proteins of the two viruses. The total radioactivity incorporated in a one-hour pulse of amino acids was similar for the two viruses, as were recoveries from the gels of the viral proteins.

A greater degree of resolution of polypeptides is afforded by electrophoresis in slab gels containing SDS (Laemmli 1970). However, the separations

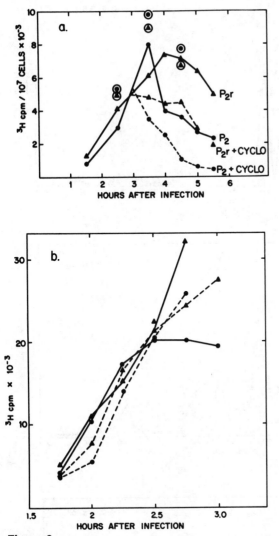

Figure 6

(*a*) RNA synthesis in HeLa cells infected by poliovirus strain P2 (●——●) or strain P2r (▲——▲).
To some cultures, cycloheximide was added at 3
hours postinfection. The cultures were labeled with
[³H]uridine as in Figure 2. Parallel cultures were
labeled for 1 hour with 0.1 μCi/ml of ³H-amino
acids, then TCA-precipitated and counted. ⊙ P2
protein synthesis; ▲ P2r protein synthesis.

(*b*) Viral RNA synthesis in HeLa cells at 35°C
or 39.5°C. Cultures were infected with P2 or P2r
poliovirus at 35°C, and after one hour, one-half the
samples were placed at 39.5°C. Thirty minutes later,
all cultures received 1 μCi/ml [³H]uridine. Every
30 minutes thereafter, some of the cultures were
treated with TCA. Accumulated counts P2,
35°C (▲——▲); P2, 39.5°C (●——●); P2r, 35°C
(●－－－●); P2r, 39.5°C (▲－－－▲).

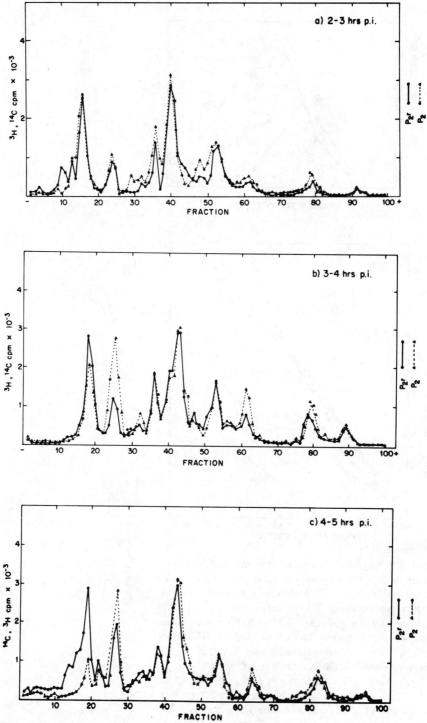

Figure 7

SDS gel electrophoresis of ³H-labeled P2 polypeptides (▲‑ ‑ ‑ ‑▲) and ¹⁴C-labeled P2r polypeptides (●——●) from HeLa cells. Labeling (5 μCi/ml) was from (*a*) 2–3 hours postinfection; (*b*) 3–4 hours postinfection; (*c*) 4–5 hours postinfection.

636

(Fig. 8) merely served to confirm the observations above. The slab gel results, which compare the cytoplasmic viral polypeptides synthesized in the third to fourth hours postinfection, indicate that $NCVP_2$ (molecular weight 75,000) of P2r virus is about four times as stable as that of P2. There is an increase in $NCVP_7$ (molecular weight 20,000) in parallel with the increase in $NCVP_4$ (molecular weight 55,000) as $NCVP_2$ is cleaved. There is another viral polypeptide, migrating slightly faster than VP_1 (perhaps $NCVP_{6b}$, Butterworth 1973), whose appearance parallels $NCVP_4$ and $NCVP_7$. Its relationship to the cleavage of $NCVP_2$ is not presently known.

Plotted in Figure 9 for ease of comparison are the rates of RNA synthesis for P2 and P2r in infected HeLa cells and the ratio for the two viruses of $NCVP_2$ to its products, $NCVP_4$ and $NCVP_7$, at various times after infection. The correlation is obvious: P2r virus, with a genetic alteration to a stable replicase, has also acquired, in parallel, a stable $NCVP_2$.

DISCUSSION

The regulation of RNA synthesis of the small RNA bacteriophages is well characterized. A phage-coded noncapsid polypeptide, whose function is to act as specific initiator, combines with host proteins to produce an RNA replicase. Later in infection, translational controls (in fact, the virus coat protein) overcome the synthesis of the replicase and inhibit further RNA synthesis (reviewed in Sugiyama, Korant and Lonberg-Holm 1972).

Polio and other picornaviruses are unable to control their replicases in a similar manner. First, they possess a nonsegmented message. In addition, initiation of translation occurs only once, at the 5' end of the message, as internal stop-start signals are lacking in eukaryotic translation machinery. And finally, all stable gene products are produced equally throughout infection (Jacobson and Baltimore 1968a; Summers, Roumiantzeff and Maizel 1971; Butterworth 1973). Thus the picornaviruses (and perhaps other mammalian viruses as well) cannot regulate translation of their mRNA, yet they do regulate several functions, including RNA synthesis.

As infection proceeds past midcycle, much as in the RNA phages, poliovirus RNA production, particularly the initiation of synthesis (Noble and Levintow 1970), declines. A logical mode of exerting control, which is supported by the results presented here, is that picornaviruses, as exemplified by poliovirus, have regulated their functions by coding for proteins that are intrinsically labile to proteases. Some of these proteins are functional, for example, in initiation of RNA synthesis.

A model for control is presented in Figure 10, which attempts to summarize diagrammatically the evidence presented here and elsewhere (Korant 1973). In this model, a host-specified serine protease (Jacobson, Asso and Baltimore 1970; Korant 1972; Summers et al. 1972) awaits the translation of the incoming viral message. The hospitable protease cleaves the viral polyprotein to several fragments, at least one of which, $NCVP_2$, is part of the replicase. If the serine protease is not functional, RNA synthesis does not ensue (see results with TLCK, Fig. 3).

Initially, and up to about 2.5 hours after infection ("early"), proteolytic

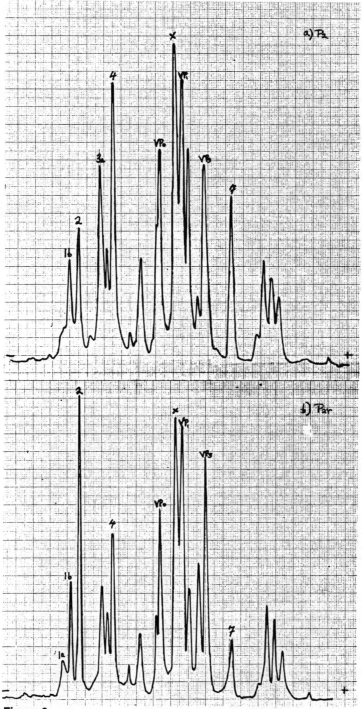

Figure 8
SDS slab gel electrophoresis (Laemmli 1970) of (*a*) P2 or (*b*) P2r polypeptides labeled with ^{14}C-amino acids from 3–4 hours post-infection. Compare with Figure 7b. The numbers above some of the peaks refer to the standard nomenclature for poliovirus cytoplasmic proteins (Summers and Maizel 1968).

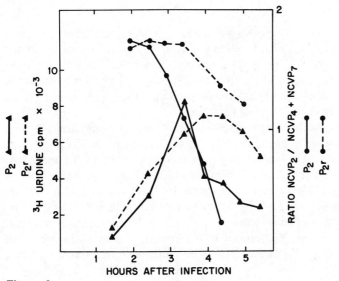

Figure 9
Comparison of rate of RNA synthesis of P2 (▲——▲) and
P2r (▲– – – –▲) viruses and ratios of NCVP₂ to its cleavage
products for P2 (●——●) and P2r (●– – – –●). These data were
obtained from infected HeLa cells.

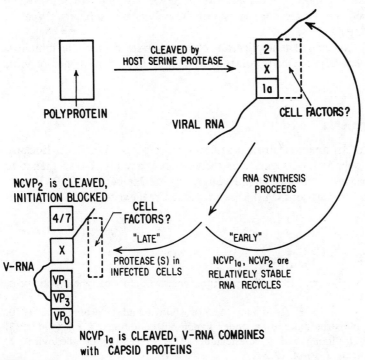

Figure 10
Model for regulation of poliovirus RNA synthesis by protein cleavage. "Cell factors" refers to hypothetical cellular proteins which may
be part of the replicase. For details and nomenclature, see text.

activities which convert precursors to capsid polypeptides (Kiehn and Holland 1970; Korant 1972, 1973; Esteban and Kerr 1974) and which inactivate $NCVP_2$ (Figs. 4a and 7) are in short supply. This results in a longer-lived replicase (Fig. 3, 4 and 9) and also less maturation of newly completed RNA strands due to a relative shortage of coat proteins. Therefore, "early" RNA can cycle again as either message or template.

But as infection proceeds, proteases are activated, or synthesized, leading to the "late" phase of replication. The capsid precursor $NCVP_{1a}$ is more rapidly cleaved to yield polypeptides which readily combine with new RNA. Also, $NCVP_2$ is rapidly cleaved (Fig. 7b,c), blocking initiation of RNA synthesis (Fig. 9). This latter reaction may be thwarted in HeLa cells by application of iodoacetamide (Fig. 3) or in monkey cells with PMSF (Fig. 4b). A virus variant may be selected (Fig. 5) which has altered nonstructural polypeptides (Scheele and Pfefferkorn 1970; Korant, in prep.) with increased stability, e.g., P2r (Fig. 7). These changes may then lead to a stable replicase (Fig. 6).

Control of RNA synthesis via proteolysis is known to occur in Bacillus, where a bacterial serine protease specifically cleaves a subunit of the DNA-dependent RNA polymerase and radically changes its specificity (Leighton et al. 1973). The situation is far less clear with mammalian viruses, much less with eukaryotic cells. However, a report of the purification of bovine and rat RNA polymerases concluded that proteolytic modifications probably accounted for structural differences between the multiple forms of the enzymes. Similar to the situation in Bacillus, the proteases (which are sensitive to PMSF) caused specific cleavage in one of the polymerase subunits (Weaver, Blatti and Rutter 1971).

It may be expected that application of appropriate proteolytic inhibitors may aid in uncovering cellular controls, as well as additional details of virus replication.

Acknowledgments

Dr. E. Knight, Jr. analyzed protein samples on slab gels. Dr. R. Z. Lockart, Jr. provided support and suggestions throughout the study. I owe thanks to several colleagues, including Drs. T. Sugiyama, S. Halperen and B. Butterworth, for their discussions. M. Parise provided excellent technical assistance.

REFERENCES

Adler, R. and D. Tershak. 1974. Temperature-sensitive defect of type 2 poliovirus. *Virology* **58**:209.

Agol, V., G. Lipskaya, E. Tolskaya, M. Voroshilova and L. Romanova. 1970. Defect of poliovirus maturation under hypotonic conditions. *Virology* **41**:533.

Anderson, C., P. Baum and R. Gesteland. 1973. Processing of adenovirus 2-induced proteins. *J. Virol.* **12**:241.

Bachrach, H., J. Swaney and G. Van de Woude. 1973. Isolation of the structural polypeptides of foot and mouth disease virus and analysis of their C-terminal sequences. *Virology* **52**:520.

Bachrach, U. and L. Benchetrit. 1974. Studies on phage internal proteins. II. Cleavage of a precursor of internal proteins during the morphogenesis of bacteriophage T4. *Virology* **59**:51.

Baltimore, D. 1969. The replication of picornaviruses. In *Biochemistry of Viruses* (ed. H. Levy), p. 101. Marcel Dekker, New York.

————. 1971. Viral genetic systems. *Trans. N. Y. Acad. Sci.* **33**:327.

Burrell, C. and P. Cooper. 1973. N-terminal aspartate, glycine and serine in poliovirus capsid protein. *J. Gen. Virol.* **21**:443.

Burrell, C., E. Martin and P. Cooper. 1970. Post translational cleavages of virus polypeptides in arbovirus-infected cells. *J. Gen. Virol.* **6**:319.

Butterworth, B. 1973. A comparison of the virus-specific polypeptides of encephalomyocarditis virus, human rhinovirus-1A and poliovirus. *Virology* **56**:439.

Butterworth, B. and B. Korant. 1974. Characterization of the large picornaviral polypeptides produced in the presence of zinc ion. *J. Virol.* **14**:282.

Butterworth, B. and R. Rueckert. 1972. Gene order of encephalomyocarditis virus as determined by studies with pactamycin. *J. Virol.* **9**:823.

Butterworth, B., L. Hall, C. Stoltzfus and R. Rueckert. 1971. Virus-specific proteins synthesized in encephalomyocarditis virus-infected HeLa cells. *Proc. Nat. Acad. Sci.* **68**:3083.

Caliguiri, L. and A. Mosser. 1971. Proteins associated with the poliovirus RNA replication complex. *Virology* **46**:375.

Cancedda, R. and M. Schlesinger. 1974. Formation of Sindbis virus capsid protein in mammalian cell-free extracts programmed with viral messenger RNA. *Proc. Nat. Acad. Sci.* **71**:1843.

Chang, C. and H. Zweerink. 1971. Fate of parental reovirus in infected cells. *Virology* **46**:544.

Chou, H., P. Black and R. Roblin. 1974. Non-selective inhibition of transformed cell growth by a protease inhibitor. *Proc. Nat. Acad. Sci.* **71**:1748.

Cooper, P. 1969. The genetic analysis of poliovirus. In *Biochemistry of Viruses* (ed. H. Levy), p. 177. Marcel Dekker, New York.

Cooper, P., D. Summers and J. Maizel. 1970. Evidence for ambiguity in the posttranslational cleavage of poliovirus proteins. *Virology* **44**:408.

Dobos, P. and E. Martin. 1972. Virus-specific polypeptides in ascites cells infected with encephalomyocarditis virus. *J. Gen. Virol.* **17**:197.

Eggen, K. and A. Shatkin. 1972. *In vitro* translation of cardiovirus ribonucleic acid by mammalian cell-free extracts. *J. Virol.* **9**:636.

Esteban, M. and I. Kerr. 1974. The synthesis of encephalomyocarditis virus polypeptides in infected L-cells and cell-free systems. *Eur. J. Biochem.* **45**:567.

Etchison, J., M. Doyle, E. Penhoet and J. Holland. 1971. Synthesis and cleavage of influenza virus proteins. *J. Virol.* **7**:155.

Fahrney, D. and A. Gold. 1963. Sulfonyl fluorides as inhibitors of esterases. I. Rates of reaction with acetylcholinesterase, chymotrypsin and trypsin. *J. Amer. Chem. Soc.* **85**:997.

Fernandez-Tomas, C. and D. Baltimore. 1973. Morphogenesis of poliovirus. III. Formation of provirion in cell-free extracts. *J. Virol.* **12**:1181.

Flanagan, J. 1966. Hydrolytic enzymes in KB cells infected with poliovirus and herpes simplex virus. *J. Bact.* **91**:789.

Friedman, T. 1974. Genetic economy of polyoma virus: Capsid polypeptides are cleavage products of the same viral gene. *Proc. Nat. Acad. Sci.* **71**:257.

Garfinkle, B. and D. Tershak. 1971. Effect of temperature on the cleavage of polypeptides during growth of LSc poliovirus. *J. Mol. Biol.* **59**:537.

Ginevskaya, V., I. Scarlat, N. Kalinina and V. Agol. 1972. Synthesis and cleavage of virus-specific proteins in Krebs II carcinoma cells infected with encephalomyocarditis virus. *Arch. Ges. Virus* **39**:98.

Goldstein, J. and S. Champe. 1974. T4-Induced activity required for specific cleavage of a bacteriophage protein *in vitro*. *J. Virol.* **13**:419.

Holland, J. and E. Kiehn. 1968. Specific cleavage of viral proteins as steps in the synthesis and maturation of enteroviruses. *Proc. Nat. Acad. Sci.* **60**:1015.

Holland, J., M. Doyle, J. Perrault, D. Kingsbury and J. Etchison. 1972. Proteinase activity in purified animal virions. *Biochem. Biophys. Res. Comm.* **46**:634.

Homma, M. 1971. Trypsin action on the growth of Sendai virus in tissue culture cells. I. Restoration of the infectivity for L cells by direct action of trypsin on L cell-borne Sendai virus. *J. Virol.* **8**:619.

Homma, M. and M. Ohuchi. 1973. Trypsin action on the growth of Sendai virus in tissue culture cells. III. Structural difference of Sendai viruses grown in eggs and tissue culture cells. *J. Virol.* **12**:1457.

Honess, R. and B. Roizman. 1973. Proteins specified by herpes simplex virus. XI. Identification and relative molar rates of synthesis of structural and nonstructural herpes virus polypeptides in the infected cell. *J. Virol.* **12**:1347.

Ishibashi, M. and J. Maizel. 1974. The polypeptides of adenovirus. V. Young virions, structural intermediates between top components and aged virions. *Virology* **57**:409.

Jacobson, M. and D. Baltimore. 1968a. Polypeptide cleavages in the formation of poliovirus proteins. *Proc. Nat. Acad. Sci.* **61**:77.

———. 1968b. Morphogenesis of poliovirus. I. Association of the viral RNA with coat protein. *J. Mol. Biol.* **33**:363.

Jacobson, M., J. Asso and D. Baltimore. 1970. Further evidence on the formation of poliovirus proteins. *J. Mol. Biol.* **49**:657.

Katz, E. and B. Moss. 1970a. Formation of a vaccinia virus structural polypeptide from a higher molecular weight precursor: Inhibition by rifampicin. *Proc. Nat. Acad. Sci.* **66**:677.

———. 1970b. Vaccinia virus structural polypeptide derived from a high molecular weight precursor: Formation and integration into virus particles. *J. Virol.* **6**:717.

Kiehn, E. and J. Holland. 1970. Synthesis and cleavage of enterovirus polypeptides in mammalian cells. *J. Virol.* **5**:358.

Klenk, H. and R. Rott. 1973. Formation of influenza virus proteins. *J. Virol.* **11**:823.

Klenk, H., C. Scholtissek and R. Rott. 1972. Inhibition of glycoprotein biosynthesis of influenza virus by D-glucosamine and 2-deoxyglucose. *Virology* **49**:723.

Korant, B. 1972. Cleavage of viral precursor proteins *in vivo* and *in vitro*. *J. Virol.* **10**:751.

———. 1973. Cleavage of poliovirus-specific polypeptide aggregates. *J. Virol.* **12**:556.

Korant, B., J. Kauer and B. Butterworth. 1974. Zinc ions inhibit replication of rhinoviruses. *Nature* **248**:588.

Korant, B., K. Lonberg-Holm, J. Noble and J. Stasny. 1972. Naturally occurring and artificially produced components of three rhinoviruses. *Virology* **48**:71.

Laemmli, U. 1970. Cleavage of structural proteins during the assembly of the head of bacteriophage T4. *Nature* **227**:680.

Laporte, J. 1969. The structure of foot-and-mouth disease virus. *J. Gen. Virol.* **4**:631.

Lazarowitz, S., R. Compans and P. Choppin. 1971. Influenza virus structural and non-structural proteins in infected cells and their plasma membranes. *Virology* **46**:830.

Lazarowitz, S., A. Goldberg and P. Choppin. 1973. Proteolytic cleavage by plasmin of the HA polypeptide of influenza virus: Host cell activation of serum plasminogen. *Virology* **56:**172.

Leighton, T., R. Doi, R. Warren and R. A. Kelln. 1973. The relationship of serine protease activity to RNA polymerase modification and sporulation in *B. subtilis*. *J. Mol. Biol.* **76:**103.

Levin, D., H. Mendelsohn, M. Schonberg, H. Klett, S. Silverstein, A. Kapuler and G. Acs. 1970. Properties of RNA transcriptase in reovirus subviral particles. *Proc. Nat. Acad. Sci.* **66:**890.

Lewenton-Kriss, S. and B. Mandel. 1972. Studies on the non-neutralizable fraction of poliovirus. *Virology* **48:**666.

Lucas-Lenard, J. 1974. Cleavage of mengovirus polyproteins *in vivo*. *J. Virol.* **14:**261.

Morser, M. and D. Burke. 1974. Cleavage of virus-specified polypeptides in cells infected with Semliki forest virus. *J. Gen. Virol.* **22:**395.

Noble, J. and L. Levintow. 1970. Dynamics of poliovirus-specific RNA synthesis and the effects of inhibitors of virus replication. *Virology* **40:**634.

Pfefferkorn, E. and M. Boyle. 1972. Selective inhibition of the synthesis of Sindbis virion proteins by an inhibitor of chymotrypsin. *J. Virol.* **9:**187.

Roumiantzeff, M., D. Summers and J. Maizel. 1971. *In vitro* protein synthetic activity of membrane-bound poliovirus polysomes. *Virology* **44:**249.

Saborio, J., S. Pong and G. Koch. 1974. Selective and reversible inhibition of initiation of protein synthesis in mammalian cells. *J. Mol. Biol.* **85:**195.

Samson, A. and C. Fox. 1973. Precursor protein for Newcastle disease virus. *J. Virol.* **12:**579.

Scheele, C. and E. Pfefferkorn. 1970. Virus-specific proteins synthesized in cells infected with RNA + temperature-sensitive mutants of Sindbis virus. *J. Virol.* **5:**329.

Scheid, A. and P. Choppin. 1974. Identification of biological activities of paramyxovirus glycoprotein. Activation of cell fusion, hemolysis, and infectivity by proteolytic cleavage of an inactive precursor protein of Sendai virus. *Virology* **57:**475.

Schlesinger, M. and S. Schlesinger. 1973. Large molecular weight precursors of Sindbis virus proteins. *J. Virol.* **11:**1013.

Schnebli, H. and M. Burger. 1972. Selective inhibition of growth of transformed cells by protease inhibitors. *Proc. Nat. Acad. Sci.* **69:**3825.

Schoellmann, G. and E. Shaw. 1963. Direct evidence for the presence of histidine in the active center of chymotrypsin. *Biochemistry* **2:**252.

Shanmugam, G., G. Vecchio, D. Attardi and M. Green. 1972. Immunological studies on viral polypeptide synthesis in cells replicating murine sarcoma-leukemia virus. *J. Virol.* **10:**447.

Shatkin, A. and J. Sipe. 1968. RNA polymerase activity in purified reovirus. *Proc. Nat. Acad. Sci.* **61:**1462.

Shaw, E., M. Mares-Guia and W. Cohen. 1965. Evidence for an active-center histidine in trypsin through use of a specific reagent, 1-chloro-3-tosyl amido-7-amino-2-heptanone, the chloromethyl ketone derived from *N*-α-tosyl-L-lysine. *Biochemistry* **4:**2219.

Silverstein, S., C. Astell, D. Levin, M. Schonberg and G. Acs. 1972. The mechanism of reovirus uncoating and gene activation *in vivo*. *Virology* **47:**797.

Simmons, D. and J. Strauss. 1974. Translation of Sindbis virus 26S RNA and 49S RNA in lysates of rabbit reticulocytes. *J. Mol. Biol.* **86:**397.

Skehel, J. 1972. Polypeptide synthesis in influenza virus infected cells. *Virology* **49:**23.

Smith, R., H. Zweerink and W. Joklik. 1969. Polypeptide components of virions, top components, and cores of reovirus type 3. *Virology* **39**:791.

Snyder, H. and T. Sreevalsan. 1973. Proteins specified by Sindbis virus in chick embryo fibroblast cells. *Biochem. Biophys. Res. Comm.* **54**:24.

———. 1974. Proteins specified by Sindbis virus in HeLa cells. *J. Virol.* **13**:541.

Strauss, J., B. Burge and J. Darnell. 1969. Sindbis virus infection of chick and hamster cells: Synthesis of virus-specified proteins. *Virology* **37**:367.

Sugiyama, T., B. Korant and K. Lonberg-Holm. 1972. RNA virus gene expression and its control. *Annu. Rev. Microbiol.* **26**:467.

Summers, D. and J. Maizel. 1968. Evidence for large precursor proteins in poliovirus synthesis. *Proc. Nat. Acad. Sci.* **59**:966.

Summers, D., J. Maizel and J. Darnell. 1965. Evidence for virus-specific noncapsid proteins in poliovirus-infected HeLa cells. *Proc. Nat. Acad. Sci.* **54**:505.

Summers, D., M. Roumiantzeff and J. Maizel. 1971. The translation and processing of poliovirus proteins. In *Ciba Foundation Symposium on Strategy of the Viral Genome* (ed. G. Wolstenholme and M. O'Connor), p. 111. Churchill Livingstone, London.

Summers, D., E. Shaw, M. Stewart and J. Maizel. 1972. Inhibition of cleavage of large poliovirus-specific precursor proteins in infected HeLa cells by inhibitors of proteolytic enzymes. *J. Virol.* **10**:880.

Taber, R., D. Rekosh and D. Baltimore. 1971. Effect of pactamycin on synthesis of poliovirus proteins: A method for genetic mapping. *J. Virol.* **8**:395.

Vogt, V. and R. Eisenman. 1973. Identification of a large polypeptide precursor of avian oncornavirus proteins. *Proc. Nat. Acad. Sci.* **70**:1734.

Weaver, R., S. Blatti and W. Rutter. 1971. Molecular structures of DNA-dependent RNA polymerases (II) from calf thymus and rat liver. *Proc. Nat. Acad. Sci.* **68**:2994.

Zweerink, H. and W. Joklik. 1970. Studies on the intracellular synthesis of reovirus-specified proteins. *Virology* **41**:501.

Activation of Cell Fusion and Infectivity by Proteolytic Cleavage of a Sendai Virus Glycoprotein

Andreas Scheid and Purnell W. Choppin

The Rockefeller University, New York, New York 10021

The paramyxovirus or parainfluenza virus group includes a large number of viruses that cause a variety of diseases in animals and man. The structure and replication of several viruses of this group, Newcastle disease virus (NDV), simian virus 5 (SV5) and Sendai virus, have been studied extensively (see reviews by Kingsbury [1973] and Choppin and Compans [1975]. The isolated virions possess several biological activities. They adsorb to host cells and agglutinate erythrocytes by binding to neuraminic acid-containing receptors on the cell surface. The site on the virion which binds to these receptors is located on a glycoprotein, the hemagglutinin, on the virion surface. Viruses that attach to cells by this mechanism also possess the enzyme neuraminidase which, by hydrolyzing the bond between neuraminic acid and carbohydrate, can inactivate the receptor on the cell surface, as well as on soluble neuraminic acid-containing glycoproteins which may act as competitive inhibitors of virus attachment. In addition, parainfluenza viruses possess the capacity to lyse erythrocytes and cause cell fusion. Cell fusion occurs in certain cells after infection and can also be observed in the absence of virus replication when sensitive cells are exposed to high concentrations of virus.

Parainfluenza viruses possess a lipoprotein envelope which is covered with spikelike projections and which encloses a helical nucleocapsid composed of a protein subunit of approximately 60,000 daltons and the single-stranded RNA genome. The envelope contains lipid, which is present in the form of a bilayer and is derived from the host cell plasma membrane during virus assembly. In contrast to the lipids, the protein content of the virion is specified by the virus genome. Figure 1 shows the polypeptides of the parainfluenza virus SV5. The functions of four proteins have been established, and the corresponding proteins have also been identified in NDV and Sendai virions. NP is the major nucleocapsid protein, and the available evidence suggests that the smallest virion protein, M, is associated with the inner surface of the lipid membrane. Two glycoproteins constitute the spikes on the surface of the virion. The basis for their designations, HN and F, will become apparent in this communication.

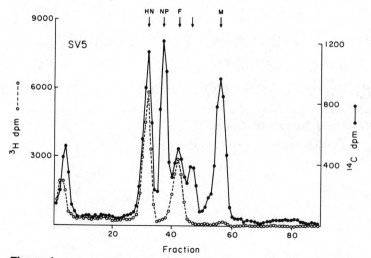

Figure 1

Polyacrylamide gel electrophoresis of the polypeptides of SV5 labeled with [14C]-amino acid mixture (●——●) and [3H]glucosamine (o– – –o). The designations of the proteins are described in the text. The protein indicated by the unlabeled arrow is closely associated with the nucleocapsid; however, its function has not been established. The origin is at the left and the anode at the right. (Modified from Chen, Compans and Choppin 1971.)

We have studied parainfluenza viruses to establish the roles of the constituents of the envelope in the various biological activities of the virion. Here we shall summarize findings which indicate that one of the two glycoproteins has both hemagglutinating and neuraminidase activities, and that the other glycoprotein, which is derived from an inactive precursor by proteolytic cleavage, is involved in cell fusion, hemolysis and the initiation of infection. These studies are described in detail elsewhere (Scheid et al. 1972; Scheid and Choppin 1973, 1974a,b,c and 1975).

Biological Activities Associated with Isolated Glycoproteins

Earlier studies indicated that the two spike glycoproteins possessed neuraminidase and hemagglutinating activities. This was based on the finding that virions which were deprived of their spikes by extensive treatment with protease were devoid of these biological activities (Chen, Compans and Choppin 1971). In order to identify the specific proteins involved in each biological activity, it was first necessary to isolate the proteins in biologically active form. This was accomplished first with SV5 (Scheid et al. 1972) and subsequently with NDV (Scheid and Choppin 1973; Seto, Becht and Rott 1973) and Sendai virus (Tozawa, Watanabe and Ishida 1973; Scheid and Choppin 1974a). The isolation procedure developed in our laboratory is based on the solubility properties of the virion proteins in nonionic detergent at varying ionic strengths. The two spike glycoproteins and the nonglycosylated membrane protein (M protein) can be solubilized by treatment of virions with

Triton X-100 and 1 M potassium chloride. The other virion proteins remain insoluble under these conditions. Removal of the salt from the envelope protein fraction yields, by precipitation, a pure preparation of the M protein. The two glycoproteins remain soluble after the removal of salt and can then be separated by rate zonal centrifugation on Triton-containing sucrose gradients. This sedimentation procedure was adequate for separating sufficient quantities to establish the biological activities, but is not convenient as a preparative procedure.

The larger isolated glycoprotein possesses both neuraminidase and hemagglutinating activity and the smaller glycoprotein neither activity. Hemagglutinating activity becomes demonstrable when the glycoprotein is allowed to aggregate in the absence of the detergent. Once it was found that only the large glycoprotein interacted with neuraminic acid-containing substances, a separation procedure was devised based on the affinity of this protein for the neuraminic acid-containing glycoprotein fetuin. Affinity chromatography using fetuin as a ligand covalently linked to Sepharose allows the separation of the two glycoproteins in preparative quantity (Scheid and Choppin 1974a,b). The analysis of the two fractions obtained from such a column is shown in Figure 2. In accordance with the earlier findings with the glycoproteins separated by velocity sedimentation, the small glycoprotein (Fig. 2a) is not

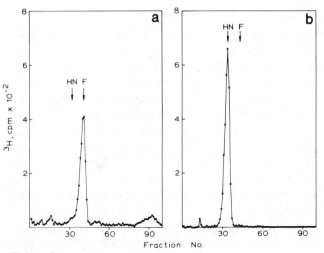

Figure 2
Polyacrylamide gel electrophoresis of the isolated SV5 glycoproteins. The glycoprotein fraction from [^{3}H]leucine-labeled virions was fractionated on fetuin-Sepharose. (*a*) Protein that does not bind to the fetuin at the initial column temperature of 0°C. (*b*) Protein that binds to fetuin at the initial column temperature of 0°C and elutes when the temperature is raised to 25°C. The arrows indicate the positions of SV5 marker glycoproteins as determined with added ^{14}C-amino acid-labeled virus polypeptides. (Reprinted, with permission, from Scheid and Choppin 1974b.)

active in the hemagglutination and neuraminidase reaction. Both activities are associated with the large glycoprotein (Fig. 2b), and this protein has thus been designated HN.

The paramyxoviruses differ from the influenza viruses, which also contain two glycoproteins, in that in influenza viruses one glycoprotein is responsible for the hemagglutinating activity of the virus and the other for the neuraminidase activity. Influenza viruses do not cause cell fusion or hemolysis. Because of the finding that in paramyxoviruses one glycoprotein possessed both hemagglutinating and neuraminidase activities and the smaller glycoprotein neither activity, it was postulated early in these studies that in parainfluenza viruses the small glycoprotein might be involved in cell fusion and hemolysis (Scheid et al. 1972; Scheid and Choppin 1973). Although the isolated, lipid-free, small glycoprotein does not hemolyze or fuse cells, this does not rule out its involvement in these activities, because there is much evidence to suggest that an orderly arrangement of protein with lipid—such as is found in the virus envelope—is required for virus-induced cell fusion. The sensitivity of this function to organic solvents has been known for a long time, and recently, cell-fusing activity could be obtained with reassembled viral glycoproteins and lipid, but not with viral glycoprotein alone (Hosaka and Shimizu 1972). Evidence that the small glycoprotein is indeed involved in cell fusion and hemolysis was subsequently obtained as described below.

The Precursor of the Small Virion Glycoprotein

Polyacrylamide gel electrophoresis patterns of Sendai virus proteins have been reported from several laboratories. Striking dissimilarities can be observed among the published patterns. An examination of the experimental conditions used suggested to us that the differences in protein composition could be explained on the basis of the growth of the virus in different host cells. That this was indeed the case has been demonstrated by the evidence described below (Scheid and Choppin 1974a,c) and by similar evidence obtained by Homma and Ohuchi (1973).

A polypeptide pattern of Sendai virus grown in chick embryo is shown in Figure 3b. Similarities of the polypeptides in Sendai, SV5 (cf. Fig. 1) and NDV virions and the functional analogies among the various viruses of the nonglycosylated polypeptides designated NP and M have been pointed out (Mountcastle, Compans and Choppin 1971; Scheid and Choppin 1974c). Two other nonglycosylated proteins are indicated by arrows in the upper row. With respect to the glycoproteins, an analogy exists between this protein pattern and those of SV5 and NDV, in that two glycoproteins, HN and F, are found in similar quantity.

The analysis of Sendai virions grown in a different host, MDBK cells, is shown in Figure 3a. These virions differ from the ones grown in chick embryo with respect to the glycoproteins. In contrast to the egg-grown virions, they contain little of the small glycoprotein (F) and large amounts of a larger glycoprotein (F_o), which is not present on egg-grown virions. The absence of F_o in egg-grown virions was established by coelectrophoresis of oppositely labeled virions from the two hosts. The largest glycoprotein, HN, is present in similar quantities in both virus preparations, and with both viruses, the HN

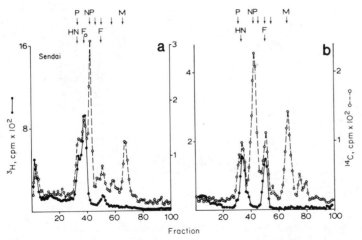

Figure 3

Polyacrylamide gel electrophoresis of the polypeptides of Sendai virus labeled with ^{14}C-amino acids (o – – –o) and [^{3}H]glucosamine (•——•). (a) Virions grown in MDBK cells. (b) Virions grown in embryonated chicken egg. The nonglycosylated proteins are indicated by the arrows in the upper row. The P protein has been associated with the RNA transcriptase (Stone, Kingsbury and Darlington 1972). The functions of the proteins indicated by unlabeled arrows have not been determined. The properties of the other virus polypeptides are described in the text. The P protein and HN frequently comigrate (a), but are sometimes resolved (b). (Modified from Scheid and Choppin 1974a,c.)

protein has been isolated and its identity as the protein with hemagglutinating and neuraminidase activities has been established. In accordance with this, virions from the two hosts do not differ significantly in their activity in the hemagglutinating and neuraminidase reactions. Thus the difference in the glycoprotein content of the virions grown in the two hosts concerns only the F and the F_0 proteins.

Because of the inverse relationship between the quantities of F and F_0 found in the virions from the two hosts, the possibility was apparent that the F_0 glycoprotein was a precursor of the F protein, and further, that complete cleavage of the F_0 protein to yield F occurred in eggs, but little cleavage occurred in MDBK cells. To investigate this directly, purified virions grown in MDBK cells were subjected to mild trypsin treatment (Fig. 4b). After trypsin treatment, the F_0 glycoprotein is lost, leaving in this region of the gel the HN protein as a single sharp peak, and the amount of the F protein is greatly increased. Thus the conversion of the precursor F to the smaller glycoprotein F_0 has been accomplished in vitro. Since the estimated molecular weight of F_0 is 65,000 daltons and that of F, 53,000 daltons, the portion removed from the precursor would be $\sim$ 12,000 daltons. Such a 12,000-dalton peptide has not been recovered. This suggests that cleavage occurs at several sites in this region.

It is apparent that the polypeptide pattern of trypsin-treated MDBK-grown

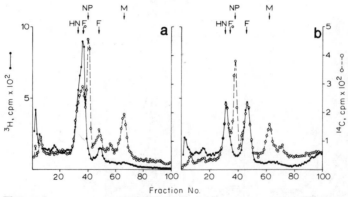

Figure 4

In vitro cleavage by trypsin of the F_0 precursor glycoprotein on Sendai virions grown in MDBK cells in the presence of ^{14}C-amino acid (o– – –o) and [^3H]glucosamine (●——●). (*a*) Control incubated without protease. (*b*) Virions treated with trypsin, 5 $\mu g/ml$ in phosphate-buffered saline pH 7.2, for 10 minutes at 37°C. (Modified from Scheid and Choppin 1974a.)

virions resembles closely that of egg-grown virions (cf. Figs. 3b and 4b) with respect to both the lack of F_0 and the abundance of F. This provides additional support for the concept that in the egg-grown virus cleavage has occurred in ovo.

Biological Activities Correlated with the Presence of the Small Glycoprotein

Sendai virions grown in MDBK cells and in the chicken embryo differ not only in their glycoprotein composition, but also in some of their biological activities; host-dependent variation in biological activities of paramyxoviruses has been observed previously (Ishida and Homma 1960; Matsumoto and Maeno 1962; Young and Ash 1970; Homma 1971, 1972; Homma and Tamagawa 1973). The egg-grown Sendai virions hemolyze and cause cell fusion, as do other parainfluenza viruses. In contrast, MDBK-grown virions do not hemolyze and do not fuse cells. Another very important difference between the Sendai virions produced by the two hosts is that the virions grown in the egg are infectious whereas those grown in MDBK cells are not. A practical consequence of this is that egg-grown Sendai virus can be used to infect other chick embryos or MDBK cells. However, the virus progeny from MDBK cells cannot be passaged to other MDBK cultures, nor does it undergo multiple cycles in the same culture. This inability to undergo multiple cycle replication provides the explanation for the inability of Sendai virus to form plaques in MDBK cells.

From these results, a correlation can be made between the biological activities of Sendai virions and their glycoprotein composition. Active hemolysis, cell fusion and infectivity are associated with the presence of the F protein, and these are properties of egg-grown virions. By contrast, MDBK-grown virions, which have little hemolyzing and cell-fusing activities and are not infectious, are deficient in the F protein, and contain a large amount of the initiation of infection, whereas the precursor F_0 is inactive.

This correlation between polypeptide composition and biological activities can also be seen in MDBK virions in which the precursor F_0 is cleaved by trypsin treatment in vitro to yield the F protein. As pointed out above, MDBK-grown virions are not infectious; however, when the virus is trypsin treated and used as an inoculum, virus production occurs. Similarly, the hemolyzing activity of MDBK-grown virus is restored by trypsin treatment. In addition, the trypsin-treated virions cause rapid fusion of BHK-21F cells (Fig. 5), whereas the untreated virions do not. Thus with trypsin treatment, the MDBK-grown virions acquire the characteristics of egg-grown virions with respect to both their glycoprotein composition and their biological activities. The F protein appears to be active in hemolysis, cell fusion and the initiation of infection, whereas the precursor F_0 is inactive.

An important clue with respect to the possible biological significance of cell fusion can be derived from the finding that cell fusion and infectivity are present or absent together. Activation of infectivity concomitant with the cleavage of the F_0 protein must involve a step beyond virus adsorbtion since

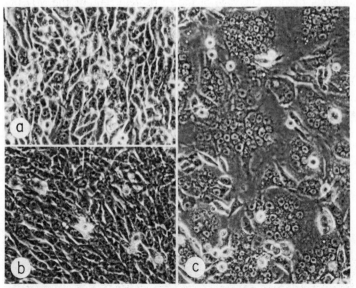

Figure 5
Fusion of BHK-21F monolayers with MDBK cell-grown Sendai virus after activation with trypsin. Virions grown in MDBK cells and concentrated to 10,240 HAU/ml were incubated with trypsin, 1 μg/ml, at 37°C for 10 minutes, or as a control, incubated with buffer. Calf serum was added to a final concentration of 1.6%, and the virus suspensions were used to inoculate monolayers of BHK-21F cells. Phase contrast micrographs were taken after 2 hours. (*a*) No virus. (*b*) Virions treated without protease (control). (*c*) Trypsin-treated virions; the cells have fused extensively and the nuclei have clustered within the syncytial cytoplasm. (Reprinted, with permission, from Scheid and Choppin 1974a.)

the protein responsible for this, the HN protein, is present and active on noninfectious MDBK-grown virions. The genome is not defective in this non-infectious virus since the virus becomes infective upon trypsin treatment. A straightforward explanation would be that the F protein is involved in a step beyond adsorption, but prior to expression of the genome, i.e., virus penetration, and that since fusion and infectivity are activated together, fusion of the virus envelope and the cell membrane is involved in the initiation of infection. Penetration of paramyxovirus nucleocapsid by fusion of viral and cell membranes was proposed by Morgan and Howe (1968) on the basis of electron microscopic studies. However, electron microscopic evidence suggesting phagocytosis as the initial step in virus penetration has also been obtained (Silverstein and Marcus 1964; Dales 1973). The present results are compatible with the view that fusion of the virus membrane with the cell membrane is involved in virus penetration. Fusion of viral and cell membranes may also be involved in hemolysis, as has been suggested by Howe and Morgan (1969).

In Vivo Activation of Sendai Virus with Trypsin

Once the nature of the defect of MDBK-grown Sendai virus was established and it was found that in vitro cleavage of the F_0 protein renders the MDBK-grown virions infectious, it also became clear that the failure of Sendai virus to undergo multiple cycle replication in MDBK cells is due to a lack of activating protease. Conversely, if protease were present in vivo to activate the progeny virus from MDBK cells, spread of virus should occur. We therefore exposed MDBK monolayers in which only a few cells had been infected with Sendai virus to trypsin. To reduce the rate of autodigestion, the N-acetylated derivative was used, and the concentration which was employed, 0.1 μg/ml, had no effect on the morphology of the monolayer. Under these conditions, and in contrast to the situation without added trypsin, all the cells in the monolayer ultimately become infected. This ability of trypsin to allow multiple cycle growth of Sendai virus in MDBK cells also allows the formation of plaques in the same cells (Fig. 6). When trypsin is included in the agar over-

Figure 6
Sendai virus plaque formation in MDBK monolayers in the presence of trypsin. Confluent monolayers in 50-mm petri dishes were inoculated with equal concentrations of Sendai virus; after the adsorption period, medium containing agar (*b*) or medium containing agar with *N*-acetyl trypsin, 0.3 μg/ml (*a*) was added. Photographs were taken after 5 days. (Reprinted, with permission, from Scheid and Choppin 1975.)

lay, virus can spread from the cell infected by the inoculum, and after several cycles of replication, the infected area becomes macroscopically visible. With several virus stocks, the plaque count was found to equal the number of egg infective doses (EID_{50}), that is, the plaquing efficiency is equal to the infectivity for the embryonated chicken egg.

In addition to its usefulness for Sendai virus titration, the plaque assay provides a convenient means for assaying proteases for their ability to substitute for trypsin in Sendai virus activation. Specifically, it was of interest to test whether plasmin is involved in the activation in an infected host, e.g., in the embryonated chicken egg or the respiratory tract of mouse or man. When serum is added to the agar overlay, no plaques develop, and serum from chicken, mouse or man inhibit the activation by trypsin. In addition, the purified plasminogens from the same species as well as the plasmins obtained by activating the plasminogens in vitro were tested. Even at concentrations as high as 5 μg/ml, plaque formation did not occur. This indicated that plasmin is not the enzyme responsible for the cleavage of the F_0 protein and the activation of the virus in the infected host, but that some other protease must be involved. This contrasts with the situation found with influenza virus, where cleavage of the hemagglutinin protein is accomplished by plasmin as well as by trypsin (Lazarowitz, Compans and Choppin 1973; Lazarowitz, Goldberg and Choppin 1973). The specificity of the F_0 cleavage is further emphasized by the fact that the other pancreatic proteases, chymotrypsin and elastase, also cannot substitute for trypsin in the activation of infectivity or hemolysis (Fig. 7). Whereas activation of hemolysis can be accomplished with trypsin at a wide range of concentrations, chymotrypsin and elastase cannot activate hemolysis at any concentration employed.

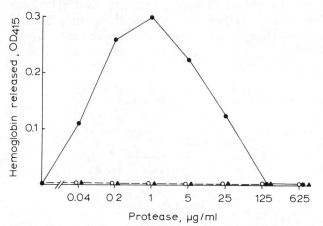

Figure 7
Susceptibility of Sendai virus hemolysis to proteolytic activation. Purified MDBK-grown virions in buffered saline, pH 7.2, were incubated for 10 minutes at 37°C with trypsin (●), chymotrypsin (○) or elastase (▲) at the concentrations indicated on the abscissa. Aliquots were then assayed for their hemolyzing activity. (Reprinted, with permission, from Scheid and Choppin 1975.)

Sendai Virus Mutants That Are Activated by Chymotrypsin

The restricted susceptibility of Sendai virus to proteolytic activation and the fact that virus spread can only occur in the presence of activating enzyme provided an excellent approach for selection of virus mutants that are activated by enzymes other than trypsin. To this end, MDBK cells were infected with virus that had been treated with sodium nitrite at pH 4.5, and chymotrypsin was added to the medium to allow the activation and spread of any mutant that would be susceptible to this enzyme. Two mutants, designated pa-c1 and pa-c2, were isolated in this way (Scheid and Choppin 1975). Infectious virus stocks can be obtained by growing the mutants in the presence of chymotrypsin. Table 1 shows the characterization of the mutants with respect to the spectrum of proteases allowing activation as determined by the plaque assay. Besides being activated by chymotrypsin, both mutants are also susceptible to activation by elastase, and, rather surprisingly, both mutants have lost their susceptibility to trypsin.

The susceptibility of these pa-c mutants was also assayed in vitro (Fig. 8). Virus grown in the absence of chymotrypsin is inactive in the hemolysis reaction, and in accordance with the findings in vivo, hemolysis activation could be accomplished with chymotrypsin and elastase, but not with trypsin. The F_o protein of the mutant viruses can be cleaved with chymotrypsin, but not with trypsin.

The results with these mutants thus provide strong support for the correlation between the virion glycoprotein composition and biological activities and indicate the specificity of the cleavage of the F_o protein that is required for activation of the virus.

One potential use of such mutants lies in the possibility to assay for the presence of activating enzymes in a host organism. Such an approach has been used to answer the question of whether in the embryonated chicken egg there is only tryptic activity available for virus activation or a broader array of activities, including chymotrypsin- or elastase-like activities. It was

Table 1
Activation of Infectivity of Sendai Virus Wild-type and Mutant pa-cl by Protease In Vivo

Virus	Protease added			
	none	trypsin	chymotrypsin	elastase
Wild type[a]	−[b]	+	−	−
Mutant pa-c1	−	−	+	+

[a] Virus stocks were grown in embryonated chicken egg. The mutant pa-cl was propagated in the presence of 30 μg chymotrypsin injected into the allantoic cavity.
[b] Infectivity was measured by plaque formation (see Fig. 6) in the presence of trypsin (0.3 μg/ml), chymotrypsin (1.0 μg/ml) or elastase (0.3 μg/ml) in the agar overlay. In the presence of activating protease, the plaque titers were $1-3 \times 10^9$ plaque-forming units per ml, whereas in the absence of protease, no plaques were seen with a 10^{-3} dilution of virus.

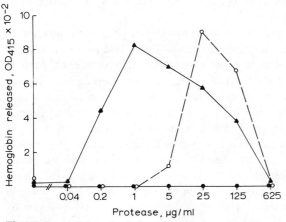

Figure 8
Susceptibility of Sendai virus mutant pa-c1 hemolysis to proteolytic activation. Purified MDBK-grown virions in buffered saline, pH 7.2, were incubated for 10 minutes at 37°C with trypsin (●), chymotrypsin (○) or elastase (▲) at the concentrations indicated on the abscissa. Aliquots were then assayed for their hemolyzing activity. (Reprinted, with permission, from Scheid and Choppin 1975.)

found that when the pa-c mutants were inoculated into eggs, they did not go through multiple cycle replication, and progeny from eggs infected at high multiplicity are inactive in hemolysis, cell fusion and infectivity and possess the F_0 precursor protein. This indicates that the enzyme activity available for virus activation in this host, the embryonated chicken egg, is rather limited. A similar approach to the identification of the activating enzyme of this virus in other hosts, such as the mouse and specifically in the bronchial epithelium, is now feasible as well. The study of additional mutants with different protease susceptibilities should refine this approach.

DISCUSSION

In certain hosts, Sendai virions are released which are defective in hemolyzing and cell-fusing activities as well as infectivity. This host-dependent variation of the virion activities has been observed repeatedly (Ishida and Homma 1960; Matsumoto and Maeno 1962; Young and Ash 1970; Homma 1971, 1972; Homma and Tamagawa 1973), and it had been assumed that host-specific inhibitors prevented the expression of the viral activities (Homma 1971, 1972). From the evidence described above, it is now clear that the variation found in Sendai virions depends on the absence or presence in the host of protease which is capable of cleaving the inactive precursor of the virion glycoprotein F_0 to yield the active protein F. This provides a biochemical basis for host-dependent variation and has allowed the identification of a protein involved in virus-induced hemolysis, cell fusion and the initiation of infection.

Because proteolytic activation must occur for Sendai virus to become infective, the nature of the proteases responsible for the activation in vivo is of particular interest. The distribution of such proteases might conceivably represent a factor responsible for the tropism of the virus for certain host species and tissues. From the studies of in vitro activation with trypsin, it is clear that the precursor protein can be cleaved on the virion. In the activation experiments in cell culture, cleavage of F_0 must occur on the mature virion or on the plasma membrane, since the added proteases act extracellularly. Another example of cleavage of a virion polypeptide after maturation and by extracellular protease has been observed with influenza virus (Lazarowitz, Compans and Choppin 1973; Lazarowitz, Goldberg and Choppin 1973), where the hemagglutinin protein may be cleaved into two subunits. In contrast to the Sendai virus F_0 protein, the influenza virus hemagglutinin may be cleaved by plasmin, and this difference is illustrated by the fact that in the same host, MDBK cells, in the presence of serum, influenza virions are produced with the hemagglutinin protein cleaved and Sendai virions with the F_0 precursor protein uncleaved. Therefore in those hosts where Sendai virus can undergo multiple cycle replication, extracellular proteases other than plasmin are presumably responsible for the activation. In addition, the possibility remains that in certain cells, cleavage may occur intracellularly, and in such a host, the F protein would be active when it is inserted into the plasma membrane.

The described Sendai virus mutants demonstrate that the susceptibility of the F_0 protein to activating protease can be changed by mutation and through the selective advantage offered by different proteases. This raises the possibility that in the natural host or during adaptation to a host in the laboratory, variants may be selected that are more susceptible to the proteases present in that host. In this context, it need be pointed out that other paramyxoviruses, which in many respects are similar to Sendai virus, appear to obtain the active F protein under quite different circumstances. Thus SV5 matures from MDBK cells with a fully active F protein. With NDV, a precursor to the F protein has been identified (Kaplan and Bratt 1973); however, this protein has only been found in infected cells and not on the virion. With NDV it is tempting to speculate that a strain-dependent variation in the susceptibility of the precursor to activating protease may account for the difference in the virulence of those strains. This possibility is supported by the fact that a protein in the position of the F_0 precursor can be detected in the published patterns (Lomniczi, Meager and Burke 1971) of proteins synthesized only in cells infected with the least virulent strains, but not in cells infected with the virulent strains.

In addition to their involvement in the activation of the F protein, proteases may be involved in another aspect of parainfluenza virus infection (Mountcastle et al. 1970, 1974). Proteases cause marked changes in the physical properties of the nucleocapsid protein of paramyxoviruses. When nucleocapsid is treated with protease in vitro, the size of this protein is reduced from ~61,000 daltons to 45,000 daltons. This cleavage renders the nucleocapsid, which in its native state is hydrophobic, water soluble. At the same time, the strength of the bonds between adjacent protein subunits increases, so that the helix changes from a loosely coiled configuration to a stiff rod. Nucleocapsid with the cleaved subunit is found under certain conditions in infected cells, but is never found in the virion. Cleavage of the subunit may prevent maturation of virions by two means. The stiff rod may be unsuitable for envelopment in the virion, which requires bending of the nucleocapsid, or the recognition site on the nucleocapsid protein subunit which is involved in the alignment of nucleocapsids beneath those areas of cell membrane which become the virus envelope may have been lost with cleavage. Thus proteolytic cleavage of the nucleocapsid could account, under certain conditions, for a block at the cell membrane of virus maturation by budding. Such a mechanism is of interest because of the ability of paramyxoviruses to cause persistent infection characterized by the accumulation of nucleocapsids within the cell. Another instance of proteolytic cleavage affecting the association behavior of the protein subunit of the helical virus nucleocapsid has been found with tobacco mosaic virus (Durham 1972). In this case, the cleavage results in the irreversible aggregation of subunits to form stacked disc rods, thereby preventing the formation of the normal single-helical structure.

SUMMARY

Parainfluenza viruses possess two glycoproteins. These are located on the outer surface of the virus envelope and have been isolated. The larger one (HN) is associated with hemagglutinating and neuraminidase activities. Direct evidence for the involvement of the small glycoprotein (F protein) in hemolysis, cell fusion and the initiation of infection has been obtained with Sendai virus. When grown in certain cells (e.g., MDBK cells), Sendai virions contain a precursor of the small glycoprotein F_o, and such particles lack hemolyzing and cell-fusing activities and are not infectious. However when the precursor is converted to the small glycoprotein (F) by proteolytic cleavage in vitro, the virions become infectious and acquire cell-fusing and hemolyzing activities. Protease that activates the virions and cleaves F_o is present in certain hosts, e.g., the embryonated chicken egg. In such hosts, the infectious virus can spread from cell to cell, whereas in cultured cells which lack the activating enzyme, virus cannot spread. However in such a system, multiple cycle replication and plaque formation are possible if trypsin is added to the medium to activate the virus.

Activation in vivo and in vitro can be accomplished with trypsin, but not with plasmin, chymotrypsin or elastase. Sendai virus mutants with altered protease susceptibility have been isolated, and two mutants have been described which are activated in vivo and in vitro by chymotrypsin and elastase, but not by trypsin. In contrast to the wild-type virus, these mutants do not spread in embryonated chicken egg, and virions from this host possess the precursor F_o.

The results demonstrate that the small glycoprotein of parainfluenza viruses is involved in hemolysis, cell fusion and the initiation of infection. The precursor of this protein is inactive and is found on virions when the host lacks activating protease. Activating protease is present only in some hosts, and this explains the host dependence of the activities of Sendai virions. On the other hand, the susceptibility of the precursor to host protease can be altered by mutation, as has been shown with the Sendai virus mutants. Thus the susceptibility of the viral F_o protein to proteases, as well as the host proteases available, determine whether infection can be established and maintained through multiple cycles. The results suggest that the presence or absence of proteolytic cleavage may affect the host range of the virus and possibly the spread and virulence of the virus within the host.

Acknowledgments

This research was supported by Research Grants AI-05600 from the National Institute of Allergy and Infectious Diseases, USPHS, and GB-43580 from the National Science Foundation.

REFERENCES

Chen, C., R. W. Compans and P. W. Choppin. 1971. Parainfluenza virus surface projections—Glycoproteins with hemagglutinin and neuraminidase activities. *J. Gen. Virol.* **11**:53.

Choppin, P. W. and R. W. Compans. 1975. Reproduction of paramyxoviruses. In *Comprehensive Virology* (ed. H. Fraenkel-Conrat and R. R. Wagner), vol. 4, p. 95. Plenum Press, New York.

Dales, S. 1973. Early events in cell-animal virus interactions. *Bact. Rev.* **37**:103.

Durham, A. C. H. 1972. The cause of irreversible polymerization of tobacco mosaic virus protein. *FEBS Letters* **25**:147.

Homma, M. 1971. Trypsin action on the growth of Sendai virus in tissue culture cells. I. Restoration of the infectivity for L cells by direct action of trypsin on L cell-borne Sendai virus. *J. Virol.* **8**:619.

———. 1972. Trypsin action on the growth of the Sendai virus in tissue culture cells. II. Restoration of the hemolytic activity of L cell-borne Sendai virus by trypsin. *J. Virol.* **9**:829.

Homma, M. and M. Ohuchi. 1973. Trypsin action on the growth of Sendai virus in tissue culture cells. III. Structural differences of Sendai viruses grown in eggs and tissue culture cells. *J. Virol.* **12**:1457.

Homma, M. and S. Tamagawa. 1973. Restoration of the fusion activity of L cell-borne Sendai virus by trypsin. *J. Gen. Virol.* **19**:423.

Hosaka, Y. and Y. K. Shimizu. 1972. Artificial assembly of envelope particles of HVJ (Sendai virus). I. Assembly of hemolytic and fusion factors from envelopes solubilized by Nonidet P40. *Virology* **49**:627.

Howe, C. and C. Morgan. 1969. Interactions between Sendai virus and human erythrocytes. *J. Virol.* **3**:70.

Ishida, N. and M. Homma. 1960. A variant Sendai virus, infectious to egg embryos but not to L cells. *Tohoku J. Exp. Med.* **73**:56.

Kaplan, J. and M. A. Bratt. 1973. Synthesis and processing of Newcastle disease virus polypeptides. *Abstr. Amer. Soc. Microbiol.*, p. 243.

Kingsbury, D. W. 1973. Paramyxovirus replication. *Curr. Topics Microbiol.* **59**:1.

Lazarowitz, S. G., R. W. Compans and P. W. Choppin. 1973. Proteolytic cleavage of the hemagglutinin polypeptide of influenza virus. Function of the uncleaved polypeptide HA. *Virology* **52**:199.

Lazarowitz, S. G., A. R. Goldberg and P. W. Choppin. 1973. Proteolytic cleavage by plasmin of the HA polypeptide of influenze virus: Host cell activation of serum plasminogen. *Virology* **56:**172.

Lomniczi, B., A. Meager and D. C. Burke. 1971. Virus RNA and protein synthesis in cells infected with different strains of Newcastle disease virus. *J. Gen. Virol.* **13:**111.

Matsumoto, T. and K. Maeno. 1962. A host-induced modification of hemagglutinating virus of Japan (HVJ, Sendai virus) in its hemolytic and cytopathic activity. *Virology* **17:**563.

Morgan, C. and C. Howe. 1968. Structure and development of viruses as observed in the electron microscope. IX. Entry of parainfluenza I (Sendai) virus. *J. Virol.* **2:**1122.

Mountcastle, W. E., R. W. Compans and P. W. Choppin. 1971. Proteins and glycoproteins of paramyxoviruses: A comparison of simian virus 5, Newcastle disease virus, and Sendai virus. *J. Virol.* **7:**47.

Mountcastle, W. E., R. W. Compans, L. A. Caliguiri and P. W. Choppin. 1970. Nucleocapsid protein subunits of simian virus 5, Newcastle disease virus, and Sendai virus. *J. Virol.* **6:**677.

Mountcastle, W. E., R. W. Compans, H. Lackland, and P. W. Choppin. 1974. Proteolytic cleavage of subunits of the nucleocapsid of the paramyxovirus simian virus 5. *J. Virol.* **14:**1253.

Scheid, A. and P. W. Choppin. 1973. Isolation and purification of the envelope proteins of Newcastle disease virus. *J. Virol.* **11:**263.

————. 1974a. Identification of biological activities of paramyxovirus glycoproteins. Activation of cell fusion, hemolysis, and infectivity by proteolytic cleavage of an inactive precursor protein of Sendai virus. *Virology* **57:**475.

————. 1974b. The hemagglutinating and neuraminidase (HN) protein of a paramyxovirus. Interaction with neuraminic acid in affinity chromatography. *Virology* **62:**125.

————. 1974c. Isolation of paramyxovirus glycoproteins and identification of their biological properties. In *Negative Strand Viruses* (ed. R. D. Barry and B. W. J. Mahy). Academic Press, New York (in press).

————. 1975. Multiple cycle replication of Sendai virus with trypsin and of Sendai virus mutants with chymotrypsin and elastase. *Virology* (in press).

Scheid, A., L. A. Caliguiri, R. W. Compans and P. W. Choppin. 1972. Isolation of paramyxovirus glycoproteins. Association of both hemagglutinating and neuraminidase activities with the larger SV5 glycoprotein. *Virology* **50:**640.

Seto, J. T., H. Becht and R. Rott. 1973. Isolation and purification of surface antigens from disrupted paramyxoviruses. *Z. Med. Mikrobiol. Immunol.* **159:**1.

Silverstein, S. C. and P. I. Marcus. 1964. Early stages of NDV-HeLa cell interaction: An electron microscopic study. *Virology* **23:**370.

Stone, H. O., D. W. Kingsbury and R. W. Darlington. 1972. Sendai virus-induced transcriptase from infected cells: Polypeptides in the transcriptive complex. *J. Virol.* **10:**1037.

Tozawa, H., M. Watanabe and N. Ishida. 1973. Structural components of Sendai virus. Serological and physicochemical characterization of hemagglutinin subunit associated with neuraminidase activity. *Virology* **55:**242.

Young, N. P. and R. J. Ash. 1970. Polykaryocyte induction by Newcastle disease virus propagated on different hosts. *J. Gen. Virol.* **7:**81.

Cleavage Associated with the Maturation of the Head of Bacteriophage T4

Ulrich K. Laemmli

Department of Biochemical Sciences, Princeton University
Princeton, New Jersey 08540

Many years ago, Crick and Watson (1956) suggested that small viruses contain a large number of identical subunits packed together in a regular manner. It also is clear that the number of efficient designs possible for biological containers constructed from identical protein molecules is very limited. The basic designs are helical tubes and icosahedral shells (Caspar and Klug 1962). Caspar and Klug's (1962) theory is based upon the idea that the protein coat of the virus is determined by the specific binding properties of identical subunits. The assembly of the virus is, therefore, a self-assembly process, which proceeds spontaneously under appropriate conditions. This concept has received ample support from the numerous demonstrations that simple viruses can be assembled in vitro from their components.

The more complex bacteriophages contain many different components, and it is evident—particularly in the bacteriophage T4—that the assembly of the major head protein, P23, is controlled by at least six gene products. The bonding specificity between P23 subunits is not stringent enough to determine the final product unambiguously. This is illustrated by the observation that P23 can assemble into a variety of structures, of different lengths and diameters, when one or another of the minor gene products is genetically eliminated. Thus control mechanisms must exist which select one assembly product from a number of energetically possible products. Moreover, many of the structural proteins of complex viruses (Korant, this volume) and bacteriophages are cleaved during the assembly process. The bonding properties of the structural subunits must change during this cleavage process, and it is unlikely that the mature particles could be reassembled following chemical disassociation into subunits.

In this paper, the major events of the assembly of the head of phage T4 are discussed, with emphasis on the possible role of the various cleavage processes associated with the maturation of the head.

Cleavage Associated with
Bacteriophage Assembly

Cleavage of structural proteins of the phage particles has been observed in phage T4 (Laemmli 1970; Dickson, Barnes and Eiserling 1970; Kellenberger and Kellenberger-Van der Kamp 1970; Hosoda and Cone 1970), λ (Murialdo and Siminovitch 1972; Hendrix and Casjens 1974), P2 (Lengyel et al. 1973) and T5 (Zweig and Cummings 1973). The first evidence for proteolytic cleavage associated with the assembly of the head of phage T4 came from work by Champe and collaborators (Champe and Eddleman 1967; Eddleman and Champe 1966). They showed that the acid-soluble internal peptides found in the head are derived from a large, acid-insoluble protein.

The proteins cleaved in phage T4 are listed in Table 1. The major head protein, P23 (MW 55,000), is cleaved into a protein called P23* (MW 45,000) during normal head maturation (Laemmli 1970; Dickson, Barnes and Eiserling 1970; Kellenberger and Kellenberger-Van der Kamp 1970; Hosoda and Cone 1970). The fragment (expected MW of about 10,000) cleaved off P23 is derived from the N-terminal end of P23 (Laemmli 1970; Celis, Smith and Brenner 1973), but apparently is not found in the mature phage, which suggests that this cleavage fragment is degraded to very small peptides or amino acids.

The product of gene 24 (P24) is a minor head protein and is cleaved during normal head formation from a 45,000- to a 43,000-dalton protein (Laemmli 1970). The protein P22 (MW 31,000) is cleaved to small fragments not detectable on polyacrylamide gels containing SDS (Laemmli 1970; Hosoda and Cone 1970). The protein P22 is the major component of the highly organized core (Laemmli, Paulson and Hitchins 1974) found in the precursor particles of the mature head and in some of the aberrant assembly products (Laemmli and Quittner 1974; Showe and Black 1973).

The fourth protein cleaved is the so-called internal protein (IPIII). This protein is basic (Black and Ahmad-Zadeh 1971) and a component of the core of the precursor particles (Laemmli and Quittner 1974). IPIII is cleaved from 23,000 to 20,000 daltons during head maturation (Laemmli

Table 1
Head Proteins Cleaved in Phage T4

Precursor	Product	Comments
P23 (MW 55,000)	P23* (MW 45,000)	coat proteins; fragment cleaved off N-terminal end
P24 (MW 45,000)	P24* (MW 43,000)	coat protein
P22 (MW 31,000)	internal peptides	core protein
IPIII (MW 23,000)	IPIII* (MW 21,000)	core protein

The molecular weights were estimated by electrophoresis in acrylamide gels containing SDS.

1970). The product IPIII* is found in the mature head and is supposedly associated with the DNA (Black and Ahmad-Zadeh 1971).

None of the small fragments cleaved off the various precursor proteins have been discovered so far, with the exception of the internal peptide II, which appears to be derived from R22 (Goldstein and Champe 1974).

Mutations in all genes (20, 21, 22, 23, 24, 31 and 40) involved in the early steps of head assembly block the cleavage of the head proteins (Laemmli 1970; Dickson, Barnes and Eiserling 1970). Many of these genes code for structural proteins of the head (Laemmli 1970) and determine the size and shape of the head, since mutation in any one of them (with the exception of gene 23) results in the accumulation of head-related assembly products (Epstein et al. 1963; Kellenberger 1966; Laemmli, Beguin and Gujer-Kellenberger 1970; Laemmli et al. 1970). We refer the reader to a recent review by Laemmli, Paulson and Hitchins (1974) for a fuller discussion of the phenotype and function of these genes. The various head-related, polymorphic structures produced when a particular gene is defective are shown in Figure 1. Amorphous aggregates ("lumps") of P23, which adhere to the cell membrane, are formed if gene 31 is defective (Kellenberger 1966; Laemmli et al. 1970). Open-ended tubular structures, called polyheads, accumulate if gene 20 or 40 is defective (Epstein et al. 1963; Kellenberger 1966; Laemmli et al. 1970). Multilayered polyheads consisting of concentric tubes of increasing diameter are observed if gene 22 is defective (Laemmli et al. 1970; Yanagida et al. 1970). In 21- or 24-defective cells (cells infected with a phage carrying a mutation in gene 21 or 24), so-called τ particles, accumulate (Epstein et al. 1963; Laemmli et al. 1970; Kellenberger, Eiserling and Boy de la Tour 1968). These particles are somewhat smaller than the mature T4 head and have a more rounded appearance (Kellenberger, Eiserling and Boy de la Tour 1968).

The fact that the cleavage of the proteins is blocked in these mutant-infected cells suggested that the head-related assembly products contain the precursor proteins (Fig. 1) (Laemmli 1970). Direct isolation of the particles has confirmed this prediction (Kellenberger, Kellenberger-Van der Kamp 1970; Laemmli and Quittner 1974; Laemmli and Johnson 1973a,b; Luftig and Lundh 1973). It is thus clear that (1) cleavage of the head proteins is not required for assembly of the various structures and occurs, as we will describe below, during the maturation of a precursor particle into the head, and (2) cleavage does not occur unless a structure is assembled with at least the gross morphology of the mature head. Furthermore, none of those precursor protein-containing particles package any detectable amount of DNA (Laemmli and Johnson 1973a,b; Luftig and Lundh 1973), which suggests that cleavage of the head proteins is necessary for the DNA-packaging process to occur.

The diagram in Figure 1 also illustrates the cooperation of the various genes in the assembly of different structures. Genes listed to the left of the arrow pointing to a particular structure are essential for the formation of this structure and genes to the right are not (Laemmli et al. 1970). This information permitted us tentatively to assign functions to some of the genes involved (Laemmli, Beguin and Gujer-Kellenberger 1970; Laemmli et al. 1970).

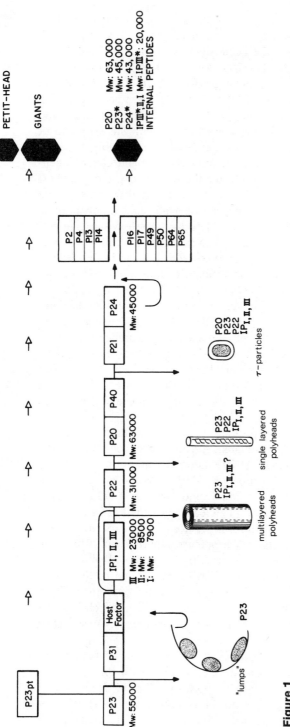

Figure 1

Cooperation of the genes in group A in the formation of the various head-related structures. This diagram illustrates the head-related structures produced when particular genes are defective. This is an updated version of an earlier presentation and includes work from different laboratories (see text). An arrow to the left of a box points to the observable result of a defect in that gene. Genes to the left of an arrow of a particular structure are essential for the formation of this structure and genes to the right are not. The internal protein genes (IPI, II and III) are the exception to this organization of the diagram, since their function is only absolutely required for multilayered polyhead formation (Showe and Black 1973). This is indicated by the line bypassing these genes. The structural components of the various structures are also listed.

Most of these structures are aberrant assembly products with the exception of the "lumps" and the τ particles found in 24-defective cells, as indicated by the arrow pointing away from these structures.

The box containing $P23_{pt}$ illustrates that mutations (pt) are known in gene 23 which result in the formation of shorter-than-normal (petit head) and giant heads (Doermann, Eiserling and Boehner 1973).

Genes 2, 4, 13, 14, 16, 17, 49, 50, 64 and 65 control later steps of head formation (see e.g., Laemmli, Paulson and Hitchins 1974).

664

In Vitro Cleavage of the Head Proteins

Compared to the elegant studies on various proteolytic enzymes reported in this book, the studies on the in vitro cleavage of the T4 head proteins are only beginning.

We have shown that partially purified polyheads, isolated from cells infected with a phage carrying a mutation in gene 20, contain the precursor proteins P23, P22 and the internal protein IPIII (Laemmli and Quittner 1974). Cleavage of the various head proteins occurs when these partially purified polyheads are incubated at 37°C (Laemmli and Quittner 1974). In a 100-minute incubation, about 60–70% of P23 (MW = 55,000) is converted to P23* (MW = 46,000), and a significant conversion of IPIII (MW = 23,000) to IPIII* (MW = 21,000) is observed. The protein P22 (MW = 31,000) disappears during this incubation and is supposedly cleaved to small fragments. An example of a cleavage of partially purified polyheads is shown in Figure 2.

The in vitro products P23* and IPIII* have the same molecular weight as the in vivo products, suggesting that the cleavage is specific (Laemmli and Quittner 1974). However, several protein fragments generated during the in vitro cleavage reaction have not been observed in vivo (Laemmli 1970). Appropriate mutant studies reveal that the products of gene 21 and 22 are

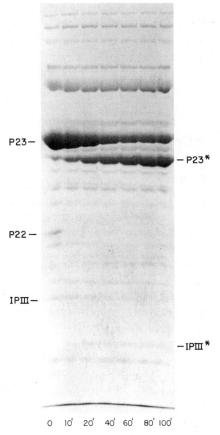

Figure 2

In vitro cleavage of the head proteins of partially purified polyheads. Partially purified polyheads were prepared from 20-defective cells as described by Laemmli and Quittner (1974). The sample of partially purified polyheads was made 10^{-3} M EDTA immediately prior to incubation at 37°C. Samples were removed at the following intervals: 0, 10, 20, 40, 60, 80, and 100 minutes, and prepared for electrophoresis on 10% acrylamide gel containing SDS.

required for the in vitro cleavage reactions. One of the genes may code for the proteolytic activity, or the absence of any of these gene products may alter the substrate (partially purified polyheads) such that the proteolytic factor cannot function.

Addition of chloroform to the reaction mixture inhibits the cleavage of partially purified polyheads (Laemmli and Quittner 1974). Chloroform extracts the core proteins from the polyheads, and the inhibitory action of chloroform could be explained by assuming that the proteolytic activity is part of the core proteins and is inactive or diluted if extracted from the polyheads. Our observation that genetic elimination of the core protein P22 also blocks the cleavage (Laemmli and Quittner 1974) is consistent with this interpretation, but by no means establishes this point. The partially purified polyheads used in this study are still heavily contaminated with bacterial and other phage proteins.

Onorato and Showe (1975) have shown that an activity which cleaves protein P22 is also associated with more highly purified particles isolated with the help of a temperature-sensitive mutant in gene 23. Interestingly, no cleavage activity was found associated with normal mature heads. They also demonstrated, in agreement with our results, that the cleavage activity is dependent on the presence of a wild-type gene 21 protein.

Goldstein and Champe (1974) also studied the in vitro cleavage of P22 in an in vitro system. They showed that the products formed in vitro include a peptide indistinguishable by several criteria from one of the internal peptides found in the mature phage. The formation of this peptide in vitro is dependent on a factor present in extracts of phage-infected cells, but missing in extracts of uninfected cells.

In summary, these experiments show that (1) protein accumulates in infected cells, coded for by gene 21, which either has proteolytic activity itself or activates a proteolytic enzyme; (2) the activity appears to be associated with particles containing the precursor proteins, but not with the mature head; and (3) no cleavage activity is found in wild-type extracts.

Pathway of Head Assembly

We have previously presented evidence (Laemmli 1970) suggesting that cleavage of the various proteins occurs following assembly of the proteins into a large structure. The observation that the various head-related structures contain the precursor proteins is certainly compatible with this result. Recently, we have been able to elucidate several steps of the maturation of the head in some detail (Laemmli and Favre 1973). The basic nature of the experiments which established this pathway is fractionation (on various gradients) of lysates of pulse-labeled wild-type and mutant-infected cells. The flow of the protein and/or DNA label through the various intermediates into the mature head is analyzed.

Our work shows that a succession of particles, prohead I (400S), prohead II (350S) and prohead III (550S), are precursors to the mature head (1100S) (Fig. 3). Proheads I and II contain no DNA; part of the DNA is packaged during the conversion of prohead II to prohead III, and the rest during the conversion of the latter particle to the mature head. The possible existence of an empty head as precursor to the full DNA-containing mature

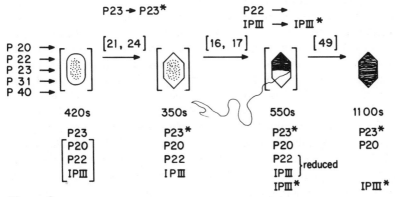

Figure 3

Tentative pathway of head maturation. The major features of the maturation of the head are outlined. The numbers over the arrows indicate the
genes which may be involved in controlling that step. The brackets indicate that the information is somewhat speculative and that we have no
hard data for the assignment of these genes to a particular step. It is
noteworthy to point out that cleavage is normal in 16⁻, 17⁻ and 49⁻
cells, whereas from the above scheme, these mutants would be expected
to block cleavage, at least of P22 and IPIII. The prefix P designates the
product of a particular gene, and the gene products found in each structure are listed. The list, however, is not complete.

head was proposed first by Luftig, Wood and Okinaka (1971) and by Simon
(1972). The cleavage of the head protein P23 to P23* largely precedes the
DNA-packaging event (Laemmli and Favre 1973), occurring during the
conversion of prohead I to prohead II, whereas cleavage of the two major
core proteins, P22 to small fragments and IPIII to IPIII*, appears to be
linked to the DNA-packaging event.

Recently, Bijlenga, Scraba and Kellenberger (1973) demonstrated that the
τ particles, which accumulate in cells infected with a temperature-sensitive
mutant in gene 24, can serve as precursors of the mature head. The τ particles
are very similar, probably identical, to prohead I particles seen in pulse-
labeled wild-type-infected cells (Laemmli and Favre 1973; Laemmli and
Johnson 1973b). The observation that the τ particles can serve as precursor
to the mature head (Bijlenga, Scraba and Kellenberger 1973) therefore
strongly supports the initial step of our pathway.

For these reasons, we have assigned gene 24 and possibly gene 21 to
control the prohead I to II conversion (Laemmli and Favre 1973). We have
also tentatively assigned genes 16 and 17 to control the prohead II to III
conversion (Laemmli and Favre 1973). Gene 49 is assigned to control the
prohead II to mature head conversion (Laemmli and Favre 1973) since this
step can be blocked reversibly with a temperature-sensitive mutant in gene 49
(Luftig, Wood and Okinaka 1971). It should be pointed out that the assignment of genes to a particular step based on the in vivo data is not a straightforward task (Laemmli, Paulson and Hitchins 1974), and therefore we consider these assignments as tentative.

Our experiment shows that prohead I and prohead II contain no DNA.

One could argue that both particles contain DNA inside the cell but lose DNA during the lysate preparation. To remove this doubt, we took advantage of the phenotype of temperature-sensitive mutants in gene 24 and 49. The τ particles (apparently identical to prohead I) and the 49-defective heads (apparently identical to prohead III) which accumulate in cells infected with the appropriate temperature-sensitive mutant in gene 24 and 49, respectively, can be converted to active phage particles if the gene products are activated by temperature shift (Bijlenga, Scraba and Kellenberger 1973; Luftig, Wood and Okinaka 1971). Both particles remain conserved and do not dissociate and reassemble during this conversion to the mature head (Luftig, Wood and Okinaka 1971; Bijlenga, Broek and Kellenberger 1974). This conversion also occurs if protein synthesis is blocked with chloramphenicol. It was therefore possible to determine the uptake of DNA synthesized following activation of P24 or P49 into particles made prior to P24 and P49 activation (Laemmli, Teaff and D'Ambrosia 1974). We found that at least 80% of the total DNA complement is packaged into τ particles following P24 activation as these particles are converted to mature heads (Laemmli, Teaff and D'Ambrosia 1974). Thus the τ particles contain very little or no DNA prior to P24 activation. The 49-defective heads package, on the average, at least 25% of the DNA complement following P49 activation (Laemmli, Teaff and D'Ambrosia 1974). This lower value was expected since the 49-defective heads are about half filled with DNA prior to P49 activation (Laemmli and Favre 1973). These experiments show that the DNA is packed into preformed heads and strongly support the pathway of head maturation (Laemmli and Favre 1973).

Structure of the Polyhead Core

The protein P23 is the major constituent of the outer coat of polyheads, whereas the proteins P22, IPIII and IPII are the constituents of the core seen in the inside of polyheads. Little information has been available about the organization of the core of the polyheads. This is due to the lability of the core, which breaks down into amorphous clumps of material or falls out of the polyhead completely when polyheads are isolated by conventional means and viewed by negative staining in the electron microscope. Some structure of low resolution is seen in the inside of polyheads examined in thin sections of cells (Kellenberger, Eiserling and Boy de la Tour 1968).

Information on the spatial organization of the core is of great importance since we know that (1) the protein P22 is involved in the initiation and size determination of the head (Laemmli et al. 1970), and (2) cleavage of the core protein P22 appears to be linked to the DNA-packaging process (Laemmli and Favre 1973).

We have developed a lysis procedure which permits the isolation of polyheads with highly organized cores (Laemmli, Paulson and Hitchins 1974) and have solved the basic structure of the core with a combination of optical diffraction (Klug and De Rosier 1966) and optical superposition methods (Markam et al. 1964).

From optical diffraction we learned that the unit cell of the surface lattice of core-containing polyheads is 110 Å, in agreement with the reported lattice of empty polyheads (Yanagida, De Rosier and Klug 1972). The core-containing polyheads show an additional feature (not seen in empty polyheads) in the form of four strong axial reflections of about 137 Å, due to the core (Laemmli, Paulson and Hitchins 1974).

In order to verify this interpretation of the optical transform and to obtain an improved picture of the polyhead core, we used the optical superposition method.

Figure 4a, shows the original image. The polyhead on the right was studied. The most prominent features of the polyhead surface are the striations spaced 55 Å apart seen along the edge of the particle. (We do not expect to see surface features over the center of the particle because the contrast provided by the surface may be obliterated by the much more heavily stained core.) We expect that a superposition image produced by translating by the surface repeat should enhance these surface features. Indeed, in Figure 4b the superposition was performed in increments corresponding to 110 Å, and the repeats along the edge are enhanced. The core has been partly suppressed (it does not show up as clearly as in Figure 4c, see below), but not completely. This is because of the small number of superpositions used in making the image and because of the high contrast in the core.

Similarly, an optical superposition image produced by translation of the original in the axial direction by the core repeat of 137 Å (Fig. 4c) should enhance the structure of the core. It is evident that the core consists of several

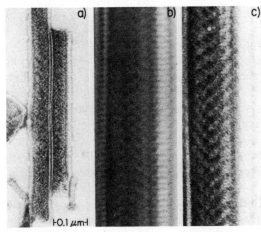

Figure 4

(a) Polyheads were prepared as described elsewhere (Laemmli, Paulson and Hitchins 1974). The polyhead on the right was used to make the superimposed polyhead images in (b) and (c). Note the cross-hatched or "herringbone" structure in the core and the striations along the edge of the particle. In (b), the image was translated by a distance corresponding to 110 Å and superimposed 10 times. In (c), the image was translated by a distance corresponding to 137 Å and superimposed 6 times.

helices spaced about 137 Å apart. The repetitious features along the edge, however, are smeared out, showing that the surface does not have this periodicity. The image of the core varies in its two sidedness; in some places, stain has contrasted both the near and far sides, giving a crosshatched appearance to the core, whereas in other places the image of one side predominates. The helices appear to be wound about a hollow center of approximately 150 Å in diameter. We have tried to determine the number of helically wound chains in the core structure, but the calculation is subject to uncertainty because we cannot determine accurately the dimensions of the core. Our best estimates indicate that there are between four and six helically wound chains (Laemmli, Paulson and Hitchins 1974).

A Possible DNA-packaging Mechanism

The core protein P22 found in the interior of preformed precursor particles is cleaved to small fragments in concert with the DNA-packaging event (Laemmli 1970; Laemmli and Favre 1973). At least one of the cleavage fragments appears to be the internal peptide VII (S. P. Champe, pers. comm.). We assume that the other internal peptide is also derived from P22, although this has not been rigorously established. These peptides were first discovered by Hershey (1957) and studied in detail by Champe and Eddleman (1967; Eddleman and Champe 1966). An extraordinary characteristic of these peptides is their amino composition. Both peptides are highly acidic; 80% of the amino acid residues are glutamic or aspartic acid for peptide II, 48% for peptide VII (Eddleman and Champe 1966). This led us to think that these acidic peptides might collapse DNA by a repulsive interaction in much the same way as polyacrylate.

That DNA can be collapsed by a repulsive interaction has been demonstrated by Lerman and his collaborators (Lerman 1971, 1973; Jordan, Lerman and Venable 1972). They have shown that in a salt solution containing a sufficient concentration of a simple polymer like polyethylene oxide or polyacrylate, high molecular weight DNA undergoes a cooperative structural transition which results in a very compact configuration. The salt is required to shield the negative charges of the DNA. Lerman's circular dichroism, X ray and birefringence show that the collapsed DNA is highly ordered and compact.

Our recent experiments established that T4 DNA collapses into fast-sedimenting structures above a critical concentration of polyglutamic acid, polyaspartic acid and of the internal peptides (Laemmli, Paulson and Hitchins 1974).

The repulsive interactions between the various polymers is nonspecific (Lerman 1973), and it is reasonable to assume that the structure of the DNA collapsed by the various polymers is similar. We have for this reason studied the structure of DNA collapsed with polyethylene oxide (PeO), a neutral polymer which is more suitable for electron microscopy (Laemmli, Paulson and Hitchins 1974).

Our studies showed that T4 DNA is collapsed in PeO into particles approaching the size of the head of the phage, in good agreement with Lerman's (1971) physiocochemical studies. Electron micrographs of these particles are shown in Figure 5. The DNA particles were negatively stained, and a 25-Å

Figure 5
Particles of polyethylene oxide-collapsed DNA. (*a, b, c*) Electron micrographs of PeO-collapsed T4 DNA; negatively stained with uranyl acetate (Laemmli, Paulson and Hitchins 1974). Magnification 145,000 ×. (*d*) Electron micrograph of a PeO-collapsed T7 DNA prepared with the cytochromic spreading technique in the presence of polyethylene oxide (details to be described elsewhere). Magnification 40,500 ×.

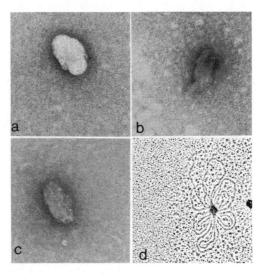

striation tangential to the particle is recognizable. These particles may be best described as a ball of string. Of course, the electron micrograph does not reveal the orientation of the DNA in the different layers.

We cannot see the individual DNA strand which might surround the particles and we have for this reason examined PeO-collapsed DNA by the cytochromic spreading technique. We found that all the collapsed DNA particles examined contain a central core surrounded by loops of DNA originating from the core (Fig. 5d). The amount of DNA in the loop varies between 10–30% of the total DNA, as determined by tracing the length of DNA in loops. No substructures are observed in the central core, but we think that the central core corresponds to the particles seen above by negative staining.

How would a repulsive interaction between the internal peptides and the DNA package the DNA into the head? In our model, we propose that one end of the DNA becomes firmly fixed to protein in the inside of the precursor head, which contains the uncleaved P22 protein (Fig. 6). Cleavage of P22

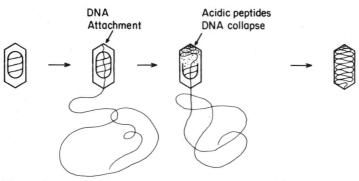

DNA
Attachment

Acidic peptides
DNA collapse

Figure 6
Model of a possible DNA-packaging mechanism. This figure tries to illustrate the model discussed in the text.

then commences, generating locally a high concentration of acidic peptides which collapse the DNA. As the DNA collapses in the interior, it will exert a pulling action on the external part of the DNA since the end is firmly attached inside the head. Thus as more and more acidic internal peptides are generated, the rest of the DNA (a head-size piece) is drawn into the head. The attractive feature of this model is that DNA is pulled into the head as a consequence of the collapse of the DNA already inside the head.

The repulsive collapse of the DNA induced by these various polymers requires a critical high concentration of salt, to shield the negative charges of the DNA. The DNA inside the head is known to be largely complexed with polyamine (Ames and Dubin 1960), which may play the role of the salt required in the vitro system.

We have discussed elsewhere whether such a model is reasonable (Laemmli, Paulson and Hitchins 1974). The major problem arises from calculation of the internal peptide concentration inside the head. The internal peptide concentration, assuming that the internal peptides have access to the total interior volume of the head, is about 10–20-fold below the critical concentration required for DNA collapse. But there are several possibilities which could increase the effective peptide concentration many times (Laemmli, Paulson and Hitchins 1974).

We have also discussed other possibilities by which the DNA could be collapsed in the interior of the head (Laemmli, Paulson and Hitchins 1974), the major ones being titration with a basic protein or sequential binding to a DNA-binding protein. In this case, structural proteins should be found which have the capacity to bind to DNA. The internal proteins are known to bind to DNA (Minagawa 1961), but recent experiments demonstrate that the internal proteins are nonessential proteins of the head (Black 1974). This appears to eliminate their possible role in the DNA-packaging process, and other DNA-binding proteins have not been found.

Nevertheless, we have studied the structure of DNA collapsed with the basic peptide polylysine in order to compare it with the PeO-collapsed DNA particles. Olins and Olins (1971) showed that DNA collapsed with polylysine appears in the same preparation as either donuts or short-term structures. Our studies confirm this (Laemmli, Paulson and Hitchins 1974), but we used negative staining and a different method to absorb the DNA to the electron microscope grid. This method permits visualization of the DNA strand in the folded complexes. Figure 7 shows the donut and stemlike structures of polylysine-collapsed DNA. The DNA, visible as a 25-Å repeating striation, is closely packed in the stem structure and is probably folded back and forth in a pleated structure since the total length of the DNA used is many times longer than that of the stem (Laemmli, Paulson and Hitchins 1974). The DNA appears to be radially distributed as a spiral in the donut structures. We estimate that both donut and stem structures are about 3–4 layers thick, assuming close packing of the DNA strand (Laemmli, Paulson and Hitchins 1974).

Recently, Richards, Williams and Calendar (1973) have obtained electron micrographs of several coliphages indicating the way the DNA is arranged within the head. These showed that the DNA appears as a set of concentric circles, as on a tightly wound spiral. Two models were proposed: in one, the

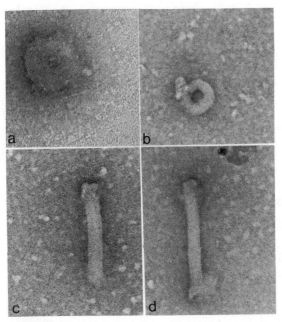

Figure 7

Donut and stem structures seen in a preparation of poly-D-lysine-collapsed DNA. (*a*) A donut structure of T4 DNA; (*c*) a donut structure of T7 DNA. (*c, d*) Representative stem structures seen in T7 DNA. The preparations were negatively stained and prepared as described by Laemmli, Paulson and Hitchins (1974). Magnification 81,000 ×.

DNA is packaged like a ball of string; in the other, it is wound coaxially like a spool.

Our electron micrographs of DNA collapsed with PeO are compatible with both models. This observation supports the idea that the DNA may be collapsed by a repulsive interaction within the head, but does not rule out a different mechanism.

Acknowledgments

This research was supported by Grant 601-617-18776 from the National Institutes of Health and Grant GB-413-40 from the National Science Foundation.

REFERENCES

Ames, B. N. and D. T. Dubin. 1960. The role of polyamines in the neutralization of bacteriophage deoxyribonucleic acid. *J. Biol. Chem.* **235**:769.

Bijlenga, R. K. L., R. Broek and E. Kellenberger. 1974. The transformation of τ-particles into T4 heads. Evidence for the conservative mode of this transformation. *J. Supramol. Struct.* **2**:45.

Bijlenga, R. K. L., D. Scraba and E. Kellenberger. 1973. Studies on the morphopoiesis of the head of T-even phage. IX. τ-Particles: Their morphology, kinetics of appearance and possible precursor function. *Virology* **56**:250.

Black, L. W. 1974. Bacteriophage T4 internal protein mutants: Isolation and properties. *Virology* **60**:166.

Black, L. W. and C. Ahmad-Zadeh. 1971. Internal proteins of bacteriophage T4D: Their characterization and relation to head structure and assembly. *J. Mol. Biol.* **57**:71.

Caspar, D. L. D. and A. Klug. 1962. Physical principles in the construction of regular viruses. *Cold Spring Harbor Symp. Quant. Biol.* **27**:1.

Celis, J. E., J. D. Smith and S. Brenner. 1973. Correlation between genetic and translational maps of gene 23 in bacteriophage T4. *Nature* **241**:130.

Champe, S. P. and H. L. Eddleman. 1967. Polypeptides associated with morphogenic defects in bacteriophage T4. In *Molecular Biology of Viruses* (ed. J. S. Colter and W. Paranchych), p. 55. Academic Press, New York.

Crick, F. H. C. and J. D. Watson. 1956. Structure of small viruses. *Nature* **177**: 473.

Dickson, R. C., S. L. Barnes and F. A. Eiserling. 1970. Structural proteins of bacteriophage T4. *J. Mol. Biol.* **53**:461.

Doermann, A. H., F. A. Eiserling and L. Boehner. 1973. Genetic control of capsid length in bacteriophage T4. I. Isolation and preliminary description of four new mutants. *J. Virol.* **12**:374.

Eddleman, H. L. and S. P. Champe. 1966. Components in T4-infected cells associated with phage assembly. *Virology* **30**:471.

Epstein, R. H., A. Bolle, C. M. Steinberg, E. Kellenberger, E. Boy de la Tour, R. Chevalley, R. S. Edgar, M. Susman, G. H. Denhardt and A. Lielausis. 1963. Physiological studies of conditional lethal mutants of bacteriophage T4D. *Cold Spring Harbor Symp. Quant. Biol.* **28**:375.

Goldstein, J. and S. P. Champe. 1974. T4 Phage induced activity required for specific cleavage of a phage protein *in vitro*. *J. Virol.* **13**:419.

Hendrix, R. W. and S. R. Casjens. 1974. Protein cleavage in bacteriophage λ tail assembly. *Virology* **61**:156.

Hershey, A. D. 1957. Some minor components of bacteriophage T2 particles. *Virology* **4**:237.

Hosoda, J. and R. Cone. 1970. Analysis of T4 proteins. I. Conversion of precursor proteins into lower molecular weight peptides during normal capsid formation. *Proc. Nat. Acad. Sci.* **66**:1275.

Jordan, C. F., L. S. Lerman and T. H. Venable, Jr. 1972. The structure and circular dichroism of DNA in concentrated polymer solutions. *Nature* **232**:67.

Kellenberger, E. 1966. Control mechanisms in bacteriophage morphopoiesis. *Ciba Foundation Symposium: Principles of Biomolecular Organization* (ed. G. E. Wolstenholme), pp. 192–226. Churchill Livingston, London.

Kellenberger, E., F. A. Eiserling and E. Boy de la Tour. 1968. Studies on the morphopoieses of the head of phage T-even. III. The cores of head-related structure. *Ultrastruct. Res.* **21**:335.

Kellenberger, E. and C. Kellenberger-Van der Kamp. 1970. On a modification of the gene product P23 according to its use as subunit of either normal capsids of phage T4 or of polyheads. *FEBS Letters* **8**:140.

Klug, A. and D. J. De Rosier. 1966. Optical filtering of electron micrographs: Reconstruction of one-sided images. *Nature* **212**:29.

Laemmli, U. K. 1970. Cleavage of structural proteins during the assembly of the head of bacteriophage T4. *Nature* **227**:680.

Laemmli, U. K. and M. Favre. 1973. Maturation of the head of bacteriophage T4. I. DNA packaging events. *J. Mol. Biol.* **80:**575.

Laemmli, U. K. and R. A. Johnson. 1973a. Maturation of the head of bacteriophage T4. II. Head-related, aberrant τ-particles. *J. Mol. Biol.* **80:**601.

————. 1973b. Bacteriophage T4 head assembly: Studies on the aberrant τ-particles. In *Virus Research: Proceedings of the 1973 I.C.N.—U.C.L.A. Symposium on Molecular Biology.* (ed. C. F. Fox), p. 279. Academic Press, New York.

Laemmli, U. K. and S. F. Quittner. 1974. Maturation of the head of bacteriophage T4. IV. The proteins of the transient head core and *in vitro* cleavage of the head protein. *Virology* **62:**483.

Laemmli, U. K., F. Beguin and G. Gujer-Kellenberger. 1970. A factor preventing the major head protein of bacteriophage T4 from random aggregation. *J. Mol. Biol.* **47:**69.

Laemmli, U. K., J. R. Paulson and V. Hitchins. 1974. Maturation of the head of bacteriophage T4. V. A possible DNA packaging mechanism: *In vitro* cleavage of the head proteins and the structure of the core of the polyheads. *J. Supramol. Struct.* **2:**276.

Laemmli, U. K., N. Teaff and J. D'Ambrosia. 1974. Maturation of the head of bacteriophage T4. III. DNA packaging into preformed heads. *J. Mol. Biol.* **88:**749.

Laemmli, U. K., E. Molbert, M. Showe and E. Kellenberger. 1970. Form-determining function of the genes required for the assembly of the head of bacteriophage T4. *J. Mol. Biol.* **49:**99.

Lengyel, J., R. Goldstein, M. Marsh, M. Sunshine and R. Calendar. 1973. Bacteriophage P2 head morphogenesis: Cleavage of the major capsid protein. *Virology* **53:**1.

Lerman, L. S. 1971. A transition to a compact form of DNA in polymer solutions (poly[vinyl pyrrolidinone]/poly[ethylene oxide]/T4phage/centrifugation). *Proc. Nat. Acad. Sci.* **68:**1886.

————. 1973. Chromosomal analogues: Long-range order in ψ-condensed DNA. *Cold Spring Harbor Symp. Quant. Biol.* **38:**59.

Luftig, R. B. and N. P. Lundh. 1973. Bacteriophage T4 head morphogenesis. Isolation, partial characterization and fate of gene 21-defective τ-particles. *Proc. Nat. Acad. Sci.* **70:**1636.

Luftig, R. B., W. B. Wood and R. Okinaka. 1971. Bacteriophage T4 head morphogenesis. I. Evidence for unfilled head membranes as intermediates. *J. Mol. Biol.* **57:**555.

Markam, R., J. H. Hitchborn, G. T. Hills and S. Freg. 1964. The anatomy of the tobacco mosaic virus. *Virology* **22:**342.

Minagawa, T. 1961. Some characteristics of the internal protein phage T2. *Virology* **13:**515.

Murialdo, H. and L. Siminovitch. 1972. The morphogenesis of bacteriophage λ. IV. Identification of gene products and control of the expression of the morphogenetic information. *Virology* **48:**785.

Olins, D. E. and A. L. Olins. 1971. Model nucleohistones: The interaction of F1 and F2a1 histones with native T7 DNA. *J. Mol. Biol.* **57:**437.

Onorato, L. and M. K. Showe. 1975. Gene 21 protein dependent proteolysis *in vitro* of purified gene 22 product of bacteriophage T4. *J. Mol. Biol.* **92:**395.

Richards, K. E., R. C. Williams and R. Calendar. 1973. Mode of DNA packing within bacteriophage heads. *J. Mol. Biol.* **78:**255.

Showe, M. and L. Black. 1973. Assembly core of bacteriophage T4: On intermediate in head formation. *Nature New Biol.* **242:**70.

Simon, L. D. 1972. Infection of *Escherichia coli* by T2 and T4 bacteriophages as seen in the electron microscope: T4 head morphogenesis. *Proc. Nat. Acad. Sci.* **69:**907.

Yanagida, M., E. Boy de la Tour, C. Alff-Steinberger and E. Kellenberger. 1970. Studies on the morphopoiesis of the head of bacteriophage T-even. VIII. Multi-layered polyheads. *J. Mol. Biol.* **50:**35.

Yanagida, M., D. J. De Rosier and A. Klug. 1972. The structure of the tubular variants of the head of bacteriophage T4 (polyheads). *J. Mol. Biol.* **65:**489.

Zweig, M. and D. Cummings. 1973. Cleavage of head and tail proteins during bacteriophage T5 assembly: Selective host involvement in the cleavage of a tail protein. *J. Mol. Biol.* **80:**505.

Introductory Overview of the Participation of Proteinases and Their Regulators in Mammalian Reproductive Physiology

H. Guy Williams-Ashman

Ben May Laboratory for Cancer Research, Department of Biochemistry,
and Department of Pharmacological and Physiological Sciences
University of Chicago, Chicago, Illinois 60637

The contributions to follow deal with proteolytic processes related to some arbitrarily selected aspects of reproduction in eutherian mammals, which exhibit distinctive modes of internal fertilization and embryo development. These presentations attempt not only to summarize what is already known about proteinases and their regulators in the tissues and secretions of male and female reproduction tracts, but also highlight certain mammalian sexual phenomena that may merit greater scrutiny from the standpoint of involvement of specific proteolytic reactions.

Since many features of the entire field could not be covered in the limited number of papers that follow, this brief introduction attempts to outline the major steps in gamete production, transport and union in which proteinases are implicated and to define certain terms. (For pertinent reviews on mammalian reproductive physiology see also Price and Williams-Ashman 1961; Mann 1964; Williams-Ashman and Hoskins 1969; Bedford 1970; Greep 1973).

Formation and Maturation of Spermatozoa and the Ejaculation of Semen

Spermatogenesis in the mammalian testis produces many millions of spermatozoa every day. It is a complex and lengthy process. The overall time normally required for a stem cell spermatogonium to differentiate into a spermatozoon in the lumen of a seminiferous tubule is rather species dependent; in man it takes roughly 70 days. A paramount intermediary process is the meiosis accomplished by spermatocytes. After meiosis is completed and under conditions where virtually all nuclear DNA replication or transcription is prohibited, the newly formed haploid spermatid gradually differentiates into a spermatozoon. The latter process of spermiogenesis involves, among other things, condensation of the chromatin of the early spermatid cell nucleus, the formation of the acrosome ("head cap"), midpiece and flagellum of the

677

spermatozoon, and the production of certain sperm-cell-specific proteins that seem to be absent from all tissues except spermatids and spermatozoa (Dixon and Smith 1968; Kistler, Geroch and Williams-Ashman 1973). There also takes place during spermiogenesis a selective breakdown of many constituents of early spermatids, notably of histones and many other proteins. The biosynthesis, posttranslational modifications and turnover of nuclear basic proteins by various cell types of the testicular germinal epithelium have been studied with regard to the replacement of histones by newly synthesized protamines during spermiogenesis in salmonoid fish (Dixon and Smith 1968; Louie and Dixon 1972). However, virtually nothing is known about the selective proteolysis of histones and other proteins during spermiogenesis in *mammalian* testes, which also entails the elaboration of arginine- and cysteine-rich proteins that eventually come in union with the DNA of mammalian sperm cell head-pieces; these mammalian sperm head-piece proteins differ markedly from the protamines of salmonoid fish (Kistler, Geroch and Williams-Ashman 1973).

Testicular spermatozoa are transported via the rete testis to the epididymis. During their slow passage through the tiny lumen of this long coiled tube, which is comprised of a number of morphologically and functionally distinct regions, the spermatozoa undergo maturational changes that are poorly understood from a mechanistic standpoint, but which, in some species, result in the acquisition by the sperm cells of the attributes of motility and the potentiality to fertilize eggs. Spermatozoa from the testis, as well as from the head or tail of the epididymis, contain a powerful proteinase of trypticlike specificity (acrosin) and apparently at least one other proteinase.

At ejaculation, contractions of the vas deferens propel spermatozoa from the epididymis into the urethra. Ejaculation also involves a concurrent squirting into the urethra of secretions of the male accessory genital glands, such as the seminal vesicles and the prostate and bulbourethral glands. In many species, the expulsion of these male genital fluids takes place during normal ejaculation in a definite order relative to the entry of sperm cells into the urethra. Male accessory genital gland secretions contain a variety of proteolytic enzymes, together with a number of proteinase inhibitors and activators. Prostatic and vesicular secretions also provide most of the bulk proteins and nonproteolytic enzymes, as well as water and other small molecules found in the fluid portion of semen or seminal plasma. The activity of proteinases in seminal plasma in some species is remarkably intense. For example, these enzymes are responsible for the fivefold or greater increase in nonprotein nitrogen and the marked changes in the electrophoretic profiles of seminal plasma bulk proteins that occur when human semen is allowed to stand for an hour or so at room or body temperature (Mann 1964). The functional significance of the multiplicity of proteolytic enzymes and their regulators in seminal plasma (Mann 1964; Williams-Ashman and Hoskins 1969) is obscure. As pointed out by Zaneveld, Polakoski and Schumacher (this volume), proteinase inhibitors associated with the acrosomes of ejaculated spermatozoa seem to be derived from seminal plasma and to be added to sperm cells when they mix prostatic and vesicular secretions during the ejaculatory process. Among various suggestions for functions of proteinases in seminal plasma of some species are (1) involvement in the postejaculatory liquefaction of normal human semen, (2) a role in the negotiation of cervical mucus by spermatozoa, and (3) formation of vasoactive small polypeptides as a result of the action of seminal plasma porteinases on bulk proteins in this fluid. The available evidence in favor of any of these hypotheses is not extremely compelling. The acrosin of the sperm-head acrosome is, however, clearly involved in the penetration of spermatozoa through the zona pellucida of mammalian eggs.

Transport of Spermatozoa to the Sites of Fertilization in the Female Genital Apparatus

Before spermatozoa deposited in the mammalian female genital tract can reach the sites of fertilization in the upper regions of the Fallopian tubes, the sperm cells must traverse a number of barriers, including the uterine cervix and the uterotubal junction. In many species, these obstacles are partly responsible for the fact that only a small fraction of the millions of spermatozoa ejaculated into the female ever rendezvous with fertilizable ova.

In some mammals, such as pigs, horses, dogs and myomorph rodents, the mechanics of coitus are such that many of the spermatozoa deposited into the female are catapulted with extreme rapidity through the cervix into the uterine lumen, so that passage of countless sperm cells along the cervical canal is not a very tortuous happening. It may be appropriate to mention here that in rats, mice and guinea pigs (as well as in certain nonhuman primates where the uterine cervix is a major barrier to sperm transport), the semen clots into a hard rubbery mass almost immediately after ejaculation. This coagulation of semen seems to be due, at least in the guinea pig, to the formation of γ-glutamyl-ε-lysyl iso-dipeptide bonds between chains of a basic protein in the seminal plasma that is derived from seminal vesicle secretions; the cross-linking reaction is apparently catalyzed by a transamidase- (transglutaminase-) like enzyme secreted by the anterior prostate (Williams-Ashman et al. 1972). In some rodent species, the postejaculatory clotting of semen gives rise to a copulation plug that occludes the vagina, thus preventing a backflow of semen from the vaginal barrel after coitus. The copulation plug later drops out of the vagina after it is loosened by phagocytic and proteolytic actions of leukocytes present in the vaginal lumen.

In other species, including rabbits, cows, sheep and humans, the semen collects at coitus as a pool in the vagina (Bedford 1970). So before the spermatozoa can enter the uterus, they must negotiate the cervix, which is packed with a mucus whose composition and consistency varies profoundly with the various phases of the estrus or menstrual cycles. Extensive investigations on human cervical mucus have disclosed that its content of glycoproteins, and many other substances including certain proteinase inhibitors, is under cyclic regulation by ovarian hormones (Schumacher 1970).

Once mammalian spermatozoa have gained access to the uterine lumen, they travel, with the aid of contractions of the myometrium and myosalpinx, to the ampullae of the Fallopian tubes (oviducts), where fertilization takes place. The uterus and oviduct secrete specific fluids which in various ways assist sperm transport and which are known to contain certain proteinases and proteinase inhibitors. In many mammalian species, freshly ejaculated spermatozoa are incapable of fertilizing recently ovulated eggs unless the sperm cells have been allowed some prior period of contact with uterine and/or oviductal secretions. The latter phenomenon is often called *capacitation* of spermatozoa. As discussed in this volume by Zaneveld, Polakoski and Schumacher and by Fritz, capacitation may involve, at least in part, a nullification by female genital secretions of the actions of "decapacitating" factor(s) derived from seminal plasma which coat the spermatozoa during ejaculation. The relationship of such "decapacitating" factor(s) to inhibitors of acrosin, or to effective catalysis by other sperm cell constituents such as the so-called corona-penetrating enzyme, is still a rather foggy subject.

Ovulation

In mammals, primordial follicles containing oocytes differentiate in the fetal ovary. Most of these follicles are lost during postnatal life by the poorly understood degenerative process known as follicular atresia. However from puberty until the end of the reproductive lifetime of the female, a small number of ovarian follicles develop during each estrus or menstrual cycle. The maturing follicles develop a fluid-filled antrum prior to ovulation. As considered by Espey (this volume), the bursting of Graafian follicles at ovulation does not seem to be evoked solely by the mounting pressure of fluid in the follicular antrum, but rather may be a more complex event involving the action of various proteinases.

Fertilization, Transport of the Early Embryo and Implantation

Eggs liberated from the Graafian follicle at ovulation are bounded by a viscous envelope called the zona pellucida, which is in turn surrounded by a layer of columnar cells known as the corona radiata, outside of which are attached the cells of the cumulus oophorus. The ova plus the latter attachments are captured by the fimbria of the oviduct and then enter the ampullae of the Fallopian tubes, where fertilization occurs if properly capacitated spermatozoa have recently arrived in that district of the female genital tract. Sperm cell hyaluronidase may be involved in dispersion of the cumulus oophorus, and a distinct corona-penetrating enzyme aids the burrowing of the spermatozoa to the outer surface of the zona pellucida. Passage of sperm cells through the zona pellucida involves the action of acrosin, and possibly other acrosomal enzymes as well.

After fertilization, the sperm cell head-piece, now inside the ovum, swells and decondenses, a process that could conceivably entail a selective breakdown of sperm head proteins (Bedford 1970). Following the formation and subsequent syngamy of male and female pronuclei, the zygote then begins to cleave as it passes down the oviduct. In many mammalian species, the zygote develops as far as a blastocyst before it reaches the endometrium of the uterus. Implantation and the formation of a placenta are initiated when the trophoblast cells of the blastocyst make contact with the endometrium of very early pregnancy. Recent experiments in the laboratories of Mintz, Denker and others (see Pinsker, Sacco and Mintz 1974) have indicated that both uterine fluid and trophoblast proteinases are required for attachment and invasion of the blastocyst into the uterine endometrium.

REFERENCES

Bedford, J. M. 1970. The saga of mammalian sperm from ejaculation to syngamy. In *Mammalian Reproduction*, 21st Mosbach Colloquium, Der Gesellschaft für Biologische Chemie (ed. H. Gibian and E. J. Plotz), pp. 124–182. Springer-Verlag, New York.

Dixon, G. H. and M. Smith. 1968. Nucleic acids and protamine in Salmon testes. *Prog. Nucleic Acid Res. Mol. Biol.* **8**:9.

Greep, R. O. 1973. Female reproductive system, parts 1 and 2. In *Handbook of Physiology, Section 7: Endocrinology* (ed. R. O. Greep and E. B. Astwood). American Physiological Society, Washington, D.C.

Kistler, W. S., M. E. Geroch and H. G. Williams-Ashman. 1973. Specific basic proteins from mammalian testes: Isolation and properties of small basic proteins from rat testes and epididymal spermatozoa. *J. Biol. Chem.* **248**:4532.

Louie, A. J. and G. H. Dixon. 1972. Trout testis cells. II. Synthesis and phosphorylation of histones and protamines in different cell types. *J. Biol. Chem.* **247**: 5498.

Mann, T. 1964. *The Biochemistry of Semen and of the Male Reproductive Tract* (2nd ed.) pp. 1–493. Wiley, New York.

Pinsker, M. C., A. G. Sacco and B. Mintz. 1974. Implantation-associated proteinase in mouse uterine fluid. *Devel. Biol.* **38**:285.

Price, D. and H. G. Williams-Ashman. 1961. The accessory reproductive glands of mammals. In *Sex and Internal Secretions* (ed. W. C. Young), 3rd ed., vol. 1, p. 366. Williams and Wilkins, Baltimore.

Schumacher, G. F. B. 1970. Biochemistry of cervical mucus. *Fertil. Steril.* **21**: 697.

Williams-Ashman, H. G. and D. D. Hoskins. 1969. Biochemistry of male accessory genital secretions in man and non-human primates. In *Urinary Infections in the Male,* p. 61. U.S. National Academy of Sciences, Washington, D.C.

Williams-Ashman, H. G., S. S. Pabalan, A. C. Notides and L. Lorand. 1972. Transamidase reactions involved in the enzymic coagulation of semen: Isolation of γ-glutamyl-ε-lysine dipeptide from clotted secretion protein of guinea pig seminal vesicle. *Proc. Nat. Acad. Sci.* **69**:2322.

The Proteolytic Enzyme Systems of Mammalian Genital Tract Secretions and Spermatozoa

Lourens J. D. Zaneveld

The Population Research Center
Medical Sciences and Engineering Division, IIT Research Institute
Chicago, Illinois 60616

Kenneth L. Polakoski

Department of Obstetrics and Gynecology, Washington University
St. Louis, Missouri 36110

Gebhard F. B. Schumacher

Laboratory for Reproductive Biochemistry and Immunology
Department of Obstetrics and Gynecology, University of Chicago
Chicago, Illinois 60637

INTRODUCTION AND HISTORICAL REVIEW

The male reproductive system consists of the following components: (1) the testis, where spermatozoa are formed within the seminiferous tubules through a process called spermatogenesis; (2) the epididymis, where the spermatozoa are stored before ejaculation; (3) the vas deferens, a tubular organ which connects the epididymis to the urethra; (4) the urethra, which is an excretory duct that extends from the bladder neck through the penis and exits exteriorly; (5) the accessory sex glands, including the ampulla of the vas deferens, the seminal vesicles, the prostate gland and the Cowper's or bulbourethral gland; and (6) the spermatozoa. Together the secretions from the accessory sex glands and from the epididymis are called the seminal plasma of the ejaculate.

Spermatozoa consist of a head and a tail, which is used for locomotion. The central portion of the head is called the nucleus and contains chromosomal material, mostly DNA and basic proteins. The anterior portion of the nucleus is covered with membranes which form a saclike structure, the acrosome. Posteriorly, the head is covered by a single membrane, called the postnuclear cap. The equatorial segment is the area of overlap between the acrosome and the postnuclear cap. The entire spermatozoon is covered by the plasma membrane.

The acrosome of the spermatozoon is formed through fusion of proacrosomal granules which are produced by the Golgi apparatus and contain a number of hydrolytic enzymes. These enzymes are quite similar to those of lysosomes, although the acrosomal enzymes which have been isolated and characterized show distinct differences from their lysosomal counterparts. One of these enzymes is a proteinase which will be discussed in detail later.

After formation, spermatozoa pass through the seminiferous tubules and are collected in the rete testis. From there they enter the epididymis where they are stored at the distal end, called the cauda epididymis. Testicular spermatozoa are not fertile and obtain the ability to fertilize during transport through the epididymis. The biochemical events involved in this maturation process are not known. At the time of ejaculation, spermatozoa are transported through the vas deferens and the urethra and mix with the accessory sex gland fluids. In man, semen is ejaculated in a coagulated form. This coagulum liquifies normally within 5 to 20 minutes, which is rather unique to the human. Other species, such as rodents, the rabbit (usually), and the monkey, form a coagulum that does not liquefy to any significant extent. Yet other species, such as the dog, do not form any seminal coagulum at all.

The female reproductive structures consist of the ovaries and the genital tract, which, depending on the species, can be subdivided into a vagina, one or two cervices, one or two uteri, two uterotubal junctions and two oviducts (Fallopian tubes). The female gamete (ovum, egg) is produced in the ovary. After ovulation, the ovum is picked up by the distal portion of the oviduct, called the fimbriae, and moves to its ampullary portion where fertilization takes place. The ovum is surrounded by three layers. The cumulus oophorus is the outermost layer. The middle layer is called the corona radiata, and the innermost layer is the zona pellucida. The first two layers consist of cells that are attached to each other through a glycoprotein matrix. The zona is non-cellular and contains mucopolysaccharides and mucoproteins. Between the zona and the egg itself, a space is present, called the perivitelline space. The egg is surrounded by a membrane called the vitelline membrane.

In various species, the ejaculate is deposited intravaginally, although many animals place their semen into the cervix or directly into the uterus. Vaginal placement requires that the spermatozoa be transported through the cervix and its mucus, a process that is possible in most species throughout the cycle. In humans, however, cervical transport is usually restricted to the ovulatory period, i.e., midcycle.

During their residence in the uterus, the spermatozoa undergo an additional ripening process, called capacitation. This requires as much as 6 hours in the rabbit, but less in other species. The requirement for capacitation is common to all mammalian species tested to date, and without this final activation process, spermatozoa are unable to penetrate the external layers of the ovum, i.e., fertilize.

In most species, all three layers surrounding the egg are intact at the time of fertilization, requiring that the spermatozoa pass through them to reach the egg. This is accomplished through the hydrolytic activity of the acrosomal enzymes and the motility of the spermatozoon. The acrosomal enzyme thought to be responsible for the penetration of the spermatozoon through the cumulus is hyaluronidase, whereas the one involved in corona passage is not well defined and has been given the name "corona-penetrating enzyme." The enzyme used to penetrate the zona is a neutral proteinase called acrosin.

Indications for a sperm enzyme, proteolytic in nature and having an effect on the layers surrounding the ovum, were reported in the early thirties (Yamane 1930, 1935). Additional reports on the sperm proteinases of various species appeared several decades later (Buruiana 1956; Waldschmidt, Hoffman and Karg 1966). It was shown that selective removal of the sperm acrosome releases this proteinase, indicating its acrosomal origin (Hartree and Srivastava 1965; Srivastava, Adams and Hartree 1965). These observations were confirmed by Stambaugh and Buckley (1968, 1969), who additionally showed that the proteinase digested the zona and had various properties in common with pancreatic trypsin. The first reports on the extensive purification and characterization of the acrosomal proteinase appeared soon thereafter (Zaneveld et al. 1970a, 1971a). At that time, the enzyme was shown to behave differently from any other known proteinase, and the name "acrosin" was suggested. This was officially accepted by the Commission on Biochemical Nomenclature in 1972 (Enzyme No. 3.4.21.10).

The association of proteinases, and various activators and inhibitors thereof, with male and female genital tract secretions has also been known for some time, but similar to the sperm acrosomal proteinase, only very recently have some of them been purified and characterized. Approximately 30 years ago, blood-clot-lysing (fibrinolytic) factors were shown to be present in seminal plasma and prostatic fluid (Huggins and Neal 1942). In the two decades following these observations, it was shown that seminal plasma contains a neutral proteinase (we will use the name "seminin" for this enzyme to prevent confusion with other seminal proteinases), proteinase inhibitors, plasminogen activator, pepsinogen, aminopeptidases, transaminases, esterases, carboxypeptidases and possibly plasminogen, plasmin and a proplasminogen activator (Harvey 1949; Oettle 1950; Lundquist 1952; Lundquist and Seedorff 1952; Tagnon, Whitmore and Shulman 1952; Tagnon et al. 1953a,b; von Kaulla and Shettles 1953, 1954; Lundquist, Thorsteinsson and Buus 1955; Karhausen and Tagnon 1955; Gotterer, Banks and Williams-Ashman 1956; Ying et al. 1956; Rasmussen, Albrechtson and Astrup 1958; Thorsteinsson 1958; Rasmussen and Albrechtsen 1960a,b; Povoa and Villela 1960). Seminin was first partially purified and characterized by Lundquist, Thorsteinsson and Buus (1955), whose results were confirmed and extended by Syner and Moghissi (1972). These same authors suggested that this enzyme may aid spermatozoa in their transport through cervical mucus (Moghissi and Syner 1970). Recently, the seminal plasminogen activator was isolated and characterized (Propping et al. 1974; Zaneveld et al. 1974), and it was shown that at least in humans, most if not all of the so-called fibrinolytic activity is due to the action of this plasminogen activator and not to proteinases such as plasmin or seminin.

The first detailed report concerning the inhibitors of male genital tract fluids appeared in the mid-sixties (Haendle et al. 1965). These were originally isolated from seminal vesicles by Fritz et al. (1967, 1970) and from seminal plasma by Zaneveld et al. (1970b) and Polakoski et al. (1971). Interest in these inhibitors increased dramatically when it appeared that they combined with spermatozoa during ejaculation (Zaneveld, Srivastava and Williams 1969, 1970) and would prevent fertilization unless removed during residence of the spermatozoon in the uterus, i.e., during capacitation (Zaneveld et al. 1971a,b). Almost at the same time it was shown that synthetic, active-site-directed reagents that inhibited acrosin could prevent fertilization and had the potential of becoming a new approach to contraception (Zaneveld, Robertson and Williams 1970).

The presence of proteinase inhibitors in female genital tract secretions was first studied by von Kaulla and Shettles (1954). These authors also described the presence of plasminogen in these secretions. Although the latter is still in question, the presence of serum inhibitors in cervical mucus was subsequently detailed initially by Schumacher, Strauss and Wied (1965) and Schumacher and Pearl (1968), and the occurrence of low molecular weight inhibitors in cervical mucus initially by Haendle, Ingrisch and Werle (1970a). Inhibitors were further shown to be present in uterine and Fallopian tube secretions by Hirschhauser et al. (1971) and Stambaugh, Seitz and Mastroianni (1974) and in amniotic fluid by Woraschk and Kressner (1962).

The following sections describe the characteristics of the components of the proteolytic enzyme systems of the mammalian reproductive tract and spermatozoa as they are known today. Due to space limitations, the discussions neither include the aminopeptidases, esterases, transaminases and carboxypeptidases of seminal plasma, nor the uterine proteinase(s) that may be involved in implantation. Otherwise, an attempt was made to present a complete overview of the field and to include all available references on the topic. It should be noted that this is a rather new area and that progress has been slow due to the difficulty in procuring enough material for study.

PROTEINASES OF SPERMATOZOA

Acrosin

The most widely studied entity of the reproductive proteolytic enzyme system is the proteinase of spermatozoa that has a pH optimum in the basic region and is called acrosin. Acrosin was demonstrated to be present in the sperm of all mammalian species tested so far, including the boar, bull, guinea pig, hamster, stallion, human, rhesus monkey, rabbit, rat and ram (Waldschmidt, Hoffman and Karg 1966; Stambaugh and Buckley 1968, 1969, 1970, 1972a; Multamaki and Niemi 1969, 1972; Garner, Salisbury and Graves 1971; Schumacher 1971; Polakoski, Zaneveld and Williams 1971; Schiessler et al. 1972; Uhlenbruck et al. 1972; Multamaki 1973; Menezo and Flechon 1973; Zaneveld, Polakoski and Williams 1973; Schill 1973, 1974; Brown and Hartree 1974). However, purification has only been attempted in the boar (Polakoski, Williams and McRorie 1972; Polakoski, McRorie and Williams 1973; Fink et al. 1972a; Schleuning, Schiessler and Fritz 1973), bull (Garner 1973; Garner and Cullison 1974), human (Zaneveld, Dragoje and Schumacher 1972; Zaneveld et al. 1974; Fritz et al. 1972a; Gilboa, Elkana and Rigbi 1973) and rabbit (Polakoski, Zaneveld and Williams 1972), with the boar having been obtained in the highest purity (Polakoski, Williams and McRorie 1972; Polakoski, McRorie and Williams 1973) and characterized in greatest detail (Polakoski, Williams and McRorie 1972; Fritz et al. 1972d; Schiessler et al. 1972; Polakoski and McRorie 1973; Schleuning and Fritz 1974).

Various procedures have been developed to remove the acrosomes from sperm, including detergent extraction (Srivastava, Zaneveld and Williams 1970; Churg, Zaneveld and Schumacher 1974), sonication and detergent treatment combined (Stambaugh and Buckley 1968, 1969; Stambaugh and Smith 1973), acidic incubation (Schleuning, Schiessler and Fritz 1973), freeze-thawing (Pedersen 1972) and other chemical treatments (Bernstein and Teichman 1973; Srivastava 1973b). In our hands, detergent extraction is the treatment of choice. Acrosin is further purified from the extracts essentially by a combination of two or all of the following procedures: ion exchange chromatography, Sephadex gel filtration or affinity chromatography. Details of the techniques used by our laboratories can be found in Polakoski, Zaneveld and Williams (1972) and Polakoski, McRorie and Williams (1973).

The highest purified fraction of rabbit acrosin prepared by our laboratories had an activity of 39 U/mg and that of boar acrosin, 271 U/mg, at 23°C when assayed on BAEE.[1] In both cases, single bands in SDS and regular disc gel electrophoresis were obtained. Other authors, using different methods of purification, reported specific activities of 44 for human acrosin (Gilboa, Elkana and Rigbi 1973) and 164 for boar acrosin (Fritz, Schleuning and Schill 1974).

Little is known about the physicochemical characterization of acrosin, primarily due to limitations of not only the purified preparations, but even of the starting material. The isoionic and isoelectric points have not been determined as yet, although one can assume that they are cationic, for acrosin is not absorbed to DEAE-cellulose at pH 8.0 and it migrates in an acidic disc gel electrophoresis system as do other basic proteins such as trypsin.

The molecular weight of rabbit acrosin was estimated as 55,000 using Sephadex gel filtration and 27,300 by SDS gel electrophoresis. It is likely, therefore, that the enzyme exists in dimeric form. A molecular weight of 30,000 was obtained using Sephadex gel filtration methods for both human (Zaneveld, Dragoje and Schumacher 1972) and boar (Polakoski, McRorie and Williams 1973) acrosin, although occasionally a second, higher molecular weight fraction was noted during the purification procedures of boar acrosin. The boar enzyme also gave a single band of 30,000 molecular weight in the SDS gel electrophoresis system. Also using SDS gel electrophoresis, Schleuning, Schiessler and Fritz (1973) and Schleuning and Fritz (1974) calculated molecular weights of 38,000 and 34,000 for the bands produced by boar acrosin, and Garner and Cullison (1974) calculated 37,000 for bull acrosin. The reported molecular weight of 75,000 for human acrosin (Gilboa, Elbana and Rigbi 1973) is difficult to explain at this time unless human acrosin also occurs in dimeric form similar to the rabbit. Multiple forms of acrosin may indeed be present, as indicated by the fact that several forms of BANA-hydrolyzing activity were separated by disc gel electrophoresis from the acrosomal extracts of bull (Garner, Salisbury and Graves 1971), boar (Polakoski, unpubl.) and human (Zaneveld, unpubl.). These forms may or may not be interconvertible. Carbohydrate has been shown to be associated with both bull (Garner and Cullison 1974) and boar (Schleuning and Fritz 1974) acrosin. An amino acid analysis of rabbit acrosin was recently published (Stambaugh and Smith 1974).

In the crude form, acrosin is very stable over a wide pH range, but gen-

[1] N-α-benzoyl-L-arginine ethyl ester. Other abbreviations to be used in the discussion to follow are: ATEE, N-acetyl-L-tyrosine ethyl ester; BANA, N-α-benzoyl-D, L-arginine-β-naphtylamide HCl; BAPA, N-α-benzoyl-D,L-arginine-p-nitroanilide; BHME, N-benzoyl-L-histidine methyl ester; BLME, N-α-benzoyl-L-lysine methyl ester; BTEE, N-benzoyl-L-tyrosine ethyl ester; DFP, diisopropylfluorophosphate; EPGB, ethyl p-guanidinobenzoate; NPGB, p-nitrophenyl-p'-guanidinobenzoate; PMSF, phenyl methyl sulfonyl fluoride; TACK, N-α-tosyl-L-arginine chloromethyl ketone; TAME, p-tosyl-L-arginine methyl ester; TLCK, N-α-p-tosyl-L-lysine chloromethyl ketone; TLME, N-α-tosyl-L-lysine methyl ester; TPCK, L-1-tosylamido-2 phenylethyl chloromethyl ketone.

erally speaking, the purer the preparation, the more labile the enzyme becomes. The highly purified acrosin is stable for several weeks if kept at 4°C at a pH of 3; however, most of the activity is irreversibly lost within minutes at room temperature at pH 8.0. This loss of activity is presumably due to autolysis. Both bovine serum albumin (BSA) and Ca^{++} protect acrosin at this pH, with the best stability obtained using a combination of BSA and Ca^{++}. Highly purified acrosin is unstable above 37°C even at pH 3.0, with about one-half of the activity destroyed in 10 minutes at 55°C, although the temperature optimum of proteinase activity against azocasein is 53°C. The preparations can be dialyzed overnight at 40°C in 0.001 M HCl, but large reductions in activity occur upon freezing or freeze-drying. This loss can be minimized in the partially purified fractions by the addition of sucrose (Schleuning, Schiessler and Fritz 1973). The pH optimum of the enzyme using an esterolytic substrate is between 7.5 and 8.5, and with azocasein is about 8.6. Acrosin is denatured by very low concentrations of urea. Experimentation with acrosin should be carried out either in plastic vials or in silicone-coated glassware since the enzyme adheres to glass (Schiessler et al. 1972).

Acrosin hydrolyzes the amine and ester derivatives of both arginine and lysine. The relative rates of hydrolysis are BAEE > BLME > TAME > BAPA > TLME, but the enzyme does not hydrolyze the chymotrypsin substrates BTEE, ATEE or BHME. The arginine-containing substrates are hydrolyzed at a faster rate than the corresponding lysine substrates. This preference for arginine residues also occurs in protein substrates, as was shown in the acrosin digests of denatured insulin, ribonuclease and lysozyme. Analyses of the digestion products from these proteins indicate that acrosin is an endoproteinase that only cleaves the carboxyl bond of arginine and lysine residues.

The possible occurrence of monomeric-dimeric forms, other molecular variants, iso-enzymes, questionable purity of the enzyme fraction, substrate activation (Schleuning and Fritz 1974), inhibition at high substrate concentration as well as product inhibition (Polakoski and McRorie 1973), and combinations of these problems have so far interfered with the obtaining of meaningful kinetic data. Some indications can be obtained, however, from the limited studies that have been performed. The apparent K_m on BAEE for boar acrosin is 4.8×10^{-5} M (Polakoski and McRorie 1973), although a value of 2.7×10^{-4} M has also been reported (Schleuning and Fritz 1974). The crude acrosomal rabbit preparation gave an apparent K_m of 5.2×10^{-6} M (Stambaugh and Buckley 1972a) and 6.4×10^{-5} M (Connors, Greenslade and Davanzo 1973).

Although Ca^{++} ions are not required for activity, they do increase the esterolytic activity of rabbit, boar and human acrosin two- to threefold. The activity of the boar enzyme is also stimulated by Mn^{++}, Ni^{++}, Mg^{++} and Co^{++} ions, while Cu^{++} ions have no apparent effect. Opposite results were obtained by Gilboa, Elkana and Rigbi (1973), who found no effect on either the activity or stability of human acrosin with Ca^{++}, Mn^{++} or Mg^{++} ions. Hg^{++} inhibits acrosin, which, together with the inhibitory properties of TLCK towards acrosin, may indicate that an active-site histidine is involved in catalysis.

Acrosin is not inhibited by EDTA or 1,10-phenanthroline, which implies that there is no absolute metal requirement. It is also not inhibited by iodoace-tate, indicating that a free -SH group does not participate in catalysis. En-zyme activity is not enhanced by cysteine, dithiothreitol or mercaptoethanol. Acrosin is inhibited by DFP and PMSF, suggesting that the enzyme is a serine proteinase. Active-site-directed reagents that inhibit acrosin are TLCK, TACK, EPGB and NPGB. The enzyme is not inhibited by TPCK.

Acrosin is also inhibited by many natural proteinase inhibitors from various sources: Kunitz and Bowman-Birk soybean trypsin inhibitors, Kunitz and Kazal pancreatic trypsin inhibitor, (chicken) ovomucoid trypsin inhibitor, trypsin-kallikrein inhibitor from bovine lung, inter-α-trypsin inhibi-tor, α_1-antitrypsin, antithrombin III, trypsin-plasmin inhibitor bdellin from leeches, leupeptin, antipain and the various low molecular weight reproduc-tive tract inhibitors (Zaneveld, Dragoje and Schumacher 1972; Zaneveld, Polakoski and Williams 1972; Fritz et al. 1972c,d,e; Fritz, Schiessler and Schleuning 1973; Fritz, Forg-Brey and Umezawa 1973; Polakoski and Mc-Rorie 1973). The Kunitz pancreatic trypsin inhibitor and α_1-antitrypsin are "progressive" inhibitors of acrosin.

Free L-arginine and various arginine analogs competitively inhibit acrosin's esterolytic hydrolysis. L-Arginine inhibits acrosin with a K_i value of 3mM when assayed on BAEE. L-Homoarginine and D-arginine also competitively inhibit hydrolysis, but L-lysine does not.

In summary, acrosin has many properties in common with trypsin and therefore can be considered to be a trypsinlike enzyme. However, acrosin is a somewhat larger molecule than trypsin, shows a higher specificity for arginine, is either not inhibited or inhibited to a different extent by some of the synthetic and natural trypsin inhibitors, and is much less stable than pancreatic trypsin. Acrosin further varies immunologically from pancreatic trypsin (Zaneveld, Schumacher and Travis 1973; Garner et al. 1975), al-though this has been argued (Stambaugh and Smith 1974). Acrosin also differs significantly from any of the other known mammalian proteinases, such as plasmin, thrombin, kallikreins and collagenase, and is therefore a unique enzyme.

Proacrosin

What appears to be a zymogen form of acrosin has recently been observed in rabbit testis and epididymal sperm (Meizel 1972; Meizel and Huang-Yang 1973) and ejaculated boar sperm (Polakoski 1974). The boar proacrosin was obtained in a highly purified form after ammonium sulfate fractionation, Sephadex G-200 gel filtration and ion exchange chromatography. It was estimated to have a molecular weight of 62,000. Sigmoidal activation curves can be obtained with either trypsin or acrosin, although proacrosin will auto-activate at pH 8.0. It is not activated by chymotrypsin, and BSA and Ca^{++} retard activation. The zymogen form of acrosin is not inactivated by TLCK.

Other Acrosomal Proteinases

Acrosomal proteinases other than acrosin were demonstrated in the crude extracts from various species. Dott and Dingle (1968) observed optimal hydrolysis with ejaculated bull sperm at pH 4.0 and 6.8, cysteine giving a markedly stimulated activity at pH 6.8. The presence of various acrosomal proteinases based on pH profiles was confirmed and extended by Allison and Hartree (1970), who used ejaculated ram sperm. They obtained maximal hydrolysis between pH 7.0–7.5 and also peaks at pH 5.5 and 3.8. The activity at all three pH's was enhanced by the addition of cysteine.

The presence of an acidic proteinase was confirmed in acrosomal extracts of ram, horse, human and rabbit (Polakoski, Williams and McRorie 1973). The boar enzyme was partially purified by ammonium sulfate fractionation and Sephadex gel filtration. Cysteine enhanced its activity, whereas iodoacetate inhibited the enzyme. DFP had no effect on the activity of the acid proteinase.

Two amidases which hydrolyze BANA maximally at pH 7.5 and 5.5 were shown in bull sperm (Garner, Salisbury and Graves 1971; Multamaki and Niemi 1972). Additionally, the naphthylamide derivatives of L-methionine, L-isoleucine, L-leucine, L-valine and L-phenylalanine are hydrolyzed at pH 8.0 by bull acrosomal extracts partially purified by Sephadex G-100 gel filtration (Meizel and Cotham 1972). This activity is stimulated by dithiothreitol and inhibited by DFP and phenanthroline. The hydrolysis of L-aspartic acid-B-naphthylamide by acrosomal extracts was not activated nor inhibited by these reagents, suggesting that different enzymes are probably involved.

A "collagenaselike" peptidase that cleaves the leucine-glycine bond of PZ-Pro-Leu-Gly-Pro-D-Arg (PZ, 4-phenylazo-benzyloxy-carbonyl) was reported to be present in human, rat and bull spermatozoa (Koran and Milkovic 1973). However, the hydrolysis of native collagen was not tested.

SEMINAL PLASMA PROTEINASES AND ACTIVATORS

Seminal Proteinases

Three proteinases have been reported to be present in seminal plasma: pepsinogen, seminin (a neutral proteinase) and plasmin(ogen). The presence of plasmin(ogen) is in doubt and has only been found by a few authors and in extremely small quantities (von Kaulla and Shettles 1953; Rasmussen and Albrechtsen 1960a,b).

Human seminal plasma contains a pepsinogen that probably originates from the seminal vesicles, has a pH optimum around 2.5, and is activated at a pH below 4 (Lundquist and Seedorff 1952). It loses activity if preincubated at pH 2.5 and subsequently at pH 8.5. Its activity in normal semen was reported to correspond to 0.002 mg pepsin per ml. The enzyme was crystallized, and the crystals appeared identical to those of gastric juice pepsin. In addition, these two enzymes hydrolyze substrates at approximately the same rate.

Seminin (also called "chymotrypsinlike enzyme," but this name is deceiving since the enzyme has but very few properties in common with chymotrypsin) has recently been studied in our laboratories and in collaboration with others (Fritz et al. 1972b; Zaneveld and Schumacher, unpubl.), and the observations confirm and extend those of previous investigators (Lundquist, Thorsteinsson and Buus 1955; Syner and Moghissi 1972). The enzyme has an approximate molecular weight of 33,000, digests gelatin very effectively, and to a lesser extent casein and hemoglobin, but not gamma globulin, albumin, transferrin, fibrin or fibrinogen. It hydrolyzes BTEE and ATEE, although poorly, but not BAEE, BANA or peptidase substrates such as the naphthylamide derivatives of alanine, leucine, phenylalanine, tyrosine or methionine. Extensive dialysis against distilled water results in a significant decrease in activity, which returns completely on addition of NaCl. Ca^{++}, Fe^{++}, Mg^{++}, Ni^{++}, Zn^{++} and Co^{++} ions and cysteine do not affect the activity of the enzyme. Heating seminin to 62°C for 10 minutes results in almost complete inactivation.

Seminin is optimally active at neutral pH (6.5–7.5) and is inactivated at a pH below 4.0 or above 9.0. Using low and high molecular weight substrates, the enzyme is not inhibited by any of the known natural or synthetic inhibitors, including the trypsin-chymotrypsin inhibitor from dog submandibular glands, the Bowman-Birk and Kunitz soybean trypsin inhibitors, the lima bean trypsin inhibitor, the Kunitz pancreatic trypsin inhibitor, α_1-antitrypsin, inter-α-trypsin inhibitor, DFP, TLCK, TPCK or the human seminal plasma proteinase inhibitors. Exceptions may be α_{1x}-antichymotrypsin and hydrocinnamic acid, which inhibit the enzyme to a small extent when a high molecular weight substrate such as gelatin is used. Rabbit bile (Hisazumi 1970) and ϵ-aminocaproic acid (EACA) (Nys et al. 1971) were also reported to possess inhibitory properties towards seminal proteolytic activity.

Seminin is probably produced by the prostate gland (Suominen, Eliasson and Niemi 1971; Tauber et al. 1973). The seminal proteinases increase vascular permeability, an effect usually associated with activity of kallikreins (Williams-Ashman and Hoskins 1969; Suominen and Niemi 1971).

Plasminogen Activators

The seminal plasminogen activator originates mostly from the prostate and Cowper's glands and the urethra (Kester 1969, 1971; Tauber et al. 1973). The activator activity of prostatic fluid has been studied most extensively (Nys et al. 1971; Tagnon and Steens-Lievens 1963; Tagnon, Whitmore and Shulman 1952; Tagnon et al. 1953a,b; Swan and Kerridge 1965; Karhausen and Tagnon 1955; Harvey 1949; Ying et al. 1956; Rasmussen and Albrechtsen 1960a,b; Rasmussen, Albrechtson and Astrup 1958; Hirschhauser and Kionke 1971a; Liedholm 1973). The seminal activator was recently isolated and characterized (Propping et al. 1974; Zaneveld et al. 1974). Purification was achieved through Sephadex gel filtration and ion exchange chromatography. Two activator fractions were obtained, both showing a single band on SDS disc gel electrophoresis. They had slightly different mobilities, corresponding to molecular weights of 70,000 (SPA-1) and 74,000 (SPA-2) (seminal plasminogen activators 1 and 2). SPA-2 is completely stable at 60°C, whereas it loses some activity at higher temperatures. SPA-1 is already unstable at 60°C. Both activators are labile above 37°C at a pH below 4 or above 9.5. EDTA, Mg^{++} and Ca^{++} ions do not affect the plasminogen activators, whereas Fe^{++} ions and cysteine completely inhibit their activity at temperatures of 60°C or above. Both activators are effectively inhibited by sera from pregnant women, although less so than urokinase. Both SPA-1 and SPA-2 show a straight inhibition curve with EACA, in contrast to urokinase which is inhibited in a biphasic fashion. Preliminary amino acid analyses of SPA-1 and SPA-2 show that they are virtually identical. Both are high in aspartic and glutamic acid, serine, glycine and leucine and low in tyrosine, methionine, isoleucine and histidine.

Urokinase antiserum inhibits the activity of its own antigen as well as that of SPA-1 and SPA-2. On immunodiffusion, only one precipitin band forms between urokinase and its antiserum, which is continuous with the precipitin bands formed between the seminal plasminogen activators, thus indicating that the seminal plasma activators are immunologically identical to urokinase. They differ in some of their biochemical properties, however, although their high stability at varying pH's and temperatures make SPA-1 and SPA-2 more similar to urokinase than to either the blood or tissue activators. Therefore, we may look upon urokinase and the two seminal plasminogen activators as "isoactivators."

It cannot be excluded that other plasminogen activators with specificities similar to the tissue activators are also present in seminal plasma (Hisazumi 1970; Rasmussen and Albrechtsen 1960a,b). The plasminogen activator activity reported to be associated with the sperm head of testicular and epididymal spermatozoa (Tympanidis and Astrup 1968) is possibly due to the activity of acrosin.

There are also indications for the presence of an inactivator of the plasminogen activator in seminal plasma since the total plasminogen activator activity consistently increases three- to fourfold on purification (Propping et al. 1974; Zaneveld et al. 1974). A proplasminogen activator that can be activated by streptokinase may also be present in semen (Rasmussen and Albrechtsen 1960a,b; Nys et al. 1971), but details concerning this protein are lacking, except that it has maximum stability at pH 7 to 11 and that streptokinase activation is almost immediate.

PROTEINASE INHIBITORS OF MALE GENITAL TRACT FLUIDS AND SPERMATOZOA

Rete Testis, Seminal Vesicle and Seminal Plasma Proteinase Inhibitors

Since they are most likely identical, the proteinase inhibitors of the male genital tract fluids and those of spermatozoa will be discussed as one group. Inhibitors were shown to be present in the rete testis, the epididymis, the seminal vesicles and seminal plasma (Haendle et al. 1965; Fink et al. 1971b; Zaneveld et al. 1971a; Suominen and Setchell 1972). Although the method of purification of the inhibitors has varied with the different authors, essentially two techniques have been applied: (1) affinity chromatography using insolubilized trypsin and subsequent final purification by one or more Sephadex gel filtration steps, and (2) TCA precipitation, Sephadex gel filtration and ion exchange chromatography. Both methods produce excellent results. The rete testis and the epididymal inhibitors are very similar to those of seminal plasma (mostly of seminal vesicle origin) and they may be identical. The rete testis inhibitor can be separated into three iso-inhibitors (Suominen, Kaufman and Setchell 1973). Each one has a molecular weight of 6500, is very stable, and inhibits acrosin and trypsin, but not other proteinases.

Two inhibitors with almost identical molecular weights were isolated from guinea pig seminal vesicles: 6600 (GP-SVI 1) and 6800 (GP-SVI 2) (Fritz et al. 1967, 1970). Both inhibitors inhibit acrosin and trypsin, but only GP-

SVI 2 inhibits plasmin. They do not inhibit chymotrypsin, thrombin or pancreatic kallikrein. GP-SVI 1 possesses 60 amino acids and is particularly high in glutamic acid, aspartic acid, glycine, proline, cysteine, leucine and arginine, whereas GP-SVI 2 is high in aspartic acid, serine, glycine and cysteine. Two forms of GP-SVI 1 were found, one (a) with an intact poly-peptide chain and one (b) with the same amino acid composition but with a cleaved Arg-X linkage. These can be interconverted by incubation with trypsin. GP-SVI 1a, but not GP-SVI 1b, is inactivated by the action of carb-oxypeptidase B or by maleic anhydride. The original residue or the Arg-X bond is therefore the active center of the molecule for complex formation with trypsin. GP-SVI 2 has an N-terminal valine residue and consists of 4 iso-inhibitors which vary by 1–5 amino acids on the N terminus. Both in-hibitors show temporary inhibition and react stoichiometrically with pro-teinases.

The inhibitors from seminal plasma have been purified from the rabbit (Zaneveld, Srivastava and Williams 1970; Zaneveld et al. 1971b), boar (Polakoski et al. 1971; Zaneveld et al. 1971a; Fink et al. 1971b, 1972a; Fritz, Schiessler and Schleuning 1973) and human (Hirschhauser and Kionke 1971b; Suominen, Multamaki and Niemi 1971; Fink et al. 1971a, 1972b; Suominen and Niemi 1972; Fritz, Schiessler and Schleuning 1973).

Using classical isolation procedures, five proteinase inhibitors have been purified from boar seminal plasma, which have molecular weights varying between 1500 and 13,000 as estimated by Sephadex gel filtration (Polakoski and Williams 1974). The inhibitor showing a molecular weight of 13,000 had previously been isolated and analyzed for its amino acid composition (Pola-koski et al. 1971; Zaneveld et al. 1971a). The smallest unit consisted of 58 amino acids, the principal ones being arginine, aspartic acid, cysteine, glu-tamic acid, glycine, lysine, phenylalanine and proline, but it lacks tryptophan and is otherwise low in valine, serine and methionine. The molecular weight, based on the amino acid analysis, was calculated as 6800, significantly less than the molecular weight of 13,000 estimated by gel filtration. This dis-crepancy was thought to be due to either the presence of carbohydrates or to dimer formation. Fink et al. (1971b) did indeed show the presence of glu-cosamine and galactosamine. These authors reported the isolation of five iso-inhibitors from boar seminal plasma, with a molecular weight of 11,600 and consisting of 99 amino acids (Fink et al. 1971b, 1972b). More recently, they revised the molecular weight to 8000 (Fritz, Schiessler and Schleuning 1973). These inhibitors inhibit acrosin, trypsin and plasmin, but not chymo-trypsin, thrombin or pancreatic kallikrein, similar to the rabbit seminal plasma inhibitor. Both the rabbit and boar inhibitors are heat and acid stable. The boar inhibitor described by Fink et al. (1971a) possesses an intact Arg-X bond at its reactive site.

Although a proteinase inhibitor was first partially purified and character-ized from human seminal plasma by Hirschhauser and Kionke (1971b), de-tailed knowledge concerning human seminal plasma inhibitors was obtained through the studies of Fink et al. (1971a,b, 1972b). Suominen, Mul-tamaki and Niemi (1971), Suominen and Niemi (1972) and Fritz, Schiessler and Schleuning (1973). Two proteinase inhibitors are present in human seminal plasma, one with a molecular weight of 5400 (HUSI-II) and the other of 12,500 (HUSI-I), based on Sephadex gel filtration experiments and amino acid analyses. HUSI-II inhibits trypsin, but not chymotrypsin, whereas HUSI-I inhibits both trypsin and chymotrypsin and to a lesser extent kallikrein. Neither inhibitor has any effect on the activity of plasmin, thrombin or urokinase, and both are heat stable. As many as eight iso-inhibitors were reported for HUSI-I (Fritz, Schiessler and Schleuning 1973). HUSI-I has 114 amino acids and is high in aspartic acid, glutamic acid, gly-cine, cysteine and lysine, whereas it is very low in methionine, isoleucine, tyrosine and phenylalanine. HUSI-II possesses 67 amino acids and is high in aspartic acid, serine, glutamic acid, glycine and cysteine, whereas it is low in alanine, proline, methionine, tyrosine and phenylalanine (Fink et al. 1972a). HUSI-II stoichiometrically inhibits acrosin, with an association constant of 5.0×10^8 M^{-1} (Zaneveld et al. 1973), whereas HUSI-I inhibits acrosin only poorly or not at all.

Based on the amino acid analyses described above and also through im-munologic comparisons (Zaneveld et al. 1974), it appears that the seminal plasma inhibitors are different from the pancreatic trypsin inhibitors.

Besides the low molecular weight inhibitors, high molecular weight inhibi-tors are also present in human seminal plasma (Schumacher 1970b; Tauber et al. 1973, 1975; Zaneveld et al. 1974). These are immunologically identical to the serum inhibitors α_1-antitrypsin and α_{1x}-antichymotrypsin. The other serum inhibitors, inter-α-trypsin inhibitor, α_2-macroglobulin, antithrombin III and the C_1-esterase inhibitor, are absent from seminal plasma, at least in humans.

Proteinase Inhibitors from Spermatozoa

It has been well established that spermatozoa also contain proteinase in-hibitors (Zaneveld, Srivastava and Williams 1969; Zaneveld, Dragoje and Schumacher 1972; Zaneveld, Polakoski and Williams 1973; Zaneveld et al. 1971a, 1974b; Haendle, Ingrisch and Werle 1970b; Polakoski, Zaneveld and Williams 1971; Hirschhauser and Baudner 1972). These inhibitors are most likely not native to the spermatozoa, but are added to them during sperm transport through the rete testis and epididymis and/or during ejaculation.

Only the inhibitor(s) from human spermatozoa has so far been purified and characterized. Zaneveld, Dragoje and Schumacher (1972) found one inhibitor with a molecular weight of 5600 that inhibits acrosin and trypsin, but not chymotrypsin, and appears identical to HUSI-II from seminal plasma. These observations were confirmed by Fritz, Schiessler and Schleuning (1973), although these investigators also found small amounts of a second proteinase inhibitor identical to HUSI-I. Preliminary studies of the inhibitor from boar spermatozoa were recently reported (Polakoski and Williams 1974).

PROTEINASE INHIBITORS OF FEMALE GENITAL TRACT FLUIDS

Similar to the proteinase inhibitors of seminal plasma, the inhibitors of the female genital tract can be divided into two general classes: inhibitors with a molecular weight below 15,000, and those with a higher molecular weight. Most of the latter are immunologically identical to serum inhibitors.

The low molecular weight inhibitor of cervical mucus was first reported to possess a molecular weight of 2200 (Haendle, Ingrisch and Werle 1970a). This was later corrected to 11,500 (Wallner and Fritz 1974). The inhibitor is acid stable and can be obtained from cervical mucus by perchloric acid extraction. It appears to be present in a free and in a masked form, probably as an enzyme-inhibitor complex. The cervical mucus inhibitor inhibits trypsin, chymotrypsin and neutral proteinases from human leukocytes, but has no effect on plasmin, thrombin, kallikrein and acrosin. The inhibitor shows a decrease during the ovulatory phase of the cycle (midcycle). Based on immunological cross-reactions, it had been reported that the inhibitor present in human cervical mucus is identical to one of the seminal plasma inhibitors (Hirschhauser, Baudner and Daume 1972). This remains to be firmly established, although the biochemical characteristics of the cervical inhibitor are very similar to those of HUSI-I.

The presence of serum inhibitors in cervical mucus has been well documented (Schumacher, Strauss and Wied 1965; Schumacher 1970a, 1973a,b; Schumacher and Zaneveld 1972, 1974). These include α_1-antitrypsin, inter-α-trypsin inhibitor, α_{1x}-antichymotrypsin, antithrombin III and the C_1-esterase inhibitor. The last two are only occasionally detectable in cervical mucus, but appear most frequently in specimens obtained from women under treatment with hormonal contraceptives. The first three inhibitors show their lowest concentrations at the ovulatory period or during estrogenic treatment, whereas they are high at other times of the cycle or under progesterone treatment. The α_1-antitrypsin levels rarely reach values that are higher than 20% of those in serum, whereas the other inhibitors reach values equal or higher than those in serum. The C_1-esterase inhibitor and the α_{1x}-antichymotrypsin inhibitor do not inhibit acrosin, whereas the other three have definite inhibitory properties (Schumacher 1971; Schumacher and Zaneveld 1972, 1974; Zaneveld, Dragoje and Schumacher 1972).

Proteinase inhibitors were also shown to be present in the uterus, Fallopian tubes and follicular fluid (von Kaulla and Shettles 1954; Hirschhauser et al. 1971; Stambaugh, Seitz and Mastroianni 1974; Hamner, pers. comm.), and even in amniotic fluid (Woraschk and Kressner 1962; Kessler, Schumacher and Zaneveld, unpubl.). Stambaugh, Seitz and Mastroianni (1974) showed that the serum inhibitors α_1-antitrypsin and α_{1x}-antichymotrypsin are present in the Fallopian tube fluids, along with some presently unidentified inhibitors that separate on electrophoresis. Similar to the ones in cervical mucus, they show a minimum concentration during the ovulatory period. No evidence was found for the presence of the other serum inhibitors.

PHYSIOLOGICAL ROLE OF PROTEOLYTIC ENZYME SYSTEMS IN THE REPRODUCTIVE PROCESS AND PRACTICAL APPLICATIONS

Recently, a multitude of reviews have been presented on the role of proteinases, activators and inhibitors in the reproductive processes (Zaneveld et al. 1971a, 1974a; Fink et al. 1972b; Williams 1973; Stambaugh 1973; Syner and Moghissi 1973; McRorie and Williams 1974; Fritz, Schleuning and Schill 1974; Zaneveld 1974, 1975). The following discussions will therefore be kept to an absolute minimum and merely summarize the conclusions of these reviews.

The location of acrosin on the spermatozoon is presently not known. It is the general opinion that at least the largest portion of acrosin is bound to the inner acrosomal membrane (Brown and Hartree 1974; Srivastava 1973a; Srivastava et al. 1974). Yanagimachi and Teichman (1972), however, show the association of a neutral proteinase (presumably acrosin) with the outer acrosomal membrane. Similar conclusions may be reached from the experiments by Stambaugh and Buckley (1972b), who showed that proteinase inhibitors bind to the outer surface of the acrosome. This may have been due, however, to a nonspecific binding of the inhibitors to the negatively charged acrosomal surface. Addition of spermatozoa to gelatin plates results in "halo" formation of digested gelatin around the anterior, acrosomal portion of the sperm head (Gaddum and Blandau 1970, 1972; Penn, Gledhill and Darzynkiewicz 1972; Benitez-Bribiesca and Velazquez-Meza 1972; Allen, Bishop and Thompson 1974), indicating that acrosin release occurs readily from this site. It is presently not known if such release normally takes place in healthy sperm or is due to sperm degradation and membrane breakdown. The ready release of acrosin under these conditions, as well as other observations (Garner et al. 1975), points to a dual location of acrosin, i.e., both on the inner and the outer acrosomal membranes (but beneath the plasma membrane).

The male reproductive tract proteinase inhibitors are most likely added to the sperm while they reside in the rete testis, epididymis and/or during ejaculation (Zaneveld, Srivastava and Williams 1969, 1970). Removal of the acrosome and subsequent release of acrosin invariably results in the occurrence of an acrosin-inhibitor complex (Zaneveld, Srivastava and Williams, 1969; Polakoski, Zaneveld and Williams 1971; Zaneveld, Polakoski and Williams 1973). Similarly, if spermatozoa are kept in solution and the supernatant is subsequently tested, 80–90% of acrosin is found complexed to inhibitor (Zaneveld, Dragoje and Schumacher 1972). The question remains, however, whether this complex is present on the spermatozoon also, or if the inhibitor is located at a different site from acrosin, for instance, on the surface of the plasma membrane. Release of the acrosin would then result in acrosin-inhibitor binding. The latter appears most plausible at the present time since some evidence has been presented for the surface location of the inhibitor (Hirschhauser and Baudner 1972). There appears to be no relationship between the concentration of inhibitor in seminal plasma and the motility or abnormal forms of spermatozoa (Ingrisch, Haendle and Werle 1970).

Acrosin has the ability of hydrolyzing the zona pellucida of the ovum. Apparently this enzyme lyses a path through this layer at the time of fertilization so that the spermatozoa can penetrate. This was definitively shown when the addition of proteinase inhibitors to spermatozoa specifically prevented their penetration through the zona (Gould 1973). Spermatozoa that reside in the uterus for some time, enough to be fully capacitated, lose all or most of their proteinase inhibitor(s) (Zaneveld, Srivastava and Williams 1969; Fritz and Schiessler, unpubl.). It was further established that the addition of inhibitors from reproductive tract secretions and from other sources to capacitated spermatozoa prevents fertilization in vivo (Zaneveld et al. 1971a,b) and in vitro (Stambaugh, Brackett and Mastroianni 1969; Suominen, Kaufman and Setchell 1973). One can conclude, therefore, that the removal of the seminal inhibitors from spermatozoa is an essential part of the capacitation process without which fertilization cannot occur. This should not be taken to imply that no other events take place during the capacitation process, since such certainly occur (Williams 1973). In addition, significant species variations are probably present.

The physiological role of the other proteolytic enzyme components of the reproductive tract is virtually unknown, although speculations are abundant. It has been suggested that seminin and the plasminogen activator are involved in the liquefaction process of human semen. This remains to be shown, and indications to date seem to negate their involvement (Tauber et al. 1973; Zaneveld et al. 1974). The inability to detect plasminogen, fibrinogen, prothrombin and factor XIII from seminal plasma (Tauber et al. 1973; Zaneveld et al. 1974) make the coagulation-liquefaction process of semen totally different from that of blood. The coagulation process of guinea pig semen has been studied in more detail, and it was shown that a transamidase (transglutaminaselike enzyme) was at least in part responsible for this process (Williams-Ashman et al. 1972). The plasminogen activator may play a role in prostatic cancer through release in the bloodstream (Tagnon, Whitmore and Shulman 1952; Tagnon et al. 1953a,b; Tagnon and Steens-Lievens 1963; Swan and Kerridge 1965).

It was suggested that sperm penetration through cervical mucus was in part accomplished by the lytic action of seminin (Moghissi and Syner 1970). Seminin hydrolyzes cervical mucoids, and sperm migration is accelerated by such hydrolyzed mucus. Addition of seminin to spermatozoa enhances their transport. However, washed spermatozoa (in the absence of seminal plasma) pass through cervical mucus at about the same rate as unwashed spermatozoa, and the role of seminin in this process is questionable. It has also been suggested that acrosin, not seminin, plays a role in cervical mucus transport. Some evidence has been presented to corroborate this (Schumacher and Zaneveld 1974), but definitive studies are lacking. It makes for an attractive hypothesis concerning the transport of spermatozoa, however, since it is known that sperm can only pass through the mucus at midcycle, the time when the proteinase inhibitor concentration of cervical mucus is the lowest. It can therefore be speculated that the cervical inhibitors play a regulatory role in the penetration of sperm through the mucus. A similar suggestion has been made in regard to the inhibitors of Fallopian tube fluid, although their effect would be to prevent sperm penetration through the ovum, or at least to inhibit excess sperm acrosin during the anovulatory periods of the cycle. Against these hypotheses stands the fact that at least in vivo the high molecular weight (serum) reproductive tract inhibitors, including α_1-antitrypsin, inter-α-trypsin inhibitor and the α_{1x}-antichymotrypsin, do not inhibit fertilization in physiological amounts when added to capacitated spermatozoa (Yang, Zaneveld and Schumacher 1975; Zaneveld 1975), and that the low molecular weight cervical mucus inhibitor does not inhibit acrosin at all.

The absolute requirement for an active acrosin to obtain sperm penetration through the zona pellucida of the ovum has led to investigations as to whether acrosin inhibitors can be used as contraceptive agents. Since natural inhibitors appear to be removed during the capacitation process, the use of natural inhibitors did not seem promising and they were indeed shown to be ineffective (Schumacher, Swartwout and Zuspan 1971). Synthetic, low molecular weight inhibitors that bind irreversibly to acrosin therefore seem to be the agents of choice. Such synthetic inhibitors effectively prevent fertilization if added to capacitated or ejaculated sperm or if deposited into the vagina before coitus (Zaneveld, Robertson and Williams 1970; Zaneveld et al. 1971a; Newell, Polakoski and Williams 1972; Migamoto and Chang 1973). The addition of such synthetic inhibitors to Delfen vaginal contraceptive creams greatly increased their antifertility properties, at least in rabbits. Since acrosin appears to be distinct from other known proteolytic enzymes of animal tissues, it should be possible to develop a synthetic inhibitor which specifically inhibits acrosin. Such an inhibitor would therefore only affect the sperm and might be administered without notable side effects. It could be given to both men and women. Another approach would be the use of acrosin antibodies, i.e., a specific immunological approach to contraception. That this can be a realistic method remains to be established, however. Other measures such as inhibiting the seminal liquefaction process or increasing the inhibitor levels in cervical mucus and Fallopian tube fluid may also be of contraceptive value.

Acknowledgments

The authors wish to thank Dr. G. Williams-Ashman for his constructive criticisms of this manuscript and all fellow investigators whose research made this comprehensive overview possible.

REFERENCES

Allen, G. J., M. W. H. Bishop and T. E. Thompson. 1974. Lysis of photographic emulsions by mammalian and chicken spermatozoa. *J. Reprod. Fert.* **36:**249.

Allison, A. C. and E. F. Hartree. 1970. Lysosomal enzymes in the acrosome and their possible role in fertilization. *J. Reprod. Fert.* **21:**501.

Benitez-Bribriesca, L. and S. Velazquez-Meza. 1972. Cytochemical demonstration of proteolytic activity of human and rat spermatozoa. *J. Reprod. Fert.* **29:**419.

Bernstein, M. H. and R. J. Teichman. 1973. A chemical procedure for extraction of the acrosomes of mammalian spermatozoa. *J. Reprod. Fert.* **33:**239.

Brown, C. R. and E. F. Hartree. 1974. Distribution of a trypsin-like proteinase in ram spermatozoa. *J. Reprod. Fert.* **36:**195.

Buruiana, L. M. 1956. Sur l'activité hyaluronidasique et trypsinique du sperme. *Naturwissenschaften* **43:**523.

Churg, A., L. J. D. Zaneveld and G. F. B. Schumacher. 1974. Fine structure of detergent treated human and rabbit spermatozoa. *Biol. Reprod.* **10:**429.

Connors, E. C., F. C. Greenslade and J. P. Davanzo. 1973. The kinetics of inhibition of rabbit sperm acrosomal proteinase by 1-chloro-3-tosylamide-7-amino-2-heptanone (TLCK). *Biol. Reprod.* **9:**57.

Dott, H. M. and J. T. Dingle. 1968. Distribution of lysosomal enzymes in the spermatozoa and cytoplasmic droplets of bull and ram. *Exp. Cell Res.* **52:**523.

Fink, E. H., Schiessler, M. Arnhold and H. Fritz. 1972a. Isolierung eines Trypsin-ahnlichen Enzymes (Akrosin) aus Eberspermien. *Hoppe-Seyler's Z. Physiol. Chem.* **353:**1633.

Fink, E., E. Jaumann, H. Fritz, H. Ingrisch and E. Werle. 1971a. Protease-Inhibitoren im menschlichen Sperma plasma. Isolierung durch Affimitatschromatographe und Hemmverhalten. *Hoppe-Seyler's Z. Physiol. Chem.* **352**:1591.

Fink, E., G. Klein, F. Hammer, G. Muller-Bardorff and H. Fritz. 1971b. Proteinproteinase inhibitors in male sex glands. In *Proceedings International Research Conference on Proteinase Inhibitors,* Munich, 1970 (ed. H. Fritz and H. Tschesche), p. 425. Walter de Gruyter, New York.

Fink, E., H. Fritz, E. Jaumann, H. Schiessler, B. Forg-Brey and E. Werle. 1972b. Protein proteinase inhibitors in male sex glands and their secretions. In *Protides of the Biological Fluids* (ed. H. Peeters), p. 425. Pergamon Press, New York.

Fritz, H., B. Forg-Brey and H. Umezawa. 1973. Leupeptin and antipain: Strong competitive inhibitors of sperm acrosomal proteinase (boar acrosin) and kallikreins from porcine organs (pancreas, submand. glands, urine). *Hoppe-Seyler's Z. Physiol. Chem.* **354**:304.

Fritz, H., H. Schiessler and W. D. Schleuning. 1973. Boar and human acrosin and acrosin inhibitors: Isolation and biochemical characterization. *Biol. Reprod.* **9**:64.

Fritz, H., W. D. Schleuning and W. B. Schill. 1974. Biochemistry and clinical significance of the trypsin-like proteinase from boar and human spermatozoa. In *Proteinase Inhibitors* (ed. H. Fritz et al.), p. 118. Springer Verlag, New York.

Fritz, H., E. Fink, R. Meister and G. Klein. 1970. Isolierung von Trypsin inhibitoren und Trypsin-Plasmin-Inhibitoren aus den Samenblasen von Meerschweinchen. *Hoppe-Seyler's Z. Physiol. Chem.* **351**:1344.

Fritz, H., B. Forg-Brey, E. Fink and H. Schiessler. 1972a. Human akrosin: Gewinnung und Eigenschaften. *Hoppe-Seyler's Z. Physiol. Chem.* **353**:1943.

Fritz, H., M. Arnhold, B. Forg-Brey, L. J. D. Zaneveld and G. F. B. Schumacher. 1972b. Verhalten der "Chymotrypsinahnlichen" Proteinase aus Humansperma gegenuber Protein-Proteinase-Inhibitoren. *Hoppe-Seyler's Z. Physiol. Chem.* **353**:1651.

Fritz, H., B. Forg-Brey, M. Meier, M. Arnhold and H. Tschesche. 1972c. Humanakrosin: Hemmbarkeit durch Protein-Proteinase-Inhibitoren. *Hoppe-Seyler's Z. Physiol. Chem.* **355**:1950.

Fritz, H., H. Schult, M. Mutzel, M. Wiedemann and E. Werle. 1967. Isolierung von Protease Inhibitoren mit Hilfe Wasser unloslicher Enzyme-Harze. *Hoppe-Seyler's Z. Physiol. Chem.* **348**:308.

Fritz, H., B. Forg-Brey, E. Fink, H. Schiessler, E. Jaumann and M. Arnhold. 1972d. Charakterisierung einer Trypsin-ahnlichen Proteinase (Akrosin) aus Eberspermien durch ihre Hemmbarkeit mit verschiedenen Protein-Proteinase Inhibitoren. I, II & III. *Hoppe-Seyler's Z. Physiol. Chem.* **353**:1007.

Fritz, H., N. Heimburger, M. Meier, M. Arnhold, L. J. D. Zaneveld and G. F. B. Schumacher 1972e. Humanakrosin: Zur Kinetik der Hemmung durch Human-Serum inhibitoren. *Hoppe-Seyler's Z. Physiol. Chem.* **353**:1953.

Gaddum, P. and R. J. Blandau. 1970. Proteolytic reaction of mammalian sperm on gelatin membranes. *Science* **170**:749.

Gaddum-Rosse, P. and R. J. Blandau. 1972. Comparative studies on the proteolysis of fixed gelatin membranes by mammalian sperm acrosomes. *Amer. J. Anat.* **134**:133.

Garner, D. L. 1973. Partial characterization of bovine acrosomal proteinase. *Biol. Reprod.* **9**:71.

Garner, D. L. and R. F. Cullison. 1974. Partial purification of bovine acrosin by affinity chromatography. *J. Chromatog.* **92**:445.

Garner, D. L., G. W. Salisbury and C. N. Graves. 1971. Electrophoretic fractionation of bovine proteins and proteinase. *Biol. Reprod.* **4**:93.

Garner, D. L., M. P. Easton, M. E. Munson and M. A. Doane. 1975. Immuno-fluorescent localization of bovine acrosin. *J. Exp. Zool.* **191**:127.

Gilboa, E., Y. Elkana and M. Rigbi. 1973. Purification and properties of human acrosin. *Eur. J. Biochem.* **39**:85.

Gotterer, G., J. Banks and H. G. Williams-Ashman. 1956. Hydrolysis of arginine esters by male accessory sexual tissues. *Proc. Soc. Exp. Biol. Med.* **92**:58.

Gould, K. G. 1973. Application of *in vitro* fertilization. *Fed. Proc.* **32**:2069.

Haendle, H., H. Ingrisch and E. Werle. 1970a. Uber einen neuen Trypsin-chymo-trypsin-Inhibitor im Cervixsekret der Frau. *Hoppe-Seyler's Z. Physiol. Chem.* **351**:545.

————. 1970b. Zur Bedeutung des Proteasen-Inhibitors des Saugespermas bei der Befruchtung und zur Frage der Identitat mit dem Dekapazitations Factor. *Klin. Woch.* **13**:824.

Haendle, H., H. Fritz, I. Trautschold and E. Werle. 1965. Uber einen hormonab-hangigen Inhibitor fur proteolytische Enzyme in mannlichen accessorischen Geschlechtsdrusen und im Sperma. *Hoppe-Seyler's Z. Physiol. Chem.* **343**:185.

Hartree, E. F. and P. N. Srivastava. 1965. Chemical composition of the acrosomes of ram spermatozoa. *J. Reprod. Fert.* **9**:47.

Harvey, C. 1949. Fibrinolysin in human semen. A method of assay and some preliminary observations. *Proc. Soc. Study Fert.* **1**:11.

Hirschhauser, C. and S. Baudner. 1972. Immunologic localization of the human seminal plasma inhibitor in human spermatozoa. *Fert. Steril.* **23**:393.

Hirschhauser, C. and M. Kionke. 1971a. Zur biochemie des Fibrinolytischen Systems in Spermaplasma. *Fortsch. Andrologie* **2**:118.

————. 1971b. Properties of a human seminal plasma inhibitor for trypsin. *Fert. Steril.* **22**:360.

Hirschhauser, C., S. Baudner and E. Daume. 1972. Immunologic identification of human seminal protease inhibitor in cervical mucus. *Fert. Steril.* **23**:630.

Hirschhauser, C., M. Kionke, E. Daume and R. Buchholz. 1971. Trypsin inhibitors in the human female genital tract. *Acta Endocrinol.* **68**:413.

Hisazumi, H. 1970. Studies on the fibrinolytic activity in human semen studied by gel filtration. *Invest. Urol.* **7**:410.

Huggins, C. B. and W. Neal. 1942. Coagulation and liquefaction of semen. Pro-teolytic enzymes and citrate in prostatic fluid. *J. Exp. Med.* **76**:527.

Ingrisch, H., H. Haendle and E. Werle. 1970. Uber die Konzentration des Trypsin-Inhibitors im Sperma von Gesunden und andrologisch Kranken und uber ihre Beziehung zu anderen Parametern des Spermas. *Andrologie* **2**:103.

Karhausen, L. and H. Tagnon. 1955. Le syndrome de fibrinolyse prostatique Nature de l'activite-proteolytique de la prostate. *Acta. Clin. Bel.* **10**:471.

Kester, R. C. 1969. Plasminogen activator in the human prostate. *J. Clin. Path.* **22**:442.

————. 1971. The distribution of plasminogen activator in the male genital tract. *J. Clin. Path.* **24**:726.

Koran, E. and S. Milkovic. 1973. "Collagenase-like" peptidase in human, rat, and bull spermatozoa. *J. Reprod. Fert.* **32**:349.

Liedholm, P. 1973. Passage of tranexamic acid (AMCA) to semen in man and its effect on the fibrinolytic activity and on migration of spermatozoa. *Fert. Steril.* **24**:517.

Lundquist, F. 1952. Studies on the biochemistry of human semen. IV. Amino acids and proteolytic enzymes. *Acta Physiol. Scand.* **25**:178.

Lundquist, F. and H. H. Seedorff. 1952. Pepsinogen in human seminal fluid. *Nature* **170**:115.

Lundquist, F., T. Thorsteinsson and O. Buus. 1955. Purification and properties of some enzymes in human seminal plasma. *Biochem. J.* **59**:69.

McRorie, R. A. and W. L. Williams. 1974. Biochemistry of mammalian fertilization. *Annu. Rev. Biochem.* **43**:777.

Meizel, S. 1972. Biochemical detection of an inactive form of a trypsin-like enzyme in rabbit testis. *J. Reprod. Fert.* **31**:459.

Meizel, S. and J. Cotham. 1972. Partial characterization of a new bull sperm arylamidase. *J. Reprod. Fert.* **28**:303.

Meizel, S. and Y. H. Huang-Yang. 1973. Further studies of an inactive form of a trypsin-like enzyme in rabbit testes. *Biochem. Biophys. Res. Comm.* **53**:1145.

Menezo, Y. and J. Flechon. 1973. Utilisation d'une microtechnique pour la determination des activities enzymatiques des spermatozoides et du plasma seminal chez le lapin et de taureau. *C. R. Acad. Sci.* **277**:1037.

Migamoto, H. and M. C. Chang. 1973. Effects of protease inhibitors on the fertilizing capacity of hamster spermatozoa. *Biol. Reprod.* **9**:533.

Moghissi, K. S. and F. N. Syner. 1970. Studies on human cervical mucus: Mucoids and their relation to sperm penetration. *Fert. Steril.* **21**:234.

Multamaki, S. 1973. Isolation of pure acrosomes by subcellular fractionation of bull spermatozoa. *Int. J. Fert.* **18**:1973.

Multamaki, S. and M. Niemi. 1969. Zona pellucida dissolving proteinase in an acrosomal preparation of the bull spermatozoa. *Scand. J. Clin. Invest.* (Suppl. 108) **23**:80.

————. 1972. Trypsin-like proteolytic activity in an acrosomal extract of bull spermatozoa. *Int. J. Fert.* **17**:43.

Newell, S. D., K. L. Polakoski and W. L. Williams. 1972. Inhibition of fertilization by proteinase inhibitors. In *Proceedings 7th International Congress on Animal Reproduction and Artificial Insemination,* vol. 3, p. 2117. Deutschen Gesellschaft für Züchtungskunde, Bonn, Germany.

Nys, M., C. Brassinne, A. Coune and H. J. Tagnon. 1971. A study of proteolytic and fibrinolytic factors in the human prostate. *Thromb. Diath. Hemorrh.* **25**:481.

Oettle, A. G. 1950. Fibrinolytic factors in human semen. *Proc. Soc. Study Fertil.* **2**:71.

Pedersen, H. 1972. The acrosomes of the human spermatozoon: A new method for its extraction, and an analysis of its trypsin-like enzyme activity. *J. Reprod. Fert.* **31**:99.

Penn, A., B. L. Gledhill and Z. Darzynkiewicz. 1972. Modification of the gelatin substrate procedure for demonstration of acrosomal proteolytic activity. *J. Histochem. Cytochem.* **20**:499.

Polakoski. K. L. 1974. Partial purification and characterization of proacrosin from boar sperm. *Fed. Proc.* **33**:1308.

Polakoski, K. L. and R. A. McRorie. 1973. Boar acrosin. II. Classification, inhibition and specificity studies of a proteinase from sperm acrosomes. *J. Biol. Chem.* **248**:8183.

Polakoski, K. L. and W. L. Williams. 1974. Isolation of proteinase inhibitors from boar sperm acrosomes and boar seminal plasma and its effect on fertilization. In *Proteinase Inhibitors* (ed. H. Fritz et al.), p. 156. Springer Verlag, New York.

Polakoski, K. L., R. A. McRorie and W. L. Williams. 1973. Boar acrosin. I. Purification and preliminary characterization of a proteinase from boar sperm acrosomes. *J. Biol. Chem.* **248**:8178.

Polakoski, K. L., W. L. Williams and R. A. McRorie. 1972. Purification and properties of acrosin, an arginyl proteinase from boar sperm acrosomes. *Fed. Proc.* **31**:278.

————. 1973. Partial purification and characterization of an acidic proteinase in sperm acrosomes. *Fed. Proc.* **32**:310.

Polakoski, K. L., L. J. D. Zaneveld and W. L. Williams. 1971. An acrosin-acrosin inhibitor complex in ejaculated boar sperm. *Biochem. Biophys. Res. Comm.* **45**: 381.

———. 1972. Purification of a proteolytic enzyme from rabbit acrosomes. *Biol. Reprod.* **6**:23.

Polakoski, K. L., L. J. D. Zaneveld, W. L. Williams and R. A. McRorie. 1971. Purification and partial characterization of a trypsin inhibitor from porcine seminal plasma. *Fed. Proc.* **30**:1077.

Povoa, H. and G. G. Villela. 1960. Transaminase in seminal plasma of man. *Experientia* **16**:629.

Propping, D., P. F. Tauber, L. J. D. Zaneveld and G. F. B. Schumacher. 1974. Purification and characterization of plasminogen activator from human seminal plasma. *Fed. Proc.* **33**:289.

Rasmussen, J. and O. K. Albrechtson. 1960a. Fibrinolytic activity in human seminal plasma. *Fert. Steril.* **11**:264.

———. 1960b. Characterization of the fibrinolytic components in the human prostate. *Scand. J. Clin. Lab. Invest.* **12**:267.

Rasmussen, J., O. K. Albrechtson and T. Astrup. 1958. The fibrinolytic activity in the human prostate and seminal fluid. *Trans. 6th Congr. Europ. Soc. Haemat.*, Copenhagen, 1957, p. 494. Karger, Basel.

Schiessler, H., H. Fritz, M. Arnhold, E. Fink and H. Tschesche. 1972. Eigenschaften des Trypsin-ahnlichen Enzymes (Akrosin) aus Eberspermien. *Hoppe-Seyler's Z. Physiol. Chem.* **353**:1638.

Schill, W-B. 1973. Acrosin activity in human spermatozoa: Methodological investigations. *Arch. Derm. Forsch.* **248**:257.

———. 1974. The influence of glycerol on the extractibility of acrosin from human spermatozoa. *Hoppe-Seyler's Z. Physiol. Chem.* **355**:229.

Schleuning, W. D. and H. Fritz. 1974. Some characteristics of highly purified boar sperm acrosin. *Hoppe-Seyler's Z. Physiol. Chem.* **355**:125.

Schleuning, W. D., H. Schiessler and H. Fritz. 1973. Highly purified acrosomal proteinase (boar acrosin): Isolation by affinity chromatography using benzamidine-cellulose and stabilization. *Hoppe-Seyler's Z. Physiol. Chem.* **354**:550.

Schumacher, G. F. B. 1970a. Biochemistry of cervical mucus. *Fert. Steril.* **21**:697.

———. 1970b. Alpha$_1$-antitrypsin in genital secretions. *J. Reprod. Med.* **5**:3.

———. 1971. Inhibition of rabbit sperm acrosomal protease by human alpha-antitrypsin and other protease inhibitors. *Contraception* **4**:67.

———. 1973a. Soluble proteins in cervical mucus. In *The Biology of the Cervix* (ed. R. J. Blandau and K. Moghissi), p. 201. University of Chicago Press, Chicago.

———. 1973b. Soluble proteins of human cervical mucus. In *Cervical Mucus in Human Reproduction, WHO Colloquium, 1972* (ed. M. Elstein, K. S. Moghissi and R. Borth), p. 93. Scripta, Copenhagen.

Schumacher, G. F. B. and M. J. Pearl. 1968. Alpha$_1$-antitrypsin in cervical mucus. *Fert. Steril.* **19**:91.

Schumacher, G. F. B. and L. J. D. Zaneveld. 1972. Inhibition of rabbit and human sperm acrosomal proteases by protease inhibitors of human origin. In *Proceedings 7th International Congress on Animal Reproduction*, vol. 3, p. 2105.

———. 1974. Proteinase inhibitors in human cervical mucus and their *in vitro* interactions with human acrosin. In *Proteinase Inhibitors* (ed. H. Fritz et al.), p. 178. Springer Verlag, New York.

Schumacher, G. F. B., E. K. Strauss and G. L. Wied. 1965. Serum proteins in cervical mucus. *Amer. J. Obstet. Gynecol.* **91**:1035.

Schumacher, G. F. B., J. R. Swartwout and F. P. Zuspan. 1971. Fertility experiments in mice and rabbits with the trypsin-kallikrein inhibitor from bovine lung. In *Proceedings International Conference on Proteinase Inhibitors,* Munich, 1970 (ed. H. Fritz and H. Tschesche), p. 247. Walter de Gruyter, New York.

Srivastava, P. N. 1973a. Location of the zona lysin. *Biol. Reprod.* **9**:84.

————. 1973b. Removal of acrosomes of ram and rabbit spermatozoa. *J. Reprod. Fert.* **33**:323.

Srivastava, P. N., C. E. Adams and E. F. Hartree. 1965. Enzymatic action of acrosomal preparations on rabbit ova *in vitro. J. Reprod. Fert.* **10**:61.

Srivastava, P. N., L. J. D. Zaneveld and W. L. Williams. 1970. Mammalian sperm acrosomal neuraminidases. *Biochem. Biophys. Res. Comm.* **39**:575.

Srivastava, P. N., J. F. Munnell, C. H. Yang and C. W. Foley. 1974. Sequential release of acrosomal membranes and acrosomal enzymes of ram spermatozoa. *J. Reprod. Fert.* **36**:363.

Stambaugh, R. 1973. Acrosomal enzymes and fertilization. In *Biology of Mammalian Fertilization and Implantation* (ed. K. Moghissi and E. S. E. Hafez), p. 187. C. C. Thomas, Springfield, Illinois.

Stambaugh, R. and J. Buckley. 1968. Zona pellucida dissociation enzymes of the rabbit sperm head. *Science* **161**:585.

————. 1969. Identification and subcellular localization of the enzymes effecting penetration of the zona pellucida by rabbit spermatozoa. *J. Reprod. Fert.* **19**:423.

————. 1970. Comparative studies of the acrosomal enzymes of rabbit, rhesus monkey and human spermatozoa. *Biol. Reprod.* **3**:275.

————. 1972a. Studies on acrosomal proteinase of rabbit spermatozoa. *Biochem. Biophys. Acta* **248**:473.

————. 1972b. Histochemical subcellular localization of the acrosomal proteinase effecting dissolution of the zona pellucida using fluorescein-labeled inhibitors. *Fert. Steril.* **23**:348.

Stambaugh, R. and M. Smith. 1973. Comparison of several extraction procedures for rabbit acrosomal enzymes. *J. Reprod. Fert.* **35**:127.

————. 1974. Amino acid content of rabbit acrosomal proteinases and its similarity to human trypsin. *Science* **186**:745.

Stambaugh, R., B. G. Brackett and L. Mastroianni. 1969. Inhibition of *in vitro* fertilization of rabbit ova by trypsin inhibitors. *Biol. Reprod.* **1**:223.

Stambaugh, R., H. M. Seitz and L. Mastroianni. 1974. Acrosomal proteinase inhibitors in rhesus monkey (*Macacca mulatta*) oviduct fluid. *Fert. Steril.* **25**:352.

Suominen, J. J. O. and M. Niemi. 1971. Influence of human seminal proteases on vascular permeability. *Nature New Biol.* **232**:1971.

————. 1972. Human seminal trypsin inhibitors. *J. Reprod. Fert.* **29**:163.

Suominen, J. J. O. and B. P. Setchell. 1972. Enzymes and trypsin inhibitor in the rete testis fluid of rams and boars. *J. Reprod. Fert.* **30**:235.

Suominen, J. J. O., R. Eliasson and M. Niemi. 1971. The relationship of the proteolytic activity of human seminal plasma to various semen characteristics. *J. Reprod. Fert.* **27**:153.

Suominen, J. J. O., M. H. Kaufman and B. P. Setchell. 1973. Prevention of fertilization *in vitro* by an acrosin inhibitor from rete testis fluid of the ram. *J. Reprod. Fert.* **34**:385.

Suominen, J. J. O., S. Multamaki and M. Niemi. 1971. Trypsin inhibitors in human semen. *Scand. J. Clin. Lab. Invest.* **27**:8.

Swan, M. T. and S. Kerridge. 1965. Fibrinolysis and carcinomia of the prostate. *J. Clin. Path.* **18**:330.

Syner, F. N. and K. S. Moghissi. 1972. Purification and properties of human seminal proteinase. *Biochem. J.* **126**:1135.

————. 1973. Properties of proteolytic enzymes and inhibitors in human semen.

In *Biology of Mammalian Fertilization and Implantation* (ed. K. S. Moghissi and E. S. E. Hafez), p. 3. C. C. Thomas, Springfield, Illinois.

Tagnon, H. J. and A. Steens-Lievens. 1963. Studies of fibrinolysis and acid phosphatase in cancer of the prostate. *Nat. Cancer Inst. Monogr.* **12**:297.

Tagnon, H. J., W. F. Whitmore and P. Shulman. 1952. Fibrinolysis in metastatic cancer of the prostate. *Cancer* **5**:9.

Tagnon, H. J., P. Shulman, W. F. Whitmore and L. A. Leone. 1953a. Prostatic fibrinolysis. Study of a case illustrating role in a hemorrhagic diathesis of cancer of the prostate. *Amer. J. Med.* **15**:875.

Tagnon, H. J., W. F. Whitmore, P. Shulman and S. C. Kravitz. 1953b. The significance of fibrinolysis occurring in patients with metastatic cancer of the prostate. *Cancer* **6**:63.

Tauber, P. F., D. Propping, L. J. D. Zaneveld and G. F. B. Schumacher. 1973. Biochemical studies on the lysis of human split ejaculates. *Biol. Reprod.* **9**:62.

Tauber, P. F., L. J. D. Zaneveld, D. Propping and G. F. B. Schumacher. 1975. Components of human split ejaculates. II. Lytic enzymes, plasminogen activator and proteinase inhibitors. *J. Reprod. Fert.* **43**:249.

Thorsteinsson, T. 1958. Proteolytic activity of human, rabbit and bull semen with special reference to peptidases in the genital tract of male rabbit. *Amer. J. Physiol.* **194**:341.

Tympanidis, K. and T. Astrup. 1968. Fibrinolytic activity of rat, rabbit and human sperm cells. *Soc. Exp. Biol. Med.* **129**:179.

Uhlenbruck, G., I. Sprenger, G. F. B. Schumacher and L. J. D. Zaneveld. 1972. Additional properties of acrosin, a proteolytic enzyme from rabbit sperm acrosomes. *Naturwissenschaften* **3**:124.

von Kaulla, K. N. and L. B. Shettles. 1953. Relationship between human seminal fluid and the fibrinolytic system. *Soc. Exp. Biol. Med.* **83**:692.

―――. 1954. Beitrag zur Kenntuis des Proteolytischen Ferment-systems in Mennschlichen Spermaplasma, Mucus Cervicalis, Tubarschleimhaut and Liquor Folliculi. *Klinische Wochenschr.* **32**:468.

Waldschmidt, M., B. Hoffman and H. Karg. 1966. Unterschungen uber die tryptische enzymaktivitat in geschlechtssekreten von bullen. *Zuchthygiene* **1**:15.

Wallner, O. and H. Fritz. 1974. Characterization of an acid-stable proteinase inhibitor in human cervical mucus. *Hoppe-Seyler's Z. Physiol. Chem.* **355**:709.

Williams, W. L. 1973. Biochemistry of capacitation of spermatozoa. In *Biology of Mammalian Fertilization and Implantation* (ed. K. Moghissi and E. S. E. Hafez), p. 19. C. C. Thomas, Springfield, Illinois.

Williams-Ashman, H. G. and D. D. Hoskins. 1969. Biochemistry of male accessory gland secretions in man and nonhuman primates. In *Urinary Infections in the Male*, p. 61. U.S. National Academy of Sciences, Washington, D.C.

Williams-Ashman, H. G., A. C. Notides, S. S. Pabalan and L. Lorand. 1972. Transamidase reactions involved in the enzymatic coagulation of semen: Isolation of glutamyl-lysine dipeptide from clotted secretion protein of guinea pig seminal vesicle. *Proc. Nat. Acad. Sci.* **69**:2322.

Woraschk, H. J. and M. Kressner. 1962. Uber die Trypsin-inhibitor-aktivitat des Fruchtwasser. *Arch. Gynakologie* **196**:622.

Yamane, J. 1930. The proteolytic action of mammalian spermatozoa and its bearing upon the second maturation division of ova. *Cytologia* **1**:394.

―――. 1935. Kausal-analytische Studien uber die Befruchtung des Kanincheneies. I. Die Dispersion der Follikelzellen und die Ablosung der Zellen der Corona radiata des Eies durch Spermatozoen. II. Die Isolierung der auf das Eizytoplasma auflosund wirkenden Substanzen aus der Spermatozoen. *Cytologia* **6**:233, 474.

Yanagimachi, R. and R. J. Teichman. 1972. Cytochemical demonstration of acrosomal proteinase in mammalian and avian spermatozoa by a silver proteinate method. *Biol. Reprod.* **6:**87.

Yang, S. L., L. J. D. Zaneveld and G. F. B. Schumacher. 1975. Effect of serum proteinase inhibitors on the fertilization capacity of rabbit spermatozoa. *Fed. Proc.* **34:**255.

Ying, S. H., E. Day, W. F. Whitmore and H. J. Tagnon. 1956. Fibrinolytic activity in human prostatic fluid and semen. *Fert. Steril.* **7:**81.

Zaneveld, L. J. D. 1974. Sperm acrosomal enzymes and their potential for contraceptive development. In *Sperm Transport, Survival and Fertilizing Ability. INSERM* **26:**435.

————. 1975. The human ejaculate and its potential for fertility control. In *Control of Male Fertility* (ed. J. J. Sciarra, C. Markland and J. J. Speidel), p. 41. Harper and Row, San Francisco.

Zaneveld, L. J. D., B. M. Dragoje and G. F. B. Schumacher. 1972. Acrosomal proteinase and proteinase inhibitor of human spermatozoa. *Science* **177:**702.

Zaneveld, L. J. D., K. L. Polakoski and W. L. Williams. 1972. Properties of a proteolytic enzyme from rabbit sperm acrosomes. *Biol. Reprod.* **6:**30.

————. 1973. A proteinase and proteinase inhibitor of mammalian sperm acrosomes. *Biol. Reprod.* **9:**219.

Zaneveld, L. J. D., R. T. Robertson and W. L. Williams. 1970. Synthetic enzyme inhibitors as antifertility agents. *FEBS Letters* **11:**345.

————. 1970. Inhibition by seminal plasma of acrosomal enzymes in intact sperm. acrosomal proteinase: Antibody inhibition and immunologic dissimilarity to human pancreatic trypsin. *Fert. Steril.* **24:**479.

Zaneveld, L. J. D., P. N. Srivastava and W. L. Williams. 1969. Relationship of a trypsin-like enzyme in rabbit spermatozoa to capacitation. *J. Reprod. Fert.* **20:** 337.

————. 1970. Inhibition by seminal plasma of acrosomal enzymes in intact sperm. *Proc. Soc. Exp. Biol. Med.* **133:**1172.

Zaneveld, L. J. D., K. L. Polakoski, R. T. Robertson and W. L. Williams. 1971a. Trypsin inhibitors and fertilization. In *Proceedings International Research Conference on Proteinase Inhibitors,* Munich, 1970 (ed. H. Fritz and H. Tschesche), p. 236. Walter de Gruyter, New York.

Zaneveld, L. J. D., K. L. Polakoski, J. Travis and W. L. Williams 1970a. Purification and properties of a trypsin-like enzyme (TLE) isolated from rabbit sperm acrosomes. In *Proceedings 3rd Annual Meeting of the Society for the Study of Reproduction,* p. 7.

Zaneveld, L. J. D., R. T. Robertson, M. Kessler and W. L. Williams. 1971b. Inhibition of fertilization *in vivo* by pancreatic and seminal plasma trypsin inhibitors. *J. Reprod. Fert.* **25:**387.

Zaneveld, L. J. D., G. F. B. Schumacher, P. F. Tauber and D. Propping. 1974a. Proteinase inhibitors and proteinases of human semen. In *Proteinase Inhibitors* (ed. H. Fritz et al.), p. 136. Springer Verlag, New York.

Zaneveld, L. J. D., L. Wagner, H. D. Schlumberger and G. F. B. Schumacher. 1974b. Immunological and biochemical studies on fractionated bull spermatozoa. *J. Reprod. Fert.* **38:**411.

Zaneveld, L. J. D., R. T. Robertson, M. Kessler, P. N. Srivastava and W. L. Williams. 1970b. Inhibition of fertilization *in vivo* by mammalian trypsin inhibitors. *Fed. Proc.* **29:**644.

Zaneveld, L. J. D., G. F. B. Schumacher, H. Fritz, E. Fink and E. Jaumann. 1973. Interaction of human sperm acrosomal proteinase with human seminal plasma proteinase inhibitors. *J. Reprod. Fert.* **32:**525.

Note Added in Proof

After this manuscript was submitted, three additional articles appeared that add to the subject matter. These are all from the same book: *Proteinase Inhibitors* (ed. H. Fritz et al.), Springer Verlag, New York, 1974. Other articles from this book had already been included in the references. The additional articles are: (1) Schiessler, H., M. Arnhold and H. Fritz: "Characterization of two proteinase inhibitors from human seminal plasma and spermatozoa," p. 147; (2) Tschesche, H., S. Kupfer, O. Lengel, R. Klauser, M. Meier and H. Fritz: "Purification, characterization and structural studies of proteinase inhibitors from boar seminal plasma and boar spermatozoa," p. 164; and (3) Polakoski, K. L. and W. L. Williams: "Studies on the purification and characterization of boar acrosin," p. 128.

Studies on the Enzymatic and Molecular Nature of Acrosomal Proteinase

Richard Stambaugh and Monica Smith

Division of Reproductive Biology
University of Pennsylvania School of Medicine
Philadelphia, Pennsylvania 19174

Acrosomal proteinase or acrosin (3.4.21.10) is contained within the acrosomes of spermatozoa (Stambaugh and Buckley 1968, 1969, 1972a; Srivastava, Adams and Hartree 1965; Zaneveld et al. 1969; Multamäki and Niemi 1972; Gaddum-Rosse and Blandau 1972; Yanagimachi and Teichman 1972), and its proteolytic activity is essential for fertilization (Stambaugh, Brackett and Mastroianni 1969; Greenslade et al. 1973; Zaneveld et al. 1971). The major role of this proteinase appears to be the digestion of a small oblique tunnel through the zona pellucida, a dense glycoprotein capsule surrounding the ovum, allowing the spermatozoon to enter the perivitelline space where fusion of the plasma membranes of the male and female gametes occurs. Possibly it may also participate in penetration of the matrix between the cells of the cumulus oophorus, a large follicular cell mass surrounding the zona pellucida and ovum at the time of ovulation. However, hyaluronidase, another acrosomal enzyme, is capable of dispensing this matrix without the aid of any proteinase, and Metz (1972) has demonstrated that hyaluronidase antibodies inhibit cumulus oophorus dispersion by spermatozoa. Dissolution of the zona pellucida by extracts of rabbit acrosomes is shown in Figure 1; trypsin inhibitors will inhibit this dissolution (Stambaugh and Buckley 1968, 1969). Figure 2 shows a rabbit spermatozoon in the process of penetrating the zona pellucida and a spermatozoon which has already reached the perivitelline space and is lying on the plasma membrane.

Acrosomal enzymes are exposed by the acrosome reaction immediately before fertilization takes place in the oviduct (Barros et al. 1967). This reaction involves fusion of the plasma membrane with the outer acrosomal membrane in a small but well-defined area on the anterior part of the spermatozoon. The acrosome then undergoes eversion, and the inner acrosomal membrane, together with its attached enzymes, becomes the leading surface of the spermatozoon as it penetrates the matrix of the cumulus oophorus and the zona pellucida. It is not yet clear whether the physiologically active forms of acrosomal proteinase and hyaluronidase are released during penetration, or whether they perform their digestive function while still attached to the

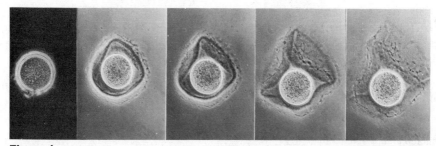

Figure 1
Dissolution of the rabbit zona pellucida by acrosomal extracts at 37.5°C.

inner acrosomal membrane. However, there seems to be little doubt that acrosomal proteinase is bound firmly to the acrosomal membrane in both ejaculated and epididymal sperm. Allison and Hartree (1970) and Zamboni (1971) have documented the lysosomal-like origin of the acrosome, but by the time the spermatozoa have matured, this organelle has evolved into a highly specialized structure for fertilization.

No attempt will be made here to review the extensive literature on the physiological relationship between this proteinase and proteinase inhibitors present in the seminal plasma, cervical mucus and oviduct fluid (Williams 1972; Fritz, Schiessler and Schleuning 1972; Schumacher 1973; Stambaugh, Seitz and Mastroianni 1974). Rather, we will review briefly some of the

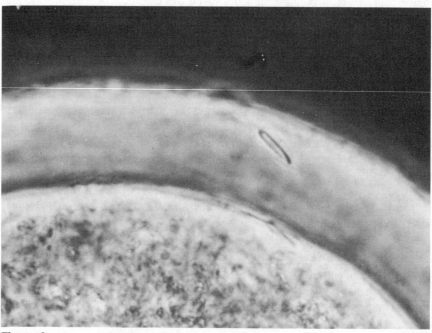

Figure 2
Rabbit ovum with one spermatozoon in the zona pellucida and a second spermatozoon in the perivitelline space lying on the egg plasma membrane.

similarities and differences that have been reported between this enzyme and the pancreatic trypsins, together with our more recent data on the molecular nature of acrosomal proteinase.

Although proteolytic activity was described in extracts of whole spermatozoa as early as 1930 by Yamane and later by Srivastava, Adams and Hartree (1965), the specific trypsinlike activity was not localized by subcellular fractionation (Stambaugh and Buckley 1968, 1969) and by cytochemistry (Stambaugh and Buckley 1972b; Yanagimachi and Teichman 1972) until later. It was established that this enzyme was essential for fertilization in vitro (Stambaugh, Brackett and Mastroianni 1969; Greenslade et al. 1973) and in vivo (Zaneveld et al. 1971), using both natural and synthetic inhibitors for acrosomal proteinase.

Acrosomal proteinase is specific for the peptide and ester bonds of lysine and arginine, it has the same pH optimum and K_m with most substrates, and many of its other enzymatic properties are quite similar to pancreatic trypsin (Table 1). The enzyme is tightly bound to the membrane, and a number of molecular weights ranging from 23,000–59,000 can be obtained depending on the extraction procedure used to solubilize the enzyme (Stambaugh and Buckley 1968, 1969; Williams 1972; Zaneveld, Dragoje and Schumacher 1972; Stambaugh and Smith 1974). However, we have recently extracted and purified this proteinase to a constant specific activity for amino acid analysis, and all of the physicochemical, enzymatic and immunological evidence indicates that only one proteinase is present in the acrosome with a molecular weight of approximately 23,000. Studies by Meizel (1972) indicate that a precursor form of the enzyme exists in testes, but by the time the spermatozoa mature, all of the proteinase exists in the active form.

The purified enzyme has an isoelectric point of 10.2, similar to that of pancreatic trypsin, but it is somewhat more heat labile than pancreatic trypsin (Table 1). However, the enzyme is stable at pH 3.0, similar to pancreatic trypsin, and will retain its activity for long periods of time at this pH and 5°C. At lower H^+ concentrations than this, the enzyme undergoes both aggregation and autodigestion, which has limited extensive physicochemical characterization of this enzyme.

Enzymatically acrosomal proteinase has the same pH optimum as pancreatic trypsin using N-α-benzoyl-L-arginine ethyl ester (BAEE) or N-α-benzoyl-D,L-arginine p-nitroanilide (BAPNA), and the K_m values are also similar for both enzymes (Table 1). Using BAEE as the substrate, the K_i values for soybean trypsin inhibitor and for benzamidine are also simliar for both enzymes (Table 1). However, acrosomal proteinase has a somewhat higher K_i value than trypsin for lima bean trypsin inhibitor, while trypsin has a higher K_i value than acrosomal proteinase for p-aminobenzamidine (Table 1). Most likely, acrosomal proteinase has a catalytic site similar to the other enzymes, since it is inactivated by DFP and TLCK and not by TPCK (Table 1), but the participation of serine and histidine in catalysis and their position in the primary structure of the enzyme has not been established.

These similarities between acrosomal proteinase and pancreatic trypsin motivated us to investigate the amino acid composition and immunological properties of this enzyme in spite of the small quantities of available enzyme. Rabbit (New Zealand White) epididymal spermatozoa from 44 male rabbits were collected and fractionated subcellularly using sonication and sucrose

Table 1
Some Physical and Enzymatic Properties of Acrosomal Proteinase and Pancreatic Trypsin

Property	Rabbit acrosomal proteinase		Bovine pancreatic trypsin
	epididymal	*ejaculated*	
pH Optimum (BAEE)[a,b]	8.2	8.2	8.2
K_m (BAEE)[a,b]	5.2×10^{-6} M	5.2×10^{-6} M	5.76×10^{-6} M
pH Optimum (BAPNA)[b]	8.2	8.2	8.2
K_m (BAPNA)[b]	10.9×10^{-3} M	10.0×10^{-3} M	6.57×10^{-3} M
K_i (soybean inhibitor)[b]	1.0×10^{-10} M	1.0×10^{-10} M	2.1×10^{-10} M
K_i (lima bean inhibitor)[b]	2.0×10^{-5} M	1.8×10^{-5} M	1.0×10^{-8} M
K_i (*p*-aminobenzamidine)[b]	8.0×10^{-6} M	8.5×10^{-6} M	3.7×10^{-5} M
K_i (benzamidine)[b]	2.6×10^{-5} M	2.3×10^{-5} M	1.8×10^{-5} M
Diisopropylfluorophosphate inhibition (DFP)[c]	+	+	+
1-Chloro-3-tosylamido-7-amino heptanone inhibition (TLCK)[c]	+	+	+
Tosylphenylalanine chloromethyl ketone inhibition (TLCK)[c]	−	−	−
$S_{20,w}$	2.7	n.d.	2.5 (bovine)[d,e] 2.68 (human)[f]
Isoelectric point[d,e]	10.2	n.d.	10.8
Half-life (50°C, pH 7.5)[b]	1.5 min	1.6 min	4.1 min

n.d. = No data. Data from [a] Stambaugh and Buckley (1968); [b] Stambaugh and Buckley (1972a); [c] Polakoski and McRorie (1973); [d] Stambaugh and Smith (1974); [e] Walsh and Neurath (1974); [f] Travis and Roberts (1969).

density gradient centrifugation as previously described (Stambaugh and Buckley 1969). After centrifugation, the extract was fractionated on sucrose gradients at 216,000g for 17 hours at 5°C. The fractions containing both hyaluronidase and a 4.3S molecular form of acrosomal proteinase were collected, dialyzed and electrofocused in a sucrose gradient (pH 7–10) for 70 hours at 3°C using an LKB 8100 apparatus and 2.0% ampholyte resins, a procedure providing good resolution of acrosomal proteinase (pI 10.2) and hyaluronidase (pI 5.9). The peak fractions of acrosomal proteinase collected from this column with a specific activity of 20,047 U/mg and an $s_{20,w}$ of 2.7S were used for amino acid analyses on a Beckman-Spinco Model 120C analyzer. This procedure was repeated four times with essentially the same amino acid analysis each time.

The results (Table 2) demonstrated a remarkable similarity in amino acid composition between rabbit acrosomal proteinase and human pancreatic trypsin (Stambaugh and Smith 1974). The glycine content was identical in both enzymes, while the structurally similar amino acid pairs, lysine and arginine, threonine and serine, and tyrosine and phenylalanine, were also

Table 2

Amino Acid Composition of Rabbit Acrosomal Proteinase Compared with Human and Bovine Trypsin

Amino acid	Acrosomal proteinase	Human trypsin[a]	Bovine trypsin[b]
Glycine	20	20	25
Lysine	10	11	14
Arginine	7	6	2
Threonine	12	10	10
Serine	22	24	33
Tyrosine	5	7	10
Phenylalanine	6	4	3
Histidine	4	3	3
Aspartic acid	22	21	22
Proline	8	9	9
Alanine	14	13	14
Methionine	2	1	2
Leucine	13	12	14
Glutamic acid	27	21	14
Valine	10	16	17
Isoleucine	8	12	15
Total	190	190	207
Half-cystine	n.d.	8	12
Tryptophan	n.d.	3	4

n.d. = No determination.
[a] Data from Travis and Roberts (1969).
[b] Data from Walsh and Neurath (1964).

present in identical numbers of residues when taken in pairs. That is, acrosomal proteinase has one less lysine, but one more arginine; it has two less serines, but two more threonines; and it has two less tryosines, but two more phenylalanines. The other amino acids all differ by only one residue, with the exception of valine, isoleucine and glutamic acid. Acrosomal proteinase has six less valine and four less isoleucine residues, but six more glutamic acid residues compared with human trypsin. Due to the small quantities of enzyme available to work with and interference in spectrophotometry by the ampholyte resins, we have not yet obtained an accurate determination of half-cystine and tryptophan residues.

Unfortunately, the amino acid composition of rabbit pancreatic trypsin has never been reported. However, we did compare the antigenic properties of these proteinases by preparing an antiserum to crystalline bovine pancreatic trypsin using Freund's adjuvant. To accomplish immunodiffusion with acrosomal proteinase, the gel must be buffered at pH 5.0. At lower hydrogen ion concentrations, acrosomal proteinase aggregates rapidly within the well of the immunodiffusion plate, and the enzyme never diffuses into the gel in sufficient quantities to form precipitation bands. At pH 5.0, distinct cross-reactions are obtained between rabbit antiserum to bovine pancreatic trypsin with human, bull, rhesus monkey and rabbit acrosomal proteinase, confirming the structural similarities between these proteinases. Three of these precipitation bands are shown in Figure 3.

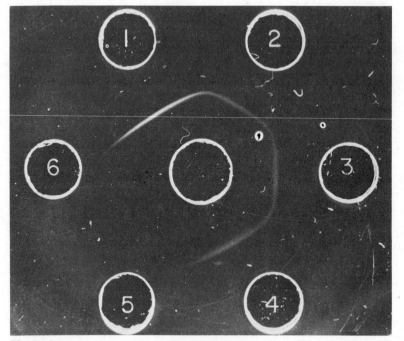

Figure 3
Immunodiffusion plate with rabbit antiserum to bovine trypsin in the center well against (1) bovine trypsin, (2) human acrosomal proteinase, (3) rabbit acrosomal proteinase, (4) bull acrosomal proteinase, (5) human thrombin and (6) bovine chymotrypsin.

These physicochemical, enzymatic, structural and antigenic properties demonstrate the striking similarity between pancreatic trypsin and acrosomal proteinase, and it will be interesting to see how similar the structures of these two enzymes are in the same species, and why one enzyme is soluble while the other is bound tightly to the acrosomal membrane.

Acknowledgments

Supported by Ford Foundation Grant No. 650-0058B and Program Project Grant NIH-HD-6274-03.

REFERENCES

Allison, A. C. and E. F. Hartree. 1970. Lysosomal enzymes in the acrosome and their possible role in fertilization. *J. Reprod. Fert.* **21**:501.

Barros, C., J. M. Bedford, L. E. Franklin and C. R. Austin. 1967. Membrane vesiculation as a feature of the mammalian acrosome reaction. *J. Cell Biol.* **34**:C1.

Fritz, H., H. Schiessler and W. Schleuning. 1972. Proteinase and proteinase inhibitors in the fertilization process: New concepts of control. In *Advances in the Biosciences* (ed. G. Raspe and S. Bernhard), vol. 10, p. 271. Pergamon Press, New York.

Gaddum-Rosse, P. and R. J. Blandau. 1972. Comparative studies on the proteolysis of fixed gelatin membranes by mammalian sperm acrosomes. *Amer. J. Anat.* **134**:133.

Greenslade, F. C., J. J. McCormack, A. F. Hirsch and J. P. Davanzo. 1973. Blockage of fertilization in *Rana pipiens* by trypsin inhibitors. *Biol. Reprod.* **8**:306.

Meizel, S. 1972. Biochemical detection and activation of an inactive form of a trypsin-like enzyme in rabbit testes. *J. Reprod. Fert.* **31**:459.

Metz, C. B. 1972. Effects of antibodies on gametes and fertilization. *Biol. Reprod.* **6**:358.

Multamäki, S. and M. Niemi. 1972. Trypsin-like proteolytic activity in an acrosomal extract of bull spermatozoa. *Int. J. Fert.* **17**:43.

Polakoski, K. L. and R. A. McRorie. 1973. Boar acrosin. *J. Biol. Chem.* **248**:8183.

Schumacher, G. F. B. 1973. Soluble proteins of human cervical mucus. In *Cervical Mucus in Human Reproduction* (ed. M. Elstein, K. S. Moghissi and R. Borth), p. 93. Scripta, Copenhagen.

Srivastava, P. N., C. E. Adams and E. F. Hartree. 1965. Enzymatic action of acrosomal preparations on the rabbit ovum in vitro. *J. Reprod. Fert.* **10**:61.

Stambaugh, R. and J. Buckley. 1968. Zona pellucida dissolution enzymes of the rabbit sperm head. *Science* **161**:585.

———. 1969. Identification and subcellular localization of the enzymes affecting penetration of the zona pellucida by rabbit spermatozoa. *J. Reprod. Fert.* **19**:423.

———. 1972a. Studies on acrosomal proteinase of rabbit spermatozoa. *Biochim. Biophys. Acta* **284**:473.

———. 1972b. Histochemical subcellular localization of the acrosomal proteinase effecting dissolution of the zona pellucida using fluorescein-labeled inhibitors. *Fert. Steril.* **23**:348.

Stambaugh, R. and M. Smith. 1974. Amino acid content of rabbit acrosomal proteinase and its similarity to pancreatic trypsin. *Science* **186**:745.

Stambaugh, R., B. G. Brackett and L. Mastroianni, Jr. 1969. Inhibition of in vitro fertilization of rabbit ova by trypsin inhibitors. *Biol. Reprod.* **1**:223.

Stambaugh, R., H. M. Seitz and L. Mastroianni. 1974. Acrosomal proteinase inhibitors in rhesus monkey (*Macaca mulatta*) oviduct fluid. *Fert. Steril.* **25**:352.

Travis, J. and R. C. Roberts. 1969. Human trypsin. Isolation and physical-chemical characterization. *Biochemistry* **8**:2884.

Walsh, K. A. and H. Neurath. 1964. Trypsinogen and chymotrypsinogen as homologous proteins. *Proc. Nat. Acad. Sci.* **52**:884.

Williams, W. L. 1972. Biochemistry of capacitation of spermatozoa. In *Biology of Mammalian Fertilization and Implantation* (ed. K. S. Moghissi and E. S. E. Hafez), p. 19. Charles C. Thomas, Springfield, Ill.

Yamane, J. 1930. The proteolytic action of mammalian spermatozoa and its bearing upon the second maturization division of ova. *Cytologia* **1**:394.

Yanagimachi, R. and R. J. Teichman. 1972. Cytochemical demonstration of acrosomal proteinase in mammalian and avian spermatozoa by a silver proteinate method. *Biol. Reprod.* **6**:87.

Zamboni, L. 1971. *Fine Morphology of Mammalian Fertilization.* Harper and Row, New York.

Zaneveld, L. J. D., B. M. Dragoje and G. F. B. Schumacher. 1972. Acrosomal proteinase and proteinase inhibitor of human spermatozoa. *Science* **177**:702.

Zaneveld, L. J. D., P. N. Srivastava and W. L. Williams. 1969. Relationship of a trypsin-like enzyme in rabbit spermatozoa to capacitation. *J. Reprod. Fert.* **20**:337.

Zaneveld, L. J. D., R. T. Robertson, M. Kessler and W. L. Williams. 1971. Inhibition of fertilization in vivo by pancreatic and seminal plasma trypsin inhibitors. *J. Reprod. Fert.* **25**:387.

Boar, Bull and Human Sperm Acrosin—Isolation, Properties and Biological Aspects

Hans Fritz, Wolf-Dieter Schleuning and Hans Schiessler

Institut für Klinische Chemie und Klinische Biochemie der Universität München
8 München 2, Germany

Wolf-Bernhard Schill

Dermatologische Klinik und Poliklinik der Universität München
Abteilung für Andrologie
8 München 2, Germany

Volker Wendt

Abteilung für Andrologie und Künstliche Besamung
Gynäkologische und Ambulatorische Tierklinik der Universität München
8 München 22, Germany

Gabriele Winkler

Organisch-Chemisches Laboratorium der Technischen Universität München
Lehrstuhl für Organische Chemie und Biochemie
8 München 2, Germany

The acrosome of the spermatozoon contains a proteolytic enzyme called acrosin which has trypsinlike specificity. The discovery, distribution, purification processes, properties and possible biological function of this proteinase have been reviewed and discussed in detail in recently published articles (Zaneveld, Polakoski and Schumacher, this volume; McRorie and Williams 1974; Fritz, Schiessler and Schleuning 1973; Williams 1972; Stambaugh 1972; Zaneveld et al. 1971a). Therefore, we will focus mainly on extraction and isolation methods developed in our laboratories and on important properties and some biological and clinical aspects of boar, bull and human acrosin elucidated in the course of our investigations.

EXTRACTION AND PURIFICATION

Isolation Procedure

Boar acrosin was isolated in preparative scale by the procedure summarized in Table 1 (Schleuning, Schiessler and Fritz 1973; Fritz, Schleuning and Schill 1974; Schleuning and Fritz 1975). This method (outlined below) is suitable for the isolation of all acid-stable acrosins irrespective of the species and yields in a few steps the highly purified proteinase.

1. Washed spermatozoa are incubated in aqueous acetic acid pH 2.0–2.7 (the extract pH depends on the stability of the acrosin: boar, pH 2.0–2.7; human, pH 2.2–2.7; bull, pH 2.7) for 15 minutes or longer, followed by centrifugation and ultrafiltration or lyophilization of the supernatant.
2. Separation of proteinase or acrosin inhibitors from acrosin is achieved by fractionation of the sperm extracts on Sephadex G-75 in acidic media pH

Table 1

Isolation of Acid-stable Acrosins

Procedure	Step	Specific activity U^a/mg	Yield %	Purification factor
Acidic extraction (H+)	1	0.05–0.10[b]		
Gel filtration (H+)	2	0.5–1.8		5–18
Neutralization, centrifugation	2a	0.6–2.2	95–100	1–2
Affinity chromatography	3	14.5	80–90	7–24
Desalting	4	14.5	95–100	1
Lyophilization	4a	12.8–13.8[c] (BAEE: 165)	92–95	

[a] Substrate: N-α-benzoyl-D,L-arginine p-nitroanilide.

[b] Approximate values (inhibitors are present!).

[c] The values calculated on the basis of protein estimations by the biuret method are given in Table 2; see also the text.

2.0–2.7 (see above). After this step, glass contact of acrosin-containing solutions should be strictly avoided (Fink et al. 1972).

3. Affinity chromatography results in the most efficient purification. The competitive acrosin inhibitor p-aminobenzamidine (Schiessler et al. 1972; Stambaugh and Buckley 1972) linked via an arm to cellulose or agarose was used as acrosin-specific adsorbent.

4. The acrosin fraction eluted in a small volume from the affinity column at pH 2.5 (cellulose) or 4.5 (agarose) is desalted on a Merckogel PGM 2000 column (equilibrated and developed with 10 mM ammonium formiate pH 3.0) and subsequently lyophilized in the presence of sucrose (1.5%, w/v).

5. Further purification of boar acrosin by ion exchange and gel chromatography has not been achieved thus far (Schleuning 1975), indicating a very high degree of purity.

Boar Acrosin, Electrophoretic Investigations

Boar acrosin isolated by the described procedure shows only one substantial protein band in acrylamide gel electrophoresis at pH 4.8 (Fig. 1). For comparison, the acidic sperm extract and the acrosin fraction obtained by gel filtration, i.e., after separation of the inhibitors (step 2), were also subjected to gel electrophoresis. Figure 2 demonstrates that the protein fraction identified in the acrylamide gel effectively hydrolyzes gelatin which had been entrapped in gel 2 and 3 and therefore should be identical with acrosin. Besides this identified acrosin fraction, other gelatin-hydrolyzing proteinases are present in the acidic sperm extract; these are removed by affinity chromatography (cf. gels 2 and 3 in Fig. 2). Further separation of the acrosin protein band could not be achieved by acrylamide gel electrophoresis in buffer solutions pH 4.8–8.3.

The results obtained in SDS-acrylamide gel electrophoresis varied with

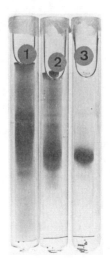

Figure 1
Acrylamide gel electrophoresis of boar acrosin preparations
of different degrees of purity. Gel 1: acidic (2% acetic acid)
sperm extract; gel 2: acrosin fraction after gel filtration
(separation of inhibitors); gel 3: lyophilized material after
affinity chromatography. Electrophoresis was performed as
described by Schleuning and Fritz (1974). Reservoir buffer:
0.05 M sodium acetate pH 4.8. Proteins were stained with
Coomassie brilliant blue.

the different batches of boar acrosin applied. Acrosin isolated previously from
spermatozoa of stored ejaculates showed three protein bands, i.e., the main
fraction with a molecular weight of about 38,000, and the two faster moving
minor fractions with molecular weights around 37,000 and 34,000, respectively (Schleuning and Fritz 1974; Fritz, Schleuning and Schill 1974). However, if freshly ejaculated spermatozoa were subjected to the extraction
procedure, mainly the 38,000 molecular weight acrosin fraction was obtained
(Fig. 3). In order to get higher yields (see below), washed spermatozoa
were shock frozen (−196°C) and stored at −25°C before extraction.

Other batches of boar acrosin were purified from spermatozoa that had

Figure 2
Determination of gelatin digestive fractions after acrylamide gel electrophoresis of boar acrosin preparations. Gel
1: pure boar acrosin, Coomassie blue-stained (cf. gel 3 in
Fig. 1); gel 2: acidic sperm extract and gel 3: pure boar
acrosin after digestion of gel-enclosed gelatin (transparent
bands). Electrophoresis was performed as described by
Schleuning and Fritz (1974) (cf. Fig. 1). Gelatin, 0.1%,
was embedded in gels 2 and 3. These gels were incubated
after electrophoresis in 0.01 M phosphate buffer pH 7.8 at
37°C for 30 minutes and subsequently stained with a mixture of 0.2% Ponceau S, 0.2% light green and 0.1%
amido black in 5% trichloroacetic acid (M. Klockow,
pers. comm.).

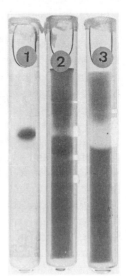

Besides the acrosin fraction (electrophoresis is performed at pH 4.8 so that the inhibitor may be partially
removed), three weaker digestion zones are to be seen in
gel 2. A relatively high amount of boar acrosin was applied
to gel 3 to demonstrate separation of the additional fractions present in gel 2; however, only a very weak, slower
moving fraction is still visible in gel 3.

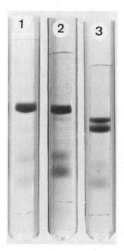

Figure 3
SDS-acrylamide gel electrophoresis of different preparations of boar acrosin. Gel 1: acrosin preparation obtained
from freshly ejaculated spermatozoa; gels 2 and 3: acrosin
preparations obtained from stored ejaculates. Reduction of
acrosin and electrophoresis were performed as described
by Schleuning and Fritz (1974).

been stored in diluted ejaculates and later shipped to us in the frozen state.
Besides the 38,000 molecular weight band, these batches contained two or
four fractions with molecular weights between 10,000 and 25,000 which did
not disappear by repurification of the active material on the affinity column
(see Fig. 3). Some batches contained two fractions with very similar molecular weights (near 38,000) in nearly equal amounts. The occurrence of
polypeptide chains of lower molecular weights than 38,000 is probably due
to the presence of active acrosin molecules in which one or more peptide
bonds are hydrolyzed in the batches under consideration. Such a limited
hydrolysis could be caused by other proteinases present in the sperm
acrosome (reviewed by Zaneveld, Polakoski and Schumacher, this volume;
Srivastava et al. 1974), because incubation of a single (38,000 MW)-band
acrosin preparation at pH 8.0 for 2–16 hours did not cause the appearance of
additional faster moving fractions. The homogeneity of the 38,000 MW
acrosin fraction was also checked by end-group analysis: the only N-terminal
amino acid residue found using the dansylation technique was alanine.

On the basis of the results given, we conclude that the molecular weight of
native boar acrosin is around 38,000 or even higher, because peptides of
lower molecular weights (<5000) are not detectable under the conditions
used in SDS gel electrophoresis. Such peptides may be separated from
acrosin molecules composed of two or more polypeptide chains of appropriate length (either in the native state or arising from internal cleavage(s)
during proacrosin activation) by reduction of the disulfide bonds. The nature
and length of the carbohydrate portion of the boar acrosin molecule
(Schleuning and Fritz 1974) may also influence the migration rate on the
SDS gel compared to the marker proteins and thus cause some uncertainty in
the molecular weight estimation.

Polakoski, McRorie and Williams (1973; Polakoski and Williams 1974)
found a molecular weight of about 30,000 for their acrosin preparation in
SDS gel electrophoresis. These authors obtained the acrosomal extracts by
incubation of ejaculated boar spermatozoa for 90 minutes in near neutral
detergent solutions. Limited enzymatic degradation of acrosin by other

acrosomal proteinases (Zaneveld, Polakoski and Williams, this volume) or proteinases probably attached to the surface of sperm head membranes (Waldschmidt, Karg and Hoffman 1964; Stambaugh and Buckley 1970) could occur under these conditions and thus be responsible for the observed relatively low molecular weight of 30,000. A more exhaustive degradation may be prevented by formation of the acrosin-inhibitor complex, in which only part of the acrosin molecule is accessible to other proteolytic enzymes, and by the shielding effect of the carbohydrate residues located at the surface of both acrosin and inhibitor (see the following paper by Fritz et al., this volume) molecules.

The observations published by Garner (1973) are of interest in this connection. He not only found a molecular weight of about 37,000 in SDS gel electrophoresis for the major component of bull acrosin, but also a fraction with a molecular weight of 44,000 (Garner and Cullison 1974). The occurrence of such multiple forms of acrosin would not be unusual if acrosin is liberated by limited proteolytic cleavage of an inactive precursor of higher molecular weight, as is indicated by more recent findings (Meizel and Huang-Yang 1973; Polakoski 1974; Meizel, Mukerji and Huang-Yang 1974; Schill and Fritz 1975a; Huang-Yang and Meizel 1975).

On the basis of the given results, we would recommend the use of freshly collected spermatozoa only, as well as short-time extraction methods, if homogeneous acrosin preparations are expected. Acidic extraction of washed spermatozoa, preferably after freeze-thawing or glycerol treatment (see below), seems to be the method of choice.

Human Acrosin, Electrophoretic Investigations

Human acrosin purified by the same procedure was separated into three gelatinolytic active fractions by polyacrylamide gel electrophoresis (Fig. 4). Chromatographic separation and molecular weight estimations of these multiple forms of human acrosin have still to be achieved. Syner (pers.

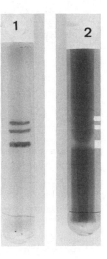

Figure 4
Determination of gelatin digestive fractions after acrylamide gel electrophoresis of human acrosin. The acrosin preparation applied was isolated like boar acrosin, including affinity chromatography, desalting and lyophilization. For electrophoresis and digestion of gelatin, see legend to Figure 2. Gel 1: human acrosin, Coomassie blue-stained; gel 2: human acrosin, digestion of gel-enclosed gelatin (transparent bands).

comm.) has also separated multiple forms of human acrosin by acrylamide gel electrophoresis.

ENZYMATIC PROPERTIES

Similarity to Trypsin

The enzymatic properties of boar and human acrosin and most probably also of the acrosins from other species (Zaneveld, Polakoski and Williams, this volume) are very similar to those of bovine or porcine pancreatic trypsin. Some characteristics common to both families of enzymes are given in Table 2. Thus far, to our knowledge, only quantitative differences have been observed but with one exception: the trypsin-chymotrypsin inhibitor HUSI-I from human seminal plasma inhibits neither boar nor human acrosin (Schiessler, Arnhold and Fritz 1974; see also the following paper by Fritz et al., this volume). Remarkable quantitative differences are: (1) the weak inhibition of acrosin by the strong trypsin-kallikrein inhibitor from bovine organs (Fritz et al. 1972a; Zaneveld et al. 1974); (2) progressive inhibition of human acrosin by α_1-antitrypsin (Fritz et al. 1972b) compared to the rapid inhibition of trypsin; (3) the relative high affinity of L-arginine to acrosin (Polakoski and Williams 1974); and (4) permanent inhibition of boar acrosin by the inhibitors from boar seminal plasma which inhibit trypsin only temporarily (Tschesche et al. 1974). Further details about the inhibitors and enzymatic properties of the boar and human acrosin isolated in our laboratory have been published elsewhere (Fritz, Schleuning and Schill 1974; Fritz, Schiessler and Schleuning 1973). Therefore, only striking differences to published data and results previously not reported will be discussed or presented here.

Table 2
Properties of Boar Acrosin and Bovine Trypsin

	Acrosin	*Trypsin*
Specific activity, U/mg		
BAPA	18.8[a]	1.2
BAEE	214[b]	36
Equal casein digestion rate by	4.5 mg	1 mg
Bimolecular velocity constant of inhibition, l/mole $\times$ min		
DFP	630	300
TLCK	23	34
Dissociation constants of complexes with inhibitors, K_i (mole/l)		
leupeptin and antipain	7×10^{-8}	2×10^{-7}
seminal acrosin inhibitors	$<1 \times 10^{-9}$	$<1 \times 10^{-9}$
Kinin liberation from kininogen	+	+

[a] Protein was estimated by the biuret method; see text.
[b] The combined test system (pH 8.7) was used; see text.

Activation by Calcium Ions

Zaneveld, Polakoski and Williams (1972) and Polakoski, McRorie and Williams (1973) reported a pronounced activation effect of calcium ions on rabbit and boar acrosin. Their highly purified acrosin preparations were extremely unstable in the neutral pH region and could not be frozen or lyophilized. In contrast, we did not find any activation of our highly purified boar and partially purified human (Fritz et al. 1972c) acrosin by calcium ions. This, as well as the lower molecular weights found for rabbit (27,300; Polakoski, Zaneveld and Williams 1972) and boar acrosin (30,000; see above) by SDS gel electrophoresis, might indicate that the acrosin molecules were degraded during detergent extraction to such an extent that their stability is reduced considerably. Thus the activating and stabilizing effect of the calcium ions could be explained by stabilization of the conformation of molecules which are labilized by limited proteolytic cleavage. Calcium ions also did not activate human acrosin purified by Gilboa, Elkana and Rigbi (1973); however, autoactivated or trypsin-activated rabbit acrosin required calcium ions for optimal activity (Huang-Yang and Meizel 1975).

Titration with NPGB and Specific Activities

p-Nitrophenyl p'-guanidinobenzoate (NPGB) was introduced by Chase and Shaw (1967, 1970) as a qualified reagent for the estimation of the number of active molecules in trypsin and trypsinlike enzyme preparations. Figure 5 shows the result of the titration of boar acrosin with NPGB. As with trypsin, deacylation proceeds very slowly after the initial burst reaction. Thus the number of active acrosin molecules can be exactly calculated from the change in extinction extrapolated to starting time zero.

Boar acrosin (14.5 U, substrate: N-α-benzoyl-D,L-arginine p-nitroanilide) and 0.77 mg (biuret) protein was applied to the test. Using the molar extinction coefficient $\epsilon_{402} = 18.3$ for p-nitrophenol, the molarity of the acrosin solution was calculated to 1.86×10^{-5} moles/l from the initial burst extinction $E_{402} = 0.340$. This means that one mole boar acrosin hydrolyzes 780 moles BAPA or 8900 moles BAEE (N-α-benzoyl-L-arginine ethyl ester) per minute under the given conditions. Referred to the amount of protein applied, which was estimated by the biuret method, a molecular weight of 41,580 is

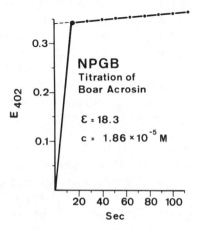

Figure 5

Titration of highly purified boar acrosin with NPGB. The titration experiment was performed according to the method of Chase and Shaw (1970). For further details, see text. Ordinate: extinction at 402 nm; abscissa: reaction time.

thus calculated for boar acrosin. (Bovine trypsin was used as reference substance for protein estimation; a multiplication factor of 32.5 was thus derived [factor for serum protein: 36.0].) This value is in relatively good agreement with the results obtained by SDS gel electrophoresis.

On the basis of the specific activity of 13.8 U (BAPA)/mg of the acrosin preparation employed, the unrealistic high molecular weight of 56,700 is calculated. Protein concentration was estimated thereby from the extinction of the acrosin solution at 278 nm (Fink et al. 1972). Obviously the molar extinction coefficient of boar acrosin differs considerably from that of the reference substance used (human γ-globulin), and therefore the specific activities previously given have to be recalculated and related to the biuret protein value given above. The specific activities of our highly purified boar acrosin preparations should thus be near 18.8 U (BAPA)/mg and 214 U (BAEE/ADH [alcohol dehydrogenase] assay, pH 8.7)/mg, respectively. The latter value cannot be directly compared with the value reported by Polakoski and Williams (1974; Zaneveld, Polakoski and Williams, this volume) for their purest acrosin preparation because of the different test systems and conditions used.

Digestion of RCM Ribonuclease

Identical molar concentrations based on NPGB titration of boar acrosin and bovine trypsin were employed for the digestion of reduced carboxymethylated (RCM) ribonuclease. Under these conditions, the optimal numbers of lysyl and arginyl bonds hydrolyzed are identical for both enzymes (Fig. 6). Polakoski and McRorie (1973) found a strong preference for the cleavage of arginyl bonds by boar acrosin. These discrepancies may be due to the different forms of acrosin and/or different enzyme to substrate ratios used by both groups.

BIOLOGICAL ASPECTS

Determination of Sperm Acrosin Activities

Potent acrosin inhibitors are present in ejaculated and epididymal spermatozoa (see the following paper by Fritz et al., this volume). This complicates the quantitative estimation of sperm acrosin activities. Zaneveld et al. (1971a; Zaneveld, Polakoski and Williams 1973; Polakoski, Zaneveld and Williams 1971) have applied acidified detergent extracts, in which the acrosin inhibitor complex is dissociated, to the substrate-containing test system. However, acrosin as well as the acrosin inhibitors may be more or less degraded by other acrosomal proteinases during prolonged detergent extraction.

Acidic extraction seems to be the method of choice if the actual state of acrosin and acrosin inhibitors in spermatozoa is to be elucidated. Enzymatic degradation or activation (if acrosin is present in an inactive precursor form, see above) processes are immediately stopped by suspending the spermatozoa in cold (4°C) acidic solution which is applied to a suitable, sensitive test system after removal of the sperm pellet (Fink et al. 1972; Fritz et al.

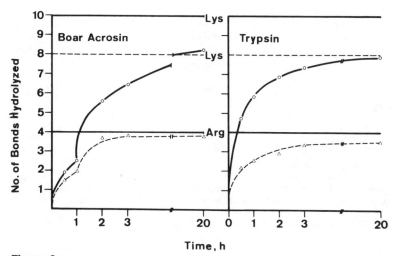

Figure 6

Digestion of RCM ribonuclease with boar acrosin and bovine trypsin. The numbers of lysyl (○——○) and arginyl (△——△) bonds hydrolyzed (ordinate) by highly purified boar acrosin and TPCK-treated bovine trypsin are given as a function of the incubation time (abscissa). The enzymatic digestion was started with a molar ratio of enzymes to substrate (7×10^{-5} M RCM ribonuclease) of 500:1 (trypsin) or 700:1 (acrosin) and continued after 1 hour with the ratio 100:1. Incubation was performed at 25°C in 0.2 M sodium phosphate pH 8.0, 0.2 mM EDTA. Further details will be given by Schiessler and Fritz (in prep.). Whereas a qualitatively similar digestion of RCM ribonuclease is observed, acrosin hydrolyzed arginyl bonds somewhat more rapidly and lysyl bonds more slowly than trypsin.

1972c; Schill 1973; Fritz, Schleuning and Schill 1974). The actual pH of the solution used for sperm extraction should be based on the following criteria: (1) the acid stability of acrosin; (2) complete dissociation of the acrosin-inhibitor complexes; and (3) the acid stability of the acrosin inhibitors. These criteria should be fulfilled in most cases between pH 2.0 and 2.7. We have successfully used the described method for the extraction of human, boar, bull and hamster sperm cells.

Sperm Acrosin Activities in Clinical Studies

Employing acidic extracts of human spermatozoa, a clear correlation was found between sperm count and sperm acrosin activity (Schill 1974a,b; Schirren et al. 1974; Schirren and Eweis 1973). Mean values of sperm acrosin activities in ejaculates with the criteria of normospermia, asthenospermia and oligospermia are given in Table 3. Whereas an average value of about 0.75 mU (substrate: BAEE) acrosin per 10^6 spermatozoa was found in normospermic and asthenospermic ejaculates, acrosin activity per sperm cell increased significantly with the decrease in the number of spermatozoa in oligospermic ejaculates (Schill 1974b; Schill, Schleuning and Fritz 1975).

724   H. Fritz et al.

Table 3

Sperm Acrosin Activities in Ejaculates with Different Semen Qualities

Semen characteristic	Normospermia	Asthenospermia[a]	Oligospermia		
No. of spermatozoa in millions/ml	40–250	40–250	20–40	10–20	<10
No. of semen specimens investigated	105	33	55	32	15
Acrosin activity in mU/10⁶ spermatozoa	0.75 ±0.27	0.70 ±0.27	0.86 ±0.47	1.07 ±0.72	1.77 ±0.88

The acidic sperm extracts were directly applied to the substrate-containing test cuvettes. The combined BAEE/ADH system was used for acrosin determination (Schiessler et al. 1972; Schill 1973). Acrosin activities (and standard deviation thereof) are given in mU/10⁶ spermatozoa. For further details, see Schill (1974b).

[a] Normal sperm count and morphology, but reduced sperm motility.

The reason for this phenomenon is still unknown. Round-headed spermatozoa, missing the acrosomal cap as a consequence of an inborn error, did not contain any measurable acrosin activity (Schill 1974b; Schirren et al. 1974).

Stimulation of Sperm Acrosin Activity

A significant increase in sperm acrosin activity is normally observed in the course of aging of human ejaculates or washed ejaculated spermatozoa (Figs. 7 and 8). The rise in sperm acrosin activity is not prevented, but in-

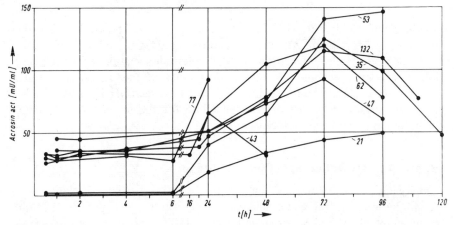

Figure 7

Change of sperm acrosin activity (substrate: BAEE, pH 8.7) during aging of human ejaculates. Samples (0.25 ml) of eight individual semen specimens (the number of sperm cells in millions per ml ejaculate is given on each curve in the figure) were incubated at 22°C for up to 120 hours. Further experimental details are given in Schill and Fritz (1975a).

Figure 8
Change of sperm acrosin activity during aging of washed human spermatozoa in 0.15 M NaCl as a function of the incubation temperature. Samples (0.25 ml) of a suspension of washed sperm cells in 0.15 M NaCl were incubated at 4, 22 and 37°C up to 144 hours. Further details are given in Schill and Fritz (1975a).

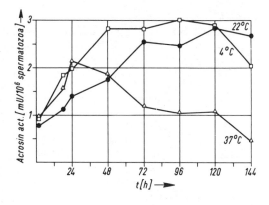

stead somewhat enhanced, in the presence of low molecular weight acrosin inhibitors in the suspension medium (Schill and Fritz 1975a). This relatively slow and temperature-dependent enhancement of the acrosin activity could be interpreted as enzymatic activation of proacrosin molecules. Besides auto-activation, another enzyme not inhibited by the acrosin inhibitors employed might be responsible for this effect. Curiously, Schill (1973) had found that in some cases an increase in sperm acrosin activity was already caused by repeated washing of the spermatozoa. The nearly total absence of an initial sperm acrosin activity in two individual cases (Fig. 7) perhaps gives us a chance to find the agent inducing the onset of the activation process.

The explanation of sperm acrosin activation during aging is complicated by the observations that a similar increase in sperm acrosin activity is also induced relatively rapidly, as a "short-time effect," by pretreatment of ejaculates with glycerol for 15 minutes only (Schill 1974c) or by freezing and thawing ejaculates or washed ejaculated spermatozoa (Table 4). However, in both cases spermatozoa of a minor number of the individual semen samples (14.3% of the glycerol-treated and 23.5% of frozen-thawed ejaculates) showed no or even a contrary response, i.e., loss of sperm acrosin activity compared to the controls (Schill 1975). This indicates that besides acrosin activation, the status of the membranes in regard to leakage and extractability of acrosin has also to be considered in such processes. The relatively rapid increase in sperm acrosin activity observed in the majority of the semen samples after glycerol treatment or freeze-thawing is more plausibly explained by structural changes of the membranes in such a way that acrosin is more easily removed.

Acrosin Activity in Individual Spermatozoa

Gaddum and Blandau (1970), and later others (Allen, Bishop and Thompson 1973; Penn, Gledhill and Darzynkiewicz 1972; Zaneveld, Polakoski and Williams, this volume; Wendt, Leidl and Fritz 1975a), demonstrated the hydrolytic effect of individual spermatozoa on gelatin layers. In order to obtain more uniform results, we developed a special modification of this

Table 4

Effects of Glycerol and Freeze-thawing on Sperm Acrosin Activity

No. of semen specimens	Control	Glycerol-treated	Frozen and thawed	Change in activity (%)	
				mean	range
68	0.76 ± 0.40	1.38 ± 0.75		+80	+8 to +282
12	0.92 ± 0.29	0.72 ± 0.29		−22	−74 to zero
102[a]	0.76 ± 0.46	1.35 ± 0.90		+78	
	0.76 ± 0.46		1.25 ± 0.91[b]	+64	
17[c]	0.94 ± 0.32	0.74 ± 0.30		−21	
	0.94 ± 0.32		0.97 ± 0.41[b]	+ 3	
28	0.78 ± 0.37	1.23 ± 0.73	1.22 ± 0.57[b]	+57	
	0.78 ± 0.37		1.13 ± 0.47[d]	+45	

Samples of individual human ejaculates were either incubated after addition of glycerol (controls: physiological saline) to a final concentration of 10% (v/v) for 15 minutes or after being frozen (−196°C) and thawed. For experimental details, see Schill (1974c, 1975). Sperm acrosin activities (and standard deviations) are given in mU (substrate: BAEE) per 10^6 spermatozoa.

[a] 85.7% from N = 119.
[b] Frozen and thawed *after* glycerol addition.
[c] 14.3% from N = 119.
[d] Frozen and thawed *without* glycerol addition.

technique (Wendt, Leidl and Fritz 1975a). Figure 9 shows the halos formed by digestion of the gelatin layer by about 90% of ejaculated bull spermatozoa after 4 hours of incubation. Figure 10 demonstrates the correlation thus found between halo area, statistically evaluated, and sperm acrosin activity. This correlation and the complete prevention of gelatin digestion after preincubation of the gelatin film with seminal or other (see below) acrosin inhibitors indicate strongly that acrosin is responsible for the observed gelatinolysis.

Obviously, gelatin digestion is possible in spite of the presence of acrosin inhibitors in the spermatozoa. In addition, preincubation of spermatozoa in media containing high amounts of seminal acrosin inhibitors, of the strong low molecular weight acrosin inhibitors leupeptin and antipain (Fritz, Förg-Brey and Umezawa 1973) and of the irreversible inhibitor TLCK (for kinetics of acrosin inhibition, see Schiessler et al. 1972; Connors, Greenslade and Davanzo 1973) did not prevent gelatinolysis at all (Wendt, Leidl and Fritz 1975b; cf. Penn, Gledhill and Darzynkiewicz 1972). Remarkably, very low amounts of pure acrosin, about 3 ng, also caused large digestion halos on the gelatin film, whereas the complex of acrosin with the seminal inhibitor as well as neutralized acidic sperm extracts showed no effect (Wendt, Leidl and Fritz 1975a).

The Actual Status of Acrosin in Spermatozoa

The speculation about the status of acrosin in spermatozoa has to be based on the following facts:

If acidic extracts of washed ejaculated spermatozoa (we used especially the

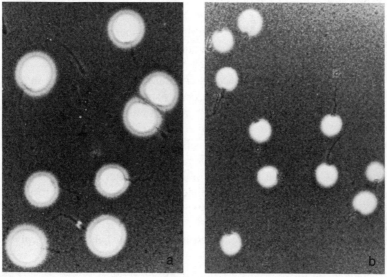

Figure 9

Lysis halos digested by ejaculated bull and boar spermatozoa on gelatin films. Bull or boar spermatozoa were applied to developed Kodak AR-10 film plates and the aducts incubated for 4 hours or 30 minutes under carefully controlled conditions (humidity, temperature). Details are given in Wendt, Leidl and Fritz (1975a). (*a*) Bull spermatozoa, 4 hours, 180 ×; (*b*) boar spermatozoa, 30 minutes, 180 ×.

Figure 10

Correlation between sperm acrosin activity and the gelatinolytic effect of ejaculated bull spermatozoa. Spermatozoa from the same ejaculate were used for the estimation of sperm acrosin activity (substrate: BAEE, pH 8.7) or applied to gelatin films. The area (ordinate) of the halos formed after 4 hours of incubation were measured. Further details are given in Wendt, Leidl and Fritz (1975a).

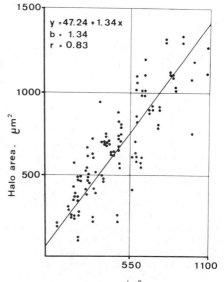

boar and human species) are neutralized, acrosin activity disappears. This is also observed if the spermatozoa are taken from glycerol-treated or aged ejaculates, i.e., after stimulation of sperm acrosin activity (see above). Similar results were obtained employing epididymal boar and hamster spermatozoa. Apparently, the amount of inhibitors present in the spermatozoa is sufficient for complete neutralization of the enzymatic potential of the sperm acrosin under suitable conditions.

On the other hand, gelatinolysis occurs even in the presence of the acrosin inhibitor(s) in the spermatozoon. This indicates strongly that acrosin and the inhibitor(s) are located each in different compartments of the sperm cell and are not present in the form of the inactive acrosin-inhibitor complex. Furthermore, preincubation of spermatozoa with the inhibitors mentioned above did not at all prevent the gelatinolytic effect; inhibition was only observed if the gelatin film was pretreated with the inhibitors, i.e., after leakage of acrosin from the sperm cell! Consequently, even low molecular weight acrosin inhibitors normally cannot penetrate the outer sperm head membranes and thus inactivate acrosin.

This view is also supported by the results shown in Table 5. The given concentrations of DFP and TLCK in the semen samples as well as the relatively long incubation times used were appointed on the basis of the results of kinetic measurements previously reported (Schiessler et al. 1972). Although free acrosin is completely inactivated under the conditions used, sperm

Table 5

Effects of DFP and TLCK on Sperm Acrosin Activity

	Controls	+DFP 0.1 µmole	+TLCK 0.4 µmole
Incubation time (hr)	1.5/4	1.5	4[a]
Acrosin activities, U (BAEE)			
ejaculate	5.5	5.6	
	8.0[b]		8.9
neutralized sperm	8.8	9.5	
extract	9.2		10.5
acrosin, inhibitor-	0.4	0	
free	0.4		0

Experimental procedure: A freshly gained boar ejaculate was divided into 1.0-ml portions. The sperm pellets of three portions were washed and subsequently extracted with acetic acid (2%, w/v); the acidic extracts were neutralized (pH 6.9) and filled up each to 1.0 ml. DFP and TLCK were added to the 1.0-ml samples (ejaculates and neutralized extracts) and a neutral solution of purified boar acrosin. The samples and appropriate controls were incubated at 25°C. Thereafter, the spermatozoa from the ejaculates were washed and the neutralized extracts dialyzed. Acrosin activity (substrate: BAEE, pH 8.7) was determined either after acidic extraction (spermatozoa), acidification (neutralized extracts) or directly (purified acrosin) as mentioned in the text. Similar results to those shown in the table were obtained if washed boar spermatozoa, human ejaculates and washed sperm cells or glycerol-treated ejaculates were employed.

[a] Chosen on the basis of the results of kinetic investigations (Schiessler et al. 1972). Obviously, the incubation time of about 15–20 minutes used in antifertility experiments is insufficient.

[b] Increase in sperm acrosin activity during aging (4 hr instead of 1.5 hr); see text.

acrosin activity was not at all diminished, i.e., both active-site-directed inhibitors cannot reach the specificity pocket of acrosin in intact spermatozoa. This also applies to neutralized acidic sperm extracts in which the inactive acrosin-inhibitor complex is present. (If the complex were to be already formed in spermatozoa, the observed gelatinolysis could not be explained.)

TLCK as well as other synthetic and naturally occurring trypsin-acrosin inhibitors have been found to be potential antifertility agents (Zaneveld, Robertson and Williams 1970; Zaneveld et al. 1971a,b; Newell, Polakoski and Williams 1972; Greenslade et al. 1973; Suominen, Kaufman and Setchell 1973; Miyamoto and Chang 1973; Palmer and Howarth, Jr. 1973; see also Zaneveld, Polakoski and Williams, this volume). However, the results of our studies indicate that neither the natural seminal acrosin inhibitors nor low molecular weight inhibitors like leupeptin, antipain, DFP and TLCK can inhibit or inactivate acrosin in ejaculated spermatozoa. Most probably, they cannot penetrate the plasma and/or outer acrosomal membranes. This seems to be true also if "capacitated" sperm cells, i.e., spermatozoa recovered from the uteri of sows 6 hours after mating, are treated with acrosin inhibitors (Wendt, Leidl and Fritz 1975b). Therefore the mechanism of inhibition of in vivo fertilization by TLCK and other trypsin-acrosin inhibitors has still to be elucidated. Whereas TLCK might have other toxic effects on spermatozoa, as discussed recently by Miyamoto and Chang (1973), other acrosin inhibitors used may be adsorbed on the surface of the sperm head membranes and later on neutralize acrosin after onset of the acrosome reaction.

The techniques now available can be applied to find out whether a correlation exists between acrosin inhibition and the antifertility effect of acrosin inhibitors. In addition, the availability of acrosin to inhibitors in epididymal, ejaculated or capacitated sperm cells can be studied with the gelatin method. Estimation of acrosin activity after acidic extraction is recommended if the effect of irreversible or active-site-directed acrosin inhibitors is to be investigated.

Localization and Absolute Amount of Acrosin in the Spermatozoon

Recently published results (Srivastava 1973; Multamäki 1973; Srivastava et al. 1974; Brown and Hartree 1974; Schill and Wolff 1974, 1975) show a relatively strong binding of the trypsinlike enzyme acrosin to acrosomal sperm membranes, most probably the inner acrosomal membrane and/or equatorial segment (see Barros, Fujimoto and Yanagimachi 1973). The latter assumption is especially supported by two observations: (1) Denuded (i.e., acrosome-free) ram spermatozoa contained nearly the total acrosin activity which could be solubilized by acid treatment or 0.1 M $CaCl_2$ solutions, but no acrosin inhibitor (Brown and Hartree 1974). (2) In the majority (76.5%) of the frozen-thawed semen samples, a significant increase in sperm acrosin activity was observed, although the plasma and outer acrosomal membranes are extensively disrupted and removed by this treatment (Schill 1975; Schill and Wolff 1974, 1975; Pedersen 1972). The detachment and solubilization of sperm-bound acrosin in Ca^{++}-containing solutions is of interest in so far as 0.05 or 0.1 M $MgCl_2$ solutions have been employed for sequential release of the plasma and outer acrosomal membranes and thus the acrosomal content which contained also the major amount of acrosin (Srivastava et al. 1974). Probably acrosin is also removed and thus solubilized from the inner acrosomal membrane by the action of Mg^{++}.

From the acrosin activity extractable from a certain number of ejaculated boar spermatozoa by acetic acid treatment (Fink et al. 1972), the specific activities of highly purified boar acrosin (Schleuning and Fritz 1975; Schiessler et al. 1972) and the results of the titration experiments with NPGB given above, the amount of acrosin present in individual boar spermatozoa is calculated at approximately 1×10^{-18} moles or 6×10^5 enzyme molecules per sperm cell.

On the basis of these figures, some interesting estimations are possible. The halos visualized on the gelatin membrane (Fig. 9) are caused by the digestive effect of about 0.04 pg (i.e., 1×10^{-18} moles) acrosin. (The lysis areas digested by the acrosin present in the spermatozoon and in the solution applied [3 ng, i.e., 1×10^{-13} moles; see above] cannot be compared on the basis of absolute amounts; very different acrosin concentrations are effective under both circumstances.) This seems to be conceivable only if the actual concentration of acrosin at the beginning of the gelatinolytic process is high enough, that is, if acrosin is bound only in certain regions at the membrane surfaces, e.g., at the inner acrosomal membrane, and especially accumulated at the equatorial segment. This assumption is in agreement with the observations of Gaddum and Blandau (1970), Gaddum-Rosse and Blandau (1972), Allen, Bishop and Thompson (1973), V. Wendt et al. (unpubl.), and Wendt, Leidl and Fritz (1975a,b) that gelatinolysis originates from the equatorial segment of the sperm cells with one exception: human spermatozoa caused uniform digestion halos over the anterior part of the sperm cells from the beginning.

The given figures are also remarkable with regard to the processes which may occur after mating. Besides the inaccessibility of acrosin to plasma proteinase inhibitors known to be present in cervical mucus (Schumacher and Zaneveld 1974), the extremely low absolute amount of acrosin in individual spermatozoa also has to be considered. Even though the dissociation constants of acrosin-inhibitor complexes are below 10^{-10} moles per liter, it is unlikely that the acrosin inhibitor concentration in cervical mucus, as well as in uterine and tubular fluids, is high enough to shift the equilibrium: acrosin + inhibitor $\rightleftharpoons$ complex in the direction of complex formation. Therefore it seems possible to us that in the female genital tract secretions (after removal of the seminal plasma), the acrosin liberated from disintegrating spermatozoa, as well as membrane-bound acrosin (after the acrosome reaction had occurred), can hydrolyze its physiological substrates without interference with acrosin inhibitors.

Effect of Acrosin on Sperm Motility and Migration

Besides the main function of acrosin, the digestion of a passage through the zona pellucida for the sperm cell (see Zaneveld, Polakoski and Williams, this volume), this enzyme may also facilitate migration of spermatozoa in the female genital tract, thereby inducing two effects: (1) a decrease in the viscosity of the cervical mucus (Schumacher and Zaneveld 1974; R. M.

Nakamura, unpubl.) and (2) an increase in sperm motility by kinin generation.

Schill and Haberland (1974) demonstrated recently that kinins can effectively stimulate sperm motility in asthenospermic ejaculates, and that the kinin-bearing natural substrate of the kallikreins, the kininogen, is present in human seminal plasma (for a review of published results see Schill and Fritz 1975b). The kinin-liberating effect of boar acrosin is quantitatively similar to that of porcine pancreatic kallikrein (S. Palm, W.–D. Schleuning and H. Fritz, unpubl.). Therefore, kinins may be liberated from kininogen, which is present in female genital fluids (H. Fritz, unpubl.), by acrosin escaping from disintegrating spermatozoa and thus enhance sperm motility and migration by "autostimulation."

Acknowledgments

Supported in part by Sonderforschungsbereich-51 München (H. F.), Deutsche Forschungsgemeinschaft (W.–B. S. and V. W.), Bayer AG Wuppertal-Elberfeld and WHO Geneva (Grant No. 2873 and H9/181/150). We are grateful to Professor W. Leidl and Dr. H. Tschesche for stimulating discussions and B. Förg-Brey and R. Hell for skillful technical assistance.

REFERENCES

Allen, G. J., M. W. H. Bishop and T. E. Thompson. 1973. Lysis of photographic emulsions by mammalian and chicken spermatozoa. *J. Reprod. Fert.* **36**:249.

Barros, C., M. Fujimoto and R. Yanagimachi. 1973. Failure of zona penetration of hamster spermatozoa after prolonged preincubation in blood serum fraction. *J. Reprod. Fert.* **35**:89.

Brown, C. R. and E. F. Hartree. 1974. Distribution of a trypsin-like proteinase in ram spermatozoa. *J. Reprod. Fert.* **36**:195.

Chase, T., Jr. and E. Shaw. 1967. *p*-Nitrophenyl *p'*-guanidino-benzoate HCl: A new reactive site titrant for trypsin. *Biochem. Biophys. Res. Comm.* **29**:508.

———. 1970. Titration of trypsin, plasmin, and thrombin with *p*-nitrophenyl *p'*-guanidino-benzoate HCl. In *Methods in Enzymology* (ed. S. Colowick and N. Kaplan) vol. 19, p. 20. Academic Press, New York.

Connors, E. C., F. C. Greenslade and J. P. Davanzo. 1973. The kinetics of inhibition of rabbit sperm acrosomal proteinase by 1-chloro-3-tosylamido-7-amino-2-heptanone (TLCK). *Biol. Reprod.* **9**:57.

Fink, E., H. Schiessler, M. Arnhold and H. Fritz. 1972. Isolierung eines Trypsin-ähnlichen Enzyms (Akrosin) aus Ebersspermien. *Z. Physiol. Chem.* **353**:1633.

Fritz, H., B. Förg-Brey and H. Umezawa. 1973. Leupeptin and antipain—Strong competitive inhibitors of sperm acrosomal proteinase (boar acrosin) and kallikreins from porcine organs (pancreas, submandibular glands, urine). *Z Physiol. Chem.* **354**:1304.

Fritz, H., H. Schiessler and W.–D. Schleuning. 1973. Proteinases and proteinase inhibitors in the fertilization process: New concepts of control? In *Advances in the Biosciences, Schering Workshop on Contraception: The Masculine Gender*, Berlin, Nov. 1972 (ed. G. Raspé and S. Bernhard), vol. 10, p. 271. Pergamon Press-Vieweg, Oxford.

Fritz, H., W.–D. Schleuning and W.–B. Schill. 1974. Biochemistry and significance of the trypsin-like proteinase acrosin from boar and human spermatozoa. In *Bayer Symposium V: Proteinase Inhibitors, Proceedings 2nd International Research Conference* (ed. H. Fritz et al.), p. 118. Springer Verlag, Berlin.

Fritz, H., B. Förg-Brey, E. Fink, H. Schiessler, E. Jaumann and M. Arnhold. 1972a. Charakterisierung einer Trypsin-ähnlichen Proteinase (Akrosin) aus Eberspermien durch ihre Hemmbarkeit mit verschiedenen Protein-Proteinase-Inhibitoren. I. Seminale Inhibitoren und Trypsin-Kallikrein-Inhibitor aus Rinderorganen. *Z. Physiol. Chem.* **353**:1007.

Fritz, H., N. Heimburger, M. Meier, H. Arnhold, L. J. D. Zaneveld and G.F.B. Schumacher. 1972b. Humanakrosin: Zur Kinetik der Hemmung durch Human-Seruminhibitoren. *Z. Physiol. Chem.* **353**:1953.

Fritz, H., B. Förg-Brey, E. Fink, M. Meier, H. Schiessler and C. Schirren. 1972c. Humanakrosin: Gewinnung und Eigenschaften. *Z. Physiol. Chem.* **353**:1943.

Gaddum, P. and R. J. Blandau. 1970. Proteolytic reaction of mammalian sperm on gelatine membranes. *Science* **170**:749.

Gaddum-Rosse, P. and R. J. Blandau. 1972. Comparative studies on the proteolysis of fixed gelatine membranes by mammalian sperm acrosomes. *Amer. J. Anat.* **134**:133.

Garner, D .L. 1973. Partial characterization of bovine acrosomal proteinase. *Biol. Reprod.* **9**:71.

Garner, L. and R. F. Cullison. 1974. Partial purification of bovine acrosin by affinity chromatography. *J. Chromatog.* **92**:445.

Gilboa, E., Y. Elkana and M. Rigbi. 1973. Purification and properties of human acrosin. *Eur. J. Biochem.* **39**:85.

Greenslade, F. C., J. J. McCormack, A. F. Hirsch and J. P. Davanzo. 1973. Blockage of fertilization in *Rana pipiens* by trypsin inhibitors. *Biol. Reprod.* **8**:306.

Huang-Yang, Y. H. J. and S. Meizel. 1975. Purification of rabbit testis proacrosin and studies of its active form. *Biol. Reprod.* **12**:232.

McRorie, R. A. and W. L. Williams. 1974. Biochemistry of mammalian fertilization. *Annu. Rev. Biochem.* **43**:777.

Meizel, S. and Y. H. J. Huang-Yang. 1973. Further studies of an inactive form of a trypsin-like enzyme in rabbit testes. *Biochem. Biophys. Res. Comm.* **53**:1145.

Meizel, S., S. K. Mukerji and Y. H. J. Huang-Yang. 1974. Biochemical studies of rabbit testes and epididymal sperm proacrosin. *J. Cell Biol.* **63**:222a.

Miyamoto, H. and M. C. Chang. 1973. Effects of proteinase inhibitors on the fertilizing capacity of hamster spermatozoa. *Biol. Reprod.* **9**:533.

Multamäki, S. 1973. Isolation of pure acrosomes by subcellular fractionation of bull spermatozoa. *Int. J. Fert.* **18**:193.

Newell, S. D., K. L. Polakoski and W. L. Williams. 1972. Inhibition of fertilization by proteinase inhibitors. *Proceedings 7th International Congress on Animal Reproduction and Artificial Insemination,* vol. 3, p. 2117.

Palmer, M. B. and B. Howarth, Jr. 1973. The requirement of a trypsin-like acrosomal enzyme for fertilization in the domestic fowl. *J. Reprod. Fert.* **35**:7.

Pedersen, H. 1972. The acrosome of the human spermatozoon: A new method for its extraction, and an analysis of its trypsin-like enzyme activity. *J. Reprod. Fert.* **31**:99.

Penn, A., B. L. Gledhill and Z. Darzynkiewiez. 1972. Modification of the gelatine substrate procedure for demonstration of acrosomal proteolytic activity. *J. Histochem. Cytochem.* **20**:499.

Polakoski, K. L. 1974. Partial purification and characterization of proacrosin from boar sperm. *Fed. Proc.* **33**:1308.

Polakoski, K. L. and R. A. McRorie. 1973. Boar acrosin. II. Classification, inhibition, and specificity studies of a proteinase from sperm acrosomes. *J. Biol. Chem.* **248**:8183.

Polakoski, K. L. and W. L. Williams. 1974. Studies on the purification and characterization of boar acrosin. In *Bayer Symposium V: Proteinase Inhibitors, Proceedings 2nd International Research Conference* (ed. H. Fritz et al.), p. 128. Springer Verlag, Berlin.

Polakoski, K. L., R. A. McRorie and W. L. Williams. 1973. Boar acrosin. I. Purification and preliminary characterization of a proteinase from boar sperm acrosomes. *J. Biol. Chem.* **248**:8178.

Polakoski, K. L., L. J. D. Zaneveld and W. L. Williams. 1971. An acrosin-acrosin inhibitor complex in ejaculated boar sperm. *Biochem. Biophys. Res. Comm.* **45**:381.

———. 1972. Purification of a proteolytic enzyme from rabbit acrosomes. *Biol. Reprod.* **6**:23.

Schiessler, H., M. Arnhold and H. Fritz. 1974. Characterization of two proteinase inhibitors from human seminal plasma and spermatozoa. In *Bayer Symposium V: Proteinase Inhibitors, Proceedings 2nd International Research Conference* (ed. H. Fritz et al.), p. 147. Springer Verlag, Berlin.

Schiessler, H., H. Fritz, M. Arnhold, E. Fink and H. Tschesche. 1972. Eigenschaften des Trypsin-ähnlichen Enzyms (Akrosin) aus Eberspermien. *Z. Physiol. Chem.* **353**:1638.

Schill, W.–B. 1973. Acrosin activity in human spermatozoa: Methodological investigations. *Arch. Derm. Forsch.* **248**:257.

———. 1974a. Methodische Untersuchungen zur Akrosinaktivität in meschlichen Spermatozoen. Korrelation zwischen Akrosin und Spermatozoendichte. In *Advances in Andrology,* vol. 3 (ed. C. Schirren), p. 80. Grosse Verlag, Berlin.

———. 1974b. Quantitative determination of acrosin activity in human spermatozoa. *Fert. Steril.* **25**:703.

———. 1974c. The influence of glycerol on the extractability of acrosin from human spermatozoa. *Z. Physiol. Chem.* **355**:225.

———. 1975. Acrosin activity of cryo-preserved human spermatozoa. *Fert. Steril.* in press).

Schill, W.–B. and H. Fritz. 1975a. N^{α}-Benzoyl-L-arginine ethyl ester-splitting activity (acrosin) in human spermatozoa and seminal plasma during aging in vitro. *Z. Physiol. Chem.* **356**:83.

———. 1975b. Kinin-induced sperm motility and fertility. In *Chemistry and Biology of the Kallikrein-Kininin System in Health and Disease, Fogarty International Center Proceedings No. 27* (ed. J. J. Pisano and K. F. Austen). U.S. Government Printing Office, Washington, D.C. (in press).

Schill, W.–B. and G. L. Haberland. 1974. Kinin-induced enhancement of sperm motility. *Z. Physiol. Chem.* **355**:229.

Schill, W.–B. and H. H. Wolff. 1974. Acrosin activity and acrosome ultrastructure of human spermatozoa after freeze preservation. *Naturwissenschaften* **61**:172.

———. 1975. Ultrastructure of human sperm acrosome and determination of acrosin activity under conditions of semen preservation. *Int. J. Fert.* **19**:217.

Schill, W.–B., W.–D. Schleuning and H. Fritz. 1975. Biochemical and clinical aspects of human sperm acrosin. In *13th World Congress on Fertility and Sterility,* Buenos Aires, Nov., 1974 (in press).

Schirren, C. and A. Eweis. 1973. Akrosinaktivität in menschlichen Spermatozoen und Spermatozoendichte im Ejakulat. *Andrologie* **5**:81.

Schirren, C., G. Laudahn, A. Eweis and I. Heinze. 1974. Morphologische und biochemische Untersuchungen im menschlichen Sperma: Akrosom-Defekte, Akrosinaktivität und Humansperma-Inhibitor-Aktivität. *Z. Hautkr.* **49**:5.

Schleuning, W.–D. 1975. Isolierung und Charakterisierung von Akrosin, einem proteolytischen Enzym aus Eberspermien. Dissertation, Medizin. Fakultät der Universität, München.

Schleuning, W.–D. and H. Fritz. 1974. Some characteristics of highly purified boar sperm acrosin. *Z. Physiol. Chem.* **355**:125.

———. 1975. Sperm acrosin. In *Methods in Enzymology* (ed. S. Colowick and N. Kaplan). Academic Press, New York (in press).

Schleuning, W.–D., H. Schiessler and H. Fritz. 1973. Highly purified acrosomal proteinase (boar acrosin): Isolation by affinity chromatography using benzamidine-cellulose and stabilization. *Z. Physiol. Chem.* **354**:550.

Schumacher, G. F. B. and L. J. D. Zaneveld. 1974. Proteinase inhibitors in human cervical mucus and their in vitro interactions with human acrosin. In *Bayer Symposium V: Proteinase Inhibitors, Proceedings 2nd International Research Conference* (ed. H. Fritz et al.), p. 178. Springer Verlag, Berlin.

Srivastava, P. N. 1973. Location of the zona lysin. *Biol. Reprod.* **9**:84.

Srivastava, P. N., J. F. Munnell, C. H. Yang and C. W. Foley. 1974. Sequential release of acrosomal membranes and acrosomal enzymes of ram spermatozoa. *J. Reprod. Fert.* **36**:363.

Stambaugh, R. L. 1972. Acrosomal enzymes and fertilization. In *Biology of Mammalian Fertilization and Implantation* (ed. K. Moghissi and E. S. E. Hafaz), p. 187. C. C. Thomas, Springfield, Illinois.

Stambaugh, R. L. and J. Buckley. 1970. Comparative studies of the acrosomal enzymes of rabbit, rhesus monkey and human spermatozoa. *Biol. Reprod.* **3**:275.

———. 1972. Studies on acrosomal proteinase of rabbit spermatozoa. *Biochim. Biophys. Acta* **284**:473.

Suominen, J., M. H. Kaufman and B. P. Setchell. 1973. Prevention of fertilization in vitro by an acrosin inhibitor from rete testis fluid of the ram. *J. Reprod. Fert.* **34**:385.

Tschesche, H., S. Kupfer, O. Lengel, R. Klauser, M. Meier and H. Fritz. 1974. Purification, characterization, and structural studies of proteinase inhibitors from boar seminal plasma and boar spermatozoa. In *Bayer Symposium V: Proteinase Inhibitors, Proceedings 2nd International Research Conference* (ed. H. Fritz et al.), p. 164. Springer Verlag, Berlin.

Waldschmidt, M., H. Karg and B. Hoffman. 1964. Proteolytische Aktivität in männlichen Geschlechtssekreten des Rindes. *Naturwissenschaften* **51**:18.

Wendt, V., W. Leidl and H. Fritz. 1975a. The lysis effect of bull spermatozoa on gelatine substrate film—Methological investigations. *Z. Physiol. Chem.* **356**:375.

———. 1975b. The influence of various proteinase inhibitors on the gelatinolytic effect of ejaculated and uterine boar spermatozoa. *Z. Physiol. Chem.* (in press).

Williams, W. L. 1972. Biochemistry of capacitation of spermatozoa. In *Biology of Mammalian Fertilization and Implantation* (ed. K. Moghissi and E. S. E. Hafez), p. 19. C. C. Thomas, Springfield, Illinois.

Zaneveld, L. J. D., K. L. Polakoski and W. L. Williams. 1972. Properties of a proteolytic enzyme from rabbit sperm acrosomes. *Biol. Reprod.* **6**:30.

———. 1973. A proteinase and proteinase inhibitor of mammalian sperm acrosomes. *Biol. Reprod.* **9**:219.

Zaneveld, L. J. D., R. T. Robertson and W. L. Williams. 1970. Synthetic enzyme inhibitors as antifertility agents. *FEBS Letters* **11**:345.

Zaneveld, L. J. D., K. L. Polakoski, R. T. Robertson and W. L. Williams. 1971a. Trypsin inhibitors and fertilization. In *Proceedings International Research Con-*

ference on Proteinase Inhibitors, Munich, 1970 (ed. H. Fritz and H. Tschesche), p. 236. Walter de Gruyter, Berlin.

Zaneveld, L. J. D., R. T. Robertson, M. Kessler and W. L. Williams. 1971b. Inhibition of fertilization in vivo by pancreatic and seminal plasma trypsin inhibitors. *J. Reprod. Fert.* **25:**387.

Zaneveld, L. J. D., G. F. B. Schumacher, P. F. Tauber and D. Propping. 1974. Proteinase inhibitors and proteinases of human semen. In *Bayer Symposium V: Proteinase Inhibitors, Proceedings 2nd International Research Conference* (ed. H. Fritz et al.), p. 136. Springer Verlag, Berlin.

Low Molecular Weight Proteinase (Acrosin) Inhibitors from Human and Boar Seminal Plasma and Spermatozoa and Human Cervical Mucus—Isolation, Properties and Biological Aspects

Hans Fritz and Hans Schiessler

Institut für Klinische Chemie und Klinische Biochemie der Universität München
D–8 München 2, Germany

Wolf-Bernhard Schill

Dermatologische Klinik und Poliklinik der Universität München, Abteilung für Andrologie
D–8 München 2, Germany

Harald Tschesche

Organisch-Chemisches Institut der Technischen Universität München
Lehrstuhl für Organische Chemie und Biochemie
D–8 München 2, Germany

Norbert Heimburger

Behringwerke AG, D–355 Marburg/Lahn, Germany

Otto Wallner

I. Frauenklinik der Universität München
D–8 München 2, Germany

Inhibition of pancreatic trypsin by human seminal plasma was first reported by Rasmussen and Albrechtsen (1960). Haendle et al. (1965), Waldschmidt, Hoffman and Karg (1966), Fritz et al. (1968) and Haendle, Ingrisch and Werle (1970) later demonstrated the presence of well-defined, acid-stable proteinase (trypsin) inhibitors of relatively low molecular weight in tissues and secretions of the male and female genital tract. The discovery and distribution, purification procedures and properties, as well as the possible biological function of these proteinase inhibitors are described and discussed in detail in recently published articles (Zaneveld, Polakoski and Schumacher, this volume; Schiessler, Arnhold and Fritz 1974; Tschesche et al. 1974; Fritz, Schleuning and Schill 1974; Zaneveld et al. 1974; Schumacher and Zaneveld 1974; Polakoski and Williams 1974a; McRorie and Williams 1974; Fritz, Schiessler and Schleuning 1973a; Fink et al. 1973; Fink et al. 1971b; Zaneveld et al. 1971a). We shall therefore confine ourselves mainly to some more recent findings obtained in our laboratories concerning the inhibitors from boar and human semen and cervical mucus: improved isolation procedures, first results of structural studies, relations between seminal and cervical inhibitors and interesting biological aspects (e.g., localization in tissues and spermatozoa, demonstration in epididymal spermatozoa and biological function).

Occurrence and Common Characteristics

Trypsin inhibitors of a protein nature were found in various glands or tissues and secretions of the male and female genital tract (Table 1). The proteinase

Table 1
Low Molecular Weight Proteinase Inhibitors in Glands or Tissues and Secretions of the Male and Female Genital Tract

Occurrence in glands, tissues, secr. (mIU/g or ml)	Inhibition of			Acid (heat) stable	MW (approx.)	Reference
	trypsin	chymo-trypsin	acrosin			
Sem. ves. (0.4–5), sem. plasma (0.2–3) of man, bull, boar, rat, mouse, hamster, guinea pig	+	−		+	6800	Haendle et al. 1965
Sem. ves. and ampullar secr. of bull	+		+	(+)		Waldschmidt, Hoffman and Karg 1966
Sem. ves. of guinea pig, mice (3.8), bull (I,II)	+	−[a]		+	5600–6800	Fritz et al. 1968
Cervical secr., human	+	+		+	2200[b]	Haendle, Ingrisch and Werle 1970
Sem. ves. of guinea pig MSSI-I[d]	+	+	−	+	11,500	Wallner and Fritz 1974
MSSI-II[d]	+	−	+[e]	+	6800	Fritz et al. 1970; Fink et al. 1971b
Sem. plasma of rabbit	+	+	+	+	7200	Zaneveld et al. 1971b
Sem. plasma of human HUSI-I	+	+	−[e]	+	10,100	Fink et al. 1971a, 1973;
HUSI-II	+	−	+[e]	+	6200	Schiessler, Arnhold and Fritz 1974
(HUSI-II)[f]	+	−		+		Hirschhäuser and Kionke 1971
(HUSI-I)[f]	+	−	−	(−)	34,000	Suominen and Niemi 1972
(HUSI-II)[f]	+	+	−	(+)	11,500	
(HUSI-I)[f]	+	−	(+)	(+)	4000	Syner and Moghissi 1972; Syner and Kurrus 1974
(HUSI-II)[f]	+	−	+	+		
Sem. plasma of boar	+			+	13,400[g]	Zaneveld et al. 1971a; Polakoski et al. 1971; Polakoski and Williams 1974a
					6800[g]	
					5800[g]	
					1600[g]	
	+	−	+[e]	+	12,000[h] 11,000[i] 5800[j]	Fink et al. 1971b; Fritz, Schiessler and Schleuning 1973b; Tschesche et al. 1974
Testes (0.04–0.22), epid. (0.05–0.4), sem. ves.	+			+		Fink et al. 1971b (from

738

Source				Max. trypsin inhibitory activity (IU)	References
(0.05–5.0), sem. plasma (0.15–3.1) of various species					Haendle 1969
Rete testes secr. (epid. and sem. ves. secr., sem. plasma) of ram (and boar)	+	−	+	6500[k]	Suominen and Setchell 1972; Suominen, Kaufman and Setchell 1973
Spermatozoa (ejaculated)					
of rabbit		+	+		Zaneveld et al. 1971a; Polakoski, Zaneveld and Williams 1971
of boar		+	+		
	+	+	+	7000[l]	Fritz, Schleuning and Schill 1974; Tschesche et al. 1974
				>7500[m]	
	+	+	+	1600[g]	Polakoski and Williams 1974a,b
				6800[g]	
				13,400[g]	
bull, boar, rabbit	+	+	+		Zaneveld, Polakoski and Williams 1973
bull	+	+	+	<13,000	D. Čechová and H. Fritz, in prep.
human	+	−	+	(5600)	Zaneveld, Dragoje and Schumacher 1972
(HUSI-II)[n]	+	+	+		Fritz et al. 1972c; Schiessler, Arnhold and Fritz 1974; Syner and Kurrus 1974
(HUSI-I)[n]	+	+	−	(10,100)	
(HUSI-II)[n]	−	−	+	(6200)	

Only those references are given in which defined substances are described. In some cases, maximal trypsin inhibitory activities in IU/g tissue or ml secretions are mentioned; for definition of IU, see legend to Table 2. Abbreviations: epid. = epididymis or epididymal; secr. = secretion(s); sem. = seminal; ves. = vesicles.

[a] Except bull inhibitor II which is identical with the basic bovine inhibitor.
[b] Corrected to about 11,500 by Wallner and Fritz (1974).
[c] Estimated later on by Fritz et al. (1972a,d).
[d] MSSI: Guinea pig seminal vesicles inhibitor I and II.
[e] Weak inhibition of acrosin reported earlier (Fritz et al. 1972a,d; Zaneveld et. al. 1973) was caused by low amounts of HUSI-II (see Fritz et al. 1972d) in the HUSI-I isoinhibitor mixture (see Schiessler, Arnhold and Fritz 1974).
[f] Probably identical with the corresponding inhibitor HUSI-I or HUSI-II.
[g] Gel filtration at pH 3.0.
[h] Gel filtration in neutral solution; in acidic solution pH 2.2: 13,500; see text.
[i] Composition and gel filtration at pH 7.6; see text.
[j] Gel filtration at pH 7.6.
[k] Inhibitor of the rete testes fluid only.
[l] Minimal value calculated from the composition (this inhibitor is soluble in neutral salt solutions; see text).
[m] Gel filtration at pH 2.6; see text.
[n] Very probably identical with HUSI-I or HUSI-II from seminal plasma (see Schiessler, Arnhold and Fritz 1974).

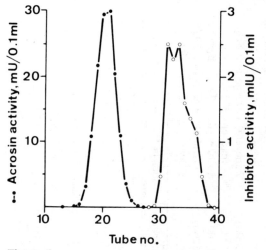

Figure 1

Fractionation of acidic extracts of epididymal hamster spermatozoa by gel filtration. The acetic acid extract pH 2.6 (Fink et al. 1972; Fritz et al. 1972c; Fritz, Schleuning and Schill 1974; Schiessler, Arnhold and Fritz 1974) of epididymal hamster spermatozoa was concentrated by ultrafiltration and subsequently applied (1.6 ml containing 2.6 U acrosin) to a Sephadex G-75 column (1.6 × 70 cm, 4°C) equilibrated and developed with acetic acid pH 2.6 at a rate of 9.5 ml/hr, 3 tubes/hr. Prior to the extraction, the suspension of the spermatozoa in phosphate buffer-glycerol (10%, v/v) solution had been frozen (for shipment), thawed, centrifuged, and the sperm pellet washed one time with physiological saline.

Acrosin activity was measured using BAEE as substrate with the combined test system (Fritz et al. 1972c; Schill 1973). Acrosin from tube 23 was employed for the estimation of the *inhibitory activity* (right ordinate) with the combined test system (the trypsin inhibition test used normally is not sensitive enough in this case). Acrosin and the inhibitory activity found in the eluted fractions cannot be compared on a quantitative basis due to the extreme low concentration of the partners under the given conditions and thus the nonproportional inhibition observed. After concentration of the inhibitor fraction by lyophilization, the inhibition of acrosin *and* trypsin could be demonstrated in the usual way (titration curves). If acidic extracts of freshly collected and washed epididymal boar spermatozoa were employed, a similar elution pattern was obtained (Weber 1975).

The activities of the eluted acrosin fractions remained unchanged after neutralization in both cases (hamster, boar), indicating a quantitative separation of the inhibitors. In contrast, acrosin activity disappeared totally if the acidic sperm extracts were neutralized prior to gel filtration.

740

inhibitors to be discussed here are characterized by their stability in acidic solutions around pH 2 and their relatively low molecular weights of about 6000 and 12,000, respectively. (The plasma proteinase inhibitors present in genital tract secretions are discussed by Zaneveld, Polakoski and Schumacher [this volume] and Schumacher and Zaneveld [1974].) Like the pancreatic Kazal-type trypsin inhibitors (Tschesche 1974; Greene, Roark and Bartelt 1974), these seminal and cervical inhibitors are also secretory proteins: they are present, e.g., in rete testes and epididymal fluid, in human cervical mucus and in especially high concentrations in the secretions of the seminal vesicles of many species (see Table 1).

The occurrence of these inhibitors in ejaculated spermatozoa is well known (Zaneveld et al. 1971a; Polakoski, Zaneveld and Williams 1971; Fink et al. 1972; Fritz et al. 1972c; Zaneveld, Dragoje and Schumacher 1972; Zaneveld, Polakoski and Williams 1973; Fritz, Scheuning and Schill 1974; Polakoski and Williams 1974a; Schiessler, Arnhold and Fritz 1974; Tschesche et al. 1974), but we were also able to establish the presence of such inhibitors directly (Fig. 1) and indirectly (see legend to Fig. 1) in epididymal boar and hamster spermatozoa. This finding is surprising in so far as one would assume from results obtained by an indirect method that acrosin inhibitors should not be present (Zaneveld et al. 1971a; Polakoski, Zaneveld and Williams 1971) or occur only in relatively low amounts (Zaneveld, Polakoski and Williams 1973) in epididymal spermatozoa compared to the ejaculated ones. This discrepancy may be due to a more or less extensive degradation of the epididymal sperm cell inhibitors during detergent extraction. This would presuppose that the inhibitors already present in epididymal and perhaps testicular spermatozoa are not identical with the inhibitors originating from the seminal vesicles which are added to the spermatozoa during ejaculation. First results support this view: antibodies produced against the inhibitors from boar seminal plasma are bound to ejaculated but not to epididymal boar sperm cells (see below).

Proteinase Inhibitors in Human Seminal Plasma, Spermatozoa and Cervical Mucus

Inhibitors in Seminal Plasma

The main steps used for the isolation of the trypsin inhibitors from seminal plasma are summarized in Table 2. The inhibitors are adsorbed to SP-Sephadex and further purified by affinity chromatography (Table 3), which works very well due to the acid stability of these substances. Gel filtration separates the HUSI-I and HUSI-II (human seminal plasma inhibitor I [trypsin-chymotrypsin inhibitor] and II [trypsin-acrosin inhibitor]) fractions on the basis of their different molecular weights (Table 4). Afterwards, fractionation of the inhibitor mixtures HUSI-I and HUSI-II by gradient elution and equilibrium chromatography is necessary to obtain homogeneous proteins. Both mixtures of multiple inhibitor forms are separated into four major fractions (Schiessler, Arnhold and Fritz 1974 and in prep.). The inhibition specificities of related inhibitors (HUSI-I$_{A-D}$, HUSI-II$_{A-D}$) are identical.

The amino acid compositions of some of the multiple inhibitor forms are given in Table 4. The molecular weights are calculated on the basis of the

Table 2
Procedure Used for Isolation of Acid-stable Trypsin Inhibitors
from Human Seminal Plasma

Purification step	Inhibitory activity (IU)	Yield (%)	Specific activity (IU/mg)	
Seminal plasma	535	(100)	0.00009	
Absorption and elution,				
SP-Sephadex C-50	486	91	0.025	
Affinity chromatography,				
trypsin-cellulose	398	87	0.9	
Fractionation on	I II		I	II
Sephadex G-75	153 209	92	0.9	1.05
Fractionation on				
SP-Sephadex C-25	I[a] A–D	90	1.73	
CM- or SP-Sephadex C-25	II[b] A–D	94		2.17

Experimental details of the isolation procedure used repeatedly are as previously published (Fink et al. 1971a; Fink et al. 1973; Schiessler, Arnhold and Fritz 1974) or will be described elsewhere (H. Schiessler, M. Arnold and H. Fritz, in prep.). The *inhibitory activity* given is expressed in IU (inhibitory units): 1 IU reduces the trypsin-catalyzed hydrolysis of BAPA by 1 μmole/min under the given conditions (Fritz, Trautschold and Werle 1974). Protein concentration was measured by absorption at 280 nm or by weight.

[a] Four major fractions are separated. In the notation used formerly (Fink et al. 1973; Schiessler, Arnhold and Fritz 1974), *all* inhibitor-containing fractions were designated by letters (A–G$_2$).

[b] Four major fractions are separated. The conditions used in SP-Sephadex C-25 gradient elution chromatography will be published elsewhere (H. Schiessler, M. Arnhold and H. Fritz, in prep.).

total numbers of residues present per molecule. Thus far, we have found no indications that carbohydrate residues are attached to the inhibitors obtained from human seminal plasma.

Of special interest are the differences in the inhibition specificities of both inhibitor groups (Table 5 and Schiessler, Arnhold and Fritz 1974). Whereas HUSI-II is a strong inhibitor of boar and human acrosin (Fig. 2), the purified multiple forms of HUSI-I do not inhibit acrosin at all. However, HUSI-I is a potent inhibitor of neutral human leukocytic proteinases, probably the elastase (J. Travis, pers. comm.). Remarkably, both HUSI-I and HUSI-II are at least weak inhibitors of human cationic and anionic trypsin (Figarella, Negri and Guy 1974). The enzymatic activities of porcine pancreatic elastase (substrate: elastin), *Aspergillus oryzae* protease, pronase and subtilisin (substrate: azo-casein) are not diminished by either inhibitor mixture I or II (H. Schiessler, M. Arnold and H. Fritz, in prep.).

Inhibitors in Spermatozoa

Two trypsin inhibitor fractions (I, II) are present in acidic extracts of washed ejaculated spermatozoa (Fritz et al. 1972c; Schiessler, Arnhold and Fritz 1974). Their inhibition specificities and molecular weights, estimated by gel filtration, are in accordance with those of the corresponding inhibitors from

Table 3

Isolation of Trypsin Inhibitors from
Human Seminal Plasma by Affinity
Chromatography Using a Trypsin-
Cellulose Column

No. Exp.	Inhibitory activity bound (pH 7.8) or eluted (pH 1.8) from the affinity column		
	IU^a bound	% eluted	in ml
1	12	83	600
2	17	100	920
3	42	86	650
4	52	84	580
5	38	89	455
6	100	100	254
7	215	94	370
8	211	93	476
9	253	100	475

The inhibitor-containing solution, 1.2–1.8
IU/ml in experiments 1–5 and 6.4–7.2 IU/ml
in experiments 6–9, was applied to the trypsin-
cellulose column (1.2 x 10 cm, 4°C; "Trypsin
polymer gebunden an CM-Cellulose," 7–10
U/mg from E. Merck, Darmstadt) which had
been equilibrated with 0.1 M triethanolamine-
HCl, 0.4 M NaCl, pH 7.8. The column was
washed with the same buffer solution until
impurities not specifically bound were removed. Subsequently, the inhibitors were dissociated from trypsin and eluted from the
column with 0.4 M KCl-HCl, pH 1.8 (flow
rate: 15 ml/hr); (see Fink et al. 1971a).

a For definition, see Table 2. One IU corresponds to the inhibition of 1 U, i.e., about 1
mg trypsin.

seminal plasma HUSI-I and HUSI-II, respectively (Table 5). This indicates
that both inhibitors are picked up by the spermatozoa from seminal fluids
either already in the epididymis or during ejaculation.

Zaneveld, Dragoje and Schumacher (1972) found only the trypsin-acrosin
inhibitor II in detergent extracts, and Syner and Kurrus (1974), besides inhibitor II, low amounts of trypsin-chymotrypsin inhibitor I. The different test
systems used might be responsible for these discrepancies: HUSI-I is a relatively weak inhibitor of trypsin so that no inhibition may be observed if very
sensitive test systems and substrates with high affinities to trypsin are employed.

Inhibitor in Cervical Mucus: Relation to HUSI-I

The proteinase inhibitor discovered in human cervical mucus by Haendle,
Ingrisch and Werle (1970) was recently characterized in more detail (Wall-

Table 4

Amino Acid and Carbohydrate Compositions of Seminal Trypsin Inhibitors in Moles/Mole

	Guinea pig seminal vesicles		Boar seminal plasma I,II		Boar sperm cells	Human seminal plasma (HUSI)				
Inhibitor:						II		I		
Fraction:	I	II	A_1^a	B^a	$-^a$	G-1	G-2^b	D	I	C^c
	$-^b$	$-^d$	A							
Asp	6	6	7			6 (5.31)		8	(7.98)	
Thr	1	4	4			3 (2.77)		4	(3.96)^d	
Ser	2	5	5			3 (2.95)	4	6	(5.83)^d	8
Glu	10	4	5			3 (2.90)		7	(6.95)	
Pro	5	2	3			5 (5.02)		12	(11.82)	
Gly	6	5	5^i			5 (5.00)		9	(8.77)^d	
Ala	1	0	2			1 (1.04)		3	(3.00)	
Cys ½	6	6	6			6 (5.69)^d		12	(11.42)^e	
Val	3	3	1			1 (0.94)		5	(4.74)^f	
Met	0	1	1			1 (1.11)		3	(2.83)^e	
Ile	4	1	2			3 (2.90)		1	(1.05)	
Leu	5	3	2			2 (1.88)		4	(3.79)^f	5
Tyr	2	4	3			3 (2.61)^d		2	(1.55)	
Phe	0	3	5			1 (0.96)		2	(1.88)	

Lys	1	4	5	4	6	4	3 (2.73)		12 (11.87)
His	2	3	3	6	6	6	2 (2.00)		0 (0.20)
Arg	6	4	5	6	6	4	6 (5.98)	5	4 (4.18)
Trp	0	0	1	0	0	4	0		0
Total	60	58	65	65	67	64	54	54	94	97
Carbohydrate residues	0	0	15[g]				0		0	0
MW	6772	6687		A: 11,000[h]			6148		10,130	

(Lys 1 for column 10; Trp 1 for column 10.)

Experimental values are given by Fritz et al. (1970) and Fink et al. (1971b) for the inhibitors from guinea pig seminal vesicles, by Tschesche et al. (1974) for the inhibitors from boar seminal plasma and spermatozoa, and by H. Schiessler, M. Arnhold and H. Fritz (in prep.) for the inhibitors from human seminal plasma (HUSI). The values calculated from 20-hour hydrolysates of HUSI-II fraction G-1 and HUSI-I fraction D, corrected for amino acid degradation or incomplete hydrolysis, are also given in the table. The amino acid compositions of HUSI-I fractions A and B are nearly identical to that of fraction D. The nomination of the various HUSI-I and HUSI-II fractions corresponds to the letters given in the elution diagrams published recently (Figs. 2 and 5 in Schiessler, Arnhold and Fritz 1974).

[a] Only those values varying from fraction A are given.
[b] Only those values varying from fraction G-1 are given.
[c] Only those values varying from fraction D are given.
[d] Corrected for degradation.
[e] After performic acid oxidation.
[f] Corrected for incomplete hydrolysis.
[g] Glucosamine, 3.40; galactosamine, 2.27; galactose, 1.98; glucose, 4.14; mannose, 1.84; (sialic acid, 0.99) (see Tschesche et al. 1974).
[h] Calculated from the composition. Two major fractions with approximate values of 5800 and 11,000 are observed in gel filtration experiments at pH 7.6; see text.
[i] Corrected recently from six to five residues on the basis of the results of the amino acid sequence.

Table 5

Properties of Seminal and Cervical Acid-stable Trypsin Inhibitors

Inhibitor source	Inhibitor fraction	Inhibition of						MW (approx.)	Multiple forms
		trypsin[a]	acrosin[b]	plasmin[c]	chymo-trypsin[a]	leukocytic proteinases[d]	other proteinases		
Guinea pig seminal vesicles	I[e]	+	+	−	−		−[f]	6600[g]	2
	II[h]	+	+	+	−		−	6700[g]	5
Boar									
seminal plasma	I	+	+	+	−	−	−[i]	11,000[j]	
spermatozoa	II	+	+	+	−	−	−[i]	6000[j]	
		+	+	+				6000[j]	
Bull									
seminal plasma	I	+	+	+	(+)			8700[k]	
	II	+	+	+	+		+[l]	6800[k]	
spermatozoa	(I,II)	+	+						
Leeches	A	+	+	+	−		−[m]	6300[g]	5
	B	+	+	+	−		−[m]	4800[g]	4
Human									
seminal plasma	I	+[n]	−	−	+	+	−[o]	10,100[g]	4
	II	+[n]	+	−	−	−	−[i,o]	6200[g]	4
spermatozoa	I	+	−		+			11,000[k]	
	II	+	−		+			6000[k]	
cervical mucus		+	−	−	+	+	−[i]	11,500[k]	

The inhibitors were isolated by affinity chromatography, except for the cervical inhibitor and the inhibitors from spermatozoa and from bull seminal plasma. For comparison, trypsin inhibitors from leeches with high affinities to acrosin are also included. For references, see text and Tables 1 and 4.

[a] Bovine.
[b] Boar and human.
[c] Porcine.
[d] Human (pH 7.65).
[e] Formerly fraction a.
[f] Porcine pancreatic kallikrein and bovine thrombin.
[g] Calculated from the amino acid composition.
[h] Formerly fraction b.
[i] Porcine pancreatic kallikrein, bovine thrombin (Haendle, Ingrisch and Werle 1970).
[j] Calculated from gel filtration experiments at pH 7.6.
[k] Derived from gel filtration experiments.
[l] Porcine and urinary kallikrein (D. Cechová and H. Fritz, in prep.).
[m] Porcine pancreatic kallikrein and subtilisin Novo.
[n] Human trypsin, too (Figarella, Negri and Guy 1974).
[o] Porcine pancreatic kallikrein and elastase, Aspergillus oryzae protease, subtilisin Novo and pronase.

Figure 2

Titration of human and boar acrosin with seminal trypsin inhibitors. Increasing amounts of the inhibitors were added to constant amounts of boar or human acrosin in 2.0 ml buffer solution pH 7.8. The remaining enzyme activity was measured using BAPA as substrate. Details of the experimental conditions used are described elsewhere (Fritz et al. 1972a,d).

On the basis of the concentrations of boar acrosin employed in these experiments ($3.5–5 \times 10^{-9}$ M; see preceding paper, Fritz et al., this volume and Fritz et al. 1972b,d) dissociation constants (K_i) below 1×10^{-10} M can be calculated from the titration curves for the acrosin-inhibitor complexes. (——•——) Human acrosin, HUSI-II$_{A-D}$; (——△———) boar acrosin, HUSI-II$_{A-D}$; (——○———) boar acrosin, inhibitor mixture I,II from boar seminal plasma (see Table 6 and Fig. 7).

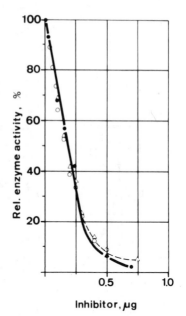

Inhibitor, μg

ner and Fritz 1974). The following properties of this inhibitor are in agreement with those reported for HUSI-I obtained from seminal plasma: inhibition of trypsin, chymotrypsin (Table 5) and neutral leukocytic proteinases (Fig. 3), but not of human acrosin, acid stability and the molecular weight of about 11,500 (estimated by gel filtration). The most striking similarity is the immunological cross-reactivity: antibodies directed against HUSI-I also produced a strong precipitation line with native or acid-treated cervical mucus (Fig. 4). These results indicate a high degree of identity between the acid-stable mucus inhibitor and HUSI-I.

Remarkably, an acid-stable inhibitor with very similar characteristics is also present in human nasal and bronchial secretions (Hochstrasser et al. 1972a; Hochstrasser et al. 1973; Hochstrasser, Feuth and Hochgesand 1974; Reichert, Hochstrasser and Conradi 1972) and the gut (Hochstrasser et al. 1972b). This inhibitor, which occurs normally in free form in bronchial mucus, is bound to leukocytic proteinases under certain pathological conditions (Hochstrasser et al. 1972a; Hochstrasser et al. 1973; Hochstrasser, Feuth and Hochgesand 1974). The acid-stable cervical mucus inhibitor is also partially liberated by acidification (Wallner and Fritz 1974) indicating the presence of an enzyme-inhibitor complex, the enzyme probably originating from leukocytes. Obviously, the acid-stable inhibitors in human mucous secretions are part of a defense mechanism directed against the lytic action of neutral leukocytic proteinases. The accumulation of leukocytes in these secretions during inflammatory processes (in cervical mucus after mating as well as the leukotatic activity of seminal plasma and spermatozoa) is well known (Wilkinson 1974; Jaszczak and Hafez 1973; Maroni, Symon and Wilkinson 1972; Maroni and Wilkinson 1971).

Hirschhäuser, Baudner and Daume (1972) claimed the occurrence of the

Figure 3

Inhibition of human leukocytic neutral proteinases by the trypsin-chymotrypsin inhibitors from human seminal plasma (HUSI-I) and cervical mucus. Increasing amounts of the inhibitors, expressed in mIU (abscissa), were incubated with constant amounts of the enzyme prepared from purulent mucous secretions (Hochstrasser et al. 1972a) for 30 minutes at 25°C in 0.5 ml 0.1 M sodium phosphate buffer pH 7.65. The remaining enzyme activities (ordinate) were estimated at 37°C with azo-casein (20 mg in 1.0 ml buffer solution) as substrate (Fritz, Trautschold and Werle 1974). The enzymatic reaction was stopped after 3 hours by addition of 1.5 ml trichloroacetic acid (5%, w/v). The trypsin inhibitory activities mIU (abscissa) were measured using BAPA as substrate. (●——●) Inhibitor from seminal plasma HUSI-I$_{A-D}$; (o——o) inhibitor from human cervical mucus. For comparison: (△——△) trypsin-kallikrein inhibitor from bovine organs; (▼——▼) inhibitor from seminal plasma HUSI-II$_{A-D}$.

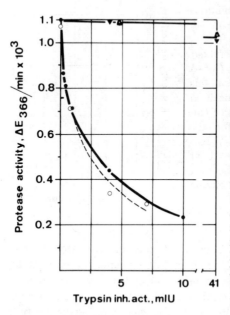

Protease activity, $\Delta E_{366}/min \times 10^3$

Trypsin inh. act., mIU

seminal plasma trypsin-acrosin inhibitor (HUSI-II) in human cervical mucus. Their conclusions were based on results obtained with antibodies which had been produced against the acid-stable trypsin inhibitor from seminal plasma: inhibition of chymotrypsin by this preparation was excluded (Hirschhäuser and Kionke 1971). Although we have not yet employed antibodies against HUSI-II, we could find no indication that an acid-stable trypsin-acrosin inhibitor similar to HUSI-II occurs in cervical mucus (Wallner and Fritz 1974). We assume that Hirschhäuser, Baudner and Daume (1972) used antibodies against a seminal plasma constituent thus far not identified.

Proteinase Inhibitors in Boar Seminal Plasma and Spermatozoa

Inhibitors in Seminal Plasma

Trypsin inhibitors were previously isolated from seminal plasma treated either with 3% perchloric acid (Fink et al. 1971b) or 2.5% trichloroacetic acid (Zaneveld et al. 1971a). The molecular weights reported for the preparations, which had been purified further by affinity chromatography, were derived from gel filtration experiments (pH 2.2: 13,500; pH 7.0: 12,000; Fink et al. 1971b) or calculated from the amino acid composition (6781, Zaneveld et al. 1971a). Recently, Polakoski and Williams (1974a) described the occurrence of inhibitor fractions with approximate molecular weights, estimated by gel

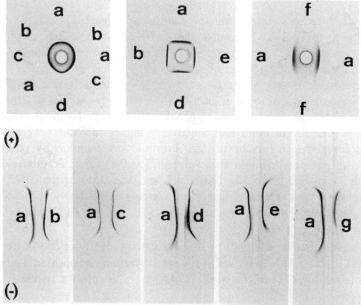

Figure 4

Immunological cross-reactivity of the trypsin-chymotrypsin inhibitors from human seminal plasma (HUSI-I) and cervical mucus. Rabbits were immunized with the trypsin-chymotrypsin inhibitor I from human seminal plasma by applying the mixture of the multiple inhibitor forms I_{A-D} (4 major fractions, see Tables 2 and 4) obtained after affinity chromatography and desalting. The antisera thus produced were employed in immunodiffusion and immunoelectrophoresis. Details of the experimental procedures will be described elsewhere.

Immunodiffusion plates: 5 μl antiserum and 4 mIU (trypsin inhibition, substrate BAPA) of each inhibitor were applied to the agarose gel (2%, w/v; 0.05 M barbital sodium pH 8.2). The precipitation lines were stained with Coomassie brilliant blue. The precipitation line obtained with the purified (after acid treatment and chromatography on CM-Sephadex) cervical mucus inhibitor was indistinguishable from the one produced by native mucus (see d).

Immunoelectrophoresis plates: 100–120 μl of the antiserum dilution (1 vol. antiserum plus 4 vols. physiological saline) and 3–4 mIU (trypsin inhibition, substrate BAPA) of each inhibitor were applied to the agarose gel (2%, w/v; 0.05 M barbital sodium pH 8.2). Electrophoresis was performed for 50 minutes in 0.05 M barbital sodium pH 8.2. The precipitation lines were stained with Coomassie brilliant blue. **a:** HUSI-I (mixture of multiple forms I_{A-D}); **b:** "HUSI-I" from washed ejaculated spermatozoa; **c:** HUSI-I_C, multiple from I_C; **d:** cervical mucus; **e:** seminal plasma; **f:** HUSI-II (mixture of multiple forms I_{A-D}), 20 mIU applied; **g:** chymotrypsin–HUSI-I_{A-D} complex (excess chymotrypsin was inhibited with Kunitz soybean trypsin inhibitor).

filtration at pH 3.0, of 13,400 (major component), 6800, 5800 and 1600 in their preparation obtained from trichloroacetic acid-treated (80°C, 10 min) seminal plasma.

The inhibitor mixture isolated from perchloric acid-treated (Fink et al. 1971b) and acidified (pH 1.0) seminal plasma by affinity chromatography was recently separated into several multiple forms by gradient elution and equilibrium chromatography (Tschesche et al. 1974). The *composition* of the three inhibitors A, B and A₁ are given in Table 4. The molecular weight of inhibitor A was calculated to be at least 11,000, which is the sum of 7600 for the protein part and 3300 for the carbohydrate portion (without sialic acid residues[1]). Keeping in mind that apparently higher molecular weights are estimated for glycoproteins by the gel filtration method than by other methods, e.g., for pancreatic kallikrein a value of 35,000 (Fritz, Trautschold and Werle 1965) instead of 26,200 (Kutzbach and Schmidt-Kastner 1972) was found, inhibitor A (or a modified form thereof) is probably identical with the 13,000-MW inhibitors described by Fink et al. (1971b) and Polakoski and Williams (1974a).

The *amino acid sequence* of inhibitor A₁, elucidated recently in one of our laboratories (H. Tschesche), is shown in Figure 5. Arginine at position 19 is the reactive-site residue (Tschesche et al. 1974) which fits into the specificity

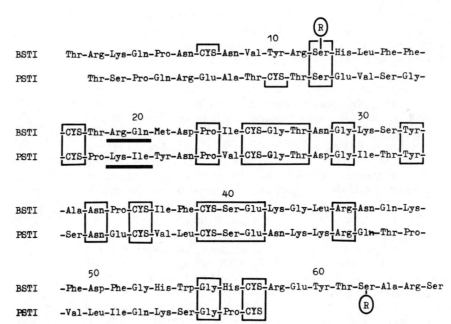

Figure 5
Structures of the boar seminal and the porcine pancreatic trypsin-acrosin inhibitor. BSTI: Boar seminal plasma trypsin inhibitor A₁ (see Table 4 and the text); PSTI: porcine pancreatic secretory trypsin inhibitor (see text); R: carbohydrate chain (the attachment sites given are based on preliminary indications).

[1] Including sialic acid and fucose found in some fractions, a molecular weight of 11,500–12,000 for the native inhibitor is very likely (H. Tschesche and R. Klauser, unpubl.).

pocket of trypsin, plasmin or acrosin. A glutamine residue (at position 20) was for the first time found next to arginine in the reactive-site bond of a trypsin inhibitor. Two carbohydrate chains are attached to serine residues which are not located in the direct vicinity of the reactive center. However, the most striking feature which is obvious from the sequence data is the structural homology of this seminal inhibitor and the pancreatic secretory trypsin inhibitors (Tschesche 1974; Greene, Roark and Bartelt 1974). Hence, the Kazal-type pancreatic trypsin inhibitors and the seminal acrosin inhibitors belong to one single class of structural homologous proteins, a finding which is especially interesting in view of a genetic relationship among the trypsin and acrosin enzyme families. Further details will be discussed elsewhere (Tschesche et al. 1975).

Like the pancreatic secretory trypsin inhibitors, the seminal inhibitors also show the phenomenon of *temporary inhibition* of trypsin (Tschesche et al. 1974; Polakoski and Williams 1974a) whereas boar acrosin is inhibited pseudo-permanently (Tschesche et al. 1974), that is, reliberation of the acrosin activity occurs much more slowly than that of trypsin.

Carbohydrate chains attached to proteins by *o*-glycosidic linkage are removed rather easily under acidic conditions. In order to see whether acidic pretreatment of seminal plasma causes cleavage of carbohydrate residues from inhibitor molecules and thus the occurrence of *multiple forms* with lower molecular weights (Polakoski and Williams 1974a; Fritz, Schleuning and Schill 1974), boar seminal plasma was fractionated directly on Sephadex G-75 in cold (8°C) salt buffer solution of pH 7.6 (Fig. 6). Freshly collected as well as stored (24 hr, 20°C), frozen-thawed and acid-treated (HCl, pH 1.0) seminal plasma showed the major trypsin inhibitor fractions I and II; sometimes also a third fraction (Ia) was separated. Fraction H contains acid-labile high molecular weight (>45,000) trypsin-acrosin inhibitors originating probably from blood plasma.

Inhibitor fractions I and II were isolated in preparative scale and twice subjected to rechromatography under identical conditions to achieve quantitative separation. Both fractions were further purified by affinity chromatography (Table 6). Inhibitor I thus obtained and a mixture of inhibitor I plus II were fractionated on Sephadex G-75; the elution diagrams are shown in Figure 7. On this basis, molecular weights of about 11,000 and 5800 were estimated for inhibitors I and II, respectively, in the salt-containing solution of pH 7.6.

In another approach, the inhibitors were directly absorbed to trypsin-cellulose from native seminal plasma and thus purified by affinity chromatography (Table 6). The activity profile obtained by gel chromatography of this inhibitor preparation (I,II) at pH 7.6 is similar to that of the inhibitor mixture I plus II (Fig. 7) and corresponds to the profile of the acid-stable inhibitors from native seminal plasma (Fig. 6).

Inhibition specificities and *immunological cross-reactivities* of inhibitors I and II were further investigated. The titration curves revealed a qualitatively identical and quantitatively very similar behavior of inhibitor I, Ia (Fig. 6) and II against the enzymes employed: boar acrosin, bovine trypsin and porcine plasmin are inhibited relatively strongly, bovine α-chymotrypsin very weakly (Table 7). Antibodies prepared against inhibitor I and inhibitors I,II

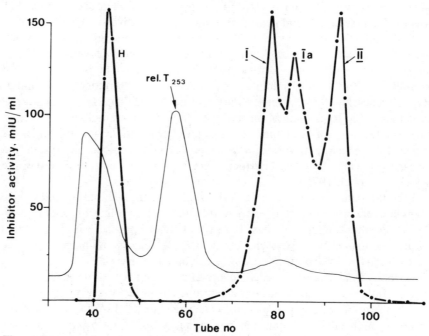

Figure 6

Fractionation of boar seminal plasma by gel filtration at pH 7.6. Boar seminal plasma, 3 ml containing 3.6 IU (trypsin inhibition, substrate BAPA), was applied to a Sephadex G-75 column (122 × 1.8 cm, 8°C) equilibrated and developed with 0.4 M NaCl, 0.1 M triethanolamine-HCl pH 7.6 at a rate of 16 ml/hr, 3.2 ml/tube. Fraction H contained acid-labile high molecular weight (>45,000) trypsin-acrosin inhibitors probably originating from serum. The acid-stable trypsin-acrosin-plasmin inhibitors are present in fractions I, Ia and II. Whereas separation of fractions I and II was achieved each time using samples from different species or after various pretreatments, fraction Ia was clearly separated in only a few cases. Ordinate: Trypsin inhibitory activity (substrate: BAPA).

(Fig. 7) produced precipitation lines of similar intensity with all of the purified inhibitor preparations (Fig. 8).

The identical inhibition specificities of inhibitors I and II and the immunological cross-reactivity among these inhibitors indicate that they are most probably derived from the same original molecule. Enzymatic modification and degradation by glycosidases and proteinases present in seminal plasma may at least partially be responsible for the occurrence of *multiple forms* in seminal plasma after ejaculation. In addition, removal of residues from the carbohydrate chains, especially sialic acid and fucose, very probably occurs during acid treatment of seminal plasma. Acid-induced cleavages may also be responsible for the presence of multiple forms of lower molecular weight (6800, 5800 and 1600) in the inhibitor preparation obtained after heating of seminal plasma with trichloroacetic acid (Polakoski and Williams 1974a). For example, the N-terminal hexapeptide and the C-terminal octapeptide (including the carbohydrate chains) outside of the disulfide bridges (see Fig. 5) can very probably be removed (resulting MW: 6100) without affecting the

Table 6

Isolation of Boar Seminal Acrosin Inhibitors by Affinity Chromatography Using Batchwise Absorption and Elution to/from Trypsin-Cellulose

Exp. no.	Inhibitory activity (IU[a]) applied	liters	bound	% Eluted[b]
	Native seminal plasma			
1	566	0.6	560	77
2	877	0.8	850	78
3	1050	1.0	995	80
4	1093	1.0	1030	90
5	1450	1.3	1273	77
9	2000	3.0	1530	74
10	2180	3.3	1868	79
11	2950	4.0	2560	83
12	1810	3.3	1698	79
13	1520	1.7	1312	88
	Inhibitor fraction I			
14	57.5	0.04	57.0	82
15	84.5	0.04	83.9	83
16	156.0	0.04	155.7	83
	Inhibitor fraction II			
17	150	0.04	147	86
18	169	0.04	157	93

Trypsin-CM-cellulose (E. Merck, Darmstadt) was equilibrated with 0.4 M NaCl, 0.1 M triethanolamine-HCl pH 7.8 and suspended in sperm-free (exps. 1–13) seminal plasma or the inhibitor solution (exps. 14–18) for from 30 minutes to 1 hour under cooling (4°C). In experiments 1–5, 9–13 and 14–18, 120, 160 and 20 ml, respectively, of the same wet trypsin-CM-cellulose was employed.

After centrifugation, the precipitate was washed five times with the cold (4°C) salt buffer solution mentioned above and one time with a 1:10 dilution of it.

The inhibitors were dissociated from trypsin and eluted from the adsorbent by suspending it four to five times in cold (4°C) 0.4 M KCl-HCl pH 2.0 (200, 400 and 20 ml in experiments 1–5, 9–13 and 14–18, respectively). The first extract was discarded. The pH of the suspension was adjusted to 2.0 with 2 N HCl during the second extraction. The major inhibitor portion was found in the second extract.

The acidic inhibitor solutions were neutralized and desalted by ultrafiltration using Diaflo membranes (UM-05 or UM-2) either directly (exps. 1–13) or after fractionation on Sephadex G-75 in salt buffer solutions pH 7.6 (inhibitor I and II). After lyophilization of the salt-free inhibitor solutions, the following specific activities (trypsin inhibition, see Table 2) were estimated: 1.9–2.2 IU/mg for the inhibitor mixtures I,II (exps. 1–13), 2.0 IU/mg for inhibitor I and 2.8 IU for inhibitor II.

[a] For definition, see Table 2: 1 IU corresponds to the inhibition of 1 U, i.e., about 1 mg trypsin.

[b] Related to the amount (activity) bound to trypsin cellulose.

754   H. Fritz et al.

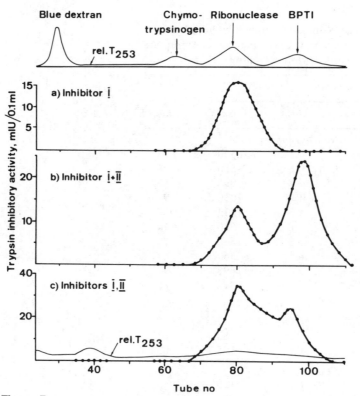

Figure 7
Gel filtration of the purified acid-stable trypsin inhibitors from boar
seminal plasma at pH 7.6. The inhibitor solution, 1–3 ml containing
2–13 IU (trypsin inhibition, substrate: BAPA), was applied to the
Sephadex G-75 column (122 × 1.6 cm, 10°C) equilibrated and
developed with 0.4 M NaCl, 0.1 M triethanolamine-HCl pH 7.6 at
a rate of 12 ml/hr, 4 ml/tube. The column was previously calibrated with marker proteins as indicated in the figure (BPTI: basic
pancreatic trypsin inhibitor). Further details are mentioned in the
text; see also Table 6.
  (*a*) Inhibitor I: Isolated by gel filtration and affinity chromatography. (*b*) Inhibitor I + II: mixture of inhibitor I and II, both isolated by gel filtration and affinity chromatography (if inhibitor II is
fractionated only, no inhibitor I is separated under the same conditions). (*c*) Inhibitors I,II: inhibitor mixture isolated by affinity
chromatography only.

basic structure of the inhibitor and, therefore, without loss of activity and
antigenicity. The possibility of the synthesis of inhibitor molecules partially
or totally devoid of carbohydrate residues or with some minor structural
alterations (see the exchange of basic amino acid residues in inhibitor A
and A_1, Table 4) due to gene mutations should also be taken into account.

Inhibitors in Spermatozoa

Gel filtration of acidic extracts of washed ejaculated spermatozoa on Sephadex G-75 at pH 2.0 resulted in the separation of acrosin from one inhibitor

Table 7

Comparison of the Inhibition of Trypsin, Acrosin, Plasmin and α-Chymotrypsin by Acid-stable Trypsin Inhibitors from Boar Seminal Plasma and Spermatozoa

Inhibitor source	Inhibition of				
	trypsin	acrosin[a]	plasmin[a]	α-chymo-trypsin[a]	kallikrein[b]
Seminal plasma					
inhibitor I	2.9	1	10	36	
inhibitor Ia	2.9	1	17	35	
inhibitor II	3.3	1	25	31	4.2
Spermatozoa[c]	3.3	1	28		

The amount (μg) of inhibitor is given, related to the acrosin value = 1, which reduced the proteolytic activity of the enzyme to 50 or 20% (α-chymotrypsin) of the original value. The activities of bovine trypsin (11.5 mU), boar acrosin (8.5 mU) and porcine plasmin (3.3 mU) were measured with N-α-benzoyl-L-arginine p-nitroanilide-HCl (0.5 mg/3 ml) as substrate. For chymotrypsin (1.5 mU) estimation, N-α-(3-carboxy-propionyl)-L-phenylalanine p-nitroanilide was used (Fritz, Trautschold and Werle 1974). Specific activities (trypsin inhibition, IU/mg) of the inhibitor preparations employed were (see text): inhibitor I, 0.63; inhibitor Ia, 0.80; inhibitor II, 2.3; inhibitor from spermatozoa, 0.93. These inhibitors have been purified only by repeated gel chromatography at pH 7.6 (see Table 6).

[a] The titration curves are not linear up to 50% inhibition.
[b] Porcine pancreatic kallikrein, 0.1 μg/3 ml.
[c] Soluble in neutral salt buffer solutions (see text).

fraction with an approximate molecular weight of 7200 (Fig. 9). After lyophilization, the portion of the inhibitor soluble in 0.4 M NaCl, 0.1 M triethanolamine-HCl pH 7.6 was fractionated on a Sephadex G-75 column equilibrated and developed with the salt buffer solution pH 7.6 mentioned above. The inhibitory activity was eluted as a symmetrical peak in a position corresponding to a molecular weight of about 6200 (at pH 7.6). The inhibitor was further purified by gradient elution chromatography (Figs. 7 and 8 in Tschesche et al. 1974; the legend of Fig. 8 in Tschesche et al. belongs to Fig. 11 and vice versa). The elution pattern thus obtained was similar to the elution profile of the inhibitors from seminal plasma (Fig. 3 in Tschesche et al. 1974). In addition, the amino acid composition of the major sperm inhibitor fraction is nearly identical with the composition of the seminal inhibitors isolated so far (Table 4). The molecular weights found by gel filtration at pH 2.0 or 7.6 would be in relatively good agreement with the value calculated from the amino acid composition if no carbohydrate chains are attached to the inhibitor molecules. Whether the carbohydrate residues are removed before adsorption of the inhibitors to the sperm cells or during contact with the sperm head membranes or in the course of extraction and purification in the acetic acid solutions used has still to be clarified.

The affinities of the sperm-associated inhibitors[2] to trypsin, acrosin and plasmin are very similar to those of the inhibitor fractions from seminal plasma (Table 7). Antibodies prepared against inhibitor I and the inhibitor

[2] The inhibitor mixture obtained after gel filtration at pH 7.6 was employed for the titration experiments.

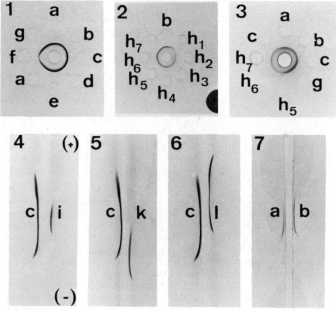

Figure 8

Immunological cross-reactivities of the trypsin-acrosin inhibitors from boar seminal plasma and spermatozoa. Rabbits were immunized either with inhibitor I or with the inhibitor mixture I,II; see *a* and *c* in Figure 7. The antibodies thus produced were employed in immunodiffusion and immunoelectrophoresis. Details of the experimental approach will be presented elsewhere.

Immunodiffusion plates: 10 or 15 (plate *1*) μl of the antibody solution $A_{I,II}$ (plates *1,2*) or A_I (plate *3*) and 5 (plate *1*), 3.5 (plate *2*) or 2 (plate *3*) mIU inhibitor (trypsin inhibition, substrate: BAPA) were applied to the agarose gel (2%, w/v; 0.05 M barbital sodium pH 8.7 in plates *1* and *2* or pH 8.2 in plate *3*). The precipitation lines were stained with Coomassie brilliant blue.

Immunoelectrophoresis: 100 μl of the antibody solution $A_{I,II}$ (plates *4–6*) or A_I (plate *7*) and 17 (plate *4*), 20 (plates *5* and *6*) and 5 (plate *7*) mIU inhibitor (complexes: calculated amount) were applied to the agarose gel (2%, w/v; 0.05 M barbital sodium pH 8.7 in plates *4–6* or pH 8.2 in plate *7*). Electrophoresis was performed for 90 minutes in the mentioned buffer solutions. The precipitation lines were stained with Coomassie brilliant blue.

a: Inhibitor I from seminal plasma; **b:** inhibitor I from seminal plasma; **c:** inhibitors I,II (*c* in Fig. 7); **d:** inhibitors I,II (another batch). **e:** boar seminal plasma; **f:** inhibitor from washed ejaculated spermatozoa (soluble at pH 2.6 but not in salt solutions pH 7.6; see text); **g:** inhibitor from washed ejaculated spermatozoa (soluble in salt solutions pH 7.6; see text); **h$_{1-7}$:** multiple forms of the inhibitors from seminal plasma (H. Tschesche and R. Klauser, unpubl.); **i:** acrosin-inhibitor complex; **k:** trypsin-inhibitor complex; **l:** inhibitor isolated from the trypsin-inhibitor complex (see k, plate *6*).

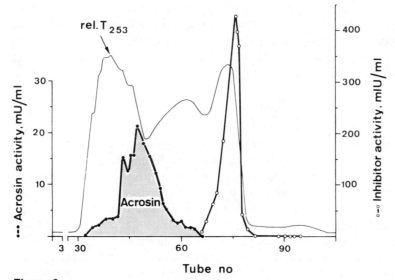

Figure 9

Gel filtration of acidic extracts of ejaculated boar spermatozoa at pH 2.0. Ejaculated boar spermatozoa were extensively washed and extracted with 2% (v/v) acetic acid adjusted to pH 2.0 by addition of 2 N HCl (see Fink et al. 1972; Fritz et al. 1972c). The acidic extract from a 30-ml sperm pellet was concentrated by ultrafiltration and applied to a Sephadex G-75 column (130 × 3.8 cm 4°C) equilibrated and developed with 2% (v/v) acetic acid, adjusted to pH 2.0 with 2 N HCl, at a rate of 24 ml/hr, 8 ml/tube. Acrosin activity was measured with N-α-benzoyl-D,L-lysine p-nitroanilide as substrate (Schiessler et al. 1972). For the estimation of the inhibition of trypsin (right ordinate), the substrate BAPA was used.

mixture I,II from seminal plasma also produced precipitation lines with the sperm-associated inhibitor (Fig. 8). These results indicate clearly that the inhibitors extracted from the sperm cells originate from seminal plasma (see earlier paper by Fritz et al., this volume).

However, besides the described sperm inhibitors which are soluble in salt buffer solution over a wide pH range (like the seminal inhibitors), other trypsin-acrosin inhibitors which are also present in appreciable amounts in the acidic sperm extracts are nearly insoluble in salt buffer solutions of pH 7.6. The purification and characterization of these inhibitors, which may originate from the spermatozoa itself or from epididymal fluids, etc., has still to be achieved.

Polakoski and Williams (1974a,b) recently demonstrated the occurrence of inhibitors of similar molecular weights (estimated by gel filtration, pH 3.0) in spermatozoa (MW 13,400, 6800, 1600) and seminal plasma. The presence of the 1600-MW fraction in the acrosomal detergent extracts is of special interest, even if it should be a degradation product of the other inhibitors. Such a molecule should have a very rigid structure representing mainly the reactive center of the inhibitor and is therefore especially interesting for studies concerned with molecular aspects of enzyme-inhibitor interactions (Huber et al. 1974; Janin, Sweet and Blow 1974). Whereas the 6800-MW inhibitor should be identical with the sperm inhibitor described by us, the 13,400-MW inhibitor could be either an association product of the 6800-MW monomer or related to the carbohydrate-containing inhibitor I from seminal plasma (see above).

Immunological Cross-reactivity and Localization

Rabbits were immunized with the inhibitor mixture I,II and inhibitor I (see Fig. 7), both purified by affinity chromatography (see Table 6). Antibodies designated $A_{I,II}$ or A_I were isolated from the appropriate antisera and used for the following investigations. Experimental details will be described elsewhere.

Cross-reactivities: The antibodies $A_{I,II}$ and A_I produced precipitation lines of similar intensity in immunodiffusion and immunoelectrophoresis with all the inhibitor fractions employed (Fig. 8). The localization of the precipitation lines in the electrophoresis plates (results to be published) indicates that the various inhibitor fractions differ in their electrophoretic mobility and thus the net charge of the molecules. This may be explained by the occurrence of multiple forms produced from the native inhibitor by removal of carbohydrate and/or amino acid residues (caused by glycosidases and exopeptidases or acid treatment), internal cleavages (caused by endopeptidases) and the synthesis of related inhibitors by mutated genes as already discussed.

In order to test the specificity of the antibodies used, the complexes of the inhibitors (mixture I,II) with trypsin and acrosin were also investigated by immunoelectrophoresis (Fig. 8). Obviously, all the immunologically reactive material is bound to trypsin or acrosin and, therefore, is inhibitor protein: the positions of the precipitation lines of the complexes are clearly different from those of the free inhibitors. Due to a higher positive net charge of the trypsin-inhibitors complex (trypsin is a strongly basic protein), the precipitation line is shifted towards the cathode. The inhibitors dissociated from the complex with trypsin migrate in the reverse direction, indicating the removal of a negatively charged portion from the inhibitor molecules.

Localization in tissues and spermatozoa: Tissue sections or extensively washed spermatozoa were treated with inhibitor antibodies $A_{I,II}$ and subsequently with fluorescein isothiocyanate-labeled goat γ-globulins directed against the inhibitor-specific rabbit γ-globulins. The detailed experimental procedures will be described elsewhere.

The epithelium cells of the seminal vesicles, the ductus deferens, the tail of the epididymis and the urethra showed an intensive fluorescence (Fig. 10). In prostate sections, the fluorescence appears in lumps, indicating the production of inhibitors in smaller cell aggregates. Besides some autoimmuno-fluorescence found in testes sections, epithelium cells of the testes and of the head and mid-piece of the epididymis showed, like the controls, no immuno-fluorescence at all. According to these results, the acrosin-trypsin inhibitors of the seminal plasma are produced mainly in the seminal vesicles, but to some extent also by the epithelium cells of the tail of the epididymis, the ductus deferens, the urethra and by special glands inside the prostate.

About 80% of washed ejaculated spermatozoa showed an intensive fluorescence of the acrosomal region, whereas epididymal spermatozoa as well

Color Plates
Figures 10 and 11

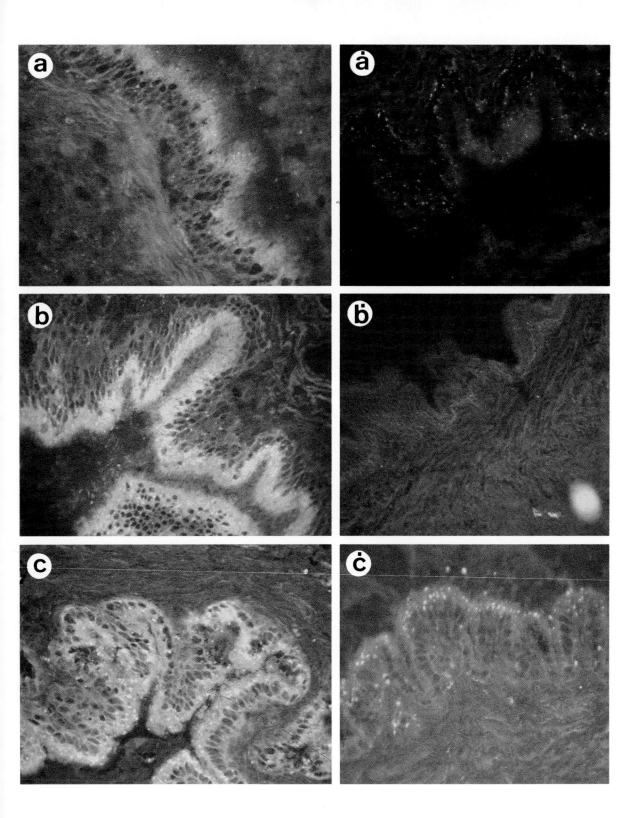

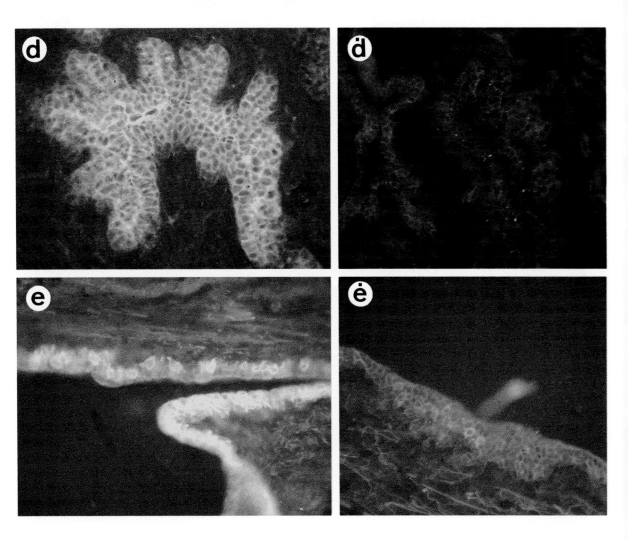

Figure 10

Immunofluorescent localization of the seminal inhibitors in the boar genital tract. Antibodies (rabbit γ-globulins) produced against the inhibitor mixture I,II (c in Fig. 7) or unspecific, non-inhibitor-directed rabbit γ-globulins (for the controls) and fluorescein isothiocyanate-labeled goat anti-rabbit γ-globulins were employed. The indirect immunofluorescent staining method was used (Nairn 1964). Details of the experimental approach will be described elsewhere. Magnification: 525X.

(a) Epididymis, tail; *(b)* ductus deferens; *(c)* seminal vesicles; *(d)* prostate; *(e)* urethra; *(à–è)* controls.

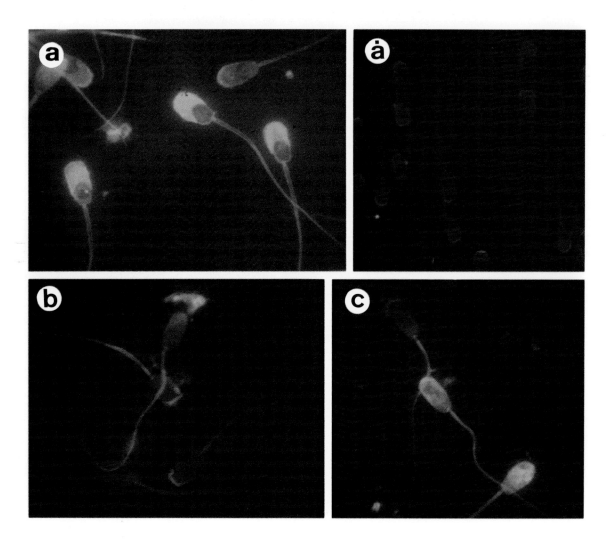

Figure 11

Immunofluorescent localization of the seminal inhibitors in boar spermatozoa. The antibodies or γ-globulins and the method were the same as described in the legend to Figure 10. Ejaculated spermatozoa were extensively washed (4–8 times) with physiological saline prior to incubation with the inhibitor-directed rabbit γ-globulins. Washed (6 times) epididymal as well as uterine spermatozoa were directly treated with the inhibitor-specific rabbit γ-globulins or, after pretreatment (15 min at 37°C, 1.5 mg inhibitor per 5 x 10^6 spermatozoa suspended in 0.25 ml physiological saline), with a highly purified inhibitor fraction (seminal plasma inhibitor A; see Table 4). Controls were run using either unspecific (non-inhibitor-directed) rabbit γ-globulins or inhibitor-specific γ-globulins which had been saturated with the highly purified inhibitor A (see Table 4) (0.4 mg inhibitor in 0.3 ml antibody solution, incubated 15 min at 37°C). Magnification: 2100X.

(a) Ejaculated spermatozoa; *(b)* uterine spermatozoa; *(c)* uterine spermatozoa, pretreated with inhibitor; *(à)* control.

as spermatozoa recovered six hours after mating from the uteri of sows did not bind inhibitor antibodies (Fig. 11). After the uterine spermatozoa had been treated with a highly purified inhibitor preparation, $A_{I,II}$ antibodies were bound again by about 40% of the spermatozoa. In contrast, epididymal spermatozoa did not bind any inhibitor antibodies after treatment with the pure inhibitor.

Obviously, the seminal plasma acrosin inhibitors are picked up by the spermatozoa during ejaculation and later on are detached from the sperm cells in the fluids of the female genital tract. However whereas uterine spermatozoa can pick up inhibitors again, the binding of inhibitors to epididymal sperm cells seems to be mediated either directly (as connecting link) or indirectly (by modification of the inhibitor or membrane surface) by a substance from the accessory glands. This view is supported by the absence of immunofluorescence in epididymal spermatozoa, although the epithelium cells of the tail of the epididymis are producing inhibitors which are cross-reacting with the seminal inhibitor antibodies.

The demonstration of the adsorption of seminal inhibitors to uterine spermatozoa is probably the basis of the results obtained in antifertility experiments: in vivo or in vitro fertilization was significantly reduced after treatment of capacitated spermatozoa with seminal acrosin inhibitors (Zaneveld et al. 1971a,b; Newell, Polakoski and Williams 1972; Suominen, Kaufman and Setchell 1973). It is unlikely, however, that the inhibitors can penetrate the outer sperm head membranes (plasma membrane, outer acrosomal membrane) and inhibit acrosin; they are most probably adsorbed only at the surface of the outer membranes (see results discussed in preceding paper, Fritz et al., this volume). However, the exact localization of the inhibitors picked up by the spermatozoa has still to be elucidated. Our preliminary conclusion that the inhibitors are only present at the surface of the outer membranes had to be reevaluated when we realized that the immunofluorescence pattern of ejaculated (as well as epididymal and testicular) spermatozoa treated with monospecific acrosin antibodies is nearly identical with that of the $A_{I,II}$-treated spermatozoa. Obviously, the outer membranes are loosened by the fixation procedure used and thus become permeable for the γ-globulin molecules.

Hirschhäuser and Baudner (1972) very probably did not use acrosin inhibitor-directed antibodies (see above) in their experiments concerned with the localization of the inhibitor substance isolated from seminal plasma in human spermatozoa. The immunologically reactive substance was easily washed off from the spermatozoa, whereas the acrosin inhibitors are not removed from boar and human spermatozoa even after the extensive washing procedure we employed.

Biological Function

Acrosin Inhibitors

Acrosin inhibitors of low molecular weight are not only added to spermatozoa during ejaculation, they are also found in *testicular* and *epididymal* tissues (Fink et al. 1971b) and fluids, for example, 12 and 24 mIU (trypsin inhibition) per ml rete testes fluid and 80 and 200 mIU/ml epididymal fluid of the boar and ram, respectively (Suominen and Setchell 1972); we found 26–250 mIU and about 40 mIU/ml epididymal fluid of the boar and hamster, respectively. These inhibitors are probably also produced by epithelium cells similar to the inhibitors from seminal vesicles. Their synthesis seems to be hormone controlled (Weber 1975).

The occurrence of acid-stable acrosin inhibitors in epididymal spermatozoa and perhaps already in testicular sperm cells is especially interesting. Whether these inhibitors are picked up by the spermatozoa from the surrounding fluids or are liberated from precursors inside the sperm cells during maturation or are synthesized as part of the sperm head membranes is still unknown.

It may be concluded from the inhibition properties and the distribution of the acrosin-trypsin inhibitors in the male genital tract that these substances are the natural antagonists of sperm acrosin. In our opinion, the *function of the acrosin inhibitors* in the male has to be seen mainly in the rapid neutralization of the enzymatic potential of acrosin whenever this enzyme is activated or liberated at the wrong time and place, i.e., before mating or capacitation of the spermatozoa in the female reproductive tract. Thus destruction of cell membranes or soluble proteins and current kinin liberation (see Fritz et al., this volume) in the male genital tract as a consequence of physiological cell death or in the course of inflammatory processes is prevented.

In this connection, it is remarkable that *complex formation* (association) between acrosin and the seminal inhibitors occurs very rapidly even if the partners are present in relatively low concentrations, e.g., each in approximately 2×10^{-9} moles per liter in the test system used for the titration experiments (Fig. 2). The equilibrium: acrosin + inhibitor $\rightleftharpoons$ complex is attained within from 1 to 5 minutes after addition of the inhibitors to the slightly alkaline (pH 7.8) acrosin solution in all cases tested so far: inhibitors I and II from guinea pig seminal vesicles, all acrosin inhibitors from boar seminal plasma and spermatozoa, inhibitor II from human seminal plasma, inhibitor B-3 from leeches (see Tables 4, 5 and Fritz et al. 1972b,d). In contrast, *dissociation* of the acrosin-inhibitor complex proceeds very slowly: we could not observe the liberation of acrosin from the complex after addition of the substrate to the test cuvettes within 30 minutes. This applies also to the BAEE/ADH (N-α-benzoyl-L-arginine ethyl ester/alcohol dehydrogenase) test system (Fritz et al. 1972c; Schill 1973) which is about ten times more sensitive than the BAPA (N-α-benzoyl-D,L-arginine p-nitroanilide) assay used for the titrations shown in Figure 2. These observations indicate long half-life times for complex dissociation even in very diluted solutions in which the concentrations of the partners are near the magnitude of the K_i values of the complexes (near or below 1×10^{-10} M; Fritz et al. 1972b and H. Schiessler, M. Arnhold and H. Fritz, in prep.).

The amount of acrosin inhibitors present in ejaculates (between 3×10^{-4} and 1×10^{-5} moles per liter) and in epididymal fluid (about 1×10^{-5} moles per liter) is sufficient to neutralize completely the acrosin which may escape from disintegrating spermatozoa. For example, an average value of 650×10^6 spermatozoa per ml boar ejaculate corresponds to 6.5×10^{-7} moles acrosin per liter. If the acrosin would be liberated from 10% of disintegrated boar spermatozoa, an inhibitor concentration of about 1–30×10^{-5} M (seminal plasma) or $1–10 \times 10^{-6}$ M (epididymal fluid) would be faced with an acrosin concentration of about 6.5×10^{-7} M if the high molecular weight acrosin inhibitors also present are not regarded.

Spermatozoa attain their fertilizing ability during migration and residence in the secretions of the female reproductive tract (Bedford 1972). Zaneveld et al. (1971a; Zaneveld, Polakoski and Schumacher, this volume) claimed that the removal of acrosin inhibitors from ejaculated spermatozoa is part of the *capacitation* process. We could now confirm directly that the acrosin inhibitors picked up by the spermatozoa from seminal plasma during ejaculation are removed during residence of the sperm cells in uterine fluid. Detachment of the seminal inhibitors may be either induced by internal factors, e.g., an alteration of the binding sites at the sperm head membranes, or external factors, e.g., inhibitor degrading or binding proteinases in uterine fluid. For the reasons discussed above, it is unlikely that a simple dissociation of the inhibitors from the complex, in which they are bound very tightly to acrosin, should be the basis of the replacement reaction. From this standpoint, the localization of acrosin and the inhibitors at different compartments of the acrosome is also logical. On the other hand, the localization of the inhibitors that are already found in epididymal spermatozoa and their fate during ejaculation and capacitation are entirely unknown.

Nevertheless, acrosin is present in free form, probably bound to the inner acrosomal membrane or equatorial segment (see Fritz et al., this volume) at the site where the actual fertilization process takes place. This view is supported by the results of experiments in which seminal acrosin inhibitors of very limited specificity could prevent fertilization in vitro and in vivo (Zaneveld et al. 1971a,b; Newell, Polakoski and Williams 1972; Suominen, Kaufman and Setchell 1973; Polakoski and Williams 1974a). We prefer the explanation that these inhibitors cannot penetrate the sperm head membranes and react with acrosin even in capacitated spermatozoa, but are relatively weakly adsorbed at the surface of the membranes and probably liberated close to acrosin during the acrosome reaction so that complex formation could occur.

In this respect, it is also interesting to note that an acid-stable acrosin inhibitor is not present in human *cervical mucus* (Wallner and Fritz 1974). The concentration of the plasmatic acrosin inhibitors (α_1-antitrypsin, inter-α-trypsin inhibitor and antithrombin III; Schumacher and Zaneveld 1974; Fritz et al. 1972e) is especially low during the time of ovulation. In addition, α_1-antitrypsin and antithrombin III react only slowly with acrosin. Apart from these considerations, it is very unlikely that the high molecular weight plasma proteinase inhibitors can penetrate the sperm head membranes.

The Human Trypsin-Chymotrypsin Inhibitor

α_1-Antitrypsin and α_2-macroglobulin are engaged in the inhibition and clearance of proteinases liberated under pathological conditions, especially of elastases and collagenases from human disintegrating leukocytes (Jeppsson and Laurell, this volume; Ohlsson 1974 and this volume). However, only α_1-antitrypsin occurs in mucous secretions of the organism (Schumacher and Zaneveld 1974; Zaneveld et al. 1974; Hochstrasser et al. 1972c, 1973). The observation of Hochstrasser et al. (1972a, 1973; Hochstrasser, Feuth and

Hochgesand 1974) of the inhibition of neutral leukocytic proteinases by the acid-stable trypsin-chymotrypsin inhibitor from bronchial and nasal secretions was, therefore, of special interest. The occurrence of inhibitors with very similar properties in seminal plasma and cervical mucus indicates that there exists besides α_1-antitrypsin an additional substance which can protect mucous membranes against the hydrolytic action of proteinases from leukocytes and perhaps also of invasive pathogens. The possible relation existing between the acid-stable trypsin-chymotrypsin inhibitor and the plasmatic inter-α-trypsin inhibitor (Hochstrasser, Feuth and Hochgesand 1974) has still to be clarified.

Acknowledgments

This work was supported by Sonderforschungsbereich-51 München (H. F., H. T.), Deutsche Forschungsgemeinschaft (W.–B. S., H. T.) and WHO, Geneva (H. F. and H. T.). We wish to thank Professor Dr. W. Leidl for valuable suggestions and discussions. We acknowledge gratefully the skillful technical assistance of Bruni Förg-Brey, Irmgard Estelmann, Sigrid Kupfer and Maria Meier. R. B. L. Gwatkin, Rahway, supplied us with epididymal hamster spermatozoa; R. Stolla, Munich, with epididymal and uterine boar spermatozoa.

REFERENCES

Bedford, J. M. 1972. Sperm transport, capacitation and fertilization. In *Reproductive Biology* (ed. H. Balin and S. Glasser), p. 338. Excerpta Medica, Amsterdam.

Figarella, C., G. A. Negri and O. Guy. 1974. Studies on inhibition of the two human trypsins. In *Proteinase Inhibitors—Second International Research Conference. Bayer Symposium V* (ed. H. Fritz et al.), p. 213. Springer Verlag, Berlin.

Fink, E., H. Schiessler, M. Arnhold and H. Fritz. 1972. Isolierung eines Trypsin-ähnlichen Enzyms (Akrosin) aus Eberspermien. *Z. Physiol. Chem.* **353**:1633.

Fink, E., E. Jaumann, H. Fritz, H. Ingrisch and E. Werle. 1971a. Protease-Inhibitoren im menschlichen Spermaplasma—Isolierung durch Affinitätschromatographie und Hemmverhalten. *Z. Physiol. Chem.* **352**:1591.

Fink, E., G. Klein, F. Hammer, G. Müller-Bardorff and H. Fritz. 1971b. Protein proteinase inhibitors in male sex glands. In *Proceedings International Research Conference on Proteinase Inhibitors* (ed. H. Fritz and H. Tschesche), p. 225. Walter de Gruyter, New York.

Fink, E., H. Fritz, E. Jaumann, H. Schiessler, B. Förg-Brey and E. Werle. 1973. Protein proteinase inhibitors in male sex glands and their secretions. In *Protides of the Biological Fluids* (ed. H. Peeters), p. 425. Pergamon Press, New York.

Fritz, H., H. Schiessler and W.–D. Schleuning. 1973a. Proteinases and proteinase inhibitors in the fertilization process: New concepts of control? In *Advances in Biosciences: Schering Workshop on Contraception—The Masculine Gender* (ed. G. Raspé and S. Bernhard), vol. 10, p. 271. Pergamon Press-Vieweg, Oxford.

———. 1973b. Boar and human acrosin and acrosin inhibitors: Isolation and biochemical characterization. *Biol. Reprod.* **9**:64.

Fritz, H., W.–D. Schleuning and W.–B. Schill. 1974. Biochemistry and clinical significance of the trypsinlike proteinase acrosin from boar and human spermatozoa. In *Proteinase Inhibitors—Second International Research Conference. Bayer Symposium V* (ed. H. Fritz et al.), p. 118. Springer Verlag, Berlin.

Fritz, H., I. Trautschold and E. Werle. 1965. Bestimmung der Molekulargewichte von neuen Trypsin-Inhibitoren mit Hilfe der Sephadex-Gelfiltration. *Z. Physiol. Chem.* **342:**253.

————. 1974. Protease-Inhibitoren. In *Methoden der enzymatischen Analyse* (ed. H. U. Bergmeyer), p. 1105. Verlag Chemie, Weinheim/Bergstr.

Fritz, H., E. Fink, R. Meister and G. Klein. 1970. Isolierung von Trypsin-Inhibitoren und Trypsin-Plasmin-Inhibitoren aus den Samenblasen von Meerschweinchen. *Z. Physiol. Chem.* **351:**1344.

Fritz, H., I. Trautschold, H. Haendle and E. Werle. 1968. Chemistry and biochemistry of proteinase inhibitors from mammalian tissues. *Ann. N.Y. Acad. Sci.* **146:**400.

Fritz, H., B. Förg-Brey, M. Meier, M. Arnhold and H. Tschesche. 1972a. Humanakrosin: Hemmbarkeit mit Protein-Proteinase Inhibitoren. *Z. Physiol. Chem.* **353:**1950.

Fritz, H., B. Förg-Brey, H. Schiessler, M. Arnhold and E. Fink. 1972b. Charakterisierung einer Trypsin-ähnlichen Proteinase (Akrosin) aus Eberspermien durch ihre Hemmbarkeit mit verschiedenen Protein-Proteinase-Inhibitoren. II. Inhibitoren aus Blutegel, Sojabohnen, Erdnüssen, Rindercolostrum und Seeanemonen. *Z. Physiol. Chem.* **353:**1010.

Fritz, H., B. Förg-Brey, E. Fink, M. Meier, H. Schiessler and C. Schirren. 1972c. Humanakrosin: Gewinnung und Eigenschaften. *Z. Physiol. Chem.* **353:**1943.

Fritz, H., B. Förg-Brey, E. Fink, H. Schiessler, E. Jaumann and M. Arnhold. 1972d. Charakterisierung einer Trypsin-ähnlichen Proteinase (Akrosin) aus Eberspermien durch ihre Hemmbarkeit mit verschiedenen Protein-Proteinase-Inhibitoren. I. Seminale Trypsin-Inhibitoren und Trypsin-Kallikrein-Inhibitor aus Rinderorganen. *Z. Physiol. Chem.* **353:**1007.

Fritz, H., N. Heimburger, M. Meier, M. Arnhold, L. J. D. Zaneveld and G. F. B. Schumacher. 1972e. Humanakrosin: Zur Kinetik der Hemmung durch Human-Seruminhibitoren. *Z. Physiol. Chem.* **353:**1953.

Greene, L. J., D. E. Roark and D. C. Bartelt. 1974. Human pancreatic secretory trypsin inhibitor. In *Proteinase Inhibitors—Second International Research Conference. Bayer Symposium V* (ed. H. Fritz et al.), p. 188. Springer Verlag, Berlin.

Haendle, H. 1969. Über neue Proteinasen-Inhibitoren menschlichen und tierischen Ursprungs. Dissertation, Medizin., Fakultät der Universität München.

Haendle, H., H. Ingrisch and E. Werle. 1970. Über einen neuen Trypsin-Chymotrypsin-Inhibitor im Cervixsekret der Frau. *Z. Physiol. Chem.* **351:**545.

Haendle, H., H. Fritz, I. Trautschold and E. Werle. 1965. Über einen hormonabhängigen Inhibitor für proteolytische Enzyme in männlichen accessorischen Geschlechtsdrüsen und im Sperma. *Z. Physiol. Chem.* **343:**185.

Hirschhäuser, C. and S. Baudner. 1972. Immunologic localization of the human seminal plasma protease inhibitor in human spermatozoa. *Fert. Steril.* **23:**393.

Hirschhäuser, C. and M. Kionke. 1971. Properties of a human seminal plasma inhibitor for trypsin. *Fert. Steril.* **22:**360.

Hirschhäuser, C., S. Baudner and E. Daume. 1972. Immunological identification of human seminal plasma protease inhibitor in cervical mucus. *Fert. Steril.* **23:**630.

Hochstrasser, K., H. Feuth and K. Hochgesand. 1974. Proteinase inhibitors of the respiratory tract: Studies on the structural relationship between acid-stable inhibitors present in the respiratory tract, plasma and urine. In *Proteinase Inhib-*

itors—Second International Research Conference. Bayer Symposium V (ed. H. Fritz et al.), p. 111. Springer Verlag, Berlin.

Hochstrasser, K., R. Reichert, E. Werle and H. Haendle. 1973. New proteinase inhibitors in the secretion of some mucous membranes of the human body. In *Protides of the Biological Fluids—Proceedings of the 20th Colloquium Brugge, 1972* (ed. H. Peeters), vol. 20, p. 417. Pergamon Press, Oxford.

Hochstrasser, K., R. Schuster, R. Reichert and N. Heimburger. 1972c. Nachweis und quantitative Bestimmung von Komplexen zwischen Leukozytenproteasen und α_1-Antitrypsin in Körpersekreten. *Z. Physiol. Chem.* **353**:1120.

Hochstrasser, K., G. Bickel, R. Reichert, H. Feuth and D. Meckl. 1972b. Nachweis von Proteaseninhibitoren im Intestinaltrakt von Mensch und Hund. *Z. Klin. Chem. Klin. Biochem.* **10**:450.

Hochstrasser, K., R. Reichert, M. Matzner, E. Werle and S. Schwarz. 1972a. Hemmbarkeit proteolytischer Enzyme in pathologischen Nasensekreten und von Leukozytenproteasen durch den natürlichen Proteaseninhibitor des Nasensekrets. *Z. Klin. Chem. Klin. Biochem.* **10**:104.

Huber, R., D. Kukla, W. Steigemann, J. Deisenhofer and A. Jones. 1974. Structure of the complex formed by bovine trypsin and bovine pancreatic trypsin inhibitor —Refinement of the crystal structure analysis. In *Proteinase Inhibitors—Second International Research Conference. Bayer Symposium V* (ed. H. Fritz et al.), p. 497. Springer Verlag, Berlin.

Janin, J., R. M. Sweet and D. M. Blow. 1974. The mode of action of soybean trypsin inhibitor as revealed by crystal structure analysis of the complex with porcine trypsin. In *Proteinase Inhibitors—Second International Research Conference. Bayer Symposium V* (ed. H. Fritz et al.), p. 513. Springer Verlag, Berlin.

Jaszczak, S. and E. S. E. Hafez. 1973. Sperm migration through the uterine cervix in the macaque during the menstrual cycle. *Amer. J. Obstet. Gynecol.* **115**:1070.

Kutzbach, C. and G. Schmidt-Kastner. 1972. Kallikrein from pig pancreas. *Z. Physiol. Chem.* **353**:1099.

Maroni, E. S. and P. C. Wilkinson. 1971. Selective chemotaxis of macrophages towards human and guinea-pig spermatozoa. *J. Reprod. Fert.* **27**:149.

Maroni, E. S., D. N. K. Symon and P. C. Wilkinson. 1972. Chemotaxis of neutrophil leucocytes towards spermatozoa and seminal fluid. *J. Reprod. Fert.* **28**:359.

McRorie, R. A. and W. L. Williams. 1974. Biochemistry of mammalian fertilization. *Annu. Rev. Biochem.* **43**:777.

Nairn, R. C., ed. 1964. *Fluorescent Protein Tracing.* E. & S. Livingstone Ltd., Edinburgh.

Newell, S. D., K. L. Polakoski and W. L. Williams. 1972. Inhibition of fertilization by proteinase inhibitors. *Proc. VII Int. Congr. Animal Reprod. Art. Insem.* **3**:2117.

Ohlsson, K. 1974. Interaction between endogenous proteases and plasma protease inhibitors *in vitro* and *in vivo.* In *Proteinase Inhibitors—Second International Research Conference. Bayer Symposium V* (ed. H. Fritz et al.), p. 96. Springer Verlag, Berlin.

Polakoski, K. L. and W. L. Williams. 1974a. Isolation of proteinase inhibitors from boar sperm acrosomes and boar seminal plasma and effect on fertilization. In *Proteinase Inhibitors—Second International Research Conference. Bayer Symposium V* (ed. H. Fritz et al.), p. 128. Springer Verlag, Berlin.

———. 1974b. Studies on purification and characterization of boar acrosin. In *Proteinase Inhibitors—Second International Research Conference. Bayer Symposium V* (ed. H. Fritz et al.), p. 128. Springer Verlag, Berlin.

Polakoski, K. L., L. J. D. Zaneveld and W. L. Williams. 1971. An acrosin-acrosin inhibitor complex in ejaculated boar spermatozoa. *Biochem. Biophys. Res. Comm.* **45**:381.

Polakoski, K. L., L. J. D. Zaneveld, W. L. Williams and R. A. McRorie. 1971. Purification and partial characterization of a trypsin inhibitor from porcine seminal plasma. *Fed. Proc.* **30**:1077.

Rasmussen, J. and O. K. Albrechtsen. 1960. Fibrinolytic activity in human seminal plasma. *Fert. Steril.* **11**:264.

Reichert, R., K. Hochstrasser and G. Conradi. 1972. Untersuchungen zur Proteasenhemmkapazität des menschlichen Bronchialsekrets. *Pneumonologie* **147**:13.

Schiessler, H., M. Arnhold and H. Fritz. 1974. Characterization of two proteinase inhibitors from human seminal plasma and spermatozoa. In *Proteinase Inhibitors—Second International Research Conference. Bayer Symposium V* (ed. H. Fritz et al.), p. 147. Springer Verlag, Berlin.

Schiessler, H., H. Fritz, M. Arnhold, E. Fink and H. Tschesche. 1972. Eigenschaften des Trypsin-ähnlichen Enzyms (Akrosin) aus Eberspermien. *Z. Physiol. Chem.* **353**:1638.

Schill, W.-B. 1973. Acrosin activity in human spermatozoa: Methodological investigations. *Arch. Derm. Forsch.* **248**:257.

Schumacher, G. F. B. and L. J. D. Zaneveld. 1974. Proteinase inhibitors in human cervical mucus and their *in vitro* interactions with human acrosin. In *Proteinase Inhibitors—Second International Research Conference. Bayer Symposium V.* (ed. H. Fritz et al.), p. 178. Springer Verlag, Berlin.

Suominen, J. J. O. and M. Niemi. 1972. Human seminal trypsin inhibitors. *J. Reprod. Fert.* **29**:163.

Suominen, J. and B. P. Setchell. 1972. Enzymes and trypsin inhibitor in the rete testes fluid of rams and boars. *J. Reprod. Fert.* **30**:235.

Suominen, J., M. H. Kaufman and B. P. Setchell. 1973. Prevention of fertilization *in vitro* by an acrosin inhibitor from rete testes fluid of the ram. *J. Reprod. Fert.* **34**:385.

Syner, F. N. and R. Kurrus. 1974. Comparison of molecular forms of trypsin acrosin isoinhibitors in human spermatozoa and seminal plasma. In *Abstracts of the 7th Annual Meeting of the Society for the Study of Reproduction,* Ottawa, p. 108. Carleton University, Ottawa, Canada.

Syner, F. N. and K. S. Moghissi. 1972. Properties of proteolytic enzymes and inhibitors in human semen. In *Biology of Mammalian Fertilization and Implantation* (ed. K. S. Moghissi and E. S. E. Hafez), p. 3. Charles C. Thomas, Springfield, Illinois.

Tschesche, H. 1974. Biochemie natürlicher Proteinase-Inhibitoren. *Angew. Chem. Int. Ed.* **13**:10.

Tschesche, H., S. Kupfer, O. Lengyel, R. Klauser and H. Fritz. 1975. Structure, biochemistry and comparative aspects of mammalian seminal plasma acrosin inhibitors. In *Protides of the Biological Fluids—Proceedings of the 23rd Colloquium Brugge, 1975* (ed. H. Peeters), vol. 23. Pergamon Press, Oxford (in press).

Tschesche, H., S. Kupfer, O. Lengel, R. Klauser, M. Meier and H. Fritz. 1974. Purification, characterization, structural studies of proteinase inhibitors from boar seminal plasma and boar spermatozoa. In *Proteinase Inhibitors—Second International Research Conference. Bayer Symposium V* (ed. H. Fritz et al.), p. 164. Springer Verlag, Berlin.

Waldschmidt, M., B. Hoffman and H. Karg. 1966. Untersuchungen über die

766   H. Fritz et al.

tryptische Enzymaktivität in Geschlechtssekreten von Bullen. *Zuchthygiene*
1:15.
Wallner, O. and H. Fritz. 1974. Characterization of an acid-stable proteinase
inhibitor in human cervical mucus. *Z. Physiol. Chem.* **355**:709.
Weber, F. 1975. Biochemische Untersuchungen über Proteinaseinhibitoren aus
Sperma und Genitalorganen vom Eber. Dissertation, Tierärztliche Fakultät der
Universität München.
Wilkinson, P. C., ed. 1974. *Chemotaxis and Inflammation.* Churchill Livingstone,
Edinburgh.
Zaneveld, L. J. D., B. M. Dragoje and G. F. B. Schumacher. 1972. Acrosomal
proteinase and proteinase inhibitor of human spermatozoa. *Science* **177**:702.
Zaneveld, L. J. D., K. L. Polakoski and W. L. Williams. 1973. A proteinase and
proteinase inhibitor of mammalian sperm acrosomes. *Biol. Reprod.* **9**:219.
Zaneveld, L. J. D., K. L. Polakoski, R. T. Robertson and W. L. Williams. 1971a.
Trypsin inhibitors and fertilization. In *Proceedings International Research Con-
ference on Proteinase Inhibitors* (ed. H. Fritz and H. Tschesche), p. 236.
Walter de Gruyter, New York.
Zaneveld, L. J. D., R. T. Robertson, M. Kessler and W. L. Williams. 1971b. In-
hibition of fertilization *in vivo* by pancreatic and seminal plasma trypsin inhib-
itors. *J. Reprod. Fert.* **25**:387.
Zaneveld, L. J. D., G. F. B. Schumacher, P. F. Tauber and D. Propping. 1974.
Proteinase inhibitors and proteinases of human semen. In *Proteinase Inhibitors
—Second International Research Conference. Bayer Symposium V* (ed. H. Fritz
et al.), p. 136. Springer Verlag, Berlin.
Zaneveld, L. J. D., G. F. B. Schumacher, H. Fritz, E. Fink and E. Jaumann. 1973.
Interaction of human sperm acrosomal proteinase with human seminal plasma
inhibitors. *J. Reprod. Fert.* **32**:525.

Evaluation of Proteolytic Activity in Mammalian Ovulation

Lawrence L. Espey

Department of Environmental Studies, Trinity University
San Antonio, Texas 78284

Mammalian ovulation is a unique biological phenomenon, unique in that it requires the rupture of healthy tissue at the surface of the ovary. This traumatic, but vital, disruption of normal tissue must occur if the ovum is to migrate into the oviduct where it can be fertilized. The conspicuous changes that take place at the site of rupture include (1) a fracture in the dense layers of collagenous tissue which encapsulate the mature ovarian follice, (2) some extravasation of blood, (3) release of the ovum, (4) subsequent formation of a blood clot, and (5) eventual wound healing.

But the histological changes are not restricted to the point of rupture: the ovulatory surge in gonadotropic hormones transforms the entire ovarian follicle into a highly secretory corpus luteum. In fact, there is reason to believe that the primary action of the gonadotropins is *luteinization,* and that rupture of the follicular surface is merely a circumstantial change that occurs as a consequence of the softening of tissue during the early stages of remodeling of the ovarian follicle into a corpus luteum.

Morphology of the Ovarian Follicle

Before going further, it would be worthwhile to briefly review the morphology of the ovarian follicle (Fig. 1). A Graafian follicle which has reached maturity protrudes above the ovarian surface and lies adjacent to the dense connective tissue sheath, the tunica albuginea, which surrounds the entire ovary. The follicle itself is encapsulated by a similar layer of collagenous tissue, the theca externa. Obviously for the egg to be released, an opening must occur in this wall of collagen. Although rupture normally takes place at the very apex of the bulge, digestion of the connective tissue appears to extend throughout the follicle, and probably into the ovarian stroma.

In looking more closely at the multilayered structure of the follicle wall (Fig. 2), the outermost layer is composed of cuboidal epithelial cells, the germinal epithelium, which is loosely attached to a diffuse basal lamina at

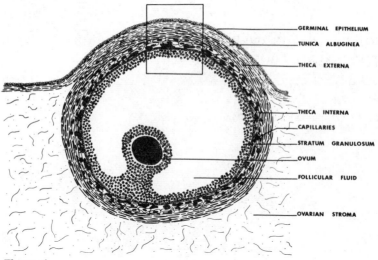

Figure 1
Structural composition of a mature ovarian follicle from the rabbit.
Boxed area indicates the region where rupture normally occurs. See
text for details.

the surface of the ovarian tunic. Beneath this is the fibrous tunica albuginea
which surrounds the ovary. This layer cannot be delineated from the collagen
and fibroblasts (the theca externa) which form the outer shell around an individual follicle. Below it is a thin layer of highly specialized fibrocytes, the
theca interna, which is very active in the secretion of female sex hormones.

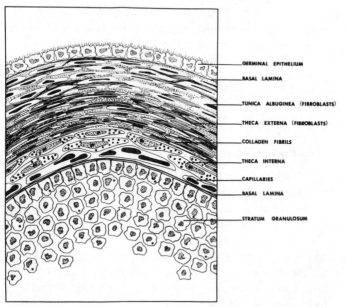

Figure 2
Close-up view of the cellular layers in the apical region of a
mature rabbit follicle. See text for details.

Adjacent to these secretory cells is an extensive network of large capillaries, essentially comprising the entire vascular supply to the follicle. These vessels do not penetrate the innermost layer of the follicle, the stratum granulosum, because of a distinct basal lamina which forms a barrier.

Under the influence of hormone stimulation, this tissue changes drastically within a matter of a few hours. All of the cells, except the germinal epithelium, become highly mitotic. In addition, the fibroblasts in the thecal layers are rapidly transformed from a resting state into motile cells. It is during the early stages of this mobilization that the extracellular matrix of collagen is decomposed (presumably by proteolytic enzymes) in a manner that leads to rupture of the follicle and extensive hemorrhage. Shortly thereafter, as the follicle begins to luteinize, the fibroblasts migrate into the interior of the follicle to lay down a network of supportive tissue for the rapidly proliferating cells of the lutein theca and the lutein granulosa which form the bulk of the developing corpus luteum. At the same time, the individual lutein cells enlarge considerably as their steroidogenic activity increases.

Experimental Animals

For specific reasons, most of our experimental tissue has been obtained from two distinctly different species of mammals. Rabbits were used because they are "induced" ovulators; that is, the ovulatory process does not occur spontaneously, but instead is induced by the act of mating. Rupture of the follicle takes place at almost exactly ten hours after coitus. Therefore, it is easy to correlate any biochemical changes in this tissue with the actual time of ovulation. The only significant disadvantage is that rabbit ovaries are relatively small. However, porcine follicles are much larger and therefore were used in addition to rabbit ovaries because they provide much greater quantities of experimental tissue. But they have a disadvantage in that hogs are spontaneous ovulators, and this makes it difficult to correlate the biochemical changes with the moment of ovulation.

The Mechanism of Ovulation

More than twelve years ago we began our studies on the mechanism of ovulation by examining the physical changes which occur in the follicle during rupture. There is now substantial evidence to show that rupture is not caused by an increase in intrafollicular pressure (Blandau and Rumery 1963; Espey and Lipner 1963; Rondell 1964), but instead is the consequence of a loosening of the collagenous connective tissue within the follicle wall. As ovulation nears, the follicle becomes distinctly more flaccid, and there is a measurable decrease in the tensile strength of the tissue (Espey 1967a).

An examination of the fine structure of the follicle has provided a close-up view of these changes (Espey 1967b). Ten hours from ovulation the thecal tissue is compact and the fibroblasts are numerous. In contrast, within minutes of ovulation the collagen fibers are dissociated and the fibroblasts appear sparse in comparison to those in follicles that are more distant from ovulation. The degeneration of the tissue is obvious. As a consequence of these changes, the tissue at the apex of the follicle begins to dissociate under the force of an hydrostatic pressure of about 20 mmHg in the follicular antrum (a pressure which is dependent upon the vascular pressure). Near the time of rupture, the fibrous outer layers of the follicle wall thin to less than one-fifth of their original width.

Proteolytic Enzymes and Ovulation

More than half a century ago an hypothesis was presented suggesting that *proteolysis* is responsible for the degradation of the follicle wall during ovulation (Schochet 1916). But it has been difficult to obtain supporting data for this hypothesis, and as recently as 1966, it was generally assumed that such enzymes were not involved in the process (Blandau 1966). However during the past decade, we have been accumulating indirect evidence that proteolytic activity probably has a significant role in this biological phenomenon. For example, the injection of small quantities of concentrated enzyme preparations directly into the antrum of a rabbit follicle can cause morphological changes similar to normal rupture (Espey and Lipner 1965). Several grades of clostridiopeptidase A, nagarse, pronase and concentrated preparations of trypsin from bovine pancreas are highly effective in causing rupture of rabbit follicles when injected in this manner. However, no response was elicited from chymotrypsin, crude peptidase, aminopeptidase, ficin, papain, lysozyme, hyase and elastase.

Also, it is now clear that a variety of proteolytic enzymes can weaken the *tensile strength* of sow follicles in vitro (Espey 1970). Their effect has been tested by incubating commercial enzyme preparations with strips of the follicle wall and then measuring the tensile strength by stretching the tissue. Under these conditions, preparations of collagenase, elastase, general protease, trypsin, α-chymotrypsin, and to a lesser extent β-chymotrypsin, were effective in reducing the tensile strength of the follicle wall.

In view of this limited information on the types of proteolytic enzymes which theoretically could be active in ovarian tissue, we assayed sow follicles and follicular fluid for a variety of enzymes. Particular attention was given to the types of enzymes which can decompose follicular connective tissue in vitro. The results of these studies have been included in a recent review article (Espey 1974), to which the reader should refer if he is interested in more details than are provided in the following summary.

The experimental follicles were collected in quantity from a local packing house and routinely staged at 20, 5 and 1 hour from ovulation. Follicular fluid was collected by syringe from follicles at each of the three stages. The fluid was centrifuged and the supernatant stored on ice until the follicle walls were also ready to be assayed. The walls from each stage of follicles were minced, homogenized in distilled water (or buffer, depending on the assay), and then assayed.

The results (Table 1) indicate that the follicle contains trypsin (in the wall, but not the fluid), cathepsin (especially at pH 3.5) and a collagenolytic enzyme which digests the synthetic substrate carbobenzoxy-glycyl-prolyl-glycyl-glycyl-prolyl-alanine (CBZ-GPGGPA). Only the enzyme that digested the synthetic hexapeptide changed significantly during ovulation. As rupture

Table 1

Hydrolytic Enzymes in Ovarian Follicles during Ovulation

Enzyme	pH	n	Relative activity in wall before rupture			Relative activity in fluid before rupture		
			20 hr	5 hr	1 hr	20 hr	5 hr	1 hr
General proteolytic	7.5	4	.01[a]	.01	.01	.01	.01	.01
Trypsin	7.8	3	.16	.18	.17	.00	.00	.00
Elastase	8.8	4	.04	.04	.03	.03	.035	.04
Cathepsin	4.7	4	.10	.09	.08	.02	.02	.02
Cathepsin	3.5	5	.36	.34	.36	.50	.51	.59
Hyase	5.3	2	.00	.00	.01	.00	.04	.01
Collagenase	8.0	7	.23	.26	.18	.18	.32	.46

[a] Since all assays involved a colorometric analysis, the values are given as optical density measurements for convenience of comparison (Espey 1974).

neared, there was a slight decrease in this activity in the follicle wall, but a simultaneous increase in activity in the follicular fluid (Espey and Rondell 1968). Realizing that this activity could reflect physiological labilization and subsequent dissipation of an enzyme that might be important in ovulation, we began an intensive effort to extract the enzyme involved.

The substrate CBZ-GPGGPA was used to monitor enzyme activity during the development of an extraction procedure. This was done in full awareness of the lack of specificity of this synthetic substrate. However, our principal interest at this stage of the investigation was to search for general collagenolytic activity, and the synthetic hexapeptide afforded a convenient, microquantitative assay. To make a very long story short, an active extract was obtained and concentrated. However when the extract was tested on strips of sow follicles in vitro, it did not decompose the tissue. Therefore, we have concluded that this "activity" *is not* important in ovulation (Espey 1974).

Current Studies

More recently, we have begun to look for collagenolytic enzymes by applying other assay methods. In particular, we have utilized the original tissue culture method, which uses a reconstituted collagen gel as the substrate (Gross and Lapiere 1962), and also the assay which is based on the detection of soluble glycine labeled with carbon-14, which is released from reconstituted collagen in a culture medium (Jeffrey, Coffey and Eisen 1971). Both of these studies are still in progress. The data from the radioactive assay are presently inconclusive and therefore will not be discussed.

Regarding the gel method, the substrate was prepared by extracting collagen from rabbit skin. The extracts were processed in a manner that led to the formation of an opaque gel of reconstituted collagen at the bottom of a culture dish. After the surface of the gel was rinsed with Dulbecco's modified medium, it was ready for tissue incubation. A positive reaction for col-

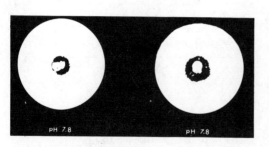

Figure 3
Demonstration of collagenolytic activity in rabbit follicles at alkaline pH. The dark halo indicates the amount of digestion of reconstituted collagen by a 1-mm^2 section of the follicle wall. Total diameter of the collagen gel is approximately 13 mm. (Photographs prepared on Kodalith to accent the area of digestion.)

lagenase was recorded if a transparent "halo" developed in the opaque gel beneath the tissue.

Tissue that was incubated at pH 7.8 caused relatively slight, but reasonably consistent, digestion of the reconstituted collagen (Fig. 3). Neither gonadotropin, cyclic AMP, progesterone nor estrogen seemed to influence the amount of lysis when these substances were added to the incubation media. Preliminary results show that the digestion which does occur at physiological pH is inhibited by EDTA, and possibly by chloroquine and Kunitz inhibitor. The activity was not clearly affected by cysteine or ceruloplasmin.

During the past few months, we have found that significantly greater digestion takes place if the tissue (and the gels) is initially rinsed in an acid buffer or in Dulbecco's medium that has been adjusted to an acid pH (Fig. 4). This activity is present over a wide pH range, with almost as much digestion occurring at pH 6.8 as at pH 5.5. However, the activity is abolished when the pH is raised to 7.0. Therefore, the hypothetical question becomes one of whether the enzyme which is involved here could be active under physiological conditions, where the extracellular pH would normally be in the alkaline range.

It may be significant that the data suggest that only a slightly acid environment is needed to initiate a large amount of enzyme activity. If there is a sufficient increase in hydrogen ions in localized sites in the follicle wall near the time of ovulation, then such an enzyme could be an active component of the degradation process. We do have some indirect evidence to support this possibility. For example, it is well known that substantial amounts of ascorbic

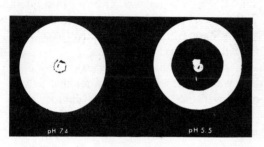

Figure 4
Demonstration of collagenolytic activity in rabbit follicles at acid pH. The halo of digested collagen is significantly larger when the tissue was incubated in an acid culture medium. (Photographs prepared on Kodalith to accent the area of digestion.)

acid are released in the ovaries in response to luteinizing hormone (Espey 1974). In addition, it may be relevant that some five years ago we noted a consistent drop from pH 7.3 to 7.1 in the follicular fluid of sow follicles near rupture (Espey 1970).

We are continuing to collect data, and although I would not be justified in drawing too many conclusions, at this stage the evidence does lead us to believe that an "acid" enzyme (if it is indeed a specific enzyme) has a major role in the deterioration of the follicle wall during ovulation. It is also possible that this follicular activity may be related to the collagenolytic enzyme that is maximally active at pH 5.5 in various rat tissues (Etherington 1973; Houck, Sharma and Carillo 1970; Shaub 1964). Also in support of this possible correlation, I should point out that several years ago we noticed that autolytic decomposition does occur in follicular tissue in vitro, with maximum activity occurring at approxmiately pH 5 (Espey 1970).

Cytological Location of Proteolysis

As a complementary experimental approach to these studies, we have used the electron microscope in an effort to localize the cytological decomposition of follicular tissue. This work has revealed intriguing multivesicular structures (Fig. 5) which protrude from the follicular fibroblasts and appear to digest the extracellular matrix in their vicinity (Espey 1971a). During the several hours preceding rupture, there is a ninefold increase in the concentration of these structures, and they are even more apparent just after ovulation when the thecal fibroblasts are proliferating into the lutein granulosa. They are especially common at the leading edges of the cytoplasmic processes which extend from the proliferating fibroblasts (Fig. 6), and this suggests that they may be important in facilitating the amoeboid movement of fibroblast through the dense collagenous tissue in the follicle wall (Espey 1971b).

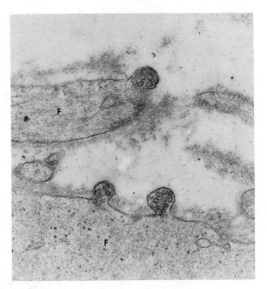

Figure 5
Electron micrograph of three conspicuous multivesicular structures which protrude from fibroblasts (F) in the thecal layer of a mature rabbit follicle. (Approx. 15,000 ×.)

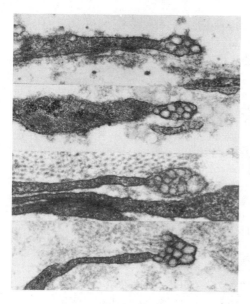

Figure 6
Plate of four electron micrographs demonstrating the common appearance of the multivesicular structures at the leading edge of cytoplasmic processes which extend from proliferating fibroblasts. (Approx. 24,000 ×.)

Further support for the hypothesis that these multivesicular structures may contain a proteolytic agent comes from our recent observation that the collagen of the relaxed symphysis pubis of the guinea pig is also digested by their contents (Fig. 7) (Chihal and Espey 1973).

Finally, I would like to conclude with a few suggestions for future studies:

1. First, and most obviously, an effort should be made to extract and characterize the acid protease in ovarian follicles.
2. It would help to know specifically how the enzyme is activated, and whether a precursor is involved.
3. What is the significance of ovarian ascorbic acid? Since it is a strong

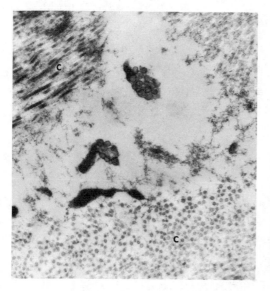

Figure 7
Electron micrograph of two conspicuous multivesicular structures which appear to be associated with the digestion of collagen (C) in the relaxed symphysis pubis of the guinea pig. (After Chihal and Espey 1973.) (Approx. 15,000 ×.)

reducing agent, there is the question of whether it could serve as an enzyme activator.

4. Is it possible to determine the hydrogen ion concentration in the extracellular spaces of the follicle wall during ovulation?
5. Does hydroxyproline and other catabolic products of connective tissue decomposition increase in the urine near the time of ovulation?
6. What is the chemical composition of the multivesicular structures which protrude from proliferating fibroblasts?
7. Are these structures present in other proliferating tissues?
8. Is there a change in the level of serum antiproteases during the menstrual cycle, or in anovulatory women, or postpartum or postmenopause?
9. Is there any correlation between the known high level of postmenopausal gonadotropins and the incidence of connective tissue diseases which occur with aging?
10. And finally, proteolysis does not last very long during the ovulatory process, and from the standpoint of fertility control, it would be interesting to know what factors turn this process off.

Acknowledgments

These studies have been supported by NIH Grant HD-06371 and NIH Contract 69-2126.

REFERENCES

Blandau, R. J. 1966. The mechanism of ovulation. In *Ovulation: Stimulation Suppression Detection* (ed., R. B. Greenblatt), p. 3. J. B. Lippincott Co., Philadelphia.

Blandau, R. J. and R. E. Rumery. 1963. Measurements of intrafollicular pressure in ovulatory and preovulatory follicles of the rat. *Fert. Steril.* **14**:330.

Chihal, H. J. and L. L. Espey. 1973. Utilization of the relaxed symphysis pubis of the guinea pig for clues to the mechanism of ovulation. *Endocrinology* **93**: 1441.

Espey, L. L. 1967a. Tenacity of porcine Graafian follicle as it approaches ovulation. *Amer. J. Physiol.* **212**:1397.

———. 1967b. Ultrastructure of the apex of the rabbit Graafian follicle during the ovulatory process. *Endocrinology* **81**:267.

———. 1970. Effect of various substances on tensile strength of sow ovarian follicles. *Amer. J. Physiol.* **219**:230.

———. 1971a. Decomposition of connective tissue in rabbit ovarian follicles by multivesicular structures of thecal fibroblasts. *Endocrinology* **88**:437.

———. 1971b. Multivesicular structures in proliferating fibroblasts of rabbit ovarian follicles during ovulation. *J. Cell Biol.* **48**:437.

———. 1974. Ovarian proteolytic enzymes and ovulation. *Biol. Reprod.* **10**:216.

Espey, L. L. and H. Lipner. 1963. Measurements of intrafollicular pressures in the rabbit ovary. *Amer. J. Physiol.* **205**:1067.

———. 1965. Enzyme-induced rupture of rabbit Graafian follicle. *Amer. J. Physiol.* **208**:208.

Espey, L. L. and P. Rondell. 1968. Collagenolytic activity in the rabbit and sow Graafian follicle during ovulation. *Amer. J. Physiol.* **214**:326.

Etherington, D. J. 1973. Collagenolytic cathepsin and acid-proteinase activities in the rat uterus during postpartum involution. *Eur. J. Biochem.* **32:**126.

Gross, J. and C. M. Lapiere. 1962. Collagenolytic activity in amphibian tissues: A tissue culture assay. *Proc. Nat. Acad. Sci.* **48:**1014.

Houck, J. C., V. K. Sharma and A. L. Carillo. 1970. Control of cutaneous collagenolysis. *Adv. Enzyme Reg.* **8:**269.

Jeffrey, J. J., R. J. Coffey and A. Z. Eisen. 1971. Studies on uterine collagenase in tissue culture. I. Relationship of enzyme production to collagen metabolism. *Biochim. Biophys. Acta* **252:**136.

Rondell, P. 1964. Follicular pressure and distensibility in ovulation. *Amer. J. Physiol.* **207:**590.

Schochet, S. S. 1916. A suggestion as to the process of ovulation and ovarian cyst formation. *Anat. Rec.* **10:**447.

Shaub, M. C. 1964. Eigenschaften und intracellulare Verteilung eines kollagenabbaienden Kathepsins. *Helv. Physiol. Pharmacol. Acta* **22:**271.

Factors Controlling the Growth of 3T3 Cells and Transformed 3T3 Cells

Robert W. Holley

The Salk Institute for Biological Studies, San Diego, California 92112

This paper will present a brief summary of our present understanding of the factors that control the growth of 3T3 cells. With this background, an attempt will be made to generalize concerning the factors that control the growth of normal cells. Finally, the loss of growth controls in transformed cells will be discussed.

CONTROL OF GROWTH OF 3T3 CELLS

Serum Factors

The growth of 3T3 cells, an established line of mouse fibroblast cells, shows an unusual dependence on serum. The cell density attained by these cells is proportional to the serum concentration in the culture medium, even up to concentrations in excess of 30% serum (Holley and Kiernan 1968). Other established cell lines and transformed cell lines show a similar proportionality of growth to the serum concentration, but usually at much lower serum concentrations, typically from 0.2–4% serum. With 10% serum, many of these cells grow to a sufficiently high density to exhaust some other constituent of the medium. The unusually high serum requirement of 3T3 cells is responsible for the low saturation density of these cells, and it is this property that has drawn attention to 3T3 cells as having a controlled, "normal" growth behavior. A high serum requirement is probably representative of only certain aspects of normal growth, but it does provide an opportunity to study serum factors that can control growth.

Studies with 3T3 cells of the growth regulatory functions of serum have indicated that the functions are complex (Holley and Kiernan 1971). At least four serum factors play roles in the initiation of DNA synthesis (Holley and Kiernan 1974a). A fifth serum factor prolongs the growth of growing cells through additional generations, but does not initiate growth in quiescent cells (Holley and Kiernan 1971 and unpubl.). Still other serum factors appear to be required for "survival" and for the cells to traverse the G_2 and M phases of the cell cycle (Paul, Lipton and Klinger 1971; Holley and Kiernan, unpubl.). Thus different serum factors control different aspects of 3T3 cell growth.

The greatest amount of attention has been given to the serum factors that are required for the initiation of DNA synthesis. These factors are of particular interest because of the key role that initiation of DNA synthesis plays in growth. The serum requirement of 3T3 cells is highest during that part of the cell cycle prior to the initiation of DNA synthesis. Lowering the serum concentration in the medium, to 0.2% for example, arrests the growth of even sparse 3T3 cells in the G_1 (or G_0) phase of the cell cycle. Raising the serum concentration reinitiates DNA synthesis and growth in the quiescent cells.

Fractionation of serum has shown that a combination of different serum fractions is required to replace serum in stimulating the initiation of DNA synthesis in quiescent 3T3 cells (Holley and Kiernan 1974a). Typical results are shown in Figure 1. Two factors in fraction A (Kaplan and Bartholomew 1972) appear to be heat labile. Two other factors, B and AS, are heat stable and though inactive by themselves, greatly increase the activity of the factors in fraction A.

The greater effectiveness of a combination of factors in initiating DNA synthesis has been confirmed by experiments in which three of the four serum fractions have been replaced by pure materials (Holley and Kiernan 1974a). The pure factors are: (1) the fibroblast growth factor (FGF) of Gospodarowicz, a new hormone isolated from pituitary glands (Gospodarowicz 1975), (2) insulin and (3) a glucocorticoid such as dexamethasone. Interactions between the pure factors are shown in Figure 2. The presence of a low concentration of one factor, for example, FGF, greatly increases the activity of a second factor, insulin, and vice versa.

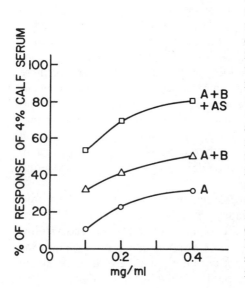

Figure 1
Initiation of DNA synthesis by fractions isolated from calf serum. The figure gives the percentage replacement of 4% calf serum achieved by (o) fraction A (the redissolved, precipitated protein fraction) from a chloroform-methanol extraction of serum; (△) fraction A plus 0.10 ml of heated (5 min at 100°C) fraction B (Kaplan and Bartholomew 1972) (the aqueous methanol-soluble fraction); (□) fraction A plus 0.10 ml each of heated fraction B and heated (5 min at 100°C) 50–70% ammonium sulfate (AS) fraction of calf serum. Fraction B and the ammonium sulfate fraction, either individually or together, gave less than 5% of the response of calf serum. (Reprinted, with permission, from Holley and Kiernan 1974a.)

Figure 2

Initiation of DNA synthesis by known factors. The figure gives the percentage replacement of 4% calf serum achieved by (•) insulin, either alone or with added dexamethasone and/or heated ammonium sulfate fraction of mouse serum; (■) varying concentrations of FGF alone; (○) varying concentrations of FGF plus 0.4 μg/ml dexamethasone; (△) FGF plus 50 ng/ml insulin plus 0.4 μg/ml dexamethasone; (□) FGF plus 50 ng/ml insulin plus 0.4 μg/ml dexamethasone plus 0.2 mg/ml protein from the heated (10 min at 100°C) 50–70% ammonium sulfate fraction of mouse serum. (Reprinted, with permission, from Holley and Kiernan 1974a.)

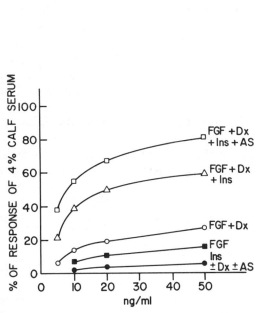

The same combination of four factors promotes the initiation of DNA synthesis both in sparse quiescent 3T3 cells and in confluent quiescent cells, the former in fresh low serum medium and the latter in high serum that has been depleted by growth of the cells (Holley and Kiernan 1974a). This evidence supports the conclusion that quiescent 3T3 cells are subject to the same growth controls whether the cells are sparse or confluent.

Requirements for the different factors appear to vary with the conditions and with the particular 3T3 cell clone. Armelin (1973) found that his 3T3 cells responded, in low serum, to pituitary extracts (which contain FGF) plus hydrocortisone. Gospodarowicz (1974) reported that serum can be replaced completely by FGF and hydrocortisone in initiating DNA synthesis in his Balb/c 3T3 cell line. In our experience, both with our 3T3 cell line and with the cell line used by Gospodarowicz, the cells appear "sick" within 24 hours in the complete absence of serum. In agreement with this, Gospodarowicz and Moran (1974) have recently reported that the cells need 0.4% serum for a good response to FGF. Thrash and Cunningham (1973) found that hydrocortisone alone initiated DNA synthesis in their confluent quiescent 3T3 cells in depleted high serum medium. Using our culture conditions, we have found that the 3T3 cell lines obtained from Dr. Cunningham and Dr. Gospodarowicz both respond to the four factors with results similar to those in Figure 2, but with quantitative differences. The Gospodarowicz 3T3 cell line showed less of a requirement for insulin, and in depleted high serum medium, this cell line gave the greatest response to dexamethasone alone, a greater response than we obtained with other 3T3 lines, including Dr. Cunningham's 3T3 cells.

Our conclusion is that 3T3 cells respond to at least four factors in serum. By varying the culture conditions, one can observe initiation of DNA synthesis by any one or any combination of these factors. The quantitative differences observed between the different 3T3 cell lines may be due to differences in the expression or affinities of hormone binding sites, to differences in the retention or destruction of hormones, or to still other causes.

The stimulation of 3T3 cells by the combination of FGF, insulin and a glucocorticoid is similar to the stimulation of growth and differentiation of mammary gland cells by prolactin, insulin and a glucocorticoid (Rivera 1964; Juergens et al. 1965; Mukherjee, Washburn and Banerjee 1973).

It should be noted that the two polypeptide hormones FGF and insulin are active in stimulating DNA synthesis in 3T3 cells at ng/ml concentrations. These concentrations are much lower than the concentrations of pronase and trypsin that have been reported to stimulate the growth of 3T3 cells (Burger 1970). This suggests that when proteases stimulate the growth of 3T3 cells (proteases are not active with our 3T3 cells under the conditions of Burger), they do so by mimicking the action of insulin and/or FGF, binding to and activating the hormone receptor sites on the cells (Kono and Barham 1971).

Low Molecular Weight Nutrients

The growth of 3T3 cells is also subject to control by certain of the common nutrients in the culture medium. We have recently shown that growth of 3T3 cells in medium with a high concentration of serum can be arrested in the G_1 (or G_0) phase of the cell cycle if the concentrations of amino acids, glucose or phosphate are reduced to low levels in the medium (Holley and Kiernan 1974b). A subsequent increase in the concentration of the nutrient that is limiting leads to the reinitiation of DNA synthesis and growth. Results in the literature (Holley 1972) suggest that a wide variety of low molecular weight nutrients have the potential of controlling the growth of mammalian cells. It is not known how important these controls are in vivo.

Mechanisms of Action of the Factors Controlling Growth

The actions of the various factors that control the growth of 3T3 cells have not been studied extensively, but some preliminary information is of interest.

Both FGF and insulin stimulate movement of 3T3 cells on the surface to which the cells are attached (Holley and Klinger, unpubl.). Combination of FGF and insulin, especially with the further addition of the other two factors required for maximum DNA synthesis, gives the greatest stimulation of movement.

FGF and insulin also stimulate the uptake of low molecular weight nutrients (Holley and Kiernan 1974b). Again a combination of factors is most active (Hilborn, unpubl.).

FGF has been reported to activate guanyl cyclase in "microsomes" prepared from 3T3 cell plasma membranes, and insulin lowers the activity of adenyl cyclase (Rudland et al. 1974). These effects are consistent with the observed rise in cyclic GMP and fall in cyclic AMP in quiescent 3T3 cells immediately after the cells are stimulated by serum (Seifert and Rudland 1974a).

It is also of interest that the concentration of cyclic GMP rises and that of cyclic AMP falls immediately after G_1-arrested cells are stimulated to grow by the addition of a nutrient. This has been observed with quiescent 3T3 cells that have been arrested in G_1 (or G_0) by limitation of phosphate or amino acids in the medium (Seifert and Rudland 1974b). Apparently growth arrest by limitation of low molecular weight nutrient also involves changes in cyclic nucleotide levels.

GROWTH CONTROL IN NORMAL CELLS

External Factors

The above results on the control of growth of 3T3 cells, as well as results in the literature with other cells (Rivera 1964; Leffert 1974), suggest that growth control of normal cells involves responses of the cell to external concentrations of hormones. Growth can also be controlled by external concentrations of common nutrients (Ley and Tobey 1972; Holley and Kiernan 1974b; Holley, Baldwin and Kiernan 1974). When growth is controlled by several factors, a combination of all the missing external factors will be required to initiate growth. That this may be a common situation in vivo is suggested by studies of the stimulation of liver cell growth (Short et al. 1972, 1973). Situations can be designed in vitro in which one factor is limiting, and growth is then dependent on that single factor. The growth of the same cell line can be made dependent on any one of several different factors. Presumably, this situation can also arise in vivo.

The results suggest that the widespread dependence of growth of cell lines on serum in vitro represents a dependence of cells on external hormones which can be supplied easily in this form. There is evidence for various insulinlike factors and for various other factors in serum (Temin, Pierson and Dulak 1972). It seems likely that many unidentified growth-controlling factors are present in serum. Serum also provides a variety of trace low molecular weight nutrients (Messmer 1973; Vallotton, Hess-Sander and Leuthardt 1965), and these control growth of cultured cells in some situations. Fractionation studies suggest that the serum factors required by different cells are different, and there are differing requirements for trace nutrients. Therefore, to understand the external factors that control growth of a cell line in vitro may require considerable investigation.

Since the response of a cell to a hormone is dependent on the expression of hormone binding sites on the cell surface (Krug, Krug and Cuatrecasas 1972; Neville 1974; Oka, Perry and Topper 1974), as well as on conditions that favor retention or destruction of the hormone, details of the relationship of growth control to external hormone concentrations remain poorly understood.

Density-dependent regulation of growth, which is so pronounced with 3T3 cells, is also not fully understood. With 3T3 cells, it is clear that density-dependent regulation is the result of an increasing requirement for serum factors as the cells become more crowded, but the cause of the increasing requirement is uncertain. It may be due simply to a diffusion boundary for serum factors (Stoker 1973), perhaps with restriction of cell surface area and increased destruction of serum factors causing the increasing serum requirement as the cells become more crowded, or it may be more complicated.

Internal Mechanisms

If one turns from a discussion of external factors that control growth to a consideration of the internal mechanisms that are responsible for the control of the initiation of DNA synthesis, present results support the view that cyclic nucleotides play a key role (Goldberg et al. 1974). One of the puzzles that remains, however, is the fact that the major changes in cyclic nucleotide levels persist for only a brief period after cells are stimulated by serum or other factors (Seifert and Rudland 1974a), yet stimulation must be continued for hours before the cells are committed to the initiation of DNA synthesis. It will be of interest to learn whether there is direct coupling of hormone-activated cyclases, i.e., through structurally associated "enzyme systems" to specific transport carriers. Such direct coupling might activate transport of specific nutrients without large changes in cellular levels of cyclic nucleotides. The major, yet transitory, changes in cyclic nucleotide concentrations might reflect the changes during the activation of various systems.

CONTROL OF GROWTH OF TRANSFORMED CELLS

Transformation of 3T3 cells by SV40 virus or by benzpyrene greatly reduces the serum requirement of the cells (Holley and Kiernan 1968, 1971; Oshiro and Di Paolo 1973). It appears that the virus-transformed cells have actually lost the requirements for all four of the serum factors that are required for the initiation of DNA synthesis by 3T3 cells, since medium that has been depleted by 3T3 cells and that is known to be deficient in all four of the factors required by 3T3 cells has full serum activity for growth of SV3T3 cells (Holley and Kiernan 1971, 1974a). Such medium, that has been depleted by 3T3 cells, is not fully active for benzpyrene-transformed 3T3 cells, however, indicating that the chemically transformed cells retain a requirement for a serum factor that is no longer required by SV3T3 cells.

The low serum requirement that remains for SV3T3 cells is for functions other than the control of the initiation of DNA synthesis. The growth rate of SV3T3 cells does decrease at low serum concentrations, but all phases of the cell cycle appear to be lengthened proportionately, and approximately half of the cells are always found in the S phase of the cell cycle (Paul, Henahan and Walter 1974). Only under very special culture conditions (Holley and Kiernan, unpubl.) is the growth of SV3T3 cells arrested in the G_1 (or G_0) phase. In contrast, benzpyrene-transformed 3T3 cells appear to arrest relatively easily in G_1 (or G_0) (Holley and Kiernan, unpubl.). The factors involved in the arrest of benzpyrene-transformed cells remain to be investigated. It is clear from these and other data (Holley, Baldwin and Kiernan 1974), however, that the loss of mechanisms that arrest cell growth in G_1 (or G_0) is not required for a cell to be malignant.

My own interpretation of present data is that the loss of growth controls in transformed cells results from changes in the cell membrane, including changes that affect transport (Holley 1972). Other hypotheses are reasonable, however. In any case, the release of proteases or other hydrolytic enzymes from the cell could play a role in causing the changes that are observed.

Acknowledgments

This research was supported in part by the American Cancer Society (#BC30A), The National Cancer Institute (CA11176) (NO1-CP-33405), and the National Science Foundation (GB 32391X). The author is an American Cancer Society Professor of Molecular Biology.

REFERENCES

Armelin, H. A. 1973. Pituitary extracts and steroid hormones in the control of 3T3 cell growth. *Proc. Nat. Acad. Sci.* **70:**2702.

Burger, M. M. 1970. Proteolytic enzymes initiating cell division and escape from contact inhibition of growth. *Nature* **227:**170.

Goldberg, N. D., M. K. Haddox, E. Dunham, C. Lopez and J. W. Hadden. 1974. The Yin Yang hypothesis of biological control: Opposing influences of cyclic GMP and cyclic AMP in the regulation of cell proliferation and other biological processes. In *Control of Proliferation in Animal Cells* (ed. B. Clarkson and R. Baserga), p. 609. Cold Spring Harbor Laboratory, Cold Spring Harbor, New York.

Gospodarowicz, D. 1974. Localisation of a fibroblast growth factor and its effect alone and with hydrocortisone on 3T3 cell growth. *Nature* **249:**123.

———. 1975. Purification of a fibroblast growth factor from bovine pituitary. *J. Biol. Chem.* **250:**2515.

Gospodarowicz, D. and J. S. Moran. 1974. Stimulation of division of sparse and confluent 3T3 cell populations by a fibroblast growth factor, dexamethasone and insulin. *Proc. Nat. Acad. Sci.* **71:**4584.

Holley, R. W. 1972. A unifying hypothesis concerning the nature of malignant growth. *Proc. Nat. Acad. Sci.* **69:**2840.

Holley, R. W. and J. A. Kiernan. 1968. "Contact inhibition" of cell division in 3T3 cells. *Proc. Nat. Acad. Sci.* **60:**300.

———. 1971. Studies of serum factors required by 3T3 and SV3T3 cells. In *Ciba Foundation Symposium: Growth Control in Cultures* (ed. G. E. W. Wolstenholme and J. Knight), p. 3. Churchill Livingstone, London.

———. 1974a. Control of the initiation of DNA synthesis in 3T3 cells: Serum factors. *Proc. Nat. Acad. Sci.* **71:**2908.

———. 1974b. Control of the initiation of DNA synthesis in 3T3 cells: Low molecular weight nutrients. *Proc. Nat. Acad. Sci.* **71:**2942.

Holley, R. W., J. H. Baldwin and J. A. Kiernan. 1974. Control of growth of a tumor cell by linoleic acid. *Proc. Nat. Acad. Sci.* **71:**3976.

Juergens, W. G., F. E. Stockdale, Y. J. Topper and J. J. Elias. 1965. Hormone-dependent differentiation of mammary gland *in vitro. Proc. Nat. Acad. Sci.* **54:**629.

Kaplan, A. E. and J. C. Bartholomew. 1972. Study of the growth response of normal and SV40-transformed 3T3 mouse fibroblasts with serum fractions obtained by use of organic solvents. *Exp. Cell Res.* **73:**262.

Kono, T. and F. W. Barham. 1971. Insulin-like effects of trypsin on fat cells. *J. Biol. Chem.* **246:**6204.

Krug, U., F. Krug and P. Cuatrecasas. 1972. Emergence of insulin receptors on human lymphocytes during *in vitro* transformation. *Proc. Nat. Acad. Sci.* **69:**2604.

Leffert, H. L. 1974. Growth control of differentiated fetal rat hepatocytes in primary monolayer culture. VII. Hormonal control of DNA synthesis and its possible significance to the problem of liver regeneration. *J. Cell Biol.* **62:**792.

Ley, K. D. and R. A. Tobey. 1970. Regulation of initiation of DNA synthesis in Chinese hamster cells. II. Induction of DNA synthesis and cell division by iso-leucine and glutamine in G_1-arrested cells in suspension culture. *J. Cell Biol.* **47:**453.

Messmer, T. O. 1973. Nature of the iron requirement for Chinese hamster V79 cells in tissue culture medium. *Exp. Cell Res.* **77:**404.

Mukherjee, A. S., L. L. Washburn and M. R. Banerjee. 1973. Role of insulin as a "permissive" hormone in mammary gland development. *Nature* **246:**159.

Neville, D. M., Jr. 1974. Receptors for polypeptide hormones: Direct studies of insulin binding to purified liver plasma membranes. *In Vitro* **9:**445.

Oka, T., J. W. Perry and Y. J. Topper. 1974. Changes in insulin responsiveness during development of mammary epithelium. *J. Cell Biol.* **62:**550.

Oshiro, Y. and J. A. Di Paolo. 1973. Loss of density-dependent regulation of multiplication of Balb/3T3 cells chemically transformed in vitro. *J. Cell Physiol.* **81:**133.

Paul, D., M. Henahan and S. Walter. 1974. Changes in growth control and growth requirements associated with neoplastic transformation *in vitro*. *J. Nat. Cancer Inst.* **53:**1499.

Paul, D., A. Lipton and I. Klinger. 1971. Serum factor requirements of normal and simian virus 40-transformed 3T3 mouse fibroblasts. *Proc. Nat. Acad. Sci.* **68:**645.

Rivera, E. M. 1964. Differential responsiveness to hormone of C3H and A mouse mammary tissues in organ cultures. *Endocrinology* **74:**853.

Rudland, P. S., D. Gospodarowicz and W. Seifert. 1974. Activation of guanyl cyclase and intracelluar cyclic GMP by fibroblast growth factor. *Nature* **250:**741.

Seifert, W. and P. S. Rudland. 1974a. Possible involvement of cyclic GMP in growth control of cultured mouse cells. *Nature* **248:**138.

--------. 1974b. Cyclic nucleotides and growth control in cultured mouse cells: Correlation of intracellular cGMP changes with a specific region of the cell cycle. *Proc. Nat. Acad. Sci.* **71:**4920.

Short, J., R. F. Brown, A. Husakova, J. R. Gilbertson, R. Zemel and I. Lieber-man. 1972. Induction of deoxyribonucleic acid synthesis in the liver of the intact animal. *J. Biol. Chem.* **247:**1757.

Short, J., N. B. Armstrong, R. Zemel and I. Lieberman. 1973. A role for amino acids in the induction of deoxyribonucleic acid synthesis in liver. *Biochem. Biophys. Res. Comm.* **50:**430.

Stoker, M. G. P. 1973. Role of diffusion boundary layer in contact inhibition of growth. *Nature* **246:**200.

Temin, H. M., R. W. Pierson, Jr. and N. C. Dulak. 1972. The role of serum in the control of multiplication of avian and mammalian cells in culture. In *Growth, Nutrition, and Metabolism of Cells in Culture* (ed. V. I. Cristafalo and G. Rothblat), pp. 50–81. Academic Press, New York.

Thrash, C. R. and D. D. Cunningham. 1973. Stimulation of division of density inhibited fibroblasts by glucocorticoids. *Nature* **242:**399.

Vallotton, M., U. Hess-Sander and F. Leuthardt. 1965. Fixation spontanée de la biotine à une proteine dans le serum-humain. *Helv. Chim. Acta* **48:**126.

The Effects of Protease Inhibitors on Cells In Vitro

Hans Peter Schnebli

Friedrich Miescher-Institut, CH-4002 Basel, Switzerland

Many tumors have been found to be associated with increased proteolytic activity when compared with their normal counterparts (Zamecnik and Stephenson 1947; Kazakova and Orekhovich 1967; Yamanishi, Dabbous and Hashimoto 1972; Bosmann and Hall 1974). Similarly, a number of in vitro transformed cells have elevated protease levels when compared with their parental cells (Bosmann and Pike 1970; Bosmann 1972; Schnebli 1972; Unkeless et al. 1973; Ossowski et al. 1973; Goldberg 1974). It has been proposed that this increased proteolytic activity is involved in the process of malignant invasion and is also responsible for the altered growth control in tumor cells.

The idea that cellular proteases may be involved in growth control was suggested further by the finding that trypsin and other proteases can release quiescent cells from density-dependent inhibition of growth (Burger 1970; Sefton and Rubin 1970; Greene, Tomita and Varon 1971; Vaheri, Ruoslahti and Hovi 1974). Protease treatment alone, however, even when it is shown to produce profound alterations in the surface structure of the cells, is not sufficient to release cells from density-dependent inhibition (Cunningham, Thrash and Glynn 1974), and the exact requirements are not known. Proteases have also been implicated in growth and development of sea urchin (Grossmann and Troll 1970; Vacquier, Espe and Douglas 1972) and starfish (Jeffrey 1972). In this context, it is interesting that esteropeptidase activity is associated with an epithelial growth factor (Jones and Ashwood-Smith 1970), a mesenchymal growth factor (Attardi, Schlesinger and Schlesinger 1967) and a nerve growth factor (Greene, Shooter and Varon 1968). The hydrolytic and growth activities, however, appear to reside on different subunits (?) of the molecules (Greene, Tomita and Varon 1971; Taylor, Cohen and Mitchell 1970), showing that they are different entities.

Troll, Klassen and Janoff (1970) and Hozumi et al. (1972) had succeeded in suppressing tumorigenesis in the skin of mice with protease inhibitors. Some three years ago when we and others set out to investigate the role(s) of proteases in the malignant transformation of cultured mouse fibroblasts, it seemed only natural that we should use protease inhibitors as tools. The meeting which resulted in this volume was a timely opportunity to critically evaluate the observed effects of protease inhibitors on cultured cells.

Handling of Protease Inhibitors

Before discussing the effects of protease inhibitors, I should like to briefly review their properties. The protease inhibitors that have been used in tissue culture work fall within three categories: N-α-Tosyl-L-arginyl methyl ester (TAME) is a substrate analog and is expected to inhibit by competing with the natural substrates of proteases and esterases (Kassell and Laskowski 1956). N-α-Tosyl-L-lysyl chloromethyl ketone (TLCK) and N-α-tosyl-L-phenylalanyl chloromethyl ketone (TPCK) are substrate analogs that react covalently and irreversibly with proteases of the trypsin and chymotrypsin family, respectively (Shaw, Mares-Guia and Cohen 1965). Ovomucoid, soybean trypsin inhibitors and the pancreatic trypsin inhibitor are macromolecules that form poorly dissociating complexes with proteases in a one-to-one molar ratio (Fraenkel-Conrat, Bean and Lineweaver 1949).

Some of the most apparent difficulties in using protease inhibitors in tissue culture stem from their toxicity and instability. There is only a narrow margin between an effective and a toxic dose of the chloromethyl ketones. With TLCK, we noted furthermore that some batches of commercial material (but not others from the same source) were highly toxic to cells, perhaps due to small amounts of impurities. Solutions of TAME, TLCK, and TPCK, particularly in tissue culture media, are very unstable. In aqueous solution, TLCK breaks down with a half-life of 1–3 hours (Shaw, Mares-Guia and Cohen 1965), and TAME has a spontaneous hydrolysis rate of 1% per hour at pH 7.5 (Roffman, Sanocka and Troll 1970). The alkylating agents (TLCK, TPCK) may also be inactivated by reaction with serum components. Increased serum concentrations have indeed been shown to antagonize the effects of TPCK (Chou, Black and Roblin 1974a). Furthermore, the sera used for tissue culture contain enzymes that rapidly hydrolyze TAME (Schnebli 1974). Ovomucoid is more stable than the small molecular weight inhibitors, but has the property of adhering progressively and tenaciously to any glass or plastic surface.

For reproducible results, it is thus essential to use only freshly made solutions of inhibitors. Because of the instability of the protease inhibitors, most investigators make daily medium changes with freshly dissolved inhibitors. It was also observed that the effect of the protease inhibitors depends on the age and density of the cell cultures: freshly seeded cells and cells at low density are most sensitive.

Effects of Protease Inhibitors

Growth Inhibition

Transformed fibroblasts differ from their normal counterparts in that in culture they continue to grow past the monolayer stage. If the growth promoting activity of a protease were responsible for this overgrowth, then inhibition of this protease should restore the growth to "normal."

The effects of four protease inhibitors on the growth of 3T3 and virally transformed 3T3 are summarized in Table 1. Since these results were obtained in different laboratories and under various experimental conditions (cell density, length of treatment, different handling of the inhibitors, etc.), they are not strictly comparable. Still, it is generally found that the growth of transformed cells is reduced at lower concentrations of inhibitors than the growth of the 3T3 cells. On the other hand, Chou, Black and Roblin (1974a) reported that they could not observe a selective inhibition of Balb/c SV3T3 by TLCK. Later, however, using Swiss 3T3 cells and treating the normal and transformed cells at similar densities, Chou, Black and Roblin (1974b) did find a difference, although "not . . . as striking a selective effect of TLCK on SV3T3 as that reported by Schnebli and Burger (1972)." The same authors also failed to observe a selective effect of TPCK on transformed cells (Chou, Black and Roblin 1974a,b). In their study, the concentrations required to get any effect were very much higher than the concentrations we used earlier (Schnebli and Burger 1972), suggesting a difference in cells or handling of the inhibitor. It should also be noted that the Balb/c 3T3 line, used earlier by Chou, Black and Roblin (1974a), has a saturation density (about 10^5 cells/cm^2) that is two to three times higher than that of the Swiss 3T3 cells used in most other laboratories.

The pattern of sensitivity towards the protease inhibitors is different for the various transformed cell lines (Schnebli and Burger 1972). SV3T3 appear to be more sensitive to TLCK, but less sensitive to TPCK and ovomucoid, than Py3T3. Hamster tumor cells (Goetz, Weinstein and Roberts 1972), like transformed 3T3, were found to be sensitive to TLCK (85% inhibition at about 45 μg/ml). The hamster cells, however, are insensitive to ovomucoid up to 2000 μg/ml (Goetz, Weinstein and Roberts 1972), but were inhibited by soybean trypsin inhibitor (50% at 1800 μg/ml). The soybean inhibitor did not affect the growth of any of the 3T3 cells (Schnebli and Burger 1972) at concentrations up to 400 μg/ml, but these experiments should perhaps be repeated with higher concentrations. Ossowski et al. (1973) reported that soybean trypsin inhibitor did not influence the growth rate of SV40-transformed hamster cells when grown in liquid medium, but reduced the colony formation in agar by 75% at 10 μg/ml.

The protease inhibitors have been found to reduce both the growth rate (Schnebli 1972; Schnebli and Burger 1972; Schnebli and Haemmerli 1974; Collard and Smets 1974; Chou, Black and Roblin 1974a, b) and the maximal cell density of the treated cultures (Schnebli and Burger 1972; Prival 1972; Chou, Black and Roblin 1974a,b). Low concentrations usually cause a decrease in growth rate, without lowering the growth plateau, while higher concentrations affect both. In contrast to this generalization, Prival (1972) found that TAME and TPCK lowered the growth plateau of SV3T3 cells without significantly reducing the growth rate. Talmadge, Noonan and Burger (1974) working with ovomucoid similarly found a much greater effect on the final cell density than on growth rate. Due to the instability of the low molecular weight inhibitors, it is likely that they act only for a short period of time before they are hydrolyzed. It is thus possible that the cultures recover

Table 1
Growth Effects of Protease Inhibitors

Inhibitor	Conc. (µg/ml)	Length of exposure (days)	% Growth inhibition			Reference
			3T3	SV3T3	Py3T3	
TAME	100	3	0		10	a
	100	4	0		8	b
	200	4	0		47	b
	500	2	15*	30*		d
	600	5	0	0		c
	1000	2	20*	36*		d
	1000	5	0	23		c
	2000	5	0	37		c
	2500	2	20*	43*		d
TLCK	10	1	0			e
	25	1	14			e
	25	4	0		20	b
	25	3	0	70	3	a
	25	3–5	7*	0*		f
	30	2		44		h
	50	3	0	87	47	a
	50	4	0		34	b
	50	2	20*	50*		d
	50	3–5	10*	17*		f
	60	2		86		h
	90	2		100		h
	100	3	15	91	63	a
	100	2	26*	70*		d
	100	3–5	12*	31*		f
TPCK	5	1	0			e
	5	3	0	15	24	a
	10	3	5	34	50	a
	20	3	8	60	73	a
	40	4	13	39		c
	50	4	59	(0)		g
	100	4	24	73		c
	100	3	~100	66		g
	150	3		78		f
	200	3	~100	~100		g
	200	4	80			c
	250	3		~100		f
Ovomucoid	200	3	0	10	30	a
	250	4	0		48	b
	300	6	(0)		49	i
	400	3	7	25	34	a
	500	2	10*	14*		d
	500	4	0		54	b
	1000	2	10*	24*		d
	2500	2	15*	34*		d

Data from [a] Schnebli and Burger (1972); [b] Schnebli (1972); [c] Prival (1972); [d] Collard and Smets (1974); [e] Sivak (1972); [f] Chou, Black and Roblin (1974a); [g] Chou, Black and Roblin (1974b); [h] Schnebli and Haemmerli (1974); [i] Talmadge, Noonan and Burger (1974). *Balb/c cells.

from the inhibition before the inhibitor is applied again after 24 hours. This would explain the inhibition of the growth rate without the lowering of the "growth plateau," but this point is not fully resolved.

Chou, Black and Roblin (1974a) have suggested that "growth plateaus" are the result of an equilibrium between limited cell proliferation and cell death. The number of dead cells floating in the medium (Goetz, Weinstein and Roberts 1972; Schnebli and Haemmerli 1974; Chou, Black and Roblin 1974b) is too small to account for the observed reduction of proliferation, but one cannot exclude the possibility of cell loss by lysis. There are, however, other reasons that make it doubtful that protease inhibitors produce the observed effects simply by killing off the cells: in treated cultures in which there is no net gain in cell number (plateau situation), the majority of cells remain attached to the culture vessel. These attached cells (e.g., TLCK treated) exclude Trypan blue and have a high plating efficiency (Schnebli and Burger 1972). That TLCK inhibits proliferation is further supported by the fact that the reduced growth rate is paralleled by a decrease in the mitotic index (Schnebli and Haemmerli 1974). Finally, TPCK-treated cells recover from the inhibition within 24 to 48 hours and regain their usual growth rate (Chou, Black and Roblin 1974b). Still, it is agreed by all investigators that the growth plateaus produced by treatment with protease inhibitors do not reflect the reacquisition of normal growth control, i.e., density-dependent inhibition of growth; the final cell density depends on the density at which the cells were treated (Chou, Black and Roblin 1974b). "Growth plateaus" can thus be generated at cell densities where the cells do not show extensive cell-to-cell contacts. Most important is the fact that TLCK- and TPCK-inhibited cells continue to synthesize DNA (Schnebli 1974; Schnebli and Haemmerli 1974; Collard and Smets 1974; Weber 1974; Chou, Black and Roblin 1974b), leading to an accumulation of cells in the G_2 phase of the cell cycle (Schnebli and Haemmerli 1974; Weber 1974; Collard and Smets 1974). TLCK has little effect on 3T3 cells, and the small reduction in growth rate in these cells is interpreted as an extension of all phases of the cell cycle (Collard and Smets 1974). In contrast to TLCK-treated transformed cells, TAME- and ovomucoid-treated SV3T3 did not accumulate in any particular part of the cell cycle, even though these inhibitors also reduced the growth rate of the transformed cells more than that of normal cells. These results strongly indicate that the various inhibitors interfere with cell proliferation in different ways, a point that will be discussed below. The G_2 arrest by TLCK and TPCK (Schnebli and Haemmerli 1974; Weber 1974; Collard and Smets 1974) shows that in contrast to our previous suggestions (Schnebli and Burger 1972; Schnebli 1973), protease inhibitors fail to induce density-dependent inhibition of growth. Density-inhibited cells are arrested in the G_1 phase of the cell cycle (Nilausen and Green 1965).

Agglutinability by Lectins

It has been known for some time that compared to normal cells, the transformed cells are more agglutinable by lectins, such as wheat germ agglutinin (Burger and Goldberg 1967) and ConA (Inbar and Sachs 1969).

TLCK- and TPCK-treated transformed cells are less agglutinable with wheat germ agglutinin (Schnebli 1974) and ConA (Schnebli and Burger 1972; Prival 1972; Collard and Smets 1974) than untreated cells. It was noted in all cases, however, that the inhibitor-treated transformed cells still agglutinated more readily than the untransformed 3T3 cells. In contrast, TAME and ovomucoid treatment of SV3T3 did not reduce their agglutinability by ConA (Collard and Smets 1974). Since TLCK- and X-ray-treated cells, but not TAME- or ovomucoid-treated cells, accumulate in the G_2 phase of the cell cycle and become less agglutinable, Collard and Smets (1974) concluded that TLCK does not reduce agglutinability directly, but only indirectly by its effect on the cell cycle.

Attachment and Morphology

Normal cells grow flatter and more firmly attached to the culture vessel than transformed cells. Attempts have been made to alter the morphology of transformed cells by treatment with protease inhibitors. TLCK causes hamster tumor cells (Goetz, Weinstein and Roberts 1972) and RSV-transformed chick embryo fibroblasts (Weber 1974) to adhere more firmly to the culture dish. This again argues against TLCK being simply toxic to the cells.

The effects of protease inhibitors on the morphology of cells are varied: TLCK causes a flattening of SV3T3 (Schnebli 1974) and RSV-chicken fibroblasts (Weber 1974), but not of hamster tumor cells (Goetz, Weinstein and Roberts 1972). The TLCK-treated RSV-chicken fibroblasts furthermore show a parallel alignment and appear to be "nearly perfect phenocopies of normal cells" (Weber 1974). Parallel alignment, but no significant flattening, of hamster sarcoma cells is obtained by treatment with basic pancreatic trypsin inhibitor (Goetz, Weinstein and Roberts 1972). "Flattening" of transformed cells suggests changes toward the normal phenotype. However, this can hardly be a specific effect of protease inhibitors since flat morphology can be induced in cultured cells by such varied agents as analogs of 3'–5' cyclic AMP (Hsie and Puck 1971; Sheppard 1971; Weber 1974), histones (Bases et al. 1973; Maciero-Coelho and Avrameas 1972), dimethylsulfoxide (Kisch et al. 1973), modified ConA (Burger and Noonan 1970), polyphloroglucin phosphate (Schnebli and Burger 1973) and sodium butyrate (Weber 1974).

Miscellaneous Effects

TLCK was found to transiently inhibit virus production in RSV T5-transformed chicken fibroblasts (Weber 1974). Weber (1974) also found that TLCK lowered the rate of 2-deoxyglucose transport in transformed chick fibroblasts. However, hexose transport was never reduced to the very low levels seen in density-inhibited cells. There was no effect on hexose transport of normal cells.

Site of Attack of Protease Inhibitors

Several groups have shown that serum pretreated with protease inhibitors (TLCK, ovomucoid) fully supports growth of transformed cells (Schnebli and Burger 1972; Collard and Smets 1974; Talmadge, Noonan and Burger 1974). From this it was concluded that the inhibitors act directly on the cells and not by inactivating serum components. This conclusion should perhaps be reconsidered in view of the finding that transformed cells activate plasmin from serum plasminogen (Unkeless et al. 1973; Ossowski et al. 1973; Goldberg 1974). Plasminogen is refractory to TLCK, but will specifically and irreversibly interact with this agent after its activation to plasmin.

Talmadge, Noonan and Burger (1974) showed that ovomucoid was effective in inhibiting growth of Py3T3 even when linked to Sepharose beads, suggesting that ovomucoid acts from the outside of the cells.

It now seems probable that the protease inhibitors utilized in in vitro studies act on different systems since the various cell lines have individual patterns of sensitivity toward protease inhibitors. The fact that TLCK and TPCK block in G_2, while TAME and ovomucoid do not, further supports this notion.

Similar growth effects of cycloheximide and TPCK led Chou, Black and Roblin (1974a) to suggest that TPCK may act by inhibiting protein synthesis. TPCK, although it is an inhibitor of bacterial protein synthesis at approximately 10^{-4} M, does not specifically affect avian or mammalian protein synthesis (Highland et al. 1974). TLCK does not interact in any specific way with protein synthetic systems. It is also noteworthy that while TLCK increased cellular attachment of transformed chick fibroblasts to the petri dish, cycloheximide did not (Weber 1974).

The experiments of Collard and Smets (1974) suggest that the chloromethyl ketones act like X rays, at least with respect to the G_2 block and the decrease in agglutinability. This is perhaps the best explanation, since TLCK and TPCK (like many radiomimetic agents) have alkylating properties. Phagocytosis appears to be higher in transformed cells than in normal cells, and this could account for the difference in sensitivity between normal and transformed cells towards these agents.

SYNOPSIS

Several protease inhibitors have been shown to affect growth, adherence, morphology, agglutinability and hexose transport of some transformed cells. The properties of inhibitor-treated cells often resemble those of normal cells, but often they differ from normal cells in important aspects. The inhibitor studies have thus not served to prove (nor to disprove) a suspected involvement of proteases in generating the transformed phenotype. Parenthetically, it should be said that neither the fact that proteases are generally increased in tumor and transformed cells nor the effects of exogenous proteases on normal cells can be taken as evidence for a role of proteases in malignancy.

The isolation of proteases that occur only, or predominantly, in transformed cells (e.g., factors that activate plasmin) is a prerequisite for a more direct approach that could involve the development of highly specific inhibitors or the use of antibodies directed against the "transformation-specific" proteases.

REFERENCES

Attardi, D. G., M. J. Schlesinger and S. Schlesinger. 1967. Submaxillary gland of the mouse: Properties of a purified protein affecting muscle tissue *in vitro*. *Science* **156**:1253.

Bases, R., F. Mendez, L. Mendez and R. Anigstein. 1973. Stimulation of HeLa cell-surface attachment by histones. *Exp. Cell Res.* **76:**441.

Bosmann, H. B. 1972. Elevated glycosidases and proteolytic enzymes in cells transformed by RNA tumor virus. *Biochim. Biophys. Acta* **264:**339.

Bosmann, H. B. and T. C. Hall. 1974. Enzyme activity in invasive tumors of human breast and colon. *Proc. Nat. Acad. Sci.* **71:**1833.

Bosmann, H. B. and G. Z. Pike. 1970. Glycoprotein synthesis and degradation: Glycoprotein:*N*-acetyl glucosamine transferase, proteolytic and glocosidase activity in normal and polyoma virus-transformed BHK cells. *Life Sci.* **9:**1433.

Burger, M. M. 1970. Proteolytic enzymes initiating cell division and escape from contact inhibition of growth. *Nature* **227:**170.

Burger, M. M. and A. R. Goldberg. 1967. Identification of a tumor-specific determinant on neoplastic cell surfaces. *Proc. Nat. Acad. Sci.* **57:**359.

Burger, M. M. and K. D. Noonan. 1970. Restoration of normal growth by covering of agglutinin sites on tumor cell surfaces. *Nature* **228:**512.

Chou, I. N., P. H. Black and R. O. Roblin. 1974a. Non-selective inhibition of transformed cell growth by a protease inhibitor. *Proc. Nat. Acad. Sci.* **71:**1748.

———. 1974b. Effects of protease inhibitors on growth of 3T3 and SV3T3 cells. In *Control of Proliferation in Animal Cells* (ed. B. Clarkson and R. Baserga), p. 339. Cold Spring Harbor Laboratory, Cold Spring Harbor, New York.

Collard, J. G. and L. A. Smets. 1974. Effect of proteolytic inhibitors in growth and surface architecture of normal and transformed cells. *Exp. Cell Res.* **86:**75.

Cunningham, D. D., C. R. Thrash and R. D. Glynn. 1974. Initiation of division of density-inhibited fibroblasts by glucocorticoids. In *Control of Proliferation in Animal Cells* (ed. B. Clarkson and R. Baserga), p. 105. Cold Spring Harbor Laboratory, Cold Spring Harbor, New York.

Fraenkel-Conrat, R., S. Bean and H. Lineweaver. 1949. Essential groups for the interaction of ovomucoid (egg white trypsin inhibitor) and trypsin, and for tryptic activity. *J. Biol. Chem.* **177:**385.

Goetz, I. E., C. Weinstein and E. Roberts. 1972. Effects of protease inhibitors on growth of hamster tumor cells. *Cancer Res.* **32:**2469.

Goldberg, A. R. 1974. Increased protease levels in transformed cells: A casein overlay assay for the detection of plasminogen activator production. *Cell* **2:**95.

Greene, L. A., E. M. Shooter and S. Varon. 1968. Enzymatic activities of mouse nerve growth factor and its subunits. *Proc. Nat. Acad. Sci.* **60:**1383.

Greene, L. A., J. T. Tomita and S. Varon. 1971. Growth stimulating activities of mouse submaxillary esteropeptidases on chick embryo fibroblasts *in vitro*. *Exp. Cell Res.* **64:**387.

Grossmann, A. and W. Troll. 1970. Tosylarginine methylester hydrolase activity in sea urchin egg membrane. *Biochim. Biophys. Acta* **312:**192.

Highland, J. H., R. L. Smith, E. Burka and J. Gordon. 1974. The effect of L-1-tosylamido-2-phenylethyl chloromethyl ketone on the activity of procaryote and eucaryote tRNA binding factors. *FEBS Letters* **39:**96.

Hozumi, M., M. Ogawa, T. Sugimura, T. Takeuchi and H. Umezawa. 1972. Inhibition of tumorigenesis in mouse skin by leupeptin, a protease inhibitor from actinomycetes. *Cancer Res.* **32:**1725.

Hsie, A. W. and T. T. Puck. 1971. Morphological transformation of Chinese hamster cells by dibutyryl adenosine cyclic 3′:5′-monophosphate and testosterone. *Proc. Nat. Acad. Sci.* **68:**358.

Inbar, M. and L. Sachs. 1969. Interaction of the carbohydrate-binding protein concanavalin A with normal and transformed cells. *Proc. Nat. Acad. Sci.* **63:**1418.

Jeffrey, W. R. 1972. Proteolytic enzyme activity during early development of starfish, *Asterias forbesii*. *Exp. Cell Res.* **72**:579.

Jones, R. O. and H. J. Ashwood-Smith. 1970. Some preliminary observations on the biochemical and biological properties of an epithelial growth factor. *Exp. Cell Res.* **59**:161.

Kassell, B. and M. Laskowski. 1956. The comparative resistance to pepsin of six naturally occurring trypsin inhibitors. *J. Biol. Chem.* **219**:203.

Kazakova, O. V. and V. N. Orekhovich. 1967. Comparative investigation of liver cathepsins of normal rats and rats with sarcoma and of cathepsins of rat sarcoma. *Bull. Exp. Biol. Med.* **64**:1207.

Kisch, A., R. O. Kelley, H. Crissman and L. Paxton. 1973. Dimethyl sulfoxide induced reversion of several features of polyoma transformed baby hamster kidney cells (BHK-21). Alterations in growth and morphology. *J. Cell Biol.* **57**:38.

Maciero-Coelho, A. and S. Avrameas. 1972. Modulation of cell behaviour *in vitro* by the substratum in fibroblastic and leukemic mouse cell lines. *Proc. Nat. Acad. Sci.* **69**:2469.

Nilausen, K. and H. Green. 1965. Reversible arrest of growth in G_1 of an established fibroblast line (3T3). *Exp. Cell Res.* **40**:166.

Ossowski, L., J. C. Unkeless, A. Tobia, J. P. Quigley, D. B. Rifkin and E. Reich. 1973. An enzymatic function associated with transformation of fibroblasts by oncogenic viruses. Mammalian fibroblast cultures transformed by DNA and RNA tumor viruses. *J. Exp. Med.* **137**:112.

Prival, J. T. 1972. Surface membrane proteins of normal and transformed mouse fibroblasts. Ph.D. thesis, Massachusetts Institute of Technology, Cambridge.

Roffman, S., H. Sanocka and W. Troll. 1970. Sensitive proteolytic enzyme assay using differential solubilities of radioactive substrates and products in biphasic systems. *Anal. Biochem.* **36**:11.

Schnebli, H. P. 1972. A protease-like activity associated with malignant cells. *Schweiz. Med. Wschr.* **102**:1194.

————. 1973. Die Rolle der Plasmamembran in der malignen Transformation. *Helv. Med. Acta* **36**:371.

————. 1974. Growth inhibition of tumor cells by protease inhibitors: Consideration of the mechanisms involved. In *Control of Proliferation in Animal Cells* (ed. B. Clarkson and R. Baserga), p. 327. Cold Spring Harbor Laboratory, Cold Spring Harbor, New York.

Schnebli H. P. and M. M. Burger. 1972. Selective inhibition of growth of transformed cells by protease inhibitors. *Proc. Nat. Acad. Sci.* **69**:3825.

————. 1973. "Flat" morphology induced in cultured transformed cells by polyphloroglucin phosphates. *Cancer Res.* **33**:3306.

Schnebli, H. P. and G. Haemmerli. 1974. Protease inhibitors do not block transformed cells in the G_1 phase of the cell cycle. *Nature* **248**:150.

Sefton, B. M. and H. Rubin. 1970. Release from density dependent growth inhibition by proteolytic enzymes. *Nature* **227**:843.

Shaw, E., M. Mares-Guia and W. Cohen. 1965. Evidence for an active center histidine in trypsin through use of a specific reagent, 1-chloro-3-tosylamide-7-amino-2-heptanone, the chloromethylketone derived from *N*-α-tosyl-L-lysine. *Biochemistry* **4**:2219.

Sheppard, J. R. 1971. Restoration of contact inhibited growth to transformed cells by dibutyl adenosine 3′:5′-cyclic monophosphate. *Proc. Nat. Acad. Sci.* **68**:1316.

Sivak, A. 1972. Induction of cell division: Role of cell membrane sites. *J. Cell Physiol.* **80**:167.

Talmadge, K. W., K. D. Noonan and M. M. Burger. 1974. The transformed cell surface: An analysis of the increased lectin agglutinability and the concept of growth control by surface proteases. In *Control of Proliferation in Animal Cells* (ed. B. Clarkson and R. Baserga), p. 313. Cold Spring Harbor Laboratory, Cold Spring Harbor, New York.

Taylor, J., S. Cohen and W. Mitchell. 1970. Epidermal growth factor (EGF): Properties of a high molecular weight species. *Fed. Proc.* **29**:670.

Troll, W., A. Klassen and A. Janoff. 1970. Tumorigenesis in mouse skin: Inhibition by synthetic inhibitors of proteases. *Science* **169**:1211.

Unkeless, J. C., A. Tobia, L. Ossowski, J. P. Quigley, D. B. Rifkin and E. Reich. 1973. An enzymatic function associated with transformation of fibroblasts by oncogenic viruses. I. Chick embryo fibroblast cultures transformed by avian RNA tumor viruses. *J. Exp. Med.* **137**:85.

Vacquier, V. D., D. Espe and L. A. Douglas. 1972. Sea urchin eggs release protease activity at fertilization. *Nature* **237**:34.

Vaheri, A., E. Ruoslahti and T. Hovi. 1974. Cell surface and growth control of chick embryo fibroblasts in culture. In *Control of Proliferation in Animal Cells* (ed. B. Clarkson and R. Baserga), p. 305. Cold Spring Harbor Laboratory, Cold Spring Harbor, New York.

Weber, M. J. 1974. Reversal of the transformed phenotype by dibutyryl-cyclic AMP and a protease inhibitor. In *Mechanisms of Virus Disease* (ed. W. S. Robinson and C. F. Fox), p. 327. W. A. Benjamin, Menlo Park, California.

Yamanishi, Y., M. K. Dabbous and K. Hashimoto. 1972. Effect of collagenolytic activity in basal cell epithelioma of the skin on reconstituted collagen and physical properties and kinetics of the crude enzyme. *Cancer Res.* **32**:2551.

Zamecnik, P. C. and M. L. Stephenson. 1947. Activity of catheptic enzymes in *p*-dimethyl-aminoazobenzene hepatomas. *Cancer Res.* **7**:326.

Effects of Added Proteases on Concanavalin A-specific Agglutinability and Proliferation of Quiescent Fibroblasts

Dennis D. Cunningham and Tsung-Shang Ho

Department of Medical Microbiology, College of Medicine
University of California, Irvine, California 92664

Several observations have led to the proposal that an alteration in the cell surface, detected by increased cell agglutinability with concanavalin A (ConA) and certain other plant lectins, brings about initiation of cell division and escape from density-dependent growth control (for review, see Burger 1973). These include observations that: (1) cells transformed by tumor viruses, chemical carcinogens and X irradiation are much more highly agglutinated by ConA than their untransformed parental cells (Inbar and Sachs 1969); (2) ConA agglutinability correlates with one expression of the transformed phenotype (lack of topoinhibition) in cells transformed by a temperature-sensitive tumor virus (Eckhart, Dulbecco and Burger 1971); (3) density-inhibited cells selected from populations of transformed cells show low agglutinability with the wheat germ lipase agglutinin (Pollack and Burger 1969); (4) selection for low agglutinability concomitantly selects for cells with density-dependent growth control (Inbar, Rabinowitz and Sachs 1969; Ozanne and Sambrook 1971); and (5) there is generally a positive correlation between agglutinability by these lectins and saturation density (Pollack and Burger 1969; Weber 1973). The possibility of a causal relationship between this membrane change and initiation of cell division was suggested by the finding that brief protease treatments at the cell surface which brought about the agglutinable state also caused density-inhibited cells to double in number (Burger 1970, 1973; Burger et al. 1972; Noonan and Burger 1973). However, our previous studies with 3T3 cells (Glynn, Thrash and Cunningham 1973) and the present experiments using secondary chick embryo fibroblasts and human diploid foreskin fibroblasts have led us to conclude that this surface change brought about by brief protease treatment is not an event that is sufficient by itself to lead to initiation of cell division and loss of density-dependent growth control.

In addition to these studies using brief protease treatments, we have also added proteases to quiescent fibroblasts and examined the effects on cell number of 24-hour to 72-hour treatments. We have confirmed the report of

795

Sefton and Rubin (1970) that a 24-hour treatment of quiescent secondary chick embryo fibroblasts with trypsin leads to a large increase in cell number. The trypsin must be present at least 4 to 8 hours to bring about a detectable increase in cell number at 24 hours. However, we have detected no significant increase in cell number after growing 3T3, human diploid foreskin or bovine embryonic trachea fibroblasts to a growth-arrested state under a variety of culture conditions and then adding proteases for periods up to three days.

Thus long-term, continuous treatment of quiescent fibroblasts with added proteases can lead to initiation of cell division in certain cases, but we do not understand the conditions necessary for the initiation. It is possible that initiation by proteases occurs more readily with avian cells, and that mammalian cells are relatively refractory to stimulation by these agents. However, it seems more likely that when different kinds of fibroblasts are grown to a quiescent state under different nutritional conditions, different factors or conditions eventually limit growth, and added proteases can initiate cell division only when certain of these become limiting.

MATERIALS AND METHODS

Swiss 3T3 fibroblasts (clone 42, obtained from Dr. George J. Todaro) were routinely cultured in Dulbecco-Vogt modified Eagle's medium containing 10% calf serum. Secondary chick embryo fibroblasts were prepared as previously described (Rein and Rubin 1968) and grown in medium 199 containing 2.0% tryptose phosphate broth and the indicated amount of chicken or calf serum. Human diploid foreskin fibroblasts (7th–10th passage) and bovine embryonic trachea fibroblasts (16th passage) were obtained from Dr. David T. Kingsbury and grown in Dulbecco-Vogt modified Eagle's medium containing the indicated amount of calf or fetal calf serum. The 3T3 cells showed no evidence of mycoplasma contamination following autoradiography with [^{3}H]-thymidine (Nardone et al. 1965). We did not check the other cells for mycoplasma contamination. All growth medium components were purchased from Grand Island Biological Company. Pronase (B grade, lot 101185), soybean trypsin inhibitor (B grade, lot 100946) and ConA (A grade, lot 210073) were obtained from Calbiochem. Trypsin (2 × crystallized from bovine pancreas, lot 123C-6860) was purchased from Sigma. Cell number was monitored by suspending cells in 0.05% trypsin, diluting with isotonic phosphate-buffered saline (PBS), and counting in a Coulter electronic particle counter. These measurements were occasionally checked by counting some samples in a hemacytometer chamber after suspending cells with 0.05% trypsin and diluting with serum-containing growth medium. ConA-specific agglutination was measured using the method of Ozanne and Sambrook (1971) with modifications. Quiescent cells were rinsed twice with PBS at 37°C and incubated for 10 minutes at 37°C in PBS or PBS containing the indicated amount of pronase or trypsin. The cells were then rinsed twice with PBS at 37°C, detached by incubating with 0.02% EDTA in PBS at 37°C, and sedimented at 120g for 10 minutes. They were then suspended in PBS containing ConA at 125 µg/ml and mixed in a Labline orbit water-bath shaker at 150 rpm for 5 minutes at 22°C. The cells were diluted by adding PBS, and a sample was placed in a hemacytometer chamber for counting. Agglutination was scored as percent of cells in clumps of three or more. Agglutination measured in the absence of ConA was less than 10%.

RESULTS

Effect of Proteases on ConA-specific Fibroblast Agglutinability

The amount of ConA-specific agglutination of quiescent human and secondary chick embryo fibroblasts after treatment with PBS or various concentrations of pronase or trypsin in PBS for 10 minutes at 37°C is shown in Figure 1. ConA-specific agglutination of control cells incubated with only PBS ranged

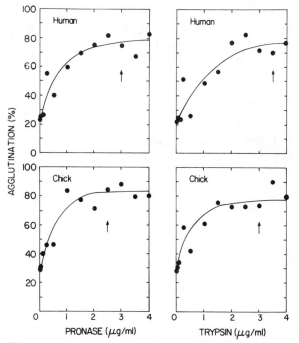

Figure 1
Effect of 10-minute protease treatments on ConA-specific agglutination of human and secondary chick embryo fibroblasts. Human fibroblasts were plated at a density of 3.0×10^4 cells/cm² in medium containing 3.0% calf serum and grown to a final density of 3.6×10^4 cells/cm² over a three-day period. Chick fibroblasts were plated at a density of 4.5×10^4 cells/cm² and grown to a final density of 1.5×10^5 cells/cm² over a three-day period. ConA-specific agglutination was measured as described in Materials and Methods. The arrows show the protease concentrations at which some rounding up of the cells was detected by phase contrast microscopy.

from 20–30%. As can be seen, treatment with pronase or trypsin increased this agglutination to plateau values of about 80%. Intermediate levels of proteases resulted in a graded increase in agglutination. The responses to pronase and trypsin were quite similar; maximal agglutination occurred after a 10-minute treatment with 2–3 µg/ml of either protease. The concentrations of pronase or trypsin which caused some rounding up of the cells (arrows in Fig. 1) were in all cases higher than those which brought about maximal ConA-specific agglutination.

Does This Surface Change Lead to Loss of Density-dependent Growth Control?

This question was answered by measuring cell number on parallel cultures of quiescent human and chick fibroblasts treated in exactly the same way with PBS or the same concentrations of pronase or trypsin in PBS for 10 minutes at 37°C. After this treatment, the cells were rinsed twice with PBS, and medium which had supported the growth of the cells to the quiescent state ("conditioned" medium) was added back to the cultures. As shown in Figures 2 and 3, there were no significant increases in cell number after treatment with

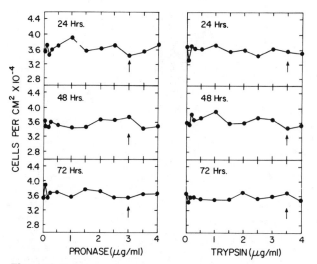

Figure 2

Effect of 10-minute protease treatments on cell density of human fibroblasts. In the same experiment described in Fig. 1, parallel cultures of quiescent cells were treated in the same way with PBS or the indicated concentration of pronase or trypsin in PBS for 10 minutes at 37°C. The cells were then rinsed twice with PBS at 37°C, and the "conditioned" medium which had supported the growth of the cells to the quiescent state was added back to the cultures. Cell number was monitored after 24, 48 and 72 hours as described in Materials and Methods. The arrows show the protease concentrations at which some rounding up of the cells occurred.

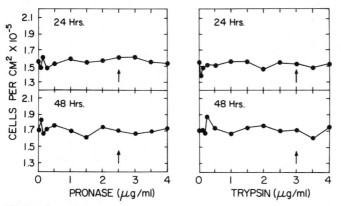

Figure 3
Effect of 10-minute protease treatments on cell density of
secondary chick embryo fibroblasts. This experiment was conducted in the same way as described in the legend to Figure 2
for human cells. Cell number was measured 24 and 48 hours
after the protease treatments.

the same concentrations of pronase or trypsin that brought about no increase to a maximal increase in ConA-specific agglutination. Thus this surface change did not lead to a loss of density-dependent growth control. We
reached this same conclusion in our earlier, similar studies with 3T3 fibroblasts
(Glynn, Thrash and Cunningham 1973).

The absence of cell division after these protease treatments could be a
result of (1) toxic materials in our protease preparations, (2) proteolytic
damage to the cells, or (3) an inability of the cells to divide in the "conditioned" medium. We ruled out these possibilities in our previous studies with
3T3 cells by showing that treatment with pronase reduced the stimulatory
response of the cells to fresh serum or cortisol by only 20% (Glynn, Thrash
and Cunningham 1973). In the present experiments, we eliminated these possibilities in another way. Cells which had been treated with 4 μg/ml of
pronase or trypsin for 10 minutes at 37°C were detached with 0.02% EDTA
in PBS and their growth in "conditioned" medium was compared to control
cells treated for 10 minutes with PBS only. These pronase or trypsin treatments did not reduce the growth of either the chick or human cells (Fig. 4).

Taken together, these results support our earlier conclusion from similar
studies using 3T3 cells (Glynn, Thrash and Cunningham 1973), that the
protease-mediated surface change measured by increased ConA-specific agglutinability is not an event sufficient by itself to lead to cell division or escape
from density-dependent growth control.

Increased Proliferation of Chick Fibroblasts
after Continuous Protease Treatment

We have confirmed the report of Sefton and Rubin (1970) that long-term,
continuous treatment of quiescent secondary chick embryo fibroblasts with
added trypsin increases cell proliferation. Figure 5 shows our experiments
with chick cells grown to quiescence in medium containing 0.7% chicken

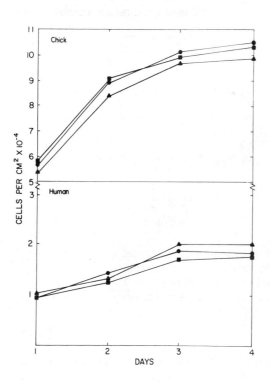

Figure 4

Effect of 10-minute pronase or trypsin teatment on subsequent growth of chick and human fibroblasts at subconfluent densities in "conditioned" medium. In the same experiment as described in the legend to Figure 1, parallel cultures of chick or human fibroblasts were detached with 0.02% EDTA in PBS after a 10-minute treatment with 4 μg/ml pronase (▲——▲), 4 μg/ml trypsin (■——■) or PBS (●——●). The cells were sedimented, resuspended in "conditioned" medium, counted, and plated at subconfluent densities (6.4 x 10^4 cells/cm² for chick cells; 1.3 x 10^4 cells/cm² for human cells). Cell number was monitored on subsequent days as described in Materials and Methods.

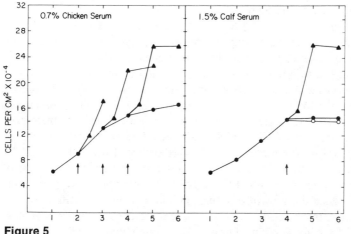

Figure 5

Effect of continuous trypsin treatment on cell density of quiescent secondary chick embryo fibroblasts. Chick fibroblasts were plated on day zero at a density of 4.5 × 10^4 cells/cm² in medium containing 0.7% chicken serum (left panel) or 1.5% calf serum (right panel). At times indicated by the arrows, trypsin in PBS was added to a final concentration of 2 μg/ml (left panel) or 20 μg/ml (right panel). Cell number was monitored as described in Materials and Methods on control cultures which received only PBS (●——●), trypsin-treated cultures (▲——▲) and cultures to which soybean trypsin inhibitor (final conc. 100 μg/ml) was added at the same time as the trypsin (○——○).

serum or 1.5% calf serum and treated with concentrations of trypsin (at times indicated by the arrows) which brought about a maximal increase in cell number. Optimal concentrations of trypsin produced an increase in cell number that was maximal after about 24 hours and that ranged from a 30 to 80% increase over control cultures in five separate experiments. This stimulation apparently depended on the proteolytic activity of trypsin, since it was prevented by the simultaneous addition of soybean trypsin inhibitor (Fig. 5, right panel). As Sefton and Rubin reported (1970), pronase also stimulates proliferation of quiescent chick cells. We found that optimal concentrations of pronase produced a response that ranged from 50 to 85% of the response brought about by optimal concentrations of trypsin.

Since our 10-minute protease treatments of quiescent chick cells led to no detectable increase in cell number, we examined the duration of trypsin treatment required to bring about an increase in cell number at 24 hours (Fig. 6). At time zero, trypsin was added to a series of quiescent chick fibroblast cultures, and at indicated times, we removed this medium, rinsed the cells, and added back "conditioned" medium containing no trypsin, which had supported the growth of parallel cultures to the quiescent state. Cell number was then monitored on all cultures at 24 hours. The trypsin had to be present for at least 4–8 hours to bring about a detectable increase in cell number at 24 hours (Fig. 6), and continued exposure to trypsin was required for maximal stimulation. We considered the possibility that the short trypsin treatments might bring about cell division at a later time, but measurements of cell number at 48 and 72 hours excluded this possibility (data not shown).

Failure of Continuous Protease Treatments to Significantly Stimulate Proliferation of Other Quiescent Fibroblasts

We have grown 3T3, human diploid foreskin and bovine embryonic trachea fibroblasts to a growth-arrested state under a variety of culture conditions,

Figure 6
Duration of trypsin treatment required to stimulate proliferation of quiescent secondary chick embryo fibroblasts. Chick fibroblasts were plated at a density of 4.5×10^4 cells/cm^2 in medium containing 0.7% chicken serum and grown to quiescence over a four-day period. At time zero, trypsin was added to a final concentration of 2 μg/ml. At indicated times, this medium was removed, cultures were rinsed twice with "conditioned" medium, and "conditioned" medium was then added back to cultures. Cell number was monitored on all cultures at 24 hours as described in Materials and Methods. Points at zero hours represent control cell counts measured at 24 hours on cultures which received only a change to "conditioned" medium at time zero.

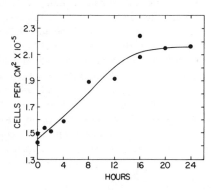

added proteases over broad ranges of concentrations, and have detected no significant increase in cell number during the following three days.

The left panel of Figure 7 shows our data for 3T3 cells grown to confluency under the usual conditions which result in "density-inhibited" cultures. We added trypsin to these cultures over a broad range of concentrations, from high concentrations which rounded up the cells (shown by arrows in Fig. 7) through twofold dilutions to concentrations that were lower by three orders of magnitude. As can be seen, none of these trypsin treatments brought about a significant increase in cell number measured at 24, 48 or 72 hours.

The absence of cell division after this continuous trypsin treatment of 3T3 fibroblasts appears not to be the result of irreversible damage to the cells by the trypsin or by an inability of the medium to support further growth. This conclusion is based on the following evidence: In the experiment shown in the

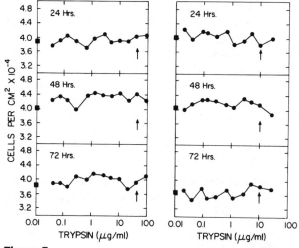

Figure 7

Effect of continuous protease treatments on cell density of quiescent 3T3 fibroblasts. 3T3 cells were plated at a density of 1.3×10^4 cells/cm^2 and grown to a final saturation density of 4.0×10^4 cells/cm^2 over a three-day period in medium containing 10% calf serum. (*Left*) Cells received no medium change before trypsin additions. (*Right*) Just before trypsin additions, the medium was changed to "conditioned" medium 199 (containing 2.0% tryptose phosphate broth and 2.0% chicken serum) which had supported the growth of chick fibroblasts for three days. Trypsin in PBS was then added to the indicated final concentrations, and cell number was monitored 24, 48 and 72 hours later as described in Materials and Methods. The solid squares on the ordinate represent cell densities of control cultures which received only PBS. The arrows represent trypsin concentrations at which some rounding up of cells was detected by phase contrast microscopy.

left panel of Figure 7, several cultures of 3T3 fibroblasts were subcultured with 0.02% EDTA in PBS after growth for 24 hours in medium containing trypsin at a final concentration of 10 µg/ml. (As shown in Fig. 7, the concentration which caused some rounding up of the cells was 40 µg/ml.) These cells were then replated at subconfluent densities in the same "conditioned" medium containing trypsin at a final concentration of 10 µg/ml. After three days, there was a 70% increase in cell number.

The stimulatory effect with chick cells, noted in the previous section, was observed with cells grown to a growth-arrested state in medium 199 containing 2.0% tryptose phosphate broth and 0.7% chicken serum. However, we observed no significant increase in cell number after adding trypsin over broad ranges of concentrations to 3T3 cells grown to a growth-arrested state in this medium supplemented with 1.5 or 7.5% chicken serum (data not presented).

To check the possibility that chick cells might release a protein-containing material into the culture medium, from which trypsin might release an active peptide, we grew 3T3 cells to the "density-inhibited" state, changed to medium which had supported the growth of chick cells for three days, added trypsin over a broad range of concentrations, and monitored cell number for three days (right, Fig. 7). As can be seen, trypsin treatments under these conditions led to no significant increase in cell number.

We also checked to see if long-term, continuous protease treatments of growth-arrested human diploid foreskin or bovine embryonic trachea fibroblasts would initiate cell division. Addition of proteases over broad ranges of concentrations to these quiescent cultures brought about no significant increase in cell number as measured at 24, 48 or 72 hours (Fig. 8).

DISCUSSION

Our previous studies with 3T3 cells led us to conclude that the protease-mediated surface change measured by increased ConA-specific cell agglutinability was not an event that was sufficient to lead to cell division and loss of density-dependent growth control (Glynn, Thrash and Cunningham 1973). We have now extended these studies to cultured cells that are not established lines and have come to the same conclusion. Ten-minute treatments with varying levels of trypsin or pronase led to graded increases in ConA-specific cell agglutinability, but after adding "conditioned" medium back to parallel protease-treated cells, there was no detectable increase in cell number. This lack of cell division was not a result of toxic materials in the proteases, extensive proteolytic damage to the cells, an inability of the cells to respond to a stimulatory signal, or extensive depletion of the medium such that it could not support further cell divisions. It is noteworthy that Chen, Teng and Buchanan (this volume) report that a 10-minute treatment of quiescent secondary chick embryo fibroblasts with 1 µg/ml trypsin leads to a 5–6-fold increase in the incorporation of [³H]thymidine into DNA when measured at 12 hours, but no detectable increase in cell number when measured at 24 hours. By extrapolation from studies on serum stimulation, they calculate that this increased incorporation should correspond to a 15% increase in cell

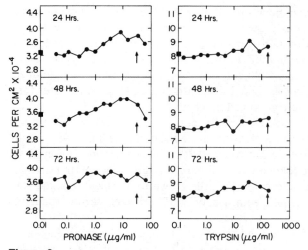

Figure 8

Effect of continuous protease treatments on cell densities of growth-arrested human and bovine embryonic trachea fibroblasts. (*Left*) Human fibroblasts were plated at a density of 3.0×10^4 cells/cm² and grown to a final saturation density of 3.4×10^4 cells/cm² over a three-day period in medium containing 3.0% calf serum. (*Right*) Bovine embryonic trachea fibroblasts were plated at a density of 3.0×10^4 cells/cm² and grown to a final saturation density of 8×10^4 cells/cm² over a three-day period in medium containing 7.5% fetal calf serum. Proteases in PBS were then added to the indicated final concentrations, and cell number was monitored as described in Materials and Methods 24, 48 and 72 hours later. The solid squares on the ordinate represent cell densities of control cultures which received only PBS. The arrows represent trypsin concentrations at which rounding up of cells was detected by phase contrast microscopy.

number, an increase that cell counting methods might not detect. We used cell number, rather than thymidine incorporation, as a measure of proliferation in our experiments, since we were concerned that small increases in incorporation might be caused by repair of DNA or changes in transport rates of thymidine or pool sizes of thymidine deoxynucleotides.

It should be pointed out that our experiments do not indicate the extent to which the cell surface alteration in *transformed* fibroblasts measured by increased ConA-specific agglutinability is involved in the loss of density-dependent growth control characteristic of these cells. The molecular changes leading to increased ConA-specific agglutinability in transformed fibroblasts might be quite different from those leading to increased agglutinability of untransformed fibroblasts after protease treatment.

Although our 10-minute protease treatments of growth-arrested fibroblasts led to no detectable increase in cell number, we found that a 24-hour treatment of quiescent secondary chick embryo fibroblasts with trypsin led to a large increase in cell number, confirming the earlier report of Sefton and Rubin (1970). Using optimal concentrations of trypsin, we observed increases in cell number that ranged from a 30 to an 80% increase over control cultures and which were maximal at about 24 hours. The increase occurred in chick fibroblasts grown to quiescence in either chicken or calf serum and was dependent on the proteolytic activity of trypsin, since the response was prevented by the simultaneous addition of soybean trypsin inhibitor. Twenty-four-hour treatments of quiescent chick fibroblasts with added pronase also led to increases in cell number, but in our experiments, as well as those of Sefton and Rubin (1970), this protease did not initiate as effectively as trypsin.

To further clarify the nature of the protease treatment required for initiation of cell division, we examined the duration of trypsin treatment required for initiation of division of quiescent secondary chick embryo fibroblasts. In our experiments, we added trypsin to a series of growth-arrested cultures and at specified times removed the trypsin-containing medium, added "conditioned" medium back to the cultures, and measured cell number on all cultures at 24 hours. We found that trypsin had to be present for at least 4–8 hours to bring about a detectable increase in cell number at 24 hours. Sefton (1972) has conducted similar experiments examining the rate of [^{3}H]thymidine incorporation into DNA at 12 hours. He also concluded that continued exposure to trypsin is required for maximal stimulation.

In contrast to these studies with quiescent chick embryo fibroblasts, we have not detected a significant increase in cell number after growing 3T3, human diploid foreskin or bovine embryonic trachea fibroblasts to a growth-arrested state under a variety of culture conditions and then adding proteases for periods up to three days. In these experiments, we added proteases over a very broad range of concentrations, from high concentrations which rounded up the cells, through twofold dilutions to concentrations that were lower by three orders of magnitude. The reasons for the differences in response between these cells and the chick cells is not clear at present. The chick cells were grown to quiescence in medium 199 containing 2.0% tryptose phosphate broth and 0.7% chicken serum. However, 3T3 cells grown to a growth-arrested state in this same medium containing 1.5 or 7.5% chicken serum showed no stimulation after adding trypsin over broad ranges of concentrations. Also, growth-arrested 3T3 cells switched to "conditioned" medium which had supported the growth of chick cells to confluency showed no increase in cell number after adding trypsin. These results indicate that the initiation of chick cell division after protease treatment does not result from trypsin releasing active peptides from protein-containing material in the tryptose phosphate broth or from materials released into the growth medium by the chick cells. It is also clear that homologous serum is not required for the response since chick cells grown to a quiescent state in medium containing calf serum increased in number after long-term treatment with trypsin.

Thus long-term treatment of quiescent fibroblasts with added proteases can lead to initiation of cell division in certain cases, but the conditions necessary for the initiation are not yet understood. It is possible that initiation by proteases occurs more readily with avian cells, and that mammalian cells are relatively refractory to stimulation by these agents. However, it seems more likely that when different kinds of fibroblasts are grown to quiescence under different nutritional conditions, different factors or conditions eventually limit growth, and added proteases can initiate division only when certain of these become limiting.

Acknowledgment

This work was supported by a grant from the National Cancer Institute of the USPHS (CA-12306).

REFERENCES

Burger, M. M. 1970. Proteolytic enzymes initiating cell division and escape from contact inhibition of growth. *Nature* **227:**170.

———. 1973. Surface changes in transformed cells detected by lectins. *Fed. Proc.* **32:**91.

Burger, M. M., B. N. Bombik, B. Breckenridge and J. R. Sheppard. 1972. Growth control and cyclic alterations of cyclic AMP in the cell cycle. *Nature New Biol.* **239:**161.

Eckhart, W., R. Dulbecco and M. Burger. 1971. Temperature-dependent surface changes in cells infected or transformed by a thermosensitive mutant of polyoma virus. *Proc. Nat. Acad. Sci.* **68:**283.

Glynn, R. D., C. R. Thrash and D. D. Cunningham. 1973. Maximal concanavalin A-specific agglutinability without loss of density-dependent growth control. *Proc. Nat. Acad. Sci.* **70:**2676.

Inbar, M. and L. Sachs. 1969. Interaction of the carbohydrate-binding protein concanavalin A with normal and transformed cells. *Proc. Nat. Acad. Sci.* **63:**1418.

Inbar, M., Z. Rabinowitz and L. Sachs. 1969. The formation of variants with a reversion of properties of transformed cells. III. Reversion of the structure of the cell surface membrane. *Int. J. Cancer* **4:**690.

Nardone, R. M., J. Todd, P. Gonzalez and E. Gaffney. 1965. Nucleoside incorporation into strain L cells: Inhibition by pleuropneumonia-like organisms. *Science* **149:**1100.

Noonan, K. D. and M. M. Burger. 1973. Induction of 3T3 cell division at the monolayer stage. *Exp. Cell Res.* **80:**405.

Ozanne, B. and J. Sambrook. 1971. Isolation of lines of cells resistant to agglutination by concanavalin A from 3T3 cells transformed by SV40. In *The Biology of Oncogenic Viruses* (ed. L. G. Silvestri), p. 248. American Elsevier, New York.

Pollack, R. E. and M. M. Burger. 1969. Surface-specific characteristics of a contact-inhibited cell line containing the SV40 viral genome. *Proc. Nat. Acad. Sci.* **62:**1074.

Rein, A., and H. Rubin. 1968. Effects of local cell concentrations upon the growth of chick embryo cells in tissue culture. *Exp. Cell Res.* **49:**666.

Sefton, B. M. 1972. Studies on the release of chick cells from growth inhibition. Ph.D. dissertation. Department of Molecular Biology, University of California, Berkeley.

Sefton, B. M. and H. Rubin. 1970. Release from density-dependent growth inhibition by proteolytic enzymes. *Nature* **227:**843.

Weber, J. 1973. Relationship between cytoagglutination and saturation density of cell growth. *J. Cell Physiol.* **81:**49.

Partial Purification and Characterization of Overgrowth Stimulating Factor

John G. Burr* and Harry Rubin

Department of Molecular Biology and Virus Laboratory
University of California, Berkeley, California 94720

Chick embryo fibroblasts transformed by infection with the Bryan strain of RSV release into their culture medium material which stimulates growth in density-inhibited populations of normal chick embryo cells (Rubin 1970a). The material responsible for the stimulation was named "overgrowth stimulating factor" (OSF). It is released into the medium in assayable quantities within 24 hours after the beginning of the morphological transformation of the culture (4–5 days after infection). The release appears to be by non-specific leakage rather than by a specific secretion process or by cell lysis (Bissell, Rubin and Hatiè 1971).

Sonic disruption of normal cells releases material that also stimulates the growth of density-inhibited cultures, but sonicates of RSV-transformed cells possess even more stimulatory activity on an OD_{280} basis soon after infection (Rubin 1970b). The cell-associated overgrowth stimulating activity in transformed cells increases until the fifth day after infection. After this time, the activity returns to near normal amounts, roughly coincident with the appearance of stimulatory activity in the culture medium.

This report describes the partial purification and characterization of OSF. In addition, the cellular overgrowth stimulating activity released by sonication of normal cells has been compared with that obtained from transformed cells.

MATERIALS AND METHODS

Cell Culture

Primary chick embryo fibroblasts were obtained by trypsinizing the body walls of ten-day-old embryos of White Leghorn strain 813 chickens (Kimber Farms, Niles, California), as described previously (Rein and Rubin 1968). Primary cells are seeded at a density of 8×10^6 cells/100-mm plastic petri

* Present address: Department of Physiology, Harvard Medical School, Boston, Massachusetts 02115

dish (Falcon Plastics) in 12.5 ml of medium 199 (Grand Island Biologicals) supplemented with 2% tryptose phosphate broth (TPB, Difco), 1% calf serum and 1% chicken serum (Microbiological Associates or Grand Island Biologicals). (The convention we will use here to designate this mixture is 199 (2:1:1), the numbers in parentheses indicating the percent TPB, calf serum and chicken serum, respectively.) Primary cultures are grown at 37°C in an atmosphere of 5% CO_2 for 3–4 days, at which time the medium is changed to fresh 199 (2:1:1). Primary fibroblasts derived from special RIF-free embryos (SPF embryos, Kimber farms) were used for cultures to be infected by RSV for the production of OSF.

Production of OSF

Secondary cultures were infected by transfer of 2×10^6 primary cells to 100-mm Falcon plastic petri dishes in 10 ml of medium 199 (2:1:1) supplemented with 0.2% glucose and containing $0.5-1 \times 10^6$ focus-forming units of the Bryan strain of RSV per ml. One-half ml of a supplement solution (70% TPB, 2% glucose and 1% $NaHCO_3$) was added to each dish one day after infection, and 1.0 ml on the second day. Three days after infection, the medium was changed to 199 (5:0:2) supplemented with 0.2% glucose. One ml of supplement was added on each of the subsequent days, and medium was collected and frozen for future assay of OSF activity on days 5, 6 and 7 after infection. This medium will be referred to as Rous conditioned medium (RCM). Virus and other particulates were removed from RCM by centrifugation at 75,000g for 2 hours at 4°C before all subsequent purification steps. Normal conditioned medium (NCM) was obtained from control, uninfected cultures by the same protocol.

OSF in Cell Sonicates

Secondary cultures were infected and grown as described above for the release of OSF into the medium. At the times described in particular experiments, the medium was removed and the cells were washed with Tris-buffered physiological saline (Tris-Saline). They were then scraped off the dish with a rubber policeman, sedimented in a clinical centrifuge, and resuspended by the addition of 0.5 ml of Tris-Saline per ml of sedimented cells. The suspended cells were then sonicated at 4°C for 5 minutes in a Raytheon sonic oscillator at 250 w and 10 hz. Cells were examined microscopically to confirm disruption, and cellular debris was removed by centrifugation at 4°C in a Sorvall angle rotor at 10,000 rpm for 20 minutes. Aliquots from the clarified supernatant thus obtained were taken for protein determination by the method of Lowry et al. (1951), and the remainder was either frozen at −60°C or used directly for chromatography.

Assay for Overgrowth Stimulating Activity

The assay for overgrowth stimulating activity was performed essentially as described previously (Rubin 1970a). Primary cells were trypsinized and seeded at 1×10^6 cells per 50-mm dish in 5 ml of medium 199 (2:0:2) and grown for two days until they were confluent. To assay for overgrowth stimulating activity, the medium was removed from these confluent cultures and replaced with 5 ml of Scherer's medium containing 5% TPB, 1% calf serum, 1% chicken serum, 0.4% Bacto-Difco agar and an aliquot of the sample whose overgrowth stimulating activity was to be measured. Eighteen hours after the cells were overlaid, the agar medium was removed, the cells were washed once with 2 ml of medium 199, and the rate of DNA synthesis in the cultures was measured by the rate of incorporation of [*methyl*-^{3}H]thymidine ([^{3}H]TdR) (5 mCi/0.055 mg, New England Nuclear) as previously described (Rubin 1970b). Overgrowth stimulating activity is expressed as the ratio of cpm incorporated into DNA by cells to which samples of RCM at various stages of purification were added in the agar overlay to the counts incorporated by cells to which either nothing or an appropriate control sample was added. Cell number was determined with a Coulter electronic counter. Mitotic activity was estimated after 6 hours of colcemide (5 × 10^{-8} M) treatment (Rubin 1970a). All points are the average of duplicate plates.

Quantitation of Recovery of Activity

The dose response curve of cells to increasing amounts of added OSF is nonlinear beyond a three- to fivefold stimulation of [^{3}H]TdR incorporation (for example, see Fig. 3). Therefore, whenever the activity of a treated sample of OSF was quantitated, it was done by comparing the dose response curve of the treated sample to the dose response curves of a dilution series of a control sample of OSF. Because of the variability from day to day of the response of assay cells to OSF, all such comparisons were made at the same time on the same set of cells.

Concentration of RCM

Concentration by diaflow ultrafiltration (Amicon) using a filter (PM-10) with a molecular weight cut-off value of 10,000 daltons was found to be both rapid and effective. Ninety percent of the OD$_{280}$ is recovered after concentration and close to 100% of the activity. A second method of concentration used was dialysis against a thick aqueous slurry of polyethylene glycol (MW 2 × 10^4) ("Aquacide III," Calbiochem) buffered with 0.05 M Tris pH 7.5. Recovery of activity was also close to 100% by this method.

Dialysis of RCM: Standard Buffer

OSF is stable to dialysis against Tris or phosphate buffers of ionic strengths ranging from 10^{-3} to 1.0 M salt; the activity is most stable to dialysis at pH 7. The presence or absence of Ca^{++} or Mg^{++} has no effect, and 0.5 mM dithiothreitol (DTT) enhances the recovery of activity. Most frequently used in the purification was 0.01 M Tris-buffered physiological saline pH 7.5 containing 0.5 mM DDT, and this will be referred to as "standard buffer."

Preparation of [^{14}C]RCM and [^{3}H]NCM

RSV-infected and control, uninfected cells were grown as described above for production of OSF. On the fourth day after infection, the medium from RSV-transformed cultures was removed and stored in the incubator. The cultures were then labeled for 1 hour in 2 ml of medium 199 containing 1/10 the usual concentration of amino acids and 50 μCi of ^{14}C-amino acids (Algal hydrolysate, New England Nuclear). The medium which had been removed was then replaced, and the cells were incubated further in this labeling medium for another 24 hours, at which time the medium was collected and frozen at −60°C. The same procedure was followed for labeling control, uninfected cultures with ^{3}H-amino acids. Before chromatography, virus was removed from RCM by centrifugation and the labeled media were dialyzed against 1 liter of standard buffer for 24 hours with one buffer change. The media were then concentrated tenfold by dialysis against polyethylene glycol.

RESULTS

Sephadex Chromatography

Rous conditioned medium, freed of virus and other particulates, was concentrated tenfold and chromatographed on Sephadex G-100. Each column

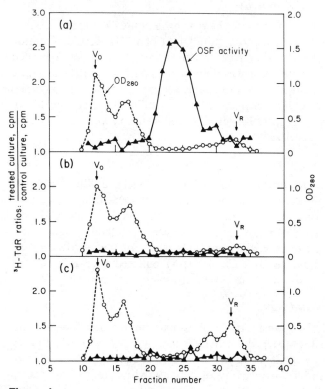

Figure 1

(a) Chromatography of RCM on Sephadex G-100. One ml of tenfold-concentrated RCM was applied to a 0.9 × 60-cm column of Sephadex equilibrated in standard buffer. The material was eluted from the column at a downward flow of 8 ml/hr. Fractions of 1.25 ml were collected, and 0.1 ml/fraction/plate of cells was assayed for overgrowth stimulating activity. (b) Chromatography of control, normal conditioned medium (NCM) on Sephadex G-100. Protocol as described for RCM. (c) Chromatography of unconditioned medium 199 (5:0:2) on Sephadex G-100. Protocol as described for RCM. (o) OD$_{280}$; (▲) overgrowth stimulating activity.

fraction was assayed for its overgrowth stimulating activity. Figure 1a shows that a single major peak of stimulatory activity was eluted from the column. A 100- to 200-fold purification was thereby achieved, but only 10–30% recovery of activity was obtained. Recombination of fractions did not result in any significant enhancement of the activity of this peak, nor did it reveal any additional activity.

Like the stimulation observed with crude RCM (Rubin 1970a), the stimulation of [³H]TdR incorporation by Sephadex-purified OSF is accompanied by an increase in mitosis and cell number (Table 1). By comparison to the elution of markers of known molecular weight, the molecular weight of this activity is estimated at 20,000 daltons.

Normal conditioned medium collected from control, uninfected cultures and medium 199 (5:0:2) which had never been exposed to cells were similarly concentrated and chromatographed. As can be seen (Fig. 1b,c), no activity was observed after chromatography. These controls rule out the possibility that the peak of activity observed in Figure 1a is simply stimulation by a serum component, unrelated to the stimulatory material responsible for the activity of RCM.

To test for the possible presence of an inhibitor of OSF activity in normal conditioned medium, an aliquot of the fractions containing maximum activity from chromatographed RCM was combined with an aliquot of the corresponding fractions from NCM. There was no diminution of activity (Fig. 2).

The activity thus purified, although more sensitive to freezing and thawing than the activity in whole RCM, is stable when simply incubated at 37°C. No loss of activity can be detected even after 24 hours at this temperature (Fig. 3).

Enzymatic Sensitivity

The sensitivity of the activity in RCM to trypsin and RNase A was examined. Since trypsin itself stimulates the growth of density-inhibited cells (Sefton

Table 1
Effect of Sephadex-purified OSF on [³H]TdR Incorporation,
Mitosis and Cell Number

	Treated cultures/control cultures		
Time after addition of OSF (hr)	[³H]TdR incorporation	mitotic index	cell number
0	1.0	1.0	1.0
12	2.0	—	1.1
18	2.2	1.8	—
25	—	2.0	—
36	4.6	—	1.25
62	4.4	—	1.5

Sephadex-purified, tenfold-concentrated RCM (0.05 ml) was added to confluent cultures in the usual agar overlay assay for OSF activity. Control cultures received the same volume of Sephadex-purified NCM.

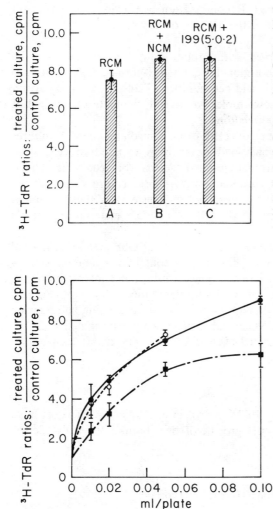

Figure 2
Absence of an inhibitor of OSF in NCM and unconditioned medium. (*A*) Activity of 0.05 ml/plate Sephadex-purified RCM. (*B*) Activity of 0.05 ml/plate Sephadex-purified RCM plus 0.05 ml/plate Sephadex-purified NCM, combined before adding to medium for agar overlay. (*C*) Activity of 0.05 ml/plate Sephadex-purified RCM plus 0.05 ml/plate Sephadex-purified unconditioned medium 199 (5: 0:2), combined before adding to medium for agar overlay. All media were concentrated tenfold before chromatography on Sephadex G-75.

Figure 3
Stability of Sephadex G-75-purified RCM. Dose response curves of Sephadex-purified OSF, incubated prior to assay for 24 hours at 37°C (●); incubated at 0°C for 24 hours (○); frozen and thawed twice (■).

and Rubin 1971), soybean trypsin inhibitor (STI) was added to the sample of RCM after trypsin digestion prior to assay. The results of this experiment are shown in Table 2. Incubation of the sample for 1 hour with trypsin at 10 μg/ml eliminated 90% of the activity. The addition of STI alone to RCM had little effect on its stimulatory activity. Incubation with RNase A at 10 μg/ml for 1 hour had no effect on the activity.

Table 2

Effect of Trypsin and Ribonuclease on OSF

Treatment	$[^3H]TdR$ ratios[a]
(1) OSF + trypsin, STI added before assay[b]	1.2
(2) OSF + STI	2.5
(3) Control, untreated OSF	2.7
(4) Trypsin + STI	0.9
(5) OSF + RNase A	2.9
(6) RNase A	1.1

[a] Ratio: Cultures receiving OSF or indicated substance/control cultures receiving only standard buffer.

[b] An aliquot of tenfold-concentrated, Sephadex-purified OSF in standard buffer was incubated (10 μg/ml) for 1 hour at 37°C. Soybean trypsin inhibitor (STI) (10 μg/ml) was then added, and the sample further incubated for 10 minutes. The sample was then assayed for overgrowth stimulating activity (0.1 ml/plate of cells). Control, untreated OSF was simultaneously incubated for 1 hour at 37°C and then assayed with (2) or without (3) STI (6 μg/ml). Trypsin (10 μg/ml) and STI (10 μg/ml) were combined (4), incubated for 10 minutes and then assayed, to control for inactivation of trypsin. OSF was similarly incubated with RNase A (10 μg/ml) (5). RNase itself was tested for overgrowth stimulating activity (6).

Absence of Measurable Proteolytic Activity

The growth stimulating activity of RCM can be mimicked by trypsin as well as several other proteases (Rubin 1970a; Sefton and Rubin 1971) and other hydrolases (Vasiliev et al. 1970). Sephadex-purified OSF was therefore examined for hydrolytic activity on several common substrates. Under the conditions shown in Table 3, no esterase, collagenase, general protease, ribonuclease or hyaluronidase activity could be detected.

Ion Exchange Chromatography

In order to examine whether the 20,000-dalton peak of activity recovered after Sephadex chromatography of RCM was due to a single component and to explore the possibilities for further purification, the elution behavior of this activity on ion exchange resins was examined. The results of chromatography on CM-Sephadex at pH 7 are shown in Figure 4. Two peaks of overgrowth stimulating activity were observed, one eluted from the column by the starting buffer, followed by a second peak eluted by 0.1 M KCl. Two peaks of activity were also observed when the Sephadex-purified material was chromatographed on DEAE-Sephadex at pH 7 (Fig. 5). In both cases, the two peaks of activity were found to be extremely labile, and neither activity could be recovered after freezing or following further chromatography.

Table 3

Absence of Measurable Hydrolytic Activity Associated with OSF

Substrate examined	Reaction conditions	Sensitivity of assay
TAME	Tris-buffered physiological saline, pH 7.5, 0.5 mM DTT	0.1 µg/ml trypsin
Yeast RNA	medium 199 pH 5, pH 7.5	0.1 µg/ml RNase A
Hide powder azure (Calbiochem)	medium 199 pH 7.5	1 ng/ml trypsin; 10 ng/ml collagenase
Azocasein (Calbiochem)	medium 199 pH 3, pH 7.5	0.2 µg/ml trypsin
^{14}C-Algal (Chlorella) protein (New England Nuclear)	medium 199 pH 4, pH 7.5, pH 10	0.1 µg/ml trypsin
Hyaluronic acid	0.01 M Tris pH 7.5	1 µg/ml hyaluronidase

Sephadex-purified OSF was assayed for hydrolysis of TAME by the method of Greene, Shooter and Varon (1968); for hydrolysis of yeast RNA by the method of Dickman, Aroskar and Kropf (1956); for hydrolysis of hyaluronidase by the method of McClean (1943); for hydrolysis of azocasein by the method of the manufacturer. ^{14}C-labeled Chlorella protein as received from NEN was TCA precipitated. The precipitate was redissolved and equilibrated in standard buffer by chromatography on Sephadex G-25 before use as a substrate. Hydrolysis was measured by rate of release of TCA-soluble cpm. OSF was present in all assays for enzymatic activity at a concentration 50–100 times that required to produce a 2–3-fold stimulation of DNA synthesis when incubated with assay cells. The sensitivity of the assay was determined by simultaneous assay of the appropriate enzyme at the concentration indicated.

$^{3}H/^{14}C$-Acrylamide Gels

SDS-acrylamide gel electrophoresis was performed on the fractions with maximum activity from Sephadex-chromatographed RCM and the corresponding fractions of chromatographed NCM. Staining with Coomassie blue revealed the presence of seven bands of protein, with no detectable differences between the number or position of proteins in the two samples (not shown). An attempt was therefore made to detect the protein(s) responsible for the activity in RCM as a band of radioactivity.

Four days after infection, RSV-transformed cultures were labeled for 24 hours with ^{14}C-amino acids, and control, uninfected cultures were simultaneously labeled with ^{3}H-amino acids. The ^{14}C-labeled RCM and ^{3}H-labeled NCM harvested on the subsequent day were then combined and cochromatographed on Sephadex G-75 (Fig. 6). The three fractions (A) immediately prior to, those containing (B) and the three subsequent (C) to the peak of OSF activity were pooled, dialyzed and concentrated tenfold.

The results of disc electrophoresis at pH 8.5 of samples A, B and C are shown in Figure 7a, b, c. In sample B, representing the fractions with maximum activity, there is a band of ^{14}C cpm (I) unaccompanied by ^{3}H label; this band is missing from the samples of both preceding and subsequent frac-

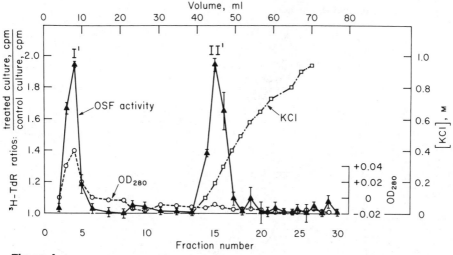

Figure 4

Chromatography of Sephadex-purified OSF on carboxymethyl-Sephadex. Five ml of tenfold-concentrated RCM were chromatographed on Sephadex G-75. The active fractions were pooled and concentrated to 1 ml and then desalted and equilibrated in 0.01 M phosphate buffer pH 7.5, plus 0.5 mM DTT, by passage over a Sephadex G-25 column. The 2 ml of material thus obtained was applied to a 0.9 × 10-cm column of CM-Sephadex equilibrated in this buffer. The column was washed with a volume of starting buffer equal to three times the volume of the column, followed by a linear gradient of KCl up to 1.0 M. Fractions containing KCl were dialyzed before assay. All fractions were assayed at 0.5 ml/plate.

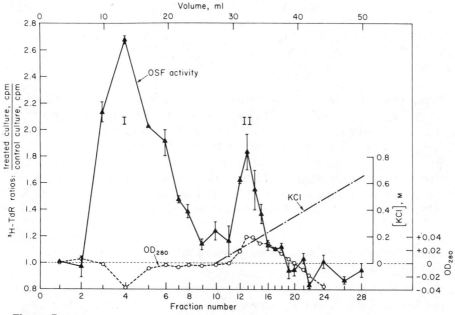

Figure 5

Chromatography of Sephadex-purified OSF on DEAE-Sephadex. Procedure and chromatography as described in Figure 4, except that sample and DEAE column were equilibrated in 0.01 M Tris, pH 7.5, plus 0.5 mM DTT.

815

816   J. G. Burr and H. Rubin

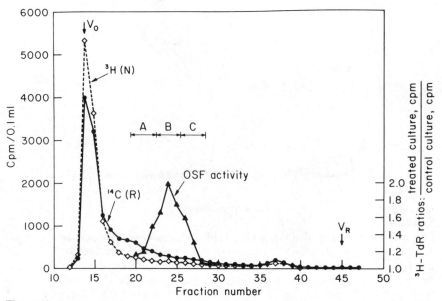

Figure 6
Chromatography on Sephadex G-75 of a pooled aliquot of ^{14}C-labeled RCM and
[3]H-labeled NCM. Dialyzed, tenfold-concentrated ^{14}C-labeled RCM (0.56 ml; 3.5×10^5 cpm) and 1.16 ml (3.5×10^5 cpm) of dialyzed, tenfold-concentrated ^{3}H-labeled NCM were combined and applied to a 1.5×90-cm column of Sephadex
G-75, equilibrated in standard buffer. Fractions of volume 3.2 ml were collected.
Fractions no. 20–30 were assayed for overgrowth stimulating activity at
0.1 ml/plate. Fractions no. 20–22 (*A*), 23–25 (*B*) and 26–28 (*C*) were pooled,
dialyzed against 0.001 M Tris pH 7.5, and then concentrated tenfold by lyophilization. Samples *A, B* and *C* were then electrophoresed (see Fig. 5). (●) ^{14}C cpm;
(◇) ^{3}H cpm; (▲) overgrowth stimulating activity.

tions. The only other peak of radioactivity in sample B, a double-labeled
band (II) running with the bromphenol blue marker, represents material
that is synthesized by both RSV-transformed and normal cells and, as confirmed in a subsequent experiment, elutes from the column at peak concentrations prior to the peak of OSF activity (near fraction no. 20 of Fig. 5).
Unfortunately, no biological activity could be recovered from the region of
the gel corresponding to band I, even when highly concentrated (100-
fold) Sephadex-purified OSF was similarly electrophoresed.

Characterization of Overgrowth Stimulating
Activity in Cell Sonicates

As discussed in the introduction, the overgrowth stimulating activity which
can be obtained by sonicating RSV-transformed cells on days 3–5 after infection increases up to fivefold above that which can be found in sonicates of
normal cells. Two questions therefore arise. (1) What is the nature of
the stimulatory activity found in sonicates of normal cells? And (2) what is
the nature of the additional stimulatory activity to be found in sonicates of

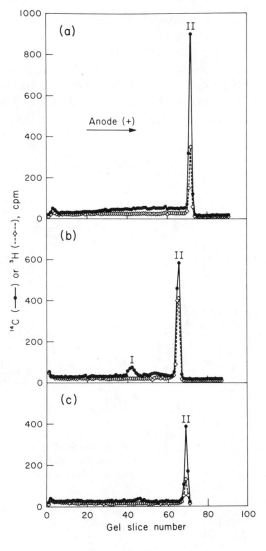

Figure 7

Electrophoresis of Sephadex-purified ¹⁴C-labeled RCM and ³H-labeled NCM. (*a*) Sample A, the fractions preceding the peak of OSF activity; (*b*) sample B, the fractions containing the peak OSF activity; (*c*) the fractions subsequent to the peak of OSF activity. Samples (0.2 ml) were applied to 7.5% acrylamide gels, pH 8.3 (Ornstein and Davis 1964).

RSV-transformed cells? These questions were approached in a preliminary way by examining the chromatographic behavior of the stimulatory activity in normal and transformed cells.

Sephadex Chromatography

RSV-infected and control, uninfected cultures were grown under the usual conditions for the production of OSF. On the fourth day, the cells were harvested and sonicated. After centrifugation, the sonicate was chromatographed on Sephadex G-75. Virtually all of the overgrowth stimulating activity from the sonicate of normal uninfected cells was excluded by Sephadex G-75 (Fig. 8). However, when an equal amount of the sap of sonicated RSV-transformed cells was chromatographed and assayed in this fashion, there was to be found (Fig. 8), in addition to the major peak of OD_{280} and stimulatory activity corresponding to that observed in the normal cells, a peak of overgrowth stimulating activity eluting at the same place on Sephadex G-75 as the activity that is later to be found in the medium. These two peaks of activity (that excluded by and that retained by the column) will henceforth be referred to as "high molecular weight cellular OSF" and "20,000-dalton cellular OSF." No 20,000-dalton cellular OSF was observed in sonicates of either density-inhibited or rapidly growing cells, even when the cell extract was two- to fourfold more concentrated than that shown in Figure 8.

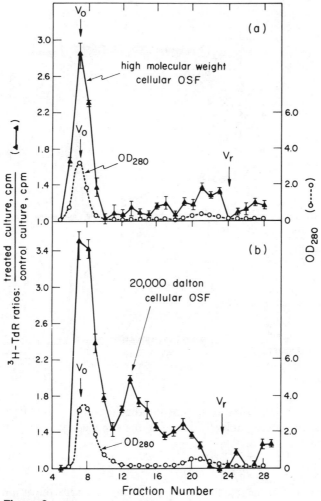

Figure 8

Sephadex G-75 chromatography of normal and RSV-transformed cell sonicates. (a) Control, normal cell sonicate. (b) RSV-transformed cell sonicate. Day-4 cultures were harvested, sonicated, and 0.5 ml (5 mg protein/ml) of the clarified sonicate was applied to a 0.6 × 25-cm column of Sephadex G-75 equilibrated in standard buffer. Fractions of 0.5 ml were collected and 0.2 ml/fraction/plate were assayed for overgrowth stimulating activity.

Chromatography on 6% Agarose

In an effort to determine the molecular weight of the material responsible for the growth stimulatory activity in the fractions excluded by Sephadex G-75, the clarified supernatants of sonicated RSV-transformed and uninfected control cells were chromatographed on molecular sieve gels of progressively larger exclusion limits. Even after chromatography on 6% agarose (exclusion limit: 6×10^6 daltons), the only activity recovered from sonicates of control normal cell cultures was excluded by the column (Fig. 9). In the case of the transformed cell sonicate (Fig. 9), there was found, in addition to this high molecular weight activity, a broad peak of stimulatory activity distributed across the fractions containing material fractionated by the column.

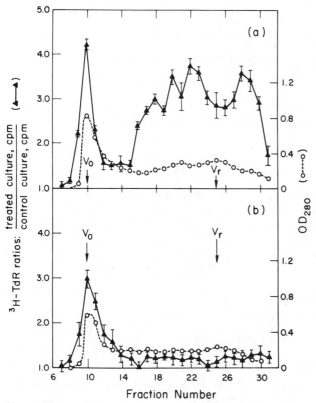

Figure 9

Six percent agarose chromatography of normal and RSV-transformed cell sonicates. (*a*) RSV-transformed cell sonicate. Day-4 RSV-transformed cells were harvested, sonicated, and 0.3 ml (10 mg protein/ml) of the clarified sonicate was applied to a 0.6×25-cm column of 6% agarose equilibrated in standard buffer. Fractions of 0.6 ml were collected, and 0.2 ml/fraction/plate was assayed for overgrowth stimulating activity. (*b*) Control, normal cell sonicate. Procedure and chromatography as described above, but using day-4 control, normal cells.

Table 4

Effects of Hydrolytic Enzymes on Cellular OSF

| | % Activity recovered ($\pm 10\%$) | |
Treatment	high molecular weight cellular OSF	20,000-dalton cellular OSF
Trypsin	100	1
Pronase	20	—
RNase	10	100
DNase	90	95
Hyaluronidase	100	100

An aliquot of the clarified supernatant of sonicated day 4 RSV-transformed cells (protein conc.: 10 mg/ml) was incubated with the indicated hydrolase for 1 hour at room temperature. Trypsin (Worthington, 1×10^4 μ/mg) was used at 100 μg/ml. Pronase (Calbiochem, 45 μ/mg) was autodigested for 1 hour at 37°C to digest nonproteolytic contaminants before addition to sonicate at 200 μg/ml. RNase A (Worthington, 3×10^3 μ/mg) was used at 30 μg/ml; DNase I (Worthington RNase-free, 2×10^3 μ/mg), at 50 μg/ml; hyaluronidase (Calbiochem, 1×10^4 μ/mg), at 60 μg/ml. Control aliquots were incubated with buffer. Treated and control aliquots were then chromatographed on 6% agarose and high molecular weight and 20,000-dalton molecular weight overgrowth stimulating activity were measured.

Enzymatic Sensitivity

In order to further characterize the high molecular weight and 20,000-dalton activities, their sensitivity to specific hydrolytic enzymes was examined (Table 4). High molecular weight cellular OSF was sensitive to pronase and ribonuclease, but not to trypsin, DNase or hyaluronidase. Like the OSF found in the medium, 20,000-dalton cellular OSF was sensitive to trypsin and resistant to RNase.

Absence of Proteolytic Activity

High molecular weight cellular OSF was tested for proteolytic activity on Hide powder azure and the uniformly ^{14}C-labeled mixture of algal (Chlorella) protein at the same time as the activity from Sephadex-purified RCM was examined (Table 3). Slight activity (equivalent to 1–2 ng/ml of trypsin) was detected with Hide powder azure as the substrate, but no activity was detected with the Chlorella protein substrate.

DISCUSSION

OSF in Rous Conditioned Medium

The stimulatory activity in Rous conditioned medium has been shown to be due to at least two proteins of molecular weight on the order of 20,000 daltons. It remains to be determined whether the two proteins are structurally unique and only coincidentally of the same molecular weight, or whether they are essentially identical in structure and function, differing only in charge as a consequence of glycosylation or some other form of modification.

Although the growth stimulatory activity of OSF can be mimicked by tryp-sin and other proteases (Sefton and Rubin 1971), no gross proteolytic activ-ity could be associated with OSF. The possibility remains, of course, that OSF is a protease, but that its action is restricted to the cleavage of only one or two bonds in the substrate protein, such that the product peptides do not become acid soluble; or that the substrate requirements are so specific that a protein representing them is not present in the collection of algal cell proteins that was tested.

Unkeless et al. (1973) have shown that chick embryo cells transformed by the Schmidt-Rupin strain of RSV produce increased amounts of a plasmino-gen activator. As a consequence, plasminogen present in serum is converted to plasmin, a potent trypsinlike enzyme. It appears that plasmin is respon-sible for several features of transformation (Ossowski et al. 1973). The ques-tion naturally arises as to the possible identity of OSF with either plasmin or plasminogen activator. OSF was not tested for its ability to convert plas-minogen to plasmin, or for fibrinolytic activity. There are, however, several differences in the physical properties of OSF and these enzymes (Table 5). On balance, it appears quite unlikely that OSF is plasmin. To the extent that the molecular weight determination by Sephadex could be in error by a factor of two, it might be possible for OSF to be plasminogen activator. In this regard, it should be mentioned that Sephadex-purified OSF stimulates DNA synthesis in density-inhibited cultures that have been washed and then incubated with OSF in the absence of added serum; also, that calf serum may be substituted for chicken serum in the usual agar overlay assay with no effect on the growth stimulation (data not shown). These observations mitigate against the proposal that OSF acts by converting plasminogen to plasmin in the assay medium.

The observation that OSF is effective in stimulating the growth of serum-starved cultures further suggests the possibility that OSF may not be a hydrolytic enzyme, but rather that it could be a serumlike growth-stimulating protein, perhaps insulinlike in activity (Pierson and Temin 1971).

Halpern and Rubin (1970) have demonstrated that chick embryo cells release prealbumin and several other proteins with antigenic properties in common with chicken serum proteins. It is unlikely that the particular pro-teins observed by Halpern and Rubin are related to the overgrowth stimulat-

Table 5

Properties of OSF, Plasminogen Activator and Plasmin

Property	OSF	Plasminogen activator	Plasmin
MW	19,000	40,000	75,000
Effect of dithiothreitol	stabilizes	inactivates	?
Effect of STI	no effect	?	inhibits
Proteolytic activity	not measurable	relatively narrow substrate specificity	relatively broad substrate specificity
Source	cellular	cellular	serum

ing activity in Rous conditioned medium since (1) they migrate in 7.5% acrylamide gels at pH 8.1 at different rates than the ^{14}C band tentatively equated with OSF and (2) they were all observed to be found in higher concentrations in medium conditioned by normal cells than in Rous conditioned medium. Nevertheless, the fact that these cells release at least some proteins similar to those in chicken serum may mean that they are capable, when transformed, of synthesizing proteins similar in activity to the growth stimulating proteins of serum (Holley and Kiernan 1971; Lipton et al. 1971; Paul, Lipton and Klinger 1971).

Also of relevance in this regard are the observations of Shodell (1972), that an L-cell line adapted to grow in protein-free medium apparently does so by virtue of the synthesis of serumlike proteins, since addition of medium conditioned by these cells supports the proliferation of serum-dependent BHK cells under otherwise protein-free conditions.

Bürke (1973; Bürke and Williams 1971) has observed an overgrowth stimulating factor of molecular weight roughly similar to OSF in medium conditioned by SV3T3 cells and a line of BHK cells transformed by SV40. This material, in addition to stimulating the growth of density-inhibited 3T3 cells, also stimulates migration of 3T3 cells in the absence of serum.

OSF in Cell Sonicates

The stimulatory activity of normal cells has been shown to be due largely to material of very high molecular weight ($>10^6$ daltons). This high molecular weight activity is sensitive to pronase and ribonuclease, and is found in roughly equal amounts in both RSV-transformed and normal cells. The additional activity observed in sonicates of RSV-transformed cells prior to the release of OSF into the medium was shown to be due in part to the presence of material of the same molecular weight as OSF; the remaining activity is associated with heterogeneous material of molecular weight intermediate to that of OSF and the very large molecular weight activity characteristic of both Rous-transformed and normal cells.

When a "wound" is made in a confluent sheet of density-inhibited fibroblasts, the cells immediately adjacent to the denuded area, and those which have migrated into it, are found to multiply at an elevated rate. The increased mitotic activity of these cells can be largely accounted for on the basis of the local decrease in population density, but Clarke et al. (1970) observed that the proportion of cells incorporating thymidine at wound edges was greater than that expected by density effects alone. It is possible that the high molecular weight cellular OSF observed in sonicates of normal cells might be released and act as a tissue culture "wound hormone" in this situation.

Role of OSF in Transformation

A number of observations suggest the possibility that OSF might be a "transformation factor" (Bürke and Williams 1971). OSF appears in assayable quantities in the medium of cultures transformed by the Bryan strain of RSV at a time roughly synchronous with the acquisition of transformed cell characteristics by the majority of nontransformed, RAV-infected cells in the culture (Rubin 1970a). When added to confluent cultures of normal cells, OSF stimulates persisting growth under agar at high cell densities. This is one of the major characteristics of RSV-transformed cells in culture and is in fact the basis of the focus-forming assay for RSV. Stimulatory material of the same apparent molecular weight as OSF is found in sonicates of transformed cells, but is not detectable in sonicates of either density-inhibited or rapidly growing normal control cells.

These observations, however, provide at best only circumstantial support for the hypothesis that OSF is directly involved in the mechanism whereby RSV transforms fibroblasts. The precariousness of any conclusions as to the presence or absence of OSF in a complex and relatively unfractionated whole cell *gemisch* on the basis of the elution profile of its biological activity on a Sephadex column is self-evident. Furthermore, although OSF does appear to increase in quantity inside transformed cells prior to its release into the medium, the increase in activity of transformed cell sonicates becomes significant only on the fourth day after infection (Rubin 1970a), which is at least two days after the RSV-infected cells have escaped density-dependent inhibition (Rubin and Colby 1968).

Of course, it is possible that concentrations of OSF too small to be detected in the medium as a whole or in sonicates of transformed cells are locally active at the surface of RSV-infected cells at this time. OSF could therefore be responsible for the escape from density-dependent inhibition by RSV-infected fibroblasts in culture. It does not follow from this that OSF is a product of the RSV genome specifically designed to stimulate host cell growth, or even that the role of OSF in cells in their in vivo environment is that of a growth-stimulating substance.

The basis for this reservation about the in vivo significance of OSF is the extreme sensitivity of chick embryo cells in monolayer culture to growth stimulation by a wide variety of agents acting at their external surface or altering their external milieu (Vasiliev et al. 1970; Rubin and Koide 1973). This sensitivity of cultured chick embryo cells to growth stimulation may not be shared by their counterparts in vivo.

Thus, for example, the "wounding" of a confluent sheet of cells in culture results in mitotic activity which is highest in the cells that have migrated into the denuded area and in the cells immediately adjacent to the free edge. In contrast, an experimentally induced skin wound in the whole animal does not lead to a stimulation of DNA synthesis and cell division in the immediately adjacent dermal fibroblasts and epithelial cells. The epithelial cells that migrate over the exposed dermis are initially mitotically inactive, and the original epidermis immediately adjacent to the wound is slower at developing an elevated rate of mitosis than that further away (Johnson and McMinn 1960). Similarly, the fibroblasts in the granulation tissue arise from the proliferation and migration of perivascular cells of the loose connective tissue; the dermal fibroblasts themselves are inactive (Grillo 1963, 1964).

Evidently, there are overriding growth-regulating influences present in these circumstances that are not in force in the tissue culture environment. It may therefore be wise to distinguish between the specific causes responsible for the release from density-dependent inhibition by cells infected by RSV in vitro (in which OSF may perhaps play a role) and those responsible for release from homeostatic regulation in vivo, even though the two mechanisms may be closely related or have a common origin in some antecedent lesion.

Acknowledgments

This investigation was supported by U.S. Public Health Service Research Grant CA05619 from the National Cancer Institute, by U.S. Public Health Service Training Grant GM01389 from the National Institute of General Medical Sciences, and by U.S. Public Health Service Grant AI01267 from the National Institute of Allergy and Infectious Diseases. The results reported here are taken from work submitted by J.G.B. in partial satisfaction of the requirements for the degree of Doctor of Philosophy. We thank Cynthia Mc-Creary for technical assistance.

REFERENCES

Bissell, M., H. Rubin and C. Hatiè. 1971. Leakage as the source of overgrowth stimulating activity in Rous sarcoma transformed cultures. *Exp. Cell Res.* **68**:404.

Bürke, R. 1973. A factor from a transformed cell line that affects cell migration. *Proc. Nat. Acad. Sci.* **70**:369.

Bürke, R. and C. Williams. 1971. Attempts to isolate SV40 transformation factor. In *Growth Control in Cell Culture* (ed. G. Wolstenholme and J. Knight), p. 107. Churchill Livingston, London.

Clarke, G., M. Stoker, A. Ludlow and M. Thornton. 1970. Requirements of serum for DNA synthesis in BHK/21 cells: Effects of density, suspension and virus transformation. *Nature* **227**:798.

Dickman, S., J. Aroskar and R. Kropf. 1956. Activation and inhibition of beef pancreas ribonuclease. *Biochim. Biophys. Acta* **21**:539.

Greene, L., E. M. Shooter and S. Varon. 1968. Enzymatic activities of mouse nerve growth factor and its subunits. *Proc. Nat. Acad. Sci.* **60**:1381.

Grillo, H. 1963. Origin of fibroblasts in wound healing: An autoradiographic study of inhibition of cellular proliferation by local X-irradiation. *Ann. Surg.* **157**:453.

———. 1964. Aspects of the origin, synthesis and evolution of fibrous tissue in repair. In *Advances in the Biology of Skin* (ed. W. Montagna and R. Billington), vol. 5, p. 128. Macmillan, New York.

Halpern, M. and H. Rubin. 1970. Proteins released from chick embryo fibroblasts in culture. *Exp. Cell Res.* **60**:95.

Holley, R. and J. Kiernan. 1971. Studies of serum factors required by 3T3 cells. In *Growth Control in Cell Culture* (ed. G. Wolstenholme and J. Knight), p. 3. Churchill Livingston, London.

Johnson, F. and R. McMinn. 1960. The cytology of wound healing of body surfaces in mammals. *Biol. Rev.* **35**:364.

Lipton, A. I. Klinger, D. Paul and R. Holley. 1971. Migration of mouse 3T3 fibroblasts in response to a serum factor. *Proc. Nat. Acad. Sci.* **68**:2799.

Lowry, O., N. Rosebrough, A. Farr and R. Randall. 1951. Protein measurement with the Folin phenol reagent. *J. Biol. Chem.* **193**:265.

McClean, D. 1943. Method of assay of hyaluronidase. *Biochem. J.* **37**:169.

Ornstein, L. and B. Davis. 1964. Disc electrophoresis. *Ann. N. Y. Acad. Sci.* **121**:321.

Ossowski, L., J. P. Quigley, G. M. Kellerman and E. Reich. 1973. Fibrinolysis associated with oncogenic transformation. *J. Exp. Med.* **138**:1056.

Paul, D., A. Lipton and I. Klinger. 1971. Serum factor requirements of normal and SV40 transformed 3T3 mouse fibroblasts. *Proc. Nat. Acad. Sci.* **68**:645.

Pierson, R. and H. Temin. 1971. The partial purification from calf serum of a fraction with multiplication stimulating activity for chicken fibroblasts in cell culture and with non-suppressible insulin-like activity. *J. Cell. Physiol.* **79**:319.

Rein, A. and H. Rubin. 1968. Effects of local cell concentration upon the growth of chick embryo cells in culture. *Exp. Cell Res.* **49**:666.

Rubin, H. 1970a. Overgrowth stimulating factor released from Rous sarcoma cells. *Science* **167**:1271.

————. 1970b. Overgrowth stimulating activity of disrupted chick embryo cells and cells infected with Rous sarcoma virus. *Proc. Nat. Acad. Sci.* **67**:1256.

Rubin, H. and C. Colby. 1968. Early release of growth inhibition in cells infected with Rous sarcoma virus. *Proc. Nat. Acad. Sci.* **60**:482.

Rubin, H. and T. Koide. 1973. Stimulation of DNA synthesis and 2-deoxy-D-glucose transport in chick embryo cultures by excessive metal concentrations and by a carcinogenic hydrocarbon. *J. Cell. Physiol.* **81**:387.

Sefton, B. and H. Rubin. 1971. Release from density dependent growth inhibition by proteolytic enzymes. *Proc. Nat. Acad. Sci.* **68**:3154.

Shodell, M. 1972. Environmental stimuli in the progress of BHK/21 cells through the cell cycle. *Proc. Nat. Acad. Sci.* **69**:1455.

Unkeless, J., A. Tobia, L. Ossowski, J. Quigley, D. Rifkin and E. Reich. 1973. An enzymatic function associated with transformation of fibroblasts by oncogenic viruses. *J. Exp. Med.* **137**:85.

Vasiliev, J., I. Gelfand, V. Guelstein and E. Fetisova. 1970. Stimulation of DNA synthesis in cultures of mouse embryo fibroblasts. *J. Cell. Physiol.* **75**:305.

Plasminogen Activator: Biochemical Characterization and Correlation with Tumorigenicity

Judith K. Christman and George Acs

Department of Pediatrics, Mount Sinai School of Medicine
New York, New York 10029

Selma Silagi

Department of Obstetrics and Gynecology, Cornell University Medical College
New York, New York 10021

Samuel C. Silverstein

Department of Cellular Physiology and Immunology
The Rockefeller University, New York, New York 10021

Reich and coworkers have described in detail their investigations of the elevated production of plasminogen activator(s) by avian and mammalian fibroblasts which have been transformed in vitro by oncogenic viruses or chemical carcinogens (Unkeless et al. 1973, 1974; Ossowski et al. 1973; Ossowski, Quigley and Reich 1974; Quigley, Ossowski and Reich 1974; Rifkin et al. 1974). Plasminogen activator(s) are produced by many transformed cells and by a few specialized normal cell lines and tissues (Rifkin et al. 1974; Unkeless, Gordon and Reich 1974). The potential importance of the expression of this enzyme to the understanding of neoplasia derives from the following observations:

1. Most malignantly transformed lines derived in the laboratory, regardless of the transforming agent, demonstrate a significantly increased level of release of plasminogen activator when compared to their normal counterparts (Unkeless et al. 1973; Rifkin et al. 1974).
2. Fibroblasts infected with temperature-sensitive Rous sarcoma virus mutants produce plasminogen activator as early as two hours after the cells are shifted to permissive temperatures (D. B. Rifkin and J. P. Bader, unpubl.), making the expression of plasminogen activator one of the earliest observable events associated with transformation.
3. Tumorigenicity and plasminogen activator disappear concurrently when a mouse melanoma line is grown in the presence of BrdU; these activities reappear together when the drug is removed (Christman et al. 1975).
4. Plasmin is a potent proteolytic enzyme capable of cleaving a wide variety of proteins; other proteases have been shown to alter the functional characteristics of the surfaces and intracellular components of animal cells in several ways: (a) Proteolysis of intact cells enhances lectin-induced agglutinability (Burger 1969) and removes or alters a variety of cell surface proteins and receptors (Glynn, Thrash and Cunningham 1973); and (b) trypsinization of nuclear components increases the

activity of DNA polymerase (Brown and Stubblefield 1974) and the extent of DNA methylation (Tosi and Scarano 1973) in cell-free systems.

Thus the presence of plasminogen activator within transformed cells (Unkeless et al. 1974; Christman and Acs 1974a) could serve a regulatory function, while the presence of this enzyme in the surrounding medium affords the transformed cell a mechanism for changing its chemical milieu, an action by which it might exert a profound effect either on its own fate or that of its normal neighbors. It remains to be determined whether plasminogen activator production is obligatory for the establishment of the transformed state or tumorigenicity, or whether it represents a function whose expression is causally unrelated to malignancy.

In order to clarify the role of plasminogen activators in neoplasia, the following would seem to be minimal requirements: (1) Plasminogen activators from several transformed lines or tumors should be purified and chemically characterized to allow comparison with one another and with known plasminogen activators produced by normal cells (e.g., urokinase); (2) the full range of natural substrates should be determined; (3) the mechanisms of synthesis and processing of both the cell-associated and released forms of the enzyme should be elucidated; and (4) specific inhibitors of the enzyme should be employed to determine the effect(s) of inactivation of plasminogen activator(s) on the processes of transformation and/or on the maintenance of malignancy.

Some progress toward these goals has been made. Several plasminogen activators have been partially or completely purified. They are all serine proteases. Their molecular weights have been determined (SV40-transformed hamster cells, 50,000 [Christman and Acs 1974b]; Rous sarcoma virus-transformed chicken fibroblasts, 39,000 [Unkeless et al. 1974]; human melanoma cells, 48,000 [Rifkin et al. 1974]). Here we review our earlier findings, report an alternative purification and amino acid analysis of the plasminogen activator released from SV40-transformed hamster cells, examine its substrate specificity, and provide evidence for the firm association of this enzyme with a membrane-enriched subcellular fraction. In addition, we have summarized our studies concerning the relationship between tumorigenicity and the presence of plasminogen activator in the BrdU-sensitive clone $B_5 59$ of mouse melanoma B16 (Christman et al. 1975).

Purification and Characterization

Figure 1 shows an analytical SDS-polyacrylamide gel electropherogram of purified plasminogen activator released by SV40-transformed hamster cells. After ion exchange chromatography on SP-C 25 Sephadex, the enzyme was isolated by electroelution from a preparative SDS-polyacrylamide gel as described in detail by Christman and Acs (1974b). Plasminogen activator purified in this way coelectrophoreses exactly with [³H]DFP-labeled plasminogen activator and with Coomassie blue-stainable material. After removing SDS (Weber and Kuter 1971), the enzyme exhibited an isoelectric point of pH 9.5.

Figure 1

Coincidence of [³H]DFP-inactivated enzyme, plasminogen activating activity and Coomassie blue-stainable protein after purification of plasminogen activator by preparative polyacrylamide gel electrophoresis (Stage 6). [³H]DFP-labeled enzyme was coelectrophoresed with active Stage 6 protein. One gel was cut into 1-mm slices; each slice was split in half. One-half was assayed directly for factor activity (o——o), and the other half was digested in 0.5 ml Soluene for 2 hours at 70°C, in order to determine [³H]-DFP radiolabel (●——●). An identical gel was fixed, stained and split longitudinally. One-half was then sliced and assayed for [³H]DFP counts. The alignment of stained bands and counts was assured by the use of fine-wire marker pins during the cutting. The

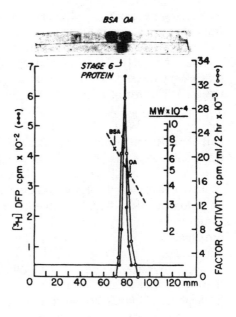

molecular weight was accurately determined by coelectrophoresing [³H]DFP-labeled factor with marker proteins and counting gel slices after staining. BSA, bovine serum albumin; OA, ovalbumin. (Reprinted, with permission, from Christman and Acs 1974a.)

One major protein band appeared at this pH. A minor protein band was found within the region of detectable enzyme activity at approximately pH 9.25. These results, taken together with the specific activity estimates derived from inactivation of the enzyme with radioactive DFP of known specific activity, indicate that the hamster-derived plasminogen activator has been purified 14,000-fold and is essentially homogeneous.

While the enzymatic function of chicken plasminogen activator is unaffected by treatment with 1% SDS for prolonged time periods (Unkeless et al. 1974), the hamster enzyme is slowly and irreversibly inactivated by exposure to this concentration of SDS. Thus during the electroelution process, careful control of the SDS concentration in the elution chamber is necessary. In addition, if the enzyme is to retain activity, SDS must be removed from the preparation, a process which leads to further reduction in yield. All in all, to isolate 30 μg of the 60–70 μg of enzyme present in 8 liters of harvest fluid required almost two weeks using these methods. For these reasons, we have developed an alternative purification scheme utilizing liquid isoelectric focusing, which overcomes some of the problems previously encountered (Christman et al., in prep.).

Plasminogen activator was purified as before by ion exchange chromatography on SP-C 25 Se′ nadex except that the column was washed with 10 mM Tris pH 8.9 prior .o elution of the enzyme. (This procedure facilitates the isofocusing by removing plasminogen and plasmin and reduces the risk of trapping the enzyme by coprecipitation with other proteins during focusing.) Before focusing, the enzyme-containing eluate was dialyzed against 0.05 M glycine buffer pH 3.5 and simultaneously concentrated at least tenfold using a Prodicon vacuum dialysis apparatus (Bio-molecular Dynamics, Beaverton, Oregon). No loss of activity occurs at this step. For the focusing, we used a 30-ml, 9-chamber, high-voltage isofocusing apparatus assembled in our laboratory by Mr. C. Whalen according to the design developed by Dr. W. D. Denckla at the Roche Institute for Molecular Biology (pat. pend.). By prefocusing a 0.5% solution of ampholines and adding the partially purified enzyme preparation to a chamber approximately 1 pH unit below the enzyme's isoelectric point, we have been able to focus the plasminogen activator at pH 9.5 in 1–2 hours. A final voltage of 13,000 was attained with wattage never exceeding 20, to avoid excess heating. All of the activity was recovered after focusing; approximately 80% of the activity was recovered in the chamber containing ampholines of the expected pH range. The contents of this chamber was immediately chromatographed on a Bio-Rad P-60 column to remove ampholines, and the excluded volume was collected directly on a 0.2–0.5-cc SP-C 25 Sephadex column. Removal of ampholine was necessary to prevent loss of enzyme activity. (The activity of the purified plasminogen activator was lost entirely after 24–48 hours exposure to pH 9.5 ampholines. This inactivation was not prevented by lowering the pH of the solution.)

When assayed immediately after elution, 100% of the plasminogen activator was recovered from the P-60 column. However, complete loss of activity occurred in 18 hours unless the enzyme was concentrated. This was achieved by binding the enzyme to SP-C 25 Sephadex. Under these conditions, the enzyme was stabilized and could be eluted in a concentrated form by passing exactly one column volume of 0.5 M $(NH_4)_2SO_4$ into the resin, allowing it to stand for 18–24 hours at 4°C, and then collecting the eluate in 250-μl fractions. Seventy-five to eighty percent of the enzymatic activity present in the isofocusing chamber, or approximately 60% of the initial SP-C 25 Sephadex eluate activity, was recovered by these methods. This compares with a 70% average recovery from the initial SP-C 25 Sephadex eluate using the preparative gel electrophoresis method (Christman and Acs 1974b). Although the maximal capacity for protein of the liquid isofocusing procedure has yet to be determined, the plasminogen activator present in 25 liters of harvest fluid can be purified in 4–5 days, allowing time for assay at all stages. At the final protein concentration obtained (40–50 μg/ml), the enzyme is stable at 4°C for at least 14 days. When examined by analytical gel electrophoresis, this material consists of one major stainable band with a molecular weight of ~50,000 and a minor band of molecular weight ~80,000. The minor band contaminant accounts for less than 5% of the protein present.

The amino acid composition of the plasminogen activator was determined using material which was first isofocused and then recovered from the 50,000 MW band after electrophoresis in a polyacrylamide gel (Table 1). Amino acid compositions of several other serine proteases and plasminogen activators are included for comparison.

The enzyme has a high cysteine content suggesting a significant degree of

Table 1

Amino Acid Composition of Plasminogen Activator from SV40-transformed Hamster Cells Compared with Several DFP-sensitive Proteases

Amino acid	Plasminogen[a] activator	Urokinase[b] S₁	S₂	α-Chymotrypsin[c]	S. griseus[d] trypsin-like	Trypsin[e]
Lys	4.4	6.5	6.7	6.9	2.7	6.3
His	3.0	3.8	4.3	1.0	.41	1.3
Arg	5.7	5.3	4.8	1.5	3.5	.9
Try	—	1.8	2.1	3.9	—	1.8
Asp	8.2	7.2	8.7	10.8	7.5	9.9
Thr	3.3	7.1	6.8	10.8	6.8	4.5
Ser	7.8	8.2	7.5	13.2	5.9	15.3
Glu	11.6	10.6	9.5	7.4	7.2	6.3
Pro	9.8	6.1	6.2	4.4	3.3	3.6
Gly	14.2	8.4	8.9	11.3	11.8	11.2
Ala	5.2	3.8	4.4	10.8	10.8	6.3
Half-Cys	4.9	3.6	4.2	4.9	2.5	5.4
Val	4.5	4.1	5.0	11.3	7.4	7.6
Met	1.4	1.7	1.7	1.0	1.1	0.9
Ile	2.9	5.8	4.4	4.9	3.3	6.7
Leu	6.9	7.7	7.3	9.4	4.6	6.3
Tyr	2.7	4.8	4.3	2.0	3.4	4.5
Phe	2.9	3.5	3.1	2.9	1.2	1.3
Total residues	(456)	267	462	204	241	223

Calculations per 100 residues.

[a] Average of two determinations. Total residues estimated on the basis of 6 Met/molecule. Unpublished data of S. Stein. Determined by fluorescamine method (Udenfriend et al. 1972).

[b] White, Barlow and Mozen (1966).

[c] Hartley and Kauffman (1966).

[d] Jurásěk, Fackre and Smillie (1969).

[e] Walsh and Neurath (1964).

intra- and interchain disulfide linkages. The presence of intrachain disulfide bonds might explain the stability of this protein's enzymatic activity in the presence of SDS and urea. The presence of interchain disulfide linkages is inferred from the capacity of β-mercaptoethanol and dithiothreitol to inactivate the enzyme, converting the active-site-containing polypeptide to a form with an apparent molecular weight of ~25,000 (Christman and Acs 1974b). We are now comparing the amino acid composition of this subunit with that of the whole enzyme to determine whether the 50,000 MW protein is composed of two identical subunits.

Substrate Specificity

As reported by Unkeless et al. (1974), the plasminogen activator released from Rous sarcoma virus-transformed chicken fibroblasts converts chicken plasminogen to plasmin in a manner analogous to the activation of human plasma by urokinase and streptokinase. That is, these plasminogen activators specifically cleave the plasminogen molecule into two polypeptide chains which remain linked together by disulfide bonds.

To determine whether zymogens other than plasminogen are activated by the hamster cell-derived plasminogen activator, chymotrypsinogen and trypsinogen were tested as substrates. No activation occurred with zymogen to activator ratios equal to or double those used for plasminogen activation employing assays where activation of less than 5% of the zymogen would have been detectable (Christman and Acs 1974a).

In another series of experiments, [^{35}S]methionine-labeled, plasma membrane-enriched fractions, prepared by the method of Atkinson and Summers (1971), were isolated from SV40-transformed hamster, SV40-transformed 3T3 and L-M cells and tested as substrates for hamster cell-derived plasminogen activator (Table 2). Less than 1% of the labeled proteins in these fractions were converted to TCA-soluble fragments after incubation with purified plasminogen activator at 37°C for 2 hours. When these

Table 2

Digestion of Plasma Membrane Proteins Derived from Normal and Transformed Cells

	% Counts released (TCA soluble)
SV40 hamster cells	
+ plasminogen activator	0.2
+ plasminogen	10
+ plasminogen and activator	15
SV40 3T3 cells	
+ plasminogen activator	0
+ plasminogen	11
+ plasminogen and activator	17
L-M cells	
+ plasminogen activator	0
+ plasminogen	1
+ plasminogen and activator	16

A plasma membrane-enriched fraction, prepared by the method of Atkinson and Summers (1971), was isolated from 5×10^7 cells which had been grown for 48 hours in medium containing 4 μCi/ml [^{35}S]methionine. The membrane fractions (containing approximately 100,-000 cpm [^{35}S]methionine) were incubated in 0.5 ml 0.1 M Tris-HCl pH 8.1 at 37°C. 4×10^{-11} moles of plasminogen and/or 10^{-12} moles of purified activator were present where indicated. After 2 hours, the reaction was terminated by bringing the mixture to 10% with TCA. Percent counts released = 100 (cpm in Millipore filtrate/cpm in filtrate + cpm in retentate). Radioactivity in filtrates was determined in Aquasol and on dried filters in Liquifluor. Background release of radioactivity by membranes incubated with no additions has been subtracted (less than 1% total).

"digested" membranes were subsequently disrupted with SDS and their proteins analyzed by chromatography on Sephadex G-100 or by electrophoresis on 5% SDS-polyacrylamide slab gels (Summers, Maizel and Darnell 1965), no breakdown of proteins was detectable. If, however, both plasminogen and plasminogen activator were added to the membrane fraction, approximately 15% of the labeled proteins were converted to acid-soluble fragments in 2 hours. Analysis of the protein pattern of these plasmin-digested membranes by Sephadex chromatography or SDS gel electrophoresis revealed that proteins of all size classes had been degraded. Thus the major action of plasminogen activator on the proteins present in these membrane-enriched fractions appears to be mediated indirectly through plasmin.

Cell-associated Plasminogen Activator

As indicated in Table 2, all of the plasma membrane-enriched fractions were alike in their susceptibility to digestion by plasmin, i.e., incubation with plasminogen plus plasminogen activator degraded approximately 15–17% of the radiolabeled proteins. These preparations, however, varied markedly when incubated with plasminogen alone. Radioactive proteins in the plasma membrane-enriched fractions from SV40-transformed hamster or SV40-transformed 3T3 cells were hydrolyzed to TCA-soluble peptides when plasminogen was added alone almost as efficiently as when plasminogen activator was added together with plasminogen. In contrast, plasma membrane-enriched fractions from L-M cells which do not express plasminogen activator were not degraded by incubation with plasminogen alone. The capacity for "self-digestion" in the presence of plasminogen correlated directly with the ability of the membrane preparations to serve as a source of plasminogen activator in the standard fibrinolytic assay (Unkeless et al. 1973) and with the ability of the cells from which they were prepared to digest fibrin when grown as monolayers on radioactive fibrin plates or to release plasminogen activator into serum-free medium (Christman and Acs 1974a).

Unkeless et al. (1974) have reported that in chicken fibroblasts transformed by Rous sarcoma virus, plasminogen activator can be localized in a subcellular fraction sedimenting between 6000 and 20,000g. This activator is stimulated 3–5-fold by addition of 0.5% Triton to the standard assay. Cell-associated activator from SV40-transformed hamster cells sedimented between 600 and 6000g (Table 3); addition of 0.5% Triton to the standard assay had no stimulatory effect. When the 6000g pellet was subfractionated by sedimentation through a sucrose density gradient, all of the recovered plasminogen activator cosedimented with membranes which form an opaque band at a density of approximately 1.17 g/cc. Electron microscopy of this material showed it to be composed mainly of smooth membranes and rough endoplasmic reticulum. There was no visible contamination with either mitochondria or lysosomes. However, about half the enzyme activity was lost during banding in sucrose, and there was no increase in the specific activity of the membrane derived from the 1.17-g/cc band as compared with the 6000g pellet. The hamster cell-associated plasminogen activator was not solubilized when these membrane-enriched fractions were treated with Triton X-100. (Similar results have been reported by Unkeless et al. [1974] for the

Table 3

Fractionation of Cell-associated Plasminogen Activator from
SV40-transformed Hamster Cells

	Total activity[a] ($\times 10^{-6}$)	Total protein (mg)	Specific activity[b] ($\times 10^{-5}$)
Lysate	18	66	2.7
Nuclei	1.6	10	1.6
Postnuclear supernatant[c]	13	56	2.3
6000g Supernatant	2	46	0.45
6000g Pellet	11.2	12.4	9

[a] (^{125}I cpm released in 2 hr at 37°C.) Total volume/aliquot volume. Conditions were chosen such that release of ^{125}I-fibrin was proportional to plasminogen activator added.
[b] Total activity/mg protein in 0.1 M Tris pH 8.1. Protein determined by the method of Lowry et al. (1951).
[c] Nuclei removed at 600g.

chicken cell-associated plasminogen activator.) After treatment of the membrane fraction with 0.5 M KCl and puromycin (Blobel and Sabatini 1971), 70–75% of the plasminogen activator remained sedimentable (10,000g for 10 min) and cosedimented with membraneous material in sucrose density gradients. The small amount of enzyme activity released by this treatment was sedimentable with higher centrifugal force and exhibited a density of more than 1.12 g/cc (D. Kyner, unpubl.). These results suggest that cell-associated plasminogen activator is bound to membranes.

Indirect evidence that the plasminogen activator is bound to the cell surface of transformed cells in such a way that its active site is exposed and functional is given in Figure 2. Plasminogen in Earle's salt solution was exposed to thoroughly washed intact cells for brief periods of time and then assayed for the presence of plasmin. Detectable amounts of plasmin were formed after a 7-minute incubation of plasminogen with intact SV40 hamster cells (Fig. 2), while no detectable plasminogen activator was released from parallel cell cultures incubated under the same conditions in the absence of plasminogen. These results should be interpreted with caution, however, since we do not know whether plasmin or plasminogen stimulate the release of activator from these cells, or whether the small amounts of activator released during this time period are unstable in the absence of other proteins.

Correlation of Tumorigenicity and Plasminogen Activator Release

To explore a possible relationship between expression of plasminogen activator and the capacity of cells to form tumors, we have studied the BrdU-sensitive clone $B_5 59$ of mouse melanoma B16 (Silagi and Bruce 1970; Silagi et al. 1972). Clone $B_5 59$ cells are highly tumorigenic in C57BL/6J mice. When $B_5 59$ cells are grown in the presence of BrdU, they rapidly change from highly melanotic cells growing in piled-up, semicolonial arrays to flat, amelanotic fibroblasts. These morphological changes are accompanied by

Figure 2

Activation of plasminogen by intact cells. Confluent monolayers of SV40 hamster cells growing on 60-mm petri dishes were washed 6 times with 10 ml/wash cold Earle's salts solution. They were then overlayered with 3 ml Earle's salts containing 4 μg/ml plasminogen and incubated at 37°C for the indicated times. The solution was removed and any cells or debris sedimented by centrifugation. One ml was then placed on an ^{125}I-fibrin-coated petri dish and incubated at 37°C for 1 hour to test for plasminogen activation (●——●). To test for release of plasminogen activator into the medium, similar cultures were incubated in Earle's salts in the absence of plasminogen. The medium was removed, centrifuged as above, and plasminogen (4 μg/ml) was added. These solutions were either incubated for 30 minutes at 37°C before assay (o– – –o) or assayed immediately (x——x).

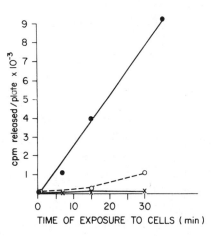

complete suppression of the tumorigenic potential of these cells (Silagi and Bruce 1970; Silagi et al. 1972; Silagi 1971; Wrathall et al. 1973). Neither the plating efficiency nor the ability of these cells to grow in culture is affected sufficiently to account for their failure to form tumors (Silagi, Newcomb and Weksler 1974). Moreover, the reduction in tumorigenicity and alterations in cell morphology induced by BrdU are reproducibly reversed by removing BrdU from the growth medium (Silagi and Bruce 1970; Silagi 1971).

Melanoma cells grown in BrdU for varying periods of time were compared for their ability to digest fibrin when plated on radioactive fibrin-coated petri dishes and for their ability to form tumors in C57BL/6J mice; the two activities were found to decrease in parallel (Table 4). On removal of BrdU from the growth medium, the cells regained both their tumorigenicity and the capability for fibrinolysis. The same pattern was observed for release of plasminogen activator into serum-free medium and the ability of intact cells to activate plasminogen (Christman et al. 1975).

A nontumorigenic clone ($C_3$471) derived from clone $B_5$59 by growth and maintenance in BrdU-containing medium (1 μg/ml) (Silagi and Bruce 1970) was tested for fibrinolytic activity in the presence of four different mammalian sera (dog, rabbit, fetal calf and mouse). $B_5$59 cells released 60% or more of the ^{125}I-labeled fibrin on an assay plate in 18–24 hours in the presence of these sera; $C_3$471 cells were either inactive in releasing ^{125}I-fibrin (dog, fetal calf, mouse sera) or released no more than 2% above background (serum from one rabbit, others proved negative). Neither the cells nor any of the subcellular fractions of $C_3$471 are capable of activating plasminogen at pH 8.1 either in the presence or absence of Triton X-100 (Christman, unpubl.), suggesting that $C_3$471 cells contain no plasminogen

Table 4

Effect of BrdU on B$_5$59 Mouse Melanoma Cells

	% Fibrinolytic activity[a]	% Tumori- genicity[b]	% Viability of cells injected
B$_5$59 melanoma—untreated control	100	100	85–98
B$_5$59—grown 24 hr with BrdU (3 μg/ml)	38–79	50	94–97
B$_5$59—grown 48 hr with BrdU (3 μg/ml)	6–30	0–30	90–97
B$_5$59—48 hr with BrdU (3 μg/ml), then replated and grown without BrdU for 5 days	90–100	100	92–93
C$_3$471—grown 1 year with BrdU (1 μg/ml) (passages 5–12)	0	0	85–98

[a] Assayed by plating cells on ^{125}I-fibrin-coated dishes as described in Table 1.

[b] Tumorigenicity of all B$_5$59 cells was tested by subcutaneous (sc) inoculation into 6-week-old C57BL/6J mice of 2×10^5 cells in all experiments. Tumorigenicity of C$_3$471 cells was tested with 10^6 cells sc. Data for tumorigenicity reflect the combination of several experiments, except those given for 24 hours, in which only 1 experiment was done with 10 mice inoculated. Mice without palpable tumors were observed for at least 70 days postinjection and dissected after being sacrificed to ascertain whether any tumor mass was present.

activator. Nonetheless, until techniques are available to detect plasminogen activator in the absence of its enzymatic activity, we cannot exclude the possibility that the observed loss of plasminogen activator in BrdU-treated mouse melanoma cells is due to the production of an inhibitor or a cell-bound masking activity, or to a conformational change in the cell membranes. However, it is already clear that B$_5$59 cells grown in BrdU do not release inhibitors of either plasmin or plasminogen activator into the serum-free medium, and that BrdU itself is not an inhibitor of either enzyme (Christman et al. 1975).

DISCUSSION

Although plasminogen activators exhibit a high level of biological activity, they represent a very small proportion of total cellular protein. Consequently, they are difficult to obtain in sufficient quantities for detailed structural analyses. We have purified the plasminogen activator released by SV40-transformed hamster cells to homogeneity, shown that it is inhibited by DFP, determined the amino acid composition of the protein, and established that it has subunits linked by disulfide bonds with the DFP-sensitive catalytic site located on the 25,000 MW subunit. Only the 50,000 MW multimer is enzymatically active. Qualitatively similar results have been obtained by Unkeless et al. (1974) in their studies of plasminogen activator from Rous sarcoma virus-transformed chicken fibroblasts. From our studies to date, only plasminogen appears to act as a substrate for the enzyme. Other zymogens, such as trypsinogen and chymotrypsinogen, are not cleaved to active enzymes by the plasminogen activator. Plasminogen activator caused no measurable release of radiolabeled peptides and no alteration in protein pattern when incubated with [^{35}S]methionine-labeled plasma membrane-enriched fractions from hamster, 3T3 or L-M cells. The observed substrate specificity of the enzyme indicates that it is not a general protease. However, our methods are not sensitive enough to detect cleavage of only a few among the many proteins in the membrane fraction. Thus we cannot be certain that the enzyme does not catalyze the cleavage of specific cellular proteins.

Cell-bound plasminogen activator is closely associated with cell membrane fractions. In contrast to the chicken cell enzyme described by Unkeless et al. (1974), the hamster enzyme is not activated by treatment of membrane fractions with the nonionic detergent Triton X-100. Cell-associated plasminogen activator bands in sucrose gradients at a density of approximately 1.17 g/cc. The enzyme is not solubilized by the treatment of particulate cellular fractions with 0.5 M KCl, by dissociation of ribosomes with puromycin, or by treatment of cell fractions with Triton X-100. For these reasons, it seems unlikely that the hamster cell plasminogen activator is stored within membrane-bound vacuoles as a soluble protein.

The capacity of washed intact cells and of the cell membrane fractions to activate plasminogen suggests that the plasminogen activator is present in an enzymatically active form on the cell surface. In this regard, it is of interest that the enzyme recovered in the medium is clearly a soluble protein, and not bound to shed cell membrane fragments, since it cannot be sedimented into a 27–60% sucrose gradient at 37,400g for 17 hours. Under these conditions, membrane fragments are sedimented into the middle of the gradient.

To explore the functional relationship between expression of plasminogen activator and tumorigenicity, we have studied the regulation of the enzyme in the BrdU-sensitive mouse melanoma described by Silagi and Bruce (1970). Our results indicate that tumorigenicity and the expression of plasminogen activator are reversibly and concurrently suppressed to approximately the same extent when these cells are grown in the presence or absence of BrdU. Although the precise mechanism by which BrdU suppresses the expression of plasminogen activator and tumorigenicity in these cells is not known, we have suggested elsewhere (Christman et al. 1975) that it may be similar to the mechanism by which this halogenated pyrimidine inhibits a variety of specialized functions in this and other cell types. Whatever the mechanism, the capacity of BrdU to reversibly inhibit the expression of plasminogen activator and tumorigenicity in B16 mouse melanoma should provide an excellent starting point for studies of the regulation of plasminogen activator synthesis and release and the relationship between the expression of this enzyme and the capacity of these cells to form melanomas in vivo.

Acknowledgments

This work was supported by Grants NP-36K from the American Cancer Society, and grants CA10095, CA16890 and AI08697 from the U.S. Public Health Service. Samuel Silverstein is an Established Investigator of the American Heart Association.

REFERENCES

Atkinson, D. H. and D. F. Summers. 1971. Purification and properties of HeLa plasma membranes. *J. Biol. Chem.* **246:**5162.

Blobel G. and D. Sabatini. 1971. Dissociation of mammalian polyribosomes into subunits by puromycin. *Proc. Nat. Acad. Sci.* **68:**390.

Brown, R. L. and E. Stubblefield. 1974. Enhancement of DNA synthesis in a mammalian cell-free system by trypsin treatment. *Proc. Nat. Acad. Sci.* **71:**2432.

Burger, M. M. 1969. A difference in the architecture of the surface membrane of normal and virally transformed cells. *Proc. Nat. Acad. Sci.* **62:**994.

Christman, J. K. and G. Acs. 1974a. Purification and biological properties of a plasminogen activator characteristic of malignantly transformed cells. In *Lipmann Symposium: Energy, Regulation and Biosynthesis in Molecular Biology* (ed. Dietmar Richter), pp. 150–164. Walter de Gruyter, Berlin/New York.

———. 1974b. Purification and characterization of a cellular fibrinolytic factor associated with oncogenic transformation: The plasminogen activator from SV-40 transformed hamster cells. *Biochim. Biophys. Acta* **340:**339.

Christman, J. K., S. Silagi, E. W. Newcomb, S. Silverstein and G. Acs. 1975. Correlated suppression by 5-bromodeoxyuridine of tumorigenicity and plasminogen activator in mouse melanoma cells. *Proc. Nat. Acad. Sci.* **72:**47.

Glynn, R. D., C. R. Thrash and D. D. Cunningham. 1973. Maximal concanavalin A specific agglutinability without loss of density dependent inhibition of growth. *Proc. Nat. Acad. Sci.* **70:**2676.

Hartley, B. S. and D. L. Kauffman. 1966. Corrections to the amino acid sequence of bovine chymotrypsinogen A. *Biochem. J.* **101:**229.

Jurásěk, L., D. Fackre and L. B. Smillie. 1969. Remarkable homology about the disulfide bridges of a trypsin-like enzyme from *Streptomyces griseus. Biochem. Biophys. Res. Comm.* **37:**99.

Lowry, O. H., N. J. Rosebrough, A. L. Farr and R. J. Randall. 1951. Protein measurement with the folin-phenol reagent. *J. Biol. Chem.* **193:**265.

Ossowski, L., J. P. Quigley and E. Reich. 1974. Fibrinolysis associated with oncogenic transformation. Morphological correlates. *J. Biol. Chem.* **249:**4312.

Ossowski, L., J. C. Unkeless, A. Tobia, J. P. Quigley, D. B. Rifkin and E. Reich. 1973. An enzymatic function associated with transformation of fibroblasts by oncogenic viruses. II. Mammalian fibroblast cultures transformed by DNA and RNA tumor viruses. *J. Exp. Med.* **137:**112.

Quigley, J. P., L. Ossowski and E. Reich. 1974. Plasminogen, the serum proenzyme activated by factors from cells transformed by oncogenic viruses. *J. Biol. Chem.* **249:**4306.

Rifkin, D. B., J. N. Loeb, G. Moore and E. Reich. 1974. Properties of plasminogen activators formed by neoplastic human cell cultures. *J. Exp. Med.* **139:**1317.

Silagi, S. 1971. Modification of malignancy by 5-bromodeoxyuridine. Studies of reversibility and immunological effects. *In Vitro* **1:**105.

Silagi, S. and S. Bruce. 1970. Suppression of malignancy and differentiation in melanotic melanoma cells. *Proc. Nat. Acad. Sci.* **66:**72.

Silagi, S., E. W. Newcomb and M. E. Weksler. 1974. Relationship of antigenicity of melanoma cell growth in 5-bromodeoxyuridine to reduced tumorigenicity. *Cancer Res.* **34:**100.

Silagi, S., P. Beju, J. Wrathall and E. Deharven. 1972. Tumorigenicity, immunogenicity and virus production in mouse melanoma cells treated with 5-bromodeoxyuridine. *Proc. Nat. Acad. Sci.* **69:**3443.

Summers, D. F., J. V. Maizel and J. E. Darnell. 1965. Evidence for virus-specific noncapsid proteins in poliovirus-infected HeLa cells. *Proc. Nat. Acad. Sci.* **54:**505.

Tosi, L. and E. Scarano. 1973. Effect of trypsin on DNA methylation in isolated nuclei from developing sea urchin embryos. *Biochem. Biophys. Res. Comm.* **55:**470.

Udenfriend, S., S. Stein, P. Böhlen, W. Dairman, W. Leimgruber and M. Weigele. 1972. A reagent for assay of amino acids, peptides, proteins, and primary amines in the picomole range. *Science* **178:**871.

Unkeless, J., S. Gordon and E. Reich. 1974. Secretion of plasminogen activator by stimulated macrophages. *J. Exp. Med.* **139:**834.

Unkeless, J., K. Danø, G. M. Kellerman and E. Reich. 1974. Fibrinolysis associated with oncogenic transformation. Partial purification and characterization of the cell factor, a plasminogen activator. *J. Biol. Chem.* **249:**4295.

Unkeless, J. C., A. Tobia, L. Ossowski, J. P. Quigley, D. B. Rifkin and E. Reich. 1973. An enzymatic function associated with transformation of fibroblasts by oncogenic viruses. I. Chick embryo fibroblast cultures transformed by avian RNA tumor viruses. *J. Exp. Med.* **137:**85.

Walsh, K. and H. Neurath. 1964. Trypsinogen and chymotrypsinogen as homologous proteins. *Proc. Nat. Acad. Sci.* **52:**884.

Weber, K. and D. Kuter. 1971. Reversible denaturation of enzymes by sodium dodecyl sulfate. *J. Biol. Chem.* **246:**4504.

White, W. F., G. H. Barlow and M. M. Mozen. 1966. The isolation and characterization of plasminogen activators (urokinase) from human urine. *Biochemistry* **5:**2160.

Wrathall, J. R., C. Oliver, S. Silagi and E. Essner. 1973. Suppression of pigmentation in mouse melanoma cells by 5-bromodeoxyuridine. *J. Cell Biol.* **57:**406.

Macromolecular Determinants
of Plasminogen Activator Synthesis

Daniel B. Rifkin, Leslie P. Beal* and E. Reich

Department of Chemical Biology, The Rockefeller University
New York, New York 10021

It has been demonstrated that enhanced production of plasminogen activator (PA) accompanies neoplastic transformation of primary or early passaged fibroblasts. Thus transformation by MSV, RSV or SV40 is associated with large (usually at least 50-fold) increases of plasminogen activator content above the level found in normal cells. This increase in PA does not occur following infection with cytocidal viruses or leukemia viruses and is therefore related to transformation rather than to virus infection or cell lysis (Unkeless et al. 1973; Ossowski et al. 1973b). The activation of plasminogen that takes place in cultures of transformed cells has been shown to influence the cellular phenotype (Ossowski et al. 1973a,b, 1974); thus growth in semisolid media, cell migration and cell morphology are reduced significantly if the formation and/or activity of plasmin are inhibited. Certain of these parameters of plasminogen activator production and cell phenotype are described elsewhere in this volume (Reich; Ossowski, Quigley and Reich; Pollack et al.; Goldberg, Wolf and Lefebvre; Christman et al.; Danø and Reich; Roblin, Chou and Black).

In view of the association between plasminogen activator synthesis and transformation, we have initiated experiments to define some of the factors that may be involved in control of activator synthesis. We have employed chick embryo fibroblasts (CEF) infected with a temperature-sensitive mutant of Rous sarcoma virus, ts 68. Cells infected with ts 68 and grown at 41°C are normal both by morphological and biochemical criteria, whereas the same cells grown at 36°C are transformed by the same criteria (Kawai and Hanafusa 1971). Because these phenotypic properties are fully reversible within a few hours following appropriate shifts in temperature, this system is a very favorable one for exploring variables that might govern activator production.

* Present Address: New York University, School of Medicine, New York, New York 10016.

842 D. B. Rifkin, L. P. Beal and E. Reich

METHODS

Secondary chick embryo fibroblasts were infected as described previously (Rifkin and Reich 1971) with virus generously provided by S. Kawai. Eagle's minimal medium supplemented with 10% fetal bovine serum was used throughout. The infected cells were subcultured and plated into 60-mm petri dishes at a density of 8×10^5 cells per 60-mm dish. The cells were incubated at either 41 or 36°C for 24 hours before the initiation of the experiment. At the beginning of each experiment, the medium was removed and replaced with fresh medium containing the appropriate drug(s) warmed to the proper temperature. The culture dishes were then incubated at the indicated temperature for the desired periods. At the end of an experiment, the medium was removed, the cell monolayer first washed twice with isotonic buffer and then removed with a policeman, and the cells collected by centrifugation at 2000g for 3 minutes. After centrifugation, the supernatant solution was removed by aspiration and the cells were dissolved in 0.3 ml of 0.50% Triton X-100 in 0.1 M Tris-HCl pH 8.1. Fifty microliters of this cell extract was added to a 35-mm petri dish containing 2 ml of Tris buffer (0.1 M, pH 8.1) supplemented with chicken plasminogen (5 μg/ml); the petri dish was coated with ^{125}I-fibrin (10 μg/cm^2; 50,000 cpm) (Unkeless et al. 1973). The values for each time point represent an average of two determinations performed on the cells removed from each of two separate petri dishes. The dishes were incubated for 3 hours at 37°C, and the supernatant fluid was then assayed for solubilized ^{125}I in a gamma-counter. The drugs were obtained from the following sources: actinomycin, Merck Sharp and Dohme, Rahway, N.J.; 5-bromotubercidin, Drs. H. B. Wood and R. Engle, Drug Development Branch, NCI, Bethesda, Md.; cycloheximide, Sigma Chemical Company, St. Louis, Mo.

RESULTS

When CEF-ts 68 cells were shifted from 41 to 36°C, increases in plasminogen activator could first be detected after 2–3 hours (Fig. 1). The intracellular activity then increased linearly for approximately 5 hours, after which it remained constant. The kinetics of increase in plasminogen activator were similar to other biochemical changes that accompany similar temperature shifts of CEF-ts 68 cells (Kawai and Hanafusa 1971). The increase in intracellular activator occurred 2–4 hours before changes in extracellular fibrinolysis (Unkeless et al. 1973); this difference is likely due to several factors, such as extracellular dilution, inhibition by components of the medium and transit time out of the cell (D. B. Rifkin, unpubl.).

Also shown in Figure 1 are the levels of plasminogen activator in cells that were shifted from the permissive to the nonpermissive temperature. The loss of activity was rapid, with a half-time of 1–2 hours. This might have been produced in several ways, including the possibility that the activator was itself temperature sensitive. Previous work had indicated that the properties of plasminogen activator were determined by the cell rather than by the transforming agent (Ossowski et al. 1973b), and a temperature-sensitive enzyme would therefore be an unexpected finding; nevertheless, an experiment was

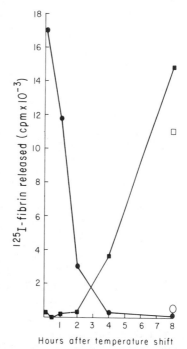

Figure 1
Intracellular levels of plasminogen activator in chick embryo fibroblasts infected with RSV ts 68. Chick embryo fibroblasts infected with RSV ts 68 were grown either at 41 or 36°C for at least one week, then trypsinized and plated at the appropriate temperature one day before initiation of the experiment. At zero time, the medium was removed from the cultures and replaced with fresh medium that had been warmed to the appropriate temperature; the cultures were then placed either at 36 or 41°C. At the indicated times, the medium was removed and the cells washed, scraped and centrifuged. The cell pellets were assayed for plasminogen activator as described in the text. (•——•) 36 → 41°C, RSV ts 68; (■——■) 41 → 36°C, RSV ts 68; (o) CEF, 41 or 36°C; (□) RSV-SR-A, 41 or 36°C.

performed (by S. T. Rohrlich) to test for this (Fig. 2). Serum-free conditioned medium was prepared from cultures infected and transformed either by wild-type Schmidt-Ruppin virus or by ts 68 (at 36°C); separate aliquots of the media were then incubated at 36 and 41°C, respectively, and the plasminogen activator activity assayed after different time intervals. The plasminogen activator activity from both cultures does not decrease significantly at either incubation temperature (Fig. 2).

To obtain additional insight into the factors that might govern the loss of plasminogen activator, the effects of several inhibitors were examined; these were added to the cultures at the time of the shift from the permissive (36°C) to the nonpermissive temperature (41°C) (Fig. 3). When cycloheximide addition accompanied the shift to 41°C, the subsequent loss of enzyme was accelerated: over one-half of the activity had disappeared within 30 minutes. In contrast, cells treated with actinomycin retained a large proportion of intracellular activator for long periods. This retention of activity was not affected by actinomycin concentration, since cells treated with higher (10 μg/ml) or lower (0.4 μg/ml) concentrations responded in a similar way (D. B. Rifkin and L. P. Beal, unpubl.). These results suggest (1) that the maintenance of intracellular enzyme depends on continuing protein synthesis, and (2) that enzyme disappearance following the temperature shift requires RNA synthesis. The rapid early drop in cellular enzyme content in the presence of actinomycin varied in different experiments, and the basis for this variation so far remains undefined.

As a further test for the presumed RNA synthesis requirement, cells were treated with 5-bromotubercidin, another inhibitor of DNA-dependent RNA synthesis. This compound is known to inhibit the synthesis of both mRNA

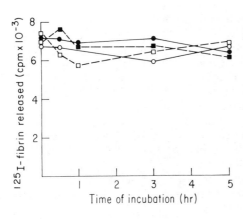

Figure 2

Temperature sensitivity of **RSV-SR-A** and RSV ts 68 induced plasminogen activators. Serum-free conditioned medium was prepared from chick embryo fibroblasts infected with either wild-type or temperature-sensitive virus at least one week before the initiation of the experiment and grown at 36°C. This serum-free conditioned medium was then incubated at either 36 or 41°C for the indicated times and assayed for plasminogen activator. (•——•) RSV-SR-A, PA incubated at 41°C; (o——o) RSV-SR-A, PA incubated at 36°C; (■---■) RSV ts 68, PA incubated at 41°C; (□---□) RSV ts 68, PA incubated at 36°C.

and rRNA, but in contrast to actinomycin, the effect of BrTu is fully reversible (Brdar, Rifkin and Reich 1973). The action of BrTu on plasminogen activator levels (Fig. 4) was very similar to that of actinomycin; enzyme levels remained high following the temperature shift up in the presence of either inhibitor. The reversibility of the BrTu effect is seen in Figure 5. In this experiment, the rapid early loss of enzyme accounted for approximately 50% of the total, and the remainder was stable in the presence of BrTu. However, removal of BrTu either at 2 or 5 hours after the temperature shift was followed by a further rapid drop in the level of activator. This is a further indication of the requirement for RNA synthesis in the disappearance of plasminogen activator that accompanies the loss of transformation.

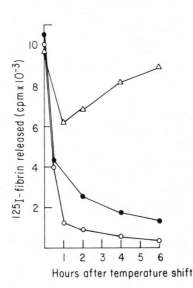

Figure 3

The effect of inhibitors of macromolecule synthesis on plasminogen activator levels in chick embryo fibroblasts infected with ts 68. Chick embryo fibroblasts infected with ts 68 were grown at 36°C. At zero time, the culture medium was removed and replaced with medium warmed to 41°C. One-third of the cultures received medium containing actinomycin (1 μl/ml); one-third, medium containing cycloheximide (20 μg/ml); and one-third, medium alone (controls). The cultures were then placed at 41°C, and at the appropriate times, the cells were scraped and assayed for plasminogen activator as described in the text. (•——•) Control; (△——△) actinomycin (1 μl/ml); (o——o) cycloheximide (20 μl/ml).

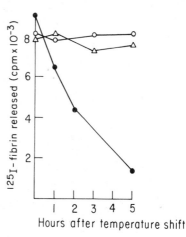

Figure 4
The effect of bromotubercidin on the intracellular level of plasminogen activator. Chick embryo fibroblasts infected with ts 68 and grown at 36°C were treated in a manner similar to those described in the legend to Figure 3, except that one set of cultures received medium containing bromotubercidin (10 μg/ml); one set, actinomycin (1 μg/ml); and the third served as control. (●——●) Control; (○——○) actinomycin; (△——△) bromotubercidin.

The maintenance of elevated levels of activator when RNA synthesis is blocked following the upward temperature shift depends on continuing protein synthesis. This is shown by cycloheximide additions to actinomycin-treated cultures in which high intracellular levels of activator normally persisted for many hours (Fig. 6). The effect of cycloheximide led to a rapid drop in cellular activator content, indicating that the enzyme level was being maintained by de novo protein synthesis. This conclusion is also implied by the observation that in addition to persisting within cells, plasminogen activator was continually secreted into the medium by actinomycin-treated cultures at 41°C (D. B. Rifkin, unpubl.).

DISCUSSION

Changes in incubation temperature of cultures infected with the temperature-sensitive virus ts 68 are known to condition the expression of the transformed

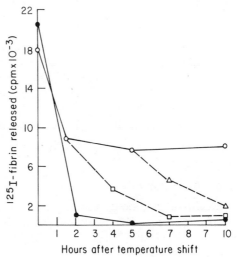

Figure 5
The reversible effect of bromotubercidin. Cultures of chick embryo fibroblasts infected with ts 68 were treated precisely as described in the experiments illustrated in Figures 3 and 4. However, at 2 and 5 hours after the temperature shift, the bromotubercidin was removed from some sets of cultures and replaced with normal medium. (●——●) Control; (○——○) bromotubercidin (10 μg/ml); (□----□) bromotubercidin washed out at 2 hours; (△----△) bromotubercidin washed out at 5 hours.

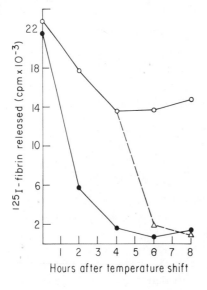

Figure 6
The effect of actinomycin and cycloheximide on the intracellular level of plasminogen activator. Cultures of chick embryo fibroblasts infected with ts 68 and grown at 36°C were treated as described in previous experiments. To one set of cultures which had received actinomycin at zero time, cycloheximide was added at 4 hours after the temperature shift. (•——•) Control; (o——o) actinomycin (10 μg/ml); (△−−−△) cycloheximide (20 μg/ml).

phenotype. As already reported elsewhere (Unkeless et al. 1973), these changes are associated with marked fluctuations in plasminogen activator production. The results presented above show that these changes occur very rapidly following temperature shifts, and the formation of plasminogen activator in such cultures is closely correlated with the expression of the viral transforming function.

Previous results (Unkeless et al. 1973a) had demonstrated the requirement for mRNA synthesis to precede plasminogen activator production following downward temperature shifts to conditions permissive for transformation. The present results show a comparable requirement for RNA synthesis to precede termination of activator formation in the converse situation, i.e., when cultures are shifted upward to temperatures that are nonpermissive for transformation, but we do not know the nature of the relevant RNA species.

The ability of actinomycin to limit the disappearance of inducible enzymes following transfer to noninducing conditions has been reported for several systems (Tomkins et al. 1972; Vilcek and Havell 1973; Whitlock and Gelboin 1973). Several interpretations have been proposed as explanations of these phenomena, including possible decreases in catabolism of specific proteins (Reel, Lee and Kenney 1970) and inhibition of synthesis of an inferred RNA species that turns over rapidly and might function as an inhibitor of translation (Tomkins et al. 1972). Our data are as yet insufficient to discriminate between these hypotheses or to eliminate other plausible explanations.

Acknowledgments

Daniel B. Rifkin holds a Faculty Research Award PR-99 from the American Cancer Society. This work was supported by grants from the National Institutes of Health, CA-13138, CA-08290-10, and from the Council for Tobacco Research.

REFERENCES

Brdar, B., D. B. Rifkin and E. Reich. 1973. Studies of Rous sarcoma virus: Effects of nucleoside analogues on virus synthesis. *J. Biol. Chem.* **248:**2397.

Kawai, S. and H. Hanafusa. 1971. The effects of reciprocal changes in temperature on the transformation state of cells infected with a Rous sarcoma virus mutant. *Virology* **46:**470.

Ossowski, L., J. P. Quigley and E. Reich. 1974. Fibrinolysis associated with oncogenic transformation. Morphological correlates. *J. Biol. Chem.* **249:**4312.

Ossowski, L., J. P. Quigley, G. M. Kellerman and E. Reich. 1973a. Fibrinolysis associated with oncogenic transformation. Requirement of plasminogen for correlated changes in cellular morphology, colony formation in agar, and cell migration. *J. Exp. Med.* **138:**1056.

Ossowski, L., J. C. Unkeless, A. Tobia, J. P. Quigley, D. B. Rifkin and E. Reich. 1973b. An enzymatic function associated with transformation of fibroblasts by oncogenic viruses. II. Mammalian fibroblast cultures transformed by DNA and RNA tumor viruses. *J. Exp. Med.* **137:**112.

Reel, J. R., K-L. Lee and F. T. Kenney. 1970. Regulation of tyrosine α-ketoglutarate transaminase in rat liver. *J. Biol. Chem.* **245:**5800.

Rifkin, D. B. and E. Reich. 1971. Selective lysis of cells transformed by Rous sarcoma virus. *Virology* **45:**172.

Tomkins, G. M., B. B. Levinson, J. D. Baxter and L. Dethlefsen. 1972. Further evidence for posttranscriptional control of inducible tyrosine aminotransferase synthesis in cultured hepatoma cells. *Nature New Biol.* **239:**9.

Unkeless, J. C., A. Tobia, L. Ossowski, J. P. Quigley, D. B. Rifkin and E. Reich. 1973. An enzymatic function associated with transformation of fibroblasts by oncogenic viruses. I. Chick embryo fibroblast cultures transformed by avian RNA tumor viruses. *J. Exp. Med.* **137:**85.

Vilcek, J. and E. A. Havell. 1973. Stabilization of interferon RNA activity by treatment of cells with metabolic inhibitors and lowering of the incubation temperatures. *Proc. Nat. Acad. Sci.* **70:**3909.

Whitlock, J. P. and H. U. Gelboin. 1973. Induction of aryl hydrocarbon (benzo [a] pyrene) hydroxylase in liver cell culture by temporary inhibition of protein synthesis. *J. Biol. Chem.* **248:**6114.

Glucocorticoid Inhibition of the Fibrinolytic Activity of Tumor Cells

Michael Wigler, John P. Ford and I. Bernard Weinstein

Institute of Cancer Research and Departments of Medicine and Microbiology
Columbia University College of Physicians and Surgeons, New York, New York 10032

We have recently discovered that glucocorticoid hormones can rapidly and virtually completely inhibit the production of plasminogen activator by certain cells (in particular, rat hepatoma cells) grown in vitro. Our studies began with the observation that dexamethasone (Dex), a potent synthetic glucocorticoid hormone, induces a phenotypic reversion in mouse L cells from a transformed, or tumor cell-like state to a more tightly regulated growth state (Ford, Wigler and Weinstein, in prep.). (In this paper, "plasminogen activator" refers to a cell factor which, in the presence of serum, leads to the lysis of fibrin in the assay system described by Reich and coworkers [Quigley, Ossowski and Reich 1974; Unkeless et al. 1974].)

METHODS

The cell lines used in the present study have been described in detail elsewhere (Weinstein et al. 1975a,b; Bomford and Weinstein 1972; Yamaguchi and Weinstein 1975; Ford, Wigler and Weinstein, in prep.). Cell cultures were grown as monolayers on plastic petri dishes. Growth media was either Dulbecco's modified minimal essential medium supplemented with 5% fetal calf serum (for 5E, W8 and ts 223 cells) or Ham's F-12 supplemented with 10% fetal calf serum (for HTC, K-16 and H-4 cells). Cells were fed every two days and passaged at low dilution upon reaching confluence.

Assays for plasminogen activator were essentially those described by Reich and coworkers (Unkeless, Gordon and Reich 1974). For measurement of intracellular activity, cell lysates were prepared as follows: Confluent or subconfluent cells in either 5- or 9-cm plastic petri dishes were washed once with phosphate-buffered saline (PBS) and then scraped into 1 ml of 0.2% Triton X-100 in water. Lysates were stored at $-20°C$ until use. Activity was assayed by incubating 0.2 ml of cell lysate, 0.8 ml of 100 mM Tris·HCl pH 8.0 and 0.025 ml of monkey serum (as plasminogen source) at 37°C in ^{125}I-fibrin-

coated 3.5-cm dishes (10 μg/cm^2, 50,000 cpm/dish). At various times thereafter, 0.2-ml aliquots of incubation mixture were sampled and the solubilized
^{125}I measured by counting in a Searle gamma-counter. Control dishes contained 0.8 ml Tris buffer, 0.2 ml 0.2% Triton X-100 in water and 0.025 ml
monkey serum. Activity was calculated as the cpm solubilized by the cell
lysate × 100, divided by the total cpm solubilized by trypsin, and expressed
as percent. Data were corrected for control dishes in which 2–6% of the cpm
was solubilized. The amount of ^{125}I solubilized was approximately a linear
function of assay time if no more than 50% of the total fibrin was digested.
In any given experiment, cell lysates were made from replicate cultures. Each
data point represents the average of two replicate incubations of two replicate
cell lysates. The data presented in this paper have not been normalized to the
amount of cell protein in lysates, and therefore the results are only semiquantitative.

RESULTS

Figure 1 illustrates the effects of Dex at 10^{-7} M on the appearance of L cells
when grown to confluence in plastic petri dishes and on the growth of these
cells in suspension in soft agar. Note that untreated monolayer cultures (Fig.

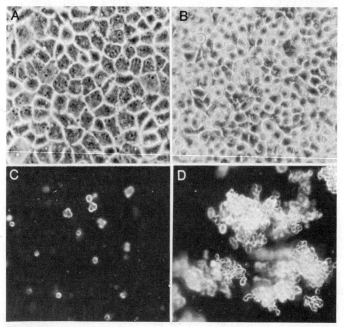

Figure 1
Effects of dexamethasone on L cell morphology and growth.
(*A,B*) Phase contrast micrographs (90 ×) of L cells grown in
monolayer to confluence: (*A*) in presence of 2 × 10^{-7} M dexamethasone; (*B*) control. Darkfield micrographs (36 ×) of L
cells grown in 0.4% agar suspension: (*C*) in presence of 2 ×
10^{-7} M dexamethasone; (*D*) control.

Table 1

Effects of Dexamethasone on L Cells

1. Round → flat cells
2. ↓ Piling up
3. ↓ Saturation density
4. ↑ Resistance to trypsinization
5. ↑ Viability at confluence
6. ↓ Growth in soft agar
7. ↓ Production of plasminogen activator

1B) contained cells that were rounded, densely packed and tended to show piling up, whereas treated cells (Fig. 1A) were flat, polygonal and less densely packed. A comparison of C and D in Figure 1 illustrates the marked inhibition of growth in soft agar suspension produced by Dex. Table 1 is a partial list of the effects of Dex that we have observed on the phenotype of L cells (Ford, Wigler and Weinstein, in prep.). We were intrigued that a marked decrease in fibrinolytic activity was found to accompany phenotypic reversion in this system and therefore have pursued this aspect further. Since L cells are not extremely high producers of plasminogen activator, we first screened other lines for a similar effect.

We examined rat hepatoma cell lines because of their well-studied responses to glucocorticoids and the general interest of our laboratory in liver cell lines. Two cell lines (HTC and H-4) derived from hepatomas induced in rats by chemical carcinogens and a rat liver cell line (5E) transformed in culture by murine sarcoma virus (MSV) (Bomford and Weinstein 1972) produced large amounts of plasminogen activator, and this activity was suppressed partially in H-4 and virtually completely in 5E and HTC by 10^{-7} M Dex. We also found that a normal epithelial rat liver cell line (K-16), a tumorigenic cell line (W-8) obtained from K-16, and a temperature-sensitive variant (ts 223) of W-8 (Yamaguchi and Weinstein 1975) all had low plasminogen activator levels even in the absence of Dex.

We chose the HTC cell line for more detailed studies on Dex inhibition of plasminogen activator production for several reasons: HTC has glucocorticoid-inducible tyrosine aminotransferase (TAT), and the response of this system to a variety of steroids has been extensively studied (Baxter and Tomkins 1971). In addition, low concentrations of Dex do not produce a gross inhibition of protein synthesis in this cell line (Thompson, Tomkins and Curran 1966), nor does prolonged culture in the presence of Dex produce growth inhibition (pers. obs.).

Table 2 lists the results when a variety of steroids were assayed for activity in suppressing plasminogen activator production in HTC cells. These results demonstrated that this effect did not occur with all steroids, but appears to be specific for glucocorticoids.

Figure 2 illustrates dose response curves comparing Dex to the less potent compound cortexolone (11 deoxycortisol). The data indicate that Dex was 20 times more potent than cortexolone on a molar basis. The dose response curves are typical for steroid hormones and presumably reflect the satura-

Table 2

Inhibition of Plasminogen Activator by Various Steroids

Addition[a]	Fibrinolytic activity (%)	Inhibition (%)
None	40	0
Testosterone (10^{-5} M)	38	0
Estradiol (10^{-5} M)	50	0
14-Hydroxy cortexolone (10^{-5} M)	23	43
Cortexolone (10^{-5} M)	3	92
Dexamethasone (10^{-7} M)	1	>97
Cortisol (10^{-6} M)	0	>97
Dibutyryl cAMP (10^{-3} M)	64	0

[a] All test agents added to replicate cultures of HTC cells 24 hours in advance of assaying for intracellular plasminogen activator. See text for details.

tion of high affinity receptors present in the cytoplasm of the HTC cells (Baxter and Tomkins 1971).

Figure 3 indicates the time course of onset of the Dex effect and compares this to untreated cells and cells treated with either actinomycin D (10 μg/ml) or cycloheximide (20 μg/ml). At zero time, media from replicate cultures were replaced with fresh media either containing no additions (control) or the indicated compound. Intracellular plasminogen activator was assayed at the indicated times. The fluctuations in the control are due to the change in the growth medium, which tends to transiently depress activator levels. Cycloheximide produced an almost immediate and rapid fall in activity, indicating

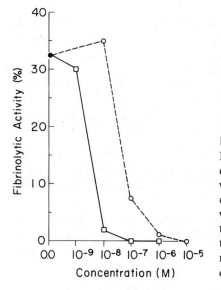

Figure 2
Dose response curves of HTC cells to dexamethasone and cortexolone. HTC cells were grown as replicate cultures, and subconfluent cultures were treated for 24 hours with either dexamethasone (□) or cortexolone (o) at the indicated concentrations before assaying for intracellular plasminogen activator. (●) Indicates activity of untreated cultures.

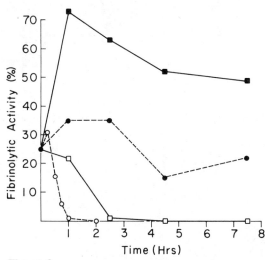

Figure 3
Time course of dexamethasone effect compared
to that of various inhibitors. At zero time, HTC
cells in replicate cultures were refed fresh growth
medium with either 10 μg/ml actinomycin D (■),
20 μg/ml cycloheximide (o), 10^{-7} M dexamethasone (□), or fresh medium without additions (•).
At the times indicated on the abscissa, cultures
were assayed for intracellular plasminogen activator.

a short intracellular half-life of preformed plasminogen activator. The response
to Dex, however, revealed a lag of about one hour followed by a very rapid
decline in activity, so that by 2.5 hours, the level was almost undetectable. As
can be seen, actinomycin D produced a 2–3-fold elevation in plasminogen
activator. Based on these results, we have tentatively concluded that Dex
exerts its inhibitory effect on plasminogen activator production at the level of
translation (or possibly posttranslation), rather than at the level of transcription.

Data presented in Table 3 indicate that the Dex effect was dependent on
de novo transcription. Replicate cultures were either untreated at 0 hours or
treated with the indicated agents at 0 or 1 hour, and the intracellular levels of
plasminogen activator were measured at 5 hours. Actinomycin D (10 μg/ml),
as noted above, and cordycepin (40 μg/ml) led to a 2–5-fold elevation in
plasminogen activator. The latter concentration of cordycepin is known to
block processing of poly(A)-containing mRNA (Darnell et al. 1971). When
either actinomycin D or cordycepin was given one hour prior to Dex, the
ability of Dex to decrease the intracellular level of plasminogen activator was
completely blocked.

Dex inhibition of plasminogen activator was reversible, as can be seen in
Figure 4. Cells were either untreated or pretreated for 4, 24 or 72 hours with
Dex. At zero time, media were removed and replaced with fresh media with-

Table 3

Effects of Drug Combinations

Addition		
0 hr	1 hr	% Activity
None	none	8
None	Dex	1
Act	none	18
Act	Dex	17
Cord	none	44
Cord	Dex	47

At the indicated times, replicate cultures of HTC cells received 10^{-7} M dexamethasone (Dex), 10 μg/ml actinomycin D (Act) or 40 μg/ml cordycepin (Cord). At 5 hours, cells were assayed for intracellular plasminogen activator.

out Dex. At 4, 12, 24 and 48 hours later, the intracellular levels of plasminogen activator were measured. It is apparent that following the removal of Dex, there was a gradual return of plasminogen activator. A curious feature is that recovery was slower in those cells pretreated with Dex for longer periods of time. The underlying mechanism for this is not known at present, but the kinetics of the return of activity are not entirely compatible with the accumulation of a single inhibitor.

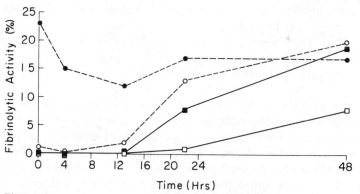

Figure 4

Recovery of activity after removal of dexamethasone. HTC cells in replicate cultures were either pretreated with 2×10^{-7} M dexamethasone for 72 (□), 24 (■) or 4 hours (○), or were untreated (●). At zero time, all cultures were washed three times with PBS and refed growth medium without dexamethasone. Cultures were refed after 24 hours. At the indicated times, cultures were assayed for intracellular plasminogen activator.

DISCUSSION

The following is our currently favored model for explaining Dex inhibition of plasminogen activator production. Dex enters the cell, combines with the cytoplasmic glucocorticoid receptor, and then passes into the nucleus where the receptor-hormone complex causes increased production of an RNA species that directly blocks (or codes for a protein that blocks) translation of the plasminogen activator mRNA. This proposal is based on the following interpretation of the effects of actinomycin D and cordycepin; namely, that the mRNA for the plasminogen activator is stable for at least 7–8 hours, a span in excess of the time needed for occurrence of the Dex effect. The possibility that actinomycin D and cordycepin both artifactually increase the longevity of plasminogen activator mRNA has not been ruled out.

The model proposed is similar to that which Tomkins and others used to explain the observation that actinomycin D blocked the deinduction of TAT synthesis in cells preinduced with Dex (Tomkins et al. 1972). Other models compatible with the present data have not been excluded. Further studies are required to elucidate possible direct or indirect effects of Dex on plasminogen activator translation, posttranslational modification or degradation, or the production of specific inhibitors. We must emphasize that the addition of Dex to the assay system itself does not inhibit fibrinolysis, and therefore, the Dex effect is exerted at the cellular level.

Our results raise the question of whether the mechanisms that maintain the lower level of plasminogen activator in certain normal cells, when compared to their transformed counterpart (Unkeless et al. 1974), are similar to those that mediate the glucocorticoid-induced decrease in plasminogen activator we have observed in transformed liver cells. The effects of actinomycin D on plasminogen activator levels in chick fibroblasts transformed by a temperature-sensitive strain of Rous sarcoma virus (observed by Rifkin, Beal and Reich, this volume) suggest that there are similarities.

Acknowledgments

The authors are indebted to Drs. D. Rifkin and E. Reich for helpful suggestions throughout the course of these studies, for sharing with us their unpublished data, and for providing the ^{125}I-fibrinogen plates used in the present assays. We thank K. Zachary for assistance in these studies. This work was supported by a National Cancer Institute contract, No. 72-3234, and research grant CA 02332.

REFERENCES

Baxter, J. D. and G. M. Tomkins. 1971. Specific cytoplasmic glucocorticoid hormone receptors in hepatoma tissue culture cells. *Proc. Nat. Acad. Sci.* **68**:932.

Bomford, R. and I. B. Weinstein. 1972. Transformation of rat epithelial-like cell line by murine sarcoma virus. *J. Nat. Cancer Inst.* **48**:379.

Darnell, J. E., L. Philipson, R. Wall and M. A. Adesnik. 1971. Polyadenylic acid sequences: Role in conversion of nuclear RNA into messenger RNA. *Science* **174**:507.

Quigley, J. P., L. Ossowski and E. Reich. 1974. Plasminogen, the serum proenyzme activated by factors from cells transformed by oncogenic viruses. *J. Biol. Chem.* **249:**4306.

Thompson, E. B., G. M. Tomkins and J. F. Curran. 1966. Induction of tyrosine α-ketoglutarate transaminase by steroid hormones in a newly established tissue culture cell line. *Proc. Nat. Acad. Sci.* **56:**296.

Tomkins, G. M., B. B. Levinson, J. D. Baxter and L. Dethlefson. 1972. Further evidence for posttranscriptional control of inducible tyrosine aminotransferase synthesis in cultured hepatoma cells. *Nature New Biol.* **239:**9.

Unkeless, J., S. Gordon and E. Reich. 1974. Secretion of plasminogen activator by stimulated macrophage. *J. Exp. Med.* **139:**835.

Unkeless, J., K. Danø, G. M. Kellerman and E. Reich. 1974. Fibrinolysis associated with oncogenic transformation: Partial purification and characterization of the cell factor, a plasminogen activator. *J. Biol. Chem.* **249:**4295.

Weinstein, I. B., R. Gebert, U. C. Stadler, J. M. Orenstein and E. M. Kaighn. 1975a. Mechanisms of chemical carcinogenesis analyzed in rat liver and hepatoma cell cultures. In *Gene Expression and Carcinogenesis in Cultured Liver* (ed. L. E. Gerschenson and E. B. Thompson). Academic Press, New York (in press).

Weinstein, I. B., J. M. Orenstein, R. Gebert, M. E. Kaighn and U. C. Stadler. 1975b. Growth and structural properties of epithelial cell cultures established from normal rat liver and chemically induced hepatomas. *Cancer Res.* **35:**253.

Yamaguchi, N. and I. B. Weinstein. 1975. Temperature-sensitive mutants of chemically transformed epithelial cells. *Proc. Nat. Acad. Sci.* **72:**214.

Plasminogen Activators
of Transformed and Normal Cells

Allan R. Goldberg, Barbara A. Wolf and Paul A. Lefebvre

The Rockefeller University, New York, New York 10021

Cells transformed by DNA or RNA oncogenic viruses are more easily agglutinated by some plant lectins than comparable normal cells (for review, see Burger 1973). Changes in agglutinability, found during transformation or upon reversion of transformed cell lines to a morphologically normal state, usually are paralleled by changes in growth properties. Enhanced agglutinability may be the result of an altered cell surface. Burger (1969) showed that normal cells can be made agglutinable by limited proteolysis of membranes. Both Burger (1970) and Sefton and Rubin (1970) demonstrated that incubation of normal, density-arrested cells with proteases stimulated cell division. Schnebli and Burger (1972) later reported that incubation of protease inhibitors with transformed cells was capable of preventing those cells from growing to high saturation densities. The corollary of this observation is that transformed cells incubated with a protease inhibitor should become markedly less agglutinable. Recent experiments in this laboratory have substantiated this prediction (Goldberg 1974a).

To determine if normal and transformed cells produce proteases that might act to alter the cell surface and if a qualitative or quantitative difference exists between normal and transformed cells in their production of proteases, we have developed and used several different assays for the detection of proteases. In particular, several different assays for the in vivo and in vitro determination of plasminogen activator activity have been employed in this laboratory.

By means of a casein-agar overlay assay, all cells growing in monolayer culture that have so far been tested appear to release proteases that are activators of plasminogen; these enzymes can convert serum plasminogen to its active form, plasmin. Chicken embryo fibroblasts transformed by any of the various strains of Rous sarcoma virus show a markedly higher level of plasminogen activator activity than do their normal, untransformed counterparts (Goldberg 1974a,b).

The cleavage of the hemagglutinin (HA) polypeptide, the largest glycoprotein of influenza virus, to polypeptides HA1 and HA2 has been used as another sensitive indicator of the production of plasminogen activators by cells (Lazarowitz, Goldberg and Choppin 1973; Goldberg and Lazarowitz 1974a,b). Comparison of the HA polypeptide of influenza virions grown in normal versus RSV-transformed chick embryo fibroblasts also has shown that plasminogen activator activity is elevated in transformed cells. In influenza virus-infected cells, the HA polypeptide appears to be cleaved while it is in association with the plasma membrane (Lazarowitz, Compans and Choppin 1971; Compans 1973), suggesting that the plasma membrane is a site of action of plasmin.

In Vivo Determination of Plasminogen Activator Production

Population analysis of plasminogen activator production by cells growing in culture has been applied to mammalian as well as avian cells (Goldberg 1974b). We have reported and consistently observed in this laboratory 10–40-fold higher levels of plasminogen activator activity in transformed

Table 1

Comparison of Plasminogen Activator Levels in Mammalian Cells Growing in Culture

Cell type	No. cells/cm² ×10⁻⁴	Human plasminogen (μg/ml)		
		1	2	5
Experiment 1				
MDBK	5.5	±	3	4
Balb/3T3	3.1	0	±	1
SV3T3	5.6	0	2	2
MSV3T3	5.9	0	±	±
Experiment 2				
MDBK	5.5	3	4	4
Balb/3T3	3.6ᵃ	±	2	3
SV3T3	7.3	±	±	1
MSV3T3	7.6	ND[b]	1	3

The overlay assay was performed as described previously (Goldberg 1974b). Cells were seeded at the indicated densities in 35-mm petri plates. Twelve hours later, 1.2 ml casein-agar overlay was added per petri plate. The overlay consisted of Dulbecco's modified Eagle's medium containing 2% commercial nonfat dried milk, 0.75% Difco agar and human plasminogen as indicated. Plasminogen activator activity was estimated indirectly by means of scoring plasmin-mediated caseinolysis 12 hours later (Exp. 1) or 24 hours later (Exp. 2). Plates were illuminated indirectly using a New Brunswick bacterial colony counter. A scale of 0 to 4 was used to indicate the extent of caseinolysis, where 0 represents no clearing, ± represents <10% clearing, 1 represents approximately 25% clearing, 2 represents approximately 50% clearing, 3 represents approximately 75% clearing and 4 represents complete clearing. (For an illustration of the varying degrees of clearing, see Fig. 1 in Goldberg 1974b.) No clearing was observed in the absence of plasminogen and/or cells.
[a] Saturation density of Balb/3T3.
[b] Not determined.

chick embryo fibroblasts in comparison with normal untransformed cells. In contrast, by means of the casein-agar overlay technique, we have observed little, if any, difference in plasminogen activator activity between contact-inhibited cultures of mouse epithelial (Balb/3T3) cells and Balb/3T3 cells transformed by either SV40 or MSV (Table 1). MDBK, an established line of bovine kidney cells, shows an even higher level of plasminogen activator activity than 3T3 cells or 3T3 transformants. Both MDBK cells and 3T3 cells are contact inhibited and do not grow in agar suspension culture, whereas SV40- or MSV-transformed 3T3 cells do not show contact inhibition of movement and do grow in agar suspension culture. In vitro assay of partially purified plasminogen activator harvested from growing cultures of 3T3 cells and MSV-transformed 3T3 cells has indicated approximately equivalent amounts of enzyme activity.

We have taken advantage of the differential rate of release of plasminogen activator by RSV-transformed chick embryo fibroblasts compared with normal fibroblasts to modify the casein-agar overlay assay so as to allow clonal analysis of cell transformation by RSV rather than just population analysis (A. Goldberg and P. Lefebvre, unpubl.). When infected monolayer cultures are overlayed with medium containing agar, casein and plasminogen, plaques develop in the casein within 2–15 hours (Fig. 1). The number of casein plaque units (CPU) is proportional to the relative virus concentration (Fig. 2) and is equal to the number of focus-forming units (FFU). Overlay at three days postinfection gives readily discernible plaques.

Optimal development of Rous sarcoma foci required seven days following infection of cells with RSV, although foci may be scored at any time between the fifth and ninth day. Therefore, the casein plaque assay allows a considerable saving in time since viral titer can be determined three days after infection.

This genetic technique also will be useful for the selection of RSV mutants which are unable to elevate the level of plasminogen activator at nonpermissive temperature. We are selecting such mutants presently and intend to

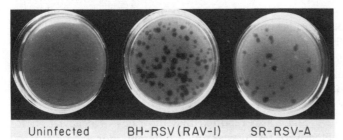

Uninfected   BH-RSV(RAV-I)   SR-RSV-A

Figure 1
Casein plaque formation. Three and one-half days following infection, 35-mm plates were overlaid with agar containing casein (1.5%) and plasminogen (5 µg/ml), as will be described elsewhere (A. Goldberg and P. Lefebvre, unpubl.). Five hours later, the plates were photographed through a blue filter against a black background. The illumination was an indirect fluorescent light.

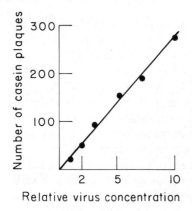

Figure 2

Relationship between virus concentration and number of casein plaques. Secondary cultures of chick embryo fibroblasts were infected with several dilutions of BH-RSV (RAV-1) and were assayed four days later by the casein-agar overlay assay as will be described elsewhere (A. Goldberg and P. Lefebvre, unpubl.).

compare them with known transformation mutants in an attempt to elucidate which RSV functions, and how many, are involved in the maintenance of transformation. By means of the casein-agar overlay technique, we can ask if all chick embryo fibroblasts infected with RSV produce plasminogen activator activity to the same level. Phrased differently, are the large and small plaque phenotypes evident in Figure 1 genetically heritable?

Effect of Protease Inhibitors on Cell Growth and Morphology

Several laboratories have reported that incubation of transformed cells growing in monolayer culture with protease inhibitors caused an inhibition of growth (Schnebli and Burger 1972; Schnebli 1974; Chou, Black and Roblin 1974). The above workers made extensive use of tosyl-lysyl-chloromethyl ketone (TLCK) and tosyl-phenylalanyl-chloromethyl ketone (TPCK) in their studies. It is likely that at the cellular level, chloromethyl ketones from single amino acid derivatives may cause side effects not related to proteolytic enzyme inhibition, such as alkylation of sulfhydryl groups. Therefore, to determine if the proper functioning of proteases is necessary for cell growth, we have investigated the effects of several different types of protease inhibitors on cells growing in culture. We have chosen to investigate several inhibitors that will inactivate either plasminogen activator, plasmin, or both. In addition, the chemical reactivity of the inhibitors used in this study is almost exclusively directed toward specific interaction with proteases.

Shaw and coworkers demonstrated that esters of *p*-guanidinobenzoate are partial substrates for trypsin (Shaw 1974). Since deacylation is remarkably slow, the protease is inactivated by such esters by complete conversion to the acyl enzyme (Fig. 3). Shaw also reported that the rates of deacylation for several proteases which were tested were quite different. For example, it required 1.5 days for plasmin to recover 50% of its activity after reaction with nitrophenyl-guanidinobenzoate (NPGB), whereas thrombin, another plasma enzyme, only required 14 minutes.

We have reported that NPGB inhibits plasmin-mediated caseinolysis (Goldberg 1974b). Using the TAME-esterase assay to measure plasminogen activator activity, we have shown NPGB to be a potent inhibitor of plasminogen activator as well as plasmin. Reich and coworkers have demonstrated

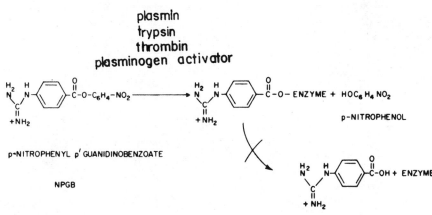

plasmin
trypsin
thrombin
plasminogen activator

p-NITROPHENYL p' GUANIDINOBENZOATE

NPGB

p-NITROPHENOL

Figure 3
Selective inactivation of proteases by stable acyl enzyme formation. (Adapted from
Shaw 1974.)

that fibrinolytic activity measured in the fibrin plate assay is inhibited by
NPGB (Unkeless et al. 1973). In addition, E. Reich and K. Danø (pers.
comm.) have reported that NPGB can block the ability of radioactive DFP to
label the active site of plasminogen activator.

Because NPGB can inhibit both plasminogen activator activity and plasmin
activity, we tested its ability to affect the growth rate of cells growing in culture. Figure 4 shows that NPGB can inhibit the growth of several established

Figure 4
Effect of NPGB on growth. Cells
were seeded at a density of 1×10^5
cells/60-mm petri plate. Twenty-
four hours later, various concentrations of NPGB were added twice
daily. Stock solutions of NPGB (10
mM) were dissolved in dimethylsulfoxide. The growth medium was
changed daily. On the third day
following treatment with NPGB,
cells were trypsinized and counted
in a hemocytometer. The percent
inhibition relative to untreated controls is plotted. The following
cell lines were used in this
study: SV3T3 = SV40-transformed
Balb/3T3; MSV3T3 = Kirsten
strain MSV-transformed Balb/3T3;
MDBK = Madin-Darby bovine kidney cells; 3T3 = Balb/3T3; Ad-HE
= 14B cells = human adenovirus
type 5-transformed hamster embryo
cells (line 9 clone).

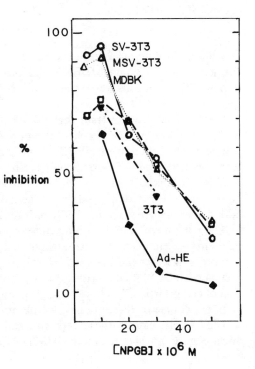

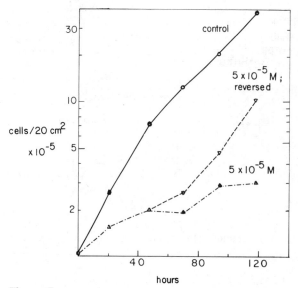

Figure 5

Reversibility of NPGB inhibition of growth of SV3T3 cells. SV3T3 cells were seeded as described in Figure 4. Twenty-four hours later, 5×10^{-5} M NPGB was added. Medium was changed daily, and NPGB was added twice daily. At 48, 72 and 96 hours, medium was removed, cells were washed twice with warm phosphate-buffered saline, and fresh medium was added either with or without 5×10^{-5} M NPGB. Duplicate plates were trypsinized, and cells were counted at the indicated times.

cell lines. No difference was observed in the degree of inhibition of untransformed cells (3T3 and MDBK) and transformed cells (SV3T3, MSV3T3 and Ad-HE). All cells were treated with NPGB while in a state of logarithmic growth.

Figures 5 and 6 illustrate that 50 μM NPGB affects the growth rate of SV3T3 cells and MSV3T3 cells. Inhibition of growth is apparent 24–48 hours following the addition of NPGB to the culture medium. When treated cells were washed at 48–72 hours and fresh medium was added daily without inhibitor, the cells resumed their normal growth rate. In contrast, when the cells were washed daily and fresh medium containing 50 μM NPGB was added, the cells grew at a markedly reduced rate.

The reaction of NPGB with a serine protease results in the formation of an acyl enzyme complex and the concomitant release of free p-nitrophenol. To determine whether the liberated hydrolysis product can affect cell growth, we incubated MSV3T3 cells with 50 μM p-nitrophenol. We observed no inhibition of cell growth.

In the course of these studies, we observed that incubation of NPGB with MSV3T3 cells caused these cells to exhibit a flatter, nonrefractile appearance. Figure 7A shows untreated MSV3T3 cells. The highly refractile cell

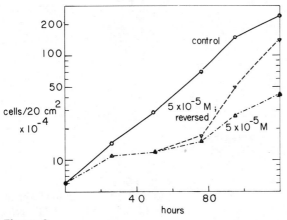

Figure 6
Reversibility of NPGB inhibition of growth of MSV-
3T3 cells. The protocol described in Figure 5 was
followed.

body and long overlapping processes are evident. In the presence of 20 μM
NPGB (Fig. 7B), the cells lose their refractility, appear to have fewer long
processes, and often align themselves parallel to one another. The effect of
NPGB on morphology was dose dependent. Figure 8 illustrates that the percentage of flat MSV3T3 cells in a culture increased with increasing concentrations of NPGB. Other workers have reported similar findings upon
incubation of transformed cells with protease inhibitors. Goetz, Weinstein and
Roberts (1972) showed that beef pancreas trypsin inhibitor promoted

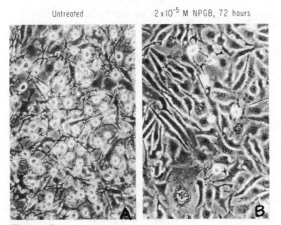

Figure 7
Morphological changes induced in MSV3T3 cells
by NPGB. (*A*) Untreated MSV3T3; (*B*)MSV3T3
cells incubated with 20 μM NPGB for 72 hours.
Cells were photographed using a Wild M-40 in-
verted phase contrast microscope. Magnification
approx. 38 ×.

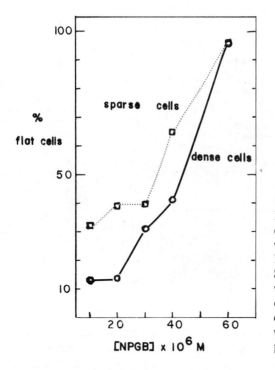

Figure 8
Effect of NPGB on the morphology of MSV3T3. MSV3T3 cells were seeded and incubated with NPGB as described for Figure 4. Seventy-two hours later, cells were photographed using phase-contrast optics. The percentage of flat cells and refractile cells was determined from Polaroid photographs.

parallel alignment of hamster tumor cells in culture. TLCK-treated SV3T3 cells (Schnebli 1974) and TLCK-treated RSV-transformed chick embryo fibroblasts (Weber, Hale and Roll, this volume) also show a flatter morphology. The rounded morphology of SV40-transformed hamster embryo fibroblasts grown in dog serum could be suppressed by the addition of soybean trypsin inhibitor to the culture medium (Ossowski et al. 1973).

Competitive inhibitors of trypsinlike enzymes also have been tested for their effect on cell growth. Pentamidine isethionate, a benzamidine analog, decreased the growth rate of all tissue culture cell lines that we tested (B. Wolf and A. Goldberg, unpubl.). It was effective at 5–40 μM.

Tripeptide chloromethyl ketones have been shown by Shaw (1974) to be more active inhibitors of trypsinlike enzymes than chloromethyl ketones derived from single amino acid residues. Experiments designed to determine the concentration of inhibitor necessary to inactivate 50% of a given amount of enzyme activity showed that alanyl-phenylalanyl-lysyl-CH_2Cl is approximately 100-fold more active than TLCK as a plasmin inhibitor. In the expectation that the use of lower concentrations of the tripeptide chloromethyl ketone would minimize undesirable side reactions, we tested Ala-Phe-Lys-CH_2Cl for its effectiveness in reducing the growth rate of MSV3T3 cells and SV3T3 cells. Figures 9 and 10 show that at concentrations between $2–3 \times 10^{-5}$ M, the growth rate was lowered markedly. Inhibition of growth was evident within one to two days after addition of the tripeptide chloromethyl ketone to the culture medium. Removal of the inhibitor from the medium reversed the growth inhibition.

Serine proteases, such as plasmin or plasminogen activator, are generally stereospecific. In interpreting the effect of a protease inhibitor in vivo, it may be helpful to examine both stereoisomers of the reagent in order to recognize

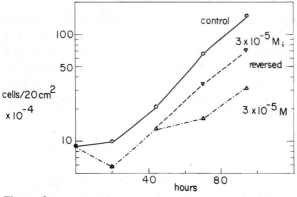

Figure 9

Reversibility of Ala-Phe-Lys-CH$_2$Cl inhibition of growth of MSV3T3. MSV3T3 cells were seeded as described for Figure 4. Twelve hours later, 3×10^{-5} M Ala-Phe-Lys-CH$_2$Cl was added. Ala-Phe-Lys-CH$_2$Cl was added twice daily and medium changed daily. At 48, 72 and 96 hours, medium was removed, cells were washed twice with warm phosphate-buffered saline, and fresh medium was added either with or without 3×10^{-5} M Ala-Phe-Lys-CH$_2$Cl. Duplicate plates were trypsinized, and cells were counted at the indicated times. Ala-Phe-Lys-CH$_2$Cl was dissolved at a concentration of 1×10^{-2} M in 10^{-3} N hydrochloric acid–50% methanol and was sterilized by filtration.

and account for nonspecific reactions. Shaw (pers. comm.) synthesized L-Ala-L-Phe-D-Lys-CH$_2$Cl, the D-stereoisomer of Ala-Phe-Lys-CH$_2$Cl. He found the L-stereoisomer to be 100-fold more effective than the D-stereoisomer as an inhibitor of plasmin in vitro. When we tested both compounds in vivo in parallel experiments, the D-stereoisomer at 2×10^{-5} M had no effect

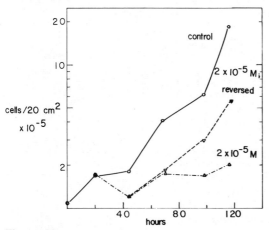

Figure 10

Reversibility of Ala-Phe-Lys-CH$_2$Cl inhibition of growth of SV3T3. The protocol was as described for Figure 9.

on cell growth, whereas the L-stereoisomer at the same concentration showed a pronounced inhibition (Fig. 11). At a tenfold higher concentration (2×10^{-4} M), the D-stereoisomer was inhibitory. At 2×10^{-4} M, TLCK is a growth inhibitor (Schnebli and Burger 1972). Since the D-stereoisomer is biologically inert toward proteases, its inhibitory effect at 2×10^{-4} M may be attributable to nonspecific alkylation of proteins, rather than to direct inactivation of trypsinlike enzymes.

CONCLUDING REMARKS

Our results demonstrate that the growth of transformed and normal cells is inhibited by incubation with protease inhibitors. We did not observe selective growth inhibition of transformed cells when they were incubated with either NPGB, pentamidine isethionate or Ala-Phe-Lys-CH$_2$Cl. Chou, Black and Roblin (1974) also showed that TPCK-mediated growth inhibition of SV3T3 was nonselective.

The effective concentration of NPGB or Ala-Phe-Lys-CH$_2$Cl may be markedly lower than the level of inhibitor that was added to the culture medium. Although these inhibitors are stable at pH 2–3, they hydrolyze spontaneously at alkaline pH. Experiments are in progress to determine the hydrolysis rate of these compounds under the conditions of pH and temperature used for the growth of tissue culture cells.

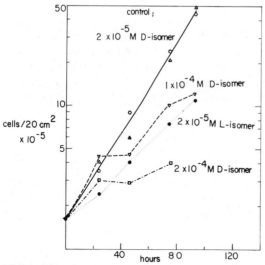

Figure 11

Effect of stereoisomers of Ala-Phe-Lys-CH$_2$Cl on the growth of SV3T3. SV3T3 cells were seeded as described for Figure 4. Twelve hours later, D- or L-Ala-Phe-Lys-CH$_2$Cl was added. Thereafter, the tripeptide chloromethyl ketones were added twice daily and medium changed daily. At the indicated times, duplicate plates were trypsinized and cells were counted.

We do not know at present if the protease inhibitors that we have added to cultures of growing cells are exerting their inhibitory effect on growth by interacting with intracellular, extracellular or plasma membrane-bound proteases. Incubation of cells with low molecular weight inhibitors that are immobilized on Sepharose beads will be useful in determining the site of action of these compounds.

Acknowledgments

We wish to thank Dr. Elliot Shaw and Mr. George Latham for their generous gift of the inhibitors used in these experiments. Our collaborative efforts on the use of these inhibitors to affect cell growth are being extended and will be published elsewhere.

Balb/3T3 and its transformants were gifts from Dr. S. Aaronson, and Ad-He cells were gifts from Dr. R. Goldman. Dr. Jim Williams isolated both the 14B cells and the line 9 clone.

These studies were supported by a research grant from the National Cancer Institute. A. R. G. is an NIH Research Career Development Awardee and a Richard King Mellon Foundation Fellow in The Rockefeller University. B. A. W. is a Public Health Service Postdoctoral Fellow of the National Cancer Institute, Fellowship No. 1F22–CA02607–01.

REFERENCES

Burger, M. M. 1969. A difference in the architecture of the surface membrane of normal and virally transformed cells. *Proc. Nat. Acad. Sci.* **62**:994.

———. 1970. Proteolytic enzymes initiating cell division and escape from contact inhibition of growth. *Nature* **227**:170.

———. 1973. Surface changes in transformed cells detected by lectins. *Fed. Proc.* **32**:91.

Chou, I.-N., P. H. Black and R. O. Roblin. 1974. Nonselective inhibition of transformed cell growth by a protease inhibitor. *Proc. Nat. Acad. Sci.* **71**:1748.

Compans, R. W. 1973. Influenza virus proteins. II. Association with components of the cytoplasm. *Virology* **51**:56.

Goetz, I. E., C. Weinstein and E. Roberts. 1972. Effects of protease inhibitors on growth of hamster tumor cells in culture. *Cancer Res.* **32**:2469.

Goldberg, A. R. 1974a. Involvement of cell and serum proteases in exposing tumor-specific agglutinin sites. *Ann. N.Y. Acad. Sci.* **234**:348.

———. 1974b. Increased protease levels in transformed cells: A casein overlay assay for the detection of plasminogen activator production. *Cell* **2**:95.

Goldberg, A. R. and S. G. Lazarowitz. 1974a. Plasminogen activators of normal and transformed cells. In *Proteinase Inhibitors* (ed. H. Fritz et al.), pp. 631–648. Springer-Verlag, Berlin.

———. 1974b. Plasminogen activators of normal and Rous sarcoma virus-transformed cells. In *Mechanisms of Virus Disease* (ed. W. S. Robinson and C. F. Fox), vol. 1, pp. 347–359. W. A. Benjamin, Menlo Park, California.

Lazarowitz, S. G., R. W. Compans and P. W. Choppin. 1971. Influenza virus structural and nonstructural proteins in infected cells and their plasma membranes. *Virology* **46**:830.

Lazarowitz, S. G., A. R. Goldberg and P. W. Choppin. 1973. Proteolytic cleavage by plasmin of the HA polypeptide of influenza virus: Host cell activation of serum plasminogen. *Virology* **56**:172.

Ossowski, L., J. C. Unkeless, A. Tobia, J. P. Quigley, D. B. Rifkin and E. Reich. 1973. An enzymatic function associated with transformation of fibroblast cultures transformed by DNA and RNA tumor viruses. *J. Exp. Med.* **137**:112.

Schnebli, H. P. 1974. Growth inhibition of tumor cells by protease inhibitors: Consideration of the mechanisms involved. In *Control of Proliferation in Animal Cells* (ed. B. Clarkson and R. Baserga), p. 327. Cold Spring Harbor Laboratory, Cold Spring Harbor, New York.

Schnebli, H. P. and M. M. Burger. 1972. Selective inhibition of growth of transformed cells by protease inhibitors. *Proc. Nat. Acad. Sci.* **69**:3825.

Sefton, B. M. and H. Rubin. 1970. Release from density dependent growth inhibition by proteolytic enzymes. *Nature* **227**:843.

Shaw, E. 1974. Progress in designing small inhibitors which discriminate among trypsin-like enzymes. In *Proteinase Inhibitors* (ed. H. Fritz et al.), pp. 531–540. Springer-Verlag, Berlin.

Unkeless, J. C., A. Tobia, L. Ossowski, J. P. Quigley, D. B. Rifkin and E. Reich. 1973. An enzymatic function associated with transformation of fibroblasts by oncogenic viruses. I. Chick embryo fibroblast cultures transformed by avian RNA tumor viruses. *J. Exp. Med.* **137**:85.

Role of Fibrinolysin T Activity in Properties of 3T3 and SV3T3 Cells

Richard Roblin, Iih-Nan Chou and Paul H. Black

Infectious Disease Unit, Massachusetts General Hospital and
Departments of Microbiology and Molecular Genetics and of Medicine
Harvard Medical School, Boston, Massachusetts 02114

Mild protease treatment of untransformed cell cultures can alter several of their phenotypic characteristics, such that growth properties (Burger 1970; Sefton and Rubin 1970), lectin agglutinability (Burger 1969), glucose transport (Sefton and Rubin 1971) and plasma membrane protein profile (Hynes 1973) resemble those of virus-transformed cell cultures. Thus the idea that enhanced activity of proteolytic enzymes in virus-transformed cells could be responsible for some of their altered phenotypic characteristics via proteolysis of membrane proteins is highly appealing. In particular, a fibrinolytic enzyme system (fibrinolysin T) which is elevated in cells transformed by Rous sarcoma virus, murine sarcoma virus or SV40 virus when compared to untransformed cells (Ossowski et al. 1973b) has recently been described. This enzyme system (plasmin) requires both a cell factor (plasminogen activator) (Unkeless et al. 1974; Christman and Acs 1974) and a serum factor (plasminogen) (Quigley, Ossowski and Reich 1974) for activity. By selective removal of plasminogen from serum, Ossowski et al. (1973a) have provided evidence suggesting that the manifestation of several cellular phenotypic characteristics (morphology, growth in agar, cell migration) are plasminogen dependent.

We have investigated the expression of fibrinolytic activity in cultures of Swiss and Balb 3T3 cells and their SV40 virus-transformed counterparts and have focused our attention upon two questions. First, is there a difference in expression of fibrinolytic activity in cultures of these two cell types, and if so, what is responsible for this difference? Second, if the fibrinolytic activity of SV3T3 cell cultures is suppressed with ϵ-aminocaproic acid (EACA), do some of the phenotypic characteristics of these virus-transformed cells revert to those of untransformed 3T3 cells? In these latter experiments, in addition to the important characteristic of saturation density, we have examined whether suppression of fibrinolytic activity affects several membrane surface properties known to be altered in SV3T3 cells compared to 3T3 cells. We report here the effect of suppressing the fibrinolytic activity on the lactoperoxidase-labeling pattern of plasma membrane proteins.

MATERIALS AND METHODS

Cells, Media and Sera

The origins and preservation of Swiss 3T3 and SV3T3 and Balb 3T3 (clone A31) and SV3T3 (clone T2) cells have been previously described (Chou, Black and Roblin 1974a,c). Swiss 3T3 and SV3T3 cells were routinely passaged in Dulbecco's medium containing 10% fetal calf serum (FCS), 250 U/ml penicillin plus 250 μg/ml streptomycin sulfate. Balb 3T3 and SV3T3 cells were routinely passaged in Eagle's minimal essential medium containing four times concentrated amino acids and vitamins (MEM X4), 10% fetal calf serum and 250 U/ml penicillin and 250 μg/ml streptomycin sulfate.

Two different lots of fetal calf serum (Gibco #A530605 and #A036620) were used in the experiments described below. Since we observed that medium containing FCS lot #A530605 supported a higher level of fibrinolytic activity than medium containing FCS lot #A036620 (see below), the fetal calf serum lot used in each experiment has been specifically identified. Unless otherwise stated, FCS lot #A530605 was used in all cell culture experiments and fibrinolysin-T assays. A single lot of dog serum (DS) (Gibco #A132814) was used for all experiments.

Assay of Fibrinolytic Activity

Fibrinolytic activity of cell culture, harvest fluids and culture medium was measured as described by Unkeless et al. (1973), except that human fibrinogen (plasminogen-free, American Red Cross National Foundation Center) was iodinated and used in place of bovine fibrinogen. Sixty-mm Falcon plastic petri dishes for cell culture experiments or 35-mm dishes for enzyme assays were coated with 10 μg/cm^2 ^{125}I-fibrinogen, dried overnight, and activated for either 2 or 18 hours by incubation in Dulbecco's medium plus 10% fetal calf serum, 250 U/ml penicillin and 250 μg/ml streptomycin sulfate. The activated dishes were then washed with TD buffer (Unkeless et al. 1973) and used for cell culture experiments or enzyme assays. In cell culture experiments, the medium was changed at 24-hour intervals, and aliquots of the culture medium were counted in a Packard gamma-counter to determine the amount of released ^{125}I-fibrinopeptides. Enzyme assays were incubated from 4–6 hours at 37°C in a CO_2 incubator and the released ^{125}I-fibrinopeptides determined by counting aliquots of the assay supernatant in a Packard gamma-counter.

EACA Inhibition of Fibrinolytic Activity

Cell culture media containing 10 mg/ml (76 mM) EACA were prepared by dissolving EACA (Sigma) in Dulbecco's medium to make a 250 mg/ml stock solution, filtering this solution through a 0.45-μ filter, and diluting it to a final EACA concentration of 10 mg/ml with Dulbecco's medium. In some cell culture experiments (Figs. 1 and 2), cells were seeded in medium without EACA, allowed to attach for 5–6 hours, and then medium containing 10 mg/ml EACA was added. To ensure continued suppression of fibrinolytic activity, media containing EACA were changed every 24 hours.

Lactoperoxidase Iodination

Cellular membrane surface proteins were labeled with ^{125}I using the lactoperoxidase/glucose oxidase technique described by Hynes (1973). Cell monolayers were washed twice with phosphate-buffered saline containing 5 mM glucose and labeled in the same buffer to which 400 μCi/ml carrier-free sodium [^{125}I]iodide (New England Nuclear), 40 μg/ml lactoperoxidase (Calbiochem, B grade, 20 IU/mg) and 0.2 U/ml glucose oxidase (Worthington, 140 IU/mg) had been added. After 10 minutes incubation at room temperature, the reaction mixture was withdrawn, and the plates were immediately covered with phosphate-buffered sodium iodide solution (PBI) containing 2 mM phenyl methyl sulfonyl fluoride (PMSF). After one additional wash with PBI-2 mM PMSF solution, the cells were scraped from one-half of the labeled plates in PBI/PMSF buffer, pelleted by centrifugation at 1000g for 10 minutes, and solubilized by addition of 0.1 ml 2% sodium dodecyl sulfate (SDS), 2 mM PMSF solution.

For electrophoretic analysis, iodinated cell samples were made 10^{-2} M dithiothreitol and heated at 100°C for 2 minutes. Fifty-μl aliquots were made 10% sucrose and 0.08% bromphenol blue and electrophoresed on cylindrical gels of 5% acrylamide.

Other cell monolayers, iodinated and washed with PBI/PMSF solution as described above, were fixed with cold 10% trichloroacetic acid for 30 minutes at room temperature, washed twice with absolute methanol, air dried, and covered with Kodak AR10 stripping film for autoradiography. After overnight exposure, the film was developed as described previously (Culp and Black 1972).

RESULTS

Expression of Fibrinolytic Activity in Cell Culture

Swiss 3T3 cells growing in fetal calf serum medium exhibited increasing fibrinolytic activity during the first four days after subculture onto ^{125}I-fibrin plates (Fig. 1) (Chou, Black and Roblin 1974b). On the next two days, the fibrinolytic activity decreased. The period of greatest expression of fibrinolytic activity corresponded to the period of active cell multiplication. As the cells became confluent, the fibrinolytic activity decreased. The reasons for this decreased fibrinolytic activity in confluent cultures will be explored in detail below.

SV3T3 cell cultures also manifest increasing fibrinolytic activity during the first four days in culture (Fig. 2). Because the cell sheet spontaneously detached from the petri dish at cell densities greater than 12×10^6 per dish, fibrinolytic activity could not be measured in untreated SV3T3 cell cultures after day 4. The reasons for detachment of the SV3T3 cell sheet at high cell densities are not known. It is unlikely that exhaustion of essential nutrients in the medium is the explanation, since the medium was changed every 24 hours. The high level of fibrinolytic activity might itself be responsible for cell detachment.

Effect of Suppression of Fibrinolytic Activity on Cell Growth

ϵ-Aminocaproic acid (EACA) is a well-known inhibitor of fibrinolysin-T (plasmin) activity in vitro (Unkeless et al. 1973; Castellino et al. 1973) and is sufficiently nontoxic to be used to suppress the fibrinolytic activity of cells

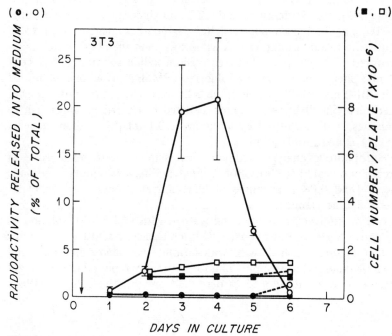

Figure 1

Effect of EACA on growth and fibrinolytic activity of Swiss 3T3 cells. At time zero, 5×10^5 Swiss 3T3 cells in 5 ml Dulbecco's medium plus 10% FCS (lot #A530605) were seeded into 60-mm Falcon plastic petri dishes. The dishes had been coated with ^{125}I-fibrinogen (10 μg/cm^2; total radioactivity 1.17×10^6 cpm/dish), which was converted to ^{125}I-fibrin by incubation for 18 hours before plating the cells. Five hours after plating the cells, the medium was removed, and 5 ml Dulbecco's medium plus 10% FCS with or without the 10 mg/ml EACA was added. At 24-hour intervals thereafter, the cell culture medium was removed, aliquots assayed for released ^{125}I-fibrinopeptides, and 5 ml fresh medium with or without EACA was added. The amount of released ^{125}I-radioactivity (1–2% of the total) on appropriate control plates without cells and with or without EACA has been subtracted. Each point represents the mean value ($\pm$S.D.) of the average of duplicate counts of medium from 4–7 replicate petri dishes. Cell growth was determined on ^{125}I-fibrin-containing plates which were prepared, seeded and treated identically to those used for assays of fibrinolytic activity. Cell counts were determined on trypsinized cell suspensions using the Cytograph laser beam cell counter. Fibrinolytic activity: (o——o) control; (●——●) EACA treated. Cell number: (□——□) control; (■——■) EACA treated. After removal of EACA, both fibrinolytic activity (●– – –o) and cell number (■– – –□) increased during the next 24 hours. (Reprinted, with permission, from Chou, Black and Roblin 1974b.)

872

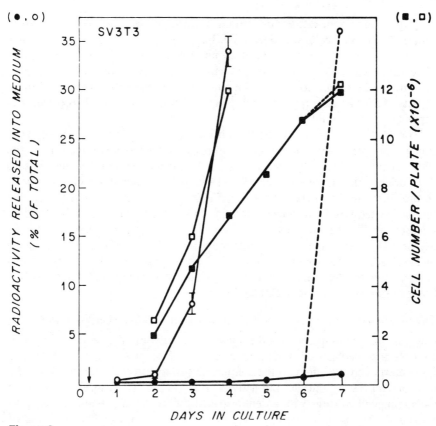

Figure 2

Effect of EACA on growth and fibrinolytic activity of Swiss SV3T3 cells. Experimental details and symbols are as described in the legend to Figure 1. Swiss SV3T3 cells were initially seeded at 5×10^5 cells/dish. (Reprinted, with permission, from Chou, Black and Roblin 1974b.)

in culture (Chou, Black and Roblin 1974b). Culturing 3T3 cells in medium containing 10 mg/ml EACA completely suppressed the fibrinolytic activity observed in the absence of the inhibitor (Fig. 1). This effect was partially reversible in the sense that removal of EACA on day 5 resulted in a small increase in fibrinolytic activity during the subsequent 24 hours.

Treatment of 3T3 cultures with EACA also decreased their saturation density, and removal of the EACA on day 5 led to a small increase in cell number by the next day (Fig. 1). We are uncertain whether this reduction in saturation density is a consequence of suppression of fibrinolytic activity or due to a side effect of the EACA. EACA treatment of 3T3 and SV3T3 cells also depresses protein synthesis (Chou, Black and Roblin 1974b), and this treatment of 3T3 cells with inhibitors of protein synthesis has been shown to cause reduction of saturation density (Chou, Black and Roblin 1974c).

Adding 10 mg/ml EACA to SV3T3 cultures (Fig. 2) suppressed their fibrinolytic activity by 83–98% when the data were expressed as cpm released/10^6 cells (Chou, Black and Roblin 1974b). Removal of EACA from suppressed SV3T3 cell cultures on day 6 led to a large burst of fibrinolytic activity during the next 24 hours (Fig. 2). Under these conditions, the growth rate of the SV3T3 cells was decreased, but they eventually reached the same cell density (12×10^6 cells/dish) as untreated cultures. Thus almost complete suppression of the fibrinolytic activity failed to restore density-dependent growth inhibition to SV3T3 cell cultures, indicating that the high plasmin activity of SV3T3 cultures is probably not responsible for their loss of density-dependent growth inhibition.

Examination of the size of the ^{125}I-fibrinopeptides suggests that the suppression of plasmin activity in SV3T3 cultures may be more complete than the 98% calculated from the data in Figure 2. Chromatography of the released ^{125}I-fibrinopeptides from 3T3 and SV3T3 cultures on Sephadex G-200 reveals several discrete size classes (Fig. 3). This is consistent with the expected limited proteolysis of fibrin by plasmin. In contrast, the ^{125}I-fibrinopeptides released by SV3T3 cultures in the presence of 10 mg/ml EACA lack the higher molecular weight components found in SV3T3 medium in the absence of EACA. Thus the small amount of ^{125}I-fibrinopeptides released by SV3T3 cultures treated with EACA may not be due to plasmin, but rather to low levels of some other proteolytic activity.

Effects of Suppression of Fibrinolytic Activity on Cell Surface Properties

One possible consequence of the plasmin activity of 3T3 and SV3T3 cell cultures is proteolysis of cellular membrane surface proteins. We have therefore determined the effect on the membrane surface protein pattern of SV3T3 cells of suppressing the plasmin activity with EACA. To ensure that we were examining cell cultures with a demonstrated difference in plasmin activity, we measured the plasmin activity of cultures seeded in parallel and treated identically to those used for lactoperoxidase iodination. The SV3T3 cells exhibited the expected increase in fibrinolytic activity (compare Fig. 4 with Fig. 2), and this increased fibrinolytic activity was completely blocked by EACA treatment. Autoradiography of cell monolayers iodinated with the lactoperoxidase technique revealed a generally uniform distribution of silver grains, with some irregularly shaped concentrations of grains at what appeared to be the cell periphery. We observed few (less than 5%) cells which were rounded and very heavily iodinated in both untreated and EACA-treated 3T3 and SV3T3 cell cultures.

Analysis of the iodinated proteins by SDS gel electrophoresis showed that although the protein profile of EACA-treated cells had more distinct peaks than the profile from untreated cells (Fig. 5), no major qualitative changes in the pattern of iodinatable membrane surface components were detected in this experiment. However, the high molecular weight protein region was not very well resolved on these 5% acrylamide gels. A more recent experiment, using a 28-cm slab gel of 9% acrylamide, suggests that EACA treatment of SV3T3 cells leads to the reappearance of 3–4 high molecular weight protein bands which are much reduced in amount in untreated SV3T3 cells (Fig. 6). These 3–4 high molecular weight protein bands are present in untreated 3T3 cells, and their presence is unaffected by treatment of 3T3 cells with 10 mg/ml EACA (Fig. 6). To date, these bands have only been detected by staining the slab gels with Coomassie blue, and we have not yet demonstrated whether any of them are membrane surface components. While

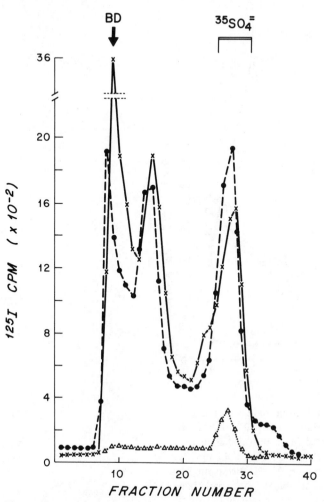

Figure 3

Sephadex G-200 chromatography of released ^{125}I-fibrinopeptides. One-ml aliquots of culture medium containing ^{125}I-fibrinopeptides were chromatographed on a 1×60-cm column of Sephadex G-200. The column was prepared and eluted with phosphate-buffered saline; fractions of 2 ml were collected and ^{125}I cpm determined in a Packard gamma-counter. BD and $^{35}SO_4$ mark the elution positions of Blue Dextran 2000 (Pharmacia) and $^{35}SO_4$ chromatographed in a separate run. (x——x) ^{125}I-Fibrinopeptides from 3T3 culture medium (day 4 of experiment shown in Fig. 1); (●– – –●) ^{125}I-fibrinopeptides from SV3T3 culture medium (day 4 of experiment shown in Fig. 2); (△– – – –△) ^{125}I-radioactivity from SV3T3 culture treated with 10 mg/ml EACA (day 4, Fig. 2).

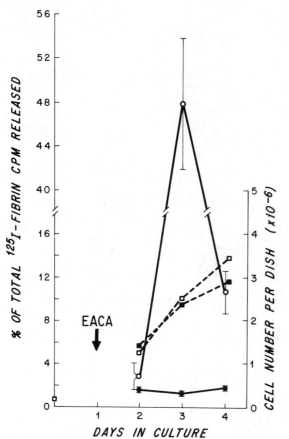

Figure 4

EACA inhibition and lactoperoxidase iodination of cell surface proteins. About 2×10^5 Swiss SV3T3 cells were seeded in 35-mm dishes in Dulbecco's medium plus 10% FCS (lot #A036620). Some of the dishes contained ^{125}I-fibrinogen (10 μg/cm^2, 9.8×10^4 cpm total) that had been converted to ^{125}I-fibrin by incubation for 2 hours in Dulbecco's medium plus 10% FCS (lot #A036620). Twenty hours after seeding, 1.5 ml of medium with or without 10 mg/ml EACA was added. Cell counts were determined with a hemocytometer on trypsinized suspensions of cells plated on dishes without ^{125}I-fibrin. Release of ^{125}I-fibrinopeptides was quantitated by counting aliquots of the culture medium in a Packard gamma-counter. No correction has been made for the release of ^{125}I-fibrinopeptides in the absence of cells. Duplicate cultures to those used for determination of cell counts and fibrinolytic activity were iodinated on day 3 using the lactoperoxidase/glucose oxidase technique. Fibrinolytic activity: (o——o) control; (•——•) EACA treated. Cell counts: (□ – – – □) control; (■ – – – ■) EACA treated.

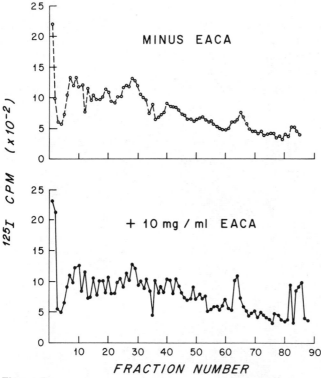

Figure 5

SDS gel electrophoresis of iodinated SV3T3 cell proteins.
Samples of iodinated SV3T3 cell proteins (Fig. 4, day 3)
were electrophoresed using 10-cm long gels of 5% acrylamide.
After fixation in cold 10% TCA, the gels were frozen and
sliced into 1-mm thick pieces which were counted in a
Packard gamma-counter. Fraction number 1 corresponds to
the top of the gel, and the position of the start of the bromphenol blue marker dye band was fraction 70.

the similarity between our results and those of Hogg (1974) suggests that at
least one of these 3–4 high molecular weight proteins will be located at the
cell surface, further work will be required to determine the location and function of these high molecular weight proteins.

Fibrinolytic Activity and Cell Growth

A decrease in the fibrinolytic activity of confluent 3T3 cell cultures, relative
to the activity of growing 3T3 cultures, was noted above (Fig. 1). In these
experiments, the decreased fibrinolytic activity of confluent cultures was probably due in part to exhaustion of the ^{125}I-fibrin substrate. In addition, however, removal of EACA inhibition in confluent 3T3 cultures led to a smaller
increase in fibrinolytic activity (Fig. 1) than did removal of inhibition in
SV3T3 cultures (Fig. 2). These results suggested a cell density-dependent
reduction of fibrinolytic activity in 3T3 cultures (Chou, Black and Roblin
1974b).

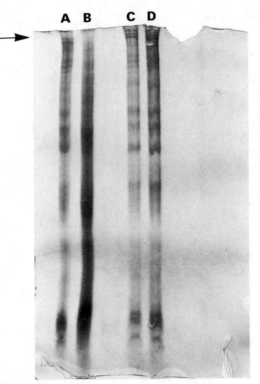

Figure 6

Effect of EACA on Swiss 3T3 and SV3T3 cell proteins. Swiss 3T3 and SV3T3 cells were plated at 3×10^5 cells/35-mm dish in Dulbecco's medium plus 10% FCS (#A036620). Twenty hours after plating, medium containing 10 mg/ml EACA was added to one-half of the cultures, while the other half received fresh medium without EACA. The plates were fluid-changed daily, and on the fifth day after plating, the cells were iodinated with the lactoperoxidase technique and prepared for electrophoresis as described in Materials and Methods. Samples were analyzed by slab gel electrophoresis (Laemmli 1970) using a 3% acrylamide stacking gel and a 9% acrylamide running gel. After electrophoresis, the gels were fixed in 30% methanol/7% acetic acid and stained with Coomassie blue. (*A*) SV3T3 cells treated with EACA; (*B*) SV3T3 cells without EACA; (*C*) 3T3 cells without EACA; (*D*) 3T3 cells treated with EACA. Left arrow indicates the high molecular weight protein region.

To test the hypothesis that decreased elaboration of plasminogen activator by confluent 3T3 cells was responsible for this effect, we prepared serum-free "harvest fluids" (HFs) from cell cultures at various times after seeding the cells. These HFs were assayed for plasmin activity on ^{125}I-fibrin plates using 10% FCS (lot #A530605) as the source of plasminogen. The plasmin activity detectable in HFs decreased when the HFs were derived from confluent 3T3 cells (Fig. 7). This was so when FCS was used as the source of plasminogen, but not when dog serum was used as the source of plasminogen. This difference is probably due to the greater concentration of plasmin inhibitors in FCS. These results indicate that elaboration of plasminogen activator decreases as 3T3 cells become confluent.

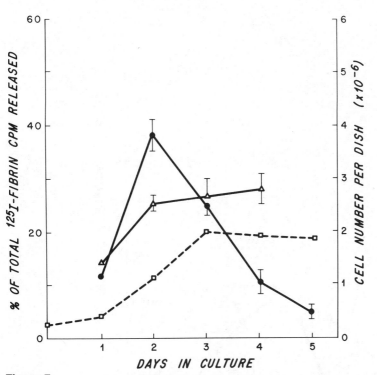

Figure 7

Fibrinolytic activity of 3T3 cell harvest fluids. Swiss SV3T3 cells 2.5×10^5 were seeded in 60-mm Falcon petri dishes in Dulbecco's medium plus 10% FCS. Twenty-four hours later, fresh Dulbecco's medium plus 10% FCS was added, and at various times thereafter, duplicate cultures were washed once with Dulbecco's medium minus serum and incubated for 18 hours in 1.5 ml of serum-free Dulbecco's medium. Nine-tenths ml of the resulting HFs and 0.1 ml of either FCS or DS was added to ^{125}I-fibrin plates, and the resulting release of ^{125}I-fibrinopeptides was measured after incubating the plates for 4 hours at 37°C in a CO_2 incubator. The fibrinolytic activity on day 1 is that of the HF made overnight starting on day 1. A blank corresponding to the activity of HF alone, without serum, has been substracted. The range and average of duplicate determinations of fibrinolytic activity are shown. Fibrinolytic activity: (●——●) with FCS; (△——△) with DS; cell count: (□----□).

A different type of experiment which led to the same conclusion is shown in Figure 8. In this case, the culture medium was not changed after plating the cells, and the fibrinolytic activity of the culture medium itself was determined. On days 3–5 after subculture, fibrinolytic activity was detectable, even though the medium contained the relatively nonactivating FCS (lot #A036620). On days 8 and 9 after subculture, plasmin activity was again undetectable. In preliminary experiments, addition of purified FCS plasminogen to confluent 3T3 culture medium failed to increase the plasmin activity, suggesting that the decreased activity after day 5 (Fig. 8) is not due to exhaustion of plasminogen. The results indicate rather that plasminogen activator is elaborated when 3T3 cells are actively growing, and that activator

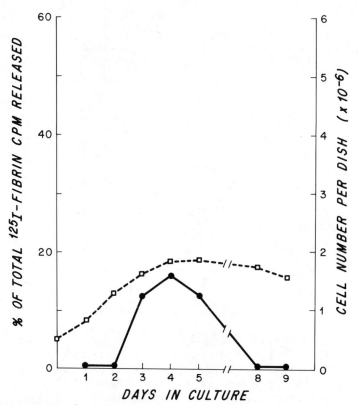

Figure 8

Fibrinolytic activity of 3T3 cell culture medium. Swiss 3T3 cells (5×10^5) were seeded in 60-mm Falcon petri dishes in Dulbecco's medium plus 10% FCS (lot #A036620) on day 0. On subsequent days, medium was removed and assayed directly on plates spread with ^{125}I-fibrinogen (10 μg/cm^2, 1.49×10^5 cpm/dish) which had been converted to ^{125}I-fibrin by a 2-hour incubation in Dulbecco's medium plus 10% FCS (lot #A036620). ^{125}I-Fibrin plates were incubated at 37°C for 4 hours. Cell counts were determined on plates seeded identically to the plates used for enzyme assay. (●——●) Fibrinolytic activity; (□––––□) cell count.

elaboration is decreased in confluent 3T3 cultures. In addition, plasmin activity is apparently slowly inactivated in 3T3 cell cultures.

Assay of the fibrinolytic activity of HFs prepared from SV3T3 cell cultures at various times after subculture is shown in Figure 9. In contrast to results obtained with HFs from 3T3 cells, the plasmin activity does not decrease as SV3T3 cells become dense. This is true regardless of whether the assays are done using FCS or dog serum as the source of plasminogen. However, this experiment was limited by an increasing tendency of these SV3T3 cells to detach from the petri dish (especially on day 4) when incubated in serum-free medium. For this reason, and for the reason that the fibrinolysis assay is not linearly related to time and enzyme concentration beyond 40% hydrolysis (Chou et al., unpubl.), the apparent slowing of the increase in fibrinolytic

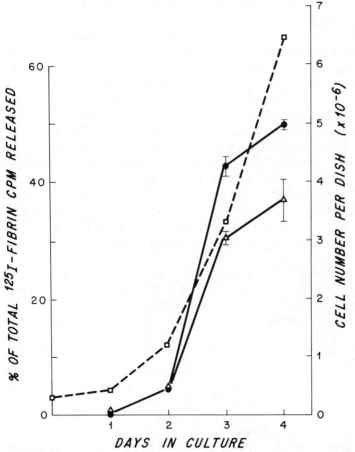

Figure 9

Fibrinolytic activity of SV3T3 cell harvest fluids. Swiss SV3T3 cells (2.5×10^5) were seeded in Dulbecco's medium plus 10% FCS (lot #A036620) on day 0. HFs were prepared and assayed and cell counts determined as described in the legend to Figure 7. Fibrinolytic activity: (●——●) with FCS; (△——△) with DS; cell count: (□ – – – □).

activity on day 4 (Fig. 9) may not accurately reflect the amount of plasminogen activator liberated by these cells. Thus dense cultures of SV3T3 cells continue to produce relatively large amounts of plasminogen activator.

DISCUSSION

Our results demonstrate that the plasmin activity of 3T3 and SV3T3 cell cultures can be essentially completely suppressed by 10 mg/ml EACA. Since the large-sized ^{125}I-fibrinopeptides characteristic of plasmin activity on fibrin are not observed in the presence of EACA (Fig. 3), the minor residual fibrinolysis in the presence of EACA is probably not due to plasmin. This high level of suppression of the plasmin activity of SV3T3 cell cultures has permitted us to determine whether their plasmin activity is responsible for loss of density-dependent growth inhibition or loss of high molecular weight membrane surface proteins.

Suppression of the plasmin activity of SV3T3 cultures fails to restore the density-dependent growth inhibition characteristic of 3T3 cells. Thus it seems likely that the high levels of plasmin activity in SV3T3 cell cultures are not the cause of loss of density-dependent growth inhibition in SV3T3 cell cultures. However at present, this conclusion must be qualified to some extent. Ambrus et al. (1968) have shown that under conditions where the action of plasmin on fibrin was almost completely eliminated by EACA, the caseinolytic activity of similar plasmin preparations was almost unaffected. Thus although the fibrinolytic activity of plasmin in the EACA-treated cultures was essentially eliminated, plasmin activity against other substrates may have remained. It should be noted, however, that Ossowski et al. (1973b) showed that SV40-transformed hamster cells grew to high density in plasminogen-depleted serum. Their data support our conclusion that the plasmin activity of SV3T3 cells is more likely to be a consequence, rather than a cause, of unrestrained cell growth.

Suppression of the plasmin activity of SV3T3 cultures does not greatly alter the overall lactoperoxidase labeling pattern (Fig. 5). However, it does appear to block the proteolysis of several high molecular weight proteins (Fig. 6). These proteins may correspond to the high molecular weight membrane surface proteins which Hogg (1974) has previously shown to be absent from SV3T3 cells. Similar losses of high molecular weight membrane surface proteins have been shown to occur in hamster cells transformed by polyoma and hamster sarcoma viruses (Hynes 1973). Our results suggest that loss of these proteins from virus-transformed hamster cells may also be due to proteolysis by plasmin.

We have also shown that the fibrinolytic activity of 3T3 cell cultures decreases as the cells become confluent. This is probably due, at least in part, to decreased elaboration of plasminogen activator by confluent 3T3 cell cultures (Fig. 7 and 8). In contrast, dense SV3T3 cultures do not show an equivalent decrease in elaboration of plasminogen activator. We find that growing 3T3 cultures elaborate relatively large amounts of plasminogen activator into the culture medium (Fig. 7). Thus the difference in fibrinolysin-T activity between untransformed 3T3 cells and their SV40 virus-transformed counterparts lies in the reduced elaboration of plasminogen activator by confluent 3T3 cells. This reduction effectively limits the fibrinolytic activity that develops in 3T3 cell cultures. Whether a similar reduction in the elaboration of plasminogen activator occurs in confluent cultures of untransformed mouse embryo fibroblasts is currently being investigated.

Although our data suggest that plasmin activity does not play a role in loss of density-dependent growth inhibition in our SV3T3 cells, it remains to be seen whether plasmin activity causes the unrestrained growth of other virus-transformed cells in vitro or "spontaneous" tumors in vivo. Because of the specificity of the action of cellular plasminogen activators on plasminogen molecules from different organisms and of the presence of inhibitors of plasmin activity in serum, it will be important to examine completely homologous serum (e.g., human tumor cells in human serum). Much more work will be required to completely evaluate the role of plasmin activity in tumor cell growth.

Acknowledgments

We thank Beverly Ash and Sara O'Donnell for excellent technical assistance. This research was supported by Grants CA10126-07 and CA15889-01 from the National Institutes of Health. I-N. C. holds an NIH postdoctoral fellowship and R. R. is a Faculty Research Associate (PRA-75) of the American Cancer Society.

REFERENCES

Ambrus, C. M., J. L. Ambrus, H. B. Lassman and I. B. Mink. 1968. Studies on the mechanism of action of inhibitors of the fibrinolysin system. *Ann. N.Y. Acad. Sci.* **146**:430.

Burger, M. M. 1969. A difference in the architecture of the surface membrane of normal and virally transformed cells. *Proc. Nat. Acad. Sci.* **62**:994.

————. 1970. Proteolytic enzymes initiating cell division and escape from contact inhibition of growth. *Nature* **227**:170.

Castellino, F. J., W. J. Brockway, J. K. Thomas, H. T. Lino and A. B. Rawitch. 1973. Rotational diffusion analysis of the conformational alterations produced in plasminogen by certain antifibrinolytic amino acids. *Biochemistry* **12**:2787.

Chou, I-N., P. H. Black and R. Roblin. 1974a. Effects of protease inhibitors on growth of 3T3 and SV3T3 cells. In *Control of Proliferation in Animal Cells* (ed. B. Clarkson and R. Baserga), p. 339. Cold Spring Harbor Laboratory, Cold Spring Harbor, New York.

————. 1974b. Suppression of fibrinolysin T activity fails to restore density-dependent growth inhibition to SV3T3 cells. *Nature* **250**:739.

————. 1974c. Non-selective inhibition of transformed cell growth by a protease inhibitor. *Proc. Nat. Acad. Sci.* **71**:1748.

Christman, J. K. and G. Acs. 1974. Purification and characterization of a cellular fibrinolytic factor associated with oncogenic transformation: The plasminogen activator of SV40-transformed hamster cells. *Biochim. Biophys. Acta* **340**:339.

Culp, L. A. and P. H. Black. 1972. Release of macromolecules from Balb/c mouse cell lines treated with chelating agents. *Biochemistry* **11**:2161.

Hogg, N. M. 1974. A comparison of membrane proteins of normal and transformed cells by lactoperoxidase labeling. *Proc. Nat. Acad. Sci.* **71**:489.

Hynes, R. O. 1973. Alteration of cell surface proteins by viral transformation and by proteolysis. *Proc. Nat. Acad. Sci.* **70**:3170.

Laemmli, U. K. 1970. Cleavage of structural proteins during the assembly of the head of bacteriophage T4. *Nature* **222**:680.

Ossowski, L., J. P. Quigley, G. M. Kellerman and E. Reich. 1973a. Fibrinolysis associated with oncogenic transformation. Requirements of plasminogen for correlated changes in cellular morphology, colony formation in agar and cell migration. *J. Exp. Med.* **138**:1056.

Ossowski, L., J. C. Unkeless, A. Tobia, J. P. Quigley, D. B. Rifkin and E. Reich. 1973b. A enzymatic function associated with transformation of fibroblasts by oncogenic viruses. II. Mammalian fibroblast cultures transformed by DNA and RNA tumor viruses. *J. Exp. Med.* **137**:112.

Quigley, J. P., L. Ossowski and E. Reich. 1974. Plasminogen, the serum proenzyme activated by factors from cells transformed by oncogenic viruses. *J. Biol. Chem.* **249**:4306.

Sefton, B. and H. Rubin. 1970. Release from density-dependent growth inhibition by proteolytic enzymes. *Nature* **227**:843.

———. 1971. Stimulation of glucose transport in cultures of density-inhibited chick embryo cells. *Proc. Nat. Acad. Sci.* **68**:3154.

Unkeless, J. C., K. Danø, G. M. Kellerman and E. Reich. 1974. Fibrinolysis associated with oncogenic transformation. Partial purification of the cell factor, a plasminogen activator. *J. Biol. Chem.* **249**:4295.

Unkeless, J. C., A. Tobia, L. Ossowski, J. P. Quigley, D. B. Rifkin and E. Reich. 1973. An enzymatic function associated with transformation by oncogenic viruses. I. Chick embryo fibroblast cultures transformed by avian RNA tumor viruses. *J. Exp. Med.* **137**:85.

Production of Plasminogen Activator and Colonial Growth in Semisolid Medium Are In Vitro Correlates of Tumorigenicity in the Immune-deficient Nude Mouse

Robert Pollack,* Rex Risser† and Susanne Conlon*

Cold Spring Harbor Laboratory, Cold Spring Harbor, New York 11724

Vicki Freedman and Seung-II Shin

Albert Einstein College of Medicine, Bronx, New York 10461

Daniel B. Rifkin

The Rockefeller University, New York, New York 10021

Multicellular animals maintain tissues at constant size by two broadly different mechanisms. In some tissues, like the skin and intestinal epithelium, cell mass is held constant, despite continuous rapid cell division, by coupling cell division to terminal differentiation of one daughter cell. In other tissues, such as the liver and the dispersed collagen-synthesizing fibroblast population, division ceases when adult cell mass is achieved and resumes only in response to injury or tissue loss. While both modes of regulation have been studied using in vitro cultured cells, fibroblasts, which regulate their division by the latter mode, have an experimental advantage in that this regulation is disrupted in vitro by oncogenic viruses. This disruption is usually called transformation.

If a normal fibroblast is sensitive to more than one signal that it should not divide, then each signal should be the basis of a separate selective assay for transformed cells. Three separate assays are in fact in current use: A normal fibroblast ceases to divide under conditions where serum is decreased or depleted, where extensive cell–cell contact is permitted, and where anchorage to a solid substrate is prevented (Fig. 1). Studies on growth control of cultured fibroblasts have used both established cell lines and primary embryo cell cultures.

Recently, we have shown that these three selective assays detect separate and distinct modes of growth control. Two lines of work led us to this conclusion. First, we showed that each selective assay could be made into a negative selective assay by adding to transformants a drug that kills growing cells while subjecting the transformants to the restrictive state (Pollack, Green and Todaro 1968; Pollack 1970; Vogel and Pollack 1974). In studies with lines

Present addresses: *SUNY, Stony Brook, New York 11790; †NIAID, NIH, Bethesda, Maryland 20014.

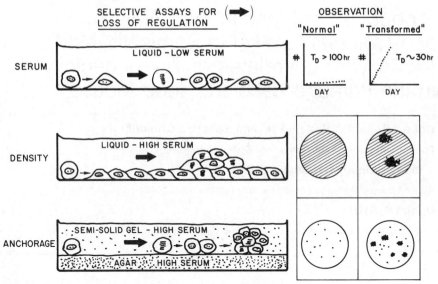

Figure 1
The three known selective assays for transformation. In each assay, transformed cells and their descendents grow, whereas untransformed cells, due to one or another restrictive environmental signals, remain alive but do not increase in number. The heavy arrow represents "transformation." Note that only the anchorage assay requires the transformant to continuously traverse the cell cycle without ever spreading out.

derived from the "normal" 3T3 mouse cell, we showed that a fully transformed cell (one that had lost sensitivity and so could grow) could, in all three assays, be the source by negative selection with BrdU (Vogel and Pollack 1973), FdU (Pollack, Green and Todaro 1968) and colchicine (Vogel, Risser and Pollack 1973) of revertants that had regained sensitivity to one, two, or all three of the different signals (Vogel, Oey and Pollack 1974).

We extended our studies on the phenotype of viral transformation by analyzing nonselectively isolated clones arising after 3T3 infection by SV40 (Risser and Pollack 1974a). Here, as with the revertants, our results showed that the three assays for transformation were not detecting the same event. We recovered clones with a spectrum of sensitivity to each of the different assays and found that a clone transformed by one assay could be normal by another (Risser and Pollack 1974a,b).

In order to determine whether the complex response of 3T3 to SV40 was intrinsic to the virus or whether it was the result of the heteroploidy and genetic instability of the established line, we examined the response of normal rat embryo cells to SV40 (Risser and Pollack 1974b; Risser, Rifkin and Pollack 1975; Pollack and Risser 1974). Here, too, we found a complex response. After a single infection, a set of clones that were fully transformed with regard to serum and density, but which differed from one another by a thousandfold in their sensitivity to the anchorage requirement, was recovered (Table 1) (Risser, Rifkin and Pollack 1975; Pollack et al. 1974).

Table 1

Expression of Parameters of Transformation by SV40-transformed Rat Embryo Cell Clones

	Serum sensitivity (doubling time [hr] in fetal calf serum)		Saturation density ([cell/cm² × 10⁴] in fetal calf serum)		Anchorage dependence ([colonies/100 cells] inoculated in methylcellulose)
	1%	10%	1%	10%	
Uninfected rat embryo cells	∞	35	—[a]	25, 30[b]	<0.001
SV40-transformed clones					
SVRE 1	25	17	6	++[c]	0.2
2	33	20	13	++	ND
3	33	13	12	++	0,2, 0.10
4	35	20	11	++	0.80
5	27	23	14	++	2.75, 2.40
6	50	20	9	++, ++	0.01
7	28	15	~10	++	ND
8	30	13	20	++	2.15
9	27	16	14	++	0.75, 1.77
10	30	18	14	++	0.30
11	38	23	≥14	++, ++	0.002
12	33	20	11	25	0.013, 0.09
13	40	20	9	++, ++	0.70

[a] No growth in 1% fetal calf serum.

[b] Each figure represents a separate experiment.

[c] Grows so dense that cell layer peels at >30 × 10⁴ cells/cm², never maintaining a stable saturation density.

METHODS

Parts of this study, including all methods used, have been published elsewhere (Risser and Pollack 1974a,b; Pollack et al. 1974; Freedman and Shin 1974).

RESULTS AND DISCUSSION

Tumorigenicity in Athymic Nude Mice

In vivo loss of growth control, measured as tumorigenicity, might be related to any of the three in vitro parameters discussed above, or any combination of them, or none at all. The different clones described above, of rat and mouse origin, served as excellent material to test this directly in the athymic nude (nu/nu) mouse (Freedman and Shin 1974). Since nu/nu mice lack a functional thymus (Pantelouris 1968), virus-induced oncogenesis in them is not complicated by cell-mediated immunorejection of virus-coded cellular antigens (Rygaard 1969; Rygaard and Povlsen 1969; Povlsen et al. 1973; Vandeputte et al. 1974).

Twenty-four different cell lines, each of whose growth controls had been assayed in all three assays, were injected into nu/nu mice. Some lines were anchorage dependent (Table 2), while others were able to grow in semisolid medium (Table 3). We chose a sufficient number of lines to provide all possible combinations of presence and absence of sensitivity to the other modes of growth control. At the present time (December, 1974), the results are quite clear. The ability to grow as a tumor in nu/nu mice is best correlated with loss of the requirement for anchorage (Table 4).

Plasminogen Activator

A major biochemical correlate of in vivo malignancy is the production by tumor cells of plasminogen activator. The evidence that tumors, and cell lines derived from them, synthesize large amounts of an activator of plasminogen, whereas most normal tissues do not, has accumulated for many years and seems to be quite general (Fischer 1946; Barnett and Baron 1959; Unkeless et al. 1973; Ossowski et al. 1973b; Ossowski et al. 1973a; Unkeless et al. 1974; Rifkin et al. 1974; Goldberg 1974; Goldhaber, Cornman and Ormsbee 1947). However, in in vitro studies it has been found that many "normal" cell lines, including BHK21 and 3T3, do synthesize appreciable amounts of activator (Rifkin et al. 1974; Goldberg 1974; D. Rifkin, L. Beal and E. Reich, this volume); and some, such as BHK21 and, under some conditions, Balb 3T3 (C. Boone, pers. comm.), are in fact tumorigenic (Sanford, Likely and Earle 1954; Defendi, Lehman and Kraemer 1963; Jarrett and Macpherson 1968). This raises a serious question about the usefulness of established "normal" cell lines for in vitro studies of oncogenic transformation.

In a recent study, we by-passed this problem by transforming diploid primary cultures of rat embryo cells (Pollack et al. 1974; Risser et al., in prep.). If any of the parameters for transformation are related to tumorigenicity, then we would expect a correlation between plasminogen activator production and expression of that parameter. With clones derived by SV40

Table 2

Anchorage-dependent Lines Tested in nu/nu Mice

Species	Cell	Serum dependent	Density dependent
Rat	primary embryo cells	yes	no
	SVRE 12	no	no
Mouse	3T3	yes	yes
	Balb 3T3	yes	yes
	revertant MSV3T3 (M22)	yes	yes
	revertant SV3T3 (F1)	no	yes
Human	embryonic fibroblasts	yes	no

Anchorage dependence = <0.01% PE in methylcellulose.

Table 3

Anchorage-independent Lines Tested in nu/nu Mice

Species	Cell	Serum dependent	Density dependent
Rat	SVRE 9	no	no
Mouse	SV3T3 (101)	no	no
	MSV3T3 (K)	no	no
	A9 (L929)	no	no
	revertant SV3T3 (Aγ)	yes	yes
	revertant SV3T3 (LS)	yes	yes
Human	D98	no	no
	HeLa	no	no

Anchorage independence = >10% PE in methylcellulose.

Table 4

Tumorigenicity in nu/nu Mice

Anchorage dependence of injected cells	No. independently isolated clones tested	Animals with tumors
yes	6	0/36
no	10	26/28

2×10^6 Cells/mouse injected sc at 4 weeks; examined for at least 4 months.

infection of primary rat embryo cells, we found that growth without anchorage was correlated with secretion of plasminogen activator (Fig. 2).

Given the correlation between plasminogen activator production and growth in semisolid medium, it is still necessary to determine whether this relationship is causal or coincidental. The most direct test, cultivation of cells

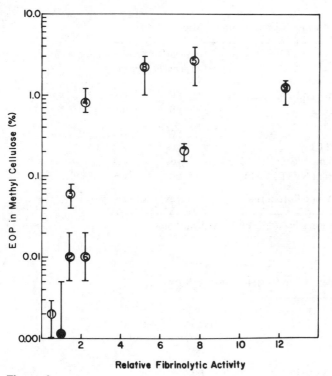

Figure 2

Anchorage dependence vs. fibrinolytic activity of normal and SV40-transformed clones of rat embryo cells. EOP = Efficiency of plating: colonies per 100 cells inoculated.

Fibrinolytic activity is that of plasminogen activated to plasmin by cellular factor. Each open circle is a separate SVRE clone; the closed circle is normal RE cells. The data for RE cells and clones 12, 3, 5 and 9 represent the averages of four separate assays of fibrinolytic activity and two assays of efficiency of plating in methylcellulose. Other lines were assayed once.

in exogenous plasmin, has not yet been accomplished, due to the lability of the protease. The following experiment, however, offers indirect evidence that plasminogen activation plays a role in anchorage-independent cell growth.

Anchorage dependence normally was assayed using methylcellulose supplemented with 10% fetal calf serum, which was the serum used to isolate the rat embryo (RE) cells and to clone the SVRE-transformed lines. Fetal calf serum supports only low levels of fibrinolysis by SVRE lines. Dog and monkey sera, however, permit high levels of fibrinolysis with rat cells (Ossowski et al. 1973b). If plasmin plays any role in growth in semisolid medium, then the SVRE lines might be expected to grow more efficiently in medium containing methylcellulose supplemented with dog or monkey serum.

To determine if this was so, we plated sister cultures of primary RE cells and three SVRE clones in regular medium on plastic dishes (Figs. 3 and 4)

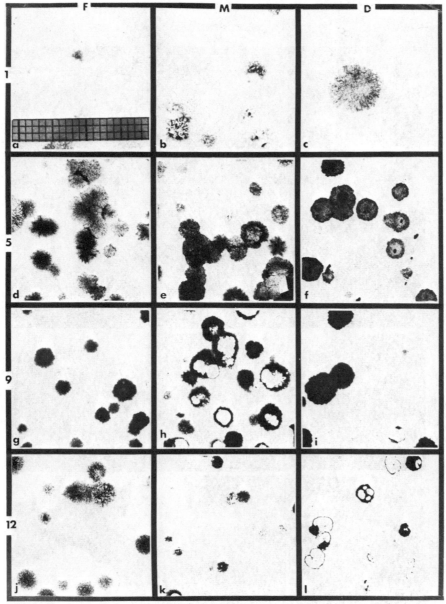

Figure 3

Four different cell populations grown on plastic in three different sera. One thousand cells each of primary rat embryo cells (1°) and SV40-transformed RE clones 5, 9 and 12 were inoculated into dishes containing Dulbecco's modified Eagle's medium supplemented with 10% fetal calf serum (F), 7.5% monkey serum plus 2.5% fetal calf serum (M), or 7.5% dog serum plus 2.5% fetal calf serum (D). After 12 days and three media changes, the plates were fixed in 3.8% formaldehyde in phosphate-buffered saline, stained with Harris hematoxylin, and photographed. The boxes in *a* are 1-mm square. (*a, b, c*) Primary RE cells; (*d, e, f*) SVRE 5 cells; (*g, h, i*) SVRE 9 cells; (*j, k, l*) SVRE 12 cells.

Note that all panels contain colonies. All three sera supported colony formation on plastic. In dog and monkey sera, the colonies of SVRE 9 and SVRE 12 grow very dense, so that their centers detach from the dish (*h, i, k* and *l*). The morphological difference is most striking in clone 12 (*j, k, l*).

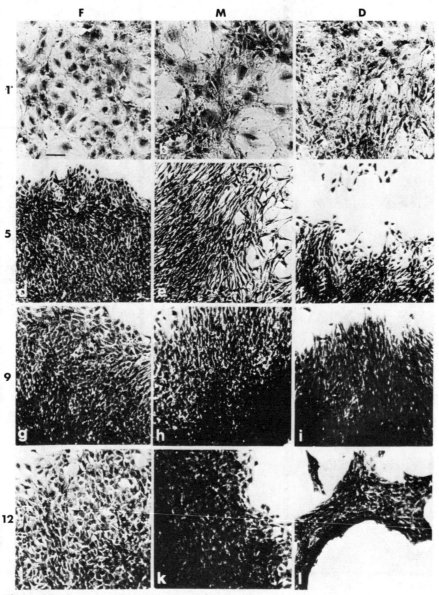

Figure 4
Same as Figure 3, but at higher magnification. Bar in *a* is 0.1 mm.

and in methylcellulose (Figs. 5 and 6) in four different serum mixtures, each making up 10% of the medium. Colony formation on plastic dishes, where anchorage was permitted, was insensitive to monkey serum for each of the three cell lines and for the primary RE cells and was greater than 10% for all cultures in fetal calf serum or monkey serum (Fig. 3). Dog serum was slightly toxic, but this toxicity was not specific for any line. In dog and monkey sera, all cultures, including primary RE cells, seemed less adherent to the dish than in fetal calf serum alone (Fig. 4).

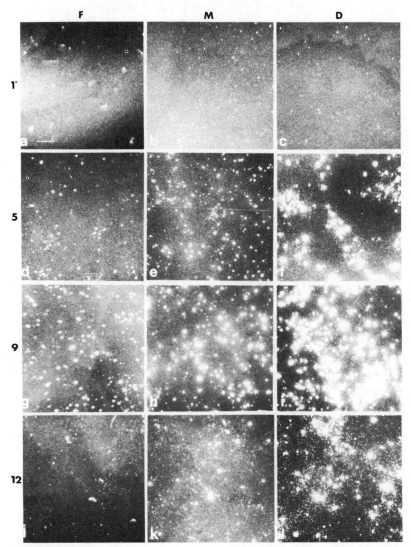

Figure 5

The panels represent the same cells, growing in the same sera, as in Figure 3. However, here the cells are suspended in methylcellulose, and the colonies that form are spherical. Primary RE cells do not form colonies in any of the sera (*a, b, c*). SVRE clones 5, 9 and 12 form a greater number of colonies in dog serum or monkey sera than in fetal calf serum (*d–l*). Growth of SVRE clone 12, in particular, is indistinguishable from that of uninfected RE cells in fetal calf serum (*a, j*) and indistinguishable from that of clones 5 and 9 in dog or monkey serum (*e, f; h, i; k, l*). The bar in *a* is 1 mm.

F   M   D

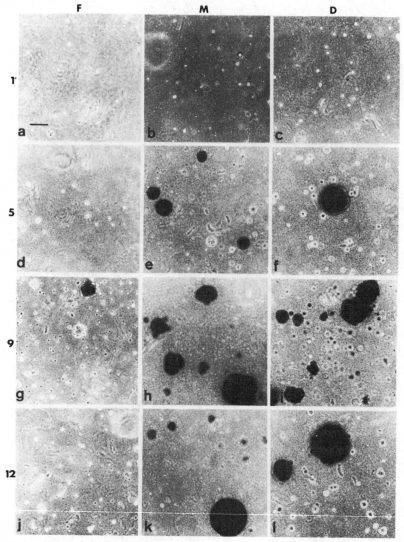

Figure 6

Same as Figure 5, but at higher magnification. Bar in *a* is 0.1 mm.

Growth in semisolid medium was strikingly dependent on the species of serum as well as on plasminogen activator production. Primary RE cells, which produced little activator, did not grow in any of the sera tested (Fig. 5). All three SVRE clones grew better in methylcellulose in media with either dog or monkey serum than in fetal calf serum alone (Fig. 5). SVRE 12 showed an especially marked increase in anchorage-independent growth (Fig. 5; Table 5).

At higher magnification, the colonies in methylcellulose are seen to be, for the most part, solid spheres ranging in size up to 0.3 mm (Fig. 6). The largest colonies contain between 10^3 and 10^4 cells. Only rarely will a colony have a

Table 5

Effect of Serum Species on Differential Specific Growth in Methylcellulose

Line	Plasminogen activator production[b]	Anchorage independence[a]			
		10% fetal	2.5% fetal 7.5% dog	5% fetal 5% dog	2.5% fetal 7.5% monkey
RE primary	1.0	≤0.01	≤0.07	≤0.01	0.04
SVRE 12	1.4	0.03	36	26	3
SVRE 5	7.6	2	27	10	14
SVRE 9	12.6	8	55	21	12

[a] Each culture was inoculated into Dulbecco's modified Eagle's medium with each of the sera given. To permit anchorage, cultures were inoculated into 60-mm dishes. Colonies were fixed in 10% formalin-PBS at 8 days after two changes of medium, stained with Harris hematoxylin, and counted. All colonies were at least 2 mm in diameter at this time. To exclude anchorage, sister cultures were inoculated in media containing the same sera, but also containing methylcellulose. Aliquots were then added to dishes precoated with agar. Colonies grew at the methylcellulose-agar interface. Cultures were fed with methylcellulose media weekly. Colonies were counted at 21 days, using a dissecting microscope and dark-field illumination. At this time, most colonies were at least 0.3 mm in diameter. Anchorage independence is calculated as [100 − 100 × (PE on plastic − PE in methylcellulose/PE on plastic)]. This is a measure of the ability of a line in a given serum to grow in the absence of anchorage corrected for its ability to form a colony on a plastic dish in that serum.

[b] To assay the ability of live cells to activate plasminogen, $2–5 \times 10^5$ cells (a subconfluent density) were plated on ^{125}I-fibrin-coated dishes (10 μg/cm^2; 4×10^4 cpm; 60-mm diameter) in 2.5% fetal calf serum and allowed to attach for 5 hours. The cultures were washed twice in serum-free medium, and then medium containing 2.5% dog serum was added. Since dog serum plasminogen is rapidly cleaved to plasmin by murine activator, the medium over cultures rich in activator becomes radioactive with time. To detect this, one-tenth of the medium was removed from each plate at 7–20 hours after switching to dog serum, mixed with 10 volumes of aquasol, and counted. Solubilization of counts was usually linear during this period, suggesting that activator production was the rate-limiting step. At 20 hours, the remaining medium was discarded, the subconfluent fibrin cultures were fixed and stained and the cells/dish calculated from counts of cells/field obtained using a Zeiss microscope (40× objective). From the radioactivity released by the cells or by an excess of trypsin and from the cell density, cellular activator was calculated as a rate: average cpm iodinated fibrin released/hr/10^5 cells/10^5 trypsinizable counts. This calculation of the rate of release of label permitted comparison of experiments by normalizing small differences from experiment to experiment in cell density and available fibrin.

looser appearance (Fig. 7). We are in the process of determining whether this difference in colony morphology is related to the extent to which plasminogen activator is secreted.

CONCLUSION

Other reports in this volume and elsewhere have shown an excellent correlation between malignancy and the production of a protease that activates the serum proenzyme plasminogen to plasmin. In order to study this protease further, it would be useful if there was an in vitro assay for tumorigenicity, and if cells that grew well in such an in vitro assay did so, at least in part, by virtue of their production of plasminogen activator.

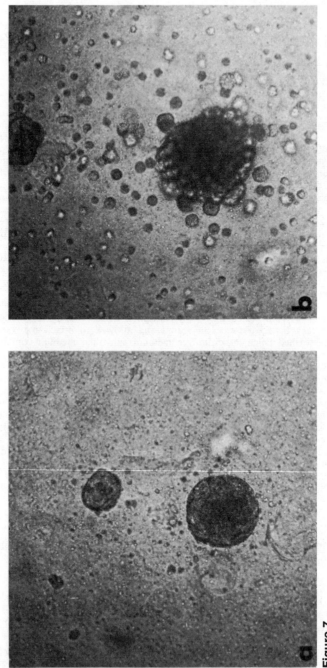

Figure 7
Selected colonies of SV40-transformed mouse cells growing in methylcellulose with 10% fetal calf serum. (*a*) Tight, spherical colony; (*b*) loose colony, possibly recruiting neighboring cells into satellite colonies.

In this paper, we present data which suggest strongly that growth in methylcellulose gel provides an in vitro assay for tumorigenicity, and that growth in methylcellulose is very likely to be dependent upon the production of plasminogen activator.

In these studies, we have used the small DNA-containing tumor virus, simian virus 40, to transform 3T3 mouse cells and primary rat embryo cells in vitro and the athymic nude mouse for tumorigenicity assay in vivo.

Anchorage and spreading of fibroblastic and epithelial cells to a solid substrate are critical prerequisites to normal cell growth. Tumor viruses cause the heritable loss of this anchorage requirement for growth, and in fact, the loss of the anchorage requirement for growth seems very well correlated with the acquisition of tumorigenicity (Weiss, Vesely and Sindelarova 1973; Eagle et al. 1970; Wiblin and Macpherson 1972; Grady and North 1974; Macpherson and Montagnier 1964; Wyke 1971; Freedman and Shin 1974). It now seems possible that both anchorage and growth control can also be diminished, at least transiently, through a mechanism involving proteases such as plasmin, collagenase and thrombin (Pollack et al. 1974; Risser, Rifkin and Pollack 1975; Muszbek and Laki 1974; Wahl et al. 1974; Chen, Teng and Buchanan, this volume).

Acknowledgments

This work was supported by grants from the National Institutes of Health, the National Science Foundation, and the American Cancer Society. We thank Gary Felsten for excellent assistance with the photography, and André Brown for assistance with nude mice.

REFERENCES

Barnett, E. and S. Baron. 1959. An activator of plasminogen produced by cell culture. *Proc. Soc. Exp. Biol. Med.* **102**:308.

Defendi, V., J. Lehman and P. Kraemer. 1963. "Morphologically normal" hamster cells with malignant properties. *Virology* **19**:592.

Eagle, H., G. E. Foley, H. Koprowski, H. Lazarus, E. M. Levine and R. A. Adams. 1970. Growth characteristics of virus-transformed cells. Maximum population density, inhibition by normal cells, serum requirement, growth in soft agar, and xenogenic transplantability. *J. Exp. Med.* **131**:863.

Fischer, A. 1946. Mechanism of the proteolytic activity of malignant tissue cells. *Nature* **157**:442.

Freedman, V. H. and S. Shin. 1974. Cellular tumorigenicity in *nude* mice: Correlation with cell growth in semi-solid medium. *Cell* **3**:355.

Goldberg, A. R. 1974. Increased protease levels in transformed cells: A casein overlay assay for the detection of plasminogen activator production. *Cell* **2**:95.

Goldhaber, P., I. Cornman and R. Ormsbee. 1947. Experimental alteration of the ability of tumor cells to lyse plasma clots *in vitro*. *Proc. Soc. Exp. Biol. Med.* **66**:590.

Grady, L. and A. North. 1974. Some effects of 5-bromodeoxyuridine on polyoma transformed mouse cells. *Exp. Cell Res.* **87**:120.

Jarrett, O. and I. Macpherson. 1968. The basis of tumorigenicity in BHK21 cells. *Int. J. Cancer* **3**:654.

Macpherson, I. and L. Montagnier. 1964. Agar suspension culture for the selective assay of cells transformed by polyoma virus. *Virology* **23**:291.

Muszbek, L. and K. Laki. 1974. Cleavage of actin by thrombin. *Proc. Nat. Acad. Sci.* **71**:2208.

Ossowski, L., J. Quigley, G. Kellerman and E. Reich. 1973a. Fibrinolysis associated with oncogenic transformation. Requirement of plasminogen for correlated changes in cellular morphology, colony-formation in agar, and cell migration. *J. Exp. Med.* **138**:1056.

Ossowski, L., J. C. Unkeless, A. Tobia, J. P. Quigley, D. B. Rifkin and E. Reich. 1973b. An enzymatic function associated with transformation of fibroblasts by oncogenic viruses. II. Mammalian fibroblast cultures transformed by DNA and RNA tumor viruses. *J. Exp. Med.* **137**:112.

Pantelouris, E. 1968. Absence of thymus in a mouse mutant. *Nature* **217**:370.

Pollack, R. 1970. Cellular and viral contributions to maintenance of the SV40-transformed state. *In Vitro* **6**:58.

Pollack, R. and R. Risser. 1974. The different stable patterns of growth control induced by SV40 infection of normal cells. In *Mechanisms of Viral Disease* (ed. W. Robinson and C. Fox), pp. 261–270. W. A. Benjamin, Menlo Park, California.

Pollack, R., H. Green and G. Todaro. 1968. Growth control in cultured cells: Selection of sublines with increased sensitivity to contact inhibition and decreased tumor-producing activity. *Proc. Nat. Acad. Sci.* **60**:126.

Pollack, R., R. Risser, S. Conlon and D. Rifkin. 1974. Plasminogen activator production accompanies loss of anchorage regulation in transformation of primary rat embryo cells by SV40 virus. *Proc. Nat. Acad. Sci.* **71**:4792.

Povlsen, C., P. Fialkow, E. Klein, G. Klein, J. Rygaard and F. Wiener. 1973. Growth and antigenic properties of a biopsy-derived Burkitt's lymphoma in thymus-less (nude) mice. *Int. J. Cancer* **11**:30.

Rifkin, D., J. Loeb, G. Moore and E. Reich. 1974. Properties of plasminogen activators formed by neoplastic human cell cultures. *J. Exp. Med.* **139**:1317.

Risser, R. and R. Pollack. 1974a. A non-selective analysis of SV40 transformation of mouse 3T3 cells. *Virology* **59**:477.

———. 1974b. Biological analysis of clones of SV40-infected mouse 3T3 cells. In *Control of Proliferation in Animal Cells* (ed. B. Clarkson and R. Baserga), p. 139. Cold Spring Harbor Laboratory, Cold Spring Harbor, New York.

Risser, R., D. Rifkin and R. Pollack. 1975. The stable classes of transformed cells induced by SV40 infection of established 3T3 cells and primary rat embryonic cells. *Cold Spring Harbor Symp. Quant. Biol.* **39**:317.

Rygaard, J. 1969. Immunobiology of the mouse mutant "nude." *Acta Path. Microbiol. Scand.* **77**:761.

Rygaard, J. and C. Povlsen. 1969. Heterotransplantation of a human malignant tumor to "nude" mice. *Acta Path. Microbiol. Scand.* **77**:758.

Sanford, K., G. Likely and W. Earle. 1954. The development of variations in transplantability and morphology within a clone of mouse fibroblasts transformed to sarcoma-producing cells *in vitro*. *J. Nat. Cancer. Inst.* **15**:215.

Unkeless, J., K. Danø, G. Kellerman and E. Reich. 1974. Fibrinolysis associated with oncogenic transformation. Partial purification and characterization of the cell factor, a plasminogen activator. *J. Biol. Chem.* **249**:4295.

Unkeless, J., A. Tobia, L. Ossowski, J. P. Quigley, D. B. Rifkin and E. Reich. 1973. An enzymatic function associated with transformation of fibroblasts by oncogenic viruses. I. Chick embryo fibroblast cultures transformed by avian RNA tumor viruses. *J. Exp. Med.* **137**:85.

Vandeputte, M., H. Eyssen, H. Sobus and P. deSomer. 1974. Induction of polyoma tumors in athymic nude mice. *Int. J. Cancer* **14:**445.

Vogel, A. and R. Pollack. 1973. Isolation and characterization of revertant cell lines. IV. Direct selection of serum-revertant sublines of SV40-transformed 3T3 mouse cells. *J. Cell. Physiol.* **82:**189.

———. 1974. Methods for obtaining revertants of transformed cells. In *Methods in Cell Biology* (ed. D. Prescott), vol. 8, p. 75. Academic Press, New York.

Vogel, A., J. Oey and R. Pollack. 1974. Two classes of revertants isolated from SV40-transformed 3T3 mouse cells. In *Control of Proliferation in Animal Cells* (ed. B. Clarkson and R. Baserga), p. 125. Cold Spring Harbor Laboratory, Cold Spring Harbor, New York.

Vogel, A., R. Risser and R. Pollack. 1973. Isolation and characterization of revertant cell lines. III. Isolation of density-revertants of SV40-transformed 3T3 cells using colchicine. *J. Cell. Physiol.* **82:**181.

Wahl, L. M., S. M. Wahl, S. E. Mergenhagen and G. R. Martin. 1974. Collagenase production by endotoxin-activated macrophages. *Proc. Nat. Acad. Sci.* **71:**3598.

Weiss, R. A., P. Vesely and J. Sindelarova. 1973. Growth regulation and tumor formation in normal and neoplastic rat cells. *Int. J. Cancer* **11:**77.

Wiblin, C. and I. Macpherson. 1972. The transformation of BHK21 hamster cells by simian virus 40. *Int. J. Cancer* **10:**296.

Wyke, J. 1971. Phenotypic variation and its control in polyoma transformed BHK21 cells. *Exp. Cell Res.* **66:**209.

Plasminogen, a Necessary Factor for Cell Migration In Vitro

Liliana Ossowski

The Rockefeller University, New York, New York 10021

James P. Quigley

SUNY Downstate Medical Center, Department of Microbiology and Immunology
Brooklyn, New York 11203

E. Reich

The Rockefeller University, New York, New York 10021

We have previously reported that oncogenic transformation in cell culture is associated with the production of high levels of plasminogen activator (Unkeless et al. 1973; Ossowski et al. 1973b; Rifkin et al. 1974). The secretion of activator and the resulting generation of the fibrinolytic and proteolytic action of plasmin have been shown to influence various phenotypic characteristics of both normal and transformed cell cultures (Ossowski et al. 1973a,b; Ossowski, Quigley and Reich 1974). One of the cellular parameters that appeared to depend on the presence of plasminogen was cell migration (Ossowski et al. 1973a).

Since cell migration is potentially of obvious significance to the invasive tendency of malignant cells, we have undertaken a more detailed study of the relationship between plasminogen and cell migration in culture. Because most fibroblasts are capable of autonomous migration when in culture, they provide a convenient model system for analyzing the factors that affect this process. Several laboratories have studied various aspects of cell migration in such cultures and have concluded that at least one, and more likely several, macromolecular factors are required for maximal migration (Clarke et al. 1970; Lipton et al. 1971; Bürk 1973).

In the present report, we document the participation of plasminogen in the migration process of various cell types. We have found that a conversion of plasminogen to plasmin is required, and that other serum components that influence the fibrinolytic system also affect the rate of migration. In addition, we have shown that the molecular requirements for the migration of transformed and normal cells may be different.

MATERIALS AND METHODS

Reagents were obtained as follows: fetal bovine serum (FBS), Reheis Chemical, Phoenix, Arizona; calf serum, Flow Laboratories, Rockville, Maryland;

and powdered culture medium, Grand Island Biological, Grand Island, N.Y. All petri dishes were disposable Falconware.

Plasminogen-depleted serum and purified plasminogen were prepared by affinity chromatography on columns of L-lysine-substituted Sepharose 4B according to procedures described in full elsewhere (Deutsch and Mertz 1970; Quigley, Ossowski and Reich 1974). Serum was depleted by two successive passages over such columns; the plasminogen, eluted with ε-amino-caproic acid, was further purified by a second cycle of affinity chromatography.

Cell Cultures

Embryonic fibroblast cultures were prepared by trypsinization of 11-day-old golden hamster embryo and 15-day-old rat embryo. 3T3, 3T3-SV40 (SV101) and SV40-transformed rat embryo fibroblasts (clone 12) were obtained from Dr. R. Pollack; B77-transformed rat embryo fibroblasts were acquired from Dr. H. Temin, and a human osteosarcoma cell line from Dr. G. Moore. Human osteosarcoma cultures were maintained in RPMI medium supplemented with 10% FBS, and all the other cells were maintained in Dulbecco's modified Eagle's medium with 5 or 10% of fetal bovine or calf serum.

Migration Assay

Transformed and normal cells were plated in 60-mm diameter gridded Falcon petri dishes and incubated for 3 days. When different serum concentrations or different sera were used, the cells were plated at correspondingly different concentrations to assure that the same cell densities would be present in the culture at the time of wounding and migration assay.

The cell monolayers were "wounded" with a razor blade as described by Bürk (1973). After wounding, the medium was aspirated and the cultures washed with fresh serum-free medium. Fresh medium, with or without serum, was then added, and the cultures were further incubated (usually for 16–20 hr), fixed with absolute methanol, and stained with Giemsa stain.

The number of cells migrating across the edge of the wound was determined by counting 6–16 grids (each grid is 1.7 mm wide and represents the assayed length unit) and averaged to yield the number of migrating cells per length unit.

Cocultivation Assay

Prior to cocultivation in Cooper dishes, 3T3 and SV101 cells were grown for at least 1 week in medium supplemented with either plasminogen-depleted serum or native serum. The 3T3 cells were trypsinized, plated onto the interior surface of a Cooper dish cover, at 1.5×10^5 cells per cover, and allowed to settle and attach for 4 hours. The bottom layers were prepared by plating 3T3 cells at 5×10^5 per plate or SV101 cells at 8×10^5 per plate. At 4 hours after plating, the covers with attached 3T3 cells were combined with the various bottom layers and incubated for 3 days before wounding. Medium was changed daily in the dishes that contained only 3T3 cells. Covers that were being cocultivated with SV101 bottom layers were transferred to freshly prepared SV101 cultures to maintain exposure to a constant concentration of transformed cells; the latter were prepared daily using EDTA (5×10^{-4}

M) for detachment of cells. At the desired time, the monolayers growing on the covers were wounded in the usual way, washed with serum-free medium, placed over an appropriate bottom layer, incubated for 18 hours, fixed and stained. Cell migration into the wound was determined in the regular way. Control experiments established that under these conditions any cells that detach from the bottom do not attach to the top of the Cooper dish.

RESULTS

The migration of cells from the edges of monolayers into "wounds" was examined using an assay system similar to that described by Bürk (1973) in which monolayers were "wounded" with a razor blade, and cell migration was monitored microscopically after fixation and staining, as a function of time. A comparison of the migration, at equal cell densities, of a transformed cell line (SV101), a nontransformed cell line (3T3) and normal secondary hamster fibroblasts is given in Table 1. When cultures were incubated in medium containing 5% FBS, the number of transformed cells migrating into the wound area was 1½-fold greater than that of normal cells. However, when the same experiment was performed in medium without serum, a much more pronounced difference was evident between the migration of transformed cells on the one hand and untransformed or normal cells on the other. Under these conditions, the number of transformed cells migrating into the wound was approximately 25-fold greater than the number of normal or untransformed cells. This finding confirms similar results reported by Lipton and Bürk (Lipton et al. 1971; Bürk 1973).

In view of previous results suggesting the requirement for plasminogen in cell migration (Ossowski et al. 1973a), experiments were performed to establish the extent to which either plasminogen or plasmin could account for the above effect of serum.

Table 1
Migration of Transformed and Normal Cells into a "Wound" in Presence and Absence of Serum

Cell type	No. cells/plate on day of wounding ($\times 10^{-6}$)	No. cells migrating into "wound"/length unit	
		in serum	no serum
SV101 (3T3-SV40)	1.7	148	25
3T3	1.6	104	1.2
Hamster (normal)	1.6	106	0.9

3T3 cells (1.8×10^6), SV101 cells (1.5×10^6) and hamster embryo fibroblasts (1.5×10^6) were plated on 60-mm, gridded petri dishes in medium containing 5% of FBS. The following day, the cells in one set of cultures were trypsinized and counted. At the same time, companion cultures were wounded and incubated in medium with 5% of FBS or in medium without serum. Fifteen hours after wounding, the cultures were fixed and stained, and the number of cells that had migrated across the edge of the wound was determined.

In a first experiment, SV101 cells were grown either in native or plasminogen-depleted fetal bovine serum until confluent monolayers of equal density were formed. The cell layers were then wounded, and the number of cells migrating into the wound in the presence of native or plasminogen-depleted serum was determined at various times thereafter. The results (Fig. 1) show that in cultures supplemented with either native or plasminogen-depleted serum, the number of cells migrating into the wound increased linearly with time. However, the reduction in plasminogen-depleted serum demonstrates a requirement for plasminogen in migration of transformed cells.

In order to define the optimal conditions for cell migration, a wide concentration range of both native and plasminogen-depleted sera was explored. Transformed SV101 cells were grown in either native or plasminogen-depleted fetal bovine serum at concentrations varying from 1–20%, and the cultures were divided into two sets. One set was first grown in medium containing the respective serum, and serum was then omitted after wounding. In the second set of cultures, the serum was present in the culture medium both before and after wounding. The results (Fig. 2) show that the number of cells migrating into a wound was insensitive to serum concentration in the range 1–5% of native serum. However, cell migration was sharply reduced in high concentrations of native serum (10–20%). When native serum was omitted from the culture following wounding, the number of cells migrating increased in proportion to the serum concentration in which the cells were grown before wounding. The optimal migration in this case occurred when cells had been grown in 10% of native serum. In contrast, incubation of cultures in increas-

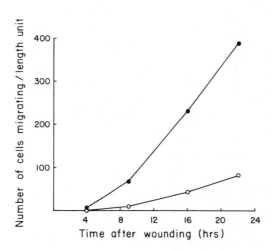

Figure 1
Time course of migration in native FBS and in plasminogen-depleted fetal bovine serum. SV101 cells grown for at least 1 week in FBS or in plasminogen-depleted FBS were trypsinized and plated at 4×10^5 cells per plate in medium with 5% of the respective serum. Three days after plating, cultures were wounded and incubated thereafter in medium with 2% of FBS or plasminogen-depleted FBS. On the day of wounding the number of cells in the two sera was $4.2–4.6 \times 10^6$/plate. The cultures were fixed, stained, and counted for migration at 4, 9, 16 and 22 hours after wounding. (•——•) Native FBS; (o——o) plasminogen-depleted FBS.

Figure 2

Effect of serum concentration (native and plasminogen-depleted) on migration of SV101 cells. SV101 cells that were incubated for at least 1 week in medium containing FBS or plasminogen-depleted FBS were trypsinized and plated at 2–5.4 × 10⁵ cells per dish in media with increasing concentrations (1–20%) of serum. The number of cells in all cultures on the day of wounding was approximately 3 × 10⁶. The cultures preincubated in 10 and 20% serum were washed twice after wounding. The cultures were fixed, stained, and migrating cells counted 15 hours after wounding. (●——●) Grown in the indicated concentration of native serum before wounding and incubated after wounding in the same concentration of native serum; (○——○) grown in the indicated concentration of native serum —no serum present after wounding; (▲——▲) grown in the indicated con-

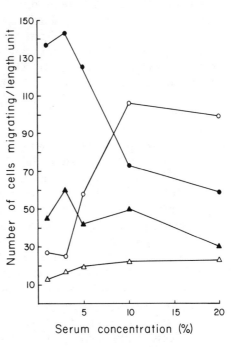

centration of plasminogen-depleted serum before wounding and incubated after wounding in the same concentration of plasminogen-depleted serum; (△——△) grown in the indicated concentration of plasminogen-depleted serum before wounding—no serum present after wounding.

ing concentration of plasminogen-depleted serum had no such pronounced effect on migration when examined under both sets of experimental conditions. In all cases, migration was reduced when cells were grown in plasminogen-depleted serum, compared with rates of migration in native serum. The maximal difference in the number of cells migrating in the two sera occurred at low serum concentrations since the inhibitory effects of high native serum concentrations obscured the differences due to plasminogen.

These results further confirm the participation of plasminogen in migration. In addition, the inhibitory effects of high serum concentration are reminiscent of comparable inhibition of fibrinolysis under similar conditions, a phenomenon which is known to be due to increasing concentration of serum plasminogen inhibitors. To test for this, the serum was first acidified and then neutralized (the serum is brought to pH 3 by dropwise addition of 1 N HCl, allowed to stand at room temperature for 2 hr, and neutralized with 1 N NaOH), a procedure that is known to destroy serum plasmin inhibitors (D. Loskutoff, unpubl.). Acid-treated and native sera were then compared in a standard migration assay.

The migration of SV101 cells was measured under standard conditions, except that the concentrations of either native or acid-treated sera were varied. Increasing concentrations of native serum produced the expected inhibition of migration (Fig. 3); in contrast, no such inhibition was observed with acid-inactivated serum, indicating that this property of native serum was abolished by the acid treatment. Thus the similar acid sensitivity of both serum protease and cell migration inhibitor(s) implies that they may belong to the same class of molecules. Although this evidence is suggestive, further purification of the serum components will be necessary to establish their identity with more certainty.

It was of interest to determine the effect of plasminogen concentration on cell migration under conditions in which a constant level of the putative serum inhibitor(s) was maintained in the culture. The following experimental approach was used: SV101 cells were preincubated in plasminogen-depleted serum which was supplemented with increasing concentrations of purified plasminogen ranging from 0–4-fold the amount present in native serum. The cultures were wounded, and the migration was monitored both in serum-supplemented and serum-free medium. The results (Fig. 4) revealed that the migration of cells incubated under both sets of conditions reflected the prior concentration of plasminogen in the medium, and that migration increased progressively with antecedent exposure to corresponding increments of plasminogen. It should be noted that cells preincubated in 3% serum containing a 4-fold excess of plasminogen, and subsequently incubated in the absence of serum, migrate as well as cells incubated in the presence of full native serum.

All of the preceding results were consistent with the interpretation that plasminogen was necessary for cell migration, at least in the case of transformed cultures. Therefore, an experiment was designed to determine whether the requirement of plasminogen in cell migration reflected its activation to plasmin or involved some other property of the zymogen itself. For this purpose, SV101 cells were grown in medium containing plasminogen-depleted

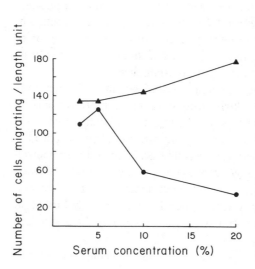

Figure 3

Effect of increasing serum concentration on migration of transformed cells: comparison of native and acid-treated sera. SV101 cells were plated in medium supplemented with 3, 5, 10 or 20% of native or acid-treated serum. On the day of wounding the number of cells in cultures incubated in native serum varied from 2.4–4.0 × 10⁶ per plate and in acid-treated serum from 2.1–2.8 × 10⁶ per plate. The cultures were fixed, stained, and counted 19 hours after wounding. (•——•) Native serum; (▲——▲) acid-treated serum.

Figure 4

Effect of plasminogen concentration on cell migration. SV101 cells grown in native or in plasminogen-depleted FBS were trypsinized and plated at 2×10^5 cells per dish in medium containing 3% of the respective serum. One day before wounding some cultures received fetal bovine plasminogen at a concentration equal to that in native FBS or 2- or 4-fold greater. On the day of wounding the number of cells in the various cultures was in the range $2.8–3.5 \times 10^6$. After wounding, some cultures were incubated in medium without serum, others in medium with the same serum equivalent to that present in preincubation medium. The cultures were fixed and stained 20 hours after wounding. The number of cells migrating in native serum per length unit was 195. (●——●) Grown in the indicated concentration of plasminogen both before and after wounding; (○——○) grown in the indicated concentration of plasminogen before wounding—no serum or plasminogen present after wounding.

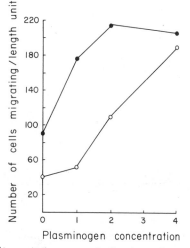

fetal bovine serum and supplemented with either purified fetal bovine plasminogen or chicken plasminogen. The molecular and catalytic properties of these two plasminogens are very similar, except that fetal bovine plasminogen is readily activated to plasmin by mammalian activators, whereas chicken plasminogen is not (Quigley, Ossowski and Reich 1974). The results shown in Figure 5 compare the effects of these plasminogens on the rate of migration. It is clear that cell migration was stimulated only in cultures supplemented with fetal bovine plasminogen; the migration of cells in medium supplemented with chicken plasminogen was no greater than in medium containing plasminogen-depleted serum alone. This result indicated that the plasminogen involvement in migration of transformed cells most likely was by way of subsequent conversion to catalytically active plasmin. This was also consistent with previous results cited showing a similar acid sensitivity of both serum plasmin and serum migration inhibitors (Fig. 3).

To determine the generality of the plasminogen/plasmin requirement for cell migration, several transformed cell lines were examined for their ability to migrate in native and in plasminogen-depleted fetal bovine serum. These included rat fibroblasts transformed by Rous virus B77 or by SV40, hamster embryo fibroblasts transformed by SV40, and a human osteosarcoma cell line. The results in Table 2 show that the removal of plasminogen from the incubation medium decreased the extent of migration of all tested cell lines.

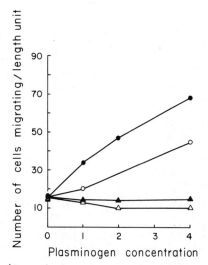

(1 equals the concentration in native serum)

Figure 5

Conversion of plasminogen to plasmin is required for optimal migration. SV101 cells grown in native or plasminogen-depleted FBS were trypsinized and plated at 2.2×10^5 cells per dish in medium containing 3% of the respective serum. One day before wounding FBS plasminogen and chicken plasminogen, respectively, were added to separate sets of cultures at a concentration equivalent to that present in native serum or 2- or 4-fold higher. On the day of wounding the number of cells in the different cultures was in the range $2.1–2.9 \times 10^6$ per dish. The wounded cultures were incubated in medium with serum (2%) supplemented with plasminogen at a concentration equivalent to that present in growth medium before wounding. One set of cultures was incubated in medium without serum. The cultures were fixed and stained 8 hours after wounding. The number of cells migrating in native serum per length unit was 59. (●——●) Plasminogen-depleted FBS supplemented with FBS plasminogen present both before and after wounding; (▲——▲) plasminogen-depleted FBS supplemented with chicken plasminogen present both before and after wounding; (○——○) plasminogen-depleted FBS supplemented with FBS plasminogen present before wounding—no serum or plasminogen present after wounding; (△——△) plasminogen-depleted FBS supplemented with chicken plasminogen present before wounding—no serum or plasminogen present after wounding.

908

Table 2

Migration of Various Transformed Cell Lines in Native and
Plasminogen-depleted Sera

Serum present		No. cells migrating/length unit in			
before wounding	after wounding	rat B77	rat SV40	hamster SV40	human tumor
Native	native	52	95	300	104
Native	none	7	26	63	7
Plasminogen- depleted	plasminogen- depleted	24	48	41	19
Plasminogen- depleted	none	6	20.5	18	5.5
Reconstituted	reconstituted	NT	95	NT	108

Rat B77 (1.5×10^5) and rat SV40 (3×10^5) were plated in Dulbecco's medium with 3% of native or plasminogen-depleted FBS. Hamster SV40 (2×10^5) was plated in Dulbecco's medium with 5% of the native or depleted serum, and human osteosarcoma (4×10^5) was plated in RPMI medium with 5% of one or the other serum. On the day of wounding, the number of cells present in the different cultures was in the range 2.3×10^6–3.0×10^6. The cultures were wounded and incubated for 15 hours in medium with 3% (hamster SV40) or 2% (rat B77, rat SV40, human osteosarcoma) of native serum, plasminogen-depleted serum or the depleted serum supplemented with plasminogen ($1 \times$ the amount present in native serum for rat SV40; $4 \times$ the amount of plasminogen present in native serum for the human line). NT = not tested.

The migration was fully restored to the level observed in native serum when depleted sera were reconstituted by addition of highly purified plasminogen.

Effect of Plasminogen on Migration of Normal Cells

Most of the preceding experiments having been performed with transformed cells, it was of interest to study the influence of plasminogen on migration of normal rat and hamster embryo fibroblasts and an untransformed 3T3 cell line. Accordingly, the respective cultures were grown in media containing either native or plasminogen-depleted serum, or plasminogen-depleted serum reconstituted with increasing concentrations of plasminogen. The decrease in the migration of normal cells incubated in plasminogen-depleted serum was comparable to that observed with transformed cells (Table 3). However in contrast to the migration of transformed cells, that of normal cells could not be fully restored by the addition of purified plasminogen to the depleted serum. Further comparison of transformed and normal cells revealed additional differences in the migration pattern of the two cell types. In contrast to the migration of transformed cells, which was inhibited by high concentrations of native serum (Fig. 2), the migration of normal cells increased with increments of serum in the incubation medium (Fig. 6). A similar response was observed when increasing concentrations of plasminogen-depleted serum were used for migration assays. However at all of the concentrations tested, the migration of normal cells in plasminogen-depleted serum was lower than that in native serum, and reconstitution with purified plasminogen gave partial restoration of migration.

Table 3

Effect of Plasminogen Depletion on Migration of 3T3 Cells and Rat and Hamster Embryo Fibroblasts

Serum present before and after wounding (FBS)	No. cells migrating/length unit in		
	3T3	rat	hamster
Native	36.6	83	191
Plasminogen-depleted	5.9 (16)[a]	40	57.3
Reconstituted with 1 × plasminogen	5.5	52	NT
Reconstituted with 2 × plasminogen	7.3	51	68.6
Reconstituted with 4 × plasminogen	7.5	55	NT

3T3 cells and hamster and rat embryo fibroblasts were grown for at least 4 days in either native or plasminogen-depleted FBS. The cultures were trypsinized and plated (3T3—1.5×10^5; hamster and rat—2×10^5) in medium with 5% of FBS or plasminogen-depleted FBS. To one set of cultures, plasminogen was added 1 day after plating. The plasminogen added was equal to the amount present in native serum (1 ×) or 2- and 4-fold greater (2 ×, 4 ×). Three days after plating, one set of cultures was trypsinized and the cells counted (3T3—3.6×10^5; rat—1.3×10^6; hamster—1.1–1.5×10^6 cell). At the same time, some cultures were wounded and incubated in medium with the appropriate serum (5%) or in medium without serum for 18 hours. NT = not tested.

[a] The value in parentheses gives the number of cells migrating in a companion culture that was grown in plasminogen-depleted FBS for 3 days before wounding and transferred to native serum after wounding.

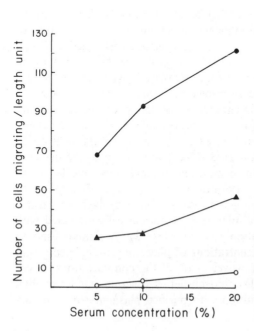

Figure 6

Effect of increasing concentration of serum on migration of 3T3 cells. 3T3 cells grown in native or in plasminogen-depleted FBS were trypsinized and plated in medium with 5, 10 or 20% of the respective serum. On the day of wounding the number of cells present in the various cultures was in the range 0.6–0.9×10^6. After wounding, the cultures were grown in medium supplemented with serum concentration equivalent to that present in preincubation mixture or in medium without serum for 18 hours. (●——●) Native serum present both before and after wounding; (▲——▲) plasminogen-depleted serum present both before and after wounding; (○——○) native serum present in preincubation medium, no serum present after wounding.

Cocultivation of 3T3 Cells with Transformed SV101 Cells

The preceding results showing (1) that increasing concentrations of plasminogen-depleted serum increased migration of normal cells, and (2) that reconstitution with plasminogen did not fully restore migration to the levels observed in native serum suggested that plasminogen might not be the only factor involved in normal cell migration. Although serum is quantitatively freed of plasminogen by affinity chromatography, this procedure also removes several other proteins, and the possibility remained that additional serum factors needed for migration of normal cells might thereby have been lost or inactivated. Furthermore, the behavior of normal cells differed clearly from that of transformed cells, since results with the latter suggested that plasminogen was both necessary and sufficient for their migration. This implied that transformed cells might synthesize and/or store the additional component(s) apparently required by normal cells. To test for this, cocultivation experiments were designed in which 3T3 cells were incubated in Cooper dishes (see Methods) over a layer of transformed SV101 cells; comparable experiments in which the bottom cell layer consisted of 3T3 cells or was cell-free served as controls. The results in Table 4 show that the migration of 3T3 cells on the dish covers was not affected by other cells in cocultivation provided that the incubation medium contained either native or plasminogen-depleted serum. However when the depleted serum was reconstituted with highly purified plasminogen, the 3T3 cells which were incubated over transformed cells showed a 2-fold increase in migration when compared either with 3T3 cells grown over a layer of 3T3 cells or grown in medium alone.

Table 4

Effect of Plasminogen on Migration of 3T3 Cells Cocultivated
with Transformed Cells (SV101)

	No. 3T3 cells migrating/length unit in		
Cells in bottom layer	serum (5%) without plasminogen	serum (5%) with plasminogen	serum (5%) native
none	13.1	10.8	31.8
3T3	14.9	11.3	31.1
SV101	11.3	24.5	32.3

The bottoms of Cooper dishes were seeded with SV101 cells (10^6), 3T3 cells (4×10^5), or incubated with medium alone. The cultures were prepared in three sets, containing 6.5 ml of media per plate, supplemented, respectively, with 5% of either native FBS, plasminogen-depleted FBS or depleted serum reconstituted with 2-fold the normal concentration of highly purified plasminogen. The covers of the Cooper dishes were prepared by plating 3T3 cells (10^5) in 0.5 ml of medium supplemented with 5% of FBS or of plasminogen-depleted FBS. The cells were allowed to settle and attach for 4 hours, and the two parts of the dishes were then combined, as indicated, for 3 days of cocultivation. Medium was changed daily in the dishes containing 3T3 or medium alone at the bottom. In order to maintain constant conditions throughout the experiment, the bottoms containing SV101 were replaced daily (10^{-4} M EDTA was used to detach cells for preparation of the fresh bottoms). The 3T3 cells growing on the tops of the Cooper dishes were wounded and further cocultivated for 20 hours before fixation, staining and counting of migrated cells.

DISCUSSION

With the reservation that only a limited number of cell types have been investigated, our findings establish that plasminogen, or, more likely, plasmin, is in some way necessary for cell migration in culture. While the responses of transformed and untransformed cells are different, the migration of both cell types is reduced in plasminogen-depleted media and at least partially restored when such media are reconstituted with the highly purified pro-enzyme. In the case of neoplastic and transformed cell lines, it seems likely that plasminogen is both necessary and sufficient for optimal migration. Thus (1) migration is reduced by up to 80% under conditions of stringent and prolonged plasminogen depletion and is fully restored by sera reconstituted with normal levels of plasminogen; and (2) preincubation with media containing elevated levels of plasminogen is sufficient to promote subsequent maximal levels of migration. This suggests that plasminogen, plasmin, or a product of plasmin action must accumulate or persist in transformed cultures for prolonged periods. (By means of ^{125}I-labeled plasminogen we have found that a significant proportion of plasminogen [at least 1% of the total available] persists in transformed, but not in normal, cell cultures for several days. This persistent fraction is greatly reduced by high concentrations of plasmin inhibitors.) (3) The serum factors that inhibit migration of transformed cells resemble known plasmin inhibitors in their sensitivity to irreversible inactivation at acid pH. (4) The reconstitution of plasminogen-depleted serum restores cell migration only if the added plasminogen is convertible to plasmin by the cellular activator in question. All of the above facts suggest that plasmin is required for optimal migration by transformed and neoplastic cells, and that it can be rate-limiting for this process in culture; there is no concrete indication concerning any mechanism by which plasmin may be exerting its effect. The possibility that plasmin may also participate in cell migration and invasiveness in animals is tantalizing to consider, but no direct evidence bearing on this question is as yet available.

The relationship between cell migration and plasminogen appears less clear for normal and untransformed cells than in the case of transformed cells. Thus high serum concentrations stimulate untransformed cell migration, whereas they inhibit migration of neoplastic cells. Furthermore, while normal cell migration is reduced by plasminogen depletion, it is not fully restored by reconstitution of media with purified plasminogen. These facts suggest that while plasminogen may be required for optimal migration by normal cells, it is not sufficient by itself, and other serum components are also necessary. Since cocultivation with transformed cells stimulates plasminogen-dependent migration by normal cells, it may be that plasminogen activator, which is present at low levels in most sera, is one of the additional serum components required for migration. However, the situation is likely to be more complex than this because stimulation of untransformed cell migration may also occur by other routes, e.g., by Bürk's factor (Bürk 1973).

Acknowledgments

This work was supported by grants from the American Cancer Society (ACS 84-E, F) and the National Institutes of Health (CA-08290).

REFERENCES

Bürk, R. R. 1973. A factor from a transformed cell line that affects migration. *Proc. Nat. Acad. Sci.* **70**:369.

Clarke, G. D., M. G. P. Stoker, A. Ludlow and M. Thornton. 1970. Requirement of serum for DNA synthesis in BHK 21 cells: Effects of density, suspension and virus transformation. *Nature* **227**:798.

Deutsch, D. G. and E. T. Mertz. 1970. Plasminogen: Purification from human plasma by affinity chromatography. *Science* **170**:109.

Lipton, A., I. Klinger, D. Paul and R. W. Holley. 1971. Migration of mouse 3T3 fibroblasts in response to serum factor. *Proc. Nat. Acad. Sci.* **68**:2799.

Ossowski, L., J. P. Quigley and E. Reich. 1974. Fibrinolysis associated with oncogenic transformation. Morphological correlates. *J. Biol. Chem.* **249**:4312.

Ossowski, L., J. P. Quigley, G. M. Kellerman and E. Reich. 1973a. Fibrinolysis associated with oncogenic transformation. Requirement of plasminogen for correlated changes in cellular morphology, colony formation in agar and cell migration. *J. Exp. Med.* **138**:1056.

Ossowski, L., J. C. Unkeless, A. Tobia, J. P. Quigley, D. B. Rifkin and E. Reich. 1973b. An enzymatic function associated with transformation of fibroblasts by oncogenic viruses. II. Mammalian fibroblast cultures transformed by DNA and RNA viruses. *J. Exp. Med.* **137**:112.

Quigley, J. P., L. Ossowski and E. Reich. 1974. Plasminogen, the serum proenzyme activated by factors from cells transformed by oncogenic viruses. *J. Biol. Chem.* **249**:4306.

Rifkin, D. B., J. N. Loeb, G. Moore and E. Reich. 1974. Properties of plasminogen activators formed by neoplastic human cell cultures. *J. Exp. Med.* **139**:1317.

Unkeless, J. C., A. Tobia, L. Ossowski, J. P. Quigley, D. B. Rifkin and E. Reich. 1973. An enzymatic function associated with transformation of fibroblasts by oncogenic viruses. I. Chick embryo fibroblast cultures transformed by Avian RNA tumor viruses. *J. Exp. Med.* **137**:85.

Role of Protease Activity in Malignant Transformation by Rous Sarcoma Virus

Michael J. Weber and Arthur H. Hale

Department of Microbiology, University of Illinois
Urbana, Illinois 61801

David E. Roll

Department of Biochemistry, University of Illinois
Urbana, Illinois 61801

We have been concerned with the question of what role proteolytic activity plays in the genesis and maintenance of the transformed phenotype. Our approach has been to determine whether some of the early cellular changes associated with malignant transformation by tumor viruses can be mimicked by treatment of normal cells with proteases or other hydrolytic enzymes, and whether some of these transformation-specific changes can be reversed by treating transformed cells with protease inhibitors. The advantage of this approach is that it allows a determination of *which* aspects of the transformed phenotype are dependent on proteolysis, gives an indication of the specificity of the proteases involved, and helps to elucidate the pathway by which viral oncogenic information modifies the cell. The major disadvantage of the approach is that the lack of specificity of some of the reagents used as probes makes cautious interpretation of results imperative.

We also have examined extracts of normal and transformed cells for changes in proteolytic activity which might be correlated with our observations on protease-dependent changes in cellular physiology.

The system we have used for these studies is chick embryo fibroblasts, either normal or transformed by Rous sarcoma virus. In addition, some experiments were performed with cells infected by RSV-T5 (Martin 1970), a mutant of Rous sarcoma virus which is temperature-conditional for transformation, but not for replication. The advantage of this system for studies on the mechanism of viral oncogenesis is that 100% of the cells can be rapidly transformed, minimizing problems associated with selection of cells whose properties have become altered as a secondary consequence of transformation.

METHODS

Adhesion Assay

Adhesion was measured by removing the growth medium from the culture dish (2 ml on a 35-mm dish) and dropping it back onto the cells from a constant height (generally 10 cm) while systematically moving the plate back and forth under the stream. The procedure was repeated a total of 5 times until the entire surface of the dish was hit by the stream. The detached cells were then transferred to 0.25% trypsin, 10^{-3} M EDTA and incubated for 0.5 hour to break up clumps. The adherent cells remaining on the dish were removed with 0.25% trypsin, 10^{-3} M EDTA. Both sets of cells were counted using a Coulter Particle Counter. The technique provides data in which duplicates agree to within 10%. In typical transformed cultures, 50–90% of the cells will be detachable, depending on the experimental conditions, whereas in an untransformed culture, between 5 and 10% of the cells are detachable. The assay does not measure the strength of adhesion, but measures the number of cells with less than a certain level of adhesiveness.

Use of TLCK

Since TLCK (tosyl-lysyl-chloromethyl ketone) is both toxic for the cells and unstable, it is worth mentioning some special precautions that must be taken in order to obtain reproducible results with the inhibitor. Cells must be healthy, proliferating, subconfluent and in culture for at least 48 hours. The TLCK should be stored as a frozen, dessicated powder, and fresh solutions made up prior to each use. The pH of the growth medium should be closely controlled and should not be allowed to rise when cells are removed from the CO_2 incubator, since the half-life of TLCK can be as little as 3–5 hours in growth medium equilibrating with the air (measured by inhibition of trypsin hydrolysis of tosyl-arginyl methyl ester). By taking these precautions, one should reproducibly be able to restore the morphology of transformed cells and their rate of hexose transport to normal and cause an increase in adherence of the cells to the dish such that 20% or fewer of the cells will be detached in the adhesion assay described above.

Protease Assays

Plasmin was assayed on plastic plates coated with ^{125}I-fibrin, as described by Unkeless et al. (1973).

Hydrolysis of ^{14}C-chlorella protein was as described by Schnebli (1972), except that BSA was added as a coprecipitant during the preparation of the substrate. Ten μl of ^{14}C-chlorella protein, containing approximately 30,000 cpm, was used for each assay. The reaction was stopped by addition of 10% TCA. Both soluble and precipitable counts were determined, and the data expressed as a percentage of counts hydrolyzed (less the background).

Cells for protease analysis were plated in Dulbecco's medium containing 2% tryptose phosphate broth, 1% calf serum and 1% heat-inactivated chicken serum on 100-mm dishes at densities of 6×10^5, 3×10^6 and 6×10^6 cells for the normal-growing, transformed and density-inhibited cultures, respectively. After 2 days in culture, extracts were prepared for analysis of protease activity as follows: Cells were washed on the dish with 12.5 ml of phosphate-buffered saline and then once quickly with RSB buffer (10 mM Tris, 10 mM NaCl, 1.5 mM $MgCl_2$, pH 7.4). They then were scraped into a small volume of RSB. After swelling, the cells were broken in a teflon glass homogenizer. More than 99% of the cells were broken as judged by microscopic examination.

Cell Culture and Biochemistry

Cell culture and transport measurements were as described previously (Weber 1973) except where indicated otherwise in the text and legends. Cells were grown in Dulbecco's medium containing 10% tryptose phosphate broth, 4% calf serum and 1% heat-inactivated chicken serum, unless otherwise noted. [^{3}H]-2-Deoxyglucose was used at 0.5 μCi/ml except for the experiments shown in Figure 1 and Table 2 where the concentration was 1.0 μCi/ml. Protein was assayed by the method of Lowry et al. (1951). Plasminogen was purified from chicken serum by the method of Deutsch and Mertz (1970) or was purchased from Sigma (porcine profibrinolysin) and was activated by urokinase (Sigma Chemical, St. Louis) as described by Robbins and Summaria (1970). Growth medium and serum were from GIBCO, Grand Island, New York. Protease inhibitors were from Sigma or Nutritional Biochemicals. Peptides were from Cyclo Chemical, Los Angeles. Isotope was from Amersham/Searle, Chicago. Hyaluronidase (Type II, 430 NFU/mg) and neuraminidase (Type V, *C. perfringens*, 0.22 U/mg) were from Sigma, and trypsin was from Miles-Seravac. Antipain, leupeptin and pepstatin were a generous gift from Drs. Aoyagi and Umezawa.

RESULTS

Do Rous-transformed Cells Contain Increased Levels of Proteases?

Transformation-specific increases in proteolytic cleavage of plasminogen to plasmin are well documented (Unkeless et al. 1973; Reich 1974). In addition, there are reports that transformed cells display increased protease activity on nonspecific substrates (Schnebli 1972; Bosmann 1972). We have assayed extracts of Rous-transformed cells and the growth medium from these cells both for activation of fibrinolysis and for generalized protease activity using either ^{125}I-fibrin or ^{14}C-chlorella protein as substrate. The data (Table 1) demonstrate that the transformed cell extracts do not display an increased ability to hydrolyze ^{14}C-chlorella protein, even though they do display an increased ability to activate serum plasminogen to plasmin. The growth medium from these cells does not display either enhanced fibrinolytic or general proteolytic activity, perhaps because proteolytic factors are not released from the cells under our growth conditions, or because inhibitors in the calf serum prevent their expression (Unkeless et al. 1973). These results demonstrate that detection of increased protease activity in Rous-transformed cells requires use of a specific substrate in order to amplify differences between normal and transformed cells.

Can the Transformed Phenotype Be Mimicked by Treatment of Normal Cells with Proteolytic or Hydrolytic Enzymes?

Among the earliest biochemical alterations to occur in cells as they become transformed is an increased rate of hexose transport (Hatanaka and Hanafusa 1970; Martin et al. 1971). Transformed cells displayed a hexose transport rate 3–5 times that seen in normal cells, growing exponentially, and 10–20 times that seen in nongrowing, density-inhibited cells (Weber 1973). It is well known that chick cells can be released from density-dependent inhibition of growth by treatment with proteolytic or other hydrolytic enzymes (Sefton

Table 1
Proteolytic Activity in Normal and Transformed Cells

	Fibrinolysis 125I-fibrin hydrolyzed cpm/8 hr	General proteolysis % 14C-chlorella protein hydrolyzed	
		10 min	8 hr
Cell extracts[a]			
growing, normal cell extract	202	4.4	5.0
growing, normal cell extract + serum[b]	209	4.3	5.4
RSV-transformed cell extract	196	3.8	5.1
RSV-transformed cell extract + serum	2376	3.2	4.2
density-inhibited cell extract	138	3.4	5.7
density-inhibited cell extract + serum	127	3.2	4.7
Growth media	cpm/mg cell protein/15 hr	% hydrolyzed/mg cell protein/15 hr	
from growing, normal cells	148	7.2	
from RSV-transformed cells	138	6.3	
from density-inhibited cells	146	5.9	

[a] Fifty-one μg of cell protein was used for all assays.
[b] Chicken serum was added as a source of plasminogen to a final concentration of 2%.

and Rubin 1970), and that accompanying this growth stimulation is an increase in the rate of hexose transport (Sefton and Rubin 1971; Vaheri, Ruoslahti and Hovi 1974). However, it has not been clear in these cases whether the hexose transport rate was increased up to the level seen in transformed cells, or whether the stimulation was only to the level characteristic of normal exponentially growing cells. Therefore we have performed a quantitative comparison of the 2-deoxyglucose transport rate of normal and transformed cells and of normal cells treated with a selection of proteolytic or other hydrolytic enzymes. It can be seen (Fig. 1) that neither trypsin, plasmin, neu-

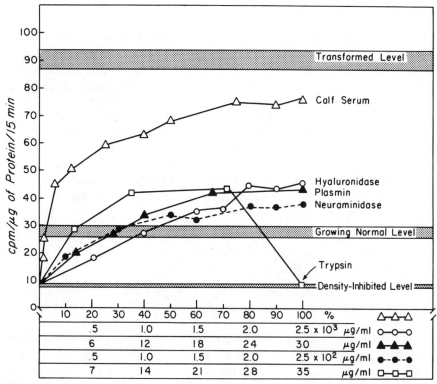

Figure 1

Stimulation of 2-deoxyglucose uptake by hydrolytic enzymes and serum. Tertiary cultures of chick embryo fibroblasts were plated at 1×10^6 cells/35-mm dish in Dulbecco's medium containing 2% tryptose phosphate broth, 1% calf serum and 1% heat-inactivated chicken serum. After 2 days in culture, the cells were stimulated by adding the specified amount of serum or enzyme directly to the culture dish. Transport determinations were made at 4 hours after stimulation, which is the point of maximum stimulation of all the agents tested. Isotope was at 1.0 μCi/ml for these experiments. Density-inhibited, normal-growing and transformed cells were always run in parallel with the experimental cultures. Porcine plasmin was used for the experiments shown, but chicken plasmin gives similar results. The hyaluronidase used contains a proteolytic activity that quantitatively accounts for all of its stimulatory properties, in agreement with the findings of Greenberg and Cunningham (1973) that the growth stimulatory ability of commercial hyaluronidase preparations does not cochromatograph with the peak of hyaluronidase activity. Neuraminidase was free of protease activity by the ^{14}C-chlorella protein assay.

Table 2
Stimulation of 2-Deoxyglucose Uptake of Growing, Normal
Cells by Proteases and Hydrolases

	Growing, normal cells	Trypsin (12.5 µg/ml)	Hyalu- ronidase (1 mg/ml)	Neuram- inidase (150 µg/ml)	RSV- transformed cells
cpm/µg/15 min	24.3	35.1	34.8	36.8	85.7

Transport measurements were made 4 hours after stimulation, which was the maximum for all the enzymes tested.

raminidase nor hyaluronidase is capable of stimulating density-inhibited cells to the hexose transport level obtained with transformed cells. Treatment of growing cells with these enzymes was no more effective (Table 2). However, high concentrations of serum could stimulate the hexose transport capacity to levels nearly equal to that displayed by the transformed cells (Fig. 1). If allowance is made for the binding of serum proteins to the cells in calculating the cpm/µg, then the stimulating ability of serum appears even better (Hale and Weber, in prep.). Thus treatment of normal cells with these enzymes does not convert them into phenocopies of transformed cells with respect to hexose transport. If hydrolysis of the cell surface plays a role in controlling the rate of hexose transport, it may be simply by altering the sensitivity of cells to regulatory factors in the serum.

Which Manifestations of the Transformed Phenotype Can Be Reversed by Protease Inhibitors?

Previous reports indicated that a variety of protease inhibitors were capable of selectively inhibiting the growth of transformed cells and of restoring to these cells a saturation density similar to that seen with normal cells (Schnebli 1974). These results are difficult to interpret, however, since the protease inhibitors do not necessarily act by restoring normal growth controls and are clearly toxic to the cells (Chou, Black and Roblin 1974). Therefore it seemed desirable to investigate this question using markers of the transformed phenotype which do not require growth assays for their detection. A decrease in adherence of cells to the dish (Weber, in prep.) and a rounding of the cell's morphology (Hatanaka and Hanafusa 1970) accompany the increase in hexose transport during transformation by Rous sarcoma virus, and thus all three of these parameters can be used as specific, early markers for the transformed phenotype. TLCK, a trypsin titrant, was tested for its ability to reverse these manifestations of the transformed state. It can be seen (Figs. 2 and 3) that TLCK was highly effective at restoring cellular morphology to normal. TLCK-treated cells became flattened and elongated, and their cell surfaces became smooth and free of microvilli. Transformed cells treated with TLCK frequently could not be distinguished from normal cells when the observations were made single-blind.

TLCK also caused the cells to adhere more tightly to the culture dish (Fig. 4). Cells transformed by the temperature-conditional mutant of Rous

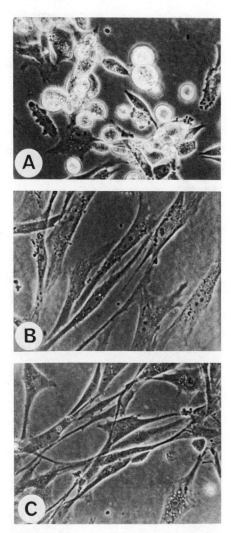

Figure 2
Phase micrographs of (*A*) Rous-transformed cells, (*B*) normal cells, and (*C*) RSV-transformed cells treated 40 hours with 50 μg/ml TLCK. Magnification, 258 ×.

sarcoma virus rapidly increased their adhesiveness when shifted to the restrictive temperature, demonstrating the close association between the decreased adhesiveness and the expression of viral oncogenic information. Addition of TLCK caused a similar, although slower, change in adhesiveness. The data demonstrate not only that the percentage of adherent cells in the TLCK-treated culture increases, but that the absolute number of detachable cells declines. Thus the protease inhibitor cannot be acting as a selective agent in this case, but must be converting transformed cells into phenotypically normal cells. The effects of TLCK on cellular adhesiveness are not always this dramatic, but by following the precautions outlined in Methods, we can routinely increase the adhesiveness of transformed cells with this inhibitor so that fewer than 20% of the cells are detachable.

Finally, TLCK caused transformed cells to lower their rate of hexose transport down to the normal level (Fig. 5). The inhibitor did not lower the transport rate below that characteristic of normal, exponentially growing

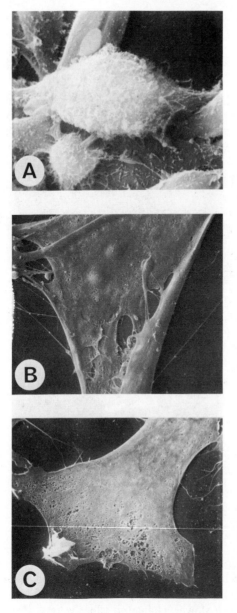

Figure 3
Scanning electron micrographs of (*A*) Rous-transformed cells, (*B*) normal cells, and (*C*) Rous-transformed cells treated 40 hours with 50 μg/ml TLCK. Magnification, 2860 ×.

cells and was without effect when added to growing, normal cells (Table 3).

Because TLCK inhibited the growth of cells, it was important to determine whether simple growth inhibition could account for its effects. Since Chou, Black and Roblin (1974) have shown that an apparently normal saturation density can be generated by selection of an appropriate concentration of cycloheximide, we asked whether cycloheximide could also mimic the effects of TLCK on cellular morphology, adhesion and hexose transport. The data in Table 4 demonstrate that inhibition of protein synthesis by cycloheximide can cause a decrease in the rate of hexose transport, presumably due to turn-

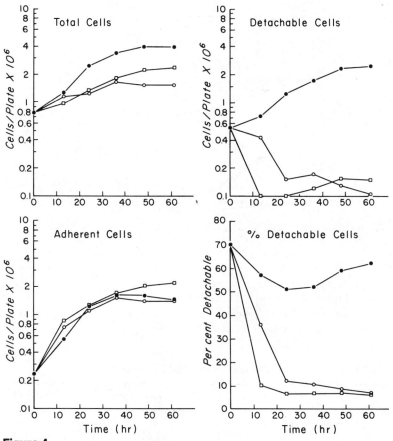

Figure 4

Growth and adhesiveness of TLCK-treated cells. Cells infected with the
temperature-conditional mutant of Rous sarcoma virus, RSV-T5, were
plated at 1.5×10^5 cells/35-mm dish at 36.5°C and allowed to remain in
culture 48 hours. At zero time, the medium was changed to fresh medium,
either with or without 50 μg/ml TLCK. One set of cultures was shifted to
41°C to "switch off" viral oncogenic information and allow the cells to
revert to a normal phenotype. Every 12 hours, fresh TLCK from a frozen,
concentrated stock was added to the TLCK-treated cultures at 25 μg/ml.
Adhesion was measured as described in Methods. (●——●) Untreated
control; (○——○) TLCK treated; (□——□) shifted to 41°C.

over of the hexose transport system. However, 0.05 μg/ml of cycloheximide,
which inhibited protein synthesis about as well as did 50 μg/ml of TLCK,
was much less effective at inhibiting hexose transport. This suggests (but
certainly does not prove) some effect of TLCK on the hexose transport system over and above its effects due to inhibition of protein synthesis. Moreover, cycloheximide did not cause a significant increase in cellular adhesiveness (Table 4). And although cells treated with 0.5 μg/ml of cycloheximide
tended to flatten out on the dish, they did not assume the elongated ap-

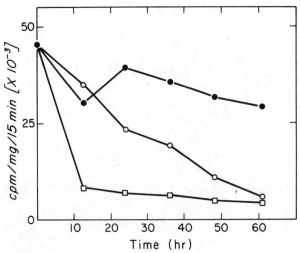

Figure 5

Inhibition of 2-deoxyglucose transport by TLCK. Procedures as in Figure 4. (●——●) Untreated control; (o——o) TLCK treated; (□——□) shifted to 41°C.

Table 3

Effect of TLCK on Hexose Transport in Normal and Transformed Cells

	2-Deoxyglucose transport (nmoles/min/mg/protein)	% Untreated control
Transformed	1.60	100
Transformed + 50 μg/ml TLCK	0.55	34
Growing, normal	0.38	24
Normal + 50 μg/ml TLCK	0.43	27
Normal + 100 μg/ml TLCK	0.52	32
Density-inhibited	0.10	6

Cells were treated with the inhibitor for 40–45 hours.

Table 4

Hexose Transport and Adhesion in Cultures Treated with TLCK or Cycloheximide

	% Inhibition protein synthesis[a]	% Inhibition 2-deoxyglucose uptake	Percentage of detachable cells[b]
50 μg/ml TLCK	60	67	22
0.05 μg/ml Cycloheximide	60	21	71
0.5 μg/ml Cycloheximide	94	75	69

Cells were treated with the inhibitors for 43 hours.

[a] Protein synthesis was measured by Lowry assay of protein increase over 43 hours.

[b] Control cultures had 76% detachable cells.

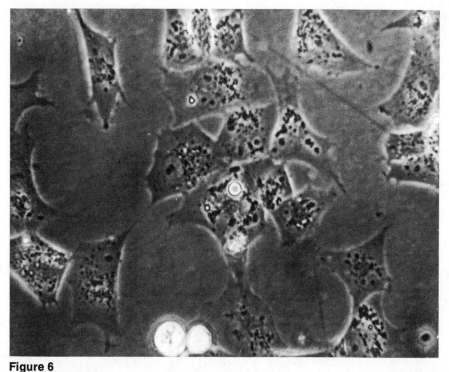

Figure 6
Phase micrograph of a RSV-transformed culture treated 43 hours with 0.5 μg/ml cycloheximide.

pearance of normal cells, but rather became stellate (Fig. 6). Cycloheximide at 0.05 μg/ml was without effect on cellular morphology. Thus we feel that simple growth inhibition does not account for the effects of TLCK on cellular morphology and adhesiveness and may not account for all of TLCK's effects on hexose transport.

The use of protease inhibitors to demonstrate the involvement of proteases in the genesis of the transformed phenotype could be made more persuasive if an effective protease inhibitor could be found which did not inhibit the growth of the cells. To this end, the inhibitors and artificial peptides listed in Table 5 were screened for their ability to cause a flattening of the cellular morphology. Only the protease inhibitors specific for trypsin were found to be effective, and only ovomucoid caused as complete a morphological change as did TLCK (Fig. 7). In addition, a synthetic dipeptide, L-lysyl-L-valine, was found to cause some flattening, and L-arginyl-L-valine, L-lysyl-L-isoleucine and L-lysyl-L-phenylalanine were partially effective. Lysine, valine acetate or o-nitrophenol were without effect. The effective inhibitors were then tested for their effects on growth, adhesion and hexose transport, with the results shown in Table 6. The data demonstrate that all of these inhibitors are capable of causing a significant increase in cellular adhesiveness, and that two of the trypsin inhibitors (ovomucoid and soybean trypsin inhibitor) did not greatly

Table 5
Protease Inhibitors and Peptides Tested for Morphological Effects on Transformed Cells

Inhibitor	Specificity	Highest conc. tested	Effective conc.
Nitrophenyl-p-guanidinobenzoate (NPGB)	trypsin	100 μg/ml	50 μg/ml
Lima bean trypsin inhibitor	trypsin	10 mg/ml[a]	none found
Soybean trypsin inhibitor	trypsin	10 mg/ml	2.5 mg/ml
Ovomucoid	trypsin	100 mg/ml	50 mg/ml
Antipain	papain, trypsin	10 mg/ml	none found
Leupeptin	papain, cathepsin B	10 mg/ml	none found
Pepstatin	pepsin, cathepsin D	10 mg/ml	none found
Phenyl methyl sulfonyl fluoride	serine proteases	1 mg/ml	none found
Tosyl-phenylalanyl-chloromethyl ketone (TPCK)	chymotrypsin	50 μg/ml[a]	none found
N-Carbobenzoxy-L-phenylalanyl-bromomethyl ketone (ZPBK)	chymotrypsin	250 μg/ml[a]	none found
N-Carbobenzoxy-L-phenylalanyl-chloromethyl ketone (ZPCK)	chymotrypsin	250 μg/ml[a]	none found
Peptide			
L-Arginyl-L-valine acetate, hemihydrate		10 mg/ml	10 mg/ml (partial)
L-Lysyl-L-valine acetate		10 mg/ml	5–10 mg/ml
L-Lysyl-L-phenylalanine diacetate		3 mg/ml	3 mg/ml (partial)
L-Lysyl-L-isoleucine diacetate		10 mg/ml	3 mg/ml (partial)
L-Lysyl-L-tyrosine, monohydrate		10 mg/ml	none found
L-Lysyl-L-alanine monohydrochloride, sesquihydrate		10 mg/ml	none found
L-Lysyl-L-tyrosyl-L-glutamic acid acetate, monohydrate		10 mg/ml	none found
L-Lysyl-L-tyrosine amide dihydrobromide, hemihydrate		10 mg/ml	none found
N-Acetyl-L-arginine methyl ester		10 mg/ml	none found

[a] Caused cell detachment.

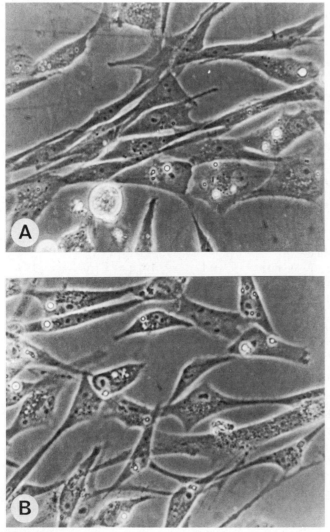

Figure 7
Phase micrographs of RSV-T5 transformed cultures treated 24 hours with 10 mg/ml lysyl-valine (*A*) or 50 mg/ml ovomucoid (*B*). Magnification, 258 ×.

inhibit cell growth. However, none of them, except TLCK, restored the rate of hexose transport to normal, including NPGB, which inhibited cell growth almost as much as TLCK.

DISCUSSION

The data presented here strongly suggest the involvement of a trypsinlike protease—presumably plasmin—in the decreased adhesiveness and rounded morphology characteristic of transformed cells (consistent with the suggestions of Ossowski et al. 1973), but not in the transformation-specific increase

Table 6

Effect of Protease Inhibitors on Adhesion and 2-Deoxyglucose
Transport by RSV-T5-transformed Cells

	Cells/plate × 10⁶			2-Deoxy-glucose transport cpm/µg/ 15 min
	detachable	total	% detachable	
Untreated control	1.5	2.1	71	34.5
Soybean trypsin inhibitor (2.5 mg/ml)	0.3	1.9	16	30.9
Ovomucoid (50 mg/ml)	0.4	1.7	23	39.7
L-Lysyl-L-valine (10 mg/ml)	0.4	1.2	33	27.1
Shifted to 41°C	0.2	1.8	11	4.5
Untreated control	1.7	3.4	50	32.7
Nitrophenyl-p-guanidino-benzoate (NPGB) (50 µg/ml)	0.1	1.8	6	33.9
Tosyl-lysyl-chloromethyl ketone (50 µg/ml, plus 25 µg/ml readded at 24 hr)	0.1	1.3	8	8.0
Shifted to 41°C	0.1	2.2	5	3.9

Cells were treated with the inhibitors for 45 hours.

in hexose transport. Treatment of normal cells with plasmin or other proteolytic and hydrolytic enzymes did not stimulate their hexose transport rate all the way up to the level that characterizes transformed cells. And treatment of transformed cells with protease inhibitors such as NPGB and soybean trypsin inhibitor, which are known to inhibit transformation-associated fibrinolysis in this system (Unkeless et al. 1973 and our unpubl. obs.), did not lower the hexose transport rate, although adhesiveness and morphology did return to normal.

The fact that the adhesive and morphological properties of the cells did not vary coordinately with changes in hexose transport indicates that these manifestations of the transformed phenotype have different proximal causes. This notion is supported by some of our other data (Weber 1974) which demonstrate that treatment of transformed cells with dibutyryl cyclic AMP restores hexose transport to normal, but has only a partial effect on adhesiveness or morphology.

The results obtained with TLCK are more complex and must be interpreted cautiously. It seems clear that TLCK, like the other effective protease inhibitors, causes transformed cells to return to normal with respect to adhesive and morphological properties. These effects of TLCK cannot be attributed to its growth inhibitory properties since growth inhibition by cycloheximide does not have similar effects. In fact, toxic doses of TLCK (in the range of 250–500 µg/ml) cause rounding of the cells and detachment from the dish. TLCK

has been reported not to inhibit fibrinolysis activated by the cell factor released from Rous-transformed chick cells (Unkeless et al. 1973), a finding
that we have confirmed. This raises the intriguing possibility that there is yet
another trypsinlike protease involved in the genesis of the transformed phenotype, separate from—and perhaps prior to—plasmin and plasminogen activator. This notion gains support from our finding that TLCK also restores the
transformed level of hexose transport to normal. However, as indicated in the
text, the evidence that this effect is due to a specific inhibition of proteases is
not conclusive. Resolution of this problem and elucidation of the interaction
of proteolytic activities in malignant transformation will require more detailed
biochemical analysis of the earliest steps in viral oncogenesis.

Acknowledgments

This work was supported by USPHS grant CA-12467. The able technical
assistance of Ms. Deborah Hohla and Ms. Shirley Yau is gratefully acknowledged.

REFERENCES

Bosmann, H. B. 1972. Elevated glycosidase and proteolytic enzymes in cells transformed by RNA tumor virus. *Biochim. Biophys. Acta* **264:**339.

Chou, I.-N., P. H. Black and R. Roblin. 1974. Effects of protease inhibitors on
growth of 3T3 and SV3T3 cells. In *Control of Proliferation in Animal Cells* (ed.
B. Clarkson and R. Baserga), p. 339. Cold Spring Harbor Laboratory, Cold
Spring Harbor, New York.

Deutsch, D. G. and E. T. Mertz. 1970. Plasminogen: Purification from human
plasma by affinity chromatography. *Science* **170:**1095.

Greenberg, D. B. and D. D. Cunningham. 1973. Does hyaluronidase initiate DNA
synthesis? *J. Cell. Physiol.* **82:**511.

Hatanaka, M. and H. Hanafusa. 1970. Analysis of a functional change in membrane in the process of cell transformation by Rous sarcoma virus alteration in
the characteristics of sugar transport. *Virology* **41:**647.

Lowry, O. H., N. J. Rosebrough, A. L. Farr and R. J. Randall. 1951. Protein
measurement with the folin phenol reagent. *J. Biol. Chem.* **246:**710.

Martin, G. S. 1970. Rous sarcoma virus: A function required for the maintenance
of the transformed state. *Nature* **227:**1021.

Martin, G. S., S. Venuta, M. J. Weber and H. Rubin. 1971. Temperature-dependent
alterations in sugar transport in cells infected by a temperature-sensitive mutant
of Rous sarcoma virus. *Proc. Nat. Acad. Sci.* **68:**2739.

Ossowski, L., J. C. Unkeless, A. Tobia, J. P. Quigley, D. B. Rifkin and E. Reich.
1973. An enzymatic function associated with transformation of fibroblasts by
oncogenic viruses. II. Mammalian fibroblast cultures transformed by DNA and
RNA tumor viruses. *J. Exp. Med.* **137:**112.

Reich, E. 1974. Tumor-associated fibrinolysis. In *Control of Proliferation in Animal Cells* (ed. B. Clarkson and B. Baserga), p. 351. Cold Spring Harbor Laboratory, Cold Spring Harbor, New York.

Robbins, K. C. and L. Summaria. 1970. Human plasminogen and plasmin. *Methods
in Enzymology* (ed. G. E. Perlmann and L. Lorand), vol. XIX, p. 184.
Academic Press, New York.

Schnebli, H. P. 1972. A protease-like activity associated with malignant cells. *Schweiz. Med. Wschr.* **102**:1194.

————. 1974. Growth inhibition of tumor cells by protease inhibitors: Consideration of the mechanisms involved. In *Control of Proliferation in Animal Cells* (ed. B. Clarkson and R. Baserga), p. 327. Cold Spring Harbor Laboratory, Cold Spring Harbor, New York.

Sefton, B. M. and H. Rubin. 1970. Release from density-dependent growth inhibition by proteolytic enzymes. *Nature* **227**:843.

————. 1971. Stimulation of glucose transport in cultures of density-inhibited chick embryo cells. *Proc. Nat. Acad. Sci.* **68**:3154.

Unkeless, J. C., A. Tobia, L. Ossowski, J. P. Quigley, D. B. Rifkin and E. Reich. 1973. An enzymatic function associated with transformation of fibroblasts by oncogenic viruses. *J. Exp. Med.* **137**:85.

Vaheri, A., E. Ruoslahti and T. Hovi. 1974. Cell surface and growth control of chick embryo fibroblasts in culture. In *Control of Proliferation in Animal Cells* (ed. B. Clarkson and R. Baserga), p. 305. Cold Spring Harbor Laboratory, Cold Spring Harbor, New York.

Weber, M. J. 1973. Hexose transport in normal and in Rous sarcoma virus-transformed cells. *J. Biol. Chem.* **248**:2978.

————. 1974. Reversal of the transformed phenotype by dibutyryl cyclic AMP and a protease inhibitor. In *Mechanisms of Virus Disease* (ed. W. S. Robinson and C. F. Fox), p. 327. W. A. Benjamin, Menlo Park, California.

Are Proteases Involved in Altering Surface Proteins during Viral Transformation?

Richard O. Hynes, John A. Wyke, Jacqueline M. Bye,
Kenneth C. Humphryes and Edward S. Pearlstein

Department of Tumour Virology, Imperial Cancer Research Fund Laboratories
Lincoln's Inn Fields, London, WC2A 3PX, England

Previous investigations of the nature of the surface proteins on cultured cells have detected differences between "normal" cells, which show density dependence of growth, movement, etc., and their virus-transformed derivatives, which do not. When lactoperoxidase-catalyzed iodination was used to label specifically those proteins exposed at the external surface, it was found that a major surface protein present on the "normal" cells was absent or considerably reduced on the transformed ones (Hynes 1973; Wickus, Branton and Robbins 1974; Hogg 1974). A similar difference was detected using galactose oxidase plus sodium borotritiide (Gahmberg and Hakomori 1973; Gahmberg, Kiehn and Hakomori 1974; Critchley 1974) and by metabolic labeling (Stone, Smith and Joklik 1974; Hynes and Humphryes 1974). Taken together, these results provide evidence for a large, external, transformation-sensitive (LETS) glycoprotein which is absent from the surfaces of virally transformed cells.

The LETS glycoprotein was also observed to be extremely sensitive to proteolytic digestion (Hynes 1973; Wickus, Branton and Robbins 1974), raising the possibility that its absence on transformed cells could be due to removal by proteases released by transformed cells (Schnebli 1972; Bosmann 1972; Unkeless et al. 1973; Ossowski et al. 1973a,b; Goldberg 1974). Proteases in general are known to produce a variety of effects in normal cells, tending to render them more like transformed cells (reviewed in Hynes 1974). It is conceivable, therefore, that the action of "transformation proteases" on the surfaces of the cells which produce them could be responsible for many of the alterations seen in transformed cells.

Given what is already known, one can propose several models to explain the absence of the LETS glycoprotein from the external surface and from purified membranes of transformed cells: (1) It is not synthesized by transformed cells; (2) it is synthesized but not inserted into the membrane because (a) it is not completed and/or processed properly, e.g., not glycosylated as normal, or not cleaved from a larger precursor, or (b) some other

931

component necessary for its correct alignment is missing; or (3) it is synthesized and inserted but is not retained by the cell because of (a) passive turnover into the medium (for reasons such as those under 2) or (b) active removal, for instance by proteolysis.

Since the LETS glycoprotein responds to growth stimuli (Hynes and Bye 1974) and is particularly sensitive to proteolysis, it is of interest to discover which of these models applies, and, in particular, whether one can make any connection between the release of proteases and altered growth controls. Such a connection is very likely to involve surface proteins. The experiments described here were designed to investigate these models.

MATERIALS AND METHODS

Cells and culture conditions were as described (Hynes 1973; Wyke and Linial 1973). The cells used were: NIL8, a normal fibroblastic cell line from hamsters; NIL8-HSV6, a transformant of NIL8 produced by treatment with hamster sarcoma virus; early passage chicken embryo fibroblasts (CEF) from C/O Brown Leghorn embryos. Rous sarcoma virus (RSV) mutants (Wyke 1973) were temperature sensitive (ts) for maintenance of transformation. Permissive and restrictive temperatures for the mutant used, ts LA 24, were 35 and 41°C, respectively. Lactoperoxidase-catalyzed iodination and SDS-polyacrylamide gel electrophoresis were as before (Hynes 1973).

Protease inhibitors used were as follows: pancreatic (Kunitz), lima bean and soybean trypsin inhibitors (Worthington Biochemicals); soybean and ovomucoid trypsin inhibitors, phenyl methyl sulfonyl fluoride (PMSF), tosyl-lysyl-chloromethyl ketone (TLCK), tosyl-arginine methyl ester (TAME) and ϵ-aminocaproic acid (Sigma, London); TLCK and TPCK (tosyl-phenylalanyl-chloromethyl ketone) (Calbiochem); Trasylol was a generous gift of Bayer Pharmaceuticals; antipain and leupeptin gifts from Dr. Brigid Hogan, and nitrophenyl-guanidinobenzoate (NPGB) a gift from Dr. J. Unkeless.

Plasminogen-depleted serum was prepared by the method of Deutsch and Mertz (1970). Diluted serum was passed at least twice over columns of lysyl-Sepharose.

RESULTS

Is the LETS Glycoprotein Synthesized by Transformed Cells?

NIL8-HSV6 cells grow as highly refractile cells, both attached to the dishes and in suspension. When cycloheximide or puromycin at concentrations which inhibited protein synthesis were added to cultures of NIL8-HSV6, the cells became flattened and most of the floating cells attached to the dish. Cells were iodinated 16–24 hours after such treatments; the results are shown in Figure 1. The cells remained impermeable to the labeling reagents, as shown by the fact that the iodination was selective. Cells treated with cycloheximide or puromycin showed iodination of a protein comigrating with the LETS glycoprotein of NIL8 cells (Fig. 1a, c–e), whereas controls did not (Fig.

Figure 1

Effect of inhibition of protein synthesis on NIL8-HSV6 cells. Autoradiograph of an SDS-polyacrylamide slab gel. Cells were iodinated 24 hours after additions as below. Normalized for equal protein. (*a*) NIL8 controls; (*b*) NIL8-HSV6 plus cycloheximide, 20 μg/ml, treated with trypsin (10 μg/ml, 10 min) before labeling; (*c*) NIL8-HSV6 plus cycloheximide, 20 μg/ml; (*d*) NIL8-HSV6 plus cycloheximide, 100 μg/ml; (*e*) NIL8-HSV6 plus puromycin, 100 μg/ml; (*f*) NIL8-HSV6 plus saline; (*g*) NIL8-HSV6 control. Arrows mark the position of the LETS protein.

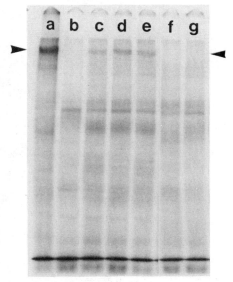

1f, g). This new iodinated band on the treated NIL8-HSV6 cells was frequently a doublet and was completely sensitive to a mild trypsin treatment (Fig. 1b) as was the LETS glycoprotein on normal NIL8 cells. Control experiments demonstrated that the doses of cycloheximide and puromycin used here inhibited protein synthesis by >95%, and that the residual synthesis did not represent a specific subclass of resistant proteins (unpubl.).

A similar experiment was performed in the chicken-RSV system. Temperature-sensitive RSV-infected CEF were cultured at 35°C, at which temperature the cells were transformed by morphological and other criteria (Wyke 1973; Wyke and Linial 1973). On shifting to 41°C, the cells flattened out within a few hours and by 16–24 hours appeared as normal CEF. This reversion to normality also occurred in the presence of inhibitory levels of cycloheximide, as observed for other ts mutants (Kawai and Hanafusa 1971; Biquard and Vigier 1972). Temperature-sensitive RSV-infected cells were iodinated at 35°C and at intervals after shift to 41°C in the presence and absence of cycloheximide. The results are shown in Figure 2. Iodination of the LETS protein increased within a few hours after shift to 41°C and this occurred also in the presence of cycloheximide. By 8 hours, the level of labeling in both control and inhibited cultures was about 30% of that observed in 41°C cells or 24 hours after shift up (unpubl.). Interestingly, addition of cycloheximide to ts RSV-infected cells at 35°C without shift up did *not* lead to reappearance of the LETS protein on the surface (see Discussion).

These experiments demonstrate that transformed hamster and chicken fibroblasts do synthesize the LETS protein, although it is not expressed in the normal manner on the cell surface. Addition of cycloheximide in NIL8-HSV6 and switch to restrictive temperature in ts RSV-infected CEF allows this latent pool to be expressed at the surface. Not only is the LETS protein undetectable by iodination on transformed cells, it is also absent or much reduced as detected by protein staining or metabolic labeling of plasma mem-

934   R. O. Hynes et al.

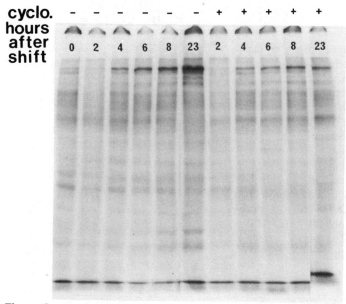

Figure 2
Shift to restrictive temperature of ts RSV-infected CEF: effect
of cycloheximide on surface proteins. Cells were iodinated at
intervals after shift from 35 to 41°C: cycloheximide at 20 μg/ml
where present. Normalized for equal protein.

branes (Stone, Smith and Joklik 1974). More important still, it is not detected
by carbohydrate labeling of whole NIL-HSV cells (Hynes and Humphryes
1974; Fig. 3). Therefore, the intracellular pool demonstrated by the cycloheximide experiments must consist either of a nonglycosylated form or of a
precursor which has a different mobility on the gels. A similar pool appears
to exist in normal NIL cells (R. O. Hynes and M. Bye, in prep.).

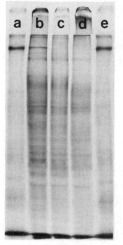

Figure 3
Glucosamine labeling of NIL8 and NIL8-HSV6 cells. (*a,e*)
Iodinated NIL8. (*b–d*) Glucosamine-labeled cells: (*b*)
NIL8, (*c*) NIL8 treated with trypsin (10 μg/ml, 10 min),
(*d*) NIL8-HSV6.

Does the LETS Protein Turn Over Faster after Transformation?

In order to test the third model, which proposes that the reason for reduction in surface levels of the LETS protein is lack of retention, ts RSV-infected CEF cultured at 41°C (at which temperature they appear normal and show high levels of iodination of the LETS protein) were iodinated and then returned to medium at 41 or 35°C. The cells shifted to 35°C transformed morphologically with the normal kinetics. At intervals, dishes were harvested and analyzed on SDS-polyacrylamide gels, and the amount of label in the LETS protein band was quantitated. Results of such an experiment are shown in Figure 4a. The radioactivity turns over at both temperatures. The rate of turnover at 35°C was either as fast as at 41°C or faster (as in Fig. 4a): the results varied from experiment to experiment for unknown reasons. However, these results should be contrasted with those in Figure 4b which show that the rate of protein *synthesis* at 35°C is only 60% of that at 41°C, as might be expected given the 6°C difference in temperature. Thus the rate of turnover of the LETS protein at 35°C is markedly faster than at 41°C when considered in relation to general metabolic activity.

Do Transformed Cells Affect Surfaces of Neighboring Normal Cells?

If transformed cells produce proteases which act upon their own surfaces, these proteases should also digest the surface proteins of normal cells cocultivated with the transformed cells. In order to test this possibility, NIL8 cells

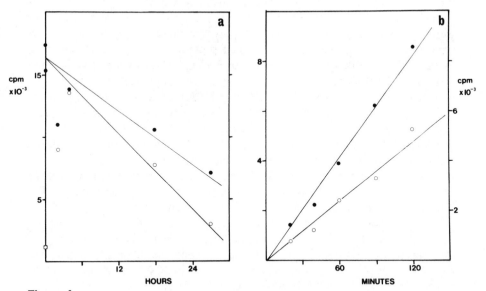

Figure 4
Effect of temperature on (*a*) rate of turnover of surface protein and (*b*) rate of protein synthesis. (*a*) Radioactivity migrating in position of the LETS protein at different times after iodination and return to culture at 35°C (-o-) or 41°C (-•-). Normalized for equal fractions of starting material. (*b*) Incorporation of [35S]-methionine at 35°C (-o-) or 41°C (-•-).

Figure 5
Cultivation of iodinated NIL8 cells alone (a,b) or with the addition of NIL8 cells (c,d) or NIL8-HSV6 cells (e,f). Cells harvested 25 hours after return to culture. Overlaid cells were in 1:1 ratio with labeled ones, i.e., 3 × 10^6 per dish. Samples represent equal proportions of starting material. (a,c,e) No added serum; (b,d,f) 10% calf serum.

were iodinated in monolayer, washed, and returned to culture medium either alone or with the addition of normal (NIL8) or transformed (NIL8-HSV6) cells. The overlaid cells were obtained in suspension by the use of EDTA; fewer than 10% of the cells were leaky to trypan blue. The added cells attached to the labeled NIL8 monolayer and spread on it. At intervals, dishes were harvested and analyzed on SDS-polyacrylamide gels; Figure 5 shows typical results. In all cases, the iodinated proteins turned over. Addition of extra NIL8 cells did not affect this turnover (Fig. 5a–d), whereas addition of NIL8-HSV6 cells accelerated it (Fig. 5e–f). A difference in turnover rate was detectable by 5 hours, and labeled LETS protein had virtually disappeared by 24 hours in the case of dishes to which NIL8-HSV6 cells had been added. This result has been confirmed using BHK and BHK-Py (unpubl.).

Although this result is consistent with the protease model discussed above, it does not prove its verity, and several other interpretations are possible (see Discussion). Further experiments to prove a role for proteases in producing this "trans" removal of LETS protein have so far been inconclusive. The "trans" removal occurs in a variety of sera (calf, hamster, chicken) and without addition of any serum (Fig. 5); it will also occur in serum depleted of plasminogen by affinity chromatography (unpubl.). These results suggest that plasminogen is not *required,* although it could be involved. In fact, one can show that plasminogen at the levels present in 10% serum can effect removal of the LETS protein from NIL8 cells (Fig. 6).

We have been unable to cause accelerated turnover of prelabeled LETS protein by addition of media conditioned by transformed cells, and addition of a variety of protease inhibitors (including several effective against plasmin) does not prevent "trans" removal in experiments such as that of Figure 5 (Table 1; Hynes and Pearlstein, in prep.).

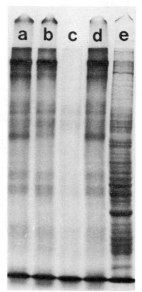

Figure 6
Effect of plasminogen on surface proteins. NIL8 cells treated (1 hr, 37°C) before iodination as follows (normalized for equal protein): (*a*) PBS; (*b*) human plasminogen, 1 casein unit/ml, 67 μg/ml; (*c*) plasminogen as (*b*) plus 100 casein units/ml streptokinase; (*d*) streptokinase alone; (*e*) [14C]leucine-labeled NIL8 to show total cellular proteins.

Do Protease Inhibitors Cause Reversion of Transformed Cells to Normal?

Several workers have reported that addition of protease inhibitors to transformed cells causes them to exhibit properties characteristic of normal cells, such as density dependence of growth (Schnebli and Burger 1972) or failure to grow in agar (Ossowski et al. 1973a). At least some of these effects now appear to be artifactual in that the growth inhibition caused by certain protease inhibitors does not lead to arrest in G_1 (Schnebli and Haemmerli 1974), is not selective for transformed cells, and is probably caused by inhibition of protein synthesis (Chou, Black and Roblin 1974a,b).

However despite these complications, it is necessary to block the putative effects of proteases in causing transformation in order to demonstrate their involvement. Accordingly, we tested a number of protease inhibitors for their ability to cause NIL8-HSV6 cells to adopt a "normal" growth pattern and/or to reexpress the LETS protein on their surfaces (Table 1). Most of them were without any effect, a few were toxic (e.g., TPCK), and only TLCK caused any signs of morphological reversion. When treated cells were iodinated, none of them showed labeling of the LETS protein, again with the possible exception of TLCK. Typical results are shown in Figure 7.

The results with TLCK were variable and depended on the density of the cells and the time at which the inhibitor was added: TLCK was more effective in causing flattening of cells and growth inhibition the earlier it was added to the cultures and the lower the cell density. Similar results have been reported by others (Chou, Black and Roblin 1974b). In most experiments, no reappearance of the LETS protein was detected by iodination (as in Fig. 7). In a few cases, trace amounts of label were observed. However, it is known that TLCK inhibits protein synthesis (Chou, Black and Roblin 1974a and pers. comm.), and we have observed that cycloheximide causes the reappearance of the LETS protein (Fig. 1).

Table 1
Effects of Protease Inhibitors

Inhibitor (µg/ml)	Tested for ability to			
	inhibit transformation of ts RSV-infected CEF	revert growth pattern of NIL8-HSV6	cause reappearance of LETS on NIL8-HSV6	inhibit "trans-" removal of LETS from NIL8 by NIL8-HSV6
Soybean trypsin inhibitor (100–500)	–	–	–	–
Lima bean trypsin inhibitor (200)		–	–	
Pancreatic trypsin inhibitor (200–500)	–	–	–	–
Ovomucoid trypsin inhibitor (200–500)		–	–	
Trasylol (100–1000 KIU/ml)	–/T	–/T		
Antipain (100–500)	–	–		
Leupeptin (100–500)	–	–		
TPCK (10–100)	T	T		
TLCK (10–100)	+/T	±/T	±/T	
PMSF (50)	–	–		
TAME (50)		–	–	
NPGB (10^{-4} M)			–	–
ε-Aminocaproic acid (1000)		–	–	–

(−) No effect; (+) some effect; (T) toxic. Fuller details, concentrations, etc., will be published elsewhere (Hynes and Wyke, in prep.; Hynes and Pearlstein, in prep.).

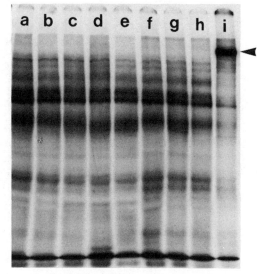

Figure 7
Iodination of NIL8-HSV6 cells grown in protease inhibitors (normalized for equal protein). (*a*) Soybean trypsin inhibitor, 100 μg/ml; (*b*) lima bean trypsin inhibitor, 100 μg/ml; (*c*) ovomucoid, 100 μg/ml; (*d*) PMSF, 50 μg/ml; (*e*) ε-aminocaproic acid, 500 μg/ml; (*f*) TLCK, 50 μg/ml; (*g*) TAME, 50 μg/ml; (*h*) control, untreated; (*i*) NIL8 marker cells, arrow marks the LETS protein.

In summary, our experiments with protease inhibitors in mammalian cell lines provide no evidence for an involvement of proteolytic enzymes in causing loss of the LETS protein. Of course, they do not rule it out.

Do Protease Inhibitors Block Transformation in ts RSV-infected Cells?

It seemed possible that it might be easier to inhibit the establishment of the transformed state than to reverse it once established. So we repeated the protease inhibitor experiments on ts RSV-infected CEF. The cells were infected and cultured at restrictive temperature. They appeared "normal" and had high levels of iodinateable LETS protein. The effects of protease inhibitors on the transformation occurring on shift to permissive temperature were then studied. Most inhibitors tested were without effect on the morphological shift and were not examined further (Table 1). TLCK at concentrations of 20–50 μg/ml largely inhibited the rounding up of the cells. This inhibition gradually wore off at times longer than 24 hours, and it could be reversed by washing away the inhibitor. One hundred μg/ml TLCK was toxic.

Cells were iodinated before and after shift to permissive temperature in the presence and absence of TLCK and after removal of the inhibitor. Results of such an experiment are shown in Figure 8. TLCK inhibited the reduction in labeling of the LETS protein that occurred in controls after the shift (Fig. 8a–e). The inhibition by 20 μg/ml could be reversed (Fig. 8f–h). In this experiment, inhibition by 40 μg/ml TLCK was not reversible (Fig. 8i), although in other experiments it was. Therefore as well as blocking the morphological transformation, TLCK blocks the loss of LETS protein associated with this transformation.

Since TLCK is known to inhibit protein synthesis, we tested its effects on CEF. Twenty and 40 μg/ml inhibited protein synthesis by 28 and 40%, respectively. Levels of cycloheximide which inhibited protein synthesis by 10–70% also occasionally blocked the transformation occurring on shift down (unpubl.). This result makes it possible that the TLCK effect is due in part

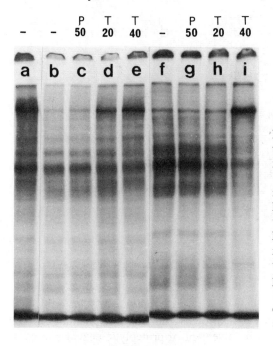

Figure 8
Effect of TLCK and PMSF on transformation of ts RSV-infected CEF. (*a*) 41 °C Control; (*b–e*) 46 hours after shift down: inhibitors as marked; (*f–i*) 46 hours after shift down: inhibitors were present for the first 21 hours. The shift in this experiment was rather slower than usual. Normalized for equal protein. Inhibitors were TLCK (T) or PMSF (P) as marked; concentrations in μg/ml as shown.

to inhibition of protein synthesis. But it remains possible that TLCK is exerting its effects by inhibiting a protease(s) released on transformation. Both the TLCK and the cycloheximide experiments were somewhat variable (cf. results with mammalian cells quoted above), and more work will be required to determine the exact nature of their inhibitory effects on transformation.

The role of plasminogen in the transformation of ts RSV-infected CEF was investigated by subculturing the cells at 41 °C into plasminogen-depleted serum and 24 hours later shifting them to 35 °C. The morphological shift proceeded as normal, as did the increase in rate of glucose transport and the disappearance of the LETS protein (data not shown). This result and the failure of leupeptin and soybean trypsin inhibitor to block the transformation suggest that plasminogen is not required for this event.

DISCUSSION

The three basic models for the absence of the LETS protein from the external surfaces and membranes of transformed cells can be summarized as: (1) lack of synthesis, (2) continued synthesis but a fault in processing, and (3) lack of retention in the membrane.

The results presented above allow one to eliminate the first model since the LETS protein can appear on the surfaces of transformed cells under conditions of protein synthesis inhibition. This proves that the protein is present in a latent form in the transformed cells (NIL8-HSV6 or ts RSV-infected CEF at permissive temperature). The reappearance during shift up in ts RSV-infected CEF can be explained merely by postulating an intracellular pool, expression of which at the surface is blocked by the functioning viral gene (model 2 or 3). Switch to restrictive temperature shuts off the viral gene, allowing expression at the surface: cycloheximide does not block this since the pool pre-exists. In the case of the reappearance on NIL8-HSV6, one has to postulate, in addition, that the positive effector necessary to block the surface expression has a shorter half-life than the pool of LETS protein. If this latter condition were not fulfilled, then mere addition of cycloheximide would not alter the surface. This condition is presumably not satisfied in ts RSV-infected CEF maintained at permissive temperature. Under our conditions, cycloheximide did not produce reappearance of the LETS protein at the surface. Only when the half-life of the (virally coded?) effector molecule was reduced by shift to restrictive temperature did reexpression occur. (Fig. 2).

As mentioned earlier, the intracellular pool demonstrated by these results must consist either of a molecule of a different size and mobility on gels or of an incompletely glycosylated precursor molecule. This follows from the fact that both transformed and trypsinized normal cells lack completely the LETS glycoprotein detected by metabolic sugar labeling (Fig. 3). Thus no glycosylated molecule of the same electrophoretic mobility can be present as an intracellular pool, certainly not to the level (at least 30% of the normal surface level) required to explain the data.

One is left, then, to decide between models 2 and 3. The experiment of Figure 4 suggests that turnover of the LETS protein is more rapid after transformation since although general metabolism as measured by overall protein synthesis is slowed at 35°C compared with 41°C (Fig. 4b), turnover of the LETS protein is either as rapid or faster (Fig. 4a). This result has also been obtained by Robbins et al. (1975) using a different ts mutant. This experiment concerns preformed LETS protein, so the accelerated turnover can only be due to indirect effects mediated by other molecules. Possibilities include alteration in other proteins or lipids essential for the retention of the LETS protein and degradation by proteases or other enzymes.

The results of the cell mixing experiments (Fig. 5) could be interpreted as evidence for proteolysis. Against this conclusion is the fact that none of the protease inhibitors tested was effective in blocking the "trans" removal of prelabeled LETS protein from NIL8 cells by NIL8-HSV6 cells. Of course, this may simply mean that the correct inhibitor has not yet been tried. Several of these inhibitors do, however, inhibit plasmin and/or plasminogen activation (Unkeless et al. 1973; Goldberg 1974), suggesting that this enzyme system is not *required*. Further evidence for this conclusion comes from the facts that no added serum was required for the "trans" removal and that it occurred equally well even when cells had been twice subcultured in plasminogen-depleted serum prior to the experiment (unpubl.). These results do not prove that plasminogen activation does not contribute to the removal of the LETS protein, as indeed it can do so (Fig. 6).

As in the case of increased turnover of the LETS protein in ts RSV-infected CEF after shift to permissive temperature, the accelerated loss of prelabeled LETS protein in the mixing experiment could be due to effects other than proteolysis. For instance, growth stimulation of the normal cells by the transformed cells by other means might well produce the same effect. It is known that G_0-arrested NIL8 cells have high levels of the LETS protein at their surfaces, whereas after stimulation with serum, this level falls (Hynes and Bye 1974). Passive release, as opposed to active cleavage by proteases, could be responsible both for this phenomenon and for the "trans" removal.

The negative results of attempts to cause reversion of NIL8-HSV6 cells to normality (Table 1, Fig. 7) are similarly inconclusive in deciding whether proteases are involved in maintenance of the transformed phenotype by these cells. Again, the failure of some of the inhibitors (e.g., soybean, pancreatic, ϵ-aminocaproic acid) to cause reversion might suggest that plasmin is not a prerequisite for transformation. Others, using different systems, have provided evidence that plasminogen activation is required for certain transformed traits (Ossowski et al. 1973b; Goldberg 1974). Conflicting results have been obtained with soybean trypsin inhibitor (Schnebli and Burger 1972; Ossowski et al. 1973a), and Schnebli and Burger (1972) have reported that ovomucoid produced density-dependent inhibition of growth in SV3T3 cells. The difficulty in interpreting results with low molecular weight inhibitors (e.g., TPCK, TLCK) have been discussed above and elsewhere (Chou, Black and Roblin 1974a,b; Schnebli and Haemmerli 1974).

The results with the ts RSV system (Fig. 8) lend some support to the protease model in that TLCK inhibits the morphological and surface protein changes associated with transformation. However, the effective doses also partially inhibit protein synthesis, and levels of cycloheximide which produce equal inhibition also occasionally block the transformation. Furthermore, if TLCK has one side effect, there is no reason why it should not have others, and it remains unproved that it is exerting its effect on transformation by inhibiting proteolysis.

In conclusion, while certain results (accelerated turnover of surface proteins after transformation in ts RSV or after addition of transformed cells to normal ones, or TLCK inhibition of transformation in ts RSV-infected CEF) are suggestive of a role for proteases in altering cell surfaces after transformation, none of them is conclusive. Alternative explanations for alterations in the display of surface proteins on transformed cells (e.g., incomplete assembly, passive turnover) have not been eliminated. It seems clear in all the systems tested that activation of plasminogen is not required for the expression of the transformed properties tested (growth, morphology, sugar transport, absence of the LETS protein). However, it remains possible that the plasminogen activator or other transformation proteases may be involved. In order to analyze more incisively the putative role for transformation proteases in affecting cell behavior via their action on cell surfaces, it will be necessary to use the purified molecules (e.g., Christman and Acs 1974) and their corresponding specific inhibitors and antibodies to investigate their effects on surface molecules and growth control.

Acknowledgment

We would like to thank Jennifer Beamand for her technical assistance.

REFERENCES

Biquard, J. M. and P. Vigier. 1972. Characteristics of a conditional mutant of Rous sarcoma virus in ability to transform cells at high temperature. *Virology* **47**:444.

Bosmann, H. B. 1972. Elevated glycosidases and proteolytic enzymes in cells transformed by RNA tumor virus. *Biochim. Biophys. Acta* **264:**339.

Chou, I., P. H. Black and R. O. Roblin. 1974a. Effects of protease inhibitors on growth of 3T3 and SV3T3 cells. In *Control of Proliferation in Animal Cells* (ed. B. Clarkson and R. Baserga), p. 339. Cold Spring Harbor Laboratory, Cold Spring Harbor, New York.

————. 1974b. Non-selective inhibition of transformed cell growth by a protease inhibitor. *Proc. Nat. Acad. Sci.* **71:**1748.

Christman, J. K. and G. Acs. 1974. Purification and characterisation of a cellular fibrinolytic factor associated with oncogenic transformation: The plasminogen activator from SV40-transformed hamster cells. *Biochim. Biophys. Acta* **340:** 339.

Critchley, D. R. 1974. Cell surface proteins of NIL1 hamster fibroblasts labelled by a galactose oxidase, tritiated borohydride method. *Cell* **3:**121.

Deutsch, D. G. and E. T. Mertz. 1970. Plasminogen: Purification from human plasma by affinity chromatography. *Science* **170:**1095.

Gahmberg, C. G. and S. Hakomori. 1973. Altered growth behavior of malignant cells associated with changes in externally labeled glycoprotein and glycolipid. *Proc. Nat. Acad. Sci.* **70:**3329.

Gahmberg, C. G., D. Kiehn and S. Hakomori. 1974. Changes in a surface-labelled galactoprotein and in glycolipid concentrations in cells transformed by a temperature-sensitive polyoma virus mutant. *Nature* **248:**413.

Goldberg, A. R. 1974. Increased protease levels in transformed cells: A casein overlay assay for the detection of plasminogen activator production. *Cell* **2:**95.

Hogg, N. M. 1974. A comparison of membrane proteins of normal and transformed cells by lactoperoxidase labeling. *Proc. Nat. Acad. Sci.* **71:**489.

Hynes, R. O. 1973. Alteration of cell-surface proteins by viral transformation and by proteolysis. *Proc. Nat. Acad. Sci.* **70:**3170.

————. 1974. Role of surface alterations in cell transformation: The importance of proteases and surface proteins. *Cell* **1:**147.

Hynes, R. O. and J. M. Bye. 1974. Density and cell cycle dependence of cell surface proteins in hamster fibroblasts. *Cell* **3:**113.

Hynes, R. O. and K. C. Humphryes. 1974. Characterization of the external proteins of hamster fibroblasts. *J. Cell Biol.* **62:**438.

Kawai, S. and H. Hanafusa. 1971. The effects of reciprocal changes in temperature on the transformed state of cells infected with a Rous sarcoma virus mutant. *Virology* **46:**470.

Ossowski, L., J. P. Quigley, G. M. Kellerman and E. Reich. 1973a. Fibrinolysis associated with oncogenic transformation. *J. Exp. Med.* **138:**1056.

Ossowski, L., J. C. Unkeless, A. Tobia, J. P. Quigley, D. B. Rifkin and E. Reich. 1973b. An enzymatic function associated with transformation of fibroblasts by oncogenic viruses. II. Mammalian fibroblast cultures transformed by DNA and RNA tumor viruses. *J. Exp. Med.* **137:**112.

Robbins, P. W., G. G. Wickus, P. E. Branton, B. J. Gaffney, C. B. Hirschberg, P. Fuchs and P. M. Blumberg. 1975. The chick fibroblast cell surface following transformation by Rous sarcoma virus. *Cold Spring Harbor Symp. Quant. Biol.* **39:**1173.

Schnebli, H. P. 1972. A protease-like activity associated with malignant cells. *Schweiz. Med. Wochenschr.* **102:**1194.

Schnebli, H. P. and M. M. Burger. 1972. Selective inhibition of growth of transformed cells by protease inhibitors. *Proc. Nat. Acad. Sci.* **96:**3825.

Schnebli, H. P. and G. Haemmerli. 1974. Protease inhibitors do not block transformed cells in the G1 phase of the cell cycle. *Nature* **248:**150.

Stone, K. R., R. E. Smith and W. K. Joklik. 1974. Changes in membrane poly-peptides that occur when chick embryo fibroblasts and NRK cells are trans-formed with avian sarcoma viruses. *Virology* **58:**86.

Unkeless, J. C., A. Tobia, L. Ossowski, J. P. Quigley, D. B. Rifkin and E. Reich. 1973. An enzymatic function associated with transformation of fibroblasts by oncogenic viruses. I. Chick embryo fibroblast cultures transformed by avian RNA tumor viruses. *J. Exp. Med.* **137:**85.

Wickus, G. G., P. E. Branton and P. W. Robbins. 1974. Rous sarcoma virus trans-formation of the chick cell surface. In *Control of Proliferation in Animal Cells* (ed. B. Clarkson and R. Baserga), p. 541. Cold Spring Harbor Laboratory, Cold Spring Harbor, New York.

Wyke, J. A. 1973. The selective isolation of temperature-sensitive mutants of Rous sarcoma virus. *Virology* **52:**587.

Wyke, J. A. and M. Linial. 1973. Temperature-sensitive avian sarcoma viruses: A physiological comparison of twenty mutants. *Virology* **53:**152.

Relation of Protease Action on the Cell Surface to Growth Control and Adhesion

Peter M. Blumberg and Phillips W. Robbins

Center for Cancer Research and Department of Biology
Massachusetts Institute of Technology, Cambridge, Massachusetts 02139

Our laboratory has been interested in the changes in cell surface structure accompanying transformation. Recent studies have focused on chick embryo fibroblasts transformed by ts 68, a mutant of the Schmidt-Ruppin strain of Rous sarcoma virus thermosensitive for transformation. In this system, transformation was observed to cause the disappearance of three proteins: Ω (MW 206,000), Δ (MW 47,000) and Z (MW 230,000–240,000), which disappeared 0–3 hours, 3–6 hours and several days, respectively, after temperature shift (Wickus and Robbins 1973; Wickus, Branton and Robbins 1974; Robbins et al. 1975). Of these proteins, Z (also referred to as 250 K protein or LETS) has been the most widely studied (Hynes 1973, 1974; Hogg 1974; Ruoslahti and Vaheri 1974; Gahmberg, Kiehn and Hakomori 1974). It is the major protein on the surface of fibroblasts, as revealed by lactoperoxidase-catalyzed iodination, and is highly sensitive to proteases such as trypsin and collagenase.

Proteases have been reported to release chick embryo fibroblasts from contact inhibition, stimulating cell division and deoxyglucose transport (Sefton and Rubin 1970, 1971; Vaheri, Ruoslahti and Hovi 1974). Moreover, transformed cells produce elevated levels of proteases, including a cell factor that converts the serum proenzyme plasminogen into the protease plasmin (Bosmann 1972; Schnebli 1972; Goldberg 1974; Unkeless et al. 1973, 1974; Unkeless, Gordon and Reich 1974; Ossowski et al. 1973a, 1973b; Ossowski, Quigley and Reich 1974; Quigley, Ossowski and Reich 1974; Reich 1974). The plasmin in turn is thought to induce some of the properties characteristic of transformed cells. It thus seemed worthwhile to examine the effect of proteases on activation of resting chick fibroblasts and on degradation of Z.

METHODS

Purified trypsin (TRL) and collagenase (CLSPA) were obtained from Worthington Biochemical; purified thrombin was the generous gift of L. Chen, and purified lactoperoxidase was the gift of G. G. Wickus. Medium was from Microbiological Associates, and serum was purchased from Microbiological Associates or Colorado Serum Company. Chick embryo fibroblasts were routinely grown in Eagle's minimum essential medium containing 2% tryptose phosphate broth and 1% heat-inactivated chicken serum (MEM-2-0-1). Where indicated, cells were grown in medium 199 containing 10% tryptose phosphate broth, 4% calf serum and 1% heat-inactivated chicken serum. Primary chick cells were transformed by ts 68, a temperature-sensitive mutant of Rous sarcoma virus, as described by Wickus and Robbins (1973). Stimulation of deoxyglucose transport was routinely measured 6 hours after protease addition using a 10-minute incubation time (Sefton and Rubin 1971). [^{3}H]Thymidine incorporation was routinely measured 18 hours after protease addition according to the procedure of Sefton and Rubin (1971).

Iodination of the chick embryo fibroblasts was by the lactoperoxidase method (Sefton, Wickus and Burge 1973). For cells preiodinated before addition of protease, the medium was removed for the iodination reaction and then returned to the plates. Labeled monolayers of cells to be examined by electrophoresis were first rinsed 2–4 times with solution A (PBS minus Ca^{++} and Mg^{++}); 0.75 ml of boiling SDS gel application buffer was added per plate, followed by phenylmethane sulfonyl fluoride (2 mM final conc.), and the samples were then subjected to SDS slab gel electrophoresis according to the procedure of Laemmli (1970). Slab gels were dried according to Maizel (1971), and radioactive bands were detected by autoradiography. For quantitation, individual bands were cut from the dried gel, eluted into 1% SDS (1 ml per sample), and counted in a liquid scintillation counter. Alternatively, the autoradiograms were scanned with a microdensitometer, and the area under the peaks determined.

RESULTS

Effects of Proteases on Z and Cell Behavior

Trypsin and Collagenase

Trypsin and collagenase at the concentrations which activate confluent chick embryo fibroblasts have little effect on the composition of the fibroblast membrane. Cells grown in MEM-2-0-1 containing only 10% of the usual concentration of methionine were labeled for 2 days with [^{35}S]methionine, 5 μCi/ml. Trypsin (2 μg/ml final conc.) or collagenase (4 μg/ml final conc.) was added to the medium. The cells were incubated for 4.5 hours. They were then washed, membranes prepared according to the method of Brunett and Till (1971), and samples subjected to SDS slab gel electrophoresis. No differences in composition could be observed, although it should be noted that the bands in the high molecular weight region of the gel (> 200,000) where Z is located were not well resolved. As previously reported for the case of somewhat more stringent conditions of proteolysis (Robbins et al. 1975), Δ was not affected.

In the case of whole cells labeled with [^{14}C]proline, trypsin and collagenase induced the disappearance of a single band on gels (Fig. 1). This proline band

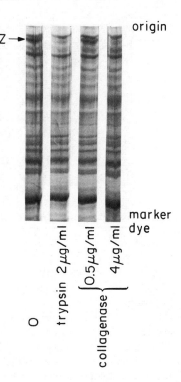

Figure 1

Effect of trypsin and collagenase on [¹⁴C]pro-line-labeled chick embryo fibroblasts. Secondary chick embryo fibroblasts were plated at a density of $1.3 \times 10^6/60$-mm plate in MEM-2-0-1 containing [¹⁴C]proline (2 μCi/ml). After 2 days, trypsin and collagenase at the final concentrations indicated were added to the medium on the plates. The plates were then washed extensively, the cells dissolved in SDS, and subjected to SDS slab gel electrophoresis on 7% gels.

is thought to be the same as Z on the basis of its identical electrophoretic mobility (Robbins et al. 1975) and its sensitivity to proteases. Similarly, treatment with either trypsin (Fig. 2) or purified collagenase (Fig. 3) followed by lactoperoxidase-catalyzed iodination of the cell surface resulted in the visible loss of only a single iodinateable band, Z. While the other iodinateable bands were retained upon treatment with trypsin under the above conditions, the bands are merely less susceptible to the action of the trypsin. High concentrations (400 μg/ml for 15 min at 39°C in solution A) led to complete release of the iodine label from the cells.

During degradation of Z by trypsin, a proteolytic fragment was frequently observed. Its molecular weight was approximately 7% (14,000) less than that of Z. The fate of the fragment upon further digestion is not known. It could either be released into the medium largely intact, or, on the other hand, it may be extensively degraded.

The effects of increasing concentrations of trypsin on Z, deoxyglucose transport and thymidine incorporation are compared in Figure 4. In this system, optimal trypsin concentrations (1–2 μg/ml) caused a 4–6-fold stimulation in deoxyglucose transport and a 4–5-fold increase in thymidine incorporation at 18 hours. Such values were associated with an increase in fibroblast cell number over 24 hours of 97% (10 separate experiments, all measurements made in duplicate). This value compares with a background increase in cell number over 24 hours of 19% (11 separate experiments, all measurements made in duplicate). The decrease in deoxyglucose transport at a trypsin concentration of 2 μg/ml in Figure 4B was caused by partial release

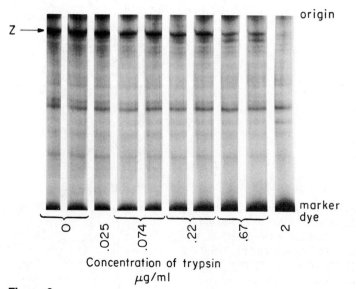

Figure 2

Effect of trypsin on iodinated chick embryo fibroblasts. Secondary chick embryo fibroblasts were plated at a density of 1.3×10^6/plate in MEM-2-0-1. After 2 days, trypsin at the final concentrations indicated was added to the medium on the plates. Iodination was performed 4.5 hours after the addition of trypsin, and samples were subjected to SDS slab gel electrophoresis on 7% gels.

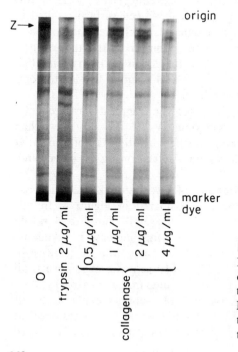

Figure 3

Effect of collagenase on iodinated chick embryo fibroblasts. Cells were treated as described in the legend to Figure 2, except that collagenase at the concentrations indicated was added to the cells.

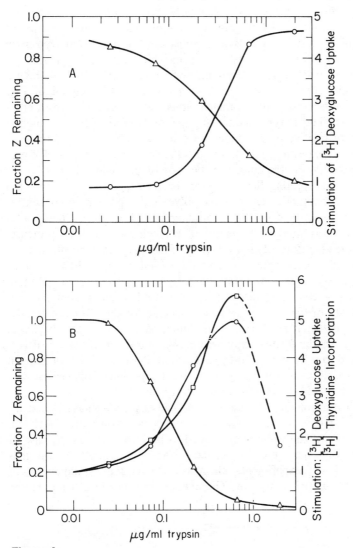

Figure 4

Effect of trypsin on Z and deoxyglucose transport. Secondary chick embryo fibroblasts were plated at a density of $1.3 \times 10^6/$ 60-mm dish in MEM-2-0-1. After incubation for 2 days, cells were treated as follows: (*A*) Z postlabeled. The indicated concentrations of trypsin were added to the medium. At 4.5 hours, the cells were iodinated. (*B*) Z prelabeled. After the cells were iodinated, trypsin at the indicated concentrations was added to the medium. At 4.5 hours, the cells were harvested. (△——△) Fraction Z remaining; (o——o) stimulation of [³H]deoxyglucose uptake; (□——□) [³H]thymidine incorporation.

of the cells from the plate. The actual concentration of trypsin inducing cell release varies a little (2–3-fold) between experiments.

The concentration of trypsin inducing 50% release of Z was quite low, 0.3–0.1 μg/ml. This degradation, moreover, took place in the presence of 1% serum, which presumably acts as a substrate for the added trypsin in competition with Z. Concentrations of trypsin which caused degradation of Z and stimulation of deoxyglucose uptake were comparable (0.3 μg/ml vs. 0.4 μg/ml, respectively, in Fig. 4A; 0.11 μg/ml vs. 0.15 μg/ml in Fig. 4B). Moreover, the results with Z labeled either before or after trypsin addition were similar. The lower residual amount of labeled Z and its slightly greater sensitivity relative to deoxyglucose transport in the case of the prelabeled cells are as expected, since in the postlabeled cells, synthesis of Z presumably continued during the course of the proteolytic treatment.

A difficulty in interpreting the results of the trypsin titrations is that the trypsin might concomitantly cleave other polypeptides present in concentrations too low to be detected on the autoradiograms. The effects on degradation of Z and on deoxyglucose transport of a much more specific protease, collagenase, were therefore of interest (Fig. 5). Here the concentration of collagenase for 50% degradation of z was 1 μg/ml, that for 50% stimulation of deoxyglucose uptake was 0.9 μg/ml. An important issue in these experiments was whether the observed effect indeed resulted from the collagenolytic activity of the enzyme, which is specific for the sequence Pro-X-Gly-Pro-Y, or whether it arose from contamination by some other protease. The evidence favors the former explanation. While the concentration of collagenase required to release Z was only severalfold higher than that of trypsin, the measured caseinolytic activity present in the purified collagenase was equivalent to only a 0.3% contamination with trypsin.

A puzzling feature of the activation of the cells by collagenase is that although degradation of Z and stimulation of deoxyglucose transport correlated, the absolute values of the stimulation tended to be lower than in the case of trypsin. Specificially, maximum deoxyglucose transport induced by collagenase averaged 72% (16 experiments, all measurements done in duplicate) of that induced by 2 μg/ml of trypsin. The average increase in cell number in

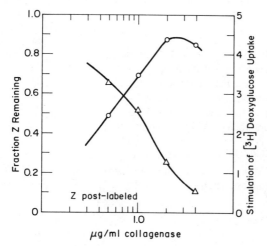

Figure 5

Effect of collagenase on Z and deoxyglucose transport (Z postlabeled). Cells were treated as described in the legend to Figure 4A, except that collagenase at the indicated concentrations was used. (△——△) Fraction Z remaining; (o——o) stimulation of [³H]deoxyglucose uptake.

24 hours was 59% (9 experiments, all measurements done in duplicate) compared to 96% for trypsin. The reason for the difference in values is not clear. It could reflect different stabilities of the enzymes in the medium, the greater tendency of collagenase to release the fibroblasts from the dish, or differing susceptibilities of the fibroblasts in the population.

Thrombin

L. Chen and J. Buchanan (in prep.) have found that brief treatment with thrombin stimulated division in resting chick embryo fibroblasts. They likewise have observed (Chen, Teng and Buchanan, this volume) that such treatments did not bring about loss of Z. We have carried out preliminary experiments to see whether these results could be confirmed under conditions where thrombin was added to the culture medium. Under such conditions, purified thrombin did, in fact, strongly stimulate deoxyglucose transport (Fig. 6). Half-maximal stimulation was effected by thrombin at a concentration of 0.1 μg/ml. In addition, the absolute level of deoxyglucose stimulation was 20% higher than it was with trypsin. The addition of thrombin did cause the level of Z to decrease (Fig. 7). However, the concentration of thrombin inducing half-maximal stimulation of deoxyglucose uptake was 10-fold lower than that causing 50% breakdown of Z. The breakdown of Z was thus not essential for activating fibroblasts under these conditions. Moreover, it should be noted that the breakdown of Z may not be a direct effect of thrombin. The thrombin could activate some other protease in the serum, e.g., plasmin, which in turn could lead to loss of Z.

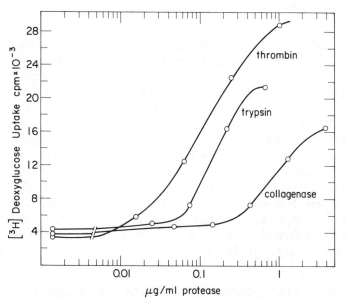

Figure 6

Stimulation of deoxyglucose transport by proteases. Secondary chick embryo fibroblasts were plated at a density of 1.3 × 10⁶/60-mm dish in MEM-2-0-1. After 2 days, proteases at the indicated final concentrations were added to the medium.

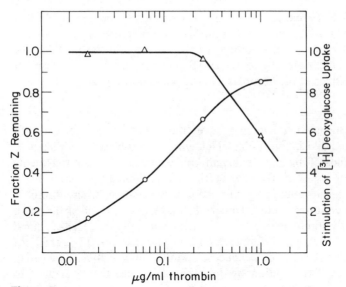

Figure 7

Effect of thrombin on Z and deoxyglucose transport (Z post-labeled). Cells were treated as in Figure 4A, except that thrombin at the indicated concentrations was used. (△——△) Fraction Z remaining; (o——o) stimulation of [³H]deoxyglucose uptake.

Stability of z in Normal and Transformed Cells

A degree of uncertainty exists about the stability of Z in normal and RSV-transformed chick embryo fibroblasts. Customarily, the transformed fibroblasts are grown in 199-10-4-1, a medium which contains 4% calf serum. Calf serum has been reported to be inhibitory ("nonpermissive") for transformation-induced fibrinolysis. Since the experiments with exogenous proteases described above were performed in MEM-2-0-1, which lacks calf serum and is not inhibitory for fibrinolysis (i.e., it is a "permissive" serum), the stability of Z in normal and transformed cells in the two media was compared (Fig. 8). Z appeared to turn over far more rapidly in the "permissive" medium MEM-2-0-1 than in the "nonpermissive" medium 199-10-4-1. The rate of turnover was not linear. In MEM-2-0-1, a large fraction of the label was lost quickly (at least half within 2 hr); the remainder disappeared more slowly. Little difference was observed in the stability of Z between normal and transformed cells. Unfortunately, undue variation in the measurements in this experiment might obscure small differences.

Effects of the Protease Inhibitor Leupeptin on Transformed Cells

Considerable interest has been devoted to the effects of protease inhibitors on transformation. An inhibitor of particular interest is leupeptin, the tripeptide derivative leucyl-leucyl-arginal (Umezawa 1972). This compound inhibits plasmin at low concentrations, 8 μg/ml causing 50% inhibition. Moreover,

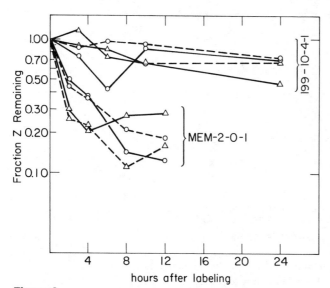

Figure 8

Stability of Z. Secondary chick embryo fibroblasts infected
as primaries with ts 68 were incubated for 24 hours at the
nonpermissive temperature, 41°C. The medium, MEM-
2-0-1 or 199-10-4-1, was then changed and the temperature
shifted where indicated to the permissive temperature,
36°C. At 16 hours, the medium was removed, the cells
iodinated, and the conditioned medium returned to the
plates. Time points were taken thereafter as indicated.
(△) Normal; (○) infected with ts 68; (———) 36°C; (– – –)
41°C.

unlike many protease inhibitors, such as TLCK (tosyl-lysine chloromethyl
ketone), it is reported not to affect protein synthesis (Rossman, Norris and
Troll 1974). For ts 68-transformed cells grown in MEM-2-0-1, concentrations up to 167 μg/ml did not affect deoxyglucose transport, cell growth or
DNA synthesis. However, the leupeptin had a clear effect on the appearance
of the cells on the plate. Since MEM-2-0-1 is a permissive serum, the ts 68-
transformed cells are released from the plate some two days after temperature
shift. During the process of release, gaps appear within the layer of cells.
These gaps enlarge until the cells assume a crochetlike pattern. The clumps
of cells then detach from the plate. Leupeptin largely, but not entirely, inhibits this process, and the cells remain more firmly attached to the plate. The
morphology of the individual cells, however, still remains that typical of
transformation. Figure 9 illustrates this effect of leupeptin on attachment.
Cells are detached by the sheer force generated by shaking the plates in a
New Brunswick gyratory shaker. Essentially all cells were released from the
plates either lacking or having low concentrations of leupeptin. The cells
remained adherent at concentrations of 50 μg/ml or above. The mechanism
by which leupeptin affects the attachment remains to be established.

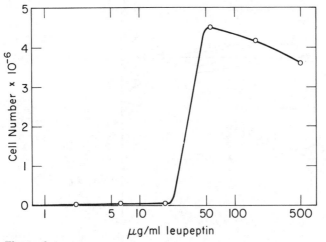

Figure 9
Effect of leupeptin on attachment of cells to plate. Secondary chick embryo fibroblasts infected with ts 68 were plated at 1.3×10^6 cells/plate in MEM-2-0-1 and incubated for 2 days at the nonpermissive temperature, 41°C. Fresh medium containing various concentrations of leupeptin was then added and the cultures shifted to 36°C. Every 12 hours the medium was replaced with fresh medium containing leupeptin. After 48 hours, plates were shaken at 130 cycles/min for 2 minutes in a New Brunswick gyratory shaker. Cells adhering to the plate were then counted.

DISCUSSION

The significance of Z in growth control is so far uncertain. The results with thrombin indicate that at least one protease, thrombin, can activate resting chick fibroblasts without degrading Z. Degradation is thus not a necessary step in activation. Is it a sufficient step? The results of the trypsin and collagenase titrations favor such a conclusion, but only with several reservations. The trypsin result is technically most persuasive. It suffers, however, from the broad specificity of the enzyme. There is little assurance that degradation of some enzyme present in amounts below the level of detectability of the gels is not responsible for the activation. Indeed, the effect of trypsin could even be on some serum component. In such cases, Z might be a single representative of an entire class of highly trypsin-sensitive proteins, one or more of which acts as a trigger for cell response. The collagenase titration renders such explanations less likely, since few proteins could be expected to be sensitive to the enzyme. However before this experiment can be persuasive, an understanding must be obtained for the failure of all cells to be induced to divide. Furthermore, since the specificity of the collagenase is quite important for this argument, it would seem desirable to carry out further controls to exclude the possibility that a contaminant in the collagenase is responsible for its activity.

The dynamics of Z synthesis and release require further study. In particular, does plasmin play a major or just a minor role in the loss of z from transformed fibroblasts? How is this loss affected by growth rate, medium and the presence of protease inhibitors such as leupeptin?

Acknowledgments

We thank Ethel Andrews and Lin Huey-Chen for expert technical assistance, and T. Sugimura for kindly providing us with leupeptin. This investigation was supported by National Institutes of Health Grant No. CA 14142-12. PMB is a postdoctoral fellow of the Helen Hay Whitney Foundation.

REFERENCES

Bosmann, H. B. 1972. Elevated glycosidases and proteolytic enzymes in cells transformed by RNA tumor viruses. *Biochim. Biophys. Acta* **264**:339.

Brunett, D. M. and J. E. Till. 1971. A rapid method for the isolation of L-cell surface membranes using an aqueous two-phase polymer system. *J. Memb. Biol.* **5**:215.

Gahmberg, C. G., D. Kiehn and S. I. Hakomori. 1974. Changes in a surface-labeled galactoprotein and in glycolipid concentrations in cells transformed by a temperature-sensitive polyoma virus mutant. *Nature* **248**:413.

Goldberg, A. R. 1974. Increased protease levels in transformed cells: A casein overlay assay for the detection of plasminogen activator production. *Cell* **2**:95.

Hogg, N. M. 1974. A comparison of membrane proteins of normal and transformed cells by lactoperoxidase labeling. *Proc. Nat. Acad. Sci.* **71**:489.

Hynes, R. O. 1973. Alteration of cell surface proteins by viral transformation and by proteolysis. *Proc. Nat. Acad. Sci.* **70**:3170.

————. 1974. Role of surface alterations in cell transformation: The importance of proteases and surface proteins. *Cell* **1**:147.

Laemmli, U. K. 1970. Cleavage of structural proteins during the assembly of the head of bacteriophage T4. *Nature* **227**:680.

Maizel, J. V. 1971. Polyacrylamide gel electrophoresis of viral proteins. In *Methods in Virology* (ed. K. Maramorosch and H. Kaprowski), vol. 5, p. 179. Academic Press, New York.

Ossowski, L., J. P. Quigley and E. Reich. 1974. Fibrinolysis associated with oncogenic transformation. Morphological correlates. *J. Biol. Chem.* **249**:4312.

Ossowski, L., J. P. Quigley, G. M. Kellerman and E. Reich. 1973a. Fibrinolysis associated with oncogenic transformation. Requirement of plasminogen for correlated changes in cellular morphology, colony formation in agar, and cell migration. *J. Exp. Med.* **138**:1056.

Ossowski, L., J. C. Unkeless, A. Tobia, J. P. Quigley, D. B. Rifkin and E. Reich. 1973b. An enzymatic function associated with transformation of fibroblasts by oncogenic viruses. II. Mammalian fibroblast cultures transformed by oncogenic viruses. *J. Exp. Med.* **137**:112.

Quigley, J. P., L. Ossowski and E. Reich. 1974. Plasminogen, the serum proenzyme activated by factors from cells transformed by oncogenic viruses. *J. Biol. Chem.* **249**:4306.

Reich, E. 1974. Tumor-associated fibrinolysis. In *Control of Proliferation in Animal Cells* (ed. B. Clarkson and R. Baserga), p. 351. Cold Spring Harbor Laboratory, Cold Spring Harbor, New York.

Robbins, P. W., G. G. Wickus, P. E. Branton, B. J. Gaffney, C. B. Hirschberg, P. Fuchs and P. M. Blumberg. 1975. The chick fibroblast cell surface after transformation by Rous sarcoma virus. *Cold Spring Harbor Symp. Quant. Biol.* **39**:1173.

Rossman, T., C. Norris and W. Troll. 1974. Inhibition of macromolecular synthesis in *Escherichia coli* by protease inhibitors. Specific reversal by glutathione of the effects of chloromethyl ketones. *J. Biol. Chem.* **249**:3412.

Ruoslahti, E. and A. Vaheri. 1974. Novel human serum protein from fibroblast plasma membrane. *Nature* **248**:789.

Schnebli, H. P. 1972. A protease-like activity associated with malignant cells. *Schweiz. Med. Wochenschr.* **102**:1194.

Sefton, B. M. and H. Rubin. 1970. Release from density dependent growth inhibition by proteolytic enzymes. *Nature* **227**:843.

————. 1971. Stimulation of glucose transport in cultures of density-inhibited chick embryo cells. *Proc. Nat. Acad. Sci.* **68**:3154.

Sefton, B. M., G. G. Wickus and B. W. Burge. 1973. Enzymatic iodination of Sindbis virus proteins. *J. Virol.* **11**:730.

Umezawa, H. 1972. Inhibitors of proteolytic enzymes. In *Enzyme Inhibitors of Microbial Origin* (ed. H. Umezawa), pp. 15–52. University Park Press, Baltimore, Maryland.

Unkeless, J. C., S. Gordon and E. Reich. 1974. Secretion of plasminogen activator by stimulated macrophages. *J. Exp. Med.* **139**:834.

Unkeless, J., K. Danø, G. M. Kellerman and E. Reich. 1974. Fibrinolysis associated with oncogenic transformation. Partial purification and characterization of the cell factor, a plasminogen activator. *J. Biol. Chem.* **249**:4295.

Unkeless, J. C., A. Tobia, L. Ossowski, J. P. Quigley, D. B. Rifkin and E. Reich. 1973. An enzymatic function associated with transformation of fibroblasts by oncogenic viruses. I. Chick embryo fibroblast cultures transformed by avian RNA tumor viruses. *J. Exp. Med.* **137**:85.

Vaheri, A., E. Ruoslahti and T. Hovi. 1974. Cell surface and growth control of chick embryo fibroblasts in culture. In *Control of Proliferation in Animal Cells* (ed. B. Clarkson and R. Baserga), p. 305. Cold Spring Harbor Laboratory, Cold Spring Harbor, New York.

Wickus, G. G. and P. W. Robbins. 1973. Plasma membrane proteins of normal and Rous sarcoma virus-transformed chick embryo fibroblasts. *Nature New Biol.* **245**:65.

Wickus, G. G., P. E. Branton and P. W. Robbins. 1974. Rous sarcoma virus transformation of the chick cell surface. In *Control of Proliferation in Animal Cells* (ed. B. Clarkson and R. Baserga), p. 541. Cold Spring Harbor Laboratory, Cold Spring Harbor, New York.

The Mitogenic Activity and Related Effects of Thrombin on Chick Embryo Fibroblasts

Lan Bo Chen, Nelson N. H. Teng and John M. Buchanan

Department of Biology, Massachusetts Institute of Technology
Cambridge, Massachusetts 02139

Trypsin and other "serine proteases" isolated from a variety of animal and plant sources are known to stimulate division of certain resting cells in culture (Burger 1970; Sefton and Rubin 1970). Our particular interest at the moment is the study of the proteases of plasma and serum in relation to their capacity to initiate cell division in chick embryo fibroblasts. Appropriate candidates include such enzymes as kallikrein, plasmin, thrombin and many other proteases discussed in this volume. This report concerns the role of thrombin as a mitogenic agent (Chen and Buchanan 1975).

MATERIALS AND METHODS

The experimental procedure used to test mitogenic activity involves the use of secondary cultures of chick embryo fibroblasts (Rein and Rubin 1968) brought to a resting state by incubation in Dulbecco's modified Eagle's medium containing 0.5% calf serum for four days with a change of medium on the second day. The protein to be tested for mitogenic activity was generally included in the culture medium without any serum supplement for the entire duration of the experiment except on special occasions to be mentioned. The incorporation of tritiated thymidine into DNA was measured at the peak of the first wave of DNA synthesis 12 hours after addition of the mitogenic agent. This measurement has been the principal method for estimating the growth-promoting activity of the agents under investigation, but, in a limited number of instances, we have also measured directly the increase in the cell number as a function of time.

Iodination of surface proteins of confluent cell cultures was carried out with slight modifications of previously described methods (Phillips and Morrison 1971; Phillips 1972; Hynes 1973; Robbins et al. 1975).

The availability of highly purified thrombin and prothrombin was absolutely essential for the interpretation of the experiments to be described.

The purity of prothrombin and thrombin has been checked by electrophoresis on SDS-polyacrylamide gels. As can be seen in Figure 1, both thrombin (A) and prothrombin (B) migrate as single bands.

RESULTS AND DISCUSSION

The Mitogenic Activity of Thrombin

Figure 2 presents the results of an experiment in which thrombin, prothrombin and calf serum are compared for their capacity to stimulate DNA synthesis 12 hours after addition to resting chick embryo fibroblasts. When added in graded amounts, serum in concentrations ranging from 0–5 mg/ml stimulates to approximately the same extent as does purified thrombin when added in concentrations of 0–10 μg/ml. Although prothrombin exhibits a low stimulation, particularly at the lower concentration, it is probable that this response results from the slow conversion of prothrombin to thrombin in the cell cultures. Thus, as one might expect, prothrombin per se has little or no mitogenic activity unless it undergoes activation to thrombin. When prothrombin is added in the presence of other serum factors, particularly factors V and X_a, as would be the case for cells stimulated by serum, one might

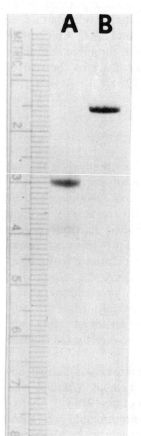

Figure 1
SDS-polyacrylamide gel electrophoretic analysis of (A) 25 μg of purified bovine thrombin and (B) 30 μg of purified bovine prothrombin. Analysis of the proteins by electrophoresis on polyacrylamide gels has been described by Weber and Osborn (1969).

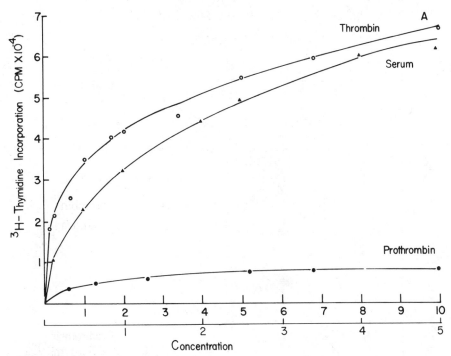

Figure 2

Dose response of mitogenic activity of bovine thrombin (0–10 μg/ml, o——o), calf serum (0–5 mg/ml, ▲——▲) and bovine prothrombin (0–10 μg/ml, •——•). Resting chick embryo fibroblasts were incubated for 12 hours with Dulbecco's modified Eagle's medium containing proteins to be tested. The cells in 1.5 ml of medium were then incubated for 1 hour with 10 μl of [³H]thymidine (0.2 mCi/ml; 6.7 Ci/mmole). Cells were then lysed with 0.1 N NaOH. DNA was precipitated with an equal volume of cold 10% trichloroacetic acid. The precipitates were washed with cold 5% trichloroacetic acid, dried, and counted. The background count of cells incubated with serum-free medium is 2400 cpm, which has been subtracted from the values reported in the figure.

anticipate a considerable utilization of prothrombin as a source of mitogenic thrombin.

In Figure 3 is shown the measurement of the increase of cell numbers over a period of 4 days when stimulation was effected by calf serum at a concentration of 5 mg/ml as compared to thrombin at a level of 2.5 μg/ml. As seen, the response of either mitogenic agent is comparable. On the other hand, the response of prothrombin is low.

Pattern of Cell Growth in Presence of Thrombin

An interesting feature of cells that have been stimulated by thrombin is the pattern of cell growth that occurs in a period of 2 to 3 days following addition of the mitogen. As shown in Figure 4A, cells grown in the presence of serum are arranged in an even and parallel manner. However, thrombin causes the random accumulation of cells in clumps sometimes with large spaces between areas of cell growth. Pictures of cells grown for 2 and 3 days

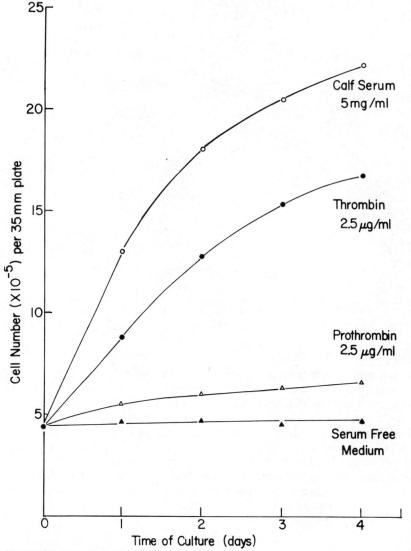

Figure 3

Stimulation of cell division of resting chick embryo fibroblasts by calf serum
(5 mg/ml, o——o), bovine thrombin (2.5 µg/ml, •——•), bovine prothrombin
(2.5 µg/ml, △——△) and serum-free medium (▲——▲). After 1, 2, 3 and
4 days, the number of cells per dish (35 mm) was counted. Each point is
the average value of three dishes.

under these conditions are shown in Figure 4B and 4C, respectively. Again,
we would like to emphasize that in the latter case there is no serum in the
medium, only thrombin and Dulbecco's modified Eagle's medium.

Effect of Thrombin on Cell Surface Proteins

Finally, we would like to discuss the use that has been made of thrombin in
the comparison of the composition of surface proteins of cells before and after

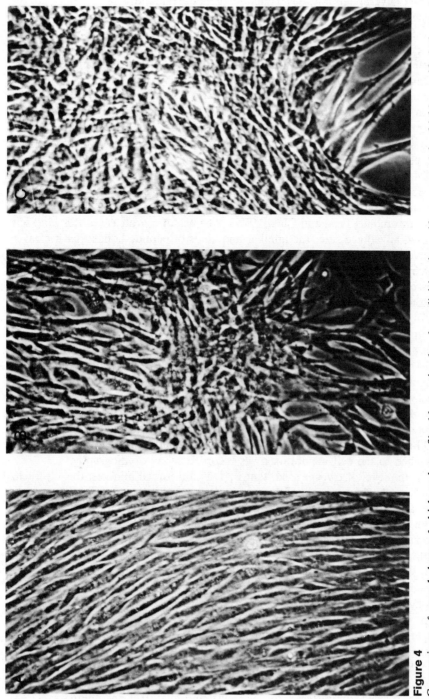

Figure 4

Comparison of morphology of chick embryo fibroblasts stimulated to division by calf serum and by highly purified thrombin. (*A*) Cells grown in calf serum (5 mg/ml); (*B*) resting cells stimulated to division by thrombin (2.5 μg/ml) for 2 days and (*C*) for 3 days.

stimulation to DNA synthesis by this protease. During the last year, investigators in several laboratories (Gahmberg and Hakomori 1974; Hogg 1974; Hynes 1973; Hynes and Humphryes 1974; Robbins et al. 1975; Ruoslahti et al. 1973; Ruoslahti and Vaheri 1974; Stone, Smith and Joklik 1974) have observed that a protein of molecular weight 210,000–250,000 (Z or "LETS" protein) present in several cell species is capable of undergoing enzymatic iodination with ^{125}I. It is readily located after electrophoresis of the cell membrane proteins on SDS-polyacrylamide gels. The interesting observation has been reported that most transformed cells lack this protein, and that when untransformed cells are stimulated to division and DNA synthesis by trypsin and some other "serine" proteases, this protein component is removed from the cell surface by proteolytic activity. This had led to the speculation (Hynes 1974) that the properties of the transformed cell and the stimulation to division of untransformed cells by proteases are in some way related to the absence or removal of this protein. We have attempted to test this hypothesis with thrombin (Teng and Chen 1975). The use of a protease of this kind has a great advantage since the number and kinds of peptide bonds that it cleaves are very limited. For example, according to present information, four arginylglycine bonds of fibrinogen are cleaved during its conversion to fibrin (Magnusson 1971). Also, a very limited number of peptide bonds of actin-containing arginine or lysine are also split by thrombin (Muszbek and Laki 1974).

In the experiment reported in Figure 5, we have compared the removal of the Z or LETS protein from the cell surface of chick embryo fibroblasts when cells were untreated (A), or treated with (B) 1 μg/ml of trypsin for 10 minutes, (C) 1 μg/ml of TPCK-trypsin for 10 minutes, (D) 4 μg/ml of α-chymotrypsin for 10 minutes, and (E) 10 μg/ml of purified thrombin for 30 minutes. The cells were then enzymatically iodinated with ^{125}I and the proteins were solubilized in SDS-buffer and subjected to electrophoresis on SDS-polyacrylamide gels. The amount of iodinated Z proteins that remained attached to the cell membranes was recorded by autoradiography.

At the level of trypsin used, that is, 1 μg/ml for 10 minutes, the Z protein was completely removed. However, the Z protein is present in approximately the same amounts in the untreated cells and in the cells treated with thrombin.

In a second experiment (Table 1), a comparison has been made of the amount of Z protein present in cells that were (1) untreated, (2) treated with 1 μg/ml of trypsin for 10 minutes, (3) treated with 50 μg of thrombin for 30 minutes, and (4) treated with 1 μg of thrombin for 12 hours. The amount of Z protein was estimated by scanning the films from the autoradiography of the gel strips. The values are recorded as percent of the untreated control. Again the trypsin removed all of the Z protein. In this particular experiment, the amount of Z protein of cells treated with thrombin or serum was essentially the same as that of the untreated control.

When the incorporation of [^{3}H]thymidine into DNA is measured, it is seen that thrombin exhibits potent mitogenic activity regardless of the manner in which it was administered to the cells, and that trypsin stimulates DNA synthesis to a considerably less extent. Undoubtedly trypsin under different conditions of concentration or time of application could have a greater mitogenic effect, but we have chosen a concentration just sufficient to cause

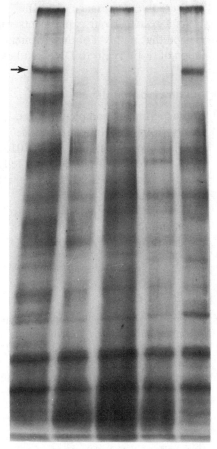

Figure 5

Autoradiograms of SDS-polyacrylamide gels of chick embryo fibroblast proteins labeled by lactoperoxidase-catalyzed iodination. Untreated control (*A*), and treatment with (*B*) 1 μg/ml trypsin (Worthington) for 10 minutes at 20°C, (*C*) 1 μg/ml TPCK-treated trypsin (Worthington, 212 U/mg) for 10 minutes at 20°C, (*D*) 4 μg/ml α-chymotrypsin (Worthington, 47 U/mg) for 10 minutes at 20°C, and (*E*) 10 μg/ml thrombin (2000 NIH U/mg) for 30 minutes at 37°C. The Z, 250 K (250,000 MW) or "LETS" protein is marked by an arrow.

Table 1

Effect of Proteases on Z Protein and DNA Synthesis

Additions to basic medium	Integrated intensity of Z protein (% of control)	[³H]Thymidine incorporation (cpm)
None (control)	—	1915
2% Calf-serum	100	53,253
Trypsin (1 μg/ml, 10 min)	0	11,265
Thrombin (50 μg/ml, 30 min)	100	45,314
Thrombin (1 μg/ml, 12 hr)	100	53,333

After the autoradiograms described in Figure 5 were prepared, the gel strips were scanned with a microdensitometer. The integrated intensity of the Z protein was measured and expressed as percent of control. In the [³H]thymidine incorporation experiment, proteases were removed at the indicated time, and the cells were washed three times with Dulbecco's modified Eagle's medium and incubated with the same medium for 12 hours. Procedures for measurement of DNA synthesis are described in the legend to Figure 2.

complete removal of the Z protein. It should be pointed out that all proteases were used in the absence of serum.

As a result of our experiments with thrombin, we have come to the conclusion that removal of the Z protein is not a necessary prerequisite for stimulation of resting cells to DNA synthesis.

Acknowledgments

The experiments completed in the laboratory of J. M. B. were supported by grants from the National Science Foundation (BMS 17669) and the National Cancer Institute (CA02015), the National Institutes of Health. N. N. H. Teng is a postdoctoral fellow of the Muscular Dystrophy Associations of America. L. B. Chen is a predoctoral fellow of the Johnson and Johnson Co.

We wish to thank Dr. P. W. Robbins for supplying chick embryo fibroblasts and Dr. D. F. Waugh and his colleagues for generous supplies of highly purified thrombin and prothrombin.

REFERENCES

Burger, M. M. 1970. Proteolytic enzymes initiating cell division and escape from contact inhibition by growth. *Nature* **227**:170.

Chen, L. B. and J. M. Buchanan. 1975. The mitogenic activity of blood components. I. Thrombin and prothrombin. *Proc. Nat. Acad. Sci.* **72**:131.

Gahmberg, C. G. and S. Hakomori. 1974. Organization of glycolipids in surface membranes: Dependency on cell cycle and on transformation. *Biochem. Biophys. Res. Comm.* **59**:283.

Hogg, N. M. 1974. A comparison of membrane proteins of normal and transformed cells in lactoperoxidase labeling. *Proc. Nat. Acad. Sci.* **71**:489.

Hynes, R. O. 1973. Alteration of cell-surface proteins by viral transformation and by proteolysis. *Proc. Nat. Acad. Sci.* **70**:3170.

———. 1974. Role of surface alterations in cell transformation: The importance of proteases and surface proteins. *Cell* **1**:147.

Hynes, R. O. and K. C. Humphryes. 1974. Characterization of the external proteins of hamster fibroblasts. *J. Cell Biol.* **62**:438.

Magnusson, S. 1971. Thrombin and prothrombin. In *The Enzymes* (ed. P. D. Boyer), vol. 3, p. 277. Academic Press, New York.

Muszbek, L. and K. Laki. 1974. Cleavage of actin by thrombin. *Proc. Nat. Acad. Sci.* **71**:2208.

Phillips, D. R. 1972. Effect of trypsin on the exposed polypeptides and glycoproteins in the human platelet membrane. *Biochemistry* **11**:4582.

Phillips, D. R. and M. Morrison. 1971. Exposed protein on the intact human erythrocyte. *Biochemistry* **10**:1877.

Rein, A. and H. Rubin. 1968. Effects of local cell concentration upon the growth of chick embryo cells in tissue culture. *Exp. Cell Res.* **49**:666.

Robbins, P. W., G. G. Wickus, P. E. Branton, B. J. Gaffney, C. B. Hirschberg, P. Fuchs and P. M. Blumberg. 1975. The chick fibroblast cell surface after transformation by Rous sarcoma virus. *Cold Spring Harbor Symp. Quant. Biol.* **39**:1173.

Ruoslahti, E. and A. Vaheri. 1974. Novel human serum protein from fibroblast plasma membrane. *Nature* **248**:789.

Ruoslahti, E., A. Vaheri, P. Kuusela and E. Linder. 1973. Fibroblast surface antigen: A new serum protein. *Biochim. Biophys. Acta* **322**:352.

Sefton, B. M. and H. Rubin. 1970. Release from density dependent growth inhibition by proteolytic enzymes. *Nature* **227**:843.

Stone, K. R., R. E. Smith and W. K. Joklik. 1974. Changes in membrane polypeptides that occur when chick embryo fibroblasts and NPK cells are transformed with avian sarcoma viruses. *Virology* **58**:86.

Teng, N. N. H. and L. B. Chen. 1975. The role of surface proteins in cell proliferation as studied with thrombin and other proteases. *Proc. Nat. Acad. Sci.* (in press).

Weber, K. and M. Osborn. 1969. The reliability of molecular weight determinations by dodecyl-polyacrylamide gel electrophoresis. *J. Biol. Chem.* **244**:4406.

Fibroblast Surface Antigen Molecules and Their Loss from Virus-transformed Cells: A Major Alteration in Cell Surface

Antti Vaheri and Erkki Ruoslahti

Departments of Virology and of Serology and Bacteriology, University of Helsinki
Haartmaninkatu 3, SF-00290 Helsinki, Finland

An alteration of growth control of cells is a distinctive feature of malignant transformation, and changes in cell surface seem to correlate with this alteration. Malignant transformation may result in the appearance of new specificities in the cell surface, or in the loss of preexisting molecules. The importance of the latter type of change lies in the fact that the lost components may be involved in the maintenance of the normal growth control.

We have recently described a major cell-type-specific glycoprotein antigen, tentatively designated as fibroblast surface (SF) antigen (Ruoslahti et al. 1973; Ruoslahti and Vaheri 1974), that is lost from virus-transformed cells (Vaheri and Ruoslahti 1974 and unpubl.). This report summarizes the information available on the molecular and biologic properties of SF antigen. The main observations are recorded in the subheadings and Tables. An attempt is also made to relate these findings to other major features in malignant transformation of fibroblasts and to consider briefly the function SF antigen molecules may have.

A Cell-type-specific Major Surface Component

Recent work in our laboratory has attempted to define the significance of cell surface changes in growth control and malignant transformation of fibroblasts (Vaheri, Ruoslahti and Hovi 1974). In order to learn more about the surface components involved in these phenomena, we used brief proteolytic digestion with matrix-bound papain to solubilize protein from the surface of fibroblasts and studied the material immunochemically. This led to the identification in cultured normal chick embryo fibroblasts of a cell-type-specific surface component (SF antigen) that is also present in chicken serum (Ruoslahti et al. 1973). A similar antigen exists in the other species that have been examined. Human SF antigen (Ruoslahti and Vaheri 1974) cross-reacts immunologically with mouse antigen (unpubl.), but only weakly with chicken SF antigen. Appropriate absorptions of the anti-SF antibody preparations remove the cross-reactions between the different mammalian species and result in species-specific reagents.

SF antigen (Table 1) is cell-type-specific. Immunofluorescence of various tissues, including liver, spleen, kidney and muscle, showed fluorescence only in loose connective tissue of chick embryo and adult chicken (Ruoslahti et al. 1973; E. Linder, J. Wartiovaara, E. Ruoslahti and A. Vaheri, unpubl.). The presence of SF antigen in circulation (serum, plasma) makes it difficult to discern whether the endothelial cells lining the vascular channels contain the antigen. Of the human cell lines and cell samples tested (Ruoslahti and Vaheri 1974 and unpubl.; A. Vaheri, E. Ruoslahti, B. Westermarck and J. Pontén, in prep.), only those of fibroblast and glial origin contained SF antigen when assayed either by immunofluorescence or other immunochemical methods. All fibroblast and glial cell lines have been positive. The fibroblast and glia SF antigens are immunologically indistinguishable. A study on the cellular and extracellular distribution of SF antigen in embryonic and adult human tissues is in progress.

It should be noted that even though SF antigen is restricted to only a few cell types, analogous, though immunologically distinct, surface molecules may well exist in other cell types.

The conclusion that the fibroblast SF antigen is a surface component is derived from the following observations: (1) Treatment of live cells with trypsin or papain removes the antigen (Ruoslahti et al. 1973; Ruoslahti and

Table 1

Biological and Molecular Properties of SF Antigen

Property	Evidence
Cell-type (fibroblast, glia)-specific	immunofluorescence, immunochemical quantitation
Associated with fibrillar structures of cell surface	immunofluorescence and scanning electron microscopy
Shed from live fibroblasts and present in circulation (serum)	immunochemical quantitation
Glycoprotein	incorporates, e.g., fucose, glucosamine, methionine and leucine; binds concanavalin A
Molecular equivalents: two polypeptides with MW 210,000 (glycosylated, accessible to lactoperoxidase, highly susceptible to trypsin) and 145,000 (uniquely rapid turnover)	gel electrophoresis, immunoadsorbents, radioimmunoassay, pulse-chase labeling experiments
Cellular SF antigen associated with a 45,000 MW polypeptide (comigrating with actin)	immunoprecipitation
Absent from transformed cells (produced but not retained at surface)	immunochemical quantitation, immunofluorescence, polypeptide analysis

For references see text.

Vaheri 1974; Wartiovaara et al. 1974). (2) Immunofluorescent staining of SF antigen in cultured fibroblasts does not require acetone treatment, but can be done in live (Ruoslahti et al. 1973) or glutaraldehyde-treated cells also (Wartiovaara et al. 1974; Vaheri and Ruoslahti, unpubl.). (3) Glutaraldehyde-treated cells will absorb anti-SF antibodies (Vaheri and Ruoslahti, unpubl.). (4) The antigen may be surface-labeled in live cells by lactoperoxidase-catalyzed iodination (J. Keski-Oja, A. Vaheri and E. Ruoslahti, unpubl.). These data indicate that SF antigen is at least partially exposed to the extracellular space.

SF antigen is a *major* surface component. Based on immunochemical data (SF antigen may be solubilized by a solution containing 8 M urea and 1% Triton X-100) and the proportion of the total protein in the gel electrophoretogram that SF antigen polypeptides account for, we have estimated that SF antigen may comprise as much as 0.5% of the total protein of cultured fibroblasts. This makes SF antigen a major component of the surface membrane.

SF Antigen Is Shed from Cultured Fibroblasts and Is Present in Circulation (Serum, Plasma)

SF antigen is synthesized by cultured fibroblasts and is found at the cell surface. Considerable quantities are shed to the extracellular medium. The following experiment may illustrate the extent of shedding: Confluent cultures of chick embryo fibroblasts (about 2×10^6 cells/20 cm²/5 ml medium 199 containing 5% calf serum) were maintained at 39°C for 24 hours. The amount of SF antigen recovered from the extracellular fluid was about 25–35% of that present in the whole culture.

In circulation, SF antigen represents a previously unrecognized type of serum (plasma) protein, i.e., it originates from the cell surface. In both chicken and human species, the serum SF antigen is a high molecular weight glycoprotein with electrophoretic mobility in the alpha region (Ruoslahti et al. 1973; Ruoslahti and Vaheri 1974). Sedimentation analysis gives a value 7–8S for the chicken SF antigen (Ruoslahti et al. 1973). Some properties of the human serum SF antigen are listed in Table 2.

Circulation of cell-surface components both from normal and malignant

Table 2
Properties of Human Serum SF Antigen

1. High molecular weight glycoprotein; S value 13.5; high molecular weight polypeptide chains
2. Immunoelectrophoretic mobility: α_2 region; altered after neuraminidase
3. Concentration in human plasma 100–200 μg/ml
4. Reaction of identity with SF antigen of human fibroblast and glia cell surface
5. Cross-reaction with mouse, but not with chicken, SF antigen
6. Binds to fibrin(ogen) at low temperature; SF antigen = so-called "cold insoluble globulin"
7. Unrelated to collagen (no hydroxyproline in amino acid analysis)

For references see text.

cells seems to be a general phenomenon. The histocompatibility antigens (Charlton and Zmijewski 1970) and certain tumor antigens (Thomson et al. 1969; Sjögren et al. 1971) also appear in circulation, although in minute concentrations. The high level of SF antigen in serum is not, perhaps, surprising considering that a very large proportion of all cells in the body are fibroblasts. As a well-defined glycoprotein, SF antigen may offer a possibility to look into the biological significance of circulating cell-surface components. It has been suggested that growth regulation (Houck, West and Sharma 1972) and the maintenance of immunological tolerance (Hellström and Hellström 1972) could be mediated by such molecules.

Molecular Equivalents of SF Antigen: Distinct High Molecular Weight Polypeptides

Polyacrylamide gel electrophoretograms of total fibroblast extracts indicated that loss of SF antigen in transformation by Rous sarcoma virus was accompanied by a loss of two polypeptides with apparent molecular weights of 210,000 and 145,000 (Vaheri and Ruoslahti 1974). The proposition that these are the molecular counterparts of chicken fibroblast SF antigen was substantiated by immunochemical purification of the antigen. The two high molecular weight bands, SF210 and SF145, were identified in electrophoretograms of cell extracts by absorption to and elution from immunoadsorbents. In addition to SF210 and SF145, immunoprecipitates of cellular SF antigen contain a 45,000 MW component (SF45). SF45 comigrates in electrophoresis (Kuusela, Ruoslahti and Vaheri 1975) with actin purified (Fine and Bray 1971) from fibroblasts. Experiments with anti-actin antibodies (Lazarides and Weber 1974) and with purified native molecules, SF antigen molecules and actin, may prove whether an association of SF antigen and actin in fact exists in normal fibroblasts. This possibility gets indirect support from the observations on the distribution of SF antigen in fibroblast surface (see below).

Immunochemical data suggest that the two polypeptide chains, SF210 and SF145, carry the same antigenic determinant. SF antigen behaves as a single antigen-antibody system, giving only one line of precipitation in immunodiffusion (Ruoslahti et al. 1973; Ruoslahti and Vaheri 1974; Vaheri and Ruoslahti 1974). Our preliminary tryptic peptide analysis of iodinated polypeptides eluted from gel electrophoresis bands (Bray and Brownlee 1973) has supported this. The peptide patterns of SF210 and SF145 appear related (unpubl.). The metabolism of SF210 and SF145 is, however, distinctly different. SF210 is glycosylated, accessible to lactoperoxidase and highly susceptible to trypsin, whereas SF145 is characterized by a uniquely rapid turnover rate in fibroblasts (Vaheri and Ruoslahti 1974; J. Keski-Oja, A. Vaheri and E. Ruoslahti, unpubl.; Kuusela, Ruoslahti and Vaheri 1975). The role of glycosylation and/or proteolytic cleavage in the relationship of the two polypeptides is being investigated at present.

Chicken serum SF antigen showed a polypeptide composition (SF210 and SF145) similar to the one from fibroblasts, except that the 45,000-dalton component was not detectable (Kuusela, Ruoslahti and Vaheri 1975). This molecular similarity reinforces the notion that the serum SF antigen originates from fibroblasts.

The human SF antigen shows similarities to that of chicken. Both the cellular and soluble serum antigen are comprised of a doublet of high molecular weight polypeptides, > 200,000 daltons, and a polypeptide of about 150,000 daltons (E. Ruoslahti and A. Vaheri, unpubl.).

An important question is the nature of the exact structural relationship between the cellular and circulating antigen (or the antigen fibroblasts shed to the culture medium). This relates to the question of how the antigen reaches the soluble state. It may be excreted as such or cleaved off from the cell surface by enzymes present in the membrane. Although the polypeptides of the cellular and serum SF antigen seem indistinguishable at the present level of resolution, the size of the polypeptides makes it difficult to exclude the possibility that they might differ by a peptide segment which could anchor the molecule to the membrane. Such a mechanism may have a precedent in the case of bacterial penicillinase. Transition of this molecule from a membrane-bound form to soluble state is considered to be mediated by proteolytic cleavage (Sargent and Lampen 1970).

SR Antigen Binds to Fibrin(ogen)— SF Antigen = "Cold Insoluble Globulin"

The concentration of SF antigen in human serum was estimated to be on the order of 100 μg/ml, and it was originally thought to represent a previously unknown serum protein (Ruoslahti and Vaheri 1974). We have now shown (Ruoslahti and Vaheri 1975) that SF antigen has properties similar to those of "cold insoluble globulin" (CIG) (Morrison, Edsall and Miller 1948; Edsall, Gilbert and Scheraga 1955; Mosesson and Umfleet 1970), and the two proteins seem to be identical.

The amount of SF antigen in serum was reduced if the blood coagulation clot was removed at a low temperature. SF antigen could be bound to Sepharose-conjugated fibrinogen and to fibrin powder at 0°C and was subsequently released when the temperature was elevated to 37°C. This procedure resulted in a tenfold enrichment of SF antigen relative to other serum proteins.

SF antigen was found to be concentrated in the cryoprecipitate fraction of human plasma and was copurified with CIG with procedures published for the latter component. SF antigen/CIG is not insoluble at low temperature as such, and its appearance in the cryoprecipitate fraction of plasma is likely to be due to its affinity to cryofibrinogen.

The biological significance of the interaction of SF antigen molecules with fibrin(ogen) is not known. It has been reported that the characteristics of transformed fibroblasts attributable to the increased fibrinolytic activity are pronounced when the cells are plated on fibrin layer (Ossowski et al. 1973). Whether the interaction of the fibroblast surface SF antigen with fibrin plays any role in these phenomena remains to be seen.

SF Antigen Is Produced but Not Retained by Transformed Cells

The amount of SF antigen was absent or greatly reduced in amount in fibroblasts transformed by five different Rous sarcoma virus (RSV) strains. Fibroblasts infected with virus mutants temperature sensitive for transformation recovered SF antigen when maintained at the nonpermissive temperature. Productive infection with a nontransforming avian type C virus did not alter the level of SF antigen characteristic of normal fibroblasts. These results were obtained either with immunochemical quantitation of SF antigen, polypeptide analysis of cell extracts (loss of SF210 and SF145 in the transformed state) (Vaheri and Ruoslahti 1974) or immunofluorescence staining (Wartiovaara et al. 1974).

The loss of SF antigen upon transformation of fibroblasts may explain the recent observations by several groups that polypeptides of high molecular weight are missing in the gel electrophoresis pattern of transformed fibroblasts or their membrane preparations from several species. There is little doubt that the external molecules found by the lactoperoxidase technique in hamster and chicken fibroblasts (Hynes 1973), in mouse (Hogg 1974), in chicken and rat (Stone, Smith and Joklik 1974) and in chicken (Wickus, Branton and Robbins 1974; Blumberg and Robbins, this volume; Yamada and Weston 1974) or by the galactose oxidase technique in hamster fibroblasts (Gahmberg and Hakomori 1973) represent SF210 molecules of the different species. Comparison of normal and RSV-transformed chick fibroblasts by Bussel and Robinson (1973) indicated that membrane preparations of transformed cells had a decreased amount of a polypeptide of an apparent molecular weight of 142,000. The relationship of this polypeptide to SF antigen (SF145 in particular) has not been established.

In accordance with these data, human fibroblasts transformed by SV40 showed a greatly reduced quantity of human SF antigen. This result was obtained both by immunodiffusion, radioimmunoassay, immunofluorescence and polypeptide analysis (unpubl.). Disappearance of a high molecular weight cell-surface protein seems to be a general phenomenon associated with transformation of fibroblasts.

The human fibroblast-SV40 system has indicated that both normal and transformed cells produce the antigen, but the latter do not retain the antigen in the surface (A. Vaheri and E. Ruoslahti, unpubl.). The phenomenon is not restricted to cell lines transformed experimentally by viruses. Our recent experiments indicate that established human tumor cell lines, gliomas and sarcomas show the same pattern as RSV- or SV40-transformed fibroblasts (A. Vaheri and E. Ruoslahti, B. Westermarck and J. Pontén, in prep.).

An Anchored Structure of Cell Surface— Possible Function of SF Antigen

Immunofluorescence and scanning electron microscopy of the same cells has indicated that SF antigen is located in fibrillar structures of the cell surface, membrane ridges and processes with a diameter of 50–200 nm. The processes extended from the periphery of the cells to the substratum and to other cells. SF antigen was detectable within 1 hour after trypsin-treated fibroblasts were reseeded in growth medium. The reappearance of SF antigen correlated with the restoration of membrane processes (Wartiovaara et al. 1974). The indirect evidence available (Kuusela, Ruoslahti and Vaheri 1975) is compatible with the possibility that the antigen is associated with actin-containing microfilaments present under the surface ridges and the membrane processes.

The uneven distribution of SF antigen implies anchoring to stable structures to prevent free lateral movement of the molecules (Frye and Edidin 1970; Taylor et al. 1971; Singer and Nicolson 1972). Whether the anchorage of fibroblast surface antigen is brought about by transmembrane restraints, such as may operate in glycophorin-spectrin interaction across red cell membrane (Singer and Nicolson 1972; Nicolson and Painter 1973), by fixed structures in or outside the membrane, or by interactions of SF antigen molecules themselves is not known at present.

In transformation (Vaheri and Ruoslahti 1974 and unpubl.), mild treatment with proteases (J. Keski-Oja, A. Vaheri and E. Ruoslahti, unpubl.) and mitosis (A. Vaheri and E. Ruoslahti, unpubl.), SF antigen, or the external lactoperoxidase-accessible polypeptide (Hynes and Bye 1974), is lost from the surface. All these conditions are associated with increased concanavalin A-effected agglutinability and loss of density-dependent inhibition of growth (Burger 1973). We know that chicken SF (A. Vaheri and E. Ruoslahti, unpubl.), as well as the analogous surface component in hamster fibroblasts, is a receptor for concanavalin A (Gahmberg and Hakomori 1975). Taken together, the data are compatible with the possibility that anchorage of SF antigen in normal fibroblasts could provide a critical restraint to membrane mobility. The anchorage of SF antigen may impede mobility of other surface components. Loss of SF antigen could result in increased structural instability of the cell surface, expressed as concanavalin A agglutinability.

It would be of considerable importance to know why transformed cells do not contain SF antigen molecules in the surface. Cellular proteases operative in the surface membrane of mitotic and tumor cells provide one of the possible mechanisms.

Acknowledgments

This work was supported by grants from the Finnish Medical Research Council, the Finnish Cancer Foundation and the Sigrid Jusélius Foundation.

REFERENCES

Bray, D. and S. M. Brownlee. 1973. Peptide mapping of proteins from acrylamide gels. *Anal. Biochem.* **55**:213.

Burger, M. M. 1973. Surface changes in transformed cells detected by lectins. *Fed. Proc.* **32**:91.

Bussel, R. H. and W. S. Robinson. 1973. Membrane proteins of uninfected and Rous sarcoma virus-transformed avian cells. *J. Virol.* **12**:320.

Charlton, R. K. and C. M. Zmijewski. 1970. Soluble HL-A7 antigen localization in beta-lipoprotein fraction of human serum. *Science* **170**:636.

Edsall, J. T., G. A. Gilbert and H. A. Scheraga. 1955. The nonclotting component of the human plasma fraction I-1 ("cold insoluble globulin"). *J. Amer. Chem. Soc.* **77**:157.

Fine, R. E. and D. Bray. 1971. Actin in growing nerve cells. *Nature New Biol.* **234**:115.

Frye, L. D. and M. Edidin. 1970. The rapid inter-mixing of cell surface antigens after formation of mouse-human heterokaryons. *J. Cell Sci.* **7**:319.

Gahmberg, C. G. and S. Hakomori. 1973. Altered growth behavior of malignant cells associated with changes in externally labeled glycoprotein and glycolipid. *Proc. Nat. Acad. Sci.* **70**:3329.

————. 1975. Surface carbohydrates of hamster fibroblasts. II. *J. Biol. Chem.* (in press).

Hellström, I. and K. E. Hellström. 1972. Can "blocking" serum factors protect against autoimmunity. *Nature* **240**:471.

Hogg, N. M. 1974. A comparison of membrane proteins of normal and transformed cells by lactoperoxidase labeling. *Proc. Nat. Acad. Sci.* **71**:489.

Houck, J. C., K. L. West and V. K. Sharma. 1972. Evidence for a fibroblast chalone. *Nature New Biol.* **340**:210.

Hynes, R. O. 1973. Alteration of cell-surface proteins by viral transformation and proteolysis. *Proc. Nat. Acad. Sci.* **70**:3170.

Hynes, R. O. and J. M. Bye. 1974. Density and cell cycle dependence of cell surface proteins in hamster fibroblasts. *Cell* **3**:113.

Kuusela, P., E. Ruoslahti and A. Vaheri. 1975. Polypeptides of a glycoprotein antigen (SF) present in serum and surface of normal but not of transformed chicken fibroblasts. *Biochim. Biophys. Acta* **379**:2954.

Lazarides, E. and K. Weber. 1974. Actin antibody: The specific visualization of actin filaments in non-muscle cells. *Proc. Nat. Acad. Sci.* **71**:2268.

Morrison, P. R., J. T. Edsall and S. G. Miller. 1948. Preparation and properties of serum and plasma proteins. XVIII. The separation of purified fibrinogen from fraction I of human plasma. *J. Amer. Chem. Soc.* **70**:3103.

Mosesson, M. W. and R. A. Umfleet. 1970. The cold insoluble globulin of human plasma. I. Purification, primary characterization and relationship to fibrinogen and other cold insoluble fraction components. *J. Biol. Chem.* **245**:5728.

Nicolson, G. L. and R. G. Painter. 1973. Anionic sites of human erythrocyte membranes. II. Anti-spectrin induced transmembrane aggregation of the binding sites for positively charged colloidal particles. *J. Cell Biol.* **59**:395.

Ossowski, L., J. P. Quigley, G. M. Kellerman and E. Reich. 1973. Fibrinolysis associated with oncogenic transformation. Requirement of plasminogen for correlated changes in cellular morphology, colony formation in agar and cell migration. *J. Exp. Med.* **138**:1056.

Ruoslahti, E. and A. Vaheri. 1974. Novel human serum protein from fibroblast plasma membrane. *Nature* **248**:789.

————. 1975. Interaction of soluble fibroblast surface (SF) antigen with fibrinogen and fibrin. Identity with cold insoluble globulin of human plasma. *J. Exp. Med.* **141**:497.

Ruoslahti, E., A. Vaheri, P. Kuusela and E. Linder. 1973. Fibroblast surface antigen: A new serum protein. *Biochim. Biophys. Acta* **322**:352.

Sargent, M. G. and J. O. Lampen. 1970. Organization of the membrane bound penicillinases of *Bacillus licheniformis*. *Arch. Biochem. Biophys.* **136**:167.

Singer, S. J. and G. L. Nicolson. 1972. The fluid mosaic model of the structure of cell membranes. *Science* **175**:720.

Sjögren, H. O., I. Hellström, S. C. Bansal and K. E. Hellström. 1971. Suggestive evidence that the "blocking antibodies" on tumor bearing individuals may be antigen-antibody complexes. *Proc. Nat. Acad. Sci.* **68**:1372.

Stone, K. R., R. E. Smith and W. K. Joklik. 1974. Changes in membrane polypeptides that occur when chick embryo fibroblasts and NRK cells are transformed with avian sarcoma viruses. *Virology* **58**:86.

Taylor, R., P. Duffus, M. Raff and S. de Petris. 1971. Redistribution and pinocytosis of lymphocyte surface immunoglobulin molecules induced by anti-immunoglobulin antibody. *Nature* **233**:1225.

Thomson, D. M. P., J. Krupey, S. O. Freedman and P. Gold. 1969. The radio-immunoassay of circulating carcinoembryonic antigen of the human digestive system. *Proc. Nat. Acad. Sci.* **64:**161.

Vaheri, A. and E. Ruoslahti. 1974. Disappearance of a major cell-type specific surface antigen (SF) after transformation of fibroblasts by Rous sarcoma virus. *Int. J. Cancer* **13:**579.

Vaheri, A., E. Ruoslahti and T. Hovi. 1974. Cell surface and growth control of chick embryo fibroblasts in culture. In *Control of Proliferation in Animal Cells* (ed. B. Clarkson and R. Baserga), p. 305. Cold Spring Harbor Laboratory, Cold Spring Harbor, New York.

Wartiovaara, J., E. Linder, E. Ruoslahti and A. Vaheri. 1974. Fibroblast surface antigen (SF). Association with fibrillar structures in normal cells. *J. Exp. Med.* **140:**1522.

Wickus, G. G., P. E. Branton and P. W. Robbins. 1974. Rous sarcoma virus transformation of the chick cell surface. In *Control of Proliferation in Animal Cells* (ed. B. Clarkson and R. Baserga), p. 541. Cold Spring Harbor Laboratory, Cold Spring Harbor, New York.

Yamada, K. M. and J. A. Weston. 1974. Isolation of a major cell surface glycoprotein from fibroblasts. *Proc. Nat. Acad. Sci.* **71:**3492.

Proteinases in Tumor Promotion and Hormone Action

Walter Troll, Toby Rossman, Joseph Katz and Mortimer Levitz

Departments of Environmental Medicine and Obstetrics & Gynecology
New York University Medical Center, New York, New York 10016

Takashi Sugimura

National Cancer Center Research Institute, Tokyo, Japan

The appearance of plasminogen activator in oncogenic transformation has rekindled interest in the study of the relationship between proteinases and tumorigenesis (Ossowski, Quigley and Reich 1974). Chemical carcinogenesis, within the framework of the two-stage theory of tumor formation and propagation, provides an excellent model for such studies (Boutwell 1974). The theory holds that a normal cell is transformed by DNA modification (initiation), examples of which are somatic mutation and activation of an oncogenic virus. Support for this concept derives from the observation that virtually all primary carcinogens tested are mutagenic in bacteria (Mukai and Troll 1969). The application of such primary carcinogens as 7,12-dimethylbenzanthracene (DMBA), β-propiolactone (BPL) or urethane to mouse skin provides a good mammalian model for the study of primary carcinogenesis. Moreover, the chronic stimulation by a promoter is essential (at least in the early stages) for sustained tumor growth.

The biological mechanism of promotion is poorly understood. Although much work has concentrated on irritants and wounding (Rous and Kidd 1941), it appears that in the mouse model, irritation is essential, but not necessarily sufficient, for promotion (Berenblum 1969). Thus phorbol-12-myristate-13-acetate (PMA), the most active principle of the skin irritant croton oil, is a more effective promoter than agents which produce the same order of irritation (Hecker 1971; Van Duuren 1969).

We have detected trypsinlike proteinase activity in the mouse ear as early as 30 minutes after the topical application of PMA. Testing the hypothesis that this proteinase activity is intimately connected with the tumor process, a variety of inhibitors were applied to the mouse ear in conjunction with PMA. The proteinase inhibitors, in addition to delaying tumor promotion, counteracted the erythema and invasion of leukocytes caused by PMA (Troll, Klassen and Janoff 1970; Janoff, Klassen and Troll 1970). The details and implications of some of these studies are presented in part of this paper.

While PMA is an interesting model promoter in mouse skin, it is unlikely that it has any role in human cancer. Sex hormones, on the other hand, have been considered possible promoters of uterine, breast and prostatic cancer in man. We have been able to demonstrate and partially characterize proteinases elaborated by rat uterus in response to estradiol and estradiol plus progesterone treatment. The enzyme shows some properties similar to the proteinase that appears in the mouse skin in response to PMA. In addition, the results of preliminary experiments suggest that the hormone-induced proteinase is capable of activating human plasminogen. The second part of this paper is concerned with details of these experiments.

PROTEINASE METHODOLOGY

The methods described in the literature were inadequate for the convenient assay of proteinases encountered in tissue extracts. Accordingly, it was imperative to develop new, sensitive methodology in order to place our initial observations in quantitative perspective.

Amino Acid Esters and Amides as Substrates

N-α-Tosyl-L-arginyl methyl ester (TAME), N-α-acetyl-L-lysine methyl ester (ALME), N-α-benzoyl-L-arginine anilide (BAA) and related substances are excellent substrates for trypsin, plasmin and thrombin (Troll, Sherry and Wachman 1954). By using substrates labeled with ^{3}H in the methanol or aniline moiety, continuous enzyme assays can be carried out (Roffman, Sanocka and Troll 1970; Roffman and Troll 1974). The reaction is carried out in a scintillation vial placed in a counter maintained at room temperature. The vial is filled with the usual toluene scintillant and a buffer phase containing the labeled substrate and enzyme. The unhydrolyzed substrate is concentrated in the water phase and as such is undetected by the phosphors. As hydrolysis proceeds, the liberated methanol or aniline passes into the toluene phase, giving rise to scintillation. The counts per minute are proportional to enzyme activity.

Protamine as Substrate Detection with Fluram

Methods used widely for the assay of proteinases employ denatured proteins such as casein and urea-denatured hemoglobin as substrates (Anson 1938; Schwabe 1973). The disadvantages reside in the lack of physical homogeneity of the substrate and in the requirement for separation of split peptides after precipitation with trichloroacetic acid. The basic protein, protamine, has attractive properties as substrate (Brown, Freedman and Troll 1973). Arginine, a target for trypsinlike enzymes, comprises more than one-half the amino acid residues. Furthermore, arginine linkage is the preferred site for plasminogen activator. Protamine contains no lysine groups, so that methods depending on the release of free amino groups for quantitation would not be complicated by high blanks. Finally, protamine (MW ~5000) exists in a single conformation, lacking tertiary structure, rendering the molecule more accessible than casein or hemoglobin to proteinases.

Amino groups can be assayed with speed and sensitivity by a fluorescence technique (Udenfriend et al. 1972). The details of the method we have developed have been previously published (Brown, Freedman and Troll 1973). Briefly, the enzyme to be quantitated is added to a solution of protamine choride in borate or phosphate buffer. At the appropriate times, a solution of 4-phenylspiro[furan-2(3H),1'-phthalan]-3,3'-dione (Fluram) in acetone is added, and the fluorescence is read at 470 mμ with activation at 390 mμ. The protamine fluorescence blank is about 15% of that obtained after complete hydrolysis and can be rendered nil by using succinylated protamine as substrate. The protamine-Fluram method permits the linear assay of 5–500 ng trypsin in a 20-minute incubation. Hydrolysis of protamine with thrombin, plasmin and urokinase has been demonstrated. Consequently, the method is applicable to the quantitation of plasminogen activation simply by preincubating the enzyme to be tested with human plasminogen for a specified time prior to the addition of protamine. The method presents distinct advantages over the fibrin plate method (Unkeless et al. 1974).

Protease Activity in Tissue in the PMA-treated Mouse

Detection of Proteinase

Groups of mice were treated with 1 μg PMA in acetone on one ear, whereas the other ear (control) received acetone alone. At specified times, mouse ears were excised, frozen in liquid nitrogen, and ground to a powder in a mortar and pestile. The powder was suspended in 0.05 M phosphate buffer, pH 7.5. The suspension was sonicated briefly, centrifuged at 10,000 RPM, and aliquots of the supernatant were assayed for activity against [³H]TAME, [³H]ALME, and protamine (Troll, Klassen and Janoff 1970).

The salient data presented in Table 1 indicate that the PMA-treated mouse ears exhibited more activities than the solvent controls. Hydrolysis of TAME was inhibited partially by N-α-tosyl-L-lysyl-chloromethane (TLCK) and N-α-tosyl-L-phenylalanyl-chloromethane (TPCK). On the other hand, protamine activity was inhibited completely by TLCK and not by TPCK, suggesting the presence of a mixture of enzymes. The enzyme hydrolysis of N-α-benzoyl-L-arginine ethyl ester in the mouse skin system is reported by Hozumi et al. (1972), who also found that acetyl-L-leucyl-L-leucyl-L-argininal (leupeptin) (Maeda et al. 1971) is inhibitory.

Table 1

Increased Esterase and Proteinase Activity in Mouse Skin
24 Hours after Treatment with PMA

	Substrate		
Skin homogenate treated with	TAME sp. act.: nmoles hydrolyzed/mg protein/hr	ALME sp. act.: nmoles hydrolyzed/mg protein/hr	protamine sp. act.: fluorescence units/mg/protein/hr
1.0 μg PMA in acetone	73	40	35
Acetone	22	21	0

Mouse ears were treated as described in text. Protein was determined with ninhydrin using crystalline bovine serum albumin as standard.

Inhibition of PMA Promotion by Proteinase Inhibitor

The importance of the proteinase appearing in response to PMA was shown by using a variety of proteinase inhibitors. In the mouse ear model, tumorigenesis was initiated with DMBA and promoted with PMA. The time of first appearance of tumor was noted, but only tumors larger than 1 mm and persisting over 30 days were scored. Studies are carried out with 1 μg and 0.1 μg PMA. Applications (10 μg) of TLCK delayed the onset of tumor 50 days with 1 μg PMA and 200 days with 0.1 μg (Fig. 1).

The relative effectiveness of 1 μg TLCK, TPCK and TAME was tested in studies where 5 μg croton oil was the promoter. Results presented in Table 2 indicate that they are all effective. TPCK virtually suppressed tumor formation for 200 days, and TLCK and TAME delayed and suppressed tumor formation below 50% of the control.

A word of caution should be injected concerning the interpretation of experiments using chloromethyl ketones (TLCK and TPCK) as proteinase inhibitors. Since they are active alkylating agents, they may react with a variety of biological substances unrelated to proteinases. For example, we have shown that TLCK and TPCK are capable of inhibiting the induction of β-galactosidase in *E. coli,* an action which was reversed by the addition of glutathione to the system. Leupeptin and ((S)-1-carboxy-2-phenylethyl) carbamoyl-L-arginyl-L-valyl-argininal (antipain) (Suda et al. 1972) had no effect on β-galactosidase even though they inhibited a trypsinlike proteinase in *E. coli.* Thus it appears that TLCK reacted with glutathione in addition to proteinases in *E. coli,* thereby producing an important biological effect (Rossman, Norris and Troll 1974).

The Hormone-treated Ovariectomized Rat

As mentioned earlier, although PMA is an interesting model promoter in mouse skin, it is unlikely that it has any role in human cancer. On the other hand, sex hormones have been considered possible promoters of uterine, breast and prostatic cancer in man. We were able to demonstrate and partially characterize a trypsinlike protease elaborated by rat uterus in response to estradiol and estradiol plus progesterone treatment. The results of preliminary experiments demonstrate that it is capable of activating human plasminogen.

In these studies, young adult ovariectomized rats were divided into three groups and treated for 4 days twice daily. The first group (U) received sesame oil only. The second group (E) received estradiol, and the third (EP) estradiol plus progesterone. On the fifth day, the uteri were excised and subcellar fractions were prepared (Levitz et al. 1974; Szego et al. 1971). Proteinases were concentrated in nuclei and the 12,000g granule.

Induction of a Nuclear Proteinase

Proteinase activity was detected in the nucleus (associated with chromatin) using degradation of histones as the indicator. Incubation of chromatin for 1 hour at 37°C prior to extraction of histones resulted in extensive degradation of the histones from EP rat uteri, little degradation of that from E rats, and no detectable breakdown of U rat histones (Fig. 2). Preincubation of the EP rat nuclei in the presence of leupeptin, antipain or sufficient protamine

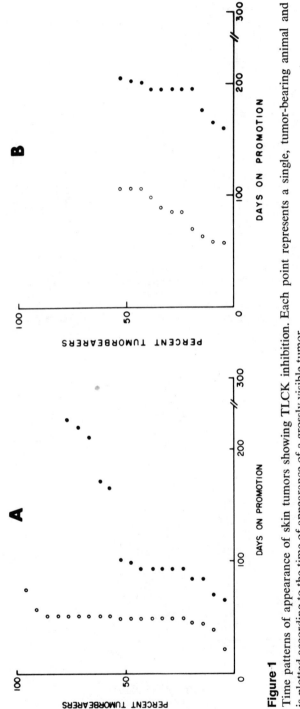

Figure 1

Time patterns of appearance of skin tumors showing TLCK inhibition. Each point represents a single, tumor-bearing animal and is plotted according to the time of appearance of a grossly visible tumor.

(*A*) Experiment 1: All animals were given an initiating treatment of 10 μg of DMBA in acetone. Three days after initiation, three applications per week of 1.0 μg of PMA in acetone was begun. (o) Control animals; (●) animals that received 10.0 μg of TLCK together with PMA. Average time of tumor appearance in controls was 51.6 days; in animals receiving TLCK it was increased to 122.4 days, a difference significant at P <0.1. The average number of tumors per animal was 2.57 for controls and 1.90 for animals receiving TLCK. This difference was significant at P <0.05 by a Student's t-test.

(*B*) Experiment 2: All animals were treated as in experiment 1, except that doses of PMA and TLCK were reduced tenfold. (o) Controls; (●) animals receiving TLCK. Average time of tumor appearance in controls was 83.9 days; in animals receiving TLCK the average time was 189.6 days. Average numbers of tumors per animal were 0.71 and 0.52 for control and treated groups, respectively, not a significant difference. The difference in average time of appearance is significant at P <0.005. (Reprinted, with permission, from Troll, Klassen and Janoff 1970.)

981

Table 2

Inhibition of Tumorigenesis by Proteinase Inhibitors

Weeks on promotion	Inhibitor treatment							
	control		TPCK		TLCK		TAME	
	T	S	T	S	T	S	T	S
10	8	19	0	21	0	21	0	21
12	10	19	0	21	0	21	0	21
14	11	19	0	21	1	21	3	21
16	11	19	0	21	4	21	5	21
18	11	19	0	21	4	21	5	21
20	11	19	0	21	4	21	5	21
22	11	19	0	21	5	21	5	21
24	11	19	1	21	5	21	5	21
30	11	19	1	21	5	21	5	21

All animals were given 10 μg of DMBA as initiator, and then 5.0 μg of croton oil in acetone was applied three times weekly as promoter. The proteinase inhibitors TLCK, TPCK and TAME were applied in DMSO three times weekly in 1.0-μg doses, 1–2 hours after applications of croton oil. All treatments were applied to ear skin of mice (the controls received DMSO alone). The average times of appearance of tumors in all three experimental groups are significantly different from the controls at $P<0.005$; T indicates the number of tumor-bearing mice; S indicates the number of survivors. (Data from Troll, Klassen and Janoff 1970.)

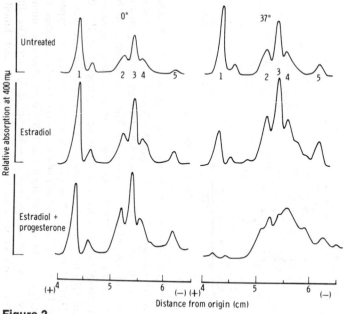

Figure 2

Polyacrylamide gel profiles of histones from uteri of hormone-treated, ovariectomized rats. Freshly prepared chromatin was incubated for 1 hour at 0–37°C, then histones were extracted and submitted to electrophoresis. (Reprinted, with permission, from Levitz et al. 1974.)

982

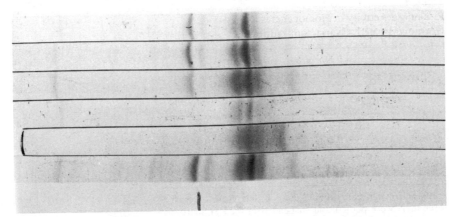

Figure 3
Photograph of polyacrylamide gels of histones from uteri of estradiol plus pro-
gesterone-treated, ovariectomized rats. Freshly prepared nuclei were incubated for
1 hour prior to extraction of histones. Histones were stained with amido black
dye. The six gels from bottom to top were from the following incubation conditions:
(1) 0°C; (2) 37°C; (3) 37°C + 100 μg/ml protamine; (4) 37°C + 500 μg/
ml protamine; (5) 37°C + 100 μg/ml antipain; (6) 37°C + 100 μg/ml leupeptin.
The origin is at the left. The vertical line below gel 1 indicates the position of the
lysine-rich histone 1. (Reprinted, with permission, from Katz et al. 1974.)

effectively protected the integrity of the histones (Fig. 3). Attempts to demonstrate proteinase activity by the protamine assay were not uniformly successful, presumably because of the tight association with histones (Bartley
and Chalkley 1970). However, upon vigorous disruption of the nuclei in a
Polytron homogenizer, such activity was detected.

Proteinase in the 12,000g Pellet

Extranuclear subcellular fractions were assayed for proteinase activity by the
protamine method. Only the 12,000g pellet was active. Activity was absent
from the U rat uterine granules and was slightly higher in the E rat uterus as
compared to the EP rat (Table 3). The specificity of the hormone-induced
proteinase activity was shown in experiments in which diaphragm and thymus
as well as uterus were examined. Only the uterus exhibited a significant
response to sex hormone treatment.

The 12,000g uterine pellets from U, E and EP rats were submitted to several conditions designed to solubilize the proteinase(s). These included
mechanical destruction, the combined application of thermal shock and
detergent action, and autolysis. Autolysis gave the best results. Incubation of
the 12,000g pellet for 1 hour consistently solubilized about 50% of the
proteinase activity. In no case was proteinase detected in either the soluble
fraction or the pellet of the U rat.

Some of the properties of the crude solubilized proteinase were determined.
The apparent K_m toward protamine substrate is about 1.0×10^6 M. The
maximum velocity is achieved by carrying out the incubations at about 45°C

Table 3

Proteinase Activity in Homogenates from
Hormone-treated, Ovariectomized Rats

	Hormone treatment		
Exp.	untreated	estradiol	estradiol + progesterone
1	3.0	33	16
2	2.5	19	13
3	5.0	30	20
4	2.5	30	19

In each experiment, the 12,000g pellets from at least three uteri were pooled and suspended in 0.25 M sucrose (0.4 ml/uterus). Values are in fluorescence units (per 100 μg DNA of tissue) determined after incubation with protamine for 4 hours and reaction with Fluram. Only values greater than 4–5 can be considered significantly different from zero.

at pH 8.5. The peptide aldehydes, antipain and leupeptin, are effective inhibitors of the solubilized proteinase in the protamine-Fluram assay.

Another property of the proteinases of the nucleus and 12,000g pellet merits comment. Although the experiments are preliminary, requiring repeating and extension, the data in Table 4 indicate that the proteinases of the 12,000g pellet and nucleus are plasminogen activators.

Table 4

Plasminogen Activation by Extracts of Subcellular Fractions of Rat Uterus

Contents of incubation	Δ Fluorescent units/2 hr
Plasminogen + protamine	1
12,000g supernate + protamine	3
12,000g supernate + plasminogen + protamine	15
Nuclear supernate + protamine	1
Nuclear supernate + plasminogen + protamine	12
Nuclear precipitate + protamine	0
Nuclear precipitate + plasminogen + protamine	6
Urokinase + protamine	0
Urokinase + plasminogen + protamine	13
Plasmin + protamine	17

Extracts are from uteri of ovariectomized rats treated with estradiol plus progesterone. The 12,000g extranuclear pellet was incubated for 1 hour at 37°C and centrifuged. Then the supernate was tested. The nuclei were disrupted vigorously in a Polytron homogenizer, centrifuged, and both the sediment and supernate were tested. In the assay, the extract was incubated with plasminogen for 1 hour, protamine was added, and at specified times the proteinase activity was determined by the Fluram reaction. In some controls, the plasminogen was omitted, whereas in other controls, urokinase was tested in the complete system and plasmin was tested against protamine. Note that levels of enzyme activity were selected such that in the absence of plasminogen, minimal proteinase activity was observed. Values presented are for 2-hour incubations with protamine.

DISCUSSION

While proteinases have been suspected of playing important roles in a wide variety of biological systems, this could not be demonstrated since their assay in tissues was difficult. The discovery that synthetic substances are good substrates for trypsinlike enzymes, coupled with the development of sensitive radioassays, presented new opportunities for these investigations. In addition, our finding that the proteolysis of the natural substrate protamine, linked with the detection of split products with Fluram, affords a highly sensitive method for measuring proteinases and activators extended these possibilities. These methods are sensitive to ng levels and have permitted us to make multiple observations on extracts from a single mouse ear or rat uterus.

Applying these assays to the mouse ear model following the application of PMA revealed the accumulation of readily detectable proteinase activity. It is important to emphasize that the appearance of proteinase is not dependent on prior initiation. However, the phenomenon appears central to the tumor process because following initiation, the coadministration of TAME or leupeptin (Hozumi et al. 1972) dramatically inhibits the onset of tumors. The technical problems of efficient extraction of these proteinases from mouse ears have so far impeded progress on their purification and further characterization.

Although there is evidence indicating that steroid sex hormones influence the course of many mammalian tumors, the mechanisms are unclear. Nevertheless, the time course of action of these hormones implicate a promotion process. A provocative finding is the localization of trypsinlike proteinases in the nucleus and the 12,000g granules in the uterus following administration of sex hormones to the ovariectomized rat. A logical course of events is synthesis of the proteinase in the cytosol, packaging in 12,000g granules (lysosomes?), and transfer to the nucleus. In the uterus, such transport is stimulated by estradiol (Szego 1974) and, according to our data, appears to be enhanced further by progesterone. Parenthetically, it is of interest that cortical steroid hormones, in particular dexamethasone, which protect the integrity of lysosomes (Weissmann et al. 1968), inhibit tumor promotion (Belman and Troll 1972). The function of proteinase transported to the nucleus may be related to the derepression of the genome by histone modification, permitting increased transcription.

It is too early to ascribe a definitive role for alkaline proteinases and plasminogen activator in tumorigenesis or endocrine action. However, the attractive possibility is presented that the judicious use of specific inhibitors such as leupeptin and antipain will pinpoint the function of these proteinases.

Acknowledgments

This work has been supported by USPH Service Grant ES00606 and Environmental Health Sciences Grant ES-00260; Grant #M74.80 from the Population Council, New York, N.Y., and Grant #CA-02071 from the National Cancer Institute, USPHS.

REFERENCES

Anson, M. L. 1938. The estimation of pepsin, trypsin, papain and cathepsin with hemoglobin. *J. Gen. Physiol.* **22**:79.

Bartley, J. and R. Chalkley. 1970. Further studies of a thymus nucleohistone-associated protease. *J. Biol. Chem.* **245**:4286.

Belman, S. and W. Troll. 1972. The inhibition of croton oil-promoted mouse skin tumorigenesis by steroid hormones. *Cancer Res.* **32**:450.

Berenblum, I. 1969. A re-evaluation of the concept of cocarcinogenesis. *Prog. Exp. Tumor Res.* **11**:21.

Boutwell, R. K. 1974. The function and mechanism of promoters of carcinogenesis. *CRC Crit. Rev. Toxicol.,* p. 419.

Brown, F., M. L. Freedman and W. Troll. 1973. Sensitive fluorescent determination of trypsin-like proteases. *Biochem. Biophys. Res. Comm.* **53**:75.

Hecker, E. 1971. Isolation and characterization of the cocarcinogenic principles from croton oil. *Meth. Cancer Res.* **6**:439.

Hozumi, M., M. Ogawa, T. Sugimura, T. Takeuchi and H. Umezawa. 1972. Inhibition of tumorigenesis in mouse skin by leupeptin a protease inhibitor from actinomycetes. *Cancer Res.* **32**:1725.

Janoff, A., A. Klassen and W. Troll. 1970. Local vascular changes induced by the cocarcinogen, phorbol myristate acetate. *Cancer Res.* **30**:2568.

Katz, J., W. Troll, J. Russo, K. Filkins and M. Levitz. 1974. Effect of protease inhibitors on *in vitro* breakdown of uterine histones from hormone-treated rats. *Endocrine Res. Comm.* **1**:331.

Levitz, M., J. Katz, P. Krone, N. N. Prochoroff and W. Troll. 1974. Hormonal influences on histones and template activity in the rat uterus. *Endocrinology* **94**: 633.

Maeda, K., K. Kawamura, S. Kondo, T. Aoyagi, T. Takeuchi and H. Umezawa. 1971. The structure and activity of leupeptins and related analogs. *J. Antibiot.* **24**:402.

Mukai, F., and W. Troll. 1969. The mutagenicity and initiating activity of some aromatic amine metabolites. *Ann. N. Y. Acad. Sci.* **163**:828.

Ossowski, L., J. P. Quigley and E. Reich. 1974. Fibrinolysis associated with oncogenic transformation: Morphological correlates. *J. Biol. Chem.* **249**:4312.

Roffman, S. and W. Troll. 1974. Microassay for proteolytic enzymes using a new radioactive anilide substrate. *Anal. Biochem.* **61**:1.

Roffman, S., U. Sanocka and W. Troll. 1970. Sensitive proteolytic enzyme assay using differential solubilities of radioactive substrates and products in biphasic systems. *Anal. Biochem.* **36**:11.

Rossman, T., C. Norris and W. Troll. 1974. Inhibition of macromolecular synthesis in *Escherichia coli* by protease inhibitors. *J. Biol. Chem.* **249**:3412.

Rous, P. and J. G. Kidd. 1941. Conditional neoplasms and subthreshold neoplastic states. A study of tar tumors in rabbits. *J. Exp. Med.* **73**:365.

Schwabe, C. 1973. A fluorescent assay for proteolytic enzymes. *Anal. Biochem.* **53**:484.

Suda, H., T. Aoyagi, M. Hamada, T. Takeuchi and H. Umezawa. 1972. Antipain: A new protease inhibitor isolated from actinomycetes. *J. Antibiot.* **25**:263.

Szego, C. M. 1974. The lysosome as a mediator of hormone action. *Recent Prog. Horm. Res.* **30**:171.

Szego, C. M., B. J. Seeler, R. A. Steadman, D. F. Hill, A. K. Kimura and J. A. Roberts. 1971. The lysosomal membrane complex: Focal point of primary steroid hormone action. *J. Biochem.* **123**:523.

Troll, W., A. Klassen, and A. Janoff. 1970. Tumorigenesis in mouse skin: Inhibition by synthetic inhibitors of proteases. *Science* **169**:1211.

Troll, W., S. Sherry and J. Wachman. 1954. The action of plasmin on synthetic substrates. *J. Biol. Chem.* **208:**85.

Udenfriend, S., S. Stein, P. Bohlen, W. Dairman, W. Leimgruber and M. Weigele. 1972. Fluorescamine: A reagent for assay of amino acids, peptides, proteins, and primary amines in the picomole range. *Science* **178:**871.

Unkeless, J., K. Danø, G. M. Kellerman and E. Reich. 1974. Fibrinolysis associated with oncogenic transformation: Partial purification and characterization of the cell factor, a plasminogen activator. *J. Biol. Chem.* **249:**4295.

Van Duuren, B. L. 1969. Tumor-promoting agents in two-stage carcinogenesis. *Prog. Exp. Tumor Res.* **11:**31.

Weissmann, G., W. Troll, B. L. Van Duuren and G. Sessa. 1968. Studies on lysosomes. X. Effects of tumor promoting agents upon biological and artificial membrane systems. *Biochem. Pharmacol.* **17:**2421.

Name Index

Italics indicate where full reference can be found; **boldface** type designates where author's article in this volume is located.

967–975, 967, 968, 969, 970, 971, 972, 973, *974, 975*
Valet, G., 232, 233, *241,* 265, *271,* 388, *404*
Vallota, E. H., 256, *271,* 277, 281, 285, 287, 288, *289, 290*
Vallotton, M., 781, *784*
van Deenen, L. L. M., *252*
Vandeputte, M., 888, *899*
van der Meer, C., *221*
Van de Woude, G., 628, *640*
Van Duuren, B. L., 977, *987*
Vane, J. R., 205, 207, 212, *214, 218*
van Eerd, J.-P., 563, *576*
Van Frank, R. M., 536, *548*
van Heyningen, R., 475
van Hooft, J. I. M., 331, *331*
Vannier, W. E., *404*
VanRietschoten, J., *547*
Varley, K. G., 562, *577*
Varon, S., 785, *792,* 814, *824*
Vasiliev, J., 813, 823, *825*
Vassalli, J. D., 338, 339
Vaughan, M., 212, *220*
Vecchio, G., *643*
Veis, A., 585, *588,* 589
Velazquez-Meza, S., 696, *698*
Velick, S. F., 552, *577*
Venable, T. H., Jr., 670, *674*
Vensel, W. H., 474, *478, 480*
Venuta, S., *929*
Vermylen, J., 69, *73,* 298, 299, *303*
Veros, A. J., 534, *545*
Verstraete, M., 291, 298, *303*
Vesell, E. S., *529, 574*
Vesely, P., 897, *899*
Vigier, P., 933, *942*
Vihert, A. M., 562, *577*
Vilcek, J., 846, *847*
Vilkas, E., 118, *122*
Vill', K., *30*
Villela, G. G., 685, *702*
Vogel, A., *353,* 885, 886, *899*
Vogel, R., 203, 209, *221*
Vogt, V., 623, *644*
Vogt, W., 204, *221,* 261, 262, 264, 266, 267, 272, 274, 287, *289,* 290
von der Mark, K., 579, 581, *588, 589, 590*
von Kaulla, K. N., 685, 690, 695, *704*
Voroshilova, M., *640*
Voynick, I. M., 41, *48, 49, 50, 453, 478*
Vreeken, J., 95, *107,* 194, *195*
Vuust, J., 585, *590*

Waaler, B. A., 67, *77*
Wachman, J., 978, *987*
Wagner, L., *705*

Wagner, R. H., 69, *73, 75*
Wahl, L. M., 897, *899*
Wahl, S. M., *899*
Waldschmidt, M., 684, 686, *704,* 719, *734,* 737, 738, *765*
Wall, R., *855*
Wallén, P., **291–303,** 291, 292, 293, 294, 296, 298, 300, *303,* 305, *310,* 311, *324*
Wallner, O., 695, *704,* **737–766,** 738, 739, 743, 747, 748, 761, *766*
Walsh, K. A., **1–11,** 1, 4, 5, *10, 11,* 16, *29,* 60, 61, *63, 64,* 68, *73, 76, 77,* 188, *189,* 539, *547,* 710, 711, *714,* 831, *839*
Walsmann, P., 370, *384*
Walter, S., 782, *784*
Walters, L., 337, *341*
Walther, P. J., 294, *303,* 305, *310,* 311, 315, *324*
Walz, D. A., *108*
Wampler, D. E., *589*
Wang, J. L., 45, *50*
Ward, P. A., 268, *270,* 274, *290*
Wardlaw, A. C., *271*
Ware, A. G., 124, *149*
Warren, L., 418, 420, *428*
Warren, R. A. J., *30, 643*
Wartiovaara, J., 968, 969, 972, *975*
Washburn, L. L., 780, *784*
Watanabe, M., 206, *221,* 646, *659*
Watanabe, S., *413*
Watnuki, N., 218
Watson, H. C., 3, 4, 5, *11,* 14, 17, *31,* 34, 35, *50,* 68, *76*
Watson, J. D., 661, *674*
Watt, S., 562, *576*
Wattiaux, R., 523, *528,* 553, *573*
Weaver, E. R., *218*
Weaver, R., 640, *644*
Webb, J. L., 459, *465*
Weber, A., 562, *577*
Weber, F., 740, 760, *766*
Weber, G., 36, *50*
Weber, J., 795, *806*
Weber, K., 260, 261, 262, 272, 312, *324,* 389, *404,* 518, *530,* 562, 563, *577,* 828, *839,* 958, *965,* 970, *974*
Weber, M. J., 789, 790, 791, *794,* **915–930,** 917, 920, 928, *929,* 930
Webster, M. E., 201, 203, 204, 206, 209, *215, 219, 221, 384, 403*
Webster, W. P., *73*
Wedgwood, R. J., *271*
Weeds, A. G., 564, *575, 577*

Wegrzynowicz, Z., *196*
Weigele, M., *839, 987*
Weinstein, C., 787, 789, 790, 792, 863, *867*
Weinstein, I. B., **849–856,** 849, 851, *855, 856*
Weinstein, M. J., *75*
Weinstock, I. M., 553, 554, *574*
Weiss, A. S., 65, *77,* 90, *93,* 210, *221*
Weiss, H. J., 69, *77*
Weiss, R. A., 897, *899*
Weiss, R. E., 325, *331*
Weissmann, G., 591, 599, *602,* 616, 618, *619,* 620, *985, 987*
Weksler, M. E., 835, *838*
Wellensiek, H. J., 266, *270*
Welsh, I. R. H., *602,* 616, *619*
Wendt, V., **715–735,** 725, 726, 727, 729, 730, *734*
Werb, Z., 472, 475, *480,* 490, *493*
Werle, E., 28, 200, 201, 202, 203, 206, *214,* 215, 216, *221,* 380, *385, 412,* 686, 694, 695, 696, 699, 700, 737, 738, 742, 743, 746, 748, 750, 755, 762, *763, 764*
Wessels, N. K., 541, *549*
Wessler, S., 72, *77,* 171, 186, *189*
West, C. D., *271*
West, D. W., 208, *220*
West, G. B., 212, *221*
West, K. L., 970, *974*
Westermarck, B., 968, 972
Westermark, B., *528*
Weston, J. A., 972, *975*
Weston, P. D., 468, 469, *478*
Westphal, O., *513, 514*
Weyers, R., *208, 221*
Whetzel, N. K., *169*
Whitaker, D. R., *10, 29*
Whitaker, J. R., 473, *480*
White, E. L., *529, 574*
White, W. F., 831, *839*
Whitfield, J. F., 211, *219*
Whitlock, J. P., 846, *847*
Whitmore, W. F., 685, 691, *704, 705*
Wiblin, C., 897, *899*
Wicher, V., 379, *386,* 410, *414*
Wickus, G. G., 931, *943, 944,* 945, 946, *956, 964,* 972, *975*
Wied, G. L., 686, 695, *702*
Wiedemann, M., *699*
Wiener, E., 485, *492*
Wiener, F., *898*
Wigler, M., **849–856,** 849, 851
Wikman-Coffelt, J., 552, *577*
Wilcox, P. E., *11, 31, 50, 149, 464*
Wilhelm, D. L., 199, *222*

Subject Index

in respiratory tract, 747
against serine/thiol proteases, 430–439
synthetic, affinity labeling of, 455–464
Protein inhibitors of plasma. *See* Plasma
protease inhibitors
Protein turnover
degradation of protein, 515–523
functions of, 523–524, 527
in myofibrils, 552–573
in viral transformation, 525–528
Proteinase(s). *See* Protease(s)
Proteolysis
in activation of Sendai virus, 652–658
in assembly of bacteriophage T4,
662–668
on cell surface, 945–954
irreversibility, 52, 55
in malignant and RSV transformation,
915–929
in plasminogen activation, 301–302
in viral replication, 621–640
Proteolytic enzymes. *See* Proteases
Prothrombin
activation of
by factor V_a, 70–71, 99–102, 112
by factor X_a 70–71, 97–102, 112,
141–147, 179–187
fragment 1·2, function of, 99–106
pathways, 103–106, 180–185
products, 96–103, 153–154
rate, 98–100, 105
regulation, 144–145
amino acid composition, 130–131
(table)
amino acid sequence, 126–129, 134,
137–141, 145
A-S fragment, 137, 141
bicarbonate incorporation, 151–154
calcium-binding properties, 100, 112–
114, 119
carbohydrate attachment sites, 137, 141
γ-carboxyglutamic acid
isolation of, 126, 135–136
position in, 145
reactivity, 118–119, 133
dicoumarol-induced, 112–116, 151
disulfide bridges, 137, 141–143, 146
phospholipid binding, 145–146
structure (schematic), 98, 114, 142–
143
vitamin K-dependent portions
comparison to others, 145–146, 152–
154
isolation, 114–116
structure, 116–119, 132
Protransglutaminase
activation by thrombin, 158, 165–166
subunit structure, 158, 160, 165

Pulmonary emphysema and α_1-antitrypsin
deficiency, 405, 410–411

Rennin, 7, 41
Reproductive process, role of proteases,
695–698
Russell's viper venom,
activation of factor X, 69, 178
inactivation of kallikreins, 203

Seminin, 685, 690–691, 697
Sendai virus(es)
glycoproteins in, 648–652
proteolytic activation of, 652–657
Serine endopeptidases, 65
Serine proteases,
activation peptides, 56–58
activation of proenzymes, 15
active site conformation, 5, 33–37, 42
amino acid residues, 57
difference spectra, 21
inactivation by chloromethyl ketones,
455–464
inhibition by leupeptins, 430–435, 438–
439
kinetics of hydrolysis, 14, 23–27, 37–
41
mechanism of activation, 59–61
in pancreas. *See* Pancreatic serine proteases
pH activity, 20–23
specificity, 3, 14–18, 27, 33–48
zymogen activity, 59–60
Serum factors and regulation of growth
control, 777–782
Serum protease inhibitors. *See* Plasma
protease inhibitors
Serum proteins C2, C4, 234–236
SF antigen,
binding to fibrin, 971
location on cell membrane, 972–973
properties, 967–971
Sialic acid, in α_1-antitrypsin, 409
Sialyltransferase activity and liver diseases, 415–426
Smooth muscle, action of anaphylatoxins
on, 274–276
Spermatozoa,
capacitation, 684–685, 697, 761
formation and maturation, 677–678
proteases of. *See* Acrosin
transport, 679–680, 684
Steroids, inhibition of plasminogen activator by, 851–852